INDEX OF TABLES

Atomic mass of light elements	836	Pressure units	261
Color scale of temperatures	772	Quality factors for several types of radiation	847
Color temperature of light sources	878	Quarks, properties of the u, d and s	923
Densities	262	Resistivity, electrical	502
Dielectric constants and breakdown strengths	484	SI base units	21
Dose equivalent, estimated annual effective	848	Sound absorption coefficients	417
Doses, recommended maximum permissible	847	Sound velocities	414
Elementary particles, partial list	918	Specific heats	306
Energy conversion table	305	Specific heats of real gases	366
Friction coefficients	111	Spectrum, colors of the visible	877
Heat of transformation	309	Surface tension	271
Indices of refraction	616	Thermal conductivities	311
Kinematic equations	218	Thermal expansion coefficients	300
Moments of inertia	174	Viscosities	280
Planets, orbits, and periods	9	Work functions	777

D0141594

MATHEMATICAL ASSISTANCE

Areas and volumes	931
Arithmetic with exponents	17–18
Conic sections	12–13
Exponential function	377–380
Field lines	148–149, 446–449, 527–530
Free body diagrams	104–107, 108–109
Logarithms	380
Problem solving	22–23
Scientific notation	17
Significant figures	16–17
Trigonometry review	87–89, 931
Unit conversion	24–25
Vectors	61–63, 67–72

CONTEMPORARY COLLEGE

PHYSICS

CONTEMPORARY COLLEGE
PHYSICS

Edwin R. Jones *and* Richard L. Childers
University of South Carolina

Addison-Wesley Publishing Company

Reading, Massachusetts • Menlo Park, California
New York • Don Mills, Ontario • Wokingham, England
Amsterdam • Bonn • Sydney • Singapore
Tokyo • Madrid • San Juan

ABOUT THE COVER

This photomontage by Marshall Henrichs suggests the breadth of physics, from the very tiny tracks of elementary particles in a bubble chamber to the macroscopic scale of everyday objects (symbolized here by the apple popularly believed to have inspired Newton's theory of universal gravitation) to the vast distances of interstellar space. Amazingly, present research in particle physics directly affects our understanding of cosmology.

Editor-in-Chief: *Robert H. Price, Jr.*
Design and Production Director: *Susan B. Trowbridge*
Sponsoring Editor: *Stuart Johnson*
Development Editor: *David M. Chelton*
Development Art Editor: *Meredith Nightingale*
Managing Editor: *Barbara Pendergast*
Production Supervisor: *Jack Casteel*
Copyeditor: *Emily Arulpragasam*
Text Designer: *Geri Davis, Quadrata, Inc.*
Layout Artist: *Geri Davis, Quadrata, Inc.*
Cover Designer: *Marshall Henrichs*
Editorial Production Services: *The Book Department, Inc.*
Illustrators: *S. T. Associates, Inc., Waldman Graphics, Inc., Boston Graphics, Inc., Gary Torrisi, Paul Wasserboehr, Susan Banta, and Jennifer DeCristoforo*
Art Consultants: *Dick Morton and Beth Anderson*
Electronic Production Consultant: *Mona Zeftel*
Manufacturing Supervisor: *Roy Logan*
Production Services Manager: *Herb Nolan*
Photo Researcher: *Laurie Sciuto*
Permissions Editor: *Mary Dyer*
Promotions Supervisor: *Julia Berkley*
Marketing Manager: *Susan Howell*
Compositor: *Waldman Graphics, Inc.*
Color Separator: *Waldman Graphics, Inc.*
Printer: *Von Hoffmann Press, Inc.*

Photo credits appear on page A-11.

Library of Congress Cataloging-in-Publication Data

Jones, Edwin R., 1938–
    Contemporary college physics / Edwin R. Jones, Richard L. Childers.
        p.   cm.
    Includes index.
    ISBN 0-201-11951-X
    1. Physics.   I. Childers, Richard L.   II. Title.
QC21.2.J66   1990
530—dc20   89-34320
                                                                            CIP

Preface

This is a comprehensive text for use in a one-year non-calculus general physics survey course. We emphasize basic principles and the unity of physics with these goals in mind: to increase student understanding of natural laws and to develop analytical skills important for the students' general and long-term education. In addition, our increased coverage of modern physics will help students better comprehend the important public policy issues facing them as citizens.

Physics is a dynamic, exciting field and we want students to see this. Our book is truly a 20th-century view of physics. Classical physics is presented from a contemporary perspective. Modern physics is treated thoroughly, as an integral part of the course. The entire book speaks to today's students, using the latest pedagogical aids.

Coverage

The coverage of topics is comprehensive, but not encyclopedic. A challenge shared by all teachers is how to cover what is needed in the time allowed. In our experience, we have found it difficult to get through classical physics and still have enough time to treat modern physics with reasonable thoroughness. Accordingly, the coverage of some topics in classical physics has been reduced. We have, however, taken care to include all topics covered on the MCAT.

We understand that different teachers will emphasize different parts of the text. To make the choices easier, we have designated some material as optional. These **sections marked with an asterisk** may be safely omitted without fear that their content is needed in subsequent sections or chapters.

Emphasis on Basic Principles

"The students can't see the forest for the trees." We have heard this over and over again from colleagues and we have made the emphasis of basic principles one of our highest priorities.

In places, this has led us to make some changes in the standard order and presentation of topics. For instance, Chapters 6 and 7 are organized to highlight the importance of conservation laws. The mathematics in these chapters is restricted to situations that can be treated using only simple equations. Chapter 8 follows with further applications of the conservation laws, including collisions where both momentum and energy are conserved and simultaneous equations are required.

Unity of Physics

Our treatment of conservation laws also illustrates another of our goals: to show that physics is not just a collection of independent ideas but is really an interconnected whole. We emphasize the connections between topics much more than do other texts. We believe this reflects the spirit of physics today and we also believe that it helps students retain more of what they've learned after they leave the course.

If you read a mystery novel all the way through in one sitting, you have at your fingertips all the clues necessary to see how the pieces fit together to solve the puzzle. However, students generally read a physics textbook in small sections and cover a group of chapters over a period of time. As a result they inevitably forget some of the clues and are less prepared to solve the puzzle—or in this case, to see the big picture and appreciate the beauty of physics. For this reason, we give frequent reminders of previously covered topics and of topics to be covered later. We also include examples and problems that require skills and information from prior chapters for their solutions.

Physics as a Scientific Endeavor

The Scientific Process: We present physics as a human activity in which new ideas are constantly being tried and in which scientific truth is never absolute. We generally introduce a new topic by describing the efforts of the scientists who made the breakthrough discoveries and advances. This "discovery" approach not only brings the material alive for the student, but also illustrates the importance of the scientific process. For example, in Chapter 1 we use the stories of Tycho Brahe and Johannes Kepler to illustrate the topics of precision and measurement in science. Other examples include how the *Principia* came to be written, Becquerel's discovery of radioactivity, and Faraday's contributions to electromagnetism. Throughout, we have emphasized physics as a way of thinking, investigating, and understanding rather than as a body of facts and theories.

Models: We introduce the concept of a model in Chapter 1 and then point out throughout the text how physicists use models as part of the scientific process. We emphasize that the first part of developing any theory is to make a model of the physical situation and state its assumptions. Then we show how later observations serve to refine the model and improve our overall understanding. Examples discussed include such fundamental models as the kinetic theory of gases, the free electron model of metals, the wave model of particles, and the quark model of matter.

Level

The text assumes that students have no previous background in physics. The basic mathematical working tools are algebra and trigonometry, and a high school course in these subjects is certainly a prerequisite. One of the challenges in teaching this course is the fact that the students' math preparation is often weaker than the teacher would like. Most students need a math refresher beyond the typical review stuck in the back of texts. To that end there are **Chapter Appendices** on key math topics in these chapters where they are first needed: basic trigonometry (Chapter 3); simultaneous equations (Chapter 8); and the exponential function (Chapter 12).

The exponential function is first used in Chapter 12 in describing the barometric formula and the distribution of molecular speeds. Subsequently, it appears in analysis of electric circuits, radioactive decay, and other topics. The addition of the exponential function to the usual mix of algebra and trigonometry affords students a better comprehension and understanding of the individual topics.

In our teaching experience, we have found that students often have difficulty separating the general features of a new mathematical technique or physical concept from the specific features shown in the initial application. We have prepared four special discussions—called **The Shape of Things to Come**—to help overcome this: (1) conic sections; (2) free-body diagrams; (3) field lines; and (4) potential energy diagrams. In each topic we preview what these ideas will look like in later applications in the book, and we refer back to these discussions when these later applications appear. In this way, students can see the features common to all applications and can focus on the important points when they appear.

Problem Solving

Solving physics problems has long been regarded by physics instructors as a key to learning physics. For this reason, there are nearly 300 worked examples in the body of the text along with a collection of thought questions and numerical problems at the end of each chapter. A short paragraph called **Hints for Solving Problems** precedes each section of problems. Answers to the odd-numbered problems are included at the back of the text.

Approximately 1,900 problems offer a mixture of complexities. They are divided into three levels of difficulty. About two-thirds of the problems are arranged according to the appropriate section. Most of these are for drill and confidence building. The remainder of the problems are not identified by section. Most of these are marked with one or two bullets to indicate their increased degree of complexity. These bulleted problems typically require the synthesis of two or more ideas for their solution and occasionally include material from previous chapters.

Illustrations

Our experience in teaching this course has taught us the value and appeal of visual illustrations for comprehension of abstract discussions of concepts. We have prepared an unprecedented program of full-color diagrams and photographs, which are pedagogically sound and visually appealing. Use of color in vector diagrams (see accompanying figure) is consistent,

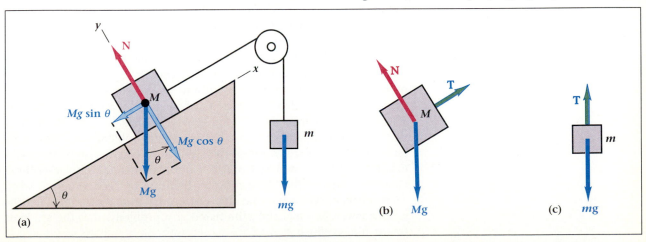

so that frictional forces, tension, and normal forces are obvious at a glance. Similarly, students can distinguish electric fields from magnetic fields, light rays from measurement lines, simply by noting color. We believe this will facilitate students' understanding of the different quantities in physics as they become more familiar with them.

A large number of photographs have been chosen for the book. We think they will reinforce abstract concepts and principles with real examples familiar to students, as well as motivate them by showing the beauty and variety of applications.

Accuracy

We have made an enormous effort to ensure accuracy.

- *Realistic Examples:* We have made a point to make the text correspond to reality. By this we mean that if we use an example of an airliner accelerating to take-off, the numerical values given for mass, take-off speed, etc. are those of a real airliner. We have tried to introduce reality by referring to real objects such as baseballs, golfballs, automobiles, and animals with realistic masses moving with realistic speeds.
- *Adherence to Nature:* We have taken care to correctly describe what is actually observed in nature as, for example, in the description of the temperature dependence of electrical resistivity of metals (Sec. 17.3), and to give correct information on friction (Sec. 4.8).
- *Fidelity to History:* We have read original papers and the current history-of-science research in order to ensure the accuracy of the presentation. Discussions of experiments correspond to what was actually done and discussions of theories correspond to what the authors actually wrote.
- *Reviewing:* Altogether, over 30 physicists have read parts or all of the manuscript in its various drafts.
- *Class-Testing:* Various drafts of the manuscript have been used in our classes at the University of South Carolina. In addition, our colleague, J.L. Safko, has class-tested the manuscript along with the Study Guide in his correspondence course.
- *Proofreading:* We were determined before we even started writing that our book would not be plagued by the typos and careless errors that have undermined many new books. To that end, we have enlisted the help of friends and family to check and recheck the proofs, and the publisher has put each stage of proof through a rigorous set of checks, by both its in-house technical staff and its outside editors.
- *Answers:* We were also determined to have correct answers to the end-of-chapter problems. Each of the problems has been worked independently by the authors. In addition, each problem was worked by either Professor Fred Watts of the College of Charleston or Professor Clyde Smith of Centenary College. On top of this, the publisher hired Craig Watkins of the Massachusetts Institute of Technology to work each problem. We hope that this process has not only confirmed the right answer, but has also eliminated any problems that the students might find confusing.

Special Features

- *Chapter Overview:* Each chapter begins with a **Word to the Student** that orients the student to the main points of the chapter, including study goals and how the chapter relates to others.
- *Chapter Summary:* Each chapter closes with a summary of important concepts and equations and a list of important terms to aid the students in their review of what has been covered. **Hints for Solving Problems** are given before the problem sets.
- *Essays:* Most chapters have one or two special topics set apart at the bottom of the page. The material in these essays is not essential for the development of later material but provides additional topics for students with differing interests. **Physics in Practice** essays generally deal with applications such as How Airplanes Fly (Chapter 9), Dipoles and Microwave Ovens (Chapter 15), and Superconductivity (Chapter 17). **Back to the Future** essays consider the origins of an idea or concept and follow through to its modern use. See, for instance, Kepler (Chapter 1), Gas Laws and Balloons (Chapter 12), and The Speed of Light (Chapter 21).
- *Additional Reading:* At the end of most chapters, students can find a list of additional readings. We have kept the lists short and concentrated on material that is current and readily available. Longer lists are included in the Instructor's Manual.

Supplements

For the Student:

- *Student Solutions Manual:* This includes solutions to the odd-numbered problems (the ones that have the answers in the book) and, like the Instructor's Manual, was written by us.
- *Student Study Guide:* This was written by Professor John Safko of the University of South Carolina. Professor Safko is an experienced study guide author. He class-tested this material in conjunction with the text manuscript.

For the Instructor:

- *Instructor's Manual:* We wrote the instructor's manual, which includes answers and solutions to all the problems in the text.
- *Test Items:* The test-item file will contain over 1,500 questions written by Professor Lewis O'Kelly of Memphis State University to accompany this text.
- *Computerized Testing:* AWTest is a test-item generating system for use on the IBM PC and compatibles. It is an algorithm-based approach that uses random numbers to generate an infinite variety of test items in ready-to-use form with answer sheets and correct answers for the instructor. Diagrams can be printed easily without need of a graphics card.
- *Overhead Transparencies:* Approximately 75 of the figures in the text have been prepared on acetate for use on an overhead projector.

Acknowledgments

We have been blessed with the strong support of our wives, Betty Jones and Sigrid Childers, who have cheerfully assisted with their careful reading of the manuscript in its many drafts. We thank them for all their encouragement and patience. We have had many profitable discussions with our colleagues at the University of South Carolina. We especially thank Chi-Kwan Au, Jim Knight, and Gary Blanpied for helpful comments on several chapters, and John Safko for suggestions on the text that arose while he was developing the Study Guide. We thank Fred Watts of the College of Charleston, Clyde Smith of Centenary College, and Craig Watkins of the Massachusetts Institute of Technology for providing independent solutions to all of the problems and for ensuring that the problems are all well stated and describe realistic conditions. Finally we acknowledge the professional staff at Addison-Wesley for their efforts in bringing this project to completion. We especially note the efforts of Stuart Johnson, David Chelton, Meredith Nightingale, Laurie Sciuto, Jack Casteel, and Geri Davis.

Reviewers

As authors we understand the importance of reviews in the development of a manuscript, and as professors we truly appreciate the outcome—a complete textbook. We would like to thank the many reviewers whose input and criticisms led us to the successful completion of our text:

John Hanneken, *Memphis State University*

Michael Lieber, *University of Arkansas*

Carl Nave, *Georgia State University*

Paul Varlashkin, *East Carolina State University*

Grant Hart, *Brigham Young University*

Lawrence Pinsky, *University of Houston*

Richard Hodson, *Montclair State College*

Lloyd Makarowitz, *State University of New York Agricultural and Technical College—Farmingdale*

Stanley Bashkin, *University of Arizona*

Paul Bender, *Washington State University*

Joseph Shinar, *Iowa State University*

Stanley Shepard, *Pennsylvania State University*

Alvin Rusk, *Southwest State University*

Robert Cole, *University of Southern California*

Roger Freedman, *University of California—Santa Barbara*

Robert Wild, *University of California—Riverside*

Frank Durham, *Tulane University*

Walter Wesley, *Moorhead State University*

Donald Franceschetti, *Memphis State University*

William Savage, *University of Iowa*

Walter Kruschwitz, *University of South Florida*

Gordon Aubrecht, *Ohio State University—Marion*

Dennis Ross, *Iowa State University*

Glen Sowell, *University of Nebraska—Lincoln*

David Markowitz, *University of Connecticut*

R. Bowen Loftin, *University of Houston*

Robert Hallock, *University of Massachusetts—Amherst*

Larry Rowen, *University of North Carolina—Chapel Hill*

Charles Brient, *Ohio State University*

George Carr, *University of Lowell*

George A Benesch, *Baylor University*

Konrad Mauersberger, *University of Minnesota*

Paul Phillipson, *University of Colorado—Boulder*

John Shelton, *College of Lake County*

Craig Watkins, *Massachusetts Institute of Technology*

Carl Babski, *Miami Dade Community College*

David Tribble, *Loyola University of Chicago*

Edward Dixon, *Iowa State University*

Greg Hussey, *Louisiana State University*

E.R. Jones
R.L. Childers

Contents

Measurement and Analysis

1

1.1 The Importance of Accurate Measurement: Tycho Brahe 3

🍎 Back to the Future: Tycho Brahe and Supernovas 6

1.2 From Data to a Model: Kepler's Laws 6

🍎 Back to the Future: Johannes Kepler 10

🍎 Shape of Things to Come: Conic Sections 12

1.3 The Measurement of Time 14

1.4 Measurements, Calculations, and Uncertainties 15

1.5 Estimates and Order-of-Magnitude Calculations 18

1.6 Units and Standards 20

1.7 Problem Solving 22

1.8 Unit Conversions 24

Motion in One Dimension

2

2.1 Speed 32

2.2 Average Velocity 34

2.3 A Graphical Interpretation
of Velocity 38

2.4 Instantaneous Velocity 40

2.5 Acceleration 43

2.6 Motion with
Constant Acceleration 46

2.7 Galileo and Free Fall 50

🍎 Back to the Future:
Galileo and Experimental Science 52

Motion in Two Dimensions

3

3.1 Addition of Vectors 61

3.2 Relative Velocity 65

3.3 Resolution of Vectors 67

3.4 Kinematics in Two Dimensions 72

3.5 Projectile Motion 75

＊**3.6** Range of a Projectile 79

🍎 Appendix: Review of Trigonometry 87

＊ Optional Sections

Force and Motion

4

4.1 The *Principia* *91*

🍎 Back to the Future: The Writing of the Principia *92*

4.2 What Is a Force? *94*

4.3 Newton's First Law—Inertia *96*

4.4 Newton's Second Law *97*

4.5 Weight, the Force of Gravity *100*

4.6 Newton's Third Law *102*

4.7 Some Applications of Newton's Laws *104*

🍎 Shape of Things to Come: Free-Body Diagrams *108*

4.8 Friction *110*

🍎 Physics in Practice:
The Friction of Automobile Tires *112*

4.9 Static Equilibrium *115*

4.10 The Laws of Motion as a Whole *119*

Uniform Circular Motion and Gravitation

5

5.1 Uniform Circular Motion *129*

5.2 Force Needed for Circular Motion *134*

5.3 The Law of Universal Gravitation *140*

5.4 The Universal Gravitational Constant *G* *143*

🍎 Back to the Future: Henry Cavendish and the Density of the Earth *144*

5.5 Gravitational Field Strength *147*

🍎 Shape of Things to Come: Field Lines *148*

Work and Energy

6

6.1 Work *157*

6.2 Energy *162*

6.3 Kinetic Energy *163*

6.4 Gravitational Potential Energy Near the Earth *165*

6.5 General Form of Gravitational Potential Energy *168*

6.6 Moment of Inertia and Rotational Kinetic Energy *172*

6.7 Conservation of Mechanical Energy *175*

🍎 Shape of Things to Come: Potential Energy Diagrams *180*

6.8 Power *182*

Linear and Angular Momentum

7

7.1 Linear Momentum *191*

7.2 Impulse *192*

7.3 Newton's Laws and the Conservation of Momentum *195*

7.4 Conservation of Momentum in One Dimension *196*

7.5 Conservation of Momentum in Two and Three Dimensions *200*

7.6 Changing Mass *203*

7.7 Torque *205*
7.8 Rotational Equilibrium *208*
7.9 Angular Momentum *211*
7.10 Conservation of Angular Momentum *213*
7.11 Angular Acceleration *217*

Combining Conservation of Energy and Momentum

8

8.1 Definition of Elastic Collisions *229*
8.2 Elastic Collisions in One Dimension *231*
8.3 Elastic Collisions in Two Dimensions *235*
8.4 Rolling Bodies *238*
8.5 Braking a Moving Car *240*
***8.6** Motion in a Gravitational Field *243*
***8.7** Escape Velocity *246*

🍎 Physics in Practice: The Earth, the Moon, and the Tides *248*

🍎 Appendix: Solving Simultaneous Equations *256*

Fluids

9

9.1 Hydrostatic Pressure *260*
9.2 Pascal's Principle *264*

🍎 Physics in Practice: Measuring Blood Pressure *266*

9.3 Archimedes' Principle *269*
***9.4** Surface Tension *271*

🍎 Physics in Practice: Surface Tension and the Lungs *272*

9.5 Streamlines and Bernoulli's Equation *274*

🍎 Physics in Practice: How Airplanes Fly *278*

***9.6** Viscosity and Poiseuille's Law *279*
***9.7** Stokes's Law *281*
***9.8** Terminal Velocity *283*
***9.9** Turbulent Flow *284*

Thermal Physics

10

10.1 Temperature and States of Matter *294*
10.2 Thermometry *295*

🍎 Back to the Future: Fahrenheit's Thermometer *298*

10.3 Thermal Expansion *300*
10.4 The Mechanical Equivalent of Heat *303*
10.5 Calorimetry *305*
10.6 Change of Phase *308*
10.7 Heat Transfer *310*

Thermodynamics

11

11.1 The First Law of Thermodynamics *320*
11.2 The Carnot Cycle and the Efficiency of Engines *326*

🍎 Physics in Practice: Gasoline Engines *330*

11.3 Heat Pumps and Refrigerators *333*
11.4 The Second Law of Thermodynamics *335*
11.5 Entropy and the Second Law *336*
*11.6 Energy and Thermal Pollution *340*

The Kinetic Theory of Gases

12

12.1 The Pressure of Air *349*
12.2 Boyle's Law *351*

🍎 Back to the Future: Gas Laws and Balloons *352*

12.3 The Law of Charles and Gay-Lussac *354*
12.4 The Ideal Gas Law *356*
12.5 The Kinetic Theory of Gases *359*
12.6 The Kinetic Theory Definition of Temperature *363*
12.7 The Specific Heats of a Gas *365*
*12.8 The van der Waals' Equation *367*
*12.9 The Barometric Formula and the Distribution of Molecular Speeds *368*

🍎 Appendix: The Exponential Function *377*

Periodic Motion

13

13.1 Hooke's Law *382*
13.2 The Simple Harmonic Oscillator *385*
13.3 Energy of a Harmonic Oscillator *389*
13.4 Period of a Harmonic Oscillator *390*
13.5 The Simple Pendulum *392*

🍎 Physics in Practice: Walking and Running *392*

13.6 Damped Harmonic Motion *395*
13.7 Forced Harmonic Motion and Resonance *397*

Wave Motion

14

14.1 Pulses on a Rope *407*
14.2 Harmonic Waves *408*
*14.3 Energy and Information Transfer by Waves *410*
14.4 Sound Waves *412*
*14.5 Measuring Sound Levels *415*

🍎 Physics in Practice: Acoustical Properties of Rooms *416*

14.6 The Doppler Effect *419*

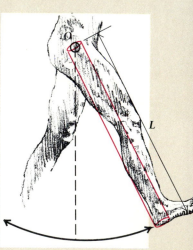

*14.7 Formation of Shock Waves *421*
14.8 Reflection of a Wave Pulse *423*
14.9 Standing Waves on a String *424*
14.10 Waves in a Vibrating Column of Air *428*
*14.11 Beats *429*

Electric Charge
and Electric Field

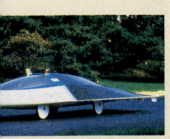

15 15.1 Electric Charge *438*
15.2 Coulomb's Law *441*
15.3 Superposition of Electric Forces *444*
15.4 The Electric Field *446*
15.5 Superposition of Electric Fields *447*
15.6 Electric Flux and Gauss's Law *450*
*15.7 A Quantitative Approach to Gauss's Law *453*

🍎 Physics in Practice: Dipoles and Microwave Ovens 456

15.8 The Electric Dipole *457*

Electric Potential
and Capacitance

16 16.1 Electric Potential *467*
16.2 The Van de Graaff Electrostatic Generator *471*
16.3 Equipotential Surfaces *473*
*16.4 Electric Potential Gradient *474*

🍎 Back to the Future: The Leyden Jar and Franklin's Kite 476

16.5 Capacitors *477*
16.6 The Parallel-Plate Capacitor *478*
16.7 Electric Field of a Parallel-Plate Capacitor *480*
16.8 Dielectrics *483*
16.9 Energy Storage in a Capacitor *486*

Electric Current
and Resistance

17 17.1 Electric Current and Electromotive Force *495*
17.2 Electric Resistance and Ohm's Law *498*

🍎 Physics in Practice: Superconductivity 500

17.3 Resistivity *501*
17.4 Power and Energy in Electric Circuits *503*
17.5 Short Circuits and Open Circuits *507*
17.6 Kirchhoff's Rules *508*
17.7 Simple Resistive Circuits *509*
17.8 Capacitors in Combination *512*
17.9 Internal Resistance of a Battery *515*

🍎 Physics in Practice: Electric Shock 516

*17.10 Home Power Distribution *517*

Magnetism

18
- **18.1** Magnets and Magnetic Fields   *526*
- **18.2** Oersted's Discovery: Electric Current Produces Magnetism   *528*
- **18.3** Magnetic Forces on Electric Currents   *530*
- **18.4** Magnetic Forces on Moving Charged Particles   *533*
- *18.5 The Cyclotron   *536*
- **18.6** Magnetic Field Due to a Current-Carrying Wire   *537*
- **18.7** Magnetic Force on a Current Loop   *541*
- **18.8** Galvanometers, Ammeters, and Voltmeters   *543*
- *18.9 Ampère's Law   *545*
- *18.10 Magnetic Materials   *547*

Electromagnetic Induction

19
- **19.1** Faraday's Law   *557*
- **19.2** Motional Emf   *561*
- **19.3** Generators and Motors   *563*
- **19.4** The Transformer   *566*
- **19.5** Inductance   *568*
- *19.6 Energy Storage in an Inductor   *570*
- **19.7** Maxwell's Equations
  and Electromagnetic Waves   *571*
- 🍎 Physics in Practice:
  Linear Accelerators for Radiation Therapy   572

Alternating-Current Circuits

20
- **20.1** The *RL* Circuit   *582*
- **20.2** The *RC* Circuit   *584*
- **20.3** Effective Values of Alternating Current   *587*
- **20.4** Reactance   *589*
- **20.5** The *RLC* Series Circuit   *592*
- **20.6** Resonant Circuits   *594*
- 🍎 Physics in Practice: AM, FM, and Stereo   596
- *20.7 Rectifiers   *598*
- *20.8 Operational Amplifiers   *601*

Geometrical Optics

21
- **21.1** Models of Light: Rays and Waves   *610*
- 🍎 Back to the Future: The Speed of Light   612
- **21.2** Reflection and Refraction   *613*
- **21.3** Total Internal Reflection   *617*
- **21.4** Thin Lenses   *620*
- **21.5** Locating Images by Ray Tracing   *622*
- **21.6** Thin Lens Equation   *624*
- **21.7** Spherical Mirrors   *628*
- *21.8 Lens Aberrations   *632*

Optical Instruments

22
22.1 The Eye *642*
22.2 The Magnifying Glass *645*
22.3 Cameras and Projectors *648*
22.4 Compound Microscopes *651*
22.5 Telescopes *653*
🍎 Back to the Future:
Development of the Telescope *654*
*22.6** Zoom Lenses *657*

Wave Optics

23
23.1 Huygens' Principle *664*
23.2 Reflection and Refraction of Light Waves *665*
23.3 Interference of Light *667*
23.4 Interference in Thin Films *671*
23.5 Diffraction by a Single Slit *674*
23.6 Multiple-Slit Diffraction and Gratings *677*
23.7 Resolution and the Rayleigh Criterion *680*
23.8 Dispersion *682*
23.9 Spectroscopes and Spectra *684*
23.10 Polarization *685*
*23.11** Scattering *689*

Relativity

24
24.1 Frames of Reference *698*
🍎 Back to the Future: Albert Einstein *700*
24.2 Einstein's Postulates of Relativity *701*
24.3 Velocity Addition *702*
24.4 Simultaneity *704*
24.5 Time Dilation *708*
🍎 Physics in Practice: The Twin Paradox *711*
24.6 Length Contraction *712*
🍎 Physics in Practice: The Appearance of Moving Objects *713*
24.7 Mass and Energy *714*
24.8 Relativistic Momentum *716*
24.9 Relativistic Kinetic Energy *719*
*24.10** The Relativistic Doppler Effect *721*
*24.11** Principle of Equivalence *723*
*24.12** General Relativity *726*

Discovery of Atomic Structure

25
- **25.1** Evidence of Atoms from Solids and Gases *735*
- **25.2** Electrolysis and the Quantization of Charge *739*
- **25.3** Avogadro's Number and the Periodic Table *740*
- **25.4** The Size of Atoms *743*
- **25.5** Crystals and X-Ray Diffraction *744*
- **25.6** Discovery of the Electron *749*
- 🍎 Physics in Practice: The Field Ion and Scanning Tunneling Microscopes *750*
- **25.7** Radioactivity *753*
- **25.8** Radioactive Decay *755*
- **25.9** Discovery of the Atomic Nucleus *757*

Origins of the Quantum Theory

26
- **26.1** Spectroscopy *767*
- 🍎 Back to the Future: Fraunhofer and the Solar Spectrum *768*
- **26.2** Balmer's Series *770*
- **26.3** Blackbody Radiation *771*
- **26.4** Planck's Hypothesis *774*
- **26.5** The Photoelectric Effect *775*
- **26.6** Bohr's Theory of the Hydrogen Atom *779*
- **26.7** Successes of the Bohr Theory *783*
- *__26.8__ Moseley and the Periodic Table *785*

Quantum Mechanics

27
- **27.1** Classical and Quantum Mechanics *793*
- **27.2** The Compton Effect *796*
- **27.3** De Broglie Waves *798*
- 🍎 Physics in Practice: Electron Microscopes *800*
- **27.4** Schrödinger's Equation *803*
- **27.5** The Uncertainty Principle *804*
- **27.6** Interpretation of the Wave Function *809*
- **27.7** The Particle in a Box *811*
- **27.8** Tunneling or Barrier Penetration *813*
- *__27.9__ Wave Theory of the Hydrogen Atom *814*
- *__27.10__ The Zeeman Effect and Space Quantization *816*
- **27.11** The Pauli Exclusion Principle *820*
- *__27.12__ Understanding the Periodic Table *821*

The Nucleus

28
- **28.1** Chadwick's Discovery of the Neutron *829*
- **28.2** Composition of the Nucleus *832*
- **28.3** Nuclear Size, Forces, and Binding Energy *833*
- **28.4** Radioactive Decay Versus Nuclear Stability *837*
- **28.5** Natural Radioactive Decay Series *839*
- **28.6** Models for Alpha, Beta, and Gamma Decay *840*

*28.7 Detectors of Radiation *844*
*28.8 Radiation Measurement and Biological Effects *846*
28.9 Induced Transmutation and Reactions *850*
28.10 Nuclear Fission *853*
28.11 Nuclear Fusion *857*

Lasers, Holography, and Color

29

29.1 Stimulated Emission of Light *864*
29.2 Lasers *865*
29.3 The Helium-Neon Laser *868*
29.4 Properties of Laser Light *869*
29.5 Holography *871*
🍎 Physics in Practice:
White-Light Holograms *874*
29.6 Light and Color *877*
29.7 Color by Addition and
Subtraction *879*

Condensed Matter

30

30.1 Types of Condensed Matter *887*
🍎 Physics in Practice: Liquid Crystal Displays *890*
30.2 The Free Electron Model of Metals *892*
30.3 Electrical Conductivity and Ohm's Law *895*
30.4 Band Theory of Solids *897*
30.5 Pure Semiconductors *900*
30.6 The Hall Effect *901*
30.7 Impure Semiconductors *903*
30.8 The *pn* Junction *904*
*30.9 Solar Cells and Light-Emitting Diodes *906*

Elementary Particle Physics

31

31.1 Particles and Antiparticles *913*
31.2 Pions and the Strong Nuclear Force *915*
31.3 More and More Particles *917*
*31.4 Accelerators and Detectors *919*
31.5 Classification of Elementary Particles *921*
31.6 The Quark Model of Matter *922*
31.7 Unified Theories *925*
31.8 Cosmology *926*

Appendices A: Formulas from Geometry and Trigonometry A-1
B: The International System of Units A-2
C: Alphabetical List of Chemical Elements A-3

Answers to Odd-Numbered Problems A-5

Index I-1

1

Measurement and Analysis

1.1 The Importance of Accurate
     Measurement: Tycho Brahe

1.2 From Data to a Model:
     Kepler's Laws

1.3 The Measurement of Time

1.4 Measurements,
     Calculations, and
     Uncertainties

1.5 Estimates and Order-of-
     Magnitude Calculations

1.6 Units and Standards

1.7 Problem Solving

1.8 Unit Conversions

A WORD TO THE STUDENT

This first chapter introduces you to, or reminds you of, several techniques and concepts you will need later. These include some ideas from elementary geometry and the proper treatment of numbers in scientific calculations. You need to know these things in this course and in your later professional career, regardless of what that career is. It is also important to understand the essential interdependence between measurement and analysis in science. We illustrate this relationship with a discussion of the works of Tycho Brahe and Johannes Kepler. In a broader context, making a planned measurement, whether measuring the positions of a planet or the motion of a runner, is called experiment. The analysis of experimental data contributes to the development of concepts that apply to a wide range of phenomena. These concepts, usually expressed in mathematical terms, are called theories. Physics is theory based on experiment, and the two are inseparable.

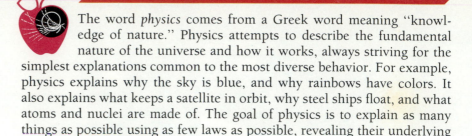

The word *physics* comes from a Greek word meaning "knowledge of nature." Physics attempts to describe the fundamental nature of the universe and how it works, always striving for the simplest explanations common to the most diverse behavior. For example, physics explains why the sky is blue, and why rainbows have colors. It also explains what keeps a satellite in orbit, why steel ships float, and what atoms and nuclei are made of. The goal of physics is to explain as many things as possible using as few laws as possible, revealing their underlying simplicity and beauty.

In achieving their goal, physicists construct models to represent the world around them. These models form conceptual frameworks that permit us to reduce complex situations into simpler, more understandable forms. For example, although we cannot see atoms, we can construct useful models of them that enable us to understand how they behave. In general, such models of physical systems take a mathematical form. It is always understood that the models are by nature incomplete and, therefore, imperfect. For instance, we can pretty well describe the throwing of a baseball if we use a model that ignores air friction. Such a model has its limitations, however, as it does not accurately describe the curved path of the ball. We can obtain better agreement with observations by using a model that includes air resistance.

Physics is an experimental science. By this we mean that the acceptance of any physical theory depends on its success in predicting and explaining reproducible observations. To understand physics, or any experimental science, we must be able to connect our theoretical description of nature with our experimental observations of nature. This connection is made through quantitative measurements. In part, a thorough understanding of physical theories rests on knowing how measurements are made and how reliable the measured information is.

Until this century, scientists assumed that a sufficiently clever observer, given enough time and money, could, in principle, measure any thing or set of things as accurately as necessary. Our understanding of the process of measurement is now more refined. We know that we cannot make a measurement that does not in some way affect the system being measured, thereby limiting the precision of our measurement. This limitation is of little or no importance in the everyday measurements with which we are most familiar—the length of a board, the speed of an automobile, one's own weight, etc. However, we will find in Chapter 27 that when we come to submolecular processes, the interaction of the observer with the measured quantity cannot be ignored.

In the first parts of this chapter we look at the work of Tycho Brahe and Johannes Kepler as an illustration of the value of careful measurement and the subsequent careful analysis of the data gathered. In the process we introduce some elementary geometry and angular measurement. In the later sections you will learn how to correctly handle numbers and units so as to maintain proper accuracy when dealing with experimental measurements and the information found in problems.

The Importance of Accurate Measurement: Tycho Brahe

1.1

In 1572 the appearance of a bright new star in the sky attracted and held the attention of a young Danish nobleman, Tycho Brahe. This type of object, given the name *nova** by Tycho, is seldom visible to the naked eye. However, when novas do appear, they often generate great excitement and interest. This particular nova was brighter than the planet Venus and was easily seen in the daytime. Tycho carefully recorded the position and brightness of the nova for a period of about two years. From his observations he concluded that the new star was not an occurrence in the earth's upper atmosphere, but was actually farther away than the moon. This idea was in complete opposition to the generally held opinion of the time that the heavens were immutable and changeless. Clearly this type of measurement cannot be made just by repeated use of a long measuring tape—so, then, how did Tycho Brahe measure such large astronomical distances?

We look at Tycho's methods in some detail because they illustrate how we can determine information about objects in the universe that are difficult to measure, such as stars, by measuring simpler quantities, such as the angle of sight of an astronomical instrument. Throughout our study of physics, we will see how physicists have managed to obtain data about events that are difficult, or even impossible, to measure directly.

Tycho's direct observation of the nova was, by necessity, a measurement of angles only (Fig. 1.1). In particular, he measured the angle between the line of sight to the nova and the line of sight to a star whose

*Nova comes from the Latin *nova stella,* which means "new star."

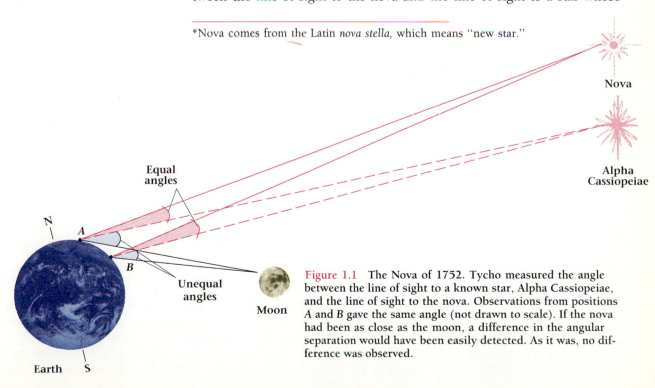

Figure 1.1 The Nova of 1752. Tycho measured the angle between the line of sight to a known star, Alpha Cassiopeiae, and the line of sight to the nova. Observations from positions A and B gave the same angle (not drawn to scale). If the nova had been as close as the moon, a difference in the angular separation would have been easily detected. As it was, no difference was observed.

position was known. Tycho noticed that the angle between the nova and the star Alpha Cassiopeiae did not change between observations made from points A and B. Tycho inferred that the nova was very far away, certainly farther than the moon and probably in the region of the fixed stars. The following example shows how we may use geometry alone to determine large or hard-to-span distances on earth.

Example 1.1

A student walking along the street notices that at one point the top of a lamp pole lies exactly on the line of sight to the top of a nearby office building (Fig. 1.2). The student knows that his eyes are 5 ft above the ground and that the height of the pole is 25 ft. He then measures the distance from his sighting location to the pole as 50 ft and the distance to the building as 120 ft. How tall is the building?

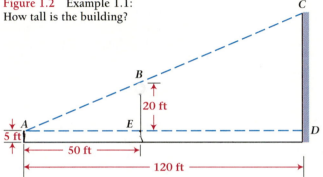

Figure 1.2   Example 1.1: How tall is the building?

Solution   One of the first things to do after reading a problem is to draw a careful diagram. (Section 1.7 discusses problem solving in more detail.) Figure 1.2 is a scale drawing of the information given. In this case the diagram itself suggests the solution.

We can write proportions between the sides of the similar triangles *ABE* and *ACD*:

$$\frac{CD}{BE} = \frac{AD}{AE}.$$

Upon multiplying both sides of the equation by *BE*, we find

$$CD = BE \times \frac{AD}{AE} = 20 \text{ ft} \times \frac{120 \text{ ft}}{50 \text{ ft}} = 48 \text{ ft}.$$

To find the height of the building, we must add the height of the eyes:

$$\text{Building height} = 48 \text{ ft} + 5 \text{ ft} = 53 \text{ ft}.$$

In a manner similar to the way that we used geometry to determine distances in Example 1.1, Tycho Brahe determined ratios of astronomical distances by measuring angles. To illustrate, let us look at one of Tycho's favorite instruments, the Great Steel Quadrant (Fig. 1.3). This device was used to determine the altitude, in degrees, of a celestial body, say a star.

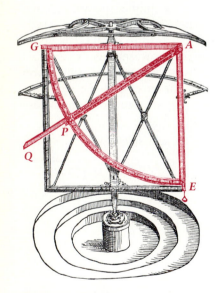

Figure 1.3   Tycho Brahe's Great Steel Quadrant. This instrument rested on a stone foundation. It was braced for maximum rigidity and leveled by means of a plumb line and screws in the base plate. A straight arm (*APQ*) with a sighting device at each end (points *A* and *P*) rotated on a pivot in the upper right-hand corner. With the right side (*AE*) held vertically, the observer used the arm to sight on a star. Then the star's angle of sight was read directly from the curved angle scale (*GPE*).

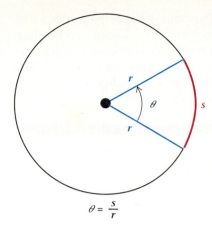

$$\theta = \frac{s}{r}$$

s = circumference
s = 2πr

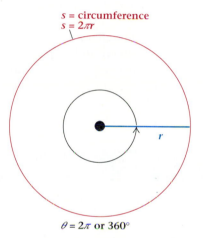

θ = 2π or 360°

Figure 1.4   Angular measure in radians. The radian measure of an angle is the ratio of the subtended arc length to the radius. If the arc length is the entire circumference of a circle, then the angle in radians is 2π.

Quadrant means "one fourth," and the quadrant was designed to measure angles over one quarter of a full circle. Since a circle is divided into 360 degrees, the scale on the quadrant could measure angles up to 90 degrees.

Let us consider one aspect of this instrument's design. For instance, how far apart would the degree marks be on the scale *GPE*? This is important if the instrument is to be read with the greatest precision. Perhaps the easiest way of determining the quadrant's scale of measurement is to introduce another unit of angular measure, the radian. If we measure the length of an arc along the circumference of a circle (Fig. 1.4), we find that the arc length s is proportional to the angle θ subtended by the arc.* For a given angle, the arc length increases in direct proportion to increases in the radius of the circle. We define the angle in **radians**, θ, to be the ratio of the arc length to the radius:

$$\theta \equiv \frac{s}{r}. \tag{1.1}$$

(Three lines instead of two in an equals sign means that the quantity on the left is defined by that equation.) Thus one radian is the angle subtended by an arc length equal to the radius of the circle.

If we let the arc length become the entire circumference of the circle, then s = 2πr. (Recall that π is just a number, equal to 3.14159 . . .) Thus the angle in radians generated in going around a complete circle is 2π. We can say that the angular change in a full rotation is either 360° or 2π radians, depending on which unit of angular measure is used.

The relationship between the radian and the degree is

$$2\pi \text{ radians} = 360°.$$

If we divide both sides by 2π, we find that

$$1 \text{ radian} = \frac{360°}{2\pi} = 57.3°.$$

We could have divided both sides by 360 and found that

$$1° = \frac{2\pi}{360} = 0.0175 \text{ radians}.$$

To return to our discussion of the Great Steel Quadrant, we multiply both sides of Eq. (1.1) by r to give

$$s = r\theta.$$

From Tycho's drawings, the radius r = AP on this instrument is known to be about 194 cm. Thus for an angle of one degree (θ = 1°),

$$s = 194 \text{ cm} \times 0.0175 \text{ radians},$$

which gives

$$s = 3.40 \text{ cm}$$

---

*A line or figure subtends the angle opposite to it in a triangle or sector of a circle.

as the distance apart for the 1° marks. This separation of more than 1 in. (1 in. = 2.54 cm) allowed for further division of the degree into 60 equal parts called minutes of arc. (One minute of arc is written 1′.) Tycho's observations were generally good to within four minutes of arc. His best measurements were good to within one minute of arc, a figure that closely approaches the limit of resolution of the human eye.

From Data to a Model: Kepler's Laws

1.2

The extensive data on planetary motion collected by Tycho would not have helped much in the search for an understanding of the heavens if they had not been analyzed. Here, *analyze* means more than tabulating or graphing information in order to examine its essential features. In this context, it means fitting the data into some broader scheme or model. Such analysis is almost always done against some background of preconceived ideas. The ideas may be modified as the analysis proceeds, but experience shows that the analysis of data outside some preconceived

BACK TO THE FUTURE

Tycho Brahe and Supernovas

How many stars can you see on a clear night with just your naked eyes? Could you identify the same stars the next night and follow their apparent path across the sky, night after night, without the aid of a telescope? That is exactly what the earliest astronomers did, with some remarkable results.

The Danish astronomer Tycho Brahe (Fig. B1.1) is generally considered to be the greatest of the "pretelescopic" astronomers. Making regular observations over a period of more than 20 years, Tycho accumulated a wealth of extremely precise data, which were to lead to one of the major advances in astronomy—Kepler's laws. Tycho's fame lies not in any astronomical theories, but in the care with which he designed and used his instruments.

Figure B1.1 Tycho Brahe (1546–1601). The family names Guldensteren and Rosenkrans appear among the names of Tycho's other noble relations. It has been suggested that Shakespeare may have taken the names Rosencrantz and Guildenstern in *Hamlet* from this source.

Several years after Tycho's astronomical observations of the new star of 1572, Frederick II, King of Norway and Denmark, gave Tycho occupancy of the island presently named Ven (about 25 km from Copenhagen), along with money for buildings and astronomical instruments. There Tycho designed and oversaw the construction of numerous instruments, like the Great Steel Quadrant, and gathered enough people to make regular observations. Though Tycho's instruments were ornate and often gilded, they were quite functional. In fact, they were probably designed with more care, and made with more precision, than any previous scientific instruments.

The death of Frederick II in 1588, and Tycho's continuing quarrels with his tenants on Ven, brought about the loss of support at court. Tycho left Ven in 1597, taking most of his data and the movable instruments with him.

conceptual framework—however vague or even erroneous—is almost unknown.

The Greek astronomer Aristarchos of Samos (ca. 270 BC) had proposed that the earth revolved around the sun. Nevertheless, by the sixteenth century the idea that the earth was the fixed center of the universe was firmly ingrained in astronomical thought and had become an article of religious faith. Nicolas Copernicus (1463–1543) published in his last year his important *De Revolutionibus Orbium Coelestium*, which outlined a workable scheme for the solar system with the sun at the center and the planets in orbits about it. His idea was initially rejected, but the work of Kepler and Galileo helped to bring about its adoption. Such a model is called a heliocentric theory, from the Greek word *helios,* for sun.

By Kepler's time, it was known that the orbits of the planets were not exact circles and that the sun was not at the exact center of the orbits. Kepler observed from Tycho's data that the planets closer to the sun moved faster than those farther away, and that a single planet moved faster when it was closer to the sun than when it was farther away. He proposed that the sun was the cause of the planet's motion and that the sun's influence might decrease with increasing distance from it. With these ideas as guidelines and the detailed observations of Tycho Brahe as his raw material,

 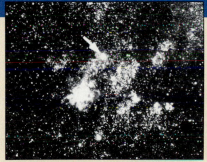

Figure B1.2 (a) The left-hand photograph was taken by Ian Shelton on the night of February 22, 1987. This area of the sky is called the large Magellanic Cloud and is visible only from the southern hemisphere. (b) The right-hand photograph of the same region of the sky was taken on the night of February 23, 1987, approximately 24 hours after the left-hand photograph. The supernova, clearly visible near the center of the photograph, is approximately 1.6×10^{18} km away.

nova, one that appeared just 32 years after Tycho's star. Yet another supernova, easily visible to the naked eye, appeared in 1987, some 383 years later. On the night of February 23 at the Carnegie Institution's observatory on Las Campas mountain in the Chilean Andes, Ian Shelton photographed the bright object shown in Fig. B1.2. The supernova remained at constant brightness for about ten days, then increased to a brightness equal to that of some of the stars in the Big Dipper.

The exploding star, officially named Supernova 1987A, created tremendous interest among astronomers and physicists because the data gathered from it contribute to an understanding of the theories that describe the evolution of stars. As we will see in Chapter 31, the observations of the large, distant supernova also provide information about the subnuclear world of the ultimate components of nature.

After some moving about and negotiating, Tycho accepted an invitation from Emperor Rudolph II to settle in Prague in 1599. He sent his son for some of his instruments from Ven and began to assemble a new group of assistants, one of whom was Kepler.

We now believe that Tycho's "new star" was a supernova occurrence in our own galaxy. A supernova is a star that flares up to hundreds of millions of times its former brightness and is thousands of times brighter than an ordinary nova. Kepler also saw a super-

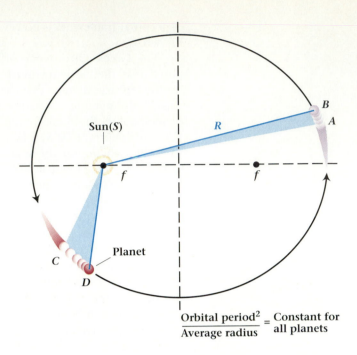

Figure 1.5 Diagram of Kepler's three laws of planetary motion. The ellipse has been exaggerated so that it has a much greater eccentricity than any actual planetary orbit. Note the sun located at one of the foci. According to the second law, the time it takes a planet to move from A to B and from C to D is the same if the areas of the pie-shaped wedges are the same. Kepler's third law says that the ratio of the orbital period squared to the cube of the average radius (R) is the same for all the planets.

Kepler constructed a model of the solar system that was compatible with both ideas.

Kepler's laws of planetary motion were among the first "laws of nature" in the modern sense. They give unambiguous predictions that are subject to verification. They are usually stated in the following way (refer to Fig. 1.5):

1. The orbit of each planet is an **ellipse*** and the sun is at one of the foci.
2. An imaginary line from the sun to a moving planet sweeps out equal areas in equal intervals of time.
3. The ratio of the square of a planet's period of revolution to the cube of its average distance from the sun is a constant.[†] This constant is the same for all planets. (The period of revolution is the time for one complete orbit.)

That Kepler worked out these laws without knowing their underlying cause is a remarkable example of perseverance. The cause of the planetary motion became clear nearly 70 years later, after Isaac Newton published the laws of gravitation and of motion. As you will see in Chapters 5 and 6, from these laws Kepler's laws can be deduced.

Kepler's laws are the first laws of nature you have encountered in this book. Like all "laws" of nature, they represent the distillation or summation of many observations. They are not "fundamental" in the same way as some other laws that we will study later; that is, they describe observations without describing the causes of these observations. How-

---

*A review of the properties of the ellipse is found on page 12.
[†]For Keplerian orbits, the average distance of the planet from the sun is the length of the semimajor axis of its orbit.

ever, Kepler's laws are similar in an important way to almost all the other laws that we will encounter: They are expressed in mathematical terms. Although they can be stated without using such terms, they are most succinct and useful when they specify the mathematical relationships involved. For example, contrast the statement that "planets move faster when closer to the sun" with Kepler's second law. The second law includes not only the information conveyed by that statement but also a basis on which to make exact predictions. Theories in the physical sciences, and physics in particular, are partly judged on their ability to predict what will occur, as well as to explain what has happened. This ability usually requires that they have a mathematical form. Likewise, compare the statement that "planetary orbits are elongated ovals" with the more precise and meaningful statement that the orbits are ellipses.

The *first law* is actually more general than the statement above. We now know that the orbit of the moon about the earth, the orbits of other moons about other planets, the orbits of comets, and the closed orbits of artificial satellites are all ellipses. Contrary to the exaggerated way in which Fig. 1.5 depicts the orbits, planetary orbits are almost circular. Table 1.1 lists the characteristics of the planetary orbits and Fig. 1.6 shows the orbits of the first four planets drawn to scale. Although we can usually approximate planetary orbits with circles, this approximation is not always acceptable for artificial satellites.

The *second law* can be used to predict the speed of a planet in one part of its orbit if we know its speed in another part. Assume, for example, that a planet takes one month to go from *A* to *B* (Fig. 1.5). If we draw pie-shaped sectors *ASB* and *CSD* so that each sector has the same area, then according to the second law, the planet will also take one month to go from *C* to *D*. Since the arc *CD* is longer than the arc *AB* and the times of travel are the same, then the speed along *CD* is greater than the speed along *AB*.

The *third law* deals with a relationship between planets, rather than predicting the behavior of a single planet alone. Stated in symbols, it

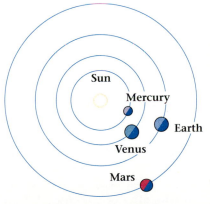

Figure 1.6 The orbits of the four innermost planets of the solar system. Relative distances from the sun are correct. On this scale the difference between the true ellipse and a circle cannot be detected. The size of the planets is exaggerated.

TABLE 1.1 The orbits and periods of the planets		
Planet	Semimajor axis of orbit in AU*	Semiminor axis of orbit in AU*	Period in years
Mercury	0.387	0.370	0.241
Venus	0.723	0.723	0.615
Earth	1.000	1.000	1.000
Mars	1.524	1.517	1.881
Jupiter	5.203	5.197	11.862
Saturn	9.529	9.515	29.458
Uranus	19.182	19.162	84.013
Neptune	30.079	30.078	164.793
Pluto	39.340	38.129	248.530

Source: The Astronomical Almanac, For the Year 1988. Washington, D.C.: U.S. Government Printing Office, 1987.
*The distances in this table are given in astronomical units, AU. One AU is a distance equal to the semimajor axis of the earth's orbit around the sun. That is, 1 AU = 1.5×10^8 km = 9.3×10^7 mi.

becomes

$$\frac{T^2}{R^3} = k, \tag{1.2}$$

where the period T is the time for one complete orbit, R is the average distance of the planet from the sun, and k is a constant that is the same for all planets.

Example 1.2

Compare the prediction of Kepler's third law for the distance of Mars from the sun with the measured value given in Table 1.1 as the semimajor axis of its orbit. Round off the value of Mars's period (a Martian year) to 1.88 years.

Solution   Since the constant k is the same for all planets, we may rewrite Eq. (1.2), using subscripts M to denote Mars and E to denote earth:

$$\frac{T_M^2}{R_M^3} = k = \frac{T_E^2}{R_E^3}$$

BACK TO THE FUTURE

Johannes Kepler

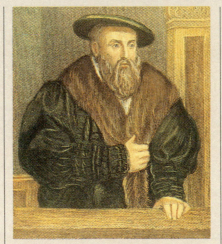

Figure B1.3 Johannes Kepler (1571–1630).

Today, children routinely learn the names of all the planets and their order from the sun, and travel throughout the solar system does not seem impossible. It's hard to imagine a time when people didn't know about the planets and how they move through space. An early step toward understanding the solar system occurred almost 400 years ago, with Johannes Kepler's analysis of planetary orbits.

Kepler (Fig. B1.3) held his first position at Graz, Austria, where his duties were both astronomical and astrological. He was struck by the fact that there existed only five regular geometrical solids and (it was thought) only five planets. Therefore he tried to explain the radii of the known planetary orbits by a scheme that was based on the geometry of the five regular geometrical solids. This mystical scheme, which accidentally gave fairly good agreement with what was then known (but ruled out the existence of any more planets), gained him considerable publicity and led to his association with Tycho Brahe in 1600. Upon Tycho's death 18 months later, Kepler took possession of his data.

Kepler spent almost ten years trying to fit Tycho's observations of the position of Mars to a circular orbit—or to some combination of circles. He even came to a point where the disagreement between his calculations and Tycho's observations was only 8 minutes of arc. This is the angle covered by a dime held on edge and viewed from a distance of about 22 in. (56 cm). But Tycho's measurements were at least twice this good, equivalent to moving the dime to 44 in. Kepler had so much faith in the precision of Tycho's observations that he knew his own analysis must be in error. He discarded his work and started over many times, arriving at last at what are now called Kepler's laws of planetary motion. Kepler discovered the first two of his three laws in the attempt to understand the orbit of the planet Mars. They ap-

or

$$R_M^3 = \frac{T_M^2 \, R_E^3}{T_E^2},$$

$$R_M = \left(\frac{T_M^2 R_E^3}{T_E^2}\right)^{\frac{1}{3}}.$$

We may use the average distance of the earth from the sun as our unit of distance. This distance is called the astronomical unit, abbreviated AU. One astronomical unit is equal to 1.50×10^{11} m. We may likewise take the year as our unit of time. From Table 1.1, the period of Mars is 1.88 years, and the period of the earth is 1 year. These values give

$$R_M = \left(\frac{(1.88 \text{ year})^2 (1 \text{ AU})^3}{(1 \text{ year})^2}\right)^{\frac{1}{3}}$$

$$= 1.52 \text{ AU}.$$

This result is in agreement with the observed value listed in Table 1.1.

peared in Kepler's *Astronomia Nova* in 1609. The third law appeared in 1619 in *Harmonices Mundi* (The Harmony of the Worlds).

Incidentally, Kepler was not the only person to benefit from the great precision of Tycho's measurements. Not only was there a general improvement in the quality of observations but in 1582 the new calendar of Pope Gregory XIII, called the Gregorian calendar, was instituted, in part because of the more accurate observations. The basic principle of this calendar system has not been changed since its introduction. The Gregorian calendar has leap years in century years divisible by 400 but not in other century years. (The year 2000 is a leap year; the year 1900 was not.) The previous calendar, with a leap year every four years, had been in use since its adoption during the reign of Julius Caesar. By the year 1580, the Julian calendar had accumulated an error of 10 days. In 1582,

Figure B1.4   A communication satellite of the type used to transmit television signals around the earth.

with the introduction of the Gregorian calendar, 10 days were dropped from October. Roman Catholic countries adopted the new calendar at once. By 1752, when Britain and its colonies made the change, the calendars were different by 11 days. The change still causes confusion. For example, George Washington was born on February 11, 1732 (old style), but we now celebrate his birthday on February 22.

Because Kepler's laws rest on more general laws of nature, including Newton's universal law of gravitation, they apply to more systems than just the planets moving about the sun. The moon in its orbit about the earth obeys Kepler's laws with an adjusted constant k. Other objects in orbit about the earth also obey the same basic laws. The orbits and periods of artificial satellites used for observations and communications can likewise be understood in terms of Kepler's laws. Figure B1.4 shows a communication satellite of the type used to distribute television programs. When boosted to the proper altitude in an equatorial orbit, these satellites appear to be stationary in the sky as they move in synchronism with the earth's rotation on its axis. (They are called geosynchronous satellites.) Television companies and home users can receive the signals by aiming their dish antennas at the satellite.

CONIC SECTIONS

Conic sections are mathematical curves that occur frequently in nature. We discuss the ellipse briefly in Chapter 1; we will encounter the parabola and hyperbola in later chapters.

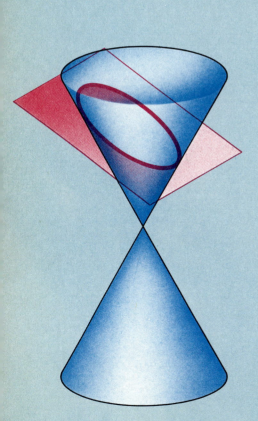

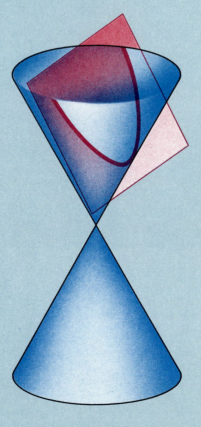

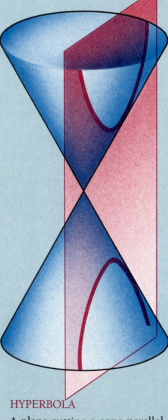

ELLIPSE
A plane cutting a cone at an angle makes an elliptical cross section.

PARABOLA
A plane cutting a cone parallel to a side makes a parabolic cross section.

HYPERBOLA
A plane cutting a cone parallel to its axis makes a hyperbolic cross section.

ELLIPSE

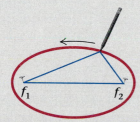

At any point on an ellipse, the sum of the distances to two fixed points (foci) is the same. An ellipse is drawn using two pins and a loop of string.

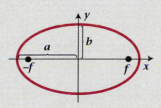

Equation:

$$\frac{x^2}{a^2} + \frac{y^2}{b^2} = 1^2$$

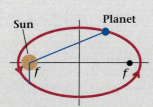

Kepler's laws: A planetary orbit is an ellipse with the sun at one focus (Chapter 1).

PARABOLA

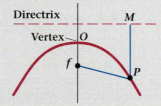

At any point on a parabola, the distance to a fixed point (focus) equals the distance to a fixed line (directrix).

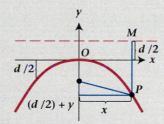

Equation:

$$y = -(1/2d)x^2$$

Projectile motion: The path of a projectile is a parabola (Chapter 3).

HYPERBOLA

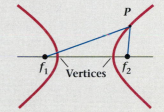

At any point on a hyperbola, the difference between the distances to two fixed points (foci) is the same.

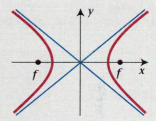

Equation:

$$\frac{x^2}{a^2} - \frac{y^2}{b^2} = 1^2$$

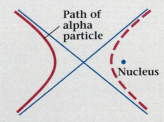

Rutherford scattering: The path of a particle deflected by a nucleus is a hyperbola (Chapter 25).

The Measurement of Time

1.3

In Tycho's day, clocks were subject to errors of between 100 and 1000 seconds per day. Though Tycho had several fine clocks, he did not make extensive use of them in his observations. His main "clock" was the regular motion of the earth itself, detected by observations of the sun and distant stars.

Not only in astronomy, but in many other fields as well, we find that the object of study changes as time passes; in fact, often the change itself is the significant observation. To some extent the usefulness of the information gathered, whether it be changes in planetary positions or the change in diameter of a tree, depends on how accurately we can determine the time interval over which the change takes place.

The improvement in timekeepers has often been tied in with ideas and developments that were in the forefront of science and technology. Advances in understanding the thermal properties of materials allowed for the development of temperature-compensated clocks. The use of electromagnets and electric switches brought about still further improvements in pendulum clocks. Studies of the electromechanical properties of quartz led eventually to crystal-controlled clocks that are many times more accurate than mechanical clocks. Figure 1.7 shows the improvement of timekeeping accuracy since the invention of the mechanical clock around the year 1300.

A better understanding of the nature of the atom has allowed us to use certain periodic properties of atoms as the characteristic that regulates

Figure 1.7 The diminishing daily error in clocks is shown as a function of the year in which the improvement was made.

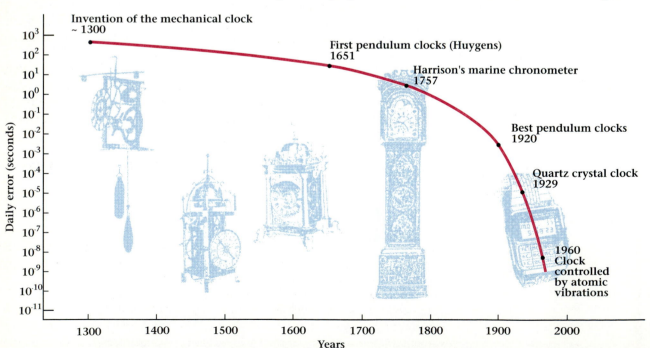

Invention of the mechanical clock ~ 1300

First pendulum clocks (Huygens) 1651

Harrison's marine chronometer 1757

Best pendulum clocks 1920

Quartz crystal clock 1929

1960 Clock controlled by atomic vibrations

Daily error (seconds)

Years

the most modern timekeepers. The present standard for timekeeping is the cesium-beam atomic clock, which can be made to a stability and accuracy of about one part in 10^{12} or even one in 10^{13}. The unit of time is the second, defined to be exactly 9 192 631 770 periods of a certain radiation from atoms of cesium 133. The symbol for the second is s.

The search continues for an even more accurate timekeeper. As scientists delve deeper into nature, we find increasingly accurate ways of keeping time. This increase, in turn, often contributes to another step forward in development and understanding in some other field.

Beyond the problem of measuring time, there is the more subtle question, What is time? Although we will discuss this question in more detail later in the text, a few comments are appropriate here. If time and time intervals are as they appear in our daily lives, surely we should have a firm grasp on their meaning as well as having ways of measuring them. Most people's intuitive feeling about time corresponds to Isaac Newton's statement that "Absolute, true, and mathematical time, of itself, and from its own nature flows equably without relation to anything external . . ." This view has been referred to as the "classical" concept of time. Later we will examine certain experimental problems and logical inconsistencies in this interpretation. However, Newton's statement provides a useful starting point. The classical way of thinking about time has proved useful in dealing with a vast range of phenomena in the physical sciences as well as in other human activities.

In practice, we measure time intervals by carrying out some specified operation. In particular, we count the number of times some regularly repeating (periodic) event occurs. For example, we measure days by the number of times the earth rotates on its axis; we measure shorter time intervals by counting the number of swings of a pendulum or the number of cycles of an alternating current. Our general concept of a clock will not depend on the details of a mechanical or electrical device, but on the operational definition of a device that counts the number of repetitions of some periodic event.

In Chapter 24 we will see that in the early part of this century, after more than 99% of humanity's recorded history had passed, our accumulated knowledge of nature required a reformulation of the classical concept of time. These changes were first made as part of Einstein's special theory of relativity.

Measurements, Calculations, and Uncertainties

1.4

Kepler relied on the precision of Tycho's data to confirm or disprove his models of planetary motion. If Tycho's data had been less precise, they could have fit Kepler's earlier model of complex arrangements of circular orbits. Kepler's use of Tycho's measurements illustrates an important difference between how we use numbers in arithmetic and algebra, and how we use them in science. This difference arises from the nature of measurement. When we say that a length is 9.2 cm, we mean that this is a length

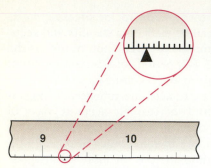

Figure 1.8 Position of a reference mark along a scale. The inset shows a magnified view, which permits greater precision in locating the position of the mark.

that has been, or can be, measured. Implicit in our statement is information about the precision with which the measurement has been made. By 9.2 cm we mean a length that is closer to 9.2 cm than to either 9.1 cm or 9.3 cm. Had we estimated between the 0.1-cm divisions, or used a ruler with finer divisions, we might have said the length was 9.23 cm (Fig. 1.8). This means a length closer to 9.23 cm than to 9.22 cm or 9.24 cm. The point is that the last figure in a quoted measurement is in some sense uncertain and does not have the same significance as the figures to its left.

The number of digits reported in a measurement, irrespective of the location of the decimal place, is called the number of **significant figures**. The number of significant figures is a reflection of how well you know a given quantity. For example, length is, in the classical sense, assumed to be infinitely divisible. This means that to specify a length "exactly" would require an infinite number of significant figures. Therefore all numbers that represent lengths are inexact; but they do carry with them information about how well those lengths are known.

Measured values are often used to calculate other quantities. For example, we can determine the circumference C of a circle from a measurement of the radius r of the circle and the expression

$$C = 2\pi r. \tag{1.3}$$

The value of π is well known and has been determined to many significant figures. To seven significant figures, $\pi = 3.141593$. If we had measured the radius of a circle to be 1.60 cm, direct application of Eq. (1.3) would give a numerical answer of

$$C = 10.053098 \text{ cm.}$$

However, this is not a physically meaningful or sensible answer. The circumference can be known only to the same precision as the radius. In this case, the precision is three significant figures. We *round off* the answer to the correct number of significant figures:

$$C = 10.1 \text{ cm.}$$

Note that the last figure has been rounded up. The rule we will follow in this book is that *if the first digit beyond the last significant figure is 5 or greater, the last significant figure is to be increased by unity (1). All other figures beyond the last significant figure are dropped. If the digit beyond the last significant figure is less than 5, the last significant figure remains unchanged.* Thus 10.05 rounds off to 10.1, but 10.04 rounds off to 10.0. (Most pocket calculators that round off automatically use this rule.)

Another numerical example will further illustrate the point. Let us calculate the area of a paperback book cover. Suppose the cover is a rectangle whose sides we measure to be 10.6 cm and 17.9 cm. To find the area A of a rectangle we multiply the length l times the width w:

$$A = l \times w.$$

The arithmetic product of 10.6 and 17.9 is 189.74. However, it is *not* correct to give the area of the cover as 189.74 cm². This would imply that the area is known to be between 189.73 cm² and 189.75 cm², a precision that is unwarranted on the basis of the precision with which the lengths of the individual sides are known.

To see what we mean, notice that the length of the first edge of the cover is between 10.55 cm and 10.64 cm, and the length of the second edge is between 17.85 cm and 17.94 cm. The product of the least of these numbers is 188.3175 and the product of the greater is 190.8816. We see, then, that even the last place *before* the decimal is uncertain, so it would be a mistake to give an answer with more than three figures. Thus we round off the computed area of 189.74 cm² to 190 cm². The general rule is that *your answer must have no more significant figures than is warranted by the least precise of your values* (that is, the value with the fewest significant figures).

Sometimes it is not clear whether the final zero or zeros in a number are significant figures or are merely needed to locate the decimal point. In the preceding example, the result of 190 cm was correct to three figures so that the zero was a significant figure. To clearly indicate the number of significant figures, we often express a number in **scientific notation**. Thus we can write 190 cm as 1.90×10^2 cm. The presence of the final zero indicates that the number is known to three significant figures. We will assume integers to be precise. Thus the answer to what is the cost of two hamburgers at $1.89 each is $3.78 and not $4.

Example 1.3

Calculate the volume of a cylindrical oil can with a diameter of 9.9 cm and a height of 13.5 cm (Fig. 1.9).

Solution   The volume of a cylinder is the product of the height h and the area of the base, which is given by

$$A = \pi \left(\frac{d}{2}\right)^2,$$

where d is the diameter. Thus the volume becomes

$$V = hA = h\pi \left(\frac{d}{2}\right)^2.$$

Upon inserting the values for h and d we get

$$V = 13.5 \text{ cm} \times \pi \times \left(\frac{9.9}{2}\right)^2 \text{ cm}^2.$$

Direct computation yields

$$V = 1039.19 \text{ cm}^3 = 1.03919 \times 10^3 \text{cm}^3.$$

However, since the diameter is known to only two significant figures, the answer must be rounded off to give

$$V = 1.0 \times 10^3 \text{cm}^3.$$

One further justification for the use of scientific notation is that it provides an easier way of writing large and small numbers and of doing arithmetic with them. To see how this is done, note that we may write the number "one thousand" in the following way:

$$1000 = 10 \times 10 \times 10 = 10^3.$$

|← 9.9 cm →|

13.5 cm

$V = h\,A = h\,\text{p}\,(d/2)^2$

Figure 1.9   Example 1.3: Calculating the volume of a cylindrical oil can.

TABLE 1.2
Powers of ten

$$10^{-3} = \frac{1}{1000} = 0.001$$

$$10^{-2} = \frac{1}{100} = 0.010$$

$$10^{-1} = \frac{1}{10} = 0.100$$

$$10^{0} = 1$$
$$10^{1} = 10$$
$$10^{2} = 100$$
$$10^{3} = 1000$$

In general we may write multiples of 10 (or 1/10) as shown in Table 1.2. With this notation, we can conveniently express very large and very small numbers using scientific notation. For example,

$$127{,}000{,}000 = 1.27 \times 100{,}000{,}000 = 1.27 \times 10^{8}$$

and

$$0.00037 = 3.7 \times 0.0001 = 3.7 \times 10^{-4}.$$

Numbers expressed in scientific notation may be multiplied and divided according to the usual rules of algebra. Remember that

$$10^{n} \times 10^{m} = 10^{n+m} \quad \text{and} \quad \frac{10^{a}}{10^{b}} = 10^{a-b}. \tag{1.4}$$

The notation is particularly useful when both large and small numbers occur in the same problem, as shown in the following example.

Example 1.4

What is the volume of a sheet of paper that is 8.60×10^{-3} cm thick, 2.16×10^{1} cm wide, and 1.40×10^{2} cm long?

Solution · The volume V of a rectangular-shaped piece of paper is given by the product of the length, width, and thickness:

$$V = \text{length} \times \text{width} \times \text{thickness},$$
$$V = (1.40 \times 10^{2} \text{ cm}) \times (2.16 \times 10^{1} \text{ cm}) \times (8.6 \times 10^{-3} \text{ cm}).$$

This may also be written

$$V = 1.40 \times 2.16 \times 8.6 \times 10^{2} \times 10^{1} \times 10^{-3} \text{ cm}^{3}.$$

Using the rules for combining exponents, we obtain

$$V = 1.40 \times 2.16 \times 8.6 \times 10^{2+1-3} \text{ cm}^{3},$$
$$V = 1.40 \times 2.16 \times 8.6 \times 10^{0} \text{ cm}^{3},$$
$$V = 1.40 \times 2.16 \times 8.6 \times 1 \text{ cm}^{3} = 26.0 \text{ cm}^{3}.$$

Estimates and Order-of-Magnitude Calculations

1.5

As we have just seen, it is important to keep track of uncertainties in measurement when calculating answers to problems. Sometimes, in everyday life as well as in science, we need to solve a problem for which we don't have enough information for a precise answer. Often we can obtain a useful answer anyway by estimating the values of the appropriate quantities. These estimates, usually made to the nearest power of ten, are called **order-of-magnitude** estimates. The resulting order-of-magnitude calculation is not exact, but usually it is accurate to within a factor of ten.

Making order-of-magnitude estimates is often easy. For example, imagine you are going to college for the first time and that you want to estimate how much money you need for books. You know that the usual load for most students is five courses and that each course requires a textbook. You can estimate the cost of a single book using the following reasoning. You know from experience that $1 is too little and that $100 too much. Even $10 is low. A reasonable estimate might be $40. Thus the estimated cost of books for one term is $5 \times \$40 = \$200$. Although the result is certainly not accurate, it is the right order of magnitude and provides a reasonable estimate to a real problem.

When making computations of this sort we often make other approximations as well. Replacing π with 3 or replacing $\sqrt{2}$ with $\frac{3}{2}$ makes little difference in the order of magnitude, but doing so greatly simplifies the calculations. The following examples illustrate the technique.

Example 1.5

Order-of-magnitude estimate: How many beans fill the jar?

A record store offers a prize to the customer with the closest guess to the correct number of jellybeans that fill a liter jar on a display counter in the store. (One liter is equal to 1000 cm^3.) Estimate what that number should be.

Solution   A careful look at the jar (Fig. 1.10) reveals several things. The jellybeans can be roughly approximated by little cylinders about 2 cm long by about 1.5 cm in diameter. Furthermore, the jellybeans are not tightly packed; perhaps only as much as 80% of the volume of the jar is filled.

The number of jellybeans is the occupied volume of the jar divided by the volume of a single jellybean,

$$\text{number of beans} = \frac{\text{volume of jar} \times 0.80}{\text{volume of 1 bean}}.$$

The volume of one jellybean is approximated by the volume of a small cylinder,

$$\text{volume of 1 bean} = h\pi \left(\frac{d}{2}\right)^2 \approx 2 \text{ cm} \times 3 \left(\frac{3 \text{ cm}}{4}\right)^2 = \frac{27}{8} \text{ cm}^3.$$

(The symbol $\approx$ means that the quantity on the left is approximately equal to the quantity on the right.) Thus the approximate number of beans in the jar is given by

$$\text{number of beans} \approx \frac{800 \text{ cm}^3}{\frac{27}{8} \text{ cm}^3} \approx 240.$$

An actual count, by one of the authors, of the jellybeans filling a quart jar (0.95 liters) gave 255 beans.

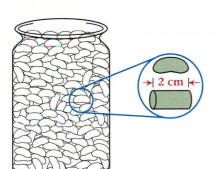

Figure 1.10   Example 1.5: How many jellybeans fill a one-liter jar?

Example 1.6

Order-of-magnitude estimate: Area of U.S. covered by cars.

Approximately what fraction of the area of the continental United States is covered by automobiles?

Solution　To answer this question we need to know three things: the area of the United States, the number of automobiles in the United States, and the average dimensions of an automobile. According to standard reference books, the area of the United States is about 9.4×10^6 km². The number of cars can be estimated from the size of the population, which is about 250 million people. We know that not every person has a car and that many families have more than one. If we estimate that there is about one car for every four people, then we get approximately 60 million automobiles. Although the sizes of automobiles vary, a reasonable approximation for the width is 2 m and for the length is 5 m, which gives an area of 10 m². The fractional area is given by

$$\text{fractional area} = \frac{\text{area covered by autos}}{\text{area of United States}}$$

$$\approx \frac{10 \text{ m}^2/\text{auto} \times 60 \times 10^6 \text{ auto}}{9.4 \times 10^6 \text{ km}^2 \times 10^6 \text{ m}^2/\text{km}^2}$$

$$\text{fractional area} \approx 6 \times 10^{-5}.$$

We estimate that automobiles cover approximately 6×10^{-5} of the area of the United States.

Units and Standards

1.6

When the distance between two objects is measured, that measurement is reported as a number. However, the particular number given depends on the size of the basic unit of measurement. Thus we can state the radius of the earth's orbit as 1 AU, 1.50×10^{11} m, 1.50×10^8 km, or 9.30×10^6 miles. In another example, suppose a surveyor needs to measure the distance between two stakes firmly stuck in the ground. If the distance is 25.4 meters, he must report both "25.4" and "meters." If he reports the distance in inches, he would say "1000 inches."

When the size of a numerical result depends on the units chosen for measurement, we say that the quantity measured has **dimensions**. The separation between the two stakes measured by the surveyor has the dimension of length. Time and mass are other familiar dimensions. In order for measurements made in different places and at different times to have meaning, we must first define **base** or **fundamental units**. Base units are sometimes referred to as standards.

In all of physics there are only seven base units, corresponding to the quantities of length, time, mass, electric current, temperature, the amount of substance, and luminous intensity. Measured quantities are expressed in terms of these units, or combinations of them. For instance, a unit of area is m² and a unit of speed is m/s. Although various systems of units have been used over the years, scientists have generally agreed to use the International System of Units, also known by its international abbreviation, SI. Table 1.3 lists the SI base units, which are largely based on the metric system. In the past, this system of units has also been called the mks system, in reference to the base units of meter, kilogram, and second.

TABLE 1.3
SI base units

Quantity	Name	Symbol
Length	meter	m
Mass	kilogram	kg
Time	second	s
Electric current	ampere	A
Thermodynamic temperature	kelvin	K
Amount of substance	mole	mol
Luminous intensity	candela	cd

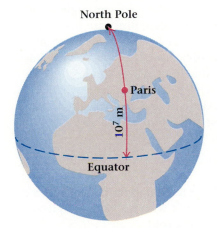

North Pole

Paris

10^7 m

Equator

Figure 1.11 The meter was originally chosen to be one ten-millionth of the distance from the equator to the North Pole along a meridian through Paris.

The basic unit of length, the meter, was approved by the first General Conference on Weights and Measures in 1889. It was defined to be the distance between two fine lines engraved near the ends of a bar of platinum-iridium alloy when the bar was maintained at 0°C. The original international prototype meter, which is still kept at the International Bureau of Weights and Measures in France, was intended to be one ten-millionth of the distance from the equator to the North Pole along a line of longitude through Paris (Fig. 1.11). In 1983 the seventeenth General Conference on Weights and Measures redefined the meter in terms of the speed of light. Details of the current definition of the meter, along with definitions of the other base units, appear at the back of this book, in Appendix B.

It is often more convenient to use the shorter symbol for a unit than to use its full name. Thus we write 10 kg instead of 10 kilograms. Table 1.3 also gives the symbols for the seven base units. Several rules govern the writing and use of these unit symbols: Roman (upright) type, generally lower case, is used for symbols of units; however, if the symbols are derived from proper names, capital roman type is used for the first letter. Unit symbols are not followed by a period and do not change in the plural.

A product of two or more units may be indicated by

$$A \cdot m \quad \text{or} \quad Am.$$

A unit may also be raised to a power, as

$$m^3 \quad \text{or} \quad cm^2.$$

Units formed from two others by division may be indicated by a horizontal line, a solidus (an oblique line, /), or a negative power; for example,

$$\frac{m}{s}, \quad m/s, \quad \text{or} \quad m \cdot s^{-1}.$$

In complicated cases, negative powers or parentheses should be used; for example,

$$m \cdot kg/(s^3 \cdot A) \quad \text{or} \quad m \cdot kg \cdot s^{-3} \cdot A^{-1}.$$

In the last section we introduced the use of scientific notation. In this notation, 1670 meters can be written as 1.67×10^3 meters. But the same quantity can also be written 1.67 kilometers—abbreviated as 1.67 km. This is an example of using a prefix to indicate a decimal multiple of a base unit. Table 1.4 gives the prefixes for other multiples or submultiples of units. Compound prefixes are not used: For example, 1 nm is correct, but not 1 mμm.

TABLE 1.4
SI prefixes

Factor	Prefix	Symbol
10^{18}	exa	E
10^{15}	peta	P
10^{12}	tera	T
10^{9}	giga	G
10^{6}	mega	M
10^{3}	kilo	k
10^{2}	hecto	h
10^{1}	deka	da
10^{-1}	deci	d
10^{-2}	centi	c
10^{-3}	milli	m
10^{-6}	micro	μ
10^{-9}	nano	n
10^{-12}	pico	p
10^{-15}	femto	f
10^{-18}	atto	a

Problem Solving

1.7

Learning physics requires more than just learning new terms and definitions and stating the concepts and laws. To find out what physics is really about, you must learn how to apply these concepts and laws to real or hypothetical situations. Experience has shown that this kind of learning cannot occur without practice. For this reason, in a physics course you may spend more time working problems than you do reading the text.

Even though the examples and problems in this text may sometimes represent an oversimplification of nature—for example, we might ignore the effects of air resistance on a moving body—they illustrate basic ideas, concepts, and techniques that you must master before going on to more complicated situations. The most helpful advice we can offer you is to use the working of problems as a method of studying. When you begin studying a chapter, go over the text examples with pencil and paper, work them out carefully in more detail than the text does, and then proceed to the problems. Although practice is the most important thing, you may find the following general rules to be helpful.

Problem-Solving Guidelines

1. Read the entire problem carefully. Then read it again with the idea of finding out what you are being told. Don't worry about the question at first; focus more on what you are being told.
2. Whenever possible, draw a diagram of the physical situation. Label the diagram with the information given in the problem. Be sure to include units, such as meters or kilograms, with the quantities. If some standard symbols have been introduced, label the parts of the diagram with them, too. For example, you might encounter R for radius, h for height, or L for distance.
3. Only after you are sure you understand what is given and after you have labeled the diagram should you tackle the question. It is often a good idea to briefly write the question, perhaps in symbols, on the paper.
4. The next step is to find a mathematical relationship between the known and unknown quantities. For example, if asked how many dimes make $1.30, you call on your knowledge of the value of a dime. Or, if asked how far an auto goes in 2 hours at 30 kilometers per hour, you call on another type of knowledge. In most cases you will need to write the relationship between the known and unknown quantities in the form of an equation, or perhaps several equations.
5. Next you should solve the equation, or equations, for the unknown quantity or quantities. This means rearranging the formulas in accord with the rules of algebra so that you have an equation with the unknown on the left-hand side of the equals sign and all of the known quantities and constants on the right-hand side.
6. Now, and only now, should you substitute numerical values into the equation. Do not substitute just "bare numbers," but substitute both

the numerical value *and* the units. As indicated in Example 1.6, units are multiplied and divided as if they were algebraic quantities. Your answer should then come out in the appropriate units. If the units do not come out correctly, you have probably made some basic error. On the other hand, if the units are correct, there is a higher probability that your work is correct. Remember that in giving your final answer you must give the proper number of significant figures.

7. As a final check you should consider whether or not your answer is reasonable. Does your result have the proper order of magnitude? You may even carry out a quick order-of-magnitude estimate as a way of confirming your work.

As an example of these ideas, let us work a sample problem.

Example 1.7

Suppose you get into your car and travel 25 km due north. Then you turn and travel an unknown distance east. Finally, you travel 32 km directly back to your starting point. How long was the eastern leg of your trip?

Solution By reading the problem carefully and drawing a diagram similar to Fig. 1.12, you will have covered steps 1, 2, and 3.

For step 4, recall the relationship between the sides of a right triangle, which is what the diagram shows. This relationship is the Pythagorean theorem: The square of the hypotenuse is equal to the sum of the squares of the other two sides. In terms of the symbols on the diagram this theorem can be written

$$c^2 = a^2 + b^2.$$

In this equation the unknown quantity is b and the known quantities are a and c. We may isolate the unknown quantity by subtracting a^2 from both sides:

$$c^2 - a^2 = a^2 + b^2 - a^2 = b^2.$$

This can also be written as

$$b^2 = c^2 - a^2.$$

To complete the solution we take the square root of both sides,

$$b = \sqrt{c^2 - a^2}.$$

The negative solution can be discarded because it is not physically meaningful.

Now, in accord with step 6, we substitute the values for the algebraic symbols:

$$b = \sqrt{(32 \text{ km})^2 - (25 \text{ km})^2}$$
$$b = \sqrt{1024 \text{ km}^2 - 625 \text{km}^2} = \sqrt{399 \text{ km}^2}$$
$$b = 20 \text{ km}.$$

Notice how the units were handled and how many significant figures were kept in the answer. The rounding off was done after all of the calculations had been made.

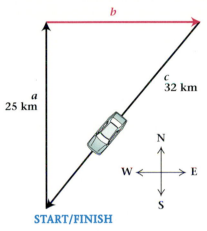

b

a
25 km

c
32 km

N

W ←——→ E

S

START/FINISH

Figure 1.12 **Example 1.7: How long is the distance b?**

Unit Conversions

1.8

Frequently you will need to change from one set of units to another. You may need to change a given number of seconds into minutes, or a given number of inches into centimeters or meters. Adopting a systematic method for doing this will almost ensure error-free conversions. Generally speaking, the rule is to write the conversion factor as a fraction whose value is equal to 1; that is, the numerator (in one set of units) equals the denominator (in the other set of units). Then multiply the quantity to be converted by this fraction. Several brief examples will help illustrate the procedure.

Example 1.8

How many centimeters are there in one foot?

Solution   First we formulate the question mathematically in terms of an unknown quantity x and then make use of the appropriate conversion factor:

$$x \text{ cm} = 1 \text{ ft} = 12 \text{ in.}$$

The foot is defined to be exactly 12 in. The inch is defined to be exactly 2.54 cm:

$$1 \text{ in.} = 2.54 \text{ cm.}$$

To find the conversion factor we divide both sides by 1 inch, obtaining

$$1 = \frac{2.54 \text{ cm}}{1 \text{ in}}.$$

Notice that the quantity on the right-hand side has the value of unity and has no dimensions, since it is equal to the left-hand side. We then return to our equation for x and multiply the right-hand side by 1 in the form of 2.54 cm/1 in.:

$$x \text{ cm} = 12 \text{ in.} \, \frac{2.54 \text{ cm}}{1 \text{ in.}} = 30.48 \text{ cm.}$$

We keep all of the significant figures here because the units are defined exactly. Notice that when you use this procedure, the correctness of the units ensures the correctness of the answer.

Example 1.9

How many minutes are there in exactly one and one-half days?

Solution   Using the outlined procedure, we obtain

$$1.5 \text{ days} = 1.5 \text{ days} \cdot \frac{24 \text{ h}}{1 \text{ day}} \cdot \frac{60 \text{ min}}{1 \text{ h}}$$

$$1.5 \text{ days} = 2160 \text{ min.}$$

Example 1.10

An automobile has a speed of 30 mi/h. What is its speed in cm/s?

Solution Following the method used in the preceding examples, we use a succession of conversion factors to change the units to the desired form:

$$30 \, \frac{mi}{h} = \frac{30 \, mi}{1 \, h} \cdot \frac{1 \, h}{60 \, min} \cdot \frac{1 \, min}{60 \, s} \cdot \frac{5280 \, ft}{1 \, mi} \cdot \frac{30.48 \, cm}{1 \, ft}$$

$$30 \, \frac{mi}{h} = 1.3 \times 10^3 \, cm/s.$$

SUMMARY

Useful Concepts

- Angles can be measured in degrees or in radians, where the angle in radians is given by

$$\theta = \frac{s}{r} \quad \text{and} \quad 1 \text{ radian} = 57.3°.$$

- Kepler's laws of planetary motion are:
  1. The orbit of each planet is an ellipse with the sun at one of the foci.
  2. An imaginary line from the sun to a moving planet sweeps out equal areas in equal intervals of time.
  3. The ratio of the square of a planet's period of revolution to the cube of its average distance from the sun is a constant. This constant is the same for all planets: $T^2/R^3 = k$.

- The results of a calculation may not have more significant figures than the least precise of the values used.
- Numbers expressed in scientific notation can be multiplied and divided in the following way:

$$(A \times 10^n) \times (B \times 10^m) = (A \times B) \times 10^{n+m}$$

$$\frac{C \times 10^n}{D \times 10^m} = \left[\frac{C}{D}\right] \times 10^{n-m}.$$

- There are seven SI base units or standards. The names of those for mass, length, and time are kilogram, meter, and second, respectively.
- Solutions to problems are best obtained by drawing a diagram, identifying the given information, and understanding the problem before attempting a solution.
- To convert from one system of units to another, first write the conversion factor as a fraction whose numerator and denominator are physically equal and then multiply this by the quantity to be converted.

Important Terms

You should be able to write the definition or meaning of each of the following:

- radian
- Kepler's laws of planetary motion
- ellipse
- significant figures
- scientific notation
- order of magnitude
- dimension
- base or fundamental unit

QUESTIONS

1.1 In Example 1.1, what assumption did we make about the way light travels?

1.2 Can you think of any reasons for dividing a full circle into 360°?

1.3 Draw a diagram showing how you would measure the angle that a distant person subtends at your eye.

1.4 Draw an ellipse by any method you choose on a piece of graph paper. Assume that a planet travels from A to B in one month, as shown in Fig. 1.5. By counting the squares on the graph paper, construct the areas *ASB* and *CSD* to be equal. Make a rough estimate of the ratio of the distances *CD/AB*.

1.5 Would you expect Kepler's second law to be valid if the line from the sun to the planet had its origin at the center of the ellipse, rather than at one of the foci? Do not attempt a formal proof but give an example to justify your answer.

1.6 Estimate the precision of your watch or a clock. You may do this by using radio or television time signals spaced 24 or 48 hours apart. Where does its precision lie on the graph in Fig. 1.7?

1.7 Discuss the difference in meaning of the three quantities 10 m, 10.0 m, and 10.00 m.

1.8 Make a list of those quantities for which you think it might be helpful to establish standard units. Why?

1.9 Explain the basic idea behind unit conversion.

1.10 Estimate the total weight of the population of the United States.

1.11 Draw an ellipse by the pin-and-string method illustrated on page 13. Do this several times and determine the eccentricity in each case. The eccentricity is defined as

$$e = \frac{f}{a} = \frac{\sqrt{a^2 - b^2}}{a}.$$

PROBLEMS

Section 1.1   The Importance of Accurate Measurement: Tycho Brahe

1.1 A rectangular field is 90 m long by 50 m wide. A man starts at one corner and walks one fourth of the way along the diagonal to the opposite corner. How far is he from the side nearest him?

1.2 Estimate the relative diameter of the sun and moon from the following information: During the total eclipse of the sun, the moon just obscures the sun. The distance from the earth to the moon is 3.84×10^8 m and the distance from the earth to the sun is 1.5×10^{11} m. (You may need to see Section 1.4.)

1.3 A path runs parallel to a brick wall. Between the path and the wall is a tree. A walk of 10.0 ft along the path causes the tree to change position against the background of the bricks by 6 bricks (Fig. 1.13). If each 8.0-in.-long brick is separated from the next by $\frac{3}{8}$ in. of mortar and the tree is 3 ft from the wall, how far is the tree from the path? (The halfway point of the 10-ft walk is the closest point on the path to the tree.)

1.4 A 12-in.-diameter phonograph record rotates about its center by one-quarter turn. (a) Through how many radians has it turned? (b) How far has a point on the rim moved?

1.5 An object 100 m away from an observer subtends a vertical angle of 0.020 radians at the observer's eye. How tall is the object?

1.6 A protractor is made so that its scale is 7.5 cm from

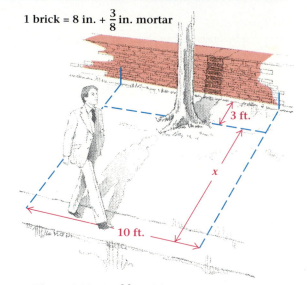

1 brick = 8 in. + $\frac{3}{8}$ in. mortar

3 ft.

x

10 ft.

Figure 1.13   Problem 1.3.

the center point. If the scale is marked in degrees, how far apart are the marks along the edge?

1.7 Phonograph turntables commonly operate at $33\frac{1}{3}$ and 45 revolutions per minute. Express these angular speeds in radians per second.

1.8 Through how many radians does the earth turn about its axis in one hour? In one day?

1.9 The moon rotates so that the same face is always toward the earth. Through how many degrees does the moon rotate about its own axis in one hour? The moon turns on its axis once every $27\frac{1}{3}$ days.

Section 1.2   From Data to a Model: Kepler's Laws

1.10 Using the pin-and-string method shown on page 13, construct an ellipse similar to the one shown in Fig.

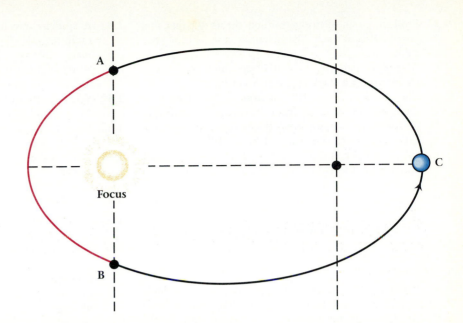

Figure 1.14 Problem 1.10.

1.14. Let this ellipse represent the orbit of some planet about the sun of another solar system. Assume that a planet goes from A to B in the figure in one interval of time; by using Kepler's first law, find how far it will travel in the same time starting from C. (*Hint:* You may want to do this on graph paper and count squares to get areas.)

1.11 What is the value of k in Eq. (1.2) for the earth if the period is expressed in years and the distance in AU? What is its value if the period is expressed in seconds and the distance expressed in meters?

1.12 Calculate the distance from the sun to Jupiter, given the information that Jupiter's period of revolution about the sun is 11.86 years.

1.13 An amateur astronomer claims that a new planet is seen beyond the orbit of Pluto and with a period of 230 years. Can this claim be true? Assume the orbit to be nearly circular.

1.14 Use the known period of $27\frac{1}{3}$ days for the motion of
• the moon about the earth and the distance from the earth to the moon of 3.84×10^8 m to calculate the radius of the orbit of an earth satellite that stays above the same point on the equator. (*Hint:* Use Kepler's third law.)

Section 1.3   The Measurement of Time

1.15 A clock loses 3 s per day. By how much will it be off at the end of one year (365 days)?

1.16 How many revolutions does the second hand of a clock make in three years? Assume no leap years in the interval.

Section 1.4   Measurements, Calculations, and Uncertainties

1.17 Calculate the volume of the rectangular board (Fig. 1.15) with height 6.5 cm, width 31.4 cm, and length 115 cm. Remember the rule regarding significant figures.

Figure 1.15 Problem 1.17.

1.18 Calculate the volume of a rectangular cereal box of height 23.5 cm, width 5.5 cm, and length 16.7 cm.

1.19 If you measure the sides of a square to be 10 cm with an accuracy of $\pm 1\%$, what is the area of the square and how many significant figures may you give in your answer?

1.20 Find the number of seconds in a year and express your answer to two significant places using scientific notation.

1.21 A rectangular file cabinet has a height of 133 cm; a width of 37.5 cm, and a length of 72.0 cm. Express its volume in scientific notation.

1.22 The area of the United States is about 9.4×10^6 km^2 and there are 2.5×10^8 people living in it. What is the population density in people per km^2?

1.23 Find the price of twelve million sheets of paper that cost 0.18 cent each. Use scientific notation for your calculation.

1.24 A sheet of paper is 8×10^{-3} cm thick. How many sheets are in a stack of paper 4 cm high?

1.25 The sun is 1.50×10^{11} m from the earth and the moon is 3.84×10^8 m from the earth. The earth–sun distance is how many times the earth–moon distance?

1.26 The distance to the sun is 1.50×10^{11} m and the speed of light is 3.0×10^8 m/s. How long does it take for light to come from the sun? (A discussion of speed is contained in Chapter 2. However, you should be able to work this problem in the same way that you find the trip time for an automobile of a given speed traveling a specified distance.)

Section 1.5   Estimates and Order-of-Magnitude Calculations

1.27 Estimate the volume of rubber worn from automobile tires each year in the United States. The average radial tire has a useful tread depth of $\frac{5}{16}$ in. and can be driven 35,000 mi before it is worn out.

1.28 How high would the stack reach if you piled one trillion dollar bills in a single stack?

1.29 Approximately how many gallons of gasoline are used by passenger cars each year in the United States?

1.30 Estimate the thickness of the pages in this book. Give your result in millimeters.

Section 1.6   Units and Standards

1.31 Express 2300 m in kilometers and in centimeters.

1.32 One liter (L) is a volume of 10^3 cm^3. How many cubic centimeters are in 2.5 milliliters?

1.33 What is the length of time in minutes of a microcentury?

1.34 How many picoseconds are there in 7.3 microseconds?

1.35 How far will light go in a vacuum in 1 nanosecond? (Speed of light $= 3 \times 10^8$ m/s)

1.36 The black grains in some types of photographic films are about 0.8 μm across. Assume that the grains have a square cross section and that they all lie in a single plane in the film. How many grains are required to completely obscure 1 square centimeter of film?

Section 1.8   Unit Conversions

1.37 What is the height in centimeters of a person who is 5' 8" tall? (1 in. $=$ 2.54 cm).

1.38 How far is 30 mi when expressed in kilometers? (1 mi $=$ 5280 ft, 1 in. $=$ 2.54 cm)

1.39 Express 120 km/h in terms of miles per hour.

1.40 When gasoline sells for $1.069 per gallon, what is the price in dollars per liter? (1 gal $=$ 3.7853 L)

1.41 A student traveling in Hamburg, Germany, finds a radio for sale. The price of the radio is given as 547 marks (DM547). If the exchange rate on that day was $1.00 $=$ DM1.6765, what is the cost of the radio in dollars and cents?

1.42 What is the area in square centimeters of an $8\frac{1}{2}'' \times 11''$ piece of paper?

1.43 How many square kilometers are there in 10 acres? (1 acre $=$ 43,560 ft^2 or 1/640 mi^2)

1.44 What is the volume in cubic yards of a cube 1 m on a side?

Additional Problems

1.45 An airplane flying at a height of 2.3 km (2300 m) takes two photographs of the ground at different points along its path. Each photo shows a tower and it is determined that the tower was directly below the flight path and midway between the points at which the photographs were taken (see Fig. 1.16). In each photograph the top of the tower is lined up with points on the ground known to be 4.0 m from the base of the tower. If the height of the tower is 20 m, how far apart were the points at which the photos were taken?

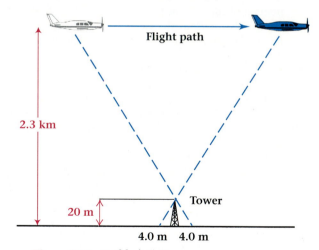

Figure 1.16   Problem 1.45.

1.46 The moon subtends an angle of 9.06×10^{-3} radians and is 3.84×10^8 m from the earth. What is the approximate diameter of the moon?

1.47 A biology student working on an experiment asks you to make a spoked wheel of 36 cm diameter. The student further specifies that the spokes be equally spaced 4.0 cm apart at the rim. Can such a wheel be made?

1.48 Eratosthenes (ca. 276–196 BC) knew that when the
• sun was directly overhead at Syene (modern Aswan)
in southern Egypt, the sun was about 7° away from
directly overhead in Alexandria (Fig. 1.17). Using the
known distance between the two observing points,
he was able to determine the circumference of the
earth. Use a figure of 770 km for the distance to
perform the same calculation.

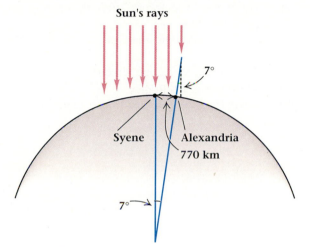

Figure 1.17   Problem 1.48.

1.49 An imaginary planet moves in an elliptical orbit with
• • semimajor axis 9 AU and eccentricity 0.7. Sketch, at
some appropriate scale, such an ellipse on graph paper. The squares of the graph paper will allow you
to determine the areas of various sectors of the ellipse
and give appropriate answers to the following questions. One full revolution is "one year" for that
planet. The eccentricity is

$$e = \frac{f}{a} = \frac{\sqrt{a^2 + b^2}}{a}.$$

(a) In terms of units of area on your sketch, what is
the rate at which the area is swept out? (b) Assume
that the planet's new year starts when the planet is
at the end of the semimajor axis farthest away from
the sun. Find the planet's position at the end of one-
third and two-thirds of a year.

1.50 Calculate the ratio of T^2/R^3 for the planets in the solar
• system. Use the data from Table 1.1. What is the
percentage difference between the highest and lowest
values?

1.51 A digital wrist watch is said to be accurate to within
15 s per month. Is this more, or less, accurate than
the best stationary mechanical clocks? Make the com-

parison by stating the accuracy of several of the timekeepers in Fig. 1.7 in the same terms as for the watch.

1.52 The speed of an automobile is said to be 103.2784
mi/h on a one-quarter-mile course. How well must
the distance be measured for this to be an accurate
statement?

1.53 How many gigaseconds are there in a century?

1.54 What fraction of a kilogram is a milligram?

1.55 In some countries the gasoline consumption of an
• automobile is expressed in liters consumed per 100
km of travel. If an automobile gets 25 mi/gal, what
is its fuel consumption in liters per 100 km?
(1 gal = 3.7853 L)

1.56 What is the volume in cubic centimeters of a cube
1.00 in. on a side?

1.57 Draw two ellipses using the technique of string and
• pins shown on page 13. Draw one that is nearly circular and one that is more oblong. Measure the axes
and determine the eccentricity of each. Recall that
the eccentricity is

$$e = \frac{f}{a} = \frac{\sqrt{a^2 + b^2}}{a}.$$

1.58 The rectangular floor of a gymnasium has sides of
• • length $x \pm \Delta x$ by $y \pm \Delta y$, where Δx and Δy are the
estimated measurement uncertainties and are small
compared to x and y. Show by direct computation
that the area of the floor and the uncertainty in that
area are given by

$$xy \pm xy \left(\frac{\Delta x}{x} + \frac{\Delta y}{y} \right)$$

when very small terms, of order $(\Delta x)^2$, are ignored.
(In most cases, this result overestimates the uncertainty in the area, because it does not take into account that the uncertainties in the lengths, Δx and
Δy, come from a series of measurements which have
a spread in their values.)

1.59 A cylindrical milkshake cup has a measured inside
• • radius of $r \pm \Delta r$ and a height of $h \pm \Delta h$. Show that
the volume of the cup is

$$V = \pi r^2 h \pm 2\pi r h \Delta r \pm \pi r^2 \Delta h$$

if very small terms, of order $(\Delta r)^2$, are ignored.

1.60 Assuming that the orbits of the planets are circular,
• • show that the product of the orbital radius with the
square of the speed of a planet is the same for all
planets. The speed is the distance traveled divided by
the time required to go that distance.

1.61 Compute the period of an artificial satellite orbiting
• • the earth at an orbit radius that is one-half the radius
of the moon's orbit. Give your answer as a fraction
of the lunar period.

ADDITIONAL READING

Astin, A. V., "Standards of Measurement." *Scientific American,* June 1968, p. 50.

Boorstin, D. J., *The Discoverers.* New York: Random House, 1983. A broad-ranging and interesting account of the origins of clocks and calendars as well as of the telescope.

Bronowski, J., *The Ascent of Man.* Boston: Little, Brown, 1973. The chapters titled "The Starry Messenger" and "The Majestic Clockwork" trace the development of our views of the universe, space, and time.

Gingerich, O., "Copernicus and Tycho." *Scientific American,* December 1973, p. 86.

Hellwid, H., K. M. Evenson, and D. J. Wineland, "Time, Frequency and Physical Measurement." *Physics Today*, December 1978, p. 23.

Wilson, C., "How Did Kepler Discover His First Two Laws?" *Scientific American,* March 1972, p. 92.

2

Motion in One Dimension

2.1   Speed

2.2   Average Velocity

2.3   A Graphical Interpretation of Velocity

2.4   Instantaneous Velocity

2.5   Acceleration

2.6   Motion with Constant Acceleration

2.7   Galileo and Free Fall

A WORD TO THE STUDENT

This chapter introduces you to a simple description of motion—one-dimensional motion, or motion along a straight line. Here we give precise mathematical meaning to several terms such as velocity and acceleration and derive the relationships between them. A thorough mastery of these ideas is necessary for understanding the following chapters, which discuss the motion of objects in two and three dimensions. Because we describe the motion of molecules and smaller bodies in much the same way we describe larger objects, the contents of this chapter are also needed in our later chapters on gases, molecules, atoms, and nuclei. As a matter of fact, most chapters in this book will make some use of the contents of this first introduction to motion.

A large part of our everyday experience, as well as our scientific experience, concerns things that move. For this reason, the study of motion is one of the most basic studies in physics. The study of motion is divided into two parts: kinematics and dynamics. The word *kinematics* is derived from the Greek word *kinema*, meaning "motion" — the same root from which we get the word *cinema*. **Kinematics** describes the positions and motions of objects in space as a function of time but does not consider the causes of motion. The study of the causes of motion is called **dynamics**.

Separating the study of motion into kinematics and dynamics is a great aid in understanding the subject. This distinction was not always present, having arisen only in the 1300s. It was one of the early steps in the development of modern **mechanics**, which is the branch of physics that deals with the kinematics and dynamics of macroscopic (large-scale) objects. In this chapter, and the next several chapters, we present the basic principles of mechanics as developed by Galileo, Newton, and others.

Kinematics provides the means for describing the motions of such varied things as planets, golf balls, and subatomic particles. Because of its precision and generality, mathematics is the natural language for kinematics. Consequently, this chapter contains many equations. However, the ideas and techniques of kinematics that you learn here are used throughout the text. The causes and effects of motion are discussed for many seemingly different situations, but the techniques used are the same. The range of these applications runs from gravity to electricity and magnetism to nuclear physics.

We begin our study of kinematics by considering motion in only one dimension. This restriction has the advantage of introducing almost all of the necessary concepts in their simplest form. In the next chapter we will learn how to apply these concepts to two- and three-dimensional motion.

Speed

2.1

Kinematics developed, in part, out of the desire to describe the apparent motion of the planets and stars. The Greek Pythagoreans were the first to give a purely geometric interpretation to the motions of the heavenly bodies. Most of the concepts they developed have been rephrased in modern mathematical expressions. The Greek ideas formed a basis for the study of motion; however, they were insufficient because they did not rely on experimental observations.

Later in this chapter we will see that the major act of linking a theory of motion with experiment was first accomplished by Galileo Galilei (1564–1642). Galileo, as he is universally known, played a uniquely pivotal role in history. His breakthrough way of linking theory with experiment influenced all subsequent scientific thought, and, to some extent, nonscientific thought as well. Galileo's work on the kinematics of falling bodies is important for at least two reasons. First, it is the study of a common example of how nature works. Second, but no less important, his work illustrates the modern view of how one should arrive at scientific conclusions. It is an example of what is sometimes called the scientific method.

Between three and four hundred years before Galileo, the works of Aristotle (384–322 BC) had been reintroduced into the Western world by Arabic writers. From that time onward, Aristotle's work was regarded as the principal source of scientific knowledge. He had written about many different topics in nature, including the observation that stones fall more quickly than feathers. (We will examine Galileo's experiments on falling bodies later, after first presenting some modern definitions and relationships.) Aristotle's entire system of knowledge was revered as correct, unchanging, and absolute. Galileo regarded this contemporary faith in Aristotle as wrong, not because Aristotle was in error—though he sometimes was—but because of the inflexible way in which the system was regarded as dogma.

Autolycus of Pitane (c. 310 BC) first defined what we now call **constant speed**. In modern terms his statement is: The speed of a body is constant when it travels equal distances in equal periods of time. It is worth emphasizing that this definition refers to neither the size nor shape nor mass nor any other property of the moving body, nor to how the body is influenced by its surroundings. The definition deals only with the motion itself. In the same way, other definitions in kinematics are restricted to properties of the motion only.

Most people are familiar with the concept of speed and know that the speed of an object is measured in units such as miles per hour, kilometers per hour, or feet per second. The speed is the ratio of the distance traveled to the time required for the travel. We define the **average speed** as the total distance, s, traveled during a particular time divided by that time interval, t:

$$\text{average speed} = \frac{\text{distance traveled}}{\text{time interval}} = \frac{s}{t}. \tag{2.1}$$

If the average speed is the same for all parts of a trip, then the speed is constant.

Example 2.1

A cheetah can run short distances at top speeds of up to 31 m/s. How far would the cheetah shown in Fig. 2.1 travel in 3.2 s when running at top speed?

Solution If we multiply both sides of the definition of average speed (Eq. 2.1) by the time, we get the distance traveled:

$$\text{distance traveled} = \text{average speed} \times \text{time}.$$

We can insert the values for average speed and time to give

$$\text{distance} = 31 \text{ m/s} \times 3.2 \text{ s}.$$

Upon carrying out the arithmetic and multiplying the units we find the

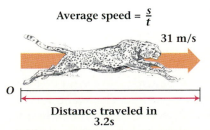

Average speed = $\frac{s}{t}$

31 m/s

O

Distance traveled in 3.2s

Figure 2.1 Example 2.1: How far can the cheetah run in 3.2 s?

distance traveled to be

$$\text{distance} = 99 \text{ m}.$$

Note that for distance to be a meaningful answer, it has to be more than just a number. It also has to have units of length—in this case, meters.

Example 2.2

Finding average speed.

On a clear October day, two students take a three-hour automobile trip to enjoy the fall foliage. In the first two hours, they travel 100 km at a constant speed. In the third hour they travel another 80 km, at a different constant speed. What is the average speed for each segment and for the entire trip?

Solution   For the first portion of the trip, the average speed is the distance traveled divided by the time:

$$\text{average speed} = \frac{\text{distance traveled}}{\text{time}}.$$

$$\text{average speed (1)} = \frac{100 \text{ km}}{2 \text{ h}} = 50 \text{ km/h}.$$

For the second portion of the trip, the average speed is given by the same formula but now with a distance of 80 km and a time of 1 h:

$$\text{average speed (2)} = \frac{80 \text{ km}}{1 \text{ h}} = 80 \text{ km/h}.$$

For the entire trip, the average speed is the total distance divided by the total time interval:

$$\text{average speed (total)} = \frac{180 \text{ km}}{3 \text{ h}} = 60 \text{ km/h}.$$

Note that the average speed for the entire trip, in this case 60 km/h, is *not*, in general, the same as the direct average of the individual speeds, which in this case is 65 km/h.

Average Velocity

2.2

As noted before, in this chapter we limit our study to motion along a line or in one dimension. This limitation makes our definition of speed and related concepts easier to understand. However, in reality, motion is usually not restricted to one dimension and we must take account of the direction as well as the speed of an object's motion. The name for the quantity that describes both the direction and speed of motion is **velocity**. We will consider two- and three-dimensional motion in the next chapter, after we discuss the general way of describing quantities with both direction and magnitude. Even though we are considering only one-dimen-

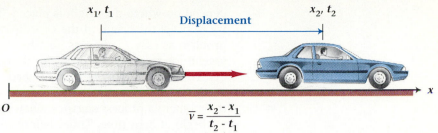

Figure 2.2   **When the car moves from point x_1 to point x_2, its displacement is given by $x_2 - x_1$.**

sional motion in this chapter, we must still take account of direction (for example, positive versus negative, or east versus west), so we will use the term velocity.

In determining a velocity, we refer to a starting point, or initial position and a destination point, or final position. That is, we set up a system of coordinates. Within this coordinate system is a zero point from which all positions are measured. Frequently, the zero point is chosen as the starting point for any motion. The change in position, including the sign of the change, is called the **displacement**. When something moves from one location to another, we say it undergoes a displacement.

Suppose a car is located at point x_1 at a time t_1, and at another point x_2 at a later time t_2 (Fig. 2.2). Then the car's **average velocity** $\bar{v}$ over the time interval is

$$\bar{v} = \frac{\text{final position} - \text{initial position}}{\text{final time} - \text{initial time}} = \frac{x_2 - x_1}{t_2 - t_1}. \tag{2.2}$$

In general, a bar over a symbol (as in $\bar{v}$) indicates the average value of that quantity, in this case the average velocity. The difference between speed and velocity is more than just an algebraic sign; it involves the difference between the total distance traveled (for speed) and the net change in position (for velocity). These two quantities are not necessarily the same. For example, a trip of 30 km away and 30 km back gives a total distance traveled of 60 km, but a net change in position of zero.

In the preceding paragraphs we used the terms initial position and final position, measuring them from the zero point of our reference system of coordinates. So far we have used a coordinate system that is just the x or y axis with the origin (zero) placed at some convenient point. Most often we have chosen the zero to be the starting point for the motion described. Notice that we can have both positive and negative positions measured from the origin. For example, in one dimension, if we define positions measured to the right of the origin to be positive, then positions measured to the left of the origin are negative. In this case, a positive velocity indicates motion to the right; a negative velocity indicates motion to the left.

Usually, when we are faced with a physical problem, we are free to choose the coordinate system and place its origin wherever we prefer. The physical locations of the fox and the rabbit shown in Fig. 2.3, for instance, do not depend on where you place the origin of the coordinate system. However, the position as *measured* within your coordinate system does depend on where the origin is placed. Two points are worth noting. First, try to place the origin of the coordinates in a position that will make the problem as easy to solve as possible. Second, having chosen a coordinate system for working the problem, do not change it during the solution.

The same argument may be made for time, the other basic quantity needed for kinematics. We often say that we "start the clock running" at the moment a process starts or when some particular event happens. Thus we begin at zero time. Times before that would be negative and all times after the beginning would be positive.

Example 2.3

Speed is not the same as velocity.

A homing pigeon is released from under the center of the arch in St. Louis. (a) What is its average velocity: (i) If it travels 50 km due east from the arch in one hour? (ii) If, instead, it travels 50 km west from the arch in one hour? (iii) If it starts 10 km east of the arch, travels to 20 km east of the arch, and then travels to 30 km west of the arch in one hour? (b) What would the speed be in each of these cases?

Solution   (a) From the statement of the problem, we see that all of the motion is confined to a straight line along the east–west direction. We begin by making a diagram to help visualize the situation (Fig. 2.4). After making the drawing we must choose which direction is positive. Here we choose east to be the positive direction and west to be negative. To find the average velocity in each case we use

$$\bar{v} = \frac{\text{final position} - \text{initial position}}{\text{final time} - \text{initial time}} = \frac{x_2 - x_1}{t_2 - t_1}.$$

Notice that the duration given by the difference between initial and final

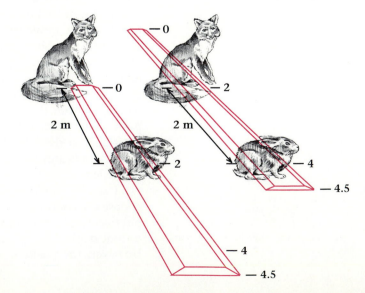

Figure 2.3   The location of the fox and the rabbit and, hence, the distance between them does not depend on the location of the origin of the coordinate system used to make the measurements.

times is the elapsed time:

$$\bar{v} = \frac{\text{final position } - \text{ initial position}}{\text{time elapsed}}.$$

So, for the three cases,

$$\text{(i) } \bar{v} = \frac{50 \text{ km} - 0 \text{ km}}{1 \text{ h}} = +50 \text{ km/h},$$

$$\text{(ii) } \bar{v} = \frac{-50 \text{ km} - 0 \text{ km}}{1 \text{ h}} = -50 \text{ km/h},$$

$$\text{(iii) } \bar{v} = \frac{-30 \text{ km} - 10 \text{ km}}{1 \text{ h}} = -40 \text{ km/h}.$$

(b) The average speed is found from

$$\text{average speed} = \frac{\text{distance traveled}}{\text{time elapsed}}.$$

For the first two cases,

$$\text{(i) average speed} = \frac{50 \text{ km}}{1 \text{ h}} = 50 \text{ km/h},$$

$$\text{(ii) average speed} = \frac{50 \text{ km}}{1 \text{ h}} = 50 \text{ km/h}.$$

For the third case, the distance traveled is 10 km east plus 50 km west, for a total distance of 60 km. Thus

$$\text{(iii) average speed} = \frac{60 \text{ km}}{1 \text{ h}} = 60 \text{ km/h}.$$

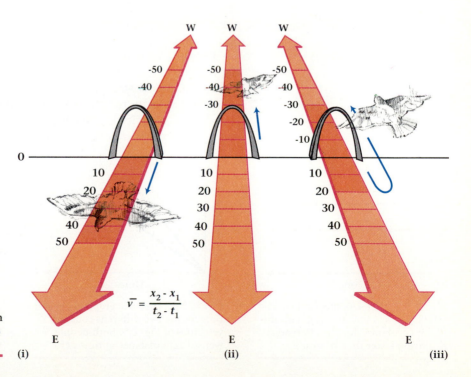

Figure 2.4 Example 2.3: Starting at the arch, (i) the pigeon flies 50 km east and (ii) the pigeon flies 50 km west. (iii) The pigeon starts 10 km east of the arch, flies first to a position 20 km east of the arch, and then flies to a position 30 km west of the arch.

Notice that the average speed is always a positive number. In situation (iii) the magnitude of the displacement and the total distance traveled are not the same. Consequently the numerical value of the average speed is not the same as that of the average velocity.

A Graphical Interpretation of Velocity

2.3

We now supplement our algebraic definition of velocity in Eq. (2.2) with a graphical interpretation of velocity. Consider two people, one running and one walking with constant velocity (Fig. 2.5). We plot the elapsed time along the horizontal axis, or **abscissa**, and the velocity along the vertical axis, or **ordinate**. (It is conventional to plot the independent variable along the abscissa and the dependent variable along the ordinate as we have done here.) The point on the graph that represents the velocity at any time traces out a smooth line as time goes by. In this case the horizontal line v_A represents the particular constant velocity of the runner. A slower-moving person, but one with constant velocity also, gives rise to the horizontal line v_B.

In Fig. 2.6, we have plotted the distance traveled against the elapsed time for the two cases in Fig 2.5. Our definitions of both velocity and speed show that the distance traversed at constant velocity or speed is directly proportional to the time. When one variable is directly proportional to the other, as *distance* is to *time* in this example, a straight line results. For this reason, such relationships are called *linear*. If we take a case in which the distance is zero at time zero, we have a line like *A*. A graph due to a lower but constant velocity gives rise to another straight line, such as the line *B*.

In Fig. 2.7 we have plotted distance against time for a case in which the velocity is not constant. Over the portion of the trip between times t_1 and t_2, the line is straight and the velocity is constant. Between times t_2 and t_3, the velocity remains the same and is still constant. Between times t_4 and t_5 the velocity is constant, but is not the same as between times t_1 and t_2 or between t_2 and t_3. The velocity is not constant for the time

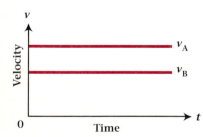

Figure 2.5 Velocity graphed against time for two people moving with different constant velocities. Person A is moving faster than person B.

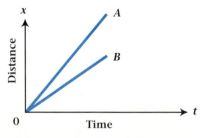

Figure 2.6 Distance graphed against time for the two different constant velocities of Fig. 2.5. Both people were at zero distance at time zero.

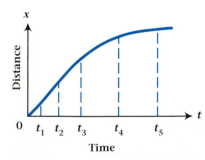

Figure 2.7 Distance graphed against time for varying velocity.

Figure 2.8   The velocity is the slope of the position–time curve. For a straight line, the slope $\Delta x/\Delta t$ is constant and is independent of the particular choice of time interval. This situation corresponds to constant velocity.

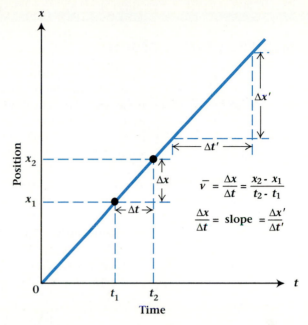

$$\bar{v} = \frac{\Delta x}{\Delta t} = \frac{x_2 - x_1}{t_2 - t_1}$$

$$\frac{\Delta x}{\Delta t} = \text{slope} = \frac{\Delta x'}{\Delta t'}$$

interval between t_3 and t_4, nor is it constant overall between times t_1 and t_5. However, we can define the average velocity for the time interval t_3 to t_4, t_1 to t_5, or any other interval.

To use our graphical method further, we need a more formal definition for the slope of a line. Suppose an object moves with constant velocity so that its position–time graph looks like Fig. 2.8. If the object was at point x_1 at time t_1, and at another point x_2 at a later time t_2, then the net change in position is given by $\Delta x = x_2 - x_1$. The symbol Δ (the Greek capital letter delta) is used to indicate a change in something—in this case, the position x. Thus Δx (delta ex) is the change in x. Similarly, the change in time is given as Δt (delta tee) $= t_2 - t_1$. The average velocity of the object during the time interval from t_1 to t_2 is then written

$$\bar{v} = \frac{x_2 - x_1}{t_2 - t_1} = \frac{\Delta x}{\Delta t}.$$

As we said earlier, the change in position $\Delta x = x_2 - x_1$ is the displacement of the moving object, including the magnitude of the change and its direction. Notice that the position of an object is the same as its displacement from the origin, or zero, of the coordinate system.

An alternative way of defining velocity is to say that velocity is the slope of the position–time curve. The **slope of a line** is defined to be the ratio of the change in the line's ordinate to the corresponding change in the abscissa. For the case of constant velocity (Fig. 2.8), the position–time curve is a straight line and the slope is $\Delta x/\Delta t$. Thus we can determine the velocity from the slope of the line. Since the distance traveled is proportional to the time, we get the same numerical value for this ratio, $\Delta x/\Delta t$, no matter what interval of time we choose to consider. Thus in Fig. 2.8 the ratio $\Delta x/\Delta t$ is equal to the ratio $\Delta x'/\Delta t'$. This behavior is characteristic of constant velocity. Notice that the slope, and therefore the velocity, has the dimensions of length/time.

Instantaneous Velocity

2.4

Suppose that a runner's velocity is not constant, but changes as illustrated in the displacement–time graph in Fig. 2.9. Now that the line is curved rather than straight, how do we find the velocity at a particular instant of time? To determine the velocity at point A, for instance, we draw a tangent to the curve at that point. Then the slope of that tangent, which is a straight line, is determined as previously described; it is given by $v_A = \Delta x_A/\Delta t_A$. The subscripts refer to the line tangent to the curve at point A. The velocity measured at any given moment (for example, at point A) is called the **instantaneous velocity**.

At the point of contact, the tangent line is parallel to the displacement–time curve. The average velocity over the time interval Δt that contains A approaches the value of the instantaneous velocity as the interval Δt becomes smaller. We can therefore define the instantaneous velocity as the limiting value of $\Delta x/\Delta t$ as Δt becomes vanishingly small. In symbols, the idea is

$$v = \frac{\Delta x}{\Delta t} \qquad \text{(where } \Delta t \text{ approaches zero).} \qquad (2.3)$$

The slopes at other points, such as B and C in Fig. 2.9, are determined in the same way. Notice that in the immediate neighborhood of point C the

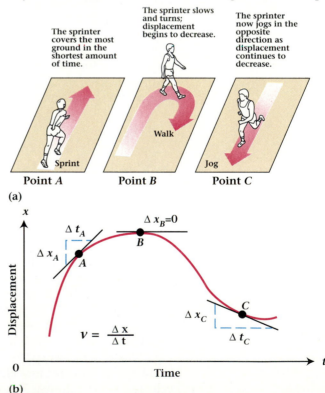

The sprinter covers the most ground in the shortest amount of time.

The sprinter slows and turns; displacement begins to decrease.

The sprinter now jogs in the opposite direction as displacement continues to decrease.

Walk

Sprint

Jog

Point A   **Point B**   **Point C**

(a)

Figure 2.9   The slope of a curved line is determined at any point by the slope of the line tangent to the curve at that point. The steepness of the slope corresponds to the magnitude of the velocity. Notice that since Δx_C is negative, the slope at point C is negative, corresponding to velocity in a negative direction (back toward the starting point).

x

Δt_A

$\Delta x_B = 0$

B

Δx_A

A

Displacement

C

Δx_C

$v = \dfrac{\Delta x}{\Delta t}$

Δt_C

0

Time

t

(b)

displacement from the starting point is decreasing as time increases. Therefore, Δx_C is negative. The slope and the instantaneous velocity are both negative at point C.

Example 2.4

Velocity of a sprinter.

At the start of a 100-m race a sprinter is poised for action. When the starter fires his gun, the sprinter pushes off the starting block and quickly reaches maximum velocity. The graph of Fig. 2.10 shows the displacement of a sprinter as a function of the time elapsed after the starting signal. Determine the instantaneous velocity of the sprinter at 0.5 s, 1.5 s, and 2.5 s.

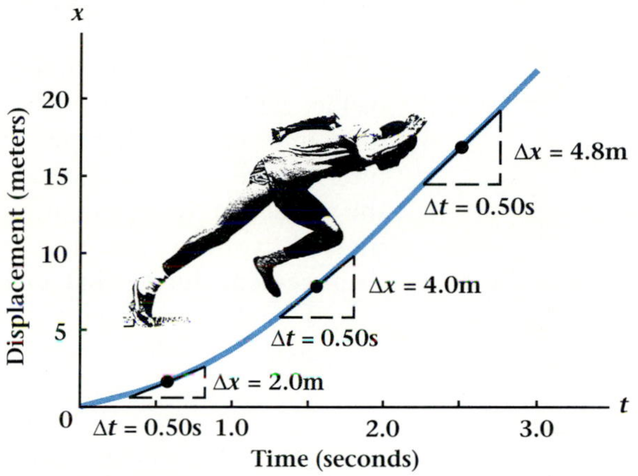

Figure 2.10   Example 2.4: Graph of displacement versus time for a sprinter. The slope of the curve at any time is measured from the line tangent to the curve at that time. The tangents shown correspond to times of 0.5, 1.5, and 2.5 s.

Solution   The results can be read directly from the graph in Fig. 2.10. We have drawn tangent lines through the points corresponding to $t =$ 0.5 s, 1.5 s, and 2.5 s.

At $t = 0.5$ s we have

$$v(t = 0.5 \text{ s}) = \frac{\Delta x}{\Delta t} = \frac{2.0 \text{ m}}{0.50 \text{ s}} = 4.0 \text{ m/s}.$$

At $t = 1.5$ s,

$$v(t = 1.5 \text{ s}) = \frac{4.0 \text{ m}}{0.50 \text{ s}} = 8.0 \text{ m/s}.$$

At $t = 2.5$ s,

$$v(t = 2.5 \text{ s}) = \frac{4.8 \text{ m}}{0.50 \text{ s}} = 9.6 \text{ m/s}.$$

In this example, we see that the velocity changes with time. The velocity increases rapidly as the runner springs from the starting block, but begins to level off as he approaches maximum velocity.

The concept of instantaneous velocity was developed by a group of scholars at Merton College, Oxford, England, in the first half of the fourteenth century. Another development of that time was the use of a graphical technique to represent kinematic quantities. In addition to being used

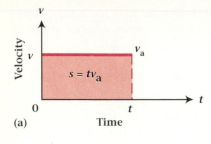

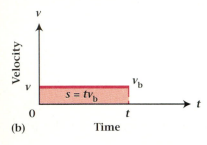

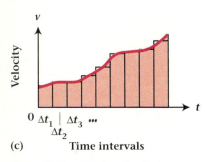

Figure 2.11 Graphs of velocity versus time for three moving objects. (a), (b) Constant velocity. (c) Time-varying velocity. The area under the curve is equivalent to the displacement.

for specific examples, graphical techniques can also help us obtain general relationships between variables. Let us consider one example of such a procedure. The example is specific to kinematics, but we will use the idea and technique many times again.

Figures 2.11(a) and 2.11(b) are velocity–time graphs for an object moving at constant velocity. The length of a vertical line from the horizontal axis to the line v_a (or v_b) is equal to the velocity. The line v_a is higher than the line v_b, indicating a greater velocity for the object represented in Fig. 2.11(a). Note that the area of the rectangle under the line v is given by tv, where t is the elapsed time and v is the velocity. However, this quantity is also the net distance traveled, s, which is the displacement. We can extend this observation to state a general principle: *The area under any velocity–time curve between two times is equivalent to the displacement during that time interval.*

To see that this last statement is true, consider the velocity–time graph of Fig. 2.11(c), for which the velocity is not constant. We can divide the total time into small time intervals of duration Δt. Over each of these intervals we may approximate the velocity by a constant value, given by the average of the initial and final velocities for the interval. Then for each interval, we have, as before, a rectangle whose area is equivalent to the displacement during that interval. The total displacement is then represented by the sum of all the areas of these rectangular strips, that is, the total area under the velocity–time curve. Thus

$$\text{total displacement} = \bar{v}_1 \Delta t_1 + \bar{v}_2 \Delta t_2 + \bar{v}_3 \Delta t_3 + \cdots,$$

where $\bar{v}_1$ is the average velocity during the first interval, $\bar{v}_2$ is the average velocity during the second interval, and so on.

If the intervals Δt are too long, or if the velocity changes too much during each interval, this approximation is unsatisfactory because we assume that the velocity is constant for each time interval Δt. However, if we make the intervals smaller and smaller, the approximation becomes more accurate. Also, as the intervals become smaller, more of them fit into a given range. The techniques of calculus, but not algebra, allow us to make the intervals as small as we want—even approaching zero time duration—and still keep track of the areas of the tremendous number of strips. We will not use calculus techniques in this text, but we will make use of two basic concepts from calculus: the meaning of the slope of a curve and that of the area under a curve.

Since the area under the velocity–time curve always represents the displacement of the moving body, we can deal with cases of nonuniform velocity. Figure 2.12(a) represents a case in which the velocity decreases from some initial value, v_0, to zero in a linear way. This graph might represent the slowing down and stopping of a car when you apply the brakes. The area under the curve, and therefore the distance traveled, is easily determined by noting that the required area is that of a right triangle. The area of the triangle is one half the height times the base, so distance traveled $= \frac{1}{2}v_0 t$.

Instead of starting with a car in motion and smoothly bringing it to rest, we could consider a car initially at rest and smoothly increase its velocity to some final value v_f (Fig. 2.12b). The distance traveled during

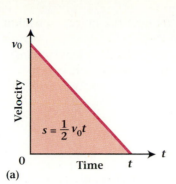

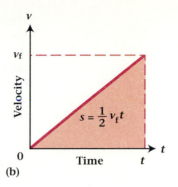

Figure 2.12 Calculating displacement from velocity–time graphs. (a) Velocity decreasing uniformly with time. (b) Velocity increasing uniformly with time.

the time the velocity is increasing is given by $s = \frac{1}{2}v_f t$. Galileo found that the motion of falling bodies follows this same behavior. We conclude that objects fall with a velocity that increases uniformly with time.

Example 2.5

Traveling on the German Autobahn in your BMW at the modest velocity of 120 km/h, you smoothly brake to a halt in 12 s. How far did you travel after first applying the brakes?

Solution   Using the result from Fig. 2.12(a), we have

$$s = \tfrac{1}{2}v_0 t = \tfrac{1}{2}(120 \text{ km/h})(12 \text{ s}),$$

$$s = \frac{120 \text{ km} \times 12 \text{ s}}{2 \text{ h}} \times \frac{1 \text{ h}}{60 \text{ min}} \times \frac{1 \text{ min}}{60 \text{ s}},$$

$$s = 0.20 \text{ km} = 200 \text{ m}.$$

Figure 2.13 The average velocity of the clock pendulum is zero; however, its average speed is not zero.

In Section 2.1 we defined average speed as the total distance traveled divided by the elapsed time. For example, if a young architect drives to the airport to pick up a client, makes a round trip of 120 km, and returns to the home office in 2 h, the average speed would be 60 km/h. On the other hand, the term *speed* is also used to mean the magnitude of the instantaneous velocity. Thus if one car travels east at 60 km/h and another travels west at 60 km/h, both have the same speed but different velocities because they are traveling in different directions. However, the average speed is not the magnitude of the average velocity. An object that oscillates back and forth through the origin, like the pendulum of the clock shown in Fig. 2.13, has an average velocity of zero because it spends equal times with equal positive and negative velocities. But its average speed is not zero.

Acceleration

2.5

In Section 2.2 we defined the average velocity of an object as its change in position divided by the time elapsed,

$$\bar{v} = \frac{\Delta x}{\Delta t}.$$

This tells us how the object's position changes with time. From our discussion in the last section, it is reasonable to define a quantity that indicates how the object's velocity changes with time. We define the **average acceleration**, $\bar{a}$, as the change in velocity divided by the time required for the change. The average acceleration can be written as

$$\bar{a} \equiv \frac{v_2 - v_1}{t_2 - t_1} = \frac{\Delta v}{\Delta t}. \tag{2.4}$$

The following two examples illustrate the meaning of average acceleration and the use of Eq. (2.4).

Example 2.6

Average acceleration.

A bicyclist starts from rest and increases her velocity at a constant rate until she reaches a speed of 2.0 m/s in 5.0 s (Fig. 2.14). What is her average acceleration?

Solution   We are given a change in velocity over a time interval, so we employ Eq. (2.4):

$$\bar{a} = \frac{2.0 \text{ m/s} - 0 \text{ m/s}}{5.0 \text{ s}} = 0.40 \frac{\text{m/s}}{\text{s}},$$

or

$$\bar{a} = 0.40 \text{ m/s}^2.$$

Notice the units of acceleration. Again we have a derived unit that is different from any previously defined quantity.

Example 2.7

A motorcyclist starts from rest and accelerates in one direction to a constant acceleration of 4.0 m/s² for 10 s (Fig. 2.15). What is the velocity of the rider at the end of the 10 s?

Solution   Here we are given a constant acceleration and a time interval, and need to find the change in velocity. We rewrite Eq. (2.4) in the form

$$\bar{v}_2 = \bar{a}(t_2 - t_1) + v_1.$$

Inserting the values of $\bar{a} = 4.0$ m/s², $t_2 - t_1 = 10$ s, and $v_1 = 0$ gives

$$v_2 = 4.0 \text{ m/s}^2 \times 10 \text{ s} + 0 = 40 \text{ m/s}.$$

Again notice the units. In working problems you should always insert the units and perform the algebraic operations on them as well as on the numbers. After all, a number without any units is just a number. But if it represents a physical quantity, it should be accompanied by the appropriate units.

A graphical approach offers further insight into the relationships between acceleration, velocity, time, and distance. From the definition of

Figure 2.14   Example 2.6: A bicyclist accelerating from rest at a constant rate reaches a speed of 2.0 m/s in 5 s.

Figure 2.15   Example 2.7: A motorcyclist accelerates at 4.0 m/s² for 10 s.

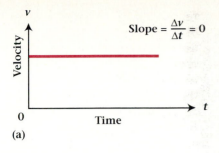

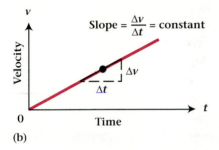

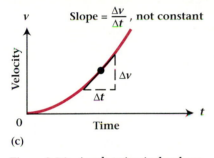

Figure 2.16 Acceleration is the slope of the curve of velocity versus time. (a) Zero acceleration. (b) Constant nonzero acceleration. (c) Variable acceleration.

average acceleration in Eq. (2.4) we see that acceleration may also be described as the slope of the curve of velocity versus time. In Fig. 2.16(a) the horizontal line (zero slope) corresponds to an object moving with constant velocity and therefore zero acceleration. In Fig. 2.16(b) the slope of the line is everywhere the same, indicating that the velocity is increasing at a constant rate. Therefore the acceleration is positive and constant. But suppose the velocity changes as shown in Fig. 2.16(c). How can we determine the acceleration at any particular instant of time? We can define the **instantaneous acceleration** in a manner similar to the way we defined the instantaneous velocity. The instantaneous acceleration is the limiting value of $\Delta v/\Delta t$ as the time interval Δt becomes vanishingly small. In symbols, the instantaneous acceleration is

$$a = \frac{\Delta v}{\Delta t} \qquad \text{(where } \Delta t \text{ approaches zero).} \qquad (2.5)$$

As the time interval Δt becomes smaller, the average acceleration approaches the value of the instantaneous acceleration. Recall that the slope of a curved line at any point is determined by the line tangent to the curve at that point. Thus the slope of a velocity–time curve at any point of time equals the instantaneous acceleration at that point.

Example 2.8

Acceleration of a sprinter.

A sprinter, at rest ($v = 0$) at the start of a race, quickly accelerates to maximum velocity (Fig. 2.17). Determine the instantaneous acceleration of the sprinter at 0.5 s, 1.5 s, and 2.5 s.

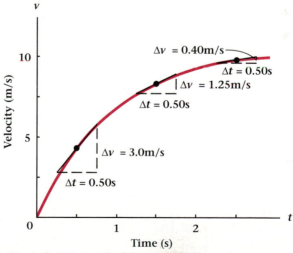

Figure 2.17 Example 2.8: Graph of the velocity of a sprinter versus time.

Solution The results can be read directly from the graph of Fig. 2.17, where we have drawn the tangent lines through the points corresponding to $t = 0.5$ s, 1.5 s, and 2.5 s.

At $t = 0.5$ s we have

$$a(t = 0.5 \text{ s}) = \frac{\Delta v}{\Delta t} = \frac{3.0 \text{ m/s}}{0.50 \text{ s}} = 6.0 \text{ m/s}^2.$$

At $t = 1.5$ s we have

$$a(t = 1.5 \text{ s}) = \frac{1.25 \text{ m/s}}{0.50 \text{ s}} = 2.5 \text{ m/s}^2.$$

At $t = 2.5$ s we have

$$a(t = 2.5 \text{ s}) = \frac{0.40 \text{ m/s}}{0.50 \text{ s}} = 0.80 \text{ m/s}^2.$$

As expected, we find that the acceleration decreases with time as the sprinter approaches maximum velocity. Compare this example with Example 2.4 and Fig. 2.10. Be sure you clearly understand how the shapes of these two curves are related.

Motion with Constant Acceleration

2.6

Many situations occur in which an object's acceleration is constant. The most common case is the motion of a freely falling object when air resistance is negligible. Because of the numerous motions that can be described by constant acceleration, we will assume constant acceleration for the remainder of this chapter. This restriction to constant acceleration allows us to develop some simple relationships among four kinematic quantities—displacement, velocity, acceleration, and time.

First, let us find an expression for the average velocity of an object moving with constant acceleration. Figure 2.18 shows a velocity–time graph of this situation. The velocity is increasing from v_1 to v_2, as discussed in Section 2.2. The displacement is the total area under the velocity–time curve, and is equal to the sum of the areas of rectangle $ABCD$ and triangle BCE. The average velocity is then

$$\bar{v} = \frac{\text{displacement}}{\text{time elapsed}} = \frac{\text{area of } ABCD + \text{ area of } BCE}{\Delta t},$$

$$\bar{v} = \frac{v_1 \Delta t + \frac{1}{2}(v_2 - v_1)\Delta t}{\Delta t} = \frac{v_1 + v_2}{2}.$$

We see that for the case of constant acceleration, and *only* for this case, the average velocity is one-half the sum of the initial and final velocities.

We can use this result, along with the definition of average velocity (Eq. 2.2) and the definition of acceleration (Eq. 2.4), to get another useful relationship. The resulting equations are simpler if we measure time in terms of elapsed time. In this case we set the initial time $t_1 = 0$, and t_2 becomes any later time t. Thus

$$t_2 - t_1 = t - 0 = t.$$

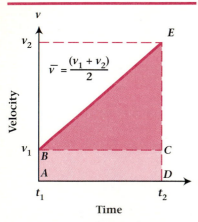

Figure 2.18   Velocity–time graph for constant acceleration, starting with an initial velocity $v_1 \neq 0$.

We also let x_0 represent the initial position (at time $t = 0$) and let x be the position at time t. Using these quantities, we can represent the distance x in terms of the average velocity (from Eq. 2.2) as

$$x - x_0 = \bar{v}t.$$

If we also let v_0 represent the initial velocity when $t = 0$ and let v be the velocity at time t, then the average velocity becomes

$$\bar{v} = \frac{v_0 + v}{2}.$$

Inserting this value of $\bar{v}$ into the equation for displacement, we find

$$x - x_0 = \tfrac{1}{2}(v_0 + v)t. \tag{2.6}$$

This equation expresses the change in displacement (distance traveled) of an object in terms of its initial and final velocities and the elapsed time of motion.

To express the distance traveled in terms of the acceleration rather than the final velocity v, we rewrite Eq. (2.4) in terms of v, v_0, and t, solving for v:

$$v = v_0 + at.$$

Then we substitute this result into Eq. (2.6), giving

$$x = x_0 + v_0t + \tfrac{1}{2}at^2. \tag{2.7}$$

You should carry out these steps to assure yourself that Eq. (2.7) is correct.

Equation (2.7) can also be obtained by examining Fig. 2.18. The term v_0t represents the area of rectangle $ABCD$ and the term $\tfrac{1}{2}at^2$ represents the area of the triangle BEC. Equation (2.7) is one of the more important and useful kinematic expressions, as we shall see in Chapter 4. The following examples illustrate the use of this relationship.

Example 2.9

Takeoff distance for an airplane.

A Boeing 747 airliner, initially at rest, undergoes a constant acceleration of 2.3 m/s² down the runway for 28 s before it lifts off (Fig. 2.19). How far does it travel down the runway before taking off?

Solution The distance the airplane travels is given by Eq. (2.7):

$$x = x_0 + v_0t + \tfrac{1}{2}at^2.$$

If we choose the coordinates so that the airliner is at rest at the origin at $t = 0$ and the direction of the acceleration is along the positive x axis, then the numerical values are $x_0 = 0$, $v_0 = 0$, $a = 2.3$ m/s², and $t = 28$ s. The distance is then

$$x = \tfrac{1}{2} \times 2.3 \text{ m/s}^2 \times (28 \text{ s})^2 = 901.6 \text{ m}.$$

Because the acceleration and time are known to only two significant figures, we round off the answer to

$$x = 900 \text{ m}.$$

Figure 2.19 Example 2.9: An airliner accelerates down the runway to gain enough velocity to become airborne.

Example 2.10

Acceleration of a drag racer.

A drag racer travels one-quarter mile in 10 s from a standing start. Calculate the acceleration in ft/s^2, under the assumption that it is constant. Choose the origin to be at the starting point and the direction of travel along the positive x axis.

Solution Setting $v_0 = 0$ in Eq. (2.7) and solving for the acceleration gives

$$a = \frac{2(x - x_0)}{t^2}.$$

Next, we insert the numerical values for the distance $x - x_0$ and t:

$$a = \frac{2 \times 0.25 \text{ mi}}{(10 \text{ s})^2} = 0.0050 \text{ mi/s}^2.$$

Since we are asked to calculate the acceleration in units of ft/s^2, we need to convert the units from miles to feet. We do so by multiplying a by the conversion factor of 5280 ft/mi:

$$a = 0.0050 \text{ mi/s}^2 \times 5280 \text{ ft/mi} = 26 \text{ ft/s}^2.$$

Example 2.11

Time for a baseball to fall.

A child leaning out a second-story window drops a baseball from rest (Fig. 2.20). If he releases the ball 4.5 m above the ground, how long does it take to fall? The downward acceleration of all falling bodies is, as we shall later see, 9.8 m/s^2.

Solution If we choose the origin of our coordinate system to be on the ground and the upward direction to be positive, then the initial position $x_0 = 4.5$ m, the final position $x = 0$, and the acceleration $a = -9.8$ m/s^2. Since the ball drops from rest, we set $v_0 = 0$ and solve Eq. (2.7) for t:

$$t^2 = \frac{2(x - x_0)}{a},$$

$$t = \pm \sqrt{\frac{2(x - x_0)}{a}} = \pm \sqrt{\frac{2 \times (-4.5 \text{ m})}{-9.8 \text{ m/s}^2}},$$

$$t = 0.96 \text{ s}.$$

We have chosen the positive value of the square root because negative time would not be meaningful.

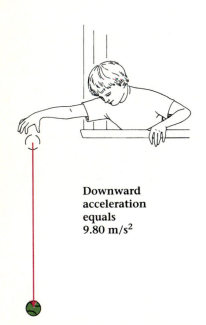

Downward
acceleration
equals
9.80 m/s^2

Figure 2.20 Example 2.11: How long does it take for the baseball to reach the ground?

We can derive another useful expression relating velocities and distances by combining Eqs. (2.4) and (2.7). Again using t to stand for elapsed time, we may rewrite Eq. (2.4) as

$$t = (v - v_0)/a.$$

$$x = x_0 + \bar{v}t$$
$$x = x_0 + \tfrac{1}{2}(v_0 + v)t$$
$$x = x_0 + v_0 t + \tfrac{1}{2}at^2$$
$$v = v_0 + at$$
$$v^2 = v_0^2 + 2a(x - x_0)$$

In these equations we have taken the initial values of time, position, and velocity to be 0, x_0, and v_0, respectively.

Inserting this into Eq. (2.6) and rearranging, we get

$$v^2 = v_0^2 + 2a(x - x_0). \tag{2.8}$$

This equation gives us a way of calculating distances, velocities, or acceleration without needing to know the elapsed time involved. It is another basic equation of kinematics that shows how our definitions of the quantities fit together consistently.

For convenience, we summarize the important kinematic equations for straight-line motion with constant acceleration in Table 2.1.

Example 2.12

Stopping distance for a car.

You are driving your new sports car at a velocity of 90 km/h, when you suddenly see a dog step into the road 50 m ahead (Fig. 2.21). You hit the brakes hard to get maximum deceleration of 7.5 m/s², that is, $a = -7.5$ m/s². How far will you go before stopping? Can you avoid hitting the dog?

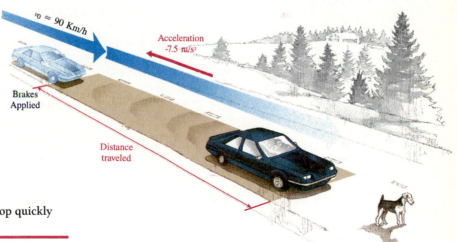

Figure 2.21   Example 2.12: Can you stop quickly enough to avoid hitting the dog?

Solution   Choose the positive x axis as the direction of travel. We are looking for distance in terms of velocity and acceleration. The values of the known quantities are $v_0 = 90$ km/h, $v = 0$, and $a = -7.5$ m/s². We can rearrange Eq. (2.8) to give the stopping distance:

$$x - x_0 = \frac{v^2 - v_0^2}{2a}.$$

Before inserting the known quantities into the equation, we should express v_0 in units of m/s:

$$v_0 = 90 \text{ km/h} \times 1000 \text{ m/km} \times 1 \text{ h}/3600 \text{ s} = 25 \text{ m/s}.$$

The value of the stopping distance can now be evaluated:

$$x - x_0 = \frac{0^2 - (25 \text{ m/s})^2}{2 \times (-7.5 \text{ m/s}^2)}$$

$$= 42 \text{ m}.$$

Fortunately, you are able to stop without hitting the dog.

Galileo and Free Fall

2.7

We now discuss in some detail a familiar and important example of constant acceleration: we examine the work by Galileo Galilei on **freely falling bodies**, objects that are moving freely under the influence of gravity. One of Galileo's earliest scientific studies of motion is illustrated in Fig. 2.22. The data for this experiment are recorded in Galileo's notes. Galileo held a ball at the top of an inclined, grooved board and marked its position. Releasing the ball, he marked its position at the end of equal intervals of time. This is much like dropping a ball from a height, except that the effect of gravity has been "diluted" by allowing the ball to roll slowly down the inclined board rather than fall straight down. The positions as measured by Galileo are given in Table 2.2. The observations show what was already known qualitatively to Galileo and others of his time—that a rolling (or falling) object picks up speed as it continues to roll (or fall). However, the debt we owe to Galileo is for his careful measurements and his quantitative (mathematical) interpretation of the data.

Galileo's object was to find a general rule describing how distances increase with increasing time of fall. After some trial and error, and with considerable insight, Galileo realized that the distance traveled was proportional to the square of the elapsed time. That is, with $x_0 = 0$,

$$x \propto t^2,$$

where the symbol means $\propto$ "is proportional to." If Galileo had possessed less imagination or a narrower point of view, he could not have come to this conclusion because the distances he measured, although close, are not exactly proportional to the square of the time. The third column in Table 2.2 lists the ratio of distance to the square of the time, which is seen to be essentially constant.

TABLE 2.2
Galileo's experiment*
Galileo's results for a ball rolled down an inclined plane.

Time t (equal intervals)	t^2	Distance x (points)	x/t^2
1	1	33	33.0
2	4	130	32.5
3	9	298	33.1
4	16	526	32.9
5	25	824	33.0
6	36	1192	33.1
7	49	1620	33.1
8	64	2104	32.9

Distances were measured in points, a unit that equals 29/30 mm.
*This experiment is described by Stillman Drake, "The Role of Music in Galileo's Experiments," *Scientific American*, June 1975, p. 98.

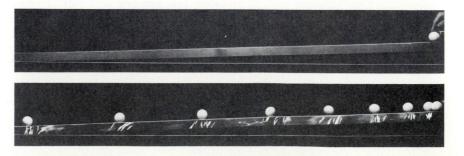

Figure 2.22   Reconstruction of Galileo's experiment on accelerated motion. The distance traveled by the ball rolling down the plane is proportional to the square of the elapsed time.

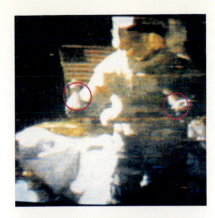

Figure 2.23 TV transmission from the moon of Apollo 15 Astronaut David R. Scott doing an elementary physics experiment. Scott dropped a hammer and a feather from waist high to illustrate that both objects are accelerated equally by the moon's gravity. Both hit the moon's surface at the same time, despite their difference in mass.

It was experiments like this, rather than the legendary Tower of Pisa experiment, which led Galileo to state "... I declare that I wish to examine the essentials of motion of a body that leaves from rest and goes with speed always increasing ... uniformly with the growth of time ... I prove the spaces passed by such a body to be in the squared ratio of the time ..."

Galileo further deduced from his observations that heavy objects fall in the same way that light objects do. Some thirty years later, Robert Boyle, in a series of experiments made possible by his new vacuum pump, showed that this observation is strictly true for bodies falling without the retarding effect of air resistance. This experiment was also demonstrated in 1971 by an astronaut on the moon. A hammer and a feather fell with the same acceleration when they were dropped from rest (Fig. 2.23).

Therefore we can write the relation between distance and time squared as an equality, where the proportionality constant k does not depend on the nature of the falling object:

$$x = kt^2.$$

In our notation, $k = \frac{1}{2}a$, and we say that all bodies fall with the same acceleration.* This acceleration is called the **acceleration of gravity** and is usually denoted by g. Its value is 9.807 m/s^2, or approximately 32 ft/s^2.

Galileo knew that there was an effect on motion due to air resistance. However, his statement, even neglecting air resistance, was in much better agreement with his observations and measurements than was the generally believed (but untested) conclusion from Aristotle, namely, that a body ten times the weight of another would fall to the ground in one-tenth the time. We see here the importance of extracting from a series of observations the essence of what is important, and setting aside or holding for later the less important aspects. It was the genius of Galileo that enabled him to see what was important and what was of secondary concern.

The measurements quoted here are not the ones Galileo referred to in *Discourses on Two New Sciences*, published in 1638, but are taken from his unpublished notes. For the published presentation of his ideas, Galileo discussed experiments in which he marked off distances in equal intervals and measured the time for those distances. He measured the short time intervals with a water clock. The result was that x/t^2 was constant, no matter which method was used. Galileo publicly described an experiment that differed from his real discovery experiment for just the same reason that scientists often do so today: to make the results more easily understood and to give the appearance that a logical mode of inquiry has been followed throughout. That a logical sequence of inquiry is often not the case is one of the important lessons in the history of science.

An extension of Galileo's experiments with inclined planes led to

---

*Strictly speaking, the magnitude of this acceleration depends on the distance from the center of the earth. However, because there is a relatively small fractional change in this distance as we move about on the earth's surface, there is negligible change in the value. The value also depends on the composition of the material underfoot, which we will also usually neglect (but see Chapter 5).

another extremely important qualitative result. Galileo positioned two inclined planes facing each other so that when a ball rolled down one, it would roll up the other. From his experiments with this setup, Galileo concluded that under ideal conditions (lack of friction and air resistance), the ball would roll up the second plane to the same height above the base as the height from which it had been released on the first plane. If the upper end of the second track were lowered, the ball would roll farther along it in order to reach the same height. As the track was lowered more, the ball would travel farther, with less deceleration each time. Thus on a level plane the ball would have no deceleration and its speed would be constant (in the absence of friction). By this reasoning, Galileo discovered the essence of what is called the law of inertia, later stated in full by Newton (see Chapter 4). Galileo realized that his experiment showed, at least for bodies on the earth, that it was not necessary to constantly apply a force in order to move an object. He concluded that rest and motion are both "natural" states of a body. This conclusion was in direct and complete opposition to most beliefs of the time. The need of a force to keep a body in motion was accepted by some as a proof of the existence of God. The conclusions from Galileo's work in mechanics were not published for many years, in part because of the contradictions they presented with Aristotle's views.

BACK TO THE FUTURE

Galileo and Experimental Science

What makes science different from other human activities? Perhaps the central distinction is the role of controlled experiments that can be reproduced by other scientists around the world. Although people had previously tested their ideas of nature with observations, Galileo (Fig. B2.1) was one of the first to make the experimental process central to the development and progress of science. His influence, which persists to the present day, is in large measure due to his literary skill in describing his theories and experiments so clearly and beautifully that quantitative methods became attractive and fashionable.

An experiment believed to have been first conducted by Galileo in 1604 was reconstructed for *Scientific American* by photographer Ben Rose according to specifications supplied by Still-

Figure B2.1   **Galileo Galilei (1564–1642).**

man Drake. The objective of this modernized test was to measure as precisely as possible the distances traveled from rest by a ball rolling down an inclined plane at the ends of eight equal times (in this case, at 0.55-s intervals). The grooved inclined plane used in the reconstruction was fitted

with a stop at the higher end, against which a 2-in. steel ball could be held.

It has sometimes been said that Galileo lived before the development of accurate clocks, and could not have measured time accurately enough to do his experiments. This statement would be true if it had been necessary to measure the elapsed time. However, he needed to determine only equal intervals of time, not total elapsed time. Galileo was especially suited to do this because of a natural ability and his musical training. (He played the lute very well, and his father was a professional musician.) Even people who have no special training in music are able to detect small differences in accurate timing.

In the modern reconstruction of Galileo's experiment the time intervals were established by singing the familiar song "Onward Christian Soldiers" at a tempo of about two beats per second. At one note the ball was released,

Figure 2.24 Example 2.13: Free fall, an experience that's like falling off a 10-story building.

Example 2.13

Free fall.

At the Six Flags amusement park near Atlanta, Ga., Free Fall riders seated four abreast in a padded gondola are taken to the top of a 10-story tower (Fig. 2.24). Then the gondola is dropped 30 m down a vertical track that curves near the bottom, where the gondola slows to a stop. How long does it take to fall from top to bottom and what maximum speed is reached?

Solution   The gondola drops from rest and falls with an acceleration of nearly 9.8 m/s². The relation between time, acceleration, and distance is Eq. (2.7) with $v_0 = 0$,

$$x - x_0 = \tfrac{1}{2}at^2.$$

This equation can be rearranged to give

$$t^2 = \frac{2(x - x_0)}{a}.$$

If we insert the values for $x - x_0$ and a, we get the time to fall:

$$t = \sqrt{\frac{2(x - x_0)}{a}} = \sqrt{\frac{2(30 \text{ m})}{9.8 \text{ m/s}^2}} = \sqrt{6.122 \text{ s}^2} = 2.5 \text{ s}.$$

and the positions of the ball at subsequent notes were marked with chalk; for comparison the exact 0.55-s positions were also captured by multiple-flash photography (see Fig. 2.22). A rubber band was then put around the plane at each chalk mark and the positions of the rubber bands were adjusted so that the audible bump made by the ball in passing each band would always come exactly at a beat of the march. The ratios of the successive measured distances were found to agree closely with a set of figures recorded by Galileo.

You may easily repeat his experiment by using a long board with rubber bands stretched around it, and tilting it at about 2° to the horizontal. Hold a heavy ball lightly in place with your finger until you are ready to release it. You will hear a slight thump as the ball rolls over each of the rubber bands. When the thumps are regular, the time intervals are of equal length. Sing a song with a good strong beat and place the rubber bands so as to keep time with the music. The results depend only on whether the beat is regular.

Today's experimental techniques are far more sophisticated than anything Galileo could have imagined. For example, the development of high-speed photography, pioneered by Dr. Harold Edgerton, has enabled us to measure time intervals as small as 10^{-6} s. These techniques have led to calculations of speed and distance not otherwise obtainable (Fig. B2.2).

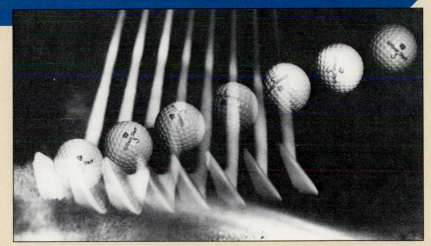

Figure B2.2   Rotation of a golf ball. Each image in this multiple-exposure high-speed photograph occurs at intervals of 1/1000 s. You can see that the ball completes one rotation in roughly four images, for a rotation speed of 4/1000 s or 250 rotations per second.

The maximum speed is reached near the bottom of the ride. We can determine the speed from Eq. (2.8), which relates the final velocity to the initial velocity, the acceleration, and the displacement. In this case, the initial velocity is zero and the equation simplifies to

$$v^2 = 2a(x - x_0),$$

or

$$v = \sqrt{2a(x - x_0)}.$$

Inserting the numerical values, we find

$$v = \sqrt{2(9.8 \text{ m/s}^2)(30 \text{ m})},$$

$$v = 24 \text{ m/s}.$$

This speed is equivalent to 54 mi/h. The riders attain this speed because the gondolas are in free fall until slowed at the bottom of the ride.

SUMMARY

Useful Concepts

- Average speed is the total distance traveled divided by the elapsed time interval.
- Average velocity is given by

$$\overline{v} = \frac{\Delta x}{\Delta t},$$

where Δx is the object's displacement and Δt is the elapsed time. Velocity has direction as well as magnitude.

- Instantaneous velocity is the limiting value of $\Delta x/\Delta t$ as Δt becomes vanishingly small. Alternatively, it is the slope of the tangent to the displacement–time curve at a point.
- The area under the velocity–time curve between two time intervals is equivalent to the displacement during that time interval.
- Average acceleration is the change in velocity divided by the time required for the change,

$$\overline{a} = \frac{\Delta v}{\Delta t}.$$

- Instantaneous acceleration is the limiting value of $\Delta v/\Delta t$

as Δt becomes vanishingly small. Alternatively, it is the slope of the tangent to the velocity–time curve at a point.
- All bodies fall with the same acceleration due to gravity, $g = 9.80 \text{ m/s}^2$. If released from rest and if air resistance is not important, they traverse the same distance in the same amount of time.
- For motion with constant acceleration, the most frequently used kinematic equations are

$$x = x_0 + \overline{v}t, \qquad\qquad v = v_0 + at,$$

$$x = x_0 + v_0t + \tfrac{1}{2}at^2, \qquad v^2 = v_0^2 + 2a(x - x_0).$$

Important Terms

You should be able to write the definition or meaning of each of the following terms:

- kinematics
- dynamics
- mechanics
- constant speed
- average speed
- velocity
- displacement
- average velocity
- abscissa
- ordinate
- slope of a line
- instantaneous velocity
- average acceleration
- instantaneous acceleration
- freely falling bodies
- acceleration of gravity

QUESTIONS

2.1 Carefully distinguish between speed and velocity.

2.2 Can an object have zero velocity but nonzero acceleration? Can it have zero acceleration and nonzero velocity? Give examples of each.

2.3 Sketch a curve of velocity versus time for the displacement–time curve of Fig. 2.10. Sketch the acceleration–time curve also.

2.4 A plot of distance against time has the shape of one-quarter of a circle. The circle is tangent to the time axis and to the distance axis. What can you conclude about the velocity and acceleration in this case?

2.5 Discuss the qualitative relationship between the total national debt, the increase per year, and the rate of change of the increase. Use a graphical method. Which would you judge to be affected more by national and world events?

2.6 The distance–time curve for a hypothetical journey has the shape of an equilateral triangle with one side along the time axis. Discuss the velocity *and* acceleration necessary to bring about such a journey. Comment on whether or not this is a realistic journey.

2.7 Use a graphical method to answer the following question. The initial population of snowy egrets on a barrier island was 100 on January 1. Each January 1 thereafter, the population was 5% more than for the previous year. What was the population at the end of 15 years? Is the rate at which the population grew a constant? (Recall the graphical meaning of constant velocity.)

2.8 In a common parlor trick, one person holds a dollar bill so that it hangs vertically. A second person places his hand so that the dollar is between but does not touch his thumb and forefinger. When the dollar is dropped, the second person tries to catch it but fails. Estimate the minimum reaction time of the second person.

2.9 Design an experiment similar to Galileo's to measure the way in which a body falls (or rolls) under the influence of gravity. You may use anything that is commonly available as a consumer item.

2.10 Summarize your concept of the scientific method as you understand it from considering the work of Galileo.

2.11 Describe another observation or experiment that shows Galileo was correct in concluding that both straight-line motion and rest are "natural states" for a body (that is, they do not require the application of an external force).

2.12 A car's speedometer is correctly calibrated for tires of a specific size. If larger-diameter tires are substituted, what will be the effect on the speedometer reading?

PROBLEMS

Hints for Solving Problems
You may find it helpful to review the steps for solving problems that were given in Chapter 1. In addition, be sure to choose the positive direction for the coordinate system used in each problem and apply the signs consistently to displacement, velocity, and acceleration. Be careful in converting units. (A conversion table is included inside the front cover.) Check your answers to see whether they are reasonable.

Section 2.1   Speed

2.1 (a) What is the speed in kilometers per hour of a car traveling at a constant speed of 55 mi/h? (b) What is the speed of the car in meters per second? (*Hint:* 1 mi = 1.609 km.)

2.2 How far will an automobile go in 3.5 h at a constant speed of 95 km/h?

2.3 What is the average speed for a trip of 137 km which requires 2.25 h?

2.4 Two bicycle riders made a 30-km trip in the same time. Cyclist A traveled continuously at an average speed of 20 km/h. Cyclist B traveled continuously except for a 20-min rest break. What was B's average speed for the time of actual riding?

2.5 How long can you afford to stop for lunch if you can drive a steady 90 km/h on the highway and must make a 200-km trip in $3\frac{1}{2}$ hours?

2.6 What is the linear speed of the earth in its orbit about the sun? Give your answer in meters per second. (Consider the earth's orbit to be a circle of radius 1.5×10^{11} m.)

2.7 A 3.0-h trip was made at an average speed of 75 km/h. For the first hour the average speed was 90 km/h. What was the average speed for the remainder of the trip?

2.8 A freight train, when it is l00 km away from the station, has a constant speed of 70 km/h toward the station. At that moment a fast bird that is perched on the locomotive takes off and flies in the direction of the station at 100 km/h with respect to the ground. When the bird gets to the station it turns around and returns to the train, where it repeats the process again. How many kilometers will the bird travel before the train reaches the station?

2.9 Your small dog gets away from you and begins running down the street. If your dog is running at a speed of 8.0 km/h and you can run 10 km/h, how many seconds head start can you afford to give the dog if you want to catch him within 30 s of when you start after him?

2.10 A jogger with a constant speed of 5 m/s runs along the edge of a square gymnasium from one corner to the corner diagonally opposite. The sides of the gym are 25 m long. With what speed would a second jogger have to run along the diagonal of the square in order to exactly match the time of the first runner's trip?

2.11 How many minutes does it take for light to go from the sun to the planet Uranus, a distance of 2.88×10^9 km? (The velocity of light is constant in free space and is 3.00×10^8 m/s.)

2.12 The speed of light is 3.0×10^8 m/s and the speed of sound is 340 m/s. Find the value of the integer n in the following statement: "If you start counting seconds when you see something happen and stop when you hear it happen, for every n seconds counted the event was about 1 km away."

Section 2.2   Average Velocity

2.13 A long-distance runner starts at a given place and runs around a circular track of 50 m radius at a speed of 6.0 m/s in the clockwise direction for 60 s. Then the runner reverses direction and runs in the counterclockwise direction at 4.0 m/s for 120 s. At the end, how far around the track is the runner from the starting point?

2.14 A delivery truck travels with a velocity of +75 km/h for 2.0 h, then with a velocity of −40 km/h for 3.0 h. How far from its origin will it be at the end of the trip?

2.15 If a greyhound runs in a straight line for 2.0 min at a velocity of 40 m/s, what must its velocity be in order to return to its starting point in 1.5 min?

2.16 Two children cross a starting line at the same time, one with a velocity of +3.5 m/s and the other with a velocity of −4.0 m/s. How far apart are they after 10 s?

2.17 An elevator in a building with floors spaced 4.0 m apart moves with a velocity of constant magnitude of 1.0 m/s. Positive velocity here indicates upward motion. The following trips were made: starting at the first floor, a trip of 1 min duration and positive velocity, followed by a trip of negative velocity lasting 20 s, and yet another trip of 8 s duration with positive velocity. On what floor of the building was the elevator at the end of the last trip?

Section 2.3   A Graphical Interpretation of Velocity

2.18 Plot the points given in the following table and sketch a curve through them. Determine the slope of the curve at $x = 5$ and at $x = 10$.

x	0	1	2	3	4	5	6	7	8	9	10	11	12	13
y	6.7	6.3	6.0	5.9	6.0	6.5	7.3	8.5	10.2	12.0	14.2	15.0	14.4	12.4

2.19 Use the following data to plot a distance–time curve. Determine the velocity (slope) at each second and plot a velocity–time graph.

Time (s)	1	2	3	4	5	6	7	8	9
Distance (m)	2	4	6	8	9.2	9.8	10	10	10

2.20 A trip was made, starting with zero speed at time zero, in such a way that the speed–time graph is approximately an isosceles triangle with the base along the time axis. The maximum speed was 25 m/s and the total elapsed time was 40 s. What distance was traveled?

2.21 You are driving a car with initial speed v_0. At time $t = 0$, you begin to increase your speed at a constant rate. Twenty seconds later you are traveling at a speed of 60 km/h. At $t = 60$ s you are traveling at a speed of 120 km/h. (a) What was your initial speed v_0? (b) What was your speed at $t = 40$ s? Use a graphical method to arrive at the answers.

Section 2.4   Instantaneous Velocity

2.22 Sketch an approximate distance–time curve for an object that has a velocity given by

$$v = (4.0 + 4.0\, t^{1/2}) \text{ m/s}$$

for the first 5 s of motion, where t is given in seconds.

2.23 The velocity–time graph of part of a cyclist's trip consists of a straight line between $v = 20$ m/s at $t = 10$ s and $v = 0$ at $t = 20$ s. What distance was covered during this portion of the trip?

2.24 The velocity–time graph of a jogger's trip is approximated by a triangle which starts at $v = 0$ at $t = 0$, rises to a maximum at $t = 6$ s, and then returns to $v = 0$ at $t = 12$ s. If the maximum speed was 6 m/s, how far did the jogger go?

Sections 2.5, 2.6   Acceleration

2.25 A motorist traveling at 90 km/h applied the brakes for 5.0 s. If the braking acceleration was -3.0 m/s^2, what was her final speed?

2.26 A motorcyclist moving with an initial velocity of 8.0 m/s undergoes a constant acceleration for 3.0 s, at which time his velocity is 17.0 m/s. (a) What is the acceleration? (b) How far does he travel during that 3.0-s interval?

2.27 A motorcycle rider moving with an initial velocity of 8.0 m/s uniformly accelerates to a speed of 17 m/s in a distance of 30 m. (a) What is the acceleration? (b) How long does this take?

2.28 The nominal stopping distances of a Nissan Sentra SE Sport Coupe are (a) 147 ft from 60.0 mi/h and (b) 264 ft from 80.0 mi/h. Determine the value of the acceleration for each case, assuming that it is constant during each event. Give your answer in units of m/s².

2.29 A BMW 535i can accelerate from 0 to 60 mi/h (26.8 m/s) in 7.9 s. What is the average acceleration during this time interval?

2.30 A small turbocharged automobile can accelerate from 0 to 48 km/h in 4.3 s and from 0 to 96 km/h in 10.4 s. In addition, under constant acceleration from rest it crosses the 0.40-km marker at a speed of 130 km/h. (a) Calculate the average acceleration needed to reach 48 km/h. (b) Calculate the average acceleration during the time it takes to go from 48 to 96 km/h. (c) What is the average acceleration over the 0.40-km course?

2.31 A spring-driven toy truck starting from rest covers a distance of 50 cm with a constant acceleration of 0.10 m/s². How long does it take to go the 50 cm?

Section 2.7   Galileo and Free Fall

2.32 How long would it take an object to fall to the ground from the top of the Leaning Tower of Pisa (height = 54.6 m)?

2.33 A cannonball is dropped from the top of a building. If the point of release is 100 ft above the ground, what is the speed of the cannonball just before it strikes the ground? Give your answer in units of ft/s and mi/h.

2.34 Use some of the data in Table 2.2 to calculate the acceleration of the ball down the track. Your answer will be in units of points/(time)².

2.35 A ball is allowed to roll from rest down an inclined • plane and the distances are marked every 2.0 s. If the second mark is made 1.60 m from the starting point, where are the first and fourth marks?

2.36 Make a table of the velocity and total distance fallen at the end of each half-second during the first 2 s for an apple dropped from rest from the top of a very tall building.

2.37 During the launch of a space shuttle, the rocket attains a speed of 17,400 mi/h 10 min after ignition. (a) Calculate the average acceleration during this time interval. (b) Compare the acceleration of the rocket with the acceleration of gravity on the earth's surface.

Additional Problems

2.38 A 9.0-h trip is made at an average speed of 50 km/h. If the first half of the distance is covered at an average speed of 45 km/h, what is the average speed for the second half of the trip?

2.39 Two automobiles travel in opposite directions from • the same starting point. If the speed of one is twice the speed of the other and they are 200 km apart at the end of one hour, what is the speed of each car?

2.40 (a) Plot the curve corresponding to $y = 3x^2 + 7$ and • find the slope at $x = 5$ by graphical means. (b) Find the area under the curve between $x = 0$ and $x = 10$.

2.41 Two cyclists start in opposite directions at the north- • ernmost point of a circular track of 25 m radius. (a) If the speed of each bike is constant and is 10 km/h for cyclist A and is 15 km/h for cyclist B, where will they meet? (b) Where will they meet the second time? Assume that they go around the track on essentially the same path, with cyclist A initially headed west and cyclist B initially headed east.

2.42 Two Winston Cup drivers race their cars on the •• 4.22-km loop track at Riverside, Calif. On one lap, as they pass the start/finish line side by side, one driver is going 180 km/h and the other is going 200 km/h. If they each maintain constant speed, how many laps will the faster car have made when it catches up with the slower one from behind?

2.43 Starting from rest, you move with a constant accel- • eration of 3.0 ft/s² for 12 s and then move with an acceleration of -3.0 ft/s² for another 12 s. (a) What is your maximum speed attained? (b) How far do you go during the whole trip? (c) What is your average speed?

2.44 Two motocross bikes start from one corner of a •• square field and go to the corner diagonally opposite in the same time t. They both start from the same place and take different routes. One travels along the diagonal with constant acceleration a and the other accelerates momentarily and then travels along the edge of the field with constant speed v. What is the relationship between a and v? Assume a negligible acceleration time for the biker traveling along the edge.

2.45 A person can throw a ball with an upward velocity •• v_0 so that it will just reach the top of a 20-m-tall building. If the same person stands on top of the building and throws the ball downward with velocity v_0, how much sooner does it reach the ground than if it were merely dropped from the same height?

2.46 A stone is dropped from rest from a height of 20 m. •• At the same time, a stone is thrown upward from the ground with a speed of 18 m/s. At what height do their paths intersect?

2.47 A ball bearing is dropped from rest at a point A. At
•• the instant it passes a mark 10 m below A, another
ball bearing is released from rest from a position
11 meters below A. (a) At what time after its release
will the second ball bearing be overtaken by the first?
(b) How far does the second bearing fall in that time?

2.48 A rock is dropped from rest from a height above a
• strange planet and a strobe-light photograph is taken.
The image is damaged in transmission to earth so that
an unknown part of the top of the picture is lost.
However, five successive images of the falling rock
can be seen. The spacing between the remaining images corresponds to 0.70, 0.90, 1.10, and 1.30 m, and
the flash rate is 4.0 flashes per second. Calculate the
acceleration of gravity on that planet.

2.49 An electric-powered model car starts from rest and
•• moves in a straight line for 4 s. The relationship between time and position is approximated by

$$x^2 + 2t^2 = 32,$$

where x is in centimeters and t is in seconds. Using
graphical methods, find the speed at $t = 1$ s and at
$t = 3$ s.

2.50 (a) A slowly moving train with 12-m-long flatcars is
• passing a station at 10 km/h. A person on the station
platform tosses rocks onto the moving flatcars at the
rate of once every second. (a) If the first rock just
hits the front edge of one car, how many rocks will
fall onto that car? (b) How many rocks will fall onto
that car if the train begins to accelerate at 0.50 m/s²
just as the first rock hits the car?

2.51 Three bugs start simultaneously at the hub of a wheel
• and crawl outward along different spokes. One crawls
with constant speed v_1, the second starts from rest
and crawls with constant acceleration a, and the
third, moving with initial speed v_2, decelerates with
acceleration $-a$. (a) Show that if all bugs reach the
rim at the same time, $v_2 = 2v_1$. (b) Find the acceleration a.

2.52 A ball dropped from the top of a tower reaches a
•• velocity v_f just before it reaches the ground. An automobile traveling toward the tower with constant
speed v_f just reaches the tower when the ball strikes
the ground. Show that at the instant the ball was
released the distance of the car from the tower was
twice the height of the tower.

2.53 A ball is thrown straight up so that it reaches a height
• of 25 m. How fast was it going when it was 5 m high?
(*Hint:* Use the symmetry of the upward and downward paths.)

2.54 An elevator at a construction site has a roof that is
•• 2.50 m above its floor. The elevator starts from the
ground and moves toward the top of a 20-m-high
building with a steady speed of 0.50 m/s. After the
elevator has been moving upward for 2.5 s, a wrench

is accidentally dropped from the top of the building
onto the roof of the elevator. At what height above
the ground is the top of the elevator when the wrench
strikes the roof?

2.55 A ball is dropped from the top of a tower at the same
•• time that one is thrown upward from the ground below. They collide at the top of the second ball's trajectory 2.0 s after the ball is thrown upward. How
tall is the tower? Neglect the height of the thrower.
(*Hint:* A ball thrown upward with speed v acquires
the same speed v when it returns again to the
ground.)

2.56 With what initial speed must you throw a ball down-
• ward from a second-story window ($h = 4.0$ m) in
order for it to reach the ground in half the time it
would have taken if it had been dropped and not
thrown?

2.57 A small parachute dropped from a 30-m-high cliff
• falls with a constant velocity of 1.2 m/s. Twenty seconds after the parachute is dropped a stone is
dropped from the cliff. Will the stone catch up with
the parachute before it reaches the ground?

2.58 The hollow cylinder shown in Fig. 2.25 is free to
• rotate about a horizontal axis. One hole is cut in the
side of the cylinder. The object of a game is to spin
the cylinder so fast that an object dropped through
the hole when it is in the uppermost position will fall
through the same hole when it has rotated to the
bottom position. If the diameter of the cylinder is
0.50 m, how many revolutions per second must it
make? (*Hint:* First calculate the time for the object to
fall the appropriate distance, then use that result to
determine the number of revolutions per second.)

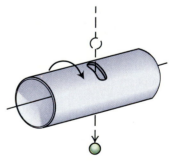

Figure 2.25   Problem 2.58.

2.59 (a) The flying time of the Concorde airplane from
New York to London (5564 km) is approximately
3 h. What is the Concorde's average speed in km/h?
(b) The Wright brothers' first sustained powered
flight at Kitty Hawk, N.C., lasted 12 s and covered
approximately 120 ft. What was their average speed
in kilometers per hour? (c) How many times faster
is the Concorde than the Wrights' first powered
flight?

2.60 According to advertisements, a Mercedes 300SE can
accelerate from 0 to 55 mi/h in 7.5 s. (a) Express
55 mi/h in meters per second. (b) Calculate the average acceleration during that 7.5 s. (c) How far does
the Mercedes travel during that time?

2.61 Figure 2.26 shows the speed curve for a Sentra SE
Sports Coupe under maximum acceleration from a
standing start. From the graph, determine the acceleration at a number of points and make your own
graph of acceleration versus time over the range from
0 to 40 s.

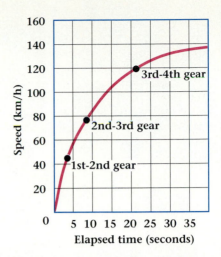

Figure 2.26   Problem 2.61: Graph of speed versus elapsed
time for a 1986 Sentra SE under maximum acceleration
from rest.

ADDITIONAL READING

Brancazio, P. J., *Sport Science*. New York: Simon and Schuster, 1984.

Casper, B. M., "Galileo and the Fall of Aristotle: A Case of Historical Injustice?"
American Journal of Physics, Vol. 45, 1977, p. 325.

Drake, S., "The Role of Music in Galileo's Experiments." *Scientific American*, June
1975, p.98.

Schroeer, D., "Brecht's Galileo: A Revisionist View." *American Journal of Physics*,
Vol. 48, 1980, p. 125.

Seeger, R. J., *Men of Physics: Galileo Galilei, His Life and His Works*. New York:
Pergamon Press, 1966.

3

Motion in Two Dimensions

3.1 Addition of Vectors

3.2 Relative Velocity

3.3 Resolution of Vectors

3.4 Kinematics in Two
     Dimensions

3.5 Projectile Motion

*3.6 Range of a Projectile

App. Review of Trigonometry

A WORD TO THE STUDENT

You should learn two principal things from this chapter; one is a technique and the other is a concept. The technique is that of using vectors to represent quantities with both direction and magnitude, such as the velocity of a runner in an open field. The elementary vector techniques presented here will be used again in almost every following chapter.

The concept to be learned is the way to treat independently the horizontal and vertical components of the motion of a body. We will use the case of a projectile as an illustration, and then extend these results to show how to solve kinematic problems in two and three dimensions. This concept will appear again in the chapters that follow, especially those on mechanics and electricity.

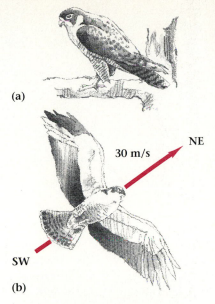

(a)

30 m/s   NE

SW

(b)

Figure 3.1 Direction and magnitude.
(a) A quantity with magnitude but
no direction, such as the falcon's
mass, is a *scalar*. (b) The speed and
direction of the falcon's flight results
in a quantity with both magnitude
and direction, called a *vector*.

Galileo's studies of motion went beyond the case of free fall in one dimension to explore the motion of projectiles, such as shells fired from a cannon. However, before we can describe the analysis of projectiles, we need to understand more about motion in two and three dimensions.

To extend the techniques developed for one-dimensional motion into two- and three-dimensional problems, it is helpful to introduce new notation. Many physical quantities can be completely specified by their magnitude alone. Such quantities are called **scalars.** Examples include such diverse things as distance, time, speed, mass, and temperature. Another physically important class of quantities is that of **vectors,** *which have direction as well as magnitude.* For example, if we say a grocery store is ten miles from home, we have not completely specified its location unless we state its direction from us—north, south, east, or west. The distance and the direction together constitute a vector called the displacement.

Many other physical quantities are vectors, in addition to displacement. These include velocity and acceleration, which were introduced in Chapter 2, as well as force and momentum, which will be defined in later chapters. Often an object (such as the falcon shown in Fig. 3.1) has both scalar properties (mass) and vector properties (velocity) at the same time.

In printed materials such as this text, we represent a vector quantity in boldface roman type, **A**. If we are referring only to the magnitude of that quantity, we use the same letter in lightface italic type. For your own purposes in working problems and taking notes, you may find that a letter with an arrow drawn over it ($\vec{A}$) is a useful symbol for a vector.

Addition of Vectors

3.1

In diagrams we frequently use an arrow to represent a vector. The arrow is drawn so that it points in the direction of the vector and so that its length is proportional to the magnitude of the vector. For instance, we might represent a displacement of 5 km east by an arrow 25 mm long (Fig. 3.2a, next page). To represent a displacement of 3 km east, we would draw a second arrow parallel to the first but shorter, in the ratio of 3 : 5, or 15 mm long. A vector representing 5 km north is directed at 90° from a vector representing 5 km east (Fig. 3.2b). Because a vector is denoted by its length and direction, it remains the same vector when translated to a new starting point (Fig. 3.2c).

Addition of scalars uses just simple math: 3 kg + 5 kg = 8 kg, for example. Addition of vectors, however, must be different to take account of the directions of the quantities. To illustrate vector addition, let us first take a simple example, after which we will state the general rule. Consider the following problem: If a woman walks 4 km north and 3 km east, how far and in what direction is she from the starting point?

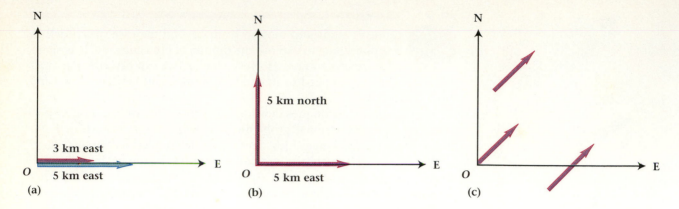

Figure 3.2   (a) The magnitude of a vector is indicated by the length of the arrow representing the vector. (b) Arrows representing displacements of 5 km east and 5 km north. The vectors have the same magnitude but different directions. (c) Three representations of the same vector.

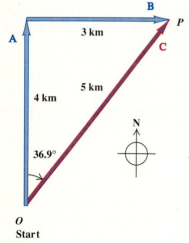

Figure 3.3   Addition of vectors. The displacement that results from walking along vectors **A** and **B** is equal to that of vector **C**. However, the sum of the magnitudes of **A** and **B** is not the same as the magnitude of **C**.

Figure 3.4   Variations in magnitude and direction. If vectors **A** and **B** are of constant magnitude but their direction changes, then the magnitude and direction of **C** will also change.

Figure 3.3 is a scale map of this walk. The initial northward walk is a displacement and hence is represented by a vector, which we have labeled **A** in the figure. The eastward journey results in a displacement represented by vector **B**. After traveling north and east, the walker arrives at point P. However, a more direct way to reach the same point would be to walk along the straight line OP. This line, which represents the resultant displacement, is represented by vector **C**, and measures 5 km at an angle of 36.9° east of north. (The distance and angle may be determined either by measuring on a scale drawing or by calculating with geometry and trigonometry.)

Whether the walker takes the indirect or the direct route, she ends up in the same place. Thus the displacement represented by the sum of the two vectors **A** + **B** equals the displacement represented by the vector **C**:

$$\mathbf{A} + \mathbf{B} = \mathbf{C}.$$

We say that we have added the two vectors **A** and **B** to get the **resultant** vector **C**.

Two important points are worth mentioning before proceeding. First, note that although the sum of the magnitudes of the individual parts of the trip is 7, this is not the magnitude of the resultant vector. The sum of two vectors depends on their directions as well as their magnitudes (Fig. 3.4). The second point is that the technique for adding vectors is the same for all vectors, whether they represent displacement or any other vector quantity—just as the rules for adding all scalars are the same, no matter whether they represent time, money, mass, or pure numbers.

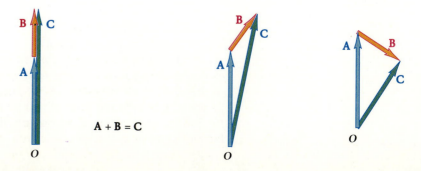

A + B = C

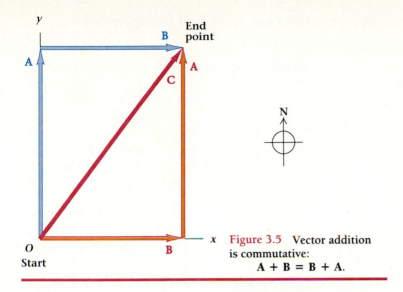

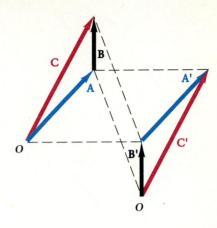

Figure 3.5   Vector addition is commutative:
$$\mathbf{A} + \mathbf{B} = \mathbf{B} + \mathbf{A}.$$

Figure 3.6   Vectors are not changed by translation.

In general, to add two vectors graphically, make a scale drawing and place the vectors "head to tail"—that is, with the tail (origin) of the second vector starting from the head (end point) of the first. Then draw the resultant vector from the origin of the first vector to the end point of the second vector. Finally, measure the length (magnitude) and direction of the resultant vector directly from the scale drawing. Alternatively, you can calculate the magnitude and direction using trigonometry. Either technique gives the same result.

In the previous example of a woman walking 4 km north and then 3 km east, she would reach the same end point if she reversed the order in which she made the two parts of the trip. Figure 3.5 shows the case in which the eastern path was traversed first (vector **B** followed by vector **A**). The final distance and direction are the same in both cases. We conclude that vector addition is commutative; that is, the order in which we add the vectors may be reversed without changing the result:

$$\mathbf{A} + \mathbf{B} = \mathbf{B} + \mathbf{A}.$$

Commutation is a general rule for the addition of any two vectors.

From our observations we can conclude that vectors may be moved (translated) without changing their value, so long as their directions and magnitudes are not changed. In Fig. 3.6 vectors **A** and **A**′ are identical. Vectors **B** and **B**′ are also identical. Thus, their resultants **C** and **C**′ are also identical in both direction and magnitude.

To add more than two vectors together, we repeat the rule by adding successive vectors head to tail (Fig. 3.7). If we wish to add three vectors **A**, **B**, and **C** we first add **A** and **B** to get a resultant **D**. Then we add **C** to **D** to give **E**. Additional vectors could also be added one at a time. It is not necessary to draw the intermediate sum **D**; we may simply add **C** by placing it next to **B** as shown in the figure.

Vectors may also be subtracted. To subtract vectors we introduce the idea of −**A**, the negative of vector **A**. By this we mean a vector of the same magnitude and parallel to **A** but pointing in the opposite direction

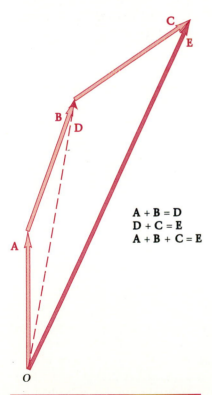

$$\mathbf{A} + \mathbf{B} = \mathbf{D}$$
$$\mathbf{D} + \mathbf{C} = \mathbf{E}$$
$$\mathbf{A} + \mathbf{B} + \mathbf{C} = \mathbf{E}$$

Figure 3.7   Addition of three vectors.

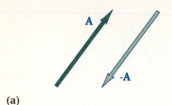

(a)

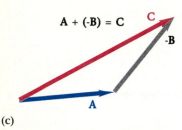

(b)

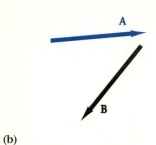

(c)

Figure 3.8 Subtraction of vectors. (a) The negative of a vector has the same magnitude but is directed in the opposite direction. (b) We can obtain the difference between two vectors by adding the first vector to the negative of the second. (c) Note how the negative of **B** has been placed head-to-tail with **A**.

(Fig. 3.8a). The subtraction of one vector from another, such as

$$\mathbf{A} - \mathbf{B} = \mathbf{C},$$

can be considered as the addition of the first vector to the negative of the second, as

$$\mathbf{A} + (-\mathbf{B}) = \mathbf{C}.$$

Figure 3.8(c) illustrates the procedure.

The result of multiplying a vector by a scalar is another vector. The vector 2**A** has a magnitude twice as great as **A** and points in the same direction as **A**. The vector −2**A** is also twice as great as **A**, but it points in the opposite direction.

The magnitude of a vector quantity is represented by the same letter used for the vector, but in lightface type instead of boldface type. An alternative notation for the magnitude is the vector symbol with vertical bars on both sides. Thus

$$\text{magnitude of } \mathbf{A} = A = |\mathbf{A}|.$$

The magnitude of a vector quantity is a scalar and is always positive.

Example 3.1

Addition of vectors.

A vector **A** has a magnitude of 7 units and a direction of 30° measured counterclockwise from the positive x axis. The vector **B** has a magnitude of 11 units and a direction of 140° measured counterclockwise from the positive x axis. What is the vector sum of **A** and **B**?

Solution   First we draw a careful scale diagram (Fig. 3.9). This diagram

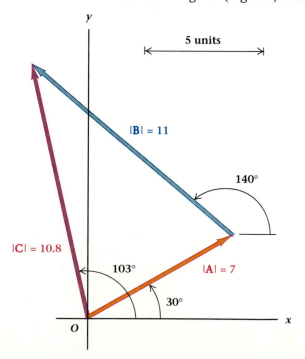

Figure 3.9   Example 3.1: Addition of vectors.

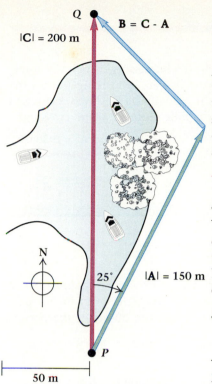

Q

B = C - A

|**C**| = 200 m

N

25°

|**A**| = 150 m

P

50 m

Figure 3.10 Example 3.2: Subtraction of vectors.

corresponds to

$$A + B = C.$$

After making the diagram, we determine the magnitude of **C** by measuring its length on the drawing and converting it to the appropriate number of units by using the scale indicated. Always indicate on the diagram the scale you use. In this case the magnitude of **C** is 10.8 units. The direction of **C** can be measured with a protractor and is 103° from the positive *x* axis.

Example 3.2

Subtraction of vectors.

Your friend standing at point *P* wishes to join you at point *Q* located a distance of 200 m due north across a lake (Fig. 3.10). Your friend decides to go around the lake by traveling along two straight-line paths **A** and **B**. If the first path **A** is 150 m at an angle of 25° east of north, what are the distance and the direction of the second path?

Solution We may formulate this vector relation as

$$B = C - A,$$

where **C** is 200 m to the north and **A** is 150 m at 25° to north. Figure 3.10 is a scale diagram from which we may determine the answer by direct measurement. The second path **B** is found to be 90 m at an angle of 45° west of north.

Relative Velocity

3.2

Velocity is not an absolute quantity but is measured relative to other objects. Thus to measure an object's velocity we must first specify the coordinate system or frame of reference in which the measurement is to be made. Ordinarily the origin of the coordinate system is fixed in some other body. For example, when describing the motion of the planets about the sun, we usually consider a frame of reference that is fixed, or stationary, with respect to the sun, even though the sun itself is in motion relative to the center of the galaxy.

When we say that an automobile is traveling 90 km/h (55 mi/h), we usually mean that it is going 90 km/h relative to the road. Imagine that you are driving down the highway at 90 km/h when another car passes you at 100 km/h. Although both cars are moving rapidly down the road, the faster car appears to overtake you very slowly. Relative to a coordinate system fixed on your car, the passing car is going only 100 km/h − 90 km/h = 10 km/h.

When two objects are traveling along the same line, the **relative velocity** of one to the other is obtained simply by subtraction. If the two velocities are not along the same line, then we need to use the techniques of vector addition to determine the relative velocity.

Example 3.3

Relative velocity on a train.

A freight train pulling several flatcars is slowly passing a highway intersection at 10 km/h. A hobo on one of the flatcars is walking toward the engine at 5 km/h (Fig. 3.11). What is the velocity of the hobo relative to an observer waiting patiently in an automobile stopped at the crossing?

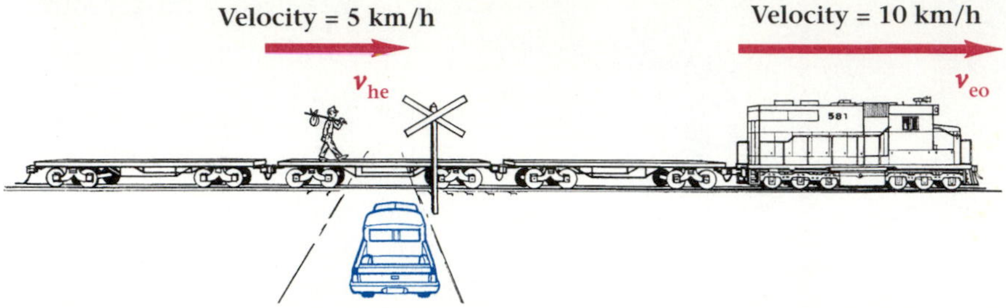

Figure 3.11   Example 3.3: A hobo walking with velocity v_{he} on a train moving with velocity v_{eo} passes a stationary observer with a velocity equal to the sum of the two velocities.

Solution   In this example both velocities are in the same direction. As the hobo walks toward the engine with a velocity $\mathbf{v}_{he}$, the train moves past the observer with velocity $\mathbf{v}_{eo}$. Thus, relative to the observer, the person on the train has a velocity

$$\mathbf{v}_{ho} = \mathbf{v}_{he} + \mathbf{v}_{eo}.$$

(Note the use of subscripts: The first refers to the object—h for hobo and e for engine—and the second is what the velocity is with respect to—o for observer and e for engine.) When we insert the numerical values we get

$$\mathbf{v}_{ho} = 5 \text{ km/h} + 10 \text{ km/h} = 15 \text{ km/h}.$$

Example 3.4

Relative velocity crossing a river.

A person who can row a boat at 5.0 km/h in still water tries to cross a river whose current moves at a rate of 3.0 km/h. The boat is pointed straight across the river, but its progress includes a downstream motion due to the river current. (a) What is the velocity of the boat with respect to the bank? (b) If the river is 200 m wide, how far downstream does the rower land?

Solution   (a) Velocity is the rate of change of displacement, which is a vector quantity. Hence velocity is also a vector and has both direction and magnitude. Thus the rule for adding relative velocities is a rule of vector addition. Figure 3.12(a) shows a sketch of the physical situation and Fig. 3.12(b) is the corresponding vector diagram. Once you have made a scale diagram, it is relatively simple to use a ruler and protractor to get the answer of 5.8 km/h at an angle of 59° with respect to the river's motion. It is equally correct to give the direction as 31° with respect to a line straight across the river. You should do this example for yourself on paper, using a scale different from the one used in the text.

   (b) The displacement of the boat is proportional to its velocity. Consequently, the ratio of the distance that the boat drifts downstream to the

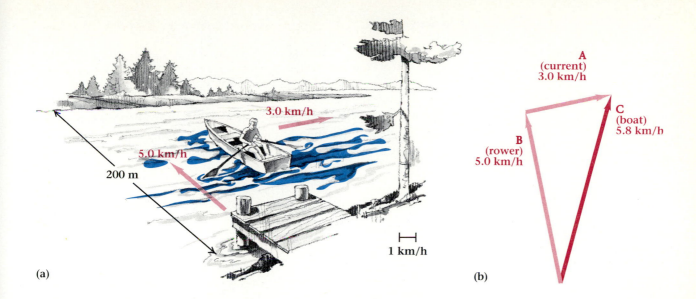

(a)

(b)

Figure 3.12 Example 3.4: (a) A boat crossing a stream with a current. (b) Vector diagram of the velocities. The boat travels downstream as well as across the river.

width across the river is the same as the ratio of the magnitude of the river's velocity to the magnitude of the velocity of the boat in still water:

$$\frac{\text{distance along bank}}{\text{width of river}} = \frac{\text{downstream current}}{\text{velocity across river}} = \frac{3 \text{ km/h}}{5 \text{ km/h}} = \frac{3}{5}.$$

Thus the boat lands downstream a distance given by

$$\text{distance} = \tfrac{3}{5} \times \text{width} = \tfrac{3}{5}(200 \text{ m})$$
$$= 120 \text{ m}.$$

Resolution of Vectors

3.3

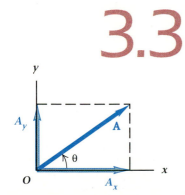

Figure 3.13 A vector A lying in the xy plane has components A_x along the x direction and A_y along the y direction. From trigonometry we see that $A_x = A \cos \theta$ and $A_y = A \sin \theta$.

It is often useful to think of a vector as the sum of two or three other vectors. We call these other vectors *components,* and they are especially convenient when we choose components at right angles to each other. In two dimensions we frequently choose these component vectors to lie along the x and y axes of a rectangular (Cartesian) coordinate system. For example, consider the vector **A** lying in the xy plane (Fig. 3.13). We can construct two component vectors by drawing lines from the end of **A** perpendicular to the x and y axes. The two vectors that lie along the x and y directions add to form **A.** When we find the magnitude of these two vectors we say that we have *resolved* the vector **A** into its x and y components.

Since the known vector and its component vectors form a right triangle, we may use trigonometry to resolve the vector into its components. (See the appendix to this chapter for a review of trigonometry.) This is why components at right angles are particularly convenient. Thus, for the vector of magnitude A that makes an angle θ with the x axis, the component A_x along the x direction and the component A_y along the y direction are given by

$$A_x = A \cos \theta,$$ (3.1)

$$A_y = A \sin \theta.$$

If we know the components A_x and A_y of a vector, then we can obtain the magnitude and direction of the vector by applying trigonometry. The magnitude A of the vector is computed from the Pythagorean theorem as

$$A = \sqrt{A_x^2 + A_y^2}.$$ (3.2)

The angle θ between the vector **A** and the x axis is determined from

$$\tan \theta = \frac{A_y}{A_x}.$$ (3.3)

Example 3.5

Resolution of a vector.

Swimming at an angle of 27° from the x axis, an angelfish has a velocity vector **v** with a magnitude of 25 cm/s (Fig. 3.14). Find the x and y components of **v**.

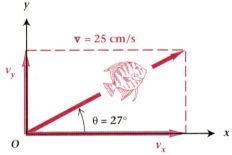

Figure 3.14   Example 3.5: Resolution of a vector into components along x and y.

Solution   To resolve vector **v** into its x and y components, we use Eq. (3.1):

$$v_x = v \cos \theta = (25 \text{ cm/s}) \cos 27° = 25 \text{ cm/s} \times 0.891 = 22 \text{ cm/s},$$

$$v_y = v \sin \theta = (25 \text{ cm/s}) \sin 27° = 25 \text{ cm/s} \times 0.454 = 11 \text{ cm/s}.$$

In many situations, the easiest way to add or subtract two vectors **A** and **B** is first to resolve each vector into components. Then we add the components along the x direction to form the x component of the new vector along that direction. Similarly, we add the y components to get the new y component. The vectors **A** and **B** add together to form a new vector **C** (Fig. 3.15a) whose components are

$$\left. \begin{array}{l} C_x = A_x + B_x, \\ C_y = A_y + B_y, \end{array} \right\} \text{ equivalent to } \mathbf{C} = \mathbf{A} + \mathbf{B}.$$

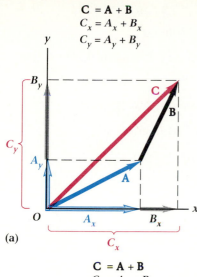

$$\mathbf{C} = \mathbf{A} + \mathbf{B}$$
$$C_x = A_x + B_x$$
$$C_y = A_y + B_y$$

(a)

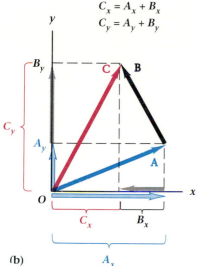

$$\mathbf{C} = \mathbf{A} + \mathbf{B}$$
$$C_x = A_x + B_x$$
$$C_y = A_y + B_y$$

(b)

Figure 3.15 Adding two vectors **A** and **B** to form a new vector **C**. (a) We add the *x* components to form C_x and the *y* components to form C_y. (b) The *x* component of **B** is directed along the negative direction, so the *x* component of **C** is less than A_x.

We can obtain the magnitude and direction of the new vector **C** from Eqs. (3.2) and (3.3) using the new components C_x and C_y. In the case illustrated in Fig. 3.15(b), the *x* component of vector **B** points in the negative direction so the sum of the *x* components is less than A_x.

If two vectors **A** and **B** are to be subtracted, we can use the same procedure of resolving each vector into components and subtracting them in proper order. Then we use the resulting components to find the magnitude and direction of the new vector **C**.

Adding or subtracting vectors by first resolving them into perpendicular components simplifies the mathematics by enabling us to add vectors in the same directions. This way we avoid having to make careful scale drawings or using trigonometry for every manipulation of vectors. The following examples illustrate the addition and subtraction of vectors by components.

Example 3.6

Adding vectors by summing components.

Vector **A** has a length of 14 cm at 60° with respect to the *x* axis and vector **B** has a length of 20 cm at 20° with respect to the *x* axis. Add the vectors by first resolving them into components and then adding the components.

Solution Figure 3.16 shows this situation. The components of the vectors are:

$$A_x = A \cos \theta_A \qquad\qquad A_y = A \sin \theta_A$$
$$= 14 \text{ cm} \cos 60° \qquad\qquad = 14 \text{ cm} \sin 60°$$
$$= 14 \text{ cm} \times 0.50 = 7.00 \text{ cm}; \qquad = 14 \text{ cm} \times 0.866 = 12.12 \text{ cm};$$
$$B_x = B \cos \theta_B \qquad\qquad B_y = B \sin \theta_B$$
$$= 20 \text{ cm} \cos 20° \qquad\qquad = 20 \text{ cm} \sin 20°$$
$$= 20 \text{ cm} \times 0.940 = 18.79 \text{ cm}; \qquad = 20 \text{ cm} \times 0.342 = 6.840 \text{ cm}.$$

The sum is given by

$$\mathbf{C} = \mathbf{A} + \mathbf{B},$$

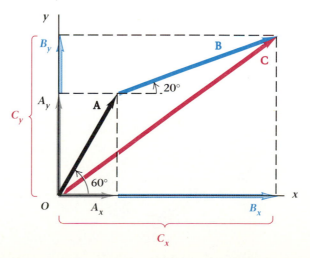

Figure 3.16 Example 3.6: Adding **A** and **B** by components.

where
$$C_x = A_x + B_x = 25.79 \text{ cm,}$$
$$C_y = A_y + B_y = 18.96 \text{ cm.}$$

The magnitude of **C** is then found from the Pythagorean theorem as

$$C = \sqrt{C_x^2 + C_y^2},$$
$$C = \sqrt{(25.79 \text{ cm})^2 + (18.96 \text{ cm})^2} = 32 \text{ cm.}$$

The angle of **C** with respect to the *x* axis is given by

$$\tan \theta = \frac{C_y}{C_x} = \frac{18.96}{25.79} = 0.735,$$
$$\theta = 36°.$$

Notice that we carried out all the computations before rounding off the answers to two significant figures. This procedure is consistent with the rules given in Chapter 1.

Example 3.7

Relative velocity by components.

A dolphin traveling at 10 km/h in still water enters a tidal current at an angle of 30° and swims in that direction. The current is moving parallel to the shore at a speed of 3.0 km/h. What is the dolphin's speed relative to the shore?

Solution Let us begin by looking at a diagram of the situation (Fig. 3.17). The velocity of the water relative to the shore is shown as $\mathbf{v}_{ws}$. The velocity of the dolphin relative to the water is $\mathbf{v}_{dw}$, which makes an angle of 30° with $\mathbf{v}_{ws}$. The vector sum of these two velocities ($\mathbf{v}_{dw} + \mathbf{v}_{ws}$) is the velocity of the dolphin relative to the shore, which we call $\mathbf{v}_{ds}$.

To calculate this quantity, we begin by resolving $\mathbf{v}_{dw}$ into components

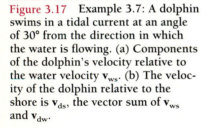

Figure 3.17 **Example 3.7: A dolphin swims in a tidal current at an angle of 30° from the direction in which the water is flowing. (a) Components of the dolphin's velocity relative to the water velocity $\mathbf{v}_{ws}$. (b) The velocity of the dolphin relative to the shore is $\mathbf{v}_{ds}$, the vector sum of $\mathbf{v}_{ws}$ and $\mathbf{v}_{dw}$.**

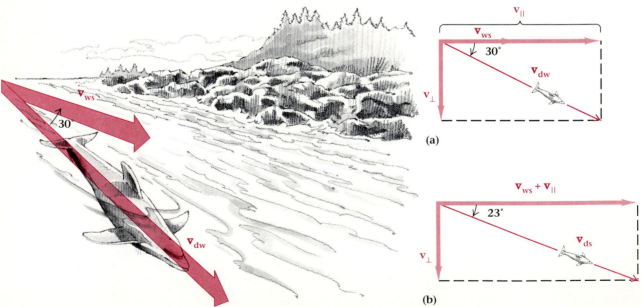

parallel and perpendicular to $\mathbf{v}_{ws}$ (Fig. 3.17a). The parallel component is

$$v_{\parallel} = v_{dw} \cos 30° = (10 \text{ km/h}) \cos 30° = 8.66 \text{ km/h}.$$

The perpendicular component is

$$v_{\perp} = v_{dw} \sin 30° = (10 \text{ km/h}) \sin 30° = 5.00 \text{ km/h}.$$

The dolphin's total speed parallel to the current is just the sum of $v_{ws} + v_{\parallel}$ or 11.66 km/h. The total speed perpendicular to the current is $v_{\perp} = 5$ km/h. The total velocity $\mathbf{v}_{ds}$ of the dolphin relative to the shore is the vector sum of these components (Fig. 3.17b).

The magnitude of $\mathbf{v}_{ds}$ is given by the Pythagorean theorem as

$$v_{ds} = \sqrt{(11.66 \text{ km/h})^2 + (5.00 \text{ km/h})^2} = 13 \text{ km/h}.$$

The magnitude of $\mathbf{v}_{ds}$ is the speed of the dolphin relative to the shore. The resulting direction is 23° away from the direction of the current. You should verify these results both by repeating the calculations yourself and by making a scale drawing and measuring with a ruler and protractor.

Example 3.8

Flying in a cross wind.

A small airplane flies with a speed relative to the ground (ground speed) of 200 km/h in a direction 18° to the east of north. If the plane is headed due north and the deviation from that direction is due to a cross wind blowing from west to east, what is the speed of the wind?

Solution Referring to Fig. 3.18, we see that the speed of the wind is

$$v_w = v \sin \theta$$
$$v_w = 200 \text{ km/h} \times \sin 18°$$
$$= 200 \text{ km/h} \times 0.309$$
$$= 61.8 \text{ km/h}.$$

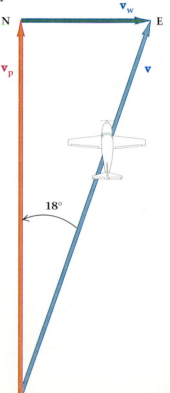

Figure 3.18 Example 3.8: An airplane headed due north is blown off course by a cross wind blowing due east.

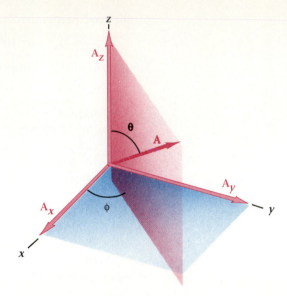

Figure 3.19 Components of a vector in three dimensions.

To this point we have considered the addition or subtraction of two vectors that lie in the same plane. Other cases occur in nature in which the vectors occupy three dimensions, rather than two. Although we will not deal with such cases very often in this text, we will state the basic principle for adding and subtracting vectors in three dimensions. The vectors to be added (or subtracted) are resolved into components along the x, y, and z axes and the components are added (or subtracted). This procedure is the same as that used for vectors lying in a plane except that there are three rather than two dimensions. Figure 3.19 shows the resolution of a vector into components in three dimensions.

Kinematics in Two Dimensions

3.4

We can now rephrase our earlier kinematic definitions in vector form. The one-dimensional kinematic equations developed in Chapter 2 for motion along a straight line are the equations for one component of a vector. For example, the average velocity vector is defined by

$$\bar{\mathbf{v}} = \frac{\Delta \mathbf{s}}{\Delta t} = \frac{\mathbf{s} - \mathbf{s}_0}{t - 0},$$
(3.4)

where $\mathbf{s}_0$ and $\mathbf{s}$ are the initial and final displacement vectors, t is the final time, and the initial time is set equal to zero.

Now we can see that the average velocity is distinct from the average speed, which depends on the total distance traveled. For example, imagine a round trip in which a bee moves a distance of 5 cm and then returns to its starting point in 1 s. During that time interval the bee travels a total distance of 10 cm and has an average speed of 10 cm/s. However, its net change in vector displacement is zero since $\mathbf{s}_0$ (the position at the start) and $\mathbf{s}$ (the position at the end) are the same. Therefore the bee has an average velocity of zero.

As in the case for one dimension, the instantaneous velocity $\mathbf{v}$ is given by the limiting value of $\Delta\mathbf{s}/\Delta t$ as Δt becomes vanishingly small. Similarly, the components of the instantaneous velocity vector are the limiting values of the component displacements divided by Δt as Δt becomes vanishingly small:

$$v_x = \frac{\Delta x}{\Delta t}, \qquad v_y = \frac{\Delta y}{\Delta t}.$$

The magnitude of the instantaneous velocity is given by

$$v = \sqrt{v_x^2 + v_y^2}.$$

The direction of motion is the direction of the instantaneous velocity, not the direction of the displacement.

The average vector acceleration is defined by

$$\bar{\mathbf{a}} \equiv \frac{\Delta\mathbf{v}}{\Delta t} = \frac{\mathbf{v} - \mathbf{v}_0}{\Delta t}. \tag{3.5}$$

Notice that an acceleration can arise from a change in the direction of the velocity as well as from a change in its magnitude, or speed. In particular, a body moving in a circle may have constant speed, yet it will be accelerating because its velocity is continuously changing direction. We will study uniform circular motion in Chapter 5.

The instantaneous acceleration vector $\mathbf{a}$ is given by the limiting value of $\Delta\mathbf{v}/\Delta t$ as Δt becomes vanishingly small. As in the case of velocity, the instantaneous acceleration can be resolved into components of acceleration a_x and a_y,

$$a_x = \frac{\Delta v_x}{\Delta t}, \qquad a_y = \frac{\Delta v_y}{\Delta t}.$$

The magnitude of the instantaneous acceleration is given by

$$a = \sqrt{a_x^2 + a_y^2}.$$

The direction of the acceleration depends on the direction of change of velocity, not on the direction of the velocity.

The kinematic equations for motion with constant acceleration can be extended to vector form and expressed in terms of their separate components. For example, Eq. (2.7) written for two dimensions becomes

$$x = x_0 + v_{x0}t + \tfrac{1}{2}a_xt^2, \tag{3.6a}$$

$$y = y_0 + v_{y0}t + \tfrac{1}{2}a_yt^2, \tag{3.6b}$$

where x_0 and y_0 are the initial positions, v_{x0} and v_{y0} are the components of initial velocity, and a_x and a_y are the components of acceleration. We can apply the same reasoning to extend these equations to the case of three dimensions.

Example 3.9

Graphical analysis of velocity and acceleration.

A particle is confined to move in a horizontal plane. Its location at time $t = 0$ is chosen as the origin of a coordinate system. The particle has an initial velocity along the x direction of 10 cm/s and is subject to a constant acceleration along the y direction of 2 cm/s^2. Determine the path of the particle by computing its position at $t = 1, 2, 3, 4,$ and 5 s and graphing the result. What is the displacement of the particle from the origin at time $t = 5$ s and what is its velocity?

Solution   We can use Eqs. (3.6) to find the x and y coordinates of the particle at the times specified:

$$x = x_0 + v_{x0}t + \tfrac{1}{2}a_x t^2 \qquad \text{and} \qquad y = y_0 + v_{y0}t + \tfrac{1}{2}a_y t^2,$$

where $x_0 = 0$, $v_{x0} = 10$ cm/s, $a_x = 0$, $y_0 = 0$, $v_{y0} = 0$, and $a_y = 2$ cm/s^2. When these values are inserted into the equations they become

$$x = 0 + (10 \text{ cm/s})t + 0 \qquad \text{and} \qquad y = 0 + 0 + \tfrac{1}{2}(2 \text{ cm/s}^2)t^2.$$

Inserting the values for the time in these equations, we can compute the coordinates x and y of the particle at each second along its path:

t	x	y
0	0	0
1	10 cm	1 cm
2	20 cm	4 cm
3	30 cm	9 cm
4	40 cm	16 cm
5	50 cm	25 cm

Figure 3.20   Example 3.9: (a) Path of the particle described. (b) Displacement of the particle at time $t = 5$ s. (c) Velocity of the particle at $t = 5$ s.

These points, plotted on the graph in Fig. 3.20(a), show the path taken by the particle.

The displacement of the particle from the origin at $t = 5$ s is the vector $\Delta\mathbf{s} = \mathbf{s} - 0$, whose coordinates are $x = 50$ cm and $y = 25$ cm.

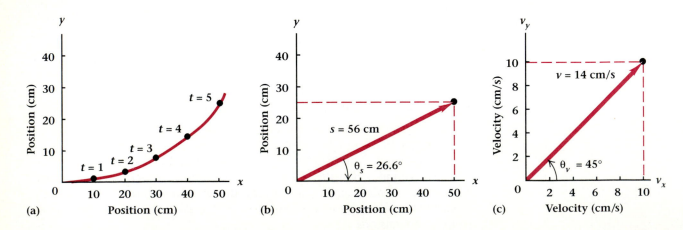

(a) Position (cm) / Position (cm)
(b) Position (cm) / Position (cm)
(c) Velocity (cm/s) / Velocity (cm/s)

Its magnitude is

$$s = \sqrt{x^2 + y^2} = \sqrt{(50)^2 + (25)^2} \text{ cm} = 56 \text{ cm}.$$

The direction of the displacement with respect to the x axis is given by the angle θ_s, where

$$\tan \theta_s = \frac{y}{x} = \frac{25}{50}.$$

This corresponds to an angle $\theta_s = 26.6°$ (Fig. 3.20b).

The equation for velocity given in Chapter 2 can be expressed in component form to give

$$v_x = v_{x0} + a_x t \quad \text{and} \quad v_y = v_{y0} + a_y t.$$

For the case at hand we are seeking v_x and v_y at $t = 5$ s. They are

$$v_x = 10 \text{ cm/s} + 0 \quad \text{and} \quad v_y = 0 + (2 \text{ cm/s}^2)(5 \text{ s})$$

$$v_x = 10 \text{ cm/s} \quad\quad\quad v_y = 10 \text{ cm/s}.$$

The velocity vector at $t = 5$ s, shown in Fig. 3.20c, has a magnitude

$$v = \sqrt{v_x^2 + v_y^2} = \sqrt{(10)^2 + (10)^2} \text{ cm/s} = 14 \text{ cm/s}.$$

The vector is directed at an angle θ_v given by

$$\tan \theta_v = v_y/v_x = 1,$$

$$\theta_v = 45°.$$

The direction in which the particle is traveling is the direction of its instantaneous velocity. Notice that this is not the same as the direction of the displacement from the origin. To help understand two-dimensional motion, you should extend this computation out to $t = 10$ s.

Projectile Motion

3.5

In the last part of his work *Discourses on Two New Sciences*, Galileo published a particularly useful idea that arose in connection with the motion of projectiles. Galileo observed that we could think of their motion as consisting of a horizontal part with constant speed and a vertical part with constant downward acceleration. Each of these motions is independent of the other, but their combination describes the motion of the projectile. For example, a ball thrown horizontally from a bridge (Fig. 3.21) falls with the same downward acceleration as a ball that is simply dropped. The downward motion is unaffected by the horizontal motion. Thus the ball thrown horizontally reaches the river at the same time as the ball dropped vertically, a result predicted by Galileo.

Galileo also used the example of a stone dropped from the mast of a ship moving with constant speed. To a shipboard observer the stone appears to travel straight down alongside the mast, falling with constant acceleration. However, to an observer on the shore, the path of the falling

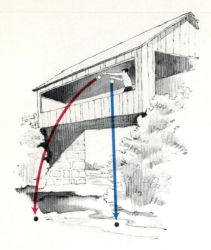

Figure 3.21   A ball thrown horizontally from a bridge has the same downward acceleration as a ball that is simply dropped.

stone on the moving ship is equivalent to projecting it horizontally from a stationary point with an initial velocity equal to the velocity of the ship.

It is important to understand that the motion seen from the shore consists of two independent parts: the downward motion with constant acceleration and the horizontal motion with constant velocity. (In this discussion we neglect the effect of air resistance.) Experiments have confirmed the independence of the horizontal and vertical components. The reason for this will become clear in Chapter 4. To clarify the idea of independent components of motion, study the following example.

Example 3.10

Throwing a ball from a height.

A ball is thrown horizontally from the Leaning Tower of Pisa (Fig. 3.22) with a velocity of 22 m/s. If the ball is thrown from a height of 49 m above ground, how far from the point directly below the launch point does it strike the ground?

Solution   We can treat the x and y motions independently and we know that the time it takes the ball to reach the ground will be the same for both motions. So, knowing the vertical height through which the ball falls, we can solve the equation of vertical motion for the time of fall and then use this value in the equations for horizontal motion to determine the horizontal distance.

The time for the ball to reach the ground depends on the height h and the vertical acceleration a. (In Section 2.7 we discussed the motion of a body falling freely under the acceleration of gravity.) We may use Eq. (3.6b), $y = y_0 + v_{y0}t + \frac{1}{2}a_y t^2$, where the initial y coordinate position is $y_0 = h$, the initial velocity in the y direction is $v_{y0} = 0$ and the acceleration in the y direction is $a_y = -g$, downward. The time for the ball to reach the ground is given by this equation with $y = 0$ and the values for y_0, v_{y0}, and a_y inserted. The equation then becomes

$$0 = h + 0 - \tfrac{1}{2}gt^2.$$

Solving for t and inserting the numerical values gives

$$t = \sqrt{\frac{2h}{g}} = \sqrt{\frac{2 \times 49 \text{ m}}{9.8 \text{ m/s}^2}},$$

$$t = 3.162 \text{ s} \approx 3.2 \text{ s}.$$

This is the elapsed time before the ball strikes the ground after it has been thrown horizontally.

The horizontal component of the motion is not accelerated and the ball continues to travel with its x component of velocity unchanged until it strikes the ground. The distance traveled in time t at constant velocity v_x is

$$x = v_x t = 22 \text{ m/s} \times 3.162 \text{ s},$$

$$x = 70 \text{ m}.$$

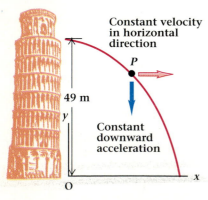

Figure 3.22   Example 3.10: A ball thrown from the Leaning Tower of Pisa.

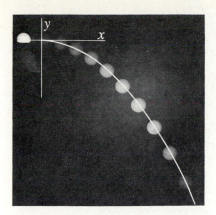

Figure 3.23 A photograph of a golf ball illuminated with a flashing strobe light and given an initial horizontal velocity. Superimposed on the photograph is a section of a parabola drawn from Eq. (3.7). Whether an object is thrown horizontally, upward, or downward, the principle remains the same: the object follows a parabolic path.

The combined effect of horizontal motion with constant velocity and vertical motion with constant acceleration is that the body moves in a parabolic path. In the following discussion we prove that the path of a projectile thrown horizontally (as in Example 3.10) is a parabola. We then extend our discussion to cases in which the body is projected at any angle, not just horizontally. If you need to review some of the properties of a parabola, refer to page 13.

Consider an object, like a golf ball, projected horizontally (Fig. 3.23). For the y (vertical) component of motion, the initial velocity is zero and the acceleration is that of gravity, giving

$$y = -\tfrac{1}{2}gt^2.$$

We use the minus sign because the acceleration of gravity is downward and we have chosen the upward direction to be positive in the figure. The projected object starts at $y = 0$ and falls to negative values of y.

The horizontal component of motion has an initial velocity but is not accelerated, so that

$$x = v_0 t.$$

Solving this equation for t and inserting the value of t into the equation for y, we get

$$y = -\frac{g}{2v_0^2} x^2. \tag{3.7}$$

Equation (3.7) has the same form as the equation for a parabola. In both cases the factor that is multiplied by x^2 on the right-hand side is a constant for a particular problem. Thus we conclude that projectile motion is parabolic.

A more detailed analysis would take account of air resistance, which causes a departure from a true parabolic path in many real situations. In this book, we will consider air resistance to be negligible unless stated otherwise. Although the assumption does not correspond to physical reality, it often gives a good approximation. The mathematical model needed to include the effects of air resistance is more complicated than we wish to analyze in this course.

Example 3.11

Hitting a falling target.

A boy aims his slingshot directly at an apple hanging in a tree. At the moment he shoots a pebble, the apple drops. Show that the pebble hits the target.

Solution For convenience, we place an xy coordinate system with its origin at the slingshot (Fig. 3.24). We first consider the motion of the projectile. Let the time the pebble takes to travel the horizontal distance D be t' and let θ be the firing angle measured from the horizontal. Then the horizontal component of the pebble's velocity is $v_0 \cos \theta$, so we may write

$$D = v_0(\cos \theta)t'$$

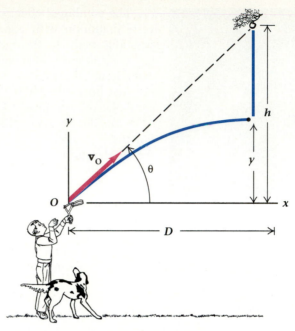

Figure 3.24   **Example 3.11: Shooting a pebble at an apple that drops at the instant the pebble is fired.**

or

$$t' = \frac{D}{v_0 \cos \theta}.$$

The vertical component of the pebble's initial velocity is $v_0 \sin \theta$, so the vertical distance traveled by the pebble during the same period of time is given by Eq. (3.6b):

$$y = v_0(\sin \theta)t' - \tfrac{1}{2}gt'^2.$$

By substituting the value of t' from the equation above, we get

$$y = D \tan \theta - \tfrac{1}{2}gt'^2.$$

Examination of the diagram shows that $D \tan \theta$ is the initial height of the apple h, so that the height reached by the pebble in time t' is

$$y = h - \tfrac{1}{2}gt'^2.$$

This height is just exactly the point to which the apple will have fallen after a time t'. So, the pebble hits the apple.

This result is independent of the speed v_0, so collision occurs for any projectile (pebble, bullet, arrow, or whatever) as long as the projectile travels a distance D before hitting the ground. For the case shown in Fig. 3.24, the projectile was still rising when it hit the target. However, the collision would still occur even if the projectile were past its maximum height. This example is often the basis for a classroom demonstration experiment.

Range of a Projectile

*3.6

Figure 3.25 depicts the path of a football projected upward that follows a parabolic arc. As usual, we have neglected air resistance. The football's initial velocity is $\mathbf{v}_0$ at an angle θ with respect to the horizontal. The horizontal component of velocity is $v_0 \cos \theta$ and is constant because there is no horizontal acceleration. The initial vertical or y component of velocity is $v_0 \sin \theta$. The vertical component of velocity changes with time as a result of gravitational acceleration. As the object rises, it slows down and its vertical component of velocity v_y at any time t is

$$v_y = v_0 \sin \theta - gt,$$

where the initial vertical velocity is $v_0 \sin \theta$ and the acceleration is $-g$.

The time for an object thrown upward from the ground to reach its maximum height and the time for it to fall to the ground again are the same, provided the initial height and final height are the same. This symmetry in the upward and downward part of the path is due to the fact that the acceleration of gravity, g, is the same for upward and downward motion at any height. The total time, T, that the object is in the air is then twice the time required for it to reach its maximum height. At the maximum height, the vertical component of velocity becomes zero; that is, the projectile stops rising before it begins to fall down. (It continues moving horizontally at this point.) Thus at the maximum height, $v_y = 0$, $t = T/2$, and the velocity equation in the vertical direction becomes

$$0 = v_0 \sin \theta - g \frac{T}{2}.$$

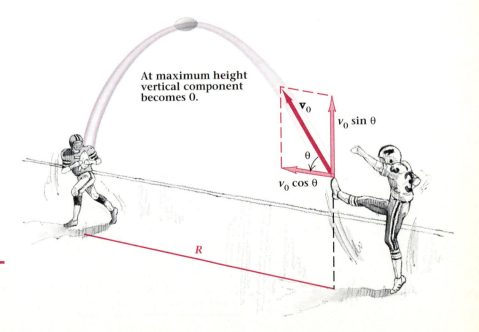

At maximum height vertical component becomes 0.

Figure 3.25 Range of a projectile (football) with initial velocity $\mathbf{v}_0$ directed at an angle θ with respect to the horizontal.

Upon rearranging, we find the time T to be

$$T = \frac{2\,v_0 \sin\theta}{g}.$$

The **range** R, indicated in Fig. 3.25, is the horizontal distance traveled in the time T with the constant horizontal velocity v_x. Notice that the range given here is the horizontal distance traveled during the time for the projectile to return to the height of its initial position. The range is given by $v_x t$, with $v_x = v_0 \cos\theta$ and $t = T$:

$$R = v_0(\cos\theta)T.$$

If the previous equation is used for the time T, the expression for the range becomes

$$R = \frac{2v_0^2}{g} \sin\theta \cos\theta.$$

Making use of the trigonometric identity $2\sin\theta\cos\theta = \sin 2\theta$, we can express the range as

$$R = \frac{v_0^2}{g} \sin 2\theta. \tag{3.8}$$

Equation (3.8) gives the range of a projectile thrown upward with an initial speed v_0 at an angle θ with respect to the horizontal. In cases where air resistance plays a role, the range is usually less than the amount given by this expression. An obvious exception is the Frisbee, a light disk that acquires a lifting force from the air passing over its surface.

Figure 3.26 shows the paths of an object projected upward at various angles and the same initial speed. Figure 3.27 is a graphical representation

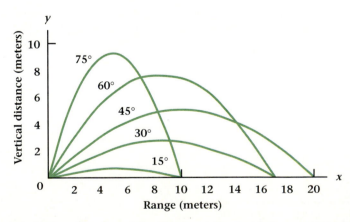

Figure 3.26   Trajectories of an object thrown upward with the same initial speed at various angles of inclination.

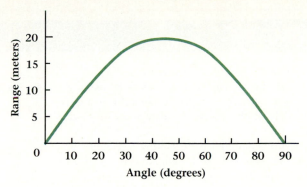

Figure 3.27 **Range versus angle with respect to the horizontal for the same initial speed.**

of Eq. (3.8). From either figure we can see that the maximum range occurs when $\theta = 45°$. (At this value of θ, sin 2θ has its maximum value of 1.) The expression for the maximum range R_{max} is then

$$R_{max} = \frac{v_0^2}{g}.$$

Example 3.12

How far is a stone's throw away?

How far will a stone travel over level ground if it is thrown upward at an angle of 30° with respect to the horizontal and with a speed of 12.0 m/s? What maximum range could be achieved with the same initial speed but with improved aim?

Solution Using Eq. (3.8), we have

$$R = \frac{v_0^2}{g} \sin 2\theta = \frac{(12 \text{ m/s})^2 \sin(2 \times 30°)}{9.80 \text{ m/s}^2} = 12.7 \text{ m.}$$

Throwing at an angle of 45°, we get the range

$$R_{max} = \frac{v_0^2}{g} = 14.7 \text{ m.}$$

We obtain an improvement of about 16%.

Because we live in a world that is more than one-dimensional, it is helpful to use vectors to properly describe the motions of objects. However, we can often separate these motions into their mutually perpendicular components and treat them separately, as we have done in this section. In this chapter we have considered only the way in which objects move. In the next and subsequent chapters we will inquire into the causes of motion and the relationships between various motions and their causes.

SUMMARY

Useful Concepts

- We can determine the sum of two vectors by using geometry, algebra, or trigonometry. However, it is a good idea, especially at the outset, to always make a scale drawing.
- Vectors are added graphically by making a scale drawing and adding the vectors head to tail. A vector **A** may be resolved into its components:

$$A_x = A \cos \theta, \qquad A_y = A \sin \theta.$$

- The angle θ is measured counterclockwise from the positive x axis.
- Vectors may be added together by adding their components. The magnitude and direction of a vector **A** in terms of its components are

$$A = \sqrt{A_x^2 + A_y^2} \qquad \text{and} \qquad \tan \theta = \frac{A_y}{A_x}.$$

- The average velocity vector is

$$\bar{\mathbf{v}} \equiv \frac{\mathbf{s} - \mathbf{s}_0}{\Delta t}.$$

- The average acceleration vector is

$$\bar{\mathbf{a}} \equiv \frac{\mathbf{v} - \mathbf{v}_0}{\Delta t}.$$

- A projectile launched with a velocity v_0 at an angle θ with the horizontal has x and y components of velocity:

$$v_x = v_0 \cos \theta \qquad \text{and} \qquad v_y = v_0 \sin \theta - gt.$$

- The path of the projectile is a parabola (neglecting air resistance), and the range of the projectile is

$$R = \frac{v_0^2 \sin 2\theta}{g}.$$

Important Terms

You should be able to write out the definitions or meanings of the following terms:

- scalar
- vector
- resultant
- relative velocity
- range of a projectile

QUESTIONS

3.1   List as many vector quantities as you can think of.

3.2   How is the meaning of the word *vectorcardiography* connected with our definition of a vector? Can you give any other compound words containing *vector* or any uses of the word outside of mathematics and physics?

3.3   What does it mean when we say that a weather map indicating temperatures represents a scalar field but that a map indicating the winds represents a vector field?

3.4   Show how you would add the following three vectors: 10 units north, 10 units south, and 10 units straight up.

3.5   Find an expression or procedure for finding the length of a vector with components along the x, y, and z axes.

3.6   In Galileo's example, if the stone had been thrown downward from the top of the mast rather than dropped from rest, what would the path look like to observers on the ship and on the shore?

3.7   If the object had been thrown horizontally from the mast of the ship in Galileo's example, what type of path would a person on the ship have observed? What type of path would a person on shore have observed? Be as explicit as you can for both cases.

3.8   So far we have considered two conic sections. In Chapter 1 we discussed the ellipse and in this chapter we used the parabola. From an encyclopedia or other reference book, find out about a hyperbola. Write down its properties and list some cases in which it occurs in connection with physical phenomena.

3.9   By inspection of Eq. (3.8) for the range of a projectile, state what must be the relationship between angles that give rise to the same range. (Also see Fig. 3.26.)

3.10   What effect does doubling the initial speed have on the range of a projectile?

PROBLEMS

Hints for Solving Problems
Draw scale diagrams carefully. Indicate your scales on the diagram. Pay careful attention to signs and apply them consistently throughout the problem. Remember that horizontal and vertical components of motion are independent of each other, but they take place during the same time interval.

Section 3.1 Addition of Vectors

3.1 If you walk six city blocks east and then eight blocks south, how many blocks are you from your starting place? What direction are you from the starting point? Give your answer as an angle measured from due east.

3.2 A motorist travels directly north for 30 km, then turns to the west and goes 18 km. How far is she from the starting point and in what direction? Give the angle with respect to the east.

3.3 Vector **A** has a magnitude of 13 units at a direction of 250°, and vector **B** has a magnitude of 27 units at a direction of 330°, both measured with respect to the positive x axis. (a) What is the vector sum of **A** and **B**? (b) What is the vector **C** = **A** − **B**? Find your answers by graphical analysis.

3.4 Add the following vectors graphically in the order given, then add them in reverse order on a separate diagram, thereby testing that vector addition is commutative: **A** = 9 units at 60° and **B** = 7 units at 180°.

3.5 A jogger runs directly east for 5.0 km, then turns and goes northwest for 7.0 km. He then travels directly west for 3.0 km. How far and in what direction is he from the starting point?

3.6 Five vectors of equal magnitude radiate outward from the center of a regular pentagon, one vector pointing toward each vertex (Fig. 3.28). Show by a graphical treatment that the sum of all five vectors is zero.

3.7 A vector **A** has a magnitude of 20 units at 200° and

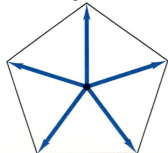

Figure 3.28 Problem 3.6.

a vector **B** has a magnitude of 37 units at 45°. What is the vector difference **A** − **B**?

3.8 The vectors diagrammed in Fig. 3.29 have the same magnitude, A. Find (a) **B** + **C**, (b) **A** + **B** + **C**, (c) **B** − **C**, (d) **A** − **B** − **C**.

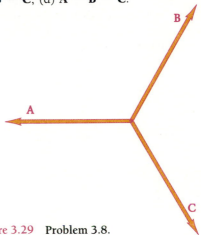

Figure 3.29 Problem 3.8.

3.9 Four ropes are tied to a stake and each is pulled toward a compass direction, N, S, E, or W. A force of 10 lb is applied to the rope pulled toward the east. Forces of 20, 30, and 40 lb are applied toward the south, west, and north, respectively. What is the net force exerted on the stake by the ropes and in what direction is it? (*Hint*: Forces are vector quantities. The net force is the vector sum of the individual forces.)

Section 3.2 Relative Velocity

3.10 A passenger rushing to catch a plane at the airport walks on a moving sidewalk at a speed of 3 km/h relative to the sidewalk in the direction that the sidewalk is moving. The sidewalk is 100 m long and moves with a steady velocity of 1 km/h. (a) How long does it take for the passenger to get from one end of the sidewalk to the other, that is, to cover the 100 m? (b) How much time does the passenger save by taking the moving sidewalk instead of just walking beside it? (c) Through what distance does the passenger walk relative to the moving sidewalk? (d) If the passenger's stride is 80 cm, how many steps are taken in going from one end of the moving sidewalk to the other?

3.11 Two cars approach each other on the highway. Car A moves north at 90 km/h, car B moves south at 70 km/h. (a) What is the velocity of car A as seen from car B? (b) What is the velocity of car B as seen from car A? (c) What are their velocities relative to car C, which is traveling north at 100 km/h?

3.12 A boat is traveling in a river with a current of 3.0 km/h. The boat is capable of traveling at 10.0 km/h in still water. How long will it take the boat to travel 7.0 km upstream? How long will it take to travel 7.0 km downstream?

3.13 A boat capable of making 9.0 km/h in still water is traveling upstream in a river flowing at 4.0 km/h (Fig. 3.30). An object lost overboard is not missed for 30 minutes. If the boat is then turned around, how long will it take to overtake the lost object? What total distance will the boat travel relative to the shore from the point of turnaround to the point of overtaking the object?

Figure 3.30   Problem 3.13.

Section 3.3   Resolution of Vectors
Solve the following vector problems algebraically.

3.14 Resolve the following vector into its x and y components: magnitude 10 directed at 30° with respect to the x direction.

3.15 A certain vector has components of 12 units along the x axis and 9.0 units along the y axis. What is the magnitude of the vector? What is the angle between the vector and the x axis?

3.16 What is the magnitude and direction of the vector with $A_x = 4.5$ and $A_y = -6.8$?

3.17 If a vector has a magnitude of 18 and an x component of -7.0, what are the two possibilities for its y component and direction?

3.18 The direction of a vector is 127° from the positive x axis, and its y component is 18.0. Find the x component and the magnitude of the vector.

3.19 (a) Find the sum **C** of the two vectors **A** and **B** given by $A_x = 3$, $A_y = 7$, and $B_x = 10$, $B_y = -9$. (b) Find the difference $\mathbf{D} = \mathbf{A} - \mathbf{B}$.

3.20 Vector **A** has a magnitude of 8.0 at an angle of 60° from the x axis. Vector **B** has a magnitude of 6.0 at an angle of $-30°$ from the x axis. Determine the vector sum $\mathbf{C} = \mathbf{A} + \mathbf{B}$ by algebraic resolution into components.

3.21 Find the difference $\mathbf{D} = \mathbf{A} - \mathbf{B}$ for the vectors given in Problem 3.20.

3.22 Find the vector $\mathbf{V} = 2\mathbf{A} - \mathbf{B}$ for **A** and **B** given in Problem 3.20.

3.23 A boat capable of making 9.0 km/h in still water is used to cross a river flowing at a speed of 4.0 km/h. (a) At what angle must the boat be directed so that

its motion will be straight across the river? (b) What is its resultant speed relative to the shore?

3.24 An airplane heading south with an air speed of 200 km/h is in a cross wind of 10 km/h blowing toward the west. How far does the airplane go in two hours and in what direction?

3.25 A vector **A** has a magnitude of two units at 30° with respect to the horizontal. A vector **B** has a magnitude of five units at 90°. Calculate $\mathbf{C} = 3\mathbf{A} + 2\mathbf{B}$.

Section 3.4 and 3.5   Kinematics in Two Dimensions and Projectile Motion

3.26 A student touring Japan accidentally drops a 500-yen coin from a height of 1.20 m above the floor of the Bullet Train, which is traveling at 250 km/h. (a) How long does it take for the coin to hit the floor? (b) Where does the coin land with respect to the floor of the train? (c) How far along the track does the train go during the time it takes for the coin to fall?

3.27 A monkey on a cliff throws a coconut horizontally
• from a height of 17 m with a speed of 2.1 m/s. If the ground below the cliff is level, how far from the base of the cliff does the coconut strike the ground?

3.28 A rock thrown horizontally from the top of a radio
• tower lands 17.0 m from the base of the tower. If the speed at which the object was projected was 9.50 m/s, how high is the tower?

3.29 A ball thrown horizontally from a 13-m-high building
• strikes the ground 5.0 m from the building. With what velocity was the ball thrown?

*Section 3.6   Range of a Projectile
Assume that air friction can be ignored for these problems.

3.30 What is the horizontal range of a ball thrown upward with a speed of 20 m/s at an angle of 30° with respect to the horizontal?

3.31 A cannonball shot at an angle of 60.0° with respect to the horizontal had a range of 196 m. What was its initial speed?

3.32 What angles give rise to a range of 100 m for a cannonball with an initial speed of 50.0 m/s?

3.33 Locusts have been observed to jump distances up to 80 cm on a level floor. Photographs of their jump show that they usually take off at an angle of about 55° from the horizontal. Calculate the initial velocity of a locust making a jump of 80 cm with a takeoff angle of 55°.

3.34 What is the maximum range for a baseball whose initial speed is 36.0 m/s?

3.35 How much farther will a golf ball with an initial speed of 75 m/s go when projected at 45° than when projected at 30°?

3.36 A rock thrown with an initial velocity of 32.5 m/s at an angle of 50° with respect to the horizontal has a range of 985 m on a certain planet. What is the acceleration of gravity on this planet?

Additional Problems

3.37 A cannon sitting on the battlement of a castle is aimed
•• away from the castle at an angle θ above the horizontal. Find an equation for the time for the cannonball to hit the level ground below in terms of the height of the battlement, the magnitude and direction of the initial velocity, and the acceleration of gravity.

3.38 A cougar leaps horizontally from the top of a cliff
• with an initial velocity of 8.25 m/s. The cliff is 6.43 m tall. Sketch the path of the cougar. What is the magnitude and direction of the velocity when the cougar is halfway to the ground?

3.39 Show that the maximum height h to which a football
•• rises when kicked from the ground at an angle θ is given by $h = 0.25\,R\tan\theta$, where R is the range.

3.40 A cannon is adjusted for maximum range R_{max} on
•• level ground. How high is the cannonball when its horizontal distance from the cannon is $\frac{3}{4}R_{max}$?

3.41 A motorcyclist rides a given distance due north, then
• drives twice that far to the west. At the end of the ride the direct distance from the starting point is 112 m. What is the length of each leg of the trip and in what direction is the end point with respect to due north?

3.42 Three vectors have the same magnitude of 14 units. They make angles of θ, 2θ, and 3θ with respect to the x axis, where $\theta = 20°$. What is their vector sum?

3.43 A salesperson made a trip starting from a certain city. The trip was made in two straight-line parts, one of which was 20 km long and to the east. The end point of the trip was 80 km from the starting point at a direction of 45° southeast. Describe the other leg of the trip. Work this as a problem in vector subtraction.

3.44 How high does a golf ball rise when projected for
• maximum range? Express your answer in terms of the range.

3.45 What spread in range (as a percent of the maximum
• range) corresponds to a spread of $\pm10°$ from the correct angle for the maximum range?

3.46 If you can throw a ball vertically upward to a height
• $h = 20$ m, what is the maximum horizontal range over which you can throw the same ball, assuming you throw it at the same initial speed?

3.47 A ladder leans against a house at an angle of 76° from the horizontal. The base of the ladder is 1.48 m from the house. (a) How long is the ladder? (b) How far up the wall does the ladder touch house?

3.48 A guy wire bracing a power pole makes an angle of 53° with the ground. How long is the guy wire if it attaches to the pole at a height of 5.50 m above the ground?

3.49 A physics student standing on the top floor of a building sights downward to a newspaper box on the ground below. The line of sight to the box makes an angle of 75° with respect to the horizontal. After coming down, the student measures the distance from the building to the box to be 40.5 m. How high above the ground were the student's eyes when he was looking at the box from the top floor?

3.50 Observations are made of a distant television tower from each end of a 100-m base line. From each position the angle between the line of sight and the base line is 85.5°. How far is the tower from the center of the line?

3.51 A point on the rim of a 15.25-cm-radius phonograph
•• disk was moved through an angle of 0.250 radians. (a) How far did the point move along the circular arc? (b) What was the straight-line distance between the starting and ending points?

3.52 A ladder that is 4.00 m long is leaning against a wall
• at an angle of 64° with respect to the ground. If the base of the ladder is moved 0.30 m farther away from the wall, how far will the top of the ladder go down?

3.53 Make a table comparing θ in radians with $\sin\theta$ for
• angles of 1°, 3°, 6°, 9°, 12°, 15°, 18°, and 21°. Comment on the range of validity of the approximation $\sin\theta \approx \theta$.

3.54 A woman traveling due north makes a detour around
•• a large lake in the following manner (Fig. 3.31). At point A she turns 46° toward the west and travels 4.91 km to B. At B she turns to the north and travels

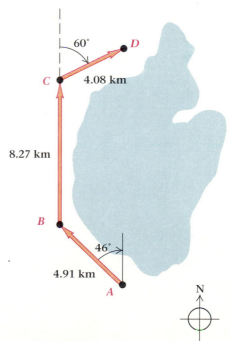

Figure 3.31 **Problem 3.54.**

a distance of 8.27 km to C. At C she turns 60° to the east and travels a distance of 4.08 km to D. At D she turns to the north and continues her trip. (a) Are the initial and final northward paths along the same line? (b) What is the distance between A and D?

3.55 A point on a horizontal plane starts at the origin and
•• moves with velocity $v_x = 10.0$ cm/s, $v_y = -5.0$ cm/s, and acceleration $a_x = 0$, $a_y = 4.0$ cm/s². (a) How far is it from the origin after 5.0 seconds? (b) In what direction is it from the origin? (c) In what direction is it moving?

3.56 Plot the path of an object in the xy plane that moves
• according to

$$x = (2 + 3t + 2t^2) \text{ cm},$$
$$y = (2t + 5t^2) \text{ cm},$$

from $t = 0$ to $t = 4$ s.

3.57 A ball is thrown upward from a platform 5.2 m high
• with a speed of 15 m/s at an angle of 40° from the horizontal. What is the magnitude of its velocity when it hits the ground?

3.58 An arrow was fired horizontally from a platform
• above the ground. Exactly 3 s after it was released, it struck the ground at an angle of 45° from the horizontal. With what speed and from what height was it launched?

3.59 A third baseman makes a throw to first base 39 m
• away. The ball leaves his hand with a speed of 38 m/s at a height of 1.5 m from the ground and making an angle of 20° with the horizontal. How high will the ball be when it gets to first base?

3.60 A third baseman makes a throw to first base 39.0 m
•• away. (a) If the ball leaves his hand traveling horizontally with a speed of 38.0 m/s at a height of 1.20 m from the ground, how far will it go before striking the ground? (b) At what angle must he throw the ball so that it reaches the first baseman's glove at a height of 1.20 m above the ground?

ADDITIONAL READING

Brancazio, P. J., "Trajectory of a Fly Ball." *The Physics Teacher*, January 1986, p. 20.

Burke, J., *The Day the Universe Changed*. Boston: Little, Brown and Company, 1985. Early attempts to understand projectile motion are described in Chapter Three.

Drake, S., "Galileo's Discovery of the Law of Free Fall." *Scientific American*, May 1973, p. 84.

Drake, S., "Galileo's Discovery of the Parabolic Trajectory." *Scientific American*, March 1975, p.102.

Gardner, M., "Mathematical Games: The Abstract Parabola Fits the Concrete World." *Scientific American*, August 1981, p. 146.

Gingerich, O., "The Galileo Affair." *Scientific American*, August 1982, p. 132.

Appendix 3
for Chapter

Review of Trigonometry

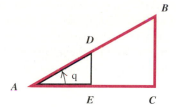

Figure A3.1 Two similar right triangles, *ABC* and *ADE*, with a common angle θ.

This brief outline of basic trigonometry, together with a few other relationships introduced later, includes most of the mathematics you need to know for this text.

It may be intuitively obvious to you (or you may remember a formal proof from geometry) that in the right triangle *ABC* of Fig. A3.1, the *ratio* of side *AC* to the hypotenuse *AB* does not depend on the physical size of the triangle, but only on the angles. For instance, in the two similar triangles *ABC* and *ADE* in Fig. A3.1, we see that

$$\frac{BC}{AB} = \frac{DE}{AD}.$$

This ratio uniquely characterizes the angle marked θ in the diagram. The ratio is called the sine of the angle θ and is written

$$\sin \theta = \frac{BC}{AB}.$$

More generally, for any right triangle,

$$\sin \theta = \frac{\text{length of the side opposite the angle } \theta}{\text{length of the hypotenuse}}.$$

Two other trigonometric functions, the cosine of the angle and the tangent of the angle, are defined as

$$\cos \theta = \frac{\text{length of the side adjacent to the angle}}{\text{length of the hypotenuse}}$$

and

$$\tan \theta = \frac{\text{length of opposite side}}{\text{length of adjacent side}}.$$

The tangent may seem somewhat superfluous, since $\tan \theta = \sin \theta / \cos \theta$,

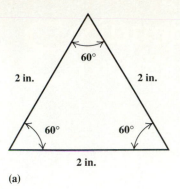

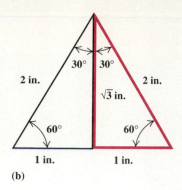

Figure A3.2   (a) An equilateral triangle with sides 2 in. long. (b) The equilateral triangle forms two right triangles, from which we can calculate some common trigonometric values.

(a)

(b)

TABLE 3.1 The values of some trigonometric functions for selected angles			
Angle θ	$\sin \theta$	$\cos \theta$	$\tan \theta$
0°	0	1	0
30°	$\dfrac{1}{2}$	$\dfrac{\sqrt{3}}{2}$	$\dfrac{1}{\sqrt{3}}$
45°	$\dfrac{1}{\sqrt{2}}$	$\dfrac{1}{\sqrt{2}}$	1
60°	$\dfrac{\sqrt{3}}{2}$	$\dfrac{1}{2}$	$\sqrt{3}$
90°	1	0	∞

Figure A3.3   Angles are measured relative to the positive x axis. The signs of the sine and cosine functions are indicated for each quadrant.

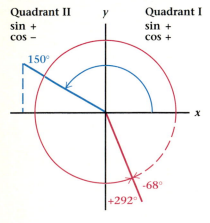

Quadrant II   y   Quadrant I
sin +   sin +
cos −   cos +

150°

x

-68°

+292°

sin −   sin −
cos −   cos +
Quadrant III   Quadrant IV

but its use is often convenient. There are other trigonometric functions, but they will not be needed in this text.

We can deduce the values of the sine and cosine for a few angles from simple geometry. Using an equilateral triangle (Fig. A3.2a), we arbitrarily assign a length of 2 in. to the sides. Then, using the Pythagorean theorem, we can determine the length of the line that bisects one of the angles (Fig. A3.2b). We then label all the known sides and angles. We can see that

$$\sin 60° = \frac{\sqrt{3}}{2} = 0.866.$$

We can use this triangle to fill in the values for the sine and cosine of 60° and 30° in Table 3.1. You should determine the values for 45° and for 0° and 90°. A scientific pocket calculator can give you the value of the sine, cosine, or tangent of any given angle.

We always measure angles in the counterclockwise direction from the positive x axis (Fig. A3.3). Angles measured in the clockwise direction are negative. Our definitions of sine, cosine, and tangent do not change for angles greater than 90°; however, the sign of the trigonometric function changes from quadrant to quadrant. For 150° the hypotenuse is positive (as always) and the side opposite the angle is positive, but the adjacent side (in the negative x direction) is negative. Thus the sine of an angle in the second quadrant (that is, an angle between 90° and 180°) is positive, but the cosine is negative. The signs of the sine and cosine functions are marked in each quadrant of Fig. A3.3.

If you know the value of the sine of an angle, you can also use your calculator to find the value of the angle. The angle θ whose sine is equal to A is written as $\theta = \sin^{-1}A$, where $\sin^{-1}$ is read as the inverse sine. The values of $\cos^{-1}$ and $\tan^{-1}$ can be obtained in a similar manner.

You can numerically verify from Table 3.1 that, for the angles listed,

$$\sin^2 \theta + \cos^2 \theta = 1.$$

We can show from Fig. A3.1 that this equation is true in general. The Pythagorean theorem gives

$$(BC)^2 + (AC)^2 = (AB)^2.$$

Dividing both sides by $(AB)^2$, we get

$$\frac{(BC)^2}{(AB)^2} + \frac{(AC)^2}{(AB)^2} = 1,$$

$$\left(\frac{BC}{AB}\right)^2 + \left(\frac{AC}{AB}\right)^2 = 1$$

or

$$\sin^2 \theta + \cos^2 \theta = 1.$$

Example 3.13

Standing on top of a tall building, you see a friend walking at a distance. The angle between your line of sight and the horizontal is 15°. How far from you is your friend? Assume the road is level and your eyes are 40.8 m above the ground.

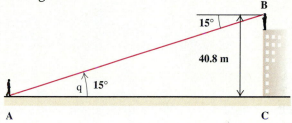

Figure A3.4 Example 3.13: How far away is your friend?

Solution Let BC be the height of the building and AB the distance from you to your friend. The angle θ in Fig. A3.4 is 15° . The sine of θ is

$$\sin \theta = \frac{BC}{AB}$$

or

$$AB = \frac{BC}{\sin \theta}.$$

Inserting the numerical values gives

$$AB = \frac{40.8 \text{ m}}{0.259} = 157 \text{ m}.$$

Example 3.14

Inverse trigonometric functions.

Three narrow strips of wood, having lengths of 30, 40, and 50 cm, are arranged on a table top to form a right triangle (Fig. A3.5). What is the angle between the 40-cm side and the 50-cm side?

Solution Using the definition of the cosine, we write

$$\cos \theta = \frac{40 \text{ cm}}{50 \text{ cm}} = 0.80.$$

The value of the angle θ is

$$\theta = \cos^{-1}(0.80) = 37°.$$

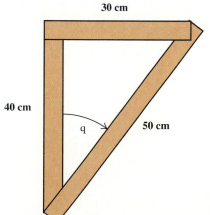

Figure A3.5 Example 3.14: Inverse trigonometric functions.

4

Force and Motion

4.1	The *Principia*
4.2	What Is a Force?
4.3	Newton's First Law— Inertia
4.4	Newton's Second Law
4.5	Weight, the Force of Gravity
4.6	Newton's Third Law
4.7	Some Applications of Newton's Laws
4.8	Friction
4.9	Static Equilibrium
4.10	The Laws of Motion as a Whole

A WORD TO THE STUDENT

In the last two chapters you learned to describe motion. In this chapter we discuss the underlying causes of motion, which are summed up in Newton's three laws. These laws enable us to describe the future (or past) motion of a body if we know the forces acting on it. We can do so regardless of the origin of the force: a muscular effort on your part, the force of gravity, or an electrical or magnetic force. Newton's laws give us the connection between cause and effect.

You will see later that the mathematical form of these laws of motion must be modified for objects that are moving extremely fast or that are very small. But for ordinary objects moving with speeds up to many thousands of kilometers per hour, these laws provide a suitable description of motion consistent with observations.

If you were to make a list of the most important books in the intellectual development of humanity, you would have to include Isaac Newton's *Philosophiae Naturalis Principia Mathematica*. Although the *Principia* was written slightly more than 300 years ago, most of the basic ideas and laws it presents, when used in the proper circumstances, are as valid today as when the *Principia* was first published. This observation is particularly significant when we consider that in the physical sciences, changes in theories and understanding are more the rule than the exception.

Why was this book written, and what did it say that was not already known? The answers to these questions constitute one of the major areas of inquiry in the history and philosophy of science. For our purposes, we will outline the historical origins only insofar as they help explain the content and importance of what is generally called **Newtonian mechanics** or classical mechanics.

The essence of Newtonian mechanics, which is widely used today to describe the motion of objects as different as golf balls and spacecraft, is found in Newton's three laws of motion. These laws, first clearly stated in the *Principia,* tell us how forces affect the motion of objects. The same laws apply whether the object is a mosquito zig-zagging through the air or a planet moving about the sun. Although Newton's laws do not apply in the domain of the very small (that is, to atoms and molecules) and the very fast (to objects moving at speeds near the speed of light), they apply to almost everything else. Thus the concepts introduced in this chapter form the basis for our understanding of much of physics and are essential to almost all of what follows. These ideas also play an important role in the development of other concepts, which, in turn, enable us to describe atoms and molecules as well as objects moving with speeds close to the speed of light.

The *Principia*

4.1

The seventeenth century included several important events that heavily influenced today's world. Twelve of the American Colonies had been established before Newton published the *Principia* in 1687. Shakespeare died early in the century; the English writers Milton, Defoe, and Swift were born in it. In 1611, the Authorized (King James) Version of the *Holy Bible* became one of the first major publications of the century.

By the last quarter of the 1600s, the works of Galileo and Kepler were known to most educated people, including Newton. Though Kepler's laws could be considered statements about the geometry and kinematics of the solar system, Kepler himself was not content with this and thought that one should look for the cause of planetary motion. Many, including Kepler, held that the planets' orbits were due to some influence of the sun.

By the 1660s and 1670s, there was some reason to believe that the sun's influence on a planet was an attractive force that depended in a simple way on the distance between them: The force was thought to be proportional to the reciprocal of the square of this distance. That is,

$$F \propto \frac{1}{r^2},$$

where r is the distance between the sun and planet and F is the attractive force. Such a force is called an **inverse-square force**. This terminology means that if the force were, say, 36 units for a distance of 6000 m, then if the distance were doubled to 12,000 m, the force would become one quarter as much, or 9 units. Similarly, if the distance were halved to 3000 m, the force would increase by a factor of 4, to 144 units. The reasons for accepting the idea of an inverse-square force for the sun–planet attraction are developed in this chapter and in Chapter 5.

One of the principal events that led to the writing of the *Principia* was Newton's deduction that an inverse-square force was responsible for the observed orbits of the planets. Further, he identified the action of the inverse-square force as the law of gravitation, and showed it to be universal by applying it to the moon, the planets, and objects on earth.

BACK TO THE FUTURE

The Writing of the *Principia*

The writing of the *Principia* came about from an argument over physics. The discussion was on the nature of gravitational force, and took place in January 1684 among three men: Christopher Wren (1632–1723), remembered today as an architect of churches and public buildings, especially St. Paul's Cathedral in London; Robert Hooke (1635–1703), who was curator of the Royal Society* at that time; and Edmund Halley (1656–1742), a young astronomer and mathematician. It was Halley (rhymes with Sally) who in 1705 predicted the return of the comet of 1682, which, upon its next appearance in 1758, was given his name.

Wren offered a prize to whichever of the other two could produce a proof

*The Royal Society of London is one of the oldest scientific societies in the world. Founded in 1660, the society is still active, publishing scientific journals.

that the force between the sun and the planets obeyed an inverse-square law. Hooke maintained that he had done so, but failed to produce evidence within a reasonable span of time. So Halley sought the help of Isaac Newton (Fig. B4.1), at that time a professor at Cambridge University. During the visit, Halley asked Newton what would be the orbit of planets attracted to the sun in an inverse-square manner. Newton replied that they would be ellipses and further said that he had produced a mathematical proof of this. Newton either could not or would not locate a copy of the proof, but promised to repeat the calculation and send it to Halley. In November, Halley received the manuscript from Newton, and it so excited him that he suggested that Newton make his results public by sending them to the Royal Society. Newton did so in the form of a tract, *De Motu,* based in part on earlier calculations. In this work he proved that if "a body revolves in an ellipse . . . the law of the centripetal force . . . is reciprocally as the square of the distance." Here was the

Figure B4.1 Isaac Newton (1642–1727). This portrait depicts Newton at age 47, two years after publication of the *Principia*. According to Newton's recollections as an old man, his peak years were 1665 and 1666. While the black plague swept through England, Newton retreated to his family's farm. There he invented calculus, completed much of the preparatory work for his theory of gravity, and discovered that white light is composed of many colored rays—all in the span of 18 months!

It would be impossible to discuss in any single book all of the significant contributions made by Newton in the *Principia*. However, several ideas stand out: the utility of the three laws of motion (which we will discuss next), the demonstration that the force law for elliptical motion was of the inverse-square type, and the universality of the law of gravitation—that is, that the law holds not only for the motion of the planets and their satellites but also for bodies on earth as well as for every other body in the universe.

Newton's ideas about the way the universe worked, as expressed in the *Principia,* transcended science and influenced Western thought in general. The idea that a mechanical universe obeyed a single set of laws suggested that all observed behavior could be explained in mechanical terms. Kepler's word "clockwork" is often used to describe this viewpoint. In economics, philosophy, law, and religion, Newtonian ideas were instrumental in bringing about the Age of Enlightenment. Leaders in these fields assumed that anything in the future could be calculated exactly, once the correct law was known, in the same way that astronomers could

solution to the physical problem of planetary orbits, the first real answer to the question, "What *must* be the force that gives rise to the observed planetary orbits if the sun is the attracting body?"

After presenting *De Motu* to the Royal Society, Halley, with the support of the society, invited Newton to write a more comprehensive version. Newton finished the complete version within 18 months, an incredibly short time for such a monumental work. The first edition of the *Principia* was available by July 5, 1687.

Newton directed the *Principia* toward a small and elite audience who understood science and mathematics. In many ways he tried to make it as difficult and inaccessible as possible. He is said to have boasted to a friend that he made the *Principia* "abstruse" so that he would not have to argue with those of lesser learning. In contrast to Galileo's use of everyday language, Newton wrote the *Principia* in Latin, the international language of learning. Newton never attempted to produce an

English version, though he lived forty more years and brought out revised editions in 1713 and 1723. The first, and only, complete English translation was published by Andrew Motte in 1729. Almost nothing is known about Motte or why he translated the third edition of the *Principia* into English.

The *Principia* was and is a difficult book, so difficult that one historian has said "it is doubtful whether any work of comparable influence can ever have been read by so few persons." Like major scientific works in other times, it had strong proponents as well as strong detractors. The German mathematician Gottfried Wilhelm Leibniz (1646–1716), who independently developed calculus at the same time that Newton did, violently opposed some of Newton's philosophy. He and Newton became bitter personal enemies, precipitating arguments among their followers for many years. However, the Newtonian concepts were admitted to the general body of scientific knowledge because of their utility and precision. The three hundredth anniversary

Figure B4.2 A British stamp commemorating the publication of the *Principia*. According to legend, Newton conceived the universal law of gravity while sitting in his garden. When he saw an apple fall from a tree, he realized that the force that caused the fruit to fall was the same as the force that holds the moon in its orbit about the earth.

of the *Principia's* publication was commemorated throughout the world (Fig. B4.2).

calculate the position of the moon at some future date. Within the last hundred years, physicists have found that it is not possible to take this deterministic viewpoint and that nature is much more subtle than it was imagined to be in Newton's day. We will discuss the contemporary viewpoint later in this book, when we get to relativity and quantum theory.

Today we still hold Newtonian mechanics in high regard because of the great range of physical properties and behavior it explains. The Newtonian approach to solving problems serves as a guide in many scientific fields. Newton's laws have not been replaced, but expanded. In contrast to some other early scientific ideas, such as the nineteenth-century theories of the nature of atoms, Newton's laws are so correct that they have not been modified in the more than 300 years since the *Principia* was first published. Though they do not correctly describe the interactions between galaxies or in the subatomic world, almost everything in between is covered. The thrust of an airplane's jet engine, the paths of baseballs and comets, how to best hit a tennis ball, and how musical instruments work, all can be understood on the basis of Newtonian mechanics.

The essential correctness of classical mechanics is reflected in part by the contents of physics textbooks. The philosophical viewpoint of textbooks in biology, and to a lesser extent chemistry, has changed during the last forty or fifty years, reflecting a change in the way biologists and chemists view their fields. Such books are different in their entirety from earlier books. This is not so much the case with physics. Although the relative importance of some topics has changed and modern discoveries have been added, almost all contemporary physics texts still begin with Newtonian mechanics.

What Is a Force?

4.2

In everyday language a **force** is a "push" or a "pull." When we push a lawnmower or a grocery cart we exert a force on it. When we pull open a drawer we exert a force on the drawer. When we drop something, it falls and we say that the force, or pull, of gravity made it fall. However, forces do not always cause motion. A book resting on a table experiences the downward force of gravity even though it does not move.

Because force has direction as well as magnitude, it is a vector quantity. If several forces act simultaneously on the same object, it is the net force that determines the motion of the object. The **net force** is the vector sum of all forces acting on the object. We often refer to the net force as the resultant force or the unbalanced force.

When you pull on the ends of a coiled spring (Fig. 4.1), the spring stretches. The harder you pull, the more the spring is extended. If you calibrate the extension of the spring, you can use the distance the spring stretches to measure the magnitude of the applied force. Spring scales that use this principle are commonly available. Once the scale is calibrated, you can use it to measure forces of different origins, such as forces resulting from muscular effort, gravitation, or magnetism.

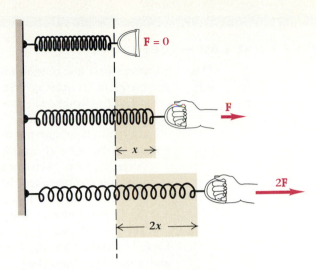

Figure 4.1 A spring attached to a fixed wall stretches through a distance proportional to the force applied to the ends. If a force F stretches the spring a distance x, then a force of $2F$ extends the spring a distance $2x$.

When you pull or push a spring you exert a *contact force*, a force in which two bodies interact directly by contact between their surfaces. The force you feel on your hand when you push a door and the force you feel on your feet when you walk are familiar contact forces. Although contact forces are common, they are not considered fundamental forces in the physical world.

A dropped stone falls because of the force of gravity. In this case the stone falls freely without being in contact with anything. The force involved is an example of action-at-a-distance; the force is exerted without actual contact between the bodies. The gravitational force is one of the *fundamental forces* in nature. The other fundamental forces are the electromagnetic force, the strong force, and the weak force. The gravitational force is responsible for the weight of bodies on the earth as well as for the motion of the planets. It acts between all masses. We will study gravitation in detail in the next chapter.

The electromagnetic force includes both electric and magnetic forces and is relatively strong. It is the electric force that you observe if you run a comb through your hair and then use the comb to pick up bits of paper. You see the magnetic force act when you pick up a pin with a magnet. The electromagnetic force is responsible for holding atoms and molecules together and for the structure of matter. Contact forces are large-scale manifestations of electromagnetic forces. We discuss electromagnetic forces in Chapters 15–20.

The strong force is the attractive force that holds together the constituents of the atomic nucleus. It is sometimes called the nuclear force. Chapters 28 and 31 include material on the strong force.

The weak force acts between all matter, but is so weak that it plays no direct part in ordinary observable behavior. It is, however, important in the interactions between subnuclear particles.

Current theories have been partially successful in unifying the basic forces of nature, so that we now understand the electric and weak force to be separate manifestations of one force, the electroweak force. Thus, according to one view, there are only three fundamental forces: the gravitational force, the strong force, and the electroweak force.

Newton's First Law—Inertia

4.3

Newton's first law of motion was translated from the Latin as: *Every body perseveres in its state of rest, or of uniform motion in a right line, unless compelled to change that state by forces impressed thereon.* In modern language, the term "right line" means "straight line." Notice that this law is a cause-and-effect statement relating force and motion. A force makes a change in the state of motion.

A statement of **Newton's first law**, also known as the law of inertia, in more concise language, is:

> **A body has a constant velocity unless there is a
> net force acting on it.**

The special case of a body at rest corresponds to a velocity of zero. Kepler believed that the inertia of bodies tended to bring them to rest, and a force was necessary to keep them in motion. (The word "inertia" is from the Latin word for "sluggish" or "inactive." In modern terms, **inertia** is the property of matter that causes objects to resist changes in motion.) Galileo, by contrast, held that certain kinds of motion could continue without the aid of a force. By the time Newton wrote the *Principia*, it had become fairly well established that the kind of motion that would continue indefinitely without any additional "push" was linear motion. The first law is a clear statement of this principle.

We readily believe that a body at rest remains at rest unless some action is taken. It is not so easily seen from everyday experience that once a body is moving with nonzero constant velocity (constant speed in a straight line), it continues to do so without any additional outside effort. Your first reaction may be that this idea goes against common sense. Nevertheless, the examples in the following paragraphs will confirm its soundness.

A paraphrase of an example given by one of Newton's acquaintances may make the first law more understandable. Suppose you exert a certain force to roll a bowling ball on an unmowed and unrolled lawn, and it rolls 20 yards. (In Newton's time, bowling took place on lawns, called bowling greens.) With the same force, you might roll it 30 yards if the grass were cut and rolled. Further smoothing of the grass surface would increase the ball's range. Since the range increases as obstacles are removed, we conclude that the removal of *all* retarding influences (forces) would allow the ball to continue rolling forever, with neither its speed nor direction changed. Once the ball is given a speed and direction by the bowler, the ball continues in its original state until acted on and retarded. In many cases, the major retarding force is friction.

Figure 4.2(a) shows a time exposure photograph of a glider moving with very little friction on an air track. While holding the glider at the left end of the track, the photographer sets off a regularly flashing light, opens the camera shutter, and gives the glider a push to the right. The constant spacing of the glider in the picture shows that the velocity of the glider remained essentially constant. This experiment suggests that even if we do not remove all friction or other forces, they may sometimes be so small that the system behaves essentially as described by the first law.

(a)

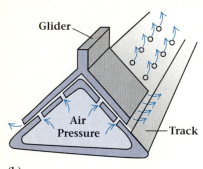

(b)

Figure 4.2   (a) A glider moving with a constant velocity on an air track. The regular spacing of the glider's successive positions indicates a constant velocity, which means that the horizontal force on it is negligible. (b) A typical linear air track is made from a hollow triangular rail that has tiny holes drilled in its upper surfaces. When pressurized, a cushion of escaping air supports a carefully fitted glider on the track. The glider floats nearly friction-free along a straight line on this air cushion.

You should consider just what we mean by uniform motion in a straight line. In the example above, the glider traveled along a straight line. However, the earth is turning on its axis, the earth goes around the sun, the solar system moves in the galaxy, and so on. Rather than answer this question now, let us temporarily assume that most of the motion discussed in this chapter can be referred to the surface of the earth. A partial answer to this question can be found in the practical working of Newtonian mechanics, but the fuller answer must be looked for within the context of relativity (Chapter 24).

Newton's Second Law

4.4

Newton's first law describes what happens when the net force acting on an object is zero. In that case the object either remains at rest or continues in motion with constant speed in a straight line. Newton's second law describes the change of motion that occurs when a nonzero net force acts on the object.

To get a sense of what Newton's second law is about, imagine that you push a subcompact car along a smooth, level street. If you push with a constant force it accelerates, moving faster as you continue to push. If a friend helps push so that the force is greater, the rate at which the speed increases is also greater; the larger force gives rise to a larger acceleration. This effect is predicted by Newton's second law, which connects force, mass, and motion.

The original translation of **Newton's second law** was: *The alteration of motion is ever proportional to the motive force impressed; and is made in the direction of the right line in which that force is impressed.* Elsewhere in the *Principia* Newton was clear that by "motion" he meant the product of the velocity and the mass. For the moment, it is sufficient to use Newton's identification of mass as the "quantity of matter." The modern name of the product of mass m and velocity $\mathbf{v}$ is **momentum**. (We consider momentum more fully in Chapter 7.) Using this term, we restate the second law:

The rate of change of momentum with time is proportional to the impressed force and is in the same direction:

$$\Delta(m\mathbf{v})/\Delta t \propto \mathbf{F}. \qquad (4.1)$$

In Eq. (4.1) $\mathbf{F}$ is the net force—that is, the vector sum of all forces acting on a body—and the change in the momentum $\Delta(m\mathbf{v})$ is in the direction of $\mathbf{F}$.

If we include a proportionality constant, the proportion becomes an equation. The value of the proportionality constant depends on the choice of units for force, mass, velocity, and time. If we choose the units appropriately, the proportionality constant can have the value 1. In that case the equation simply becomes

$$\frac{\Delta(m\mathbf{v})}{\Delta t} = \mathbf{F}. \qquad (4.2)$$

In the majority of real situations, the mass of an object does not change, so that the change in momentum is just the mass times the change in velocity. Then

$$\frac{\Delta(m\mathbf{v})}{\Delta t} = m\frac{\Delta\mathbf{v}}{\Delta t} = m\mathbf{a}.$$

Thus the second law may be expressed as

$$\mathbf{a} = \frac{\mathbf{F}}{m}. \tag{4.3}$$

Rearranging, we have the most common statement of the second law in the form:

$$\mathbf{F} = m\mathbf{a}. \tag{4.4}$$

Although this last equation is mathematically correct, you should use care in interpreting it. It is the force that causes the acceleration, not vice versa.

The ability to predict the motion of a body from a knowledge of the forces acting on it is one of the most useful accomplishments of physics. Not only is it useful for predicting the motion of bodies under mechanical or gravitational forces, but we can also calculate the motion when the force is due to other causes, such as magnetism or electricity. In other cases the force is not known and may be inferred from a knowledge of the motion.

Before going on to examples of the second law, it is worth noting that Newton's meaning of the second law is properly represented by Eq. (4.1). Equation (4.4), or even $\mathbf{F} \propto \mathbf{a}$, as it is sometimes stated, is a restricted special case for objects with constant mass. In this case, the acceleration is proportional to the force, and the direction of the acceleration is the same as the direction of the force.

Compare the glider moving with no horizontal force on it (Fig. 4.2) to the photo in Fig. 4.3(a). In the latter case, a constant horizontal force has been applied to the same glider. The increased spacings between strobe flashes indicate that the glider accelerates as it moves to the right. The force was applied to the glider in the manner shown in Fig. 4.3(b).

Measurements taken from the photograph are listed in Table 4.1. They show that over each interval of time the velocity increases by the same amount; that is, the acceleration is constant. In this demonstration, a constant force has given rise to a constant acceleration. In another experiment, similar to this one, the horizontal force on the glider was changed by hanging another mass alongside the first. Measurements on the photograph again showed motion with constant acceleration.

Indeed, by using a range of accelerating forces and a number of different gliders, we always observe that for a given total mass (glider plus hanging mass), the acceleration is proportional to the force. Rearrangement of the mathematical statement of Newton's second law gives

$$m = F/a.$$

Then we can call on common experience and this equation to tell us that

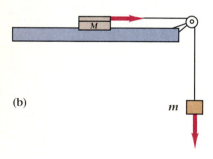

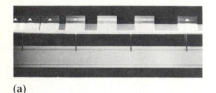

(a)

(b)

M

m

Figure 4.3 (a) An air-track glider moving under the influence of a small hanging mass. The photograph was made by the same technique as in Fig. 4.2. The increased spacing between the positions of the glider for successive strobe flashes indicates acceleration. (b) A simplified drawing of the air track showing the hanging mass. As the mass falls, it pulls the glider along with a constant force.

TABLE 4.1			
The motion of an air-track glider subjected to a constant force			
Time	x (cm)	$\Delta x/\Delta t$ (cm/time interval)	$\Delta(\Delta x)/(\Delta t)^2$ (cm/(time interval)^2)
0	131.4		
		8.4	
1	139.8		3.0
		11.4	
2	151.2		3.2
		14.6	
3	165.8		3.1
		17.7	
4	183.5		3.0
		20.7	
5	204.2		3.1
		23.8	
6	228.0		

Time intervals were uniformly spaced. The position given is x, the location of the front of the glider. There is some uncertainty in the fourth digit. The change in x is Δx, which is proportional to the velocity. The acceleration is proportional to the change in Δx and is given by $\Delta(\Delta x)/(\Delta t)^2$. From the table we see that the acceleration is constant.

what we have called **mass** is a quantitative measure of the inertia of a body. Finally, we say that the more massive a body, the larger the force necessary to give it a particular acceleration.

Example 4.1

Safety regulations require that all cars traveling at a given speed be able to stop within a given distance. What must be the relationship between the minimum braking forces allowed by law for two cars whose masses stand in the ratio of 3 to 2?

Solution The requirement for stopping in equal distances from the same initial speeds for both cars means they must both have the same deceleration. We can write the two braking forces F_1 and F_2 as

$$F_1 = m_1 a_1,$$
$$F_2 = m_2 a_2.$$

However, we require that

$$a_1 = a_2,$$

so

$$F_1/F_2 = m_1/m_2 = 3/2.$$

In the SI system of units, the unit of mass is the kilogram. The present standard of mass is a platinum-iridium cylinder whose mass is defined to be, by international agreement, one kilogram. This standard is kept at the International Bureau of Weights and Measures near Paris. Other bodies are assigned their mass value by a comparison with the standard. If, when using Eq. (4.4), the unit of mass is the kilogram and the units of acceleration are meter/second^2, then the units of force become kilogram-meter/second^2. This combination of units is given the name **newton** (N). We say: **A force of one newton acting on a one-kilogram mass gives it an acceleration of one meter/second^2.**

Example 4.2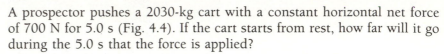

What force is necessary to give a 0.80-kg air-track glider a horizontal acceleration of 1.5 m/s²?

Solution Again we apply the second law:

$$F = ma = 0.80 \text{ kg} \times 1.5 \text{ m/s}^2 = 1.2 \text{ N}.$$

Example 4.3

A prospector pushes a 2030-kg cart with a constant horizontal net force of 700 N for 5.0 s (Fig. 4.4). If the cart starts from rest, how far will it go during the 5.0 s that the force is applied?

Solution We first determine the acceleration. From Newton's second law we get

$$a = \frac{F}{m} = \frac{700 \text{ N}}{2030 \text{ kg}} = \frac{700 \text{ kg m/s}^2}{2030 \text{ kg}} = \frac{700}{2030} \text{ m/s}^2.$$

Using this acceleration and our knowledge of kinematics, we can find the distance. The appropriate expression is:

$$x = v_0 t + \tfrac{1}{2} a t^2.$$

Since the cart starts from rest, $v_0 = 0$. Inserting the values for a and t, we get

$$x = \frac{1}{2} \left(\frac{700}{2030} \text{ m/s}^2 \right) \times (5.0 \text{ s})^2 = \frac{350}{2030} \times 25 \text{ m},$$

$$x = 4.3 \text{ m}.$$

Figure 4.4 Example 4.3: A prospector pushes a cart loaded with ore.

Weight, the Force of Gravity

4.5

One important example of Newton's second law is the expression for an object's weight. The **weight** of an object on earth is the gravitational force exerted on it by the earth. In Galileo's time most scholars held Aristotle's belief that heavy (weightier) objects fall faster than light ones, just because they are heavier. As we saw in Chapter 2, Galileo showed the reasoning to be incorrect. Some objects do fall less swiftly than others because of air resistance, which slows the fall of objects such as feathers and leaves. More compact objects such as stones fall faster because, for a given mass, they offer a smaller area to the air, thus reducing the effect of air friction to a negligible amount. In a vacuum, all objects fall with the same acceleration, regardless of their weight.

As a result, we know that near the earth's surface, when we neglect air resistance, the acceleration is the same for all falling bodies. This constant acceleration is known as the acceleration of gravity, **g**, and has the standard value 9.807 m/s². * When an object is dropped near the earth's surface, it is accelerated by the gravitational force (equal to its weight) with an acceleration **g**. Thus by Newton's second law the weight becomes

$$\mathbf{w} = m\mathbf{g}. \tag{4.5}$$

We see in this equation the relation between mass and weight: *Weight is a force proportional to the mass of a body and g is the constant of proportionality.* If we change the force of gravity on an object (say, by taking it to the moon), its weight will change, even though its mass remains constant.

When an object such as a brick rests on the ground, the gravitational force continues to act on the brick, even though it is not accelerating. According to Newton's second law, the net force on the brick at rest must be zero. There must be another force acting on the brick that opposes the gravitational force. This force is provided by the ground. The force provided by the ground is perpendicular to the surface of contact and is known as the **normal force.** (In geometry, the word "normal" means "perpendicular.") The normal force is simply the resistance of the ground to the motion of the brick acted upon by gravity. It is this normal force that keeps the brick from sinking into the ground.

Example 4.4

The weight of a book.

What is the weight of a textbook whose mass is 0.80 kg?

Solution   From Newton's second law,

$$F = ma = mg,$$

where the numerical value of g is 9.80 m/s². The force becomes

$$F = 0.80 \text{ kg} \times 9.80 \text{ m/s}^2 = 7.8 \text{ kg m/s}^2 = 7.8 \text{ N}.$$

The weight of the book is 7.8 N.

Example 4.5

An inclined plane.

A furniture van has a smooth ramp for making deliveries. The ramp makes an angle θ with the horizontal. A large crate of mass m is placed at the top of the ramp. Assuming the ramp to be a frictionless plane, what is the acceleration of the crate?

Solution   To find the acceleration of the crate we must first find the net force acting on it. The gravitational force $m\mathbf{g}$ acts downward on the block

---

*Although the value of g is the same for all objects in a given locality, it is measurably different at other localities. There is about 0.5% variation with latitude, with g being smaller at the equator than at the poles. An additional variation depends on altitude and on the underlying geological formation. For example, the value of g in Denver, Colo., is 9.796 m/s², while the value in Greenwich, England, is 9.812 m/s².

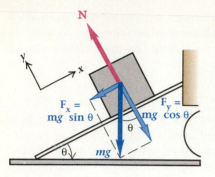

Figure 4.5   Example 4.5: A large crate slides down a slippery incline.

(Fig. 4.5). In this case it is convenient to choose a coordinate system in which one axis is parallel to the surface of the inclined plane. We can resolve the gravitational force vector into components parallel and perpendicular to the surface of the plane. We choose the positive x direction to be upward along the plane, and the positive y direction perpendicular to it as indicated in the figure. The resulting components of the gravitational force are

$$F_x = -mg \sin \theta$$

parallel to the surface of the plane and

$$F_y = -mg \cos \theta$$

perpendicular to the plane.

The plane supports the crate with a normal force N perpendicular to the plane. We assume that the ramp is rigid and that the crate can slide freely over the surface. There is no acceleration perpendicular to the plane, so the net force in the y direction must be 0:

$$F_{y \text{ net}} = N - mg \cos \theta = 0.$$

The only force in the x direction is the gravitational force,

$$F_{x \text{ net}} = -mg \sin \theta.$$

Because there are no other forces along the x direction, this force is the net force along x that causes the crate to move down the plane. The acceleration is given by Newton's second law:

$$a = \frac{F_{x \text{ net}}}{m} = \frac{-mg \sin \theta}{m} = -g \sin \theta.$$

The negative sign indicates that the crate accelerates to the left in the figure. Notice that the acceleration of the crate depends only on the acceleration of gravity (g) and on the angle of the inclined plane, and not on the mass of the crate. The crate moves with the same acceleration if it contains heavy furniture or if it is empty.

Newton's Third Law

4.6

Newton's **third law** of motion is: *To every action there is always opposed an equal reaction; or the mutual actions of two bodies upon each other are always equal, and directed to contrary parts.* The statement of the third law is sometimes shortened to:

For every action there is an equal and opposite reaction.

This is perhaps the easiest to remember and hardest to use of all the laws of motion.

Newton's own words help us understand the third law a little better. "Whatever draws or presses another is as much drawn or pressed by that other. If you press a stone with your finger, the finger is also pressed by the stone. If a horse draws a stone tied to a rope, the horse will be equally

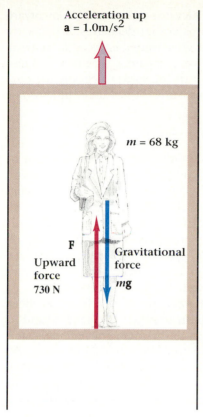

Acceleration up
a = 1.0m/s^2

m = 68 kg

F
Upward
force
730 N

Gravitational
force
m**g**

Figure 4.6 Example 4.6: A passenger in an accelerating elevator has an apparent weight different from her real weight.

drawn back towards the stone; for the distended rope, by the same endeavor to relax or unbend itself, will draw the horse as much toward the stone as it does the stone toward the horse . . ."

Restating the third law in more quantitative terms, we have:

If body A exerts a force **F**$_{AB}$ on body B, then B exerts a force **F**$_{BA}$ on A, so that **F**$_{AB}$ = − **F**$_{BA}$.

The action and reaction forces of Newton's third law are equal in magnitude, opposite in direction, and *act upon different bodies.*

We should point out that when applying the second law we consider only one of these forces at a time. Body A is subjected to the applied force **F**$_{BA}$. The acceleration of A is thus proportional to **F**$_{BA}$ and inversely related to its own mass m_A. Similarly, the body B is subjected to the force **F**$_{AB}$ and is accelerated by it.

Example 4.6

Weight in an elevator.

A 68-kg passenger rides in an elevator that is accelerating upward at 1.0 m/s^2 due to external forces. What is the force exerted by the passenger on the floor of the elevator?

Solution The situation is illustrated in Fig. 4.6. Two forces act on the passenger: the gravitational force m**g** pulling down and the force **F** exerted by the floor, pushing up. The difference between these forces provides the upward acceleration of the passenger, which in this case is the same as the acceleration of the elevator. By Newton's second law we have

$$F_{\text{net}} = F - mg = ma,$$

or by rearranging,

$$F = ma + mg = m(a + g),$$
$$F = 68 \text{ kg } (1.0 + 9.8) \text{ m/s}^2 = 730 \text{ N}.$$

By the third law we know that the passenger pushes on the floor with a force equal and opposite to the force of the floor on the passenger. So the passenger exerts a downward force of 730 N on the floor of the elevator. If the passenger were standing on a scale placed on the elevator floor, the resultant force would be the force read on the scale. This force is the passenger's apparent weight. If the elevator were at rest or moving with a constant speed, the scale would read

$$F = mg = 68 \text{ kg} \times 9.8 \text{ m/s}^2 = 670 \text{ N}.$$

If the elevator had been accelerating downward, the apparent weight of the passenger would be less than normal:

$$F = m(g - a).$$

You can feel this change in apparent weight when riding in an elevator. If you walk around in the elevator, when it accelerates as it starts to ascend or comes to a stop during a descent, you will notice how much more effort is needed to walk naturally. On the other hand, as the elevator begins to descend (or comes to a stop when moving upward) you will feel lighter and sense an additional spring in your step as you walk.

Figure 4.7 Two astronauts demonstrate the effects of weightlessness.

According to Example 4.6, when an elevator moves with a downward acceleration a, the apparent weight of a passenger decreases. If you were riding in the elevator you would say that your weight is less than normal. If the downward acceleration increased, your apparent weight would also decrease. If the downward acceleration of the elevator reaches g, the condition of free fall, your apparent weight measured by a scale would become zero:

$$F = m(g - a) = m(g - g) = 0.$$

This condition is called **weightlessness**. Notice that your mass would be unchanged. The gravitational force on you, and thus your real weight, would also be unchanged, even though your apparent weight becomes zero. Because you and the elevator are both falling with the same acceleration, you are free to float within the elevator. A slight push against the floor sends you to the ceiling. Because there is no net force between you and the floor, the concepts of "up" and "down" with respect to the elevator no longer apply. You are just as comfortable with your head toward the floor or lying on your side as with your head toward the ceiling.

Weightlessness is experienced by the occupants of orbiting artificial satellites like the space shuttle or Skylab. In that case, the satellite and its occupants are both in free fall and subject to the same gravitational force, just as in the case of the accelerating elevator. During weightlessness the occupants of artificial satellites move around freely (Fig 4.7). Remember, the explanation is that the satellite and its occupants are undergoing the same acceleration. There is no force acting between them, but the force of gravity still acts on all of them.

Some Applications of Newton's Laws

4.7

A useful device for illustrating Newton's second law was devised by George Atwood (1746–1807), nearly a century after the *Principia* appeared (Fig. 4.8a). His original device had a pendulum to measure time and elaborate wheel work to reduce frictional forces. In the simplified diagram (Fig. 4.8b), a flexible cord connecting two masses, m_1 and m_2, runs over a pulley that is free to turn without friction. Since the masses of the pulley and the cord are very small in comparison with the masses m_1 and m_2, we can disregard them in our analysis.

We include the force provided by the cord, which we have labeled with the symbol T. The magnitude of this force is called the **tension** in the cord. (Recall Newton's example of a horse pulling a stone with a rope, mentioned in the previous section.) We begin by making a separate vector diagram for each body, showing all the forces that act on that body. Such a diagram is called a **free-body diagram** (see page 108). For example, in Fig. 4.9 we see that two forces act upon mass m_2: a downward force m_2g due to gravity and an upward force T provided by the cord. We choose the downward motion of mass m_2 as the positive direction (that is, we

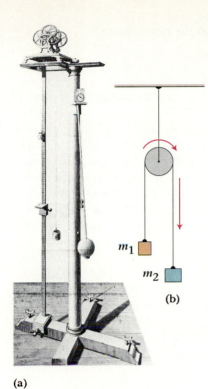

(a)

Figure 4.8 (a) Atwood's machine as illustrated in his 1784 book on mechanics. (b) A simplified version of an Atwood machine in which two masses are suspended by a cord passing over a frictionless pulley.

assume $m_2 > m_1$). The net downward force on m_2 is therefore

$$F_2 = m_2 g - T = m_2 a.$$

The acceleration is positive (downward) when $m_2 g$ exceeds T. Since m_1 must move up when m_2 moves down, the corresponding positive direction for m_1 is upward. The net upward force on m_1 (as seen in Fig. 4.9b) is

$$F_1 = T - m_1 g = m_1 a.$$

The acceleration a must be the same for m_1 and m_2, since they are joined by the cord and we assume that the cord does not stretch.

The tension T in the cord is the same at each end if we neglect all effects of the pulley. Upon rearranging the equations we get

$$T = m_2 g - m_2 a = m_1 g + m_1 a.$$

This equation may be solved to find the acceleration,

$$a = \frac{m_2 - m_1}{m_2 + m_1} g.$$

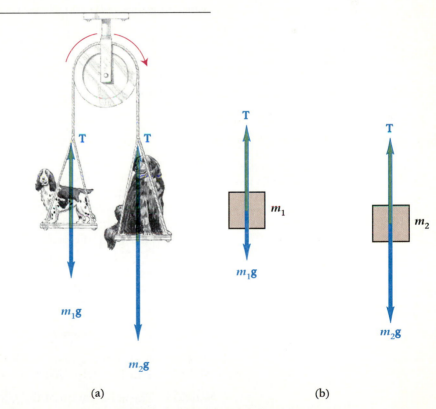

(a)   **(b)**

Figure 4.9 (a) A simple pulley problem, showing the upward forces due to tension (T) in the cord. (b) Free-body diagrams, showing the forces acting on the two single masses, m_1 and m_2.

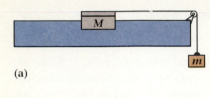

(a)

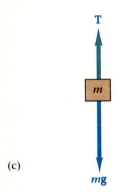

(b)

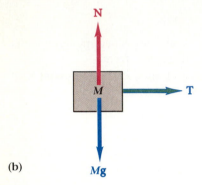

(c)

Figure 4.10 (a) An air-track glider of mass M connected to a small mass m by a light cord passed over a frictionless pulley. (b) A free-body diagram for the glider. The forces are the downward force of gravity Mg, the upward support N due to the air track (equal and opposite to Mg), and the horizontal force T due to the tension in the cord. (c) Free-body diagram for the small mass m.

We can now compute the tension T from the acceleration,

$$T = \frac{2m_1 m_2}{m_1 + m_2} g.$$

Notice that when $m_1 = m_2$, the tension T equals the weight mg and there is no acceleration.

As another example of analysis using free-body diagrams, let's consider an air-track glider of large mass M moving on a frictionless air track (Fig. 4.10). A small mass m is attached to M by a light cord that passes over a frictionless pulley of negligible mass. The forces on M are shown in the free-body diagram of Fig. 4.10(b). The gravitational force Mg is equal and opposite to the supporting force N provided by the air track because there is no vertical acceleration. An unbalanced horizontal force T, exerted by the string, accelerates the mass to the right:

$$T = Ma.$$

The forces acting on the smaller mass m (Fig. 4.10c) are a downward gravitational force mg and the upward force T, the tension in the string:

$$ma = mg - T.$$

Combining these two equations, we find that

$$a = \frac{m}{m + M} g.$$

We can find the tension T by substituting for a in either of the force equations above,

$$T = \frac{mM}{m + M} g.$$

The tension T is always less than the downward force of gravity on the hanging mass m and the acceleration is always less than g.

In complicated situations, the application of free-body diagrams separates the problem into manageable pieces so that we can find the appropriate equations. In simpler situations, including the preceding one, we can sometimes grasp the problem at once in an intuitive fashion. Thus for the air-track glider and hanging mass, the net unbalanced force on the system is mg. Since the masses move together, the total mass $m + M$ is to be used in the second law. Thus $F_{net} = mg = (m + M)a$, which gives the same acceleration found above.

Example 4.7

Suppose an air-track glider of 1.000 kg mass is connected to a mass of 0.015 kg as in Fig. 4.10. What is the acceleration of the glider?

Solution   The acceleration of the glider is obtained from Newton's second law,

$$F_{net} = m_{total}a,$$

where F_{net} is given by mg and the total mass is the sum of the masses,

$m + M$. Thus

$$mg = (m + M)a.$$

Upon rearranging we find the acceleration to be

$$a = \frac{mg}{m + M}$$

$$a = \frac{0.015 \text{ kg} \times 9.8 \text{ m/s}^2}{1.015 \text{ kg}} = 0.14 \text{ m/s}^2.$$

Example 4.8

Combining a pulley and an inclined plane.

Suppose that a block of mass M on an inclined plane is joined to a mass m by a cord over a pulley (Fig. 4.11). The block slides on a frictionless surface and the effects of the pulley are negligible. What are the magnitude and direction of the acceleration of the block if the surface is inclined at 20° and $m = \frac{1}{2}M$?

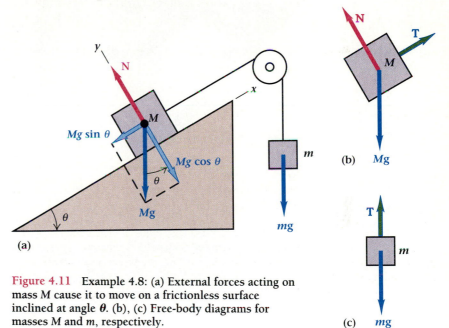

Figure 4.11   Example 4.8: (a) External forces acting on mass M cause it to move on a frictionless surface inclined at angle $\boldsymbol{\theta}$. (b), (c) Free-body diagrams for masses M and m, respectively.

Solution   We resolve the gravitational force on block M into components parallel and perpendicular to the surface of the inclined plane. The perpendicular component of the weight is balanced by the normal force N of the plane supporting the block:

$$N - Mg \cos \theta = 0.$$

Two forces act on the block in the direction parallel to the plane: the gravitational component $-Mg \sin \theta$ and the tension force T due to the cord, where we choose the direction up the ramp to be positive. Applying Newton's second law to the mass of the block, M, we find the net force

THE SHAPE OF THINGS TO COME

FREE-BODY DIAGRAMS

Free-body diagrams show all forces acting on an object. They are a useful first step for analyzing the motion of a physical situation. We will use free-body diagrams throughout this book.

PHYSICAL SITUATION

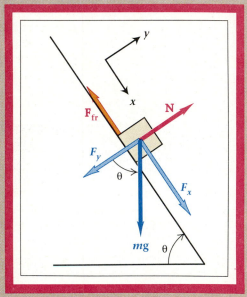

FREE-BODY DIAGRAM

Always keep your free-body diagrams simple, and as uncluttered as possible. Draw all forces as vector arrows, in approximately correct size and direction, and show bodies as simple particles or blocks. Show only the geometry important to solving the problem. Indicate the positive direction of distance, velocity, or acceleration, depending on the problem.

to be

$$F_{\text{net(block)}} = T - Mg \sin \theta = Ma.$$

From the free-body diagram of the hanging mass (Fig. 4.11c) the net force is

$$F_{\text{net(mass m)}} = mg - T = ma,$$

where a positive net force corresponds to the mass m accelerating downward and M accelerating up the incline. We can add these two equations to eliminate T, resulting in an equation for the acceleration in terms of the masses and the gravitational acceleration g:

$$mg - Mg \sin \theta = (m + M)a.$$

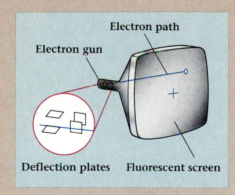

FREE-FALL WITH AIR RESISTANCE

A sky diver in free fall reaches a constant velocity due to the exact balance between the upward forces of buoyancy and air resistance and the downward force of his weight. In the free-body diagram, we show the diver as a particle and show all the forces acting on him. The velocity **v** is downward and constant. (See Ch. 9)

ELECTROSTATIC FORCES

In a cathode-ray tube, as in your television picture tube, electrons are accelerated toward the screen at the large end of the tube. The path of an electron depends on the force due to the electric field between two parallel plates near the electron gun, as shown in the free-body diagam. We can control the electron's path by controlling the strength of these forces. (See Ch. 16)

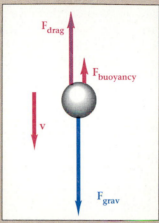

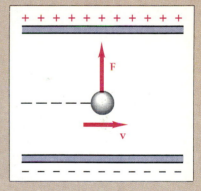

The acceleration is then

$$a = \frac{(m - M \sin \theta)g}{m + M}.$$

If we insert the values $m = \frac{1}{2}M$ and $\theta = 20°$, we get

$$a = \frac{Mg(\frac{1}{2} - \sin 20°)}{\frac{3}{2}M},$$

or

$$a = \tfrac{2}{3}g(0.5 - 0.34) = 0.11g.$$

The direction of the acceleration is up the incline.

Friction

4.8

So far we have been careful to ignore the effects of friction. However, when one object slides over the surface of another, its motion is always opposed by a retarding force that resists this motion. This force is called **friction**. Frictional forces are especially important to us in our everyday lives, for without them we could not walk or hold things with our hands; cars would be unable to start or stop; nails and screws would be useless. Because frictional forces are so common, we will introduce them here. We first examine frictional forces and show how to work problems that include friction. Then we present a description of the origins of friction.

Frictional forces are not fundamental forces like gravity or electromagnetism, but arise as reaction to other applied forces. Consider the motion of a solid object in contact with a horizontal surface. The object might be a brick on the floor or a telephone on a tabletop (Fig. 4.12). Initially the telephone of weight **w** rests on the horizontal surface. If we apply a small horizontal force **T** to the telephone by its cord, the telephone does not slide; instead it remains at rest. Clearly, according to Newton's second law, there must be another force acting on the telephone that is equal and opposite to **T** so that the net force is zero and the telephone remains stationary. This force is the frictional force $\mathbf{F}_{fr}$ exerted on the telephone by the surface. If **T** is made smaller, the frictional force must also decrease so that the two remain equal but opposite. When the applied force **T** becomes large enough, the telephone begins to slide in the direction of the applied force. The frictional force is still present and directed opposite to the applied force, but is no longer equal in magnitude to **T**. The net difference in these forces causes the telephone to accelerate.

The general observations of frictional behavior have been known for nearly 500 years. (1) For objects in relative motion, that is, sliding or rolling, the force of friction always acts in a direction opposite to the direction of motion. (2) The frictional force is proportional to the perpendicular (or normal) force between the two surfaces in contact. (3) For solid objects, the frictional force is approximately independent of the area of contact between the surfaces. (4) The frictional force depends on the

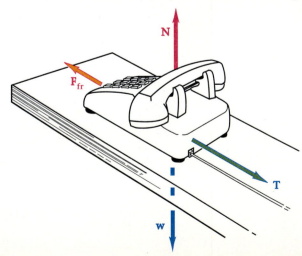

Figure 4.12   A telephone of weight **w** rests on a horizontal surface. When a horizontal force **T** is applied through its cord, the telephone's motion is opposed by a frictional force $\mathbf{F}_{fr}$.

Figure 4.13 On level ground, the horses pull against the frictional force between the sled and the ground. As more weight is added to the sled, the horses have to pull against the increased friction.

particular materials that make up the surfaces. These empirical rules usually hold to a good approximation. For instance, observation (2) is borne out by experiences such as that illustrated in Fig. 4.13, in which the sled is harder to pull when its load is increased. However, in some situations the usual rules of friction do not apply. We consider one such situation on p. 112.

Let's put these statements about friction in quantitative form. For the static case, in which there is no relative motion between the surfaces, the magnitude of the frictional force is

$$F_{\text{fr}} \leq \mu N \qquad \text{(static friction)}, \qquad (4.6\text{a})$$

where N is the magnitude of the normal force and the proportionality constant μ is the (dimensionless) **coefficient of friction**. The value of μ depends on the objects involved and on the condition of their surfaces. Table 4.2 lists typical values of μ for a few cases.

Equation (4.6a) actually sets an upper limit for the frictional force $\mathbf{F}_{\text{fr}}$ since, as we have seen, when an object is at rest the frictional force must be equal and opposite to the applied tangential force. If no tangential force is applied, there is no opposing frictional force. As the tangential force increases, the frictional force increases to oppose it. Ultimately, the frictional force reaches the maximum value expressed by Eq. (4.6a). If a still greater external force is applied, the object no longer remains at rest but begins to slide. For this case, in which the surfaces slide against each other, the magnitude of the frictional force is

$$F_{\text{fr}} = \mu N \qquad \text{(kinetic friction)}. \qquad (4.6\text{b})$$

Notice that the expression of Eq. (4.6b) is an equality and not an inequality as given in Eq. (4.6a). Although the coefficient of friction is not truly constant, Eq. (4.6) is a good empirical rule for approximating the force needed in many practical situations.

The magnitude of the frictional force depends on whether the two surfaces are in relative motion. For identical surface conditions and constant pressures, the coefficient of friction generally decreases slowly with

TABLE 4.2
Coefficients of friction*

Materials	Conditions	μ
Glass on glass	Clean	0.9–1.0
Wood on wood	Clean and dry	0.25–0.5
Wood on wood	Wet	0.2
Steel on steel	Clean	0.58
Steel on steel	Motor oil lubricant	0.2
Rubber on solids	Dry	1–4
Teflon on steel	Clean	0.04
Waxed hickory on dry snow		0.03–0.06
Brass on ice		0.02–0.08

*Approximate values. Frictional coefficients vary with surface conditions and cleanliness.

increasing relative speed.* If the normal force or the speed becomes too large, Eq. (4.6) no longer applies. It is important to realize that an empirical law such as this has its limitations, beyond which it does not work.

Sometimes two frictional coefficients are given: μ_s for static friction and μ_k for kinetic, or sliding, friction. However, because the coefficient of friction depends on speed, and because it varies greatly as a result of conditions such as surface moisture, cleanliness, and wear, the differences in the two are poorly known and not very reproducible. For these reasons we have not drawn a distinction between static and kinetic coefficients in the examples and in the problems. You should bear in mind that the values of the coefficients given in the table are only approximate. Ordinarily, for identical surface conditions, μ is slightly greater for static friction than it is for sliding friction. The situation is very complicated. As Richard Feynman observed,

> Many people believe that the friction to be overcome to get something started (static friction) exceeds the force required to keep it sliding (sliding friction), but with dry metals it is very hard to show any

*H. L. Armstrong, "How Dry Friction Really Behaves," *American Journal of Physics,* September 1985, p. 910; I. V. Kragelskii, *Friction and Wear* (translated from the Russian by L. Rosen), Washington: Butterworths, 1965; E. Rabinowicz, "Resource Letter F-1 on Friction," *American Journal of Physics,* December 1963, p. 897.

PHYSICS IN PRACTICE

The Friction of Automobile Tires

How do tires affect your safety when you drive your car along the highway? What factors help to prevent skidding and allow you to control your car when turning and stopping? What does friction have to do with this?

The tread pattern of rubber tires plays a major role in determining their friction, or skid resistance. Under dry conditions on paved roads, a smooth tire gives better traction than a grooved or patterned tread because a larger area of contact is available to develop the frictional forces. For this reason, the tires used for auto racing on the tracks at Darlington, Indianapolis, Talladega, and elsewhere have a smooth surface with no tread design (Fig. B4.3). Unfortunately, a smooth tire develops very little traction under wet condi-

Figure B4.3 Race cars driven on the superspeedways are equipped with wide, smooth tires known as "racing slicks."

tions because the frictional mechanism is reduced by a lubricating film of water between the tire and the road. A patterned tire provides grooves or channels into which the water can squeeze as the tire rolls along the road, thus again providing a region of direct contact between tire and road (Fig. B4.4). A patterned tire gives typical dry and wet frictional coefficients of about 0.7 and 0.4, respectively. These values represent a compromise between the extreme values of about 0.9 (dry) and 0.1 (wet) obtained with a smooth tire.

Classical friction theory must be modified for tires because of their structural flexibility and the stretch of the tread rubber. Instead of depending solely on the coefficient of friction at the tire–road interface (which is determined by the nature of the road surface and the tread rubber compound), maximum stopping ability also depends on the resistance of the tread to tearing

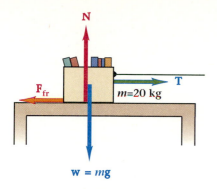

Figure 4.14 Example 4.9: A box of books moves only if the applied force **T** is greater than the force of friction between the books and the table.

difference. The opinion probably arises from experiences where small bits of oil or lubricant are present, or where blocks, for example, are supported by springs or other flexible supports so that they appear to bind.*

Example 4.9

A horizontal force of 100 N is applied to a box of books of mass 20 kg resting on a wooden table (Fig. 4.14). Does the box slide if the coefficient of friction of the box on the table is 0.40? Assuming that the box moves, find its acceleration.

Solution The normal force N between the box and the table is just equal to the weight of the box, $mg = 196$ N. The maximum frictional force is

$$F_{fr} = \mu N = 0.40 \times 196 \text{ N} = 78 \text{ N}.$$

This force is less than the applied force, so the box slides in the direction of T.

The box is accelerated by the net force $T - F_{fr}$. Thus from Newton's

*R. P. Feynman, R. P. Leighton, and M. Sands, *The Feynman Lectures on Physics.* Reading, Mass.: Addison-Wesley, 1964, p. 12–15.

Figure B4.4 Tires used on racing cars driven only on dry tracks have a solid contact area, like that of the Indianapolis 500 race tire shown (top left). Grooved tires designed for general use provide traction under wet conditions by channeling water away from the tire. Because it has no similar tread pattern, the racing tire cannot be driven on a wet track.

under the forces that occur during braking.

When a car is braked to a hard stop on a dry road, the maximum frictional force developed can be greater than the strength of the tread. The result is that instead of the tire merely sliding along the road, rubber is torn off the tread at the tire–road interface. Undoubtedly the tread resistance to this tearing is a combination of the rubber strength and the grooves and slots that make up the tread design.

The weight of the car is unevenly distributed over the tire–road contact area, creating areas of high and low pressure. (This is much like what you feel when you step on a pebble while walking in thin-soled shoes.) The resistance of the tread to tearing increases in the areas of higher pressure, where the tread is more compressed, causing an effective increase in traction.

Further, the size of the contact area is very important in car tires because the traction is dynamic rather than static; that is, it changes as the tire rolls along. The maximum coefficient of friction can occur anywhere in the contact area, so that the greater the area, the greater the likelihood of maximum traction. Thus, under identical load and on the same dry surface, the wider tire has a greater contact area and develops higher traction, resulting in greater stopping ability.

Next time you need to buy tires, think about what kind of climate you live in, what kind of roads you drive on, and what speeds you drive. If you live in a region with good paved roads, you may not need tires with extra tread. If you drive in areas with mud or snow, you need a tread designed for those conditions.

(We discuss braking a moving car in more detail in Section 8.5.)

second law we get

$$ma = F_{\text{net}} = T - F_{\text{fr}},$$

and the acceleration is

$$a = \frac{(T - F_{\text{fr}})}{m}$$

$$a = \frac{(100 - 78)\ \text{N}}{20\ \text{kg}} = 1.1\ \text{m/s}^2.$$

What are the causes of frictional forces and how do we know about them? It is often supposed that frictional effects originate in the roughness of the surfaces in contact with one another. In fact, experiments have shown that friction does not generally increase with roughness. For example, two pieces of smooth, flat glass show much more frictional drag than two pieces of rough, ground glass. The frictional forces arise primarily from molecular forces in the regions of real contact. Thus friction is determined not so much by the effect of the roughness or smoothness of the surfaces as by the molecular forces in the area of actual contact.

Studies of blocks sliding on blocks show that friction is generally independent of the contact area of the blocks. This area, which is determined by multiplying length times width, is more properly thought of as the apparent area of coverage. Because objects that appear smooth may be microscopically rough, the apparent area of coverage is usually much larger than the actual area of contact. As the normal force increases, the actual contact area also increases because of deformations of the two surfaces at their interface. It is this increase in contact area with increasing load that gives the apparent connection between friction and normal force.

In dealing with surfaces that are easily deformed, an increase in apparent area may closely approximate an increase in the actual area of contact. For this reason the width (and hence area) of tires can make a significant difference in how an automobile drives. This effect of dependence on area stands in sharp contrast to experiments with metal blocks sliding over metal surfaces, where changes in the apparent area of the block make little difference in the observed friction.

Whenever you measure the friction of one metal block sliding against another block of the same metal, it is not really a case of pure metal sliding on pure metal. The surfaces of each block contain oxides and other impurities. If the surfaces are carefully cleaned in a high vacuum and are touched together, they will stick, forming a cold weld. This surprising result happens when microscopically clean surfaces touch, since the bonding at the interface is then the same as it is anywhere else within the metal. Thus the friction that we normally observe is due principally to the surface layer of contaminants that is always present, which serves to reduce the intermolecular forces at the interface.

A thin layer of oil is often placed between surfaces to make motion easier. This idea is not new; lubrication with fats and oils has been common for thousands of years. The principal effect of a lubricating film is to

Figure 4.15 Hovercraft vehicles use powerful blowers to maintain a cushion of air beneath them to float over land and sea with little frictional drag.

diminish the attractive forces of the sliding surfaces and thereby reduce their cohesion. Even an invisible layer of oil can reduce the friction of dry surfaces by several orders of magnitude.

A different effect occurs when the lubricating film separating two sliding surfaces is thick compared to the dimensions of molecules. Then the friction depends on the properties of the film rather than on the surfaces. A striking example of this is the use of a thin layer of air to support heavy objects. Because the friction of the air is so small, the objects appear to glide along almost without friction. This effect has been put to use in the air track of Fig. 4.2. Its application to air bearings has made possible the modern high-speed drills used by dentists. Hovercraft vehicles that float on a cushion of air are in regular service across the English channel (Fig. 4.15).

Static Equilibrium

4.9

We have seen that a body subjected to a net force has an acceleration proportional to that force. But what if the vector sum of the forces acting on the body is zero? This is the condition of translational **equilibrium**, a state of motion in which the velocity of the body is constant. If the velocity of the body is zero, then the body is at rest and is said to be in **static equilibrium**. If the body is in motion with constant velocity, then we say it is in **dynamic equilibrium**.

In addition to translational motion, a body can also have rotational motion. However, these two types of motion can be separated and treated independently. We shall defer discussion of rotational motion and rotational equilibrium until Chapter 7. Our immediate concern is with static equilibrium of translational motion.

We begin by examining the forces on a refrigerator resting on a horizontal floor (Fig. 4.16). We know that for a refrigerator of mass m, a gravitational force $m\mathbf{g}$ pulls down on it, as illustrated in the figure. But from observation and Newton's laws, we have come to expect that the refrigerator remains at rest. Therefore, another force must be present, equal and opposite to the gravitational force. If this were not so, a net unbalanced force would be acting on the refrigerator and it would accelerate. This equal and opposite force is provided by the floor, which pushes up on the refrigerator with a normal force $\mathbf{N}$.

The condition of translational equilibrium for any object is given mathematically by the statement that the vector sum of all forces acting on that object must be zero. That is,

$$\mathbf{F}_{net} = \mathbf{F}_1 + \mathbf{F}_2 + \mathbf{F}_3 + \cdots = 0.$$

We can represent this summation more compactly as

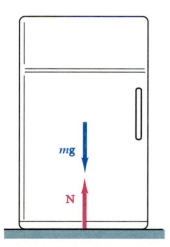

Figure 4.16 A refrigerator rests in static equilibrium on a horizontal floor. Therefore its weight must be balanced by an upward normal force.

$$\mathbf{F}_{net} = \sum_i \mathbf{F}_i = 0, \tag{4.7}$$

where i stands for the indices 1, 2, 3, 4, . . . and the symbol Σ_i represents the sum over all values of i. Here the forces $\mathbf{F}_i$ all act on the same object. This study of objects and forces in equilibrium is a special case of dynamics called **statics**.

The vector equation (4.7) can be easily handled by resolving the forces into components along the vertical (y) direction and the horizontal (x) direction. If the forces are in equilibrium, then their x components and their y components must also be in equilibrium separately. The resulting equilibrium conditions are

$$\sum_i F_{ix} = 0 \quad \text{and} \quad \sum_i F_{iy} = 0.$$

Some examples follow.

Example 4.10

A child sits on a toboggan that rests on a snow-covered hill making an angle θ with the horizontal. If the coefficient of friction is 0.10, what is the maximum angle at which the toboggan remains at rest? (Assume the hill can be approximated by an inclined plane.)

Solution   We can resolve the weight of the toboggan plus child, $m\mathbf{g}$, into components parallel and perpendicular to the incline (Fig. 4.17). The perpendicular component of the weight is equal and opposite to the normal force $\mathbf{N}$. The parallel component F_1 down the incline is opposed by the frictional force $\mathbf{F}_{fr}$. From Fig. 4.17 we see that the relations between $\mathbf{N}$, F_1, and $m\mathbf{g}$ are

$$N = mg \cos \theta$$

and

$$F_1 = mg \sin \theta.$$

For the toboggan to be at rest, the condition of equilibrium requires that the vector sum of the forces must be zero in each direction. Thus the magnitude of F_1 must be equal to the magnitude of the force of friction F_{fr}. The maximum value for F_{fr} is μN. Thus

$$F_1 = \mu N$$

at the maximum angle, known as the angle of repose. We can determine

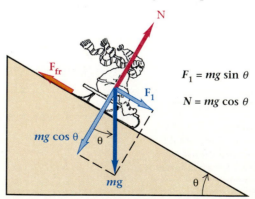

Figure 4.17   Example 4.10: How steep can the hill be before the toboggan slides down?

this angle by inserting the relations for F_1 and N into this equation. The result is

$$F_1 = mg \sin \theta = \mu N = \mu mg \cos \theta,$$

or

$$\mu = \tan \theta.$$

When $\mu = 0.10$, $\theta = 5.7°$.

For angles less than $5.7°$, the toboggan remains stationary on the incline, regardless of its weight. For angles greater than $5.7°$, the toboggan slides down.

Example 4.11

Hanging a lantern.

A lantern of mass m is suspended by a string that is joined to two other strings as shown in Fig. 4.18(a). What is the tension in each string if they make equal angles of $45°$ from the support beam as shown? Ignore the mass of the string.

Solution Figure 4.18(b) shows a free-body diagram of the mass m, which is acted on by a downward gravitational force mg and by an upward force of tension in the string T_1. For the lantern to be in static equilibrium,

$$T_1 = mg.$$

This equation gives the tension in one string, but how do we find the relationships for the other strings? The answer is to make a free-body diagram for the knot (Fig. 4.18c). At the knot, the tension $\mathbf{T}_1$ in the lower string is downward. The tensions in the other two strings are $\mathbf{T}_2$ and $\mathbf{T}_3$ in the directions indicated. According to Eq. (4.7), the condition for the equilibrium can be expressed as

$$\mathbf{T}_1 + \mathbf{T}_2 + \mathbf{T}_3 = 0.$$

Treating the horizontal components first, we get

$$\sum T_x = T_{2x} + T_{3x} = 0.$$

Taking the positive direction to the right, we see that

$$T_{3x} = T_3 \cos 45° = \frac{T_3}{\sqrt{2}}$$

and

$$T_{2x} = -T_2 \cos 45° = -\frac{T_2}{\sqrt{2}}.$$

Upon inserting these values in the equilibrium equation, we find that $T_2 = T_3$, something we might have suspected from the symmetry of the situation.

We now have enough information to evaluate T_2 (and T_3) in terms of T_1 by summing the vertical components in equilibrium:

$$\sum T_y = T_{1y} + T_{2y} + T_{3y} = 0.$$

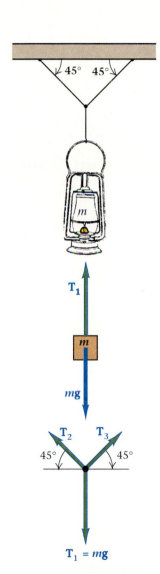

Figure 4.18 Example 4.11: (a) A lantern of mass m suspended by one string joined to two other strings. (b) A free-body diagram for the mass m. (c) A free-body diagram for the knot joining the strings.

If we take the positive direction to be upward, then

$$T_{3y} = T_3 \sin 45° = \frac{T_3}{\sqrt{2}},$$

$$T_{2y} = T_2 \sin 45° = \frac{T_2}{\sqrt{2}} = \frac{T_3}{\sqrt{2}},$$

and

$$T_1 = -mg,$$

so

$$\sum T_y = -mg + \frac{T_3}{\sqrt{2}} + \frac{T_3}{\sqrt{2}} = 0.$$

This gives

$$T_3 = T_2 = \frac{mg}{\sqrt{2}}.$$

In summary, the three tensions are $T_1 = mg$, $T_2 = mg/\sqrt{2}$, and $T_3 = mg/\sqrt{2}$.

Example 4.12

Equilibrium of a child's swing.

A child of mass M sits in a light swing suspended by a rope of negligible mass. The child's sister pushes him forward by a horizontal force until the rope makes an angle θ with the vertical (Fig. 4.19). What is the tension in the rope and how much horizontal force is required to hold the child in that position?

Solution   We solve the problem by essentially the same method used in Example 4.11. First we sum the vertical components of force,

$$\sum F_y = T \cos \theta - Mg = 0.$$

There are only two vertical components, since the force F is purely horizontal. We obtain the tension immediately as

$$T = \frac{Mg}{\cos \theta}.$$

The condition of equilibrium for the horizontal forces gives

$$\sum F_x = F - T \sin \theta = 0$$

or

$$F = T \sin \theta.$$

The horizontal force F can also be found in terms of Mg by inserting the value for T found above:

$$F = \frac{Mg \sin \theta}{\cos \theta} = Mg \tan \theta.$$

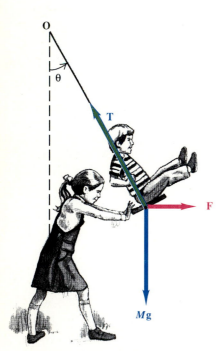

Figure 4.19   Example 4.12: A horizontal force **F** pushes a child in a swing until the rope makes an angle θ with the vertical.

Notice that for $\theta = 0$, $F = 0$. For $\theta = 45°$, the force becomes $F = Mg$. As θ approaches 90°, F must increase toward infinity; that is, a purely horizontal force can never make the rope become perfectly horizontal as long as it has to support the downward weight.

The Laws of Motion as a Whole

4.10

We saw earlier that a system of coordinate axes fixed to a particular body is called a **reference frame**. Our description of motion is always in terms of motion with respect to some particular reference frame. For example, we usually describe the motion of autos, rockets, baseballs, and other ordinary objects with respect to the earth. On the other hand, we usually describe the motion of the planets with respect to a reference frame attached to the sun. Motions viewed in one reference frame may not appear the same when viewed from another. A passenger who drops a pencil while riding in a bus moving with constant velocity sees it fall straight down. To an observer standing by the road, however, the path of the pencil is a parabola as the pencil moves both forward and downward.

When you place a ball on a flat floor, you expect it to remain in place in accordance with your experience and Newton's laws. However, if you were on a carousel turning with a constant rate and placed a ball on the floor, it would begin to roll toward the outer edge. The law of inertia does not hold in the reference frame fixed on the rotating carousel. Similarly, if you placed a ball on the floor of an accelerating truck, the ball would begin to roll in the direction opposite to the direction of the truck's acceleration. Reference frames such as these, in which the law of inertia (Newton's first law) does not hold, are called *noninertial* reference frames.

Inertial reference frames are reference frames in which the law of inertia does hold. In an inertial reference frame an object at rest will remain at rest if no net force acts on it. We usually consider Newton's laws to be valid on the face of the earth, even though the earth is not truly an inertial reference frame because it rotates. The effects of the rotation are not large and for most purposes Newton's laws apply. The effects of rotation can be calculated by using Newton's laws and an inertial reference frame. Such calculations help explain the motion of the air responsible for the formation of hurricanes and tornadoes.

We have introduced Newton's laws and some of their limitations. But the real content of these laws is that forces have some independent properties in addition to that expressed in the law $F = ma$. Not only can we use the laws to calculate acceleration when a given force acts on a mass, but we can use Newton's laws to investigate the forces observed in nature. By studying accelerations we can find how forces depend on other quantities, such as distance. In this way Newton's laws become a tool for understanding nature.

In Newton's *Principia* the laws of motion are in a section called "Axioms, or Laws of Motion." As with axioms in geometry, we may postulate

Figure 4.20 A speed–distance diagram illustrating the range of applicability of Newtonian mechanics. (Note that the scales chosen are not linear, but logarithmic.) The laws of classical physics (center area) are consistent with observations of everyday life. However, when dealing with very small or very large distances, or with very high speeds, these laws do not adequately describe what we observe. In those regions, we need the laws of quantum mechanics and relativity instead of Newton's laws to describe and predict the physical observations.

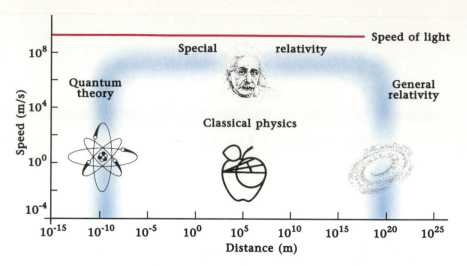

a set of axioms for dynamics that give rise to an abstractly satisfactory scheme of dynamics, just so long as the axioms are not contradictory. However, Newton wanted to explain nature as it was observed. This desire determined his choice of axioms. Strictly speaking, the success of Newtonian dynamics rests, not on verification of individual axioms, but on the success of the entire scheme in predicting what we observe.

In this chapter, "what we observe" has meant "ordinary-sized things moving with ordinary velocities." If, instead, we want to explain observations that include a much wider range of sizes and velocities, alternative or modified axioms are needed. In particular, if the range of observed velocities is to include those comparable to the velocity of light, then we need the postulates of Einstein's relativity (Chapter 24). If we go to small dimensions comparable to atomic sizes, then the postulates of quantum mechanics are required (Chapter 27). Figure 4.20 illustrates the domain of Newtonian physics. The overall success of Newtonian dynamics in its proper domain is evident from the great variety of situations covered and the enormous range of sizes, from the motion of the solar system to the motion of helium atoms in a gas.

SUMMARY

Useful Concepts

- Newton's laws tell us that a body in motion stays in motion at constant velocity along a straight line unless acted upon by an external force; that a net force applied to a body causes it to change its motion by accelerating, and similarly, an accelerating body does so because a net force is applied; and that every action has an equal and opposite reaction, applied to different bodies. In terms of dynamical quantities, the mathematical statements of Newton's three laws of motion are usually given as follows:

1. A body has a constant velocity unless there is a net force acting on it.
2. The rate of change of momentum is proportional to the impressed force and is in the same direction,

$$\mathbf{F} = \Delta(m\mathbf{v})/\Delta t.$$

For the case of constant mass, Newton's second law is

$$\mathbf{F} = m\mathbf{a}.$$

3. If a body A exerts a force $\mathbf{F}_{AB}$ on a body B, then B exerts a force $\mathbf{F}_{BA}$ on A, so that $\mathbf{F}_{AB} = -\mathbf{F}_{BA}$.

- The weight of an object is the gravitational force on that object,

$$\mathbf{w} = m\mathbf{g}.$$

- The frictional force of one object sliding on another is

$$F \leq \mu N \quad \text{(static)}; \qquad F = \mu N \quad \text{(kinetic)}.$$

- For a body to be in translational equilibrium, the sum of all the forces acting on it must be zero:

$$\sum \mathbf{F}_i = 0.$$

Important Terms

You should be able to write the definition or meaning of each of the following terms:

- Newtonian mechanics
- inverse-square force
- force
- net force
- inertia
- momentum
- mass
- newton
- weight
- normal force
- weightlessness
- tension
- free-body diagram
- friction
- coefficient of friction
- equilibrium
- static equilibrium
- dynamic equilibrium
- inertial reference frame

QUESTIONS

4.1 Why do packages slide off the seat of a car that is braked hard?

4.2 Describe the difference between mass and weight.

4.3 A tennis ball thrown against the wall bounces back toward the thrower. Where does the force come from that sends it back?

4.4 Describe some of your everyday activities that would be seriously hindered, if not impossible, if there were no friction.

4.5 A child sits in a swing that is not moving. Describe all forces present and identify all action-reaction pairs.

4.6 Imagine a large adult and a small child on roller skates. Describe their motions if they push off each other.

4.7 A book rests motionless on a table. Does that mean there are no forces on it?

4.8 Suspend a heavy weight from a light string and attach a similar string below it (Fig. 4.21). If you pull on the lower string with steadily increasing force, the upper string will break; if you pull the lower string with a jerk, the lower string breaks. Explain both cases.

Figure 4.21 Question 4.8.

4.9 Given the existence of a standard mass, devise a way of dynamically comparing other masses with the standard.

4.10 Two teams of children are playing tug-of-war. The rope passes through a small hole in a high fence that separates the two teams. Neither team can see the other. Both teams pull mightily, but neither budges. As lunchtime approaches, the members of one team decide to tie their end of the rope to a stout tree while they take a lunch break. Can the other team tell that the first is not pulling on the motionless rope? Analyze the forces in this problem.

4.11 Identify all of the forces present in the situation of Fig. 4.22. Which forces are action-reaction pairs?

Figure 4.22 Question 4.11.

4.12 A spring scale is used to weigh beans on an elevator. How will the readings for a given amount of beans change when the elevator is (a) going down with constant velocity? (b) moving with a constant downward acceleration less than g? (c) moving upward with a constant velocity? (d) accelerating upward with an acceleration a?

4.13 A man goes over Niagara Falls in a barrel with windows in the side. During the descent, the man takes out an apple, holds it up in front of his face and releases it. Describe what is seen by (a) an observer on the bank looking through the window and (b) the man in the barrel.

4.14 A person on an upward-moving elevator is throwing darts at a target fixed to the wall of the elevator. In order to hit the target, how should she aim the dart if the elevator has (a) constant velocity, (b) constant upward acceleration, (c) constant downward acceleration?

PROBLEMS

Hints for Working Problems

To become skillful at calculating the answers to motion problems, you must have considerable experience. Thus many practice problems are included in this and other chapters. You should work numerous problems in order to develop your understanding of the laws of motion. If all you can do is repeat Newton's three laws, you are like a musician who is able only to repeat the names of the notes in the music: In both cases the ability may be desirable, but it is of no real value unless you can also perform.

An object's acceleration is proportional to the net external applied force, so forces should be added vectorially. Remember that action-reaction pairs act on different objects. In inclined-plane problems, it is often helpful to resolve forces into components parallel and perpendicular to the plane. In static equilibrium problems, the equilibrium condition applies independently in each direction.

Section 4.4   Newton's Second Law

4.1 (a) Find the force that produces an acceleration of 4.0 m/s^2 for a 0.50-kg cantaloupe. (b) If the same force is applied to a 20-kg watermelon, what will its acceleration be?

4.2 Three coplanar forces act on a 7.0-kg mass: 14 N directed at 0°, 15 N at 138°, and 16 N at 275°. What is the magnitude and direction of the acceleration?

4.3 What steady forward force must be exerted on a 1.20×10^6-kg train initially at rest in order to give it a velocity of 30.0 km/h in 5.00 minutes? (Neglect friction.)

4.4 A 400-kg ice boat moves on runners on essentially frictionless ice. A steady wind blows, applying a constant force to the sail. At the end of an 8.0-s run, the acceleration is 0.50 m/s^2. (a) What was the acceleration at the beginning of the run? (b) What was the force due to the wind? (c) What retarding force must be applied at the end of 4.0 s to bring the ice boat to rest by the end of the next 4 s? (The wind is still blowing. Assume the boat was at rest at time $t = 0$.)

4.5 The total horizontal force exerted between the tires of a 1500-kg automobile and the ground is 980 N. If the car starts from rest, how far will it go in 5.0 s?

4.6 A toy truck of 1.0 kg is at rest on a horizontal frictionless surface. At time $t = 0$ there are no horizontal forces acting on the truck. At $t = 1.0$ s the truck is suddenly acted upon by a force $F = 1.0$ N. This force is maintained until $t = 2.0$ s, when the force becomes 2.0 N. At $t = 3.0$ s the force becomes 3.0 N, etc. Construct a graph of the velocity as a function of time. From this make a graph of displacement versus time.

Section 4.5   Weight, the Force of Gravity

4.7 A force of 1 newton is equal to 0.2248 lb. (a) Compute the weight in newtons of a 150-lb man. (b) Compute the mass of the man.

4.8 What is the gravitational force on a 1.00-kg mass?

4.9 What is the weight of a 9.00-kg mass?

4.10 What is the mass in kilograms of an object that weighs 1.000 lb? *Hint:* 1 N = 0.2248 lb.

4.11 A body of mass m resting on a frictionless surface is given a horizontal acceleration of 4.5 m/s^2 by a force of 8.7 N. (a) What is the mass of the body? (b) What is the weight (force due to gravity) of the body?

4.12 The downward acceleration of a falling body on earth is about 9.80 m/s^2. On the moon the same quantity is 1.96 m/s^2. An astronaut in a space suit has a mass of 145 kg. (a) What is the astronaut's weight on earth? (b) On the moon? (c) What is the astronaut's mass on the moon?

4.13 A standard kilogram mass was prepared in Paris, where $g = 9.81$ m/s^2, and sent to Washington, where $g = 9.80$ m/s^2. What was the percentage change in the weight of the standard?

4.14 A Saturn-Apollo launch rocket has a mass of 5.40×10^5 kg. What is the initial acceleration of the rocket if the thrust at lift-off is 7.40×10^6 N?

Section 4.6   Newton's Third Law

4.15 An 80-kg man stands in an elevator. What force does he exert on the floor of the elevator under the following conditions? (a) The elevator is stationary.

(b) The elevator accelerates upward at 2.0 m/s^2. (c) The elevator rises with constant velocity of 4.0 m/s. (d) While going up, the elevator decelerates at 1.5 m/s^2. (e) The elevator goes down with constant velocity of 7.0 m/s.

4.16 An unabridged dictionary of mass m rests on a weak table. (a) What is the reaction force to the force of the book on the table? (b) What is the reaction force to the force of gravity on the book? (c) The table top collapses under the book's weight. Answer (a) and (b) for these conditions.

4.17 A bird is placed on a perch in a large closed box that is sitting on a spring scale. What will happen to the scale reading when (a) the bird jumps off the perch, (b) the bird is flying, (c) the bird lights on the perch again?

4.18 What upward force must be exerted at A to hold the basket shown in Fig. 4.23?

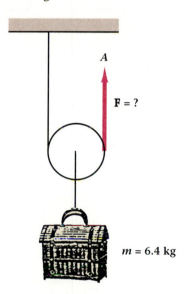

Figure 4.23   Problem 4.18.

4.19 A person weighing 650 N stands on spring scales in an elevator that is moving downward with constant speed of 3.6 m/s. The brakes suddenly grab, bringing the elevator to a stop in 1.8 s. Describe the scale readings from just before the brakes grab until after the elevator is at rest.

4.20 A large 500-kg magnet exerts a constant force of 3.00 N on a 0.250-kg bar magnet. What force does the bar magnet exert on the big magnet?

Section 4.7   Some Applications of Newton's Laws

4.21 Figure 4.24 shows a graph of distance along a straight course as a function of time for a 1000-kg dragster. Construct a graph of the horizontal force between the tires and ground (no slipping) as a function of time.

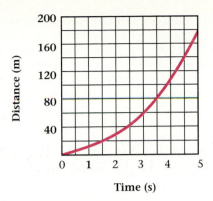

Figure 4.24   Problem 4.21.

(*Hint*: First construct a plot of velocity, taking points, say, every 0.5 s.)

4.22 Two air-track gliders m_1 and m_2 are joined together with a light string (Fig. 4.25). A constant horizontal force of 5.0 N to the right is applied to mass m_2. (a) If $m_1 = 2.0$ kg and $m_2 = 0.50$ kg, what is the acceleration of the gliders? (b) What is the tension in the cord joining them?

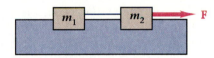

Figure 4.25   Problem 4.22.

4.23 Find the accelerations and the tension for the situation of Problem 4.22, given that $m_2 = 2.0$ kg and $m_1 = 0.50$ kg.

4.24 (a) A 20.0-kg bucket is lowered by a rope with constant velocity of 0.50 m/s. What is the tension in the rope? (b) A 20.0-kg bucket is lowered with a constant downward acceleration of 1.00 m/s^2. What is the tension in the rope? (c) A 15.0-kg bucket is raised with a constant upward acceleration of 1.00 m/s^2. What is the tension in the rope?

4.25 A .0050-kg bullet traveling with a speed of 200 m/s
• penetrates a large wooden fence post to a depth of 0.030 m. What was the average resisting force exerted on the bullet?

Section 4.8   Friction

4.26 A 10-kg lead brick rests on a wooden table. If a force of 46 N is required to slide the brick across the table at a constant speed, what is the coefficient of friction?

4.27 What is the minimum force required to pull a 50.0-kg bag of fertilizer across the floor if the coefficient of friction is 0.37?

4.28 A 1.00-kg weight, initially at rest, is pulled across a table with a force of 3.00 N. (a) Will it move if the coefficient of friction is 0.45? (b) What is the coefficient of friction if it just begins to move? (c) What is its acceleration if the coefficient of friction is 0.20?

4.29 A 5.00-kg concrete block rests on a table. The coefficient of friction between the block and the table is 0.55. A 4.00-kg weight is attached to the block by a string of negligible mass passed over a light frictionless pulley (Fig. 4.26). What is the acceleration of the block when the 4.00-kg weight is released?

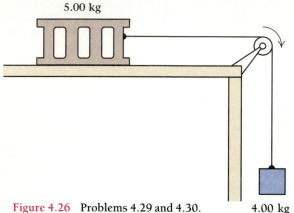

5.00 kg

Figure 4.26   Problems 4.29 and 4.30.   4.00 kg

4.30 A 5.00-kg concrete block rests on a table. A 4.00-kg mass is attached to the block by a string passing over a light, frictionless pulley (Fig. 4.26). If the acceleration of the block is measured to be 1.00 m/s², what is the coefficient of friction between the block and the table?

4.31 A 4.0-kg wooden block rests on a table. The coefficient of friction between the block and the table is 0.40. A 5.0-kg mass is attached to the block by a horizontal string passed over a frictionless pulley of negligible mass. (a) What is the acceleration of the block when the 5.0-kg mass is released? (b) What is the tension in the string during the acceleration?

4.32 Determine the acceleration and the tension for the situation in Problem 4.31 when the coefficient of friction between the block and the table is 0.25.

4.33 What is the coefficient of friction between a sled and a plane inclined at 30° from the horizontal if the sled just slides without accelerating when given an initial push?

Section 4.9   Static Equilibrium

4.34 Calculate the angle of repose for a glass cube on a glass incline, given that the coefficient of friction is 0.95. (*Hint:* see Example 4.10.)

4.35 Calculate the angle of repose of a cubic steel paperweight on a clean dry steel incline. (*Hint:* see Example 4.10.)

4.36 A sign is hung from cables as shown in Fig. 4.27. What is the tension in each cable if the sign weighs 150 N? (*Hint:* Ignore the weight of the cable.)

Figure 4.27   Problem 4.36.

4.37 A fish of mass *m* is suspended by a string as shown in Fig. 4.28. The string is fastened securely at point *C* but will pull loose from the wall at *A* when the string tension exceeds 10 N. What is the maximum mass of the fish that can be supported by the string?

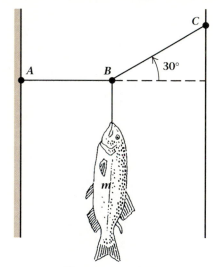

Figure 4.28   Problems 4.37 and 4.38.

4.38 What is the force exerted by the string on the wall at point *A* in Fig. 4.28 if the suspended fish has mass *m* = 0.15 kg?

4.39 A physicist finds that her car is stuck in the sand and cannot be driven out. Unable to push it out, she ties a stout rope from the front of the car to a large tree 30 m away and directly in front of the car. She then pushes on the middle of the rope with a force of 400 N in a direction perpendicular to the length of the rope. If the midpoint of the rope is displaced by 3.0 m, what is the force applied to the car?

4.40 Three equal masses are suspended from frictionless pulleys as shown in Fig. 4.29. What are the angles of the strings with respect to the horizontal when the system comes to equilibrium?

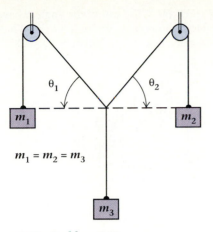

Figure 4.29 Problem 4.40.

4.41 Suppose that the masses in Problem 4.40 are not identical. What are the angles θ_1 and θ_2 if $m_1 = m_2 = 2.0$ kg and $m_3 = 3.0$ kg?

4.42 A 60-lb child is seated in a swing of negligible mass. How much horizontal force is required to pull the child and swing aside so that the support rope makes an angle of 30° with the vertical?

4.43 Assume that the Atwood machine of Fig. 4.8 is in static equilibrium. Make a diagram showing all forces. What differences would appear if the mass $m_1 \neq m_2$?

Section 4.10 The Laws of Motion as a Whole

4.44 A monkey clinging to one end of a rope that passes over a frictionless pulley is balanced by an exactly equal weight of bananas on the other end of the rope (Fig. 4.30). The monkey begins to climb the rope. (a) What will happen? If you need additional information, make a specific assumption and answer the question in that light. (b) After the monkey has traversed 3.0 m of rope, he stops. How high has he climbed?

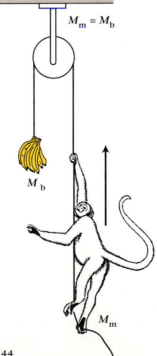

Figure 4.30 Problem 4.44.

4.45 The starship Enterprise is on a mission into deep space, where no man has gone before. It encounters an alien spacecraft, and Captain Kirk orders activation of the tractor beam to pull the alien craft to the starship. The alien ship is made of a super-dense alloy so that its mass is 8 times the mass of the starship. Assuming the two craft have zero relative velocity at the time the tractor beam is activated, describe their motions.

4.46 A truck loaded with heavy cartons is forced to stop suddenly with a deceleration of 5.0 m/s². Calculate the minimum coefficient of friction between the cartons and the truck bed, given that the cartons do not slide.

Additional Problems

4.47 What minimum force is required to drag a carton of
• books across the floor if the force is applied at an angle of 45° to the horizontal (Fig. 4.31)? Take the mass of the carton as 40 kg and the coefficient of friction as 0.60.

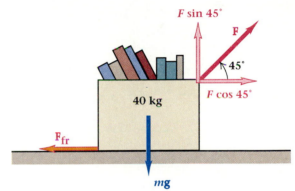

Figure 4.31 Problem 4.47.

4.48 A 5000-kg cable-car is moving up a 15% grade, that
• is, one that rises 15 ft in a horizontal distance of 100 ft. Find the force that must be exerted on it to keep it moving with constant velocity. Assume that effects of friction may be neglected.

4.49 A 500-kg trailer being pulled behind a car is subject
• to a 100-N retarding force due to friction. What force must the car exert on the trailer if: (a) the trailer is to move forward at a constant speed of 25 km/h? (b) the trailer is to move forward with an acceleration of 2.0 m/s²? (c) starting from rest, the trailer and car are to travel 150 m in 10 s?

4.50 (a) What is the minimum time in which one can hoist
• a 1.00-kg rock a height of 10.0 m if the string used to pull the rock up has a breaking strength of 10.8 N? Assume the rock to be initially at rest. (b) If the string is replaced by one that is 50% stronger, by what

percentage will the minimum time for the hoist be reduced?

4.51 A 10-kg block is placed on a frictionless table and connected to a 5.0-kg block by a string that extends horizontally across the table over a pulley and down to the 5-kg block. What is the acceleration of the blocks? Ignore friction in the pulley.

4.52 A simple Atwood machine composed of a single pul-
•• ley and two masses m_1 and m_2 is on an elevator. When $m_1 = 44.7$ kg and $m_2 = 45.3$ kg it takes 5.00 s for mass m_2 to descend exactly one meter from rest relative to the elevator. What is the elevator's motion? (That is, is it moving with constant velocity or accelerating up or down?)

4.53 A classroom demonstration is done with an Atwood
•• machine. The masses are $m_1 = 1.00$ kg and $m_2 = 1.10$ kg (Fig. 4.32). If the larger mass descends a distance of 3.00 m from rest in 3.6 seconds, what is the acceleration of gravity at that place? (Ignore effects of pulley mass and friction.)

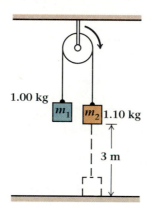

Figure 4.32   Problem 4.53.

4.54 For the accompanying diagram (Fig. 4.33), describe
• what will happen and what the spring scale S will read if (a) $m_1 = m_2 = 1.00$ kg, (b) $m_1 = 1.20$, $m_2 = 1.00$ kg. (*Hint:* Ignore the mass of the scale.)

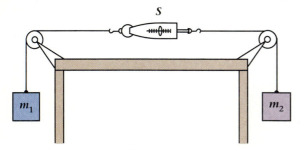

Figure 4.33   Problem 4.54.

4.55 A 10.0-kg block is placed on a frictionless inclined
•• plane and connected to a 5.0-kg block as shown in

Fig. 4.34. (a) What would the angle θ have to be for the blocks to remain motionless? (b) What would be the acceleration of the blocks if $\theta = 37°$?

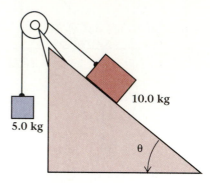

Figure 4.34   Problem 4.55.

4.56 A block of mass 4.7 kg slides from rest down a 20-
• m-long inclined plane making an angle of 30° with the horizontal. If the block takes 10 s to slide down the plane, what is the retarding force due to friction? The frictional force is parallel to the plane and opposite to the direction of motion of the block.

4.57 Two clay pots joined together by a light string rest
•• on a table (Fig. 4.35). The frictional coefficient between the pots and the table is 0.35. The pots are also joined to a 4.0-kg mass by a string of negligible mass passed over an ideal pulley as shown in the figure. Calculate the acceleration of the system when the 4.0-kg mass is released. Also find the tensions T_1 and T_2 in the strings during acceleration.

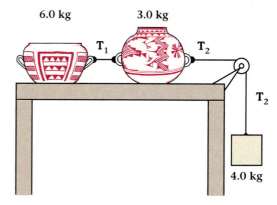

Figure 4.35   Problems 4.57 and 4.58.

4.58 Consider the situation shown in Fig. 4.35 but with
•• the 4.0-kg mass replaced by a 6.5-kg mass. Find the acceleration of the system and the tensions T_1 and T_2, assuming that the frictional coefficient is 0.35.

4.59 A plant is hung from strings as shown in Fig. 4.36.
• What is the tension in each cable if the plant weighs 20.0 N? (*Hint*: Ignore the weight of the string.)

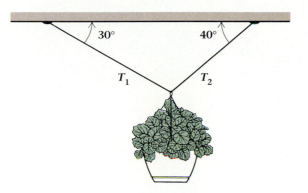

Figure 4.36   Problem 4.59.

4.60 Suppose that the weight w_2 in Fig. 4.37 is 400 N.
• What must be the values of the weights w_1 and w_3?

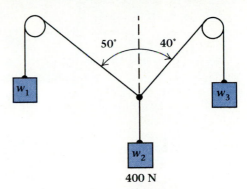

Figure 4.37   Problem 4.60.

4.61 A 65-kg skier goes down a 30° slope. (a) What would
• be the acceleration of the skier if friction could be neglected? (b) A skier continues to accelerate until gravitational force is balanced by the normal force and by the frictional forces due to the skis on the snow and the air resistance. What must be the frictional forces if the skier is no longer accelerating?

ADDITIONAL READING

Armstrong, H. L., "How Dry Friction Really Behaves." *American Journal of Physics*, Vol. 53, September 1985, p. 910.

Badger, N. W., "The Royal Institution." *The Physics Teacher*, Vol. 14, October 1976, p. 402.

Cohen, I. B., "Newton's Discovery of Gravity." *Scientific American*, Vol. 244, March 1981, p. 166.

Feynman, R. P., R. B. Leighton, and M. Sands, *The Feynman Lectures on Physics*. Reading, Mass.: Addison-Wesley, 1963. Chapter 12 contains a discussion of the nature of forces.

Franklin, A., "Inertia in the Middle Ages." *The Physics Teacher*, Vol. 16, April 1978, p. 201.

Manuel, F. E., *A Portrait of Isaac Newton*. Cambridge, Mass.: Harvard University Press, 1968.

Moore, D. F., *The Friction of Pneumatic Tyres*. Amsterdam: Elsevier, 1975.

5

Uniform Circular Motion and Gravitation

5.1 Uniform Circular Motion

5.2 Force Needed for Circular Motion

5.3 The Law of Universal Gravitation

5.4 The Universal Gravitational Constant G

5.5 Gravitational Field Strength

A WORD TO THE STUDENT

Up to this point we have studied one-dimensional motion along a straight line and projectile motion, which is two-dimensional. We begin this chapter by describing uniform circular motion, which is also two-dimensional. Some of the terminology in this discussion will be useful later for describing other kinds of motion, such as rotation and oscillation. We then proceed to examine gravitation and the motion resulting from gravitational forces, which includes uniform circular motion. An understanding of the principle of gravitation allows us to explain the motion of the earth, the moon, and artificial satellites. It also gives us greater insight into the meaning of weight and the behavior of freely falling bodies. Finally we present a brief introduction to a concept whose importance will grow later on in this book: the field.

We saw in the preceding chapter how we can use Newton's laws to calculate the motion of an object once we know the forces acting on it. However, Newton's laws are very general, and do not depend on what kind of force acts on a particular object.

Newton's law of universal gravitation was the first example of expressing a fundamental force of nature in mathematical form; the second force to be described mathematically was the electrostatic force, almost 100 years later (see Chapter 15). This process of mathematical development continues today for nuclear and subnuclear forces (Chapters 28 and 31).

Planets moving under the influence of gravity move in elliptical paths about the sun. In many cases these ellipses are almost circles. Therefore, to understand the motion of planets and satellites in their orbits, we first examine the dynamics of circular motion. Circular motion is quite common in nature, and the ideas and descriptions presented in this chapter will come up again frequently in later chapters.

Uniform Circular Motion

5.1

Consider a point particle moving along a circular path of radius r with constant speed v. A particle moving in this manner is said to undergo **uniform circular motion**. By a "particle" we mean an object of negligible size and constant mass. We use the idea of a particle as a way of creating a simplified model of a real physical situation, so our description of circular motion is not too difficult mathematically. However, the resulting equations apply with relative accuracy to real situations, such as a dot of paint on the rim of a phonograph record rotating on a turntable at a steady rate. To a good approximation, we can extend our description of uniform circular motion to a merry-go-round ride or to a satellite revolving around the earth.

As we learned in our study of kinematics, a particle's speed is determined by measuring the distance traveled along its path and dividing by the elapsed time. Although the dot moves along its circular path with a constant speed v, its instantaneous velocity vector is constantly changing because the direction of its motion is constantly changing. At any instant of time, the instantaneous velocity is tangential to the circle (Fig. 5.1). In a time interval Δt the object moves along the circular path from one point, say P in Fig. 5.1(a), to another point, Q. The instantaneous velocity vectors at points P and Q, given by $\mathbf{v}_P$ and $\mathbf{v}_Q$, respectively, have the same magnitude but differ in direction. Each instantaneous velocity is tangent to the circle and perpendicular to the radius r at the point in question.

Because the velocity is constantly changing, there must be an acceleration. (Remember, acceleration occurs whenever the velocity changes in magnitude *or direction*.) During the time interval Δt, the velocity changes by an amount $\Delta \mathbf{v} = \mathbf{v}_Q - \mathbf{v}_P$. From the definition of average acceleration we have

$$\overline{\mathbf{a}} \equiv \frac{\Delta \mathbf{v}}{\Delta t}. \tag{5.1}$$

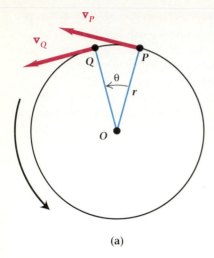

(a)

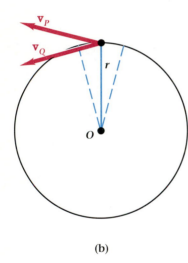

(b)

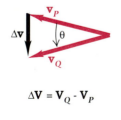

$$\Delta \mathbf{V} = \mathbf{V}_Q - \mathbf{V}_P$$

(c)

Figure 5.1 (a) The instantaneous velocity vectors at points P and Q for a particle moving in a circular path. (b) Vectors $\mathbf{v}_P$ and $\mathbf{v}_Q$ are translated to a common origin for comparison. (c) The vector $\Delta \mathbf{v} = \mathbf{v}_Q - \mathbf{v}_P$. As $\boldsymbol{\theta}$ becomes very small, $\Delta \mathbf{v}$ becomes perpendicular to both $\mathbf{v}_Q$ and $\mathbf{v}_P$.

To evaluate the acceleration we translate the vectors $\mathbf{v}_P$ and $\mathbf{v}_Q$ to a common origin (Fig. 5.lb). The change in velocity $\Delta \mathbf{v}$ is shown in Fig. 5.1(c). As the time interval Δt is made smaller, points P and Q are found closer together, reducing the angle θ. From the figure you can see that this angle is also the angle between the two velocity vectors. Eventually the angle θ becomes so small that $\mathbf{v}_P$ and $\mathbf{v}_Q$ are almost parallel and their difference $\Delta \mathbf{v}$ is almost perpendicular to both of them. In the limit where Δt goes to zero, $\Delta \mathbf{v}$ is exactly perpendicular to $\mathbf{v}$. Hence the instantaneous acceleration, which is in the same direction as $\Delta \mathbf{v}$, is directed radially toward the center of the circular path. Therefore *a particle moving with constant speed around a circle is always accelerated toward the center*. In this special case, the particle has uniform circular motion, and its acceleration is always perpendicular to the velocity. This acceleration is called the **centripetal** (center-seeking) **acceleration**.

The centripetal acceleration can be readily evaluated. The angle θ between $\mathbf{v}_P$ and $\mathbf{v}_Q$, when measured in radians, is the ratio of the arc length to the radius, as defined in Eq. (1.1) of Chapter 1. The arc length is the product of the speed and the time interval, so the angle becomes

$$\theta = \frac{\text{arc length}}{\text{radius}} = \frac{v\,\Delta t}{r}. \qquad (5.2)$$

From the geometry of Fig. 5.1(c) we see that as θ gets smaller, the magnitude of $\Delta \mathbf{v}$ (indicated by $|\Delta \mathbf{v}|$) is approximately the same as the arc length made by moving a vector of magnitude $|\mathbf{v}_P|$ through the angle θ. Thus θ may also be expressed as the ratio

$$\theta \approx \frac{|\Delta \mathbf{v}|}{|\mathbf{v}_P|}.$$

Since the magnitude of the velocity is constant, $|\mathbf{v}_P| = v$ and we can write

$$\theta \frac{|\Delta \mathbf{v}|}{v}.$$

The magnitude of the average acceleration becomes

$$|\bar{\mathbf{a}}| = \frac{|\Delta \mathbf{v}|}{\Delta t} \approx \frac{v\theta}{\Delta t}.$$

Upon substituting Eq. (5.2) for θ we find that, in the limit of very small angles, we have

$$a_c = \frac{v^2}{r}, \qquad (5.3)$$

where the subscript c denotes centripetal acceleration. Note that at high speeds, the velocity vector changes direction rapidly (large centripetal acceleration). Also, if a particle travels in a circle of large radius, the velocity changes direction more slowly (small centripetal acceleration). Nevertheless, the velocity is always tangential to the circle and the acceleration always points to the center of the circle. A numerical example of calculating centripetal acceleration follows.

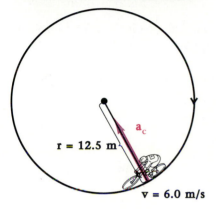

r = 12.5 m

a_c

v = 6.0 m/s

Figure 5.2 **Example 5.1:** Bicycle racer on a circular track.

Example 5.1

Riding a bike in a circle.

A bicycle racer rides with constant speed around a circular track 25 m in diameter (Fig. 5.2). What is the acceleration of the bicycle toward the center of the track if its speed is 6.0 m/s?

Solution Since the speed around the circle is constant, we can compute the acceleration directly from Eq. (5.3):

$$a_c = \frac{v^2}{r},$$

$$a_c = \frac{(6.0 \text{ m/s})^2}{12.5 \text{ m}} = \frac{36 \ (\text{m/s})^2}{12.5 \text{ m}} = 2.9 \text{ m/s}^2.$$

Remember that this acceleration is directed toward the center of the circle as the bicycle moves at a constant speed around the circular track.

Sometimes it is more convenient to describe circular motion in terms of other quantities. For example, suppose that we know an object's **period** *T*, which is the time it takes the object to complete one revolution around its circular path. During this time, the object travels with constant speed *v* along a distance equal to the circumference of the circle,

$$vT = 2\pi r.$$

Upon rearranging, we get

$$v = \frac{2\pi r}{T}.$$

We can insert this expression into Eq. (5.3) to get the centripetal acceleration in terms of *r* and *T*,

$$a_c = \frac{4\pi^2 r}{T^2}. \tag{5.4}$$

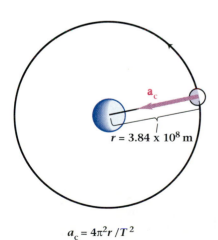

a_c

$r = 3.84 \times 10^8$ m

$a_c = 4\pi^2 r / T^2$
$T = 27.3$ days

Figure 5.3 **Example 5.2:** The moon orbiting the earth.

Example 5.2

Centripetal acceleration of the moon.

Determine the centripetal acceleration of the moon toward the earth and compare that acceleration with the acceleration of bodies falling on the earth.

Solution According to Newton's first law, the moon would move with constant velocity in a straight line unless it were acted on by a force. We can infer the presence of a force from the fact that the moon moves with approximately uniform circular motion about the earth (Fig. 5.3). The acceleration toward the earth can be calculated using a mean earth–moon distance of 3.84×10^8 m and a period of 27.3 days:

$$a_c = \frac{4\pi^2 r}{T^2} = \frac{4\pi^2 (3.84 \times 10^8 \text{ m})}{(27.3 \text{ days} \times 24 \text{ h/day} \times 3600 \text{ s/h})^2},$$

$$a_c = 2.72 \times 10^{-3} \text{ m/s}^2.$$

The ratio of the center-to-center earth–moon distance to the earth's radius is about 60. The ratio of the acceleration of falling bodies on the earth's surface to the acceleration of the moon toward the earth is

$$\frac{9.8 \text{ m/s}^2}{2.72 \times 10^{-3} \text{ m/s}^2} = 3600.$$

This is approximately the same as the square of the ratio of the distances. Considerations of this kind helped convince Newton and others of his time of the inverse-square nature of the gravitational force.

The **frequency** f is the number of complete revolutions, or cycles, an object makes per unit of time. The frequency f is the reciprocal of the period T. If an object takes a time T to complete one revolution around the circle, then the number of revolutions per unit time, the frequency, is

$$f = \frac{1}{T}. \tag{5.5}$$

In general, frequency has the dimension of inverse time, that is,

$$\frac{1}{\text{time}}.$$

The SI unit of frequency is the **hertz**, abbreviated Hz:

$$1 \text{ Hz} = 1 \text{ s}^{-1}.$$

When Eq. (5.5) is inserted into Eq. (5.4), we get an expression for the centripetal acceleration in terms of the frequency:

$$a_c = 4\pi^2 f^2 r.$$

Example 5.3

Linear speed of the rim of a rotating wheel.

An industrial grinding wheel with a 25.4-cm diameter spins at a rate of 1910 revolutions per minute. What is the linear speed of a point on the rim?

Solution   The given rate is the frequency f. The speed of a point on the rim, a distance r from the axis of rotation, is

$$v = \frac{2\pi r}{T} = 2\pi r f.$$

Inserting the values of r and f we get

$$v = (2\pi) \left(\frac{25.4 \text{ cm}}{2} \right) \left(\frac{1910}{\text{min}} \right) \left(\frac{1 \text{ min}}{60 \text{ s}} \right),$$

$$v = 2540 \text{ cm/s} = 25.4 \text{ m/s}.$$

The position of a point in circular motion with tangential velocity v at a constant radial distance r from the center of the circle is given by the

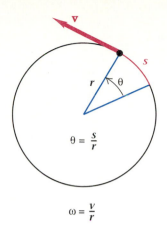

$\theta = \dfrac{s}{r}$

$\omega = \dfrac{v}{r}$

Figure 5.4 A point moving with a constant speed v along a circle of radius r is described in terms of its angular position θ. Its angular velocity is $\omega = v/r$.

angle θ shown in Fig. 5.4. This angle is defined in terms of the arc length s and the radius r ($\theta = s/r$), as defined earlier. The rate of change of this angle is the **angular velocity** ω. The average angular velocity $\overline{\omega}$ is

$$\overline{\omega} \equiv \frac{\Delta\theta}{\Delta t} = \frac{\Delta s}{\Delta t} \cdot \frac{1}{r}.$$

When the time interval is very small, $\Delta s/\Delta t$ becomes the instantaneous speed v. Then the instantaneous angular velocity ω becomes

$$\omega = \frac{v}{r}. \tag{5.6}$$

We can express the centripetal acceleration in terms of the angular velocity as

$$a_c = \omega^2 r.$$

The dimension of ω is the reciprocal of time (that is, time^{-1}) and the units are radians per second (rad/s). Remember that, since the angle in radians was defined as a ratio of lengths, it has no dimension. The unit of radian is carried as a reminder that the angles are measured in radians and not in degrees.

When the change in time Δt is one period, the change in angle corresponds to one complete revolution or 2π rad. Thus we can also express the angular velocity as

$$\omega = \frac{\Delta\theta}{\Delta t} = \frac{2\pi}{T},$$

so that

$$\omega = 2\pi f. \tag{5.7}$$

Since ω is directly proportional to f and has the dimension of inverse time, it is often called the **angular frequency**. The two names for ω may be used interchangeably. We will use these terms again in describing rotations and oscillations.

Example 5.4

At the Six Flags amusement park near Atlanta, the Wheelie carries passengers in a circular path with a radius of 7.7 m (Fig. 5.5). If the ride makes a complete rotation every 4.0 s, what acceleration does a passenger experience due to the circular motion?

Solution Because the riders travel in a circle, they undergo a centripetal acceleration. We start with

$$a_c = v^2/r$$

and use the relation $v = 2\pi r/T$ to generate the form

$$a_c = \frac{4\pi^2 r}{T^2} = \frac{4\pi^2(7.7 \text{ m})}{(4.0 \text{ s})^2} = 19 \text{ m/s}^2.$$

Notice that this is almost twice the acceleration of a body in free fall.

Figure 5.5 Example 5.4: The Wheelie.

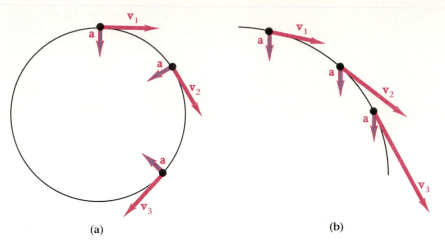

Figure 5.6 (a) In uniform circular motion, the acceleration is always perpendicular to the velocity, so that only the direction of the velocity changes. (b) In other two-dimensional motions, such as projectile motion, the acceleration is neither perpendicular nor parallel to the velocity. Then both direction and magnitude of the velocity change.

(a)   (b)

Before we leave this section, let's gather together the various formulas we have used for centripetal acceleration:

$$a_c = \frac{v^2}{r}, \tag{5.8a}$$

$$a_c = \frac{4\pi^2 r}{T^2}, \tag{5.8b}$$

$$a_c = 4\pi^2 f^2 r, \tag{5.8c}$$

$$a_c = \omega^2 r. \tag{5.8d}$$

These four equations all convey the same meaning. Each can be converted to the others with the aid of equations such as Eqs. (5.6) and (5.7), which connect the quantities v, r, ω, f, and T. Thus the four equations are really the same, and knowledge of one is equivalent to knowledge of all if you understand the meanings of the various terms. You do not need to remember them all. Instead, remember one of them, such as Eq. (5.8a), and the definitions of f, T, and ω. Then you can generate the other versions when you need them.

Uniform circular motion represents the special case of two-dimensional motion in which the acceleration is always perpendicular to the velocity. In that case, the acceleration changes only the direction of the velocity, and not its magnitude (speed). We found earlier that for projectile motion the acceleration is generally neither parallel nor perpendicular to the velocity, so that both magnitude and direction of the velocity change. Figure 5.6 depicts these two cases.

Force Needed for Circular Motion

5.2

We have just seen that an object of mass m moving in a circular path with a uniform speed v is accelerated because the direction of its instantaneous velocity is continuously changing. For example, a toy airplane whirled in a circle by a string is accelerated. By Newton's second law, the net force acting on the object is in the same direction as the observed acceleration. This net force is given the special name **centripetal force** because it is directed toward the center of the circle. It is this net force that causes the motion of the object to be circular; without the centripetal force, the object would travel in a straight line and not in a circle. The magnitude of the centripetal force is obtained from the second law as

$$F_c = ma_c,$$

$$F_c = \frac{mv^2}{r}. \tag{5.9}$$

Substituting any other expression for centripetal acceleration into the definition of centripetal force also gives a valid equation. For example, centripetal force is also given by $m\omega^2 r$.

As we have seen, circular motion requires acceleration and the acceleration is the result of a net force. Thus any object undergoing circular motion necessarily experiences a force that causes the object to move in a circular path. It is this force that we call the centripetal force. A word of caution here is appropriate. Centripetal force is not a fundamental force in the same sense that gravity is a fundamental force. It is just the name we give to the net force—whatever its origin—that causes an object to move in a circle. Computations of centripetal force are illustrated by the following examples.

Example 5.5

The force of the earth on the moon.

Approximately how much force does the earth exert on the moon?

Solution   Assume the moon's orbit to be circular about a stationary earth. Use the data given in Fig. 5.3. The mass of the moon is 7.36×10^{22} kg. Then

$$F = ma_c = m \frac{4\pi^2 r}{T^2},$$

$$F = \frac{(7.36 \times 10^{22} \text{ kg}) \, 4\pi^2 \, (3.84 \times 10^8 \text{ m})}{(27.3 \text{ days} \times 8.64 \times 10^4 \text{ s/day})^2},$$

$$F = 2.01 \times 10^{20} \text{ N}.$$

Example 5.6

A student ties a 0.060-kg lead fishing weight to the end of a piece of string and whirls it around in a horizontal circle. If the radius of the circle is 0.30 m and the object moves with a speed of 2.0 m/s, what is the horizontal component of force that directs the lead weight toward the center of the circle (Fig. 5.7a)? What is the tension in the string?

Solution   To begin, we draw the free-body diagram of Fig. 5.7(b). The tension **F** along the string provides both a horizontal and a vertical force. The centripetal acceleration is provided by the horizontal component F_c.

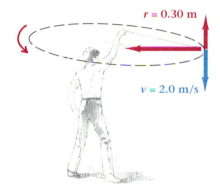

$r = 0.30$ m

$v = 2.0$ m/s

(a)

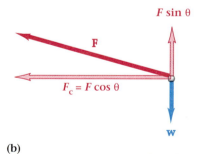

$F \sin\theta$

F

$F_c = F \cos\theta$

w

(b)

Figure 5.7   Example 5.6: (a) A fishing weight is tied to a string and whirled in a horizontal circle. (b) The free-body diagram for (a). The force **F** along the string must provide both the horizontal centripetal force and the upward vertical force that opposes the gravitational force **w**.

The weight will move up or down until the vertical component of **F** is equal and opposite to the gravitational force **w**. The horizontal component is

$$F_c = ma_c = \frac{mv^2}{r},$$

$$F_c = \frac{(0.060 \text{ kg})(2.0 \text{ m/s})^2}{0.30 \text{ m}} = 0.80 \text{ N}.$$

The tension in the string is the vector sum of F_c and the vertical force opposing **w**. Its magnitude is

$$F = \sqrt{F_c^2 + w^2},$$

$$F = \sqrt{(0.80 \text{ N})^2 + (0.060 \text{ kg} \times 9.80 \text{ m/s}^2)^2} = 0.99 \text{ N}.$$

A car moving in a circle with constant speed must be acted on by a force in order to execute circular rather than straight-line motion. This centripetal force is provided by the friction between the tires and the road. Passengers inside the car must also be subject to a centripetal force or they will not travel in the same path as the car. They experience forces exerted by the seat or the door of the car that cause them to move along the same path. This description is in accord with what is seen by an observer at rest outside the car.

Inside the car, the driver may actually believe she experiences an outward force that pushes her against the car door. In reality, however, it is the car seat and door that press inward on the driver, causing her to move in a circular path (Fig. 5.8). The occupant of the car exerts a force on the car that is equal and opposite to the centripetal force. This outward force, which is experienced only by the person in the car moving in a curve (a rotating reference frame), is called a centrifugal force. In a non-rotating reference frame, such as that of the observer at rest, there is no

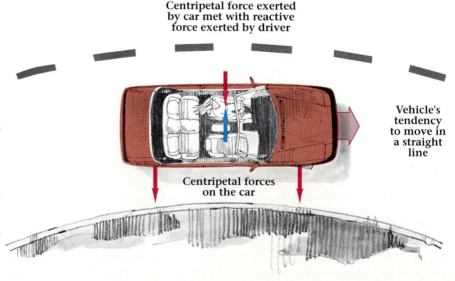

Centripetal force exerted by car met with reactive force exerted by driver

Vehicle's tendency to move in a straight line

Centripetal forces on the car

Figure 5.8 Centripetal forces cause both driver and car to move together in a circular path. Friction between the tires and the road exerts an inward force on the car; the car seat and door exert an inward force on the driver. By Newton's third law, the driver at rest inside the car also exerts an outward force on the car seat, equal in magnitude and opposite in direction to the centripetal force.

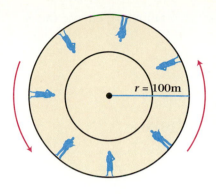

$$mg = \frac{m4\pi r^2}{T^2}$$

Figure 5.9 Example 5.7: A rotating space station. Occupants are kept on the outer wall by an "artificial gravity" produced by the rotational motion.

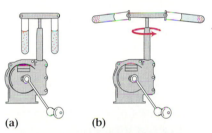

(a)   (b)

(c)

Figure 5.10 (a) A simple centrifuge. At rest the tubes hang down as shown. (b) As the shaft rotates at high speed, the tubes swing out. (c) Commercial laboratory centrifuges with a fixed tube angle commonly rotate at speeds of 3400 rpm.

centrifugal force. The words *centripetal* and *centrifugal* were used in their present-day context in the *Principia*.

Example 5.7

A uniformly rotating space station.

Imagine a giant donut-shaped space station located so far from all heavenly bodies that the force of gravity may be neglected. To enable the occupants to live a "normal" life, the donut rotates and the inhabitants live on the part of the donut farthest from the center (Fig. 5.9). If the outside diameter of the space station is 200 m, what must be its period of rotation so that the passengers at the periphery will perceive an artificial gravity equal to the normal gravity at the earth's surface?

Solution The weight of a person of mass m on the earth is a force

$$F = mg.$$

The centripetal force required to carry the person around a circle of radius r is

$$F = ma_c = \frac{m4\pi^2 r}{T^2}.$$

We may equate these two force expressions and solve for the period T:

$$mg = \frac{m4\pi^2 r}{T^2},$$

$$T = 2\pi\sqrt{\frac{r}{g}} = 2\pi\sqrt{\frac{100 \text{ m}}{9.8 \text{ m/s}^2}} = 20 \text{ s}.$$

A common laboratory tool that operates by the principle of centripetal force is the centrifuge. This device is primarily used to increase the sedimentation rate of small particles suspended in a liquid or to separate slightly dissimilar liquids. One type of centrifuge consists of a balanced pair of tubes attached to a shaft that can rotate at high speed (Fig. 5.10). The tubes are mounted on pivots so that they hang vertically when at rest. When the rotor spins, the tubes swing outward until at very high speeds they are virtually horizontal. In this position the liquid is unable to exert the centripetal force required to keep the small suspended particles moving in a circle. Consequently, the particles move outward toward the end (bottom) of the tubes. At high speeds, the resulting forces on the particles may be many times greater than the gravitational force, so that the small particles collect at the bottom of the tubes more quickly than if left to settle by gravity alone. Further details will be found in the discussion of Stokes's law in Chapter 9.

Example 5.8

Calculate the acceleration at the center of a centrifuge tube 8.0 cm from the axis of rotation, given that the centrifuge rotates at 55 rotations per second.

Solution   Since we are given a frequency of rotation, we choose the acceleration formula

$$a_c = \omega^2 r,$$

where the angular velocity is

$$\omega = 2\pi f.$$

So

$$a_c = (2\pi f)^2 r = (2\pi 55/\text{s})^2\, 0.080 \text{ m},$$

$$a_c = 9.6 \times 10^3 \text{ m/s}^2.$$

We may compare this acceleration with the acceleration of gravity, $g = 9.8 \text{ m/s}^2$, to get

$$a_c = 970\, g.$$

On a flat curve a car is turned by the frictional force exerted on the tires by the road. If the frictional force is not large enough, the car does not travel around the proper curve, but instead skids toward the outside. For high-speed turns on highways and race tracks, it is common practice to bank the curves to compensate for the tendency of vehicles to skid outward. Banking the curves also reduces the sideways force on the passengers and thus makes them more comfortable.

Figure 5.11(a) shows the force vectors acting on a car rounding an unbanked curve. The weight of the car is shown as the downward vector *m***g**. The normal forces $\mathbf{F}_N$ of the road supporting the car act on the tires. The sum of these normal forces just equals the weight of the car. The frictional forces $\mathbf{F}_{fr}$ acting on the tires provide the unbalanced force that gives the car its centripetal acceleration and causes it to turn.

If the curved road is banked at an angle θ, the normal force has a component toward the center of the circle (Fig. 5.11b). When the car is moving, this inward component of the normal force still points to the center of the circle. At one particular speed, this force component provides just the necessary force for turning the car without skidding, even if there are no sideways frictional forces between the tires and the road. At this speed the car will not skid even on an ice-covered road.

At this nonskidding speed, all of the forces between the car and the road are perpendicular or normal to the road surface. The road exerts a normal force **N** on the car. The vertical component $N \cos \theta$ must be equal and opposite to the gravitational force *m***g**. The horizontal component of **N** provides the centripetal force that turns the car. Thus we get

$$N \cos \theta = mg$$

and

$$N \sin \theta = \frac{mv^2}{r}.$$

We can divide the second equation by the first to get

$$\tan \theta = \frac{v^2}{gr}. \tag{5.10}$$

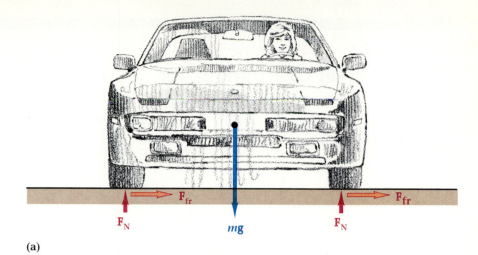

(a)

Figure 5.11   (a) A car rounding a
flat curve. The car's weight is
balanced by the sum of the normal
forces $\mathbf{F}_N$. Frictional forces $\mathbf{F}_{fr}$
between the tires and the road pro-
vide the centripetal acceleration.
(b) On a banked road the normal force
may be resolved into a vertical component and a horizontal component
directed toward the center of the
curve. For each bank angle there is
one particular speed for which the
normal force provides the necessary
centripetal force.

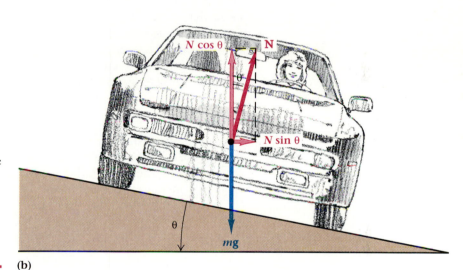

(b)

This equation gives the banking angle θ for a curve of a given radius r to
be negotiated at a speed v without tending to slide out or in, away from
the circular path.

Example 5.9

Bank angle of a race track.

A race track designed for average speeds of 240 km/h (66.7 m/s) is to
have a turn with a radius of 975 m. To what angle must the track be
banked so that cars traveling 240 km/h have no tendency to slip sideways?

Solution   We can determine θ from Eq. (5.10):

$$\tan \theta = \frac{v^2}{gr} = \frac{(66.7 \text{ m/s})^2}{(975 \text{ m})(9.8 \text{ m/s}^2)} = 0.466,$$

$$\theta = 25°.$$

The Law of Universal Gravitation

5.3

Now that we have examined uniform circular motion, let us return to the subject that prompted Newton's writing of the *Principia*: gravitation. In the *Principia* Newton showed that the gravitational force acting on a body moving in an elliptical orbit—as a planet does—is inversely proportional to the square of the distance from the body to the center of force; that is, $F \propto 1/r^2$. We have seen in Example 5.2 why this was a reasonable conjecture, at least for a circular orbit.

The gravitational force between objects depends not only on the distance between them, but also on their masses. We have seen that the gravitational force on an object near the earth's surface is directly proportional to that object's mass. Furthermore, from Newton's third law we know that the same object exerts an equal and opposite force on the earth. From such reasoning, Newton proposed that the magnitude of the gravitational force between two objects is proportional to *both* their masses. Thus the force between any two bodies with masses m_1 and m_2 has the form

$$F \propto \frac{m_1 m_2}{r^2},$$

where r is the distance between them. Strictly speaking, the law applies to what we call point masses, which are objects that have no size. However, for the sun, planets, and other bodies with spherical symmetry, the distance r is measured as the distance between their respective centers.

When we insert a constant of proportionality, this statement becomes

$$F = \frac{Gm_1 m_2}{r^2}. \tag{5.11}$$

Equation (5.11) expresses Newton's **law of universal gravitation**. It states that every particle in the universe attracts every other particle with a force that is directly proportional to the product of their masses and inversely proportional to the square of the distance between them. Note that there is an attractive force between *any* pair of objects, whether they be the sun and the earth, the earth and you, or a leaf and a grain of sand.

Newton had some difficulty in showing to his own satisfaction that the law of universal gravitation holds when applied to extended spherical bodies of uniform density*—the planets are approximately this—if r is measured from the center of one sphere to the center of the other (Fig. 5.12). Newton's mathematical work on this problem eventually led to his discovery of calculus, with which he proved that the mass of a symmetrical object of uniform density behaves under the law of universal gravitation

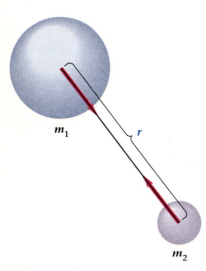

Figure 5.12   For spherical bodies of uniform density, the gravitational force between them is found from Newton's law with the distance r measured from the center of one to the center of the other.

---

*The **density**, ρ, of an object is defined as its mass per unit volume: $\rho = m/V$, where m is the mass of the object whose volume is V.

exactly as if it were concentrated at the point of the object's center of symmetry. Such a point is called the **center of gravity**.

During the next two centuries, astronomers showed that the law of universal gravitation accounts for the motion of the entire solar system with great precision. The one exception was the motion of Mercury, a problem solved only in the twentieth century by the introduction of Einstein's theory of general relativity. Today's astronomers view gravitation as the force shaping the structure of the universe. No other fundamental force acts over such enormous distances and is always an attractive force, never balanced by repulsive forces. We believe gravitational forces are responsible for the formation of stars from clouds of gas and the formation of galaxies from millions of stars. When huge masses like these are involved, gravitational forces can be awesome in magnitude.

The **universal gravitational constant** G must be determined by experiment. Its numerical value depends on the system of units in which we make the measurement. Before describing the determination of G, let us consider two examples that allow us to use the law of gravitation without knowing the value of G. The first gives a result for things here on the surface of the earth; the second makes a prediction about the solar system.

Example 5.10

Calculating g in terms of G.

Consider a mass m falling near the earth's surface. Find its acceleration g in terms of the universal gravitational constant G and draw some conclusions from the form of the answer.

Solution   The gravitational force on the body is

$$F = \frac{GmM_E}{r^2},$$

where M_E is the mass of the earth and r, the distance of the mass from the center of the earth, is essentially the earth's radius (Fig. 5.13).

We have already noted that the gravitational force on a body at the earth's surface is

$$F = mg.$$

Setting the two expressions for the gravitational force on m equal to each other, we get

$$mg = \frac{GmM_E}{r^2},$$

or

$$g = \frac{GM_E}{r^2}.$$

Both G and M_E are constant, and r does not change significantly for small variations in height near the surface of the earth. Thus the right-hand side of this equation does not change appreciably with position on the earth's surface. For this reason we may replace r with the average radius of the

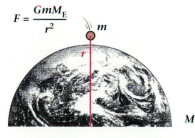

$$F = \frac{GmM_E}{r^2}$$

m

r

M_E

Figure 5.13   Example 5.10: A small object near the earth's surface is attracted by a gravitational force $F = GmM_E/r^2$.

earth R_E to get

$$g = \frac{GM_E}{R_E^2}.$$

As a result, the law of gravitation predicts that the acceleration due to gravity of an object at the earth's surface is approximately constant and does not depend on the mass of the object. Experimentally we know that g does not vary appreciably from one place to another. This constancy of g is just what Galileo found. Thus the law of universal gravitation not only describes the forces that hold the planets in their orbits, but also describes the forces on objects close to the earth.

Example 5.11

Derivation of Kepler's third law.

Show that Kepler's third law follows from the law of universal gravitation. [Recall that Kepler's third law states that for all planets the ratio (period)^2/(distance from sun)^3 is the same.]

Solution   We make the approximation that the orbits of the planets are circles. We have seen in Chapter 1 that this is essentially correct for those planets visible to the unaided eye. It is also approximately true that the orbital speed is constant.

Using the symbols in Fig. 5.14, we can say that the sun's gravitational force on any planet is

$$F = \frac{GmM}{r^2}.$$

Because the mass of the sun is so much larger than the mass of the planet, we can assume, as Kepler did, that the sun lies at the center of the planetary orbit. The circular orbit implies a centripetal force

$$F_c = \frac{4\pi^2 mr}{T^2}.$$

This net force for circular motion is provided by the gravitational force. Equating these two forces, we get

$$\frac{GmM}{r^2} = \frac{4\pi^2 mr}{T^2}.$$

Rearranging gives

$$\frac{T^2}{r^3} = \frac{4\pi^2}{GM}.$$

The right-hand side of this equation depends on the mass of the sun and on the universal gravitational constant, but not on any property of the planet. Therefore the ratio on the left-hand side is identical for all planets, just as Kepler observed. However, now we see that this result is not an independent physical law, but a consequence of a more fundamental law—the law of universal gravitation. We have derived Kepler's third law only for uniform circular motion of the planet, but the result is true for elliptical orbits if we use the average distance from the sun for r.

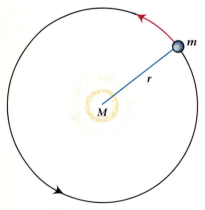

Figure 5.14   Example 5.11: A planet of mass m makes a circular orbit of radius r about the sun (mass M) with a period T. (Figure is not drawn to scale.)

The Universal Gravitational Constant *G*

5.4

After publication of Newton's *Principia*, experimenters tried to make some independent test of the law of universal gravitation, despite its obvious success in explaining the main features of planetary and lunar motion. Perhaps the most satisfying type of experiment would be one in which known masses are placed a known distance apart and the attractive force between them is measured. Under such conditions the validity of the law could be checked and *G* determined in a straightforward manner from Eq. (5.11). However, common experience tells us that the gravitational force between ordinary bodies is extremely small. To measure this slight force, a sensitive device, called a torsion balance, was invented independently by Charles Coulomb in France and John Michell in England in the 1770s. Coulomb used his balance to measure the force between electrical charges. (See Chapter 14.) Michell, a professor of geology at Cambridge, designed his balance to measure gravitational forces and so to "weigh the earth." He did not complete the work before his death, and the apparatus eventually passed into the hands of Henry Cavendish, who refined and made use of it.

In 1798, 71 years after Newton's death, Cavendish first measured the force between small masses on earth. His interest, like that of many of his contemporaries, was in finding the earth's density. Only much later were his experimental results interpreted by others to give a value for *G*. Thus, although Cavendish is often remembered for determining the value of *G*, in fact he never did so.

Since the time of Cavendish, the value of *G* has been determined in a number of ways. The presently accepted value is

$$G = 6.673 \times 10^{-11}\,\text{N} \cdot \text{m}^2/\text{kg}^2.$$

Using this value of *G*, we can apply the law of universal gravitation to many different situations. The examples that follow illustrate a few of these.

Example 5.12

Use the law of universal gravitation and the measured value of the acceleration of gravity *g* to determine the average density of the earth.

Solution We start with the result of Example 5.10 that

$$g = \frac{GM_E}{R_E^2},$$

where M_E is the mass of the earth, R_E is its radius, and *g* is the acceleration of gravity. Let ρ be the average density of the earth, defined as the ratio of the earth's mass to its volume, $\rho = M_E/V$. If we take the earth to be a sphere of radius R_E, then

$$\rho = \frac{M_E}{\frac{4}{3}\pi R_E^3}.$$

The equation for g can then be rewritten in terms of the density as

$$g = \frac{G\ (\frac{4}{3}\pi R_{\rm E}^3 \rho)}{R_{\rm E}^2} = \tfrac{4}{3}G\pi R_{\rm E}\rho.$$

Upon rearranging, we find the density to be

$$\rho = \frac{3g}{4\pi R_{\rm E}G}.$$

BACK TO THE FUTURE

Henry Cavendish and the Density of the Earth

Like most eighteenth-century English scientists, Henry Cavendish (1731–1810) was influenced by the questions in Newton's *Principia* and *Optics*. This influence led him to investigate gravitational forces. Because gravitational forces between ordinary objects are so very small, Cavendish had to use a special balance, based on Michell's design, to measure them.

Cavendish's 1798 paper, *Experiment to Determine the Density of the Earth*, contained a drawing of the torsion balance used in his experiment (Fig. B5.1). In Cavendish's own words, "The apparatus is very simple; it consists of a wooden arm, 6 feet long, made so as to unite great strength with little weight. This arm is suspended in a horizontal position, by a slender wire 40 inches long, and to each extremity is hung a leaden ball about 2 inches in diameter; and the whole is enclosed in

a narrow wooden case, to defend it from the wind.

"As no more force is required to make this turn round on its center, than what is necessary to twist the suspending wire, it is plain, that if the wire is sufficiently slender, the most minute force, such as the attraction of a leaden weight a few inches in diameter, will be sufficient to draw the arm sensibly aside. The weights which Mr. Michell intended to use were 8 inches in diameter. One of these was to be placed on one side of the case, opposite to one

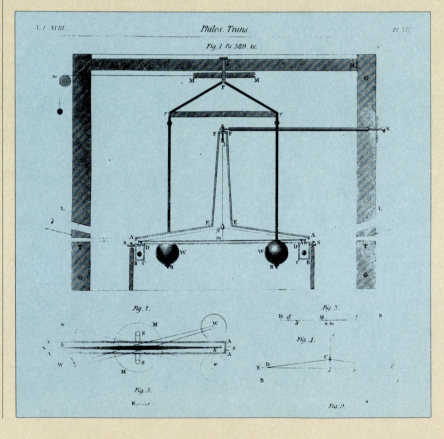

Figure B5.1 The Cavendish balance: A torsional balance constructed of two small masses balanced on a light rod suspended by a thin fiber. Two large masses (lead spheres) are arranged symmetrically on either side of the rod so as to give it a rotational force. The restoring force is supplied by the fiber.

Inserting the numerical values, we get

$$\rho = \frac{3(9.8 \text{ m/s}^2)}{4\pi(6.38 \times 10^6 \text{ m})(6.67 \times 10^{-11} \text{ N} \cdot \text{m}^2 \text{ kg}^{-2})},$$

$$\rho = 5.5 \times 10^3 \text{ kg/m}^3.$$

This is the average density of the entire earth and is 5.5 times the density of water.

---

of the balls, and as near it as could conveniently be done, and the other on the other side, opposite to the other ball, so that the attraction of both the weights would conspire in drawing the arm aside; and when its position, as affected by these weights, was ascertained, the weights were to be removed to the other side of the case, so as to draw the arm the contrary way, and the position of the arm was to be again determined; and, consequently, half the difference of these positions would show how much the arm was drawn aside by the attraction of the weights."

By improving on Michell's apparatus and by using the utmost care, Cavendish performed this delicate experiment in a series of 17 trials whose results clustered around an average value of 5.48 for the density of the earth as compared with water. (See Example 5.12.) Recall that the density of any material is the ratio of its mass to its volume. Cavendish reckoned that his measurement was correct to within about 7%. His result corresponds to a value for *G* of $(6.70 \pm 0.48) \times 10^{-11}$ N · m²/kg², which differs little from the modern value of

$G = 6.673 \times 10^{-11}$ N · m²/kg².

Cavendish's interest in the density of the earth has its counterpart today. Local variations in the density of the earth's crust can yield information about mineral and oil deposits. Consequently, geologists have developed instruments to measure the acceleration of gravity with great precision. One type of gravity meter uses a very sensitive spring scale in which a mass is pulled down more where the force of gravity (and therefore *g*) is larger. Another type uses the connection between the period of a pendulum and the acceleration of gravity, a topic discussed in Chapter 13.

These small variations from place to place can be used to map the underlying geological structure. Modern gravity meters can easily measure variations in g as small as 10^{-6} m/s². Figure B5.2 is a map showing the variations in the acceleration of gravity in South Carolina. The contours for constant values of g are spaced at intervals of 5×10^{-5} m/s². A trained geologist can use such a map to help infer information about the mineral content and seismic structure of the underlying terrain.

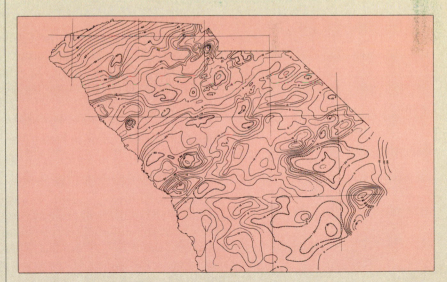

Figure B5.2 Details from a gravity map of South Carolina. The contour lines are spaced 5 mgal apart. The *gal*, a unit named after Galileo, is 1 cm/s². Thus 1 mgal is approximately 10^{-6} g.

Example 5.13

Finding the period of the moon's orbit.

Calculate the period of the moon's orbit about the earth, assuming a constant distance $r = 3.84 \times 10^8$ m.

Solution   The gravitational force between the earth and moon is

$$F = G \frac{M_E m}{r^2},$$

where M_E is the mass of the earth (5.98×10^{24} kg), m is the mass of the moon, and r is the earth–moon distance. For any case of uniform circular motion, the magnitude of the attractive force, whatever its nature, must equal the centripetal force given by

$$F_c = \frac{4\pi^2 m r}{T^2},$$

where we have used Eq. (5.8c) for the centripetal acceleration. The centripetal force is provided by the gravitational force, so that

$$\frac{GM_E m}{r^2} = \frac{4\pi^2 m r}{T^2}.$$

Solving for T gives

$$T = \sqrt{\frac{4\pi^2 r^3}{GM_E}} = \sqrt{\frac{4\pi^2 (3.84 \times 10^8 \text{ m})^3}{(6.67 \times 10^{-11} \text{ N} \cdot \text{m}^2/\text{kg}^2)(5.98 \times 10^{24} \text{ kg})}},$$

$$T = 2.37 \times 10^6 \text{ s},$$

or

$$T = 27.4 \text{ days}.$$

The result obtained here is slightly greater than the observed period of 27.3 days. The discrepancy occurs because we assumed the moon to orbit around a stationary earth. In actuality they both move about a common point that is near, but not at, the center of the earth.

Example 5.14

Period of a low earth-orbiting satellite.

Estimate the period of an artificial earth-orbiting satellite that passes just above the earth's surface (Fig. 5.15).

Solution   In reality, a satellite's orbit must be high enough so that the satellite is above most of the earth's atmosphere. However, the height at which low-orbiting satellites operate is small compared with the radius of the earth (about 6.4×10^6 m). Therefore, for estimating purposes, we approximate the radius of the orbit by saying that it's the same as the radius of the earth.

We set the force required to give a circular orbit—the centripetal force—equal to the gravitational force, as in Example 5.13. We let the mass of the satellite be m, the mass of the earth M_E, the radius of the orbit

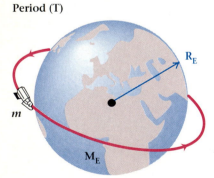

Period (T)

Figure 5.15   Example 5.14: Finding the period of an earth-orbiting satellite.

R_E, and the satellite's period T. Then

$$\frac{m4\pi^2R_E}{T^2} = \frac{GmM_E}{R_E^2},$$

or

$$T = \sqrt{\frac{4\pi^2R_E^3}{GM_E}}.$$

Rather than substituting the numerical values now, we use the result of Example 5.10 that $gR_E^2 = GM_E$, so that the above expression for the period becomes

$$T = \sqrt{\frac{4\pi^2R_E^3}{gR_E^2}} = \sqrt{\frac{4\pi^2R_E}{g}}.$$

Notice that the period depends only on the radius of the earth and the acceleration of gravity. To complete our estimate we insert the approximate values of $\pi^2 \approx 10$, $g \approx 10$ m/s^2, and $R_E \approx 6.4 \times 10^6$ m:

$$T \approx \sqrt{\frac{4(10)(6.4 \times 10^6 \text{ m})}{10 \text{ m/s}^2}} \approx 5100 \text{ s} \approx 85 \text{ min}.$$

This value compares favorably with the observed periods of low-orbiting satellites.

Gravitational Field Strength

5.5

We have already used the idea of gravitational force. In doing so we considered a property of matter (its mass) and the effects of that property (gravitational force) throughout space. Now we examine gravitation from a different but related point of view: gravitational field strength, or more simply, the gravitational field. We will use the general idea of a field later when studying electricity and magnetism.

We define the **gravitational field strength** *at any point in space to be the gravitational force per unit mass on a test mass m_0.* Thus, at a point in space where a test mass m_0 experiences a gravitational force **F**, the gravitational field strength is

$$\boldsymbol{\Gamma} \equiv \frac{\mathbf{F}}{m_0}. \tag{5.12}$$

Note that the gravitational field strength is just the acceleration that a unit mass would experience at that point in space. Later, when we examine the electric field, we will find that its field strength is not an acceleration.

THE SHAPE OF THINGS TO COME

FIELD LINES

Field lines show us graphically the direction and magnitude of the field at any point of space. The direction of the field is in the direction of the line, and the magnitude of the field is proportional to the density of lines.

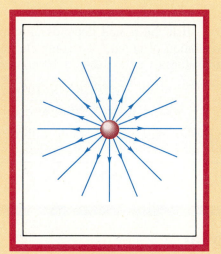

Inverse-Square Field
(Point Source)

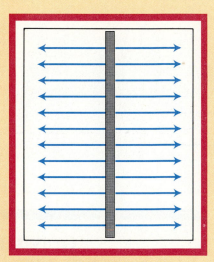

Uniform Field
(Infinite Plane Source)

The field due to a point mass or charge is an inverse-square field, in which field lines converge to or diverge from the source in all directions. The lines are closer together (stronger field) at small distances from the source in accord with a $1/r^2$ field strength. In a uniform (constant) field, due to an infinite plane of mass or charge, field lines are parallel and equally spaced apart.

In spite of this, the concepts of field and field strength apply in both cases, allowing us to draw useful analogies between gravitation and electricity when they are considered from the standpoint of fields.

Since the gravitational force is a vector, the field must also be a vector, having both magnitude and direction. If the gravitational force arises from the attraction of the test mass by a mass M located a distance r from the test mass, then the magnitude of the field strength is

$$\Gamma = \frac{F}{m_0} = \frac{GMm_0}{r^2 m_0} = \frac{GM}{r^2}.$$

ELECTROSTATIC FIELD

A large static charge causes each hair on this person's head to repel each other hair.

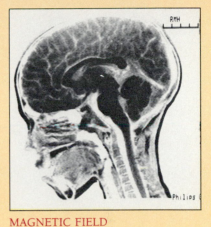

MAGNETIC FIELD

Magnetic image resulting from behavior of magnetic dipoles in a magnetic field.

Two unlike charges attract. A tiny positive charge between the source charges experiences a large force (Chapter 15).

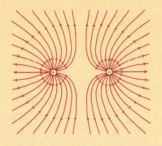

Two like charges repel. A tiny positive charge midway between the source charges experiences no force (Chapter 15).

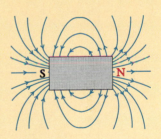

A tiny test magnet (like a compass needle) will align itself with the field lines from north pole to south pole (Chapter 18).

The field vector lies along the line from M to m and is directed toward the mass M. For example, the gravitational field at the earth's surface is a vector directed toward the center of the earth with magnitude 9.8 m/s².

Once the gravitational field at any point in space has been determined (either by measurement or by calculation), then we can compute the force on any other mass m placed at that point g by using $\mathbf{F} = \boldsymbol{\Gamma}m$. We may, in fact, say that the gravitational field is a property of space. Instead of focusing on the mass of an object and how the gravitational force depends on distance, we focus on the space itself and how a property of the space

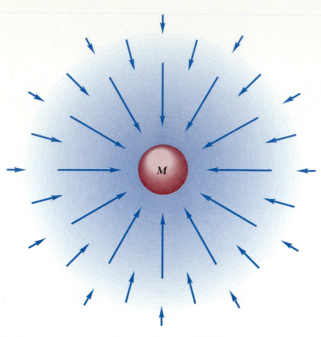

Figure 5.16   A representation of the gravitational field around a point mass *M*. Each arrow indicates the magnitude and direction of the field at the base of the arrow.

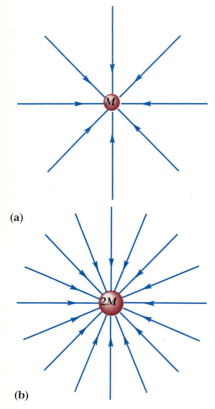

Figure 5.17   Lines of force (a) around a mass *M* and (b) around a mass 2*M*.

(the field) is affected by the presence of objects near and far. Masses can then be treated as sources of the gravitational field, and the force on some particular mass is determined by the field present at the location of that mass. We can represent this field visually with the aid of arrows representing the direction and magnitude of the field at different points in space. This representation is shown in Fig. 5.16 to illustrate the field around a point mass *M*. Each arrow represents the field at the base of the arrow. The lengths of the arrows are proportional to the magnitude of the field at each point. (We have shown a two-dimensional drawing, but the gravitational field itself extends outward from the source in all directions in three-dimensional space.)

Another way to help visualize the gravitational field is to diagram **lines of force**. These continuous lines are drawn in the direction of the force on a test mass (Fig. 5.17). The relative number of lines is proportional to the strength of the force and hence proportional to the field. Such a representation helps show that the field strength diminishes as the distance from the mass increases. The farther away from the source of the gravitational field, the farther apart the lines are, and the weaker the field becomes.

We will use the field concept again in the next chapter when we discuss the gravitational field. We will also draw lines of force later, in the chapters on electricity and magnetism, as an aid in understanding the forces encountered there. The field concept was introduced by Michael Faraday (Chapter 19) in connection with his experiments on electricity and has become the contemporary way of describing the effects and interactions of the fundamental forces of nature.

SUMMARY

Useful Concepts

- Uniform circular motion can be described by an angular velocity and a centripetal (center-seeking) acceleration. The centripetal acceleration always points to the center of the circular path of motion and is perpendicular to the linear velocity:

$$a_c = \frac{v^2}{r}.$$

- The frequency of revolution is the reciprocal of the period, which is the time required for one revolution:

$$f = \frac{1}{T}.$$

- The angular velocity (also called angular frequency) is

$$\omega = \frac{v}{r}.$$

- It is proportional to the rotational frequency,

$$\omega = 2\pi f.$$

- The centripetal acceleration used in Newton's second law gives the centripetal force,

$$F_c = \frac{mv^2}{r}.$$

- The centripetal force is not a fundamental force, but describes the effect of any force applied to a body, which keeps it moving in uniform circular motion.
- Newton's law of universal gravitation gives the force between *any* two point masses, no matter what their mass and no matter where they are located,

$$F = \frac{Gm_1m_2}{r^2}.$$

- The value of the universal gravitational constant G is

$$G = 6.673 \times 10^{-11}\ \text{N} \cdot \text{m}^2/\text{kg}^2.$$

- The gravitational field strength is the gravitational force per unit mass on a test mass m_0 at a point in space,

$$\mathbf{\Gamma} \equiv \frac{\mathbf{F}}{m}.$$

- We can visualize the gravitational field by drawing continuous lines of force showing the direction of the gravitational force at any point.

Important Terms

You should be able to write the definition or meaning of each of the following terms:

- uniform circular motion
- centripetal acceleration
- period
- frequency
- hertz
- angular velocity (angular frequency)
- centripetal force
- law of universal gravitation
- density
- center of gravity
- universal gravitational constant
- gravitational field strength
- lines of force

QUESTIONS

5.1 Describe the forces acting on a passenger riding on a Ferris wheel. Explain any difference between the behavior near the top of the ride and that near the bottom.

5.2 When a fenderless bicycle is ridden along a wet street, the rider gets a wet stripe down his back from water droplets thrown off the rear wheel. Explain this observation.

5.3 Explain the action of the spin cycle in removing water from clothes in an automatic washing machine.

5.4 What complications would you encounter in playing catch with a baseball if you and your partner were standing on a rotating carousel?

5.5 How can an object in uniform circular motion be moving at a constant speed and at the same time be constantly accelerated?

5.6 Make an order-of-magnitude estimate of the linear speed of a point on the earth's equator due to the earth's rotational motion. Compare this speed with the linear speed of the earth's orbital motion.

5.7 What are the restrictions on the orbit of a communications satellite if it is to appear motionless in the sky as viewed from a location on the earth?

5.8 If the average distance between the earth and the moon were suddenly cut in half, how would the lunar period be affected?

5.9 What is universal about the law of universal gravitation?

5.10 Assume that you are inside a train moving at constant speed. All of the window shades are drawn so that you cannot see out. You have with you a sensitive spring balance with a mass hanging from it and a finely ruled protractor. How can you use these instruments to tell whether the train is traveling along a straight track or going around a curve?

PROBLEMS

Hints for Solving Problems

In uniform circular motion, the speed is constant and the centripetal acceleration, which is perpendicular to the velocity, points to the center of the circle. The formulas for centripetal force describe the force necessary to make an object move in a circle. In this chapter the applied force is often gravity. Remember that objects in circular motion obey Newton's three laws. Also, remember to draw and label a free-body diagram before attempting a problem.

Section 5.1   Uniform Circular Motion

5.1 The period of a stone swung in a horizontal circle on a 2-m cord is 1 s. (a) What is its angular velocity in rad/s? (b) What is its linear speed in m/s? (c) What is its radial acceleration in m/s²?

5.2 The earth is 1.5×10^{11} m from the sun and has a period of about 365 days. Assume the earth's orbit to be circular and determine the magnitude and direction of its radial acceleration in km/s².

5.3 What is the centripetal acceleration of an automobile driving at 40 km/h on a circular track of radius 20 m?

5.4 A marker on a movie reel turning with constant speed moves through a circle of 8.50 cm radius in 1.00 s. What is the centripetal acceleration of the marker?

5.5 (a) Compute the angular velocity of a phonograph turntable rotating at $33\frac{1}{3}$ rev/min. (b) Calculate the angular velocity at 45 rev/min.

5.6 Compute the centripetal acceleration of a spider sitting on the rim of a 12-in. diameter phonograph record rotating at $33\frac{1}{3}$ rev/min.

5.7 A phonograph record rotates at the rate of 45 rev/min. (a) Express its frequency in hertz. (b) What is its period?

5.8 An automobile tire is 68 cm in diameter. (a) At what frequency f does the tire rotate when the automobile is traveling at a speed of 50 km/h? (b) What is the angular frequency ω?

5.9 A bicycle tire is 66 cm in diameter. (a) At what frequency f does the tire rotate when the bicycle is traveling at a speed of 30 km/h? (b) What is the angular frequency ω?

5.10 Jupiter's moon Europa has an average orbital radius of 6.67×10^8 m and a period of 85.2 h. Calculate the magnitude of (a) the tangential velocity, (b) the angular velocity, and (c) the centripetal acceleration of Europa.

5.11 Jupiter's moon Io has an average orbital radius of 4.22×10^8 m and a period of 42.5 h. Calculate the magnitude of (a) the tangential velocity, (b) the angular velocity, and (c) the centripetal acceleration of Io.

Section 5.2   Force Needed for Circular Motion

5.12 A stunt pilot in an airplane diving vertically downward at a speed of 220 km/h turns vertically upward by following an approximately semicircular path with a radius of 180 m (Fig. 5.18). (a) How many g's does the pilot experience due to his motion alone? (b) By what factor does the pilot's weight appear to increase at the bottom of the dive?

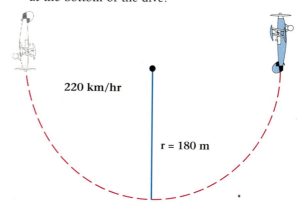

220 km/hr

r = 180 m

Figure 5.18   Problem 5.12.

5.13 (a) The basket in an automatic clothes dryer rotates about a horizontal axis. The basket is rotated so that the force exerted by the basket on clothes located at the basket's edge is zero at the top of the path. If the radius of the basket is 0.65 m, how fast must the basket turn to accomplish this? (b) Draw vectors whose lengths are proportional to the force exerted by the basket at the lowest point and halfway up.

5.14 A 0.20-kg toy whistle can be whirled in a horizontal circle of 1.00 m radius at a maximum of 3.00 rev/s before the string breaks. What is the breaking strength of the string?

5.15 An aluminum can is half filled with water and whirled about in a horizontal circle on the end of a string 1.00 m long. If the total mass of the can and its contents is 0.115 kg and its tangential speed is 10.0 m/s, what is the tension in the string?

5.16 Calculate the centripetal force on a 2000-kg automobile rounding a curve of 175 m radius at a speed of 50 km/h.

5.17 A space station 400 m in diameter rotates fast enough so that the artificial gravity at the outer edge is 1.5g. (a) What is the frequency of rotation? (b) What is its period? (c) At what distance from the center will the artificial gravity be 0.75g?

5.18 Calculate the angle of banking required for a curve of 200 m radius so that a car rounding the curve at 80 km/h would have no tendency to sway outward or inward. Assume the surface is frictionless.

5.19 A race track curve has a radius of 100 m and is banked at an angle of 68°. For what speed was the curve designed?

Section 5.3 The Law of Universal Gravitation

5.20 An apple ($m = 0.20$ kg) falls to the earth. Determine (a) the apple's acceleration toward the earth; (b) the earth's acceleration toward the apple. (c) Discuss the appropriate reference frames in which to determine the accelerations in (a) and (b).

5.21 At what fraction of the distance from the earth to the moon will their opposing gravitational forces on a spaceship traveling between them be equal in magnitude? (*Hint:* The mass of the moon is 0.0123 times the mass of the earth.)

5.22 What is the ratio of the acceleration of gravity on the surface of the moon to the acceleration of gravity on the surface of the earth? (*Hint:* The radius of the moon is 0.273 times the radius of the earth, and the mass of the moon is 0.0123 times the mass of the earth.)

Section 5.4 The Universal Gravitational Constant G

5.23 Gold has a density of 1.93×10^4 kg/m³. (a) Calculate the volume of 0.500 kg of gold. (b) If this amount were shaped into a cube, what would be the length of one edge of that cube?

5.24 Magnesium has a density of 1.74×10^3 kg/m³. (a) Calculate the volume of 0.500 kg of magnesium. (b) If this amount were shaped into a cube, what would be the length of one edge of that cube?

5.25 Calculate the mass of a lead brick of dimensions 5.0 cm × 10 cm × 30 cm, given that the density of lead is 1.13×10^4 kg/m³.

5.26 A block of copper 5.0 cm × 5.0 cm × 20 cm has a mass of 4.45 kg. Use these data to compute the density of copper.

5.27 Suppose the earth and moon were held at rest at their present separation and then released to move under their mutual gravitational attraction. What would be the initial acceleration of each body? Be specific about your choice of coordinate frames.

5.28 In Henry Cavendish's famous experiment, we can determine that the force between the ball and weight in each pair was about 1.53×10^{-7} N when the torsional balance passed through the equilibrium position. From the data listed in Fig. 5.19, calculate the universal constant of gravitation.

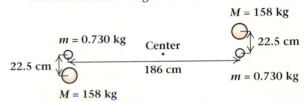

Figure 5.19 Problem 5.28: Diagram of a torsion pendulum in its equilibrium position as viewed from above.

5.29 Compute the gravitational force between the sun ($M = 1.99 \times 10^{30}$ kg) and the planet Uranus ($m = 14.5\,M_E$, $R = 19.2$ A.U.).

5.30 (a) What is the radius of the orbit of a communications relay satellite that always remains above one point on the earth's surface? Such an orbit is called a geosynchronous orbit. (b) Can such a satellite be placed in geosynchronous orbit over *any* point on the earth's surface? Why?

5.31 What is the period of an artificial satellite whose orbital radius is one half that of a geosynchronous satellite?

5.32 What is the period of an earth satellite whose orbital radius is one third that of the distance from the earth to the moon?

5.33 Calculate the orbital period of Venus from knowledge of G, the mass of the sun ($M = 1.99 \times 10^{30}$ kg), and the orbital radius of 1.08×10^{11} m.

5.34 (a) Calculate the mass of the sun from the radius of the earth's orbit (1.5×10^{11} m), the earth's period in its orbit, and the gravitational constant G. (b) What is the density of the sun and how does it compare with the density of the earth? (*Hint:* The sun's radius is 6.96×10^8 m.)

5.35 Calculate the mass of Jupiter from the knowledge that its satellite Io orbits at an average distance of 4.22×10^5 km from its center with an orbital period of 42.5 h.

5.36 Jupiter's mass is about 318 times that of the earth and its radius is 11 times the earth's radius. (a) Calculate the acceleration of gravity at the surface of

Jupiter. Express your answer in g's. (b) Also calculate the average density of Jupiter.

5.37 Saturn's mass is 95 times that of earth and its radius is 9 times that of earth. (a) Calculate the acceleration of gravity at the surface of Saturn. Express your answer in g's. (b) Also calculate the average density of Saturn.

5.38 An astronaut weighing 700 N on Earth travels to the planet Mars. What does the astronaut weigh on Mars? (*Hint:* The mass of Mars is 0.107 of the earth's mass. The radius of Mars is 0.53 of the earth's radius.)

5.39 A 25-kg fuel tank is taken to the moon. What is the weight of the fuel tank on the lunar surface? (*Hint:* See Problem 5.22.)

Section 5.5   Gravitational Field Strength

5.40 What is the gravitational field strength on the surface of the earth due to the earth alone? In your answer give magnitude, direction, and proper units to agree with Eq. (5.12).

5.41 Two 1.0-kg masses are located with one at each of two corners of an equilateral triangle with sides 1.0 m long. Determine the magnitude and direction of the gravitational field at the third corner due to these masses. Neglect the effects of all other masses, including the earth.

5.42 Compare the magnitude of the gravitational field at the surface of the earth due to the moon with that due to the sun. (*Hint:* The mass of the sun is 1.99×10^{30} kg, the mass of the moon is 7.36×10^{22} kg, and the distance from the surface of the earth to the moon is $3.78 \ 10^8$ m.)

Additional Problems

5.43 You drive an automobile of mass 2.10×10^3 kg from
• sea level to the top of a mountain 2.05 km high. What is its change in weight?

5.44 A popular amusement park ride consists of a broad
• short cylinder arranged so that it rotates around its vertical axis (Fig. 5.20). People stand inside the cylinder with their backs to the outer wall and "feel pushed back" when the cylinder rotates. When the cylinder is rotating fast enough, it is tipped so that its axis of rotation is almost horizontal. If the radius of the cylinder is 4.5 m, how fast must it rotate so that the riders do not fall away from the walls at the topmost position? Give your answer in hertz.

5.45 A belt passes over a pulley having a 14.0-cm radius. (a) If the linear speed of the belt is 300 cm/s, what is the angular velocity of the pulley? (b) What is the rotational frequency of the pulley?

5.46 A pulley of radius 15 cm is connected with a belt to another pulley having a radius of 5.0 cm. (a) If the large pulley is turning at a rate of 10 Hz, what is the

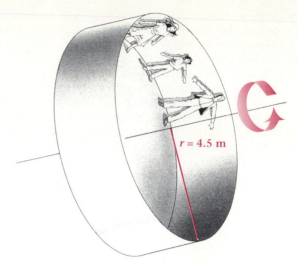

Figure 5.20   Problem 5.44.

rate of the smaller pulley? (b) What is the linear speed of the belt?

5.47 You swing a bucket of water in a vertical circle at arm's length (0.70 m). What is the minimum number of revolutions per second you must maintain to keep the water from spilling out of the bucket?

5.48 A bob of mass m is whirled in a circular path on the
• end of a string 1.0 m long. If the string makes an angle of 20° with the vertical (Fig. 5.21), what is the tangential speed of the ball?

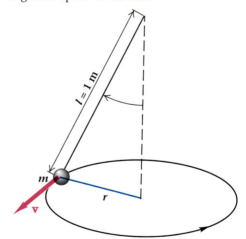

Figure 5.21   Problem 5.48.

5.49 Before Cavendish, Robert Hooke attempted to
•• measure G by weighing the same object at heights differing by 300 ft (91.4 m). (a) What fractional change in weight would be predicted by the law of universal gravitation? (b) What difference in height would he have had to use to detect a change in weight of one grain in one pound troy? (One pound troy equals 5760 grains.)

5.50 If an astronaut dropped a small rock near the surface of Mars, how far would the rock fall in 1.00 s? (*Hint:* The mass of Mars is 0.107 the mass of earth, and the radius of Mars is 0.530 the radius of the earth.)

5.51 The average density of the planet Mercury is 5.61×10^3 kg/m^3. The acceleration of gravity on its surface is 3.92 m/s^2. Calculate the average radius of Mercury.

5.52 (a) If all the physical dimensions in the solar system were scaled up by a factor of 2 and the masses of the sun, planets, and moons scaled up by a factor of 2, what would the moon's period be in terms of present years? (The moon's present period is 0.075 of the present year.) (b) What would be its period in terms of "new years"?

5.53 Assuming the earth to be an oblate spheroid, so that the distance from the center to the equator is 27 mi (43.5 km) greater than the distance from the center to the poles, calculate the approximate percent change in a person's apparent weight at the poles and at the equator. Take into account the fact that the earth is turning.

5.54 Plot a graph of the gravitational force on a rocket ship of mass m moving along a line between the earth and the moon as a function of the distance from the earth. Include the gravitational effects of both the moon and the earth. The mass of the moon is 0.0123 times the mass of the earth.

5.55 Passengers riding in the Great Six Flags Air Racer are spun around a tall steel tower (Fig. 5.22). At top speed the planes fly at a 56° bank approximately 46 m from the tower. In this position the support chains make an angle of 56° with the vertical. Calculate the speed of the planes.

Figure 5.22 Problem 5.55.

5.56 Look up the orbital period of a space shuttle flight and estimate its average height above the surface of the earth.

5.57 (a) Compute the mass of the earth from knowledge of the earth–moon distance (3.84×10^8 m) and of the lunar period (27.3 days). (b) Then calculate the average density of the earth. The average radius of the earth is 6.38×10^6 m.

5.58 The gravitational attraction due to a spherical mass is the same as that of a point mass of the same magnitude located at the center of the sphere. Use this fact to calculate the size of a lead sphere that would attract you with a force 10^{-3} times your own weight if you stood right next to it. (*Hint:* The density of lead is 1.13×10^4 kg/m^3.)

5.59 A highway curve with a radius of 750 m is banked properly for a car traveling 120 km/h. If a 1590-kg Porshe 928S rounds the curve at 230 km/h, how much sideways force must the tires exert against the road if the car does not skid?

5.60 A long string tied to an overhead support has a weight hanger attached to its lower end. The string can support a maximum weight Mg without breaking. A piece of the same string is doubled and a mass M is attached to one end of the doubled string. The mass is then swung in a vertical circle of 0.75 m radius. What is the maximum frequency of rotation that can be maintained without breaking the string?

5.61 The Wheelie, described in Example 5.4 and shown in Fig. 5.5, can be tilted until its plane of rotation makes an angle of 89° with the horizontal. Show that when it is in this position, the force exerted by the seats on the riders at the top is equivalent to only 0.94 g, while at the bottom it is 2.94 g.

5.62 An automatic tumble dryer has a 0.65-m-diameter basket that rotates about a horizontal axis. As the basket turns, the clothes fall away from the basket's edge and tumble over. If the clothes fall away from the basket at a point 60° from the vertical (Fig. 5.23), what is the rate of rotation in units of revolutions per minute?

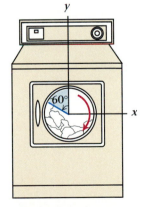

Figure 5.23 Problem 5.62.

5.63 An electron with mass 9.11×10^{-31} kg moves with a speed of 2.00×10^6 m/s in a circle of 2.85 cm radius under the influence of a magnetic field. A proton of mass 1.67×10^{-27} kg, moving in the same plane with the same speed, experiences the same centripetal force. What is the radius of the proton's orbit?

6

Work and Energy

6.1 Work

6.2 Energy

6.3 Kinetic Energy

6.4 Gravitational Potential Energy Near the Earth

6.5 General Form of Gravitational Potential Energy

6.6 Moment of Inertia and Rotational Kinetic Energy

6.7 Conservation of Mechanical Energy

6.8 Power

A WORD TO THE STUDENT

This chapter introduces several very important concepts. Chief among these are work, energy, and power. Especially important is the section on conservation of mechanical energy, since in later chapters we will show how the ideas presented here apply to other forms of energy. The law of energy conservation is one of the most powerful tools available for analyzing physical situations, enabling us to solve many problems that would otherwise be too difficult to attempt. Even those problems that can be solved by other methods are often made simpler and clearer when approached from the standpoint of energy conservation.

In our presentation of the law of conservation of energy we include the important contribution due to the energy of rotational motion. We will build upon this material, along with our definitions of angular kinematics from Section 5.1, when we discuss rotational dynamics in Chapter 7.

Perhaps the most generally useful idea in all of science is the concept of energy and its conservation. Energy is a vital part of our daily lives. The food we eat gives our bodies energy for movement; electrical energy lights our homes and streets; oil and gas propel our cars and keep us warm. These are all examples of using energy. In this chapter we define work and mechanical energy, and arrive at quantitative relationships between them. We will extend and apply these principles in the following chapters.

The terms *force* and *energy* were not always clearly defined. Before the mid-nineteenth century, they were often used interchangeably. However, progress in mechanics and in thermal physics helped clarify these ideas and the distinction between them. In 1807 the English scientist Thomas Young (1773–1829) introduced the word *energy* to denote the quantity of work that a system can do. Later, the Scottish engineer and thermal physicist W. J. M. Rankine (1820–1872) popularized this definition and coined the terms *potential energy* and *conservation of energy*. As is often the case in science, this clarification of terms and definitions led to greater insight and understanding of natural laws and their consequences.

Today the principle of conservation of energy is part of the framework of physical theory. Our faith in this principle is based on years of experience. We will encounter this idea in one form or another throughout the rest of the book and use it to derive other results. Indeed, we will see that many laws in various areas of physics are simply alternative versions of the law of conservation of energy, stated in different terms.

An important part of mechanical energy is the energy due to motion, including rotational as well as linear motion. We discuss rotational energy in this chapter, using some of the kinematic terms introduced with uniform circular motion in the previous chapter. We will examine the dynamics of rotation in the next chapter, when we study additional conservation laws in mechanics.

Work

6.1

The word *work* means many different things to us in our daily lives. We do work when we rake the yard, or buy groceries, or drive a truck. Some of us would even say we do work when we study physics. But in the physical sciences, the meaning of *work* is more precise and restricted than in everyday usage. If we exert a constant force **F** on an object (Fig. 6.1), causing it to move a distance x parallel to **F**, then the **work** W done by the force is defined to be the product of the magnitude of the force times the distance through which it acts.

There are two important conditions in our definition of work. First, the force must be exerted through a distance. In other words, *the force must move the object.* Consider the ancient Greek myth about Atlas, who held up the sky. Atlas is often depicted as a stooped but powerful figure, bearing the earth on his shoulders. Atlas soon tired of his terrible burden.

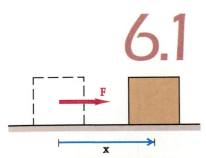

Figure 6.1 A force **F** acting in a direction parallel to the displacement **x** of an object does an amount of work $W = Fx$.

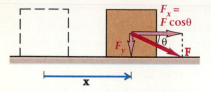

Figure 6.2 A force **F** making an angle θ with the horizontal pushes a box through the displacement **x** and does an amount of work $W = Fx \cos \theta$.

But, according to our definition, as long as he held the earth stationary, he actually did no work on it. In a similar fashion, you could push with all your might against a stationary wall until your muscles ached with the effort. Nevertheless, if the wall did not move, you would not have done any work on the wall.

Second, for work to be done, the force must have a component parallel to the direction of motion. If an applied force is not along the direction of motion, we can resolve it into components parallel to and perpendicular to the displacement (Fig. 6.2). *Only the component of force that is parallel to the displacement contributes to the work.* Thus, if the force **F** makes an angle θ with the line of motion, chosen as the x direction in the figure, then the component of force that contributes to the work is $F_x = F \cos \theta$. Mathematically, the work done is defined to be

$$W \equiv F_x x = Fx \cos \theta. \tag{6.1}$$

When **F** is along the direction of **x** (as shown in Fig. 6.1), then $\theta = 0$ and $\cos \theta = 1$. For that special case the work becomes

$$W = Fx.$$

Notice that, although the work done depends on the force, work itself is a scalar, not a vector, quantity. It has magnitude but no direction. We can add amounts of work directly, just like any other scalar quantities. When the component of the force is in the same direction as the displacement, the work done on the object is positive. When the component of the force is opposite to the displacement, the work done on the object is negative. If the force is perpendicular to the displacement, the work is zero (Fig. 6.3).

The SI unit for work is the newton-meter or $k \cdot gm^2/s^2$. This combination unit has also been given the name **joule** (J), in honor of James Prescott Joule (1818–1889), one of the great contributors to our understanding of energy (see Chapter 10):

$$1 \text{ joule (J)} = 1 \text{ N} \cdot \text{m} = 1 \text{ kg} \cdot \text{m}^2/\text{s}^2.$$

Figure 6.3 (a) A person does positive work by lifting the weight with an applied force **F** in the same direction as the displacement **x**. (b) The frictional force F_{fr} does negative work by acting in a direction opposite the displacement. (c) When the applied force is perpendicular to the displacement, zero work is done.

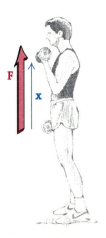

(a)

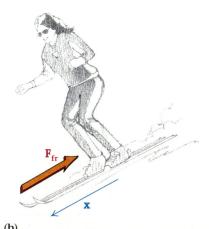

(b)

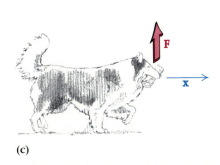

(c)

In the British system, the unit of work is the foot-pound:

$$1 \text{ ft-lb} = 1.356 \text{ J}.$$

You can get some feeling for the size of the joule by looking ahead at Table 6.2.

Example 6.1

Pulling an object parallel to the floor.

A dock worker pulls a crate 3.2 m along the floor with a force of constant magnitude of 250 N. The force is applied parallel to the floor as shown in Fig. 6.4. How much work is accomplished?

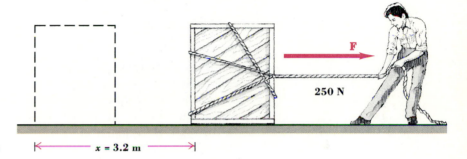

Figure 6.4 Example 6.1: A dock worker pulls a crate with a force of 250 N for a distance of 3.2 m in a direction parallel to the force.

Solution The angle between the force and the direction of motion is zero and so $\cos \theta = \cos 0 = 1$. Therefore

$$W = Fx \cos 0,$$
$$W = (250 \text{ N})(3.2 \text{ m})(1) = 800 \text{ N·m} = 800 \text{ J}.$$

Example 6.2

Pulling an object at an angle.

A child pulls a toy 2.0 m across the floor by a string, applying a force of constant magnitude 0.80 N (Fig. 6.5). During the first meter the string is parallel to the floor. During the second meter the string makes an angle of 30° with the horizontal direction. What is the total work done by the child on the toy?

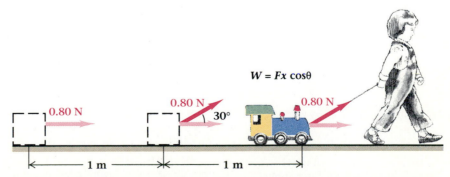

Figure 6.5 Example 6.2: A child pulls a toy, first with the string parallel to the motion and then at an angle of 30°.

Solution   We must calculate the work separately for the first and second parts of the motion and add them together. For the first part, the work W_1 is

$$W_1 = F_1 x_1 \cos \theta_1, \qquad \text{where } \theta_1 = 0;$$

$$W_1 = (0.80 \text{ N})(1.0 \text{ m})(1) = 0.80 \text{ J}.$$

For the second part of the motion, the work W_2 is

$$W_2 = F_2 x_2 \cos \theta_2$$

$$= (0.80 \text{ N})(1.0 \text{ m})(\cos 30°) = (0.80 \text{ N})(1.0 \text{ m})(0.87),$$

$$W_2 = 0.69 \text{ J}.$$

The total work W is then

$$W = W_1 + W_2 = 0.80 \text{ J} + 0.69 \text{ J} = 1.5 \text{ J}.$$

If the force exerted on a moving object is constant, then we can calculate the work by the simple application of Eq. (6.1), as Example 6.1 illustrates. In Example 6.2, the force changed after the first meter of displacement, but then remained constant for the remainder of the motion. In this case we applied Eq. (6.1) to each part of the motion separately. However, often the force exerted on an object changes continuously, and in these situations we cannot apply Eq. (6.1) to calculate the work done on the object. Instead, we must calculate the small amount of work ΔW for small displacements Δx, over which the force is constant or approximately constant:

$$\Delta W = F \, \Delta x \cos \theta.$$

Then we can add together these amounts of work ΔW to give the total work for the entire process.

An important example of this idea is a spring that obeys what is called Hooke's law. We know that the more force we apply, the more a spring stretches. For a spring that obeys Hooke's law the extension of the spring is proportional to the applied force. In Fig. 6.6, the extension or displacement of the spring from its equilibrium position is denoted by x. The applied force F_{ext} required to extend it that far is given by

$$F_{ext} = kx, \tag{6.2}$$

where k is a constant whose value is determined for each particular spring.* Note that the units of k must be N/m.

Let us calculate how much work is done in stretching a spring a total distance x. We use the graphical technique introduced in Section 2.4, where we found that the displacement is given by the area under the velocity–time curve. Here we find that the work is given by the area under the force–displacement curve. Figure 6.7 is a graph of the applied force

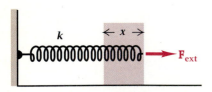

k

Spring in equilibrium

k

$\leftarrow x \rightarrow$

F_{ext}

Figure 6.6   A spring that obeys Hooke's law. F_{ext} is the force applied to the spring to give an extension x.

---

*Further discussion of Hooke's law is found in Chapter 13. Usually, Hooke's law is written as $F = -kx$, where F is the force exerted *by* the spring and is in the opposite direction from the spring extension x.

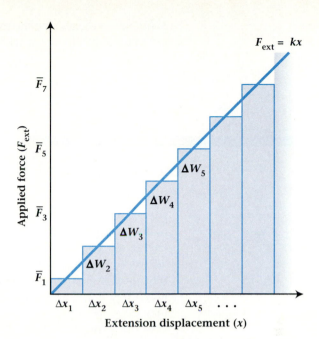

Figure 6.7 The force F_{ext} plotted against the extension of the spring of Figure 6.6. The work done in extending the spring equals the area under the force–displacement curve.

plotted against the spring extension. (The extension of the spring is a displacement.) We mark off small displacements Δx along the abscissa of the graph and draw vertical lines from them up to the force–displacement curve. Then much as we did in Chapter 2, we form a series of rectangles whose width is Δx and whose height is halfway between the applied force at the beginning and at the end of Δx. This height corresponds to the average applied force $\overline{F}$ over each displacement Δx.

We wish to calculate the small amounts of work ΔW corresponding to the small displacements Δx and then add them together:

$$W = \Delta W_1 + \Delta W_2 + \Delta W_3 + \cdots$$

In each small displacement Δx, the work equals the product of Δx with the average force over that displacement. This is just the area of each rectangular strip in Fig. 6.7. The total work then corresponds to the total area of all of the strips. As we decrease the size of each displacement Δx, thus increasing the number of strips, their total area approaches the area of the triangle with base x and altitude F_{ext}. Therefore

$$W = \tfrac{1}{2}xF_{ext} = \tfrac{1}{2}x(kx)$$

or

$$\boxed{W = \tfrac{1}{2}kx^2.} \tag{6.3}$$

The work done in stretching a spring by an amount x is given by $\tfrac{1}{2}kx^2$.

Example 6.3

Work required to stretch a spring.

How much work is required to extend an exercise spring by 15 cm if the spring constant k has the value 800 N/m?

Solution   We can calculate the work by substituting directly into Eq. (6.3), the expression for the work done in stretching a spring. We find

$$W = \tfrac{1}{2}kx^2 = \tfrac{1}{2}(800 \text{ N/m})(0.15 \text{ m})^2,$$
$$W = 9.0 \text{ J}.$$

Energy

6.2

Now that we understand the concept of work, we can use it to define energy. Because energy appears in so many different forms, it is difficult to give a single, brief definition. Thus we begin by considering mechanical energy. As a start, we can define **energy** by saying: *Energy is the ability to do work*. A compressed spring has energy because it may do work in returning to its uncompressed state. A falling body has energy because it may drive a stake into the ground upon striking it. Gunpowder, which may do work on exploding, has energy. An electrical battery has energy because it can turn an electric motor that does work.

An ocean wave or a falling rock has mechanical energy (Fig. 6.8), whereas gunpowder and gasoline have chemical energy, and steam has thermal energy. Energy has many forms and may be transformed from one form to another. Chemical energy changes to mechanical energy when an automobile engine burns gasoline; mechanical energy changes to electrical energy when water from a dam turns the turbine of an electric generator; and mechanical energy turns into thermal energy when you rub your hands together to warm them.

Energy, in all its forms, is measured in the same units as work. Other energy units in common use, in addition to the joule and the foot-pound, arise out of convenience in different situations. The British thermal unit

Figure 6.8   An ocean wave possesses energy associated with its motion. This is a form of mechanical energy called kinetic energy.

TABLE 6.1 The values of some common energy units in joules	
1 ft-lb	= 1.356 J
1 Btu	= 1.055×10^3 J
1 kWh	= 3.600×10^6 J
1 calorie*	= 4.187 J

*The unit ordinarily used in nutrition is the Calorie or kilocalorie, which is equal to 10^3 calories. This unit is also called a "large calorie." Note that the larger unit is distinguished from the smaller unit by being written with a capital letter.

TABLE 6.2 Approximate energy values	
Source	Approximate energy (in J)
Kinetic energy of moon in its orbit	1.16×10^{28}
Total U.S. use in one year (from all sources, 1988 est.)	8.4×10^{19}
Generated by Hoover Dam in one year	4×10^{16}
Obtained by burning one ton of coal	30×10^9
Obtained by burning 1000 ft^3 of natural gas	1×10^9
Obtained by burning 1 gallon of gasoline	2×10^8
Kinetic energy of a car at 60 mph	1×10^6
1 Btu	1×10^3
1 calorie	4.187
Released by fission of one atom of uranium	1.8×10^{-11}
Kinetic energy of a molecule of air	6×10^{-21}

(Btu) and calorie (cal) are often useful in discussing thermal energy, and the kilowatt-hour (kWh) is most frequently used in the case of electrical energy. These units will be discussed again and are mentioned here only to emphasize that they all measure the same thing. Their values in joules are listed in Table 6.1. Table 6.2 gives the energy values in joules of a range of phenomena.

Kinetic Energy

6.3

A body in motion possesses energy associated with its motion because it can do work upon impact with another object. This energy of motion is called **kinetic energy**.

Suppose we consider a free body (that is, one that is not restrained nor acted upon by other forces) and subject it to a constant force **F**. This force may or may not be gravitational in origin; we only require that it be constant. Under the influence of this constant force, the body moves a distance x. The work done on the body is $W = Fx$. We can then use Newton's second law to replace the force by the product of mass times acceleration, obtaining

$$W = max.$$

If the object is initially moving in the direction of **F** with a speed v_1, then after moving through the distance x it will have a speed v_2 given by the kinematic expression from Chapter 2:

$$v_2^2 = v_1^2 + 2ax.$$

If we rearrange this last expression and multiply by $m/2$, we get

$$max = \tfrac{1}{2}mv_2^2 - \tfrac{1}{2}mv_1^2$$

or

$$W = \tfrac{1}{2}mv_2^2 - \tfrac{1}{2}mv_1^2 \tag{6.4}$$

In applying a force to the object, we performed an amount of work $W = max$. The effect of the work done on the object has been to change its motion. The quantity $\frac{1}{2}mv^2$ is given the name *kinetic energy* (KE). More specifically, this quantity is called the **translational kinetic energy**. A body of mass m moving with a speed v possesses a kinetic energy due to its translational motion that is given by

$$\text{KE} \equiv \tfrac{1}{2}mv^2. \qquad (6.5)$$

Thus the right-hand side of Eq. (6.4) is the difference between the final and initial kinetic energies. The SI units for kinetic energy are $\text{kg·m}^2/\text{s}^2$ or joules, the same as the units for work.

Equation (6.4) is known as the **work–energy theorem:** *The work done on a body by the* net force *acting on it is equal to the change in kinetic energy of the body.* The left-hand term represents the work done on the body. The terms on the right of the equals sign represent a change in its kinetic energy. If work is done on the body, then its kinetic energy changes. Conversely, if the kinetic energy of the body is observed to change because its velocity has changed, then we know that work was done on the body. The work–energy theorem emphasizes that work, or equivalently energy, is needed to set a body in motion. The theorem is valid even for a force that is not constant.

Example 6.4

Kinetic energy of a thrown baseball.

A baseball player throws a 0.17-kg baseball at a speed of 36 m/s. Find its kinetic energy.

Solution   From the definition of kinetic energy we have

$$\text{KE} = \tfrac{1}{2}mv^2,$$
$$\text{KE} = \tfrac{1}{2}(0.17 \text{ kg})(36 \text{ m/s})^2 = 110 \text{ J}.$$

Example 6.5

Speed of an alpha particle.

Sometimes we can measure the energy of the particles of nuclear physics by determining how far they travel in matter before stopping. Using this technique, a physicist determines that an alpha particle had an initial kinetic energy of 8.0×10^{-14} J. The mass of an alpha particle is known to be 6.65×10^{-27} kg. What was the initial speed of the alpha particle in m/s? What was the speed when expressed as a fraction of the speed of light ($c = 3 \times 10^8$ m/s)?

Solution   We can calculate the particle's speed from our definition of kinetic energy even if we don't know what an alpha particle is. (We will describe it later, in Chapter 25.) From our definition of kinetic energy, we have

$$\text{KE} = \tfrac{1}{2}mv^2,$$

which we can solve for v to give

$$v = \sqrt{\frac{2KE}{m}}.$$

Inserting the numerical values gives

$$v = \sqrt{\frac{2(8.0 \times 10^{-14} \text{ J})}{6.65 \times 10^{-27} \text{ kg}}},$$
$$v = 4.9 \times 10^6 \text{ m/s}.$$

We may compare the speed of the alpha particle to the speed of light c:

$$\frac{v}{c} = \frac{4.9 \times 10^6 \text{ m/s}}{3 \times 10^8 \text{ m/s}} = 0.016,$$

or

$$v = 0.016c.$$

Example 6.6

Work done in accelerating a car from rest.

(a) How much work is done to move a 1400-kg Saab 9000 Turbo automobile from rest to 25 m/s (55 mi/h) on a level road? (b) If this takes place over a distance of 130 m, what is the average net force?

Solution   (a) We can use the work–energy theorem to find the work with v_1 set equal to zero, $v_2 = 25$ m/s, and m = 1400 kg:

$$W = \tfrac{1}{2}mv_2^2 - \tfrac{1}{2}mv_1^2,$$
$$W = \tfrac{1}{2}(1400 \text{ kg})(25 \text{ m/s})^2 = 4.4 \times 10^5 \text{ J}.$$

(b) The average net force can be found from

$$W = Fx,$$
$$F = \frac{W}{x} = \frac{4.4 \times 10^5 \text{ J}}{130 \text{ m}} = 3.4 \times 10^3 \text{ N}.$$

Note that we could also compute the car's average acceleration and then use Newton's second law to find the force. However, the present method of solution uses only scalar quantities, rather than vectors, and for more complicated situations is the easier method to use.

Gravitational Potential Energy Near the Earth

6.4

We have seen that an object in motion has kinetic energy. Energy also occurs in other forms. When a spring is stretched, it acquires energy called potential energy. For example, when you do work to wind the spring of a toy car, you give the spring potential energy; it has the ability to move

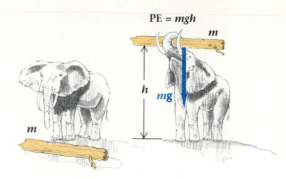

Figure 6.9   The gravitational potential energy of a log of mass m raised to a height h above the earth's surface is given by mgh.

the car when it unwinds and therefore to do work. Similar examples include a jack-in-the-box and the compressed gas in an aerosol spray can. A mass lifted from the ground also gains potential energy. If it is dropped, it loses potential energy and gains kinetic energy. If it strikes a stake on the ground, the mass can do work by driving the stake farther into the ground.

In all of these examples, work must be done to increase an object's potential energy, and in all cases the object acquires the capability of doing work when released. We can think of potential energy as stored energy. These examples illustrate mechanical potential energy; other forms of potential energy include chemical and electrical potential energy, which we will discuss in later chapters.

Let us examine gravitational potential energy, one of the most familiar forms of potential energy. Figure 6.9 shows a log of mass m initially at rest on the ground. It is then lifted slowly at a uniform velocity with a constant upward force just strong enough to equal the downward force of gravity mg. If the log is raised from the ground to a height h, the work done on the log is the net lifting force times the distance traveled, or

$$W = Fh = mgh.$$

If the log is released, it will fall. As it falls it accelerates, gaining velocity and kinetic energy, and thereby the ability to do work. Because the log at height h is capable of doing work if it is released, we say it has potential energy due to its position. More specifically, we say in this case that it has **gravitational potential energy**. The change in the log's gravitational potential energy is equal to the work required to raise the log to that height.

Thus near the earth's surface an object's gravitational potential energy with respect to some reference level is

$$PE = mgh, \tag{6.6}$$

where h is the height above the reference level. Again, the units are joules. Note specifically that the gravitational potential energy depends on the position of the body and is not a property of the body alone.

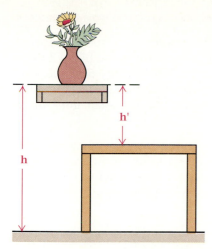

Figure 6.10 The potential energy of a vase with respect to the table is less than its potential energy with respect to the floor.

Example 6.7

How much potential energy does a 7.5-kg ceiling fan have with respect to the floor when it is 3.0 m above it?

Solution Using Eq. (6.6), we see that

$$PE = mgh,$$
$$PE = (7.5 \text{ kg})(9.8 \text{ m/s}^2)(3.0 \text{ m}) = 220 \text{ J}.$$

Gravitational potential energy must always be referred to a specific reference level. In Fig. 6.10, the potential energy of the vase with respect to the floor is greater than its potential energy with respect to the table. Yet, there is only one physical situation, not two. This example does not imply any ambiguity in the concept of potential energy; it only points out that the reference level is arbitrary. However, once you have chosen the reference level from which to measure potential energy in a given physical situation, you must keep that same reference throughout your analysis and calculation of that situation. Remember, *the physically important quantity is the difference in potential energy between two levels.*

Figure 6.11 shows several different paths by which a stack of boards can be lifted from point *A* to point *B*. Though some pathways are longer than others, *the same amount of work is done in each case.* Remember that only the displacement in the direction of the force contributes to the work done on an object. In this case the direction of the gravitational force is vertical and the net vertical displacement is the same for each pathway. Thus we conclude that the gravitational potential energy depends only on the difference between the heights of *A* and *B*. We will discuss this further in Section 6.7.

The term *potential energy* does not mean that the energy is not real energy. Rather, it means that the energy is stored and is available to be converted into work or some other form of energy.

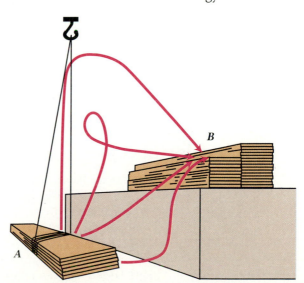

Figure 6.11 The work done in lifting a load between two heights in a uniform gravitational field is independent of the path traveled by the load.

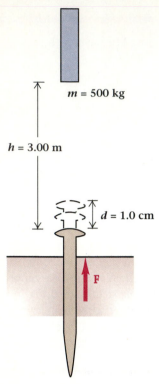

Figure 6.12   Example 6.8: A pile driver drops a 500-kg mass onto a piling, driving it deeper into the ground.

Example 6.8

In Fig. 6.12, the 500-kg mass of a pile driver is dropped from a height of 3.00 m onto a piling in the ground. The impact drives the piling 1.0 cm deeper into the ground. If all the original potential energy of the mass is converted into work in driving the piling into the ground, what was the frictional force acting on the piling? Assume the frictional force to be constant over the 1.0-cm travel.

Solution   We can solve this problem by using the concept of energy. This method has the advantage that we do not need to know the exact nature of the interaction, only that all of the energy is transferred.

The initial potential energy relative to the final position is

$$PE = mgh.$$

The driver falls through a distance h, striking the piling and driving it into the ground. We assume all of the potential energy goes into causing this motion. In doing so, the driver must overcome the frictional force between the ground and the piling. The work needed to drive the piling through a distance d is

$$W = Fd.$$

We equate the work done and the change in potential energy to give

$$Fd = mgh$$

$$F = \frac{mgh}{d}$$

$$F = \frac{(500 \text{ kg})(9.8 \text{ m/s})(3.00 \text{ m})}{0.010 \text{ m}}$$

$$= 1.5 \times 10^6 \text{ N}.$$

In reality, some of the original potential energy would go into radiated sound and into heating of the point of impact. Most of the original energy goes into moving the stake against the frictional force, energy that is eventually dissipated as heat. (We will discuss this in more detail in Chapter 10.)

General Form of Gravitational Potential Energy

6.5

So far we have considered gravitational potential energy near the surface of the earth. If we wish to determine the potential energy of an object that is far removed from the earth's surface, then Eq. (6.6) is inadequate. We derived Eq. (6.6) under the assumption that the gravitational force on an object is constant. This assumption is approximately true for bodies near the earth's surface, but the approximation fails as the distances involved get larger. We can't expect Eq. (6.6) to be appropriate for describing the potential energy of the earth–moon system or even of an artificial satellite orbiting the earth.

Suppose we wish to calculate the potential energy difference between points A and B, where A is at a distance r_A from the earth's center and B is at a distance r_B from the earth's center (Fig. 6.13). We can determine this energy difference by computing the work required to move an object of mass m from A to B. First we move the object along the path of constant radius from point A to a point C. Remember, work is done only when we exert a force through a distance to overcome the gravitational force. Since the direction of the gravitational force is radial, no work is done in going from point A to point C. To move from C to B, we must apply a force to move the mass outward. This force opposes the gravitational force given by Newton's law of gravitation,

$$F = \frac{GM_E m}{r^2},$$

where M_E is the mass of the earth.

The total work expended in going from C to B in the earth's gravitational field is the product of the average force exerted over the distance times the distance moved. That is,

$$W_{CB} = W_{AB} = \overline{F}_{AB}(r_B - r_A).$$

We have used r_A here since the magnitudes of r_A and r_C are equal. Because of the inverse-square nature of the force, the appropriate average turns out to be the geometric average, or geometric mean. (This result can be shown with the aid of calculus.) The proper average for other forces may be different. However, here we have

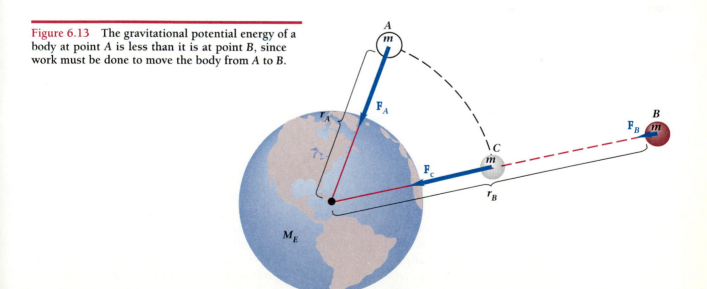

Figure 6.13 The gravitational potential energy of a body at point A is less than it is at point B, since work must be done to move the body from A to B.

Inserting this force into the equation for the work done, we get

$$W_{AB} = \frac{GM_Em}{r_Ar_B}(r_B - r_A),$$

which can also be expressed as

$$W_{AB} = \frac{GM_Em}{r_A} - \frac{GM_Em}{r_B}.$$

The work done equals the change in potential energy ΔPE_{AB} that results when the mass m is moved from A to B. We would like to express this as the difference between the potential energy at B and the potential energy at A,

$$\Delta PE_{AB} = PE_B - PE_A.$$

We can do so if we rearrange the expression for W_{AB} to get

$$W_{AB} = \Delta PE_{AB} = \frac{-GM_Em}{r_B} - \frac{-GM_Em}{r_A}, \tag{6.7}$$

where the potential energies at r_A and r_B are given by

$$PE_A = \frac{-GM_Em}{r_A}$$

and

$$PE_B = \frac{-GM_Em}{r_B}.$$

When we considered potential energy in Section 6.4 it was necessary to choose a reference, or zero. The choice of reference was based on convenience, the important physical quantity being the difference in potential energy between two positions. When the difference in gravitational potential energy is given by Eq. (6.7), we can again choose a reference on the basis of convenience. We will choose $r = \infty$ (infinity) to be the zero reference. This choice has the advantage of making the potential energy at the reference distance equal to zero. Then the potential energy at any distance is

$$PE = \frac{-GM_Em}{r}. \tag{6.8}$$

Equation (6.8) gives the potential energy at a distance r from the earth's center and is in agreement with our previous observation that objects have more potential energy as they are moved away from the earth. According to Eq. (6.8), a body's potential energy is negative near the earth's surface and becomes less negative (i.e., greater) as it moves away from the earth. The maximum potential energy is zero, the value obtained at an infinite distance from the earth's center. Figure 6.14 shows a graph of the potential energy of a 1-kg mass as a function of distance from the center of the earth.

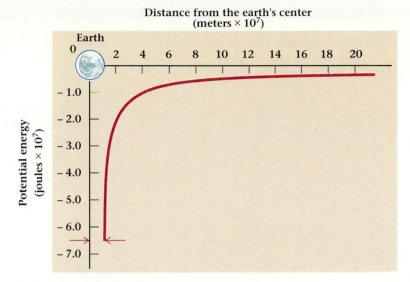

Figure 6.14 The potential energy of a 1-kg mass as a function of distance from the earth's center. The graph starts at the earth's surface.

It might seem as if we have two different expressions for the change in potential energy of a body lifted from the earth's surface. Certainly, Eqs. (6.6) and (6.8) do not appear to be the same. We can show, however, that they give essentially the same answer for small changes in distance from the earth's surface. Let us imagine that a mass m is initially at rest on the surface of the earth. We then lift it to a height h. What is its change in potential energy? If R_E is the radius of the earth, then Eq. (6.7) gives

$$\Delta\text{PE} = \frac{-GM_Em}{R_E + h} - \frac{-GM_Em}{R_E},$$

$$\Delta\text{PE} = GM_Em\left(\frac{1}{R_E} - \frac{1}{R_E + h}\right),$$

$$\Delta\text{PE} = GM_Em\left(\frac{h}{R_E(R_E + h)}\right).$$

If R_E is much greater than h, then the quantity in the brackets becomes approximately h/R_E^2 and the change in potential energy is given by*

$$\Delta\text{PE} = \frac{GM_Emh}{R_E^2}.$$

In Chapter 5 we found that the acceleration of gravity g is given by $g = GM_E/R_E^2$, so that the above equation can be written as

$$\Delta\text{PE} = mgh.$$

This equation is the same as Eq. (6.6). Thus for changes in position near the earth's surface, Eqs. (6.6) and (6.8) both give the same answer. For motion over distances that are not small compared with the earth's radius, you must use Eq. (6.8).

---

*If h is 1 km above the earth's surface, the approximation of $h/R_E(R_E + h)$ by h/R_E^2 gives an error of only 0.015%.

Moment of Inertia and Rotational Kinetic Energy

6.6

An important part of a moving object's total kinetic energy is its energy due to rotation. In fact, it can be shown that any general motion of an object, no matter how complicated, is equivalent to a combination of translational and rotational motion, so rotational effects are extremely common in nature. Sometimes an object's rotational kinetic energy even exceeds its translational kinetic energy. In order to present a more complete picture of conservation of energy right from the start, we discuss rotational kinetic energy here, using several kinematic quantities introduced in Section 5.1 for uniform circular motion. These quantities can describe rotational motion as well as circular motion; for example, a point on an object rotating uniformly about an axis traverses a circular path. The angular velocity of the object is the same as the angular velocity of that point. However, rotational motion is not always uniform, having tangential acceleration as well as radial (centripetal) acceleration. Also, in this section we introduce a quantity associated with the rotation of solid objects: the moment of inertia. In Chapter 7 we will discuss the dynamics of an object in rotational motion. Then in Chapter 8 we will examine these topics together and see how they operate simultaneously in several common situations.

When an applied force sets an object in rotation, it does work on the object. For example, you do work when you start a top or gyroscope spinning. The rotating body has kinetic energy due to its rotary motion, and this kinetic energy equals the work done in causing the rotation. We now calculate this work for a particularly simple case: a single point mass that is constrained to rotate about an axis O at a fixed distance r from that axis (Fig. 6.15).

Suppose the mass is initially moving with tangential speed v_1 and is subject to a force that is always perpendicular to the fixed length r. The work done moving the mass through a small arc s about the axis is

$$W = Fs.$$

Using Newton's second law to represent the force as ma we get

$$W = mas.$$

In the same way that we found the work in Eq. (6.4), we find the work done here is

$$W = \tfrac{1}{2}mv_2^2 - \tfrac{1}{2}mv_1^2,$$

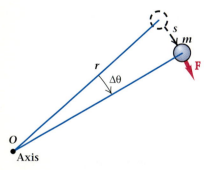

Figure 6.15   If a point mass moving at a set distance r about a fixed axis is subjected to a force parallel to the direction of its motion, an increase in its kinetic energy results. This rotational kinetic energy is $\tfrac{1}{2}mr^2\omega^2$.

where v_1 and v_2 are the initial and final tangential velocities. We may relate these tangential velocities to the angular velocities through $v = r\omega$ (Section 5.1). Thus we can express the work as

$$W = \tfrac{1}{2}mr^2\omega_2^2 - \tfrac{1}{2}mr^2\omega_1^2,$$

where r is the radial distance from the axis of rotation to the mass m and ω is the angular velocity in radians per second.

The work done is equal to the change of rotational kinetic energy, where the **rotational kinetic energy** is defined by

$$KE_{rot} \equiv \tfrac{1}{2}mr^2\omega^2.$$

This is equal to the work done upon the body to set it rotating with angular velocity ω. The rotational kinetic energy may also be written as

$$KE_{rot} = \tfrac{1}{2}I\omega^2, \tag{6.9}$$

where I, the **moment of inertia** of the point mass at radius r, is given by

$$I = mr^2.$$

In this form the rotational kinetic energy is similar to the expression for the translational kinetic energy. Note that both expressions are one half the product of a term depending only on the properties of the body times the square of a velocity term.

We have just computed the rotational kinetic energy for a single point mass moving in a circle. However, it can be shown that Eq. (6.9) is more general and may be used to find the rotational kinetic energy of any extended rigid body provided we use the proper moment of inertia I. The moment of inertia plays the same role in the formulas for rotational motion that mass does in the ones for linear motion. (We will see this again in Chapter 7.) However, the moment of inertia depends both on the mass and on the geometry, or shape, of the body. We can calculate the moment of inertia of an extended body by summing the moments of inertia of each small element of the body, obtaining the moment of inertia of the whole. Except for a few simple cases, the techniques of calculus are needed to perform the summation. Table 6.3 (p. 174) gives the moments of inertia for several bodies about specific axes. Inserting these expressions into Eq. (6.9), we can readily compute the rotational kinetic energy.

Example 6.9

Kinetic energy of a rotating record.

How much energy is required to set a 12-in. phonograph record into rotation at $33\tfrac{1}{3}$ revolutions per minute? The record has a mass of 0.115 kg.

Solution From Table 6.3, the moment of inertia of a disk of mass m and radius R rotating about an axis through its center is $I = \tfrac{1}{2}mR^2$. Substituting this value for the moment of inertia I into the equation for rotational kinetic energy, we get

$$KE_{rot} = \tfrac{1}{2}I\omega^2 = \tfrac{1}{2}(\tfrac{1}{2}mR^2)\omega^2 = \tfrac{1}{4}mR^2\omega^2.$$

We now insert the numerical values, noting that

$$R = 6 \text{ in.} = (6 \text{ in.})(0.0254 \text{ m/in.}) = 0.152 \text{ m}$$

and

$$\omega = 2\pi f = 2\pi(33\tfrac{1}{3} \text{ rev/min})(1 \text{ min/60 s}) = 3.49 \text{ rad/s}.$$

TABLE 6.3
Moments of inertia for some regular objects of mass *m*

Dumbbell about axis through center perpendicular to length. Connecting rod of negligible mass.

Thin ring about axis through center

Disk about axis through center

Solid sphere about any diameter

Thin rod about axis through one end perpendicular to length

Thin rod about axis through center perpendicular to length

Disk about axis through one edge parallel to the symmetry axis

Thus

$$KE_{rot} = \tfrac{1}{4}(0.115 \text{ kg})(0.152 \text{ m})^2(3.49 \text{ rad/s})^2,$$

$$KE_{rot} = 8.1 \times 10^{-3} \text{ J}.$$

We have included rotational motion in this section to give you an idea of the different kinds of mechanical energy. However, when we discuss conservation of mechanical energy in the next section, we will assume that all the kinetic energy is due to translational motion. We do this to keep the mathematics simple when introducing this new idea. We will return to rotational kinetic energy in the examples of energy conservation discussed in Chapter 8.

Conservation of Mechanical Energy

6.7

When you apply a force to a spring and stretch it, the spring returns to its original length when released. On the other hand, if you apply a force to a book and push it across a table against the frictional force, the book does not return to its original position when you release it. In both cases you do work against a resisting force, but the nature of the forces is different. The spring force represents what is called a conservative force; friction is a dissipative, or nonconservative, force.

If the work done by a force on an object depends only on the initial and final positions of the object, then that force is a **conservative force**. The work is independent of the path taken. If you move an object against a conservative force and return it to the starting place, the total work done is zero. For example, if you do an amount of work to lift a barbell from the floor, the same amount of work is done on you if you lower the barbell to its initial location (Fig. 6.16). (This does not mean that a weight lifter does no work in lifting and lowering a barbell. Energy losses occur within the body.)

As we saw earlier, the gravitational force is a conservative force. The expression for the work given in Eq. (6.7) depends on the distances r_A and r_B (Fig. 6.13), not on the path taken between them. A spring with a restoring force proportional to its extension is another example of a conservative force. We conclude, from the requirement that the work done be independent of the path taken, that conservative forces must be forces that depend on position, rather than forces that may vary with time, speed of the object, or some other parameter.

The work done against a frictional force depends on the path. If you push a coin from A to B (Fig. 6.17) with constant speed, the work done is different for different paths. Friction is classified as a nonconservative force. The resistance of air to the motion of a body and the resistance of a liquid, such as water, to motion through it, are other examples of nonconservative forces.

An important distinction between conservative and nonconservative forces is that we can write an expression for the potential energy for conservative

Figure 6.16 The work done in raising a barbell is equal to the work done on the weight lifter by the barbell if it is lowered to its initial position.

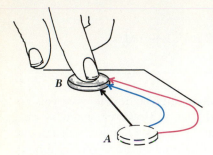

Figure 6.17   The work done to push a coin from A to B depends on the path taken.

forces. You have seen this for gravitation, and should realize that the energy needed to compress a spring that obeys Hooke's law is the stored, or potential, energy. No such expression is possible for frictional forces. The term "conservative" force is appropriate because a conservative force corresponds to the conservation, or constancy, of the kinetic energy plus potential energy.

Let us return to the case of a body that is raised to a small height h above the ground (Fig. 6.18). Over this distance the earth's gravitational field is constant. The potential energy of the body is PE $= mgh$. If the rope breaks, the body is released from rest and falls. As it falls it gains speed as a result of the acceleration of gravity, and at any instant has a kinetic energy corresponding to its instantaneous speed, given by

$$KE = \tfrac{1}{2}mv^2.$$

If the speed at height h_1 is v_1, then we can determine the speed v_2 at height h_2 by using the kinematic expression from Chapter 2 with g inserted for the acceleration:

$$v_2^2 = v_1^2 + 2gs.$$

The kinetic energy at h_2 is then

$$KE(h_2) = \tfrac{1}{2}mv_2^2 = \tfrac{1}{2}mv_1^2 + mgs.$$

The distance s is the difference in the two heights, $s = h_1 - h_2$. When this value of s is inserted into the above equation, we find

$$\tfrac{1}{2}mv_2^2 = \tfrac{1}{2}mv_1^2 + mgh_1 - mgh_2.$$

This may be rearranged to give

$$\tfrac{1}{2}mv_2^2 + mgh_2 = \tfrac{1}{2}mv_1^2 + mgh_1. \qquad (6.10a)$$

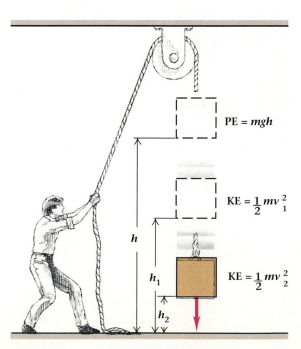

Figure 6.18   A body is released from height h above the ground. It falls to height h_1, where its speed is v_1, and to height h_2, where it has a new speed v_2. Its total kinetic and potential energy stays the same.

The sum of the body's kinetic energy and potential energy is its total **mechanical energy**. According to Eq. (6.10a), the mechanical energy at height h_2 is equal to the mechanical energy at height h_1. Thus an object's total mechanical energy is constant for motion in a constant gravitational field provided that no other forces are introduced. For this special case we say that the mechanical energy is *conserved*. The total energy is the same at the top (h) as it is at any other height. The total mechanical energy E can be written in the form

$$E = \text{KE} + \text{PE}. \qquad (6.10b)$$

The value of the potential energy (PE) and kinetic energy (KE) may change, but their sum, the total energy (E), is a constant and does not change.

Equations (6.10a) and (6.10b), along with additional observations and experiments, lead to the generalization that *if the forces are all conservative, the sum of the kinetic and potential energy is a constant*. This statement is the **law of conservation of mechanical energy**.

The principle of conservation of energy is extremely useful in a wide variety of cases. In many cases the effects of friction are small enough so that we can ignore them and use the law of conservation of mechanical energy. In later chapters we will study other forms of energy, such as thermal energy and electrical energy. We will find that energy can be transformed from one form to another and that we can extend the principle of conservation of energy to include those other forms. Example 6.10, which follows, demonstrates an application of conservation of mechanical energy.

Those laws of nature that state that some quantity is the same before and after an event or interaction are called **conservation laws**. They reflect one of our most basic ways of describing nature. The law of conservation of mechanical energy is, in some sense, a more fundamental principle than Newton's mechanics, which we used to derive it here. Because conservation laws allow us to consider quantities, such as energy, that do not change during an event, we do not need to know the details of the interaction. We simply deal with the value of the conserved quantity before and after the interaction. Moreover, the fact that some quantities are conserved, and some are not, tells us something about nature itself.

In all of physics there are only a relatively small number of conservation laws. We will introduce two more in the next chapter and only a few in the remainder of the text. Other conserved quantities include momentum (Chapter 7) and electric charge (Chapter 15). It can be argued that conservation laws represent the most powerful and, at the same time, simplest view of nature. They allow us not only to understand how much electrical energy it takes to keep our house warm or cool, but also to discover new particles and new behavior on the atomic and subatomic scale.

Example 6.10

A monkey in a tree drops a coconut from rest to the ground 7.5 m below. Use the principle of conservation of energy to determine its speed just before it strikes the ground.

Solution   The total mechanical energy E of the coconut is a constant. It consists of two parts, KE and PE, which are not individually constant, but whose sum is constant:

$$E = KE + PE.$$

Using subscripts T for top and 0 for ground, we equate the total energy at the top to the total energy at the bottom.

$$KE_T + PE_T = KE_0 + PE_0.$$

But $KE_T = 0$ because the initial speed is zero, and $PE_0 = 0$ because $h = 0$. Thus we have

$$PE_T = KE_0$$

or

$$mgh = \tfrac{1}{2}mv^2.$$

Upon solving for v we get

$$v = \sqrt{2gh},$$
$$v = \sqrt{2(9.8 \text{ m/s}^2)(7.5 \text{ m})} = 12 \text{ m/s}.$$

Note that we could have obtained the same result using the methods described in Chapter 2. However, the principle of energy conservation is important because it can be applied in much more complicated situations.

Let us consider what happens to a roller coaster as it moves along a track (Fig. 6.19a). The height of the track above some reference level determines the potential energy at that point. In part (a), a car of mass m leaves point A with no initial velocity ($v_A = 0$). We can determine its speed at any other location B using the principle of conservation of energy:

$$KE_A + PE_A = KE_B + PE_B,$$
$$0 + mgh_A = \tfrac{1}{2}mv_B^2 + mgh_B,$$
$$\tfrac{1}{2}mv_B^2 = mgh_A - mgh_B,$$

or

$$v_B = \sqrt{2g(h_A - h_B)}.$$

This result suggests a diagrammatic way of viewing a situation in which energy is conserved. In Fig. 6.19(b) we have plotted the potential energy of the car as a function of its horizontal displacement x. Since the potential energy is directly proportional to the height h, the potential energy curve has the same shape as the track itself. Such a graph of the potential energy against the displacement is called a **potential-energy diagram**. The total energy is represented here by E_1 and is constant. In this case, the total energy E_1 is equal to the initial potential energy PE_A. On the diagram you can see that the kinetic energy is given by the distance between the total-energy curve and the potential-energy curve. Thus the kinetic energy at point B is

$$KE_B = E_1 - PE_B.$$

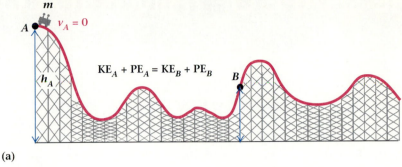

(a)

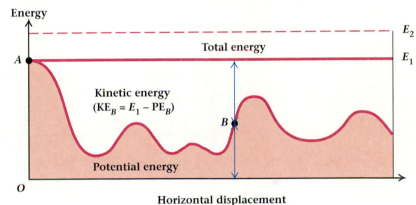

(b)

Figure 6.19   (a) A roller coaster on a track. The total mechanical energy $E = KE + PE$ is conserved. (b) The potential-energy diagram for the roller coaster on the track. We choose the zero energy level to be ground level. The kinetic energy is the difference between total energy E_1 and the potential energy. (c) The potential-energy diagram where the zero energy level is chosen to be the energy of point A. The kinetic energy is still the difference between the total energy $E = 0$ and the potential energy.

(c)

When these energies are expressed in terms of position and speed we get

$$\tfrac{1}{2}mv_B^2 = E_1 - mgh_B,$$

or

$$v_B = \sqrt{2\left(\frac{E_1}{m} - gh_B\right)}.$$

However, since for this case the total energy E_1 is equal to the initial potential energy mgh_A, this expression becomes

$$v_B = \sqrt{2g(h_A - h_B)}.$$

This result is the same as our earlier finding.

POTENTIAL-ENERGY DIAGRAMS

Potential-energy diagrams plot energy on the vertical axis against distance from a reference point on the horizontal axis. They are a useful tool for seeing how an object's energy changes as it moves within a field.

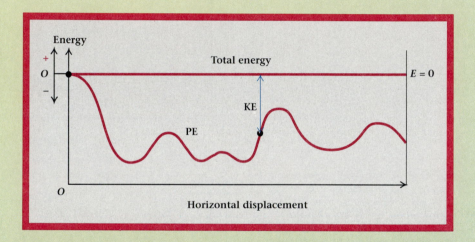

An object's total energy consists of kinetic energy (always positive) plus potential energy. The shape of the potential-energy diagram tells us how much energy an object has at any distance from the source of the field.

If the roller coaster in part (a) had been moving at point A, it would have had an initial kinetic energy, as well as an initial potential energy. Then the total energy would have been greater than E_1, as shown by the dotted line E_2 in part (b). In this case the speed at any point B would be

$$v_B = \sqrt{2\left(\frac{E_2}{m} - gh_B\right)}.$$

This is just the same as the previous expression, with E_1 replaced by E_2. In either case the value of the total energy would be known from the initial conditions.

We emphasize that the roller coaster's kinetic energy, and therefore its speed, is determined by the difference between the total-energy line, or curve, and the potential-energy curve. As indicated by the potential-

GRAVITATIONAL POTENTIAL ENERGY

The earth as seen from the moon.

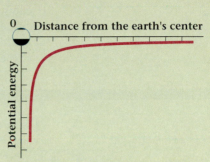

0 **Distance from the earth's center**

Potential energy

Gravitational potential energy is always negative, indicating the constant attractive force of gravitation (Chapter 8).

MOLECULAR POTENTIAL ENERGY

A crystalline solid.

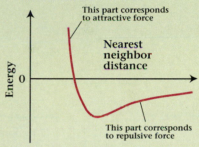

This part corresponds to attractive force

Nearest neighbor distance

Energy

0

This part corresponds to repulsive force

The potential energy of two atoms or molecules interacting shows an attractive force at large distances and a repulsive force at much smaller distances. The minimum potential energy occurs at the stable interatomic spacing of the solid (Chapter 30).

energy diagram, at a point where the potential energy is smaller, the kinetic energy, and therefore the speed, is larger. From such a diagram you can determine the speed if you know the position.

The conservation relationship among total energy, potential energy, and kinetic energy remains the same regardless of the reference level selected. Thus if we choose a new reference level from which to measure the potential energy of the roller coaster car, the kinetic energy is the same as it was before, at any specific location. For example, in Fig. 6.19(c) we choose the level of A to have zero energy. At B, the potential energy is now negative relative to our new reference level. Using primes to distinguish this reference level from the one in part (b), we have

$$0 = KE_B' + PE_B',$$
$$0 = \tfrac{1}{2}mv_B^2 - mgh.$$

From the diagram we see that $h = h_A - h_B$, so we have

$$v_B = \sqrt{2g(h_A - h_B)},$$

exactly the same as before.

If we had some other initial total energy, E_2', then we could make the analysis in the same way. We emphasize that the particular reference level chosen for the zero of potential energy does not change the results. We are free to choose the reference zero as a matter of convenience, but once chosen, it must not be changed during a problem.

Power

6.8

In many cases it is useful to know not just the total amount of work being done, but how rapidly work is being done. For instance, if you have a motor that can provide only a certain amount of work in one day, and you wish to accomplish twice that much work, then you must either take two days for the job or get an additional motor. We define **power** as the time rate of doing work; that is,

$$P \equiv \frac{\Delta W}{\Delta t},$$ (6.11)

where ΔW is the amount of work done in the time interval Δt. In SI units, work is measured in joules and time in seconds. The unit of power is the joule/second, a combination that has been given the name **watt** (abbreviated W):

$$1 \text{ joule/second} = 1 \text{ watt.}$$

The measurement of power grew out of the need of early steam-engine builders to specify the properties of their engines. James Watt (1736–1819), the most inventive of these engine builders, developed the steam engine into an efficient, versatile engine that could be used to drive machinery. As part of his efforts to measure the power of his steam engine, Watt made systematic measurements on the work a horse could perform in a given time. From these measurements he defined a horsepower as 550 ft-lb/s. The relationship between the watt and the horsepower (hp) is

$$1 \text{ hp} = 746 \text{ W} = 0.746 \text{ kW.}$$

Table 6.4 gives the approximate power produced or consumed in a number of situations.

Example 6.11

A jogger's power output.

Using a calibrated treadmill, a jogger measures her energy output to be 4.0×10^5 J for an hour's run. What average output power did she develop?

TABLE 6.4 Approximate power production and consumption	
Device	**Approximate power (watts)**
Hoover Dam	1.34×10^9
Automobile use of chemical energy at 40 mi/h	7×10^4
Electric stove	1.2×10^4
Average per capita use of electricity (U.S.)	1×10^3
Refrigerator-freezer	615
Solid-state color TV	200
Available solar power per square meter (ave. U.S. over 24-h day)	180
Pocket calculator (LCD display)	7.5×10^{-4}

Solution   From the definition of power as the rate of doing work, we have

$$P = \frac{\Delta W}{\Delta t} = \frac{4.0 \times 10^5 \, \text{J}}{3600 \, \text{s}},$$

$$P = 110 \, \text{W}.$$

Compare this with the power consumption of a household electric lamp.

Example 6.12

Power used running upstairs.

A 70-kg person runs up a staircase 3.0 m high in 3.5 s. How much power does he develop?

Solution   If we assume that the work is done at a constant rate, then the definition of Eq. (6.11) becomes

$$P = \frac{W}{t}.$$

In this case, the work done is the change in gravitational potential energy, mgh, so that the power becomes

$$P = \frac{mgh}{t} = \frac{70 \, \text{kg} \times 9.8 \, \text{m/s} \times 3.0 \, \text{m}}{3.5 \, \text{s}} = 590 \, \text{W}.$$

Note that we have actually computed the average power, which is the total work done divided by the total time.

The power may also be given in units of horsepower as

$$P = 590 \, \text{W} \times \frac{1 \, \text{hp}}{746 \, \text{W}} = 0.79 \, \text{hp}.$$

This result is consistent with the general observation that humans are capable of power outputs in the range of 0.5 to 1 hp for 30 s. For longer periods of time, human power output is much decreased. For steady work over 8 h, human power is of the order of 0.1 or 0.2 hp.

Our definition of power in Eq. (6.11) applies to all types of work, whether mechanical, electrical, or thermal. However, we can rewrite the definition in a special way for mechanical work by simply rearranging terms. When a force acts on an object so that it moves with a speed v, we can calculate the power from the force and the speed. If we consider the force to be constant, the change in work is $\Delta W = F \Delta x$. Then power becomes

$$P = F \frac{\Delta x}{\Delta t},$$

or

$$P = Fv. \tag{6.12}$$

It can be shown that Eq. (6.12) is true for instantaneous power even when the force is not constant.

Example 6.13

Power used pushing a car.

A motorist whose car is out of gas pushes his car at a steady speed of 0.75 m/s with a force of 200 N. What power does he deliver to the car?

Solution   The power is given by

$$P = Fv,$$
$$P = (200 \text{ N})(0.75 \text{ m/s}) = 150 \text{ W} = 0.20 \text{ hp}.$$

SUMMARY

Useful Concepts

- The work done on an object by a force F acting over a distance x at an angle θ to the displacement is

$$W = Fx \cos \theta.$$

- Energy is the ability to do work.
- The force required to extend a spring a distance x is

$$F_{\text{ext}} = kx.$$

- The work required to stretch a spring a distance x, and the energy stored in it, is

$$W = \tfrac{1}{2}kx^2.$$

- An object's kinetic energy of translation is

$$KE = \tfrac{1}{2}mv^2.$$

- According to the work–energy theorem, the work done on a body by the net force acting on it is equal to the change in kinetic energy of the body.

- The level from which you choose to measure an object's gravitational potential energy is arbitrary, but you must not change it while working a problem. The gravitational potential energy of a mass m at a height h near the earth's surface is given by

$$PE = mgh.$$

- The gravitational potential energy of a mass m at any distance r from the center of the earth and above the earth's surface is

$$PE = \frac{-GM_E m}{r}.$$

- The work done by a conservative force acting on an object depends only on the initial and final positions of the object, not on the path taken. The sum of kinetic energy and potential energy is a constant if the forces are conservative. This is the law of conservation of mechanical energy.

- The kinetic energy of rotation for an object with moment of inertia I rotating with an angular velocity ω is

$$KE_{rot} = \tfrac{1}{2}I\omega^2.$$

- Power is defined as the rate of doing work:

$$P \equiv \frac{\Delta W}{\Delta t}.$$

Important Terms

You should be able to write the definition or meaning of each of the following terms:

- work
- joule
- energy
- kinetic energy
- translational kinetic energy
- work–energy theorem
- gravitational potential energy
- rotational kinetic energy
- moment of inertia
- conservative force
- mechanical energy
- conservation of mechanical energy
- conservation laws
- potential-energy diagram
- power
- watt

QUESTIONS

6.1 Why is no work done on an object when a force acting on the object does not move it?

6.2 What happens to the work done in stretching a spring?

6.3 A baseball and a Ping-Pong ball are thrown with the same velocity. Which one has the greater kinetic energy? Why?

6.4 Explain how a pile driver works.

6.5 Do you do the same work to lift a 1-kg weight through a vertical height of 1 m everywhere on the face of the earth?

6.6 A ball rolls across the floor. Is it possible for its translational and rotational kinetic energies to be the same?

6.7 Explain the basic ideas that govern the design and operation of a roller coaster. Under what conditions can successive hills be as high as or higher than the initial one?

6.8 Time yourself to see how long it takes you to do 10 deep knee bends. Then estimate how much work you do in lifting yourself back up each time. From these numbers, estimate the rate at which you were doing work.

6.9 Why do you shift to a lower gear to pedal a multi-speed bicycle uphill? Do you save any energy in doing so?

6.10 Explain the observation that smaller cars generally get better fuel mileage than larger cars.

6.11 An engineer desires to store energy in a rotating flywheel of a given mass and radius. Should he select a flywheel in the shape of a uniform solid disk or one with most of the mass on the rim? What is the ratio of the energies that can be stored in these two cases if both wheels rotate with the same angular frequency ω?

PROBLEMS

Hints for Solving Problems

You should be careful to distinguish the initial and final states of a situation involving energy. Determine the potential energy and kinetic energy for each state. Then equate the initial mechanical energy to the final mechanical energy. Be careful about the signs: Work done on an object is positive, work done by the object is negative. A force in the direction of displacement does positive work; a force opposite to the direction of displacement does negative work; a force perpendicular to the displacement does zero work. Be careful about the reference level for gravitational potential energy. Once you choose a reference level, do not change it while solving the problem.

Section 6.1   Work

6.1 How much work does a 55-kg woman do against gravity when climbing from the bottom to the top of a 3.0-m-high staircase?

6.2 A steady force of 20 N is applied to a pushcart that moves parallel to the direction of the force. If 75 J of work are done, how far does the cart move?

6.3 A gardener pushes a box of tools across a driveway by applying a 30-N force that pushes downward at an angle of 20° with respect to the horizontal. How much work is done if the box is moved 2.0 m in the horizontal direction?

6.4 A boy uses 75 J to push a sled a distance of 3.0 m across the snow, applying the force in a horizontal direction. (a) How much force is needed to push the same sled through the same distance if the force is applied in a downward direction of 45° with respect to the horizontal? (b) How much force is required if it is applied in an upward direction of 45°? Assume that the work required is unchanged.

6.5 A worker does 300 J of work against a frictional retarding force of 15 N in pushing a power sweeper across a floor in 3.0 s. If the sweeper moves with constant speed, how fast is it going?

6.6 A force of 20 N is needed to hold a spring extended 5.0 cm from its equilibrium position. How much work is done in extending the spring?

6.7 A person finds that he can stretch a spring exercise device 1.3 times as far as a friend can. (a) What is the ratio of the force they can apply? (b) What is the ratio of the work they each do in stretching the spring?

6.8 A robot arm expends 15 J of work to extend a spring 10 cm from its equilibrium position. How much force is required to hold it in that position?

Section 6.2   Energy

6.9 How many Btu are equivalent to 1 kWh? (*Hint:* See Table 6.1.)

6.10 (a) Approximately how many kilowatt-hours of electrical energy does one ton of coal produce if one third of the energy available is converted to electrical energy? (b) How much gasoline would be required to produce the same energy? (*Hint:* Refer to the data in Table 6.2.)

6.11 How many electricity-generating plants the size of Hoover Dam would be required to supply the total energy consumption of the United States? (Use the data of Table 6.2.)

6.12 How many gallons of gasoline would it take to produce the electrical energy equal to that generated by Hoover Dam in one year? Assume that one third of the energy of the gasoline is converted to electrical energy and two thirds is lost.

Section 6.3   Kinetic Energy

6.13 (a) What is the kinetic energy of an 1800-kg car moving at 20 m/s? (b) At 88 km/h?

6.14 What is the kinetic energy of a 5.0×10^5-kg locomotive moving with a speed of 40 km/h?

6.15 A 2.5-g Ping-Pong ball at rest is set in motion by the use of 1.8 J of energy. If all of the energy goes into the motion of the ball, what is the ball's maximum speed?

6.16 What is the ratio of the kinetic energy of a car traveling at 60 mi/h to its kinetic energy at 20 mi/h?

6.17 With what speed would a 50-kg girl have to run to have the same kinetic energy as a 1000-kg automobile traveling at 2.0 km/h?

6.18 How far does a 1.0-kg stone with a kinetic energy of 3.0 J go in 2.0 s if it is moving in a straight line?

Section 6.4   Gravitational Potential Energy Near the Earth

6.19 How much potential energy with respect to ground level does a 10.0-kg lead weight have when it is 2.00 m above the surface of the ground? Give your answer in (a) joules and (b) ft-lb.

6.20 Two blocks of 6.00 kg and 3.00 kg are hung over a pulley on a rope, with the 6-kg block resting on the floor. What is the change in the potential energy of the system if the 6-kg block is raised 0.800 m?

6.21 By calculating the work done along each part of the path and adding the amounts (taking account of the signs), show that the work done in moving a 1-kg mass to a point 1 m over from and 1 m above its initial point is the same for all four paths shown in Fig. 6.20.

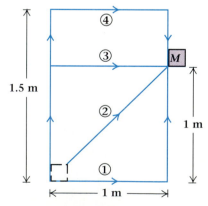

Figure 6.20   Problem 6.21.

6.22 A 0.25-kg textbook rests on a shelf 0.75 m above a desk top, which in turn is 0.80 m above the floor. (a) What is the potential energy of the book with respect to the desk top? (b) With respect to the floor?

Section 6.5 General Form of Gravitational Potential Energy

6.23 What is the value of the gravitational potential energy of a 1.00-kg mass on the surface of the earth if the zero of potential energy is taken at $r = \infty$?

6.24 What is the gravitational potential energy of the moon with respect to the earth if the zero of potential energy is taken at $r = \infty$?

6.25 What is the change in gravitational potential energy of a 1.00-kg mass that is carried from the surface of the earth to a distance of one earth radius above the surface?

Section 6.6 Moment of Inertia and Rotational Kinetic Energy

6.26 What is the kinetic energy of a 0.125-kg 12-in. phonograph record when rotated at 45 rev/min? (Refer to Table 6.3 for the moment of inertia.)

6.27 A solid iron cylinder has a radius of 0.250 m and a mass of 154 kg. Calculate the kinetic energy of this cylinder when it is rotating about its axis at a rate of 12.5 rev/s. (Refer to Table 6.3 for the moment of inertia.)

6.28 A solid bowling ball with a radius of 10.9 cm and a mass of 7.0 kg rolls along a bowling alley at a linear speed of 2.0 m/s. (a) What is its translational kinetic energy and (b) what is its rotational kinetic energy?

6.29 What is the approximate rotational kinetic energy of a 66-cm diameter bicycle wheel of mass 4.0 kg when the bicycle is traveling at 15 km/h?

Section 6.7 Conservation of Mechanical Energy

6.30 With what speed does a hammer dropped from a height of 23 m strike the ground? Neglect air friction.

6.31 A 30-kg suitcase falls from an airplane at a height of 2000 m. (a) If it loses 90% of its energy through friction with the air, what energy does it have just before it strikes the ground? (b) What speed does it have just before it strikes the ground?

6.32 A 4.0-kg box of books is pulled up a 4.0-m-long in-
 • clined plank to a height of 1.5 m. A force of 20 N parallel to the incline is needed to pull the box up the plank. (a) How much energy is used and (b) what percent of that is lost to friction?

6.33 An egg falls from a nest at a height of 3.0 m. What speed will it have when it is 0.50 m from the ground?

6.34 A ball is thrown upward with sufficient velocity to send it to a height of 9.0 m. What is its speed when it has reached a height of 8.0 m?

6.35 A block slides on a semicircular frictionless track (Fig. 6.21). If it starts from rest at position A, what is its speed at the point marked B?

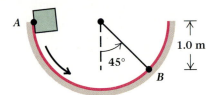

Figure 6.21   **Problem 6.35.**

Section 6.8 Power

6.36 What is the horsepower of a 250-W light bulb?

6.37 About how many storage batteries would an electric car need in order to develop 80 hp? (Assume conventional electric batteries can deliver energy on a continuous basis at a rate of about 300 W.)

6.38 How many watts are used to push a loaded cart with a force of 40.0 N through a horizontal distance of 10.0 m in 3.00 s?

6.39 How many horsepower are developed when a 2.25-kg book is lifted 0.520 m in 2.00 s?

6.40 An electric motor that can develop 1.0 hp is used to lift a mass of 25 kg through a distance of 10 m. What is the minimum time in which it can do this?

6.41 An elevator is powered by a 10-hp motor. What is the maximum mass it can raise with uniform speed through a distance of 40 m in 18 s?

6.42 If electricity costs $0.08 per kilowatt-hour, how much does it cost to use a 250-W light bulb for 8 h?

6.43 How much power is developed when a constant force of 200 N moves a canoe at a speed of 5.0 m/s?

6.44 An automobile engine develops 30 hp in moving the automobile at a constant speed of 50 mi/h. What is the average retarding force due to such things as wind resistance, internal friction, tire friction, etc.?

Additional Problems

6.45 (a) A 50-kg gymnast stretches a vertical spring by
 • 0.50 m when she hangs from it. How much energy is stored in the spring? (b) The spring is cut into two equal lengths and the gymnast hangs from one section. In this case the spring stretches by 0.25 m. How much energy is stored in the spring this time?

6.46 (a) How much work is needed to push a 100-kg pack-
 • ing crate a distance of 2.0 m up a frictionless inclined plane that makes an angle of 20° with the horizontal? (b) How much work would be required if the coefficient of friction were 0.20?

6.47 (a) How much work is done to move an 8.0-kg block 10 cm to the right if the spring is initially relaxed (Fig. 6.22)? The spring constant is 20 N/m and the coefficient of friction between the block and the floor is 0.50.

6.48 A force is given by $F = kx^2$, where x is in meters and $k = 10$ N/m². What is the work done by this force when it acts from $x = 0$ to $x = 0.1$ m? (*Hint:* Use a graphical technique.)

6.49 The area of the United States (excluding Alaska and Hawaii) is approximately 3.5×10^6 mi². (a) Use the data in Table 6.4 to estimate the total solar power falling on this part of the United States. (b) Assuming a population of approximately 2.5×10^8 people, what area would have to be devoted to solar collectors with a 10% conversion efficiency in order to get all of the needed power from solar power? (c) What percent of the area of the United States would be covered with solar collectors? (d) How does the answer to (b) compare with the area of the state in which you live?

6.50 Sam can throw a baseball twice as fast as can his little brother, Bill. How many times as much kinetic energy can Sam give the baseball as Bill can?

6.51 (a) How much work is required to increase the speed of a 1200-kg automobile from 10 km/h to 30 km/h? (b) How much work is required to further increase the speed by the same amount, this time from 30 km/h to 50 km/h? Neglect the effects of friction.

6.52 How high from the surface of the earth must an object be raised so that the increase in potential energy as given by PE $= mgh$ and by PE $= -GM_{E}m/r$ will differ by 2%? Express this distance as a multiple of the earth's radius.

6.53 Calculate the total energy of the moon with respect to the earth. (*Hint:* Add the kinetic and potential energies.)

6.54 The block on the loop-the-loop in Fig. 6.23 slides without friction. From what height must it start at A so that it presses against the track at B with a net upward force equal to its own weight?

6.55 (a) A solid 4.0-kg wheel of radius 0.23 m is initially at rest. How much work is required to make it rotate at 3.0 rev/s about its axis? (b) If the energy of the rotating wheel is doubled, how many revolutions per second will it make?

6.56 A 5.0-kg object slides down an inclined board from a height of 2.0 m to the ground, losing 10% of its energy to friction. What is its speed when it reaches the ground?

6.57 A 40-kg child sits in a swing suspended with 2.5-m-long ropes. The swing is held aside so that the ropes make an angle of 15° with the vertical. If let go, what speed will the child have at the bottom of the arc? (*Hint:* Use conservation of energy.)

6.58 From what height must a car be dropped to give it the same kinetic energy just before impact that it has when traveling at 60 km/h?

6.59 A 0.50-kg grapefruit depresses the pan of a spring balance 1.0 cm when resting upon it. If the grapefruit is dropped from a height of 0.20 m, how far will the pan be depressed if all the energy of the grapefruit goes into compressing the spring? (*Hint:* The energy in a compressed or stretched spring is the same for the same displacement.)

6.60 The 5.0-kg mass in Fig. 6.24 is released from rest 1.0 m above the floor. If the coefficient of friction between the 2.0-kg mass and the table is 0.28, what is the speed of the 5.0-kg mass just before it strikes the floor? (*Hint:* Use conservation of energy.)

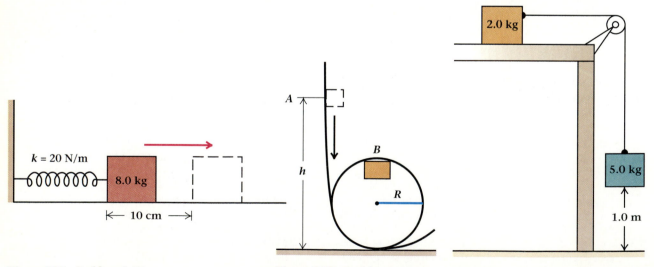

Figure 6.22   Problem 6.47.

Figure 6.23   Problem 6.54.

Figure 6.24   Problem 6.60.

6.61 Suppose the roller coaster car in Fig. 6.25 starts from rest at point *A* and moves without friction. How fast is it going at points *B*, *C*, and *D*? What constant deceleration must be applied at *D* to have it stop at *E*?

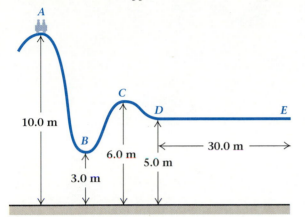

Figure 6.25 Problem 6.61.

6.62 Niagara Falls is about 48 m high. An estimated 10^4
• kg of water pass over the falls every second. If this energy were usefully employed, what power could be produced?

6.63 Show that the height *h* to which an animal of mass
•• *m* can jump is given approximately by

$$h = \frac{1}{2g} \left(\frac{2sP}{m} \right)^{2/3},$$

where *P* is the power expended and *s* is the distance over which the animal accelerates.

6.64 The power *P* delivered by a windmill whose blades
•• sweep in a circle of diameter *D* by a wind of speed *v* is proportional to the square of the diameter and the cube of the wind speed. Show that this dependence on diameter and wind speed is what you would expect in the case of 100% efficient energy transfer by

showing that the energy transferred per unit time is equal to

$$P = \frac{\pi}{8} \rho D^2 v^3.$$

(*Hint:* First calculate the kinetic energy in a "tube" of air of density ρ moving with a speed *v*, then determine the rate at which this energy is transferred to the windmill.)

6.65 Show for a satellite moving in a circle about a center
• of gravitational attraction, such as the sun or Earth, that

$$R \times KE = \text{constant},$$

where *R* is the radius of the orbit and KE is the translational kinetic energy of the satellite.

6.66 A block in Fig. 6.26 is initially at rest on an inclined
•• plane at the equilibrium position that it would have if there were no friction between the block and the plane. How much work is required to move the block 10 cm down the plane (a) if the frictional coefficient is $\mu = 0$, and (b) if the frictional coefficient is $\mu = 0.17$?

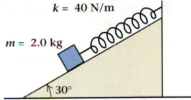

Figure 6.26 Problem 6.66.

6.67 Instruments used by geologists can detect changes in
•• the earth's gravitational field strength as small as 10^{-7} m/s². To what approximate change in height above the earth's surface does this change in field strength correspond? (*Hint:* Use a mathematical procedure similar to that used to determine the difference in gravitational potential energy in Section 6.5.)

ADDITIONAL READING

Alexander, R. M., "Walking and Running." *American Scientist,* July 1984, p. 348.

Brooks, A. N., A. V. Abbott, and D. Gordon, "Human Powered Watercraft." *Scientific American,* December 1986, p. 120.

Gray, C. L., Jr., and F. von Hipple, "The Fuel Economy of Light Vehicles." *Scientific American,* May 1981, p. 48.

Nadel, E. R., and S. R. Bussolari, "The Daedalus Project: Physiological Problems and Solutions." *American Scientist,* July 1988, p. 351.

Scientific American, Issue on Energy and Power, September 1981.

Wilson, D. G., "A Short History of Human-Powered Vehicles." *American Scientist,* July–August 1986, p. 350.

7

Linear and Angular Momentum

7.1 Linear Momentum

7.2 Impulse

7.3 Newton's Laws and the Conservation of Momentum

7.4 Conservation of Momentum in One-Dimensional Collisions

7.5 Conservation of Momentum in Two- and Three-Dimensional Collisions

7.6 Changing Mass

7.7 Torque

7.8 Rotational Equilibrium

7.9 Angular Momentum

7.10 Conservation of Angular Momentum

7.11 Angular Acceleration

A WORD TO THE STUDENT

We first introduced the concept of momentum in our discussion of Newton's second law. Now we take a closer look at momentum, examining both linear and angular momentum. Each of these quantities is the basis of an independent conservation law and, together with conservation of mechanical energy, is often the starting point for analysis of motion. We illustrate the operation of these laws with several examples, including simple collisions. However, we postpone analyzing more complicated collisions, involving the solution of two or more equations simultaneously, until the next chapter.

As in our discussion of rotational kinetic energy in Chapter 6, here we emphasize the similarities between the equations of translational and rotational motion. We will use these concepts of motion and of momentum conservation in discussing many different topics in physics.

Nearly two decades before Newton's *Principia* was published, the Royal Society of London issued a request for experimental clarification of collision phenomena. Responses were received from several of Newton's contemporaries, including Sir Christopher Wren and Christiaan Huygens (1629–1695). Their observations led to the discovery of laws governing the exchange of momentum and energy between two colliding objects. These ideas were known to Newton and influenced his work. Their most important result was the law of conservation of linear momentum. According to this law, the total momentum after a collision is the same as the total momentum before the collision. This law made a key contribution to the growing understanding of mechanics.

The momentum that we have encountered in Newton's second law is often called linear momentum to distinguish it from the angular momentum associated with rotary motion. Just as linear momentum is often a conserved quantity, so is angular momentum. The independent laws concerning the conservation of energy, of linear momentum, and of angular momentum are among the most basic laws in contemporary physics. Although we will derive these laws from Newton's laws of motion, in some respects they are even more fundamental and far-reaching than Newton's laws. For example, even in situations where Newton's laws do not apply, such as speeds approaching the speed of light or dimensions on atomic scales, these conservation laws are still valid. The use of conservation laws is one of the most fundamental ways of describing nature.

In our discussion we point out the analogies between translation and rotation. As we stated earlier, any general motion of an object can be analyzed as an independent combination of translation and rotation, so these concepts underlie our description of all kinds of different physical behavior.

For simplicity, we focus on one conservation law at a time. In the next chapter we examine the interplay of energy conservation and momentum conservation, as well as the combined effects of translation and rotation.

Linear Momentum

7.1

The concept of momentum is extremely important in physics. Whenever we examine a moving object, we must consider both its mass and its velocity. The **linear momentum** of a body with mass m, traveling with velocity $\mathbf{v}$, is defined to be the product of the mass and the velocity. Momentum is associated with an object's translational motion. Since mass is a scalar quantity and velocity is a vector quantity, their product, momentum (which we designate with the letter $\mathbf{p}$), is a vector quantity:

$$\mathbf{p} \equiv m\mathbf{v}. \tag{7.1}$$

The word *momentum* (pl. *momenta*) is a Latin word meaning "movement" or "moving power."

Most of us are intuitively aware of the importance of momentum in understanding the behavior of moving objects. For example, consider the difference between being hit by a bicycle traveling at 10 m/s (22 mi/h) and being hit by a locomotive traveling at the same velocity. We see that the mass of an object is an important consideration! Velocity is also an important part of the object's motion; just think of the difference between being hit by a baseball thrown by a child and being hit by a baseball thrown by a major league pitcher.

Impulse

7.2

When a baseball hits a bat (Fig. 7.1), or when two billiard balls collide, they exert forces on each other over a very short time interval. Forces of this type, which exist only over a very short time, are called impulsive forces. We will examine such forces to see how they can be related to momentum.

In Chapter 4, we introduced Newton's second law in the form

$$\mathbf{F} = \frac{\Delta m\mathbf{v}}{\Delta t} = \frac{\Delta \mathbf{p}}{\Delta t},$$

where **F** is the net force applied to an object and $\Delta \mathbf{p}$ is its change in momentum during a time Δt. If we multiply both sides of this equation by the time interval Δt we get

$$\mathbf{F}\,\Delta t = \Delta \mathbf{p}. \qquad (7.2)$$

Figure 7.1   A high-speed photograph of a bat striking a baseball. The interaction takes place over a very small time interval.

The quantity on the left, $\mathbf{F}\,\Delta t$, is called the **impulse**. It is the product of the force **F** and the time interval Δt over which the force acts. Even though the force occurs very briefly, the force is not usually constant over the time interval (Fig. 7.2). For this reason, we often replace the force in Eq.

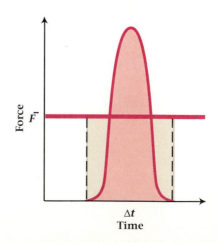

Figure 7.2   Average and instantaneous force during a typical brief collision between two moving bodies. The area under the curve of force versus time is equal to the impulse. Since the area of the rectangle whose height is the average force equals the area under the curve, we can replace the instantaneous force by the average force to obtain the impulse.

(7.2) with the average force $\overline{\mathbf{F}}$ over the time of interaction Δt:

$$\overline{\mathbf{F}} \, \Delta t = \Delta \mathbf{p}. \qquad (7.3)$$

The usefulness of impulse is illustrated in the two following examples.

Example 7.1

Force on a ball hit by a bat.

Using the following data, determine the force on a baseball hit by a bat. The baseball has a mass of 0.14 kg and an initial speed of 30 m/s. It rebounds from the bat with a speed of 40 m/s in the opposite direction and is in contact with the bat for 0.0020 s. (High-speed photographs such as Fig. 7.1 can be used to determine the contact time.)

Solution Since the mass of the ball is constant, we can rewrite Eq. (7.3) as

$$\overline{F} = \frac{\Delta p}{\Delta t} = \frac{\Delta mv}{\Delta t} = \frac{m \, \Delta v}{\Delta t}.$$

We choose the direction of v_{final} as positive; then v_{initial} must be negative. The change in momentum becomes

$$m \, \Delta v = m(v_{\text{final}} - v_{\text{initial}}) = 0.14 \text{ kg } [40 \text{ m/s} - (-30 \text{ m/s})]$$
$$= 0.14 \text{ kg} \times 70 \text{ m/s}.$$

The average force is then

$$\overline{F} = \frac{(0.14 \text{ kg})(70 \text{ m/s})}{0.0020 \text{ s}} = 4900 \text{ kg·m/s}^2 = 4900 \text{ N}.$$

Typically, the maximum force is much greater than the average force, as shown in Fig. 7.2.

Example 7.2

A 45-kg teenager jumps to the ground from a platform 1.0 m high (Fig. 7.3). If she bends her knees slightly on landing, lowering herself by only 2.0 cm, what is the average force with which her feet hit the ground?

Solution To determine the girl's momentum, we need to know the girl's speed just before striking the ground. We obtain it from the kinematic relation

$$v^2 = v_0^2 + 2gh,$$

where g is the gravitational acceleration, h is the height of the platform, v is the speed just before striking the ground, and v_0 is the initial speed. Since $v_0 = 0$, the speed v becomes $v = \sqrt{2gh}$.

The force of landing acts over the time during which she bends her knees to absorb the shock. If we assume the deceleration to be constant, then the slowing-down distance d is given by the product of the average speed and the time:

$$d = \frac{(v_\text{i} + v_\text{f})}{2} \, \Delta t.$$

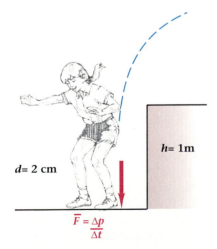

d= 2 cm

h= 1m

$$\overline{F} = \frac{\Delta p}{\Delta t}$$

Figure 7.3 Example 7.2: A person jumping from a platform to the ground bends her knees to absorb the shock.

Here the initial speed is $v_i = v$ and the final speed v_f is zero, so

$$d = \frac{v}{2}\,\Delta t.$$

The time over which the slowing-down, or knee-bending, process takes place is

$$\Delta t = \frac{2d}{v}.$$

The girl's momentum changes from its initial value at the moment her feet strike the ground to a value of zero at the end of the knee flexing. So we may write the change of momentum as

$$\Delta p = p(\text{final}) - p(\text{initial}),$$
$$\Delta p = 0 - mv.$$

Then the average impulsive force is

$$\overline{F} = \frac{\Delta p}{\Delta t}.$$

Upon substituting $-mv$ for Δp and $2d/v$ for Δt, we get

$$\overline{F} = \frac{-mv}{2d/v}$$

$$\overline{F} = \frac{-mv^2}{2d}.$$

Using $v = \sqrt{2gh}$, we find

$$\overline{F} = \frac{-m(\sqrt{2gh})^2}{2d}$$

$$\overline{F} = -\frac{mgh}{d}.$$

Inserting the numerical values, including $g = 9.8$ m/s^2, we get

$$\overline{F} = \frac{-(45\text{ kg})(9.8\text{ m/s}^2)(1.0\text{ m})}{0.020\text{ m}},$$

$$\overline{F} = -22{,}000\text{ N (about 5000 lb)}.$$

Notice that the impulsive force is inversely proportional to the knee-bending distance d. A greater amount of bending reduces the average force, while a smaller amount increases the force. If the girl had landed with her knees locked, the force would have been so large that she might have suffered one or more broken bones. The force that we have just calculated is the impulsive force that changes her momentum. In addition, there is a contribution due to her weight (even when she is standing still), but in this case it is small enough for us to neglect when compared with the impulsive force.

Newton's Laws and the Conservation of Momentum

7.3

We now introduce a fundamental principle known as the conservation of momentum. To do so we first consider the motion of a single particle or object. Then we examine the motion of a system of interacting objects.

If a single object of mass m is subject to zero net force, Newton's second law states that the rate of change of momentum with time is zero. This is equivalent to saying that *the object's momentum remains constant when the net force acting on it is zero.* That is, if

$$\frac{\Delta \mathbf{p}}{\Delta t} = \mathbf{F} = 0,$$

then

$$\mathbf{p} = m\mathbf{v} = \text{constant.}$$

Most interesting physical systems consist of two or more objects interacting with one another. For such systems we can apply Newton's laws to each individual object. In doing so we find that while the momenta of the individual objects may change, *the total momentum of the system is constant whenever the net external force on the system is zero.* The effect of the internal forces of interaction among the objects is to exchange momentum among them in such a way that the total momentum is conserved.

To see how this result arises, consider the behavior of the two interacting bodies in Fig. 7.4. We are free to consider each of them independently (part a) or to regard them together as a system (part b). We can write the total force on one body as the sum of the forces exerted from outside the system plus the force exerted on it by the other body inside the system. Let us write the total force on one body, say body A, as

$$\mathbf{F}_A(\text{total}) = \mathbf{F}_A(\text{internal}) + \mathbf{F}_A(\text{external}).$$

The force on the other body is

$$\mathbf{F}_B(\text{total}) = \mathbf{F}_B(\text{internal}) + \mathbf{F}_B(\text{external}).$$

The internal forces make up an action-reaction pair and, by Newton's third law, are equal in magnitude but opposite in direction. As a result, the sum of the internal forces is zero. Consequently, if we add up these two equations we find that the net force on the system is just the sum of the external forces on the two bodies:

$$\mathbf{F}(\text{total}) = \mathbf{F}_A(\text{external}) + \mathbf{F}_B(\text{external}).$$

The total rate of change of the system's momentum is equal to the sum of the external forces and is independent of the internal forces. We can express this result as

$$\mathbf{F}_{\text{net}} = \frac{\Delta \mathbf{p}}{\Delta t}, \tag{7.4}$$

where $\mathbf{F}_{\text{net}}$ is the net external force and $\mathbf{p}$ is the total momentum of the

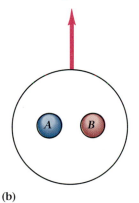

$\mathbf{F}_A$ (external) $\mathbf{F}_B$ (external)

$\mathbf{F}_A$ (internal) $\mathbf{F}_B$ (internal)

(a)

$\mathbf{F}$ (total) = $\mathbf{F}_A$ (external) + $\mathbf{F}_B$ (external)

(b)

Figure 7.4 A system of two interacting bodies. (a) Considered individually, each body is subject to a net force comprised of two forces, one external, coming from outside the system, and one internal, due to the other body. (b) Considered as a system, the net force on the two bodies is just the sum of the external forces.

system. Equation (7.4) is the extension of Newton's second law to a system of two bodies. By similar reasoning we can extend it to include systems of three, four, or any number of bodies. In all cases, **when the net external force on a system is zero, the total momentum of that system is constant**. This is a statement of the law of **conservation of linear momentum**.

As we said earlier, conservation of linear momentum is one of the most useful laws in physics, enabling us to determine the momentum of a system after an interaction without needing to know all the details of the interaction. We will illustrate the application of momentum conservation with examples in the next few sections.

Conservation of Momentum in One-Dimensional Collisions

7.4

As we said at the beginning of this chapter, Christiaan Huygens and others knew of the concept of momentum conservation even before Newton published the *Principia*. The results of Huygens's investigations of collisions independently led to the following statement of the law of conservation of linear momentum: *Provided there are no external forces acting on a system, the total momentum before collision equals the total momentum after collision.* For a collision involving two bodies, the conservation law can be expressed symbolically as

$$m_1\mathbf{v}_1 + m_2\mathbf{v}_2 = m_1\mathbf{v}_1' + m_2\mathbf{v}_2'. \qquad (7.5)$$

The subscripts indicate which of the two bodies is referred to; unprimed quantities stand for values before the collision and primed quantities stand for values after the collision. Although we have derived this general statement from theoretical considerations, it has been shown to be true experimentally as well.

Equation (7.5) is a statement of conservation of linear momentum that follows directly from Eq. (7.4). We will find it an especially useful version in problems involving collisions. Note that we do not need to know anything about the details of the collision mechanism itself. The rule holds for collisions between hard elastic bodies, such as billiard balls or glass spheres, as well as for collisions between soft bodies that do not "bounce" upon colliding, such as blobs of putty. In fact, cases in which the two objects stick together are an important class of collisions, known as **perfectly inelastic collisions**. A collision between two railroad cars that couple together upon impact is another example of a perfectly inelastic collision. Let us now analyze conservation of momentum in inelastic collisions. In the next chapter we will examine the slightly more complicated case of elastic collisions.

First, let's simplify our discussion to just head-on, or one-dimensional, collisions. Suppose, for example, that we consider the perfectly inelastic collision of two gliders on a linear air track (Fig. 7.5). The air track allows us to control the gliders' motions so that they move only in one direction.

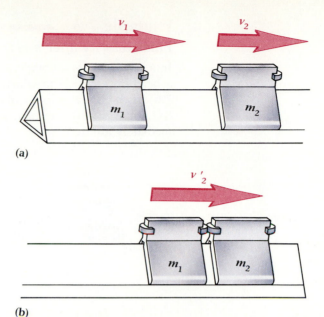

Figure 7.5 A one-dimensional collision of two gliders on an air track. (a) Before collision. (b) If putty or sticky tape is attached to their bumpers, the two gliders stick together. The resulting collision is completely inelastic.

This means that the vector equation for momentum conservation (Eq. 7.5) reduces to a single one-dimensional algebraic equation:

$$m_1 v_1 + m_2 v_2 = m_1 v_1' + m_2 v_2'.$$

If the collision is perfectly inelastic, the two bodies stick together and $v_1' = v_2'$. We can immediately determine the final velocity in terms of the masses and the initial velocities.

Suppose that the glider of mass m_1 is given an initial velocity v_1 and that glider m_2 has an initial velocity v_2 in the same direction. If $v_1 > v_2$, a collision occurs when glider m_1 overtakes glider m_2. If the gliders stick together after the collision, the final velocities v_1' and v_2' are identical. Then the momentum equation becomes

$$m_1 v_1 + m_2 v_2 = (m_1 + m_2) v_2'.$$

Solving for the final velocity v_2', we get

$$v_2' = \frac{m_1 v_1 + m_2 v_2}{m_1 + m_2},$$

$$v_2' = \frac{m_1}{m_1 + m_2} v_1 + \frac{m_2}{m_1 + m_2} v_2.$$

This last equation reveals the significance of the restriction to perfectly inelastic collisions. If we know the masses and initial velocities, we can compute the final velocity readily from momentum considerations alone. If the collision is not perfectly inelastic, so that the two bodies do not stick together and therefore have different velocities after collision, then the more general equation (Eq. 7.5) applies. In collisions that are not perfectly inelastic, knowing the masses and initial velocities is not enough to determine the final velocities—we need more information. We consider this point again in Chapter 8.

Example 7.3

Final velocity after a perfectly inelastic collision.

In a safety test of automobile equipment, two cars of equal mass undergo a head-on collision in which they stick together after the collision (Fig. 7.6). If the initial velocity of one was 5.0 km/h and that of the other was 8.0 km/h toward the first, what is the velocity of the combination immediately after collision?

$v_1 = 5.0$ km/h $v_2 = -8.0$ km/h

$$p = mv$$

v'

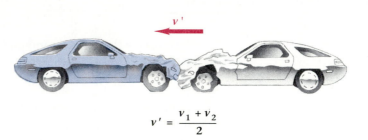

Figure 7.6   Example 7.3: A head-on collision between two cars of equal mass. Because momentum is the product of mass and velocity, after the collision the vehicles travel in the direction of the faster-moving car.

$$v' = \frac{v_1 + v_2}{2}$$

Solution   Because the cars are traveling in opposite directions, their velocities have opposite signs. We arbitrarily call the direction of the first car the positive direction, $v_1 = 5.0$ km/h. Then the direction on the second car is negative, $v_2 = -8.0$ km/h.

The masses are equal and can be canceled from the equation of momentum conservation, which then becomes

$$v_1 + v_2 = v_1' + v_2'.$$

After impact the cars stick together and thus travel together. As a result, the final velocity is the same for each car, so $v_1' = v_2' = v'$. We may then write

$$v_1 + v_2 = 2v',$$

or

$$v' = \frac{v_1 + v_2}{2},$$

$$v' = \frac{+5.0 \text{ km/h} - 8.0 \text{ km/h}}{2} = -1.5 \text{ km/h}.$$

The negative sign indicates that the final motion is in the same direction as the initial motion of the second car.

Example 7.4

Recoil velocity of an ice skater.

A 60-kg ice skater is standing at rest on a frozen lake. The friction between his skates and the surface of the ice is negligible. If he throws a 2.0-kg chunk of ice horizontally with a velocity of 12 m/s, what is his recoil velocity?

Solution This situation is essentially that of a perfectly inelastic collision run in reverse; that is, a system separates into two bodies with no net force acting on the system as a whole. (We can neglect the force of gravity because it acts along a direction perpendicular to the direction of separation.) So we can apply the law of conservation of momentum along the direction of motion.

The total momentum before the interaction is zero. Therefore the total momentum afterward must also be zero. Writing the equation for momentum conservation, we get

$$0 = m_1 v_1' + m_2 v_2'.$$

If $m_1 v_1'$ is the momentum of the ice, then the skater recoils with a velocity

$$v_2' = -\frac{m_1 v_1'}{m_2},$$

$$v_2' = -\frac{(2.0 \text{ kg} \times 12 \text{ m/s})}{60 \text{ kg}} = -0.40 \text{ m/s}.$$

The negative sign indicates that the skater's motion is in the direction opposite to that in which the ice was thrown.

Example 7.5

Momentum of a rocket and its payload.

A booster rocket and its payload are traveling at a speed of 900 m/s. An explosion separates the booster from the payload with a relative speed of 100 m/s, and the payload is thrown forward along the initial direction of motion. Find the velocity of the payload and of the booster immediately after they separate. Assume that the mass of the booster is four times that of the payload and that the effects of gravity are negligible.

Solution The payload is thrown forward, so momentum transfer to the booster must be in the backward direction. We can solve this problem by applying the principle of momentum conservation. If the mass of the payload is M, the mass of the booster is $4M$. Before the collision, both pieces are attached and travel with the same velocity $v_i = 900$ m/s. Thus the initial momentum is

$$\text{momentum before} = (M + 4M)v_i = (5M)v_i.$$

After the separation the momentum is

$$\text{momentum after} = Mv_p + (4M)v_b,$$

where v_p = payload velocity and v_b = booster velocity. Conservation of momentum tells us that

$$\text{momentum before} = \text{momentum after}$$

or

$$(5M)v_i = Mv_p + 4Mv_b.$$

We have two unknown quantities, v_b and v_p, so we need another equation relating them. This is the statement of relative speed:

$$\text{relative speed} = 100 \text{ m/s} = v_p - v_b.$$

Solving for v_p and inserting the result into the equation of momentum conservation, we get

$$5Mv_i = M(v_b + 100 \text{ m/s}) + 4Mv_b,$$

$$= 5Mv_b + 100M \text{ m/s},$$

which may be solved for the booster velocity:

$$v_b = v_i - 20 \text{ m/s} = 880 \text{ m/s}.$$

The payload velocity is 100 m/s faster, or

$$v_p = 980 \text{ m/s}.$$

Conservation of Momentum in Two- and Three-Dimensional Collisions

7.5

In the previous examples we limited ourselves to one-dimensional situations. However, in many common situations, such as billiard ball collisions, the objects move in different directions after the collision and it is necessary to consider two or three dimensions. Then the vector aspect of momentum becomes important. We first write out the general equations, simplify to two dimensions, and then consider a specific example. As before, provided that the net external force on the system is zero, we write the conservation rule as: *The momentum before collision equals the momentum after collision.*

In the general case, we could resolve each velocity vector in Eq. (7.5) into three mutually perpendicular components (v_x, v_y, v_z). For the vector equation to be true, the component equations must also be satisfied separately. As a consequence, the vector equation contains three independent statements, corresponding to its three components:

$$x \text{ component:} \quad m_1 v_{1x} + m_2 v_{2x} = m_1 v'_{1x} + m_2 v'_{2x}, \quad (7.6a)$$

$$y \text{ component:} \quad m_1 v_{1y} + m_2 v_{2y} = m_1 v'_{1y} + m_2 v'_{2y}, \quad (7.6b)$$

$$z \text{ component:} \quad m_1 v_{1z} + m_2 v_{2z} = m_1 v'_{1z} + m_2 v'_{2z}. \quad (7.6c)$$

We have used two subscripts on each velocity component. The first subscript indicates whether the velocity is for mass 1 or mass 2 and the second subscript indicates which component (x, y, or z) of velocity is being considered.

Consider the situation shown in Fig. 7.7. An object m_1 with an initial velocity $\mathbf{v}_1$ strikes an object m_2 initially at rest. These objects could be billiard balls, subatomic particles, whatever you like. After the collision, we observe the object that was initially in motion to be traveling with a velocity $\mathbf{v}'_1$ at an angle θ_1 with respect to the forward direction, which in this case is the x direction. Can we determine the speed and direction of the second object?

The directions of the velocity vectors $\mathbf{v}_1$ and $\mathbf{v}'_1$ define a plane, which we may choose as the xy plane. There is no initial momentum in the z direction and $\mathbf{v}'_1$ has no z component, so there can be no z component of velocity associated with $\mathbf{v}'_2$. Thus $\mathbf{v}'_2$ must also lie in the xy plane. Because all the motion is in the xy plane—that is, over a flat surface—the problem is two-dimensional only. If we choose the x direction as the initial direc-

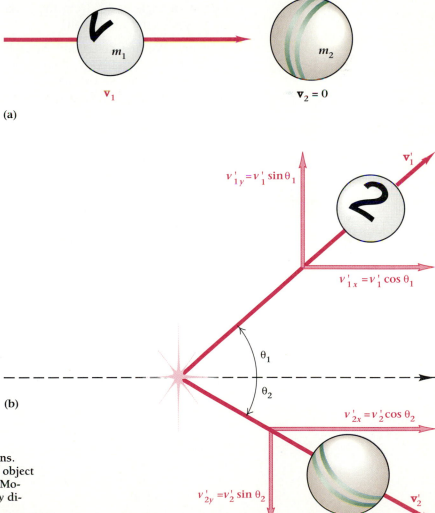

(a)

(b)

Figure 7.7   A collision in two dimensions.
(a) Object m_1 moving in the x direction, object m_2 at rest. (b) Velocities after collision. Momentum is conserved in both the x and y directions.

tion of $\mathbf{v}_1$, then the equations of momentum conservation reduce to

$$(p_x) \qquad m_1 v_{1x} = m_1 v'_{1x} + m_2 v'_{2x} \qquad (7.7a)$$

and

$$(p_y) \qquad 0 = m_1 v'_{1y} + m_2 v'_{2y}. \qquad (7.7b)$$

Because these two equations must hold at the same time, they are called simultaneous equations. The appendix to Chapter 8 contains a review of methods for solving simultaneous equations. An application of these equations to momentum conservation in two dimensions is illustrated in the following example.

Example 7.6

A billiard ball moving at 10.0 m/s along the positive x axis collides with a second billiard ball at rest. The balls have identical masses. After the collision, the incoming (or incident) ball moves on with a speed of 7.7 m/s at an angle of 40° from the x axis. What is the speed and direction of motion of the struck ball?

Solution We first determine the individual components of the final velocity of the second body, then calculate its direction of motion. In this example $m_1 = m_2$, so the mass may be eliminated in Eq. (7.7). Referring to the diagram in Fig. 7.7(b), we see that the x component of velocity of the incident ball after collision is

$$v'_{1x} = v'_1 \cos \theta_1.$$

Substituting this equation into Eq. (7.7a), we obtain the x component of velocity of the struck ball,

$$v'_{2x} = (v_1 - v'_1 \cos \theta_1).$$

The y component of motion of the second ball is determined in a like manner. First, from the diagram we have

$$v'_{1y} = v'_1 \sin \theta_1,$$

which gives, upon substitution into Eq. (7.7b),

$$v'_{2y} = -v'_1 \sin \theta_1.$$

The minus sign indicates that the motion is in the negative y direction. The final speed and direction of the second body are now

$$v'_2 = \sqrt{v'^2_{2x} + v'^2_{2y}} = \sqrt{(v_1 - v'_1 \cos \theta_1)^2 + (-v'_1 \sin \theta_1)^2},$$

$$\tan \theta_2 = \frac{v'_{2y}}{v'_{2x}} = \frac{-v'_1 \sin \theta_1}{v_1 - v'_1 \cos \theta_1}.$$

We can evaluate the speed and direction of mass 2 when the values for v_1, v'_1, and θ_1 are inserted into these equations. They give $v'_2 = 6.8$ m/s at $\theta_2 = 50°$ below the x axis. Note that we cannot determine the final velocities and angles from a knowledge of the initial conditions only, although it can be done in some special cases, as we will point out in Chapter 8.

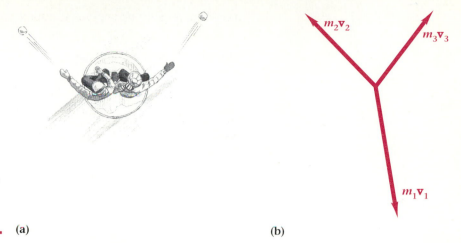

(a)   **(b)**

Figure 7.8   (a) Two occupants of a stationary ice sled simultaneously throw objects from the sled in different directions. (b) Momentum diagram for the situation of (a) after the objects have been thrown.

For a situation involving more than two bodies, we still have the same general principles. Figure 7.8 shows one such case, which is two-dimensional and involves the movement of three bodies. Two occupants of an ice sled simultaneously throw objects horizontally from the sled in different directions. The accompanying vector diagram shows the final momenta of the objects and the sled. To analyze the situation we would write x and y components for the momenta as before. Only now we would have three bodies in the equations, which would be extensions of Eqs. (7.6).

Changing Mass

7.6

In Chapter 4 we considered the dynamics of bodies with constant mass. Because the mass was constant it was easy, when given the force, to find the acceleration and then the subsequent motion of the body. However, in many practical cases the mass is not constant.

An example of a system with changing mass is the railroad car of Fig. 7.9. The car is in motion while taking on sand at a constant rate. For simplicity, we assume that the car has some initial velocity, v_1, and rolls freely without any friction. It is not connected to a locomotive.

The initial momentum p_1 is the product of the mass of the car M and

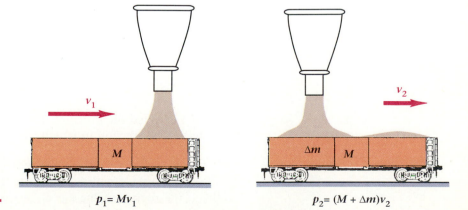

Figure 7.9   A railroad car in motion takes on sand from a stationary hopper. Momentum in the horizontal direction is conserved despite the system's changing mass.

$p_1 = Mv_1$ $p_2 = (M + \Delta m)v_2$

Figure 7.10 The basic principle of rocket propulsion: Gas expelled from the rear of the rocket causes it to move forward because of the conservation of momentum.

its initial velocity v_1,

$$p_1 = Mv_1.$$

Although the falling sand has zero initial speed in the horizontal direction, it is accelerated horizontally when it hits the moving car. The resulting action-reaction pair speeds up the sand while slowing down the car. After a time Δt, an amount of sand Δm has been added and the system (car plus sand) has a new velocity v_2. The corresponding momentum of the system is then

$$p_2 = (M + \Delta m)v_2.$$

The change in momentum is

$$p_2 - p_1 = Mv_2 + \Delta m v_2 - Mv_1 = M(v_2 - v_1) + \Delta m v_2,$$

which can be written as

$$\Delta p = M \, \Delta v + v_2 \, \Delta m.$$

If we divide through by Δt, we find the system's change in momentum per unit time is

$$\frac{\Delta p}{\Delta t} = M \frac{\Delta v}{\Delta t} + v_2 \frac{\Delta m}{\Delta t}.$$

But the change in momentum per unit time is equal to the net external force applied. Since there is no external force along the direction of motion, $\Delta p / \Delta t = 0$ and

$$M \frac{\Delta v}{\Delta t} = -v_2 \frac{\Delta m}{\Delta t}.$$

We see that $\Delta v / \Delta t$ is negative, as expected; the car is slowing down. If we wanted to keep the car moving with constant speed so that $v_1 = v_2$, then we would have to supply an external force $F = v_1 \, \Delta m / \Delta t$.

A second example of motion with changing mass is a rocket, whose mass diminishes as it consumes its fuel. Figure 7.10 represents the basic principle of rocket propulsion. Gases are expelled at high velocity from the rear of the rocket. Conservation of momentum for the rocket–gas system then requires a forward velocity of the rocket.

Because the force of gravity acts on an earthbound rocket, we cannot properly apply the principle of conservation of momentum without considering the earth–rocket system. For simplicity, let us imagine a rocket so far from the earth that we can neglect external forces. The rocket engine ejects gases at the rate of $\Delta m / \Delta t$ (where m is the mass of the fuel) at a velocity v_r with respect to the rocket. The velocity v_r is taken to be constant here. Thus the reaction force on the rocket is

$$F_R = v_r \frac{\Delta m}{\Delta t}. \tag{7.8}$$

This force is called the *thrust* of the rocket motor. The direction of $\mathbf{F}_R$ is opposite to the direction in which the gases are expelled.

Example 7.7

A fully fueled rocket with a total mass of 5000 kg is set to be fired vertically. If the rocket engine ejects its exhaust gases at a speed of 3.0 ×

10^3 m/s and burns fuel at the rate of 50 kg/s, what is the rocket's initial upward acceleration? Don't forget to include the effects of gravity.

Solution   The thrust F_R generated by the rocket engine is the burn rate times the exhaust velocity:

$$F_R = v_r \, \Delta m / \Delta t = 3.0 \times 10^3 \text{ m/s} \times 50 \text{ kg/s} = 1.5 \times 10^5 \text{ N}.$$

The rocket is acted upon by two forces: an upward force, F_R, given by the rocket engine, and a downward force, mg, due to gravity. The initial acceleration of the rocket is given by the net force divided by the initial mass:

$$a = \frac{F_{net}}{m} = \frac{F_R - mg}{m} = \frac{F_R}{m} - g = \frac{1.5 \times 10^5 \text{ N}}{5 \times 10^3 \text{ kg}} - 9.8 \text{ m/s}^2,$$

$$a = 20 \text{ m/s}^2 \text{ upward.}$$

7.7 Torque

Conservation of linear momentum applies only to translational motion of an object. However, there is an analogous conservation law that applies to rotational motion. To understand this conservation principle, let's first examine the dynamics of rotational motion.

A door hinged at one edge is a familiar object. We can pull on it in several ways (Fig. 7.11). However, the most effective way to open the door

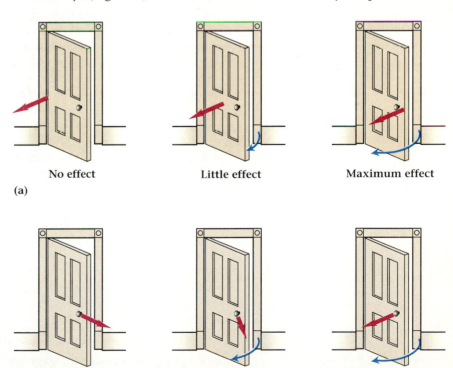

Figure 7.11   A door free to rotate about its hinges. (a) The force applied farthest from the hinges produces the greatest torque. (b) The force applied at right angles to the door produces the greatest effect.

(a)   No effect   Little effect   Maximum effect

(b)   No effect   Some effect   Maximum effect

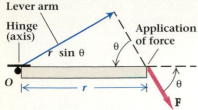

Figure 7.12   A force **F** applied at a distance r causes a torque $\tau = rF \sin \theta$ about the point O. The effective lever arm is $r \sin \boldsymbol{\theta}$.

is to grab the edge of the door farthest from the hinges and pull at right angles to the door. Intuitively, we realize that the distance from the hinge to the point of application of the force, as well as the magnitude and direction of the force, are important factors affecting the tendency of the door to rotate.

The quantity measuring how effectively a force causes rotation is called **torque.** The greater the distance from the axis of rotation (door hinges) to the point where we apply the force (door handle), the greater the torque. Also, maximum torque occurs when the direction of the applied force is perpendicular to a line drawn between the axis and the point where the force is applied. By contrast, when the line and the force are in the same direction, there is no torque.

Figure 7.12 illustrates the application of a force causing a door to rotate about an axis at point O, looking down from above the door. The torque τ about the point O is defined as

$$\tau \equiv rF \sin \theta, \tag{7.9}$$

where r is the magnitude of the displacement from the axis to the point of application of the force **F** and θ is the angle between the direction of the force and the direction of **r**. The maximum torque occurs when θ is 90°, that is, when **r** and **F** are perpendicular. When **r** and **F** are in the same direction, θ becomes zero and there is no torque.

We see from the figure that applying a torque is equivalent to applying a force perpendicular to a lever. The lever arm is defined as the perpendicular distance from the axis of rotation to the line of action of the force. From the geometry, we see that the lever arm is $r \sin \theta$. Therefore saying that the torque is the product of force times the lever arm ($r \sin \theta$) is consistent with Eq. (7.9). Thinking of torque as the product of a force and a lever arm often makes a problem easier to analyze.

Figure 7.13 shows two forces generating torques. In both cases the forces are at right angles to **r**. Although the resulting torques have the same magnitude, they tend to cause rotations in opposite directions. Thus torque is really a vector quantity, with both magnitude and direction. The magnitude is given by Eq. (7.9) and the direction is along the axis of rotation. This direction is perpendicular to the plane containing both the line of application of the force and the line from the axis to the point of application (Fig. 7.14a). The direction of the torque points along the direction a right-handed screw will move if **r** is rotated by **F**. Figure 7.14(b) shows an alternative way of establishing this direction.

The units of torque are the units of length multiplied by the units of force. Thus the SI units for torque are the newton-meter (N·m). In the British system, the units for torque are foot-pounds (ft-lb). The units of torque are the same as those for work and energy; however, torque and work represent very different physical quantities. Moreover, remember that torque is a vector, while work is a scalar.

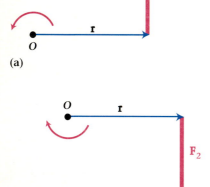

Figure 7.13   (a) Force $\mathbf{F_1}$ produces a torque causing a counterclockwise rotation about O. (b) Force $\mathbf{F_2}$ produces a clockwise rotation about O.

Example 7.8

Force needed to produce a given torque.

The instructions for replacing the head gasket on an automobile engine say that the bolts should be "torqued down" to 90 N·m. If you use a

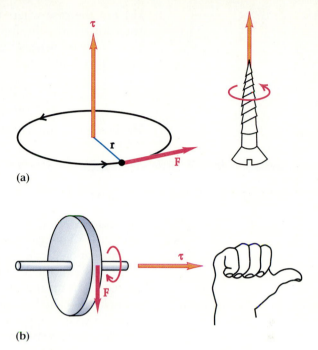

(a)

(b)

wrench that is 45 cm long, how much force must you apply in a direction
perpendicular to the wrench handle to accomplish this? (Too much torque
will pinch the gasket so that it will not seal properly.)

Solution   The physical situation is shown in Fig. 7.15. We are told that
the angle between the direction of the force and the line from the axis to
the point of application is 90°. Thus Eq. (7.10) becomes

$$\tau = rF$$

and

$$F = \frac{\tau}{r},$$

$$F = \frac{90 \text{ N·m}}{45 \text{ cm}} \times \frac{100 \text{ cm}}{1 \text{ m}} = 200 \text{ N}.$$

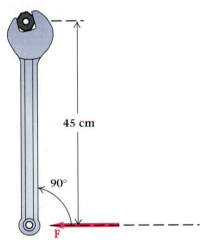

45 cm

90°

F

Figure 7.15   Example 7.8: Using a
wrench to tighten the nut on a bolt.
A force **F** is applied near the end of
the handle.

Example 7.9

Torque on a bicycle pedal.

The crank arm of a bicycle pedal is 16.5 cm long. If a 52.0-kg woman
puts all her weight on one pedal, how much torque is developed (a) when
the crank is horizontal and (b) when the pedal is 15° from the top?

Solution   (a) For the case where the crank arm is horizontal, the angle
between the downward force of the woman's weight and the crank arm
is 90°. The lever arm is the full extent of the crank arm, and the torque
is

$$\tau = rF = r(mg) = 16.5 \text{ cm } (52.0 \text{ kg} \times 9.81 \text{ m/s}^2) \ 10^{-2} \text{ m/cm},$$

$$\tau = 84.2 \text{ N·m}.$$

(b) For the case where the pedal is 15° from the top, the angle between the direction of the force and the direction of the crank arm is 15°, so that the lever arm is 16.5 cm × sin 15° = 4.27 cm = 0.0427 m. The resulting torque is

$$\tau = (\text{lever arm})(\text{force})$$
$$= (0.0427 \text{ m})(52.0 \text{ kg} \times 9.80 \text{ m/s}^2) = 22.1 \text{ N·m}.$$

Have you ever noticed this difference when pedaling a bicycle up a hill?

Rotational Equilibrium

7.8

We saw in Section 4.9 that an object acted on by two forces, equal in magnitude but opposite in direction, is not accelerated. We say that it is in translational equilibrium. However, an object in translational equilibrium may still rotate. A pair of forces, such as $\mathbf{F}_1$ and $\mathbf{F}_2$ of Fig. 7.16, that are equal in magnitude but opposite in direction *and not lying along the same line* is called a **couple**.

The couple applies a torque about O equal to the sum of the torques due to the individual forces. Note that although the forces are in opposite directions, they tend to rotate the body in the same direction about O. If the distance between the lines of application of the forces is s, the torque produced by the two equal forces is

$$\tau = \frac{s}{2} F_1 + \frac{s}{2} F_2.$$

However, since $F_1 = F_2 = F$, we may write

$$\tau = sF.$$

If we want to keep the object from rotating, we must subject it to another torque of the same magnitude but in the opposite direction. We may thus conclude that for a body to be in **rotational equilibrium** the sum of the torques must be zero, or

$$\sum_{i=1}^{N} \tau_i = 0, \tag{7.10}$$

where there are N individual torques τ_i.

The idea expressed in Eq. (7.10) is sometimes called the *second condition for equilibrium*. The first condition for equilibrium was that the vector sum of the forces on a body be equal to zero (Section 4.9). This condition ensures that there is no translational acceleration. The second condition ensures that there is no rotational, or angular, acceleration. (Angular acceleration is just the rotational analogue of linear acceleration; we will discuss it in more detail in Section 7.11.) An object satisfying both conditions of equilibrium is said to be in equilibrium. If such an object is stationary, it is in *static equilibrium;* if it is moving with neither translational nor rotational acceleration, it is in *dynamic equilibrium.*

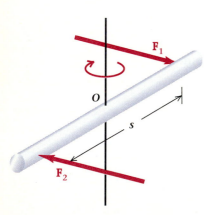

Figure 7.16 A couple. Two forces of equal magnitude act opposite and parallel to each other, but not along the same line. The rod is in translational equilibrium, but can still rotate.

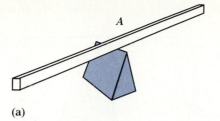

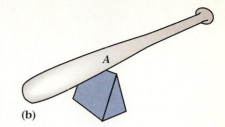

(a)   **(b)**

Figure 7.17 (a) A uniform rod balanced at its center of mass. (b) A nonuniform rod balanced at its center of mass.

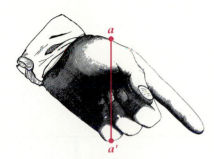

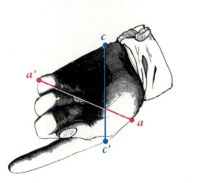

Figure 7.18 Locating the center of mass of a two-dimensional object.

Figure 7.17(a) shows a uniform rod, such as a meter stick, placed on a fulcrum. It balances at its midpoint, marked by A. A nonuniform rod, like a baseball bat (Fig. 7.17b), will also balance at some point, though not at its midpoint. In each case the upward force due to the fulcrum is equal to the entire weight of the object, as required by the first condition for equilibrium. In addition, the sum of the torques about the fulcrum due to gravity equals zero if the rod is in rotational equilibrium. In each case we can say that all the weight of the bar acts downward at A. The object behaves as if its mass were concentrated at a point that lies directly above A. This point is called the **center of mass** of the body.*

We can determine the center of mass of an extended two-dimensional object by the method shown in Fig. 7.18. The sheet metal figure is freely suspended from one point, a, and when it hangs motionless, we draw a vertical line, aa', directly downward from the suspension point. The center of mass of the body must lie somewhere along the line aa'; if it did not, the gravitational force would produce a torque about the point of suspension, causing a rotation. When the figure is suspended from another point on its edge (Fig. 7.18b), we repeat the procedure. The center of mass must lie somewhere along the new vertical line cc'. The center of mass is then located at the intersection of the lines aa' and cc'.

If a body is in static equilibrium, not only does it not rotate about some particular axis, but it does not rotate about any axis at all. We are free, therefore, to choose any possible rotation axis for the purpose of computing torques. Proper choice of the rotation axis can usually simplify the computations, since any force acting through the axis of rotation produces zero torque. Examples 7.10 and 7.11 illustrate this point.

Example 7.10

Summing torques in static equilibrium.

A 5.0-kg mass (m_2) and an unknown mass (m_3) hang from a 1.0-m rod of 2.0-kg mass (m_1), as shown in Fig. 7.19. The rod is supported on a knife-edge fulcrum at a distance 35 cm from one end. How large is the mass m_3 if the rod and masses are to balance on the knife-edge?

Solution As we said above, if the rod is in static equilibrium, not only does it not rotate about some particular axis, but it does not rotate about any axis at all. We are free to pick any point for the axis to go through.

*Strictly speaking, the *center of gravity* is the point at which the force of gravity can be considered to act. For spherical bodies, like the planets, that point is their geometric center. Frequently the center of gravity and the center of mass are the same. They are different only when the external gravitational field is not uniform over the body.

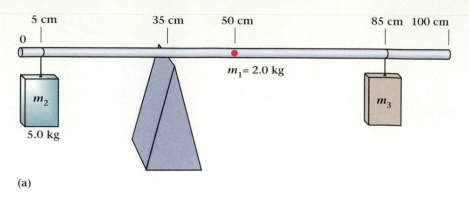

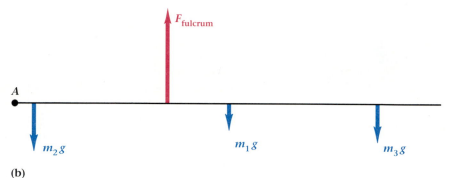

Figure 7.19   Example 7.10: A rod and two other masses in static equilibrium. (a) Physical situation. (b) A force diagram for the rod.

For this example we pick the left-hand end of the rod (point A on Fig. 7.19b).

We may consider all the weight of the rod to act downward at its center of mass, which in this case is at 50 cm. We then proceed as if a single mass m_1 equal to 2.0 kg (the mass of the rod) were hung from the rod at that point. Forces act at four points along the bar: the downward forces $m_1 g$, $m_2 g$, and $m_3 g$ at the points indicated and an upward force F at the fulcrum. The upward force must equal the total downward force in order to have translational equilibrium:

$$F_{\text{fulcrum}} = m_1 g + m_2 g + m_3 g.$$

We can utilize Eq. (7.10) by calling torques that tend to rotate things in a counterclockwise direction positive and those that tend to rotate things in a clockwise direction negative. The mass $m_1 = 5.0$ kg, $m_2 = 2.0$ kg, and m_3 is unknown. According to Eq. (7.10),

$$\sum \tau = \sum_i r_i F_i = 0.$$

Therefore the sum of the torques about point A is as follows:

$$-r_2 m_2 g + r_F F_{\text{fulcrum}} - r_1 m_1 g - r_3 m_3 g = 0,$$

$$-r_2 m_2 g + r_F (m_1 g + m_2 g + m_3 g) - r_1 m_1 g - r_3 m_3 g = 0.$$

Since g is a common factor, it can be divided out to give

$$-r_2 m_2 + r_F (m_1 + m_2 + m_3) - r_1 m_1 - r_3 m_3 = 0.$$

Next we insert the appropriate numerical values to get

$$-(0.05 \text{ m})(5.0 \text{ kg}) + (0.35 \text{ m})(5.0 \text{ kg} + 2.0 \text{ kg} + m_3)$$
$$- (0.50 \text{ m})(2.0 \text{ kg}) - (0.85 \text{ kg})(m_3) = 0,$$

$$-0.25 \text{ m·kg} + 2.45 \text{ m·kg} + 0.35 \text{ m}(m_3)$$
$$- 1.0 \text{ m·kg} - 0.85 \text{ m}(m_3) = 0.$$

Gathering terms, we get

$$(0.50 \text{ m})(m_3) - 1.20 \text{ m} \cdot \text{kg} = 0.$$

So

$$m_3 = 2.4 \text{ kg}.$$

Example 7.11

Torques about a different axis.

Calculate the mass m_3 in Example 7.10 by computing torques about the fulcrum.

Solution   If we choose the axis of rotation to be about the fulcrum, then we have only three torques to consider: the counterclockwise torque due to m_2 and the clockwise torques due to m_1 and m_3. The torque due to the fulcrum force becomes zero because the moment arm is now zero. The sum of the torques becomes

$$-r_2' m_2 g + r_1' m_1 g + r_3' m_3 g = 0.$$

Upon rearranging we find

$$m_3 = \frac{r_2' m_2 - r_1' m_1}{r_3'}.$$

From Fig. 7.19 we see that $r_2' = 30$ cm, $r_1' = 15$ cm, and $r_3' = 50$ cm. When the numerical values are inserted into the equation for m_3 we get

$$m_3 = \frac{30 \text{ cm} \times 5.0 \text{ kg} - 15 \text{ cm} \times 2.0 \text{ kg}}{50 \text{ cm}},$$

$$m_3 = 2.4 \text{ kg}.$$

As expected, we get the same answer as in Example 7.10. This time the equation was simpler and the computation reduced because we only had to consider three torques. The force acting through the point of rotation produced no torque and did not have to be considered.

Angular Momentum

7.9

Suppose you have a wheel mounted on an axle on which it is free to turn. To set the wheel turning you need to apply a torque to it about its axle to overcome the wheel's inertia. Once the wheel is set in motion, it continues to rotate at constant angular velocity until another torque is applied.

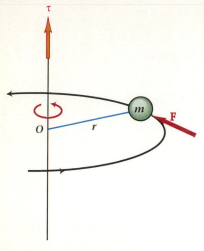

Figure 7.20   A mass constrained to move about a fixed point O at a distance r is subjected to a force $\mathbf{F}$. The resulting torque τ changes the angular velocity of the mass.

(We neglect the retarding effects of a torque due to friction.) In both cases, a torque causes the angular velocity to change. This situation is analogous to the application of a force to a body to change its linear velocity. In this section we develop this analogy further.

Consider a single particle of mass m that is free to move in a plane, at a fixed distance about the origin O (Fig. 7.20). This model might represent the physical situation of a ball bearing glued to one end of a soda straw of negligible mass, which is fitted to a friction-free pivot at the other end. A force applied to the ball bearing in a direction perpendicular to the straw causes the bearing to accelerate. The corresponding torque is

$$\tau = rF = r\left(\frac{\Delta p}{\Delta t}\right) = rm\,\frac{\Delta v}{\Delta t}.$$

However, because r and m are constant, we can write this as

$$\tau = \frac{\Delta(rmv)}{\Delta t}.$$

Thus the application of a given torque causes a proportional change in the quantity rmv. This equation is the rotational equivalent to Newton's second law, that the rate of change of momentum is proportional to the force applied.

We can rewrite the equation for the torque as the rate of change of a new quantity called the angular momentum L:

$$\tau = \frac{\Delta L}{\Delta t}. \qquad (7.11)$$

The **angular momentum** is the product of the linear momentum mv and the radius r:

$$L \equiv rmv. \qquad (7.12)$$

The quantity $L \equiv rmv$ is the magnitude of the angular momentum vector $\mathbf{L}$. Its direction is defined in a manner similar to the way in which we defined the direction of the torque in Fig. 7.14. That is, the angular momentum vector is at right angles to the plane containing $\mathbf{r}$ and $\mathbf{v}$. The direction of the angular momentum is given by the right-hand rule and is directed to the right for the wheel in Fig. 7.21.

We may use the definition of angular velocity, $\omega = v/r$ (Chapter 5), to write the magnitude of the angular momentum in terms of the angular

Figure 7.21   A wheel rotating on an axle. The direction of the angular momentum $\mathbf{L}$ is found from the right-hand rule. The fingers are curled so that they point along the direction of rotation; then the thumb points along the direction of the angular momentum.

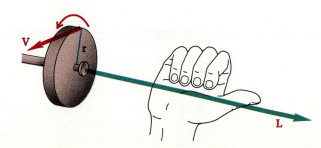

velocity ω:

$$L = mr^2\omega. \tag{7.13}$$

Then we can separate the angular momentum into two parts: one term that depends on the properties of the body—that is, its mass and shape—and another term that is the angular velocity. For a single point mass m at a fixed distance r from the axis of rotation, we can write

$$L = I\omega, \tag{7.14}$$

where I is the moment of inertia (defined in Section 6.6). For more general objects, Eq. (7.14) still holds with the appropriate moment of inertia $I = \Sigma mr^2$ (Table 6.3). Thus we see that angular momentum is the product of moment of inertia and angular velocity, in the same way that linear momentum is the product of mass and linear velocity.

Conservation of Angular Momentum

7.10

If the torque applied to a body is zero, the change in its angular momentum with respect to time is also zero. Thus the angular momentum is constant. This situation is analogous to the conservation of linear momentum. That is, if

$$\tau = \frac{\Delta L}{\Delta t} = 0,$$

then

$$\Delta L = 0,$$

and

$$L = I_i\omega_i = I_f\omega_f = \text{constant},$$

where the subscripts i and f stand for initial and final values, respectively.

When the net applied torque on an object is zero, its angular momentum is conserved. This statement of the law of **conservation of angular momentum**, like the conservation of linear momentum, expresses an extremely important physical principle. It applies not only to large-scale phenomena, but also to atomic and nuclear phenomena. It applies to objects moving in a curved path or simply spinning about an axis.

If an object rotates with a large angular momentum, its axis of rotation remains relatively stationary in space unless a large torque is applied transverse to the initial rotation axis. The upright stability of a spinning top is a consequence of the fact that the torque due to gravity about the support point produces a changing angular momentum at right angles to the initial angular momentum. The vector sum of these two momenta causes the vertical axis of the top to move in a nearly circular path. As the top slows down, its angular momentum becomes smaller and the torque causes it to fall over.

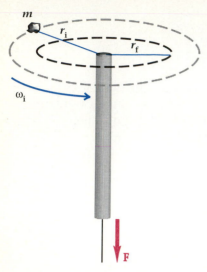

Figure 7.22 **Example 7.12:** A stone is whirled around in a horizontal plane.

Example 7.12

Conservation of angular momentum.

A stone attached to a string is whirled around in a horizontal circle (Fig. 7.22). If the stone is originally moving at a rate of 0.5 rad/s, what will its rate of revolution (i.e., its angular velocity) become if the radius of the circle is halved?

Solution   When the string is pulled in, the force acts through the axis of rotation. Thus there is no torque applied to the stone and the angular momentum is unchanged. The equation for conservation of angular momentum, with subscripts i and f standing for initial and final values, respectively, becomes

$$m r_i^2 \omega_i = m r_f^2 \omega_f,$$

or

$$\omega_f = \frac{r_i^2}{r_f^2} \omega_i.$$

Substituting the initial angular velocity of 0.5 rad/s and the final radius $\frac{1}{2} r_i$ into the equation gives

$$\omega_f = \left(\frac{r_i}{0.5 \, r_i} \right)^2 0.5 \text{ rad/s},$$

which reduces to

$$\omega_f = 2 \text{ rad/s}.$$

Note that when angular momentum is constant, decreasing the radius increases the angular velocity.

In many situations, an object's moment of inertia I may change. This is especially true in some sports activities. For example, in the case of a high diver, the only external force is due to gravity, which acts on the diver's center of mass. Thus there is no external torque acting on the diver and angular momentum is conserved. Since the product $I\omega$ remains constant, if I is made smaller, ω must increase. From the definition of I as $I = \Sigma mr^2$, we can see that the moment of inertia decreases when an object's mass is brought closer to the axis of rotation (smaller r). The diver does this by "tucking in" (Fig. 7.23). The diver's change in moment of inertia is accompanied by a corresponding increase in angular speed, allowing him to flip rapidly. The diver comes out of the tucked position to slow down his angular speed and enter the water vertically.

The motion of the diver combines rotational motion with translational motion. Careful observation shows that the diver's center of mass travels in a parabolic arc. This is the same path it would follow if it were a point mass. This is so because the net force (due to gravity in this case) acting on the diver can be considered to act on his center of mass. The motion of the diver about the center of mass is further complicated because location of the body's center of mass is affected by the positions of the limbs.

Figure 7.23 Motion of a diver.
(a) In the tucked position, a diver has a smaller moment of inertia and thus a greater angular velocity than at any other point in the dive. (b) The center of mass of the diver follows a parabolic path, even though the location of the center of mass relative to the body changes as the positions of the arms and legs change.

(a)

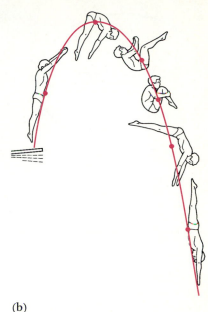

(b)

This effect can be seen in Fig. 7.23(b), where the center of mass is represented by the dot drawn on the diver.

Example 7.13

Conservation of angular momentum of an ice skater.

An ice skater starts spinning at a rate of 1.5 rev/s with arms extended. She then pulls her arms in close to her body, resulting in a decrease of her moment of inertia to three quarters of the initial value. What is the skater's final angular velocity?

Solution The skater's motion of her arms produces no external torque. Therefore, if we use the subscript i to denote initial values and f to denote final values, we have, from conservation of angular momentum,

$$I_i \omega_i = I_f \omega_f,$$

or

$$\omega_f = \frac{I_i}{I_f}\, \omega_i.$$

We are told that $I_f = \frac{3}{4}I_i$ and that the skater's initial frequency is $f = 1.5$ rev/s. Since the angular velocity is $\omega = 2\pi f$, we have

$$\omega_f = \frac{I_i \times 2\pi \times 1.5 \text{ rad/s}}{\frac{3}{4}I_i},$$

$$\omega_f = 4\pi \text{ rad/s},$$

which corresponds to a rotational rate of two rotations per second.

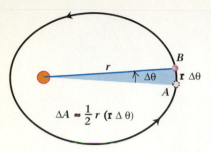

$$\Delta A \approx \tfrac{1}{2} r\,(r\,\Delta\theta)$$

Figure 7.24 **Example 7.14: A portion of the orbit of a planet moving about the sun. During a time Δt the planet moves, making an angle $\Delta\theta$ with the sun.**

Example 7.14

Show that Kepler's law of equal areas is equivalent to the law of conservation of angular momentum. (This result was, in fact, obtained by Newton in the early part of the *Principia*.)

Solution   Figure 7.24 shows a planet of mass m in an orbit about the sun. We do not need to know anything about the force on this planet other than that it is directed along the line joining the planet and the sun, as required by the law of universal gravitation. As the planet moves from point A to point B in a time Δt, it sweeps out an area ΔA, which is approximately triangular. The area A of a triangle is given by

$$A = \tfrac{1}{2}(\text{base})(\text{altitude}).$$

For the triangle of Fig. 7.24 we have

$$\Delta A \approx \tfrac{1}{2}(r)(r\,\Delta\theta),$$

or

$$\Delta A \approx \tfrac{1}{2}r^2\,\Delta\theta.$$

The rate at which the area is swept out, $\Delta A/\Delta t$, is then given by

$$\frac{\Delta A}{\Delta t} \approx \frac{1}{2}\,r^2\,\frac{\Delta\theta}{\Delta t},$$

which becomes an equality in the limit of small Δt. In this limit, the quantity $\Delta\theta/\Delta t$ becomes the angular velocity ω. Thus

$$\frac{\Delta A}{\Delta t} = \frac{1}{2}\,r^2\omega,$$

which may be rewritten in the form

$$\frac{\Delta A}{\Delta t} = \frac{mr^2\omega}{2m}.$$

Since $mr^2\omega = L$, the angular momentum, we have

$$\frac{\Delta A}{\Delta t} = \frac{L}{2m}.$$

If L is constant, then the rate at which area is swept out is constant.

The dominant force on each planet is the gravitational attraction of the sun, which is a radially directed force that produces no torque. Forces between the planets are very much smaller. Thus the torque on the planets in their orbits is essentially zero and the angular momentum L of any planet is constant. Our result predicts that for a given planet, $\Delta A/\Delta t$ is a constant; that is, the rate at which the planet sweeps out an area is constant. We have just derived Kepler's second law (equal areas in equal time) from the law of conservation of angular momentum.

An additional observation may be made. The conservation of angular momentum not only means that the magnitude of the angular momentum is constant, but also that its direction in space is constant. The constancy of direction leads to the conclusion that the plane of a planet's orbit is fixed in space.

Angular Acceleration

7.11

We have already seen that the application of a net torque causes a change in an object's rotational motion. To set a wheel into motion, for example, or to change its rotational velocity, a torque is required. We described the effects of torque earlier with Eq. (7.11), which related torque to the rate of change of angular momentum. The angular momentum, in turn, can be expressed as the product of the moment of inertia with the angular velocity, which leads to

$$\tau = \frac{\Delta(I\omega)}{\Delta t}.$$

If the moment of inertia is constant, as it is for a wheel spinning about its axis, then the equation for the torque becomes

$$\tau = I\frac{\Delta\omega}{\Delta t}.$$

The quantity $\Delta\omega/\Delta t$, the rate of change of the angular velocity with time, is the **average angular acceleration**, $\overline{\alpha}$:

$$\overline{\alpha} \equiv \frac{\Delta\omega}{\Delta t}. \tag{7.15}$$

As with instantaneous linear acceleration, the instantaneous angular acceleration α is found from $\Delta\omega/\Delta t$ in the limit as Δt becomes vanishingly small. Substituting α for $\Delta\omega/\Delta t$ in the equation for torque above, we can write

$$\tau = I\alpha. \tag{7.16}$$

Equation (7.16) is the rotational equivalent of Newton's second law. Notice the similarity between it and the relation $F = ma$. Equation (7.16) may be expressed in words as follows: An object's angular acceleration is proportional to the applied torque and the proportionality constant is the moment of inertia of the object. The torque corresponds to the force in Newton's law, the angular acceleration corresponds to the linear acceleration, and the moment of inertia plays the same role as the mass. When the angular acceleration is constant, we can derive a set of equations involving angular displacement and angular velocity that correspond exactly to the kinematic equations developed in Chapter 2 for the case of constant linear acceleration. Table 7.1 lists these analogous linear and rotational equations.

Imagine a point on a rotating object, say a wheel, that is located a distance r from the rotation axis. It has an instantaneous tangential velocity $v = r\omega$ and a centripetal acceleration $a_c = \omega^2 r$. If a torque acts on the wheel, there will be an angular acceleration α, causing a change in the wheel's angular velocity. Since there is a corresponding change in the tangential velocity of the point, there must also be an acceleration tangent to the path of the point. This tangential acceleration is

$$a_t = r\alpha.$$

TABLE 7.1
Summary of the linear and rotational equations

Linear		Rotational	
Kinematics			
$\bar{v} = \Delta x/\Delta t$	(Ch. 2)	$\bar{\omega} = \Delta\theta/\Delta t$	(Ch. 5)
$\bar{a} = \Delta v/\Delta t$	(Ch. 2)	$\bar{\alpha} = \Delta\omega/\Delta t$	(Ch. 7)
$v^2 = v_0^2 + 2a(x - x_0)$	(Ch. 2)	$\omega^2 = \omega_0^2 + 2\alpha(\theta - \theta_0)$	
$v = v_0 + at$	(Ch. 2)	$\omega = \omega_0 + \alpha t$	
$x = x_0 + \frac{1}{2}(v_0 + v)t$	(Ch. 2)	$\theta = \theta_0 + \frac{1}{2}(\omega + \omega_0)t$	
$x = x_0 + v_0 t + \frac{1}{2}at^2$	(Ch. 2)	$\theta = \theta_0 + \omega_0 t + \frac{1}{2}\alpha t^2$	
Conditions for equilibrium			
$\sum_i \mathbf{F}_i = 0$	(Ch. 4)	$\sum_i \tau_i = 0$	(Ch. 7)
Momentum			
$\mathbf{p} = m\mathbf{v}$	(Ch. 4)	$L = I\omega$	(Ch. 7)
Kinetic energy			
$KE = \frac{1}{2}mv^2$	(Ch. 6)	$KE_{rot} = \frac{1}{2}I\omega^2$	(Ch. 6)

This table contains analogous linear (translational) and rotational equations. They were first introduced in the chapters indicated. Some expressions that were not derived in the text have been included because you may find them useful in other applications.

The instantaneous acceleration of the point is the vector sum of the radial acceleration a_c and the tangential acceleration a_t.

Example 7.15

A cylindrical winch of radius R and moment of inertia I is free to rotate without friction about an axis (Fig 7.25a). A cord of negligible mass is wrapped about the shaft and attached to a bucket of mass m. When the bucket is released, it accelerates downward as a result of gravitational attraction. Find the acceleration of the bucket.

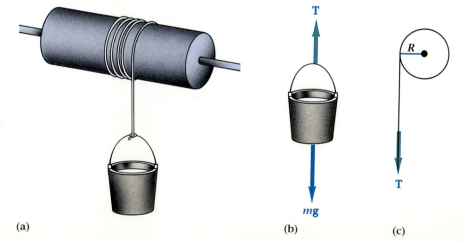

Figure 7.25 Example 7.15: (a) A cylinder rotates about its symmetry axis. A cord is wound about the cylinder and a bucket is suspended from the end of the cord. (b) Free-body diagram for the bucket of mass m. (c) Free-body diagram for the torque about the axis of the cylinder.

(a)

(b)

(c)

Solution   We first examine the forces on the bucket, shown as a free-body diagram in Fig. 7.25(b). The downward gravitational force is *mg* and the upward force *T* is the tension in the cord. If we choose the downward direction for the positive acceleration, then by Newton's second law

$$ma = mg - T,$$

and *a* is positive when *mg* exceeds *T*.

The cylinder is subject to a torque, given by *RT* (Fig. 7.25c). Its angular acceleration is given by

$$\tau = RT = I\alpha.$$

When the bucket moves down a distance *s*, a point on the circumference of the cylinder must also move through a distance *s*. Similarly, if the bucket accelerates an amount *a*, so does a point on the edge of the cylinder. The tangential acceleration *a* of a point on the circumference of the cylinder and the angular acceleration of the cylinder are related in the same way as the tangential velocity and the angular velocity are related. That is,

$$a = R\alpha.$$

We may use this relationship to eliminate α in the torque equation:

$$RT = \frac{Ia}{R},$$

or

$$T = \frac{Ia}{R^2}.$$

We may insert this last expression into the equation for the acceleration of the bucket. In that case

$$ma = mg - \frac{Ia}{R^2}.$$

Upon rearranging we get

$$\left(m + \frac{I}{R^2}\right)a = mg$$

or

$$a = \frac{mg}{m + \dfrac{I}{R^2}} = \left(\frac{1}{1 + \dfrac{I}{mR^2}}\right)g.$$

This last equation gives the acceleration of the bucket. The effect of the cylinder's inertia is to reduce the acceleration of the bucket. When *I* becomes very large, *a* becomes very small. When *I* becomes small, *a* approaches *g*. The acceleration reverts to that for free fall in the limit that the inertia of the cylinder approaches zero.

SUMMARY

Useful Concepts

This chapter introduces two fundamental conservation laws of physics:

- *Conservation of linear momentum:* The total linear momentum of a system is constant whenever the net external force on the system is zero.
- *Conservation of angular momentum:* The total angular momentum of a system is constant whenever the net external torque on the system is zero.
- Linear momentum is given by

$$\mathbf{p} = m\mathbf{v}.$$

- An impulsive force is related to the change in momentum by

$$\mathbf{F}\,\Delta t = \Delta\mathbf{p}.$$

- For two bodies in collision, the statement of conservation of momentum becomes

$$m_1\mathbf{v}_1 + m_2\mathbf{v}_2 = m_1\mathbf{v}_1' + m_2\mathbf{v}_2'.$$

- In the case of a rocket engine with a changing mass m of burning fuel, the thrust is

$$F_{\mathrm{R}} = v_{\mathrm{r}}\frac{\Delta m}{\Delta t}.$$

This is a form of Newton's second law with variable mass, $F = \Delta(mv)/\Delta t$.

- The torque about a point is

$$\tau = rF\sin\theta.$$

- Counterclockwise torques are positive, clockwise torques are negative.
- The conditions for equilibrium are:

1. The vector sum of all the external forces must be zero:

$$\Sigma\,\mathbf{F}_{\mathrm{i}} = 0.$$

2. The sum of all the external torques must be zero:

$$\Sigma\,\tau_{\mathrm{i}} = 0.$$

- If the object in equilibrium is at rest, neither translating nor rotating, the equilibrium is static. If the object is in motion, but not accelerating linearly or rotationally, the equilibrium is dynamic.

- A torque produces a change in an object's angular momentum $L = rmv$,

$$\tau = \frac{\Delta L}{\Delta t}.$$

- The angular momentum of a point mass m moving in a circle of radius r with angular velocity ω is

$$L = mr^2\omega.$$

- For an extended body the angular momentum is

$$L = I\omega,$$

where I is the moment of inertia, $I = \Sigma\,mr^2$.

- Angular acceleration is the time rate of change of angular velocity:

$$\overline{\alpha} = \Delta\omega/\Delta t.$$

- For rotation about a fixed axis of a body with a constant moment of inertia, the torque is related to the angular acceleration by

$$\tau = I\alpha.$$

- The equations developed for rotational motion can be obtained from the corresponding equations for linear motion if we replace x by θ, v by ω, a by α, F by τ, and m by I.

Important Terms

You should be able to write the definition or meaning of each of the following terms:

- linear momentum
- impulse
- conservation of linear momentum
- perfectly inelastic collision
- torque
- couple
- rotational equilibrium
- center of mass
- angular momentum
- conservation of angular momentum
- angular acceleration

QUESTIONS

7.1   Discuss the usage of the word *momentum* in everyday conversation and compare it with the physical definition of momentum.

7.2   Explain the conservation of momentum that occurs when a fast-moving baseball is struck by a bat.

7.3   A golf club strikes a stationary golf ball and sends it flying. Which has the greater momentum change? Why?

7.4   Under what conditions can a tiny compact car have the same momentum as a large station wagon?

7.5   A golf ball is thrown hard at a brick wall. A lump of soft clay with the same mass as the golf ball is thrown against the wall with the same initial velocity. Which of these events delivers the greater impulse to the wall?

7.6   What is the function of seat belts and air bags in automobiles? How do they reduce injuries?

7.7   When a balloon is blown up and released, it flies about as the air escapes. What makes it go?

7.8   Carefully distinguish between force and torque. Give examples of forces without torques and forces that produce torques.

7.9   Why do car owners go to the trouble to balance automobile tires? What happens when car wheels are unbalanced? Why?

7.10 Can a diver pull into a tuck and rotate while diving if he leaves the diving board with no angular velocity? Why?

7.11 What would happen to the planets if the gravitational force had a tangential component as well as a radial component?

7.12 The velocity of a bullet fired from a rifle held against the shooter's shoulder is measured very carefully. The rifle is then clamped to a massive bench so that it has no measurable recoil. How does that affect the velocity of the bullet?

7.13 Explain how ice skaters can quickly go from a slow to a fast spin and vice versa. What happens to their angular velocity and their moment of inertia? Is an external torque required?

7.14 Explain how a yo-yo works.

7.15 A cat held upside down and dropped can right itself before it hits the floor (Fig. 7.26). Explain how the cat does so, since there are no external torques present. (*Hint:* Study the figure.)

7.16 How does a balancing pole help a tight-rope walker?

7.17 Suppose you are holding the far end of a long garden hose of uniform diameter which extends straight from a faucet. Do you feel any force when the water is turned on? Would it make any difference if you bent the end of the hose through a 90° angle?

7.18 The gondola of a hot-air balloon carries a heavy flywheel mounted with its axis of rotation perpendicular to the earth's surface. Before lifting off, the wheel is at rest with respect to the gondola. After lifting off, the wheel is set into motion by an electric motor. Would an observer on the ground be able to tell that the wheel was set into motion?

Figure 7.26   Question 7.15.

PROBLEMS

Hints for Solving Problems
Collisions can be treated in terms of the final and initial states. When there is no net force, the final momentum equals the initial momentum. When there is no net torque, the final angular momentum equals the initial angular momentum. If you choose counterclockwise torques to be positive, then clockwise torques are negative. A net torque causes a change in angular momentum with time, which is equivalent to angular acceleration times the moment of inertia. Remember the correspondence between linear quantities and rotational quantities.

Section 7.1   Linear Momentum

7.1   What is the momentum of an 1800-kg automobile traveling at 35.0 km/h?

7.2   What is the momentum of an ocean liner of mass 3.0×10^7 kg when it is approaching a dock at a speed of 0.50 cm/s?

7.3   What is the momentum of a proton traveling at one hundredth the velocity of light? (The mass of a proton is 1.67×10^{-27} kg; the velocity of light is 3.0×10^8 m/s.)

7.4   (a) What is the momentum of a 100-kg football player running at a top speed of 10 m/s? (b) What is the momentum of a 2.36-g rifle bullet traveling at 382 m/s?

Section 7.2   Impulse

7.5   A 0.14-kg baseball was hit with an average force of 4500 N. It was thrown with a speed of 25 m/s and had a speed of 32 m/s after it was hit. How long was it in contact with the bat?

7.6   A 0.046-kg golf ball is hit in such a way that its speed is 60 m/s immediately after being struck. If the ball is in contact with the club head for 8.0×10^{-4} s, what is the average force on the ball?

7.7   A 0.14-kg baseball with an initial speed of 28 m/s rebounds with a speed of 34 m/s after being struck with a bat. If the duration of contact between ball and bat was 2.1 ms, what was the average force between ball and bat?

Section 7.4   Conservation of Momentum in One-Dimensional Collisions

7.8   A 78-kg ice hockey player standing on a frictionless sheet of ice throws a 6.0-kg bowling ball horizontally with a speed of 3.0 m/s. With what speed does the hockey player recoil?

7.9   A 3.5-kg rifle fires a 10-g bullet with a velocity of 850 m/s. What is the recoil velocity of the rifle?

7.10   A child playing marbles shoots a marble directly at another marble at rest. The first marble stops and the second marble continues in a straight line with the same speed that the first marble had initially. What is the ratio of the masses of the two marbles?

7.11   A 2.0-kg mass with a speed of 0.50 m/s collides head-on with a 1.5-kg mass moving with a speed of 0.30 m/s toward the first mass. After the collision the 2.0-kg mass stops. What is the speed of the second mass after the collision?

7.12   A 0.20-kg model railroad car moving with a speed of 0.24 m/s is struck from behind by an 0.42-kg model locomotive moving along the same line with a speed of 0.52 m/s. If they stick together after the collision, what is their velocity?

7.13   A baseball is thrown against a target that is free to move and rebounds from it with a speed 0.80 of its initial speed. The mass of the target is 20 times that of the baseball. If the baseball had an initial speed of 28 m/s, how far will the target move in 0.10 s?

7.14   Hans Brinker, with a mass of 59 kg, is standing in the middle of a frozen lake of frictionless ice. He wants to reach the shore in the shortest possible time without skating. Hans has two 3-kg snowballs. He can throw snowballs with a relative speed of 10 m/s, regardless of their mass. Determine whether Hans should throw both masses at one time or throw one mass and then throw the other mass one second later. (If this is not a realistic question, state what is wrong with the assumptions made.)

Section 7.5   Conservation of Momentum in Two- and Three-Dimensional Collisions

7.15   Two cars approach each other along streets that meet at a right angle. They collide at the intersection. After the collision they stick together. If one car has a mass of 1300 kg and an initial speed of 2.25 m/s and the other has a mass of 1800 kg and an initial speed of 4.50 m/s, what will be their speed and direction immediately after impact?

7.16   Two cars approaching each other at right angles collide and stick together. One car has a mass of 1200 kg and a speed of 30 km/h in the positive x direction before the collision. The second car has a mass of 1500 kg and was traveling in the positive y direction. After the collision the two cars move off at an angle of 64° to the x axis. What was the speed of the heavier auto?

7.17 A railroad track lies alongside a frozen lake. A railroad train moves along the track with a constant speed of 7.0 m/s. A boy on a frictionless ice sled is initially moving parallel to the train track with the same speed as the train and 10 m away from it. The boy and sled together weigh 100 kg and the boy carries a 5.0-kg bag of sugar with him. At some time the boy tosses the sugar to a girl on the train with a velocity of 2.0 m/s perpendicular to the direction of his motion. The girl on the train catches the bag of sugar and immediately throws it back to the boy, who catches it. The girl on the train throws the bag of sugar perpendicular to her motion with the same speed at which she caught it. What is the final direction and speed of the boy and sled?

Section 7.6   Changing Mass

7.18 A 24-ton freight car rolls along a straight track with a speed of 10 mi/h. A 2.0-ton automobile falls vertically and lands on it from above and then falls off after a few moments. What is the final speed of the railroad car?

7.19 What is the thrust of a rocket that burns fuel at a rate of 1.3×10^4 kg/s if the exhaust gases have a velocity of 2.5×10^3 m/s with respect to the rocket?

7.20 The initial mass of a rocket is 2.6×10^6 kg. A fuel-burning rate of 1.0×10^4 kg/s gives an initial acceleration of 1.5 m/s². What is the velocity of the exhaust gases? Assume the rocket is in free space.

Section 7.7   Torque

7.21 If a force of 4 N is needed to open a 1-m wide door when applied at the edge opposite the hinges, what force must be applied to open the door if you push against the door 10 cm from the hinged side?

7.22 If a person can apply a maximum force of 50 lb, what is the minimum length of a wrench needed to apply a 35-ft-lb torque to the bolts on a motorcycle engine?

7.23 A clock has a second hand whose tip rubs against the inside of the glass cover. If the frictional force between the glass cover and the tip of the second hand is .0020 N and the length of the hand is 8.0 cm, what is the minimum torque that must be applied to the second hand if the clock is not to be stopped?

Section 7.8   Rotational Equilibrium

7.24 A claw hammer is used to pry up a nail. Approximately how much force is applied to the nail when a force of 100 N is applied to the handle as shown in Fig. 7.27?

7.25 A person with upper arm vertical and forearm horizontal holds a 4.5-kg iron cannon ball (Fig. 7.28). Assume the mass of the forearm and hand is 1.5 kg with a center of mass 15 cm from the elbow. The

center of the cannon ball is 32 cm from the elbow and the force of the biceps is applied 5.0 cm from the elbow. (a) What force is exerted on the forearm by the biceps muscle? (b) What force is exerted on the upper arm at the elbow contact?

7.26 A bar balances 30.0 cm from one end. When a 0.75-kg mass is hung from that end, the balance point moves 8.0 cm toward that end. What is the mass of the bar?

7.27 When the front wheels of a truck are run onto a platform scale the scale reads W_1. When the rear wheels are run onto the scale so that the front wheels are off, it reads W_2. (a) Prove that the total weight of

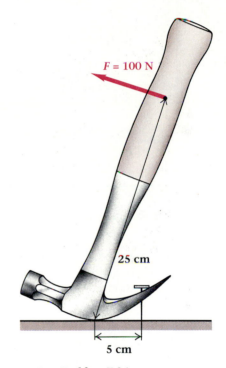

F = 100 N

25 cm

5 cm

Figure 7.27   Problem 7.24.

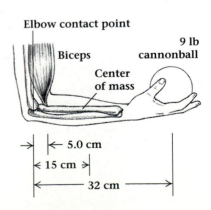

Elbow contact point

Biceps

Center of mass

9 lb cannonball

5.0 cm

15 cm

32 cm

Figure 7.28   Problem 7.25.

the truck is $W_1 + W_2$. (b) Prove that if the truck is loaded so that its center of gravity is halfway between the front and rear wheels, the total weight is $2W_1$.

7.28 Two children are playing on a balanced seesaw. One child with a mass of 42 kg sits 1.5 m from the center. Where on the other side must the second child sit if her mass is 34 kg?

7.29 A 4.0-m-long iron bar of uniform cross section is held perpendicular to the wall by a wire joined from the end of the bar to the wall (Fig. 7.29). What is the tension in the wire if the iron bar weighs 400 N and the wire is 5.0 m long?

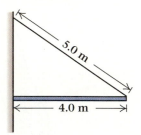

Figure 7.29 Problem 7.29.

Section 7.10 Conservation of Angular Momentum

7.30 A 0.5-kg flashlight is swung at the end of a string in a horizontal circle of 0.80-m radius with a constant angular speed. If no torque is applied, what must the radius become if the angular speed of the flashlight is to be halved?

7.31 A toy airplane moves at the end of a string about a fixed point in a horizontal circle of 1.0 m radius. If the linear speed is 5.0 m/s, what will the speed become if the string is pulled up to give a radius of 0.75 m? Assume there is no torque.

7.32 A 12-in. record is dropped on a large 12-in. phonograph turntable that is freely turning at $33\frac{1}{3}$ rev/min. The mass of the record is 50 g and the mass of the turntable is 1.00 kg. What is the final speed of the turntable in revolutions per minute? (*Hint:* In this case you may add the angular momenta in a manner analogous to adding linear momenta.)

Section 7.11 Angular Acceleration

7.33 A torque of 12 N·m is applied to a heavy wheel whose moment of inertia is $I = 30$ kg·m². (a) What is the angular acceleration of the wheel? (b) If the wheel was initially at rest and the torque is applied for 10 s, what would be the rotational frequency ω of the wheel at the end of the 10 s?

7.34 An iron disk has a radius of 0.500 m and a mass of 300 kg. The disk is mounted on its axis so that it is free to spin. (a) What torque is required to give it an angular acceleration $\alpha = 1.00$ rad/s²? (b) If the

torque is applied at the edge of the disk, how much force is required? (c) How much force is required if the force is applied at a distance 5 cm from the axis of rotation?

7.35 A wheel whose moment of inertia is 30 k·gm² is subjected to a torque of 10 Nm. If the wheel is initially moving with an angular velocity $\omega = 6$ rad/s when the torque is applied, what will be its angular velocity if the torque is applied for 6 s?

7.36 A wheel of radius R and moment of inertia I is subjected to a torque τ for a time t. If the wheel was initially at rest at $t = 0$, what is its angular velocity at time t? Also calculate the tangential speed of the outer edge of the wheel.

7.37 A wheel starts from rest and rotates about its axis with constant angular acceleration. After 6.0 s have elapsed it has rotated through an angle of 25 radians. (a) What is the angular acceleration of the wheel? (b) What is the angular velocity when the time $t = 6.0$ s? (c) What is the centripetal acceleration of a point on the wheel a distance $r = 0.45$ m from the axis at $t = 6.0$ s?

Additional Problems

7.38 A 0.046-kg golf ball was hit with an impulsive force
• that averaged 8000 N. If the ball was in contact with the club head for 5.0×10^{-4} s, what was the speed of the ball immediately after impact?

7.39 Two toy locomotives approach each other along the
• same line, and upon collision both stop dead still. If one locomotive has three times the speed of the other and the sum of their masses is 2.88 kg, what is the mass of each locomotive?

7.40 A bowling ball collides with another bowling ball at
•• rest. The first ball rebounds with a speed one tenth of its original speed. The second ball moves in the same direction as the first and with a speed one fifth of the original speed of the first ball. What is the ratio of the masses of the balls?

7.41 A 1.0-kg toy car moves freely along a model railroad
•• track with a speed of 0.50 m/s. An additional 0.20-kg mass is carried on the car in a device that projects the mass horizontally away from the car with a speed of 0.20 m/s relative to the initial motion of the car. Calculate the final velocity of the car along the track when the 0.20-kg mass is projected (a) in the same direction as the car's motion, (b) in the opposite direction from the car's motion, and (c) at right angles to the car's motion.

7.42 A 3000-kg rocket and its 500-kg payload are traveling
• at a speed of 2000 m/s. The payload is thrown forward by an explosion that separates it from the rocket with a relative velocity of 140 m/s. (a) What is the velocity of the payload after separation? (b) What is the velocity of the rocket?

7.43 A bowling ball with initial velocity of v_0 strikes a
• stationary bowling ball a glancing blow. The first ball
goes off in a direction of 30° from the initial direction
with a speed of 4.0 m/s. The second ball recoils in a
direction of 45° from the initial direction with a speed
of 3.0 m/s (Fig. 7.30). What is the ratio of the mass
of the struck ball to that of the incident ball?

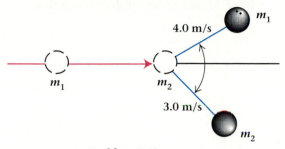

Figure 7.30 Problem 7.43.

7.44 A 900-kg car collides with a 1400-kg car that was
• initially at rest at the origin of an xy coordinate system. After the collision the lighter car moves at 20
km/h in a direction of 40° with respect to the positive
x axis. The heavier car moves at 10 km/h along the
positive x axis. What was the initial speed and direction of the lighter car?

7.45 A 1000-kg car collides with a 1200-kg car that was
•• initially at rest at the origin of an xy coordinate system (Fig. 7.31). After the collision the lighter car
moves at 20 km/h in a direction of 30° with respect
to the positive x axis. The heavier car moves at 12
km/h at $-44°$ with respect to the positive x axis.
What was the initial speed and direction of the lighter
car?

7.46 A rocket with initial mass of 8.0×10^3 kg is fired
•• vertically. Its exhaust gases have a relative velocity of
2.5×10^3 m/s and are ejected at a rate of 40 kg/s.
(a) What is the initial acceleration of the rocket?
(b) What is its acceleration after 20 s have elapsed?

7.47 A rocket with initial mass of 8.0×10^3 kg is fired in
•• the vertical direction. Its exhaust gases are ejected at
the rate of 45 kg/s with a relative velocity of $2.4 \times$
10^3 m/s. What is the initial acceleration of the
rocket? What is the acceleration after 25 s have
elapsed?

7.48 A rocket with initial mass of 5.0×10^3 kg is fired
•• vertically. Its exhaust gases are ejected at the rate
of 30 kg/s with a relative velocity of 3.0×10^3 m/s.
(a) Find the initial acceleration of the rocket. (b) How
high does the rocket climb in 10 s? (*Hint:* Ignore the
change in mass to perform this calculation.) (c) How
would the answer to part (b) change if you included
the change in mass of the rocket?

7.49 In Example 7.10, determine the mass m_3 by calculat-
• ing torques about a point 0.50 m from the left-hand
end of the bar.

7.50 In Example 7.10, determine the mass m_3 by calculat-
• ing torques about a point 0.10 m to the left of the
right-hand end of the bar.

7.51 Figure 7.32(a) shows a person's outstretched arm
•• holding a 5.0-kg dumbbell. The deltoid muscle is attached so that the force is applied at an angle of 17°
from the horizontal at a point halfway between the
shoulder joint and the center of mass of the arm (Fig.
7.32b). If the mass of the arm is 7.0 kg, what must
be the tension in the deltoid muscle?

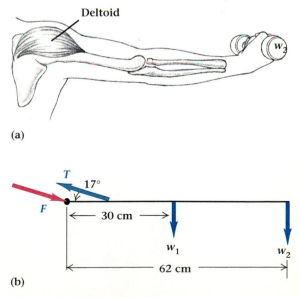

Figure 7.32 Problem 7.51.

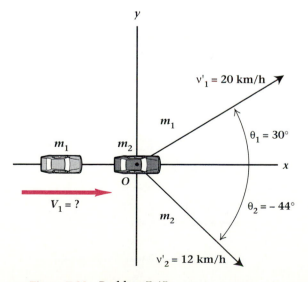

Figure 7.31 Problem 7.45.

7.52 Figure 7.33 is a diagram of the jaw. Chewing is accomplished by the force of the masseter muscle closing the jaw about the fulcrum A. The distance $x_2 = 3x_1$. If the muscle exerts a force of 400 N, what force is applied by the front teeth?

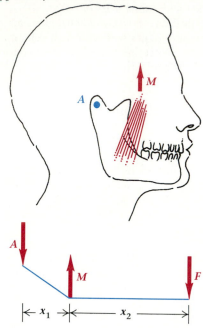

Figure 7.33   Problem 7.52.

7.53 A sign weighing 400 N is suspended at the end of a uniform rod 4.00 m long weighing 500 N (Fig. 7.34). (a) What is the tension in the support cable if it makes an angle $\theta = 40°$ with the rod? (b) What would be the tension if the cable were moved so that $\theta = 55°$?

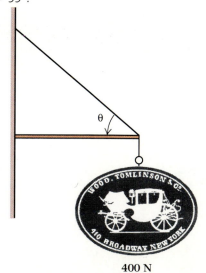

400 N

Figure 7.34   Problem 7.53.

7.54 Suppose the sign in Fig. 7.34 is suspended from the center of the boom rather than at the end. The cable is still connected to the end of the boom. What would be the tension in the cable if it makes an angle $\theta = 40°$ with the bar?

7.55 Find the vertical and horizontal components of force exerted by the wall on the bar suspended as shown in Fig. 7.35. The weight of the bar is 300 N.

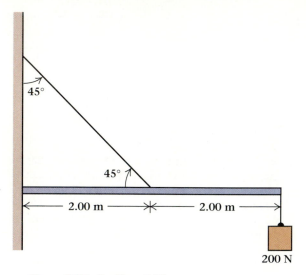

Figure 7.35   Problem 7.55.

7.56 A car engine idles at 800 rev/min. With the car in neutral, you depress the accelerator. After 1.2 s the tachometer indicates an engine speed of 3400 rev/min. (a) What are the initial and final angular velocities in rad/s? (b) What is the average angular acceleration during the 1.20-s interval? (c) How many revolutions does the engine make during that 1.20 s?

7.57 A yo-yo resting on a horizontal table is free to roll (Fig. 7.36). When the string is pulled horizontally to the left, the yo-yo rolls to the left. When the string is pulled vertically, the yo-yo rolls to the right. Prove that the yo-yo will slide without rolling when the string is pulled at an angle θ given by $\sin \theta = r/R$.

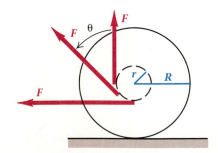

Figure 7.36   Problem 7.57.

7.58 A 75-kg marine does pushups as shown in Fig. 7.37.
• What are the forces on his hands and feet?

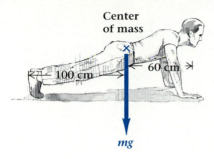

Center
of mass

← 100 cm → 60 cm →

mg

Figure 7.37   Problem 7.58.

7.59 In a student experiment, a torque of 2.0×10^{-2}
• N·m is applied to a rigid aluminum pipe, causing it to move about an axis through its center and perpendicular to its length with an acceleration $\alpha = 0.43$ rad/s^2. (a) What is the moment of inertia of the pipe? (b) The rigid pipe is made from light tubing with a mass of 0.20 kg inserted in each end. How far apart are the two masses? (*Hint:* Refer to Table 6.3.)

7.60 An unpowered flywheel is slowed by a constant fric-
• tional torque. At time $t = 0$ it has an angular velocity of 200 rad/s. Ten seconds later its velocity has decreased by 15%. What is its angular velocity at (a) time $t = 50$ s and (b) $t = 100$ s?

7.61 A playground merry-go-round has a disk-shaped plat-
•• form that rotates with negligible friction about a vertical axis. The disk has a mass of 200 kg and a radius of 1.8 m. A 36-kg child rides at the center of the merry-go-round while a playmate sets it turning at

0.25 rev/s. If the child then walks along a radius to the outer edge of the disk, how fast will the disk be turning?

7.62 Two masses, one of value m and the other of $2m$,
•• hang from light strings wrapped around a uniform solid cylinder of mass M and radius R that is free to rotate about a horizontal axis (Fig. 7.38). Find the acceleration of the masses when the cylinder is released from rest. Neglect effects of friction and the mass of the string.

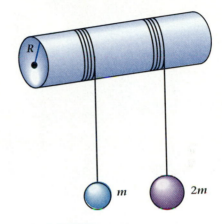

Figure 7.38   Problem 7.62.

7.63 (a) Show that the kinetic energy of a satellite in
•• circular orbit of radius r about the earth is one half the magnitude of its gravitational potential energy. (b) What is its total mechanical energy?

ADDITIONAL READINGS

Brody, H., "The Moment of Inertia of a Tennis Racket." *The Physics Teacher,* April 1985, p. 312.

Burger, W., "The Yo-Yo: A Toy Flywheel." *American Scientist,* March 1984, p. 137.

Edge, R. D., "Murphy's Law or Jelly-Side Down." *The Physics Teacher,* September 1988, p. 392.

Frohlich, C., "The Physics of Somersaulting and Twisting." *Scientific American,* March 1980, p. 154.

Laws, K., "Physics and Dance." *American Scientist,* September 1985, p. 426.

Walker, J., "The Amateur Scientist: The Physics of the Follow, the Draw, and the Massé (in Billiards and Pool)." *Scientific American,* July 1983, p. 124.

8

Combining Conservation of Energy and Momentum

8.1   Definition of Elastic Collisions

8.2   Elastic Collisions in One Dimension

8.3   Elastic Collisions in Two Dimensions

8.4   Rolling Bodies

*8.5   Braking a Moving Car

*8.6   Motion in a Gravitational Field

*8.7   Escape Velocity

App.   Solving Simultaneous Equations

A WORD TO THE STUDENT

In this chapter we bring together the laws of conservation of energy, linear momentum, and angular momentum. You are already familiar with the significant ideas themselves; this chapter shows you how to use them in combination and extend their range of usefulness. We have included several topics that could have been covered in earlier chapters, such as collision problems and motion in a gravitational potential. However, we have gathered these topics here to emphasize the usefulness and unity of the conservation laws. In some cases we apply these laws individually, but in others we must apply them simultaneously to solve particular problems. Even though the concepts are not difficult, when both laws are required for a problem the amount of algebra needed to get a final answer can require a lot of work. Even simple collisions between billiard balls can give rise to a great deal of algebraic manipulation. For this reason, you should be particularly careful to study and fill in all the mathematical steps in the examples. In that way you will be able to handle the assigned problems.

We have emphasized several times in the past two chapters that the application of conservation laws can solve an immense range of physical problems. Indeed, their importance becomes evident as you see how many different situations from all areas of natural science are governed by the same conservation laws. The collision of two subatomic particles and the orbiting of the sun by a comet obey the same conservation rules for energy and momentum, even though the important force is different in the two cases. The transformation of energy between living cells and the transformation of the chemical energy of fuel to warm your home obey similar principles, though we will have to extend our ideas of energy to understand thermal processes (Chapters 14 and 15).

In this chapter we analyze in some detail several different physical situations. The common threads that bind these analyses together are the principles of conservation of energy and conservation of momentum. Because we frequently apply these principles simultaneously, we often have to solve simultaneous equations and manipulate these equations algebraically. However, we never go beyond routine algebra and elementary trigonometry. The mathematics may sometimes become tedious, but there are no principles that you have not already used.

The type of analysis we show here is similar in some respects to how physicists approach a new problem. The starting point is a simplified model of the physical situation (with assumptions and restrictions stated). Conserved quantities, such as energy or momentum, are carefully noted. The conservation laws are then used to get a solution. In fact, this type of analysis from conservation laws has led to the discovery of new subatomic particles, as we will see in Chapters 28 and 31.

Definition of Elastic Collisions

8.1

We have already encountered examples of momentum conservation in Chapter 7. There, for simplicity, we considered perfectly inelastic collisions, in which two colliding objects stick together after impact. We now wish to broaden our understanding to include collisions in which the colliding objects rebound from each other.

Consider a ball of mass m dropped straight down onto a hard surface so that it rebounds straight up (Fig. 8.1, p. 230). Let us label the initial height from which it was dropped as h, and the height to which it rebounds as h'. We show that the ratio of the magnitude of the ball's velocity just after impact to its magnitude immediately before impact can be expressed solely in terms of h' and h.

If the ball is dropped from rest, its total energy immediately before impact with the surface is equal to the total energy it had at the top, provided that we can neglect air resistance:

$$PE_{bottom} + KE_{bottom} = PE_{top} + KE_{top}.$$

The kinetic energy at the top is zero because the ball starts from rest. The kinetic energy at the bottom is then equal to the difference in potential energy between the top and the bottom:

$$KE_{bottom} = PE_{top} - PE_{bottom}.$$

If we choose the hard surface to be the zero level of potential energy, then the potential energy at the bottom is zero and the potential energy at the point of release—that is, at the height h—is mgh. The kinetic energy at the bottom thus becomes

$$\tfrac{1}{2}mv^2 = mgh.$$

Rearranging this equation, we obtain

$$v = \sqrt{2gh}.$$

The speed v is the speed immediately before impact.

Similar analysis for the upward path gives the speed v' immediately after impact in terms of the maximum rebound height h' as

$$v' = \sqrt{2gh'}.$$

The ratio of the speed immediately after impact to the speed just before impact is

$$\frac{v'}{v} = \sqrt{\frac{h'}{h}}.$$

If the ball were to rebound to its initial height (that is, to $h' = h$), then we would say that the collision is elastic. In this case v' would be the same as v and the kinetic energy immediately after the collision would be the same as the kinetic energy immediately before the collision. We call a collision **elastic** when the kinetic energy is conserved. If the kinetic energy is not conserved, then the collision is called **inelastic**. Note that inelastic collisions do not have to be perfectly inelastic. The colliding objects may have some recoil and loss of kinetic energy at the same time. In inelastic collisions the total energy is still conserved, but some of the initial kinetic energy is converted into other forms, such as thermal energy.

When we studied momentum conservation in Chapter 7 we considered perfectly inelastic collisions, in which the colliding objects did not rebound at all, but stuck together after the collision. Momentum was conserved in these inelastic collisions, but kinetic energy was not. At the other extreme are the elastic collisions, in which *both* kinetic energy and momentum are conserved simultaneously. Most collisions are neither elastic nor perfectly inelastic. Car crashes and bouncing balls are prime examples of such collisions (Fig. 8.2). However, many situations can be approximated as either elastic or perfectly inelastic. The collisions of hard spheres, such as billiard balls, are nearly elastic. A blob of modeling clay thrown against the wall is an example of a perfectly inelastic collision.

Christiaan Huygens observed in 1669 that when hard bodies collided, the sum of the quantity mv^2 for each body before the collision was the same as the sum after the collision. He noted this conservation rule at the same time that he observed conservation of momentum. The conservation of kinetic energy upon collision of hard spheres was one of a series of

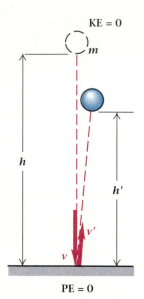

Figure 8.1   A ball of mass m dropped onto a hard surface from a height h rebounds to a height h'. If $h' = h$, the collision is elastic and kinetic energy is conserved.

Figure 8.2   Successive bounces of a golf ball on a hard surface, showing the loss of energy with each bounce.

conclusions that were drawn from observation of actual behavior of a particular class of colliding bodies.

The laws of conservation of energy and momentum have been validated countless times over the years. Today they have become such a fundamental part of our belief about nature that they are used as analytical tools in experiments, rather than being the subjects of experiments to test their validity. However, conservation of kinetic energy alone is not always sufficient to give an answer. Kinetic energy may be exchanged for potential energy or, as we shall see in Chapter 24, for mass. The rule for the conservation of total energy includes energy in all its forms. Because we believe that energy and momentum are conserved, we can often obtain information about a collision or other interaction without ever observing it directly. Instead, we measure energy and momentum of objects before and after the interaction and determine facts about the nature of the interaction itself.

Elastic Collisions in One Dimension

8.2

We begin our analysis of elastic collisions by considering a collision in one dimension and applying the laws of conservation of momentum and conservation of kinetic energy to this situation. We will expand our analysis to two dimensions in the next section.

Figure 8.3 shows two masses constrained to move along the x axis only. The first object, of mass m_1, is traveling in such a way that it will collide head-on with the second object, of mass m_2, which may or may not be at rest. We choose the positive x direction to be the direction of travel of the first body. For interactions in one dimension, we can use the law of conservation of momentum for two bodies to write

$$m_1 v_1 + m_2 v_2 = m_1 v_1' + m_2 v_2'. \qquad (8.1)$$

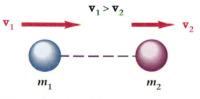

Figure 8.3 An object of mass m_1 travels with velocity $\mathbf{v}_1$ prior to a head-on collision with an object of mass m_2 and velocity $\mathbf{v}_2 < \mathbf{v}_1$.

Here v_1 and v_2 are the initial velocities of the first and second objects, respectively, while v_1' and v_2' are their velocities immediately after the collision. If the collision is elastic we can also use the conservation of kinetic energy to get

$$\tfrac{1}{2}m_1 v_1^2 + \tfrac{1}{2}m_2 v_2^2 = \tfrac{1}{2}m_1 v_1'^2 + \tfrac{1}{2}m_2 v_2'^2 \qquad (8.2)$$

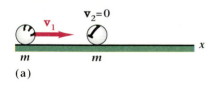

(a)

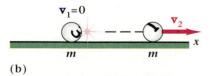

(b)

Figure 8.4 A head-on collision of two billiard balls. (a) Before impact; (b) after impact.

We can then use Eqs. (8.1) and (8.2) to predict the results of any straight-line, elastic collision between two bodies.

A case of special interest, and one well known to billiard players, is the one-dimensional elastic collision between objects of equal mass in which one object (m_2) is at rest before the collision (Fig. 8.4). Because collisions between billiard balls are very nearly elastic, we describe them by assuming elastic conditions. For simplicity, we also neglect the effects of rotations. For the situation $m_1 = m_2$ and $v_2 = 0$, the momentum conservation equation simplifies to

$$v_1 = v_1' + v_2', \tag{8.3}$$

and the kinetic energy equation becomes

$$v_1^2 = v_1'^2 + v_2'^2. \tag{8.4}$$

We can solve these two equations simultaneously by substitution, giving the final velocities v_1' and v_2' in terms of the initial velocity. You should show for yourself that one result is

$$v_1' = 0,$$

$$v_2' = v_1.$$

This result says that the first ball stops and the second ball moves with the same velocity that the first ball had initially. Equations (8.3) and (8.4) are also satisfied by the values $v_1' = v_1$ and $v_2' = 0$. These values are not useful, however, because they are the same as the initial conditions and do not correspond to a collision.

It is important to understand that these results depend on the *simultaneous* application of the law of conservation of momentum (Eq. 8.1) and the law of conservation of energy (Eq. 8.2). Neither law alone is sufficient to predict the results of a scattering process. (Because a body is usually scattered from its initial direction by a collision, we often refer to collisions as scattering.) The conservation laws also hold for elastic collisions in which the masses are unequal and one or both bodies are initially in motion. The following examples illustrate the application of these conservation laws.

Example 8.1

An elastic collision.

Two railroad boxcars are initially in motion to the right along the same straight line, which we denote as the x axis with the positive direction to the right (Fig. 8.5). The ratio of the masses of the two cars is $m_1/m_2 = 1/2$. The rearmost boxcar m_1 has a velocity twice as great as the forward boxcar m_2. The faster-moving car m_1 overtakes the other car and collides with it. After the collision, what is the velocity of each boxcar if the collision is elastic?

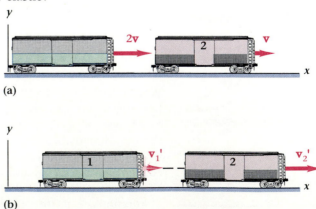

Figure 8.5   Example 8.1: Elastic collision between two boxcars of unequal mass.

Solution   This example illustrates a common situation. We will find that we get two sets of answers that satisfy our mathematical formulation of the laws of conservation of energy and momentum. We must then take a careful look at the situation in order to select the answer that corresponds to physical reality. We first obtain the mathematical solutions and then consider their physical meaning. You should fill in the missing algebraic steps for yourself.

If the initial velocity of the second body m_2 is v, then the initial velocity of the first body m_1 is $2v$. If we write $m_1 = m$ and $m_2 = 2m$, the law of conservation of momentum becomes

$$(m)(2v) + (2m)(v) = mv_1' + (2m)v_2'.$$

This equation may be simplified as

$$4v = v_1' + 2v_2'.$$

Conservation of kinetic energy gives

$$\tfrac{1}{2}(m)(2v)^2 + \tfrac{1}{2}(2m)(v)^2 = \tfrac{1}{2}mv_1'^2 + \tfrac{1}{2}(2m)v_2'^2,$$

or

$$6v^2 = v_1'^2 + 2v_2'^2.$$

We can rearrange the momentum equation to give v_1' in terms of the other velocities. Then we substitute v_1' into the energy equation to give

$$3v_2'^2 - 8vv_2' + 5v^2 = 0.$$

We now have a quadratic equation in v_2'. (The initial velocity v is a constant.) If we write the general form of a quadratic equation in x as

$$ax^2 + bx + c = 0,$$

where a, b, and c are constants, then the general solutions are given by the quadratic formula:

$$x = \frac{-b \pm \sqrt{b^2 - 4ac}}{2a}.$$

The $\pm$ sign indicates that there are two solutions: one for the plus sign and one for the minus sign. Both solutions must be investigated when we apply the quadratic formula to the equation for v_2'. In this case $a = 3$, $b = -8v$, and $c = 5v^2$. The solutions are

$$v_2' = \frac{+8v \pm \sqrt{(-8v)^2 - 4(3)(5v^2)}}{2(3)} = \frac{8v \pm 2v}{6}.$$

If we choose the plus sign we find

$$v_2' = \tfrac{5}{3}v.$$

If we choose the minus sign we find

$$v_2' = v.$$

Both of these values cannot represent the physical situation, since only one event actually happens. However, we must keep both until we can eliminate one of them by physical considerations. Upon substituting $v_2' =$

v into the momentum equation we get

$$v_1' = 2v.$$

Upon substitution of $v_2' = \frac{5}{3}v$ into the momentum equation we have

$$v_1' = \frac{3}{2}v.$$

Thus from the purely algebraic standpoint we have two pairs of possible answers:

$$\{v_1' = \tfrac{2}{3}v \quad \text{and} \quad v_2' = \tfrac{5}{3}v\}$$

or

$$\{v_1' = 2v \quad \text{and} \quad v_2' = v\},$$

both of which satisfy the conservation equations.

The resolution of the problem comes from a careful look at what the answers mean. Remember that this is a straight-line collision and that boxcar 1 is the rearmost car. Physically it must always remain the rearmost car because it cannot pass through the other car. This means that after the collision, its velocity in its initial direction must be less than the velocity of car 2. Therefore the correct answer is $v_1' = \frac{2}{3}v$ and $v_2' = \frac{5}{3}v$. Note that the other answer corresponding to both cars maintaining their initial velocities, but requiring boxcar 1 to pass through boxcar 2 without collision, is unphysical. Because the equation for conservation of kinetic energy is quadratic, we always obtain two sets of answers; however, one set corresponds to the initial conditions and can be discarded.

Example 8.2

Elastic collisions of two unequal air-track gliders.

In an experiment on a linear air track, a glider of mass m moving with initial speed v_0 to the right collides in an elastic collision with a glider of mass $2m$ initially at rest. What are the final velocities of the two gliders?

Solution   You should sketch a diagram for yourself before proceeding.

We apply conservation of momentum to get

$$mv_0 = mv_1' + 2mv_2',$$

where v_1' is the final velocity of the light glider and v_2' is the final velocity of the heavy glider. Conservation of energy gives

$$\tfrac{1}{2}(m)v_0^2 = \tfrac{1}{2}(m)v_1'^2 + \tfrac{1}{2}(2m)v_2'^2.$$

We may eliminate the common factors of $\frac{1}{2}m$ to get two equations that must be solved simultaneously for the two unknown velocities:

$$v_0 - v_1' = 2v_2' \quad \text{(momentum)},$$
$$v_0^2 - v_1'^2 = 2v_2'^2 \quad \text{(energy)}.$$

The momentum equation may be squared to get

$$(v_0^2 - 2v_0v_1' + v_1'^2) = 4v_2'^2.$$

Now we multiply the energy equation by 2 and subtract the above equation from it, obtaining

$$v_0^2 + 2v_0v_1' - 3v_1'^2 = 0.$$

This equation may be factored, giving us

$$(v_0 - v_1')(v_0 + 3v_1') = 0.$$

As before, we have two solutions,

$$v_1' = v_0$$

and

$$v_1' = -\tfrac{1}{3}v_0.$$

The solution $v_1' = v_0$ indicates that the body of mass m has the same velocity after the collision that it had before the collision; that is, it implies that there was no collision. We choose instead the physically meaningful solution of $v_1' = -\tfrac{1}{3}v_0$. The negative sign indicates that the light glider recoils in the direction opposite to its initial motion.

If we insert $v_1' = -\tfrac{1}{3}v_0$ into the momentum equation, we obtain

$$v_2' = \tfrac{2}{3}v_0.$$

You may wish to insert these values into the energy equation to verify their correctness.

Elastic Collisions in Two Dimensions

8.3

When two billiard balls collide on the surface of a billiard table, the general result is that they move off in different directions. Since all velocities lie in the plane of the surface, the collision is two-dimensional. We can analyze elastic collisions in two dimensions by using the fact that momentum is a vector quantity. There are two equations for conservation of momentum, one for the x components and one for the y components of momentum, and one equation for conservation of energy. (Remember that energy is a scalar quantity.) However, four quantities must be determined to specify completely the final outcome of a two-dimensional collision: the two velocity components of each body, or equivalently, the magnitudes and directions of the two vector velocities. These four quantities cannot be determined from three equations, even if we know all of the initial velocity components. However, if one of these quantities can be specified, either from observation or from some physical consideration, then the other three quantities can be uniquely determined.

Figure 8.6 (p. 236) depicts an elastic collision between two spheres of mass m_1 and m_2. One of the spheres, m_2, is initially at rest. Using the law of conservation of momentum, we obtain an equation for the x component of momentum:

$$m_1 v_{1x} = m_1 v_{1x}' + m_2 v_{2x}' \qquad \text{(momentum along } x\text{)}. \qquad (8.5)$$

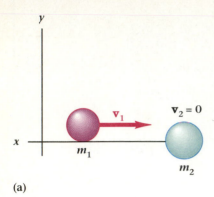

(a)

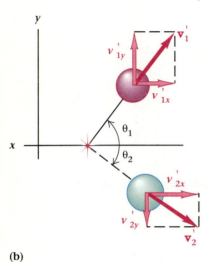

(b)

Figure 8.6 A glancing collision in two dimensions, viewed from above. (a) Before collision. (b) After the collision the two spheres move off in different directions.

Conservation of the y component of momentum gives

$$0 = m_1 v'_{1y} + m_2 v'_{2y} \quad \text{(momentum along } y\text{)}. \quad (8.6)$$

To simplify our work, we have deliberately chosen a coordinate system in which the initial velocity is along the x axis. This choice guarantees that there is no initial momentum in the y direction. Conservation of energy gives

$$\tfrac{1}{2}m_1 v_1^2 = \tfrac{1}{2}m_1 v'^2_1 + \tfrac{1}{2}m_2 v'^2_2. \quad \text{(kinetic energy)}. \quad (8.7)$$

Simultaneous solution of these three equations is necessary to determine the subsequent motion of the spheres. However, we still require prior knowledge of one postcollision velocity component or angle before we can find the other quantities. The following example illustrates the use of conservation laws in two dimensions.

Example 8.3

Elastic collision of two billiard balls.

Consider two billiard balls of equal mass that collide elastically. One of them has an initial speed v_1 and the other is at rest. After the collision, the ball that was initially in motion is deflected at an observed angle θ_1 with respect to its original direction. Find the final speeds of each ball and the direction in which the other ball is moving.

Solution   Referring to Fig. 8.6, we can rewrite the equation for the x component of momentum (Eq. 8.5) in terms of the angles θ_1 and θ_2,

$$v_1 = v'_1 \cos \theta_1 + v'_2 \cos \theta_2 \quad (x \text{ momentum}),$$

where θ_1 and θ_2 are both defined as positive in the diagram. Notice that since both balls have the same mass, m is a common factor to all terms and has been eliminated from the equation. Similarly, the equation for conservation of the y component of momentum (Eq. 8.6) may be expressed as

$$v'_1 \sin \theta_1 = v'_2 \sin \theta_2 \quad (y \text{ momentum}),$$

and the kinetic energy conservation equation (Eq. 8.7) becomes

$$v_1^2 = v'^2_1 + v'^2_2 \quad \text{(kinetic energy)}.$$

The x component momentum equation may be rearranged and squared to give

$$v_1^2 - 2v_1 v'_1 \cos \theta_1 + v'^2_1 \cos^2 \theta_1 = v'^2_2 \cos^2 \theta_2.$$

The y momentum equation may also be squared to give

$$v'^2_1 \sin^2 \theta_1 = v'^2_2 \sin^2 \theta_2.$$

When these two equations are added we have

$$v_1^2 - 2v_1 v'_1 \cos \theta_1 + v'^2_1 = v'^2_2,$$

where we have made use of the trigonometric identity $\cos^2 \theta + \sin^2 \theta = 1$.

At this point we can combine the kinetic energy equation and the last equation to eliminate v_2'. The result expresses v_1' in terms of the initial speed v_1 and the angle θ_1:

$$v_1^2 - 2v_1v_1' \cos \theta_1 + v_1'^2 = v_1^2 - v_1'^2,$$

$$v_1'^2 = 2v_1v_1' \cos \theta_1 - v_1'^2,$$

$$v_1'^2 = v_1v_1' \cos \theta_1.$$

One solution corresponds to $v_1' = 0$, but that is a head-on collision. The other solution, corresponding to ball 1 emerging at angle θ_1, is

$$v_1' = v_1 \cos \theta_1.$$

This value of v_1' may, in turn, be inserted into the kinetic energy equation to yield v_2':

$$v_2' = v_1 \sin \theta_1.$$

(To get this result you need to use the trigonometric identity $\cos^2 \theta + \sin^2 \theta = 1$.) Finally, the recoil angle of the second ball can be obtained from the y momentum equation as

$$\sin \theta_2 = \frac{v_1'}{v_2'} \sin \theta_1 = \frac{v_1'}{v_1}.$$

We see that knowledge of the initial speed v_1 and the recoil angle θ_1 is sufficient to determine the magnitude of the other speeds v_1' and v_2' as well as the angle θ_2.

Example 8.4

Scattering angles in elastic collisions of equal masses.

A ball of mass m moving with a speed of 2.00 m/s collides with a stationary ball of the same mass in an elastic collision. The incident ball is scattered at an angle of 20° from its original direction. Find the final speed of each ball and the direction of recoil of the struck ball.

Solution This problem is a numerical example of the collision described in Example 8.3. We are given that $v_1 = 2.00$ m/s and $\theta_1 = 20°$. We may use the equation developed in Example 8.3 to find the speed v_1' of the scattered ball:

$$v_1' = v_1 \cos \theta_1 = (2.00 \text{ m/s}) \cos 20° = 1.88 \text{ m/s}.$$

We obtain the speed of the struck ball from the kinetic energy equation as

$$v_2' = \sqrt{v_1^2 - v_1'^2} = \sqrt{2.00^2 - 1.88^2} = 0.682 \text{ m/s}.$$

Finally, the recoil angle of the second ball is obtained from

$$\sin \theta_2 = v_1'/v_1 = 0.940,$$

$$\theta_2 = 70°.$$

In Example 8.4 the two balls of equal mass move off at right angles

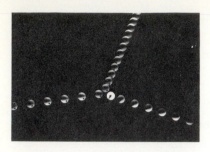

Figure 8.7 A multiflash photograph of an elastic collision between two balls of equal mass. After the collision the two balls move off at a right angle to each other.

to each other. This motion is characteristic of elastic collisions between two bodies of equal mass. No matter what the scattering angle of the incoming ball, the recoil direction of the struck ball is at right angles to it (Fig. 8.7). This behavior may be shown formally by examining the equations developed for v_1', v_2', and $\sin \theta_2$. They may be combined to show that $\sin \theta_2 = \cos \theta_1$. Thus the two angles are complementary and their sum always adds to 90°.

The preceding examples, in which we arbitrarily restricted ourselves to two dimensions, are actually applicable to three-dimensional problems. Conservation of momentum in collisions involving only two particles ensures that the trajectories of the particles both before and after the collision all lie in one plane. Therefore many real three-dimensional problems may be reduced to the equivalent two-dimensional problem by the appropriate selection of coordinate systems.

Rolling Bodies

8.4

In Section 6.6 we introduced rotational kinetic energy, but for the sake of keeping the mathematics simple, we did not discuss any examples that included both translational and rotational kinetic energy. Now we extend the use of the law of conservation of mechanical energy to include, at the same time, translational as well as rotational kinetic energy. Both considerations need to be taken into account when dealing with such things as rolling tires and hoops. They are even important when describing the motion of molecules in a gas.

Figure 8.8 shows a disk of mass m and radius r at the top of an inclined plane. The axis of the disk is parallel to the top edge of the plane so that, when released, the disk rolls straight down the plane. If the frictional force is great enough, there is no sliding and the disk rolls without slipping. The thickness of the disk is not important here, except that for a given material and radius, the total mass of the disk depends upon its thickness.

At the top of the plane the disk has a potential energy mgh relative to its position at the bottom. Here h is the vertical distance through which the center of mass moves from the top to the bottom of the plane. If the disk rolls down to the bottom of the plane without slipping, then all of the initial potential energy is completely transformed into kinetic energies of rotation and translation at the bottom. Because the disk rolls without slipping, we can neglect energy loss due to friction. Then we may extend the idea of conservation of mechanical energy to include rotational as well as translational kinetic energy. Consequently,

$$\Delta PE + \Delta KE_{trans} + \Delta KE_{rot} = 0,$$

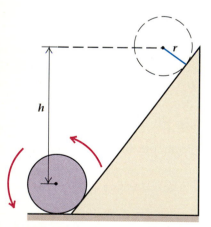

Figure 8.8 A disk rolling down an inclined plane. Its potential energy at the top is transformed into translational and rotational kinetic energy at the bottom.

or

$$-\Delta PE = \Delta KE_{trans} + \Delta KE_{rot}.$$

Here KE_{trans} is the kinetic energy due to translation of the center of mass of the disk and KE_{rot} is the kinetic energy of rotation about its center of

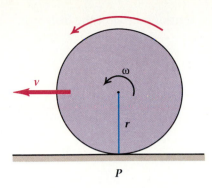

Figure 8.9 A disk rolling with speed v rotates with an angular velocity $\omega = v/r$.

mass. The sum of these two terms is the total kinetic energy of the rolling disk:

$$\mathrm{KE}_{\mathrm{tot}} = \mathrm{KE}_{\mathrm{trans}} + \mathrm{KE}_{\mathrm{rot}} = \tfrac{1}{2}mv^2 + \tfrac{1}{2}I\omega^2. \qquad (8.8)$$

We see that the magnitude of the decrease in potential energy is equal to the gain in total kinetic energy, or

$$mgh = \tfrac{1}{2}mv^2 + \tfrac{1}{2}I\omega^2, \qquad (8.9)$$

where v and ω are the linear and angular speeds of the disk when it reaches the bottom of the plane and I is the moment of inertia of the disk about its center of mass. The expression for the kinetic energy of rotation is taken from Chapter 6, Eq. (6.9). Table 6.3 (p. 174) lists moments of inertia for several objects of different shapes. Equation (8.9) is valid for any round object, whether it be a disk, a hoop, or a wheel. However, you must use the correct moment of inertia for the case at hand.

Because the disk rolls without slipping, there is a direct relationship between its linear and angular motions. If the disk is rolling with a speed v, then the instantaneous motion of the center of mass about the point of contact P (Fig. 8.9) is a rotation with angular velocity $\omega = v/r$. Since the center of mass is moving steadily in a straight line, a point on the rim must be in rotation about it at the same angular velocity. Thus the angular velocity and the linear speed are related through $v = r\omega$. Then we can express Eq. (8.9) solely in terms of either v or ω. Let us choose to eliminate ω, obtaining

$$mgh = \tfrac{1}{2}mv^2 + \tfrac{1}{2}I\frac{v^2}{r^2}.$$

We can determine the speed of the disk at the bottom of the incline by solving this equation for v. The result is that after the body has rolled through a vertical height h, its speed is

$$v = \sqrt{\frac{2gh}{1 + \dfrac{I}{mr^2}}}. \qquad (8.10)$$

We see that the body's speed at the bottom of the plane depends on the height of the plane and on the moment of inertia of the disk or other round object. Since the moment of inertia always depends linearly on the mass, the term I/mr^2 is independent of m and depends only on the geometry of the body. The following example illustrates a consequence of that point.

Example 8.5

Which rolls faster, a hoop or a disk?

A uniform solid disk of radius R and mass m and a hoop of the same radius and mass are released from rest from the top of an incline (Fig. 8.10). Which object is moving more rapidly at the bottom?

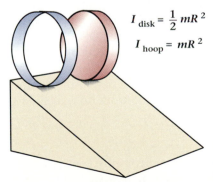

$$I_{\mathrm{disk}} = \tfrac{1}{2}mR^2$$

$$I_{\mathrm{hoop}} = mR^2$$

Figure 8.10 Example 8.5: A solid disk and a hoop of the same mass and radius on an inclined plane. Which one rolls to the bottom faster?

Solution   From Table 6.3 we find the moment of inertia of the disk to be

$$I_{disk} = \tfrac{1}{2}mR^2.$$

The moment of inertia of the hoop is

$$I_{hoop} = mR^2.$$

Inserting these values for the moment of inertia into Eq. (8.10), we get

$$v_{disk} = \sqrt{\frac{2gh}{1 + \tfrac{1}{2}}} = \sqrt{\frac{4gh}{3}}$$

and

$$v_{hoop} = \sqrt{\frac{2gh}{1 + 1}} = \sqrt{gh}.$$

The disk, because of its smaller moment of inertia, has a greater speed than the hoop at any point along the inclined plane, including the bottom. Consequently, if they are released simultaneously, the disk reaches the bottom first. Notice that this result is independent of the mass of each object.

Braking a Moving Car

*8.5

How do energy considerations affect the stopping distance of a car? If we know the frictional coefficient between the appropriate surfaces and the initial speed, we can calculate the distance required to stop a moving object. For example, consider an automobile traveling along a level (horizontal) road. Normally, in a controlled stop, the kinetic energy of the car is dissipated in the friction of the brakes. However, tests have shown that for some conditions the fastest way to stop on a dry road is to lock the wheels by applying the brakes as fast as possible and holding the wheels locked until the car skids to a stop.* This is not true on a wet or icy road. In this section we assume the wheels are locked. (The antilock brakes now being used on some cars are designed to prevent the loss of control that occurs when the wheels do not lock evenly, causing the car to spin. These devices bring the car to a stop along a controlled path.)

If the brakes are applied hard enough to lock the wheels, then the kinetic energy of the moving car is dissipated by the friction of the tires against the road. If the frictional force is constant, the work done in sliding a distance s is:

$$W = Fs,$$

where F is the magnitude of the frictional force. From the law of conservation of energy, the work done in braking the car to a stop must equal

*J. Walker, "The Amateur Scientist: In an Emergency Stop, Should a Car's Wheels Be Locked or Should the Braking be Controlled?" *Scientific American,* February 1989, p. 104.

the initial kinetic energy:

$$Fs = \tfrac{1}{2}mv^2.$$

The frictional force is given by μN, where μ is the coefficient of friction and N is the normal force (Fig. 8.11). For a level road, the total normal force on the wheels is just the weight, mg. Substituting these terms into the equation above, we find that the stopping distance s in which one skids to a stop is

$$s = \frac{v^2}{2\mu g}. \tag{8.11}$$

Equation (8.11) shows that when the frictional coefficient μ decreases (less friction), the stopping distance s increases, as we would expect. It also shows that s increases when the speed increases, but the increase is not linear; s increases as the square of v. For example, a car traveling 50 km/h requires more than twice the stopping distance of a car traveling 35 km/h. Similarly, it takes four times as far to stop from a speed of 70 km/h as it does from 35 km/h.

To use Eq. (8.11) to calculate stopping distances, we must use a consistent set of units. If speed is given in units of meters per second, then g should be in meters per second squared. The distance would then be in units of meters. However, by a conversion of units, Eq. (8.11) can be rewritten in an approximate but handy form in which speeds are given in kilometers per hour and distances in meters. The stopping distance becomes

$$s = \frac{v^2}{2\mu g} = \frac{[v \ (\text{km/h})]^2}{2\mu \times 9.8 \ \text{m/s}^2} \left(\frac{1000 \ \text{m/km}}{3600 \ \text{s/h}} \right)^2.$$

Thus, when speed is given in kilometers per hour, the stopping distance in meters is given by

$$s \ (\text{m}) \approx \frac{[v \ (\text{km/h})]^2}{250\mu}.$$

A similar formula can be derived for speeds in miles per hour and distances in feet, or indeed for any desired combination.

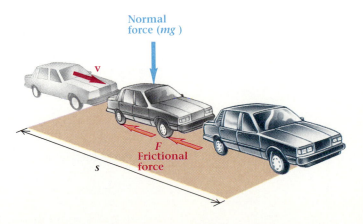

Figure 8.11 A car brought to a sliding stop. The work done by the frictional force equals the change in kinetic energy.

Example 8.6

Stopping distance from 90 km/h on a dry road.

You are driving a car at 90 km/h on a dry road and suddenly brake to a stop. If the coefficient of friction between the tires and the road is 0.63, what is the minimum stopping distance?

Solution   By equating the work done in braking to the initial kinetic energy, we get the relation of Eq. (8.11):

$$s = \frac{v^2}{2\mu g}.$$

For speed v in km/h, the distance s is given in meters by:

$$s\ (m) \approx \frac{[v\ (km/h)]^2}{250\mu} = \frac{(90\ km/h)^2}{250 \times 0.63} = 51\ m.$$

Example 8.7

Compare the stopping distances for two identical cars initially traveling 27 m/s, one along a level road and the other down a 10° incline. Take the frictional coefficient to be $\mu = 0.60$.

Solution   It is harder to stop the car going downhill than it is to stop the one moving on the level road because of two contributing effects. First, the frictional force is reduced because the normal force is reduced. Second, the *net* stopping force is reduced because the gravitational force is pulling the car downhill.

    For the car on the level road, we can simply apply Eq. (8.11),

$$s = \frac{v^2}{2\mu g} = \frac{(27\ m/s)^2}{2 \times 0.60 \times 9.8\ m/s^2} = 62\ m.$$

    For the car on the hill (Fig. 8.12), we first observe that the normal forces on the tires add up to a total normal force opposing the component of the car's weight perpendicular to the hill. In the direction parallel to the hill, the net decelerating force F is the difference between the frictional force μN and the tangential component F_t of the gravitational force:

$$F = \mu N - F_t.$$

Upon substituting for the normal and tangential forces, we find that

$$F = \mu mg \cos \theta - mg \sin \theta.$$

From the energy conservation equation we see that

$$Fs = \tfrac{1}{2}mv^2,$$

$$s = \frac{\tfrac{1}{2}mv^2}{F} = \frac{v^2}{2g(\mu \cos \theta - \sin \theta)}.$$

When the values for v, g, μ, and θ are inserted, we get

$$s = \frac{(27\ m/s^2)^2}{2 \times 9.8\ m/s^2\ (0.60 \cos 10° - \sin 10°)},$$

$$s = 89\ m.$$

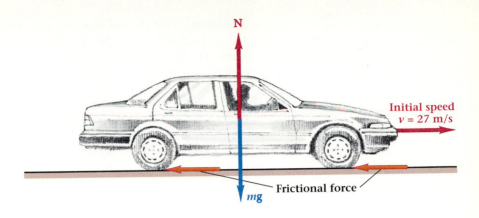

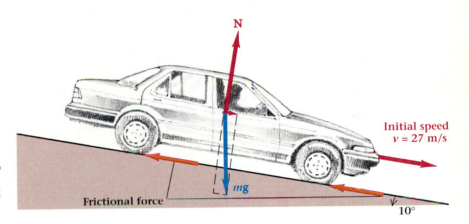

Figure 8.12   **Example 8.7: Cars brought to a stop on level ground and on a hill.**

Thus, when going down a 10° hill, it takes nearly 44% greater distance to stop a car traveling 27 m/s than it would if the car were traveling on level ground.

As the hill becomes steeper, the frictional force decreases while the gravitational force tending to pull the car downhill increases. These two forces become equal when $\mu = \tan\theta$. For steeper inclines it would not be possible to stop the car from sliding down the hill. For icy conditions with $\mu = 0.1$, the forces become equal on a hill inclined at only 5.7°.

Motion in a Gravitational Field

*8.6

In this and the next section we discuss some aspects of the motion of objects, such as spacecraft, that are projected from the earth and orbit around it. These topics are included here because they demonstrate the power of conservation laws and because the mathematical techniques are similar to those we have already used in this chapter. Here we use only the laws of universal gravitation and of conservation of energy. Although

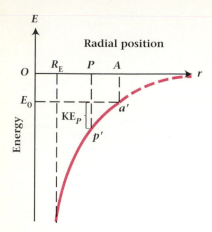

Figure 8.13   Potential-energy diagram of a body thrown straight up. Energy is shown along the vertical axis and radial position from the center of the earth along the horizontal axis.

a complete treatment of satellite and spacecraft motion requires the conservation of angular momentum as well, we have chosen to restrict ourselves to situations where angular momentum need not be included, so that the mathematical analyses are simplified.

An object thrown straight up reaches a maximum height h that depends on its initial upward velocity v_0. We can determine the relation between h and v_0 from the principle of energy conservation by equating the potential energy at height h to the kinetic energy at height 0 (Example 6.10). The result is $v_0 = \sqrt{2gh}$, where g is the gravitational acceleration. In arriving at this relationship, we assumed that the distance h is small enough to allow us to consider the gravitational force to be constant. If we consider objects such as rockets and satellites, which are projected to great distances above the earth's surface, the assumption of a uniform gravitational force is no longer acceptable. It is possible, though tedious, to derive the correct relationship between speed and position from a consideration of the actual forces involved. However, it is much easier to analyze such problems from the standpoint of energy. This approach has the additional advantage of introducing some important techniques that are of general usefulness.

The gravitational potential energy of an object of mass m at a distance r from the center of the earth was given earlier (Section 6.5). For the case in which the potential energy is set equal to zero at $r = \infty$, it is

$$PE = \frac{-GM_E m}{r},$$

where G is the universal gravitational constant and M_E is the earth's mass. (We have neglected any effects due to the earth's rotation.) The potential-energy diagram for this case is the curve in Fig. 8.13, which plots potential energy against radial distance from the center of the earth. The potential energy of the object is negative for any finite distance r.

If the object is thrown straight upward, it has a total energy E_0 that is the sum of its kinetic and potential energies. Its total energy remains constant as it moves in the earth's gravitational field if we neglect air resistance. The kinetic and potential energies at any point P are related by

$$KE_P + PE_P = E_0.$$

If the total energy E_0 is known, then the kinetic energy at any point is the difference between the total energy and the potential energy. The total energy E_0 is indicated in Fig. 8.13 for a particular case in which E_0 is negative. For any radial position $r < A$, where A depends on the total energy E_0, the object has a positive kinetic energy, as indicated at P. As the object moves away from the earth toward A, its kinetic energy decreases and its potential energy increases. When it reaches A, its kinetic energy becomes zero. At this point, the body stops and falls back toward the earth. When it passes the point P on its way back to the earth's surface, it has the same kinetic energy that it had at that same point on the way up, and is moving with the same speed but now headed down instead of up.

A simple way to visualize the object's motion is to imagine it represented by a bead sliding without friction along a wire bent into the shape of the potential-energy curve. An upward push sends the bead along the wire from r'_E to a', where it stops and slides back down to r'_E. The bead loses speed as it goes from r'_E to a' and regains speed as it returns. This analogy can be quite useful, but you must remember that the actual motion of the object is along a straight line directed radially away from the earth.

Let us return to the question of finding the relationship between the object's initial upward speed v_0 and the maximum height h to which it can ascend. We take the initial location to correspond to the earth's surface at $r = R_E$. Applying the rule of conservation of energy gives

$$KE(R_E) + PE(R_E) = KE(r) + PE(r).$$

At maximum height h ($r = R_E + h$), the kinetic energy of the object is zero and the energy equation becomes

$$\tfrac{1}{2}mv_0^2 - GM_Em/R_E = 0 - GM_Em/r.$$

We can rearrange this to give

$$v_0 = \sqrt{2GM_E\left(\frac{1}{R_E} - \frac{1}{r}\right)}.$$

This equation gives the initial upward speed required for a body at the earth's surface if it is to rise to a distance r from the center of the earth.

Compare this equation for v_0 with the simpler form given earlier for a uniform gravitational field, $v_0 = \sqrt{2gh}$. To make the comparison easier, we express the gravitational constant G in terms of the acceleration g (see Example 5.10):

$$GM_E = gR_E^2.$$

The result of inserting this into the above equation for speed is

$$v_0 = \sqrt{2gR_E^2\left(\frac{1}{R_E} - \frac{1}{r}\right)}.$$

We may also replace r by $R_E + h$ to get

$$v_0 = \sqrt{\frac{2ghR_E}{R_E + h}}. \tag{8.12}$$

If h is small, so that $R_E + h \approx R_E$, then Eq. (8.12) reverts to the simpler form of the case for constant gravitational force. However, Eq. (8.12) is correct for all heights. The value of g remains the value measured at the earth's surface.

Example 8.8

Initial speed for a rocket reaching height 0.1 R_E.

A rocket is projected upward from the earth's surface to a height of 0.0100 R_E, where R_E is the mean radius of the earth. (a) What initial speed is necessary? (b) What fractional error would occur if the calculation were made under the assumption that the earth's gravitational field were uniform at all heights?

Solution   (a) For the correct answer we use

$$v_0 = \sqrt{\frac{2ghR_E}{R_E + h}}$$

$$= \sqrt{\frac{2g(0.0100R_E)(R_E)}{R_E + 0.0100R_E}} = \sqrt{2gR_E(0.00990)}$$

$$= \sqrt{2(9.80 \text{ m/s}^2)(6.38 \times 10^6 \text{ m})(0.00990)}$$

$$= 1100 \text{ m/s} \approx 2400 \text{ mi/h.}$$

(b) Let us take the ratio of the correct to the approximate expressions for the initial speed:

$$\frac{v_0 \text{ (correct)}}{v_0 \text{ (approximate)}} = \frac{\sqrt{\dfrac{2ghR_E}{R_E + h}}}{\sqrt{2gh}} = \sqrt{\frac{R_E}{R_E + h}} = \sqrt{\frac{1.00}{1.00 + 0.010}},$$

$$\frac{v_0 \text{ (correct)}}{v_0 \text{ (approximate)}} = 0.995.$$

As we expect, since the strength of the gravitational field decreases instead of remaining constant with increasing height, the correct initial speed is less than the one calculated from the approximation. Nevertheless, for a distance of nearly 600 km, the difference in the calculations is only 0.5%.

Escape Velocity

*8.7

Look at the potential energy of a body thrown straight up (Fig. 8.13) and remember the analogy of the bead on the wire. We see that for a projectile to escape the earth—that is, for the bead not to slide back down again—its initial energy must be enough to move the point a' to infinity. Thus the total initial energy must be at least zero ($E_0 \geq 0$). For zero total energy, the projectile will "reach infinity" with zero speed. Rockets (or other objects) with energy less than zero cannot escape the earth. Bodies with total energy greater than zero, like E_1 in Fig. 8.14, are unbound. For a projectile that is just able to escape from the earth's gravitational field, the total energy is zero.

A rocket that is fired for only a short time behaves essentially like a projectile that is given an initial upward velocity. The minimum initial upward speed necessary for the rocket not to fall back to earth again is called the **escape velocity** v_{esc}, and is found by setting the projectile's total energy equal to zero:

$$\tfrac{1}{2}mv_{esc}^2 - \frac{GM_E m}{R_E} = 0.$$

If we again use $GM_E = gR_E^2$ we find that

$$v_{esc} = \sqrt{2gR_E}.$$

(8.13)

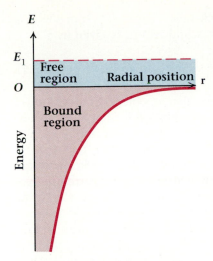

Figure 8.14 Potential-energy diagram for a mass attracted to the earth. Objects with total energy E less than zero are bound to the earth. Those with E greater than zero are free to escape from it. Particles with $E = 0$ are just barely free and approach $r = \infty$ with zero velocity.

Note that the escape velocity does not depend on the mass of the rocket. The value of the escape velocity for any object on the earth can be calculated to be 11.2 km/s. It is interesting to note that the speed needed to completely escape from the earth's gravitational field is only $\sqrt{2}$ times as much as the speed of a body in orbit just above the earth's surface.

Example 8.9

What is the escape velocity for a rocket on the surface of Mars?

Solution We can obtain the escape velocity from the surface of Mars from the law of conservation of energy in the same way that we obtained Eq. (8.13):

$$\tfrac{1}{2}mv_{esc}^2 - \frac{GM_M m}{R_M} = 0.$$

The escape velocity is then

$$v_{esc} = \sqrt{\frac{2GM_M}{R_M}}.$$

When the mass of Mars, 6.58×10^{23} kg, and its radius, 3.38×10^6 m, are inserted into the equation for v_{esc}, we get

$$v_{esc} = \sqrt{\frac{2 \times 6.67 \times 10^{-11} \,\text{N·m}^2/\text{kg}^2 \times 6.58 \times 10^{+23} \,\text{kg}}{3.38 \times 10^6 \,\text{m}}},$$

$$v_{esc} = 5.10 \times 10^3 \,\text{m/s}.$$

The escape velocity from the surface of Mars is less than one half of the escape velocity from the earth's surface.

The expression for the escape velocity in the previous example leads to an interesting consequence that is outside the realm of Newtonian mechanics. Objects with greater mass or smaller radius than the earth (that is to say, denser objects) have greater escape velocities. For stars, the escape velocity is quite high. If a star is dense enough, the escape velocity approaches the speed of light c, the limiting speed for all motion. Therefore when

$$\sqrt{\frac{2GM}{R}} = c,$$

nothing, not even light itself, can escape. Because no light can escape from such a dense object, it is called a *black hole*.

The idea of a black hole was first proposed by P. S. Laplace (1749–1827) in 1796, before much of our present understanding of the nature of light or the stars was known. Current astronomical theories allow for the existence of black holes, although very different from Laplace's original conception. We know that stars are not permanent objects, but evolve and change over long periods of time. At the end of the lifetime of some extremely massive stars the core can undergo a collapse; in this case it is possible for the central density to become large enough for a black hole to form. Although a proper understanding of black holes requires astronomy, general relativity, and quantum mechanics, the numerical result is the same as Laplace obtained.

We can obtain the radius of a black hole of mass M from the last equation. This radius, called the Schwarzschild radius, is given by

$$R_S = \frac{2GM}{c^2}. \tag{8.14}$$

The Schwarzschild radius indicates the size that an object of mass M must have if it is to have the enormous density of a black hole. For a black hole with the mass of the sun, the Schwarzschild radius is only about 3 km.

Even though black holes give off no light from within, they can be detected by their gravitational effects. For example, a black hole and a star can form an orbiting system, whose combined motions conform approximately to Kepler's laws. If we can infer the motion of the star from astronomical observations, then we can detect the presence and mass of the black hole. The motion of at least one star, Cygnus X-1, is most simply explained by saying that it is a member of a system containing a black hole. The interaction between the black hole and the star is complex, but calculations indicate that the extremely strong gravitational field of the black hole causes it to pull in matter from the star (Fig. 8.15). The observed x-ray emissions from the Cygnus X-1 system support these calculations.

PHYSICS IN PRACTICE

The Earth, the Moon, and the Tides

The periodic rise and fall of the ocean on the beach—the tides—are familiar to everyone who has spent time at the seashore (Fig. B8.1). In the open ocean the tides are approximately half a meter high. As the tides approach the shore, the geographic features of the shoreline often channel the water so that typical shore tides are about two meters. These tides vary from place to place. In some areas they are smaller, while in a few locations they are much greater. In Canada's Bay of Fundy the tidal level varies by as much as 15 m.

People have often dreamed of harnessing the motion of the tides to produce electricity. However, the possibility of doing so is restricted to those few places where the tidal variations are sufficiently large and where a dam can be constructed across the channel. At the present time, the expense of build-

Figure B8.1   Ocean tides.

ing such facilities has rendered them impractical in comparison with other means of generating electric power.

The tides are primarily caused by the gravitational pull of the moon.* In addition to the ocean tides, the moon also causes tides in the solid body of the earth, but these earth tides are harder to observe. As the moon moves in its orbit, the earth also moves, because each moves about the center of mass of the earth–moon system. Due to the inverse-square nature of the gravitational force, the water on the side of the earth near the moon is pulled toward the moon with a greater-than-average force, while the water on the far side is pulled with a less-than-average force. Moreover, the motion of the earth about the center of mass also helps raise a tidal bulge on the side away from the moon. As a result, two bulges appear in the water, on opposite sides of the earth.

*The sun also produces a tidal effect, but it is less than half that of the moon.

Figure 8.15   An artistic representation of a black hole and a star in an orbiting system.

Because the rotation of the earth about its axis is faster than the motion of the moon about the earth, and because of the frictional forces between the ocean currents and the sea floor, the earth drags the tidal bulges ahead of the position they would otherwise have (Fig. B8.2). This asymmetrical position of the bulges relative to the line joining the centers of the earth and the moon produces a net torque on the moon. This torque acts to increase the moon's angular momentum. By Newton's third law a torque of equal magnitude acts to slow the rotation of the earth.

Although the total angular momentum of the earth–moon system is conserved, angular momentum is transferred from the earth to the moon. The total mechanical energy decreases as a result of the frictional losses of the tides. Consequently, the length of the day steadily increases as the earth's rate of rotation slows, and the length of the month decreases as the moon speeds up. Because of this increase in speed, and therefore energy, the distance of the moon from the earth also increases. These effects have been measured; the length of the day is gradually increasing at a rate of about 20 μs per year and the moon is slowly moving away at approximately 3 cm per year. Calculations show that the moon will continue to move away from the earth until it reaches a distance of about 75 earth radii. Then the length of the day will equal the length of the month and the motion of the earth and moon will be synchronized. The earth will then keep the same face toward the moon, just as the moon now keeps the same face toward the earth.

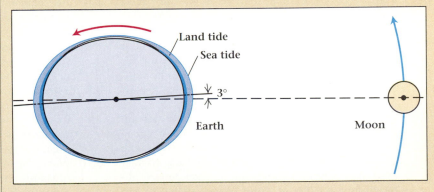

Figure B8.2   Tidal bulges occur 3° ahead of the line between the centers of the earth and the moon because of the earth's rotation. The view is from above the polar axis. (The sizes of the tides are greatly exaggerated for the sake of clarity.)

SUMMARY

Useful Concepts

In this chapter we applied the laws of conservation of energy and momentum to analyze elastic collisions, rolling bodies, and motion in a gravitational field. These examples indicate the wide variety of problems solvable by conservation laws.

- In one dimension the law of conservation of momentum for collision of two objects is

$$m_1 v_1 + m_2 v_2 = m_1 v_1' + m_2 v_2'.$$

- In an elastic collision, kinetic energy is conserved, leading to the equation

$$\tfrac{1}{2}m_1 v_1^2 + \tfrac{1}{2}m_2 v_2^2 = \tfrac{1}{2}m_1 v_1'^2 + \tfrac{1}{2}m_2 v_2'^2.$$

- For collisions in two dimensions, we take account of the vector nature of momentum, determining separate conservation equations in the x and y directions.
- The total kinetic energy of a rolling body separates into the translational kinetic energy of the center of mass

and rotational kinetic energy about the center of mass,

$$KE_{tot} = KE_{trans} + KE_{rot} = \tfrac{1}{2}mv^2 + \tfrac{1}{2}I\omega^2.$$

- The gravitational potential of an object of mass m at a distance r from the center of the earth is

$$PE = -\frac{GM_E m}{r}.$$

We can use this formula and the conservation of energy to determine the escape velocity from the earth,

$$v_{esc} = \sqrt{2gR_E}.$$

Important Terms

You should be able to write the definition or meaning of each of the following terms:

- elastic collision
- inelastic collision
- escape velocity

QUESTIONS

8.1 A stationary firecracker is hung by a thread from a tree limb and then explodes into three pieces. Do the paths of the three pieces lie in a plane? Explain your answer with care.

8.2 Give some examples of elastic, inelastic, and perfectly inelastic collisions.

8.3 Two bodies collide elastically in midair. The only restriction on their initial motions is that they must collide. Explain how this situation can be viewed as only a two-dimensional problem.

8.4 A wooden disk and an iron ring have the same mass and radius. Suppose they are both placed at the top of an inclined plane and released at the same time. Which object reaches the bottom of the incline first? Explain your answer.

8.5 A cosmonaut wishes to dock his spacecraft with another craft several hundred meters ahead of him in the same orbit. The two spacecraft are moving with the same speed at the same radius in the same circular orbit. The cosmonaut can use thruster rockets directed fore, aft, up, or down. Which should he use and in what order? Describe the subsequent motion of his craft.

8.6 A projectile fired from the earth's surface needs to be given an upward speed equal to the escape velocity v_{esc} in order to escape from the earth. Would a rocket fired from the earth need that same speed in order to escape? Why?

8.7 Can an object have, at the same time, more kinetic energy but less momentum than another object?

8.8 Are there any combinations of masses and initial velocities in a one-dimensional elastic collision so that the velocity of one of the bodies is unchanged by the collision? Are there any conditions under which the speed is the same after the collision as it was before? In either case, explain your answer.

8.9 The large wheels in Fig. 8.16 have the same radius and mass, and they turn without friction. If the weight W, which is suspended by cords wrapped around the wheels, is allowed to fall from rest, which angle will decrease more rapidly, θ_1 or θ_2?

8.10 When the supporting stick S is jerked out from the apparatus shown in Fig. 8.17, the board falls down about the hinged end H. The ball B is caught by the cup C. Explain how the cup C can reach the ground before the ball, even though the ball is in free fall.

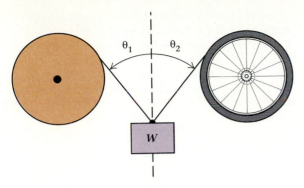

Figure 8.16   Question 8.9.

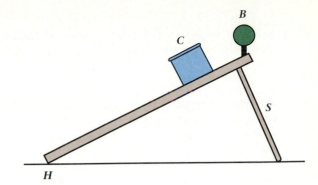

Figure 8.17   Question 8.10.

PROBLEMS

Hints for Solving Problems
Remember that for elastic collisions, both kinetic energy and momentum are conserved. You will find it helpful to draw two diagrams of a collision, one before and one after impact. In a gravitational field, the potential energy is always negative if the reference level is zero at infinity.

Section 8.1   Definition of Elastic Collisions

8.1 A glass marble is dropped onto a steel slab from a height of 2.0 m. If the marble rebounds to a height of 1.6 m, what fraction of its initial energy was lost? Where has the energy gone?

8.2 A ball bearing is dropped from a height of 2.0 m onto a steel slab. If the ball bounces to a height of 1.4 m, what is its upward speed as it passes the 1.0-m mark?

Section 8.2   Elastic Collisions in One Dimension

8.3 Two billiard balls travel toward each other along a straight line with the same speed. What are their speeds and directions after an elastic collision? Assume the balls do not rotate.

8.4 Two air-track gliders of equal mass are initially moving in the same direction along a straight line. The rearmost glider has an initial speed of 3.0 m/s and the forward glider has a speed of 2.0 m/s. If the collision is elastic, what are the speeds and directions of the gliders after the collision?

8.5 An air-track glider with an initial speed of 4.0 m/s has a head-on collision with another glider at rest, which is three times as massive. What are the final speeds and directions of the gliders if the collision is elastic?

8.6 Two glass marbles moving along a straight line toward each other undergo an elastic collision. The speed of one marble is $2v$ and its mass is m; the speed of the other marble is v and its mass is $2m$. What are the speeds and directions of the marbles after the collision?

8.7 After an elastic collision between two pool balls of equal mass, one is observed to have a speed of 3.0 m/s along the positive x axis and the other has a speed of 2.0 m/s along the negative x axis. What were the original speeds and directions of the two balls?

8.8 A 0.400-kg toy truck moving at an initial speed of 0.100 m/s collides head-on with a 0.300-kg toy car at rest. The collision is perfectly inelastic, so that the two toys stick together. (a) Find their final speed and (b) calculate how much kinetic energy was lost in the collision.

8.9 A 30-caliber rifle with a mass of 2.95 kg fires a 7.1-g bullet with a speed of 600 m/s with respect to the ground. What kinetic energy is released by the explosion of the gunpowder that fires the bullet?

8.10 A 1000-kg automobile going 10 m/s collides head-on with a 1200-kg automobile traveling in the opposite direction at 4.0 m/s. (a) What is the maximum kinetic energy that can be dissipated in damaging the two cars? (b) What limits this amount of energy?

8.11 An air-track glider with an initial speed of 4.0 m/s collides head-on with another glider of equal mass initially at rest. The struck glider has a ball of wax on its bumper that deforms during the collision. What fraction of the initial energy is used in deforming the wax if the final speed of the struck glider is 3.0 m/s?

Section 8.3   Elastic Collisions in Two Dimensions

8.12 A ball of mass m moving with a speed v_0 collides elastically with a stationary ball of the same mass. The incident ball is scattered at an angle of 60° from its incident direction. What fraction of the initial kinetic energy is imparted to the struck ball?

8.13 An elastic collision occurs between two air hockey pucks in which one puck is at rest and the other is moving with a speed of 0.50 m/s. After the collision the puck initially in motion makes an angle of 20° with its original direction and the struck puck makes an angle of 70° with the same direction. What are the final speeds of each puck?

8.14 A proton traveling with speed v_0 collides elastically with another proton initially at rest. After the collision the protons move off, making angles of 45° with respect to the direction of motion of the incident proton. What are the final speeds of the protons after the collision in terms of the initial speed v_0?

8.15 A deuteron collides elastically with another deuteron initially at rest. After the collision one deuteron is observed to have half of the original kinetic energy. In what direction must it be moving? What is the direction of motion of the struck deuteron?

8.16 After colliding elastically with a stationary puck of the same mass, an air hockey puck is observed to have a momentum whose magnitude is only half the magnitude of its original momentum. What angle does it make with its original direction?

Section 8.4   Rolling Bodies

8.17 A hoop and a disk, both of 0.50-m radius and 2.0-kg mass, are released from the top of an inclined plane 3.0 m high and 8.0 m long. What is the speed of each when it reaches the bottom? Assume that they both roll without slipping.

8.18 You simultaneously release a 1.0-kg hoop of 0.50-m radius and a 1.0-kg disk of 0.25 m radius from the upper end of an inclined plane of 2.5 m height. Calculate the speed of each at the bottom of the plane, assuming that they roll without slipping.

8.19 A hoop released from the top of an inclined plane has a speed of 15 m/s at the bottom. How high is the upper end of the inclined plane? Assume that the hoop rolled without slipping.

8.20 In a demonstration an iron hoop rolls without slipping down an inclined plane from a height h to the bottom. In another demonstration the plane is lubricated and the hoop slides down the plane from the same height without rolling. What is the ratio of the speeds at the bottom?

8.21 A sphere is released from rest at the top of an inclined plane. Derive an expression for the speed of the sphere at a point a distance h below its starting point. Assume that the sphere rolls without slipping.

8.22 A hoop of mass m is released from rest and rolls without slipping down a hill to a point that is a distance h lower than the starting point. Show that at this time the hoop will be rotating with an angular velocity

$$\omega = \sqrt{\frac{gh}{r^2}}.$$

*Section 8.5   Braking a Moving Car

8.23 Show that for speed in miles per hour, the stopping distance in feet is given by

$$s \text{ (ft)} \approx \frac{[v \text{ (mph)}]^2}{30\mu}.$$

8.24 Calculate the stopping distance for a car traveling at a speed of 120 km/h on a snow-packed road where the frictional coefficient is 0.45.

8.25 Calculate the stopping distance for a car traveling at a speed of 90 km/h on a wet asphalt road where the frictional coefficient is 0.55.

8.26 (a) Calculate the stopping distance for a car traveling at a speed of 40 mi/h on a dry concrete road where the frictional coefficient is 0.70. (b) How much farther would it take to stop from the same speed on smooth wet ice with a frictional coefficient of 0.10?

8.27 (a) Calculate the stopping distances for a car traveling at speeds of 25, 35, 45, and 55 mi/h on a dry concrete road where the frictional coefficient is 0.70. (b) Plot the stopping distance as a function of speed from zero to 55 mi/h.

8.28 A car is braked hard to a stop from a speed of 30 m/s. If the frictional coefficient is 0.70, what is the deceleration force in g's?

8.29 A car travels 5.0 m/s on smooth ice with $\mu = 0.10$. How far will it take to stop the car after the brakes are applied? How far will it take to stop the car if it is headed down a 5° hill?

*Section 8.6   Motion in a Gravitational Field

8.30 With what upward speed must a rocket be projected to reach a height above the earth equal to the earth's radius? (Ignore air friction.)

8.31 How fast must you project an object for it to reach a height equal to the moon's distance from the earth?

8.32 A rocket is projected upward from the earth with an initial speed v_0 that carries it to a distance $R = \frac{3}{2}R_E$ from the center of the earth. What is the value of v_0?

8.33 Imagine that a meteor headed straight for the earth has an approach speed of 8.0×10^3 m/s at a distance of $3R_E$ from the center of the earth. What will its speed be when it reaches the atmosphere at $R \approx R_E$?

8.34 Imagine that a meteor headed straight for the earth has an approach speed of 8.0×10^3 m/s at a distance of $4R_E$ from the center of the earth. What will its speed be when it falls to a distance of $1.5R_E$ from the center of the earth?

*Section 8.7 Escape Velocity

8.35 A distant planet has a mass of $0.82M_E$ and a radius of $0.95R_E$. What is the ratio of the escape velocity from this planet to the escape velocity from the earth?

8.36 The escape velocity from a distant planet is 35 km/s. If the acceleration of gravity on the planet surface is 11.1 m/s^2, what is the radius of the planet?

8.37 The radius of Uranus is approximately 3.69 times the radius of the earth. If the escape velocity from Uranus is 22 km/s, what is the ratio of the acceleration of gravity on Uranus to its value on earth?

8.38 Compute the escape velocity from the surface of Venus, which has a radius of 6.31×10^6 m and a mass of 4.82×10^{24} kg.

Additional Problems

8.39 A tennis ball dropped from a height of 3.00 m loses 50% of its mechanical energy at each bounce. To what height does it rise after the third bounce?

8.40 A 58-kg girl skis down a slope from a height of 9.0 m above the bottom of a hill. At the bottom she plows into a snowdrift that stops her in 2.0 s. What average force does she exert on the snow? How far does she go into the snowdrift?

8.41 A 1200-kg spacecraft is separated from its 4800-kg booster stage by an explosion (Fig. 8.18). The two parts move away from each other with a relative speed of 100 m/s. Calculate the energy imparted to the two masses by the explosion.

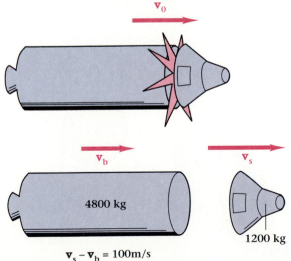

Figure 8.18 Problem 8.41.

8.42 A glass marble is dropped down an elevator shaft and
•• hits a thick glass plate on top of an elevator that is

descending at a speed of 2.0 m/s. The marble hits the glass plate 3.0 m below the point from which it was dropped. If the collision is elastic, how high will the marble rise, relative to the point from which it was dropped?

8.43 (a) A ball of mass $2m$ is projected upward with speed
•• v_0 from the floor (Fig. 8.19). Another ball of mass m is hung from the ceiling by a light string at a height h directly above the first ball, so that the projected ball collides with it. Derive an expression for the height above the floor to which the second ball will rise as a function of v_0, h, and g, assuming that the collision is elastic.

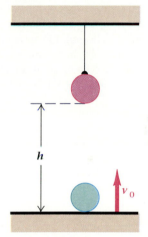

Figure 8.19 Problem 8.43.

8.44 A rocket moving at 1000 m/s consists of a 1200-kg
•• space capsule and a 6000-kg booster stage. An explosion that separates them throws the capsule forward so that it has a speed of 1100 m/s. How much energy was released in the explosion?

8.45 A 1500-kg car moving at an unknown speed strikes
• a 1000-kg car initially at rest. After the collision, the two cars stick together and skid to a stop on asphalt, where the friction coefficient is 0.70. The police measure the skid marks to be 33 m long. Determine whether or not the car was exceeding the speed limit of 55 mi/h (90 km/h).

8.46 A parked car was struck by a moving car, which tried
• to stop but could not. At the scene of the accident, skid marks on dry asphalt leading up to the point of impact were measured to be 40 m long. From the final positions of the cars, insurance investigators determined that the oncoming car was traveling 32 km/h just prior to impact. Estimate the minimum speed of the oncoming car prior to the beginning of its skid. Assume a friction coefficient of 0.70.

8.47 An object is shot directly upward from the earth's
•• surface with an initial speed v_0 sufficient for it to reach a height of 50 km. (a) Do not assume the

earth's gravitational field to be constant and show that the initial speed is given by the approximate formula $v_0 \approx \sqrt{2gh}$. (b) Calculate the initial speed v_0 and express your answer as a fraction of the escape velocity from earth.

8.48 A tennis ball collides with an identical ball at rest.
•• The collision occurs in such a way that one fourth of the initial kinetic energy is lost in deformation of the balls. The outgoing balls leave the impact point, making equal angles with respect to the direction of the original ball. Determine the angle θ between the direction of one of the balls and the initial direction of the first ball.

8.49 Two identical bumper cars collide elastically at right
•• angles, with one going twice as fast as the other. If the faster car is deflected 45° away from its original direction of travel, by how much is the slower car deflected away from its original direction of travel?

8.50 A physics teacher stands on a freely rotating platform
• (Fig. 8.20). He holds a dumbbell in each hand of his outstretched arms while a student gives him a push until his angular velocity reaches 1.5 rad/s. When the freely spinning professor pulls his hands in close to his body, his angular velocity increases to 5.0 rad/s. What is the ratio of his final kinetic energy to his initial kinetic energy? How do you account for this result?

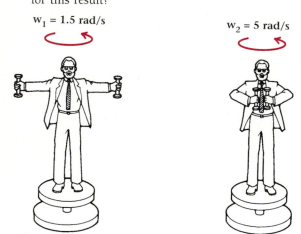

Figure 8.20   Problem 8.50.

8.51 A 1.0-m-radius flywheel is to be made from steel in
•• the form of a solid disk. If the flywheel, when turning at 60 rev/min, is to store as much energy as a 100-watt lamp uses in 1.0 min, how thick must the flywheel be? (*Hint:* The density of steel is 7.8 g cm^{-3}.)

8.52 A piece of metal is pressed against the rim of a 1.6-
•• kg, 19-cm-diameter grinding wheel that is turning at 2400 rev/min. The metal has a coefficient of friction of 0.85 with respect to the wheel. When the motor

is cut off, with how much force must you press to stop the wheel in 20 s?

8.53 Include rotational energy in the equation for the con-
• servation of energy for collisions in one dimension (Section 8.2). Assume the collisions to be between billiard balls that roll without slipping and that do not transfer any rotation upon contact. Derive a pair of expressions corresponding to Eqs. (8.3) and (8.4) for the case in which $m_1 = m_2$ and $v_2 = 0$.

8.54 A solid cylinder rolls without slipping on a horizontal
•• surface. It collides with a resting hollow cylinder of the same mass and radius in a collision in which kinetic energy and momentum are conserved. If the initial speed of the rolling cylinder is 0.25 m/s, what is the final speed of the struck cylinder if it also rolls without slipping?

8.55 When fitted over a narrow inclined plane and re-
•• leased, a yo-yo rolls slowly down the plane and then speeds up when it hits the floor (Fig. 8.21). The diameter of the center rod of the yo-yo is one fifth of the diameter of the outer disks. (a) Show that the speed at the bottom of the plane is

$$v = \sqrt{\frac{4gh}{27}},$$

where h is the vertical distance through which the center of mass descends. (b) What is the speed on the floor? Neglect the mass of the rod that joins the disks.

8.56 If the 0.70-kg block in Fig. 8.22 is released from rest,
•• what speed will it have just before it hits the floor if there is no friction at the wheel's axis? (*Hint:* Use conservation of energy and consider both translational and rotational kinetic energy.)

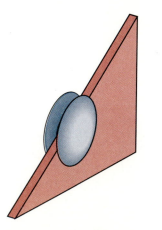

Figure 8.21   Problem 8.55.

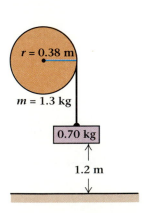

Figure 8.22   Problem 8.56.

8.57 Assume that the axis of the wheel in Problem 8.61 is
•• not frictionless. What is the frictional torque if the speed just before the 0.70-kg block hits the floor is half of the speed it would have without friction?

8.58 A force of 5.0 N is required to pull a 0.30-kg toy car
•• across the floor at constant speed. The car is placed at the bottom of a ramp (Fig. 8.23) and a wooden ball of 0.05 m radius and the same mass as the car is allowed to roll down the plane from a height of 0.54 m and strike the car. How far will the car go if the collision is elastic?

0.54 m

Figure 8.23   Problem 8.58.

8.59 A 5.00-g glass marble moving at 1.2 m/s collides
•• head-on with an identical stationary marble. After the collision, both marbles then move in the direction of the oncoming marble. The struck marble moves with a speed 100/99 of the initial speed of the oncoming marble. After the collision a small piece of the struck marble is found directly below the point of impact. What is the mass of the small piece? (*Hint:* This problem is a one-dimensional classical mechanics example of a technique used in nuclear and elementary par-

ticle physics—determining the mass of an unobserved particle by measuring the initial and final kinematic quantities and then using the conservation laws to find the missing mass. Assume in this case that momentum and kinetic energy are conserved.)

8.60 (a) What is the Schwarzschild radius of a body with
• the mass of the earth? (b) What would be the average density of that body if its radius were the same as its Schwarzschild radius?

8.61 A 2.5-kg block collides with a horizontal spring of
•• negligible mass and spring constant $k = 320$ N/m. The block compresses the spring by 8.5 cm from its rest position (Fig. 8.24). How fast was the block going when it hit the spring? The frictional coefficient between the block and the horizontal surface is 0.40.

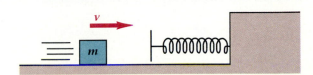

v

m

Figure 8.24   Problem 8.61.

8.62 What is the potential energy of a 1.0-kg mass at a distance three quarters of the way from the earth to the moon? Ignore effects of the sun.

8.63 At what fraction of the distance between Jupiter $(1.90 \times 10^{27}$ kg) and the sun $(1.99 \times 10^{30}$ kg) is the gravitational attraction due to the planet equal to the attraction due to the sun?

ADDITIONAL READINGS

Brancazio, P. J., "The Physics of Kicking a Football." *The Physics Teacher*, October 1985, p. 403.

Damask, A. C., "Forensic Physics of Vehicle Accidents." *Physics Today*, March 1987, p. 36.

Frolich, C., ed., *Physics of Sports: Selected Reprints*. College Park, Md.: American Association of Physics Teachers, 1986.

Hall, D. E., "The Hazards of Encountering a Black Hole." *The Physics Teacher*, December 1985, p. 540.

Sherfinski, J., "Coupled Systems with Constrained Accelerations." *The Physics Teacher*, October 1988, p. 454.

Walker, J., "The Amateur Scientist: Drop Two Stacked Balls from Waist Height; The Top Ball May Bounce Up to the Ceiling." *Scientific American*, October 1988, p. 140.

Appendix for Chapter 8

Solving Simultaneous Equations

Physics problems often lead to two or more equations that need to be solved at the same time to get the required answers. We call these equations simultaneous equations. The following problem illustrates how we solve simultaneous equations: A bicyclist heads north at a steady speed of 15 km/h from a point 3 km from a park. Three hours before the cyclist departed, a walker started walking north from the park along the same path at 5 km/h. When and where does the bicyclist overtake the walker?

Let y be the distance from the park where the bicyclist overtakes the walker. Then for the bicyclist

$$y = (15 \text{ km/hr}) \, t + 3 \text{ km},$$

where t is the cyclist's elapsed time in hours. The walker, having traveled for $t + 3$ h, goes the same distance,

$$y = (t + 3) \, 5 \text{ km/hr} = (5 \text{ km/hr}) \, t + 15 \text{ km}.$$

Each of these equations contains two unknown quantities, y and t. Neither equation alone can give us the answers, but we can solve them simultaneously to obtain both unknowns.

There are several techniques for solving simultaneous equations, all using only simple algebra and all giving the same result for a given problem. When you use these techniques to solve physics problems, be sure you take proper care of the units. However, because our intent here is to review the mathematical techniques, we will simplify them by not including the units.

To solve the two equations just developed, let's write them again without units. We can eliminate the variable y by subtracting the second equation from the first. They become

$$
\begin{array}{rl}
y = 15t + 3 & \qquad (1) \\
- \quad y = 5t + 15 & \qquad (2) \\
\hline
0 = 10t - 12. &
\end{array}
$$

This new equation involves only t and may be rearranged to give

$$10t = 12.$$

Thus

$$t = \frac{12}{10} = 1.2.$$

Now that we have a value for t, it may be substituted into either of the original equations to give

$$y = 15t + 3 = 15(1.2) + 3 = 21,$$
$$y = 5t + 15 = 5(1.2) + 15 = 21.$$

Notice that we get the same answer in both cases. Only one is necessary, but you may want to compute both as a check on your work. The physical answer to the original problem is that the cyclist overtakes the walker 1.2 h after the cyclist started. They meet at a position 21 km from the park.

We can solve the same pair of equations by eliminating t first. To do so we multiply the first equation by 5 and the second equation by 15. This procedure gives two new equations with the same coefficient for t. If we subtract the lower equation from the upper one, we get

$$
\begin{array}{r}
5y = 75t + 15 \\
- \quad 15y = 75t + 225 \\
\hline
-10y = \qquad -210.
\end{array}
$$

Thus

$$y = 21.$$

This is the same result as before. If you substitute this value of y into either original equation and solve for t, you will again get $t = 1.2$.

These examples are called the *addition and subtraction* method. Strictly speaking we used "subtraction" in both examples. Suppose the physical process had given equations of the type

$$z = 12x + 20$$

and

$$3z = -2x + 7.$$

Then we could solve them by multiplying the second equation by 6 and adding the two equations:

$$
\begin{array}{r}
z = 12x + 20 \\
+ \quad 6z = -12x + 42 \\
\hline
7z = \qquad 62.
\end{array}
$$

The unknown quantity z is then

$$z = \frac{62}{7} = 8.86.$$

The value for x can be found by substituting z into one of the original equations.

Another technique for solving simultaneous equations makes use of substitution. For example, in the problem of the cyclist and walker, we could rewrite the first equation to express t in terms of y:

$$t = \tfrac{1}{15}y - \tfrac{1}{5}.$$

We then substitute this value of t into the second equation to get

$$y = 5(\tfrac{1}{15}y - \tfrac{1}{5}) + 15,$$
$$y = \tfrac{1}{3}y - 1 + 15,$$
$$y\,(1 - \tfrac{1}{3}) = \tfrac{2}{3}y = 14,$$
$$y = \frac{14}{2/3} = 21.$$

As before, we can substitute this value of y into either original equation to get the value for t.

If your simultaneous equations include the square of an unknown quantity, you should use these same techniques to eliminate all but one variable. If the resulting equation still contains a squared term, it may be necessary to use the quadratic formula to find the solution. Such a problem is worked out in detail in Example 8.1. In some cases, for instance the two-dimensional elastic collision of two objects, you will have three simultaneous equations to be solved for three unknown quantities. (Example 8.3 is one such case.) By applying any of the procedures we have just described, you can sequentially eliminate variables and thus reduce the number of equations to a single equation in one variable. The resulting solution can then be used to determine the other unknowns. In general, the number of independent equations you will need must be at least as great as the number of unknowns.

9

Fluids

9.1 Hydrostatic Pressure

9.2 Pascal's Principle

9.3 Archimedes' Principle

*9.4 Surface Tension

9.5 Streamlines and Bernoulli's Equation

*9.6 Viscosity and Poiseuille's Law

*9.7 Stokes's Law

*9.8 Terminal Velocity

*9.9 Turbulent Flow

A WORD TO THE STUDENT

We have now completed our presentation of basic Newtonian mechanics, with applications to motions and interactions of individual particles and objects. Our next step is to discuss the behavior of large collections of particles, such as liquids and gases.

This chapter on the behavior of fluids is largely self-contained, although we assume a familiarity with mechanics as presented in the preceding chapters, especially conservation of energy. We first discuss fluids at rest, and in the later sections we treat fluids in motion. Since the category of fluids includes such common and important substances as air, water, and blood, the ideas in this chapter find application in all areas of science. We have chosen examples from several fields, including biology and engineering. After studying this chapter, you should be able to recognize the proper conditions for applying the various laws of fluid behavior, particularly those of fluid flow. These laws are valid only for certain types of fluids and particular types of flow. They cannot give reliable results beyond their proper domain of applicability.

The study of fluids dates back to some of the earliest discoveries in physics. Many of the principles we examine in this chapter are associated with great scientists of the past, such as Archimedes (third century B.C.), Pascal (seventeenth century), Bernoulli (eighteenth century), and Stokes (nineteenth century). However, the study of fluid flow remains an active area of research. For example, the principles of fluid flow are used to minimize the aerodynamic resistance of a moving car or plane. Weather forecasters use computer simulation methods to model the fluid flow of our atmosphere, trying to understand the origin of hurricanes as well as of stable air currents like the jet stream. The occurrence of stable patterns of fluid flow from seemingly random initial conditions has become an exciting area of contemporary research in physical model-building and computer science.

In this chapter we discuss the fundamental properties of fluids at rest (the study of *hydrostatics*) and in motion (the study of *hydrodynamics*). The treatment here is somewhat different from that of other chapters in that we present most results without derivation. Instead, we rely on physical intuition to show that these results are reasonable. We do this because in many cases the derivations are prohibitively long. However, these ideas are all developed from classical mechanics, primarily Newton's laws and conservation of energy. Although we introduce only the basic ideas of fluid behavior, we touch on a wide range of applications from medicine to engineering to sports.

A **fluid** is any substance that cannot maintain its own shape; in other words, it is a substance that has no rigidity. It can flow and alter its shape to conform to the outlines of its container. This definition includes liquids, such as water; gases, such as air; very slowly flowing substances, such as tar and some plastics; and even some mixtures of solids and liquids that can flow, such as mud. Gases are easily compressible and have no natural volume; that is, they expand to uniformly fill the container in which they are held. Liquids, on the other hand, are practically incompressible, and a given mass of liquid has a characteristic volume. If the liquid volume is less than that of its container, the liquid will have a well-defined surface bounding its volume.

Unless otherwise stated, we deal in this chapter with fluids that cannot be compressed, that is, with liquids. However, we consider fluids to have viscosity, which is the friction that resists the motion of objects through the fluid.

Hydrostatic Pressure

9.1

Consider an upright cylinder containing a liquid (Fig. 9.1). A force acts on the bottom of the cylinder as a result of the weight of the liquid inside. We define the **pressure** P as the *force per unit area*:

$$P \equiv \frac{\text{force}}{\text{area}}. \tag{9.1}$$

TABLE 9.1 Some common units of pressure	
Name	Value $(N/m^2 = Pa)$
1 pascal (Pa)	1
1 bar	1.00×10^5
1 atmosphere (atm)	1.01×10^5
1 mm Hg	1.33×10^2
1 torr	1.33×10^2
1 lb/in² (psi)	6.89×10^3

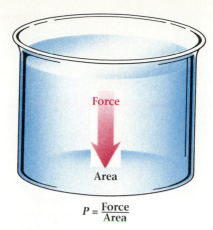

$$P = \frac{\text{Force}}{\text{Area}}$$

Figure 9.1 A fluid exerts a pressure on the bottom of its cylindrical container equal to the total weight of the fluid divided by the area of the bottom of the container.

If we insert the weight of the liquid for the force in Eq. (9.1) we find that

$$P = \frac{\text{weight}}{\text{area}} = \frac{mg}{A},$$

where m is the mass of the liquid, A is the area of the bottom of the cylinder, and g is the acceleration of gravity. Note that although force is a vector quantity, pressure is a scalar.

The different conditions under which pressure measurements are made have led to the development of a variety of commonly used units. The SI unit of pressure, N/m^2, is given the name **pascal** (Pa). Table 9.1 lists several other common units of pressure, along with their relationship to the pascal.

A tire gauge is a familiar pressure-measuring device (Fig. 9.2). One form consists of a hollow cylinder fitted with a spring-loaded piston. When you press the gauge against a tire valve stem, the pressurized air from the tire enters the cylinder. The air exerts a force on the piston that is equal to the pressure in the cylinder times the area of the face of the piston ($F = P \cdot A$). If the spring obeys Hooke's law, the piston is pushed a distance that is proportional to the force and therefore to the pressure. The rod on the end of the piston can be calibrated to read the pressure directly. We will discuss some other pressure-measuring devices later.

It is common to calibrate pressure gauges, such as the tire gauge, so that they indicate only the pressure in excess of atmospheric pressure. This pressure is called *gauge pressure*. As we will discuss in more detail in Chapter 12, the atmosphere exerts a pressure of about 1.013×10^5 N/m^2 (101.3 kPa or 14.7 lb/in²) at sea level and this value must be added to the gauge pressure to give the total pressure.

It is convenient to use the concept of density when discussing pressure. The density of a substance, ρ, was defined in Chapter 5 as its mass per unit volume, so we can write the total mass of an object as the product of its density and its volume. Table 9.2 gives the densities of several gases, liquids, and solids. Equation (9.1) can be written in terms of the density

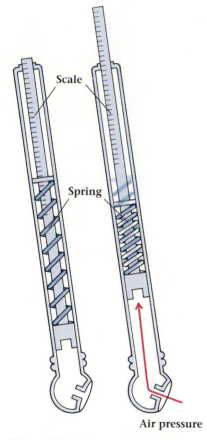

Scale

Spring

Air pressure

Figure 9.2 A tire-pressure gauge. The extension of the scale is proportional to the force on the spring (Hooke's law), which is, in turn, proportional to the air pressure in the tire. When the gauge is removed from the tire, the scale, which is not attached to the spring, remains extended, held in position by friction with the sleeve that holds it.

TABLE 9.2
Densities of some common materials

Material	Temperature (°C)	Density (kg/m³)
Gases*		
Hydrogen	0	0.09
Helium	0	0.18
Nitrogen	0	1.25
Air	0	1.29
Oxygen	0	1.43
Carbon dioxide	0	1.98
Liquids		
Gasoline	20	0.68×10^3
Methyl alcohol	20	0.791×10^3
Water	20	0.998×10^3
Seawater	20	1.03×10^3
Glycerin	20	1.26×10^3
Mercury	20	13.6×10^3
Solids		
Wood, balsa	—	$(0.12–0.20) \times 10^3$
Wood, pine	—	$(0.37–0.64) \times 10^3$
Wood, oak	—	$(0.67–0.79) \times 10^3$
Butter	—	$(0.86–0.87) \times 10^3$
Ice	20	0.92×10^3
Brick	—	$(1.4–2.2) \times 10^3$
Glass, common	—	$(2.4–2.8) \times 10^3$
Aluminum	20	2.70×10^3
Iron	20	7.87×10^3
Lead	20	11.4×10^3
Uranium	20	18.95×10^3
Gold	20	19.3×10^3

*The density given here is for a pressure P_0 of one standard atmosphere = 1.01325×10^5 Pa. The density ρ of a gas at other temperatures and pressures is given by

$$\rho = \rho_0 \left(\frac{P \times 273.15}{P_0 \times (T + 273.15)} \right),$$

where P is the pressure and T is the temperature in degrees Celsius.

as

$$P = \frac{mg}{A} = \frac{\rho V g}{A} = \frac{\rho h A g}{A}$$

or

$$P = \rho g h, \tag{9.2}$$

where ρ is the density and h is the height of the liquid (Fig. 9.3).

We can now see that pressure is directly proportional to both the density and the depth of the liquid. Thus separate containers of different size, holding identical liquids of uniform density, have equal pressure at equal depth. If the two containers are filled to the same height, they have

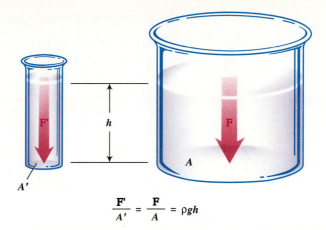

$$\frac{\textbf{F}'}{\textbf{A}'} = \frac{\textbf{F}}{\textbf{A}} = \rho g h$$

Figure 9.3 Containers of different size, filled to the same height with identical fluids of uniform density, have equal pressure at the bottom.

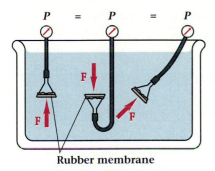

Rubber membrane

Figure 9.4 At constant depth, the pressure exerted by the fluid is the same in all directions.

equal pressure at the bottom, even though the total force on the bottom surface due to the liquid is greater at the bottom of the larger container (Fig. 9.3). The pressure depends on the depth and not on the cross section. A thin upright tube filled to a height of 3 m with water has the same pressure at its bottom as does a large lake that is 3 m deep.

A force also is exerted on the sides of the container, and a corresponding pressure is present there. Moreover, a pressure exists at any point within the body of the liquid. If the liquid is at rest, this pressure is independent of direction. Thus the pressure on the rubber membrane in Fig. 9.4 is the same regardless of the orientation of its surface, provided that the center of the membrane is kept at a constant depth. If the pressures in opposite directions were not the same, a pressure difference would arise, resulting in an unbalanced force acting on the liquid. This would cause the liquid to flow, which contradicts our original assumption that the liquid is at rest. So the pressure must depend only on the height of the liquid above the point in question, according to Eq. (9.2). Thus the difference in pressure ΔP between two points that differ in depth by Δh is

$$\Delta P = \rho g \, \Delta h. \tag{9.3}$$

Example 9.1

Pressure at the bottom of a tank.

A tank is filled with water to a depth of 1.5 m. What is the pressure at the bottom of the tank due to the water alone?

Solution We can compute the pressure from Eq. (9.2). The density of water is approximately 10^3 kg/m^3 and the height h is 1.5 m. So

$$P = (10^3 \text{ kg/m}^3)(9.8 \text{ m/s}^2)(1.5 \text{ m}) = 1.5 \times 10^4 \text{ N/m}^2.$$

The combination of units N/m^2 is the pascal (Pa). Thus the pressure at the bottom of the tank due to the water is 1.5×10^4 Pa, or 15 kPa. Note that the pressure due to the water is completely independent of the size and shape of the tank. Note also that, if we wish to know the total pressure on the bottom of the tank, we must add atmospheric pressure to our answer here.

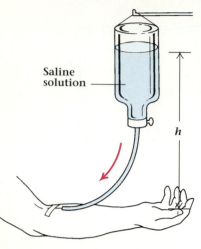

Figure 9.5   Example 9.2: A fluid is fed into a patient's arm from a suspended container. The pressure of the fluid must exceed the pressure in the arm.

Example 9.2

Pressure for an intravenous infusion.

A nurse administers medication in a saline solution to a patient by infusion into a vein in the patient's arm (Fig. 9.5). The density of the solution is 1.0×10^3 kg/m^3 and the gauge pressure inside the vein is 2.4×10^3 Pa. How high above the insertion point must the container be hung so that there is sufficient pressure to force the fluid into the patient?

Solution   The container must be hung high enough that the gauge pressure due to the liquid in the tube and container is at least as great as the gauge pressure inside the vein:

$$P_{\text{liquid}} = \rho g h = 2.4 \times 10^3 \text{ Pa}.$$

Solving for the height h, we get

$$h = \frac{2.4 \times 10^3 \text{ Pa}}{\rho g} = \frac{2.4 \times 10^3 \text{ Pa}}{9.8 \text{ m/s}^2 \times 1.0 \times 10^3 \text{ kg/m}^3},$$

$$h = 0.24 \text{ m} = 24 \text{ cm}.$$

To actually establish a flow through the needle, the container would need to be higher than this result.

Pascal's Principle

9.2

It is often desirable to know the pressure at one point in a fluid when we know it at another. The pressure at any point in the liquid in Fig. 9.6 depends on its depth below the surface h and on any additional pressure (such as atmospheric pressure, P_{atm}) exerted on the liquid above that point. If we know the pressure on the surface of the liquid (P_{atm}) and wish to determine the pressure at point B, our equation reads

$$P_B = P_{\text{atm}} + \rho g h. \tag{9.4}$$

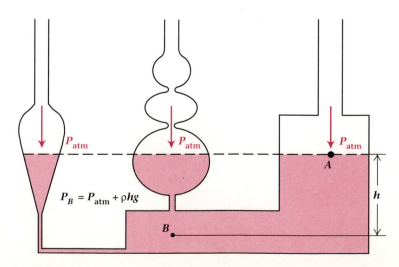

Figure 9.6   Pascal's principle. The pressure at point B is the pressure due to the liquid plus the atmospheric pressure P_{atm} on the surface.

If the pressure at A (that is, P_{atm}) is increased, then the pressure at B is correspondingly increased by the same amount. This fact was recognized by Blaise Pascal (1623–1662) and is embodied in the statement known as **Pascal's principle:** *The pressure applied at one point in an enclosed fluid is transmitted undiminished to every part of the fluid and to the walls of the container.* Pascal's principle holds for gases as well as for liquids, with some minor modifications due to the change in volume of a gas when the pressure is changed.

Example 9.3

You can make a simple hydraulic lift by fitting a piston attached to a handle into a 3-cm-diameter cylinder, which is connected to a larger cylinder of 24 cm diameter (Fig. 9.7). If a 50-kg (110-lb) woman puts all her weight on the handle of the smaller piston, how much weight can be lifted by the larger one?

Solution   By Pascal's principle, the same applied pressure is transmitted everywhere within the enclosed liquid system. In particular, if the heights of the pistons a and b are the same, the pressure on the pistons must be the same, including the pressures due to the applied force. Using subscripts a and b to denote the quantities at each place, we can write

$$P_a = P_b.$$

But the pressure is the force per unit area:

$$\frac{F_a}{A_a} = \frac{F_b}{A_b}.$$

Figure 9.7   Example 9.3: A woman pushes down on the piston a. The pressure is transmitted undiminished to piston b.

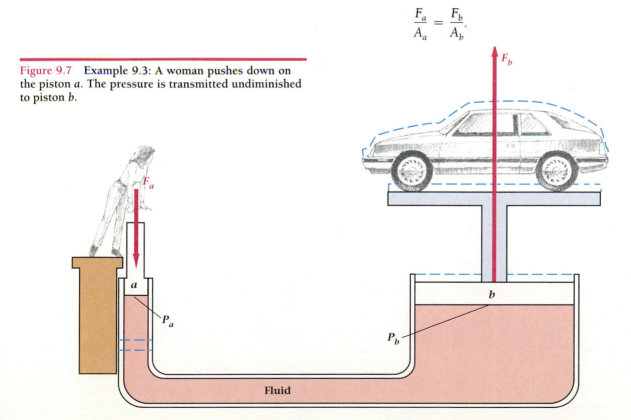

The area of the circular pistons is πr^2, so that

$$\frac{F_a}{\pi r_a^2} = \frac{F_b}{\pi r_b^2}.$$

Solving for F_b gives

$$F_b = F_a \left(\frac{r_b}{r_a}\right)^2.$$

We see that the force is multiplied by the square of the ratio of the radii of the cylinders. In this case, the only force we need to consider is the applied force $F_a = mg$, so the answer is

$$F_b = 50 \text{ kg} \times 9.8 \text{ m/s}^2 \left(\frac{12 \text{ cm}}{1.5 \text{ cm}}\right)^2,$$

$$F_b = 3.14 \times 10^4 \text{ N} \approx 7000 \text{ lb} \approx 3.5 \text{ tons}.$$

This is enough force to lift two Nissan Stanzas, which together weigh 2.6×10^4 N.

PHYSICS IN PRACTICE

Measuring Blood Pressure

Almost any time you get a medical checkup, someone measures your blood pressure. The procedure is one of the most common in medicine: Someone wraps a cuff around your arm, inflates the cuff until it's tight, then listens through a stethoscope held to your arm while letting the cuff slowly deflate. What is happening during this procedure? The answer is that the person is measuring the pressure in a fluid, your blood.

The heart is a large muscle, responsible for pumping oxygen-supplying blood to all parts of the body. It works like two pumps: The blood returns from the body through the veins to the right side of the heart, which pumps the blood to the lungs. The lungs remove carbon dioxide from the blood and add oxygen. The left side of the heart receives the oxygenated blood from the lungs and pumps it throughout the body by way of the arteries. The blood flows from the arteries to the veins through capillary beds.

Two pressures in the heart's action are of particular medical interest: the systolic pressure, when the heart is contracted, and the diastolic pressure, when the heart is relaxed between beats. Normal heart action causes arterial blood pressure to oscillate between these two values. Abnormally high or low arterial blood pressure can sometimes indicate physical and mental conditions of varying degrees of seriousness.

The most direct way of measuring blood pressure is to insert a fluid-filled tube into the artery and connect it to a pressure gauge. Though this is sometimes done, it is neither comfortable nor convenient enough for routine physical examinations.

The commonly used indirect method involves a device called a sphygmomanometer. A nonelastic cuff that has an inflatable bag within it is placed around the upper arm, roughly at the same vertical level as the heart. The cuff is connected directly to some pressure gauge, such as a manometer (Fig. B9.1). When the cuff is inflated, the tissue in the arm is compressed; if sufficient pressure is applied, the flow

of arterial blood in the arm stops. If the cuff is long enough and if it is applied snugly, the pressure in the tissues in the arm is the same as the pressure in the inflated part of the cuff, and is also the same as the pressure in the artery. In effect, Pascal's principle holds for the system composed of cuff, arm, and artery.

After the blood flow has been cut off, the pressure in the cuff is reduced by releasing some of the air. The falling pressure corresponds to the dashed line in Fig. B9.1. At some point the maximum arterial pressure slightly exceeds the pressure in the surrounding tissue and cuff, allowing the blood to resume flowing. The acceleration of the blood through the arteries gives rise to a characteristic sound, which can be identified by means of a stethoscope. When this sound occurs, the manometer indicates the maximum, or systolic, pressure. As the pressure in the cuff falls further, a second change in the sound is heard, characteristic of the drop below diastolic pressure. The readings shown in Fig. B9.1 correspond to the two pressures and are reported as "110 over 80," which is a

Example 9.4

Pressure on a scuba diver.

A scuba diver searches for treasure at a depth of 20.0 m below the surface of the sea. At what pressure must the scuba (self-contained underwater breathing apparatus) device deliver air to the diver?

Solution The pressure at the diver's depth is greater than atmospheric pressure because of the water above the diver. If the air breathed in is not at the same pressure as the external pressure on the diver's chest, the excess pressure will collapse the chest. Thus the breathing apparatus must deliver air to the diver at the pressure of the surrounding water.

The pressure on the diver is

$$P = P_{atm} + \rho g h,$$

where P_{atm} is the pressure due to the atmosphere pressing down on the sea (101.3 kPa), ρ is the density of sea water (approximately 1030 kg/m^3), and h is the depth of the diver below the surface. When these values are

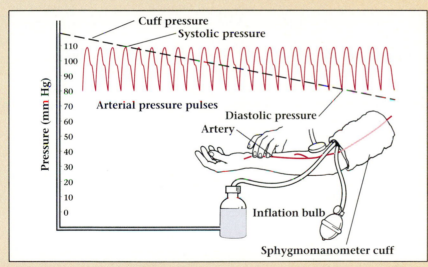

Figure B9.1 Measuring blood pressure with a sphygmomanometer. Identifiable sounds occur in the arm when the cuff pressure falls below the systolic and diastolic pressures.

ments have been used for decades in human medicine, where repeatability and practicality are more important than accuracy. Even though we know that the best cuff size for accurate measurement of systolic pressure is different from the best size for accurate measurement of diastolic pressure, most medical practitioners use a single cuff size for each patient.

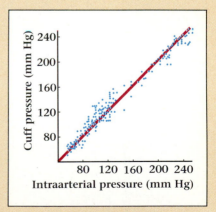

Figure B9.2 Comparison of blood pressure measurements by sphygmomanometry with direct measurements of arterial pressure.

typical value of the pressures for a healthy person.

The measurements made by this technique are subject to some variation because of the need to fit the cuff properly and the need to estimate reliably the point at which the sound changes. Other factors, such as the condition of the manometer, the size of the arm, and the rate at which the cuff is inflated and deflated, can have an effect. Figure B9.2 shows a comparison between pressures measured directly in the artery and pressures measured indirectly by a type of sphygmomanometer.

Indirect blood pressure measure-

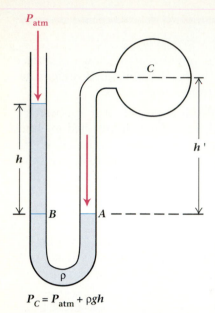

P_{atm}

C

h'

h

B A

ρ

$$P_C = P_{atm} + \rho gh$$

Figure 9.8 An open U-tube manometer. The difference in the heights of the liquid indicates the pressure in the chamber C.

inserted into the equation, we get

$$P = 101.3 \times 10^3 \text{ Pa} + 1030 \text{ kg/m}^3 \times 9.80 \text{ m/s}^2 \times 20.0 \text{ m},$$
$$P = 101.3 \times 10^3 \text{ Pa} + 202 \times 10^3 \text{ Pa} = 303 \text{ kPa}.$$

When expressed in terms of atmospheres, the pressure becomes

$$P = 303 \text{ kPa} \left(\frac{1 \text{ atm}}{101.3 \text{ kPa}} \right) = 2.99 \text{ atm} \approx 3 \text{ atm}.$$

In other words, at a depth of 20.0 m, the pressure is three times the pressure at the surface.

In addition to the tire gauge mentioned earlier, there are several other pressure-measuring devices. One of these is the *manometer*, which measures a pressure by balancing a force due to the pressure with the weight of a column of liquid (Fig. 9.8). The pressure to be measured by this open-tube manometer is the pressure in the chamber C. If the density of air is neglected, or the height h' is not too great, then Pascal's principle says that the pressure at A is the same as at C. Pascal's principle further says that the pressure at B is the same as at A. Therefore we can determine the gauge pressure at C by measuring the difference in the levels (heights) of the liquid h. This, as we saw in Section 9.1, is ρgh, where ρ is the density of the liquid. In some cases, it is common to state the pressure in terms of the height h. Thus blood pressures are given as so many millimeters of mercury (mm Hg).

A barometer is a type of manometer used for measuring atmospheric pressure. A long tube is filled with mercury and inverted in a bowl of mercury (Fig. 9.9a). The mercury drops in the tube until it is about 76 cm above the level of the mercury in the bowl. The atmosphere exerts a pressure on the mercury in the bowl that just balances the column of mercury. Small changes in atmospheric pressure, which are associated with changes in the weather, can be read from the barometer. Aneroid

Figure 9.9 (a) Mercury barometer. (b) Aneroid barometer.

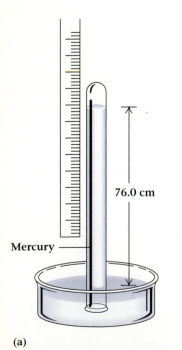

76.0 cm

Mercury

(a)

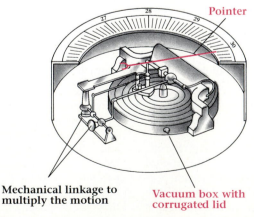

Pointer

Mechanical linkage to multiply the motion

Vacuum box with corrugated lid

(b)

barometers are also commonly used to measure atmospheric pressure. The basic component of an aneroid (without liquid) barometer is a partially evacuated, sealed chamber made of thin metal with corrugated sides (Fig. 9.9b). The chamber expands and contracts in response to changes in atmospheric pressure. This movement of the chamber walls is transmitted to a pointer by a mechanical linkage.

Archimedes' Principle

9.3

The story has been told that Archimedes (287–212 B.C.) conceived of the principle that bears his name after King Hiero asked him to determine the actual composition of his crown, which was alleged to be pure gold. Archimedes was ordered to do so without damaging the crown. According to legend, the Greek scientist's inspiration came to him as he lay partially submerged in his bath. On getting into the tub he observed that the more his body sank into the tub, the more water ran out over the top. He immediately jumped out of the tub and rushed through the streets naked, shouting excitedly in a loud voice "Eureka" ("I have found it"). **Archimedes' principle** says that: *A body, whether completely or partially submerged in a fluid, is buoyed upward by a force that is equal to the weight of the displaced fluid.* How this principle allowed Archimedes to solve the problem of the king's crown is seen in Problem 9.23.

Archimedes' principle can be obtained by a simple nonmathematical argument. Consider an object of arbitrary but known shape, such as an aluminum cube, that hangs completely submerged in a fluid, such as water (Fig. 9.10a). Now imagine that you remove the cube and replace it with a volume of water having the same size and shape as the "missing cube." (This is the amount of fluid displaced by the object.) Since this volume of fluid does not sink to the bottom of the container, the surrounding fluid must provide a net upward force, which counteracts the downward force due to the weight of fluid in the volume (Fig. 9.10b).

This upward force due to the surrounding water is the **buoyant force** and is equal to the weight of fluid in the volume displaced by the object. If you replace the aluminum cube with a lead cube of the same size, the

Figure 9.10 (a) An aluminum cube suspended in water. (b) An imaginary volume in the water, equal to the volume of the cube. The difference between the upward force and the downward force on the volume is equal to the weight of the water in the volume. (c) If the string is cut, the cube sinks because it is denser than water.

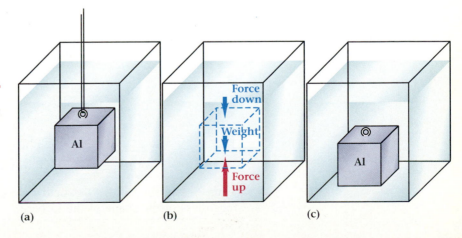

(a)   (b)   (c)

Figure 9.11 The Goodyear airship *Europa* above the British research vessel *Eye of the Wind*. Although both the airship and the boat contain materials of high density, their average density is less than or equal to that of the fluid in which they float.

buoyant force remains the same: It depends only on the volume of the submerged object, not on its mass or density. Of course, since aluminum is denser than water, the buoyant force alone cannot support the cube, which sinks to the bottom of the container if the supporting string is cut (Fig. 9.10c). If the fluid were mercury, which has a density greater than that of aluminum, the aluminum cube would float. We can generalize to a case in which the body is partially submerged, as in the following Example 9.5, of an iceberg.

Blimps, hot-air balloons, and other lighter-than-air craft furnish another example of Archimedes' principle. They float in the air just as a submerged fish floats in the water. The blimp in Fig. 9.11 obviously has individual parts that will not float in air. However, the *average* density of the whole craft, including passengers, must be less than the density of air to accomplish an unpowered takeoff. To meet this requirement in a practical way, the blimp contains a large volume of helium. The same type of observation can be made about large ships, which float even though they are made of steel and carry dense objects. The explanation is that the average density of the entire ship, including all of the air spaces, is less than the density of water.

Example 9.5

Buoyant force on an iceberg.

Icebergs are made of fresh-water ice, which has a density ρ_i of 0.92×10^3 kg/m³ at 0°C. Ocean water, largely because of the dissolved salt, has a density ρ_w of about 1.03×10^3 kg/m³. What fraction of an iceberg lies below the surface?

Solution For simplicity, assume the iceberg to be a cube of volume L^3. The downward force acting on it is due to gravity:

$$F_{\text{grav}} = mg = \text{volume} \times \text{density} \times g = L^3 \times \rho_i \times g.$$

The buoyant force, according to Archimedes' principle, depends on the volume of ice submerged in the water. If the bottom of the floating cube is a depth d below the surface of the water, then the buoyant force is

$$F_{\text{buoyancy}} = d \times L^2 \times \rho_w \times g.$$

These forces are equal when the iceberg is floating in equilibrium. Therefore

$$F_{\text{buoyancy}} = F_{\text{grav}}$$

or

$$dL^2 g\rho_w = L^3 g\rho_i$$

and

$$\frac{d}{L} = \frac{\rho_i}{\rho_w} = \frac{0.92}{1.03} = 0.89.$$

Thus 89% of the iceberg lies below the surface. This turns out to be true regardless of the shape of the iceberg.

Surface Tension

*9.4

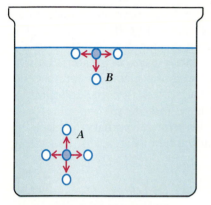

Figure 9.12 A molecule inside a liquid (point *A*) is attracted to other molecules on all sides. A molecule on the surface (point *B*) is attracted only to other molecules in and below the surface. As a result, the surface acts like a membrane under tension.

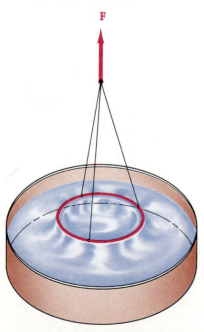

Figure 9.13 A force is required to lift a ring out of the liquid.

Most people have observed the phenomenon called surface tension. You have seen spherical water drops hanging from a spider's web; or objects denser than water, like a needle or a razor blade, floating on the surface; or a water bug skittering over a pond. In each case, the liquid surface exhibits some of the characteristics of a stretched membrane under tension.

The molecules that compose a liquid attract one another; otherwise the liquid would not have a definite volume. Most liquids are nearly incompressible because the intermolecular forces are repulsive at distances less than their normal separations. At any point *A* inside a liquid at rest, the net molecular force on a molecule of the liquid is zero because the molecules surrounding it provide opposing forces in balance (Fig. 9.12). However, as shown, at a point *B* on the surface of the liquid the situation changes. A molecule experiences attractive forces only in the direction of the other molecules of the liquid, and since at the surface these forces are not balanced by attractive forces in the outward direction, there is a net inward force. This inward force makes the surface act like a stretched drumhead. The surface of a liquid resists any effort to increase its area.

We can determine values of surface tension by measuring the force necessary to lift a ring out of the liquid (Fig. 9.13). The required force is proportional to *C*, the circumference of the ring. As the ring rises, a thin film of liquid clings to it until the weight of the film exceeds the attractive forces holding the film together. The film has two surfaces, one inside and one outside of the ring. Thus the total length along which the lifting force acts is 2*C*. The **surface tension** γ is the ratio of the surface force to the length along which it acts:

$$\gamma = F/2C. \tag{9.5}$$

Notice that the surface tension is not simply a force but is a force divided by a length. The units of surface tension are N/m. Table 9.3 lists the surface tensions of some common liquids.

An equivalent way of thinking about surface tension is to think of it as the energy per unit area of the surface. That this is reasonable can be seen by writing the units N/m as N·m/m², which is J/m². The equilibrium

TABLE 9.3 Measured values of surface tension		
Liquid in contact with air	Temperature (°C)	Surface tension (10^{-3} N/m)
Water	0	75.6
Water	25	72.0
Water	80	62.6
Ethyl alcohol	20	22.8
Acetone	20	23.7
Glycerin	20	63.4
Mercury	20	435

Figure 9.14 Close-up of the splash of a milk drop falling on a hard surface.

configuration of a surface corresponds to its lowest possible energy. One consequence is that liquids try to minimize their surface area. As a result, liquids tend to assume the shape of a sphere, which is the shape with the smallest surface area (and therefore energy) for a given volume. Figure 9.14 is a view of a splash of milk, showing the detached spheres in the process of separating from the body of liquid. For the same reason, soap bubbles tend to take the shape of spheres.

The attractive force between like molecules that acts to hold a liquid together is called cohesion. The attractive force between unlike molecules—say, between those of water and glass—is called adhesion. The adhesive forces may be as great as, or even greater than, cohesive forces. For example, the adhesion of water to clean glass is greater than the cohesion of water to itself, so the water wets the glass. By contrast, the adhesion of water to a newly waxed surface is less than the cohesion, so water does not wet the surface, but beads up into drops.

If a tube of relatively small internal diameter (a capillary) is placed in a liquid that wets it, the liquid rises up in the tube (Fig. 9.15). This effect,

PHYSICS IN PRACTICE

Surface Tension and the Lungs

Take a deep breath. You've probably never thought about it, but some interesting physics goes on in your lungs every time you breathe. It all takes place without your having to think about it, but the interplay is fascinating.

The air passages into the two lungs branch and branch again until they end in tiny air sacs called *alveoli* (Fig. B9.3). It is here that the exchange of gases with the blood takes place. An adult's lungs contain on the order of 600 million alveoli, each with an average radius of 60 μm. The surface tension of the material that coats the inside of the alveoli governs several important functions of the lungs.

First, consider two soap bubbles connected to a pipe that contains a valve (Fig. B9.4). This arrangement is sometimes shown to a class and the question asked: "What will happen if the valve is opened?" When the valve

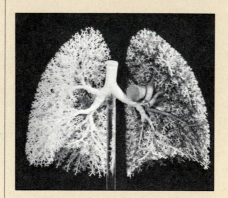

Figure B9.3 A plastic cast of the air passages in the lungs. The passages end in tiny sacs called alveoli.

is opened, the larger bubble grows and the smaller bubble becomes smaller until it completely goes away. We can understand this observation from the standpoint of energy. Just as a free drop of liquid takes on a spherical shape to minimize its total surface energy, the bubbles change relative sizes in such a way as to minimize their combined surface area, and therefore the surface energy. The surface area of a single bubble is about 30% less than the surface area of two smaller bubbles of equal size that together have the same volume as the larger bubble. So, when the valve is opened, the two bub-

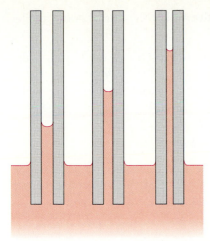

Figure 9.15 The internal diameters of the three tubes decrease from left to right.

known as *capillary action,* is due to the cohesive forces of surface tension and the adhesive forces between the liquid and the glass tube. The liquid rises until the upward force due to surface tension is equaled by the weight of the liquid in the tube.

The surface tension of a liquid, such as water or oil, may be greatly reduced by the addition of certain chemicals called surfactants, or surface-active agents. Detergents are one type of surface-active agent. The lowering of the surface tension of water by detergents increases the ability of water to "wet" a surface. This improved wetting allows more water molecules to penetrate cloth fibers and wash away particles of soil.

Example 9.6

Capillary action.

Show that the height to which a liquid rises in a narrow tube is given by $h = 2\gamma \cos\theta / r\rho g$, where r is the radius of the column of liquid, ρ is its density, g is the acceleration of gravity, γ is the surface tension, and θ is

---

bles become one system and form a single bubble.

If the effect of large bubbles absorbing smaller ones were to take place in

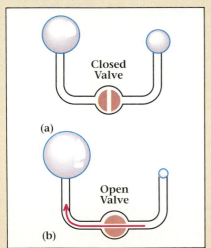

Figure B9.4 (a) Two soap bubbles of different radius are connected through a pipe. (b) When the valve is opened, the larger bubble grows as the smaller one shrinks.

the lungs, the smaller alveoli would all collapse and the larger ones would grow. This does not happen because of the pulmonary surfactant, the material that coats the inside of the lungs. Experiments have shown that the surface tension of the pulmonary surfactant *increases* with its area, in contrast with the behavior of water and most other liquids. This means that the surface energy of larger alveoli, in spite of the smaller surface-to-volume ratio, can be the same as that of the smaller alveoli. Therefore the larger and smaller air sacs can exist in equilibrium.

The experimental results that show the variation of surface tension with area also explain another observation about the lungs. If you take a big breath and then relax your chest muscles, the air is expelled from your lungs. Part of the reason that the air flows out is that the lungs have been blown up, much like a balloon, and the elasticity of the tissues causes them to contract and force the air out. However, tests have shown that the elasticity of the lung

tissue itself is not sufficient to completely explain what happens. A large part of the effect is due to the surface tension in the alveoli, which causes them to contract and expel the air. Furthermore, the effect depends on a material for which the surface tension increases with area, or else the tendency of the lungs to expel air when inflated would be much less than it is.

Normal breathing is possible only when the pulmonary surfactant is present in the proper amount and has the proper surface tension. If the surface tension is greater than normal, the lungs tend to collapse and breathing is difficult. If the surface tension is lower than normal, the tendency of the lungs to expel air when inflated decreases. Some newborn babies, especially premature ones, do not have enough surfactant, making lung expansion difficult. Unless they receive immediate treatment, these infants die soon after birth because of inadequate ventilation. This condition is known as respiratory distress syndrome.

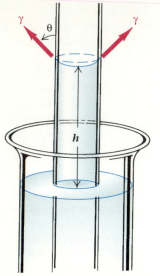

Figure 9.16 Example 9.6: Analysis of the height to which a liquid rises in a capillary tube.

the angle the liquid surface makes with the wall at the line of contact (Fig. 9.16).

Solution   First note that, for equilibrium, the weight of the liquid in the tube must be equal to the upward component of the adhesive force due to surface tension. The weight, or downward force F_d, is given by the density of the liquid times the volume of the column times the acceleration of gravity:

$$F_d = \rho g V = \rho g h \pi r^2.$$

The upward force F_u, due to the surface tension acting all around the inside of the cylindrical surface, is the product of the upward force per unit length ($\gamma \cos \theta$) times the inner circumference of the cylinder:

$$F_u = 2\pi r \gamma \cos \theta.$$

Equating the upward and downward forces gives

$$\rho g h \pi r^2 = 2\pi r \gamma \cos \theta,$$

or

$$h = 2\gamma \cos \theta / r \rho g.$$

Streamlines and Bernoulli's Equation

9.5

Let us now turn our attention to fluids in motion. Consider an incompressible fluid flowing steadily through a tube from a region of cross-sectional area A_1 to a region of area A_2 (Fig. 9.17). Because the fluid is incompressible, the same amount of it leaves each region toward the right per unit time as enters from the left per unit time. The volume of fluid that flows into the tube across A_1 in a time interval Δt is $\Delta V_1 = A_1 v_1 \, \Delta t$, where v_1 is the velocity of the fluid at A_1. If the density of the fluid is ρ, then the mass of fluid that flows into the tube in time Δt is $\rho A_1 v_1 \, \Delta t$. Similarly, the mass of fluid that flows out of the tube through A_2 in the same time Δt is $\rho A_2 v_2 \, \Delta t$. Since the mass of fluid entering is the same as

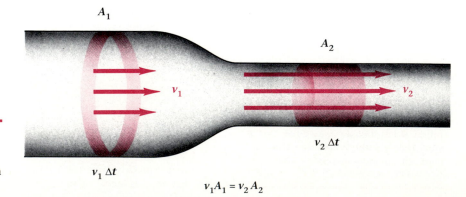

Figure 9.17 Fluid passing cross section A_1 with speed v_1 passes through cross section A_2 with a new speed v_2, as required by the equation of continuity.

the mass leaving, we get

$$\rho A_1 v_1 \, \Delta t = \rho A_2 v_2 \, \Delta t.$$

We can divide out the density because it is constant for an incompressible fluid. Then we get

$$v_1 A_1 = v_2 A_2. \tag{9.6}$$

This equation is called the **equation of continuity** and will be useful throughout our discussion of fluids in motion. It states that the flow of material (mass) through a tube of changing cross section is constant when the density of the fluid does not change. That is, the equation of continuity is a statement of conservation of mass. Notice that the product Av is the volume rate of flow, that is, the volume of fluid passing a given cross section per unit time. In SI units the volume rate of flow is measured in m^3/s.

Example 9.7

Calculating velocity change from the equation of continuity.

A horizontal pipe of 25 cm^2 cross section carries water at a velocity of 3.0 m/s. The pipe feeds into a smaller pipe with a cross section of only 15 cm^2. What is the velocity of water in the smaller pipe?

Solution   We can determine the velocity of water in the smaller pipe from the equation of continuity. The velocity and area in the large pipe are $v_1 = 3.0$ m/s and $A_1 = 25$ cm^2. The area of the smaller pipe is $A_2 = 15$ cm^2. Thus

$$v_2 = \frac{v_1 A_1}{A_2} = \frac{3.0 \text{ m/s} \times 25 \text{ cm}^2}{15 \text{ cm}^2} = 5.0 \text{ m/s}.$$

We wish to consider first the flow of a fluid that not only is incompressible but has no internal friction, or viscosity. (We will introduce the effects of viscosity in Section 9.6.) Figure 9.18 shows a fluid flowing from region A to region B. If the neighboring layers of fluid move past each other smoothly, each small element of fluid follows a path called a **streamline**. This smooth streamline flow is known as **laminar** flow. If we draw all of the streamlines from the boundary of region A to some later position B, we outline a tube of flow. The equation of continuity (Eq. 9.6) applies to a tube of flow. When the fluid exceeds a certain critical velocity, the

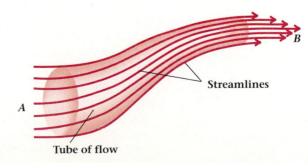

Figure 9.18   Fluid passing through area A later passes through area B. The streamlines mark the paths of small elements of the fluid. The tube that connects A to B along the streamlines is the tube of flow.

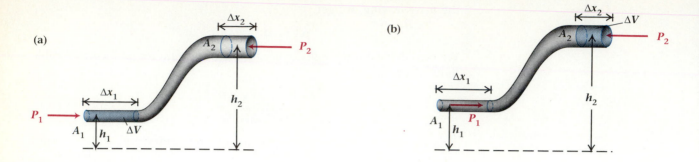

Figure 9.19   A volume ΔV of incompressible fluid is moved along a tube of flow from (a) region 1 to (b) region 2.

flow is no longer laminar but becomes **turbulent** and is characterized by an irregular, complex motion. (Turbulent flow will be considered in Section 9.9.)

We wish to find a relationship among the variables that describe laminar flow. To do so we make use of the principle of conservation of energy. In particular we calculate the work done on a small element of fluid moving along a tube of flow and then use the work–energy theorem to equate the change in kinetic energy to this work. To move a small element of fluid through a distance of Δx_1 at region 1 (Fig. 9.19) requires an amount of work $P_1 A_1 \, \Delta x_1$. At the same time, the same amount of fluid (given by $A_2 \, \Delta x_2$) moves a distance Δx_2 at region 2. The work in this case is $-P_2 A_2 \, \Delta x_2$. The negative sign indicates that the element of fluid at region 2 moves against the force due to the pressure of the fluid to its right.

Because the fluid is not compressible, the volume of the fluid displaced at region 1 is equal to the volume of the fluid displaced at region 2; $A_1 \, \Delta x_1 = A_2 \, \Delta x_2$. The work done against gravity in the net motion of fluid ($m = \rho A_1 \, \Delta x_1 = \rho A_2 \, \Delta x_2 = \rho \Delta V$) from region 1 to region 2 is $g\rho \, \Delta V \, (h_2 - h_1)$. So the total work done is

$$P_1 \, \Delta V - P_2 \, \Delta V + \rho g \, \Delta V h_2 - \rho g \, \Delta V h_1.$$

According to the work–energy theorem, the change in kinetic energy of the mass $\rho \, \Delta V$ is equal to this work, so that

$$\tfrac{1}{2}\rho \, \Delta V v_1^2 - \tfrac{1}{2}\rho \, \Delta V v_2^2 = P_1 \, \Delta V - P_2 \, \Delta V + \rho g \, \Delta V h_2 - \rho g \, \Delta V h_1.$$

Dividing through by ΔV and rearranging terms gives

$$\boxed{P_1 + \rho g h_1 + \tfrac{1}{2}\rho v_1^2 = P_2 + \rho g h_2 + \tfrac{1}{2}\rho v_2^2 = \text{constant.}} \qquad (9.7)$$

This expression is called **Bernoulli's equation**.* It describes the relationship of a fluid's pressure, velocity, and height as it moves along a pipe or other tube of flow. But remember, Bernoulli's equation is valid only for incompressible fluids of negligible viscosity in laminar flow. Under these conditions, Bernoulli's equation expresses conservation of energy in a moving fluid.

---

*This problem was first solved by the Swiss mathematician and physicist Daniel Bernoulli (1700–1782), who published it in a book on fluid flow in 1738.

If we consider a horizontal pipe ($h_1 = h_2$), the equation of continuity tells us that the fluid flows more rapidly in a constricted region of the pipe. If we combine the equation of continuity [$v_1 = (A_2/A_1)v_2$] with Bernoulli's equation we get

$$P_2 = P_1 + \frac{\rho v_2^2 (A_2^2 - A_1^2)}{2A_1^2}. \tag{9.8}$$

The term in parentheses on the right-hand side is negative if A_1 is greater than A_2. In that case, the pressure P_2 is less than the pressure P_1. This result, which arises from the principles of conservation of energy and mass, tells us that the pressure is less in the constricted region of the pipe. Similarly, if A_2 is greater than A_1, then P_2 is larger than P_1. In other words, when a moving fluid enters a narrower section of pipe (or artery), its speed increases but the pressure on the fluid decreases.

Equation (9.7) holds strictly only for incompressible nonviscous fluids. But the general qualitative conclusion above applies to both gases and liquids. If we consider an incompressible but viscous fluid, such as water or blood, we get

$$P_2 < P_1 + \frac{\rho v_2^2 (A_2^2 - A_1^2)}{2A_1^2}.$$

The inequality arises from noticing that in a viscous fluid, some of the work done is dissipated by the internal frictional forces in the liquid.

Example 9.8

Calculating pressure differences with Bernoulli's equation.

Determine the pressure change that occurs on going from the larger-diameter pipe to the smaller pipe for the conditions of Example 9.7; that is, take $A_1 = 25$ cm^2, $A_2 = 15$ cm^2, and $v_2 = 5.0$ m/s.

Solution   Since the pipe is horizontal, we can find the pressure change $P_2 - P_1$ from Eq. (9.8). It is

$$\Delta P = P_2 - P_1 = \rho v_2^2 \frac{(A_2^2 - A_1^2)}{2A_1^2}.$$

Here ρ is the density of water $= 10^3$ kg/m^3. Therefore

$$\Delta P = (10^3 \text{ kg/m}^3)(5.0 \text{ m/s})^2 \left(\frac{(15 \text{ cm}^2)^2 - (25 \text{ cm}^2)^2}{2(25 \text{ cm}^2)^2} \right),$$

$$\Delta P = -8.0 \times 10^3 \text{ Pa}.$$

Note that the pressure is smaller in the smaller pipe.

Let us now consider an example of Bernoulli's equation. Note that the effects are based on the relationship expressed in Bernoulli's equation between fluid velocity and pressure.

The speed of a fluid flowing in a pipe can be measured with a device called a *Venturi meter* (Fig. 9.20). If the fluid in the pipe is flowing, the pressure at B is lower than the pressure at A. The difference in pressures

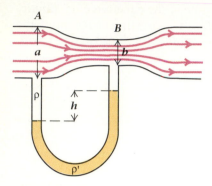

Figure 9.20   A Venturi meter enables us to measure fluid speed by observing height differences in a U-tube.

is a function of the rate at which the fluid flows. The Venturi meter measures this difference in pressure with the U-shaped tube. Using Bernoulli's equation (Eq. 9.7) and the equation of continuity (Eq. 9.6), we can derive the following relationship for the velocity of the fluid at position A:

$$v = b \left(\frac{2(\rho' - \rho)gh}{\rho(a^2 - b^2)} \right)^{1/2},$$

where a and b are the cross-sectional areas at A and B, respectively, ρ is the density of the fluid flowing in the pipe, and ρ' is the density of the liquid in the U-tube. The difference in height of the liquid in the two arms of the U-tube is h and the gravitational acceleration is denoted by g. Thus the Venturi meter enables us to determine the flow velocity from a measurement of the height difference h.

PHYSICS IN PRACTICE

How Airplanes Fly

Air travel is one of the great triumphs of the twentieth century. Every day hundreds of thousands of people are carried through the air to destinations all around the world. In every case the flight of heavier-than-air craft results from the flow of air around their wings.

Before their first powered flight in December 1903, the Wright brothers tested many different wing shapes in a wind tunnel to find the shape that produced the most lifting force. This shape, often called an airfoil, is shown in Fig. B9.5 along with the streamlines of the air moving past. The fluid moving over the top travels a greater distance than that moving just under the bottom of the wing. Consequently, the fluid moving over the top must travel faster in order to conform with the shape of the wing and still maintain the natural streamline. The shape of the wing also crowds the streamlines together above the wing, just as in the case of a constricting pipe. The result is that the region immediately above the wing experiences reduced pressure relative to the region immediately below the wing. Because the downward force on the top of the wing is less than the upward force on the bottom, a net upward force, or lift, arises from the air flow. (Beyond the airfoil the flowing air has a downward component of velocity. By Newton's third law, the reaction force to the net downward force exerted on the air is the lift.) Note that for lift to occur, a flow of air is required relative to the wing. The lift occurs equally well for a wing moving through stationary air or for air moving past a stationary wing.

You can demonstrate this effect with a small piece of paper, about 10 × 15 cm. Hold the short edge close below your lower lip and blow vigorously across the top of the paper (Fig. B9.6). The motion of the air above the paper will cause it to rise. This same effect helps to lift a plane into the air.

In addition, the angle of attack, or tilt of the wing relative to the air flow, can be changed to get additional lift from the deflection of the air stream (see Fig. B9.5). If the leading edge of the wing is higher than the trailing edge, the force of air against the underside of the wing is greater than its force against the upper side. In this case lift

Figure B9.5   An airfoil in a moving fluid.

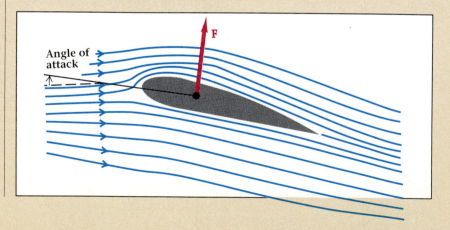

Viscosity and Poiseuille's Law

*9.6

Bernoulli's equation predicts that for a horizontal pipe of uniform cross section, the pressure in a moving fluid is constant. If there were no fluid friction, or viscosity, this would indeed be the case. However, for a viscous fluid flowing through a horizontal pipe of uniform cross section, the fluid pressure decreases with distance along the direction of flow.

Viscosity is that property of a fluid that indicates its internal friction. The more viscous a fluid, the greater the force required to cause one layer of fluid to slide past another. Viscosity is what prevents objects from moving freely through a fluid, or a fluid from flowing freely in a pipe. The viscosity of gases is less than that of liquids, and the viscosity of water

occurs even for a flat wing. However, if the angle of attack becomes too great, the streamline flow gives way to turbulence and the pressure difference is no longer maintained. If the turbulence is great enough, the lift diminishes and the plane stalls.

In general, as the flow of air past the wing increases, both the lift force and the drag force (the resistance to forward motion) increase. Aircraft wings are designed so that pilots can change the wing shape during flight, produc-

ing greater lift for the slower speeds of takeoff and landing and producing less drag at cruising speeds. During takeoff and landing, flaps are extended backward and downward from the trailing edge of the wing (Fig. B9.7), increasing lift by imparting a greater downward velocity to the air. On some planes, extending the flaps increases the wing area as much as 25%, resulting in a much increased drag. At the same time, the leading edge of the wing may be moved forward, creating a slot that di-

Figure B9.7   Flaps are extended during takeoff and landing to increase the lift.

rects a high-speed layer of air over the top surface of the wing to reduce turbulence and increase lift. At higher speeds, the pilot closes the slot and retracts the flaps to reduce the drag forces. Passengers in commercial aircraft can easily see these changes in the wing during flight.

We should point out that lift is not in strict accord with Bernoulli's equation. The reason is that the Bernoulli equation holds exactly only for incompressible nonviscous fluids, yet air is both compressible and viscous. However, the pressure difference, and hence the lift, does occur in air, even if the amount is not in exact agreement with Eq. (9.7).

Figure B9.6   A demonstration of lift. Blowing across the paper causes it to rise.

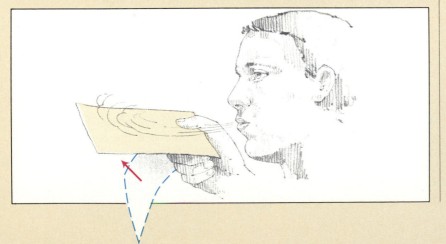

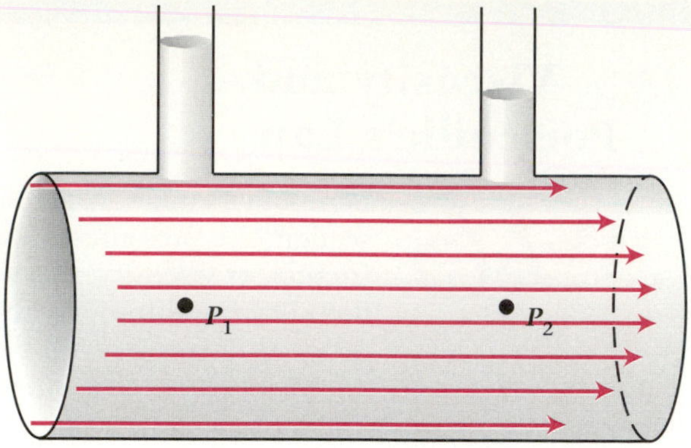

Figure 9.21 The pressure in a viscous fluid along a horizontal pipe of uniform cross section diminishes with distance along the direction of flow. Thus pressure P_2 is less than pressure P_1. Note, too, that fluid flows fastest in the center of the pipe.

and light oils is less than that of molasses and heavy oils. Your experience with fluids such as motor oils and syrups tells you that viscosity increases with decreasing temperature. Thus, it is hard to start a car engine in subzero weather when the oil is thick and flows slowly, but it is easy to start the same car on a hot summer day when the oil is warm and flows readily.

An analogy is often used to help explain the microscopic mechanism of viscosity. Consider two railroad trains traveling at different speeds in the same direction on parallel tracks. Each train carries a large number of identical sandbags. As the faster train passes the slower train, the bags on each train are thrown to the other train. The bags from the slower train have less momentum than equivalent bags on the faster train and therefore slow the faster train down when they land. Conversely, the bags from the faster train tend to speed up the slower train. If this exchange continued indefinitely, the trains would come to the same speed, just as if there were a frictional or viscous force between them.

Let us return to the situation of a fluid moving through a horizontal pipe. The walls of the pipe exert a resistive force, or drag, on the adjacent layers of fluid. These layers, in turn, slow down the next adjacent layers, and so on. As a result, the rate of flow is slowest near the pipe walls and fastest in the center of the pipe. Therefore, for a given rate of flow the pressure difference between two points along the length of the pipe depends on the radius of the pipe (Fig. 9.21). The pressure difference between the two points is also related to a quantity known as the coefficient of viscosity or simply the viscosity of the fluid. The exact relation is given by the following equation, called **Poiseuille's law**:

$$P_1 - P_2 = 8\,\frac{Q\eta L}{\pi R^4}, \tag{9.9}$$

where Q is the flow rate in m^3/s, η is the coefficient of viscosity, R is the radius of the pipe, and L is the separation between the test points. If R and L are given in meters and the pressure is given in pascals, the unit of the coefficient of viscosity is the pascal-second (Pa·s). This equation is often used experimentally to determine the coefficient of viscosity of a liquid. Table 9.4 lists the coefficients of viscosity for a number of fluids.

TABLE 9.4
Viscosities of some common fluids

Fluid	Temperature (°C)	Viscosity (10^{-3} Pa·s)
Air	20	0.018
Water	40	0.656
Water	20	1.005
Motor oil (SAE 10)	30	200
Glycerin	20	1490
Castor oil	20	986
Mercury	20	1550

Example 9.9

Blood flow in arterioles.

Blood in the extremities of the body is carried by arterioles, small vessels with an average diameter of about 0.1 mm. The muscles in the walls can contract and change the diameter of the "pipe," thereby decreasing the flow of the blood, a viscous fluid. Sometimes great strain or shock causes a severe reduction in blood flow. Approximately by what amount would the arterioles have to contract to reduce the blood flow to 30% of its former value if we assume that the pressure drop remains constant?

Solution   We use Poiseuille's law, writing subscripts 1 for the normal condition and 2 for the reduced-flow case. Then, because the pressure drop is constant, we have

$$8 \, \frac{Q_1 \eta_1 L}{\pi R_1^4} = 8 \, \frac{Q_2 \eta_2 L}{\pi R_2^4}.$$

Though the viscosity of the blood is not independent of its velocity or of the diameter of the tube through which it flows, we assume it to be constant over the range considered here. Then we have

$$\frac{R_2^4}{R_1^4} = \frac{Q_2}{Q_1}, \quad \text{or} \quad \frac{R_2}{R_1} = \left(\frac{Q_2}{Q_1}\right)^{1/4}.$$

If the flow is reduced to 30% of its initial value, then $Q_2/Q_1 = 0.30$. Inserting this in the above equation, we find that

$$R_2/R_1 = (0.30)^{1/4} = 0.74.$$

The arteriole has constricted to 74% of its original diameter. For example, if the diameter of the arteriole were 0.10 mm originally, then its diameter after constriction would be 0.074 mm. The relative change in flow is greater than the relative change in the diameter of the tube.

In general, the diameter of blood vessels, unlike that of glass or metal tubes, increases as the internal pressure increases, because the blood vessels are distensible. Moreover, in blood vessels as small as capillaries, the viscosity of whole blood is as little as half of what it is in large vessels. This effect is due to the alignment of the red blood cells as they pass through the narrow vessels.

Stokes's Law

*9.7

An object moving through a fluid experiences a resistive force, or drag, that is proportional to the viscosity of the fluid. If the object is moving slowly enough, the drag force is proportional to its speed v. If the object is a sphere of radius r, the force is

$$F = 6\pi\eta r v, \tag{9.10}$$

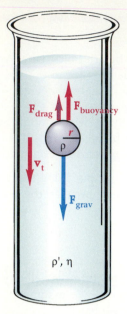

Figure 9.22 A sphere falling in a viscous liquid reaches a terminal velocity v_t that depends on the radius and density of the sphere and the density and viscosity of the liquid.

where η is again the coefficient of viscosity. This equation is known as **Stokes's law,** after Sir George Stokes (1819–1903), who first conceived it in 1845. Stokes's law can be used to relate the speed of a sphere falling in a liquid to the viscosity of that liquid.

Consider a solid sphere of radius r dropped into the top of a column of liquid (Fig. 9.22). At the top of the column the sphere accelerates downward under the influence of gravity. However, there are two additional forces, both acting upward: the constant buoyant force and a velocity-dependent retarding force given by Stokes's law. When the sum of the upward forces is equal to the gravitational force, the sphere travels with a constant speed v_t, called the **terminal velocity**. To determine this speed, we write the equation for the equilibrium of forces:

$$F_{grav} = F_{bouyancy} + F_{drag}.$$

We can express the gravitational force in terms of the density ρ of the sphere, its volume $\frac{4}{3}\pi r^3$, and g:

$$F_{grav} = \frac{4}{3}\pi r^3 \rho g.$$

The buoyant force is equal to the weight of the displaced liquid, which has a density ρ':

$$F_{bouyancy} = \frac{4}{3}\pi r^3 \rho' g.$$

The retarding force is expressed by Stokes's law with the speed v_t:

$$F_{drag} = 6\pi\eta r v_t.$$

Combining these equations, we get an expression for the terminal velocity:

$$v_t = \frac{2r^2 g}{9\eta}(\rho - \rho'). \tag{9.11}$$

The terminal velocity is also called the sedimentation velocity by biologists and geologists.

Example 9.10

Finding terminal velocity.

An aluminum sphere of radius 1.0 mm is dropped into a bottle of glycerin at 20°C. What is the terminal velocity of the sphere?

Solution Using Eq. (9.11), we calculate the terminal velocity directly as

$$v_t = \frac{2r^2 g}{9\eta}(\rho - \rho').$$

The radius in meters is 1.0×10^{-3} m. The densities, from Table 9.2, are $\rho = 2.7 \times 10^3$ kg/m^3 and $\rho' = 1.26 \times 10^3$ kg/m^3. The viscosity, from Table 9.4, is 1.49 Pa·s. Thus

$$v_t = \left(\frac{2 \times 1.0 \times 10^{-6} \text{ m}^2 \times 9.8 \text{ ms}^{-2}}{9 \times 1.49 \text{ Nm}^{-2}\text{s}}\right)(2.7 - 1.26) \times 10^3 \text{ kg m}^{-3},$$

$$v_t = 2.1 \times 10^{-3} \text{ m}^2\cdot\text{s}^{-2}\cdot\text{kg/N}\cdot\text{s}.$$

But the units of N are kg·m·s^{-2}, so the velocity is in m/s:

$$v_t = 2.1 \times 10^{-3} \text{ m/s or 2.1 mm/s.}$$

Stokes's law suggests a method of measuring the size of small particles. If the rate at which material settles from a suspension is measured and the other parameters are known, then Eq. (9.11) gives the particle size. In the case of particles whose radius is less than about 5×10^{-6} m, the settling rate is prohibitively slow, even when the densities of the particles and fluid are quite different. However, we can increase the rate by using a centrifuge (see Section 5.2). The centripetal force per unit mass in an ordinary centrifuge may be 100g, while that in a modern ultracentrifuge may go as high as $5 \times 10^5 g$.

Stokes's law is also useful in the consideration of geological processes in which the rate of sedimentation is important. Modifications are made to take account of nonspherical particles.

Terminal Velocity

*9.8

An important class of effects occurs whenever an object moves more rapidly than we assume it does for the case in which Stokes's law holds. The retarding force of air on a falling raindrop, a skydiver, and a moving car illustrates this behavior. As long as the speed is low enough so that the fluid flows smoothly around the object, the retarding force, or drag, is proportional to the speed. But when the speed becomes so great that the streamline flow breaks up and the fluid swirls around in eddies, the drag becomes more nearly proportional to the square of the velocity. The retarding force can then be represented as a sum of two terms: one proportional to the speed v and the other proportional to v^2. The magnitude of the force becomes

$$F = bv + cv^2,$$

where b and c are constants determined for each case being considered.

An object falling from rest through the air begins its fall with an acceleration equal to the acceleration of gravity. The retarding force increases as the object's speed increases, so that the acceleration decreases and eventually becomes zero. The resulting constant velocity is the terminal velocity v_t at which the downward force of gravity is opposed by a retarding force of equal magnitude due to the air resistance:

$$mg = bv_t + cv_t^2.$$

In many cases of practical interest, the constant b is so small that the term bv_t can be neglected. In that case the terminal velocity may be written as

$$v_t = \sqrt{\frac{mg}{c}}.$$

Elementary analysis (see Problem 9.71) shows that the constant c depends

TABLE 9.5 Aerodynamic drag coefficients	
Shape	C_D
1988 Buick Regal	0.305
1988 BMW 735i	0.32
Typical 1970 U.S. auto	0.50
Typical 1970 U.S. station wagon	0.60
Small truck	0.70

on the density ρ of the air and the area A of the body presented to the air flow. Then the equation for the terminal velocity is

$$v_t = \sqrt{\frac{mg}{C_D \frac{\rho}{2} A}},$$

where C_D is called the drag coefficient. This equation also holds for objects moving horizontally through the air at any speed if mg is replaced by the retarding, or drag, force on the object. Thus the aerodynamic drag on a moving object, such as a car, is approximately

$$F_{drag} = 0.65 C_D A v^2, \tag{9.12}$$

where F_{drag} is in newtons when the variables are in SI units.

One of the objects of contemporary automobile design is to reduce the drag coefficient in order to improve fuel economy. (Some representative drag coefficients are given in Table 9.5.) On the other hand, the design objective for a parachute is to have a large value of both C_D and A so that the descent is slow (Fig. 9.23).

For a skydiver falling through air, the terminal velocity is approximately 120 mi/h (or about 60 m/s); for a feather it may be as small as 0.1 m/s. For a raindrop 2 mm in diameter, the terminal velocity is about 7 m/s. Under the assumptions of free fall, such a raindrop would reach a speed of 7 m/s in less than three quarters of a second, while it fell a distance of only 2.5 m. In this case the effects of air resistance become very important.

In our discussion of the range of a projectile in Chapter 3, we used a simplified model of reality in which we neglected the effects of air resistance. The prediction given by our model there was that a thrown baseball would follow a parabolic path. In a more realistic model that includes the drag force due to the air, the path of the ball is nonparabolic and the ball lands short of the range predicted with the simple model. Furthermore, the launch angle for maximum range is less than 45° and depends on the initial speed of the ball. If the ball spins, additional forces arise from the resulting turbulence that can also affect its path. These additional forces are described in the next section.

Figure 9.23 A parachutist falling with terminal velocity. Parachutes are designed for large drag coefficients and large areas.

Turbulent Flow

*9.9

When you move your finger slowly through a liquid, such as water, you feel only a moderate force of resistance. This resistance arises from two sources: an inertial resistance to the acceleration of the water being displaced and a viscous drag force. As you move your finger faster, the resistance becomes larger because you are moving more water in the same time. The ratio of the inertial force to the viscous force, called the **Reynolds number**, is a useful parameter for describing fluid flow and for

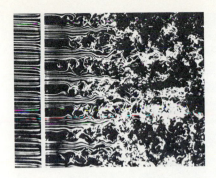

Figure 9.24 An example of turbulent flow. A laminar flow of smoke trails passes through a grid, which induces turbulent flow.

determining the onset of turbulence. It is given by

$$Re = \frac{\rho v L}{\eta}, \tag{9.13}$$

where ρ and η are the density and viscosity of the fluid, v is the velocity of the object, and L is a length characteristic of the object. In this case, L is the length of your finger.

A large value of Re indicates large motion of the fluid. When motion through the fluid exceeds a certain critical speed, the laminar flow of the fluid around the object becomes turbulent and is characterized by an irregular, complex motion. It is the motion of the object relative to the fluid that is important; that is, a stationary object and a moving fluid give the same results. Similar behavior is observed in the flow of a fluid through a pipe. Laminar and turbulent flow are illustrated in Fig. 9.24.

In 1883 the Irish-born engineer Osborne Reynolds (1842–1912) discovered that laminar flow can become turbulent if the speed is sufficiently large. For flow through a tube the Reynolds number becomes

$$Re = \rho v D / \eta, \tag{9.14}$$

where D is the diameter of the tube. Experiments showed that the Reynolds number is of primary importance in determining whether the flow is turbulent or laminar. The Reynolds number is dimensionless and has the same value in any consistent system of units. Observations show that for flow through a pipe, Reynolds numbers of less than about 2000 correspond to laminar flow and Reynolds numbers greater than 3000 correspond to turbulent flow. At values between 2000 and 3000 the flow is unstable and may change back and forth from one type of flow to another.

The frictional forces in turbulent flow are much greater than in non-turbulent flow. It is therefore often desirable to maintain laminar flow, whether in the case of water in pipes or blood in arteries and veins. In rigid pipes this is accomplished primarily through using large pipes and low velocities. Blood vessels, however, are flexible and may expand with an increase in pressure. Turbulence in pipes can also be reduced by the appropriate placement of deflecting vanes and wires.

A related effect of turbulent air flow, this time around a spinning ball, causes the curve of a baseball or golf ball. To see this effect, imagine a stationary ball subjected to a flow of air (Fig. 9.25a). The motion of the air past the ball departs from streamline flow at the velocities normally attained by a thrown baseball and some turbulence occurs, as shown. However, because of the symmetry of the ball, there is no force on the ball perpendicular to the direction of flow.

When the ball is spinning, the friction between the ball's surface and the air drags a layer of air around with the ball (Fig. 9.25b). The pattern of turbulence becomes asymmetric and the streamlines become more crowded at the top of the figure than at the bottom. The net result is a lowering of the pressure at the top and a force transverse to both the direction of flow and the axis of spin. This transverse force is the force that causes the ball to curve. By Newton's third law, there must also be a

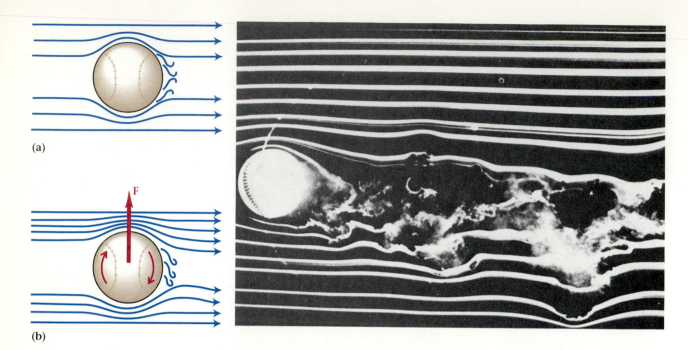

(a)

(b)

Figure 9.25 (a) Air flow about a baseball that is not spinning. (b) Air flow about a spinning baseball. (c) Photograph of smoke trails around a spinning baseball.

net force acting on the air whose effect is to deflect the air downward (Fig. 9.25c). (You can also analyze the behavior of the ball with Bernoulli's equation.)

In a golf stroke the impact with the slanted head of the club gives a spin to the ball, similar to the illustration of Fig. 9.25(c), where the ball is traveling from right to left. The force resulting from the spin gives an additional lift to the ball that keeps it aloft longer, thus allowing it to travel a greater horizontal distance. Similarly, when the face of the club does not strike the ball properly, additional spin about a vertical axis is introduced that causes the ball to "hook" or "slice" off to one side. The surface of a golf ball is intentionally roughened by the addition of dimples to the surface during manufacturing. These dimples play a significant role in providing proper friction to enhance the ball's time of flight.

SUMMARY

Useful Concepts

- Pressure is defined to be the force per unit area,

$$P = F/A.$$

The gauge pressure due to the weight of a fluid of density ρ at a depth h is

$$P = \rho g h.$$

- Pascal's principle: The pressure applied at one point in an enclosed fluid is transmitted undiminished to every part of the fluid and to the walls of the container.

- Archimedes' principle: A body, whether completely or partially submerged in a fluid, is buoyed upward by a force that is equal to the weight of the displaced liquid.
- The equation of continuity,

$$v_1 A_1 = v_2 A_2,$$

is a statement of the principle of conservation of mass for fluids of constant density.

- For an incompressible nonviscous fluid in laminar flow, Bernoulli's equation is

$$P + \rho g h + \tfrac{1}{2}\rho v^2 = \text{constant}.$$

- For a viscous fluid flowing in a horizontal pipe, Poiseuille's law gives

$$P_1 - P_2 = 8Q\eta L/\pi R^4,$$

where Q is the flow rate, L is the distance along the direction of flow, η is the coefficient of viscosity, and R is the radius of the pipe. The force on a sphere moving in a viscous liquid is given by Stokes's law:

$$F = 6\pi\eta r v.$$

- Aerodynamic drag is given by

$$F_{\text{drag}} = 0.65 C_D A v^2,$$

where C_D is the drag coefficient.

- The Reynolds number is used to characterize turbulent flow through a tube and is given by

$$Re = \rho v D/\eta.$$

Important Terms

You should be able to write out the definitions of each of the following terms:

- fluid
- pressure
- pascal
- Pascal's principle
- Archimedes' principle
- buoyant force
- surface tension

- equation of continuity
- streamline
- laminar flow
- turbulent flow
- viscosity
- terminal velocity
- Reynolds number

QUESTIONS

9.1 You are floating on a rubber raft in a small swimming pool in which the water level has been carefully measured. You throw overboard some wooden blocks that had been on the raft and watch the blocks float on the water. What happens to the water level as measured on the edge of the pool? Would the water level behave differently if the blocks were concrete that sank to the bottom of the pool?

9.2 Explain how a siphon works.

9.3 Three containers have the same base area but not the same shape. According to Pascal's law, when they are filled to the same height with water, the same force acts on the base of each one. Yet when they are weighed on a scale, they do not weigh the same. Explain this apparent contradiction, which is sometimes called the *hydrostatic paradox*.

9.4 A block of wood floats half-submerged in a container of water. If the same container were in an earth-orbiting satellite, how would the block float? Explain your reasoning.

9.5 The port of Hamburg is about 60 mi inland from the North Sea on the river Elbe. When a container ship left the port a sailor noticed that there was a paint spot just at the water line. Where was the paint spot when the ship reached the open sea? Explain.

9.6 A sealed hollow glass tube is weighted at one end so that it floats upright when placed in a liquid. The level at which the tube floats depends on its weight and the density of the liquid. When calibrated, such a device can be used to measure fluid densities and is called a hydrometer. Discuss the effect of adhesive and cohesive forces on the level at which the tube floats.

9.7 A thin piece of wood is cut into the shape shown in Fig. 9.26 and a small piece of soap is placed at the point marked *A*. When the "boat" is placed in a pan of water, it moves in the direction indicated by the arrow. Explain.

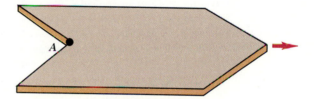

Figure 9.26 **Question 9.7.**

9.8 A Ping-Pong ball can be suspended in a vertical stream of air, such as that from the exhaust pipe of a vacuum cleaner. If the ball is given a small impulse to the side, it will return to the center of the stream rather than being ejected from the stream. Explain.

9.9 The card in the diagram in Fig. 9.27 will not fall away

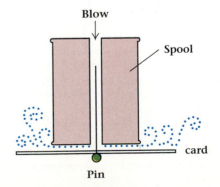

Figure 9.27 **Question 9.9.**

from the spool as long as one blows through the hole. Explain this effect. The pin in the center is only to eliminate sideways motion and does not exert any vertical force.

9.10 The viscosity of liquids decreases with increasing temperature. Give as many examples as you can where this effect can be encountered in everyday experience.

9.11 List some sports in which it is desirable to reduce turbulent fluid flow, and give some of the ways in which this is done.

9.12 How does the external design of a car affect its high-speed behavior? Why do race cars have spoilers on the back?

9.13 The explanation often given of how airplanes fly de-

pends on the fact that the cross section of the wings is usually not symmetric but is curved on the upper surface and flat on the lower surface. How, then, do stunt flyers manage to fly upside down for considerable distances?

9.14 When syrup or oil is slowly poured from a container, the diameter of the stream decreases for a distance below the point at which it leaves the container. Explain this observation.

9.15 (a) If your mass is 62 kg and you float in water with only a negligible amount of your body above the surface, what is your approximate volume? (b) Determine your own volume by using this technique, or an improvement on it that takes account of how deep you really float.

PROBLEMS

Hints for Solving Problems

In using Pascal's principle at some point in a fluid, make sure you know the pressure due to depth at that point before you consider the effect of an applied pressure. Remember that Bernoulli's principle applies to incompressible, nonviscous fluids in laminar flow. Be especially careful with units, since many of the quantities used in fluid flow involve derived units rather than fundamental units.

Section 9.1   Hydrostatic Pressure

9.1 An automobile tire is properly inflated at a pressure of 35 psi. What is its pressure expressed in kPa?

9.2 A woman wearing high-heeled shoes places about 50% of her full weight on a single heel when walking. (a) Assuming the woman to weigh 530 N, what is the pressure on the ground under one heel if the area of contact is 6.5 cm^2? 1.0 cm^2 ? (b) How does this compare with the pressure underneath an elephant's foot? For computation, assume that a full-grown elephant weighs 37000 N and is standing evenly on four feet. Approximate the feet as circles 38 cm in diameter.

9.3 A 1350-kg automobile is supported by four tires inflated to a gauge pressure of 180 kPa. Ignoring the effects of tread thickness, calculate the area of contact between the tires and the road.

9.4 A water tower provides pressure for a water supply system. What is the maximum water pressure available at the bottom of the tower, given that the water level is 30 m above the place where the pressure is to be measured?

9.5 At the base of the Hoover Dam on the Colorado River,

the depth of the water is 726 ft. What is the pressure at the base of the dam? Neglect the pressure due to the atmosphere.

9.6 A rectangular fish tank measures 30 cm by 60 cm by 40 cm high. (a) If the tank is filled with water to a depth of 38 cm, what is the pressure at the bottom due to the water? (b) What is the total force of the water on the bottom?

9.7 What is the pressure at the bottom of a 2.0-km-deep oil well filled with oil of 8.6×10^2-kg/m^3 density?

9.8 A Rolex Sea-Diver wrist watch is guaranteed to be water resistant down to a depth of 4000 ft below sea level. A special version of this model ran after having been submerged to a depth of 35,000 ft in the Marianas Trench in the Pacific Ocean. To what pressures do these depths correspond? (Take the density of sea water to be 1.026×10^3 kg/m^3.)

9.9 Organisms have been found living in the oceans where the pressure is as high as 1000 atm. To what depth does this pressure correspond? (Take the density of sea water to be 1.026×10^3 kg/m^3.)

9.10 The gauge pressure at the bottom of a reservoir is four times what it is at a depth of 1.2 m. How deep is the reservoir?

9.11 A 1.00-m-tall pipe is filled to the halfway level with glycerin and then to the top with water. What is the gauge pressure at the bottom?

Section 9.2   Pascal's Principle

9.12 The column of mercury in a barometer stands 76.0 cm high. How tall would the barometer have to be if the mercury were replaced by water?

9.13 A hydraulic jack is made with a small piston 1.0 cm in diameter that is used to move a large piston 10 cm in diameter. If a man can exert a force of 100 N on the small piston, how heavy a load can he lift with the jack?

9.14 A hydraulic press has a large piston with a cross-sectional area of 400 cm² and a small piston with a cross-sectional area of 5.0 cm². What is the force on the large piston when a force of 1500 N is applied to the small piston?

9.15 A hydraulic lift of the type shown in Fig. 9.28 is used to raise a car weighing 15,000 N. The piston that supports the car has a diameter of 36 cm. What pressure of air within the system is required to just hold the car in place?

9.16 An air compressor maintains a pressure of 700 kPa over the hydraulic fluid in a tank (Fig. 9.28). The large piston that lifts the car has a cross-sectional area of 0.280 m². What is the maximum weight that it can lift?

9.17 A U-tube is partially filled with equal volumes of water and mercury (Fig. 9.29). If each liquid fills a 20-cm-long section of the tube, what will be the difference in the levels of the upper surfaces?

Section 9.3   Archimedes' Principle

9.18 A block of pine wood (ρ = 0.40 g/cm³) is floating on a pond. The block is 10 cm × 40 cm × 5 cm thick. (a) How much of the block protrudes above the water? (b) If the block is made to carry a load by placing additional mass on top of it, how much mass must be added to just submerge the block?

9.19 A wooden block 20 cm × 20 cm × 10 cm has a density of 0.60 g/cm³. (a) How much iron (ρ = 7.86 g/cm³) can be placed on top of the block if the top of the block is to be level with the water around it? (b) If iron were attached to the bottom of the block instead, how much iron would it take to bring the top of the wooden block down to the level of the water? Why?

9.20 A solid cube of unknown composition is seen floating upright in water with 70% of it below the surface. Calculate the density of the material.

9.21 (a) A block of iron is suspended from one end of an equal-arm balance by a thin wire. To balance the scales, 2.35 kg are needed on the scale pan at the other end. What is the volume of the block? (b) Next, a beaker of water is placed so that the iron block, suspended as in part (a), is submerged in the beaker but not touching the bottom (Fig. 9.30). What mass is now necessary to balance the scales?

9.22 A Goodyear blimp has a volume of 5750 m³ and a
• mass of 4300 kg when empty. What additional load is it able to lift when the entire volume is filled with

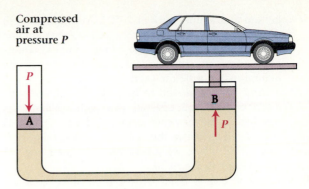

Figure 9.28 Problems 9.15 and 9.16. An air compressor provides a pressure at the piston *A*. The pressure is transmitted through the hydraulic fluid to the piston *B*.

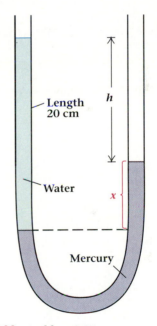

Figure 9.29 Problem 9.17.

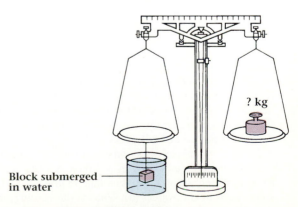

Figure 9.30 Problem 9.21.

helium at a temperature of 20°C? (Your answer will differ from the actual load because the entire volume is not filled with helium, the helium is not at atmospheric pressure, and the temperature of the air and helium may not be the same.)

9.23 A king's crown is said to be solid gold but may be
  • made of lead and covered with gold. When it is weighed in air the scale reads 0.475 kg. When it is submerged in water the scale reads 0.437 kg. Is it solid gold? If not, what percentage by mass is gold? (*Hint:* Refer to Fig. 9.30.)

9.24 A plastic bag is filled with helium at atmospheric pressure and 21°C. How large a volume of helium is required to lift a 50-kg girl off her feet? Assume that the mass of the bag is negligible. (*Hint:* See Table 9.2.)

9.25 If, in Problem 9.24, hot air is used instead of helium, what is the required volume for the balloon if the air inside can be maintained at a temperature of 44°C? Assume the outside air is at 21°C. (*Hint:* See the footnote in Table 9.2.)

9.26 A wooden dowel is placed in a test tube containing water. The dowel floats with 60% of its length below the water surface. How much of the dowel is submerged when it floats in methyl alcohol?

9.27 A glass disk *G* is held tightly against the end of a vertically held cylindrical tube *C* while the tube is lowered 20 cm under the water (Fig. 9.31). What is the maximum thickness the disk can have and not fall away from the cylinder? (*Hint:* The density of the glass is $\rho = 2.5 \times 10^3$ kg/m^3.)

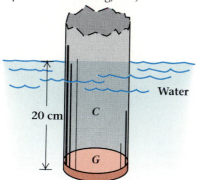

Figure 9.31   Problem 9.27.

9.28 A cylinder of solid uranium weighs 9.34 kg in air, 8.84 kg in water, and 2.54 kg in another liquid. (a) What is the volume of the cylinder? (b) What is the density of the uranium cylinder? (c) What is the density of the liquid? (d) Identify the liquid.

9.29 A 1.0-kg container of water sits on a scale *A*. A piece of aluminum 10 cm × 10 cm × 10 cm is suspended from a spring scale *B* so that half of the block is submerged in the water. (a) What is the reading on the spring scale? (b) What is the reading on scale *A*?

*Section 9.4   Surface Tension

9.30 A 5.0-cm-diameter ring is used to determine the surface tension of a liquid. What is the liquid's surface tension if, in addition to the ring's weight, a force of 2.3×10^{-2} N is required to lift the ring from the liquid?

9.31 What force is required to overcome surface tension when raising a horizontal 10-cm-diameter ring out of water at 25°C?

9.32 How high will water rise in a capillary tube with an inside diameter of 0.01 mm? Assume the contact angle to be $\theta = 0°$. Take the temperature to be 25°C.

9.33 By what factor is the surface energy of a 2.0-cm-diameter soap bubble increased when it is blown up to a diameter of 4.0 cm?

9.34 A wire frame with a sliding crosspiece is dipped into a soap solution and held vertically (Fig. 9.32). The length of the crosspiece is 5.0 cm and it has a mass of 0.265 g. What is the value of the surface tension of the soap solution if the weight of the crosspiece is just balanced by the surface tension force?

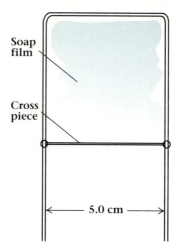

Figure 9.32   Problem 9.34.

9.35 Water is poured into an upright U-tube in which the legs have different internal diameters. If the diameter of one is 0.6 mm and the diameter of the other is 1.2 mm, what will be the difference in the height of the water in the two legs? (*Hint:* Take the contact angle to be 0°.)

Section 9.5   Streamlines and Bernoulli's Equation

9.36 A large storage tank is filled with water. Neglecting viscosity, show that the speed of water emerging through a hole in the side a distance *h* below the surface is $v = \sqrt{2gh}$. (*Hint:* Use Bernoulli's equation.)

9.37 Water is flowing in a horizontal pipe of variable cross section. Where the cross-sectional area is 1.0×10^{-2} m^2, the pressure is 5.0×10^5 Pa and the velocity is 0.50 m/s. In a constricted region where the area is 4.0×10^{-4} m^2, what are the pressure and velocity?

9.38 Water is flowing in a horizontal pipe of varying cross section. At one point where the cross-sectional area is 1.0×10^{-2} m^2, the velocity of the water is 2.0 m/s and the pressure is 15 kPa. In another region of the pipe the velocity is 3.0 m/s. What is the cross-sectional area at the second position and what is the pressure there?

9.39 What is the pressure change in water going from a 5.0-cm-diameter pipe to a 2.0-cm-diameter pipe if the velocity in the smaller pipe is 3.0 m/s?

9.40 What is the pressure change in methyl alcohol flowing from a 4.0-cm-diameter pipe to a 2.0-cm-diameter pipe if the velocity in the larger pipe is 0.40 m/s?

*Section 9.6 Viscosity and Poiseuille's Law

9.41 By what fraction would the blood flow be reduced if an arteriole were reduced to 0.95 of its former diameter? (Assume the pressure and viscosity to be constant.)

9.42 A horizontal garden hose 15 m long with an interior diameter of 1.25 cm is used to deliver water at the rate of 40 cm^3/s. What is the pressure drop from one end of the hose to the other? (*Hint:* Assume a temperature of 20°C.)

9.43 Mercury flows through a horizontal pipe 4.0 cm in diameter and 0.50 m long. If the pressure drop from one end of the pipe to the other is 1.0×10^4 Pa (about $\frac{1}{10}$ atm), what is the rate of flow through the tube?

9.44 How high above the point of injection must a container of blood plasma be, if the plasma is to enter the patient's arm at a rate of 3.0 cm^3/min through a needle that is 50 mm long and has an inside diameter of 0.55 mm? Assume the pressure in the vein to be 15 mm Hg. (Assume also that the density of blood plasma is 1.05 g/cm^3 and its viscosity is about 1.5×10^{-3} Pa·s.)

*Section 9.7 Stokes's Law

9.45 A steel ball bearing 8.0 mm in diameter is dropped into a cylinder of glycerin. The densities of steel and glycerin are 7.8×10^3 and 1.26×10^3 kg/m^3, respectively. What is the terminal velocity of the ball bearing?

9.46 Compare the sedimentation rates for a mixture of spherical particles that are all of the same material but have diameters that differ in the ratio of 1 : 2 : 3.

9.47 (a) A bottle of light corn syrup is taken from a refrigerator ($T = 5°C$) and a glass marble of density 2.5×10^3 kg/m^3 is dropped into it. The marble takes 45 s to sink to the bottom. The diameter of the marble is 1.57 cm, the depth of the liquid is 12.1 cm, and its density is 1.2×10^3 kg/m^3. What is the viscosity of the syrup at that temperature? (b) If the bottle is kept out of the refrigerator for several hours and the experiment done again, the marble takes 5.0 s to fall through the liquid. What is the viscosity of the syrup at room temperature?

9.48 A glass marble of density 2.5×10^3 kg/m^3 and 5.0 mm diameter is dropped into a cylinder containing castor oil of density 900 kg/m^3. What is the terminal velocity of the marble?

*Section 9.8 Terminal Velocity

9.49 A ball moving through air with a speed v experiences a drag force in newtons of $F = 6.20 \times 10^{-6}\, v + 3.5 \times 10^{-4}\, v^2$, where v is in m/s. Calculate the speed at which the linear term and the square term make equal contributions to the drag.

9.50 A small balloon is inflated to a diameter of 20 cm and has a total mass of 0.40 g. When it is allowed to fall in air the balloon has a drag force predominantly due to the v^2 term in the drag equation. Calculate its terminal velocity, given that the coefficient c equals 9.0×10^{-3} kg/m.

9.51 When the engines of a jet airliner develop 1.00×10^5 N of thrust (driving force), it reaches an air speed of 750 km/h. Calculate the thrust required for speeds of 800 km/h and 600 km/h. What does your result suggest about the relationship between fuel consumption and speed if the thrust is proportional to fuel consumption? (*Hint:* Neglect the linear term in the drag equation.)

9.52 The speed of an automobile increases from 90 km/h (56 mi/h) to 115 km/h (71 mi/h). What is the ratio of the drag forces at the two speeds? Ignore the linear term in the drag equation.

9.53 Two spherical objects have the same size and same surface roughness. One of them is heavier than the other. Show that if both objects are simultaneously released from rest from the same height, the heavier one strikes the ground first.

*Section 9.9 Turbulent Flow

9.54 (a) Calculate the terminal velocity for a steel ball of radius 0.50 cm and density 7.8×10^3 kg/m^3 falling through water. Assume that Stokes's law applies. (b) Calculate the Reynolds number by setting L in Eq. (9.13) equal to the diameter of the ball. (c) Is the flow really laminar? What does this suggest about the value for the terminal velocity? (*Hint:* Assume a temperature of 20°C.)

9.55 If the flow of a liquid in a 2-cm pipe is just barely laminar, what size pipe would be needed to maintain laminar flow if the flow rate were to be twice as much as in the first pipe?

9.56 Show that the Reynolds number is dimensionless.

9.57 What is the order of magnitude of the lowest speed that a Ping-Pong ball can have in air for the flow to remain turbulent?

Additional Problems

9.58 The deep research vessel Alvin can dive to depths of
• 4000 m. What is the force of the water on one of the submarine's 30.5-cm-diameter circular ports? The pressure inside the titanium hull is one atmosphere.

9.59 A swimming pool is 50 m long by 23 m wide, and is less deep at one end than at the other. The depth at the shallow end is 1.22 m, and the depth at the deep end is 4.35 m. What is the difference in pressure on the bottom at opposite ends of the pool?

9.60 A dense liquid is poured into a 1-m-deep container
•• and a less dense liquid carefully poured on top of it, so as to form two layers. After many days the liquids have become mixed, but not thoroughly so. At the bottom of the container the density is still that of the denser liquid and at the top the density is that of the lighter liquid. Tests show that the density is given by

$$\rho = (1 + 0.26x^2) \times 10^3 \text{ kg/m}^3,$$

where x is the distance below the surface in meters. What is the pressure in this liquid mixture at a depth of 0.5 m below the surface? (*Hint:* Use the graphical technique of Chapters 3 and 6. Add up the weight of individual thin layers whose density is nearly constant. Using graph paper will help.)

9.61 If the internal volume of a hot-air balloon is 2180 m³,
• at what temperature must the air be to keep a 475-kg balloon and loaded gondola in the air when the outside temperature is 20°C? (*Hint:* See the footnote in Table 9.2.)

9.62 An object of density ρ and mass m is submerged in a
• liquid with a smaller density ρ_0. Show that the effective weight of the submerged object is

$$w_{\text{eff}} = mg(1 - \rho_0/\rho).$$

9.63 Suppose you blow air at a speed of 10 m/s across
• one end of a U-shaped tube containing water. What will be the difference in the heights of the water surfaces on the two sides? Which one will be higher?

9.64 A Venturi meter of the type shown in Fig. 9.20 has
• mercury in the U-shaped tube. The meter is used to measure the flow speed of water. At A the pipe diameter is 10.0 cm; at B it is 5.0 cm. What is the speed of the water at A if the differential height h is 5.0 cm?

9.65 A garden hose has an interior diameter of 0.95 cm and a nozzle with a 0.40-cm-diameter opening. (a) If 30 liters per minute flow through the hose, what is the speed of the water emerging from the nozzle? (b) If the nozzle is held horizontally 1.1 m above level ground, how far will the stream go before it hits the ground?

9.66 Use the results of Problem 9.36 to show that you
•• obtain the maximum range for water leaving a hole in the side of a tank resting on the ground when the hole is halfway between the top and bottom surfaces of the liquid.

9.67 On each heartbeat about 70 cm³ of blood is forced
• from the heart at an average pressure of approximately 105 mm Hg. What is the average power output if the heart beats 70 times each minute?

9.68 Use the equation of continuity and Bernoulli's law to
•• derive the equation for the speed measured by a Venturi meter. (*Hint:* The answer is given in the text.)

9.69 A geological sample from a river bed forms sediment
• at the rate of 1.0 g/day. How many revolutions per second would a centrifuge have to achieve to increase the sedimentation rate to 3.0 g/h? Assume that the sample is placed 5.0 cm from the axis of rotation of the centrifuge.

9.70 A ball moving through the air experiences a drag
• force of

$$F \text{ (newtons)} = 6.20 \times 10^{-6} v + 3.5 \times 10^{-4} v^2,$$

where v is in m/s. What is the terminal velocity for the ball if it falls from a great height? Assume that the ball has a mass of 5.0×10^{-2} kg.

9.71 Show that the terminal velocity of an object falling
• in air can be estimated by

$$v_t = \sqrt{\frac{mg}{kA\rho}},$$

where m is the mass of the object, A is its cross-sectional area, ρ is the density of air, and k is a dimensionless constant whose value is 1 or less and depends on the shape of the object. (*Hint:* During its fall the object "sweeps out" a vertical tube of air. Assume that the drag force is proportional to the rate at which the falling object transfers momentum to this tube of air, and that the rate of change of momentum is the product of the velocity and the rate at which the mass of the air in the tube is displaced.)

9.72 How much water per hour can be delivered by a
• ¾-in. pipe in which laminar flow is maintained? (Assume a temperature of 20°C.)

9.73 What is the minimum diameter of a pipe through
• which 1.00 m³ of glycerin at 20°C can be made to flow per hour if the flow is to be laminar?

10

Thermal Physics

10.1 Temperature and States of Matter

10.2 Thermometry

10.3 Thermal Expansion

10.4 The Mechanical Equivalent of Heat

10.5 Calorimetry

10.6 Change of Phase

10.7 Heat Transfer

A WORD TO THE STUDENT

The study of temperature effects and the flow of energy (heat) will occupy us for the next three chapters. These chapters cover the topics of thermal physics, thermodynamics, and kinetic theory of gases. In mechanics we focused on the behavior and interactions of one or two individual objects. Here we will focus on the behavior of a collective system of objects, such as a gas composed of many molecules. We will find that although one of our main tools of analysis continues to be the law of conservation of energy, we will need an additional energy law.

The first few sections of this chapter deal with the measurement of temperature and heat, topics of both scientific and economic interest. We first establish the connection between heat and energy, then extend the principle of energy conservation to the exchange of heat between substances at different temperatures. Finally, we consider thermal conductivity and insulation. A thorough understanding of these ideas is needed before you study the laws of thermodynamics in Chapter 11.

The development of thermal physics was greatly influenced by the early developments in chemistry (see Chapter 26), and by the practical concerns and search for efficiency that characterized the Industrial Revolution in the eighteenth and nineteenth centuries. By that time, mechanics was relatively well developed, but electricity did not become of practical and commercial importance until the end of the nineteenth century. As a consequence, the disciplines of mechanics, heat, and electricity all evolved along different paths. As recently as the late eighteenth century, the study of heat was not related to the study of mechanics. As a result, the definitions and units of measurement for temperature and heat were developed independently of the definitions and units for work and mechanical energy. Not until the mid-nineteenth century did James Prescott Joule quantitatively connect the unit of thermal energy with the unit of mechanical work, allowing us to see mechanics and thermal physics as parts of a greater whole.

Today we think of heat as a form of energy transfer, as mentioned in Chapter 6 and described in more detail in Chapter 11. The effects of heat and temperature changes are a fundamental aspect of many physical situations, from the study of star formation to research in lasers.

Temperature and States of Matter

10.1

Figure 10.1   Matter principally exists in three phases: solid (icebergs), liquid (water), and gas (air and water vapor).

It will help in our discussion of thermal physics if we first briefly outline what we mean by different states of matter. As an example, water can have the form of ice (solid), water (liquid), or steam (gas). Many other materials can exist as solids, liquids, or gases. Such distinct forms, or states, of matter are called *phases* (Fig. 10.1). The change from one state, or phase, to another, such as the melting of ice, is usually caused by a transfer of heat.

The molecules of a gas move about freely, except when they collide with other gas molecules or the walls of the container. The average separation between molecules is large compared with their own size and, as a result, a gas has no definite volume. Consequently we can compress or expand a gas to fill a container of any shape or size. In a liquid, the average separation between molecules is comparable to their own diameters. Individual molecules are free to move about, but because of the forces between them they move so that the average separation between near neighbors remains constant. As a result, a liquid is virtually incompressible and has a definite volume, although its shape can change to match the shape of its container. In solids the separations are comparable to the separations in liquids, but the binding forces are so strong that the atoms in a solid are not free to move about. Instead, the atoms of a solid are confined to small oscillations about fixed positions. Thus a solid has not only a definite volume but a definite shape as well.

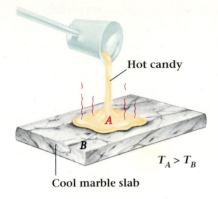

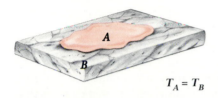

Figure 10.2 The hot, molten candy and the cool marble slab reach thermal equilibrium after a period of time. At thermal equilibrium, their temperatures are equal.

Energy is associated with the motion of molecules in any state of matter. In fact, as we will see, changes in temperature and changes from one state to another are simply large-scale manifestations of changes in the energy of the atoms and molecules that compose the material.

The concept of temperature originated in human sensory perception of the environment. When you touch an object, you say it is relatively "hot" or "cold." This response early led to attempts to describe the feeling in terms of the objects; for example, this rock is warmer than that rock. The desire to quantify and measure such differences in warmth culminated in the idea of **temperature**: the number assigned to an object as an indication of its warmth. Devices used to measure temperature are called **thermometers**. They are used not only as quantitative indicators to measure what the hand can feel, but also to extend the range of measurements far beyond the sense of touch. The range of thermometers extends from temperatures low enough to freeze the gases of the air to the enormous temperatures at the interiors of stars.

Our sensory perception allows us to define another useful term. Place an object A, which feels hot to your hand, in contact with an object B, which feels cold to your hand. After a period of time has passed, they will give the same sensation to your hand. Objects A and B are said to be in **thermal equilibrium** and their temperatures are equal (Fig. 10.2). We can extend this idea to say that two objects that are not touching are in thermal equilibrium if, upon being placed in contact, their temperature would not change.

Thermometry

10.2

The development and calibration of thermometers and the establishment of temperature scales are the essence of *thermometry,* the science of temperature and its measurement. The basis of thermometry is that some physical properties vary with temperature in a quantitative and repeatable fashion. Some of these thermometric properties are the volume of a gas or liquid, the length of a metallic strip, the electrical resistance of a conductor, and the light-transmitting properties of a crystal. Any physical system whose properties change with variations in temperature can be used as a thermometer. The choice of a particular thermometer depends primarily on the range of temperatures to be studied. We measure the change in some property, say the length of a column of liquid, and then associate a change in temperature with our measurement of the change in length.

We choose to discuss the liquid-in-glass thermometer here because of its simplicity and great familiarity. However, we emphasize that it is only one of many possible types of thermometers. For example, some thermometers determine temperature by measuring the amount of light transmitted by a crystal (Fig. 10.3a). The amount of light transmitted by the crystal depends on the temperature and is reproducible. This type of thermometer is used in medicine for applications that require a small, remotely readable thermometer that does not conduct electricity.

0.6 mm

Light out
to detector   Light in

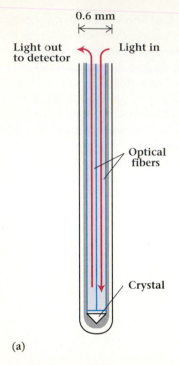

Optical
fibers

Crystal

(a)

(b)

Figure 10.3   (a) A thermometer based on the temperature dependence of the transmission of infrared light through a gallium arsenide crystal. The thermometer has no metal or other electrically conducting parts and can be calibrated to within about 0.2°C. (b) A mercury-in-glass fever thermometer.

The liquid-in-glass thermometer was invented around 1650 by the Grand Duke of Tuscany, Ferdinand II, one of Galileo's fellow countrymen and a patron of science. This thermometer makes use of thermal expansion, the effect you use to loosen the metal cap on a glass bottle by holding it under a stream of hot water. (We discuss thermal expansion further in the next section.) In this thermometer a liquid indicator is sealed into a glass capillary tube having a bulb at one end. When the temperature increases, thermal expansion causes both the volume of the glass bulb and the volume of the liquid to increase. If they both expanded at the same rate, we would observe no change. But since the liquid expands at a greater rate than the glass does, the liquid is forced to expand into the tube as the temperature increases. By using a relatively large bulb and a narrow tube, it is possible to make a thermometer that we can read easily from a scale scribed on the glass. The common fever thermometer is made this way (Fig. 10.3b).

Any thermometer, whether liquid-in-glass or one that depends on other thermal properties, must be calibrated to make it a useful instrument, capable of quantitative and reproducible measurements. For example, a liquid-in-glass thermometer can be calibrated by marking on the glass the position of the liquid column at a set of reference temperatures or standards. Temperatures between the reference marks are interpreted as proportional to the length of the liquid column. In fact, this is just how the **Celsius temperature scale** is defined. The two fixed points of reference are the ice point and the steam point. The ice point is defined as the equilibrium temperature of a mixture of ice and water at a pressure of one atmosphere; the steam point is defined as the equilibrium temperature of water and steam at a pressure of one atmosphere. The numbers assigned to these two points in the Celsius scale are arbitrarily chosen as 0 for the ice point and 100 for the steam point.

Assuming that the cross section of the thermometer capillary is uniform and that the rate of expansion of the liquid with changes in temperature is constant, we then can mark the distance between the ice and steam points into 100 equal parts. We can easily compare the level of the liquid to the nearest mark, called a degree.* We thus measure temperature in units of degrees Celsius, abbreviated as °C. (This scale was originally known as the centigrade scale because it has one hundred divisions between the principal reference marks. The present name was adopted to honor the Swedish astronomer Anders Celsius, who popularized the scale in 1742.)

Figure 10.4 shows a Celsius scale thermometer at a temperature T, at which the liquid is extended a distance L beyond the zero position. We may calculate the temperature by

$$T \; (°C) = \left(\frac{L}{L_0}\right) \times 100,$$

where L_0 is the distance between the 0° and 100° marks. Here we have defined the temperature scale to be a linear function of the length L of the liquid column. (There is no fundamental reason for doing this; we

*Some early thermometers were marked in 360 divisions, like the parts of a circle; thus the term "degree" became applied to temperature.

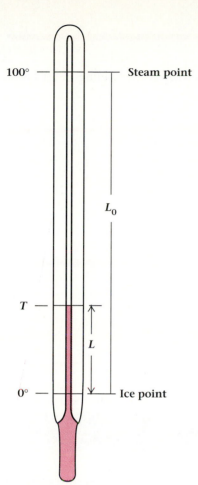

Figure 10.4 A liquid-in-glass thermometer with a Celsius scale. The temperature is proportional to the ratio of the distance L to the reference distance L_0.

could define other functions equally well. However, the linear relationship is the simplest.) When the temperature scale is defined this way, other important physical properties turn out to be approximately independent of temperature.

Although the Celsius temperature scale is widely used, there is nothing fundamental about choosing the ice point to be 0° and the steam point as 100°. The **Fahrenheit temperature scale** assigns a value of 32° to the ice point and 212° to the steam point, a difference of exactly 180° (Fig. 10.5). It is easy to transform temperatures in one system into temperatures in the other system. For example, let us find the relationship that transforms temperature in °C to temperature in °F. Just remember that 0°C is equivalent to 32°F, and that a range of 180° on the Fahrenheit scale is 100° on the Celsius scale. Therefore one Fahrenheit degree is equivalent to $\frac{5}{9}$ of one Celsius degree. We can then write

$$T \, (^\circ\text{F}) = \tfrac{9}{5} T \, (^\circ\text{C}) + 32. \tag{10.1}$$

You can convince yourself that this is correct by substituting the Celsius values for the freezing and boiling temperatures of water and seeing whether you get the correct Fahrenheit values. You can use this same substitution to check whether you have remembered the formula correctly when you must recall it without the book.

Equation (10.1) gives the Fahrenheit temperature if the Celsius temperature is known. It can easily be rearranged to express the Celsius temperature in terms of Fahrenheit. Two examples of these transformations follow.

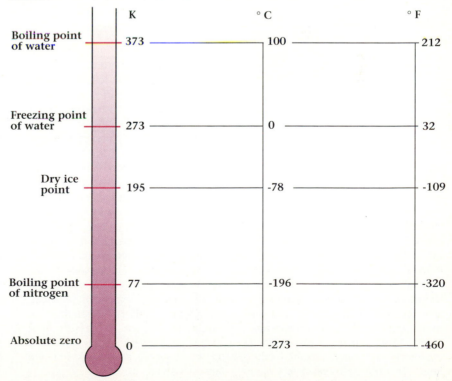

Figure 10.5 A comparison of the Fahrenheit, Celsius, and Kelvin scales of temperature.

Example 10.1

Converting Fahrenheit to Celsius.

On a day when the temperature is 86°F, what is the reading of a Celsius thermometer?

Solution   By using Eq. (10.1) we have

$$T \; (°F) = \tfrac{9}{5} T \; (°C) + 32,$$

which can be rearranged to give

$$T \; (°C) = \tfrac{5}{9}[T \; (°F) - 32],$$
$$T \; (°C) = \tfrac{5}{9}(86 - 32) = \tfrac{5}{9}(54) = 30°C.$$

Example 10.2

Converting Celsius to Fahrenheit.

What Fahrenheit temperature is equivalent to 37.0°C?

Solution   Application of Eq. (10.1) yields

$$T \; (°F) = \tfrac{9}{5}(37.0) + 32 = 66.6 + 32 = 98.6°F.$$

BACK TO THE FUTURE

Fahrenheit's Thermometer

Although the Celsius scale is becoming increasingly common in the United States, most people in the United States still think in terms of Fahrenheit temperatures when deciding what to wear outside. Have you ever wondered why the freezing temperature is 32°F? Why not 0° or 100°? What's so special about the numbers in 32°F, 212°F, or 98.6°F?

At the beginning of the eighteenth century, the Danish astronomer Ole Roemer (famous for the first measurements that showed that the velocity of light is finite) devised a temperature scale of his own for use with the alcohol-in-glass thermometers that he constructed. His thermometers attracted the attention of Gabriel Fahrenheit (1686–1736), a manufacturer of meteorological instruments in the Netherlands. In 1708 he traveled to Copen-

hagen to meet Roemer and see his thermometers, which were based on two reference points. For one reference Roemer used a mixture of ice, water, and salt to reach the lowest temperatures then attainable in the laboratory, which he called zero. His other reference was the boiling point of water, which he arbitrarily designated as 60 degrees.

Fahrenheit returned home to make thermometers like Roemer's. In 1714 he overcame technical difficulties with alcohol thermometers by substituting mercury as the expanding liquid. The use of mercury extended the range of temperature measurements from well below Roemer's zero to well beyond the boiling point of water. Furthermore, mercury expanded and contracted more uniformly than the other liquids then in use. As a result, Fahrenheit could mark his mercury thermometers more accurately and with finer divisions.

By 1724 Fahrenheit had adopted a new scale, similar to Roemer's but with much finer divisions. For the zero point he chose the same reference as Roemer. However, since his thermometer was intended for meteorological observations, he wanted a second reference point that would be nearer the maximum observed temperature for weather. He chose the normal temperature of the human body as the upper reference point, which he called 96°. (Body temperature had earlier been used as a reference point by Newton, who regarded it as absolutely constant for a healthy person.) Fahrenheit gave no reason for his choice of 96, but it may have been due to his desire for a finer scale and because 96 is evenly divisible by 2, 3, 4, 8, and 12.

Why didn't Fahrenheit choose the freezing point of water for his zero reference, as Newton had done before him and as Celsius did later on? Perhaps Fahrenheit was influenced by

In both the Fahrenheit and Celsius temperature scales, the assignment of the zero point is arbitrary. We can readily achieve temperatures below these zero points. However, one temperature scale has a more fundamental choice of zero. This scale was proposed in 1848 by William Thomson, Lord Kelvin (1824–1907) and arose from the study of gases. Kelvin's scale uses intervals equal to those of the Celsius degree, but with zero set at the lowest theoretical temperature that a gas can reach. The scale is based on the fact that a gas at 0°C will lose 1/273.15 of its volume for a 1°C drop in temperature. If this reduction in volume were to continue with decreasing temperature, and if the gas did not liquefy, the volume would become zero at −273.15°C. This temperature is called **absolute zero** and is the temperature at which all molecular motion would stop. The temperature scale based on this zero is the **Kelvin temperature scale**. We will discuss the physical meaning of this observation in more detail in Chapter 12. In this chapter you can just consider the Kelvin scale to be a temperature scale with degree intervals of the same size as the Celsius degree, but with the zero point at −273.15°C.

The conversion between Celsius and Kelvin temperatures is a simple one,

$$T(K) = T\,(°C) + 273.15. \tag{10.2}$$

Roemer, or he may have wanted to avoid the inconvenience of repeatedly using negative temperatures during winter. Also, we should point out that in the early 1700s it was widely believed that water did not always freeze at the same temperature. Soon, using his newly calibrated thermometers, Fahrenheit learned that water always froze at 32° on his scale. He immediately added this third reference point to his instruments.

Fahrenheit also made other thermometers, with scales of different lengths depending on their intended use. His instruments for measuring the temperatures of boiling liquors began at 0 and extended 600 degrees to nearly the boiling point of mercury. A report of his thermometers was published in the *Philosophical Transactions* in 1724. Almost at once his scale was adopted in Great Britain and the Netherlands and gained wide acceptance throughout the English-speaking countries.

The Fahrenheit scale in use today differs slightly from the original. The two fixed points are the ice point, assigned a value of 32°F, and the steam point, assigned a value of 212°F. On

Figure B10.1 An infrared pyrometer is used to measure the temperature of molten steel.

this scale the normal human body temperature is 98.6°F, slightly higher than the 96° originally chosen by Fahrenheit.

Today the Celsius scale and the Kelvin scale have replaced the Fahrenheit scale for scientific work. Also, the range of temperatures that can now be measured has been extended by many orders of magnitude since Fahrenheit's time. Modern thermometry uses many different physical properties to indicate temperatures, spanning a range from the extreme lows near 10^{-6} K to the temperature of the stars at about 10^4 K. The choice of thermometer depends on the temperature to be measured. For example, an infrared pyrometer, which uses the infrared radiation from hot matter to measure temperature, can be used to find the temperature of molten steel (Fig. B10.1). Pyrometers like this measure temperatures ranging from −30°C to 3000°C.

The unit of absolute temperature is the kelvin (K). It is not written with a degree sign. A temperature of 0°C is simply 273.15 K. A comparison of the various scales is shown in Fig. 10.5.

Thermal Expansion

10.3

In the preceding section we described the liquid-in-glass thermometer, which is based on the difference in the rates of thermal expansion of the indicating liquid and of the glass envelope. We can calibrate such a thermometer without knowing the individual expansion rates. However, once a temperature scale is established, we can show that, to a good approximation, most solid objects change length in direct proportion to a change in temperature. Also, the change in length is proportional to the initial length of the object. Thus, for the same increase in temperature, a long copper bar expands more than a shorter one, but the ratios of change in length to initial length are the same for both bars (Fig. 10.6). We call this behavior **linear thermal expansion,** that is, expansion in one dimension. The reason for this expansion is that the increase in temperature causes greater amplitudes of vibration of the atoms in the solid, giving a greater average distance between them.

Let us describe linear thermal expansion mathematically. For an object of initial length L_0, the change in length ΔL due to a change in temperature ΔT can be expressed as

$$\Delta L = L_0 \alpha \, \Delta T. \tag{10.3}$$

The proportionality constant α, called the **thermal expansion coefficient**, has the dimension of inverse temperature, or $°C^{-1}$. Table 10.1 lists thermal expansion coefficients for a number of common materials. The coefficients themselves have some slight temperature dependence, and are given here

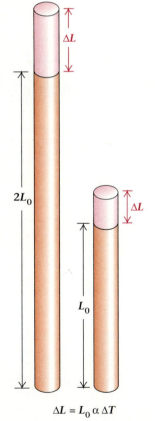

$$\Delta L = L_0 \, \alpha \, \Delta T$$

Figure 10.6 A copper bar that is twice as long as a shorter copper bar undergoes twice as much expansion for the same temperature change.

TABLE 10.1
Coefficients of thermal expansion at 20°C*

Material	Linear coefficient α $(10^{-6} \, °C^{-1})$	Volume coefficient β $(10^{-6} \, °C^{-1})$
Aluminum	24	72
Brass	19	57
Brick and concrete	10–12	30–36
Copper	17	51
Glass (ordinary)	9	27
Glass (Pyrex)	3	9
Invar	0.7	2.1
Iron and steel	12	36
Lead	29	87
Ice	51	153 ($-20°C$ to $-1°C$)
Gasoline		950
Mercury		180
Water		210

*The values given are approximate. They vary with the composition of alloys, glasses, and composite materials, and with temperature.

for 20°C, but you may take them as constant for the purposes of your study.

For an increase in temperature ΔT, a rod of initial length L_0 expands to a new length L. The change in length is

$$L - L_0 = \Delta L.$$

If we combine this equation with Eq. (10.3), we can express the length L at the new temperature T in terms of the initial length L_0 at the initial temperature T_0:

$$L = L_0 (1 + \alpha \Delta T)$$

or

$$L = L_0 [1 + \alpha(T - T_0)]. \tag{10.4}$$

Figure 10.7 Example 10.3: The Verrazano-Narrows Bridge. Changes in temperature cause the bridge components to expand or contract.

Example 10.3

Thermal expansion of a steel bridge.

The Verrazano-Narrows Bridge between Brooklyn and Staten Island in New York City is one of the world's longest suspension bridges, with a center span of 1300 m (Fig. 10.7). Because the temperature variation over a year may be quite large, allowance must be made for thermal expansion and contraction of its materials. Assuming that the bridge is steel and, for safety, allowing for a temperature range of 120°C, how much thermal expansion must be allowed for in the center span? (This allowance is made by using expansion joints and components that can move with respect to each other.)

Solution We can obtain the change in length of the center span from our definition of thermal expansion:

$$\Delta L = L_0 \, \alpha \, \Delta T.$$

Upon inserting the numerical values, including α from Table 10.1, we get

$$\Delta L = 1300 \text{ m} \times (12 \times 10^{-6} \, °C^{-1}) \times (120°C) = 1.87 \text{ m}.$$

The allowance for expansion must be 1.87 m.

Example 10.4

Change in length of a hot-water pipe.

A copper hot-water pipe is 10.0 m long when cut and installed in a building on a day when the temperature is 10°C. How long is the pipe when it carries hot water at 60°C if the pipe is free to expand?

Solution We can use Eq. (10.4):

$$L = L_0[1 + \alpha(T - T_0)],$$
$$L = 10.0 \text{ m}[1 + (17 \times 10^{-6} \, °C^{-1})(60°C - 10°C)],$$
$$L = 10.0 \text{ m}[1.00085] = 10.0085 \text{ m} = 1000.85 \text{ cm}.$$

The pipe is 0.85 cm longer.

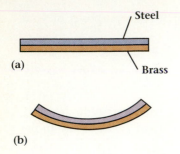

(a)

(b)

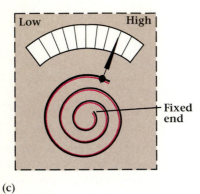

(c)

Figure 10.8 (a) A bimetallic strip of steel and brass at room temperature. (b) At a higher temperature, the brass expands more than the steel, bending the strip. (c) A thermometer made with a bimetallic strip.

A common application of linear thermal expansion is the bimetallic strip, which is made by joining two strips of metal with different thermal expansion coefficients (Fig. 10.8). The unequal expansion or contraction of the two metals with change in temperature causes the bimetallic strip to bend or curl. Bimetallic strips are used to make thermometers and sensing elements in thermostats.

In some cases we do not observe the expansion anticipated with increasing temperature because the object is clamped or otherwise held fixed. This is the case with modern continuous-welded railroad track. Formerly rail was laid in short sections containing gaps called expansion joints between them. The rails were free to slide back and forth on the ties as the temperature changed. Continuous-welded rail is laid in unbroken segments of any length, with a practical limit of 25 mi imposed by the need for electrical breaks for signal purposes. Spikes clamp the rail firmly to the ties, which distribute the forces caused by temperature changes to the ballast (the small rocks and gravel packed around the railroad ties), and thereby to the earth. The large forces are distributed because the rails are clamped down at closely spaced intervals.

Suppose we consider the expansion of a rectangular metal plate as its temperature changes by ΔT (Fig. 10.9). A straight line drawn on the plate in any direction would expand with the linear expansion coefficient of the metal. Along the horizontal (x) direction the plate would expand with a coefficient α. If the material is homogeneous, the plate expands with the same α in the vertical (y) direction or, indeed, in any other direction. Thus the plate enlarges horizontally by an amount $\Delta x = x_0 \alpha \, \Delta T$ and vertically by an amount $\Delta y = y_0 \alpha \, \Delta T$, giving an increase in area of approximately $2\alpha \, \Delta T$ (see Problem 10.53).

If we consider the thickness of the plate, it, too, increases with increasing temperature. If the temperature change is not too great, the change in volume ΔV of a homogeneous material is also proportional to the change in temperature ΔT and to the original volume V_0, so that we have

$$\Delta V = \beta V_0 \, \Delta T, \tag{10.5}$$

where β is the volume coefficient of thermal expansion. The units of β are also $°C^{-1}$. The volume coefficient of thermal expansion β is approximately three times the value of the linear coefficient of thermal expansion α. Values for β are given in Table 10.1.

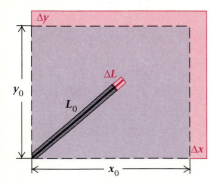

Figure 10.9 A rectangular plate expands in all directions with the same linear thermal expansion coefficient.

Example 10.5

Volume expansion of a glass bottle of water.

A 1.00-liter glass bottle is filled to the brim with water at a room temperature of 20°C. The temperature of the bottle and the water are then raised to 95°C. Does the water spill over, or does the level go down, and by how much?

Solution Think of the glass bottle as the "skin" of a solid piece of glass, all of which expands uniformly. Then the change in volume of the inside of the bottle is just the same as the change in volume of the solid interior.

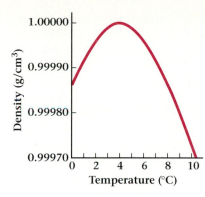

Figure 10.10 The density of water as a function of temperature.

We may write the change in volume ΔV_{glass} for the bottle as

$$\Delta V_{glass} = \beta V_0 \, \Delta T,$$

$$\Delta V_{glass} = (27 \times 10^{-6} \, °C^{-1})(1.00 \times 10^{-3} \, m^3)(95°C - 10°C),$$

$$\Delta V_{glass} = 2.30 \times 10^{-6} \, m^3 = 2.30 \, cm^3.$$

For the water the change in volume ΔV_{water} is

$$\Delta V_{water} = \beta V_0 \, \Delta T,$$

$$\Delta V_{water} = (210 \times 10^{-6} \, °C^{-1})(1.00 \times 10^{-3} \, m^3)(95°C - 10°C),$$

$$\Delta V_{water} = 17.9 \times 10^{-6} \, m^3 = 17.9 \, cm^3.$$

The expansion of the water is greater than the expansion of the bottle. The amount of water that will run over the edge is

$$\Delta V_{water} - \Delta V_{glass} = 17.9 \, cm^3 - 2.30 \, cm^3 = 15.6 \, cm^3.$$

Most liquids expand smoothly with increasing temperature. Alternatively, we can say that they become less dense with increasing temperature. Water, however, is an exception (Fig. 10.10). Its density is greatest at 4°C and is less for both higher and lower temperatures. Water also becomes less dense on freezing, again in contrast to most liquids. This effect has important consequences for aquatic life. When a body of water, such as a lake, cools, the cooler water at the top flows to the bottom because of its greater density. When the lake reaches 4°, this flow stops because the colder top of the lake is less dense as the temperature drops below 4°. As a consequence the top of the lake freezes first, while the lower depths remain at 4°. So water freezes from the top down. If water behaved like other substances, lakes would freeze from the bottom up, and the continuous circulation of warmer water to the top would cause more efficient freezing. Under those conditions lakes would freeze solid more frequently than they actually do. But as it is, lakes do not frequently freeze solid, even in the coldest climates. This effect is aided by the fact that the ice layer acts as an insulating blanket over the water. Fish survive by staying on the bottom, where the temperature is at least 4°C.

The Mechanical Equivalent of Heat

10.4

Before the mid-eighteenth century, the distinction between temperature and heat was not clear and the two were often confused. At that time it was generally thought that heat was some kind of fluid, called *caloric,* which could be added to, or taken away from, a substance to make it hot or cold. We now know that **heat flow** is a form of energy transfer that occurs when there is a temperature difference between objects. An example of the distinction between heat and temperature is sometimes given by comparing a flaming match and a warm hot-water radiator in the same cool room. The flaming match is at a much higher temperature than the radiator, but you do not expect it to appreciably warm the room. On the

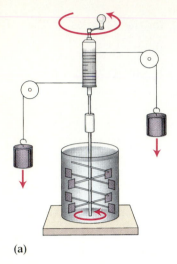

(a)

View from above

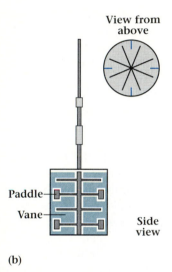

Paddle—
Vane—

Side view

(b)

Figure 10.11   (a) Sketch of Joule's apparatus. Falling weights turn a set of paddles in a water-filled container. (b) A cross-sectional view showing the paddles and the stationary vanes. Joule's apparatus measured the conversion of mechanical energy into thermal energy.

other hand, although the radiator is at a lower temperature than the match, enough heat flows from it to keep you warm. In both cases the heat flow is from an object at a higher temperature to surroundings at a lower temperature.

The first evidence for the connection between heat and energy transfer came when the American-born Benjamin Thompson, Count Rumford (1753–1814), was serving as minister of war in Bavaria. While supervising the boring of cannon he became curious about the tremendous amount of heat generated. His interest led to some detailed experiments on the nature of heat and heat capacities. He concluded that the increase in temperature was due to the work done in the boring process. Despite the implications of Rumford's work, the popular notion of heat as the fluid caloric still persisted, since that theory explained all the results in which people were generally interested.

In 1842 Julius Mayer (1814–1878) suggested that heat and mechanical work were equivalent and that one could be transformed into the other. He even went so far as to show that the temperature of water could be raised 1°C by mechanical agitation alone. However, he failed to determine the amount of work required for such a change.

The connection between heat flow and work was conclusively demonstrated a year later, in 1843, by James Prescott Joule (1818–1889). Joule devised an experiment in which the change of potential energy of falling weights was used to churn the water in an insulated container (Fig. 10.11) This famous apparatus contained paddles for stirring the water and stationary vanes to break up the flow, so that the water was not merely set into rotational motion (kinetic energy). The frictional drag of the water caused the weights to fall very slowly, so that their kinetic energy was quite small. The potential energy lost by the falling weights was imparted to the water and was detected as a change in temperature. In this way Joule showed that the temperature of one pound of water could be raised one degree Fahrenheit by the expenditure of 772 ft-lb of mechanical work.* He proved the direct conversion of mechanical energy into thermal energy (heat) and measured the numerical factor relating mechanical units to heat units.

As mentioned earlier, the definitions and units for mechanical energy developed independently of the definitions and units of heat, which grew out of the study of the properties of water. Two separate units for measuring heat were devised. In Britain the primary unit was the British thermal unit (Btu), the amount of heat required to raise the temperature of one pound of water one degree Fahrenheit. In Europe, where the metric system was in use, the calorie (cal) was defined as the heat required to raise the temperature of one gram of water by one degree Celsius. The Calorie (spelled with a capital C) used in discussing diet and nutrition is a kilo-calorie (10^3 cal).

In honor of Joule's contribution to science, his name was given to the common unit of energy. The joule equals one newton-meter and is roughly one fourth the size of the calorie:

$$1 \text{ calorie} = 4.1868 \text{ joules.}$$

*Better measurements place this value as 778 ft-lb, less than 1% greater than Joule's carefully determined number.

	J	cal	kcal	Btu	kWh
TABLE 10.2 Conversion table for some common energy units					
1 J = 1	1	0.239	2.39×10^{-4}	9.48×10^{-4}	2.78×10^{-7}
1 cal = 4.19		1	10^{-3}	3.97×10^{-3}	1.16×10^{-6}
1 kcal = 4190		1000	1	3.97	1.16×10^{-3}
1 Btu = 1060		252	2.52×10^{-1}	1	2.93×10^{-4}
1 kWh = 3.60×10^6		8.60×10^5	8.60×10^2	3.41×10^3	1

The relationships between several energy units are given for reference in Table 10.2.

The calorie is not recognized as an SI unit. The appropriate SI unit for energy is the joule. However, the calorie is still used in many practical applications and in several fields of research.* Therefore we will use both the joule and the calorie in our examples. You should be able to work with either unit.

Example 10.6

When you do work against a frictional force, mechanical energy is transformed into thermal energy. Suppose you push a 2.4-kg textbook 0.85 m across a level table at a constant speed. If the coefficient of friction between the book and the table is 0.25, how much energy is dissipated? Give your answer in calories.

Solution   For a constant speed, the force you apply equals the resisting frictional force, which is the product of the coefficient of friction μ and the normal force mg. The work done to move the book through a distance d is

$$W = \mu mgd = (0.25)(2.4 \text{ kg})(9.8 \text{ m/s}^2)(0.85 \text{ m}) = 5.0 \text{ J}.$$

This expenditure of energy corresponds to

$$5.0 \text{ J} \frac{1 \text{ cal}}{4.1868 \text{ J}} = 1.2 \text{ cal}.$$

Calorimetry

10.5

The measurement of quantities of heat exchanged, a process known as *calorimetry*, was introduced in the late 1700s. Chemists of the time found that when a hot object, such as a brass block, was immersed in a water bath, the resulting change in temperature of the water bath depended on both the mass and the initial temperature of the block. The temperature change was interpreted by them as a measure of the heat contained in the

---

*There are several definitions of the calorie that result in slightly different numerical values when expressed in joules. The differences are small, occurring in the fourth significant figure.

object. Further observations showed that, when two similar brass blocks at the same initial temperature were immersed in identical water baths, the more massive block caused a greater temperature change. Similarly, for two identical blocks at different temperatures, the hotter block gave rise to a greater temperature change in the bath. Finally, for blocks of the same mass and initial temperature, but of different composition, the change in temperature was different for different materials.

We can synthesize these observations by describing the objects in terms of their **heat capacity**, which is the amount of heat required to change an object's temperature by 1°C. Blocks of the same material but of different masses have heat capacities proportional to their mass. Thus we define an intrinsic quantity peculiar to each material, called its *specific heat capacity*, the ratio of the heat capacity to the mass. The specific heat capacity, or simply the **specific heat**, as it is usually called, is **the heat required per unit mass to change the temperature of a substance by one degree**. A material with a high specific heat, like water, requires a lot of heat to change its temperature, while a material with a low specific heat, like silver, requires little heat to change its temperature.

The amount of heat Q required to warm an object of mass m by raising its temperature ΔT is given by

$$Q = mc\, \Delta T, \tag{10.6}$$

where c is the specific heat of the material from which the object is made. If the object cools, then the temperature change is negative and the heat Q is given off by the object. The units of specific heat are cal/g·°C, J/kg·°C, or Btu/lb·°F. A list of specific heats is given in Table 10.3.

We have said that heat is a form of energy transfer. Therefore we can predict temperature changes in systems of two or more substances in thermal contact by applying the principle of conservation of energy: The heat (or energy) lost by the cooling objects must equal the heat (or energy)

TABLE 10.3
Specific heat for some common materials at 25°C*

Substance	Specific heat (J/kg·°C)	Specific heat (cal/g·°C) or (kcal/kg·K)
Water (0°C to 100°C)	4190	1.00
Ice (−10°C to 0°C)	2090	0.50
Steam (100°C)	2010	0.48
Wood	1700	0.4
Aluminum	900	0.215
Marble	860	0.21
Glass	840	0.200
Iron	448	0.107
Copper	390	0.0920
Zinc	386	0.0922
Silver	236	0.0564
Lead	128	0.0305

*The specific heat of most materials varies slightly with temperature; however, you may take them to be constant.

gained by the substances being warmed. We take the quantity of heat *added* to a body to be positive, and the heat *lost* by a body to be negative. Then we say that the sum of all the heat flows to all bodies in thermal contact is equal to zero. When we use this convention, ΔT is always $T_{final} - T_{initial}$. A negative value of ΔT means that heat has left the body. Example 10.7 illustrates this principle using a Styrofoam cup as an insulating, low-heat-capacity container. This experiment can easily be done and the measured values compared with the calculations.

Example 10.7

A Styrofoam cup of negligible heat capacity contains 150 g of water at 10°C. If you add 100 g of water at a temperature of 85°C, what is the final temperature of the mixture after it has been thoroughly mixed?

Solution   By the principle of conservation of energy, we expect that:

$$\text{heat gained (positive)} + \text{heat lost (negative)} = 0.$$

We also expect that the final temperature T of the mixture will be between 10° and 85°C.
   The heat gained by the cooler water is

$$\text{heat gained} = m_1 c \, \Delta T_1 = m_1 c(T_{final} - T_{initial}) = (150 \text{ g})c(T - 10°C).$$

The heat lost by the hotter water is

$$\text{heat lost} = m_2 c \, \Delta T_2 = m_2 c(T_{final} - T_{initial}) = (100 \text{ g})c(T - 85°C).$$

When the heat lost plus the heat gained is set equal to zero, the resulting expression determines a unique value for the final temperature T:

$$150(T - 10°C) + 100(T - 85°C) = 0,$$
$$(150 + 100)T = (8500 + 1500)°C,$$
$$T = \frac{10,000°C}{250} = 40°C.$$

   Note that in this particular problem there was no necessity for knowing the specific heat because both components of the mixture were of the same material: water.

Example 10.8

Using a calorimeter to determine specific heat.

A small metal block (mass of 75 g) is heated in an oven to 90°C. It is then taken from the oven and immediately placed in a calorimeter, a thermally insulated container designed for measurements of heat. The calorimeter contains 300 g of water at 10°C. The heat capacity of the calorimeter is negligible and the final temperature is 14°C. Identify the composition of the block from the following list: aluminum, iron, silver, or zinc.

Solution   Again we apply the rule of energy conservation: Heat lost plus heat gained equals zero. The heat lost by the block was

$$Q_1 = m_1 c_1 \, \Delta T_1 = (75 \times 10^{-3} \text{ kg})c_1(14°C - 90°C),$$
$$Q_1 = -5.70c_1(\text{kg·°C}).$$

The heat gained by the water was

$$Q_2 = m_2 c_2 \, \Delta T_2 = (300 \times 10^{-3} \text{ kg})(4190 \text{ J/kg·°C})(14°C - 10°C),$$

$$Q_2 = 5028 \text{ J}.$$

Conservation of energy gives

$$\text{heat lost} + \text{heat gained} = 0,$$

$$-5.70c_1 \text{ kg·°C} + 5028 \text{ J} = 0.$$

We rearrange to find

$$5.700c_1 \text{ kg·°C} = 5028 \text{ J},$$

or

$$c_1 = \frac{5028 \text{ J}}{5.70 \text{ kg·°C}} = 880 \text{ J/kg·°C}.$$

By comparing this value for c with those in Table 10.3, we see that the metal block in this example is probably made of aluminum.

Change of Phase

10.6

We know from experience that when heat is supplied to ice, it melts into water and that steam, when cooled, condenses into water. The transformation from one physical state to another (for instance, from solid to liquid, or from gas to liquid) takes place *with no change in temperature* and is called a change of phase. If we perform a careful calorimetric measurement during a phase change, we find that our previous description of heat exchange is incomplete. In addition to the heat absorbed or released in proportion to changes in temperature, there is also an amount of heat associated with a phase change (Fig. 10.12). This quantity is called the **heat of transformation** (sometimes the *latent heat of transformation*) L, defined as the ratio of the amount of heat Q absorbed (or released) to the mass m of material undergoing the phase change:

$$L \equiv Q/m.$$

We can express the heat absorbed (or released) in terms of L:

$$Q = mL. \tag{10.7}$$

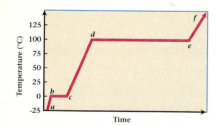

Figure 10.12 Temperature as a function of time for heat applied to water at a constant rate. The temperature remains constant during a change of phase. Ice is warmed to 0°C (*a* to *b*), melts (*b* to *c*), the resulting water is heated (*c* to *d*) and boils (*d* to *e*). The final steam can obtain higher temperatures (*e* to *f*).

Heats of transformation are expressed in units of J/kg or cal/g. The term "latent heat" goes back to the early days of the study of heat, when it referred to the absorption of heat without an accompanying change in temperature. In more recent times, people refer simply to the heat of transformation. If the phase change is from the solid to the liquid phase (or from the liquid to the solid), we refer to the **heat of fusion**; for the liquid–vapor phase change we use the term **heat of vaporization**. The energy added (or removed) in the form of the heat of transformation goes into rearranging the internal structure of the substance. For example,

TABLE 10.4
Heat of transformation of various substances at atmospheric pressure

Substance	Melting point (K)	Heat of fusion (J/kg)	Heat of fusion (cal/g)	Boiling point (K)	Heat of vaporization (J/kg)	Heat of vaporization (cal/g)
Water	273	3.34×10^5	79.7	373	22.6×10^5	539
Iron	1808	2.89×10^5	69.1	3023	63.4×10^5	1520
Copper	1356	2.05×10^5	48.9	2840	48.0×10^5	1150
Oxygen	54.4	0.14×10^5	3.3	90.2	2.13×10^5	50.9
Nitrogen	63.3	0.26×10^5	6.1	77.3	2.01×10^5	48.0

when a solid becomes a liquid, energy is required to overcome the forces that keep the material in the solid state.

Table 10.4 lists heats of transformation for several substances. We see from the table that the heat of fusion for ice is 3.34×10^5 J/kg (79.7 cal/g). This means that 3.34×10^5 J are required to melt each kilogram of ice. Conversely, 3.34×10^5 J are released by each kilogram of water that freezes.

Example 10.9

A 100-g copper calorimeter contains 300 g of water at room temperature ($T = 23°C$). If 50 g of ice at 0°C is added to the calorimeter, what is the final temperature of the system?

Solution Again we solve the problem by applying the principle of conservation of energy:

$$\text{heat lost} + \text{heat gained} = 0.$$

However, in this case heat is gained by the ice in melting, as well as by the melted ice in being warmed from 0°C to the final temperature T:

$$Q(\text{gained}) = m_{ice}L + m_{ice}c_{water}\,\Delta T,$$
$$Q(\text{gained}) = (50\text{ g})(80\text{ cal/g}) + (50\text{ g})(1\text{ cal/g·°C})(T - 0°C),$$
$$Q(\text{gained}) = 4000\text{ cal} + 50T\text{ cal/°C}.$$

The heat lost is given up by the calorimeter (subscript c) and the original water (subscript w):

$$Q(\text{lost}) = m_w c_w\,\Delta T + m_c c_c\,\Delta T = (m_w c_w + m_c c_c)(T_f - T_i)$$
$$= [(300\text{ g})(1\text{ cal/g·°C})$$
$$+ (100\text{ g})(0.092\text{ cal/g·°C})](T - 23°C),$$
$$Q(\text{lost}) = (309\text{ cal/°C})(T - 23°C),$$
$$Q(\text{lost}) = 309T\text{ cal/°C} - 7110\text{ cal}.$$

When the heat gained is added to the heat lost, we find

$$4000\text{ cal} + 50T\text{ cal/°C} + 309T\text{ cal/°C} - 7110\text{ cal} = 0.$$

Upon rearranging, we find

$$359T\text{ cal/°C} = 3110\text{ cal},$$
$$T = 8.7°C.$$

Example 10.10

Repeat the calculations for Example 10.9, but this time use 100 g of ice. What happens now?

Solution   The expression for heat lost is the same as before,

$$Q(\text{lost}) = 309T \text{ cal/}°\text{C} - 7110 \text{ cal},$$

and is a maximum of -7110 cal when the final temperature is $0°\text{C}$. For the heat gained we find

$$Q(\text{gained}) = 8000 \text{ cal} + 100T \text{ cal/}°\text{C},$$

which is a minimum of 8000 cal at $T = 0°\text{C}$. When we apply the conservation equation, we arrive at a negative value for T, an obvious error. (If T were negative all the water would turn to ice. In addition, we expect the final temperature to lie in the range between $0°\text{C}$ and $23°\text{C}$.) What went wrong?

We see that the maximum amount of heat available from the water and calorimeter is 7110 cal. Therefore not all of the ice will melt, since that would require 8000 cal, an amount that exceeds the heat available. The amount of ice melted is

$$\frac{7110 \text{ cal}}{80 \text{ cal/g}} = 89 \text{ g}.$$

The remaining mixture of water and 11 g of ice has a temperature of $0°\text{C}$. (In time, that ice will also melt as a result of the slow leakage of heat through the insulation of the calorimeter.)

The home ice cream churn provides a useful application of heat of fusion. The mixture to be turned into ice cream is placed in a metal can (with good heat-conducting properties) surrounded by ice. Rock salt is then poured onto the ice. At the ice–salt interface, the ice melts because there is a chemical interaction with the salt. The salt solution has a freezing point much lower than that of pure water. The solution can thus provide the energy for melting the rest of the ice. The melting ice absorbs heat from the salt solution, making the solution much colder than $0°\text{C}$. Even though the temperature of the ice is also lowered, it continues to melt at its surface because of the effects of the salt. The salt solution, in turn, absorbs heat from the ice cream mixture through the walls of the metal can, allowing the mixture to cool enough to become firm. Without the addition of salt to the ice, the temperature of the mixture would never become low enough to form ice cream.

Heat Transfer

10.7

Heat transfer takes place in three ways: conduction, convection, and radiation. When one end of a metal rod is heated, the other end gets warm. This is an example of **conduction**, in which thermal energy is transferred without any net movement of the material itself. Conduction is a relatively slow process. A more rapid process of heat transfer is accomplished through the mass motion or flow of some fluid, such as air or water, and is called **convection**. This transfer takes place when warm air flows about a room and when hot and cold liquids are poured together. A still more rapid transfer of thermal energy is accomplished by **radiation**, a process that requires neither contact nor mass flow. The energy from the sun comes to us by radiation. We also feel radiation from warm stoves, fires, and radiators. In this section we will principally discuss some aspects of conduction and radiation. Because of the mathematical complexities, the details of convection will be left to more advanced work.

An object's usefulness as a thermal conductor (or insulator) depends on a number of things, including its thickness and the nature of the material from which it is made. For example, no sensible person would try to use a good heat conductor like aluminum as a protective layer when picking up a hot pan. On the other hand, a padded cloth potholder is a good insulator and works quite well.

Suppose we imagine a wall of uniform material, such as plasterboard, that separates a warm room from a cold one. After a period of time, a steady temperature change occurs across the wall and a steady flow of heat goes from the warmer room to the cooler one (Fig. 10.13). Experiments show that the time rate at which heat flows ($\Delta Q/\Delta t$) through the wall is proportional to the area A, proportional to the temperature difference ($T_2 - T_1$), and inversely proportional to the thickness L of the wall. This information is contained in the heat flow equation:

$$\frac{\Delta Q}{\Delta t} = KA \frac{(T_2 - T_1)}{L}. \tag{10.8}$$

The constant K is called the **thermal conductivity** and is characteristic of the material making up the wall. The SI units of heat flow are J/s or W, so the SI units of K are J/(s·m·°C) or W/m·°C.

A high thermal conductivity indicates a good heat conductor; a low thermal conductivity indicates a good heat insulator. Some representative values of K for common materials are tabulated in Table 10.5. To have a good insulator, such as a potholder or the outside walls of a house, the first requirement is to choose a material with a small thermal conductivity, so the heat flow in Eq. (10.8) is small. In addition, by minimizing the area of contact A and making the path length L as long as possible, we can further reduce the heat flow.

The effectiveness of insulation is rated by another quantity, called thermal resistance, or R value. The **R value** is the ratio of a material's thickness to its thermal conductivity:

$$R \equiv L/K. \tag{10.9}$$

For the outside walls of your home you want a good heat insulator; that is, you want materials and insulation with a high R value. In the United States, the R value is given in the British system units of ft²·h·°F/Btu. R values are useful because you can simply add them to obtain the R value

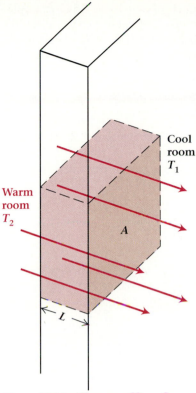

Figure 10.13 The rate of heat flow by conduction through a wall of area A depends on the temperature difference divided by the thickness, $(T_2 - T_1)/L$, as well as the thermal conductivity K of the wall's material.

TABLE 10.5
Thermal conductivities of some common materials at 27°C

Material	K(W/m·K)
Aluminum	237
Copper	398
Iron	80.3
Silver	427
Tungsten	178
Brick	0.4–0.8
Water	0.61
Air	0.026
Asbestos	0.083
Wood (pine)	0.11–0.14
Glass	0.72–0.86
Fiberglass	0.040
Polystyrene foam	0.033
Polyurethane foam	0.020

resulting from multiple layers of insulation (Problem 10.61). For example, a 6-in.-thick layer of fiberglass insulation has an R value of 19. Insulation made from two layers of 6-in. fiberglass has an R value of 38. For a given material, the R value is directly proportional to the thickness, which is represented by L in Eq. (10.9).

Example 10.11

Heat conduction from an oven.

A small oven has a surface area of 0.20 m². The insulated walls are 1.5 cm thick with an average thermal conductivity of 4.0×10^{-2} W/m·°C. What is the rate of heat loss if the temperature inside the oven is maintained at 220°C and the outside temperature is 20°C?

Solution   The rate of heat loss is computed from

$$\frac{\Delta Q}{\Delta t} = KA \frac{(T_2 - T_1)}{L},$$

$$\frac{\Delta Q}{\Delta t} = \frac{4.0 \times 10^{-2} \text{ W/m·°C} \times 0.20 \text{ m}^2 (220°C - 20°C)}{0.015 \text{ m}},$$

$$\frac{\Delta Q}{\Delta t} = 110 \text{ W}.$$

A hot object also loses energy by radiation. This radiation, which is known as electromagnetic radiation, is similar to light (see Chapter 19), and can pass through empty space (a vacuum). The warmth you feel when you warm yourself by a fire is due to this radiation. If the object is hot enough, some of the radiation is visible and can indeed be seen.

The rate at which an object radiates energy is proportional to its surface area A and to the fourth power of its absolute temperature T. The total energy radiated from an object per unit time is found experimentally to be

$$P = \sigma e A T^4, \tag{10.10}$$

where σ is the Stefan-Boltzmann constant, which has the value $\sigma = 5.67 \times 10^{-8}$ W·m^{-2}·K^{-4}, and e is a constant called the *emissivity*. The emissivity is a dimensionless number between 0 and 1 that describes the nature of the emitting surface. The emissivity is larger for dark, rough surfaces and smaller for smooth, shiny ones. Equation (10.10) is known as the **Stefan-Boltzmann** law.

According to the Stefan-Boltzmann law, all objects radiate energy, no matter what their temperature happens to be. Why then do they not lose all their thermal energy by radiation and cool down to 0 K? The answer is that they also absorb radiation from surrounding objects and eventually come to thermal equilibrium with their environment. The book in your hand is radiating, but it is also absorbing radiation from its surroundings. If the book (or other object) is at a temperature T and its surroundings are at a different temperature T_s, the net energy gained (or lost) by the book is given by

$$P_{\text{net}} = \sigma e A (T^4 - T_s^4), \tag{10.11}$$

where A is the surface area of the book. Notice that we have used the same emissivity for absorption as for radiation. This must be correct, since the net heat exchange must go to zero when $T = T_s$. Thus a good radiator is also a good absorber.

Because of the T^4 term in Eq. (10.10), the total power radiated grows rapidly as the temperature increases. For example, an object at a temperature of 273°C (546 K) radiates 16 times more power than it does at 0°C (273 K). The distribution of the radiation, which is composed of many different wavelengths, is also a function of temperature. It is the change in this distribution, along with the increase in radiant energy with an increase in temperature, that accounts for the onset of the glow and the change of color of a hot object as its temperature is raised. We will discuss these issues further when we describe blackbody radiation in Chapter 26.

Example 10.12

A patient waiting to be seen by his physician is asked to remove all his clothes in an examination room that is at 16°C. Calculate the rate of heat loss by radiation from the patient, given that his skin temperature is 34°C and his surface area is 1.6 m². Assume an emissivity of 0.80.

Solution   From Eq. (10.11), the rate of heat loss by radiation is

$$P_{net} = \sigma e A (T^4 - T_s^4),$$

$$P_{net} = (5.67 \times 10^{-8}\,\text{W·m}^{-2}\text{·K}^{-4})(0.80)(1.6\,\text{m}^2) \\ [(307\,\text{K})^4 - (289\,\text{K})^4],$$

$$P_{net} = 140\,\text{W}.$$

The problem of choosing good insulation is not always merely that of finding a poor thermal conductor. Air is a poor conductor, yet a hot object left exposed in the air cools rapidly as a result of convection currents in the air, which continually bring cool air in contact with the object. These convection currents are caused by the expansion of the air as it is warmed. The warmer air is lighter than the cooler surrounding air and rises as a result of buoyancy. Energy is also lost by direct radiation. To reduce these effects and still capitalize on the low conductivity of air, we may use an insulating material that contains many tiny pockets of air so that convection is reduced to nearly zero. Pockets of trapped air account for the good insulating qualities of Styrofoam, fiberglass, down, felt, and woolen clothing. In addition, the many surfaces of the insulating material help reduce the radiant loss by reflection and by radiation back toward the object.

The vacuum flask, used so effectively to keep hot foods hot and cold foods cold, has an inner glass container surrounded by an outer one (Fig. 10.14). The space between is evacuated and sealed. The vacuum between the containers offers little heat loss through either conduction or convection. Radiation losses are minimized by coating the wall of the vacuum space with a highly reflecting layer of silver. Thus the principal means of heat leakage is through the plug at the mouth of the container and through the glass joining the inner and outer containers. This type of flask is called a Dewar flask after the Scottish scientist Sir James Dewar, who first used it in 1892. It is also known by the trade name, Thermos bottle.

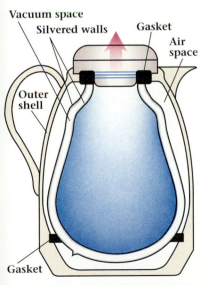

Figure 10.14   Cross-sectional view of a vacuum flask. The vacuum space reduces heat loss by conduction or convection and the reflecting surfaces reduce heat loss by radiation.

SUMMARY

Useful Concepts

- Heat flow is a form of energy transfer between two objects; temperature is a measure of an object's warmth.
- The relationship between Fahrenheit and Celsius temperatures is

$$T \text{ (F°)} = \tfrac{9}{5} T \text{ (°C)} + 32.$$

- The relationship between the Kelvin and Celsius temperature scales is

$$T \text{ (K)} = T \text{ (°C)} + 273.15.$$

- A body expands or contracts when the temperature changes. The length at a new temperature T in terms of the initial length L_0 at T_0 is

$$L = L_0[1 + \alpha(T - T_0)],$$

where α is the thermal expansion coefficient.
- Joule showed experimentally that heat and mechanical work are equivalent, determining the relationship

$$1 \text{ calorie} = 4.1868 \text{ joules.}$$

- Calorimetry is based on the idea of the conservation of thermal energy. In a thermally isolated system,

    heat gained (positive) + heat lost (negative) = 0.

- The heat Q required to change the temperature of a body of mass m is

$$Q = mc \, \Delta T,$$

where c is the specific heat capacity. The heat required to change the phase of a mass m of material with latent heat of transformation L is

$$Q = mL.$$

- Heat of transformation is absorbed or given off during a substance's change of phase; however, the temperature remains constant during the phase change.

- The heat conduction equation is

$$\frac{\Delta Q}{\Delta t} = \frac{KA(T_2 - T_1)}{L}.$$

- The R value of insulation is the ratio of thickness to thermal conductivity:

$$R \equiv L/K.$$

- Energy is radiated from all objects. The rate at which the energy is radiated is given by the Stefan-Boltzmann law:

$$P = \sigma e A T^4.$$

Important Terms

You should be able to write the definition or meaning of each of the following terms:

- temperature
- thermometer
- thermal equilibrium
- Celsius temperature scale
- Fahrenheit temperature scale
- absolute zero
- Kelvin temperature scale
- linear thermal expansion
- thermal expansion coefficient
- heat flow
- heat capacity
- specific heat
- heat of transformation
- heat of fusion
- heat of vaporization
- conduction
- convection
- radiation
- thermal conductivity
- R value

QUESTIONS

10.1 Why do some materials, such as glass and metal, usually feel cold and some other materials, such as cloth, usually feel warm? (Refer to Tables 10.3 and 10.5.)

10.2 A square brass plate has a large circular hole cut in its center. If the plate is heated, it will expand. Will the diameter of the hole expand or contract? Explain your answer.

10.3 How does thermal expansion affect the precision of a pendulum clock?

10.4 What differences are there between a fever thermometer that measures from 35 to 42°C and a laboratory thermometer that measures from −10 to 110°C?

10.5 What happens to the heat of transformation absorbed during a phase transition?

10.6 Devise a thermometer that relies on some property other than thermal expansion to indicate changes in temperature.

10.7 Why are concrete highways made in short sections with tar-filled gaps between them?

10.8 How does the thickness of a pot or frying pan affect the way it cooks? What effect does the pot's composition (e.g., aluminum, steel, or ceramic) have on the way it cooks?

10.9 A sports car going 30 km/h is braked to a stop without skidding. What happens to its kinetic energy?

10.10 Can you warm up a cup of coffee by stirring vigorously?

10.11 People in hot arid regions frequently store water in canvas bags through which some of the water can seep. What is the purpose of doing this?

10.12 Why is the climate of coastal cities milder than that of cities in the midst of large land areas?

10.13 Is there any insulating advantage for windows made of three panes of glass separated by two air spaces over those made of the equivalent total thickness of material in two panes of glass separated by one air space? Explain your reasoning.

PROBLEMS

Hints for Solving Problems

Be especially careful to use a consistent set of units, based on the same temperature scale. In an isolated system, the heat lost plus the heat gained is zero, where heat gained is positive and heat lost is negative. Heat exchanged includes $\Delta Q = mc \, \Delta T$, due to temperature change, and $\Delta Q = mL$, due to phase change. The final temperature of a mixture of hot and cold liquids will be intermediate between their initial temperatures.

Section 10.2   Thermometry

10.1 What is generally regarded as the record high temperature of 57.8°C occurred in Tripoli in Northern Africa in 1922. The record low of −89.2°C was recorded at the Soviet Antarctica station Vostok in 1983. Convert these temperatures to the Fahrenheit scale.

10.2 Human body temperature is about 98.6°F. Convert this to the Celsius scale.

10.3 At what temperature do the Fahrenheit and Celsius scales have the same numerical value?

10.4 Express the following Kelvin temperatures in degrees Celsius and in degrees Fahrenheit: 10 K, 300 K, and 450 K.

10.5 Find a relationship expressing the temperature in degrees Fahrenheit in terms of the temperature in kelvins.

10.6 At atmospheric pressure the boiling point of helium is 4.2 K. What is the boiling point of helium on the Celsius scale? On the Fahrenheit scale?

10.7 At atmospheric pressure the boiling point of nitrogen is −195.8°C. What is the boiling point of nitrogen on the Kelvin scale? On the Fahrenheit scale?

10.8 An approximate way of converting from the Fahrenheit scale to the Celsius scale is to subtract 32 from the Fahrenheit temperature and divide the result by two. How much error in °C does this give for Fahrenheit temperatures of 80°, 40°, 10°, and −10°?

Section 10.3   Thermal Expansion

10.9 An aluminum wire 100.0 cm long at 10°C is observed to be 100.2 cm long at a temperature of 80°C. What is its coefficient of linear expansion?

10.10 A steel plate with a circular hole of 2.01 cm radius and a copper ball with a radius of 2.00 cm are initially at 15°C. If their temperature is raised to 275°C, will the ball still fit through the hole?

10.11 In the days of horse-drawn wagons, iron tires used to be placed on wooden wagon wheels by heating the iron rims and slipping them over the wheels before they cooled. If an iron tire is made to fit tightly around a 1.50-m-diameter wheel at 15°C, what diameter will the tire have when it is heated to 800°C?

10.12 A brass rod is measured to be 3.00 m long at a temperature of 15°C. How much does it expand when heated to a temperature of 150°C?

10.13 A motorist fills the 60-liter tank of his automobile with gasoline at 60°F. The automobile is left in the sun and the temperature of the gasoline rises to 110°F. Approximately how much gasoline is lost to overflow caused by thermal expansion? (*Hint:* Ignore the expansion of the tank.)

10.14 A concrete highway has expansion joints at intervals of 18 m. How wide must the expansion joints be to allow for thermal changes over the temperature range from −10°C to 40°C? (*Hint:* Use $\alpha = 10 \times 10^{-6} \, °C^{-1}$.)

10.15 A building with a steel framework is 50 m high. How much taller is it on a summer day when the temperature is 30°C than on a −5°C winter day?

10.16 Show that the density of an object ρ with respect to its density ρ_0 at some reference temperature is given approximately by $\rho = \rho_0(1 - \beta \, \Delta T)$, where β is the thermal coefficient of volume expansion and ΔT is the difference between the object's temperature and the reference temperature.

10.17 A framework of rods is made as shown in Fig 10.15. How far and in what direction will the point A move when the temperature is increased by 100°C?

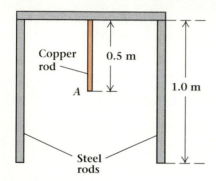

Figure 10.15   Problem 10.17.

10.18 An iron bar is exactly one meter long at 15°C. A brass bar is 0.5 mm shorter than the iron bar at 15°C. At what temperature will they be the same length if they are heated to the same temperature?

10.19 What is the increase in volume of an aluminum sphere with a radius of 10.0 cm when it is heated from 0°C to 100°C?

Section 10.4   The Mechanical Equivalent of Heat

10.20 The food energy of a jelly donut is about 250 kcal. Show that the jelly donut could be used as a unit of energy where 1 jelly donut $\approx$ 1 MJ.

10.21 The Btu is defined to be the amount of heat needed to raise the temperature of one pound of water by 1°F. Show that 1 Btu = 252 cal.

10.22 (a) How many joules are required to raise 1.0 kg of water from room temperature of 22°C to its boiling point? (b) If this work were used to lift a 50-kg boy instead, how high could he be lifted?

10.23 A 1500-W heater is submerged into one kilogram of water that is well below 100°C. At what rate, in °C/s, does the temperature rise when the heater is operating at its rated power?

10.24 What would be the maximum possible increase in temperature of water as it fell from the top to the bottom of a 23-m-high waterfall?

10.25 Two hundred grams of lead shot is placed in a 1.5-m-long cardboard tube, which is closed at both ends. If the tube is in a vertical position and then quickly inverted, the shot falls through the length of the tube. If this is done 50 times in succession, what is the maximum increase in temperature of the shot?

10.26 In an experiment similar to the one done by Joule (Fig. 10.11), a 10-kg weight is allowed to fall through a distance of 1.6 m. If there is 2.0 kg of water in the container, what will its temperature rise be?

10.27 How many calories are released in stopping a car that has a mass of 1780 kg and is traveling at 60 km/h?

Section 10.5   Calorimetry

10.28 Suppose that 250 g of water at 85°C is mixed with 80 g of water at 15°C in an insulated container of negligible heat capacity. What is the final temperature?

10.29 Assume that the specific heats of coffee and water are the same. How much cool water at 59°F must be mixed into a 150-g cup of coffee at 187°F to cool it to 135°F? Ignore the mass of the cup.

10.30 A cup of negligible heat capacity holds 180 g of hot tea at 80°C. A 100-g silver spoon at 25°C is immersed in the tea and comes to thermal equilibrium. By how much does the temperature of the tea decrease? (*Hint:* Assume that the specific heat of tea is the same as that of water.)

10.31 You pour 150 g of hot coffee at 86°C into a 200-g glass cup at 22°C. If they come to thermal equilibrium quickly, what is the final temperature?

10.32 A cook pours 400 g of hot water at 98°C into a 235-g aluminum pot initially at 15°C. If they come to thermal equilibrium quickly, what is the final temperature?

10.33 A 2.0-cal/s heater is submerged in a 1.0-liter beaker full of water for 10 min. What is the temperature rise if all of the heat goes into raising the temperature of the water?

10.34 If 1.0 kg of lead shot at 150°C is poured into 1.0 kg of water at 23°C, what is the final temperature? Neglect the effect of the container, and assume that no steam escapes.

Section 10.6   Change of Phase

10.35 How many joules are required to change one kilogram of ice at $-10°C$ to water at 15°C?

10.36 From your experience, estimate the final temperature that will result if a single ice cube is placed in a cup of hot coffee that has just been boiled. Make reasonable estimates of the size of the ice cube and the volume of coffee.

10.37 How many calories are required to change 500 g of ice at $-14°C$ to steam at 110°C?

10.38 How many grams of ice at 0°C can be melted by the heat released when one gram of steam condenses to water at 100°C?

10.39 How many joules are required to change 1.0 kg of solid iron at 18°C to a liquid at 2500°C? (Take the melting temperature of iron to be 1535°C, the specific heat of both solid and molten iron to be the same, and the heat of fusion to be 64 cal/g.)

10.40 If 20 g of steam at 100°C is mixed into 80 g of water at 20°C, what will be the final temperature if no heat is lost?

10.41 If the energy that goes into evaporating one kilogram of water from Lake Michigan at 20°C were used instead to raise that amount of water above the surface of the lake, how high would it be lifted?

10.42 A container of negligible heat capacity is filled with 5.0 kg of crushed ice and then placed on a hot plate that supplies heat to the ice at a rate of 30 W. What volume per second of water is produced?

Section 10.7 Heat Transfer

10.43 A large window of ordinary plate glass is 1 m wide, 1.5 m high, and 3.0 mm thick. Use the heat flow equation to calculate the rate of heat loss through the window on a day when the inside temperature is 22°C (72°F) and the outside temperature is 0°C. (In practice, the actual rate of heat loss through the window is much smaller than the value calculated on the basis of the heat flow equation with $T_1 =$ inside temperature and $T_2 =$ outside temperature. The layers of air on both sides of the glass act as additional insulation, so that the temperature difference across the glass is substantially reduced.)

10.44 A wall is insulated with glass wool of thermal conductivity $K = 0.040$ W/cm·°C. What is the rate of heat loss through an area 1.0 m wide by 1.5 m high, insulated with a layer of glass wool 15 cm thick, if the temperature difference across the layer is 20°C?

10.45 An outside wall of a room is 2.44 m (8 ft) high by
  • 5.0 m (16.4 ft) wide. (a) Calculate the rate of heat loss through the wall, assuming there are no windows and the wall has an average R value of 15. Take the temperature difference across the wall to be 25°C. (b) What is the rate of heat loss through the wall if it has a single glass window with an R value of 1, an area of 1 m² = 10.76 ft², and a thickness of 2.54 mm?

10.46 A Styrofoam cooler has a surface area of 0.50 m² and an average thickness of 2.0 cm. How long will it take for 1.5 kg of ice to melt in the cooler if the outside temperature is 30°C? (The thermal conductivity of the Styrofoam is 0.030 W/m·°C.)

10.47 What is the rate of heat flow along a copper bar 1.0 m long having a cross section of 1.0 cm² if one end of the bar is at 0°C and the other one is at 100°C?

10.48 A certain lamp is designed to operate at a temperature of 3200 K. If the lamp is operated at a higher voltage that raises its temperature to 3400 K, what will be the fractional increase in radiant energy?

10.49 What is the rate of energy radiated per unit area from a blackbody with emissivity = 1 at temperatures of 300 K, 1000 K, 3000 K?

10.50 The rate of radiation from the sun is determined to be 6.25×10^7 W/m². Use this value to compute the effective temperature of the sun. Assume an emissivity of 1.

10.51 The radiation from the sun is received from all parts of the sun's disk, including the less luminous outer edges. One estimate placed the radiation rate at the center to be 16% greater than the average value of 6.25×10^7 W/m². Use this estimate to determine the temperature of the sun near the center of the sun's disk. Assume an emissivity of 1.

Additional Problems

10.52 A physicist defines a new temperature scale that has
  • its zero at a room temperature of 70°F and its 10° mark at approximately body temperature (say 98°F). (a) Derive an expression for converting from the Fahrenheit scale to this new scale. (b) Derive an expression for converting from the Celsius scale to this new scale.

10.53 A flat plate with area A_0 at temperature T_0 is char-
  • acterized by a linear expansion coefficient α. Show that to a good approximation the area A of the plate at temperature T is given by

$$A = A_0[1 + 2\alpha(T - T_0)].$$

10.54 A certain material has a linear expansion coefficient
  • α. Show that, to a good approximation, the volume expansion coefficient is $\beta = 3\alpha$.

10.55 A clock with a brass pendulum keeps correct time
  • at 20°C. (a) What is the fractional change in pendulum length when the temperature rises to 38°C? (Hint: The thermal expansion coefficient is $\alpha = 18.5 \times 10^{-6}$/°C.) (b) The period of a pendulum varies as the square root of its length. By how many seconds will the clock be in error after running 24 h at 38°C?

10.56 An automobile having a mass of 1900 kg and trav-
  • eling at a speed of 30 m/s is braked smoothly to a stop without skidding in 15 s. (a) How much energy is dissipated in the brakes? (b) What is the average power delivered to the brakes during stopping? (c) If the total heat capacity of the braking systems (shoes, drums, etc.) is 0.75 kcal/°C, what is the temperature rise of the brakes during the stop?

10.57 A 175-g copper block at 90°C is dropped into an
  • aluminum calorimeter cup initially at 20°C. The calorimeter cup has a mass of 400 g and contains 430 g of water, also at 20°C. What is the final temperature of the system?

10.58 (a) If 120 g of hot coffee at 85°C is poured into an
• insulated cup containing 200 g of ice, how many
grams of liquid will there be when the system
reaches thermal equilibrium? (b) How much ice will
remain?

10.59 What is the R value of a wall that loses heat at the
• rate of 10 W/m² when the temperature difference
across the wall is 20°C? Express your answer in
units of ft²·h·°F/Btu.

10.60 An iron bar 50 cm long is welded to a copper bar
• 50 cm long and of the same diameter. The free end
of the iron bar is kept at 0°C while the free end of
the copper bar is at 100°C. (a) What is the temperature of the junction point? (b) What is the heat
flow down the rod if its cross-sectional area is
1.5 cm²?

10.61 Show that the effect of using two different adjacent
• pieces of insulating material, with R values of R_1 and
R_2, is to give an equivalent R value of $R_1 + R_2$.

10.62 A person's metabolic rate can be measured using
• what is called a flow calorimeter. The person is
placed in a large insulated container through which
water can flow. The flowing water carries away the
heat produced by the body. If a resting person is
known to have a thermal power output of 85 W,
what will the temperature difference between the
intake and outflow water be when the flow rate is
1.0 liter each 5.0 min?

10.63 A liquid of unknown specific heat at a temperature
• • of 20°C was mixed with water at 80°C in a well-
insulated container. The final temperature was
measured to be 50°C and the combined mass of the
two liquids was measured to be 1000 g. In a second
experiment with both liquids at the same initial temperature, 20 g less of the liquid of unknown specific
heat was poured into the same amount of water as
before. This time the equilibrium temperature was
found to be 52°C. Determine the specific heat.

10.64 If you remove 1000 cal from 1.5 g of steam at 100°C,
• what will you have left?

10.65 If you start with a container of 500 g of water at
• 20°C and add 300 g of ice, how many grams of steam
at 100°C will you have to condense into the water,
by bubbling it through a tube from the bottom, in
order to return the mixture to its original temperature?

10.66 A 50-W electrical heating element is placed in a
• well-insulated container into which has just been
placed 500 g of water at 20°C and 300 g of ice at
0°C. How long will it take before all of the contents
are evaporated?

10.67 A 0.50-kg piece of glowing-hot iron at a temperature
• • of 1000°C is put into an insulated container holding
one kilogram of water at 23°C. How much water
will be left after equilibrium has been reached? Assume that no steam escapes.

10.68 A 50-g piece of iron and a 40-g piece of copper at
• • the same temperature of 80°C are put into an insulated container that has 400 g of water and 100 g
of ice, all at 0°C. What is the equilibrium temperature?

10.69 If a one cubic meter sealed container of air at 0°C
• and one atmosphere pressure is heated by supplying
heat at the rate of 10.0 W for three minutes, by how
much will the temperature rise? Assume the container to be perfectly insulated. Under these conditions the density of air is 1.293 kg/m³ and its specific heat is 804 J·kg⁻¹·K⁻¹.

ADDITIONAL READING

Brown, S. C., "Benjamin Thompson, Count Rumford." *The Physics Teacher*, May 1976, p. 263.

Donnely, R. J., "Leo Dana: Cryogenic Science and Technology." *Physics Today*, April 1987, p. 38.

Jones, E. R., "Fahrenheit and Celsius, A History." *The Physics Teacher*, November 1980, p. 594.

Rosenfeld, A. H., and D. Hafemeister, "Energy-Efficient Buildings." *Scientific Amer-ican,* April 1988, p. 78.

Ward, D. S., "Advances in Solar Heating and Cooling Systems." *The Physics Teacher,* April 1976, p. 199.

Wineland, D.J., and W. M. Itano, "Laser Cooling." *Physics Today*, June 1987, p. 34.

11

Thermodynamics

11.1   The First Law of
       Thermodynamics

11.2   The Carnot Cycle and
       the Efficiency of Engines

11.3   Heat Pumps and
       Refrigerators

11.4   The Second Law of
       Thermodynamics

11.5   Entropy and the
       Second Law

*11.6 Energy and Thermal
       Pollution

A WORD TO THE STUDENT

In this chapter we introduce the fundamental laws of thermodynamics. These laws are as basic to our understanding of heat and energy flow as the laws of conservation of mechanical energy and momentum are to our understanding of motion. The laws of thermodynamics are founded on many experimental observations and apply to a broad range of different physical behavior. In fact, almost all motion, whether in thunderstorms or in jet engines, is ultimately due to heat.

The first law of thermodynamics is an expression of the law of conservation of energy that explicitly includes heat energy. The second law sets limits on the efficiency of machines and processes that change one form of energy into another. The second law has particular importance for all practical systems, including air conditioners, heat pumps, refrigerators, electrical generators, steam turbines, and internal combustion engines. Together these two laws govern the flow of thermal energy in the universe.

The area of physics concerned with the relationships between heat and work is called **thermodynamics**. Our recognition of heat as a form of energy and our application of energy conservation in calorimetry provide introductory glimpses into this science. But the concerns of thermodynamics are much broader and more elegant than simply the measurement of heat. The underlying basis of thermodynamics is found in two general laws of nature abstracted from our universal observations and experience. The first law states that you can't get more energy out of a system than you put into it, in all forms. While this sounds like a straightforward statement, its implications are quite profound and important. The second law says that the transfer of energy by heat flow has a direction; in other words, not all processes in nature are reversible. If a polar bear lies down in the snow, heat from its body will melt the snow; but the bear can't extract energy from the snow to warm itself. Thus the flow of heat has a direction—from hot to cold. By logical extension of these laws, we can correlate many measurable properties of matter with one another. Thermodynamic formulas predict many relationships between properties of matter and have the same general validity as the two laws on which they are based.

The development of thermodynamics grew out of a very practical concern with the operation of steam engines in the early nineteenth century. James Watt (1736–1819), a Scottish engineer and inventor, made steam power practical by markedly improving the efficiency of steam engines. His improvements, which resulted from his keen physical insight into thermal processes, were of such magnitude that he is often spoken of as the inventor of the steam engine.

The principles of thermodynamics that were developed in the eighteenth and nineteenth centuries led to tremendous advances in the power and efficiency of engines, from Watt's early steam engines to today's steam-driven electric power plants. The search for new and more efficient sources of energy continues. But we recognize that all potential advances must still follow the laws of thermodynamics; no matter what new sources we tap, we can't get something for nothing.

The First Law of Thermodynamics

11.1

We saw in the previous chapter that two objects in thermal contact can exchange heat as long as they are at different temperatures. The warmer object cools as the cooler one warms until they reach a common temperature at which no further changes take place. Two objects in this condition are said to be in a state of thermal equilibrium. In this state, energy ceases to flow from one object to the other, and they are at the same temperature throughout.

What happens if we introduce a third object? For example, suppose

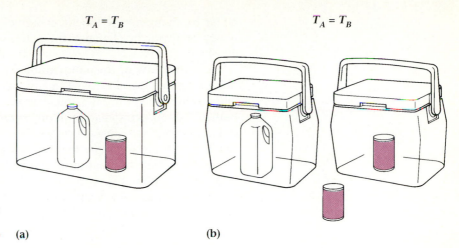

$T_A = T_B$ $T_A = T_B$

(a)   (b)

Figure 11.1 (a) Objects *A* and *B* at different temperatures come to thermal equilibrium when placed in thermal contact. (b) They remain in thermal equilibrium even when separated if no heat is exchanged with their environment. Then if object *C* is in thermal equilibrium with *A*, is it also in equilibrium with *B*?

we take a bottle of milk *A* from the refrigerator and place it in an ideal insulating chest (Fig. 11.1a), where it exchanges heat with a can of orange juice *B* until they come to thermal equilibrium. (The chest insulates them from the outside world.) They remain in thermal equilibrium if they are separated. That is, if we place the juice in another insulated chest, it is still in equilibrium with the milk. Now, if a second can of juice *C* is in thermal equilibrium with the milk *A* (Fig. 11.1b), we may ask the question: What is the relationship of *B* to *C*? Experiments show that *B* and *C* are also in equilibrium. This result may be stated as: **Two systems, each in thermal equilibrium with a third system, are in thermal equilibrium with each other.** This rule is known as the **zeroth law of thermodynamics.** It is called the "zeroth" law because it logically precedes the statements of the first and second laws of thermodynamics, but was not recognized as an important and fundamental law of nature until after these other laws had been stated and named.

In the study of thermodynamics we introduce the concept of a system and consider the transfer of energy into or out of the system by heat or work. The system could be any physical system, such as a machine (an automobile engine), or it could be a chemical system (a burning log), or a biological system (you). A **thermodynamic system** is any collection of objects considered together; the rest of the universe is the environment of the system. A thermodynamic system interacts with its surrounding environment by heat transfer and/or work. As a result of this energy exchange with the environment, the system's internal energy may change. By **internal energy** we mean the total kinetic and potential energy associated with the internal state of the atoms composing the system. In addition, the system may also have kinetic and potential energies due to its collective motion and outside forces, such as the force of gravity.

Be sure you understand the differences among *temperature, internal energy,* and *heat.* Temperature is a measure of the warmth of an object; as we will see in Chapter 12, on an atomic level it is determined by the average random kinetic energy of the object's atoms. Internal energy is the sum of the kinetic and potential energies of the internal motion of all the atoms in the object. Heat is the transfer of energy to or from an object, either by changing the kinetic energy of the atoms (changing an object's temperature) or by changing the potential energy of the atoms (changing an object's phase).

The **first law of thermodynamics** is based on the idea that energy is neither created nor destroyed in any thermodynamic system. As is true of many other scientific laws, there is no absolute proof for the first law. Rather, it is an extrapolation of our experience and has no known exception. However, we must be sure we know all the forms in which energy can occur before we apply the first law to some specific system.

In the usual formulation of the first law, we consider the transfer of heat into a system, the work performed by the system, and the change in the system's internal energy. If we let Q be the net amount of heat flowing *into* a system during some process and W be the net work done *by* the system, then conservation of energy gives

$$Q = W + \Delta U,$$

where ΔU is the change in the system's internal energy. Upon rearranging, we find

$$\Delta U = Q - W. \qquad (11.1)$$

The meaning of two of the terms in this equation is clear from prior chapters: Work was encountered early in our study of mechanics; and heat, as a form of energy transfer, was treated in the previous chapter. A negative value of Q means that heat is given out by the system instead of being added to the system. Similarly, a negative value of W means that work is being performed on the system rather than being done by the system. The internal energy U of the system can take a variety of forms, as we will see later, and depends on the temperature of the system. Equation (11.1) is the usual mathematical statement of **the first law of thermodynamics**. In words, **the change in internal energy of a system equals the difference between the heat taken in by the system and the work done by the system**.

Since the first law is a statement about energy, to apply it we need to specify, or measure, the energy or energy change of a system. For a purely mechanical system this is fairly easy to do because we can measure the masses and determine their velocities or positions. But with heat, things are different. There are no perfect heat insulators to keep energy confined to a certain place. For that reason, among others, we must carefully identify the system under consideration and separate "the system" from "the rest of the universe" or what is usually called the *environment*.

A thermodynamic system, no matter what its composition, may undergo several special kinds of processes involving energy. If no heat enters or leaves the system during some process, then the system is said to be perfectly isolated from its environment and the process is called **adiabatic**. A good approximation to an adiabatic process is anything that happens so rapidly that heat does not have time to flow in or out of the system, as in the rapid compression of air in a tire pump. If $Q = 0$ in the first law (Eq. 11.1), we are left with

$$\Delta U = -W \qquad \text{(adiabatic process)}.$$

Thus in an adiabatic process, the system does not exchange heat with the environment, and the change in internal energy is the negative of the work done.

If the temperature of a system does not change during a process, the process is said to be **isothermal.** An approximately isothermal process proceeds so slowly that the rate of change in temperature is negligible. Slow compression of air in a tire pump is an example of such a process, provided that air is not allowed to flow out of the cylinder. Because the kinetic energy of the air molecules is proportional to the temperature, the internal energy remains constant during an isothermal process that involves no change of phase or chemical change. Therefore, from the first law, the heat absorbed by the system must equal the work done by the system:

$$Q = W \qquad \text{(isothermal process)}.$$

Most processes in nature are neither strictly adiabatic nor strictly isothermal, but we can approximate many processes by treating them as one or the other, or as one followed by the other.

A process in which the volume of the system does not change is called isovolumetric or **isochoric.** Heating a gas in a tightly sealed container is an example of such a process. In a process that goes forward at constant volume, no displacement can take place, so the work done by the system is zero. (Remember, work is force times displacement.) Then, from the first law, we have only two terms,

$$Q = \Delta U \qquad \text{(isochoric process)}.$$

That is, in an isochoric process, no work is done by the system, and any heat added to the system goes into increasing its internal energy.

If the pressure does not change during a process, the process is called **isobaric.** One example of an isobaric process is the boiling of water in an open container. Since the container is open, the process occurs at constant atmospheric pressure. At the boiling point, the temperature of the water no longer increases with the addition of heat; instead there is a change of phase from water to steam.

Many of the thermodynamic systems we will use to illustrate new principles consist of a fluid, often a gas. So we often find it convenient to express work in terms of pressure rather than force. In practice, a thermodynamic system might be the water and steam system in a steam engine, or the system of gasoline and air in an internal combustion engine.

Figure 11.2 shows a cylinder of gas fitted with a piston. If the gas pushes the piston out an amount Δx, the increment of work done $F\,\Delta x$ can be written in terms of the pressure P and the change in volume ΔV:

$$\Delta W = F\,\Delta x = \frac{F}{A}\,(\Delta x A) = P\,\Delta V. \qquad (11.2)$$

The total work done by the gas is the sum of these small increments, taking proper account of the relationship between pressure and volume in the case being investigated. A gas-filled cylinder fitted with a piston is the simplest form of what is called a *heat engine.* Such a heat engine is used to model more complicated systems, from automobile engines to biological systems.

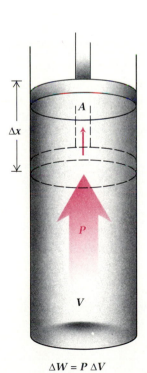

$\Delta W = P\,\Delta V$

Figure 11.2 The increment of work ΔW done by an expanding gas is $P\,\Delta V$.

As we noted, the change in a system's internal energy can take many different forms, including changes in temperature, changes of phase, and chemical changes. In the following examples we will consider a change in internal energy corresponding to a change in temperature. However, this is not the only possibility.

Example 11.1

Internal energy change in an isochoric process.

Show that in an isochoric process, the change in temperature is proportional to the change in internal energy.

Solution   As we have seen, in an isochoric process, $Q = \Delta U$. From Chapter 10 we also know that the heat added to a mass m of a substance is related to the temperature change ΔT and the specific heat c by

$$Q = mc\,\Delta T.$$

(We assume that no change in phase occurs in this process.) Upon equating these two expressions for Q, we obtain

$$\Delta U = mc\,\Delta T.$$

If the system were a gas, we would need to use the appropriate specific heat, that is, the specific heat for constant volume. (See Chapter 12.)

What we have shown in this special case is generally true: Temperature is associated with the internal energy of a system.

Example 11.2

Internal energy change in an isobaric process.

Gas confined by a piston (Fig. 11.2) in a heat engine expands against a constant pressure of 100 kPa (nearly one atmosphere). When 2×10^4 J of heat are absorbed by the system, the volume of the gas expands from 0.15 m³ to 0.25 m³. (a) What is the work done by the system during this process? (b) What is the change in internal energy of the system?

Solution   (a) The work done at constant pressure is

$$W = P\,\Delta V = P(V_{\text{final}} - V_{\text{initial}}),$$
$$W = 100 \times 10^3 \text{ N/m}^2 \, (0.25 \text{ m}^3 - 0.15 \text{ m}^3) = 1 \times 10^4 \text{J}.$$

(b) The change in internal energy is obtained from the first law of thermodynamics:

$$\Delta U = Q - W.$$

In this case Q is 2×10^4 J. The work W done by the gas was just found to be 1×10^4 J. Thus the change in internal energy is

$$\Delta U = 2 \times 10^4 \text{J} - 1 \times 10^4 \text{J} = 1 \times 10^4 \text{J}.$$

The first law of thermodynamics relates two measurable quantities that pertain to changes in a system: the heat added to or given out by the system and the work performed on or done by the system. We can measure the thermodynamic properties of systems in many ways: the length of a

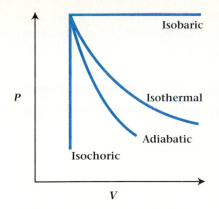

Figure 11.3 *PV* diagrams for isochoric, adiabatic, isothermal, and isobaric processes. The lines indicate the thermodynamic state of a system during the process, as measured by the two state variables pressure and volume.

mercury column in a thermometer; the volume, pressure, or temperature of a sample of gas; or the resistance of a resistor. All of these properties are called **state variables** or state coordinates. A state variable is a physical property that characterizes the state of a system independently of how that particular state is reached. For example, a cup of tea that is at room temperature has the same temperature whether it cools from boiling temperature or is heated from freezing temperature; temperature is a state variable. Under some conditions, two state coordinates completely specify the thermodynamic state of a system. Then we can represent any possible state by a point on a two-dimensional plot (Fig. 11.3). The amount of heat added to or released by a thermodynamic system is not a state variable, and although the work done by a system often does involve state variables, such as "*P ΔV* work" done by a gas, work itself is not a state variable. An interesting aspect of the first law is that it asserts that the internal energy of a system is a state variable; that is, the energy difference between the heat into a system and the work done by that system does not depend on the details of how that heat was added or how that work was done. In this case, the difference between two quantities that are not state variables is itself a state variable.

An important goal of thermodynamic studies is to consider systems in different thermodynamic states and follow the changes in the state variables as the systems undergo changes. However, thermodynamic state variables are meaningful only for systems in equilibrium. For example, when a large sample of gas undergoes a rapid expansion, the temperature and pressure may fluctuate from place to place within the gas. Thus, describing the temperature or pressure of the sample while it changes is not meaningful. To determine clearly defined values of state variables, we must consider reversible processes. A **reversible process** is one in which the system is very nearly in equilibrium all the time. A reversal of the controlling factors causes the system to exactly retrace its path in the opposite direction, back to its initial state. Alternatively, we can say that a reversal of the controlling factors causes a reversal of the energy transformation. There is no wasted energy.

Reversible processes are allowed by the first law of thermodynamics and we will use them as examples of how to think about thermodynamic processes. However, a reversible process cannot take place if friction or any other form of energy loss is present. The situation is somewhat like learning basic mechanics without including friction in everything from the start. We will see later that the second law of thermodynamics shows that completely reversible processes do not occur in nature. However, they are still useful models for analyzing changes in the state of a system.

Because many machines operate in approximately reversible cycles, it is important to examine the implications of the first law for a cyclic process in which a system begins with an internal energy U at an initial temperature T, exchanges heat and work with its surroundings, and returns to its initial state characterized by U and T. In any number of complete cycles, $\Delta U = 0$ so that $Q = W$. Thus the first law tells us that it is impossible for a machine (or a system) in any number of complete cycles to put out more energy in the form of work than it takes in as heat. A machine that could do this would be called a perpetual motion machine of the first kind. The first law is sometimes stated in terms of such a machine as, *A perpetual motion machine of the first kind is impossible*; or, to put it more succinctly, *You can't win.*

The Carnot Cycle and the Efficiency of Engines

11.2

Figure 11.4 shows an early steam engine built by the firm of Boulton and Watt. Steam came from the boiler (B) at the left and entered a condensing cylinder near the large upright cylinder (C). Here the steam was cooled and condensed to water, creating a partial vacuum in the large cylinder that caused a piston (P) to descend. This motion was transmitted to the vertical rods (R) on the right through the large reciprocating beam (RB) at the top. The upward motion of the right-hand rods pumped water from a mine. Rotary motion and the use of the pressure of expanding steam for the driving force did not come until later.

Naturally it was important to regulate an engine to develop its maximum power output and to determine what that output was. Watt accomplished this with a technique he developed and used privately for many years. In this method, Watt attached a device called an indicator to the engine. The mechanism included a card connected to a piston so that it could move back and forth as the piston moved within the cylinder (Fig. 11.5). The displacement of this main piston was proportional to the volume of the main cylinder. Another, smaller cylinder with a spring-loaded piston was connected to the main cylinder. The position of this small piston indicated the pressure. It also moved a lever with a pencil on its

Figure 11.4 An early steam engine built by the firm of Boulton and Watt. Its operation is described in the text.

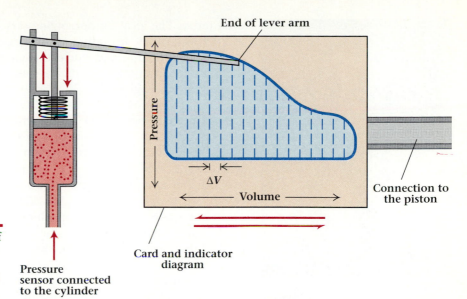

Figure 11.5 A schematic drawing of a steam engine indicator mechanism showing an indicator diagram. The area within the curve is proportional to the work done in each cycle.

far end. As the card and small cylinder moved, the pencil traced out a curve of pressure versus volume. The resulting diagram was called an indicator diagram. This term is still used for measurements made on internal combustion engines. The indicator diagram is a cyclic *PV* diagram, similar to the *PV* diagrams we looked at earlier for adiabatic, isothermal, or isochoric processes.

The area inside the curve of an indicator diagram is proportional to the work done in each cycle. To see this, suppose that we divide the area inside the indicator diagram into narrow vertical strips. The width of each strip is proportional to the change in volume ΔV, and the height of each strip indicates the pressure change *P*. The small area of each strip $P\,\Delta V$ represents an increment of work. The sum of all these incremental values of work, which is the total work per cycle, corresponds to the area inside the curve. Because the time for each cycle is essentially constant, the area inside an indicator diagram also measures the power of an engine.

Though steam engines were considerably improved by Watt and others, the basis for understanding the general principles of heat engines did not come until 1824, when the French engineer Sadi Carnot (1796–1832) published a treatise on this subject. In doing so, Carnot formulated the basic ideas of thermodynamics. He said that *all* movements were ultimately due to heat. It made no difference whether they occurred in natural phenomena, such as rain, storms, earthquakes, and volcanos, or in mechanical devices such as steam engines (Fig. 11.6). In view of modern knowledge, Carnot's vision of nature was slightly simplified, but his understanding of heat as the generator of motive power was essentially correct. His work was unappreciated except by his closest friends until, sixteen years after Carnot's death, Lord Kelvin pointed out its fundamental theoretical and practical importance.

Although we will discuss Carnot's ideas in terms of an ideal engine that cannot actually be built, the ideas have great practical importance

Figure 11.6 **The motions of torna-dos, volcanos, and jet engines are due to heat. Thus their behavior is governed by the laws of thermodynamics.**

even today. The ideal Carnot engine sets an upper limit on the efficiency of all real engines, including steam engines, Diesel and gasoline (Otto) engines, jet engines, and nuclear reactors. Furthermore, studies of the theoretical Carnot engine indicate some of the factors that affect the efficiency of real engines.

Carnot's genius saw beyond all of the complicated mechanisms of various steam engines to the fundamental process underlying the working of all engines—the transformation of one type of energy (heat) into another (mechanical work). He recognized that work could be done only when heat flowed from a higher temperature to a lower one. So Carnot proposed an *ideal* heat engine that operates cyclically and reversibly between two temperatures. This so-called Carnot engine is not 100% efficient, but is as efficient as any machine could be in transforming heat into work. Carnot analyzed the transformation of energy during one complete cycle of this engine's performance and determined the conditions for maximum efficiency. Only later was Watt's indicator diagram proposed as a basis for a mathematical discussion of Carnot's ideas.

In our example, the working substance of the engine is an ideal gas* confined within a cylinder by means of a frictionless piston (Fig. 11.7). We use an ideal gas for mathematical simplicity; however, the results would be the same for any working substance in the Carnot cycle.

Figure 11.7 shows diagrammatically the other components of a Carnot engine, in addition to the cylinder containing the working substance. A hot body of infinite thermal capacity, called a *heat reservoir*, supplies heat without lowering its own temperature. (This reservoir can be approximated by any source of heat, such as the sun, that is much larger than the needs of the engine.) An insulating platform, together with the sides of the cylinder and piston, acts as a perfect insulator against the flow of heat. The cold body, called a *heat sink*, is also of infinite thermal capacity so that it can absorb heat without raising its own temperature. (This sink can be approximated by any large body, such as the ocean, that can absorb much more heat than the engine can generate.) Finally, there is a second insulating platform. Operation of the Carnot engine consists of moving the cylinder in a prescribed manner from one of these platforms to the other and then repeating the cycle.

The **Carnot cycle** consists of four reversible processes, two isothermal and two adiabatic:

Step 1.   We start the cycle with the cylinder in contact with the heat reservoir, where the working substance (gas) takes in an amount of heat Q_H at a high temperature T_H. Because the system absorbs heat in a reversible process, its temperature is the same as the reservoir's; that is, this is an isothermal process. As the heat is absorbed, the gas expands. This expansion is represented in Fig. 11.8 by going from A to B along an isothermal curve. During this isothermal process, the system's internal energy does not change, so according to the first law the work done by the system is equal to the heat input.

---

*Chapter 12 describes an ideal gas. Use of an ideal gas allows us to calculate the exact shape of the curves for the cyclic process shown in Fig. 11.8.

Figure 11.7 A diagrammatic version of a Carnot engine using an ideal gas as the working fluid. Heat is taken in isothermally at A, followed by: adiabatic expansion at B, isothermal compression at C (where heat is expelled), and finally, adiabatic compression at D.

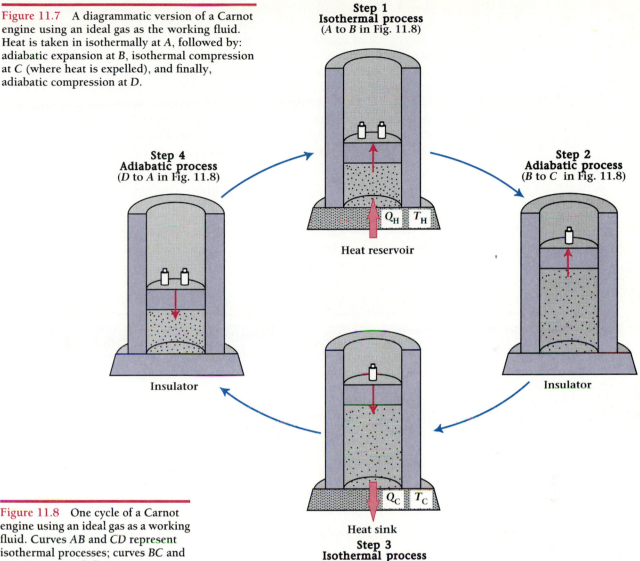

Step 1
Isothermal process
(A to B in Fig. 11.8)

Heat reservoir

Step 4
Adiabatic process
(D to A in Fig. 11.8)

Step 2
Adiabatic process
(B to C in Fig. 11.8)

Insulator

Insulator

Step 3
Isothermal process
(C to D in Fig. 11.8)

Heat sink

Figure 11.8 One cycle of a Carnot engine using an ideal gas as a working fluid. Curves AB and CD represent isothermal processes; curves BC and DA represent adiabatic processes. Compare with Figure 11.7.

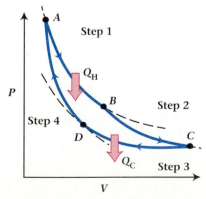

Step 2. The cylinder is then moved to the insulating body, where the heat input is zero. The load on the piston is reduced and the gas is allowed to expand, this time along an adiabatic curve (B to C). As the gas continues to do work by expanding, its internal energy must go down. This expansion is accompanied by a decrease in temperature along the curve BC until the cylinder reaches the temperature of the heat sink.

Step 3. Next the cylinder is moved to the heat sink. Here the gas undergoes an isothermal contraction (C to D) in which an amount of heat Q_C is expelled to the cold reservoir at temperature T_C. As in the previous isothermal process, the heat intake equals the work done. However, in this case, since heat is exhausted, the work is negative; that is, work is done on the system.

Step 4. In the final step of the Carnot cycle, the cylinder is moved back to the insulating body. The load on the piston is increased and the gas undergoes an adiabatic compression (*D* to *A*). Again the heat exchange is zero, and because the volume is decreasing (work being done on the system), the internal energy and the temperature increase. When the temperature of the gas again reaches that of the heat reservoir, the cylinder is transferred to the heat reservoir and the cycle starts again. In this way the working substance returns to the same internal energy that it had at the start of the complete cycle. Thus, by the first law of thermodynamics, the work done must equal the net heat flow into the cylinder:

$$W = Q_H - Q_C,$$

where Q_H and Q_C are taken to be positive quantities. The process is shown schematically in Fig. 11.9.

You should go over Figs. 11.7 and 11.8 again to be sure you understand how they correspond.

PHYSICS IN PRACTICE

Gasoline Engines

Ads for new cars often stress the increased efficiency of the new models compared with what you're driving now. In fact, the last few years have seen real improvements in engine efficiency. But how far can this continue? Let's find out by analyzing a simple model of a typical car engine.

Internal combustion engines form a special class of heat engines that generate the input heat by the combustion of fuel within the engine itself. Examples of internal combustion engines include gasoline engines, Diesel engines, and gas turbines. Here we consider the gasoline engine as a representative example of internal combustion engines.

The operating cycle of the gasoline engine used in most cars is a four-stroke cycle (Fig. B11.1). In the *intake stroke* a mixture of air and gasoline vapor is drawn through the intake valve into the cylinder by the downward motion of the piston. The valve closes and the fuel–air mixture is compressed. At the top of this *compression stroke* the gases are ignited by an elec-

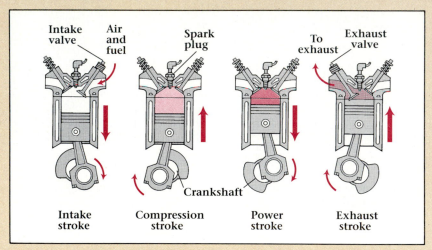

Figure B11.1 The operating cycle of a four-stroke gasoline engine. The piston goes up and down twice during each cycle.

tric spark from the spark plug, raising the temperature and pressure of the gases. The hot gases then expand against the piston in the *power stroke*, delivering energy to the crankshaft. The exhaust valve opens as the piston moves upward again, expelling the burned gases in the *exhaust stroke*. The exhaust valve closes, the intake valve opens, and the cycle is ready to repeat.

Analysis of an indicator diagram of a real gasoline engine is very difficult (Fig. B11.2a). For this reason, the gasoline engine is usually analyzed with a simplified model of the cycle called the Otto cycle, after its developer, Nicholas Otto (1832–1891). The Otto cycle begins at point *A* on the *PV* diagram of Fig. B11.2(b). The volume expands at constant pressure to point *B* as the pis-

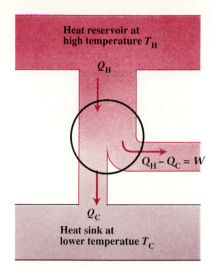

Figure 11.9   A schematic representation of a heat engine. The work done equals the net heat flow into the system, $Q_H - Q_C$.

ton moves down during the intake stroke. During the compression stroke the gases are compressed adiabatically to point C. Ignition of the gas by the spark causes an isochoric change to point D at higher temperature and pressure, which is followed by an adiabatic expansion to point E during the power stroke. Opening the exhaust valve causes the pressure to drop isochorically to point B, and this drop is followed by a decrease in volume at constant pressure as the piston moves through the exhaust stroke.

The work done by the gasoline engine is found from the area enclosed in the curve on the PV diagram, which for the idealized Otto cycle is the loop B-C-D-E-B. Comparison of the Otto cycle with a Carnot cycle operating between the same two temperatures shows that the efficiency of the Carnot cycle is more than that of the Otto cycle. The Otto cycle is, in turn, considerably more efficient than the actual gasoline engine cycle that it represents. Real gasoline engines achieve thermodynamic efficiencies of 20 to 25%, roughly half the value predicted from the simplified Otto model.

In recent years, manufacturers have been designing and building more efficient cars. Electronic sensors have been installed to monitor exhaust emissions, while computer control of air–fuel mixtures is now common. Other advances such as lean-burn engines, turbocharging, multiple valves, and cast-aluminum engine blocks have been used to make cars more fuel-efficient. Future developments will undoubtedly include even more computer control of the combustion process and electronically controlled transmissions to provide the optimum gearing between the engine and the wheels for every situation.

Figure B11.2   Indicator diagrams for (a) a real gasoline engine and (b) an idealized Otto cycle.

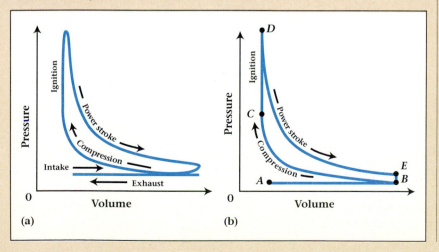

TABLE 11.1 Practical efficiencies of real engines	
Type of engine	Efficiency (%)
Gasoline engine	20–25
Diesel engine	26–38
Steam turbine	
Nuclear-powered	35
Coal-fired	40

We define the **thermal efficiency** of any system, such as a machine, to be the ratio of the work done to the heat input:

$$\text{thermal efficiency} = \frac{W}{Q_H} = \frac{Q_H - Q_C}{Q_H}. \tag{11.3}$$

In Chapter 12 we show that for an ideal gas the internal energy is proportional to the Kelvin temperature. From that fact and from a detailed examination of the Carnot cycle for an ideal gas, Kelvin showed that

$$\frac{Q_C}{Q_H} = \frac{T_C}{T_H},$$

where the temperatures are the absolute thermodynamic temperatures measured on the Kelvin scale. The thermal efficiency of an ideal engine is thus

$$\text{thermal efficiency} = 1 - \frac{T_C}{T_H} \quad \text{(ideal).} \tag{11.4}$$

It can be shown that *all* reversible engines operating in cycles between the same two heat reservoirs have the same efficiency regardless of the operating fluid. Moreover, no heat engine of any kind, operating in cycles between the same two reservoirs, can have an efficiency greater than that of a reversible Carnot engine. Thus, even if there were no losses due to friction and heat leakage, the absolute maximum efficiency of a heat engine is given by Eq. (11.4). The efficiency of any real engine is certain to be less than that of the ideal engine. Table 11.1 gives some examples of typical efficiencies.

Example 11.3

Maximum efficiency of a steam engine.

What is the maximum possible thermal efficiency of a steam engine that takes in steam at 100°C and exhausts it at 27°C?

Solution   We can calculate the efficiency using Eq. (11.4), after converting the temperatures to the Kelvin scale:

$$T_H = 100°C = 373 \text{ K,}$$

$$T_C = 27°C = 300 \text{ K,}$$

$$\text{efficiency} = 1 - T_C/T_H = 1 - 300 \text{ K}/373 \text{ K} = 0.20.$$

The theoretical efficiency of this engine is only 20%. Note that this is true regardless of the details of the engine's operation.

Example 11.4

Maximum efficiency of a power plant.

Calculate the maximum theoretical efficiency of a power plant that has a high-temperature reservoir at 500°C and a low-temperature exhaust (cold reservoir) at 50°C.

Solution   Again we must convert to absolute temperatures:

$$T_H = 500°C = 773 \text{ K},$$

$$T_C = 50°C = 323 \text{ K},$$

$$\text{efficiency} = 1 - T_C/T_H = 1 - 323 \text{ K}/773 \text{ K} = 0.58.$$

The maximum theoretical efficiency is 58%. The increase above that of Example 11.3 is due to the increase in T_H. Because real power plants do not reach ideal efficiency, a realistic value of the efficiency of a power plant is closer to 40%.

Heat Pumps and Refrigerators

11.3

By reversing the direction of the Carnot cycle, we can put work into the system and transfer heat from a low temperature to a higher one. A system operated in this manner is called a **refrigerator** (Fig. 11.10). In this case the ratio of the heat extracted from the cold reservoir to the work supplied is similar to an efficiency and is called the **coefficient of performance** (c.p.):

$$\text{c.p.(refrigerator)} = \frac{Q_C}{W} = \frac{Q_C}{Q_H - Q_C}.$$

With ideal gas as the working substance, this becomes

$$\text{c.p.(refrigerator)} = \frac{T_C}{T_H - T_C}. \tag{11.5}$$

Notice that the amount of work required to run a refrigerator increases with the temperature difference between T_H and T_C.

Figure 11.11 shows a diagram of the thermal part of a freon-cycle refrigerator, a type found in many homes. The typical refrigerator compresses a refrigerant gas (freon) to a pressure of several atmospheres. The resulting hot gas is forced through a heat exchanger (the condenser) external to or on the side walls of the refrigerator, where the gas is cooled to near room temperature and thereby condensed into a liquid. The cool liquid then flows at high pressure through a narrow tube to a much larger tube (the evaporator), which is a region of much lower pressure. In this region the liquid evaporates because of the reduced pressure, absorbing heat from the contents of the refrigerator. The resulting cold gas is then drawn through the low-pressure tube back to the compressor, where the cycle starts over.

Compare the diagram of Fig. 11.11 with the schematic of Fig. 11.10. The work input is provided by the compressor; the heat input is absorbed by the gas at the evaporator at low temperature; and the heat exhausted is given up at higher temperature through the condenser. Thus a refrig-

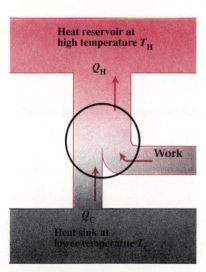

Figure 11.10   A schematic representation of a refrigerator. We have to put work into the system to transfer heat from lower temperature to higher temperature.

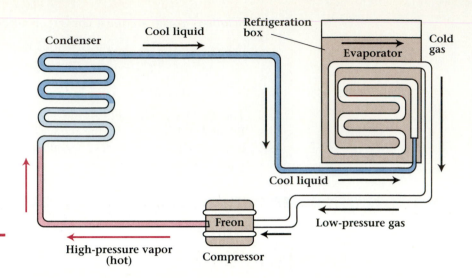

Figure 11.11   A diagram of a common freon-cycle refrigerator.

erator transfers heat energy from a cooler body to a warmer one at the expense of the work supplied.

An air conditioner is a refrigerator that is designed to take heat from within a house and exhaust it to the outdoors. When such a system is reversed so that it cools the outdoors and delivers heat to the inside of the house, it is called a **heat pump**. The coefficient of performance of a heat pump is defined to be the ratio of the heat delivered inside the house (the high temperature reservoir) to the work supplied:

$$\text{c.p.(heat pump)} = \frac{Q_H}{W} = \frac{Q_H}{Q_H - Q_C}.$$

In terms of temperatures, the coefficient of performance becomes

$$\text{c.p.(heat pump)} = \frac{T_H}{T_H - T_C}. \tag{11.6}$$

Example 11.5

Temperature inside a refrigerator.

A household refrigerator has a coefficient of performance of 6.0. If the room temperature outside the refrigerator is 30°C, what is the lowest temperature that can be obtained inside the refrigerator?

Solution   The coefficient of performance, c.p., is

$$\text{c.p.(refrigerator)} = \frac{T_C}{T_H - T_C}.$$

We can rearrange this equation to give

$$T_C = \frac{(\text{c.p.})T_H}{1 + (\text{c.p.})}.$$

Then we obtain T_C by inserting the numerical values for the coefficient of performance and the temperature T_H. Remember that T_H must be measured on the Kelvin scale:

$$T_H = 30°C = (30 + 273) \text{ K} = 303 \text{ K},$$

$$T_C = \frac{6.0 \times 303 \text{ K}}{1 + 6.0} = 260 \text{ K}.$$

Upon converting T_C to the Celsius temperature, we see that the coldest temperature attainable inside the refrigerator is

$$T_C = -13°C.$$

The Second Law of Thermodynamics

11.4

It might seem to you, as it did to those who followed immediately after Carnot, that there was either some outright contradiction or at least a lack of clarity in his ideas about heat. On the one hand we have the concept of the mechanical equivalent of heat, which Joule's experiments had confirmed. On the other hand we have Carnot's result that even the most efficient heat engine conceivable could not convert all of its heat input into mechanical output. Didn't this result contradict the conservation of energy?

The key to resolving this problem lies in seeing that Carnot's results refer to the amount of work available for use as output, and not to whether the total energy is conserved. The situation is somewhat analogous to the concept of gravitational potential. In that case, work can be done only when a body goes from one height to a lower one. The greater the difference in heights, the greater the amount of work that can be done. If the body cannot fall to a lower height, then no work can be done, no matter how much potential energy the body originally had. Similarly, for a heat engine, no work can be done unless heat can be taken in at one temperature and exhausted at a lower temperature.

Out of this seeming contradiction came the formulation of the **second law of thermodynamics**. It was first expressed mathematically by the German theoretical physicist Rudolf Clausius (1822–1888) and shortly afterwards by Lord Kelvin. Although these two formulations appear to be different, both were an outgrowth of Carnot's ideas. The two statements can be shown to be equivalent.

1. **Clausius statement of the second law:** Heat cannot, by itself, pass from a colder to a warmer body.
2. **Kelvin-Planck* statement of the second law:** It is impossible for any system to undergo a cyclic process whose *sole* result is the absorption of heat from a single reservoir at a single temperature and the performance of an equivalent amount of work.

---

*Planck's thermodynamic studies led to ideas that revolutionized our notions of thermal radiation. (See Chapter 26.)

Let us look into some of the implications of these statements. Remember that they express in general ways the results of experimenting with and observing the behavior of heat. The Clausius statement of the second law of thermodynamics is consistent with our experience. If an ice cube and a cup of hot chocolate are placed in contact, heat will flow from the hot chocolate to the ice cube until they come to the same temperature. The principle of conservation of energy—the first law of thermodynamics—does not tell us anything about how this process proceeds. It would not be a violation of the first law of thermodynamics if heat were to flow from the ice cube to the hot chocolate.

The second law of thermodynamics is different from the laws of mechanics. It does not describe the interactions between individual particles, but instead describes the overall behavior of collections of many particles. The second law of thermodynamics says something about the sequence, or order, in which events naturally take place. In mechanics, individual events are always reversible; we say that they are symmetric in time. If we make a movie of the collision between two air-track gliders and then look at the movie, the collision that we see satisfies all the laws of mechanics, regardless of whether we show the movie forward or backward. The collision has time-reversal symmetry. However, if we make a movie of an egg frying and then show the movie backward, the result violates all our previous experience. It is in this sense that the second law tells us which way is forward in time and which way is backward.

Entropy and the Second Law

11.5

We can gain additional insight into the meaning of the second law of thermodynamics by considering it from a standpoint first introduced by Clausius in 1850. He introduced a new thermodynamic state variable called *entropy,* which has two Greek roots and means much the same as "turning into." **Entropy** is a measure of how much energy or heat is unavailable for conversion into work.

When a system at Kelvin temperature T undergoes a *reversible* process by absorbing an amount of heat Q, its increase in entropy ΔS is

$$\Delta S \equiv \frac{Q}{T}.$$

(11.7)

Notice that we are defining entropy for a reversible process, which does not occur in nature. However, for our purposes, we can use this definition for processes that are approximately reversible.

Before we examine the meaning of this state variable further, let's see an example of calculating changes in entropy. We will find that, just as in problems involving potential energy or heat, the changes are the significant quantity. In this example the temperatures are not constant. Ideally, we should use the techniques of calculus to add up the increments of entropy change over many small intervals of almost constant temperature. For this example and for the problems at the end of the chapter, the difference between the exact answer from calculus and the approximate answer obtained by using the average temperature in Eq. (11.7) is less than 1% in all cases. However, you should keep in mind that this method is only approximate and is not always valid. Problem 11.54 is concerned with the difference between the exact and the approximate solutions.

Example 11.6

Entropy change in a mixture of hot and cold water.

A student mixes 100 g of water at 60°C (sample 1) with 200 g of water at 40°C (sample 2). Determine the change in entropy of the system.

Solution   Using the methods of the last chapter, we first determine the final temperature and the heat gained or lost by each sample of water. We use the equation

$$\text{heat lost} + \text{heat gained} = 0,$$

where heat gained or lost equals $mc\,\Delta T$. The heat lost for sample 1 is $100\text{ g} \times 1\text{ cal/g·°C} \times (T - 60°\text{C})$. The heat gained by sample 2 is $200\text{ g} \times 1\text{ cal/g·°C} \times (T - 40°\text{C})$. When these are added and set equal to zero, we obtain a final temperature of 46.7°C. The heat gained or lost is found to be

$$Q_1 = m_1 c\,\Delta T = -1.33\text{ kcal}$$

and

$$Q_2 = m_2 c\,\Delta T = +1.33\text{ kcal}.$$

As expected, the heat lost by one sample is equal in magnitude to the heat gained by the other.

The average temperature of the first sample is 53.3°C, or 326 K. The average temperature of the second sample is 43.3°C, or 316 K. The change in entropy of the initially warm water (sample 1) is then approximately

$$\Delta S_1 = \frac{Q_1}{T_1} = \frac{-1.33\text{ kcal}}{326\text{ K}} = -4.08 \times 10^{-3}\text{ kcal/K}.$$

The change in the entropy of the initially cool water (sample 2) is

$$\Delta S_2 = \frac{Q_2}{T_2} = \frac{+1.33\text{ kcal}}{316\text{ K}} = +4.21 \times 10^{-3}\text{ kcal/K}.$$

The change in the entropy of the system is the sum of the changes in entropy of the component parts,

$$\Delta S = \Delta S_1 + \Delta S_2 = 0.13 \times 10^{-3}\text{ kcal/K}.$$

Notice that the change in the entropy of the system is positive. Also notice that no matter what initial temperatures we start with and no matter what the mass of each sample we choose, the change *still* will be positive. You can see this from the fact that the heat energy gained always equals the heat energy lost and that the term for the initially warmer sample has both a minus sign and the larger denominator.

Example 11.7

Does the entropy of the universe change as a result of the operation of the power plant in Example 11.4?

Solution   If we consider the power plant to be a reversible Carnot engine, and if we assume that the surroundings interact reversibly, the entropy of the surroundings (the universe) is unchanged. This is so because the entropy lost by the high-temperature reservoir is exactly matched by the entropy gained by the low-temperature reservoir. The amount of entropy lost by the high-temperature reservoir when it gives up an amount of heat Q_H is

$$\Delta S_1 = \frac{Q_H}{T_H}.$$

The entropy gained by the low-temperature reservoir when it absorbs an amount of heat Q_C is

$$\Delta S_2 = \frac{Q_C}{T_C}.$$

But, as we have seen, for a Carnot engine,

$$\frac{Q_H}{T_H} = \frac{Q_C}{T_C}.$$

Consequently, the net change in entropy of the surroundings is zero.

In a real power plant, however, the efficiency is always less than that of an ideal Carnot engine. As a result, the heat Q_C delivered to the low-temperature reservoir is greater than the amount that would be delivered by a Carnot engine. This means that the increase in entropy of the low-temperature reservoir is greater than the decrease in entropy of the high-temperature reservoir, so that there is a net increase in the entropy of the universe.

Examples 11.6 and 11.7 are special cases of a much broader principle, discovered by Clausius: *In any process the entropy of the universe increases or remains constant.* (This is another way of expressing the second law.) Entropy remains constant only in the case of reversible processes, which do not occur naturally. So the entropy principle predicts that the entropy of the natural universe always increases. It is extremely important to remember that this does *not* mean that the entropy of a local segment cannot decrease—the entropy did decrease for the warm water in Example 11.6— but the total entropy of a system and its surroundings always increases. All observations and calculations indicate that if entropy decreases in one place, it simultaneously increases by an equal or larger amount somewhere else. Thus entropy is quite different from such concepts as energy, momentum, and angular momentum, which we have previously encountered, because entropy is not conserved. In fact, the opposite is true. Entropy can be created and in natural processes entropy always increases if all systems taking part in the process are considered. We can examine the meaning of the entropy principle from two equivalent viewpoints, which we briefly outline in the remainder of this section.

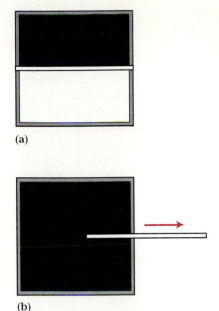

(a)

(b)

Figure 11.12   (a) A box in which all the gas molecules are confined to one side. (b) The same box after removing the dividing partition. The gas is now less ordered, and has increased entropy.

If we have two bodies at different temperatures—say, a hot stove and a block of ice—we can connect a heat engine between them and extract useful work. If, instead, we place the two bodies in direct thermal contact, they will come to thermal equilibrium. In agreement with the first law of thermodynamics, the total energy content of the stove and the ice is the same before contact as that of the stove and water (melted ice) after they have been placed in contact. However, once they reach equilibrium, we cannot separate them again and expect to extract work from them with a heat engine. Something has changed, even though the total energy has not. What has changed is the availability of the energy to do work. An increase in entropy means a decrease in the energy available to do work, not a decrease in the total energy.

Another viewpoint connects entropy with probability and statistics. This insight is due to Ludwig Boltzmann (1844–1906), who showed that an increase in the entropy of a system or substance corresponds to an increased degree of disorder in the atoms or molecules composing the substance. The most probable—that is, the most statistically favored— arrangement of molecules is the one with the most molecular disorder. For example, suppose you have a box with a partition dividing it in two, with a gas on one side of the partition and the other side evacuated (Fig. 11.12). With all of the molecules of the gas in one side you have a highly ordered situation. If the partition is removed, however, the gas molecules will soon distribute themselves throughout the box and be moving in random directions—a less ordered situation. We can calculate that this change of order corresponds to an increase in the entropy of the gas. The probability that the molecules will all return to their original corner position at the same time is vanishingly small. J. W. Gibbs (1839–1903), the first great theoretical physicist in the United States, once called entropy a measure of "mixed-upness."

The statistical view is represented by an example that originated with Sir Arthur S. Eddington (1882–1944). A new deck of 52 cards comes in a preestablished order. As you shuffle the deck and shuffle it again—an action corresponding to the occurrence of thermodynamic processes—the order of the cards becomes randomized. No matter how many times you shuffle the deck, you do not expect it to return to its original order. Though this event is certainly a possibility, its *probability* is so low that it is not worth considering. Similarly, systems undergoing some physical process proceed from order to disorder because disorder is so much more probable. The first law does not say that systems will not, of their own accord, become ordered again; but the second law says that the probability of their doing so is, in practical terms, zero. The reason is that any physical system is made up of so many molecules that the probability of its going back to its ordered state is infinitely smaller than the probability that a deck of 52 cards will be reordered upon repeated shufflings.

An advanced area of physics called *statistical mechanics* relates entropy to collections of large numbers of particles. Though we will not go into statistical mechanics in this text, it does provide a mathematical basis for calculating entropy as a measure of disorder, as well as for calculating other thermodynamic state variables and properties. The kinetic theory model described in Chapter 12 is a simple form of statistical mechanics.

The concept of increasing disorder applies to the entire universe as well as to systems here on earth. Current understanding of the early history of the universe is that it began as a highly compressed, hot "fireball," which has been expanding for something on the order of 15 billion years. This expansion corresponds to going from a more ordered to a disordered state. During the expansion, the temperature of the universe decreases and its entropy increases. (Most of the universe, except for the stars and their associated planets, is quite cold, the average temperature being about 3 K.)

Just as in the case of the two liquids mixed in Example 11.6, the entropy and disorder of the universe increase as hot bodies cool and cold bodies warm. If the entropy principle is true throughout the universe, we can envision some time in the far distant future when everything in the universe will have reached a uniform temperature. No heat could flow, no work could be done, and no change in energy or motion could take place. Neither engines nor plants nor animals would be able to extract energy. This possible occurrence is often called the *heat death of the universe*.

The heat death of the universe may happen in billions of years or it may not happen at all. If, at some point, the universe begins to contract again into a fireball, then heat death will not occur. At the present time, it is not known which of the two cases, contraction or endless expansion, will occur.

Clausius summarized the laws of thermodynamics by saying that (1) the energy content of the universe is constant and (2) the entropy of the universe always increases. However, because of the complexity of the problem, we cannot determine in advance at what point the entropy will have reached its maximum value and the available energy will have decreased to zero.

The first law expressed the idea of conservation of energy. We stated it succinctly as *You can't win*. The second law expresses the idea that all available energy cannot be converted into useful work. So not only can you not win, but *You can't break even*.

Energy and Thermal Pollution

*11.6

We obtain most of the necessities and comforts of life from of the generation of power, in one way or another. The word *generation* should perhaps be replaced by the word *conversion*. The burning of fuel, as in a car engine, converts chemical potential energy in the fuel into heat energy; the use of water power, as in a hydroelectric power station, converts gravitational potential energy of the water into electrical energy; and so on. The conversion and utilization of energy from falling water, burning fossil fuels, nuclear reactors, solar converters, or any other source cannot take place without the cost of some heat being discharged somewhere.

The conversion involved in producing and using energy is always accomplished by machines or organisms whose efficiency is less than that of a Carnot engine. Thus an inevitable consequence of energy conversion is the discharge of heat. We have no option in this respect. The unwanted release of thermal energy into the environment is known as *thermal pollution*. It can be controlled and spread over a large area to minimize local changes in temperature, but it cannot be eliminated.

During the early 1980s several independent estimates of the rate of energy generated by humans on the earth gave an estimate of about 5×10^{12} W. It is possible that through conservation and other measures, this figure need not increase as rapidly as once thought, but it is unlikely to decrease in the foreseeable future. Overall heat production from human devices is less than 3×10^{-4} of the rate of energy absorption from the sun, and therefore would not have an appreciable influence if it were distributed uniformly over the earth. But, of course, it is not. In heavily populated regions, the heat production rate may be several percent of the rate of energy absorption from the sun.

Approximately 75% of the electrical power in the United States is generated in fossil fuel steam plants and most of the resulting waste heat is deposited in streams and rivers. This concentrated dumping of energy can raise the temperature of the water by several degrees. Such a small change is insignificant to humans and most warm-blooded animals, but can drastically influence which type of fish will predominate in rivers and thereby modify the entire local food chain. Other effects can be even more unpleasant, since higher temperatures often provide a more hospitable environment for pathogenic organisms.

There are several approaches to solving, or at least lessening, the problem of thermal pollution. One approach in which active research is going on is the development of alternative energy sources. The techniques being studied include the harnessing of wind power, tidal power, solar power, ocean temperature differences, and a host of other ingenious ideas (Fig. 11.13). A second approach to reducing thermal pollution involves

Figure 11.13 Alternative energy sources. (a) The *Sunraycer*, a solar-powered car. (b) A solar collector for generating power in France. (c) Some of the nearly 16,000 wind turbines operating in California. (d) Hoover Dam, one of the world's largest generators of hydroelectric power.

(a)

(b)

(c)

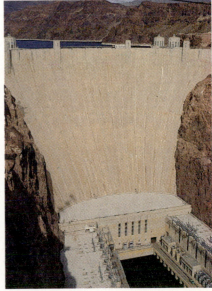

(d)

Figure 11.14 **Cooling tower at the Paradise power plant.**

measures for saving energy, from better building insulation to smaller, more efficient cars. However, the fact remains that a large amount of energy will continue to be discharged into the environment.

Another possible remedy is to extract further energy from heat exhausts. For example, exhaust heat can be used to warm buildings, resulting in lower direct-exhaust temperature and less additional fuel needed to heat the buildings. Another solution is to distribute heat exhaust more widely by depositing as much as possible into the atmosphere rather than into bodies of water. Figure 11.14 is a photograph of a cooling tower that is designed for this purpose.

There is no single solution to the problems of thermal pollution, but perhaps the combination of alternative sources, conservation, and wide distribution of waste heat can make the problem manageable.

SUMMARY

Useful Concepts

- Thermodynamics is the branch of physics concerned with the relationships between heat and work. These relationships govern the operation of all engines and devices that convert energy from one form to another, including processes that take place in living organisms.
- The zeroth law of thermodynamics states that when two systems are each in thermal equilibrium with a third system, they are in thermal equilibrium with each other.
- The first law of thermodynamics is a statement of conservation of energy,

$$\Delta U = Q - W,$$

where ΔU is the change in a system's internal energy, Q is the heat added to the system, and W is the work done by the system.

- The thermal efficiency of a Carnot engine is

$$\text{thermal efficiency} = 1 - \frac{T_C}{T_H}.$$

- The coefficient of performance of a refrigerator is

$$\text{c.p.(refrigerator)} = \frac{T_C}{T_H - T_C}.$$

- The coefficient of performance of a heat pump is

$$\text{c.p.(heat pump)} = \frac{T_H}{T_H - T_C}.$$

- The second law of thermodynamics can be stated in several ways. Two of them are:

    Clausius statement of the second law: Heat cannot, by itself, pass from a colder to a warmer body.

    Kelvin-Planck statement of the second law: It is impossible for any system to undergo a cyclic process whose *sole* result is the absorption of heat from a single reservoir at a single temperature and the performance of an equivalent amount of work.

- When a system at temperature T absorbs an amount of heat Q in a reversible process, the change in entropy is

$$\Delta S \equiv \frac{Q}{T}.$$

- Entropy is a measure of the disorder of a system. In any process the entropy of the universe increases or remains constant.
- Thermal pollution is the unwanted release of thermal energy into the environment.

Important Terms

You should be able to write the definition or meaning of each of the following terms:

- thermodynamics
- zeroth law of thermodynamics
- thermodynamic system
- internal energy
- first law of thermodynamics
- adiabatic process
- isothermal process
- isochoric process
- isobaric process
- state variable
- reversible process
- Carnot cycle
- thermal efficiency
- refrigerator
- coefficient of performance
- heat pump
- second law of thermodynamics
- entropy

QUESTIONS

11.1 Apple 1 is in thermal equilibrium with apple 2, apple 2 is in thermal equilibrium with apple 3, and so on to apple 6. Are apples 1 and 6 in thermal equilibrium? Can you prove your answer using the zeroth law of thermodynamics, or are you simply using your physical intuition (which is probably correct here)?

11.2 Assume that in Joule's paddle-wheel experiment (Chapter 10), the thermodynamic system is the water and it is well insulated from its surroundings. Has heat been added? Has work been done? Has the internal energy changed?

11.3 As you bend a wire back and forth, it becomes warm, then hot, and finally breaks. Discuss what is happening from the standpoint of the first law of thermodynamics.

11.4 According to the first law of thermodynamics, would you expect the temperature at which a substance melts or freezes to change if the substance is subject to a constant pressure?

11.5 A weight is placed on top of a vertically held brass rod that has its lower end on a fixed support. Discuss what happens from an energy standpoint as the rod is heated.

11.6 If we place a hot brick in thermal contact with a cold brick, does the first law of thermodynamics allow the hot brick to become hotter and the cold brick to become colder? What is allowed by the second law?

11.7 A circular rod fits into a circular hole in a metal block. The interface is oiled to make it easier to turn the rod. When the rod is turned rapidly for a long time, the block and rod become warm. Extremely precise measurements show that the mechanical energy delivered in turning the rod is more than the heat energy developed. Discuss this result in terms of the first law of thermodynamics. (*Hint:* Can any changes occur in the molecular structure of oil?)

11.8 Can you cool a kitchen by leaving the refrigerator door open? What will happen?

11.9 A heat engine can extract energy from any system maintaining a temperature difference between two bodies. List as many cases as you can where this is actually done, starting with the steam engine. Give some cases in nature where a temperature difference is maintained but in which we do not, for economic or engineering reasons, extract energy.

11.10 Explain how a heat pump can add more energy to a house than is used by the electric motor driving the pump.

11.11 Discuss and give specific examples of the effects that prevent real engines from reaching their Carnot efficiency.

11.12 Assume that the exhaust from a conventional fossil fuel power plant is at a temperature of 50°C. (a) Plot the maximum theoretical efficiency as a function of the high, or input, temperature. (b) Does the shape of the curve suggest any practical considerations about the usefulness of increasing the input temperature?

11.13 A sealed container is divided into two equal parts by a removable partition. One side is filled with pure oxygen and the other side is filled with air at the same pressure. (a) If the partition is removed, does the entropy of the system increase, remain constant, or decrease? (b) What would the answer be if both sides were originally filled with oxygen?

11.14 Discuss the statement that the eventual destiny of the physical universe is stagnation.

11.15 The earth is in approximate thermal equilibrium with its environment, absorbing radiant energy from the sun and reradiating energy at a lower temperature to its environment at the same rate. Recalling that processes on the earth increase its entropy, make an educated guess about the overall entropy change of the earth.

PROBLEMS

Hints for Solving Problems
In using the first law of thermodynamics, you need to pay careful attention to the signs: Work done on a system is negative, work done by a system is positive, whereas heat into a system is positive, heat out of a system is negative.

In formulas for thermal efficiency and coefficient of performance, the temperatures are always in kelvins. Always be clear about what is the system you are studying and what are its surroundings.

Section 11.1   The First Law of Thermodynamics

11.1   An insulated system takes in 5.00 kcal of heat and has 2000 J of work done on it. What is the change in internal energy of the system?

11.2   One thousand joules of mechanical work is done by an insulated system to which 5.00 kcal of heat is added. What is the change in the internal energy of the system?

11.3   (a) What is the change in internal energy of 100 g of ice as it melts to water at 0°C if we assume the density of water and ice to be the same? (b) Would the answer be larger or smaller if you took into account the fact that ice is less dense than water at 0°C? Explain your answer.

11.4   In a laboratory experiment, a heat engine takes in 50 cal of heat and does 50 J of work in a reversible process. What is the change in internal energy of the engine during this process?

11.5   An inventor tells you he has designed a new type of electrical generator. He says that a working model produces 1000 W while using 75 Btu of fuel per minute. He mentions that an interesting side effect is that the apparatus does not heat up its environment. Is it possible for such a device to exist?

11.6   Show that the work done during an isobaric process is $W = P(V_2 - V_1)$.

11.7   A 2.0-g piece of copper undergoes isochoric heating, which causes a temperature increase of 10°C. By how many joules does the internal energy change? (See Table 10.3.)

11.8   Gas confined by the piston in a particular heat engine expands against a constant pressure of 3.0×10^5 N/m². When 10.0 kcal of heat is added to the system, the volume expands from 0.10 m³ to 0.20 m³. What is the change in internal energy of the system?

11.9   (a) Consider a system that goes from a to b via the path acb on the PV diagram of Fig. 11.15. If the heat absorbed is 400 J and the system does 150 J of work, what is the change in internal energy of the system? (b) If the system does 50 J of work along path adb, how much heat energy must the system absorb?

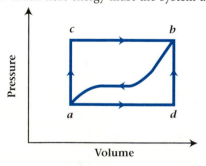

Figure 11.15   Problems 11.9 and 11.10.

11.10   When a system is taken from state a to b along the path acb in Fig. 11.15, 75 J of heat flows into the system while 25 J of work is done by the system. (a) How much heat flows into the system when it goes from a to b along the path adb if the work done is 10 J? (b) When the system returns to a along the curved path, the system performs 15 J of work. Does the system absorb or give out heat during that process? (c) How much heat is exchanged in part (b)?

11.11   A container is filled with 0.010 kg of carbon dioxide and sealed at atmospheric pressure and 20°C. It is then placed in a bath of ice water and allowed to come to thermal equilibrium. The container is then immersed in a water bath at 60°C. What is the difference in internal energy of the gas between the initial state, when the container was first filled, and the final state? The specific heat of carbon dioxide under these conditions is 638 J·kg^{-1}·K^{-1}.

11.12   The working fluid in a heat engine is initially at pressure P_0 and volume V_0. It is operated in a cycle shown in Fig. 11.16. What is the net work done by the engine in each cycle?

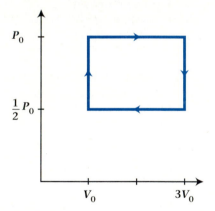

Figure 11.16   Problem 11.12.

11.13   A heat engine undergoes a process in which its internal energy decreases by 400 J while it is doing 250 J of work. How much heat is taken in (or given out) by the engine during this process?

Section 11.2   The Carnot Cycle and the Efficiency of Engines

11.14   A hypothetical indicator diagram is traced out in which the pressure rises linearly with increasing volume to some value P_1 at volume V_1, then falls to zero immediately and remains there while the volume decreases to some volume V_0, at which point the cycle starts again. What is the work per cycle represented by this indicator diagram?

11.15   From the standpoint of efficiency, show whether it is better to increase the highest temperature of a Carnot engine by ΔT or to decrease the lowest temperature by the same amount.

11.16 A Carnot heat engine takes in 100 J at a temperature of 100°C. At the end of the cycle, 50 J are exhausted. What is the temperature of the exhaust?

11.17 At the end of a cycle, 17 J are exhausted from a Carnot engine operating between 375°C and 150°C. (a) How many joules were taken in during each cycle? (b) How much work was done?

11.18 What is the temperature at which 45 J per cycle are added to a Carnot engine if 38 J are exhausted at a temperature of 100°C?

11.19 How much work is done per cycle by a Carnot engine operating between +10°C and −20°C if 12.0 J of heat are taken in at the higher temperature during each cycle?

11.20 An engine is measured to have an actual efficiency of 60%, which is 15% less than the maximum theoretical efficiency. If the temperature difference between the input and the output is 300°C, what is the exhaust temperature?

11.21 What is the maximum efficiency of an engine for which the ratio of the energy exhausted to the work done per cycle is 0.25?

11.22 A Carnot engine operating at 30% efficiency exhausts heat into a heat sink at 283 K. By how much would the temperature of the heat reservoir need to be raised to increase the efficiency of the engine to 45%?

11.23 What is the maximum efficiency of a heat engine that takes in heat at a temperature of 173°C and has an exhaust temperature of 75°C?

11.24 In the summer the temperature of water at the lower depths of Lake Geneva is 14°C. The surface temperature of the lake is about 20°C. What would the maximum efficiency be for any power-generating device that took advantage of this temperature difference?

11.25 (a) What is the efficiency of an ideal engine operating between 275°C and 0°C? (b) What is the efficiency if the engine operates over the same temperature difference, but with the input and output temperature both raised by 100°C?

11.26 An inventor claims to have devised a steam turbine that is 75% efficient. The steam enters the turbine at a temperature of 300°C and exhausts at the temperature of the surrounding atmosphere. Do you believe his claim?

11.27 An engine does 2000 J of work and exhausts 2.25 kcal each cycle while operating between 700°C and 400°C. What is its actual efficiency? How does it compare with the maximum theoretical efficiency?

11.28 A nuclear power plant in South Carolina produces electrical energy at the rate of 900 MW. The plant has a thermal efficiency of 30%. At what rate is energy released by the nuclear reactor? What is the rate of release of heat to the surrounding environment?

11.29 A steam-electric power plant delivers 900 MW of electric power. The surplus heat is exhausted into a river with a flow of 5.0×10^4 kg/s, causing a change in temperature of 12°C. (a) What is the efficiency of the power plant? (b) What is the rate of the thermal source?

Section 11.3   Heat Pumps and Refrigerators

11.30 What is the coefficient of performance of a refrigerator that maintains an inside temperature of 0°C in a room where the temperature is 25°C?

11.31 Show that the difference between the theoretical coefficient of performance for a heat pump and the same device running as a refrigerator is unity.

11.32 What maximum temperature can be maintained inside a house with a heat pump whose coefficient of performance is 8 if the outside temperature is 0°C?

11.33 What is the minimum theoretical temperature that can be maintained in a refrigerator with a coefficient of performance of 5 when operated in a room where the temperature is 16°C?

11.34 Show that the efficiency e of a Carnot engine and the coefficient of performance c.p.refrig of a Carnot refrigerator operating between the same temperatures are related by

$$e = 1/(1 + \text{c.p.refrig}).$$

11.35 (a) Calculate the coefficient of performance of a heat pump used to maintain an inside temperature of 21°C (70°F) on a day when the outside temperature is 7°C (45°F). (b) What is the coefficient of performance when the outside temperature drops to −12°C (10°F)?

11.36 (a) A Carnot engine operates between heat reservoirs at 420 K and 300 K. If it absorbs 500 J of heat from the hot reservoir in each cycle, how much work is delivered by the engine in each cycle? (b) If the engine is operated in reverse as a refrigerator acting between the two reservoirs, how much work input is required per cycle to remove 500 J of heat from the low-temperature reservoir?

Section 11.5   Entropy and the Second Law

11.37 What is the change in entropy when 100 g of ice melts to water at 0°C?

11.38 Determine the change in entropy when 50 g of cream at 14°C are put into 350 g of coffee at 80°C. Assume that the specific heat of both cream and coffee is essentially the same as that of water.

11.39 What is the change in entropy of the water when 100 g of water are converted to steam at 100°C?

11.40 What is the change in entropy when enough heat is supplied so that 80 g of ice initially at 0°C melts, warms up to 100°C, and changes to steam at 100°C?

11.41 What is the change in entropy when 50 g of water at 10.0°C are mixed with 50 g of water at 12.0°C?

11.42 An exhaust tube from a steam generator is submerged in a large container of cool water. When 10.0 g of steam are condensed into water at 100°C by bubbling it through the cold water, what is the change of entropy of the steam?

11.43 What is the change in entropy per second of the river water for the power plant in Problem 11.29?

11.44 A 1.0-kg lead block slides slowly down an inclined plane and comes to rest at the bottom. If the initial height of the block was 0.45 m above its final resting place and if the temperature of the block, the table, and the surrounding air remains at 27°C, by how much does the entropy of the universe change as a result of this process?

11.45 A student takes a 2.5-kg block of ice at 0°C, places it on a large rock outcropping, and watches the ice melt. (a) What is the entropy change of the ice (water)? (b) If the source of heat (the rock) is very massive and remains at a constant 21°C, what is the entropy change of the rock? (c) What is the total entropy change?

Additional Problems

11.46 When you drink ice water, it is warmed by your body to body temperature (37°C). Energy is required to warm the water; therefore you could conceivably lose weight by consuming lots of ice water to help you burn off your food energy. (a) How much ice water at 0°C would it take to offset eating two jelly donuts? (See Problem 10.20.) (b) Estimate the change in entropy and state any assumptions made.

11.47 A fluid is heated so that its original volume V_0 is doubled. The pressure is observed to vary linearly with the volume according to $P = P_0(1 + bV)$, where b and P_0 are constants. Find an expression for the work done by the expanding fluid.

11.48 Suppose that water is put into the refrigerator of Problem 11.30 at room temperature of 25°C. What is the cost of making 100 lb (45 kg) of ice if electricity costs $0.08 per kilowatt hour?

11.49 A Carnot engine has an exhaust temperature of 120°C and exhausts 80 J per cycle. If the input temperature is 325°C, how much work does the engine do per cycle?

11.50 An ideal engine always takes in heat at the same input temperature T_H. When the exhaust temperature is 120°C, the efficiency is 30%. (a) What is the input temperature? (b) If the exhaust temperature is changed so that the efficiency rises to 37%, what is the new exhaust temperature?

11.51 In order to increase the output of some heat engines a two-step process is used. The input and output temperatures of the first engine are T_1 and T_2. The exhaust of the first engine provides the input to the second engine at temperature T_2 and with a final exhaust temperature of T_3. Show that the overall efficiency is the same as that of a single engine operating between temperatures T_1 and T_3.

11.52 The ignition temperature of a typical automobile engine is 17,000°C and the spent gas is exhausted from the engine at 120°C. (a) What is the maximum possible efficiency of such an engine? (b) The energy per unit volume released by combustion of gasoline is about 0.35×10^8 J/L. If the power required to overcome wind friction is 22 kW when the car is driven at 100 km/h and if the fuel consumption is 8.5 km/L, what is the actual efficiency when driving at 100 km/h? How does your answer compare with the maximum theoretical value?

11.53 Heat is added uniformly to a 1.0-kg sample of water at a constant rate of 35 W. The water is initially at a temperature of 10°C. (a) What is the change in entropy of the water over the temperature interval between 10°C and 11°C? (b) What is the change in entropy over the temperature interval from 98°C to 99°C?

11.54 (a) Using the technique of adding up the area under a curve, find the total change in entropy of a 1.0-kg sample of water as it is heated from 20°C to 40°C. You may want to divide the temperature range into 2° intervals. (b) Compute the entropy change as the total heat added, divided by the average temperature. (c) How do these two results compare?

11.55 Consider a large steam-electric plant that produces 1000 MW of electric power. (a) If the plant is 33.3% efficient, what must be the energy rate of the thermal source used? (b) What is the rate at which thermal energy is exhausted to the environment? (c) If this energy were dumped into a river whose flow was 6×10^4 kg/s, what would be the average rise in the temperature of the river?

11.56 (a) Show that when a Carnot cycle is plotted as a temperature-versus-entropy diagram, the resulting graph is a rectangle. Calculate (b) the heat absorbed, (c) the heat exhausted, and (d) the work done by the system whose Carnot cycle is shown in Fig. 11.17.

11.57 A typical person consumes 2000 kcal of food per day and converts most of that energy into heat. (a) Under the assumption that all of the food energy is released as heat, calculate the rate of heat released in watts. (b) If all of this heat is transferred to the surrounding air, which remains at an essentially constant temperature of 24 °C, what is the rate at which entropy is being delivered to the universe?

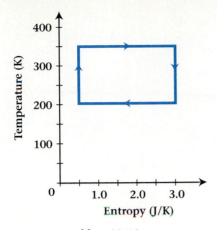

Figure 11.17   Problem 11.56.

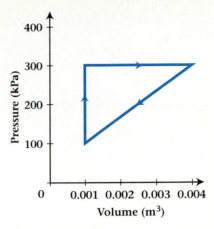

Figure 11.18   Problem 11.60.

11.58 A copper rod 30 cm long having a cross-sectional
•• area of 4.0 cm² is used to connect a boiler maintained at 100°C to a large block of ice at 0°C.
(a) What is the rate of heat flow through the rod?
What is the rate of change in entropy of (b) the
boiler, (c) the rod, (d) the ice, and (e) the universe
due to the heat flow through the bar?

11.59 Two similar coal-fired electric power plants generate
• electric energy at the same rate. One plant operates
with an efficiency of 30%, while its newer companion operates with an efficiency of 40%. How much
more waste heat is released to the environment by
the first plant than by the newer one?

11.60 The gas in a heat engine undergoes the process
• shown in the *PV* diagram of Fig. 11.18. (a) What is
the work done during one complete cycle? (b) What
is the net heat added to the system during one complete cycle?

ADDITIONAL READING

Bennett, C. H., "Demons, Engines, and the Second Law." *Scientific American,*
November 1987, p. 108.

Cropper, W. H., "Carnot's Function: Origins of the Thermodynamic Concept of
Temperature." *American Journal of Physics*, Vol. 55, February 1987, p. 120.

Crutchfield, J. P., J. D. Farmer, and N. H. Packard, "Chaos." *Scientific American,*
December 1986, p. 46.

Hafemeister, D., "Science and Society Test X: Energy Conservation." *American
Journal of Physics*, April 1987, p. 307.

Ord-Hume, A. W. J. G., *Perpetual Motion, The History of an Obsession.* New York:
St. Martin's Press, Inc., 1977.

Wilson, S. S., "Sadi Carnot." *Scientific American*, August 1981, p. 134.

12

The Kinetic Theory of Gases

12.1	The Pressure of Air
12.2	Boyle's Law
12.3	The Law of Charles and Gay-Lussac
12.4	The Ideal Gas Law
12.5	The Kinetic Theory of Gases
12.6	The Kinetic-Theory Definition of Temperature
12.7	The Specific Heats of a Gas
*12.8	The van der Waals Equation
*12.9	The Barometric Formula and the Distribution of Molecular Speeds
App.	The Exponential Function

A WORD TO THE STUDENT

We now examine the behavior of gases and consider theoretical models that allow us to understand this behavior. We use concepts from the previous two chapters in our discussion, as well as basic ideas from mechanics. This chapter illustrates some of the things we can learn by applying a simple model to matter. In fact, we have referred to gases several times in our development of thermal physics and thermodynamics. Now that we have these concepts available, we can present a complete description of gases.

The initial sections of this chapter sum up the observed behavior of gases in terms of empirical laws. In subsequent sections we develop these observations into a model that can explain and predict gas behavior. We also present, in an appendix to this chapter, a discussion of the exponential function. If you are not already familiar with this function you should learn it now; we will use it again to describe electric circuits, radioactive decay, and atomic structure.

In this chapter we examine for the first time the concept of atoms. The atomic concept is that matter consists of enormous numbers of tiny components called atoms. The properties and interactions of atoms, or of the molecules into which they may combine, are responsible for the observed properties and behavior of matter in bulk form. Up to now we have dealt only with objects large enough to see and measure. Here we apply the ideas from mechanics to predict how the atoms or molecules in a gas will behave.

Because we can neither see nor measure individual atoms directly, we must make some assumptions about what they are like. As we have discussed before, this process is part of making a theoretical model. In addition, because we clearly cannot follow the motions of every individual atom, we must decide how to evaluate average behavior. Our model will not be perfect—that is, it will not correspond to reality in every way—but the model will allow us to make predictions about the large-scale thermal behavior of gases. We can then test those predictions experimentally. The value of our model, called the kinetic theory of gases, depends on how close it comes to correctly describing the experimental observations.

Before considering the construction and use of a model of a gas, we briefly review the gas laws as they were determined over the years by observations and experiments. An understanding of the nature of air, and subsequently of other gases, could only come with the development of new devices and measuring instruments. The seventeenth century saw the discovery of such new instruments as the telescope, microscope, thermometer, barometer, pendulum clock, and air pump. These developments, as well as the founding of the first scientific societies, gave impetus to many kinds of scientific study, including the investigation of gases.

Today active research continues into the behavior of gases, using the latest computer simulation techniques and other modern methods of investigation. Much of this research has been spurred on by major environmental problems connected with the atmosphere, like the depletion of the earth's ozone layer. Solutions to such problems depend on understanding the behavior of gases.

The Pressure of Air

12.1

Galileo was the first person to record a fact that others had surely observed: that a lift pump (Fig. 12.1) could raise water only as high as 10.4 m (34 ft). In 1634, Evangelista Torricelli (1608–1649), who was briefly associated with Galileo, provided an explanation for this fact. Previous thinking, from Aristotle onward, had assumed that the entire universe must be filled with something material. It was said that "nature abhorred a vacuum." People thought the lift pump worked because the water rose up into the chamber to avoid the formation of a vacuum at the top. But Torricelli believed that the water was forced up the pipe by the pressure of the

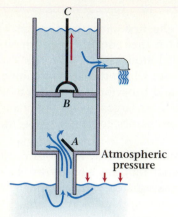

Figure 12.1   A lift pump. When rod C is lifted, valve B closes and valve A opens. Then water below the piston flows into the chamber and the water above flows out at the same time. When rod C is pushed down, A closes and B opens, allowing water to flow above the piston.

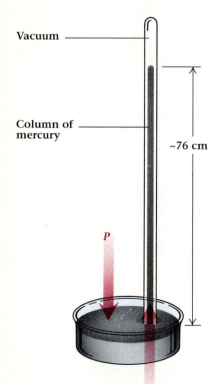

Figure 12.2   In a Torricellian tube, atmospheric pressure supports a column of mercury approximately 76 cm tall.

atmosphere on the surface of the water at the bottom of the pump. He reasoned that since mercury is about 14 times as dense as water, then if he were correct, atmospheric pressure should support a column of mercury approximately 1/14 as high as the maximum water column:

$$10.4 \text{ m water} \left(\frac{\text{density of water}}{\text{density of mercury}}\right) = 10.4 \text{ m} \left(\frac{1}{13.6}\right)$$

$$= 0.76 \text{ m mercury.}$$

Figure 12.2 shows a *Torricellian tube,* the forerunner of today's barometer. If a glass tube, closed at one end, is completely filled with mercury and then inverted into a bowl of mercury, the column of mercury in the tube drops until it reaches a height of about 76 cm above the lower surface, just as Torricelli predicted. In accord with Pascal's principle (Chapter 9), the pressure of the atmosphere on the surface of the mercury in the bowl is equal to the pressure due to the weight of the mercury in the tube. If this were not so, the mercury would flow about because it would not be in static equilibrium. The space between the top of the liquid and the end of the tube contains no air and was named the *Torricellian vacuum.*

Example 12.1

Determine the pressure due to the atmosphere, using Torricelli's results.

Solution   We could make immediate use of the results contained in Chapter 9 on fluids, but it is more instructive to consider the problem from a more fundamental standpoint. The weight of mercury in the tube in Fig. 12.2 is the product of its density ρ times the volume times the acceleration of gravity:

$$\text{weight} = \rho \cdot h A \cdot g,$$

where h is the height and A the area of the mercury column. For the fluid to be in static equilibrium, an upward force must be present. In this case it comes from the atmospheric pressure acting downward on the mercury in the bowl and then, according to Pascal's principle, being transmitted equally to all parts of the fluid. The upward force on the bottom of the tube is therefore

$$F = PA,$$

where P is the pressure due to the atmosphere. Because the fluid is in equilibrium, we can set the two equations equal to each other:

$$\rho \cdot h A \cdot g = PA$$

or

$$P = \rho h g.$$

This is the same expression as we found earlier, in Section 9.1. Inserting the numbers for g, h, and the density of mercury, we have

$$P = (13.6 \times 10^3 \text{ kg/m}^3)(0.760 \text{ m})(9.80 \text{ m/s}^2),$$

$$P = 101 \times 10^3 \text{ N/m}^2,$$

$$P = 1.01 \times 10^5 \text{ N/m}^2 = 1.01 \times 10^5 \text{ Pa.}$$

Figure 12.3 Otto von Guericke's experiment in Magdeburg, Germany, utilized an early vacuum pump. Two teams of eight horses each could not separate the sealed and evacuated hemispheres against the force of atmospheric pressure.

This pressure is atmospheric pressure. An atmosphere is a unit of pressure equal to the pressure of the earth's atmosphere at sea level. It is defined to be 1.01325×10^5 N/m^2.

Within a few years of Torricelli's experiments, Pascal suggested that the atmosphere is like an ocean of air, in which the pressure is greater at the bottom than at higher altitudes. This suggestion was soon confirmed when a Torricellian tube was carried from sea level to a mountain top and the mercury column was observed to be shorter at the higher elevations.

A dramatic demonstration of atmospheric pressure was provided by Otto von Guericke (1602–1686), who used his own invention, the air pump, in the famous demonstration of the Magdeburg hemispheres (Fig. 12.3). Two hemispheres, each about 55 cm in diameter, were fitted together and most of the inside air removed. The hemispheres were held together by the pressure of the surrounding air. Two teams of eight horses each, when hitched to the hemispheres, could not pull them apart. Problem 12.47 asks you to find the force holding them together.

Boyle's Law

12.2

Robert Boyle (1627–1691) advanced the study of gases using an air pump made by Robert Hooke that was greatly improved over that of von Guericke.* The result for which Boyle is best known today is the relationship between the pressure and the volume of an enclosed gas. **Boyle's law states that the pressure exerted by a gas is inversely proportional to the volume in which it is enclosed.** Boyle's law is usually written

$$PV = \text{constant,}$$

where P is the gas pressure, V its volume, and the value of the constant depends on the initial conditions. A complete statement of Boyle's law includes the condition that both the temperature and the amount of gas must be held constant.

Alternatively, Boyle's law may be written

$$P_1V_1 = P_2V_2, \tag{12.1}$$

where the subscripts 1 and 2 refer to different physical states of the same sample of gas with the temperature held constant. Figure 12.4 illustrates Boyle's law. A quantitative example of the application of Boyle's law follows.

We should point out that while Boyle's law is applicable over a wide range of pressures, it does not always apply. For example, if the temperature is low enough, a sample of gas will condense to a liquid at sufficiently high pressure. For carbon dioxide at 31°C, this pressure is about 7.38 ×

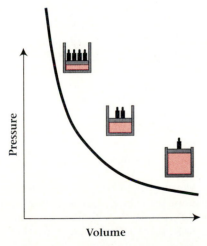

Figure 12.4 A graph of pressure versus volume for a gas enclosed in a cylinder at constant temperature. According to Boyle's law, PV is constant.

*Boyle was one of the founders of the Royal Society, chartered in 1662, which published Newton's first papers and of which Newton was later president. Among the interests of this early group were pneumatic experiments, which were often carried out as a diversion.

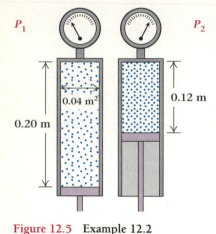

Figure 12.5   Example 12.2

10^6 N/m² or 72.9 atm. Other ways in which the behavior of real gases differs from the predictions of the gas laws will be discussed in Section 12.8, on the van der Waals equation.

Example 12.2

Boyle's law for the gas in a cylinder.

A cylinder with a height of 0.20 m and a cross-sectional area of 0.040 m² has a close-fitting piston that may be moved to change the internal volume of the cylinder (Fig. 12.5). Air at atmospheric pressure fills the cylinder. If the piston is pushed in until it is within 0.12 m of the end of the cylinder, what is the new pressure of the air? Assume that the temperature of the gas remains constant and that the volume of gas in the gauge is small compared with the volume of the cylinder.

BACK TO THE FUTURE

Gas Laws and Balloons

The first human flight occurred in the western suburbs of Paris on November 21, 1783, when two passengers made a 25-minute flight in a hot-air balloon designed by Joseph and Etienne Montgolfier (Fig. B12.1). The brothers had been experimenting with balloons for several years. Benjamin Franklin, then serving as ambassador to France, was one of the official observers of this first manned flight. When asked what was the use of flying he is said to have replied, "What use is a newborn baby?"

The speculation and rumor that had accompanied earlier experiments by the Montgolfiers gave rise to efforts by others. A few days after the Montgolfier flight J. A. C. Charles, in a balloon of his own design, made an ascent with a companion. At the end of the flight his companion got out of the gondola and Charles alone soared to a height of 9000 ft in about ten minutes, making temperature and pressure measurements along the way.

According to Archimedes' principle, a balloon rises if its overall density is less than the density of the air it displaces. Thus a balloon needs very-low-density gas to carry people through the air. Although the Montgolfier brothers knew about the discovery of hydrogen by Henry Cavendish in 1766, they elected to use hot air. Their considera-

Figure B12.1   An early hot-air balloon similar to the one designed by the Montgolfier brothers.

tions were primarily practical and economic. Professor Charles, who had received encouragement from Franklin, selected hydrogen instead. (Hydrogen is the least dense of all the gases.) For balloons of the same size, a hydrogen-filled balloon has several times more lifting force than a hot-air balloon. The first few years of ballooning were filled with controversy over the relative merits of hydrogen versus hot air. The laws of gas behavior were discovered at this time by people who were often associated with ballooning. For instance, the first qualitative work on the thermal expansion of gases was probably done by Charles in about 1787, several years after his first hydrogen balloon flight. However, his results went unpublished.

Another balloonist, J. L. Gay-Lussac, was one of the first to make ascents for scientific purposes. He was an active chemist and made two important discoveries about gases. He independently studied the thermal expansion of gases and published his results in 1802, fifteen years after Charles's work. In his paper he referred to the

Solution Let subscripts 1 and 2 indicate the situation before and after the piston is pushed in. Then, according to Boyle's law, we can write

$$P_1V_1 = P_2V_2,$$

which can be written

$$P_2 = \frac{P_1V_1}{V_2}.$$

Then we have

$$P_2 = \frac{1.0 \times 10^5 \text{ N/m}^2 \, (0.20 \text{ m} \times 0.040 \text{ m}^2)}{0.12 \text{ m} \times 0.040 \text{ m}^2},$$

$$P_2 = 1.7 \times 10^5 \text{ N/m}^2.$$

The new pressure is 1.7×10^5 N/m^2 or 1.7×10^5 Pa.

work of Charles, noting that Charles had not actually obtained the correct results for wet gases. The law of thermal expansion of gases is given the names of both Gay-Lussac and Charles.

During the next hundred years, ballooning evolved into sporting, military, and commercial applications, culminating in the invention of the rigid-frame airship by count Ferdinand von Zeppelin in the last part of the nineteenth century. Zeppelin's airships became the luxury liners of the sky. The Zeppelin was made of a light, rigid metal framework covered with fabric. Inside were sealed bags of hydrogen gas and compartments for passengers. The cabin that hung beneath the Zeppelin contained the bridge and navigation rooms. The Hindenburg, a Zeppelin built in 1936, was the largest flying machine ever made, with space to carry 72 passengers at 80 mph for over 8000 mi. The Hindenburg made more than 50 successful flights, including 36 across the Atlantic, before it exploded on landing at Lakehurst, New Jersey, on its first transatlantic flight of 1937. It was the last of more than 100 Zeppelins built. For reasons that were both technical and political, the Hindenburg disaster marked the

Figure B12.2 A modern hot-air balloon.

end of the era of rigid lighter-than-air ships.

The form in which ballooning is most frequently seen today is that of the colorful and beautiful hot-air sports balloon (Fig. B12.2). From shortly after the time of the Montgolfiers until the 1950s, most balloons used hydrogen or helium, the two lightest gases. The expense of large quantities of gas effectively kept private individuals from participating. On October 10, 1960, the age of the modern hot-air sports balloon was born when Ed Youst flew a hot-air balloon. To improve the performance, he built propane burners of his own design. These heat sources, which are at the heart of modern sports ballooning, are capable of delivering several million Btu per hour (10 million Btu/h = 2.9 MW), giving temperatures near 100°C inside the top of a rising balloon. Hot-air balloons must be relatively large because even very hot air is only slightly less dense than the atmosphere. A typical three- or four-person balloon has a volume of about 2200 m^3.

The Law of Charles and Gay-Lussac

12.3

Boyle's law relates the pressure and the volume of a gas at constant temperature. Figure 12.6(a) shows an apparatus used to investigate the effect of temperature change on the volume of a gas at constant pressure. We introduce some particular gas, say gas A, into the cylinder and place a weight on the piston. This arrangement creates an enclosed variable volume in which the pressure (due to the weight) remains constant. Then we bring the gas to thermal equilibrium at several different temperatures. We record these temperatures and the corresponding volumes and graph the data (Fig. 12.6b, line A). If we repeat this procedure with another gas, say B, with the same initial volume and temperature, we get the same results (line B). However, if we use gas B with a different initial volume, we obtain data producing a new line, marked B' in Fig. 12.6(b). Whatever gas we use, the results are always the same: The plotted data lie along a straight line.

To see what this result means, we may write the equation of any of these straight lines in the form

$$V = V_0(1 + \beta T),$$

where V is the volume at temperature T and V_0 and β are constants. The experimental result is that β is the same for all gases. If T is measured in degrees Celsius, β has the experimental value of $1/273.15°C$.

If a gas is brought to some temperature T_1, then its volume V_1 is given by

$$V_1 = V_0(1 + \beta T_1).$$

If the gas is brought to a different temperature T_2, its volume V_2 is

$$V_2 = V_0(1 + \beta T_2).$$

Dividing one of these equations by the other and writing in the numerical value for β gives

$$\frac{V_1}{V_2} = \frac{273.15 + T_1}{273.15 + T_2}.$$

It is evident that, at least so far as gases are concerned, it would be convenient to introduce a new temperature scale in which the temperature T is related to the Celsius temperature, T_C, by

$$T = 273.15 + T_C. \tag{12.2}$$

As we learned in Chapter 10, this temperature scale is called the Kelvin scale. The scale was introduced in 1848 by Lord Kelvin, who observed that if the lines for all gases in Fig. 12.6(b) were extrapolated to lower temperatures, they would intersect at $-273.15°C$ and zero volume. That does not mean that a gas would vanish at this temperature, even if it could be cooled so low before liquefying. It means, as we will see in more detail in Section 12.6, that the molecules of the gas would have a minimum

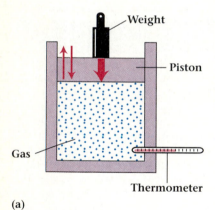

(a)

(b)

Figure 12.6 (a) Apparatus for measuring the volume of a gas as a function of temperature at constant pressure. (b) Typical graphical representation. The volume is directly proportional to the temperature.

amount of energy at that temperature. The scale is an absolute scale in the sense that its zero ($-273.15°C$) is the lower limit for temperatures as defined by macroscopic thermometers. *We will use the Kelvin scale throughout the remainder of this chapter.*

If the temperature is expressed in kelvins, we find that, when the pressure is held constant, the volume is proportional to the temperature. This statement is the **law of Charles and Gay-Lussac,** which can be expressed mathematically as

$$\frac{V_1}{V_2} = \frac{T_1}{T_2} \qquad \text{or} \qquad \frac{V}{T} = \text{constant.} \tag{12.3}$$

As with Boyle's law, the amount of gas also must be held constant for Eq. (12.3) to be valid.

Example 12.3

To what temperature would the air in a hot-air balloon have to be heated so that its mass would be 0.98 times that of an equal volume of air at a temperature of 25°C?

Solution   As the air in a hot-air balloon is heated, it expands and some of it escapes through the hole at the bottom in order to remain at constant atmospheric pressure. The mass of air remaining inside the fixed volume of the bag is less than the mass of an equivalent volume of the surrounding cooler air. Thus the density of the air inside the bag is reduced and the balloon floats in accord with Archimedes' principle. The mass of cool gas m_1 at temperature T_1 that was originally inside the volume V_1 of the bag expands to a new volume V_2 upon heating to temperature T_2. Since the volume of the bag is constant, the amount of hot gas m_2 that remains inside is proportional to the original mass and to the ratio of V_1/V_2. Thus

$$m_2 = \frac{V_1}{V_2} m_1,$$

which can be rearranged as

$$\frac{V_2}{V_1} = \frac{m_1}{m_2}.$$

The law of Charles and Gay-Lussac, Eq. (12.3), can be written as

$$T_2 = \frac{V_2}{V_1} T_1 = \frac{m_1}{m_2} T_1.$$

The ratio of m_1/m_2 is

$$\frac{m_1}{m_2} = \frac{1}{0.98}.$$

Therefore, upon putting the numbers into the law of Charles and Gay-Lussac, we get

$$T_2 = \frac{1}{0.98} (273 + 25) \text{ K} = 304 \text{ K.}$$

Upon converting back to degrees Celsius, we get

$$T_2 = 31°C.$$

12.4 The Ideal Gas Law

Boyle's law and the law of Charles and Gay-Lussac are special cases of a more general expression called the **ideal gas law**.* It can be inferred from them and is usually written

$$PV = nRT. \tag{12.4}$$

Here, as before, P, V, and T stand for pressure, volume, and temperature, respectively; R is a constant that is the same for all gases, and so is called the **universal gas constant**. If pressure is measured in the SI unit of pascals, volume in cubic meters, and temperature in kelvins, then R has the value 8.314 joule/mole·K.

The quantity of gas, measured in moles, is given by n. A **mole** (abbreviated mol) is the amount of material whose mass in grams is numerically equal to the molecular mass of the substance. For example, the molecular mass of oxygen gas is 32; a mole of oxygen gas is 32 g. (Oxygen molecules are diatomic; that is, each molecule consists of two atoms.) Avogadro's principle, discussed in Chapter 25, leads to the conclusion that a mole of any gas contains the same number of molecules. This number is called **Avogadro's number**, N_A, and is $N_A = 6.02 \times 10^{23}$ molecules/mole. The modern definition of the mole encompasses the concept of isotopes, discussed in Chapter 28. *The mole is the amount of substance of a system that contains as many elementary entities as there are atoms in 0.012 kg of carbon-12.* Table 12.1 lists the molecular masses of a few common gases. The dependence of the ideal gas law on n, the number of moles, can be seen from the observation that doubling an amount of gas at constant temperature and volume will double its pressure. One mole of nitrogen gas at 273.15 K and 101 kPa pressure occupies a volume of 22.4 L; two moles of nitrogen gas at the same temperature and volume have a pressure of 202 kPa.

An equation that links the pressure, volume, and temperature of a sample of matter is called an **equation of state**. Equation (12.4) is one such equation. It is the equation of state for an ideal gas, which is described in detail in the next section.

In ordinary situations, the ideal gas law predicts the behavior of many gases quite well. The agreement between the observed and predicted behavior of real gases is good near atmospheric pressure and ordinary temperatures. Deviations from the predictions are greatest when the pressure is very great or the temperature is very low. At the extremes of these conditions, the gases condense to liquids. For the examples and problems

---

*The definition of an ideal gas is given in Section 12.5. In brief, it is a gas that obeys Eq. (12.4).

TABLE 12.1
Approximate molecular masses of some common gases

Substance	Symbol	Molecular mass (g/mol)
Molecular hydrogen	H_2	2
Helium	He	4
Water vapor	H_2O	18
Neon	Ne	20
Molecular nitrogen	N_2	28
Molecular oxygen	O_2	32
Argon	Ar	40
Carbon dioxide	CO_2	44

given in this chapter, the deviations from the ideal gas law are small. Exceptions are discussed in Section 12.8, where we describe a modification of the ideal gas law that improves the agreement between prediction and observation.

Example 12.4

Using the ideal gas law.

During a chemistry laboratory experiment, a sample of hydrogen gas is collected into a 0.50-L flask at room temperature (23°C) and pressure (1.00 atm). It is cooled to 5°C and transferred to a container whose volume is 0.12 L. What pressure does the gas exert on the walls of the final container?

Solution Using the subscript 1 to indicate the initial conditions, we have

$$P_1V_1 = nRT_1.$$

Using the subscript 2 to denote conditions in the new container, we write

$$P_2V_2 = nRT_2.$$

Because the number of moles of gas is constant, we can combine the equations of state for the two situations to get

$$\frac{P_1V_1}{T_1} = \frac{P_2V_2}{T_2} = nR$$

or

$$P_2 = \frac{V_1 T_2 P_1}{V_2 T_1}.$$

Substituting the data gives

$$P_2 = \frac{0.50 \text{ L } (273 + 5) \text{ K } (1.00 \text{ atm})}{0.12 \text{ L } (273 + 23) \text{ K}} = 3.9 \text{ atm.}$$

Notice that quantities that enter only as a ratio (such as volumes, V_1/V_2) can be given in any system of units as long as they are the same. Temperatures, however, must be given in kelvins.

Example 12.5

Calculating density with the ideal gas law.

What is the density of carbon dioxide gas at a temperature of 23°C and atmospheric pressure?

Solution   The mass m of a sample of matter is related to the number of moles by

$$m = nM,$$

where n is the number of moles and M is the gram molecular mass. For carbon dioxide, CO_2, the gram molecular mass is $M_{carbon} + 2M_{oxygen} = (12.0 + 2 \times 16.0) = 44.0 \text{ g/mol}$.

The density is given by

$$\rho = \frac{m}{V} = \frac{nM}{nRT/P} = \frac{MP}{RT}.$$

Substituting for the molecular mass of CO_2 in the units of kg/mol and for the other variables we have

$$\rho = \frac{(44 \times 10^{-3} \text{ kg/mol})(1.01 \times 10^5 \text{ N/m}^2)}{(8.31 \text{ J/mol·K})[(273 + 23) \text{ K}]},$$

$$\rho = 1.81 \text{ kg/m}^3.$$

This value compares well with the experimentally measured one of 1.84 kg/m³.

Example 12.6

Air pressure in an automobile tire.

An automobile tire is filled to a gauge pressure of 240 kPa (35 lb/in²) early in the morning when the temperature is 15°C. After the car is driven all day over hot roads, the tire temperature is 70°C. Estimate the new gauge pressure. (Remember, gauge pressure is the pressure in excess of atmospheric pressure.)

Solution   We assume the volume of the tire to be approximately constant, so that we have

$$\frac{P_1}{T_1} = \frac{P_2}{T_2},$$

$$P_2 = P_1 \frac{T_2}{T_1},$$

$$P_2 = \frac{(240 \text{ kPa} + 101 \text{ kPa})(273 + 70)}{(273 + 15)} = \frac{341 \times 343}{288},$$

$$P_2 = 406 \text{ kPa}.$$

But this is the absolute pressure, so

gauge pressure = 406 kPa − 101 kPa = 305 kPa   or   44 lb/in².

The Kinetic Theory of Gases

12.5

The gas laws describe what happens to a gas under various conditions, but they say nothing about why gases act this way. Explaining the "why" behind observed behavior of matter has always been a fundamental motivation in physics. As early as the time of Boyle and Newton (that is, late seventeenth century) there were efforts to explain the observed behavior of gases on a fundamental basis. Newton proposed that a gas might consist of tiny particles called molecules, which exert a repulsive force on each other. Such an idea can lead to Boyle's law, though Newton and Boyle did not claim this to be the only possible explanation.

The development of a successful theory of gas behavior did not take place until the middle of the nineteenth century. Many people contributed to this development, and the theory has not been named after any particular scientist. However, since the theory assumes that a gas consists of particles in motion, it is usually called the **kinetic theory of gases**.

In the kinetic theory of gases, we assume that a gas consists of many particles. Here "many" means so numerous that we cannot hope to trace out their individual paths. In fact, it may not be desirable to do this even if we could. The things we want to determine from a model of a gas are the things that determine its behavior—an equation of state and the thermal and mechanical properties of a gas—not the directions and speeds of a large number of individual particles. If we do not treat the molecules individually, then we must determine their average behavior. This implies that we need to develop a statistical theory that includes the rules of probability.

We will first construct a model of a gas by making detailed assumptions about its nature. Then we will calculate how this gas should behave, using our knowledge of Newton's laws and of the conservation of momentum and mechanical energy, and by taking the appropriate averages. For instance, we will derive a relationship between the pressure and volume of this model gas. One test of the validity of our assumptions about the gas model will be how closely our predictions agree with what we observe for a real gas.

This example of model making is typical of many theories in modern science. We cannot, in the ordinary sense, "see" individual molecules, atoms, or their constituent parts, so our explanations of what happens in many biological, chemical, and physical phenomena must be made in terms of an abstract model. Such models should not be taken as absolute replicas of reality in all cases. For example, we will see in Chapter 27 that the "solar system" model of an atom, though convenient for representing some aspects of atomic behavior, does not literally correspond to reality. Furthermore, we will see that in many cases, as we examine smaller and smaller physical constituents of a system, and therefore consider more of them, we are forced to calculate the observed behavior as statistical averages. This will be the case with a gas.

Our rules for formulating a model are:

1. We will not make any assumptions that are not specifically needed in our derivation.
2. The assumptions made will be as simple as possible.
3. The assumptions will, when possible, have some basis in physical experience.

We will first list the *principal assumptions of the kinetic theory of gases,* and then use them in deriving a relationship between the pressure and the volume of a gas. These assumptions constitute the definition of an ideal gas. Any gas that obeys the relationships derived from these assumptions at all temperatures and pressures is called an **ideal gas.**

1. *A sample of gas consists of many identical molecules. In this context "many" means so many that one could not hope to trace out their individual paths.*
2. *The molecules are very far apart in comparison to their size; that is, the total volume of the molecules is negligible when compared with the size of their container.*
3. *The direction of motion of any molecule is random; on the average, no direction is preferred above another and the molecules move with a variety of speeds.*
4. *The molecules are treated as if they were hard spheres. This assumption, which gives rise to the name "billiard ball model," means that there are no forces acting between molecules except when the molecules collide and that the collisions are elastic. In addition, we treat collisions with the walls of the container as elastic collisions.*
5. *The molecules obey Newton's laws of motion.*

These assumptions are to some extent based on everyday experience. For instance, we know that molecules of air cannot be seen by the unaided eye. This means they must be much smaller than spheres with a radius of about 0.1 mm. The assumption of random directions seems reasonable because if the molecules of a gas had a preferred direction, the gas would have a net flow in that direction.

The assumption that molecules of a gas are like hard spheres is not the only, or even the most realistic, assumption we could make. However, it is the simplest assumption and therefore, lacking specific information to the contrary, the most appropriate in this case. We have listed the use of Newtonian mechanics in order to emphasize that the results of our calculations with this model depend not only on the physical properties of the model, but also on the mathematical techniques used to obtain its predictions.

Now that we have stated the assumptions of the kinetic theory of gases, let us analyze their consequences and compare these with observable properties of a particular gas. We will do this by deriving mathematical expressions for those physical quantities of importance in mechanics, such as momentum, force, and kinetic energy. We start by considering an ideal gas confined to a cubical box with sides of length L. The gas molecules move in a manner determined by Newton's three laws and the laws of conservation of energy and momentum. We assume that the number den-

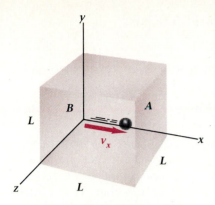

Figure 12.7 A model for calculating the pressure of a gas from kinetic theory. The gas molecule is confined to a box of edge length L as shown.

sity, that is, the number of molecules per unit volume, is on the average the same throughout the box. For convenience we choose a coordinate system aligned with the edges of the box (Fig. 12.7).

Consider a molecule of mass m moving parallel to the x axis in the positive direction. For the moment, we neglect collisions between molecules and consider only collisions between the molecules and the walls of the box. If the molecule has a velocity v_x it will rebound from wall A with a velocity $-v_x$ because we assumed that the collisions are elastic, and the walls are essentially immobile. The molecule's change in velocity is $-2v_x$ and its change in momentum is $-2mv_x$. The momentum transferred to the wall is $+2mv_x$.

If the molecule encounters no other molecules, but moves until it strikes wall B and rebounds again to strike wall A, the time between collisions with face A is

$$\Delta t = \frac{2L}{v_x}.$$

The average momentum change per unit time at face A is

$$\frac{\Delta p}{\Delta t} = \frac{2mv_x^2}{2L}.$$

We can interpret this expression as the average force on the wall due to any molecule that has an x component of velocity given by v_x. The total force due to N molecules is

$$F_{total} = N\,\frac{\overline{\Delta p}}{\Delta t} = \frac{Nm\overline{v_x^2}}{L},$$

where the bar indicates, as before, the average of that quantity.

By dividing both sides of the equation by the area of the face L^2, we have the pressure P,

$$P = \frac{F}{A} = \frac{F}{L^2} = \frac{Nm\overline{v_x^2}}{L^3} = \frac{Nm\overline{v_x^2}}{V}$$

or

$$PV = Nm\overline{v_x^2}, \tag{12.5}$$

where P is the pressure on the wall and V is the volume of the box.

The square of the total velocity of a molecule is the sum of the squares of the three component velocities:

$$v^2 = v_x^2 + v_y^2 + v_z^2.$$

For a large number of molecules, the assumption of no preferred direction (assumption 3) leads to an average squared velocity of

$$\overline{v^2} = \overline{v_x^2 + v_y^2 + v_z^2} = \overline{v_x^2} + \overline{v_y^2} + \overline{v_z^2}.$$

The average values of the components are identical because the x, y, and z directions are equivalent and independent. Therefore

$$\overline{v^2} = 3\overline{v_x^2}.$$

Thus $\overline{v_x^2}$ in Eq. (12.5) can be replaced by $\overline{v^2}/3$ to give

$$PV = \tfrac{1}{3}Nm\overline{v^2}.$$ (12.6a)

Since $\overline{KE} = \tfrac{1}{2}m\overline{v^2}$, where $\overline{KE}$ is the average kinetic energy per molecule, we have

$$PV = \tfrac{2}{3}N\,\overline{KE}.$$ (12.6b)

The quantities on the left-hand side of Eq. (12.6) are macroscopic (large-scale) quantities. The quantities on the right-hand side are microscopic (molecular-scale) variables. We have derived an expression linking the microscopic, and generally unobservable, properties of molecules, such as their masses and speeds, with the easily observed large-scale properties, such as pressure and volume.

In deriving Eq. (12.6) we assumed that there were no collisions between the particles. However, if we include elastic collisions, the result is unchanged. Because of the perfect exchange of velocities in elastic collisions between identical particles, the component of momentum mv_x of a particle leaving face B is still carried to face A of the box in Fig. 12.7 by some other particle. The time duration of elastic collisions is negligible compared with the time between collisions. So the absence of collisions in our derivation does not affect its validity.

Example 12.7

Average speed of a gas molecule.

Estimate the average speed of oxygen molecules at the standard temperature and pressure of 0°C and one atmosphere. Assume that oxygen can be treated as an ideal gas.

Solution   Equation (12.6a) can be rewritten as

$$\overline{v^2} = \frac{3PV}{Nm}.$$

But since Nm is the total mass of the gas, we can write

$$\overline{v^2} = \frac{3P}{\rho},$$

where $\rho = Nm/V$ is the mass density. This expression represents the average of the speed squared or, as it is also called, the mean-square speed.

The quantity $\sqrt{\overline{v^2}}$ is called the root-mean-square speed, abbreviated as rms speed:

$$v_{rms} = \sqrt{\overline{v^2}} = \sqrt{\frac{3P}{\rho}}.$$

Inserting a pressure of one atmosphere and the density of oxygen at one atmosphere and 0°C, we get

$$v_{rms} = \sqrt{\frac{3 \times 1.01 \times 10^5 \text{ N/m}^2}{1.43 \text{ kg/m}^3}},$$

$$v_{rms} = 460 \text{ m/s} \approx 1700 \text{ km/h } (1000 \text{ mi/h}).$$

This result for the rms speed of oxygen molecules is the same order of magnitude as the speed of sound (317 m/s) observed in oxygen at this temperature and pressure. This similarity is to be expected, since sound waves are transmitted by the motion of air molecules. According to our model, the molecules of a gas in a closed container are moving with great speeds. The average speed of the particles is slightly smaller than the rms speed. Calculation of the average speed requires a detailed knowledge of the distribution of speeds among the particles of the gas; we shall discuss this later in the chapter.

The Kinetic-Theory Definition of Temperature

12.6

According to Eq. (12.6b), the product of pressure and volume of an ideal gas can be expressed in terms of the average kinetic energy per molecule, $\overline{KE}$:

$$PV = \tfrac{2}{3} N \, \overline{KE}.$$

The product $N \, \overline{KE}$ is the total kinetic energy of the gas. However, since we have not considered rotations or internal vibrations of the molecules in our model of a gas, this kinetic energy represents the total internal energy of the gas, U. Thus we can write

$$PV = \tfrac{2}{3} U.$$

This equation is the same as Boyle's law if we assume that keeping the temperature constant and keeping the internal energy constant are the same thing.

By combining this equation with the ideal gas law (Eq. 12.4), we can relate the internal energy to the temperature through

$$nRT = \tfrac{2}{3} U.$$

We can rearrange to find

$$U = \tfrac{3}{2} nRT. \tag{12.7}$$

This result for the internal energy of an ideal gas allows us to derive expressions for observable properties of the gas. In the following section we will use the first law of thermodynamics and the internal energy derived here to obtain an equation for the specific heat of a gas, which can be measured and compared with our predictions.

Rewriting the ideal gas law in terms of the average kinetic energy per molecule yields

$$nRT = \tfrac{2}{3} N \, \overline{KE}.$$

The number of molecules N is just the number of moles n times Avogadro's number ($N = n N_A$). Thus

$$\overline{KE} = \frac{3}{2} \frac{n}{N} RT = \frac{3}{2} \frac{RT}{N_A},$$

or

$$\boxed{\overline{KE} = \tfrac{3}{2}kT.}$$

(12.8a)

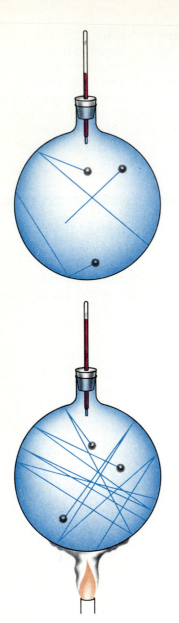

Figure 12.8 The average kinetic energy of a gas depends only on the temperature of the gas. The greater the temperature, the greater the average kinetic energy of the molecules.

The constant $k = R/N_A$ is called the **Boltzmann constant** and has a value of 1.3807×10^{-23} J/K. It occurs frequently in the physics of gases and is sometimes called the gas constant per molecule. Equation (12.8a) provides a new definition of temperature in terms of the microscopic mechanical properties of a gas. Specifically, temperature is a measure of the average kinetic energy of the molecules of a gas (Fig. 12.8),

$$T = \frac{2}{3} \frac{\overline{KE}}{k}.$$

(12.8b)

Furthermore, we see that the average kinetic energy of a molecule in an ideal gas depends only on the temperature, not on the pressure or type of gas.

The connection between temperature and the average kinetic energy of the molecules is an important result. It tells us what is happening on a molecular scale. An increase in temperature corresponds to an increase in the average speed of the molecules; a decrease in temperature corresponds to a slowing down of the molecules. Neither phenomenon is cause nor effect; instead, temperature is a large-scale manifestation of motion at the molecular level of gases, solids, and liquids. Knowing this helps us to better understand the relationships among temperature, heat flow, and energy that we discussed in Chapter 10.

Example 12.8

Determining molecular speed from temperature.

What is the rms speed of a nitrogen molecule at a temperature of 300 K? Assume that nitrogen behaves as an ideal gas. The mass of a nitrogen molecule is $m = 4.65 \times 10^{-26}$ kg (nitrogen is another diatomic molecule).

Solution   From Eq. (12.8) we have

$$\overline{KE} = \tfrac{1}{2}m\overline{v^2} = \tfrac{3}{2}kT,$$

so

$$\overline{v^2} = 3kT/m,$$

and

$$v_{rms} = \sqrt{\overline{v^2}} = \sqrt{\frac{3kT}{m}}.$$

Upon inserting the values for k, T, and m, we find that

$$v_{rms} = \sqrt{\frac{3 \times 1.38 \times 10^{-23} \text{ J·K}^{-1} \times 300 \text{ K}}{4.65 \times 10^{-26} \text{ kg}}},$$

$$v_{rms} = 517 \text{ m/s}.$$

The Specific Heats of a Gas

12.7

In the previous sections we have seen that the equation of state of an ideal gas can be derived from the set of defining assumptions about gases. We can use this same model to determine other properties of an ideal gas, such as those relating to viscosity and diffusion. In this section we will use the model to calculate the specific heats of a gas and then compare our results with experimental observations.

In Chapter 10 we saw that the relationship between the change in temperature ΔT of a mass m with specific heat c and the heat added Q is

$$Q = mc\,\Delta T.$$

In the case of gases, it is convenient to use a molar specific heat (sometimes called the molar heat capacity), which is the heat needed to raise one mole of a substance through one degree Celsius. However, for a gas, we must define two specific heats. The amount of heat needed to raise the temperature of a gas by one degree depends on whether the volume of the gas is kept constant or allowed to expand. We therefore introduce C_v, the **molar specific heat at constant volume**, and C_p, the **molar specific heat at constant pressure**.

Using the first law of thermodynamics, let us consider what happens when we heat an ideal gas at constant volume. From Chapter 11, the first law of thermodynamics is

$$Q = \Delta U + W,$$

where Q is the heat absorbed by the gas, ΔU is the change in internal energy, and W is the work done by the gas. If the volume does not change, then no mechanical work is done by the gas and all the heat goes into changing the internal energy. Thus

$$Q = \Delta U.$$

From Chapter 10, the heat added is related to the temperature change by

$$Q = nC_v\,\Delta T,$$

where n is the number of moles and C_v is the molar specific heat. Comparison of these two equations gives

$$\Delta U = nC_v\,\Delta T. \tag{12.9}$$

For an ideal gas, the internal energy is due solely to the kinetic energy of the molecules. Thus the change in internal energy may be obtained from Eq. (12.7):

$$\tfrac{3}{2}nR\,\Delta T = nC_v\,\Delta T.$$

This equation can be simplified to give

$$\boxed{C_v = \tfrac{3}{2}R} = 12.47 \text{ J/mol·K.} \tag{12.10}$$

We have derived a numerical value for the molar specific heat of an ideal gas at constant volume, using the kinetic theory of gases and the first law of thermodynamics. Note that our model assumes elastic collisions and does not consider rotations nor vibrations of the molecules.

We can also find the molar specific heat of an ideal gas at constant pressure with the help of the first law of thermodynamics. If the expansion is carried out at constant pressure, the work done is

$$\Delta W = P \, \Delta V.$$

For constant pressure, we can calculate $P \, \Delta V$ from the equation of state of an ideal gas (Eq. 12.4),

$$P \, \Delta V = nR \, \Delta T.$$

The heat flow Q for a process at constant pressure is, by definition,

$$Q = nC_p \, \Delta T.$$

Upon combining these equations with the first law, we find that

$$\Delta U = Q - W,$$
$$nC_v \, \Delta T = nC_p \, \Delta T - nR \, \Delta T,$$

where we have used Eq. (12.9) for the change in internal energy. Upon simplification we see that

$$C_p - C_v = R. \tag{12.11}$$

For an ideal gas, we can use Eq. (12.10) to see that

$$C_p = \tfrac{5}{2}R = 20.79 \text{ J/mol·K}.$$

Table 12.2 gives the specific heats of some real gases. Note that the agreement between the predictions of kinetic theory and the experimentally measured values is excellent for monatomic gases (gases whose molecules are single atoms), but less good for gases with more complicated molecules. One reason for this is that more complicated molecules may vibrate and rotate, thus having other forms of energy in addition to translational kinetic energy. It is possible to take account of this additional

---

TABLE 12.2
Observed molar specific heats of some real gases

Gas	Type	Temperature (°C)	C_p	C_v	$C_p - C_v$
			\multicolumn{3}{c}{$(\text{J·mol}^{-1}\text{·K}^{-1})$}		
Argon (Ar)	Monatomic	15°C	20.95	12.56	8.39
Neon (Ne)	Monatomic	15°C	20.78	12.67	8.11
Nitrogen (N_2)	Diatomic	15°C	29.04	20.68	8.36
Oxygen (O_2)	Diatomic	15°C	29.17	20.82	8.35
Carbon dioxide (CO_2)	Triatomic	15°C	36.64	27.88	8.76
Steam (H_2O)	Triatomic	100°C	36.35	27.45	8.90
Ethane (C_2H_2)	Polyatomic	15°C	48.60	39.84	8.76

$$R = 8.31 \text{ J·mol}^{-1}\text{·K}^{-1}$$

energy using concepts from classical, or Newtonian, mechanics. Doing so greatly improves agreement, but is still not entirely satisfactory. The fact is that, on an atomic level, Newton's laws must be modified. The appropriate laws for that scale are those of quantum mechanics, a theory that will be introduced in another context in later chapters.

Example 12.9

Specific heat of an ideal gas.

How much heat is required to raise the temperature of one mole of an ideal gas by 10 K if the gas is maintained at constant pressure?

Solution At constant pressure, the molar heat capacity is given by $C_p = \frac{5}{2}R$. The heat required is given by

$$Q = nC_p \, \Delta T.$$

Here $n = 1$, $\Delta T = 10$ K, and $R = 8.3$ J/mol·K. Thus

$$Q = (1 \text{ mol})(\tfrac{5}{2}R) \, \Delta T = 1 \text{ mol} \times \tfrac{5}{2}(8.3 \text{ J/mol·K}) \, 10 \text{ K},$$
$$Q = 208 \text{ J}.$$

The van der Waals Equation

*12.8

So far in this chapter we have dealt only with ideal gases. We have also seen that in many cases, the behavior of real gases corresponds closely to the predictions of a model based on noninteracting spheres of negligible volume. Yet we do not believe that this model represents even our best guess at physical reality. Molecules do have size and their collisions are not the same as billiard-ball collisions.

We can take account of certain types of molecular interactions and of the nonzero size of the molecules by making modifications to the equation of state of an ideal gas. The first major step in this direction was taken in 1873, when the Dutch physicist J. D. van der Waals (1837–1923) proposed that the equation of state of a gas be written as

$$(P + an^2/V^2)(V - nb) = nRT. \tag{12.12}$$

In this equation, called the **van der Waals equation**, the constant b is called the covolume and nb is of the order of magnitude of the volume actually occupied by the molecules. Its effect is to reduce the available volume of a gas in calculations with the equation of state. The interaction between the molecules is taken into account in a very general way by the constant a. Both a and b must be determined experimentally and are different for different gases.

Figure 12.9(a) is the PV diagram for one mole of an ideal gas. The lines of constant temperature are called isotherms. Figure 12.9(b) shows the isotherms for a gas that obeys the van der Waals equation, using values of a and b determined for carbon dioxide. Note that at high temperatures, the results are almost the same as those for an ideal gas. The behavior of

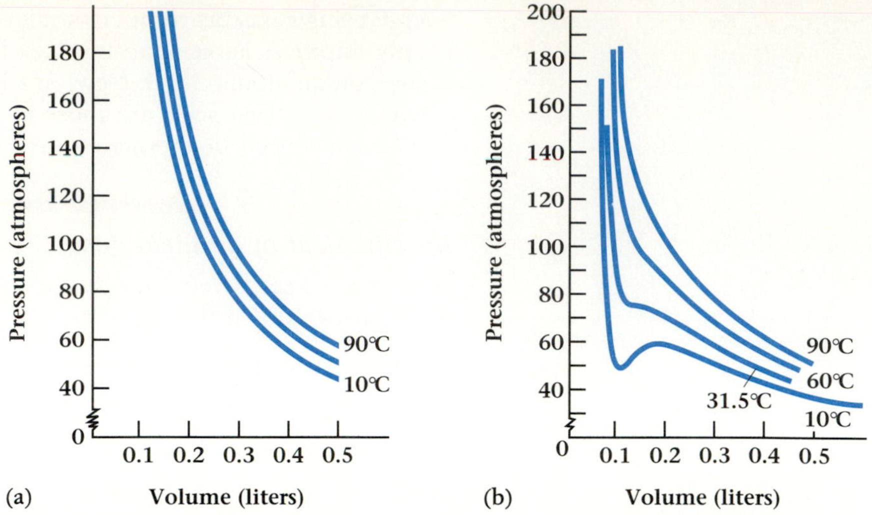

Figure 12.9 (a) Isotherms for one mole of an ideal gas. (b) Isotherms of a van der Waals gas plotted for one mole of carbon dioxide. At high temperatures, the isotherms are almost the same as those for an ideal gas.

(a)   Volume (liters)

(b)   Volume (liters)

the van der Waals isotherms at lower temperatures does not represent the behavior of any real gases. These temperatures are ones at which the actual gas may be a liquid.

Numerous attempts have been made to improve on the van der Waals equation. Many of these have brought better agreement, but not necessarily better understanding. The difficulty of accurately representing the intramolecular forces is one of the principal difficulties. There is also the basic problem, mentioned before, that the laws of classical physics are not entirely correct for such small dimensions.

The Barometric Formula and the Distribution of Molecular Speeds

*12.9

So far in this chapter, we have considered only the average speeds of molecules. However, there are times when we need to know the distribution of the molecular speeds in a gas. By distribution we mean a mathematical expression that tells us what fraction of the molecules have speeds in a given range.

The problem of the distribution of molecular speeds was first solved by James Clerk Maxwell (1831–1879) in 1860. Maxwell's significant contribution was the introduction of statistical ideas into classical mechanics. We will first derive a formula for pressure as a function of height in the atmosphere as a way of introducing you to the exponential function. Then we will present Maxwell's results, which are expressed with this mathematical function. (We discuss the exponential function in detail in an appendix to this chapter.)

Consider a tall column of air, which might represent a cross section of the atmosphere (Fig. 12.10). We will assume the temperature to be the same at all points in this atmosphere. Let us now examine the gas between two horizontal planes at altitudes z and $z + \Delta z$. The pressure at height

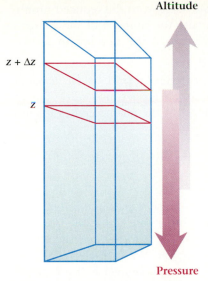

Altitude

$z + \Delta z$

z

Pressure

Figure 12.10 A simplified model of the atmosphere at constant temperature. Pressure is greater at lower altitudes.

z is greater than the pressure at height $z + \Delta z$ because of the weight of the air contained between z and $z + \Delta z$. Although the atmosphere does not have a well-defined height, we are able to write the difference in pressures because it depends on the difference in heights:

$$P_{z+\Delta z} - P_z = \Delta P = -\rho g\, \Delta z,$$

where ρ is the mass density of the gas. The minus sign indicates that the pressure decreases as the altitude increases. If we assume that the density of the air does not vary over a small change in height, we may use the ideal gas law to express this density. The total mass M of the gas is the number of moles n times the number of molecules in a mole N_A times the mass of one molecule m. The density is then

$$\rho = \frac{\text{mass}}{\text{volume}} = \frac{M}{V} = \frac{nN_Am}{nRT/P},$$

$$\rho = \frac{m}{kT}\,P,$$

where k is the Boltzmann constant. We can insert this value for ρ into the equation for ΔP to get

$$\frac{\Delta P}{\Delta z} = -\rho g = -\frac{mg}{kT}\,P. \tag{12.13}$$

This equation is a relationship between changes in pressure and changes in height. We wish to find an expression for pressure as a function of height alone. If we interpret Eq. (12.13) graphically, we see that we are looking for a function P whose slope $\Delta P/\Delta z$ is proportional to the value of P. The result is that

$$P(z) = P_0 e^{-mgz/kT}, \tag{12.14}$$

where P_0 is the pressure at $z = 0$ and e is an irrational number with a value approximately equal to 2.718. (If you are not already familiar with the exponential function e, see the chapter appendix.) Although we would not expect full agreement between the prediction of Eq. (12.14) and the observed pressure in the earth's atmosphere because T is not constant, there is rather good general accord.

We may also express Eq. (12.14) in terms of the number density n, the number of molecules per unit volume. We can do so because at constant temperature, the pressure and density of an ideal gas are proportional. Thus

$$n = n_0 e^{-mgz/kT}. \tag{12.15}$$

This equation is called the *barometric formula*. It gives the number of molecules per unit volume as a function of height z in our idealized atmosphere.

Example 12.10

Relative density of air on Pike's Peak.

People who have been to high elevations on mountains know that it is more difficult to breathe at higher altitudes than at lower ones. This effect is due to the reduced air pressure or, equivalently, to the reduced number of air molecules per unit volume. Estimate the number of molecules of air per unit volume on Pike's Peak relative to the number in Denver, Colorado.

Solution   The atmosphere is about 21% oxygen (at 5.31×10^{-26} kg per molecule) and 79% nitrogen (at 4.67×10^{-26} kg per molecule). For our computations we will use the average mass of the air molecules for m in Eq. (12.15). The average mass is

$$m = (0.21)(5.31 \times 10^{-26} \text{ kg}) + (0.79)(4.67 \times 10^{-26} \text{ kg}),$$

$$m = 4.80 \times 10^{-26} \text{ kg}.$$

The temperature is not uniform, but because we use the Kelvin scale, only a small error is introduced by assuming a constant value of $0°C = 273$ K. The altitude of Denver is 1610 m above sea level and the altitude of Pike's Peak is 2690 m, some 1080 m higher. We can use Eq. (12.15) if we let n_0 represent the density of molecules at Denver and n represent the density at Pike's Peak. Then z is the difference in their altitudes, 1080 m. When these and the other values are inserted into the equation we find

$$\frac{n}{n_0} = e^{-mgz/kT} = e^{-\frac{4.80 \times 10^{-26} \text{ kg} \times 9.8 \text{ m/s}^2 \times 1080 \text{ m}}{1.38 \times 10^{-23} \text{ J/K} \times 273 \text{ K}}},$$

$$\frac{n}{n_0} = 0.87.$$

The difficulty of breathing on Pike's Peak is understandable because the number of molecules per unit volume is only 87% as great as in Denver. How does the density of air molecules on Pike's Peak compare with the density where you live?

By continuing to analyze gases in the same way and by using the results of the kinetic theory, we can extend the derivation of the barometric formula to get an expression for the fraction of particles in a gas within a range of speeds. This result was obtained (though by another method) by Maxwell, who showed that the distribution in speeds is given by

$$f(v) \ \Delta v = 4\pi(m/2\pi kT)^{3/2} \ v^2 e^{-mv2/2kT} \ \Delta v, \qquad (12.16)$$

where $f(v)$ is the fraction of molecules that have speeds between v and $v + \Delta v$. Equation (12.16) is called the **Maxwell-Boltzmann distribution** function. Figure 12.11 displays this function, computed for oxygen, neon, and helium gases at a temperature of 295 K. Figure 12.12 shows the distribution for oxygen at 295 K and 1000 K. Notice that the Maxwell-Boltzmann distribution function is not an exponential curve, although the equation does contain an exponential function. We do not expect you to memorize the details of this formula, but you should become familiar in a general way with the factors that determine the distribution and their

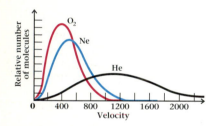

Figure 12.11   Maxwell-Boltzmann molecular-speed distribution for three different gases at $T = 295$ K. The differences are due to the different molecular masses.

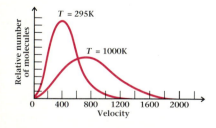

Figure 12.12   The distribution of molecular speeds for oxygen gas according to the Maxwell-Boltzmann distribution function at two different temperatures.

significance. In particular, the shift of the curve with temperature and the factor in the exponent of kinetic energy $\frac{1}{2}mv^2$ divided by kT are worth noting.

The peak of each curve represents the most probable molecular speed for that temperature and gas. That is, the curve peaks at that speed exhibited by the largest fraction of molecules in the gas. Though an appreciable number of molecules have speeds near the peak, the overall range of speeds is large in every case. More massive molecules have lower most probable speeds, in agreement with the predictions of kinetic theory for a given temperature. For a given gas, the most probable molecular speed becomes greater with increasing temperature. In particular, more molecules have high speeds and fewer molecules have low speeds, throughout the range of speeds. This ties in with our earlier results from kinetic theory that higher temperatures mean higher molecular kinetic energies.

In 1955, Miller and Kusch conducted a high-precision experiment to compare a measured speed distribution with the Maxwell-Boltzmann equation. They placed a container A, filled with a known gas, and a molecule detector B at opposite ends of a cylinder C of length L (Fig. 12.13). A large number of helical grooves were cut into the surface of the cylinder, only one of which is shown in the diagram. The cylinder rotated about

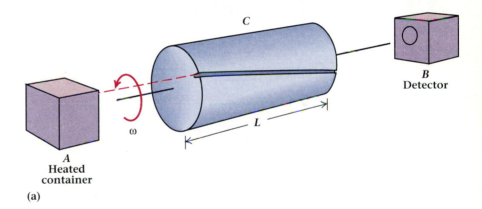

(a)

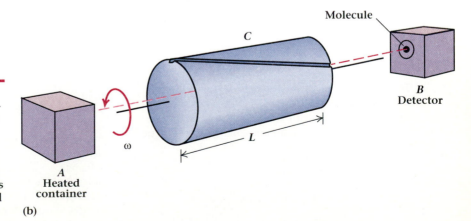

(b)

Figure 12.13 Apparatus for measuring the distribution of molecular speeds. As the grooved cylinder rotates, a molecule entering the groove at one end at just the right speed (a) can continue in straight-line motion and emerge from the far end. (b) Molecules with other speeds will strike the edge of the groove and be scattered.

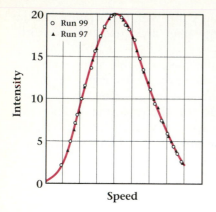

Figure 12.14

Distribution of molecular speeds in a gas. The points represent experimental observations; the solid line represents the Maxwell-Boltzmann distribution. (From R. C. Miller, and P. Kusch, "Velocity Distributions in Potassium and Thallium Atomic Beams." *Physical Review*, Vol. 99, August 15, 1955, p. 1314.)

its axis at a constant angular velocity ω. As the cylinder rotated, it was bombarded by molecules coming out of an aperture in the heated container *A*. A molecule from *A* would reach detector *B* only if its velocity matched the cylinder's speed of rotation. The velocity of those molecules that reached *B* was then recorded as

$$v = \frac{\omega L}{\Phi},$$

where Φ is the angle through which one end of the groove was turned with respect to the other. Miller and Kusch recorded the number of molecules detected (beam intensity) as a function of ω. This distribution of beam velocities was then used to give the distribution of molecular speeds in the container (Fig. 12.14). For other gases, the results were similar. In all cases, the distribution was consistent with the predictions of the Maxwell-Boltzmann distribution function.

SUMMARY

Useful Concepts

- Boyle's law states that if the temperature and the amount of a gas are held constant, then

$$P_1V_1 = P_2V_2 \quad \text{or} \quad PV = \text{constant}.$$

- The law of Charles and Gay-Lussac states that for a gas at constant pressure,

$$\frac{V}{T} = \text{constant}.$$

- We can predict the behavior of gases using the ideal gas law. The agreement between prediction and observation is best when temperatures and pressures are near ordinary values. The ideal gas law is

$$PV = nRT.$$

- The kinetic theory of gases shows how it is possible to connect the macroscopic properties of a gas, such as pressure and volume, with its microscopic properties, such as average molecular kinetic energy. The kinetic theory gives, for an ideal gas,

$$PV = \tfrac{2}{3}N\,\overline{\text{KE}}.$$

- The kinetic-theory definition of temperature is

$$\overline{\text{KE}} = \tfrac{3}{2}kT.$$

- From kinetic theory we get the results that

$$C_{\text{v}} = \tfrac{3}{2}R \quad \text{and} \quad C_{\text{p}} - C_{\text{v}} = R.$$

- One equation of state for a nonideal gas is the van der Waals equation,

$$(P + an^2/V^2)(V - nb) = nRT.$$

- The pressure at a given height in the atmosphere is given by

$$P(z) = P_0 e^{-mgz/kT}.$$

Important Terms

You should be able to write the definition or meaning of each of the following:

- Boyle's law
- law of Charles and Gay-Lussac
- ideal gas law
- universal gas constant
- mole
- Avogadro's number
- equation of state
- kinetic theory of gases
- ideal gas
- Boltzmann constant
- molar specific heat at constant volume
- molar specific heat at constant pressure
- van der Waals equation
- Maxwell-Boltzmann distribution

QUESTIONS

12.1 What would happen if you took a barometer to the bottom of a swimming pool? What would happen if you took it to the moon?

12.2 A barometer of the type shown in Fig 12.2 is carried on an earth-orbiting satellite in which the cabin is pressurized to one atmosphere. Describe what an astronaut would observe if the apparatus were first laid on its side and then turned upside down.

12.3 If the atmosphere were made of only oxygen, instead of being primarily nitrogen, would barometers read higher or lower than they do now?

12.4 A car owner's manual says to measure the tire pressure before driving. What difference would it make if the pressure were measured after driving several miles at highway speeds?

12.5 Explain in words why the pressure of a gas increases when its volume is reduced.

12.6 In the kinetic model of a gas we assumed elastic collisions between the molecules. How would the model change if inelastic collisions were allowed?

12.7 How would the results of our kinetic model be modified if we assumed that the size of the molecules is not negligible in comparison to the distance between them?

12.8 Give an explanation for the difference between heat and temperature, based on the kinetic theory.

12.9 Explain how the molar specific heats of argon and helium can be the same, even though the atomic mass of argon is ten times greater than that of helium.

12.10 At high altitude, the ratio of nitrogen to oxygen in the atmosphere increases above the ratio at sea level. Why?

12.11 Is the total kinetic energy of the molecules of air in a warm room greater than the total kinetic energy of the molecules of air in the same room when it is cool? Why? (*Hint:* The pressure is unchanged.)

12.12 An attempt is made to construct a barometer as indicated in Fig. 12.15. The barometer differs from the ordinary Torricellian tube only in that two liquids are employed as shown. (a) Can such a barometer be made if liquid A is mercury and liquid B is water? (b) Can such a barometer be made if liquid A is water and liquid B is mercury? Explain your reasoning in each case.

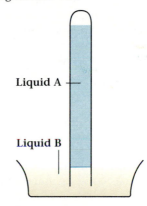

Figure 12.15   Question 12.12.

PROBLEMS

Hints for Solving Problems
Remember to convert temperatures to kelvins. It is often helpful to establish ratios of old and new state variables (P, V, T) when calculating changes in a gas.

Section 12.1   The Pressure of Air

12.1 How high will the liquid stand in a barometer filled with glycerol if the atmospheric pressure is 101 kPa and the temperature is 20°C? (*Hint:* The density of glycerol is 1.26 g/cm³.)

12.2 A mercury barometer is tilted at an angle of 45° from the vertical. How far up the tube is the end of the mercury column, as measured along the tube? Assume that the temperature is 20°C and the atmospheric pressure is 101 kPa.

12.3 Estimate the "total weight" of the atmosphere, using your knowledge of barometers and given the radius of the earth.

12.4 What is the total downward atmospheric force on the top of a 0.21-m × 0.28-m book lying on a table? Why doesn't this force break the table?

12.5 The air-conditioning system of a department store is designed to maintain a pressure of 0.20 in. of water above atmospheric pressure. What is the net force due to pressure on a rectangular display window that is 3.0 m × 4.0 m? (*Hint:* First convert the pressure of 0.20 in. of water to pascals.)

Section 12.2   Boyle's Law

12.6 The cylinder of a bicycle pump has an interior diameter of 2.0 cm and a length of 25 cm. It is used to put air into a tire where the pressure is already 240 kPa. How far down must you push the piston before air begins flowing into the tire?

12.7 Air at a pressure of one atmosphere is confined to a volume V_0. If it is compressed isothermally to a volume of $\frac{1}{3}V_0$, what is the resulting pressure?

12.8 A cylinder of 10.0 cm² cross-sectional area is fitted with a movable piston. Air is introduced into the cylinder at room temperature of 20°C. The temperature is held constant as the volume is compressed to half its initial volume. How much force must be applied to the piston to maintain it in its new position?

12.9 If you can use your abdominal and chest muscles to decrease the volume of your lungs by 20%, what pressure can you develop by this method alone?

12.10 A spherical bubble rises from the bottom of a lake. If the lake temperature is uniform and the bubble doubles its volume by the time it reaches the surface, how deep is the lake?

Section 12.3   The Law of Charles and Gay-Lussac

12.11 A 1.00-L sample of an ideal gas at room temperature (23°C) is taken outside on a cool day (4°C) in a constant-pressure container similar to the one in Fig. 12.6(a). What is the new volume?

12.12 (a) By what fraction of its initial volume does the volume of a gas decrease at constant pressure if the temperature is lowered from 100°C to 0°C? (b) What will be the fractional change if the temperature is lowered from 0°C to −100°C?

12.13 A one-meter-long glass tube is sealed at one end. A drop of mercury large enough to close off the tube is placed at the midpoint when the temperature is 0°C. Where will the mercury be when the closed end of the tube is immersed in boiling water?

12.14 A column of dry air was sealed off from the atmosphere by closing one end of a glass tube and placing a drop of mercury in the tube 0.50 m from the closed end. This was done at some unknown temperature. The tube was then placed into a freezer where the temperature was known to be −10°C. After the tube reached the temperature of the freezer, the mercury slug was found to be 42 cm from the closed end. What was the original temperature?

12.15 We wish to double the volume of a gas held at constant pressure. To what temperature must we heat it if its original temperature was (a) 0°C, (b) 100°C, (c) 1000°C?

Section 12.4   The Ideal Gas Law

12.16 Calculate the volume occupied by 20 g of carbon dioxide gas at a pressure of one atmosphere and a temperature of 25°C. Assume that the ideal gas law is obeyed.

12.17 The temperature of a 1.00-L sample of gas at atmospheric pressure is 37°C. How many moles of gas are in the sample?

12.18 A cylindrical propane tank used for portable cooking stoves has an interior length of 16.5 cm and a diameter of 6.5 cm. The mass of the gas inside is 400 g. Approximately how much pressure is exerted on the walls of the container when the temperature is 75°F? (*Hint:* Propane has a molecular mass of 44 g/mol.)

12.19 An unknown gas has a density of 1.784 kg/m³ at STP. (STP, for standard temperature and pressure, means a temperature of 0°C and a pressure of 101 kPa.) The gas does not burn, support combustion, or react strongly with polished metal surfaces. (a) What is the gram molecular mass? (b) What is the most probable identity of the gas?

12.20 A cylinder closed at both ends has a piston in between that is free to slide. At 20°C, the piston is exactly in the center when hydrogen gas is in one end and helium in the other (Fig. 12.16). (a) What is the ratio of the number of hydrogen molecules to the number of helium molecules? (b) Where will the piston be if the temperature is raised to 57°C?

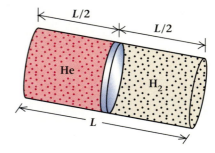

Figure 12.16   Problem 12.20.

12.21 One cubic centimeter of water is converted to steam at atmospheric pressure and 100°C. What is the volume of the resulting steam?

Section 12.5 and 12.6   The Kinetic Theory of Gases and the Kinetic-Theory Definition of Temperature

12.22 Show, using the numbers 1, 3, 7, and 8, that the root-mean-square value and the average value are not the same.

12.23 What is the rms speed of hydrogen molecules at a temperature of 23°C? (*Hint:* The mass of a hydrogen molecule is 3.34×10^{-27} kg.)

12.24 (a) At what temperature is the rms speed of a molecule of hydrogen gas equal to 2200 m/s? (b) If you wished to reduce the rms speed of the molecules in hydrogen gas to 1100 m/s, what temperature would be required?

12.25 What is the density of a gas, at a pressure of one atmosphere, in which the molecules have an rms speed of 450 m/s?

12.26 Hydrogen (H_2) and helium (He) gases are mixed in a container. What is the ratio of the rms velocity of the heavier gas to that of the lighter gas? The mass of a helium molecule is approximately twice the mass of a hydrogen molecule.

12.27 The velocity of sound in a gas is proportional to the rms velocity of the molecules. What is the ratio of the speed of sound at 27°C to that at 0°C?

12.28 Show that the speed of sound in an ideal gas depends on the temperature and molecular mass, but not on the density. (*Hint:* See Problem 12.27.)

12.29 What is the kinetic energy of an oxygen molecule (O_2) at a temperature of 300 K? Assume oxygen to be an ideal gas.

12.30 Find the kinetic energy of a nitrogen molecule at a temperature of 500 K. Assume ideal gas behavior.

12.31 (a) If the root-mean-square velocity of a molecule in a gas at 300 K were 100 m/s, what would be its mass? (b) Is this a realistic example?

12.32 A student plans to build a cubical box to hold a gas at a temperature of 17°C and a pressure of 101 kPa. He wants the number of gas molecules inside to equal the population of the United States (250 million). What length are the edges of the box?

12.33 Calculate the internal energy of one mole of helium gas at a temperature of 300 K. Assume ideal gas behavior.

12.34 Calculate the total internal energy of an ideal gas confined to a volume of 10 L at a pressure of 2.0 atm and a temperature of 300 K.

Section 12.7 The Specific Heats of a Gas

12.35 How many joules of heat are required to raise the temperature of one mole of an ideal gas by 150°C at (a) constant pressure and (b) constant volume?

12.36 How many joules are needed to change the temperature of 0.040 kg of hydrogen from 20°C to 60°C at constant pressure and then from 60°C to 100°C at constant volume? Assume hydrogen to be an ideal gas.

12.37 If 820 J are supplied to 2.5 mol of an ideal gas, by how much will the temperature change if the process is carried out (a) at constant volume, (b) at constant pressure?

12.38 What is the ratio of the rms velocity in one mole of an ideal gas initially at STP after 50 J are added at constant pressure to the rms velocity if the 50 J are added at constant volume?

*Section 12.8 The van der Waals Equation

12.39 (a) What is the ratio of the pressures in a one-cubic-meter volume of a gas at temperatures of 300 K and 400 K if the gas is an ideal gas? (b) What is this ratio if the gas is a van der Waals gas with $a = 0.45$ Pa·m^6/mol^2 and $b = 5.0 \times 10^{-5}$ m^3/mol? Assume that the volume is 1.0 m^3 and $n = 40$.

12.40 (a) Calculate the ratio of pressures in a 0.75-m^3 volume of a gas at temperatures of 300 K and 900 K, assuming that the gas is an ideal gas. (b) What is the ratio if the gas is a van der Waals gas with $a = 0.50$ Pa·m^6·mol^{-2} and $b = 5 \times 10^{-5}$ m^3·mol^{-1}? Assume that the volume is 1 m^3 and $n = 50$.

*Section 12.9 The Barometric Formula and the Distribution of Molecular Speeds

12.41 Plot a graph of the number density versus height for an atmosphere consisting of oxygen (0_2) only at a temperature of 23°C. Assume a unit density ($n = 1$) at $z = 0$. You may use Table A12.2 for the evaluation of the exponential function. Plot your graph from $z = 0$ to $z = 15$ km.

12.42 Calculate the height at which atmospheric pressure is half of the sea-level pressure P_0 for an atmosphere consisting solely of nitrogen (N_2) at a uniform temperature of 300 K.

12.43 Calculate the atmospheric pressure at a height of 5900 m for air with average molecular mass of 29 g/mol and uniform temperature of 290 K, given that the sea level pressure is P_0.

12.44 The velocity of sound in air is given by $v = \sqrt{1.4P/\rho}$, where P is the pressure and ρ the density of the air. What is the ratio of the rms velocity of the air molecules to the velocity of sound?

12.45 Use the barometric formula to estimate the ratio of the atmospheric pressure at Myrtle Beach, South Carolina (sea-level elevation), when the air temperature is 27°C, to the pressure at Denver, Colorado (1610 m above sea level), when the temperature is 10°C. (*Hint:* Use the mass given in Example 12.10.)

12.46 (a) At what height above sea level is the atmospheric pressure half of the pressure at sea level? Assume that the temperature is constant. (b) How high must you go for the pressure to drop to one fourth of the pressure at sea level? (*Hint:* Use the mass given in Example 12.10.)

Additional Problems

12.47 (a) How much force would the teams of horses have had to apply to pull the Magdeburg hemispheres apart? Recall that the diameter of the hemispheres was 55 cm. A rough estimate is that 90% of the air was removed from the sphere by the vacuum pump. (b) Is there anything surprising in this result?

12.48 An upright cylinder, 1.00 m tall and closed at its
• lower end, is fitted with a light piston that is free to slide (Fig. 12.17). Initially the piston is in the center. A cuplike cavity is formed by the top of the piston and the upper cylinder walls. Water is poured into the cavity until it is full. At what fraction of the total height of the cylinder will the piston be when the cavity is full? Assume that the lower portion of the cylinder contains an ideal gas at constant temperature.

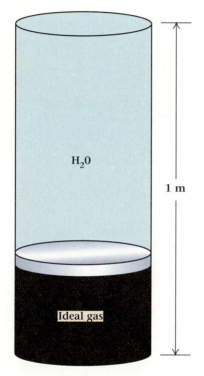

H₂0

1 m

Ideal gas

Figure 12.17   Problem 12.48.

12.49 (a) Approximately how many molecules of air at
•• STP (0°C and 101 kPa) are there in a room measuring 10 ft × 12 ft × 8 ft? (b) How much do the molecules weigh if the average molecular mass for air is 28.8 g/mol? (c) What is their average separation? (d) If they were all compacted so that the

material had the same density as water, what volume would they occupy?

12.50 Nitrogen gas (N_2) is held in a 1.0-L container at
• a pressure of 15 atm and a temperature of 18°C. (a) How many moles of nitrogen are inside the container? (b) What is the total mass of nitrogen in the container?

12.51 A bubble of 1.0 cm diameter is released at the bot-
• tom of a lake that is 30 m deep. The temperature at the bottom is 5°C, and near the surface it is 17°C. What is the diameter of the bubble when it reaches the surface?

12.52 (a) How many molecules are there in a cubic meter
• of argon at a temperature of 23°C and a pressure of one atmosphere? (b) What is the total kinetic energy of these molecules? (c) How fast would someone have to throw a baseball (0.17 kg) in order for it to have the same kinetic energy as the total kinetic energy in part (b)?

12.53 A sample of an ideal gas at 100°C is heated until
• both the pressure and the volume are doubled. Calculate the heat needed to do this by each of the following two-step processes. (a) The gas is first heated at constant pressure until its volume is doubled, then heated at constant volume until the pressure is doubled. (b) The gas is first heated at constant volume, then is heated at constant pressure.

12.54 The maximum value of the Maxwell-Boltzmann
• distribution function occurs for a speed $v_m = \sqrt{2kT/m}$. Find an expression for the magnitude of the distribution function at this speed.

12.55 What is the ratio of the probability that a molecule
• has the most probable speed to the probability that it has a speed of twice the most probable speed? (See Problem 12.54.)

12.56 (a) At what temperature is the rms speed of nitrogen
•• gas equal to the escape velocity from the earth? (b) At what temperature is the rms speed of oxygen gas equal to the escape velocity? (c) How would this result affect the composition of the air in the upper atmosphere, where the temperature is about 1000 K? (*Hint:* The masses of oxygen and nitrogen molecules can be found in Example 12.10.)

12.57 A 1.0-L container of an ideal gas at 300 kPa and
•• 20°C is connected through a small tube to a 10-L container of the same gas at 100 kPa and 25°C. After the gas has come to equilibrium, its temperature is 22°C. What is the pressure of the gas?

Appendix for Chapter 12

The Exponential Function

In many cases in the physical and biological sciences, the rate of change of a variable is proportional to the variable itself. The growth rate of a biological cell, the rate of growth of a population (whether people, plants, or bacteria), the rate of decay of a radioactive material, the rate of cooling of a hot object—all these examples represent rates of change that are proportional to the amount of material present. The function that reflects this characteristic is called the *exponential function*.

Let us examine some of the properties of the exponential function. We call our function, the dependent variable, y and call the independent variable x. The variable x can be displacement, time, temperature, or any other quantity. In Fig. A12.1 we have drawn curve a, whose slope $\Delta y/\Delta x$ increases as x increases. The curve b has a constant slope and curve c has a slope that decreases with increasing x. The relationship between x and y in the exponential function has a graphical form that looks more like curve a than like the other curves in Fig. A12.1. For the moment, we are considering only increasing rates of change.

Now let us look for a specific relationship between x and y that satisfies our requirements. Table A12.1 shows the powers-of-ten notation that we have frequently used earlier in this book. Powers of other numbers can be treated in the same way. The table gives several examples and shows that neither the base numbers nor the powers to which they are raised need to be integers. We display these relationships graphically in Figs.

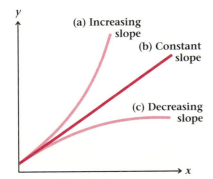

Figure A12.1 Curves with various slopes. The slope increases with increasing x for curve (a). The slope is constant for curve (b). The slope decreases with increasing x for curve (c).

(a) Increasing slope

(b) Constant slope

(c) Decreasing slope

TABLE A12.1 Several base numbers raised to various powers		
$10^0 = 1$	$1.5^0 = 1$	$2^0 = 1$
$10^1 = 10$	$1.5^1 = 1.5$	$2^1 = 2$
$10^2 = 100$	$1.5^2 = 2.3$	$2^2 = 4$
$10^3 = 1000$	$1.5^3 = 3.4$	$2^3 = 8$
$10^4 = 10,000$	$1.5^4 = 5.1$	$2^4 = 16$
$10^{4.5} = 31,622$	$1.5^{4.5} = 6.2$	$2^{4.5} = 22.6$
$10^5 = 100,000$	$1.5^5 = 7.6$	$2^5 = 32$

377

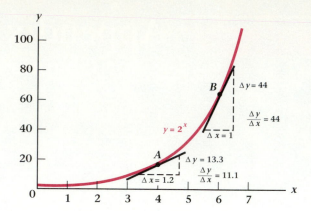

Figure A12.2   A graph of $y = 2^x$. The tangents at A and at B are the slopes of the curve at those points.

A12.2 and A12.3. Notice that the slope increases with increasing x for all of these curves. In Fig. A12.2 we have drawn a tangent to the curve at two points, A and B, and have determined the slope of the curve at these particular values of x. We can plot the slope $\Delta y/\Delta x$ of the curve as a function of x. The gray lines drawn in Fig. A12.3 are the slopes of the respective colored curves.

After studying Fig. A12.3, we can conclude that there is only one curve for which the slope and the original curve are identical. Such a curve corresponds to a base number we call e, whose magnitude is such that, for every point x,

$$\text{slope } (e^x) = \frac{\Delta(e^x)}{\Delta x} = e^x.$$

You should be able to guess from the figure that the number e must be between 2 and 3. In fact, it is an irrational number with a value of approximately 2.718. We see, then, that there is a function of x for which the slope of that function is just equal to the function itself. This function is the exponential function e^x. Table A12.2 gives values of e^x for some values of x.

Figure A12.3   Graphs of the form $y = a^x$ and their slopes. The gray lines give the values of y; the colored lines give the slopes of the gray lines.

In the equation for pressure in the atmosphere (Eq. 12.13), we required only that the slope of the curve be proportional to the curve, not

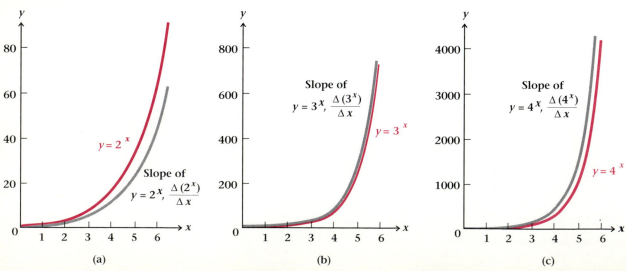

(a)   (b)   (c)

TABLE A12.2 Values of e^x			
x	e^x	x	e^x
0.0	1.000	4.0	54.60
0.2	1.221	4.2	66.69
0.4	1.492	4.4	81.45
0.6	1.822	4.6	99.48
0.8	2.226	4.8	121.5
1.0	2.718	5.0	148.4
1.2	3.320	5.2	181.3
1.4	4.055	5.4	221.4
1.6	4.953	5.6	270.4
1.8	6.050	5.8	330.3
2.0	7.389	6.0	403.4
2.2	9.025	6.2	492.7
2.4	11.02	6.4	601.8
2.6	13.46	6.6	735.1
2.8	16.44	6.8	897.8
3.0	20.09	7.0	1097
3.2	24.53	7.2	1339
3.4	29.96	7.4	1636
3.6	36.60	7.6	1998
3.8	44.70	7.8	2441

Note: $e^{-x} = 1/e^x$.

equal to it. This requirement, too, is satisfied by an exponential function. For example, consider the number e raised to the power ax, where a is a constant. From the above equation we have

$$\text{slope}\,(e^{ax}) = \frac{\Delta(e^{ax})}{\Delta ax} = e^{ax},$$

where we have replaced x with ax. Since a is constant, $\Delta(ax) = a(\Delta x)$ and the equation becomes

$$\frac{\Delta(e^{ax})}{a\,\Delta x} = e^{ax},$$

$$\frac{\Delta(e^{ax})}{\Delta x} = ae^{ax}.$$

The exponential function e^x closely represents many processes observed in nature. It is characterized not only by its rapid increase with increasing x, but also by the increase in the rate of change with increasing x.

To get a feeling for how rapidly the values of $y = e^x$ grow with increasing x, think of graphing the function on a large blackboard, with the axes scaled in centimeters. At $x = 1$ cm, the graph is $y = e^1 \approx 3$ cm above the x axis. At $x = 6$ cm, the graph is $y = e^6 \approx 403$ cm ≈ 4 m high (it is about to go through the ceiling if it hasn't done so already). At $x = 10$ cm, the graph is $e^{10} \approx 22,026$ cm ≈ 220 m high, higher than most buildings. At $x = 24$ cm, the graph is more than halfway to the moon, and at $x = 43$ cm from the origin, the graph is high enough to reach past the nearest star, Proxima Centauri:

$$e^{43}\text{ cm} \approx 4.7 \times 10^{18}\text{ cm} = 4.7 \times 10^{13}\text{ km}$$
$$= 1.57 \times 10^8\text{ light seconds (light}$$
$$\text{travels at 300,000 km/s in a vacuum)}$$
$$= 5.0\text{ light years.}$$

The distance to Proxima Centauri is about 4.3 light years. Yet, with $x = 43$ cm from the origin, the y component of the graph is still less than 2 feet to the right of the y axis.*

If we have a case in which the rate of change in y decreases in proportion to increasing x, then we have a result of the form $y = e^{-x}$ (Fig. A12.4), as is the case for the barometric formula of Eq. (12.15). Such a situation is called exponential decay. We will encounter equations of this form when we study radioactive decay in Chapter 25.

In the previous paragraphs we have discussed the exponential function $y = e^x$. Table A12.2 gives values of y corresponding to a range of values of x. However, in some cases we may already know y but need to know x. We can find x by taking the logarithm of both sides of this equation. In general, if $N = a^b$, a is called the base. The logarithm of N with respect to the base a is the power to which the base must be raised to give N. In practice, both $e = 2.718\ldots$ and 10 are commonly used as bases for

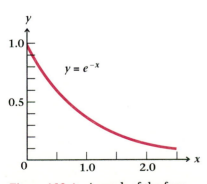

Figure A12.4 A graph of the function $y = e^{-x}$.

*From G. B. Thomas, Jr., and R. L. Finney, *Calculus and Analytic Geometry*, 7th ed. (Reading, Mass.: Addison-Wesley, 1987), p. 422.

logarithms. When the base e is used, we refer to the logarithm as the natural logarithm and use the symbol ln. When base 10 is used, the symbol for the logarithm is log.

When we take the natural logarithm of both sides of the equation $y = e^x$, we get $\ln y = x$. We can find the natural logarithm with a pocket calculator or with the aid of a table. For example, suppose we know that $7.00 = e^x$ and want to find x. In this case, we find $x = \ln 7.00 = 1.95$. Be careful in your computations that you use the natural logarithm (ln) and not the logarithm to the base 10 (log).

The following relationships for logarithms are very useful:

$$\log (xy) = \log x + \log y,$$
$$\log (x/y) = \log x - \log y,$$
$$\log (x^n) = n \log x.$$

Although we have written these relations with log (base 10), they are valid for any base, including base e.

PRACTICE PROBLEMS

A12.1 Plot the curve $y = e^{-0.1x}$ from $x = 0$ to $x = 10$. Determine the slope at $x = 5$ by calculation and by graphical techniques.

A12.2 Plot the curve $y = e^{-0.02x}$ from $x = 0$ to $x = 100$. Determine the slope at $x = 40$ by calculation and by graphical techniques.

A12.3 Plot a graph of $y = e^x$ from $x = -2$ to $x = +2$ in steps of $\Delta x = 0.2$.

A12.4 The intensity of a beam of light traveling in a glass fiber decreases with distance x according to

$$I = I_0 e^{-kx},$$

where I_0 is the intensity at $x = 0$ and k is the absorption coefficient. What is the absorption coefficient if the intensity decreases to $0.50\, I_0$ in a distance of 3.5 km?

A12.5 •• The rate R_0 at which a sample of radioactive material emits radiation is measured to be 1200 particles per minute at time $t = 0$. The rate R at a later time t is given by $R = R_0^{-0.693t/T}$, where T equals 26 min and is called the half-life of the material. What is the rate of the radiation in particles per minute at $t = 2.0$ h?

A12.6 •• In a biology experiment, the number of cells N in a particular population is given by $N = N_0 e^{at}$, where N_0 is the number of cells at time $t = 0$, and $a = 0.30/h$. By what factor will the population increase in 24 h?

A12.7 •• To what power does the base e have to be raised to give a number equal to the population of the United States (250×10^6)?

13

Periodic Motion

13.1 Hooke's Law

13.2 The Simple Harmonic
      Oscillator

13.3 Energy of a Harmonic
      Oscillator

13.4 Period of a Harmonic
      Oscillator

13.5 The Simple Pendulum

13.6 Damped Harmonic Motion

13.7 Forced Harmonic Motion
      and Resonance

A WORD TO THE STUDENT

So far we have studied translational and rotational motion, and we have used them in examining several areas of physics. There is another important kind of motion that we introduce now and use throughout the rest of the book: oscillatory, or vibratory, motion. Such motion repeats itself, like a clock pendulum moving back and forth, so it is also known as periodic motion. The simplest kind of periodic motion is called simple harmonic motion, and you should learn its general characteristics and the conditions that lead to it.

Objects and systems that oscillate are important in themselves, but their study also serves to introduce wave motion (Chapter 14). To describe vibratory motion we will use the laws of mechanics that you have already learned. We will then extend this general description to the specific cases of vibrating springs and the oscillating pendulum. We will end by considering two practical situations: a vibrating system that includes friction, and a vibrating system that moves in response to forces applied at different frequencies. Later, we will find that these two cases are relevant to our study of electronic circuits and atomic structure.

In physics any behavior that repeats itself regularly is called **periodic**. Almost any physical object or system can be made to undergo periodic, or oscillatory, motion. Objects that display such motion abound in everyday life: a child's swing, a guitar string, a bell, or the quartz crystal in an electronic wrist watch. Even the earth itself undergoes oscillatory motion due to seismic activity. Many oscillating systems produce waves. For example, vibrating mechanical bodies generate sound waves, and electrical oscillations generate electromagnetic waves that make radio and television broadcasting possible. In the next chapter we will consider the details of wave motion.

For a mechanical system to be capable of self-sustaining oscillations, it must meet two requirements. The system must have inertia, or mass, and there must also be a force that acts to restore the system to its equilibrium, or lowest-energy, state. A stretched spring or a pendulum pulled to one side exhibits these properties. On the other hand, a light piece of thread held suspended from one end and a kicked pile of sand do not, and they also do not exhibit oscillatory behavior to any great extent.

There are many types of periodic motion, from the complicated, but repeating, pattern of the electrical impulses of a human heart (Fig. 13.1a), to the simple repeating pattern of the end of a vibrating tuning fork (Fig. 13.1b). The latter type of vibratory motion, which is both common and simple in its mathematical description, is called simple harmonic motion. In this motion the displacement varies sinusoidally with time; that is, the graph of the displacement versus time is a sine curve as shown in Fig. 13.1(b). Another example, which we will discuss in detail, is the oscillation of the pendulum of a grandfather clock. In this chapter we restrict ourselves to simple harmonic motion. Even if the motion of a periodic system is more complex, its behavior has features in common with simple harmonic motion. Furthermore, as we will see in Chapter 14, we can analyze complex periodic motion as a combination of a number of simple harmonic motions.

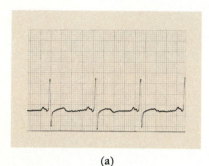

(a)

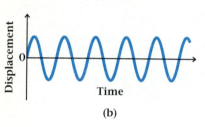

(b)

Figure 13.1 The repeating pattern of an electrocardiogram (a) is more complicated than the smooth, sinusoidal motion of the end of a tuning fork (b). However, both patterns are periodic.

Hooke's Law

13.1

We saw in Chapter 6 that the force required to stretch a spring is, to a good approximation, proportional to the distance by which the spring is extended. This relationship between an applied force and the change in the length of a spring is known as **Hooke's law**, in honor of Robert Hooke,* who first enunciated it in 1678. The elastic behavior of many materials, including some woods, bone, and steel, can be described by Hooke's law if the materials are not stretched too far. Further stretching leads to a nonlinear relationship between force and extension and, finally, to breaking or rupture.

---

*Hooke is also well known for his work in biology. His book *Micrographia*, published in 1665, contained the first drawings of tiny objects seen with the aid of a microscope; in fact Hooke was the one who coined the word *cell* as it is used in biology.

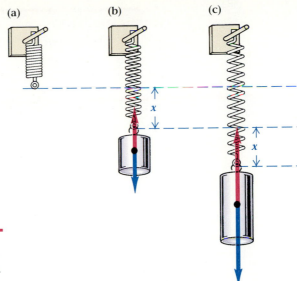

Figure 13.2   The extension of a spring is proportional to the applied force. (a) Spring with no load. (b) Spring extended a distance x by 0.5 N load. (c) Spring extended a distance $2x$ by 1.0 N load.

If we hang different weights on a spring that obeys Hooke's law, the elongation is proportional to the applied force (Fig. 13.2). When the spring reaches equilibrium, the gravitational force acting downward on the mass must be balanced by an upward force due to the spring. This so-called **restoring force** acts in a direction opposite to the direction of the displacement of the end of the spring. Although we have illustrated the case of stretching a spring, the situation of compressing a spring is exactly the same. As the spring is compressed beyond equilibrium, a restoring force acts to resist the compression.

Figure 13.3 shows a block of mass m attached to a horizontal coil spring. The block is at rest, but is free to move along a frictionless surface. We choose this arrangement so that we can avoid the changes in gravitational potential energy that occur in a situation like the one in Fig. 13.2.

What happens when we pull the block a little to the right? Let's analyze this situation mathematically. In Fig. 13.3(a) the block is at rest at the equilibrium position $x = 0$. When the block is displaced a distance x to the right, the spring exerts a restoring force F to the left (Fig. 13.3b). The relation between displacement and restoring force is

$$F = -kx. \tag{13.1}$$

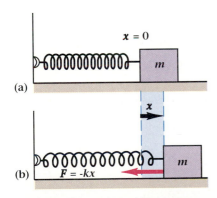

Figure 13.3   (a) A mass m attached to a spring rests on a frictionless surface. (b) When mass m is displaced by an amount **x**, the spring exerts a restoring force **F** in a direction opposite to **x**.

If k is a constant, Eq. (13.1) is the mathematical form of Hooke's law. For many materials, k is constant if the displacement x is not too large. This proportionality constant k is called the **spring constant**. It is also known as the stiffness constant. The negative sign in the Hooke's law equation indicates that the restoring force due to the spring is in the opposite direction from the displacement. Since we are considering only one-dimensional motion of an object at the end of a spring, we do not need to use vector notation in our equation.

TABLE 13.1 Measurements on a spring		
Mass m applied (kg)	Force = mg (N)	Length of spring (cm)
0.00	0.00	16.2
0.050	0.49	21.4
0.100	0.98	26.5
0.150	1.47	31.7
0.200	1.96	36.9
0.250	2.45	42.0
0.300	2.94	47.2
0.350	3.43	52.4

Example 13.1

We would like to use a coil spring as a scale. To do so we must determine whether its extension is proportional to the weight hung from it. Using a setup similar to the one in Fig. 13.2, we measure the length of the spring for different weights hung on one end. The measurements are given in Table 13.1. Determine whether the spring obeys Hooke's law and, if it does, find the spring constant.

Solution   To help visualize the behavior of the spring, we make a graph of spring length versus applied force (Fig. 13.4). Note that the measurements are not the displacement of the spring from equilibrium, but rather its total length. The graph is linear, which means that the spring does obey Hooke's law over the range used.

To determine the spring constant, note that in Eq. (13.1), x is defined as the displacement from equilibrium. We choose the positive direction to be downward from the equilibrium position. The total displacement is the change in the spring's length, 52.4 cm − 16.2 cm = 36.2 cm = 0.362 m. At equilibrium the force applied by the spring is equal in magnitude and opposite in direction to the applied gravitational force; that is, the force in the Hooke's law equation is −3.43 N. The spring constant k is

$$k = -\frac{F}{x} = -\frac{-3.43 \text{ N}}{0.362 \text{ m}} = 9.48 \text{ N/m}.$$

We can find the spring constant in another way. The graph of Fig. 13.4 is a plot of the length of the spring versus the applied force and is represented by the straight-line equation

$$x = \left(\frac{1}{k}\right) F.$$

The slope of this straight line is the reciprocal of the spring constant k; that is,

$$\text{slope} = \frac{\Delta x}{\Delta F} = \frac{1}{k}.$$

First draw the line that best fits the data. Then calculate the slope of that line and take its reciprocal to find the spring constant k.

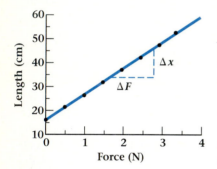

Figure 13.4   **Example 13.1: A graph of spring length versus applied force.**

If the block in Fig 13.3 is displaced an amount x from its equilibrium position, an external force equal and opposite to the restoring force is required to keep it there. If this external force is suddenly released, the unbalanced spring force accelerates the block. The acceleration is found by combining Newton's second law with Hooke's law:

$$a = \frac{F}{m} = -\frac{k}{m}x. \tag{13.2}$$

The acceleration is proportional to the displacement and in the opposite direction. We will see that this same proportionality between acceleration and displacement occurs for children's swings, clock pendulums, and many types of vibrating bodies, as well as for some quantities that describe the behavior of electrical circuits. In the next section we will see that whenever this relationship occurs we get an especially simple type of oscillatory behavior.

Example 13.2

Acceleration of a mass on a spring.

The spring and a block of mass 0.350 kg in Example 13.1 are placed horizontally, as in Fig. 13.3. What is the initial acceleration of the block if the spring is compressed 6.0 cm and released from rest?

Solution   If we choose our coordinate axis so that x increases to the right, compressing the spring corresponds to $x = -6.0/\text{cm}$. The acceleration is then

$$a = \frac{-kx}{m},$$

$$a = \frac{-9.48 \text{ N/m} \times -0.060 \text{ m}}{0.350 \text{ kg}} = 1.6 \text{ m/s}^2.$$

The acceleration is positive (and hence must be in the positive x direction, or to the right), a result that agrees with your intuition about which way the block will initially move.

The Simple Harmonic Oscillator

13.2

We want to find a general description of motion for which the acceleration is proportional to the negative of the displacement, as it is for Hooke's law springs and for many other systems. The motion of any system whose acceleration is proportional to the negative displacement is **simple harmonic motion**. As we will see, the projection (or shadow) of uniform circular motion onto a straight line is an example of simple harmonic motion.

   We already know that circular motion is periodic. When we view, from above, a toy train engine going around a circular track, we see cir-

Shadow of the engine

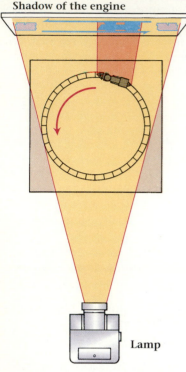

Figure 13.5 A simple arrangement illustrating the connection between uniform circular motion and simple harmonic motion. As the toy train engine moves around the circular track, it casts a shadow on the wall that moves with simple harmonic motion.

cular motion (Fig. 13.5). But if we look at the shadow of the train cast by a lamp edge-on to the track, the engine's shadow appears to oscillate back and forth. We will show this type of motion to be the same simple harmonic motion as the oscillations of a mass on a spring.

We begin by investigating the projection onto the x axis of the uniform circular motion of a point (Fig. 13.6). If P is a point moving with constant speed v_0 in a circle of radius x_0, then the projection, or shadow, of P on the x axis oscillates back and forth between $+x_0$ and $-x_0$. The position of the point along the x axis is

$$x = x_0 \cos \theta.$$

Since $\cos \theta$ is always between -1 and $+1$, we see that x_0 is the maximum displacement of the projection of the point.

As we saw in Chapter 5, any object moving in a circle with constant speed is accelerated radially toward the center. The projection of this acceleration along the x axis is

$$a_x = -a_0 \cos \theta,$$

where a_0 is constant and positive and is the maximum acceleration of the point. The minus sign occurs because the acceleration vector is in the direction opposite to the position vector directed from the origin to the point P.

For uniform circular motion, the angle θ must increase steadily with time; that is, θ is proportional to t. As we saw in Chapters 1 and 5, the time required to make one complete revolution ($\theta = 2\pi$) is the period T. Thus the relationship between angle and time is

$$\theta = 2\pi \frac{t}{T}.$$

The projection along the x axis of the circular motion (the x coordinate

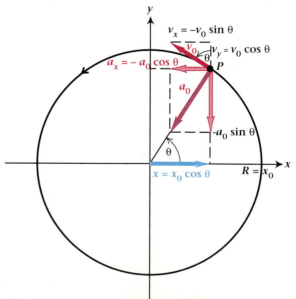

Figure 13.6 A point P moving around a circular path with a constant tangential speed v_0. The projection along the x axis of the radius vector to P is $x_0 \cos \theta$. The projection of the acceleration vector $\mathbf{a}_0$ along the x axis is $-a_0 \cos \theta$.

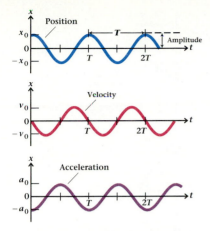

Figure 13.7 (a) The graph of x position as a function of time for the moving point of Fig. 13.6. This is also the graph of displacement as a function of time for the mass at the end of the spring of Fig. 13.3. The time interval between successive maxima is the period T. (b) Velocity along the x direction as a function of time. (c) Acceleration along the x direction as a function of time.

of the point) is therefore

$$x = x_0 \cos 2\pi \frac{t}{T}. \tag{13.3}$$

The acceleration is also a function of time, given by

$$a_x = -a_0 \cos 2\pi \frac{t}{T}. \tag{13.4}$$

Upon solving Eq. (13.3) for $\cos(2\pi t/T)$ and substituting into Eq. (13.4), we have

$$a_x = -(a_0/x_0)x. \tag{13.5}$$

This says that for the oscillating projection of uniform circular motion on the x axis, the acceleration is proportional to the negative of the displacement. This is the same relationship between acceleration and displacement that we had for Hooke's law. Therefore Eqs. (13.3) and (13.4) also describe the position and acceleration of a mass hung from a spring.

The velocity is also a function of time, and from Fig. 13.6 we find that it is proportional to $-\sin \theta$:

$$v_x = -v_0 \sin 2\pi \frac{t}{T}. \tag{13.6}$$

The maximum value x_0 is called the **amplitude** of the displacement about zero. Similarly, v_0 and a_0 are the amplitudes of the velocity and acceleration, respectively.

When the position of a harmonic oscillator is plotted as a function of time, we obtain the graph in Fig. 13.7(a). The time required for the system to go through one complete oscillation is the period T. This is illustrated in the figure as the interval between two successive crests. More generally, the period is the interval between any two successive points along the curve that have the same magnitude and slope. In the earlier example of the toy train engine, the time for the engine to make one complete revolution—the period—is the same as the time for one complete back-and-forth oscillation of the engine's shadow. The amplitude corresponds to the radius of the circle in circular motion, in this case the radius of the track.

If a block attached to a spring that obeys Hooke's law is displaced by an initial amount x_0 and released from rest, its motion is described by the curves in Fig. 13.7. We release it at time $t = 0$ with no initial velocity, but with an initial displacement and an initial acceleration opposite in direction to the displacement (Fig. 13.8). The block moves toward the equilibrium position $x = 0$, gaining speed as it moves. As it reaches the position of zero displacement, its momentum keeps it moving, even though the restoring force is zero at that point. The block is then displaced in the other direction. A restoring force proportional to this displacement

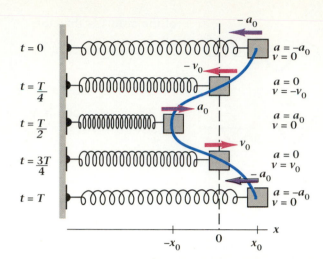

Figure 13.8 Relationship between velocity and acceleration for a mass moving with simple harmonic motion at the end of a spring. The velocity is zero when the acceleration is maximum, and vice versa.

gradually slows the block to a stop and then accelerates it in the opposite direction, back toward the initial position. In this way the block oscillates periodically with time. If there were no frictional forces to slow (damp) the motion, the block would oscillate indefinitely. We call this system a **simple harmonic oscillator** and its motion is sinusoidal.

In the previous paragraphs we described the motion by $x = x_0 \cos 2\pi t/T$ and showed this motion in a number of graphs. In all of these cases we assumed that the displacement was x_0 at the time we called $t = 0$. However, the important general property of simple harmonic motion is not the displacement at time $t = 0$, but the shape of the curves that describe displacement, velocity, and acceleration. Notice that all of the following equations have the same behavior as a function of time but differ in magnitude at the time we choose to call $t = 0$ (Fig. 13.9):

$$x = x_0 \cos(2\pi t/T + \phi),$$

$$x = x_0 \cos(2\pi t/T) \qquad\qquad (\phi = 0),$$

$$x = x_0 \cos(2\pi t/T + \pi/2) = x_0 \sin(2\pi t/T) \qquad (\phi = \pi/2).$$

The angle ϕ is called the **phase angle** and can be selected so that our choice of starting time and initial displacement are in agreement.

From the preceding discussion it seems reasonable that a block on a spring oscillates in a sinusoidal manner. However, we want to know if any other type of motion is possible when the acceleration is proportional to the negative of the displacement. The answer is no, for the following

Figure 13.9 Displacement curves for a harmonic oscillator corresponding to different values of the phase angle ϕ: curve A, $\phi = 0$; curve B, $\phi = \phi$; and curve C, $\phi = \pi/2$. The amplitude $x_0 = 1$.

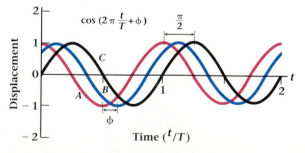

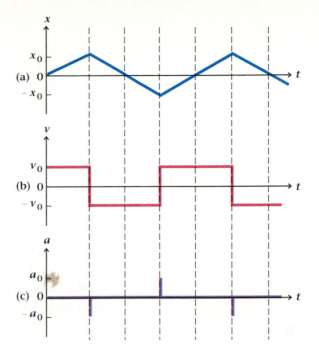

Figure 13.10   (a) A triangular displacement curve. (b) Velocity curve representing the slope of the displacement curve, $v = \Delta x/\Delta t$ as Δt approaches 0. (c) The acceleration curve $a = \Delta v/\Delta t$. Clearly the acceleration is not proportional to the displacement.

reason. The velocity is the time rate of change of the displacement, and the acceleration is the time rate of change of the velocity. That is, the acceleration is given by the slope of the velocity curve, which in turn is given by the slope of the displacement curve. Curves of oscillations other than sinusoidal all give the result that the acceleration is not proportional to the displacement. We show this result explicitly for the triangular waveform of Fig. 13.10. The acceleration calculated from this displacement curve bears no resemblance to the initial curve. However, the acceleration curve of Fig. 13.7(c) is just the negative of the displacement curve of Fig. 13.7(a). Sinusoidal motion is indeed special.

Energy of a Harmonic Oscillator

13.3

Work must be done to stretch a spring. As we saw in Chapter 6, the increase in the potential energy of a Hooke's law spring that is due to extending the spring from $x = 0$ to $x = x_0$ is

$$PE = \tfrac{1}{2}kx_0^2. \tag{13.7}$$

If we release the spring from rest at x_0, its potential energy changes to kinetic energy as it moves toward the equilibrium position. As the end of the spring passes through $x = 0$, all its energy is kinetic and the magnitude of its velocity is a maximum. By applying the principle of energy conservation, we can find the relation between the maximum displacement x_0 and the maximum velocity v_0:*

$$\tfrac{1}{2}mv_0^2 = \tfrac{1}{2}kx_0^2,$$

or

$$v_0 = x_0\sqrt{\frac{k}{m}}.$$

The total energy is the sum of the kinetic and potential energies at any time and is a constant if we neglect dissipative forces in the spring. We can write the total energy as

$$E = \tfrac{1}{2}mv^2 + \tfrac{1}{2}kx^2,$$

where the instantaneous values of displacement and velocity are given by Eqs. (13.3) and (13.5).

Example 13.3

Calculating the speed of a harmonic oscillator.

A 3.0-kg ball is attached to a spring of negligible mass and with a spring constant $k = 40$ N/m. The ball is displaced 0.10 m from equilibrium and released from rest. What is the maximum speed of the ball as it undergoes simple harmonic motion?

Solution We can compute the maximum speed from the energy of the system. The maximum speed occurs at $x = 0$, when the kinetic energy is maximum and is equal to the initial potential energy. Then

$$\tfrac{1}{2}mv_0^2 = \tfrac{1}{2}kx_0^2.$$

So

$$v_0 = x_0\sqrt{\frac{k}{m}} = (0.10 \text{ m})\sqrt{\frac{40 \text{ N/m}}{3.0 \text{ kg}}} = 0.37 \text{ m/s}.$$

Period of a Harmonic Oscillator

13.4

We have seen that the velocity of a harmonic oscillator depends on the spring constant and on the mass. Therefore the period, or the time required to complete one cycle of the motion, also depends on them. We can obtain the exact relationship by analyzing the circular motion of the particle in Fig. 13.6. The particle moves with constant speed v_0. In one period T it traverses a circular path of length $2\pi x_0$; that is,

$$v_0 T = 2\pi x_0.$$

*Note that the maximum velocity and maximum displacement do not occur at the same time. The velocity reaches a maximum v_0 when $x = 0$. Similarly, the displacement reaches a maximum x_0 when the velocity is zero.

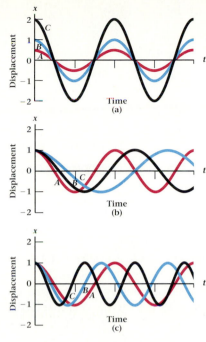

Figure 13.11 Displacement plotted against time for a simple harmonic oscillator. (a) The values of k and m are held constant while x_0 increases from curve A to B to C. (b) The values of k and x_0 are held constant while m increases from curve A to B to C. (c) The values of x_0 and m are held constant while k increases from curve A to B to C.

The period is then

$$T = \frac{2\pi x_0}{v_0}.$$

We have already seen that $v_0 = x_0 \sqrt{k/m}$, so the period may be expressed as

$$T = 2\pi \sqrt{\frac{m}{k}}. \tag{13.8}$$

To show that this equation is consistent with our intuition, we have plotted the motion of several spring systems, each of which contains the same oscillator plus two others for which one of the parameters has been changed. Figure 13.11(a) shows the effect of displacing the block by different initial amounts. The amplitude of the oscillations is larger for larger initial displacements, but because m and k are unchanged, the period is the same. The period of the system is completely determined by the mass m and the spring constant k; how far it moves is influenced by what you do to it. In Fig. 13.11(b) we have used the same spring constant and the same initial displacement from the equilibrium position, but we have increased the mass for succeeding curves. As you would expect, the more massive systems are slower to respond and have longer periods. Figure 13.11(c) corresponds to springs of increasing k, or increasing stiffness. As expected, the stiffer the spring, the shorter the period. These curves emphasize that to change the period, you have to change the physical characteristics of the system, that is, m or k. You cannot vary the period by simply pulling the mass down farther before you release it. The period of a freely oscillating system depends on the properties of the system and is independent of the way in which the oscillation is initiated. This period is called the **natural period** of the oscillator.

As we saw in Chapter 5, the reciprocal of the period is the frequency f, the number of complete cycles per unit time:

$$f = 1/T. \tag{13.9}$$

Recall that the dimension of frequency is inverse time and its unit is the hertz (Hz). When convenient, we can express the displacement, velocity, and acceleration of an oscillator in terms of the frequency rather than the period. The frequency associated with the natural period is the **natural frequency** of an oscillator. It is the frequency of an oscillator that has been set into motion and then left to oscillate freely. Examples include masses on springs, pendulums, and guitar strings. We will see in Section 13.7 that the natural frequency is the frequency at which energy is most easily fed into an oscillating system.

Example 13.4

Period and frequency of oscillation.

A block of 2.0 kg oscillates on a spring of constant $k = 20$ N/m. What is the period of one complete oscillation? What is the natural frequency of the oscillations?

Solution   The period of oscillation is given by Eq. (13.8):

$$T = 2\pi\sqrt{\frac{m}{k}} = 2\pi\sqrt{\frac{2.0\text{ kg}}{20\text{ N/m}}} = \frac{2\pi}{\sqrt{10}}\text{ s} = 2.0\text{ s.}$$

The frequency is the reciprocal of the period:

$$f = \frac{1}{T} = \frac{1}{2.0\text{ s}} = 0.50\text{ Hz.}$$

The Simple Pendulum

13.5

A simple pendulum behaves as a simple harmonic oscillator when the amplitude of its motion is small. This is approximately the case in a grandfather clock, for example. Let us examine the pendulum and find out what determines its period.

A **simple pendulum** consists of a mass suspended by a light string of constant length L attached to a rigid support. When the mass is displaced to the side through an angle θ with the vertical (Fig. 13.12), the gravita-

PHYSICS IN PRACTICE

Walking and Running

To get some idea of the fundamental mechanical principles of walking and running let us approximate the leg by a long thin rod of uniform cross section. We could more closely approximate a real leg with a more complicated model, but the nature of our conclusions would not change.*

Figure B13.1 shows a rod of length L supported at its upper end (point O) and free to swing. The period of a freely swinging rod supported at its upper end is

$$T = 2\pi\sqrt{\frac{2L}{3g}}.$$

This expression depends on the length of the rod and the acceleration of grav-

*This discussion of walking and running is adapted by permission of the author from P. Davidovits, *Physics in Biology and Medicine*, Englewood Cliffs, N. J.: Prentice-Hall, Inc., 1975.

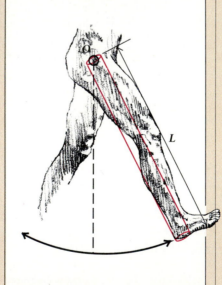

Figure B13.1   We can approximate the shape of a freely swinging leg by a uniform rod of length L that pivots about its upper end.

ity in the same way as does the period of a simple pendulum. The factor of 2/3 is a result of the mass being distributed uniformly along the rod, rather than being concentrated at its lower end.

This expression gives an approximate value for the period of a freely swinging leg. You may wish to satisfy yourself that it is a reasonable approximation by doing the following simple experiment. Stand up and swing your leg back and forth as freely as you can. Do not use muscular effort to hold it back or to swing it rapidly. After you feel you can swing your leg freely, count the number of complete swings in ten seconds and calculate the period. Compare the period you observe with the prediction from the equation. To determine the length L, measure your leg length from the hip socket.

We can use this model to estimate a person's natural gait. Let us assume that a natural gait is the one involving

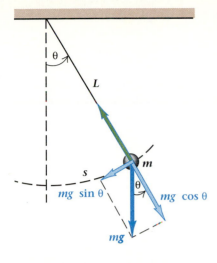

Figure 13.12 A simple pendulum consisting of a mass *m* at the end of a string of length *L*. The magnitude of the restoring force is *mg* sin *θ*.

tional force **mg** has a component along the direction of the string and a component perpendicular to the string. This perpendicular component provides the restoring force:

$$\text{restoring force} = -mg\sin\theta.$$

The displacement *s* is the length along the arc through which the pendulum swings, given by

$$s = L\theta.$$

When the angle *θ* is sufficiently small, sin *θ* may be approximated by the angle *θ* in radians:*

$$\sin\theta \approx \theta, \qquad \text{for small } \theta.$$

---

*How much error do we introduce by replacing the sine of an angle with the angle in radians for an angle of 10°? The sine of 10° is 0.1736. The angle measured in radians is 0.1745, just slightly larger. The relative error is

$$\frac{\sin\theta - \theta}{\sin\theta} = \frac{-0.0009}{0.1736} \approx -5 \times 10^{-3}.$$

That is, the error is about one half of one percent for *θ* = 10°. For smaller angles, the relative error is even less.

---

the least muscular effort—the gait with the period found above. To a first approximation, we assume that the length of a stride is proportional to the length of the leg. The time for a single stride is one half of the period given above. The walking speed *v* thus depends on the leg length:

$$v_{\text{walk}} \propto \frac{L}{T/2} \propto \sqrt{L}.$$

This equation predicts that people with longer legs have a more rapid natural walking gait. The prediction is made on the basis of a model that assumes minimum energy expenditure and utilizes an oversimplified description of the leg. However, the prediction is borne out by common experience.

When a person runs, an important change must be made in our model. During running, the leg does not swing freely but is subject to a torque about *O*. The torque is a result of the force *F* applied by the muscles. This force is roughly proportional to the cross-sectional area of the muscles involved. If we assume that, for people of different size, the relative proportions of the leg are the same, then the cross-sectional area, and therefore the force *F*, depends on the square of the length *L*. The torque is then proportional to the product of *F* with *L*:

$$\tau \propto FL \propto L^2 \cdot L \propto L^3.$$

The moment of inertia *I* is proportional to the mass times the square of the length (Chapter 6). Again we assume that all legs have essentially the same proportions; that is, width and thickness are proportional to length. Then the mass varies as the cube of the length:

$$I \propto mL^2 \propto L^5.$$

It can be shown that the period *T* of a long rod oscillating about one end and subject to a torque depends on the maximum torque and moment of inertia:

$$T \propto \sqrt{\frac{I}{\tau}}.$$

Upon substituting for *I* and *τ*, we find

$$T \propto \sqrt{\frac{L^5}{L^3}} \propto L.$$

The speed of running is proportional to the frequency of taking steps times the length of a step:

$$v_{\text{run}} \propto fL \propto \frac{L}{T} \propto \frac{L}{L} = 1.$$

So we have the prediction that, for animals with similarly shaped legs, the speed of running does not depend on the leg length. This prediction is not, of course, strictly true. However, the model does offer an explanation for the observation that the ordinary walking rate of people with long legs is greater than that of people with short legs, whereas the rate at which they can run is not always appreciably greater.

Then the restoring force is proportional to the displacement and we have the condition of a harmonic oscillator:

$$F = -mg\theta = -(mg/L)s. \tag{13.10}$$

This equation has the same form as the spring equation (13.1) except that here the spring constant k is replaced by mg/L. The period is given by Eq. (13.8) with this substitution made for k. Thus

$$T = 2\pi\sqrt{\frac{m}{k}} = 2\pi\sqrt{\frac{m}{mg/L}},$$

$$T = 2\pi\sqrt{\frac{L}{g}}. \tag{13.11}$$

This is an interesting result. The period of the simple pendulum is independent of the mass of the pendulum bob. The period is also independent of the amplitude of the motion (provided it is relatively small). However, the period of the pendulum does depend on its length.

The constancy of the period of a pendulum was discovered in the sixteenth century by Galileo, who is said to have observed the swinging of the chandeliers in the Cathedral of Pisa. Using his pulse as a reference timer, he noticed that the chandeliers' periods of swing were independent of their amplitudes. This observation led to the invention by Christiaan Huygens of the pendulum clock, which became the standard timekeeper for nearly 300 years. Pendulum clocks have a weight near the end of the pendulum which is used to adjust the period. Lowering the weight increases the effective length of the pendulum to make the period longer.

Notice that when the period and length of a pendulum are well known, Eq. (13.11) provides a method for measuring the gravitational acceleration g. It is also appropriate to note that this simple result for the pendulum is valid only for small-amplitude oscillations. At larger amplitudes the period is no longer independent of the amplitude and the motion is no longer simple harmonic, though it is still periodic.

Example 13.5

The seconds pendulum.

A grandfather clock has a pendulum 1.0 m long. What is its period?

Solution   The period is given by Eq. (13.11):

$$T = 2\pi\sqrt{\frac{L}{g}} = 2\pi\sqrt{\frac{1.0\text{ m}}{9.8\text{ m/s}^2}} = 2.0\text{ s}.$$

This length pendulum is frequently used in grandfather clocks. It is sometimes called a "seconds pendulum" because it makes a swing from one side to the other in one second, making the "tick" and the "tock" one second apart.

Damped Harmonic Motion

13.6

Up to now, we have not considered the effects of friction on harmonic oscillators. Ignoring friction simplified the introduction to the topic and, as we have seen in both examples and problems, is justified because our results closely describe what actually happens when the frictional forces are weak or when we do not consider many oscillations. Yet friction is always present and should be included in order to describe those cases where friction plays an important role. To be more complete, both Eq. (13.1) for the spring and Eq. (13.10) for the simple pendulum should have an additional term that describes the frictional force. This frictional force for an oscillator is often called **damping.**

Figure 13.13(a) shows an oscillator with damping provided by the resistance of the water in the jar to the motion of the vanes on the light rod attached to the bottom of the mass m. According to the principle of energy conservation, the amplitude of the oscillations of a damped system decreases as time goes by, unless energy is constantly supplied to replace the energy lost by friction. Thus a spring oscillator that receives an initial displacement and is then left alone will gradually run down. For the particular system shown in Fig. 13.13, observations show that the damping is approximately proportional to the velocity. Including such a term in Eq. (13.1) leads to the result that if the mass is lifted up a distance y_0 and released, the displacement as a function of time is approximately

$$y = y_0 e^{-\gamma t} \cos(2\pi f t), \tag{13.12}$$

where γ depends on the frictional force and the mass. If the damping is not too large, f is approximately the same as the natural frequency of the

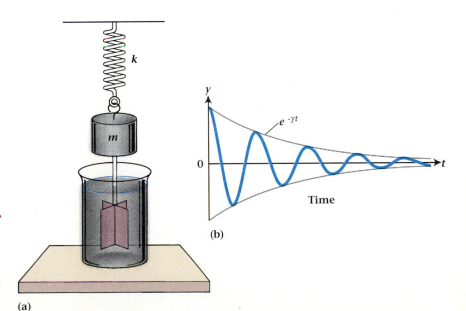

Figure 13.13 (a) A simple harmonic oscillator with damping provided by viscous (frictional) forces. (b) Displacement of an underdamped harmonic oscillator as a function of time. The amplitude of the motion decreases exponentially with time.

(a)

(b)

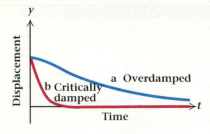

Figure 13.14 The displacement of (a) an overdamped harmonic oscillator and (b) a critically damped oscillator as a function of time.

system.* The motion is shown by the solid curve in Fig. 13.13(b). We can look on this behavior as an oscillation at almost the natural frequency of the system but with an amplitude that decreases exponentially with time. This is shown by the dashed lines in Fig. 13.13(b), which graphs the factor $e^{-\gamma t}$.

Though not all harmonic oscillators have damping that is proportional to the velocity, Eq. (13.12) closely represents what happens in many real physical situations. As an extreme example, if the friction is constant but independent of velocity, a graph of the motion will be similar to Fig. 13.13(b), but the curves that form the upper and lower boundaries of the oscillatory motion will be replaced by straight lines.

If the harmonic oscillator is a pendulum, rather than a spring, Eq. (13.12) still describes the motion, but the linear displacement x is now replaced by the angular displacement of the pendulum. The same basic mathematical form holds for other damped oscillatory systems, whether they are mechanical, electrical, or acoustical systems.

Equation (13.12) and Fig. 13.13 describe the case of relatively weak damping and correspond to what is called *underdamped* motion. If the frictional force is too strong, the oscillator will not move at all. For damping that is strong, but yet allows the system to move, the motion is called *overdamped,* shown by curve (a) in Fig. 13.14. A particularly important case lies between the underdamped and overdamped cases. This case is called *critical damping* and corresponds to the case in which the damping coefficient γ is equal to the natural frequency of the system. In this case a damped spring that is displaced and released returns as quickly as possible to the equilibrium position without overshooting (Fig. 13.14b). If we apply a constant force to a critically damped system, it moves to a new equilibrium position in the minimum time with no overshoot. The design of pointers on meters, shock absorbers, hydraulic and pneumatic door closers, and other practical devices is based on the proper design of damped harmonic oscillators.

Example 13.6

A swing with 2.50-m ropes is pulled aside and let go. After 20 oscillations the maximum amplitude is one half of its original value. Assume that the swing behaves like the damped harmonic oscillator described in Eq. (13.12). (a) What is the value of γ? (b) Approximately what amplitude, relative to its original amplitude, will the swing have after 35 oscillations?

Solution   (a) We assume that the swing oscillates at its natural frequency, so that the period T is

$$T = 2\pi\sqrt{\frac{L}{g}} = 2\pi\sqrt{\frac{2.50 \text{ m}}{9.80 \text{ m/s}^2}} = 3.173 \text{ s}.$$

Twenty oscillations will then take

$$20T = 63.47 \text{ s}.$$

---

*The frequency f is related to the natural frequency f_0 by $f^2 = f_0^2 - (\gamma/2\pi)^2$. The complete expression for the displacement under the conditions just described is $y = y_0 e^{-\gamma t}[\cos 2\pi ft + (\gamma/\omega) \sin 2\pi ft]$. For lightly damped systems this last term is small.

Because the amplitude decreases exponentially to 0.5 of the original amplitude in time $t = 20T$, we can write

$$0.5 = e^{-\gamma t} = e^{-\gamma 63.47}.$$

We may now determine γ by taking the natural logarithm of both sides of the equation. You can find the logarithm of 0.5 with a calculator:

$$\ln(0.5) = -0.6931.$$

The logarithm of the exponential is just the value of the exponent:

$$\ln(e^{-\gamma 63.47}) = -\gamma 63.47.$$

Thus

$$0.6931 = \gamma 63.47$$

and

$$\gamma = 0.0109/s.$$

(b) Thirty-five oscillations will take

$$35T = 111.1 \text{ s},$$

so the amplitude, as a fraction f of the original amplitude, is

$$f = e^{-\gamma t} = e^{-(0.0109)(111.1)} = 0.298.$$

After 35 oscillations the swing will have about one third of its original amplitude.

Forced Harmonic Motion and Resonance

13.7

Before continuing, do the following simple experiment. Make an oscillator by joining several rubber bands together and hanging an unopened soft drink can from the end (Fig. 13.15). You will find it easy to attach the can to the rubber band by looping the bottom rubber band under the pull-up tab. You may try some other combination of elastic elements and mass; the idea is to make an oscillator with a natural period of about one-half to one second. The components are not critical, but you should make a system with a period long enough so that you can easily observe what happens. Get a rough idea of the natural frequency of the system by looping one end of the rubber band around the tip of a finger, pulling the can down, and observing what happens when you release it.

Next, suspend the system from your finger and move your finger up and down slowly. You will find that the can moves up and down in phase with your finger, and with about the same amplitude. Increase the rate of your finger's motion and observe the change that occurs as the frequency of your finger comes closer to the natural frequency of the system. The amplitude of the can's motion increases greatly and, at the natural frequency of the system, a very small motion of your finger makes the can move up and down with a large amplitude. At this point energy is being

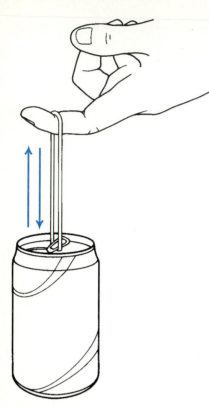

Figure 13.15 A soft drink can suspended from rubber bands can be used to demonstrate resonance.

fed into the system at the natural frequency of the system, a condition called **resonance.** The natural frequency is also known as the **resonant frequency.** As we will see in the next chapter, natural resonant systems usually have more than one resonant frequency.

As you continue to increase the frequency with which you move your finger up and down, you will find that the amplitude of the can's motion decreases. Eventually you will reach a frequency at which the can stays relatively motionless, even though you move your finger through quite a large distance.

What you have done is to observe the response of a lightly damped harmonic oscillator (the rubber bands and can) to an applied force as a function of the frequency of that force. Actually, rubber bands do not obey Hooke's law strictly, but the behavior of the system is close enough to a harmonic oscillator for us to study the general principles.

A graph of the amplitude of a forced damped harmonic oscillator as a function of the forcing frequency shows the behavior we have just described (Fig. 13.16). Two curves have been plotted, corresponding to different amounts of damping. As you would expect, large damping in the system causes the maximum amplitude at resonance to be lower than otherwise.

The idea of resonance is extremely important for mechanical and electrical systems because energy is most effectively transferred when it is supplied at the natural frequency of the system. When you push a swing to make it go higher, you push at its natural frequency. When you tune a radio or TV set, you are adjusting the natural, or resonant, frequency of the electrical circuit to match the frequency of the radio or TV station you want to receive. We will make additional use of the idea of resonance in Chapter 14 on waves.

Resonance effects can also have undesirable or even destructive effects. The rattle in a car's body or an annoying buzz in a stereo speaker is often due to resonance. Most people have heard that a powerful singer can shatter a glass by singing at the right frequency. Equally famous is the warning that a group of people should not march in step across a bridge, lest the frequency of the footsteps match some natural frequency of the bridge. These are all examples of resonance.

Frequently we need a system that does not transfer energy efficiently. An example is a mechanism for isolating sensitive apparatus from vibrations. A common solution to a problem of vibration is to fasten the source of vibration on an elastic mounting to cushion and absorb the shock. What may not be obvious is that the selection of the wrong elastic mounting can make the problem worse. Isolation is accomplished by decreasing the natural frequency of the system relative to the frequency of the vibration source. The reason this technique works can be seen from Fig. 13.16, and from recalling the experiment with the rubber band and the can. The least energy is transferred when the frequency of the driving force is much higher than the natural frequency of the system.

We have based our discussion of resonance, and the response of a system to forcing, entirely on the behavior of a mass attached to a spring that obeys Hooke's law. However, the same general principles and results apply to other oscillating systems, whether they be mechanical, electrical, or otherwise.

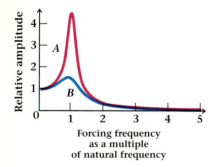

Figure 13.16 The response of a damped harmonic oscillator plotted as a function of the frequency of the driving force. Curve *B* corresponds to greater damping than curve *A*.

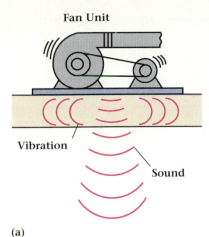

Fan Unit

Vibration

Sound

(a)

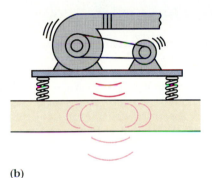

(b)

Figure 13.17 (a) A fan bolted directly to the roof generates vibrations throughout the building. (b) Shock-mounting the fan on springs reduces the vibrations transmitted to the building, provided we choose the proper spring constants.

Example 13.7

Minimizing the transmission of vibrations.

The fan unit for a building's heating and air-conditioning system is rigidly mounted on the roof and runs continuously (Fig. 13.17a). The vibrations are transmitted throughout the structure of the building and generate unacceptable vibration levels. To reduce the vibration felt below, the system is to be attached to a spring-mounted slab. The fan shaft turns at 1800 rpm (revolutions per minute) and the combined mass of the unit and the mounting slab (Fig. 13.17b) is 576 kg. What is the proper stiffness constant for the springs used to support the slab? Assume four springs, one at each corner.

Solution The oscillation system in this case consists of the motor, the fan, the mounting platform, and the springs. One rule of thumb sometimes used is that the driving, or disturbing, frequency should be at least three times the natural frequency of the system. For many cases a factor of five is adequate and for critical conditions a factor of twelve or more is appropriate. We can achieve these factors by lowering the natural frequency of the system. If we choose a one-to-five ratio, which corresponds to a reduction in the force of the vibrations to the building of about 96%, the desired natural frequency of the system is

$$\frac{1}{5} \times 1800 \text{ rpm} \times \frac{1}{60 \text{ s/min}} = 6.0 \text{ Hz}.$$

The proper springs can be selected by using

$$f = \frac{1}{T} = \frac{1}{2\pi}\sqrt{\frac{k}{m}}.$$

Solving for the spring constant k gives us

$$k = m(2\pi f)^2 = 576 \text{ kg}(2\pi 6.0/\text{s})^2 = 8.19 \times 10^5 \text{ N/m}.$$

This would be the largest desirable spring constant if all of the mass were supported by one spring. Since there are four springs, one at each corner of the mounting slab, then each one of these four springs will have a spring constant, or stiffness, of $\frac{1}{4} \times 8.19 \times 10^5 \text{ N/m} = 2.05 \times 10^5 \text{ N/m}$.

SUMMARY

Useful Concepts.

- Many phenomena can be described by an equation such as Hooke's law. These phenomena are examples of simple harmonic motion and occur in electrical as well as mechanical systems.
- Hooke's law is

$$F = -kx.$$

- For simple harmonic motion

$$x = x_0 \cos 2\pi t/T,$$

$$v = -v_0 \sin 2\pi t/T = -\sqrt{\frac{k}{m}} x_0 \sin 2\pi t/T,$$

$$a = -a_0 \cos 2\pi t/T = -\frac{k}{m} x_0 \cos 2\pi t/T.$$

- The frequency is the reciprocal of the period,

$$f = \frac{1}{T}.$$

- The potential energy of a stretched spring is

$$PE = \tfrac{1}{2}kx_0^2.$$

- The period of a mass oscillating on a spring is

$$T = 2\pi\sqrt{\frac{m}{k}}.$$

- The period of a simple pendulum is

$$T = 2\pi\sqrt{\frac{L}{g}}.$$

- The displacement of a damped harmonic oscillator is the product of an exponentially decreasing term and an oscillating term:

$$y = y_0 e^{-\gamma t}\cos(2\pi ft).$$

- Energy is transferred most effectively at the natural frequency of the system, a condition known as resonance.

Important Terms

You should be able to write the definition or meaning of each of the following terms:

- periodic
- Hooke's law
- restoring force
- spring constant
- simple harmonic motion
- amplitude
- simple harmonic oscillator
- phase angle
- natural period
- natural frequency
- simple pendulum
- damping
- resonance
- resonant frequency

QUESTIONS

13.1 A simple pendulum suspended from the ceiling of an elevator has a period T_0 when the elevator is not moving. What is the effect on the period when the elevator is (a) accelerated upward, (b) accelerated downward, (c) moving with a constant speed?

13.2 A glider on a horizontal air track is connected to the end of the track by a spring and placed on an elevator. What is the effect on the period of its oscillations when the elevator is (a) accelerated upward, (b) accelerated downward, (c) moving with a constant speed?

13.3 A hollow plastic sphere is held well below the surface of a swimming pool by a spring attached to the bottom. Do the vertical oscillations of the sphere show damped simple harmonic motion?

13.4 The pendulum on the clock that strikes the hours on the bell "Big Ben" in London was particularly noted for its accuracy. Formerly the period was regulated by placing a penny on top of the bob of the 3.98-m long pendulum. Will placing a penny there make the clock run faster or slower? Explain.

13.5 Consider the motion of a meter stick swinging from one end and that of a one-meter-long simple pendulum of the same mass. Are their periods the same? Give a physical reason for your answer.

13.6 You can consider an automobile suspension to be a forced damped harmonic oscillator, consisting of the springs and shock absorbers, which is forced by the wheels' up-and-down motion over bumps and ruts. What are some of the important design criteria? Should there be much or little damping?

13.7 Determine the effective spring constant of a bathroom scale by estimating the amount the scale compresses when you stand on it.

13.8 If the outer case of an air conditioner rattles when the unit runs, what can you do that will probably make it stop?

13.9 Why is it that sometimes you hear rattles when you drive a car at one speed, but they go away when you drive faster or slower?

13.10 A lump of clay is dropped onto an upright spring. What determines the maximum compression? (*Hint:* Consider conservation of energy.)

PROBLEMS

Hints for Solving Problems

Simple harmonic motion occurs when the acceleration is proportional to the negative of the displacement. In simple harmonic motion, maximum acceleration occurs at maximum displacement and zero velocity. The period of a simple harmonic oscillator made with a spring depends only on the mass and the spring constant or, for a simple pendulum, on the pendulum's length and the acceleration of gravity.

Section 13.1   Hooke's Law

13.1 A 10.0-N bag of fruit extends a grocer's spring scale 10.0 cm. What is the spring constant of the spring?

13.2 A coil spring has a spring constant of 45 N/m. If the full length of the spring is 35 cm when a 1.0-kg mass is hung from it, what is the equilibrium length of the spring when the 1.0-kg mass is removed?

13.3 A fisherman's spring scale is extended to a total length of 0.18 m when a 6.12-kg fish is suspended from it. If the spring constant is 1000 N/m, what is the total length of the spring when an 11.4-kg fish is suspended from it?

13.4 A coil spring is extended by 2.50 cm when it supports a 1.00-N weight. If an identical spring is joined to the end of the first one, what is the extension of the combined spring when it supports the same 1.00-N weight? Assume the spring masses to be negligible.

13.5 Two identical coil springs are mounted side by side so that they jointly support a weight hanger of 2.00 N. When an additional 4.00 N is added to the hanger it is displaced downward by 3.00 cm. What is the force constant of the individual springs?

13.6 Determine whether the following two sets of data were taken from springs that obey Hooke's law. In each case objects of increasing mass were hung from a suspended spring and the length of the spring measured. If Hooke's law is obeyed, determine the spring constant. If not, decide whether there is any part of the range over which Hooke's law is obeyed, and determine the spring constant for that part.

Mass (kg)	Length of spring A (cm)	Length of spring B (cm)
0.0	15.7	8.4
1.0	16.5	15.6
2.0	17.8	20.5
3.0	19.3	21.4
4.0	20.4	21.6
5.0	21.3	22.1
6.0	22.8	22.5
7.0	24.1	22.6
8.0	25.0	23.1
9.0	26.6	25.8
10.0	27.6	28.2

13.7 Two springs have the same spring constant. One of them is 0.40 m long and the other 0.25 m long. If they are connected end-to-end between rigid supports 1.30 m apart, where will the connection point of the two springs lie relative to the supported end of the short spring?

13.8 A spring gun consists of a horizontal spring with $k = 50$ N/m that is compressed 17 cm when the gun is cocked. The gun fires a 150-g projectile. What is its initial acceleration?

13.9 A child's pail is filled with sand and hung from a vertical spring, extending the spring 10 cm. If the pail is pulled down an additional 20 cm and released from rest, what is the pail's initial upward acceleration?

Section 13.2   The Simple Harmonic Oscillator

13.10 The position of a harmonic oscillator is described by $x = x_0 \cos 2\pi t/T$, where the displacement amplitude is $x_0 = 10$ cm and the period T is 0.25 s. Calculate the displacement at $t = 0.75$ s, at $t = 2.0$ s, and at $t = 2.3$ s.

13.11 A harmonic oscillator with an amplitude of 30 cm has a displacement of 30 cm at $t = 0$. At $t = 0.20$ s it has a displacement of 27 cm, without having passed through zero displacement. What is the period of the oscillator's motion?

13.12 A block moving with simple harmonic motion has a period of 2.0 s. The maximum displacement from equilibrium in any direction is 5.0 cm. (a) Write an equation to describe the displacement, given that at time $t = 0$ the displacement is zero and the velocity is positive. (b) Write an equation for the displacement, given that at time $t = 0$ the displacement is a maximum and the velocity is zero.

13.13 A 0.50-kg air-track glider is attached to the end of the track by a horizontal coil spring of $k = 20.0$ N/m. The glider is displaced 15.0 cm from its equilibrium position and released, so that it oscillates back and forth on the track. (a) What is the maximum acceleration of the glider? (b) What is its acceleration at a time equal to 1/8 of the oscillator's period? (c) What is its position at a time equal to 1/8 of the oscillator's period?

13.14 Show that for a harmonic oscillator $v_0 = 2\pi x_0/T$, where x_0 is the amplitude of the displacement, v_0 is the amplitude of the velocity, and T is the period. (*Hint*: Use Fig. 13.6.)

13.15 A can of beans hung on a vertical spring is pulled down from its equilibrium position and released. It travels up and then back down to the position from which it was released in 0.75 s, traveling a total distance of 0.75 m. How far from its release position was the can 0.12 s after it was released?

13.16 A harmonic oscillator has a period of 0.314 s and an amplitude of 7.0 cm. At $t = 0$ it is at $x = 0$. (a) How far does it travel between $t = 0$ and $t = 0.063$ s? (b) How far does it travel between $t = 0.283$ s and $t = 0.345$ s?

Section 13.3 Energy of a Harmonic Oscillator

13.17 Two joules are required to extend a spring by 0.25 m. What is the spring constant?

13.18 The potential energy stored in the compressed spring of a toy dartgun is 0.35 J. The spring constant is 28 N/m. By how much is the spring compressed?

13.19 The spring described in Example 13.2 is stretched to 5.0 cm beyond its equilibrium length. How much work is required to stretch the spring?

13.20 A 4.0-kg block oscillates with an amplitude of 0.080 m on a light spring of constant $k = 15$ N/m. (a) What is the maximum value of the block's velocity? (b) What is its maximum acceleration?

13.21 A 0.50-kg plush toy oscillates at the end of a long light spring of constant $k = 10.0$ N/m and negligible mass. The maximum speed of the toy as it oscillates is 0.30 m/s. (a) What is the amplitude of the oscillation? (b) How much energy is stored in the spring at its greatest extension?

13.22 A harmonic oscillator of frequency $f = 0.50$ Hz has an amplitude of 10.0 cm. (a) What is the maximum speed of its motion? (b) What is the maximum acceleration?

Section 13.4 Period of a Harmonic Oscillator

13.23 A 0.500-kg puppet oscillates at the end of a light spring of constant $k = 2.00$ N/m. (a) What is the frequency of the oscillations? (b) What is the period?

13.24 A 0.400-kg brass block is attached to a spring of negligible mass. What is the value of the spring constant if the vibrational frequency is 0.500 Hz?

13.25 A 2.0-kg fish is attached to a spring of constant $k = 50$ N/m and of negligible mass. The spring is stretched 8.0 cm from equilibrium and released from rest. (a) What is the maximum speed of the fish as it oscillates? (b) What is the frequency of the oscillation?

13.26 A 500-g lead weight is attached to a spring of negligible mass. When the weight is set into motion it oscillates with a period of 2.0 s. (a) What is the spring constant k? (b) What force is required to stretch the spring by 2.0 cm?

13.27 A circus performer bobs up and down at the end of a long elastic rope at a rate of once every two seconds. By how much is the rope extended beyond its unloaded length when the performer hangs at rest?

13.28 The prongs of a 440-Hz tuning fork vibrate with simple harmonic motion. A scratch on the end of one of the prongs travels through a total distance of 1.00 mm as the prong moves from one extreme position to the other. (a) What is the maximum speed of the scratch? (b) What is the maximum acceleration of the scratch?

Section 13.5 The Simple Pendulum

13.29 A simple pendulum makes 93 complete swings in one minute. (a) What is its frequency? (b) What is its period of oscillation?

13.30 A 30-kg child is playing on a swing of negligible mass. The child swings so high that at the peak of the motion her center of mass is 1.0 m above where it is at the bottom of the swing. How fast is the child moving when she passes through the minimum point of her swing?

13.31 A clock pendulum has a period of 0.750 s. (a) How long is the pendulum? (b) How long must it be to have a period of 2.00 s?

13.32 A 21-kg child swings on a playground swing 3.0 m long. (a) What is the period of her motion? (b) If her older brother, who weighs twice as much as she does, rides in the swing instead, what is the period of the motion? Assume the center of mass to be at the position of the seat.

13.33 A child swings on a playground swing attached to chains 4.0 m long. (a) Calculate the period of the swing for small-amplitude oscillations. (b) What would be the new period if the seat height were raised 1.0 m? Assume the center of mass to be at the position of the seat.

13.34 Until 1987, a 22.25-m long pendulum hung in the Smithsonian Institute, National Museum of American History. (a) Calculate the period of the pendulum for small-amplitude oscillations. (b) What would be the new period if the pendulum were reinstalled with a length one third of its original length?

13.35 Students in an elementary physics laboratory measured the period of a simple pendulum to be 2.475 s. Careful measurement showed the length of the pendulum to be 151.8 cm. What was the acceleration of gravity in their laboratory?

13.36 A 800-kg wrecking ball hangs from the end of a crane by a 30.0-m cable. If the crane operator quickly moves the end of the crane 3.00 m to the left, how many seconds pass before the wrecking ball passes below the end of the crane?

13.37 The lengths of pendulums with periods of two seconds are given for several locations. Find the distance through which an object will fall in one-half second at these locations.

St. Thomas	99.11 cm
New York	99.32 cm
London	99.41 cm
Spitzbergen	99.61 cm

13.38 A rule sometimes used by clock makers is that a pendulum oscillating with V oscillations per minute has a length L in inches given by $L = (187.6/V)^2$. Verify this expression.

Section 13.6   Damped Harmonic Motion

13.39 Graph the motion of a damped harmonic oscillator described by Eq. (13.12) for which $\gamma = 0.04/s$ and $2\pi f = 6.0/s$. Extend your graph out to $t = 12$ s.

13.40 A 1.26-m pendulum is pulled aside and released. Air resistance gives a damping coefficient of $\gamma = 0.020/s$. The support wire makes an angle of 3.30° with the vertical direction 1.50 s after the release. What was the angle from the vertical of the original displacement?

13.41 A 94-cm simple pendulum is pulled aside a small distance and released. After 120 oscillations the amplitude is one half of its starting value. The damping is proportional to the speed of the pendulum bob. (a) What is the value of the damping coefficient γ if the behavior is described by Eq. (13.12)? (b) What fraction of the original amplitude will remain after 15 min?

Section 13.7   Forced Harmonic Motion and Resonance

13.42 A motor that turns at 12,000 rpm is suspended from above by a single spring to isolate the building from its vibration. If the natural frequency of the system is to be one fourth of the vibration frequency, by how many centimeters should the spring extend beyond its unloaded length when the motor is hung on it?

13.43 Repeat the experiment with the rubber bands and drink can described in Section 13.7. This time determine the percent difference between your observed frequency of finger motion at the resonant frequency and the frequency of the system calculated from the spring constant and the mass of the can. Because of the assumptions, you should expect only approximate agreement. Determine the spring constant by measuring the extension of the rubber bands when the can is hung from them. (Assume that Hooke's law holds, neglect the mass of the metal can compared with the mass of the liquid, and if necessary assume the density of soft drinks to be the same as the density of water.)

13.44 What should be the length of the pendulum in Fig. 13.18 to give the 1.0-kg air-track glider the maximum amplitude if the spring constant is $k = 120$ N/m?

Additional Problems

13.45 Two 35-cm springs have different spring constants.
• If the springs are connected end-to-end and the outer ends connected to supports that are 100 cm apart, the connection point is 45 cm from the nearest support. What is the ratio of the spring constants?

13.46 When a physics textbook is hung from a spring attached to the ceiling the spring stretches 4.0 cm.
• The book is next hung from three springs exactly like the first one (Fig. 13.19). What is the total extension of the spring system?

13.47 When a 3.00-kg cube of metal is hung from a spring,
• the spring stretches by 7.50 cm. The same block is placed on a frictionless horizontal surface and the spring connected to a support at the same level as the block. The block is pulled 3.00 cm away from its equilibrium position and released. What is the initial acceleration of the block?

13.48 A physics student wants to build a spring gun that
• when aimed horizontally will give a 20-g ball an initial acceleration of g by releasing a spring that has been compressed 15 cm. (a) What spring constant is necessary? (b) What is the velocity of the ball as it emerges from the gun?

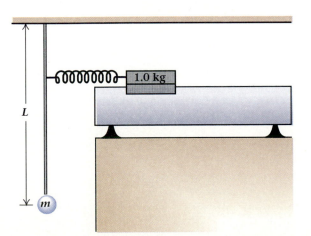

Figure 13.18   Problem 13.44.

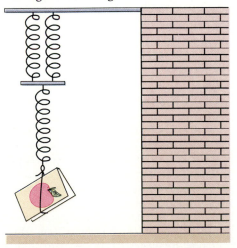

Figure 13.19   Problem 13.46.

13.49 Show that for a harmonic oscillator the displace-
•• ment x and velocity v are related by

$$x = x_0 \sqrt{1 - \left(\frac{v}{v_0}\right)^2},$$

where x_0 is the maximum displacement and v_0 is the maximum velocity.

13.50 A person whirls a 0.250-kg stone in a 1.00-m-radius
•• horizontal circle at the end of a string. The horizontal component of the force with which the person pulls on the string is 5.00 N. Write expressions for the projections of the displacement, velocity, and acceleration on a diameter of the circle.

13.51 It is claimed that if a hole could be bored straight
•• through the earth, and if an object were dropped into the hole, the object would oscillate back and forth along this diameter with simple harmonic motion. Test this claim by assuming that the earth is of uniform density, getting an expression for the force on the object, and comparing the expression with Eq. (13.1). (*Hint:* A small object at a distance R from the center of a mass with spherical symmetry will experience a gravitational force toward the center that depends only on the mass inside an imaginary sphere of radius R. This is true whether the object is outside or inside the spherical mass.)

13.52 By substituting for v and x as functions of time in
• the expression for the total energy of a harmonic oscillator, show that the total energy is conserved; that is, show that the total energy is independent of time.

13.53 A solid plastic block lies on a smooth floor. It is
• connected to the wall by a long horizontal coil spring with a spring constant of 86 N/m. If the block is pulled aside 37 cm and released, how much heat will have been generated before the block comes to rest?

13.54 Plot on the same graph the potential energy and the
• total energy of a harmonic oscillator as a function of time if $m = 0.10$ kg, $k = 20$ N/m, and $x_0 = 5.0$ cm. Plot for t between 0 and 2 s.

13.55 A mass m is attached to a vertical spring and re-
•• leased. The mass oscillates but eventually comes to rest a distance h below its initial position. Find expressions for the change in gravitational potential energy of the mass and for the energy stored in the spring as a function of m, h, and the acceleration of gravity. Explain why the two expressions are, or are not, the same.

13.56 When a box of cookies is hung on the end of a
• spring, the spring stretches by 6.7 cm. What is the oscillation frequency of the system if the box is pulled down and released?

13.57 A U-shaped tube is partially filled with water.
•• (a) Show that if the water is displaced from equilibrium it executes simple harmonic motion. (b) Find an expression for the period of the oscillations. (Let the diameter of the tube be d and the length of the water column be L.)

13.58 Assume that the length of the water column in a
•• very long upright U-tube is 50 cm. If the damping coefficient γ is 0.30/s, how many oscillations must occur for the amplitude to decrease to one fourth of its original value?

13.59 A long 3.0-cm-diameter cylinder of ice is weighted
• at the bottom so that it floats upright in a container of water at 0°C. The mass of the cylinder with the weight is 300 g. When the cylinder is pushed down and released, it bobs up and down. Show that this motion is simple harmonic motion and find its frequency. (*Hint:* You will need to know the density of water at 0°C.)

13.60 Tarzan is across the river from Jane, who is in dan-
• ger of being blown up by a time-bomb set for 6.9 s. Tarzan can save Jane only by swinging across the river on a 21-m vine hung directly over the center of the river. Because Tarzan is clinging to another vine, he can swing over only with the force of gravity and is unable to push off. (a) Will he arrive in time to save Jane? If so, by what margin? (b) Where are Tarzan and Jane when the bomb goes off?

13.61 A clock pendulum that "ticks seconds" (has a pe-
•• riod of two seconds) on earth is set up on the moon. What is the pendulum's period on the moon? ($M_m = 7.36 \times 10^{22}$ kg, $R_m = 1.74 \times 10^6$ m.

13.62 A pendulum bob on a string of length L is arranged
•• as shown (Fig. 13.20). The point marked P is a smooth peg and is at a distance $L/2$ below the sup-

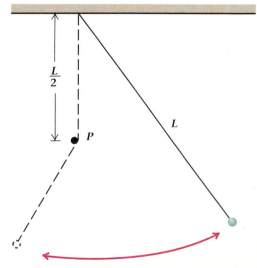

Figure 13.20   Problem 13.62.

port. Write an expression for the period for small oscillations of this system.

13.63 A swing is hung from the limb of a tree. When a
•• young child gets into the swing it settles 2.0 cm. Because of the flexibility of the tree limb, the swing bobs up and down as well as swinging back and forth. When the swing is gently bobbed up and down with no swinging motion, the frequency is 10 times the frequency of the swing when it is gently swung so as not to excite bobbing motion. How long is the swing?

13.64 An oscillator with a 0.46-s period is made from a
• mass suspended from a spring. The mass is placed on a frictionless surface that makes a 45° angle with the horizontal and the spring is attached at the top of the incline (Fig. 13.21). What is the new period of the oscillator?

13.65 The apparatus shown in Fig. 13.22 consists of a
•• mass m attached to the end of a light rod (negligible mass) of length L and hinged at one end. A spring of spring constant k is joined to the rod a distance a from the hinged end. Show that the frequency of oscillation of this apparatus is

$$f = \frac{a}{2\pi L}\sqrt{\frac{k}{m}}$$

by first showing that it is equivalent to a mass hanging from a spring of constant $k(a/L)^2$.

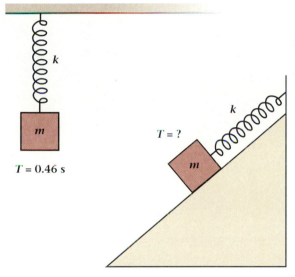

Figure 13.21   Problem 13.64.

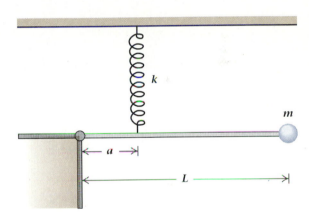

Figure 13.22   Problem 13.65.

ADDITIONAL READING

Egan, D. M., *Concepts in Architectural Acoustics.* New York: McGraw-Hill, 1988. A handy reference for solving a broad range of acoustical and vibrational problems.

Jones, G.E., and J. L. Ferguson, "Easy Displacement Versus Time Graphs for a Vibrating String: Tuning a Guitar by Television." *American Journal of Physics,* Vol. 48, May 1980, p. 362.

Rossing, T. D., "The Physics of Kettledrums." *Scientific American,* November 1982, p. 172.

Walker, J., "The Amateur Scientist: Strange Things Happen When Two Pendulums Interact Through a Variety of Interconnections." *Scientific American,* October 1985, p. 176.

Yin, T. P., "The Control of Vibration and Noise." *Scientific American,* January 1969, p. 98.

14

Wave Motion

14.1	Pulses on a Rope
14.2	Harmonic Waves
*14.3	Energy and Information Transfer by Waves
14.4	Sound Waves
*14.5	Measuring Sound Levels
14.6	The Doppler Effect
*14.7	Formation of Shock Waves
14.8	Reflection of a Wave Pulse
14.9	Standing Waves on a String
14.10	Waves in a Vibrating Column of Air
*14.11	Beats

A WORD TO THE STUDENT

In this chapter we use our knowledge of oscillatory motion to describe the motion of waves. We put these methods to immediate use in examining sound and other mechanical waves. The principles in this chapter will also be important for studying the behavior of light waves (Chapter 23). Furthermore, the basic principles of wave behavior are essential for understanding the behavior of matter on the atomic and subatomic scale (Chapter 26). Thus the concepts developed in this chapter will also be necessary for learning the quantum theory (Chapter 27).

The study of periodic vibrations and waves is one of the oldest scientific studies. Some four centuries before Christ, the Pythagoreans first established the connection between musical sounds and mathematics. This study continued through the middle ages, and was later developed by Huygens and Galileo. In the eighteenth century an interest in musical instruments prompted the mathematical study of vibrating bodies and of the propagation of sound through air. By the nineteenth century the study of sound had become an integral part of physics and was known as *acoustics*.

Studies of wave motion also advanced through research into the behavior of light, the discipline known as *optics*. Robert Hooke and Christiaan Huygens, contemporaries of Newton, believed light to be a form of wave motion, and Huygens developed a mathematical model for light waves. However, not until the early 1800s did Thomas Young firmly establish the wave nature of light. By the end of that century Heinrich Hertz had established that what we now call radio waves were of the same nature as light waves. In the early twentieth century wave concepts were used to develop the theory of quantum mechanics, which is sometimes called wave mechanics.

In our daily lives we see many examples of waves and oscillations: radio and television waves reach around the world; musical sounds are created and modified by computers, as well as by traditional instruments; and ultrasound and x rays are commonly used in medical diagnosis and treatment.

There are similarities between vibratory motion and wave motion. We use mechanical concepts such as displacement, velocity, acceleration, and energy in describing them both. However, you must take care to distinguish between these two phenomena. As we develop the wave concept we will point out both the similarities and the differences.

Pulses on a Rope

14.1

The simplest way to begin the study of waves is to think of the propagation of a pulse along a rope. If one end of a tautly stretched rope is suddenly snapped up and back down to its initial position, the action generates a wave pulse that travels along the rope to the other end (Fig. 14.1). As you can see from the figure, the displacement of the rope due to this wave pulse is at right angles to the direction in which the pulse travels. For this reason we call this type of motion a **transverse wave pulse**. If we watch any particular point of the rope very closely, we see that the motion of that point along the direction of the rope is very slight compared with its motion perpendicular to the rope. Each section of the rope makes a single oscillation up and down. But the oscillations are not independent, for each section of the rope is connected to the adjacent sections. Thus the propagation of a transverse wave pulse along the rope is a collective motion of the whole rope, not simply the isolated behavior of any one section.

In addition to distinguishing between the motion of a pulse, or wave,

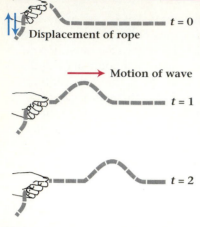

Displacement of rope

$t = 0$

Motion of wave

$t = 1$

$t = 2$

Figure 14.1   A wave pulse traveling along a rope. The displacement of the rope is vertical, whereas the motion of the wave pulse is horizontal.

and the motion of the medium (here, the rope), we must distinguish the kind of motions involved. For example, the velocity of a wave on a rope may be constant, while the transverse velocity of a point on the rope may be sinusoidal in time. In this case the particles of the medium oscillate back and forth about an equilibrium position, while the wave propagates along the rope at its own speed. We will see later, in Chapter 19, that electromagnetic waves, such as light and radio waves, do not require a medium for their propagation. Their behavior is unlike that of waves on a rope, sound waves, and water waves, which do require a medium for their propagation.

The wave pulse we have just described is called a **traveling wave**. The pulse occurs at one place at one time, and at another place at a later time. The distance the pulse travels is proportional to the elapsed time. We have here a very general description of a wave: A **wave** is a disturbance that transfers energy from one point to another without imparting net motion to the medium through which it propagates. This description is quite general and, as you can see, does not require the wave to be repetitive.

In our initial example, the physical action was the bending of the rope into a distorted shape, which propagated along the length of the rope. This is a very special case: a one-dimensional, transverse wave pulse. We shall continue to examine special waves in order to give simple and concrete examples of wave properties. However, you should remember that the wave concept is very broad and very powerful, and applies to a wide range of phenomena from water waves to sound waves to light.

Harmonic Waves

14.2

Let us now consider another special one-dimensional wave: a wave generated by the simple harmonic motion of one end of a long extended rope. In such a **harmonic wave,** the distance between successive maxima, or wave crests, is the **wavelength** λ (Fig. 14.2). As the end of the rope moves up and down with harmonic motion, it generates a sinusoidal wave. (Remember, a sinusoidal wave has the shape of a sine curve.) Note that while any given point on the *rope* moves up and down, the point does not undergo any displacement along the x axis. By contrast, any given point on the *wave*—a crest, say—moves one wavelength λ along the x axis in one period.

The displacement of a harmonic wave is a function of both position and time. At the instant of time $t = 0$ (Fig. 14.2) the displacement of the wave $y(x, t)$ is described by a sinusoidal function of position x,

$$y(x, t)\big|_{t=0} = y_0 \sin \frac{2\pi x}{\lambda},$$

where λ is the wavelength of the wave and y_0 is its amplitude. (That is, y_0 is the maximum vertical displacement of the rope from its equilibrium position.) However, if we look only at the position $x = 0$, the motion of

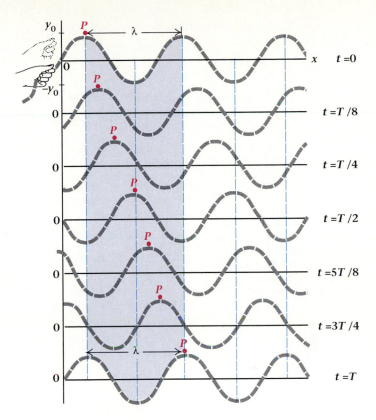

Figure 14.2 Harmonic waves traveling along a rope are generated by a sinusoidal motion of one end. A point on the crest of the wave, such as P, moves one wavelength λ in one period T.

the rope as a function of time is described by

$$y(x, t)\big|_{x=0} = -y_0 \sin \frac{2\pi t}{T},$$

where T is the period of the wave. (Compare this with Section 13.2.) The general mathematical form of the wave, including both position and time, is given by

$$y(x, t) = y_0 \sin 2\pi \left(\frac{x}{\lambda} - \frac{t}{T} \right). \tag{14.1}$$

The function $y(x, t)$ describes the displacement of the rope at any position x and any time t. Equation (14.1) is the mathematical description of a *traveling harmonic wave* moving in the positive x direction. We could use a cosine function for this equation just as well, but we choose the sine function so that $y = 0$ when $x = 0$ and $t = 0$.

If we follow the motion of a single wave crest, we see that in one complete period it travels one full wavelength. The wave, therefore, is traveling with a speed v given by

$$v = \frac{\lambda}{T}.$$

Remembering that the frequency is the reciprocal of the period, we see

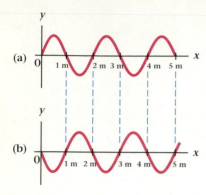

Figure 14.3   Example 14.1: A diagram of two snapshots of a wave on a rope. Picture (b) was taken $\frac{1}{10}$ s after picture (a). The scale is shown in meters. The amplitude of the wave is exaggerated to make it easier to see.

that

$$v = \lambda f. \qquad (14.2)$$

This relation between a wave's speed, frequency, and wavelength is very important and we shall frequently make use of it throughout the book.

Example 14.1

Speed of a harmonic wave.

Figure 14.3 represents two snapshots of a wave on a rope. The snapshots were taken $\frac{1}{10}$ s apart. We know that the wave was traveling to the right and that it moved by less than one wavelength between pictures. What is its (a) wavelength, (b) wave speed, and (c) frequency?

Solution   (a) Examining the figure, we see that the distance between two successive crests, or the wavelength, is 2 m.

(b) During the $\frac{1}{10}$-s interval the wave moved to the right a distance of half a wavelength, or 1 m. The wave speed is the distance traveled divided by the time interval, or 10 m/s to the right.

(c) Since we now know both wavelength and speed, we can obtain the frequency from

$$f = \frac{v}{\lambda} = \frac{10 \text{ m/s}}{2 \text{ m}} = 5 \text{ s}^{-1} = 5 \text{ Hz}.$$

Energy and Information Transfer by Waves

*14.3

We experience the energy transferred by waves in many situations: We feel the force of an ocean wave, our skin is warmed by the light waves from the sun, we hear sound waves . Furthermore, most of the information that we receive comes to us by waves. Speech and music are transmitted by sound waves. Radio and television are transmitted by electromagnetic waves. The reflected light by which you read this page is a wave. How does the energy (and, hence, the information) transmitted by the wave depend on the properties of the wave? We answer this question by first considering how energy is transferred by a single pulse. Then we extend our results to get an expression for the energy of a harmonic wave.

At time $t = 0$, a small segment of the rope around point P in Fig. 14.4, with mass Δm and length Δl, is at rest and has no kinetic energy. The up-and-down hand motion provides the energy required to start the pulse along the rope. As the leading edge of the pulse reaches P, the segment Δl begins to move upward. As the wave crest passes the segment Δl, the segment moves to its highest position and then starts down again, possessing kinetic energy while it is in motion. When the entire pulse has passed P, the segment Δl returns to rest and again has no kinetic energy. The progress of the pulse along the rope corresponds to the flow of energy

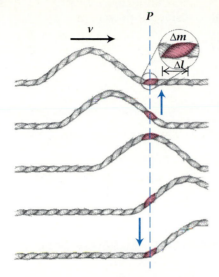

Figure 14.4 An element of mass Δm at point P is given kinetic energy as the wave pulse passes by with a speed v.

along the rope. Any other type of pulse, including a pulse traveling through the air, would transfer energy along the direction of propagation in a similar manner.

How much energy has been transferred past P during a time t? For a traveling harmonic wave on a rope, each point moves with simple harmonic motion in the transverse (y) direction. As we saw in Section 13.3, Eq. (13.7), in the absence of damping, the total energy of a harmonic oscillator is equal to its potential energy at maximum displacement y_0, that is, $\frac{1}{2}ky_0^2$. Also in Chapter 13, the relationship between mass, spring constant, and frequency was found to be $f = (1/2\pi)\sqrt{k/m}$. If we treat the segment of the rope as a harmonic oscillator with mass Δm moving at frequency f, we can find an effective spring constant $k = (2\pi f)^2\,\Delta m$. The energy associated with the motion of this segment of the rope is then

$$\Delta E = 2\pi^2\,\Delta m f^2 y_0^2.$$

We now have an important result: The energy of a wave depends on the square of the amplitude of the wave. Thus a wave with twice the amplitude of an otherwise equivalent wave (same frequency, same medium) will have four times as much energy.

To find the rate of energy flow, we observe that Δm can be written as $\rho A\,\Delta l$, where ρ is the density, A the cross-sectional area, and Δl the length of the rope segment. In a time Δt the wave with speed v passes a length $\Delta l = v\,\Delta t$, so that we can substitute $\Delta m = \rho A v\,\Delta t$ into the equation for ΔE. We obtain an expression for the energy transported in time Δt:

$$\Delta E = 2\pi^2 A\rho v f^2 y_0^2\,\Delta t.$$

The rate at which energy propagates along the rope is the power P,

$$P = \frac{\Delta E}{\Delta t} = 2\pi^2 A\rho v f^2 y_0^2.$$

The more generally useful parameter is the **intensity** I, defined to be the power flowing through unit area. For the case at hand, the intensity in watts per square meter is

$$I = \frac{P}{A} = 2\pi^2\rho v f^2 y_0^2. \qquad (14.3)$$

Although we have derived this result for the specific case of waves on a rope, it does give the correct dependence of the intensity on the density of the medium, the wave velocity, the frequency, and the amplitude appropriate for any traveling harmonic wave.

Example 14.2

Power transmitted by waves.

Waves of the same frequency and velocity travel along two identical ropes. The power transmitted down rope 1 is 0.30 mW. If the waves in rope 2 have an amplitude 1.6 times those in the first rope, how much power is transmitted along rope 2?

Solution   We may write the ratio of the transmitted powers as

$$\frac{P_2}{P_1} = \frac{2\pi^2 A\rho v f^2 y_{02}^2}{2\pi^2 A\rho v f^2 y_{01}^2},$$

which becomes

$$\frac{P_2}{P_1} = \frac{y_{02}^2}{y_{01}^2}.$$

Thus

$$P_2 = P_1 \frac{y_{02}^2}{y_{01}^2} = 0.30 \text{ mW} \frac{1.6^2}{1.0^2} = 0.77 \text{ mW}.$$

Sound Waves

14.4

So far we have been discussing **transverse waves,** such as a wave pulse along a rope, in which the particles of the rope move at right angles to the direction of propagation of the wave. However, waves also occur in which the particles of the wave medium move back and forth along the direction of propagation. Such waves are known as **longitudinal waves.** For example, if a stretched spring is alternately expanded and compressed, a compressional oscillation travels along the spring (Fig. 14.5). The sections of the spring where the coils are tightly packed are called *compressions* and the sections where the coils are farther apart are called *rarefactions.* The terms compression and rarefaction are also used to describe relative density for other types of longitudinal waves.

Sound waves in air are longitudinal waves. The vibrations of a drumhead or loudspeaker exert varying pressure on the air. As the pressure is applied, the gas molecules crowd together and push against adjacent molecules. These molecules, in turn, strike their neighbors. The resulting pulse of compressed air moves away from the pressure source. As the compression passes, the individual gas molecules move back to their original positions. Thus while the wave travels in a longitudinal fashion, the molecules themselves only vibrate.

Sound waves travel out in all directions from an isolated source and are therefore three-dimensional waves. The outward-moving compressions

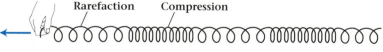

Direction of vibration

Figure 14.5 Example of a longitudinal wave in a spring. The direction of vibration is parallel to the direction of the wave.

Direction of wave

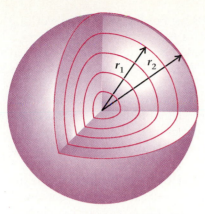

Figure 14.6 Waves expanding from a point source have spherical wavefronts.

(or rarefactions) correspond to expanding spherical shells called **wavefronts** (Fig. 14.6). The radius of any shell increases at the speed of sound. If we are far away from the source, a small piece of the spherical wavefront will be approximately flat, or planar. We call these plane wavefronts, or simply **plane waves**.

As the spherical wave expands, energy is conserved if the damping is negligible, as it is over short ranges in air. Therefore the power passing through a spherical shell of one radius is the same as that passing through a shell of any other radius. The power passing through the shell of radius r is the product of the intensity I with the area of the shell, $4\pi r^2$. For two different radii, r_1 and r_2, we can write

$$I_1 4\pi r_1^2 = I_2 4\pi r_2^2.$$

Upon rearranging we get

$$\frac{I_2}{I_1} = \frac{r_1^2}{r_2^2}.$$

Thus the intensity of a spherical wave decreases inversely with the square of its distance from the source:

$$I = \frac{I_0}{r^2}. \qquad (14.4)$$

Example 14.3

Change of sound intensity with distance.

A loudspeaker on a tall pole standing in a field of tall grass generates a sound of high frequency at an intensity of 1.0×10^{-5} W/m² at the position of the ears of a person standing 8.0 m directly below it. If the person walks away from the pole so that she is 24 m from the loudspeaker, what is the sound intensity at the new position of her ears?

Solution The grass will almost completely absorb the high-frequency sound that reaches it, so the expanding sound wave reaching the listener's ears will be a section of a spherical wave with no additional effects from reflected waves. We can then compare the intensities at the two positions:

$$\frac{I_2}{I_1} = \frac{r_1^2}{r_2^2},$$

$$I_2 = \frac{r_1^2}{r_2^2} I_1 = \frac{(8.0 \text{ m})^2}{(24 \text{ m})^2} \times 1.0 \times 10^{-5} \text{ W/m}^2,$$

$$I_2 = 0.11 \times 10^{-5} \text{ W/m}^2.$$

The speed with which sound waves propagate in air depends on atmospheric pressure, temperature, and humidity. At standard sea-level pressure and 0°C, the speed of sound in dry air is 331.5 m/s. The speed of sound at other temperatures is adequately represented by the expression

$$v(T) = (331.5 + 0.6T) \text{ m/s},$$

TABLE 14.1. Speed of sound in some representative materials at 15°C	
Material	Speed (m/s)
Air	340
Polyethylene	920
Helium	961
Water	1500
Marble	3810
Wood (oak) along the fiber	3850
Aluminum	5000
Iron	5120

where the temperature T is in degrees Celsius. For the purpose of working problems in this text, it will be convenient to use the value of 340 m/s for the speed of sound. This value corresponds to the speed of sound in air at a temperature of approximately 15°C. (In other units it is 1100 ft/s or 760 mi/h.)

Sound travels slowly enough to make us aware of its finite speed. The delay between a flash of lightning and the crash of thunder occurs because the speed of sound is much slower than that of light. For the same reason, a baseball outfielder can see the batter's swing before he hears the sound of the bat hitting the ball.

The vibration of a drumhead or a guitar body pushes against the air to produce pressure waves that propagate to our ears and are perceived as sound. The response of human ears is limited to a range of frequencies from about 20 Hz to about 20,000 Hz. In general, as we grow older the upper limit of audible frequencies drops. It is not uncommon to find people with adequate hearing whose range of audible frequencies extends to only 10,000 or 15,000 Hz. Those frequencies between 20 Hz and 20,000 Hz are usually referred to as *audio*, or *sonic*, frequencies. Vibrational frequencies above 20,000 Hz are beyond our hearing and are referred to as *ultrasonic* frequencies. Similarly, extremely low frequencies are called *infrasonic* frequencies. Ultrasonic vibrations generate soundlike waves, but we do not hear them because of the limitations of our ears. It is well known that many animals, including dogs, can hear frequencies well above those audible to people.

The *pitch* of a sound or musical note is a subjective judgment of its highness or lowness. It is determined primarily by its frequency; a high pitch corresponds to a high frequency. However, the loudness of a sound at a given frequency can influence its apparent pitch.

Sound waves can propagate in solids and liquids, as well as in gases. The velocity of sound is much greater in denser solids and liquids than in air. The sound waves in liquids are longitudinal, just as in air. However, vibrations can propagate in solids as both longitudinal and transverse waves, and the corresponding wave speeds may be different. A list of representative sound speeds is given in Table 14.1.

There are many applications of ultrasonic waves. Because of their relatively short wavelengths, they can be focused into narrow beams and directed more easily than the longer waves of audible sound. (We will see this effect again in the discussion of wave optics in Chapter 23.) Figure 14.7 illustrates the use of ultrasonic waves to measure distances. A pulse of waves of frequency between 25 and 40 kHz is emitted from the ranging device, which automatically measures the time for the echo of the pulse to return. The distance to the wall or other object is computed from the travel time of the pulse and the speed of sound in air, taking into account the dependence of the speed on temperature and humidity. The computation is made by the microcircuits within the ranging device and the result is displayed on a numerical panel. These instruments can measure with a precision of about 0.6 cm. Similar devices are used on some automatically focusing cameras to determine the proper position for the lens.

When the ultrasonic waves strike the wall in Fig. 14.7, they are partly reflected and partly transmitted. The transmitted wave is also partly re-

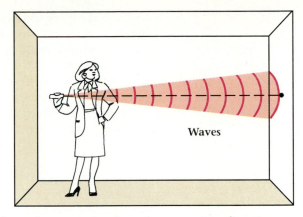

Waves

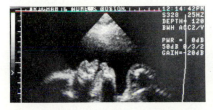

Figure 14.8   **An ultrasonic image of a human fetus inside the mother's body.**

flected and partly transmitted at the next interface it encounters. The amount of the reflection depends on the relative speed of sound on each side of the interface. (This, too, is a general property of waves and not just a property of ultrasonic waves.) The returning signals from each interface can be used to generate an image of the obstacles encountered by the waves. Ultrasonic imaging is routinely used by physicians to "look" inside the human body (Fig. 14.8). The visual image is reconstructed from the information contained in the amplitude and phase of the reflected waves. Ultrasonic imaging is generally considered a safer technique than x-ray imaging.

Example 14.4

What is the wavelength of the sound waves when someone sings a "standard A" ($f = 440$ Hz)?

Solution   Assume the room temperature to be about 15°C, so that the speed of sound is 340 m/s. We may use Eq. (14.2) to get

$$\lambda = \frac{v}{f} = \frac{340 \text{ m/s}}{440 \text{ s}^{-1}} = 0.77 \text{ m}.$$

Measuring Sound Levels

*14.5

The human ear is an extremely sensitive detector, capable of hearing sounds over an extremely large range of intensities. For example, a passing train generates sound intensities that may be 10^4 to 10^6 times greater than the sound intensity due to a buzzing mosquito, yet we can hear both sounds clearly. Such a range of pressure responsiveness is remarkable. Although our ears are sensitive to this enormous range of sound intensities, our subjective judgment of loudness does not directly correspond to the magnitude of the sound intensity. Experiments have shown that, to a good approximation, people perceive a sound to be about twice as loud as a reference sound when its intensity is ten times as large as that of the reference. A sound perceived to be four times as loud as a reference requires an increase in sound intensity by a factor of 100. This relationship is approximately logarithmic; that is, loudness is proportional to the logarithm of the sound intensity. Thus it is natural to employ a scale of measurement that is also logarithmic.

The unit of sound intensity measurement is the **decibel,** dB. The decibel unit is one-tenth the size of the bel (a unit named in honor of the inventor of the telephone, Alexander Graham Bell). The **intensity level** IL in decibels is defined to be 10 times the logarithm of the ratio of two intensities I and I_0; that is,

$$\text{IL (in decibels)} = 10 \log_{10} (I/I_0), \qquad (14.5)$$

where I_0 is the reference intensity. For example, if the intensity I exceeds

---

PHYSICS IN PRACTICE

Acoustical Properties of Rooms

Have you ever noticed that the same music group sounds better in one auditorium than in another, or that the same lecturer is easier to understand in one room than in another? What you hear depends not only on the source of the sound but also on the acoustical properties of the room. Until the end of the nineteenth century auditoriums for music and lectures were good only by chance, or because they were built in imitation of good examples.

About 1895, Professor Wallace Sabin of Harvard University began the first systematic study of room acoustics. He discovered that one of the more important acoustical properties of a room is its reverberation time, which is the time for the sound pressure level to decrease by 60 dB. The reverberation time in a large stone-walled cathedral is several seconds, while a small, heavily carpeted and draped room has a much shorter reverberation time. Experience has shown that the optimum reverberation times depend on the size of the room and its intended use (Fig. B14.1).

The reverberation time in a given room is determined by the amount and type of sound-absorbing material present. The sound absorption a of a surface is the product of the actual surface area and the sound-absorption coefficient of the surface material (Table B14.1). When the area is in square meters, the absorption unit is named the *metric sabin*. The total sound absorption in a room is the sum of the absorptions for the individual surfaces.

After many experiments, Sabin developed an empirical relationship between the reverberation time T_{60}, the volume V of the room, and the sound absorption a in the room. When distances are measured in meters, the re-

Figure B14.1 **Optimum reverberation times at 500 Hz for rooms of different size and use.**

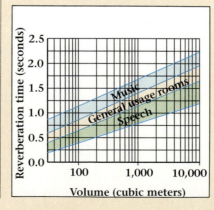

verberation time in seconds is

$$T_{60} = 0.16 \frac{V}{a}.$$

For example, a room 9 m wide × 12 m long × 4 m high has a volume of 432 m³. The total area of the walls is 168 m², and the areas of the ceiling and of the floor are 108 m² each. If the walls are plywood paneling, the floor wood, and the ceiling gypsum board (dry wall), the total sound absorption at 500 Hz is:

$$
\begin{aligned}
a_{\text{walls}} &= 168 \text{ m}^2 \times 0.17 \\
&= 28.6 \text{ metric sabins} \\
a_{\text{floor}} &= 108 \text{ m}^2 \times 0.10 \\
&= 10.8 \text{ metric sabins} \\
a_{\text{ceiling}} &= 108 \text{ m}^2 \times 0.05 \\
&= 5.4 \text{ metric sabins} \\
\hline
a_{\text{total}} &= 44.8 \text{ metric sabins}
\end{aligned}
$$

Therefore the predicted reverberation time of the empty room is

$$
\begin{aligned}
T_{60} &= 0.16 \frac{V}{a} \\
&= 0.16 \frac{432 \text{ m}^3}{44.8 \text{ metric sabins}} = 1.5 \text{ s}.
\end{aligned}
$$

This value is somewhat longer than is optimum for general use in a room of this size (Fig. B14.1). However, with seats and people in the room, the reverberation time is shortened. As an

the reference intensity I_0 by a factor of 4, the intensity level of I is 6 decibels (6 dB) above I_0:

$$\text{IL} = 10 \log 4 = 10(0.6) = 6 \text{ dB}.$$

Since the measure of sound intensity level in decibels is really a comparison of two intensities, a given sound level cannot be stated in units of decibels unless the reference level is known. The standard reference level of sound intensity is 10^{-12} W/m², which is approximately the intensity that can just barely be heard by a person with good hearing. This intensity is called the threshold of hearing. The intensity level of a very quiet recording studio is about 20 dB, corresponding to a sound intensity 100 times greater than the quietest sound you can hear. Figure 14.9 (p. 418) shows a comparison of sound intensity levels due to various sources.

TABLE B14.1
Typical sound absorption coefficients*

Material	Absorption coefficients at		
	125 Hz	500 Hz	2000 Hz
Acoustical ceiling tile ($\frac{3}{4}$ in.)	0.76	0.83	0.99
Brick, unglazed	0.02	0.03	0.05
Carpet, heavy, on concrete	0.02	0.14	0.60
Concrete block, painted	0.10	0.06	0.09
Glass, ordinary window glass	0.35	0.18	0.07
Heavy drapes (to half area)	0.14	0.55	0.70
Floors: Concrete or terrazzo	0.01	0.02	0.02
Linoleum or asphalt tile	0.02	0.03	0.03
Wood	0.15	0.10	0.06
Gypsum board, $\frac{1}{2}$ in.	0.29	0.05	0.07
Plaster on lath	0.14	0.06	0.04
Plywood paneling $\frac{3}{8}$ in.	0.28	0.17	0.10
Seating: Metal or wood chairs	0.15	0.22	0.38
Upholstered chairs	0.19	0.56	0.61
Person in an upholstered chair	0.39	0.80	0.92
Students in tablet-arm chairs	0.30	0.49	0.87

*From M. David Egan, *Architectural Acoustics* (McGraw-Hill, 1988)

dependence of the reverberation time. For example, a *warm* sound is the result of a longer reverberation time in the lower, or bass, frequencies.

Reverberation time is only one of the important acoustical properties of a room. Careful design is needed if the sound quality is to be the same at each seat. The direction from which the sound appears to come and the apparent distance of the hearer from the source are other important considerations. Frequently overhead reflectors and other splayed surfaces near the performers or speakers are used to increase the sense of intimacy and to enhance the balance and blend of the sound.

Modern auditorium design not only employs modified versions of Sabin's equation and techniques for calculating other acoustical properties, but also includes computer simulation and scale models. The most successful designs result from the work of a skilled and experienced person using the best of technology, research, and artistry. Even then, the desired result is not always achieved for large halls. Electronic reinforcement, with amplification and delays, can help adjust an auditorium's acoustical properties, but it cannot usually turn a bad auditorium into a good one.

example, if the floor were three-quarters covered with students in tablet-arm chairs, approximately 30 metric sabins would be added. The new reverberation time would be less than a second, which is in the acceptable range. Such calculations are not expected to give the precise value of the actual reverberation time, but they do indicate the approximate value expected under specific conditions.

The overall acoustical quality of a room is influenced by the reverberation time at all frequencies, not just 500 Hz. Subjective descriptions of an auditorium's sound, such as *warm*, *live*, and *brilliant*, depend not only on the length but also on the frequency

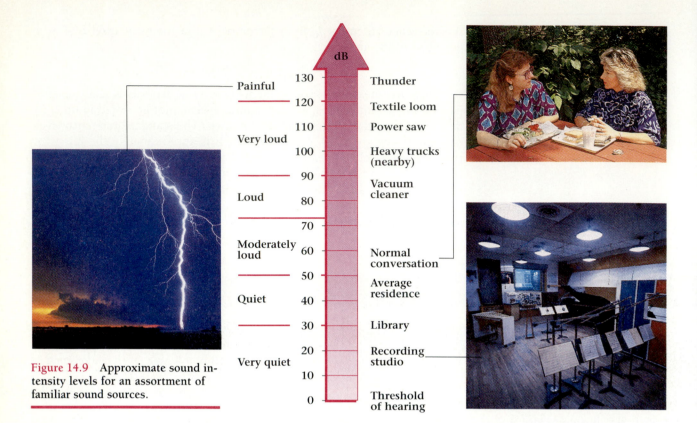

dB

Painful	130 — Thunder
	120 — Textile loom
	110 — Power saw
Very loud	100 — Heavy trucks (nearby)
	90 — Vacuum cleaner
Loud	80 —
	70 —
Moderately loud	60 — Normal conversation
	50 — Average residence
Quiet	40 —
	30 — Library
	20 — Recording studio
Very quiet	10 —
	0 — Threshold of hearing

Figure 14.9   Approximate sound intensity levels for an assortment of familiar sound sources.

Listening to loud sounds over a period of time causes a temporary shift of the hearing threshold to higher intensity levels. The effect is a temporary hearing loss known as hearing fatigue. The duration and magnitude of the fatigue depend on both the intensity of the sound causing it and the length of exposure time. If the exposure time and intensity are great enough, the hearing loss may become permanent instead of temporary. Permanent hearing impairment has been found to occur in persons exposed to steady noise levels in excess of 90 dB over very long periods. Greater losses are suffered by people exposed to greater noise levels.

The ear does not respond equally well to all frequencies in the audio range. It is considerably more sensitive to frequencies between 2000 and 5000 Hz than to either higher or lower frequencies. For this reason, sound level meters are often designed to have a frequency response matching that of the human ear. A meter that does so is said to be A-weighted, and measurements are often given in A-weighted decibels or dBA.

Example 14.5

Measuring sound level change in decibels.

At a party, a stereo tape deck is playing at maximum volume. Suddenly a dancer trips over a wire, and one speaker stops playing. What is the reduction in sound intensity level, measured in dB?

Solution   Let's assume that the two speakers are balanced, that is, that an equal sound level is produced by each. When one fails, the resulting sound intensity is one half of the initial intensity. This can be expressed in dB as

$$\text{intensity change in dB} = 10 \log \tfrac{1}{2} = -3 \text{ dB}.$$

The reduction in sound level due to failure of one half of the sound system is only 3 dB. Thus a 3-dB reduction in sound level corresponds to reducing the intensity by a factor of one half. Conversely, a 3-dB increase in sound level corresponds to twice as much sound intensity incident on the ear.

The Doppler Effect

14.6

Most of us are familiar with the rise and subsequent drop in pitch of an automobile horn as it approaches and then passes. As the moving car approaches a stationary listener, the sound waves crowd together, causing an increase in the frequency of the sound heard. After the car has passed and is moving away from the listener, the waves spread out and the observed frequency is lower. This change in frequency associated with the relative motion between a sound source and a listener (observer) is called the **Doppler effect**, after Christian Doppler (1803–1853), the Austrian physicist who first explained the phenomenon.

To understand the cause of the Doppler effect, study Fig. 14.10, which shows a source of sound waves moving to the right. The source radiates spherical waves, shown here as circles. Each wave crest moves out as an expanding sphere, but since the source is moving, it emits each successive wave at a different location. As a result, the waves moving in the same direction as the source are crowded together, while those moving in the opposite direction are spread farther apart. The wave speed is constant whether the source moves or not. Thus where the wavelength is shortened, the frequency is increased, and where the wavelength is lengthened, the frequency is reduced.

In the case of a source of frequency f in motion toward the observer, we can readily determine the new frequency f' heard by the observer. During one period T of the source, the wave moves out a distance vT, where v represents the speed of sound. During this same time interval, the source, moving at a speed v_S, has moved a distance $v_S T$. The difference between these two distances is the new wavelength λ':

$$\lambda' = (v - v_S)T.$$

The frequency is obtained from the wavelength in the usual way:

$$f' = \frac{v}{\lambda'} = \frac{v}{(v - v_S)T},$$

or, using the fact that $f = 1/T$,

$$f' = \frac{f}{(1 - v_S/v)} \qquad \text{(source approaching)}.$$

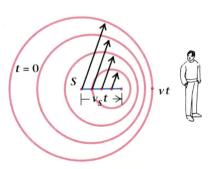

Figure 14.10 The Doppler effect: As a sound source moves toward an observer, the wavelength decreases, causing the observer to hear a higher frequency. During a time t, the wave advances a distance vt while the source advances a distance $v_S t$.

If the source is moving away from the observer, the wavelength reaching the observer is increased instead of shortened. Using the same argument as above, we find the new frequency by the same expression with a plus sign in front of the v_S/v term. The two cases can be written as one:

$$f' = \frac{f}{(1 \mp v_S/v)},\qquad(14.6)$$

where the upper sign means that the source is approaching the observer and the lower sign means that the source is moving away.

Example 14.6

Doppler shift of the sound from a moving police car.

A police car horn emits a 250-Hz tone when sitting still. What frequency does a stationary observer hear if the police car sounds its horn while approaching at a speed of 27.0 m/s (60 mi/h)? What frequency is heard if the horn is sounded as the car is leaving at 27.0 m/s?

Solution   The apparent frequency for the approaching car is found from Eq. (14.6), where we use the negative sign and the speed of sound is taken as 340 m/s:

$$f' = \frac{250 \text{ Hz}}{1 - \dfrac{27.0 \text{ m/s}}{340 \text{ m/s}}} = 272 \text{ Hz}.$$

When the car passes, the frequency is lowered according to Eq. (14.6) with the positive sign, and f' becomes

$$f' = \frac{250 \text{ Hz}}{1 + \dfrac{27.0 \text{ m/s}}{340 \text{ m/s}}} = 232 \text{ Hz}.$$

For waves transmitted by a medium, such as sound waves in air, a slightly different Doppler formula results when the observer, rather than the source, is in motion. If a stationary observer hears a frequency f, an observer O moving with a speed v_O toward the source would hear a higher frequency f', because he would encounter wavefronts more rapidly as a result of his motion toward the source. In a time t, the stationary observer receives ft wave crests. In the same time t, the moving observer travels a distance $v_O t$, which corresponds to $v_O t/\lambda$ wavelengths. The total number of wave crests encountered is

$$\text{number of wave crests} = ft + v_O t/\lambda.$$

The resulting frequency is the number of waves divided by the time,

$$f' = f + \frac{v_O}{\lambda} = f + \frac{v_O f}{v},$$

or $$f' = f\left(1 + \frac{v_O}{v}\right)\qquad \text{(observer approaching)}.$$

If the observer were moving away from the source, fewer waves would reach him per second. The general result for a moving observer is

$$f' = f\left(1 \pm \frac{v_O}{v}\right),$$
(14.7)

where the plus sign corresponds to an observer approaching the sound source and the minus sign corresponds to the observer moving away. Although this is similar to Eq. (14.6), the two equations are not quite the same. However, when v_S and v_O are much less than v, the two sets of formulas give essentially the same results.

The Doppler effect is not limited to sound waves alone but applies to other kinds of waves as well. It even applies to electromagnetic waves such as light and radar, although the formulas derived here for sound waves differ from the formulas that apply to light and radio waves. The correct equations to use for light are found in Chapter 24. In astronomy, the measured Doppler shifts in the light received from stars have been the principal sources of information on stellar motions. In more familiar situations, radar waves bounced off a moving target are Doppler-shifted as a result of the motion of the reflecting object. By measuring the shift in frequency of a radar wave beamed at a moving object, we can measure the speed of the object precisely. This technique has been widely adapted for measuring the speed of cars, planes, thunderstorms, and even baseballs.

Example 14.7

Doppler shift heard by a moving observer.

If the police car in Example 14.6 were sounding its horn while stationary, what frequency would be heard by an observer who was approaching it at a speed of 27.0 m/s (60 mi/h)?

Solution   Here we have the case of an observer moving toward a stationary source. Equation (14.7) with the plus sign is the appropriate expression:

$$f' = f\left(1 + \frac{v_O}{v}\right),$$

$$f' = 250 \text{ Hz}\left(1 + \frac{27.0 \text{ m/s}}{340 \text{ m/s}}\right) = 270 \text{ Hz.}$$

Notice that this is not the same answer as in Example 14.6.

Formation of Shock Waves

*14.7

We have seen in the Doppler effect that the wavefronts produced by a moving source of sound are crowded together in the direction toward which the source is traveling. As the speed of the source increases, the crowding becomes more pronounced. What happens when the speed of the source becomes greater than the wave speed? In this case, the source moves faster than the waves and the arguments used to describe the Doppler effect no longer apply. Instead, the spherical waves expanding from the source at subsequent positions along the path of the source all combine, forming a single conical wavefront known as a **shock wave** (Fig. 14.11). Because the shock wave is composed of many wavefronts acting together, it has a large amplitude.

At time $t = 0$ the source emits a wave from point O. At a later time t, the wavefront has expanded to a radius $r = vt$ and the source has traveled a distance $v_s t$ to reach point S. Subsequent wavefronts also expand, so that at time t they just reach the tangent line drawn from S to the wavefront centered at O. The resulting envelope of wavefronts forms a cone of half-angle θ given by

$$\sin \theta = \frac{v}{v_S}. \tag{14.8}$$

The ratio v_S/v, called the Mach number, is often used to give the speed in terms of the speed of sound. Thus a speed 1.5 times the speed of sound is referred to as Mach 1.5.

When the shock wave is produced by an airplane moving at a speed greater than the speed of sound, that is, at supersonic speed, the shock wave is known as a *sonic boom*. Figure 14.12 shows the shock wave produced in air by a supersonic projectile moving at Mach 1.70. Notice that in addition to the shock wave produced at the front end, lesser shock waves appear at the rear of the projectile. High-speed aircraft often produce two or more shock waves, which are associated with the nose, the tail, and other projections on the aircraft.

Figure 14.11 Shock waves are generated when the source of sound moves faster than the speed of sound waves in the medium. The waves combine to produce a conical wavefront.

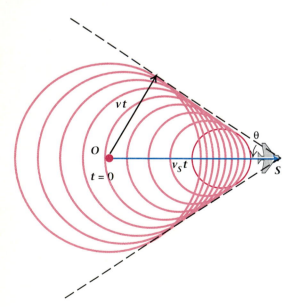

Figure 14.12 A shock wave generated by a projectile traveling at Mach 1.7.

Reflection of a Wave Pulse

14.8

When a wave traveling along a string reaches the end of the string, it is reflected. The exact way in which it is reflected depends on whether the end of the string is fixed or free to move. Let us examine the two cases separately.

When a wave pulse reaches the far end of a string that is fixed to a wall at that end, the wave does not suddenly stop, but is reflected. If no energy is dissipated at the far end of the string, the reflected wave has a magnitude equal to that of the incident wave; however, the direction of the displacement will be reversed (Fig. 14.13a). This reversal happens because as the pulse encounters the wall, the upward force of the pulse on the end of the string pulls upward on the wall. As a result, according to Newton's third law the wall pulls downward on the string. This reaction

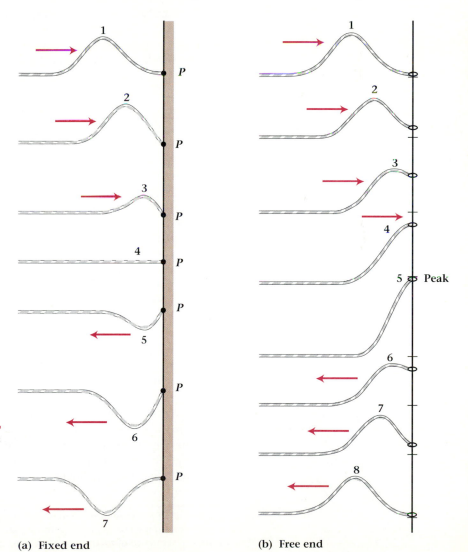

Figure 14.13 (a) Reflection of a wave pulse from a fixed end point. The direction of displacement of the reflected wave is inverted. (b) Reflection of a wave pulse from an end that is free to move. The direction of displacement of the reflected wave is unchanged. (Numbers indicate the sequence of motion.)

(a) Fixed end

(b) Free end

force causes the string to snap downward, initiating a reflected pulse that moves off with an inverted (or negative) amplitude.

What happens, now, if the string is free to move at its far end? Again, a wave pulse traveling along the string is reflected when it reaches that end (Fig. 14.13b). But in this case we see that the reflected wave has the same direction of displacement as the incident wave. As the pulse reaches the end of the string, the string moves up in response to the pulse. As the end of the string begins to return to its initial position, it starts a pulse back along the string, just as if the end motion were due to some outside force. The result is a pulse exactly like the incident wave pulse, except that its direction of travel is reversed.

Standing Waves on a String

14.9

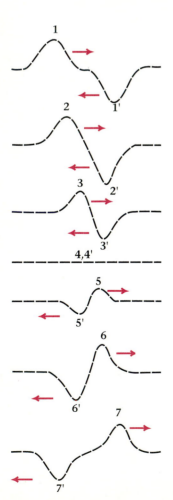

If two wave pulses are started along a rope from opposite ends, the waves will meet, pass through each other, and continue their motion as though nothing had happened. The resultant wave observed during the time the two individual waves overlap is the algebraic sum of the individual amplitudes (Fig. 14.14). The same behavior is displayed by sound waves: When two or three people are talking in the same room, you can distinguish the individual voices even if they speak simultaneously. The loudness of the sound does vary, however, depending on how many voices are heard simultaneously. This effect is an example of superposition. The **principle of superposition** says that at any instant, the resultant combination wave is the algebraic sum of all the component waves. The principle of superposition applies to many types of wave motion, and so has enormous impact—it allows us to analyze combinations of waves. Superposition is one of the major differences between a particle model and a wave model: Particles collide, but waves superpose.

Let us consider the vibrational behavior of a taut, but flexible, string held fixed at each end. A guitar string is a good example. If the string is pulled aside at some point and then released, it begins to vibrate. We can obtain a description of its motion by analyzing it in terms of oppositely directed wave pulses originating at the point where the initial displacement occurred. These waves travel back and forth along the string, undergoing reflection each time they reach the ends of the string. The resulting motion of the string is then the superposition of the amplitudes of all these pulses.

Imagine that the guitar string oscillates with harmonic motion. For a given string, the speed of waves along the string depends on the tension in the string. If the tension is constant, the speed of the waves is also constant. Now, suppose a wave propagates along the string and back, and reaches the origination point at just the right time so that the displacement of the reflected wave is in the same direction as the displacement of the next wave. Then the two displacements add together and the resulting

Figure 14.14   Two wave pulses moving in opposite directions pass through each other. The resultant wave is the algebraic sum of the individual waves. (Numbers indicate the sequence of motion for each wave pulse.)

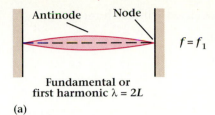

$f = f_1$

Fundamental or
first harmonic $\lambda = 2L$

(a)

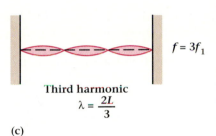

$f = 2f_1$

Second harmonic
$\lambda = L$

(b)

$f = 3f_1$

Third harmonic
$\lambda = \dfrac{2L}{3}$

(c)

Figure 14.15 Standing waves on a string. The points labeled as nodes do not move. The points that vibrate with maximum amplitude are called antinodes.

wave has a larger amplitude. The two waves are said to be *in phase* and the resulting build-up of amplitude is called **constructive interference**. At many other frequencies (and wavelengths) the displacements of the waves are in opposing directions. These waves are said to be *out of phase* and their amplitudes tend to cancel. This effect is called **destructive interference.** Thus waves of these frequencies are suppressed.

If the string is driven periodically, it responds by vibrating at the frequency of the driving force. However, the amplitude of the vibration will be much greater if the string is driven at a resonance frequency. On the other hand, if the string is struck sharply and thereafter allowed to vibrate on its own, only the resonant frequencies will persist.

The lowest resonant frequency of vibration of the string (or other object) is called its **fundamental frequency**. Resonant frequencies that are integer multiples of the fundamental frequency are called **harmonic frequencies**. For our string, the fundamental frequency is the first harmonic; the frequency that is double this value is the second harmonic, and so on. All resonant frequencies higher than the fundamental, whether they are integer multiples of the fundamental or not, are called **overtones**. For example, the overtones of an idealized drum head are not harmonic but stand in the frequency ratio of 1.0 : 1.6 : 2.1 : 2.3.

Each resonant frequency corresponds to an oscillation of the entire string. For the present case, the ends of the string are fixed. The vibrational motion with lowest frequency (first harmonic) is shown in Fig. 14.15(a); this vibration corresponds to a sinusoidal motion of the entire string moving up and down between the supports. At any instant the displacement of the string is described by a sine curve; but this curve does not travel along the string, as we have seen for traveling waves. Instead, the whole string moves up and down. We have a **standing wave**.

Since each end of the string does not move, the end points must be special. They are **nodes**—points of zero vibrational amplitude—and the distance between adjacent nodes is always one half of the wavelength. Halfway between each pair of nodes is an **antinode**, a point on the string that vibrates with the greatest amplitude. Thus the fundamental vibration of a string with fixed end points has a node at each end and a single antinode in between. The length of the string is half a wavelength. The fundamental frequency of the string depends on the speed of waves along the string and on the length of the string. The lowest frequency of vibration of the string is

$$f = \frac{v}{\lambda} = \frac{v}{2L},$$

where L is the length of the string and v is the speed of the wave along the string. Other resonant frequencies also occur. Since the two end points are nodes, the length of the string must be an integral number of half wavelengths. That is, for a string of length L, the resonant wavelengths must be consistent with

$$L = n\frac{\lambda}{2}, \qquad \text{where } n = 1, 2, 3, \ldots$$

For fixed L, there is a wavelength λ_n associated with each integer n,

$$\lambda_n = \frac{2L}{n}.$$

The frequency is obtained from Eq. (14.2),

$$f_n = \frac{v}{\lambda_n} = n\,\frac{v}{2L}. \tag{14.9}$$

For flexible strings the wave speed is given in meters per second by

$$v = \sqrt{\frac{T}{m/L}}, \tag{14.10}$$

where T is the tension in the string in newtons and (m/L) is the linear mass density in kilograms per meter. The resonant frequencies of a stretched flexible string are then given by

$$f_n = \frac{n}{2L}\sqrt{\frac{T}{m/L}}, \qquad n = 1, 2, 3 \ldots \tag{14.11}$$

Increasing the tension in the string raises the speed of waves along it and thus raises the natural vibrational frequencies. When you hear a stringed instrument being tuned, the musician is adjusting the tension in the strings. In addition, by pressing the string down at different points on the fingerboard, it is possible to shorten the vibrational length of the string— and thus increase the fundamental frequency—by varying amounts. That is how a musician is able to play many different notes with one string.

Recall from our earlier discussion that the harmonic frequencies of a vibrating string are integer multiples of the fundamental frequency. A frequency of twice the fundamental frequency is called the second harmonic or first overtone, one of three times the fundamental is the third harmonic or second overtone, and so on. Not all harmonics are present in all vibrating systems. We will see an example in Section 14.10.

A string that is plucked vibrates in a complicated way, with many of its harmonics contributing to its motion. In fact, it is this combination of harmonics that gives a vibrating string its distinctive sound. The same is true for the human voice. You recognize someone's voice not just because of its pitch and the way that person enunciates, but also because of that person's particular combination of overtones.

Example 14.8

Fundamental frequency of a guitar string.

What is the fundamental frequency of a guitar string whose tension is adjusted to give a wave speed of 143 m/s if the length of string free to vibrate is 0.65 m?

Solution   The fundamental wavelength of a string fixed on both ends is twice the length of the string. The frequency is found from the ratio of sound speed to wavelength, as in Eq. (14.9), where we set $n = 1$ for the fundamental frequency:

$$f = \frac{v}{\lambda_n} = \frac{v}{2L} = \frac{143 \text{ m/s}}{2(0.65 \text{ m})} = 110 \text{ Hz.}$$

The sound from many musical instruments, such as pianos, guitars, and violins, originates in a set of strings, each vibrating with a specific fundamental frequency. In music the term frequency is not usually employed, but other terms are used to describe essentially the same thing. When the frequency of two tones is in the ratio of 1 : 2, they are said to be one *octave* apart. Between these two tones, each octave contains intermediate tones, arranged in a sequence that is repeated in every octave. Specifically, in Western music the octave is divided into twelve semitones, of which seven are selected to form a major or minor scale. A scale is just a regular series of tones arranged in order of frequency. A scale using the full sequence of twelve semitones is called a chromatic scale.

To Western ears, pleasing combinations of tones are those with frequencies that stand in whole-number ratios, such as 1 : 2, 4 : 5, and even 4 : 5 : 6. Without going into detail, we will say that difficulties arise when this idea is used to fill an octave with twelve tones. Therefore several scales have been developed that almost satisfy the requirements. The scale to which most musical instruments are tuned today is called the *equal-tempered scale*. In this scale each of the twelve tones in an octave has a frequency $\sqrt[12]{2}$ times the frequency of the previous one. As you can see (Table 14.2), the frequency ratios are not exactly whole numbers, although they are fairly close.

Although we refer to an acoustic guitar or a violin as a "stringed

TABLE 14.2
Frequencies of the notes in an octave beginning with middle C

Name	Equal-tempered frequency	Ratio to C_4	Nearest whole-number ratio to C_4	
C_4	261.63	1.0000	1.0000	
$C_4^\#$	277.18	1.0594		
D_4	293.66	1.1224	1.1250	(9/8)
$D_4^\#$	311.13	1.1892		
E_4	329.63	1.2599	1.2656	(81/64)
F_4	349.23	1.3348	1.3333	(4/3)
$F_4^\#$	369.99	1.4142		
G_4	392.00	1.4983	1.5000	(3/2)
$G_4^\#$	415.30	1.5874		
A_4	440.00	1.6818	1.6875	(27/16)
$A_4^\#$	466.16	1.7818		
B_4	493.88	1.8877	1.8984	(243/128)
C_5	523.25	2.0000	2.0000	

Figure 14.16 A guitar player changes the length of the vibrating string by pressing the string against the neck of the instrument with his fingers.

instrument," the sound does not come directly from the string. A plucked guitar string that is stretched between two stone columns can be heard only by placing your ear close to the string. The string has too little surface area and insufficient amplitude to set much air in motion. Instead, the sound you hear from a guitar or violin comes from the vibrating body of the instrument. The frequency that is heard depends on the length of the particular string plucked or bowed. As we noted earlier, this length is determined by the performer, who chooses where to press the string to the fingerboard (Fig. 14.16).

The quality of the tone you hear, say from a violin, results primarily from the way the body of the violin is made. That is, the construction of the instrument determines the relative strength of the overtones for each tone that the body radiates. The maker must choose exactly which pieces of wood to use, their thickness at every point, the way in which the parts are joined, the finishing process, and many other details of the construction. Although it is now fairly well known what acoustical properties a violin must have to give it a good tone over its entire range, to actually build a good instrument requires knowledge, skill, and practice.

Waves in a Vibrating Column of Air

14.10

Hollow pipes have long been used for making musical sounds. Flutes, organ pipes, and children's whistles produce sounds in similar ways. To understand how they work, let us examine the behavior of air in a hollow pipe that is open at both ends. If you blow air across one end, the disturbance due to the moving air at that end propagates along the pipe to the far end. When it reaches the far end, part of the wave is reflected, in analogy with a wave reflected along a string whose end point is free to move. Since there is motion of the air at that end, the end point is an antinode with respect to the flow of air (Fig. 14.17). All harmonic frequencies are possible, just as for a string. And, as with the string, the fundamental wavelength is twice the length of the pipe.

If one end of the pipe is closed off, the air is not free to move any further in that direction and the closed end becomes a node. The resonant behavior of the pipe is completely changed. Since one end is a node and the other is an antinode, the lowest frequency (longest wavelength) vibration has no other nodes or antinodes between the ends. Thus the fundamental wavelength is four times the length of the pipe (Fig. 14.18). The fundamental frequency is therefore much lower than for a pipe of the same length but open on both ends.

Above the fundamental frequency, the next resonant vibrational mode

Figure 14.17 Standing waves in an open pipe. The ends of the pipe are antinodes. The fundamental resonant wavelength is twice the length of the pipe.

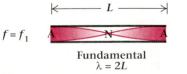

$f = f_1$

Fundamental
$\lambda = 2L$

$f = 2f_1$

Second harmonic
$\lambda = L$

$f = 3f_1$

Third harmonic
$\lambda = \dfrac{2L}{3}$

Fundamental
$\lambda = 4L$

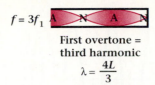

$f = 3f_1$

**First overtone =
third harmonic**
$$\lambda = \frac{4L}{3}$$

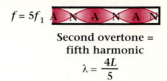

$f = 5f_1$

**Second overtone =
fifth harmonic**
$$\lambda = \frac{4L}{5}$$

Figure 14.18 Standing waves in a pipe closed at one end. The fundamental resonant wavelength is four times the length of the pipe.

in our closed pipe has both an antinode and a node between the ends. Since this mode represents the next higher frequency, it is called the first overtone. But it is not the first harmonic. Its wavelength is one third of the fundamental wavelength and its frequency is three times the fundamental, so it is the third harmonic. Similarly, the next higher frequency, the second overtone, is the fifth harmonic. Thus for a pipe closed at one end, not all harmonics are possible; only odd-numbered harmonics can be excited.

What harmonics are excited also depends on the initiating disturbance. For example, if you blow gently across the top of a pop bottle, it resonates softly at its fundamental frequency. But if you blow much harder, the higher pitch of an overtone is heard because the faster airstream creates higher frequencies in the exciting disturbance. This same effect can also be achieved by increasing the air pressure to an organ pipe.

Beats

*14.11

When two sound sources that have almost the same frequency are sounded together, an interesting effect occurs. You hear a sound with a frequency that is the average of the two. However, the loudness of this sound repeatedly grows and then decays, rather than being constant. Such repeated variations in amplitude are called **beats**, and the occurrence of beats is a general characteristic of waves.

If the frequency of one of the wave sources is changed, there is a corresponding change in the rate at which the amplitude varies. This rate is called the beat frequency. As the frequencies come closer together, the beat frequency becomes slower. Thus a musician can tune a guitar to another sound source by listening for the beats while increasing or decreasing the tension in each string. Eventually the beats become so slow that they effectively vanish, and the two sources are then in tune.

Beats are easily explained by considering two sinusoidal waves y_1 and y_2 of the same amplitude y_0, but of different frequencies f_1 and f_2. The superposition principle says the combined amplitude y is the algebraic sum of the individual amplitudes:

$$y = y_1 + y_2,$$
$$y = y_0 \sin 2\pi f_1 t + y_0 \sin 2\pi f_2 t,$$
$$y = y_0 (\sin 2\pi f_1 t + \sin 2\pi f_2 t).$$

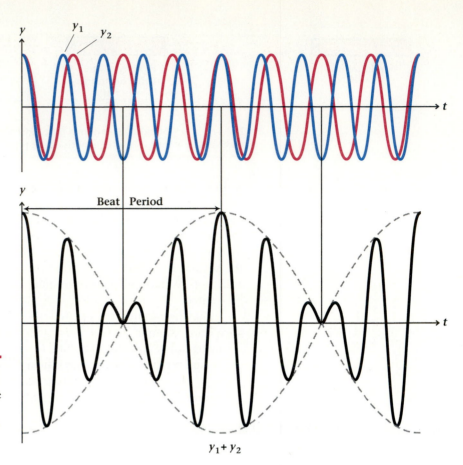

Figure 14.19 Two waves of slightly different frequencies add to produce a wave that oscillates with the average of their frequencies. The resulting amplitude also oscillates, with a beat frequency equal to the difference between the source frequencies.

Using the trigonometric identity for the sum of the sines of two angles, we have*

$$y = \left[2y_0 \cos 2\pi \left(\frac{f_1 - f_2}{2} \right) t \right] \sin 2\pi \left(\frac{f_1 + f_2}{2} \right) t. \qquad (14.12)$$

This equation represents the amplitude of a sound wave with a frequency that is the average of the two source frequencies (the sine term) and with a slowly varying amplitude (the cosine term) (Fig. 14.19). Our sensation of loudness depends on the intensity, which is proportional to the square of the amplitude. The rate at which the loudness maxima occur is the beat frequency. Its magnitude is the difference in the two source frequencies, $|f_1 - f_2|$. Beats can easily be heard up to frequencies of about 10 Hz. Beyond that, they are hard to distinguish.

---

*The following relationship is derived from the definition of the trigonometric functions:

$$\sin A + \sin B = 2 \cos \frac{A - B}{2} \cdot \sin \frac{A + B}{2}.$$

SUMMARY

Useful Concepts

- The general expression for a one-dimensional traveling wave is

$$y(x,t) = y_0 \sin 2\pi \left(\frac{x}{\lambda} - \frac{t}{T} \right).$$

- The relationship between the speed, the frequency, and the wavelength of a wave is

$$v = \lambda f.$$

- The intensity of a wave on a rope is the power per unit area transmitted by the wave,

$$I = 2\pi^2 \rho v f^2 y_0^2,$$

where ρ is the density of the medium.

- For a spherically expanding wave, the intensity decreases with the square of the distance from the source,

$$I = \frac{I_0}{r^2}.$$

- The intensity level IL of a sound is measured in decibels relative to a a reference sound intensity I_0:

$$\text{IL (in dB)} = 10 \log_{10}(I/I_0).$$

- The observed frequency change due to motion of the source or observer is called the Doppler shift. The frequency heard is given by

$$f' = \frac{f}{(1 \mp v_s/v)}, \qquad \text{moving source,}$$

$$f' = f(1 \pm v_O/v), \qquad \text{moving observer.}$$

The upper sign in these equations corresponds to approaching motion and the lower sign corresponds to leaving.

- If the speed of the source exceeds the speed of the wave, a shock wave is formed. The shock wavefront makes an angle θ with the direction of motion of the source, given by

$$\sin \theta = \frac{v}{v_S}.$$

- The frequency of standing waves in a string is

$$f_n = \frac{n}{2L} \sqrt{\frac{T}{m/L}}, \qquad n = 1, 2, 3 \ldots$$

- When two sounds of almost the same frequency are heard at the same time, the sensation is of a single tone whose amplitude is rising and falling. These variations in amplitude are called beats. The rate at which they occur, called the beat frequency, is

$$|f_1 - f_2|.$$

Important Terms

You should be able to write the definition or meaning of each of the following terms:

- transverse wave pulse
- traveling wave
- wave
- harmonic wave
- wavelength
- intensity
- transverse wave
- longitudinal wave
- wavefront
- plane wave
- decibel
- intensity level
- Doppler effect
- shock wave
- principle of superposition
- constructive interference
- destructive interference
- fundamental frequency
- harmonic frequencies
- overtone
- standing wave
- node
- antinode
- beats

QUESTIONS

14.1 A stereo amplifier is designed to have a channel separation of 40 dB. Explain what this means.

14.2 You can estimate the distance to a lightning flash by counting the seconds that elapse between the time you see the flash and when you hear the accompanying thunder. (a) The approximate distance in miles is obtained by dividing the time in seconds by 5. Explain why this is so. (b) Estimate the distance to a lightning flash that you see 10 s before you hear the thunder.

14.3 The speed of sound in helium is much faster than in air at the same temperature and pressure. Use this fact to explain the observation that if you fill your lungs with helium and try to talk, your voice will be higher pitched than normal.

14.4 If an organ pipe is placed in a large chamber in which the air pressure is reduced to slightly below atmospheric pressure, will the frequency of the sounded note change?

14.5 If the whine of a bullet were 1000 Hz as it passed through the air, what would you hear if the bullet were traveling away from you with the speed of sound?

14.6 Discuss why the lower-pitched strings in most musical instruments are wrapped with a thin coil of wire.

14.7 How could you get a stretched string to vibrate in only its second harmonic mode?

14.8 Describe what you will hear in the following situation: You swing a constant frequency source attached to a string around your head in a 1-m-diameter horizontal circle. You are standing near a brick wall that reflects the sound.

14.9 If you moisten your finger and rub it around the rim of a thin-stemmed glass, you often produce a sound. How is this tone produced? On what does it depend? Why must you wet your finger? What changes will take place if the glass is half filled with water?

14.10 What does it mean to say that musical scales are logarithmic?

PROBLEMS

Hints for Solving Problems

Assume the speed of sound to be 340 m/s corresponding to a temperature of about 15°C and ignore variations due to temperature unless specifically stated otherwise in the problem. For harmonic waves, assume that $y = 0$ at $x = 0$ when $t = 0$ (corresponding to a sine wave). Remember to use the correct sign of v_S or v_O when calculating Doppler shifts. Also be sure to determine whether a vibrating string or column of air has nodes or antinodes at its ends.

Section 14.2   Harmonic Waves

14.1 Harmonic waves are sent along a rope at a speed of 10 m/s. What is the frequency of the wave if successive wave crests are 1.5 m apart?

14.2 Write the general expression for a traveling harmonic wave of amplitude y_0 moving in the negative x direction with wavelength λ and period T.

14.3 Rewrite the general expression for a traveling harmonic wave moving in the positive x direction in terms of the angular frequency ω and the wave number $k = 2\pi/\lambda$.

14.4 A harmonic wave moving in the positive x direction has an amplitude of 3.0 cm, a speed of 40 cm/s, and a wavelength of 40 cm. Calculate the displacement due to the wave at: (a) $x = 0.0$ cm, $t = 2.0$ s; and (b) $x = 10$ cm, and $t = 20$ s.

14.5 Radio waves travel with the speed of light, approximately 3.00×10^8 m/s in both vacuum and air. What are the wavelengths of the radio signals from the following stations? (a) An FM station broadcasting at 92.9 MHz. (b) A standard broadcast AM station operating at 536 kHz. (c) A shortwave station operating at 11,900 kHz?

14.6 A shortwave receiver has "90-m band" written on one end of the dial and "11-m band" on the other. What range of frequencies does this radio receive? (*Hint:* Take the speed of radio waves to be 3.00×10^8 m/s.)

14.7 The human eye is sensitive to light of wavelengths from about 400 to 700 nm. What range of frequencies does this correspond to? (*Hint:* Take the speed of light to be 3.00×10^8 m/s.)

*Section 14.3   Energy and Information Transfer by Waves

14.8 Show that the expression for the power transmitted by a harmonic wave on a rope can also be written as

$$P = \tfrac{1}{2}\mu v \omega^2 y_0^2,$$

where μ is the mass per unit length of the rope and $\omega = 2\pi f$.

14.9 A child sends traveling waves along a rope that passes through the narrow slits in a picket fence. The waves are transmitting 0.30 W. Another child puts some boards on the fence to restrict the amplitude of the waves that can pass, so they become only one third of the original amplitude. What must the first child do to maintain the same level of power transmission?

14.10 A 6.0-m-long string with a circular cross section of 2.0 mm² has a mass of 8.5 g. Transverse waves propagate along the string with a speed of 189 m/s. One end of the string is forced to oscillate at 120 Hz with an amplitude of 0.53 cm. What power is transmitted along the string?

Section 14.4   Sound Waves

14.11 Calculate the speed of sound in dry air at a temperature of 35°C.

14.12 (a) What is the wavelength of a 2000-Hz sound wave in air? (b) What is the wavelength of a 2000-Hz sound wave in water?

14.13 (a) Calculate the frequencies of the pressure waves in air having the following wavelengths: 3.4 cm, 34 cm, 340 cm, and 34 m. (b) Which of these frequencies can be heard by human ears?

14.14 What is the wavelength of the waves produced in air by an ultrasonic distance measuring device if the frequency is 40 kHz?

14.15 What are the shortest and longest wavelengths produced by a stereo system that has a useful range of 50 to 17,000 Hz?

14.16 You are 10 m from a loudspeaker, where you hear a sound intensity I_0. How far must you be from the speaker to reduce the sound intensity that you hear to $I_0/3$? (*Hint:* Assume spherical waves.)

14.17 Show that for spherical harmonic traveling waves the amplitudes y_i at distances r_i from the source are related through

$$\frac{y_1}{y_2} = \frac{r_2}{r_1},$$

where the subscripts 1 and 2 refer to different locations.

14.18 When sounded, two identical car horns produce sound intensities at your ear in the ratio of 6 : 1. If the nearer car is 8.0 m away from you, how far away is the other car? (*Hint:* Assume spherical wavefronts.)

14.19 A loudspeaker sounds a 500-Hz tone. What is the difference in the wavelength of the sound waves produced in air at −3°C and at 38°C? (*Hint:* Use the expression for the speed of sound as a function of temperature.)

14.20 Reference is sometimes made to a singer "hitting high C." What is the wavelength of this sound at 12°C? (The frequency of high C, more precisely called C_6, is 1047 Hz.)

*Section 14.5   Measuring Sound Levels

14.21 What is the ratio of the sound intensity near a power saw to that in a recording studio?

14.22 An audio amplifier has a power gain of 100,000. Express this gain in units of decibels.

14.23 A sound is 5000 times as intense as the minimum threshold of hearing. What is the intensity level of this sound in decibels?

14.24 If one sound is 25 dB greater than another sound, what is the ratio of their intensities?

14.25 In a good FM radio receiver, the radio signal detected may be as much as 65 dB greater than the noise signal. What is the ratio of signal intensity to noise intensity?

14.26 The upper limit of the intensity level of sound waves is about 194 dB. This intensity corresponds to 100% modulation of the air; that is, the amplitude of the pressure oscillations corresponds to that of atmospheric pressure. What intensity (in watts per square meter) would be required to generate this level of sound?

14.27 When listening to ordinary conversation at 68 dB, what is the intensity of the sound at your ears? Express your answer in watts per square meter.

Section 14.6   The Doppler Effect

14.28 A train approaching a station at a speed of 40 m/s sounds a 2000-Hz whistle. (a) What is the apparent frequency heard by an observer standing at the station? (b) What is the drop in frequency heard as the train passes by?

14.29 A train going 40 m/s approaches a crossing bell whose frequency is 820 Hz. (a) What frequency is heard by passengers on the train? (b) What frequency is heard after the train passes the bell?

14.30 You are standing by the railroad track when a train sounding a 750-Hz whistle passes you at 80 km/h. What is the difference in the frequency you hear from the train approaching and departing?

14.31 A 440-Hz source has been sounding in air for a long time. (a) What frequency will you hear if you move away from it at 0.90 times the speed of sound? (b) What frequency will you hear if you move away from it at the speed of sound?

*Section 14.7   Formation of Shock Waves

14.32 A rocket model in a wind tunnel generates a shock wave with a half-angle of 65° when the gas is flowing at 400 m/s. What is the speed of sound in the gas?

14.33 A bullet traveling in helium at 15°C creates a shock wave with a half-angle of 45.0°. What is the half-angle of the shock wave if the bullet travels with the same velocity through air at the same temperature?

14.34 You hear the sonic boom of a high-speed jet plane exactly 3.0 s after it passes directly overhead in level flight. At the time you hear the boom, you see the plane at an angle of 20° above the horizon. How fast is the plane traveling? Assume that the speed of sound at the altitude of the plane is 325 m/s.

14.35 What is the half-angle for the shock wave generated by the Concorde aircraft when it is flying at Mach 2.1?

14.36 What is the speed of the air flow around a model airplane in a wind tunnel if the half-angle of the shock wave is 37°? Give your answer as a Mach number.

14.37 A projectile moving at a speed of 418 m/s through an atmosphere of pure nitrogen creates a shock wave with a half-angle of 53°. What is the speed of sound in the nitrogen?

Section 14.9   Standing Waves on a String

14.38 A guitar string is stretched between supports 0.65 m apart. When the string is plucked the fundamental frequency heard is 440 Hz. What is the speed of sound in the string?

14.39 A nylon string is stretched between supports 1.20 m apart. Given that the speed of sound in the string is 800 m/s, find the frequency of the fundamental vibration and the first two overtones.

14.40 A steel wire is stretched taut between supports one meter apart. (a) What is the fundamental wavelength of vibration of the wire? (b) What is the fundamental frequency if the speed of sound in the wire is 2050 m/s? (c) What is the wavelength of the fundamental frequency in air?

14.41 The vibrations from an 800-Hz tuning fork set up standing waves in a string clamped at both ends. The wave speed in the string is known to be 400 m/s for the tension used. The standing wave is observed to have four antinodes and an amplitude of 2.0 mm. How long is the string?

14.42 A 5.00-kg mass is suspended from the ceiling by a 20.0-g wire 1.60 m long. What is the fundamental frequency of the wire?

14.43 Two adjacent strings on an unusual stringed instrument have the same mass per unit length and are subjected to the same tension. One of the strings is 1.44 times as long the other. What is the ratio of the fundamental frequency of the longer string to the second harmonic of the other one?

14.44 Use graphical techniques to draw the wave that is the superposition of two harmonic traveling waves, each with amplitude of 2.0 cm, one with wavelength of 3.0 cm and the other with wavelength of 5.0 cm. Sketch the wave at time $t = 0$.

14.45 An experiment shows that a pulse propagates with a speed of 260 m/s along a nylon cord subject to a tension of 68.0 N. What is the mass per unit length of the cord?

14.46 A wire of mass 0.030 kg is stretched between fixed supports a distance 1.50 m apart. (a) If the tension in the wire is 850 N, what is the speed of a wave along the wire? (b) How does the wave speed compare with the speed of sound in air?

14.47 What is the frequency ratio of two notes that are two octaves apart?

14.48 What is the ratio of frequencies of a higher note to a lower note that is 16 semitones below it on the equal-tempered scale?

14.49 A string stretched between two supports sets up standing waves with two nodes between the ends when driven at a frequency of 230 Hz. (a) At what frequency will it have three nodes? (b) What order harmonic is such a wave? (c) What is the frequency of the fundamental?

Section 14.10   Waves in a Vibrating Column of Air

14.50 Calculate the fundamental frequency and the first three overtones of a hollow pipe 30 cm long and open at both ends.

14.51 (a) What is the fundamental frequency of a hollow tube 50 cm long and open at both ends? (b) What would be the frequency if the tube were closed at one end?

14.52 Calculate the fundamental and first three overtones of a hollow pipe 25 cm long and closed at one end.

14.53 Low-frequency standing waves can sometimes be generated in tall cylindrical structures, such as unused smokestacks. What would be the frequency of the fundamental acoustical vibration in a 50-m-tall smokestack if it were closed at the bottom?

*Section 14.11   Beats

14.54 Two tuning forks are sounded simultaneously and a beat frequency of 2.0 Hz is heard. If the frequency of the higher-pitched fork is 262 Hz, what is the frequency of the other one?

14.55 Two identical guitar strings are stretched between supports that are not the same distance apart. The fundamental frequency of the higher-pitched string is 400 Hz and the speed of sound in both wires is 150 m/s. How much longer is the lower-pitched string if the beat frequency is 2.0 Hz?

14.56 How many beats are heard when two organ pipes, each open at both ends, are sounded together if one pipe is 50 cm long and the other is 52 cm long?

14.57 Piano tuners use overtones and beats as an aid in tuning a piano. If a tuner hears a beat frequency of 1.0 Hz between the second harmonic of a string whose fundamental frequency is 440 Hz and the third harmonic of another note, what is the other note? (*Hint:* Assume that the tuner has already determined that the second harmonic of 440 Hz is the higher of the two frequencies.)

Additional Problems

14.58 A uniform string of linear density 10 g/m is tied to
• the ceiling of an elevator. A 5.0-kg mass is hung from the other end of the string at a point 1.00 m from the ceiling. If the elevator accelerates upward at 0.10g, what is the fundamental frequency of the string?

14.59 A wire stretched between supports 50 cm apart has
• a fundamental frequency of 400 Hz. Can it be reso-

nantly excited by an object vibrating at a frequency of (a) 2000 Hz? (b) 3000 Hz?

14.60 A strip of paper is pulled along at 20 cm/s at a right
• angle to the plane of a 0.49-m-long pendulum. The pendulum has a small, continuously running ink-jet device attached to the bob, which is arranged so that it marks the position of the pendulum on the paper (Fig. 14.20). The bob is held 8.0 cm to one side and released at $t = 0$. Write the equation for the wave drawn on the paper.

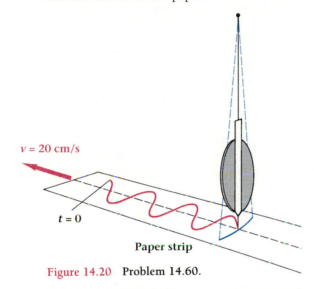

$v = 20$ cm/s

$t = 0$

Paper strip

Figure 14.20 Problem 14.60.

14.61 A cart moves with a constant speed of 0.30 m/s in
• the x direction. A 2.3-kg lead brick is hung from a support on the cart with a spring of spring constant $k = 20$ N/m. The brick is pulled down 6.0 cm below the equilibrium point and released at $t = 0$. Write an expression for the y position of a red dot painted on the center of the brick as a function of x and t. Assume that $x = 0$ when $t = 0$. You may wish to call the equilibrium position $y = 0$.

14.62 A car moving with a speed of 20 m/s sounds a horn
• of frequency 1000 Hz. (a) What is the frequency heard by a stationary observer positioned ahead of the car? (b) What frequency does the observer hear if the wind is blowing at a speed of 5 m/s from the car toward the observer?

14.63 A car traveling 30 m/s overtakes another car going
• only 25 m/s. When the faster car is still behind the slower one, it sounds a horn of frequency 1500 Hz. What is the frequency heard by the driver of the slower car?

14.64 A trailer truck traveling east at 30 m/s sounds
• • a 1000-Hz horn. (a) What frequency is heard by an approaching driver headed west at 40 m/s? (b) What frequency is heard if the approaching driver is also headed east?

14.65 A student whirls an electronic buzzer around her
• • head. The buzzer, which makes an 800-Hz tone when at rest, is swung on a string in a 0.75-m-radius horizontal circle at three revolutions per second. Derive an expression for the frequency as a function of time of the sound heard by a distant observer.

14.66 An automobile passes by on the street blowing its
• horn. A music student standing on the sidewalk observes that the pitch of the horn drops by a musical fifth; i.e., the frequency heard after it passes is only three quarters of the frequency heard when it was approaching. (a) Calculate the speed of the car in meters per second. (b) Also give your answer in units of kilometers per hour.

14.67 A copper and a steel pipe are of the same diameter
• • and length at 8°C. The fundamental frequency of the air resonance of both is 180 Hz at this temperature when open at both ends. How many beats per second will be heard at 27°C?

14.68 At $x = 15.0$ cm and $t = 2.00$ s the displacement
• of a traveling wave is 8.66 cm. If the amplitude of the wave is 10.0 cm and its wavelength is 8.00 cm, what is its period? (*Hint:* Assume that at $t = 0$, $y = 0$ at $x = 0$.)

14.69 A traveling wave has a wavelength of 0.25 m and a
• • period of 0.10 s. (a) If you stand in one place, how many crests pass you per second? (b) If you travel in the same direction as the wave at a speed of 1.0 m/s, how many crests pass you in one minute? (c) If you travel in the same direction as the wave at a speed of 3.0 m/s, how many crests do you intercept in one minute? (d) If you go in the opposite direction from the direction of the wave motion at 0.50 m/s and start from a crest at $t = 0$, how far away from you will that same crest be at $t = 32$ s?

14.70 Rewrite the expression for the speed of sound as a function of temperature so that the speed is in feet per second and the temperature in degrees Fahrenheit.

14.71 Show that the change in wavelength $\Delta\lambda$ of the
• • sound from a source due to a change in the temperature of the air ΔT is approximately

$$\Delta\lambda \approx (1.8 \times 10^{-3})\lambda\,\Delta T,$$

where $\Delta\lambda$ and λ are in meters and ΔT is in degrees Celsius.

14.72 In Chapter 12 we stated that the speed of sound in
• • a gas is proportional to v_{rms} of the gas molecules and therefore is proportional to the square root of the temperature in Kelvins. In this chapter we asserted that the speed of sound is well represented by a constant plus a term that is proportional to the temperature in degrees Celsius. Compare these two statements by plotting both predictions on the same

graph from 0°C to 100°C and determining the maximum percentage difference between the two predictions in the specified temperature range. (*Hint:* The expression from Chapter 12 may be written in the form $v = 331.5\sqrt{T/273.15}$, where T is in kelvins.

14.73 By how many decibels do you reduce the sound
• intensity level due to a source of sound if you triple your distance from it? (Assume that the waves expand spherically.)

14.74 If you detect a frequency shift of 30 Hz in the 6000-
• Hz bell of a bicycle as it approaches and then leaves you, how fast is the bicycle going?

14.75 (a) Derive an expression for the frequency as a func-
• • tion of time that you would hear if you dropped a source of frequency f_0 from a tall tower. (b) Derive an expression for the frequency heard by an observer on the ground.

14.76 If you move at 15 m/s toward a 2000-Hz source that
• is moving toward you with a ground speed of 40 m/s, what frequency do you hear?

14.77 Show that the expression for a standing wave may
• • be derived by adding together two traveling waves that are of the same amplitude and frequency but that travel in opposite directions.

14.78 The six strings of a classical guitar are all 65.5 cm
• long and are tuned to frequencies of 82, 110, 147, 196, 247, and 330 Hz. The mass per unit length of the set of strings is, starting with the one of lowest frequency, 5.05×10^{-3}, 3.71×10^{-3}, $2.21 \times$ 10^{-3}, 1.01×10^{-3}, 0.58×10^{-3}, and 0.44×10^{-3} kg/m. (a) What is the tension in each string? (b) What is the total string force?

14.79 Two strings are each 1.00 m long with a mass of
• • 0.10 g. Both are subjected to the same tension and are located in an elevator. One string is stretched between fixed supports. One end of the other string is fixed to the ceiling of the elevator while the other end supports a 6.00-kg mass. When the elevator is at rest, both strings vibrate with the same fundamental frequency. When the elevator accelerates upward, the notes are different and a beat frequency of 2 Hz is heard. What is the acceleration of the elevator?

14.80 Singing in the shower is a habit with some people,
• • even those who do not normally sing elsewhere. To get some insight into the reason for this, calculate the first four harmonics associated with each of the linear dimensions of a 1.65 m × 2.30 m × 2.43 m room and put them in ascending order. Assume that $v = 340$ m/s.

14.81 Five identical looms operating in a textile mill pro-
• • duce an average sound intensity level of 85 dB. If two additional looms are put into operation in the same room, what is the new sound intensity level?

14.82 The sound intensity level in a factory is 93 dB when
• • seven identical metal stamping machines are all in operation. If three of the machines are shut down for maintenance, what is the sound intensity level due to the remaining machines?

ADDITIONAL READING

Fletcher, N. H., and S. Thwaites, "The Physics of Organ Pipes." *Scientific American,* January 1983, p. 94.

Mathews, M. V., and J. R. Pierce, "The Computer as a Musical Instrument." *Scientific American,* February 1987, p. 126.

Rossing, T. D., *The Science of Sound.* Reading, Mass: Addison-Wesley, 1982. This book covers a broad range of topics from musical instruments to high-fidelity electronics to the acoustical properties of auditoriums. It contains much material not previously available in one book. The mathematical level of Rossing's book is about the same as in this text.

Schroeder, M. R., "Toward Better Acoustics for Concert Halls." *Physics Today,* October 1980, p. 24.

15

Electric Charge and Electric Field

15.1 Electric Charge

15.2 Coulomb's Law

15.3 Superposition of Electric Forces

15.4 The Electric Field

15.5 Superposition of Electric Fields

15.6 Electric Flux and Gauss's Law

*15.7 A Quantitative Approach to Gauss's Law

15.8 The Electric Dipole

A WORD TO THE STUDENT

You are already familiar with many electrical phenomena. Lightning is one; another is the shock you get upon touching a door knob after walking across a rug on a dry day. In each case you see a brief spark, but the effect does not persist. Such events are due to what we call static electricity.

Electricity and magnetism are different aspects of a single fundamental force of nature, called electromagnetism. We will study the far-reaching effects of this force in the next six chapters. We begin by considering the forces that arise from static electricity. As you will see, the law describing the electrostatic force has the same mathematical form as the law of universal gravitation. We then define the electric field in a manner similar to the way we defined the gravitational field, and we describe the behavior of electrified bodies in terms of their interactions with an electric field. Because the field concept plays a prominent role in the theory of electricity, you must give particular attention to the field and its applications in order to understand the chapters that follow. The laws of electricity given here apply both to large-scale events and to atomic and nuclear interactions, as you will see in later chapters.

Human beings have known about electrical effects for thousands of years. As far back as the ninth century B.C., people searched the seashores for a yellowish-brown fossil resin called amber, which they carved and polished into beads and jewelry. By 600 B.C. the Greeks had discovered that when amber was rubbed briskly with a cloth, it would attract and pick up small, light objects such as bits of feathers or straw, or even thin scraps of metal. The Greek word for amber is *elektron,* and it is from this root word that we get our word *electricity.* Lightning is another example of an electrical effect in nature (Fig. 15.1). However, the relationship between lightning and the behavior of amber went unrecognized for centuries. Similarly, early sailors often watched the brush of light called St. Elmo's fire that danced atop their ship's mast during storms, but it, too, was looked upon as a completely separate phenomenon from lightning. Still another effect observed in nature is the ability of some fish and eels to shock their prey. All of these seemingly different effects are examples of electricity.

As we will see, the basic element of electricity is electric charge, which is a fundamental property of matter. The Englishman William Gilbert (1544–1603) was an early investigator of what we now call **electrostatics**, the study of electric charges at rest. Gilbert used a simple but sensitive test instrument based on a lightweight stick pivoted about its center on a needle point. Gilbert found that if he rubbed a piece of amber and held it near the end of the stick, the stick moved noticeably about its pivot, indicating an attraction by the amber. He tried rubbing other objects and found that many of them produced similar effects. In this way Gilbert found a force that was not present before the amber or other substance was rubbed. This force was due, not to gravitation, but to a different source: electricity.

As we study electricity, we still use the concepts of mechanics, especially such ideas as the conservation of energy and the relationship between force and acceleration. In fact, the nongravitational forces that we studied in mechanics, such as friction and Hooke's law, are actually due to electrical forces between molecules and atoms. The common contact forces of daily life are electrical in origin. Consequently, understanding the electrical nature of matter helps us understand mechanics at the molecular level.

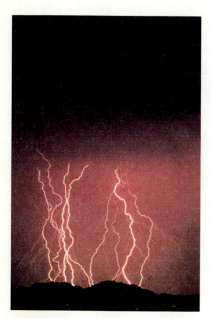

Figure 15.1 Lightning, a dramatic example of electricity.

Electric Charge

15.1

The realization that the key to making electricity was the rubbing of material led to the development of frictional electric machines. The first true frictional electric machine was developed about 1705 by Francis Hauksbee. Hauksbee's machine consisted of a glass sphere turned by a crank (Fig. 15.2). When rubbed, the sphere was capable of making sparks and causing light threads to stand straight out, thereby demonstrating the ability of electricity to repel as well as to attract. An object such as Hauksbee's globe is said to become "charged" when rubbed, meaning that it possesses a net

Glass sphere

Figure 15.2 Hauksbee's electrostatic machine. The glass sphere was turned by a crank, becoming charged by friction when rubbed.

electric charge. It is this net charge that is the source of the electric force of attraction or repulsion. Charge, like mass, is a fundamental property of matter.

In 1731 Stephen Gray announced the discovery of what is today called conduction of electricity. Through a series of experiments, Gray found that he could transmit electric charge over a great distance through a metal wire supported by silk cords. This observation led to the classification of materials into conductors and insulators. A **conductor** is a material through which charge may flow easily; an **insulator** is a material through which charge flows poorly or not at all.*

Two years later, in 1733, Charles-François du Fay, superintendent of gardens to the king of France, discovered that rubbing various objects together could generate two different kinds of electricity. Du Fay found that two pieces of rubbed glass repelled each other, but rubbed amber was strongly attracted to rubbed glass (Fig. 15.3). His experiments provided evidence for two kinds of electric charge and for a basic rule describing their behavior: **Like charges exert a repulsive force on each other** and **unlike charges exert an attractive force on each other**.

To explain this observed behavior of electricity, Benjamin Franklin (1706–1790) proposed a "single fluid" theory of electricity. According to his theory only one kind of charge is mobile. The two types of electric behavior—that found on rubbed amber and that found on rubbed glass— are due to a deficiency or an excess of the more mobile kind of charge, transferred during the rubbing. Franklin designated the effect produced by rubbing glass as positive electricity and that produced by rubbing amber as negative electricity. We still use these terms today.

In solids the mobile charges are negative **electrons**, which are discussed in detail in later chapters. When electrons are removed from an object, positive charges remain. These consist of the nuclei of the atoms that make up the material. In solids the positive charges are not mobile. In liquids and gases, however, they, too, are free to move.

*Modern electronic devices are made from semiconductors, which are materials that are neither good conductors nor good insulators, but show a behavior intermediate between the two. The nature of semiconductors is described in Chapter 30.

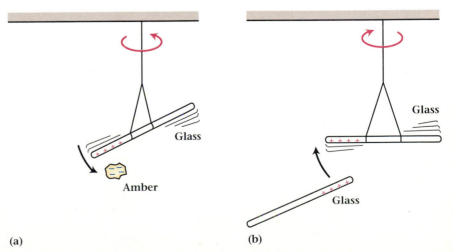

Figure 15.3 (a) A rubbed glass rod and a rubbed piece of amber attract each other. (b) Two rubbed glass rods repel each other.

Glass

Amber

(a)

Glass

Glass

(b)

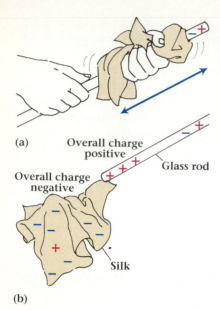

(a)

(b)

Figure 15.4   (a) Rubbing a glass rod with silk transfers electric charge. (b) The glass becomes positively charged and the silk becomes negatively charged.

Figure 15.5   Charging an electroscope by induction. Step 1: An electroscope without any excess charge. Steps 2 through 5 use a negatively charged rod to induce a positive charge on the electroscope.

Macroscopic objects are nearly always electrically neutral; that is, they have neither excess negative nor excess positive charge. However, we can give them a negative charge by adding extra electrons. Conversely, we can make them positive by removing electrons, so as to leave them deficient in negative charge (and thus positive). No charge is removed from any object except by being transferred to another object. For example, in rubbing glass with silk (Fig. 15.4), electrons are removed from the glass, leaving it positive. At the same time, the transfer of electrons to the cloth makes the cloth negative. The excess positive charge left on the glass is equal in amount (but opposite in type) to the excess negative charge transferred to the cloth. This equality illustrates a basic principle of physics known as the **law of conservation of charge**, which states that the *total amount of electric charge in the universe remains constant.* That is, *single charges can be neither created nor destroyed.* This principle is one of the fundamental observations of nature, equal in importance to the conservation laws for energy, momentum, and angular momentum. No violations of this principle have ever been observed. Charges can be created (and destroyed) only in pairs of equal magnitude and opposite sign.

The presence of electric charge can be detected by an *electroscope* (Fig. 15.5). It consists of a conducting metal rod, *A*, with a pair of flexible metallic leaves, *B*, attached to its lower part. The rod is insulated from the case, *C*. When charge is present on the assembly of rod and leaves, the leaves swing out because of repulsion between the like charges on each leaf. By making the leaves very light, the electroscope can be made quite sensitive to small amounts of charge.

We can illustrate the nature of electric charge by charging an electroscope by induction, as shown in Fig. 15.5. Step 1: The electroscope is initially uncharged; that is, it contains equal amounts of positive and negative charge. Step 2: When we bring a negatively charged rod near to (but not touching) the metal knob on top of rod *A*, negative charges in the electroscope rod are repelled to the lower part, leaving the knob positively charged. The negative charges go to the two leaves, which move apart

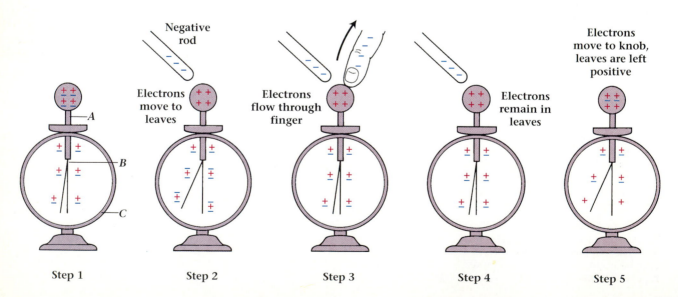

Step 1   Step 2   Step 3   Step 4   Step 5

because of the repulsive force between the negative charges. Step 3: When we touch the rod with a finger, the negative charges, still repelled by the negatively charged rod, flow out of the electroscope through our body to the earth. This procedure is called *grounding*. The leaves on the electroscope then collapse. Step 4: Removing the finger does not change the position of the leaves. Step 5: When the negatively charged rod is removed, the entire electroscope is left with a deficiency of negative charge, that is, with an excess of positive charge. The leaves again diverge, this time from the repulsive force between two positive charges.

We can explain what happens during the charging process in terms of moving positive and negative charges or in terms of moving negative charges only. Notice that the charge remaining on the electroscope is opposite to that of the charging rod. Thus we have induced a charge on the electroscope without ever touching it with a charged object. This technique is known as **charging by induction**.

Coulomb's Law

15.2

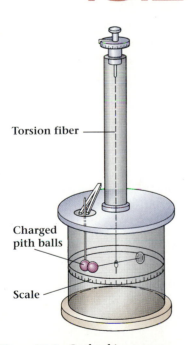

Torsion fiber

Charged pith balls

Scale

Figure 15.6 Coulomb's apparatus. The torsion balance measured the electric force on the charged spheres in the same way that Cavendish's balance measured the gravitational force.

Isaac Newton's influence on those who came after him can be seen in the work of the French engineer and physicist Charles Coulomb (1736–1806). He attacked the problem of quantifying the attractive and repulsive forces in static electricity with little concern for the nature of charge itself. Coulomb said that the important thing was to get a mathematical expression for the force.

To measure the relatively small electrostatic forces, Coulomb used a torsional balance (Fig. 15.6). Coulomb's balance was similar to the Cavendish balance used later to measure the gravitational force between masses (discussed in Chapter 5). In his experiments on electrical forces, Coulomb charged two small pith balls (pith is a light, spongy material from the center of plant stems) and measured the force between them, in much the same way that Cavendish later measured the force of gravitational attraction. In Coulomb's words the first results were that "The repulsive force between two small spheres electrified with the same type of electricity is inversely proportional to the square of the distance between the centers of the two spheres."* He also verified the inverse-square distance law for the attractive force between unlike charges.

The electrostatic force also depends on the product of the charges. Thus the magnitude of the force between two point charges is

$$F = k\frac{q_1 q_2}{r^2}, \tag{15.1}$$

where q_1 and q_2 are the values of the two charges, r is the separation between them, and k is a proportionality constant. This relationship is

*C. Coulomb, *Mémoires de l'Académie des Sciences,* 1785, p. 569 ff., translated in *Great Experiments in Physics.* Edited by M. H. Shamos, New York: Holt, Rinehart and Winston, 1959, p. 63.

$F = k q_1 q_2 / r^2$

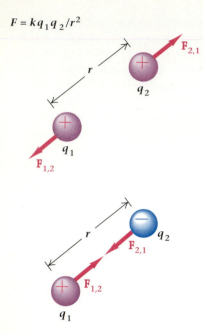

Figure 15.7 The force $\mathbf{F}_{1,2}$ exerted on charge q_1 by charge q_2 is equal in magnitude and opposite in direction to the force $\mathbf{F}_{2,1}$ exerted on charge q_2 by charge q_1.

known as **Coulomb's law**. The force is directed along the line joining the two charges. If both q_1 and q_2 are positive, their product is positive and so F is positive, indicating a repulsive force. Similarly, if both charges are negative, F is again positive and the force is again repulsive. However, if one charge is positive and the other negative, then F is negative, indicating an attractive force.

Notice that there are two forces, one acting on one charge and another acting on the other charge. These two forces form an action-reaction pair, in accord with Newton's third law. Thus the force $\mathbf{F}_{1,2}$ exerted on charge q_1 by charge q_2 is equal in magnitude and opposite in direction to the force $\mathbf{F}_{2,1}$ exerted on charge q_2 by charge q_1 (Fig. 15.7).

Coulomb did not actually prove that the force generally depends on the product of the charges, although he did test it for a few cases. The analogy with Newton's formula for the law of gravitation most likely influenced Coulomb's choice of what to measure as well as his interpretation of the results.

The magnitude of the proportionality constant k in Eq. (15.1) depends on the system of units we choose for measuring the charges and their separation. If we choose k to be a dimensionless constant, then Eq. (15.1) could be the defining equation for electrical charge. In that case the dimension of charge would be length times the square root of force. An entire system of electrical units could be built up on the basis of this choice. However, present practice favors the SI system, in which the base unit is the unit for electric current, the ampere (defined in Chapter 17). The SI unit for charge, the **coulomb** (C), is derived from the ampere. If we measure the amount of charge in coulombs and the separation in meters, then Eq. (15.1) correctly gives the force in newtons if

$$k = 8.988 \times 10^9 \ \text{N·m}^2/\text{C}^2 \approx 9.0 \times 10^9 \ \text{N·m}^2/\text{C}^2.$$

It is often convenient to define the constant in Coulomb's law to be

$$k = \frac{1}{4\pi\epsilon_0}.$$

While this appears to unnecessarily complicate the force equation, it does eliminate a factor of 4π from other equations we shall encounter later. The quantity ϵ_0 is known as the **permittivity of free space** and has the value

$$\epsilon_0 = \frac{1}{4\pi k} = 8.854 \times 10^{-12} \ \frac{\text{C}^2}{\text{N·m}^2}.$$

When the permittivity constant is used, Coulomb's law takes the form

$$F = \frac{1}{4\pi\epsilon_0} \frac{q_1 q_2}{r^2}.$$

The most fundamental unit of charge in nature is the charge on an electron or proton. This quantity is denoted by e, the elementary charge, which has the value

$$e = 1.602 \times 10^{-19} \ \text{C}.$$

Thus it takes about 6.24×10^{18} electrons to make one coulomb of charge.

TABLE 15.1 Approximate charge on some objects*	
Object	**Charge (C)**
Single electron	$\approx 10^{-19}$
Aerosols and ink-jet printer drops	$\approx 10^{-15}$
Person on insulating stand	$\approx 10^{-6}$
Charge transferred between storm cloud and earth	$\approx 10^{2}-10^{4}$

*From Joseph M. Crowley, *Fundamentals of Applied Electrostatics* (John Wiley, 1986)

The electrostatic force is so great that separate charges of coulomb magnitude are rarely encountered. A more typical range of charge encountered in electrostatic experiments is from 10^{-9} to 10^{-6} C. Table 15.1 lists the charge on several objects. Experimentally, it has been found that electric charge always occurs in multiples of the elementary charge e. For this reason we say that charge is *quantized*; that is, it occurs only in integer multiples of e.* Two examples of Coulomb's law and the magnitude of electrostatic forces are given below.

Example 15.1

Calculating electrostatic force.

Calculate the repulsive force between a pair of like charges, each of one microcoulomb, separated by a distance of one centimeter.

Solution We can calculate the force from Eq. (15.1), inserting $q_1 = q_2 = 1 \times 10^{-6}$ C, $k = 9 \times 10^{9}$ N·m²/C², and $r = 10^{-2}$ m. Then

$$F = k \frac{q_1 q_2}{r^2} = 9 \times 10^{9} \frac{\text{N·m}^2}{\text{C}^2} \frac{(10^{-6}\ \text{C})(10^{-6}\ \text{C})}{(10^{-2}\ \text{m})^2},$$

$$F = 90\ \text{N}.$$

Example 15.2

Comparing electrostatic and gravitational forces.

Compare the electrostatic repulsive force between two electrons with the attractive force due to gravitation. (The properties of the electron are described in later chapters; however, some of its properties are listed in the table in the back endpapers of this book.)

Solution The electrostatic force is given by Coulomb's law and the gravitational force is given by Newton's law of gravitation. Both of these forces depend on $1/r^2$. The ratio of these forces is then independent of the separation r between the electrons.

The electrical force is

$$F_e = k \frac{q^2}{r^2} \qquad \text{(repulsive)}$$

and the gravitational force is

$$F_g = G \frac{m^2}{r^2} \qquad \text{(attractive)}.$$

The ratio of these two forces is

$$\frac{F_e}{F_g} = \frac{k}{G} \frac{q^2}{m^2}.$$

*Current models of nuclear particles such as the proton postulate the existence of component particles called quarks with fractional values of the electron charge. Most interpretations of these models indicate that isolated quarks, and hence their fractional charges, are not observable in the ordinary sense. So far, no isolated quarks have been detected. (See Chapter 31.)

The electric charge of the electron is -1.6×10^{-19} C and its mass is 9.1×10^{-31} kg. The ratio of the forces is

$$\frac{F_e}{F_g} = \frac{9.0 \times 10^9 \text{ N·m}^2/\text{C}^2}{6.67 \times 10^{-11} \text{ N·m}^2/\text{kg}^2} \frac{(1.6 \times 10^{-19} \text{ C})^2}{(9.1 \times 10^{-31} \text{ kg})^2},$$

$$\frac{F_e}{F_g} = 4.2 \times 10^{42}.$$

The electrostatic force between electrons is immensely greater than the gravitational force between them. Thus, when describing the behavior of electrons, it is usually sufficient to consider only the electrical forces and neglect the gravitational effects.

The magnitude of the force between electric charges is in large part the reason why doing static electric experiments is so difficult. Because of the huge force of attraction between unlike charges, it is very hard to separate a large number of charges and keep them separated.

Superposition of Electric Forces

15.3

As we have seen, electrical forces are slightly more complicated than gravitational forces because they can be either attractive or repulsive. We deal with this complexity by adopting a convention for describing the direction of the force. Consider the two charges in Fig. 15.8. The force $\mathbf{F}_{2,1}$ acting on the charge q_2 due to charge q_1 is directed along the line extending from q_1 to q_2. When the values of q_1 and q_2 are inserted into Eq. (15.1), we must be careful to include their algebraic sign. If both charges are positive or if both are negative, their product is positive. Thus $\mathbf{F}_{2,1}$ would be a vector with magnitude $|kq_1q_2/r^2|$ pointing in the direction from q_1 to q_2. If, however, one of the charges is negative and the other is positive, then their product is negative. In this case the force on q_2 is a vector with magnitude $|kq_1q_2/r^2|$ in the direction from q_2 toward q_1. Thus the magnitude of the force is the same, but the direction of the force is determined from the signs of the charges.

If several charges interact with one another, we can determine the net force on any one charge by summing up the individual contributions to the force due to each of the other charges considered independently. This independence of the forces is known as the law of superposition. For example, in Fig. 15.9 the repulsive force on a positive charge q_3 due to the positive charges q_1 and q_2 is found by determining the vector sum of the two individual forces that act on q_3. The force $\mathbf{F}_{3,1}$ exerted on q_3 by the charge q_1 is independent of the presence of the other charge, q_2. Similarly, the force $\mathbf{F}_{3,2}$ exerted on q_3 by the charge q_2 is independent of the presence of q_1.

Superposition is not a self-evident principle for all types of forces; each case must be examined separately. However, superposition is valid for electrical forces. An example follows.

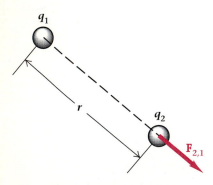

Figure 15.8   The direction of the force $\mathbf{F}_{2,1}$ on a charge q_2 due to another charge q_1 is along the line between the charges and depends on the signs of both charges.

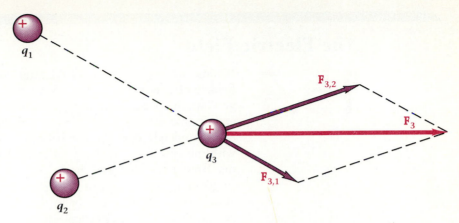

Figure 15.9 The force $\mathbf{F}_3$ on charge q_3 is the vector sum of the forces due to q_1 and q_2, considered independently.

Example 15.3

Electrostatic force on three charges in a line.

Calculate the force on the charge q_3 due to the other two charges located as shown in Fig. 15.10(a). Take the magnitudes of the charges to be $q_1 = -4\ \mu C$, $q_2 = +1\ \mu C$, and $q_3 = +1\ \mu C$.

Solution Before making any calculations, we observe from the signs of the charges that the force between q_2 and q_3 is repulsive, while that between q_1 and q_3 is attractive. Thus, in the free-body diagram in Fig. 15.10(b), the force $\mathbf{F}_{3,2}$ is directed to the right and $\mathbf{F}_{3,1}$ is directed to the left. The total force on q_3 is the difference between the magnitudes of $\mathbf{F}_{3,2}$ and $\mathbf{F}_{3,1}$. Thus

$$F_3 = \left| k\frac{q_2 q_3}{(r_{2,3})^2} \right| - \left| k\frac{q_1 q_3}{(r_{1,3})^2} \right|,$$

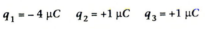

and is directed to the right if positive and to the left if negative. The vertical bars indicate the magnitude, or absolute value, of a quantity. The expression for F_3 can be manipulated to give

$$F_3 = kq_3 \left\{ \frac{|q_2|}{(r_{2,3})^2} - \frac{|q_1|}{(r_{1,3})^2} \right\}.$$

Since the magnitude of $q_1 = 4q_2$ and $r_{1,3} = 2r_{2,3}$, we get

$$F_3 = kq_3 \left\{ \frac{|q_2|}{(r_{2,3})^2} - \frac{|4q_2|}{(2r_{2,3})^2} \right\}$$

$$F_3 = kq_3 \frac{|q_2|}{(r_{2,3})^2} \left\{ \frac{1}{1} - \frac{4}{2^2} \right\} = 0.$$

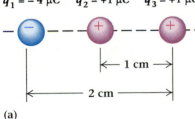

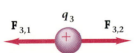

(b)

Figure 15.10 Example 15.3: (a) Three charges arranged along a line. (b) Free-body diagram for charge q_3. The two forces $\mathbf{F}_{3,1}$ and $\mathbf{F}_{3,2}$ are equal but oppositely directed. The magnitude of each force is calculated independently of the other.

The net force on the charge q_3 is zero.

Notice that the magnitude of force $\mathbf{F}_{3,1}$ did not change because charge q_2 is located between charges q_1 and q_3. This is the power and the beauty of the principle of superposition of electrostatic forces due to individual charges. The force on one charge is found by considering the effects of other charges one at a time and then summing them. Of course, we have to remember that the sum is a vector sum because force is a vector.

The Electric Field

15.4

In describing gravitation in Chapter 5 we introduced the concept of a field, which is a physical quantity defined at all points in space. In the gravitational case, field strength is the gravitational force per unit mass acting on a test mass. The sources of that force are other masses, but we can completely describe the behavior of the test mass by its interaction with the *field*. We can also predict the behavior of any other known mass, just from a knowledge of the field in the region where the mass is located. In general, the concept of field provides insight into many physical situations. It is useful not only in describing gravitational interactions, but also in dealing with electrical interactions.

We define the **electric field E** at a point in space as the force **F** per unit charge exerted on a small positive test charge q_0 placed at that point,

$$E = \frac{F}{q_0}.$$

(15.2)

This field is caused by other electric charges distributed about the test charge. Thus Eq. (15.2) defines the field due to this distribution of charge, not the field caused by the test charge. We choose the test charge q_0 to be very small so that its presence does not distort the field being measured.

The electric field is a vector. Its magnitude is determined by dividing the magnitude of the force exerted on a test charge by that charge. Thus the units of electric field are newtons per coulomb, N/C. The direction of the field vector is the direction of the force on a positive test charge. Once we know the electric field at a point, we can readily compute the force on any electric charge placed at that point. Since we can determine the forces, we can calculate the motion of the charge.

It is easy to find the field of a single point charge Q that is isolated from other charges. We first locate Q at the origin and then use Coulomb's law to determine the force on a positive test charge q_0 at a displacement **r** from the origin. The magnitude of **F** is

$$F = k\frac{Qq_0}{r^2}.$$

The magnitude of the electric field at point r is

$$E = \frac{F}{q_0} = k\frac{Q}{r^2}.$$

(15.3)

If Q is a positive charge, then the field is directed radially away from it. If Q is negative, the direction of the field is toward it.

The direction of the electric field of an isolated point charge is illustrated in Fig. 15.11. Here we have used *lines of force* (also called *field lines*) to represent the electric field, just as we did earlier for the gravitational field. (Remember that the relative number of lines of force is proportional to the magnitude of the force and hence to the strength of the field.) In an electrostatic field, lines of force always begin at positive charges and

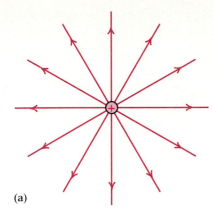

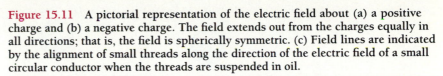

Figure 15.11 A pictorial representation of the electric field about (a) a positive charge and (b) a negative charge. The field extends out from the charges equally in all directions; that is, the field is spherically symmetric. (c) Field lines are indicated by the alignment of small threads along the direction of the electric field of a small circular conductor when the threads are suspended in oil.

end at negative charges. They are closer together where the field is stronger and farther apart where the field is weaker. The patterns of lines of force for gravitational and electrostatic fields are similar, which is not surprising because both fields have a $1/r^2$ behavior; that is, they decrease inversely as the square of the distance from the source.

(a)

Example 15.4

What is the magnitude and direction of the electric field 0.50 m from a positive point charge of 1.2×10^{-10} C?

Solution   We can compute the magnitude of the electric field directly from Eq. (15.3):

$$E = \frac{kQ}{r^2} = \frac{9.0 \times 10^9 \text{ N·m}^2/\text{C}^2 \times 1.2 \times 10^{-10} \text{ C}}{(0.50 \text{ m})^2} = 4.3 \text{ N/C}.$$

The direction of the field is radially outward from the point charge because the charge is positive.

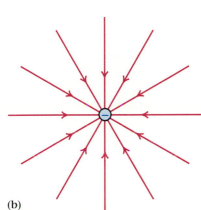

Example 15.5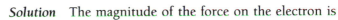

An electron (charge $e = -1.6 \times 10^{-19}$ C and mass 9.1×10^{-31} kg) is injected into a region of uniform electric field of magnitude $E = 1.0 \times 10^5$ N/C. What is the acceleration of the electron?

Solution   The magnitude of the force on the electron is

$$F = |\mathbf{F}| = |qE| = |-e||E|.$$

The acceleration is the force divided by the mass,

(b)

$$a = \frac{F}{m} = \frac{1.6 \times 10^{-19} \text{ C} \times 1.0 \times 10^5 \text{ N/C}}{9.1 \times 10^{-31} \text{ kg}} = 1.8 \times 10^{16} \text{ m/s}^2.$$

The direction of the acceleration is along the direction of the force, which is *opposite* to the direction of the electric field because the charge on the electron is negative. Note that Newton's laws of motion apply to electric forces just as they do to mechanical forces.

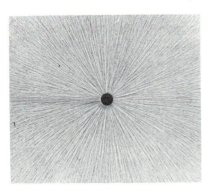

(c)

Superposition of Electric Fields

15.5

We have just examined the electric field due to a single isolated point charge. What is the field due to several source charges? For example, suppose we have three point charges, arranged as shown in Fig. 15.12. The force on a test charge q_0 at point P is the vector sum of the forces

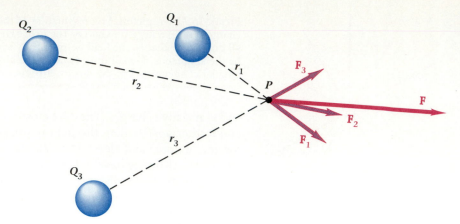

Figure 15.12   Three charges produce an electric field in space. The field measured at point P is the vector sum of the individual fields.

due to each source charge calculated separately. As we described earlier, this is nothing more than a superposition of the individual forces. Thus

$$\mathbf{F} = \mathbf{F}_1 + \mathbf{F}_2 + \mathbf{F}_3.$$

The electric field at P is found by dividing the force by the q_0, obtaining

$$\mathbf{E} = \frac{\mathbf{F}}{q_0} = \frac{\mathbf{F}_1}{q_0} + \frac{\mathbf{F}_2}{q_0} + \frac{\mathbf{F}_3}{q_0},$$

or

$$\mathbf{E} = \mathbf{E}_1 + \mathbf{E}_2 + \mathbf{E}_3.$$

In other words, the electric field resulting from several point charges is just the superposition of their individual fields. Thus we have arrived at a rule for superposition of fields from the rule for superposition of forces.

Figures 15.13 and 15.14 illustrate the fields of two equal positive charges and of two equal but unlike charges. Although these figures show only the field in a plane containing the two charges, the field is actually

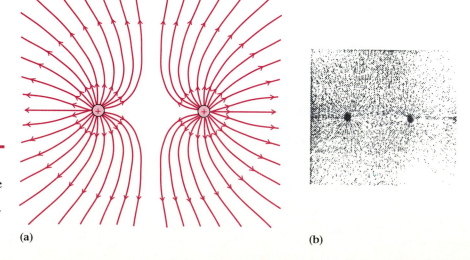

Figure 15.13   (a) Lines of electric field near two identical charges. The field is weakest directly between the charges, as is shown by the absence of field lines. (b) Field lines are indicated by small threads suspended in oil.

(a)

(b)

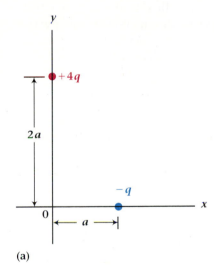

Figure 15.14 (a) Lines of electric field near two equal but opposite charges. The field is strongest directly between the charges, as is shown by the high concentration of field lines. (b) Field lines are indicated by small threads suspended in oil.

(a)   (b)

three-dimensional. It is symmetric about the axis joining the two charges, so that if you imagined rotating the charges about that axis, the shape of the field would be unchanged.

Example 15.6

Electric field of two charges.

A tiny pith ball with charge $-q$ is placed on the x axis a distance a from the origin. A second pith ball with charge $+4q$ is placed on the y axis a distance $2a$ from the origin. What is the electric field at the origin?

Solution   Figure 15.15(a) illustrates the situation. The field at O due to the charge $-q$ is directed along the positive x axis with magnitude

$$E_x = k\frac{q}{a^2}.$$

The field at O due to the charge $+4q$ is directed along the negative y axis with magnitude

$$E_y = k\frac{4q}{(2a)^2}$$

$$E_y = k\frac{q}{a^2}.$$

The two components of the field are equal in magnitude and directed at right angles. Consequently, the resulting field vector has a magnitude

$$E = \sqrt{2\left(k\frac{q}{a^2}\right)^2}$$

$$E = \sqrt{2}\,k\frac{q}{a^2},$$

and is directed at an angle of $-45°$ from the positive x axis (Fig. 15.15b).

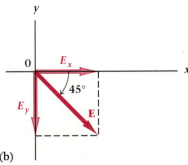

(a)

(b)

Figure 15.15 Example 15.6: (a) Position of the charges. (b) Calculated x and y components and resultant electric field at the origin.

Electric Flux and Gauss's Law

15.6

We now wish to examine a law of electrostatics that relates the electric field to the charge that generates it. In principle, we can always calculate an electric field using Coulomb's law, the definition of the field, and the principle of superposition. However, an alternative way of looking at the same physical situation leads us to a new law that is more powerful than Coulomb's law and that provides more insight into the connection between field and charge. Our first step is to define a quantity, related to the electric field, that we call the electric flux. Then we show that the net amount of electric flux that passes through an arbitrary closed surface is directly related to the amount of electric charge inside that surface. This relationship is called Gauss's law. In the following section we will analyze this relationship mathematically and show that it's true for any electrostatic field.

First we need to define flux. In general, flux is the rate at which something passes through a surface. For example, the rate at which water moves through a cross section of a pipe may be described by its flux. In this case, the flux would be the amount of water passing through the pipe per unit time.

We define electric flux in a similar manner. To begin, consider a uniform electric field passing through a small area of a surface (Fig. 15.16a). Equal spaces between the lines in the figure indicate that the field is uniform. The number of lines passing through the surface represents

Figure 15.16 (a) A small surface area oriented perpendicular to a uniform electric field. The number of lines of force that pass through the surface is proportional to the electric flux. (b) In the same field, more lines pass through a larger surface than through the surface in (a). (c) More field lines pass through the same surface in a stronger field than in a weaker one. (d) When the surface is not perpendicular to the direction of the field, fewer lines pass through. (e) If the surface is parallel to the field direction, no lines pass through it and the flux is zero.

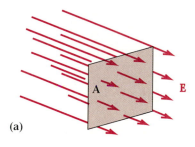

(a)

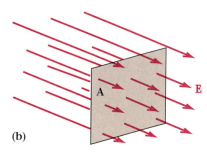

(b)

Larger area, larger flux

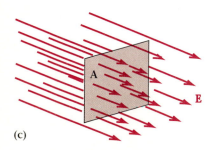

(c)

Larger field, larger flux

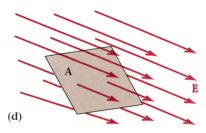

(d)

Area not perpendicular to field, smaller flux

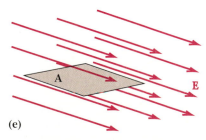

(e)

Area parallel to field, zero flux

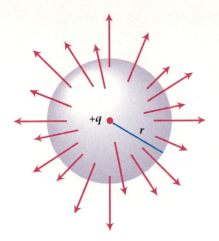

Figure 15.17 The lines of electric field emerging from a positive point charge $+q$ pass through a spherical surface centered about the charge. The number of field lines passing through the surface is independent of the radius of the surface.

the electric flux. If we have a larger surface (Fig. 15.16b), more field lines pass through, corresponding to a larger flux. If the electric field were stronger, we would indicate it by drawing the field lines closer together, as described in Section 15.4. In that case more lines would pass through our surface, corresponding to an even larger flux (Fig. 15.16c). Because the magnitude of the electric field is proportional to the number of lines of force in a given area, we define the **electric flux** to be the number of field lines that pass through a given surface. Notice that when the surface area is turned so that it is no longer perpendicular to the electric field, fewer electric field lines pass through (Fig. 15.16d). In this case, the flux is reduced, even though the area is the same. In fact, when the surface area is turned parallel to the direction of the field, no field lines pass through at all and the flux is zero (Fig. 15.16e).

We now use our definition of flux and our knowledge of point charges to arrive at Gauss's law. Imagine that we surround an isolated point charge $+q$ with an imaginary spherical surface of radius r centered about the point charge (Fig. 15.17). The electric flux that passes through this closed surface is represented by the number of field lines that cross the surface. If the field lines are directed outward through the surface, the flux is positive; if the field lines are directed inward, the flux is negative. The total number of field lines passing through our closed spherical surface is proportional to the magnitude of the charge and is independent of the radius r. Thus the electric flux through the surface is directly proportional to the charge enclosed within the surface. If we replace the charge with $-q$, then the direction of field lines is reversed and the flux becomes negative. If we combine several charges at the central point of the sphere, the sign of the net charge will determine the sign of the flux. We are led to **Gauss's law for electrostatics**, which says that **the net electric flux through any (real or imaginary) closed surface is directly proportional to the net electric charge enclosed within that surface.**

We have not proved Gauss's law here. Instead we have described a special case, but one that clearly conforms to Gauss's law. However, Gauss's law is more general. It is valid regardless of the number of charges involved, their positions inside or outside the closed surface, or the shape of the surface itself. Gauss's law stands as one of the fundamental laws of electricity. The (usually imaginary) surface used for application of Gauss's law is often called a **Gaussian surface**.

We can qualitatively verify Gauss's law by examining Fig. 15.18. Look at the various Gaussian surfaces drawn in cross section in the figure and think about the net flux through each one. The field lines that emerge through surface A surrounding the positive charge q are all directed outward and correspond to a net positive flux. The field lines that enter surface B surrounding the negative charge $(-q)$ are all directed inward and correspond to a net negative flux. The field lines that enter one part of surface C, which surrounds no charge, emerge from the other side. Thus the net flux is zero. Finally, the large surface D surrounding both charges contains no net charge. The lines of force that originate on $+q$ pass out of surface D at one end and then enter the surface on the other end. The net flux is zero. Each of these situations is consistent with Gauss's law.

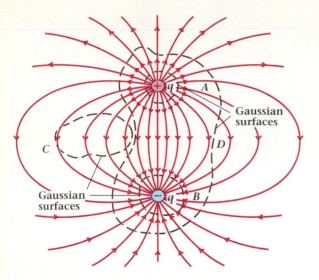

Figure 15.18 Four Gaussian surfaces, shown in cross section, in the region of two point charges of the same magnitude but opposite sign. Surface *A* surrounds a positive charge and has a net positive flux. Surface *B* surrounds a negative charge and has a net negative flux. Surfaces *C* and *D* surround zero net charge and thus have zero flux.

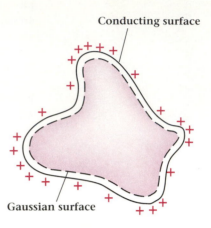

Figure 15.19 The dashed line represents a Gaussian surface just inside the surface of a charged conductor, shown here in cross section. We can use Gauss's law to show that when a charged conductor has reached electrical equilibrium, any excess charge on the conductor must reside on its outer surface.

Figure 15.20 A man inside a Faraday cage is shielded from an electrical discharge.

We can use Gauss's law to prove an important assertion: **An excess charge placed on a conductor resides entirely on its outer surface**. This effect was observed experimentally by Benjamin Franklin long before Gauss's law was known. For example, if an excess charge is placed on an isolated conductor of any arbitrary shape (Fig. 15.19), the charge sets up a field within the conductor. This field causes the mobile charges within the conductor to move about until the internal field reduces to zero. When the internal field reaches zero, the charge stops moving and the system is in static equilibrium. Thus for a charged conductor *in equilibrium,* the electric field inside the conductor must be zero. Now, since the internal electric field is zero, the flux through an imaginary surface just inside the actual surface of the conductor is also zero. Such a Gaussian surface is indicated by the dashed line of Fig. 15.19. Therefore, there is no net or excess charge within the interior of the conductor. The charge must necessarily reside on the outer surface.

If excess charge resides on the surface of a conductor, then even a thin covering of conducting material can be used to surround a volume and shield it from external static charges. (Such an arrangement is often called a Faraday cage, after Michael Faraday, who made extensive experiments on shielding.) The volume can even be shielded from the effects of external electric fields. Thus the occupants of a room surrounded by conducting material are safe from external electrical effects as long as they remain inside the room. Even a wire grid Faraday cage (Fig. 15.20) can shield the occupant within from electrical discharges, including lightning.

A Quantitative Approach to Gauss's Law

*15.7

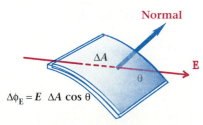

$\Delta\phi_E = E \; \Delta A \cos\theta$

Figure 15.21 Electric field passing through a small surface of area **ΔA**. The normal to the surface makes an angle **θ** with the direction of the electric field. The electric flux **Δφ_E** is equal to the product of the field strength E, the area **ΔA**, and the cosine of the angle between the direction of the field and the surface normal.

In the previous section, when we discussed electric flux and Gauss's law, we reached the important conclusion that for static cases, any excess electric charge on a conductor resides on its surface. In this section we take a more quantitative approach to Gauss's law and show how we can use it to find the value of an electric field.

First let's consider a uniform electric field passing through a small area of an arbitrary surface ΔA (Fig. 15.21). We wish to describe the electric flux in terms of the electric field **E** that exists in the region of this small area. In the previous section we defined the flux as the number of field lines that pass through the surface. Since we expressed the magnitude of electric field E in terms of the number of field lines through a given area, we can also define the electric flux as the amount of electric field that passes through a given surface area multiplied by the area. As we have just seen, when a surface is perpendicular to the direction of the field, the maximum number of field lines can pass through and we have the maximum flux. As the surface tips so that the angle between the plane of the surface and the field gets smaller, the effective area becomes smaller and fewer field lines can pass through. For this reason, our mathematical formula for flux must include the effect of the angle between the surface and the field lines. In practice, we usually define the orientation of a surface as the direction of a line perpendicular to the surface (the *normal*). So, in defining flux we consider the angle between the field direction and the normal.

The electric flux through a small portion of surface area ΔA is the product of the magnitude of the electric field $|\mathbf{E}|$, the magnitude of the surface area ΔA, and the cosine of the angle θ between the direction of the field and the direction of the normal to the surface (Fig. 15.21). The electric flux $\Delta\phi_E$ is then

$$\Delta\phi_E = E \; \Delta A \cos\theta.$$

Now let's consider the surface to be a sphere of radius r centered about a source charge Q (Fig. 15.22). At the surface of the sphere the field has magnitude

$$E = k\frac{Q}{r^2}$$

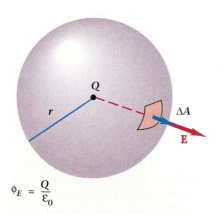

$\phi_E = \dfrac{Q}{\varepsilon_0}$

Figure 15.22 A spherical surface symmetrically surrounding a point charge **Q**. The surface normal is parallel to the electric field at all points of the surface.

and is perpendicular to the surface at all points. Since **E** is perpendicular to the surface everywhere, the angle θ between the field and the normal at any point on the surface is zero and $\cos\theta = \cos 0 = 1$. The flux through an area ΔA then becomes

$$\Delta\phi_E = E \; \Delta A$$

or

$$\Delta\phi_E = k\frac{Q \; \Delta A}{r^2}.$$

The total flux ϕ_E through the entire surface is the sum of all contributions $\Delta\phi_E$ over the surface of the sphere:

$$\phi_E = (\Delta\phi_E)_1 + (\Delta\phi_E)_2 + (\Delta\phi_E)_3 + \cdots$$

$$= k\frac{Q}{r^2}\,\Delta A_1 + k\frac{Q}{r^2}\,\Delta A_2 + k\frac{Q}{r^2}\,\Delta A_3 + \cdots$$

We can remove the factor kQ/r^2 from the sum because it has the same constant value everywhere on the surface of the sphere:

$$\phi_E = k\frac{Q}{r^2}\,(\Delta A_1 + \Delta A_2 + \Delta A_3 + \cdots).$$

We are left with the sum of all the individual areas of surface, which is just the total surface area of the sphere, $4\pi r^2$. The total flux thus becomes

$$\phi_E = k\frac{Q}{r^2} \times 4\pi r^2 = 4\pi k Q.$$

If we replace k by $1/4\pi\epsilon_0$, we find that

$$\boxed{\phi_E = \frac{Q}{\epsilon_0}.} \tag{15.4}$$

Equation (15.4) is the mathematical statement of Gauss's law. This is one place where introducing the permittivity ϵ_0 results in a simpler equation. Notice that the total flux through the surface depends only on the amount of charge Q contained *within* the surface. (The amount of charge *outside* the surface has no net effect on the flux.) If Q is positive, then the flux is positive, meaning that the field is directed outward from the charge; if Q is negative, the flux is negative, so the field is directed inward toward the charge; and if Q is zero, the flux is also zero, and there is no net field through the surface.

Although we have derived Gauss's law for a spherical surface, it is valid regardless of the shape of the surface involved. Because of this, we can choose a Gaussian surface of any shape around a charge. Proper choice of this surface can often simplify the flux calculation. Gauss's law holds regardless of where the charge is located within the surface. It even holds for the case where the charge is outside the closed surface. In that case the charge inside is zero, so the net flux is also zero.

In the general case, as shown in Fig. 15.18, the net electric flux through any (real or imaginary) closed surface is directly proportional to the net electric charge enclosed within that surface, regardless of the number of charges involved, their positions inside or outside the surface, or the shape of the surface itself. Two examples of using Gauss's law follow.

Example 15.7

Gauss's law in a uniform field.

Verify that Gauss's law correctly describes the flux through a closed cylindrical surface in a region of uniform electric field **E** in which there are no charges.

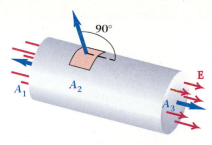

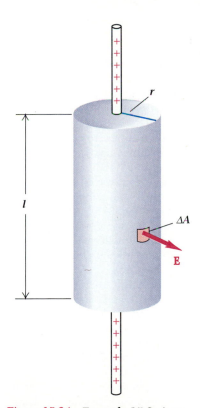

Figure 15.23 Example 15.7: A cylindrical Gaussian surface in a uniform electric field. The net charge enclosed by the surface is zero, so the net flux is zero. (As many field lines pass into the surface as pass out of the surface.)

Solution   Let the cylindrical surface be oriented with its axis parallel to the direction of the electric field (Fig. 15.23). We choose this orientation to simplify the flux calculations. Let us designate the area of the left-end surface of the cylinder as A_1, the area of the side of the cylinder as A_2, and that of the right-end surface as A_3. The flux is then

$$\phi_E = EA_1 \cos \theta_1 + EA_2 \cos \theta_2 + EA_3 \cos \theta_3.$$

Since the field lines are directed perpendicularly inward through the surface A_1, $\theta_1 = 180°$ and $\cos \theta_1 = -1$. The surface A_2 is everywhere at right angles to **E**, so $\theta_2 = 90°$ and $\cos \theta_2 = 0$. The field lines pass perpendicularly outward through surface A_3, so $\theta_3 = 0°$ and $\cos \theta_3 = +1$. Thus

$$\phi_E = EA_1(-1) + EA_2(0) + EA_3(1).$$

But A_1 and A_3 are equal, so

$$\phi_E = E(-A_1 + A_3) = 0.$$

This result is exactly what we expect from Gauss's law because no charges are located within the cylindrical surface. Notice that there is electric field within the cylinder, but the net flux is zero because the inward flux and outward flux are equal.

Example 15.8

Field of an infinite line of charge.

Figure 15.24 shows a section of an infinitely long charged plastic rod. The rod is uniformly charged with a constant linear charge density λ (the charge per unit length). Use Gauss's law to show that the electric field due to this uniform line of charge is proportional to $1/r$, where r is the distance from the line.

Solution   The electric field near the uniformly charged rod must be radially directed because of the symmetry of the situation. The field must have cylindrical symmetry because the appearance of the charged rod is unchanged by rotating the rod about its axis. The field must also be independent of position along the rod because the distance to either end is infinite, so that any part of the rod is exactly the same as any other part.

We can apply Gauss's law to an imaginary cylindrical surface centered about the line of charge (Fig. 15.24). Again, we choose the shape of the surface to simplify the calculations through symmetry considerations. The flux is

$$\phi_E = \sum E\,\Delta A \cos \theta = \sum_{\text{ends}} E\,\Delta A \cos \theta_1 + \sum_{\text{cylinder}} E\,\Delta A \cos \theta_2.$$

The field is parallel to the plane of the end surfaces, so $\theta_1 = 90°$ and

$$\sum_{\text{ends}} E\,\Delta A \cos \theta_1 = 0.$$

Since the cylinder is symmetrically positioned about the line, the magnitude of the field E is constant over the cylindrical surface and the angle θ_2 between **E** and the normal is zero. Thus

$$\phi_E = \sum_{\text{cylinder}} E\,\Delta A \cos 0 = E \sum \Delta A,$$

Figure 15.24 Example 15.8: A cylindrical Gaussian surface of radius r and length l centered about an infinite line of charge. This situation approximates the field around a uniformly charged long thin rod.

where the summation is over the cylindrical surface. The surface area of a cylinder of radius r and length l is $2\pi rl$, so

$$\phi_E = 2\pi rlE.$$

By Gauss's law the flux must be

$$\phi_E = \frac{Q}{\epsilon_0} = \frac{\lambda l}{\epsilon_0},$$

since λl is the charge enclosed by the surface. Upon combining these two expressions for the flux we see that

$$E = \frac{\lambda}{2\pi\epsilon_0 r}.$$

The magnitude of the field is proportional to $1/r$.

PHYSICS IN PRACTICE

Dipoles and Microwave Ovens

Although microwave ovens are a relatively recent invention, they have become widespread across the world. Many people don't give it a thought when they pop leftover meals into the microwave oven to heat them, or when they use the oven for fast and easy cooking. But why is microwave cooking so fast: How does it work and how safe is it?

All substances consist of molecules, which are made up of atoms. For example, a water molecule is made up of two atoms of hydrogen and one atom of oxygen. Because atoms contain positive charges (protons) and negative charges (electrons) in equal number, molecules themselves are not charged. However, as we have seen, in some cases the charge is not distributed symmetrically within the molecule, giving rise to dipole moments. For example, water has a permanent dipole moment (Fig. B15.1).

An electrical force exists between two polar molecules, even though they are both uncharged. Some of the bonding between molecules is due to this dipole–dipole interaction. In a uniform electric field a polar molecule experiences a torque that tends to align its dipole moment with the field. If we reverse the direction of the field, a new torque arises that tends to rotate the molecule through 180°. If the molecule is isolated, it can rotate freely. However, if the molecule is in a substance and bound to other molecules, it encounters retarding "friction" to the applied torque. This friction is due to the disruption of the bonds between mole-

Figure B15.1   Water molecules have a permanent dipole moment directed from the negative (oxygen) end to the positive (hydrogen) end.

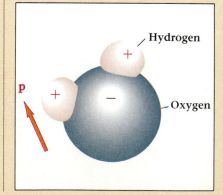

cules. If the direction of the field is changed rapidly, the energy used against friction appears as heat in the substance itself.

Microwaves, which are high-frequency radio waves, can provide such a rapidly changing electric field. In North America, microwave ovens designed for home use operate at a frequency of 2450 MHz (wavelength of 12.2 cm). Microwaves are produced and confined within the small volume of the oven (Fig. B15.2). These waves, like sound waves or light waves, may be transmitted, reflected, or absorbed. Microwaves are easily transmitted through air, glass, paper, and many types of plastics. Microwaves are reflected from metal, but absorbed by water, fat, and sugar. In most foods the microwaves penetrate one to two inches. As they penetrate, they are absorbed by the food, which generates heat primarily as a result of the disruption of the intermolecular bonds of the water molecules as they rapidly change direction.

The rapidity with which microwave ovens heat and cook is due not so much to the total power delivered into the oven as to the fact that the microwaves penetrate the outer layers of the

The Electric Dipole

15.8

Many common electrical situations involve two equal but opposite charges separated by a fixed distance. This combination of charges is called an **electric dipole**.* For example, the hydrogen chloride (HCl) molecule is negative at one end, the Cl end, and positive at the other end, the H end (Fig. 15.25, p. 458). Many other molecules also have some asymmetry of charge and act as electric dipoles, including water (H_2O), sulfur dioxide (SO_2), and chloroform ($CHCl_3$). Spherically symmetric molecules such as methane (CH_4) and carbon tetrachloride (CCl_4) do not act as dipoles.

In a uniform electric field a dipole does not experience any net force

---

*The term dipole implies that there are two poles or charges.

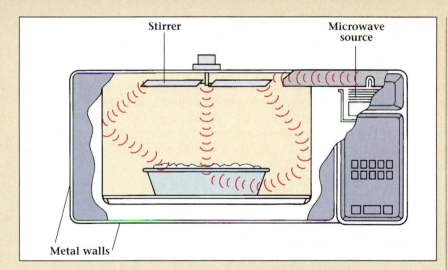

Figure B15.2   A typical microwave oven. Microwaves from the source are directed into the oven cavity, where they are contained by metal walls. The metal stirrer helps distribute the microwave energy evenly throughout the oven.

food and deliver their energy within the food itself. In a conventional oven, energy is transferred to the surface of the food by convection through the low-thermal-conductivity air. In both conventional and microwave ovens, food cooks from the outside in.

Because microwaves reflect from the inside walls of the oven, it is possible to set up standing waves within the oven, similar to the way standing sound waves are set up in a pipe. The nodes and antinodes correspond to cold and hot spots, which lead to uneven cooking. This effect is reduced by rotating metal fans (stirrers) intended to break up standing waves and by a rotating cooking platform. Food en-

closed within metal containers does not cook because the metal shields the food from the microwaves.

Caution is appropriate when using microwave ovens. The microwave power output of typical household ovens is about 500 W, a level high enough to cause burns. However, the level of microwave power need not be high enough to cause cooking for serious damage to occur. Therefore, standards have been set to limit how much radiation may escape from the oven. The 1985 U. S. federal standard is that the microwave power two inches from the door shall not exceed 1 mW/cm^2 from a new oven. Studies have indicated that levels of from 0.1 to 1.0 mW/cm^2 have no apparent harmful effect over an eight-hour period. For safety, a conducting plastic seal ensures that microwave energy does not leak out around the door. For additional safety, an electrical interlock shuts off the oven when you open the door. To allow you to see inside the oven with the door shut, most microwave ovens have windows with a metal screen or mesh. This mesh is opaque to the long wavelengths of the microwaves but easily lets pass the shorter waves of visible light.

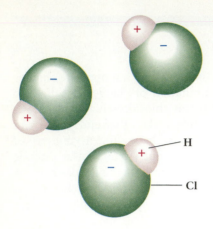

Figure 15.25 The HCl molecule is an electric dipole because its charge is not spherically symmetric.

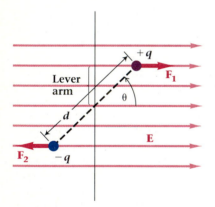

Figure 15.26 The forces acting on an electric dipole in a uniform electric field cause it to rotate.

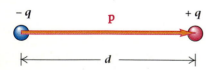

Figure 15.27 Direction of the dipole moment **p** of two opposite charges of magnitude q separated by a distance d.

because the force on the positive charge is equal and opposite to the force on the negative charge. As a result, there is no net translational acceleration. However, this does not imply that nothing happens. On the contrary, a net *torque* arises that causes the dipole to rotate so as to align along the direction of the field. The situation is illustrated in Fig. 15.26, where the clockwise torque generated by F_1 about the center of the dipole is

$$\tau_1 = F_1 \frac{d}{2} \sin \theta.$$

Here θ is the angle between the dipole axis and the electric field and d is the distance between the charges. Similarly, the torque resulting from the force on the negative charge is also clockwise and is given by

$$\tau_2 = F_2 \frac{d}{2} \sin \theta.$$

The magnitudes of the forces are equal because the magnitudes of the charges are identical. So the total torque, which is the sum of the individual torques, is

$$\tau = \tau_1 + \tau_2 = Fd \sin \theta.$$

The force F is given by qE. This leads to

$$\tau = qdE \sin \theta.$$

The torque depends on the magnitude of the field (E), the charge (q), the separation between the charges (d), and the angle θ.

The product qd is usually given the name **dipole moment** and for electric dipoles is often represented by the letter p. If we define the dipole moment as a vector of magnitude p with direction from the negative to the positive charge (Fig. 15.27), then the torque is the product of the dipole moment p, the field E, and the sine of the angle between them:

$$\boxed{\tau = pE \sin \theta.} \tag{15.5}$$

The direction of the rotation produced by the torque rotates **p** into alignment with **E**.

Example 15.9

What is the maximum torque on a molecule of HCl (dipole moment $p = 3.6 \times 10^{-30}$ C·m) in a region of electric field $E = 1.0 \times 10^4$ N/C?

Solution The maximum torque occurs when the direction of the dipole moment vector is at right angles to the direction of the electric field vector. In that case the torque is

$$\tau = pE \sin 90° = pE = 3.6 \times 10^{-30}\,\text{C·m} \times 1.0 \times 10^4\,\text{N/C},$$
$$\tau = 3.6 \times 10^{-26}\,\text{N·m}.$$

Although the torque seems very small, you must remember that the mo-

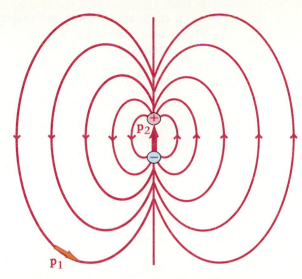

Figure 15.28 When dipole $\mathbf{p}_1$, free to rotate, is brought near a fixed dipole $\mathbf{p}_2$, the effect of the field due to $\mathbf{p}_2$ is to rotate the free dipole $\mathbf{p}_1$ so that its dipole moment aligns with the field.

ment of inertia is also very small (about 2.7×10^{-47} kg·m^2), so the molecule rotates rapidly.

We have shown that when a dipole is placed in an external electric field, a torque acts to align it with the field. For example, consider the behavior of a free dipole interacting with the field of a second dipole fixed in space (Fig. 15.28). The free dipole then rotates into alignment with the field of the fixed dipole. It follows that we must do work to rotate the dipole to some position other than alignment along the field direction. We can, therefore, associate a potential energy with the orientation of the dipole in the field (Fig. 15.29). If we consider the position in which the dipole is perpendicular to the field as the reference position, then the potential energy decreases when the dipole rotates into the direction of the field and increases when the dipole rotates in the opposite direction. The potential energy is

$$U = -pE \cos \theta. \tag{15.6}$$

Although it is helpful to visualize the motion of a dipole in terms of the torques acting on it, usually the physically important quantity is the potential energy associated with the dipole's orientation. We shall examine the magnetic dipole in Chapter 18. Then in later chapters we will show

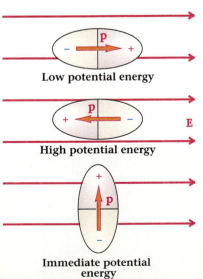

Low potential energy

High potential energy

Immediate potential energy

Figure 15.29 A dipole in an electric field has a potential energy that depends on the orientation of the dipole moment vector relative to the direction of the field.

how the electric and magnetic dipole energies enter into the theory of atoms and how we may use them to explain the magnetic behavior of solids.

Example 15.10

Energy of a molecular dipole.

A water molecule has an electric dipole moment $p = 6.2 \times 10^{-30}$ C·m. If the molecule is in an electric field of 1.0×10^3 N/C, how much energy does it take to rotate the dipole from parallel alignment to antiparallel alignment with the field?

Solution   The energy required is equal to the change in potential energy:

$$\Delta U = U_{\text{antiparallel}} - U_{\text{parallel}} = -pE \cos 180° - (-pE \cos 0°),$$

$$\Delta U = pE + pE = 2pE = 2 \times 6.2 \times 10^{-30} \text{ C·m} \times 1.0 \times 10^3 \text{ N/C},$$

$$\Delta U = 1.2 \times 10^{-26} \text{ J}.$$

SUMMARY

Useful Concepts

- Single charges can be neither created nor destroyed (the law of conservation of charge).
- Like charges repel and unlike charges attract.
- Coulomb's law for the force between charges is

$$F = k\frac{q_1 q_2}{r^2},$$

where r is the distance between charges q_1 and q_2. If q_1 and q_2 are measured in coulombs and r in meters, then $k = 9.0 \times 10^9$ N·m²·C⁻². The direction of the force is along the line between the two charges.

- The principle of superposition of electrical forces states that the combined effect of many forces acting simultaneously is the same as the vector sum of the individual forces.
- The electric field is defined as the electrical force per unit positive charge:

$$\mathbf{E} = \frac{\mathbf{F}}{q_0}.$$

- The principle of superposition of electrical fields states that the combined effect of many fields acting simultaneously is the same as the vector sum of the individual fields.
- Gauss's law states that the net electric flux through any closed surface is directly proportional to the net electric charge enclosed within that surface. The mathematical

statement of Gauss's law is

$$\phi_E = \frac{Q}{\epsilon_0}.$$

- The excess charge on a conductor lies on its exterior surface.
- The torque on a dipole in an electric field is

$$\tau = pE \sin \theta,$$

where p is the magnitude of the dipole moment and θ is the angle between the direction of the field and the direction of the dipole moment vector.

- The potential energy of a dipole of moment p in an electric field E is

$$U = -pE \cos \theta.$$

Important Terms

You should be able to write the meaning or definition of each of the following terms:

- electrostatics
- electric charge
- conductor
- insulator
- electrons
- law of conservation of charge
- charging by induction
- Coulomb's law
- coulomb (C)
- permittivity of free space
- electric field
- electric flux
- Gauss's law
- Gaussian surface
- electric dipole
- dipole moment

QUESTIONS

15.1 It is common to observe a static spark when you touch the door knob after shuffling across a rug in the winter. Why is it uncommon to get a similar spark in the summer?

15.2 Gasoline tank trucks can become charged as they travel. How can this happen and how can it be prevented?

15.3 Why does your hair tend to stand on end when combed on a cold dry day?

15.4 Where does the charge come from when an object is charged by friction?

15.5 When tiny scraps of paper are placed between two charged plates, they bounce back and forth between the plates. Explain this occurrence.

15.6 When charging an electroscope by induction, why is it necessary to remove your finger before removing the charging rod?

15.7 What electrostatic experiment can you do to tell whether two charges have the same magnitude?

15.8 Discuss the similarities and differences between the gravitational force between masses and the electric force between charges.

15.9 Tall buildings made of concrete and steel are frequently struck by lightning without any apparent damage. Can you explain why?

15.10 Two equal negative charges are each placed at a corner of an equilateral triangle. A positive charge of equal magnitude is placed at the third corner. Sketch the field lines for this situation.

15.11 Are the Coulomb forces between electrical charges conservative or non-conservative forces? Explain your answer.

15.12 What are the practical consequences of the observation that the electric field inside a hollow conductor is zero regardless of how much electric charge is placed on its outer surface? What does this imply regarding the safety of a person inside an automobile in a thunderstorm?

PROBLEMS

Hints for Solving Problems
In applying Coulomb's law and the definition of the electric field, you should express all quantities in SI units. Remember that electric field lines are directed from positive charges to negative charges, also that Gauss's law applies to the net flux and to the net charge enclosed within a Gaussian surface.

Section 15.2 Coulomb's Law

15.1 A point charge of $+2.0\ \mu C$ is located 0.15 m from a second point charge of $-4.0\ \mu C$. What are the magnitude and the direction of the force on each charge?

15.2 Two equal point charges are located one cm apart. If the electrostatic force between them is $1.0 \times 10^{-2}\ N$, what must be the magnitude of the charges?

15.3 Two charges of equal magnitude exert an attractive force of $4.0 \times 10^{-4}\ N$ on each other. If the magnitude of each charge is $2.0\ \mu C$, how far apart are the charges?

15.4 A charge of $+3.0\ \mu C$ exerts a force of $3.2 \times 10^{-2}\ N$ on a charge of $-1.6\ \mu C$. How far apart are the charges?

15.5 Two small charged pieces of amber experience a force when they are separated by a distance R. If the charge on each of the bodies is tripled and the separation between them is halved, what is the ratio of the new force to the original force between them?

15.6 Two charged styrofoam balls are moved so that the force between them becomes ten times greater than it was originally. What is the ratio of their new separation to their original separation?

15.7 Two conducting spheres of the same size have charges $q_1 = -4.0 \times 10^{-6}\ C$ and $q_2 = +8.0 \times 10^{-6}\ C$ and are separated by 3.0 cm. (a) What is the force between them? (b) What would be the force if the spheres were touched together and again separated by the same distance?

15.8 The nucleus of an atom contains protons of charge $+1.60 \times 10^{-19}\ C$ and neutrons, which are uncharged. (a) What is the ratio of the electrical force to the gravitational force between two protons? (b) Can the attractive force of gravity be responsible for holding the nucleus together? (*Hint:* Proton mass $= 1.67 \times 10^{-27}$ kg.)

15.9 A model of the hydrogen atom consists of a proton at the center with an electron in a surrounding orbit. The charges of the proton and the electron have opposite signs and equal magnitude of $q = 1.6 \times 10^{-19}\ C$. The average separation between the electron and the proton is approximately 5.3×10^{-11} m. What is the Coulomb force between them?

15.10 Suppose that the attraction between the moon and the earth were due to Coulomb forces rather than gravitational force. What would be the magnitude of the charge required if equal but opposite charges resided on both earth and moon? (*Hint:* Mass of earth $= 5.98 \times 10^{24}$ kg; mass of moon $= 7.36 \times 10^{22}$ kg; earth–moon distance $= 3.84 \times 10^{8}$ m.)

15.11 Two equal charges of 2.6 μC are placed a distance x apart. What must be the value of x if the force between the charges is 2.0×10^{-8} N?

15.12 Two small, positively charged spheres experience a mutual repulsive force of 1.52 N when they are 0.200 m apart. The sum of the charges on the two spheres is 6.00 μC. What is the charge on each sphere?

15.13 Modern particle accelerators can be used to produce particles that are not usually observed by other means. One such particle is the antiproton, a particle with the same mass as a proton and with charge of the same magnitude but opposite sign. When a proton and an antiproton are simultaneously produced, what is the Coulomb force between them when they are 2.50×10^{-10} m apart? (*Hint:* The charge on the proton is $+1.60 \times 10^{-19}$ C.)

Section 15.3   Superposition of Electric Forces

15.14 Two point charges of magnitude $+q$ are separated by a distance r. A third charge of equal magnitude and opposite sign is placed midway between the two positive charges. What is the net electrostatic force on the end charges? What is the net electrostatic force on the middle charge?

15.15 Four equal point charges $+q$ are placed at the corners of a square of edge length L. Find the force on any one of the charges.

15.16 Four point charges of 5.0 μC are placed at the corners of a square that is 30 cm on a side. Two charges, diagonally opposite each other, are positive and the other two are negative. What are the magnitude and the direction of the force on one of the charges?

15.17 Three equal charges, each of $+6.0$ μC, are spaced along a line. The end charges are each 2.0 m from the central charge. What are the magnitude and the direction of the force on each charge?

15.18 Three equal charges are placed at the corners of an equilateral triangle 0.50 m on a side. What are the magnitude and the direction of the force on each charge if the charges are each -3.0 nC?

15.19 Four charges are spaced along the x axis at 0.50-m intervals, beginning at the origin. Starting with the charge at the origin, the charges have values of $+5.0$ μC, -3.0 μC, -4.0 μC, and $+5.0$ μC. What is the net force on the -3.0-μC charge?

Section 15.4   The Electric Field

15.20 What is the electric field strength E at a point 0.200 m from a point charge of $+1.30$ μC?

15.21 Calculate the electric field 10.0 cm from a point charge of 1.35 μC.

15.22 A spherical copper shell has a charge q uniformly spread over its surface. What is the electric field at the center of the shell?

15.23 A point charge q produces an electric field of magnitude 90.5 N/C at a distance of 1.56 m. If the field is directed toward the charge, what is the value of q?

15.24 (a) Sketch the pattern of the electric field around an isolated positive charge Q. (b) Sketch the pattern of the electric field around an isolated negative charge of magnitude $2Q$.

15.25 Sketch the pattern of the electric field around two otherwise isolated charges separated by a small distance d. Assume that both charges are positive and that the magnitude of one of the charges is twice that of the other.

15.26 A positive charge of 1.00 μC is located in a uniform field of 1.00×10^{5} N/C. A negative charge of -0.100 μC is brought near enough to the positive charge that the attractive force between the charges just equals the force on the positive charge due to the field. How close are the two charges?

15.27 A tiny styrofoam ball of mass 0.500 g is suspended by a light thread of negligible mass. The ball is electrically charged. Then the ball is placed in a uniform horizontal electric field of 400 N/C. What is the charge q on the ball when it is deflected by 15° (Fig. 15.30)?

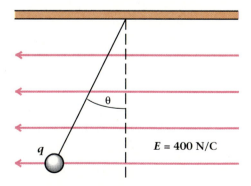

Figure 15.30   Problem 15.27.

15.28 A tiny ball of mass 0.500 g is suspended by a thread of negligible mass in a horizontal electric field of

400 N/C. If the ball has a charge $q = 9.00 \, \mu C$, what will be the deflection angle of the string measured from the vertical?

15.29 An electron (mass $= 9.11 \times 10^{-31}$ kg and charge $= 1.60 \times 10^{-19}$ C) is initially at rest just to the right of the leftmost plate in Fig. 15.31. The separation between the two plates is 4.00 cm. A uniform electric field of 5000 N/C directed to the left is suddenly applied between the two plates. (a) What is the force on the electron? (b) What kinetic energy will the electron have when it strikes the other plate?

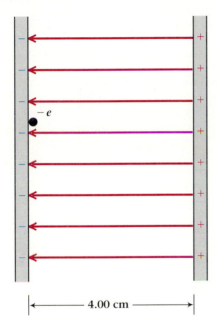

Figure 15.31 Problem 15.29.

15.30 Show that the kinetic energy acquired in a time t by a particle of mass m and charge q released from rest in a uniform electric field E is given by

$$KE = \frac{1}{2m}(Eqt)^2.$$

15.31 An electric charge $Q = 1.50 \, \mu C$ is in a region of electric field with components $E_x = 6000$ N/C, $E_y = 8000$ N/C, and $E_z = 0$. What are the magnitude and the direction of the force on the charge Q?

Section 15.5   Superposition of Electric Fields

15.32 An electric field of 12.6 N/C directed due north is superposed on an existing electric field of 17.4 N/C directed due east. What are the magnitude and the direction of the resultant field?

15.33 Three equal positive point charges are located at the corners of an equilateral triangle. Sketch the pattern of the electric field in the plane of the triangle.

15.34 Three equal positive point charges of magnitude Q are located at three corners of a square. A negative charge $-3Q$ is placed on the fourth corner. (a) Sketch the pattern of the electric field in the plane of the triangle. (b) At the position of the negative charge, what is the magnitude of the electric field due to the three positive charges? (c) What is the force on the negative charge if the edge length of the square is d?

*Section 15.7   A Quantitative Approach to Gauss's Law

15.35 A hollow spherical shell is uniformly charged with a total charge Q. Show that the electric field outside the shell is everywhere the same as the field due to a point charge Q located at the center of the shell.

15.36 A point charge Q is at the center of a conducting spherical shell of radius R. The total charge of the shell is $-Q$. (a) What is the field in the region between the point charge and the shell? (b) What is the field outside the shell? (*Hint:* Use Gauss's law and remember that the charge that contributes is the total charge within the surface.)

15.37 A large insulating sheet is given a uniform static charge of surface charge density σ. Show that the electric field near the sheet is given by $E = \sigma/2\epsilon_0$. (The surface charge density is the charge per unit area of the surface.)

15.38 Use Gauss's law to determine the electric field inside a hollow conducting sphere that carries a total charge Q. Extend your answer to a conductor of any shape.

15.39 A narrow hole is drilled along the diameter of a solid
•   sphere. The sphere carries a positive electric charge distributed uniformly throughout with a charge per unit volume ρ. A negative charge q is placed within the narrow hole. (a) Show that the charge q will be attracted to the center of the sphere with a force proportional to the distance from q to the center of the sphere. (b) Find an expression for the motion of the charged particle as a function of time.

Section 15.8   The Electric Dipole

15.40 A sulfur dioxide molecule has an electric dipole moment $p = 5.3 \times 10^{-30}$ C·m. If the molecule is in an electric field of 1.0×10^{4} V/m, what is the change in potential energy if it rotates from antiparallel to parallel alignment of its electric dipole moment with the direction of the electric field?

15.41 An electric dipole $p = 3.2 \times 10^{-30}$ C·m is in a region of uniform electric field of $2.5 \, 10^{3}$ V/m. Calculate the energy required to rotate the dipole from parallel alignment to antiparallel alignment with the field.

15.42 A water molecule of dipole moment $p = 6.1 \times 10^{-30}$ C·m is in a region of electric field $E = 1.4 \times 10^3$ V/m. What is the maximum torque on the dipole?

15.43 A dipole composed of two charges of magnitude 3.00×10^{-6} C experiences a maximum torque of 1.30×10^{-5} N·m when placed in a region of electric field of 1.00×10^3 N/C. What is the separation between the two charges that make up the dipole?

15.44 A dipole of $+q$ and $-q$ separated by distance $2a$ is surrounded by an imaginary Gaussian surface. (a) What is the electric flux through that surface? (b) Can Gauss's law be used to determine the electric field at the surface? Explain.

15.45 A water molecule, $p = 6.2 \times 10^{-30}$ C·m, is oriented along the direction of an electric field E. If the potential energy of the dipole in the field is -6.2×10^{-28} J, what is the magnitude of the electric field?

15.46 An experiment is done with a substance whose molecular dipole moment is not known. If the substance is subjected to a field of 1.50×10^4 N/C, each molecule releases an average energy of 2.28×10^{-25} J when rotating from antiparallel to parallel to the applied field. What is the molecular dipole moment of this material?

Additional Problems

15.47 Two equally charged insulating balls each weigh
• 0.10 g and hang from a common point by identical threads 30 cm long. The balls repel each other so that the separation between them is 8.0 cm. What is the magnitude of the charge on each ball?

15.48 If, after two hours, the spheres of Problem 15.47 are
• found to be only 4.0 cm apart, what is the average rate at which charge has leaked off the spheres?

15.49 How far apart must two protons be if the Coulomb
• force between them is equal to the weight of a single proton at the earth's surface? Assume that the two protons are far removed from any other charges. The charge of the proton is 1.60×10^{-19} C and its mass is 1.67×10^{-27} kg.

15.50 Imagine that the effects of gravity have been "turned
•• off" and that you want to place a 50-kg satellite into a synchronous circular orbit with the same radius that it would obtain if gravity were "turned on." How much electric charge would be required if the satellite were held in orbit by Coulomb forces? For simplicity, assume that the magnitude of the net charge on the satellite and on the earth are the same and that the charge on the earth is uniformly spread across the earth's surface.

15.51 The hydrogen atom may be described as a positively
•• charged proton surrounded by an orbiting, negatively charged electron. In a simple model, the electron is in a circular orbit of radius $r = 5.29 \times 10^{-11}$ m about a stationary proton. The mass of the proton is $m_p = 1.67 \times 10^{-27}$ kg and the mass of the electron is $m_e = 9.11 \times 10^{-31}$ kg. The charge on the proton is $+e$ and the charge on the electron is $-e$, where $e = 1.60 \times 10^{-19}$ C. (a) What is the Coulomb force between the proton and the electron? (b) What is the linear speed of the electron in its orbit? (c) What is the orbital frequency of the electron about the proton?

15.52 Four equal charges of magnitude Q are arranged
• on the corners of a square whose edge length is a. (a) What is the force on a test charge q_0 placed at the center of the square? (b) What are the magnitude and the direction of the force on the test charge q_0 if it is placed at the midpoint of one of the edges?

15.53 A charge of $+3.0 \times 10^{-6}$ C is located at the point
• $x = +0.10$ m, $y = 0$. A second charge of -3.0×10^{-6} C is located at $x = -0.10$ m, $y = 0$. (a) What are the magnitude and the direction of the electric field at the point $x = 0$, $y = 0$ and (b) at the point $x = 0$, $y = +0.30$ m?

15.54 Four equal point charges of magnitude Q are ar-
• ranged on the corners of a square of edge length 1.0 cm. Evaluate the electric field at (a) the center and (b) the midpoint of one edge, given that $Q = 1.0 \times 10^{-9}$ C.

15.55 A tiny plastic ball of mass m carries an electric
•• charge q. The ball is attached to a charged nonconducting sheet with uniform surface charge density σ by means of a nonconducting string of negligible mass (Fig. 15.32). Derive a formula for the angle θ between the string and the vertically oriented sheet in terms of m, q, σ, and g.

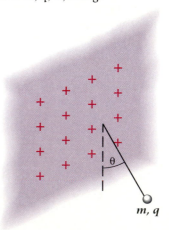

Figure 15.32   Problem 15.55.

15.56 Show that the electric field along the axis of a dipole
•• of dipole moment p is given by $E = p/2\pi\epsilon_0 r^3$ when the distance r from the center of the dipole is large compared with the separation between the charges.

15.57 A solid insulating sphere of radius R has an electric
•• charge uniformly distributed throughout its volume. The charge per unit volume is ρ. (a) Show that outside the sphere the electric field is given by $E = \rho R^3/3\epsilon_0 r^2$, where r is the distance from the center of the sphere. (b) Show that the field inside the sphere is given by $\rho r/3\epsilon_0$. (c) Draw a graph of the field from $r = 0$ to $r = 3R$. (*Hint:* Use Gauss's law and remember that it is the total charge within the Gaussian surface that contributes to the electric field at that surface.)

15.58 A spherical shell with inner radius a and outer ra-
•• dius b has a uniform charge Q distributed throughout its volume. Calculate the electric field for $r < a$, $a < r < b$, and $b < r$.

15.59 Show that Coulomb's law can be obtained from
•• Gauss's law.

15.60 Four equal charges of $+3.0 \times 10^{-7}$ C are placed
•• on the corners of one face of a cube of edge length 10.0 cm. A charge of -3.0×10^{-7} C is placed at the center of the cube. What are the magnitude and the direction of the force on the charge at the center of the cube?

15.61 A proton accelerates from rest to 3.00×10^6 m/s
• in 1.00×10^{-6} s in a uniform electric field **E**. What is the magnitude of the electric field? (*Hint:* The proton has a mass $m_p = 1.67 \times 10^{-27}$ kg and a charge $+e = 1.60 \times 10^{-19}$ C.)

15.62 An electron is injected horizontally into a vertical
•• electric field. Show that the path of the electron is parabolic. (*Hint:* Neglect the effects of gravity and compare this case with the case of projectile motion in a gravitational field in Chapter 3.)

15.63 A small nonconducting sphere of mass m falls a dis-
• tance d from one metal plate to another directly beneath it. If the sphere possesses a charge q, show that it will return to its original position in the same time that it took to fall when a uniform electric field $E = 2mg/q$ is present between the plates.

15.64 A small droplet of oil with mass of 2.00×10^{-15}
• kg is held suspended in a region of uniform electric field directed upward with a magnitude of 6125 N/C. (a) Is the excess charge on the droplet positive or negative? (b) How many excess elementary charges reside on the droplet? (The elementary charge is $e = 1.6 \times 10^{-19}$ C.)

ADDITIONAL READING

Burland, D. M., and L. B. Schein, "Physics of Electrophotography." *Physics Today,* May 1986, p. 46.

Edge, R. D., "String and Sticky Tape Experiments, Electrostatics with Soft-drink Cans." *The Physics Teacher,* September 1984, p. 396.

Greenslade, T. B., Jr., "Electrostatic Toys." *The Physics Teacher,* November 1982, p. 552.

Hackmann, W. D., *Electricity from Glass: The History of the Frictional Electrical Machine 1600–1850.* Alkphen aan den Rijn, The Netherlands: Sijthoff & Noordhoff, 1978. A thorough account of the discoveries of electrostatic phenomena and the development of electrostatic generators.

Walker, J., "The Amateur Scientist: How to Capture on Film the Faint Glow Emitted When Sticky Tape Is Peeled off a Surface." *Scientific American,* December 1987, p. 138.

Walker, J., " The Amateur Scientist: The Secret of a Microwave Oven's Cooking Action Is Disclosed." *Scientific American,* February 1987, p. 134.

Williams, E. R., "The Electrification of Thunderstorms." *Scientific American,* November 1988, p. 88.

16

Electric Potential and Capacitance

16.1   Electric Potential

16.2   The Van de Graaff Electrostatic Generator

16.3   Equipotential Surfaces

*16.4   Electric Potential Gradient

16.5   Capacitors

16.6   The Parallel-Plate Capacitor

16.7   Electric Field of a Parallel-Plate Capacitor

16.8   Dielectrics

16.9   Energy Storage in a Capacitor

A WORD TO THE STUDENT

Having studied the electrostatic field of a charged body, we now turn our attention to electric potential. We define potential as the electric potential energy per unit charge. This concept enables us to calculate electrical effects by using a scalar quantity instead of a vector, much as we did in Chapter 8 for gravitation. The advantage of this approach will be evident in the simpler calculations in this chapter compared with those in the preceding chapter.

With the fundamental ideas of charge, field, and potential, we can begin to examine some of the basic components of electric circuits—in particular, the capacitor. The classic model of the capacitor is two parallel conducting plates that are spaced close to each other but do not touch. When the two plates are oppositely charged, the charges stay in place because of Coulomb forces. Thus we can use capacitors to store charge and hence energy. Capacitors are widely found in electrical and electronic equipment, as we will see in Chapter 20.

As we have seen, the early theories and experiments in electricity concentrated on the properties of charge. For example, electrostatic generators like Hauksbee's glass globe were eventually analyzed in terms of separating charges by applying frictional force. However, by the late eighteenth century gravitation had been successfully treated in terms of gravitational fields, and a similar approach began to carry over to electricity. Understanding the nature of electric potential allowed people to develop larger electrostatic generators, capable of separating even greater amounts of charge. In fact, the idea of electric potential—or voltage, as it is often called—is still one of the key concepts in studying electric circuits and electronic devices.

The basic ideas of work and energy are the same when applied to electrical forces as when applied to gravitational or any other forces. We develop the idea of electric potential by analyzing the work that is done and the change in energy that occurs when a charge is moved in an electric field. Then, to illustrate what a useful idea potential can be, we discuss an important application of electrostatics: the Van de Graaff electrostatic generator. We go on to generalize the idea of potential, which gives us a background for understanding a simple electric circuit device, the capacitor. Capacitors, which evolved from early devices for storing charge, are part of most common electronic circuits, from the tuning circuit in your radio to the ignition system in your car. Their operation can be understood by building on the ideas of electric charge and electric potential.

Electric Potential

16.1

We learned in Chapter 15 that according to Coulomb's law, a positive test charge q_0 located a distance r_A away from a positive charge Q (Fig. 16.1a, p. 468) experiences an electrostatic force of magnitude

$$F_A = k\frac{Qq_0}{r_A^2}$$

in the direction away from Q. If q_0 is moved closer to Q, say to a point r_B as in Fig. 16.1(b), it then experiences a force of magnitude

$$F_B = k\frac{Qq_0}{r_B^2},$$

still directed away from Q. Because r_B is less than r_A, the force F_B is greater than F_A.

To move charge q_0 from r_A to r_B, we must apply a force in a direction opposite to the electrostatic force. The work done on the test charge by this external force is

$$W = \overline{F}(r_A - r_B),$$

where $\overline{F}$ is the average force supplied over the interval from A to B. The

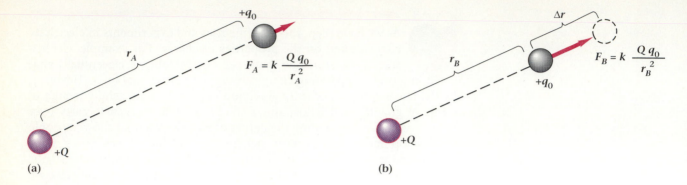

Figure 16.1 (a) A test charge q_0 located a distance r_A from a source charge Q. (b) The test charge is moved a distance Δr closer to source charge Q, which results in a change in the potential energy.

proper value of the average force is*

$$\overline{F} = k\frac{Qq_0}{r_A r_B},$$

so the work becomes

$$W = kQq_0\frac{r_A - r_B}{r_A r_B},$$

or

$$W = kQq_0\left(\frac{1}{r_B} - \frac{1}{r_A}\right). \tag{16.1}$$

The work done against the field in moving the test charge to r_B can be regained by removing the applied force. This would allow the Coulomb force to repel the test charge back to r_A. If the test charge is released from rest at r_B, it will be in motion as it passes r_A and will therefore have kinetic energy. From conservation of energy, we know that this kinetic energy must come from somewhere, which means that the work given by Eq. (16.1) corresponds to an increase in potential energy of the test charge. Since the interaction is electrical we call this *electrical potential energy*. The electrostatic force is a conservative force, just like the gravitational force, so the work done does not depend on the path of the charge, but only on its starting and ending positions.

Equation (16.1) gives us the *difference* in electrical potential energy between two points in space due to a point charge. In establishing a zero of potential energy, we are free to pick any absolute reference potential that suits our convenience. Traditionally, we choose the zero of potential to correspond to the case in which the charges are infinitely separated, that is, for $r_A = \infty$. With this choice, the electrical potential energy of the test charge q_0 located a finite distance r away from a source charge Q is

$$PE_{elec} = W = \frac{kQq_0}{r}. \tag{16.2}$$

There is a strong similarity between the equation for electrical potential energy and the equation for gravitational potential energy, given in

---

*The quantity $r_A r_B$ is intermediate in value between r_A^2 and r_B^2 and is called the geometric mean of the two.

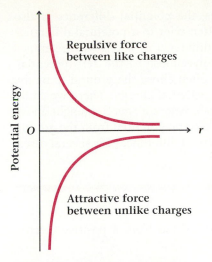

Figure 16.2 The electrostatic force between two charges corresponds to either of two potential energy curves, depending on whether the force is attractive or repulsive.

Chapter 6 as

$$PE_{grav} = -G\frac{Mm}{r}.$$

In both cases we take the zero reference point to be at $r = \infty$. In the gravitational case the force is attractive and the potential is always negative. For a positive charge Q, the electrical force on a positive charge q_0 is repulsive and the potential energy is positive. However, if Q is negative, resulting in an attractive force between Q and q_0, the electrical potential energy is negative. We can draw a potential-energy diagram for q_0 in the electric field of Q similar to the gravitational potential-energy diagrams shown in Chapter 6. However, in this case, when the charges have the same sign the potential energy is positive and when the charges have unlike signs the potential energy is negative (Fig. 16.2).

We define the **electric potential** V at a point in an electric field as the electrical potential energy divided by the magnitude of the test charge q_0:

$$V \equiv \frac{PE_{elec}}{q_0}. \tag{16.3}$$

The electric potential due to a point charge Q is obtained from the potential energy of Eq. (16.2) by dividing by q_0,

$$V = k\frac{Q}{r} = \frac{Q}{4\pi\epsilon_0 r}. \tag{16.4}$$

If the charge Q is positive, the potential is positive, and if Q is negative, the potential is negative. The unit of electric potential is the **volt**:

1 volt = 1 joule/coulomb.

The electric potential at any point in space is defined to be the work per unit charge required to bring a charge from infinity to that point. The electric potential due to two or more point charges is easily obtained. From the principle of superposition of forces, the work done in bringing a test charge from infinity can be separated into parts associated with the individual forces due to each single charge. The total work is the sum of these individual contributions. Because electric potential is a scalar, not a vector quantity, the potential at a given point in space is just the algebraic sum of the potential due to the separate source charges.

Frequently, just as for the case of gravitation, the quantity of interest is not the absolute potential, but the potential difference between two points. The **electric potential difference** is the ratio of the work required to move a charge from one point to another to the magnitude of the charge. The potential difference between points A and B is

$$V_{AB} = V_B - V_A = \frac{W_{AB}}{q}. \tag{16.5}$$

TABLE 16.1
Electric potential differences found in common situations.*

Situation	Potential difference (volts)
pn-junction in a transistor	10^{-1}
Biological cell	10^{-1}
Single cell battery	1
Household electric outlet	10^2
TV sets	10^4
Thunderclouds	10^8

*From Joseph M. Crowley, *Fundamentals of Applied Electrostatics* (John Wiley, 1986)

Since the potential is measured in volts, the potential difference is also measured in volts. For this reason we often refer to a potential difference as a voltage. Many times, the potential difference is taken with reference to the *ground* (earth), which we generally choose as the zero potential. We do this for the same reason that we often chose the ground to be the reference when describing gravitational potential energy. There we could calculate a change in potential energy *mgh*, where *h* was the height above the ground. Here we want the change in electric potential from the ground to some other point. Table 16.1 lists potential differences for several cases.

Example 16.1

Potential due to a point charge.

What is the electric potential at a point 5.0 cm from a positive electric point charge of 10 nC?

Solution   We assume that no other charges are present and that the potential is zero at infinity. Then according to Eq. (16.4), the potential is

$$V = k\frac{Q}{r} = 9.0 \times 10^9 \text{ N·m}^2\text{·C}^{-2} \times \frac{10 \times 10^{-9} \text{ C}}{5.0 \times 10^{-2} \text{ m}},$$
$$V = 1.8 \times 10^3 \text{ N·m/C} = 1.8 \times 10^3 \text{ V} = 1.8 \text{ kV}.$$

Because the charge *Q* is positive, the potential is positive.

Example 16.2

Potential due to several point charges.

What is the electric potential at the center of the square shown in Fig. 16.3? The values of the charges are $q_1 = 1.0$ nC, $q_2 = -2.0$ nC, $q_3 = +3.0$ nC, and $q_4 = -4.0$ nC. Assume the square to have an edge length $d = 1.0$ m.

Solution   The center of the square is equidistant from all four charges, a distance of $d/\sqrt{2}$. Let's call this distance *r*. Then we can calculate the potential by adding up the potentials due to the individual charges,

$$V = V_1 + V_2 + V_3 + V_4 = \frac{kq_1}{r} + \frac{kq_2}{r} + \frac{kq_3}{r} + \frac{kq_4}{r},$$
$$V = \frac{k}{r}(q_1 + q_2 + q_3 + q_4),$$
$$V = \frac{9.0 \times 10^9 \text{ N·m}^2/\text{C}^2 \times (1.0 \; -2.0 \; +3.0 \; -4.0) \times 10^{-9} \text{ C}}{1.0 \text{ m}/\sqrt{2}},$$
$$V = -25 \text{ N·m/C} = -25 \text{ V}.$$

The electric potential at the center of the square is thus 25 V below the zero level reference potential, which corresponds to charges being infinitely far away. This means you would have to do work on a positive test charge to move it from the center of the square to infinity.

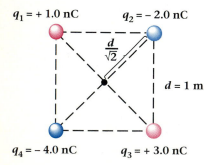

$q_1 = +1.0$ nC $q_2 = -2.0$ nC

$\frac{d}{\sqrt{2}}$ $d = 1$ m

$q_4 = -4.0$ nC $q_3 = +3.0$ nC

Figure 16.3   Example 16.2: Four point charges arranged in a square of edge *d*.

The Van de Graaff
Electrostatic Generator

16.2

The electrostatic generator is a machine for producing very large electric potential differences, of the order of millions of volts. The modern form of the electrostatic generator was developed in 1931 by the American physicist Robert J. Van de Graaff (1901–1967). Since the chief application of these electrostatic generators is to accelerate charged particles to very high kinetic energies, they are often referred to as accelerators.

The operation of a **Van de Graaff generator** can be explained in terms of the principles described in the preceding sections. These same principles apply to the research accelerators as well as to the classroom demonstration Van de Graaff generator shown in Fig. 16.4(a). A motor-driven pulley drives a belt of insulating material within a column of the machine (Fig. 16.4b). At the base of the column, the belt passes the sharp points of a charged metal "comb," which removes negative charges (electrons) from the belt, leaving the belt positively charged. The charged belt moves upward into a large, hollow, conducting dome. Here a second conducting wire, connected to a comb of sharp points near the top of the belt, supplies negative electrons from the dome to the positively charged belt, leaving positive charge on the dome.

Since the dome is a conductor, its excess charge resides only on the outside surface, as was shown in Section 15.6. By Gauss's law, there is no

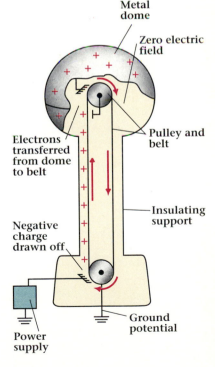

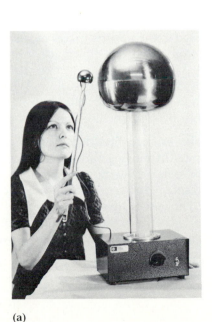

Figure 16.4   (a) A small demonstration Van de Graaff generator.
(b) Schematic representation of a Van de Graaff generator. A moving belt removes electrons from the hollow dome, leaving a large positive charge on the dome.

(a)

(b)

Figure 16.5 A research Van de Graaff generator used for producing beams of energetic particles.

field inside the dome due to these outer charges. However, there is an electric field inside the dome due to the positively charged belt. This field draws negative charges from the dome regardless of the dome's potential. As a result, a large positive charge builds up on the dome, accompanied by a large increase in voltage, which is proportional to the total charge.

There are limits to the potential difference that can be maintained between the dome and surrounding objects. Eventually the potential difference becomes so great that the air can no longer act as an insulator, and the dome discharges with a spark of miniature lightning. The motion of ions in the air can also cause the dome to lose charge. In addition, if the generator is operated in air when the humidity is high, charge leaks off because of moisture on the insulating surfaces and on the belt. Consequently, the maximum potential is reduced. For this reason the electrostatic generators used in research are usually enclosed in a sealed container, where they can be surrounded by a controlled, dry atmosphere of air, nitrogen, or other gas. Figure 16.5 shows a research Van de Graaff generator with the outer enclosure removed.

If the dome is charged positively to several million volts above ground potential, positive charges can be accelerated away from the dome. By placing a source of charged particles inside the dome, we can generate a stream of highly energetic particles. The energy of these particles is determined by the potential of the dome. In practice, Van de Graaff accelerators are used to produce beams of high-energy particles for use in nuclear physics and nuclear medicine.

Example 16.3

Speed of an accelerated ion.

In the manufacture of some integrated circuit chips, which are the hearts of computers, watches, and other electronic equipment, boron atoms are implanted in the upper surfaces of silicon to control its electrical behavior. The boron atoms, which acquire a high velocity in passing through the potential difference of an electrostatic accelerator, bury themselves in the silicon when they hit the surface. What is the speed of a singly charged boron ion (an atom with one electron removed) that has been accelerated from rest by a potential difference of 86.0 kV?

Solution   We can find the speed by using the principle of conservation of energy. In moving through the potential difference the boron ion of mass m_B gains a kinetic energy $\frac{1}{2}m_B v^2$, where v is the final velocity. This kinetic energy is a result of the work done on the ion by the electric field as the ion moves through the electric potential difference. The work is $W = qV$, where q is the charge on the ion and V is the potential difference. Then from the work–energy theorem we get

$$\tfrac{1}{2}m_B v^2 = qV,$$

or

$$v = \sqrt{\frac{2qV}{m_B}}.$$

The charge q is equal to the magnitude of the electron charge, 1.60×10^{-19} C, and the mass of the boron ion is 1.83×10^{-26} kg. Inserting these values into the equation for v, we find that

$$v = \sqrt{\frac{2 \times 1.60 \times 10^{-19} \text{ C} \times 86.0 \times 10^{3} \text{ V}}{1.83 \times 10^{-26} \text{ kg}}},$$

$$v = 1.23 \times 10^{6} \text{ m/s}.$$

Equipotential Surfaces

16.3

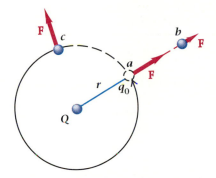

Figure 16.6 A test charge q_0 changes potential energy when moving along a radial path away from a fixed charge Q, but not when moving in a circular path about Q. The circular path is an equipotential line.

Figure 16.7 The equipotential surfaces around a point charge have the form of a nested set of spherical shells. The surface closest to the charge has the largest value of potential.

We have seen pictorial representations of electric fields in Figs. 15.11, 15.13, and 15.14. These representations were based on the concept of electrostatic force, a vector. We now wish to introduce a pictorial representation of electric potential based on the concept of energy, a scalar. You may find it helpful to review the ideas used in Chapter 6 for making potential-energy diagrams.

Figure 16.6 shows a test charge q_0 initially located a distance r from a charge Q whose position is fixed. If the test charge q_0 moves from point a to point b along a radial path, the force due to the electric field of Q does work on the test charge. As a result, the potential energy of q_0 is changed (Fig. 16.2). On the other hand, if the charge moves from point a to point c along the circular path of constant radius, its motion is perpendicular to the electric force and no work is done by (or against) the field. Therefore the potential energy of q_0 is the same at point a as at point c. As long as the charge moves at right angles to the electric field (and hence to the electric force), its potential energy remains unchanged. In this case, q_0 could move completely around Q along the circular path without any work being done by the electric field. The potential energy of the test charge is the same everywhere on the circle, so the electric potential must also be constant along the circle. For this reason, we call any circle about the source charge Q an equipotential line.

As we noted in Chapter 15, lines of force about a charge extend symmetrically in three dimensions. Similarly, the two-dimensional equipotential circle around a point charge is actually a three-dimensional spherical surface. Any surface for which all points are at the same potential is called an **equipotential surface**. Figure 16.7 depicts the equipotential surfaces around a point charge as a set of nested spherical shells. Each equipotential surface represents a different potential, decreasing in magnitude with increasing distance from the source charge. The choice of specific potentials is arbitrary; an equipotential surface can be drawn through any point in the field.

Equipotential surfaces may be drawn corresponding to any given configuration of electric field. We construct the contours of the equipotential surfaces by making them everywhere perpendicular to the electric field vector (or lines of force). Figure 16.8 (p. 474) shows a cross section of the equipotential surfaces for (a) a pair of like charges and (b) a pair of unlike charges. Notice that equipotential lines are always perpendicular to field lines. A cross section of the equipotential lines around four charges (a quadrupole of $+q$, $+q$, $-q$, and $-q$) is shown in Fig. 16.8(c).

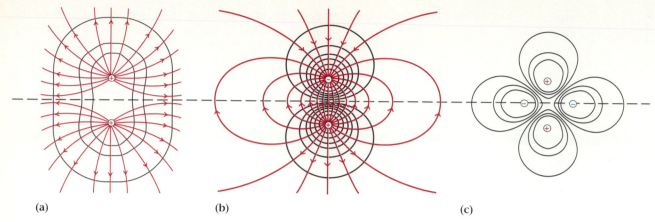

(a)   (b)   (c)

Figure 16.8   Cross sections of equipotential surfaces around (a) a pair of like charges, (b) a pair of unlike charges, and (c) two pairs of unlike charges. The black lines represent the equipotential surfaces and the red lines indicate the electric field.

In Chapter 15 we found that any excess charge placed on a conductor must reside on its surface. If the charge is at rest, whatever electric field is present must be perpendicular to the surface, since any field parallel to the surface would cause the charges to move. Since the field is perpendicular to the surface, all points along the surface must be at the same potential. We conclude that **when all charges have come to rest, the surface of a conductor is always an equipotential surface**. Also, as we saw in Chapter 15, at equilibrium there is no field inside a conductor, so all interior points have the same potential as the surface. Thus it takes no work to move a charge around anywhere within a conductor.

Electric Potential Gradient

*16.4

For any configuration of source charges, a correspondence exists between the electric field produced and the associated equipotential surfaces. We have already seen that we can construct equipotential surfaces from a knowledge of the field lines. The converse is also true; namely, we can determine the electric field from a knowledge of the potentials.

For example, suppose we wish to move a test charge q_0 from one equipotential surface to another along a path Δs (Fig. 16.9). If the potential increases by ΔV, we must do an amount of work

$$\Delta W = q_0 \, \Delta V.$$

If the path Δs is perpendicular to the equipotential surfaces, then it must lie along the direction of the electric field present in that region of space. So, we can also calculate the work done in terms of the force due to that field:

$$\Delta W = F \, \Delta s = q_0 E \, \Delta s.$$

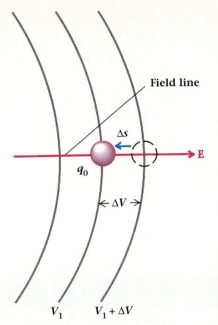

Field line

Δs

q_0

$\leftarrow \Delta V \rightarrow$

V_1 $V_1 + \Delta V$

Figure 16.9 An amount of work $\Delta W = q_0\,\Delta V$ is required to move a test charge q_0 along a distance Δs from one equipotential surface to another.

Upon comparing these two equations, we find that

$$E = -\frac{\Delta V}{\Delta s},$$

(16.6)

where we included the negative sign to give the correct direction of the field. An electric field exerts a force on a positive test charge to move it in the direction of the field, at the same time going from higher potential to lower potential. So, if the potential increases as we move a positive displacement of magnitude Δs, then the electric field must point in the opposite direction. Thus, in Fig. 16.9, if ΔV is positive, the field is directed opposite to the displacement.

The quantity $\Delta V/\Delta s$ is called the **electric potential gradient**. It is the rate of change of the electric potential with distance, that is, the change in potential with unit distance. The potential gradient is a vector whose magnitude is equal to the magnitude of the electric field and whose direction is opposite to the direction of the electric field. Thus the electric field is equal to the negative of the potential gradient.

It is apparent from Eq. (16.6) that the dimensions of the electric field are the same as that of electric potential divided by distance. In SI units the electric field is given in volts per meter. Indeed, the most common way to measure E is in units of potential divided by a convenient length unit, such as V/m, V/mm, kV/m, etc.

Example 16.4

Dimensional analysis of potential gradient.

Show that the units of newtons per coulomb, which we first used for electric field, are equivalent to volts per meter.

Solution We start by recalling the definition of the volt:

$$\text{volt} = \text{joule/coulomb}.$$

The joule was defined as a newton·meter, so that

$$\text{volt} = \text{newton·meter/coulomb}$$

and

$$\text{volt/meter} = \text{newton/coulomb}.$$

Example 16.5

Calculation of electric field.

The electric potential difference between the parallel deflection plates in an oscilloscope is 300 V. If the potential drops uniformly when going from one plate to the other and if the distance between the plates is 0.75 cm, what is the magnitude of the electric field between them and in which direction does it point?

Solution   Let us choose the positive direction of Δs to be in the direction of increasing potential, that is, from the lower-voltage plate to the higher-voltage one. We can then apply Eq. (16.6):

$$E = -\frac{\Delta V}{\Delta s}.$$

Here $\Delta V = +300$ V when $\Delta s = +0.75$ cm $= 0.75 \times 10^{-2}$ m. Thus

$$E = -\frac{300 \text{ V}}{0.75 \times 10^{-2} \text{ m}} = -40{,}000 \text{ V/m}.$$

The negative value of E tells us that E is directed opposite to Δs. Thus E is directed from the higher-voltage plate toward the lower-voltage one.

BACK TO THE FUTURE

The Leyden Jar and Franklin's Kite

Early experimenters in electrostatics faced a very difficult problem. The magnitude of the force between electric charges makes it difficult to separate large amounts of charge and keep them separated. Even with Hauksbee's electrostatic machine and others developed later, it was most difficult to store isolated charge.

In 1746 the Dutch physicist Pieter van Musschenbroek (1692–1761) was pursuing experiments at the University of Leyden to build up and store an extremely large charge. These efforts culminated in a device that became widely known as the *Leyden jar* (Fig. B16.1). This device is a glass jar covered inside and out with metal foil. It has a wooden lid with a metal rod running through it. A metal chain hangs down from the rod to touch the metal foil lining the inside of the jar. With these Leyden jars electric charge was collected in quantities so large that it was used to shock hundreds of people.

The mechanism of charging a Leyden jar is easily understood. First the outer metal coating of the jar is grounded, say by a metal wire running

into the ground. The center conductor is brought near an electrostatic machine, which, for illustration, we consider to be positively charged. When the center conductor is brought into contact with the machine, the mobile negative charges (electrons) in the center conductor drain off onto the

machine. As a result, a positive charge is left on the inner foil liner of the jar. This inner charge attracts excess negative charge to the outer foil, the negative charge coming from the earth or ground by means of the ground wire.

The charging process can be repeated several times to build up an

Figure B16.1   The Leyden jar, consisting of a glass insulator between two metal foils.

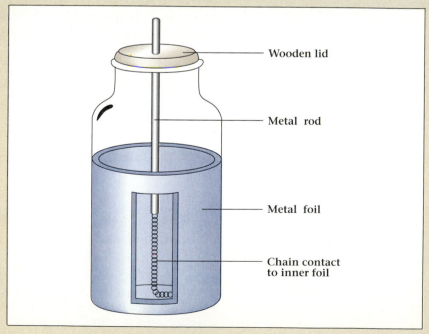

Wooden lid

Metal rod

Metal foil

Chain contact to inner foil

Capacitors

16.5

Any system of two conductors placed near each other but separated by an insulator, such as the metal foils on a Leyden jar (see p. 476), can store more charge than can an isolated conductor. It was said that the Leyden jar concentrated, or condensed, the charge, and the name *condenser* has persisted to the present day. However, since the important measure of these devices is their capacity to store charge, in time they became known as capacitors. A **capacitor** is a device for storing charge.

Capacitors come in many shapes and sizes, made from a variety of materials. The only thing they all have in common is the close placement of two or more conducting sheets or plates separated by a thin layer of

---

extremely large charge on the jar. After the charging is completed, the ground is disconnected. Because the two conductors are well insulated, the jar is able to store a surprisingly large amount of charge for a long time. The glass acts as an insulator to keep the charges apart. If it were not for the leakage of charge into the air and across the insulating surfaces of jar and lid, the charge would remain indefinitely. As it is, the charge can remain for long periods of time if the jar is clean and dry and the air is dry.

If a conductor attached to the outer coating of a charged Leyden jar is brought near the center conductor, the jar will discharge with a large spark. In Musschenbroek's laboratory, people acting as experimental subjects were knocked about and stunned by the terrifying shock they received from the jar. If one person held the jar while another person touched the insulated conductor, both felt a shock when they joined their free hands together. In some experiments, several people holding hands made a human chain through which the charge could pass. The end persons primarily felt the shock if the others clasped their hands firmly. Human chains exceeding 200 persons in length were formed as the

experimenters demonstrated their new tool.

Benjamin Franklin also experimented with Leyden jars and wondered if electricity might be the same as lightning. In 1752 he charged a Leyden jar with the electricity obtained from a kite that he flew during a thunderstorm (Fig. B16.2). For this proof that lightning and electricity were one and the same, Franklin was made a member of England's scientific Royal Society.

Franklin's experiment was extremely hazardous; he was fortunate to carry it out unscathed. Indeed, several

Figure B16.2   Benjamin Franklin's experiment.

people who tried to duplicate his experiment were struck by lightning and killed.

Franklin noted that a charged object could be discharged by bringing a metal sphere close to the object. The discharge could take place through the air—perhaps even with a spark. However, if a sharply pointed metal rod was brought close to the charged object, it discharged when the rod was considerably farther away, and the discharge was accomplished without a spark. Today we would say that the potential gradient, and hence the field, was greater near the sharp point than near a rounded surface. Thus charge was drained off the tip of the pointed rod by the field, causing the discharge of the object.

Franklin reasoned that if a pointed metal rod were placed on the roof of a house and grounded with a heavy wire, it could discharge thunderclouds harmlessly. In addition, in case of a lightning strike, this "lightning rod" would provide a safe electrical path to the ground for the lightning's energy, thus protecting the house from damage. Indeed, lightning rods are very effective and are in common use in America, especially on tall buildings and in rural areas.

(a)

(b)

Figure 16.10   (a) Some typical capacitors. (b) The circuit symbol for a capacitor.

insulation. In some capacitors the conductors are metal foil and the insulator is paper; other capacitors contain alternating layers of metal plates and mica. Still others in modern integrated circuits consist of thin strips of silicon oxide deposited on silicon chips and covered with a layer of metal. Figure 16.10(a) shows some typical capacitors.

Capacitors play an important role in many electrical and electronic circuits. For example, in computers and television sets, the insulating layers of capacitors can be used to isolate portions of a circuit from unwanted electric potentials. Capacitors are also used to store charge, and hence energy, to be given up when needed. In this sense, a charged capacitor is similar to a compressed spring, for both store energy that can be released when needed. The circuit symbol for a capacitor (Fig. 16.10b) suggests an interrupted conducting path due to the insulating layer.

The **capacitance** C of a capacitor is defined as the ratio of the quantity of charge stored on it to the potential difference V between the conductors.

$$C \equiv \frac{q}{V}.$$ (16.7)

The magnitude of the capacitance is a constant for each particular capacitor and depends only on the geometrical shape of the capacitor and on the nature of the insulating material between the conductors. When the charge is measured in coulombs and the potential in volts, the capacitance is measured in **farads** (F):

$$1 \text{ farad} = 1 \text{ coulomb/volt.}$$

The unit for capacitance is named in honor of the English physicist and chemist Michael Faraday (1791–1867), who was renowned for his work in electrochemistry and electricity. The farad is a large unit; many practical capacitors have values of microfarads, nanofarads, or even picofarads.

Because, for a given capacitor, the value of the capacitance is fixed, we can rearrange Eq. (16.7) to show the linear relationship between the charge stored and the voltage on the capacitor:

$$q = CV.$$

Thus, if the potential difference between the plates is increased, the charge stored on the capacitor is also increased. Similarly, if you need to store a large charge with only a small applied voltage, then you need a large capacitance.

The Parallel-Plate Capacitor

16.6

In its simplest form a capacitor consists of a pair of parallel metal plates separated by a small distance d (Fig. 16.11). The air between the plates is the insulator. When a charge $+q$ is stored on one plate and a charge $-q$ on the other, an electric potential difference exists between the plates. For

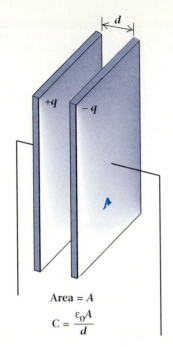

Area = A

$$C = \frac{\epsilon_0 A}{d}$$

Figure 16.11 Two oppositely charged parallel plates form a capacitor. The capacitance depends on the area A of the plates and inversely on the distance d between them.

a given separation between the plates we expect that larger plates can hold more charge and thus contribute to a larger capacitance. Similarly, for a fixed area we expect the capacitance to decrease as the separation increases because the influence of each plate on the other is reduced; that is, the force of attraction between the plates is reduced. The result for two parallel plates of area A that is large compared with their separation d is a capacitance

$$C = \frac{\epsilon_0 A}{d}. \tag{16.8}$$

We will derive this equation formally in the next section. Equation (16.8) is appropriate when the plates are in a vacuum. When the plates are in air, the capacitance is changed only slightly, as we shall see in Section 16.8. The symbol ϵ_0 is the permittivity of free space, as given in Chapter 15.

Example 16.6

Area of a 1-F parallel-plate capacitor.

A parallel-plate capacitor is designed to have a capacitance of 1.00 F when the plates are separated by 1.00 mm in a vacuum. What must be the area of the plates?

Solution   We use Eq. (16.8), which relates capacitance, area, and separation:

$$C = \frac{\epsilon_0 A}{d},$$

or

$$A = \frac{Cd}{\epsilon_0}.$$

We next insert the values for C, d, and ϵ_0 to get

$$A = \frac{1.00 \text{ F} \times 1.00 \times 10^{-3} \text{ m}}{8.85 \times 10^{-12} \text{ C}^2/\text{N·m}^2} = 1.13 \times 10^8 \text{ F·N·m}^3/\text{C}^2.$$

We expect that the area should be in square meters. Let's examine the units to be sure. Remember that

$$1 \text{ farad} = 1 \text{ coulomb/volt},$$
$$1 \text{ joule} = 1 \text{ newton·meter},$$

and

$$1 \text{ volt} = 1 \text{ joule/coulomb}.$$

So

$$\text{F·N·m}^3/\text{C}^2 = \frac{\text{C}}{\text{V}} \frac{\text{J}}{\text{m}} \frac{\text{m}^3}{\text{C}^2} = \frac{1}{\text{V}} \frac{\text{J}}{\text{C}} \text{m}^2 = \text{m}^2.$$

The area of a 1.00 F parallel-plate capacitor with a 1.00-mm separation is $1.13 \times 10^8 \text{ m}^2$, almost the size of the city of San Francisco!

Example 16.7

Calculate the capacitance of a parallel-plate capacitor with plate area of 0.50 m^2 separated by 0.10 mm.

Solution   We use Eq. (16.8) to calculate capacitance:

$$C = \frac{\epsilon_0 A}{d} = \frac{8.85 \times 10^{-12} \frac{\text{C}^2}{\text{N·m}^2} \times 0.50 \text{ m}^2}{0.10 \times 10^{-3} \text{ m}}$$

$$= 4.4 \times 10^{-8} \text{ C/V} = 4.4 \times 10^{-8} \text{ F}.$$

The capacitance is very small, 44 nF.

Example 16.8

Dimensional analysis of ϵ_0.

Show that the units of the permittivity constant ϵ_0 can be expressed as farads per meter.

Solution   The units of ϵ_0 were given as $\text{C}^2/(\text{N·m}^2)$. From Eq. (16.8) we see that ϵ_0 must have dimensions of farads per meter. We can rearrange the units to put them in the desired form as follows:

$$\frac{\text{C}^2}{\text{N·m}^2} = \frac{\text{C}^2}{\text{J·m}} = \frac{\text{C}}{\frac{\text{J}}{\text{C}} \text{ m}},$$

where we use the fact that one joule = 1 newton·meter. We further observe that joule/coulomb = volt and coulomb/volt = farad, so

$$\text{C}^2/(\text{N·m}^2) = \text{F/m}.$$

We will usually find it convenient to express ϵ_0 in these units of F/m.

Electric Field of a Parallel-Plate Capacitor

16.7

One very important situation in electrostatics is that of a uniform electric field, that is, a region in which E is everywhere the same. Suppose we place two large identical flat conducting plates parallel to each other and close together to form a parallel-plate capacitor. We charge one of the plates uniformly with positive charge and give the other an identical amount of negative charge. From the symmetry of this configuration, the field between the plates will be very nearly uniform and directed from the positive plate to the negative plate (Fig. 16.12). Some fringing of the field occurs at the edges, where the field lines become distorted and the magnitude rapidly diminishes, but that can be made arbitrarily small by increasing the area of the plates and reducing their separation.

Except for positions near the edges, the field is constant everywhere between the plates and by symmetry must be perpendicular to the surfaces

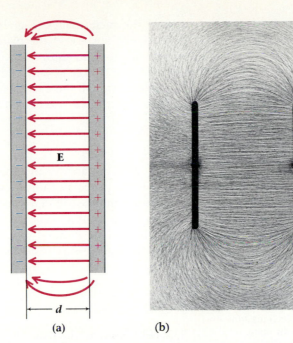

E

d

(a) (b)

Figure 16.12 The electric field due to a pair of oppositely charged parallel plates is uniform except near the edges. (a) Schematic representation of the field. (b) Field lines as indicated by small threads suspended in oil between oppositely charged plates.

of the plates. We can use Gauss's law to show that the field is indeed constant and does not depend on the distance between the plates. Imagine a small cylindrical Gaussian surface placed between the plates and oriented with its axis perpendicular to the plates (Fig. 16.13). The electric field is parallel to the curved surface C, so the flux through that portion is zero. Since no net charge is contained within the surface of the cylinder, the net flux through it must be zero. In other words, the flux entering at the upper end must exactly equal the flux leaving through the lower end. Since the areas A of the top and bottom surfaces are identical, the field at the top must be equal to the field at the bottom. Because we did not require the Gaussian surface to be any particular size, we conclude that the electric field $\mathbf{E}$ must be constant everywhere between the plates (except near the edges).

We can also use Gauss's law to find the magnitude of the electric field. Imagine another Gaussian surface, this time a cylinder with one end inside

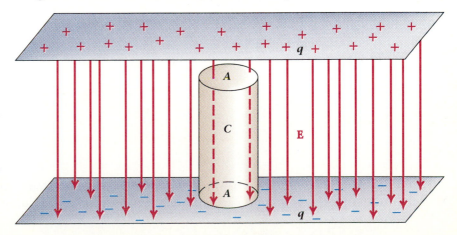

Figure 16.13 A cylindrical Gaussian surface between two parallel charged plates, oriented with its axis perpendicular to the plates.

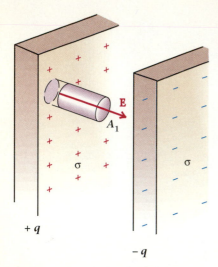

Figure 16.14 A cylindrical Gaussian surface oriented with its axis perpendicular to two parallel charged plates. One end of the Gaussian surface is inside the metal plate.

one of the metal plates (Fig. 16.14). Under the conditions we have described, the surface of the plate is an equipotential and **E** must be perpendicular to it as shown. According to Gauss's law, the flux is proportional to the charge q_1 enclosed by the cylinder. For this alignment there is no flux through the curved side of the cylinder. Thus all the flux must emerge through the end of area A_1. The statement of Gauss's law becomes

$$\phi_{net} = EA_1 = \frac{q_1}{\epsilon_0},$$

where A_1 is the area of the end of the cylinder, q_1 is the charge within, and ϵ_0 is the proportionality constant. For a flat plate we expect the charge to be uniformly distributed over the surface. In that case the ratio of charge to area is constant and can be taken as equal to the total charge q on the plate divided by its total surface area A. The ratio of q/A is called the surface charge density, usually denoted by σ. Thus the magnitude of the field is

$$E = \frac{q_1}{\epsilon_0 A_1} = \frac{q}{\epsilon_0 A} = \frac{\sigma}{\epsilon_0}. \tag{16.9}$$

The field between the plates is constant, independent of the distance from either plate.

A test charge q placed in this constant field experiences a force $\mathbf{F} = q\mathbf{E}$. If the charge moves through the distance d from the positive to the negative plate, the work done by the field is $W = Fd = qEd$. This amount of work is equal to the loss of electric potential energy by the test charge. The electric potential difference between capacitor plates is then given by

$$V = \frac{W}{q} = \frac{qEd}{q} = Ed. \tag{16.10}$$

We can now calculate the capacitance of the parallel-plate capacitor. We start with the definition of capacitance,

$$C = \frac{q}{V}.$$

We replace the potential V with Ed, according to Eq. (16.10). Thus

$$C = \frac{q}{Ed}.$$

We then substitute for E using Eq. (16.9) to get

$$C = \frac{q}{qd/\epsilon_0 A},$$

or

$$C = \frac{\epsilon_0 A}{d}.$$

This equation is the same expression for the capacitance of a parallel-plate capacitor that was given earlier as Eq. (16.8).

Example 16.9

Force on a charge placed between the plates of a capacitor.

A parallel-plate capacitor whose plates are 2.5 cm apart is charged to a potential difference of 100 V. What would be the force on a test charge of 1 μC placed between the plates?

Solution The force on a charge q is

$$F = qE.$$

The electric field E is given by

$$E = \frac{V}{d},$$

where V = potential difference and d = separation between the plates. Combining these two expressions, we get

$$F = \frac{qV}{d} = \frac{1 \times 10^{-6} \text{ C} \times 100 \text{ V}}{2.5 \times 10^{-2} \text{ m}},$$

$$F = 4 \times 10^{-3} \text{ N}.$$

Dielectrics

16.8

The capacitance given by Eq. (16.8) for a parallel-plate capacitor is appropriate only when there is a vacuum between the capacitor plates. When a nonconducting (electrically insulating) material is inserted between the plates, the capacitance increases, even though the area and separation remain constant. The ratio of the new capacitance to the capacitance in a vacuum is called the **dielectric constant** κ (Greek letter kappa):

$$\kappa = \frac{C \text{ (dielectric layer)}}{C \text{ (vacuum layer)}}. \tag{16.11}$$

Materials that do not conduct electric charge are also known as **dielectrics**. Table 16.2 (p. 484) lists the dielectric constants for several materials commonly used in capacitors.

For a parallel-plate capacitor, the effect of adding an insulating layer of dielectric constant κ that fills the space between the plates is to modify Eq. (16.8) so that it becomes

$$C = \frac{\kappa \epsilon_0 A}{d}. \tag{16.12}$$

Thus the value of a capacitor depends on the area of the plates, the dielectric constant of the insulating material, and the thickness of the dielectric layer.

Material	Dielectric constant κ	Dielectric strength (10^6 V/m)
Vacuum	1.0000	∞
Air (dry at 1 atm)	1.0006	3
Teflon	2.1	60
Polyethylene	2.6	25
Mylar	3.1	240
Paper	3.5	14
Mica	3–6	160
Glass	5–10	13
Aluminum oxide (amorphous)	8–9	700–900
Polyvinyldienedifluoride (PVDF)	11	200
Tantalum pentoxide	25	526

TABLE 16.2
Representative dielectric constants and breakdown strengths

We can gain some insight into the behavior of dielectric materials by imagining the effect on a piece of dielectric placed between the plates of a capacitor (Fig. 16.15). When the plates are charged, an electrical force acts on the dielectric to pull any negative charges toward the positively charged plate and push any positive charges toward the negatively charged plate. Because the dielectric material is nonconducting, the charges within the dielectric are not mobile. Still, a slight reorientation takes place, resulting in an induced charge that appears on the surface of the dielectric. In some cases the dielectric consists of polar molecules whose dipole moments become oriented by the external electric field (Fig. 16.16). Within

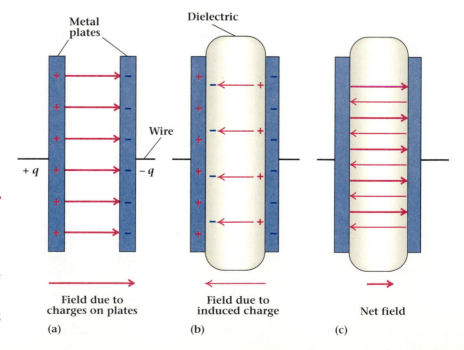

Figure 16.15   (a) The field of a parallel-plate capacitor without any material between the plates. (b) An induced charge on the surfaces of an insulating material placed between the plates generates a field within the material opposite in direction to the original field. (c) The net field between the plates is reduced, leading to an increased capacitance.

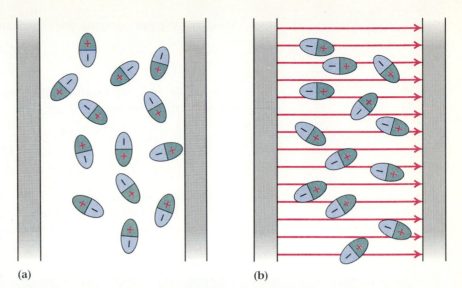

Figure 16.16 (a) Polar molecules are randomly oriented in the absence of an electric field. (b) In the presence of an electric field the dipoles become partially aligned along the direction of the field.

(a)   (b)

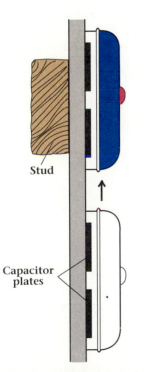

Figure 16.17 An electronic stud finder uses a change in capacitance to locate the studs hidden behind a wall.

the material these dipolar charges tend to cancel each other, but at the surfaces they induce a charge. The induced charge is bound and cannot be removed from the dielectric. Because the polarity of the induced charge is opposite to the charge on the plates, its presence serves to reduce the electric field in the region between the plates. When the field is reduced, the potential difference between the plates is also reduced according to the relation $V = Ed$. Thus for the same charge residing on the plates, the presence of the dielectric reduces the potential difference and thereby increases the capacitance.

An electronic stud finder (Fig. 16.17) uses a variation in dielectric constant to locate the studs hidden behind a wall. The stud finder has a capacitor located on the back with plates spaced wide apart so that the electric field between them penetrates the wall. The presence of a stud is detected by a change in capacitance due to the dielectric material of the stud.

Another important characteristic of a dielectric is its **dielectric strength**, which is the maximum applied potential per unit thickness that the material can withstand before breaking down. If the dielectric strength is exceeded, a spark occurs, along with a catastrophic failure of the material. The electric sparks you encounter from shuffling across the rug on a dry day result from a breakdown of the insulating layer of air. Lightning bolts are also sparks that occur when the dielectric strength of the air is exceeded. Table 16.2 includes the dielectric strengths of the materials listed.

Example 16.10

Maximum voltage across a capacitor.

A capacitor is made with a dielectric layer of Mylar film that is 12 μm thick. The effective area of the film and conducting plates is 0.1 m². What is the capacitance of the capacitor and what voltage can it withstand?

Solution   We can compute the capacitance from Eq. (16.12),

$$C = \frac{\kappa\epsilon_0 A}{d}.$$

From Table 16.2, we find the dielectric constant κ for Mylar to be 3.1. Upon inserting the values we get

$$C = \frac{3.1 \times 8.85 \times 10^{-12}\ \text{F·m}^{-1} \times 0.1\ \text{m}^2}{12 \times 10^{-6}\ \text{m}},$$

$$C = 0.23 \times 10^{-6}\ \text{F} = 0.23\ \mu\text{F}.$$

The dielectric strength of Mylar is also given in Table 16.2 as 240×10^6 V/m. If we multiply the dielectric strength by the film thickness, we get the maximum voltage that the capacitor can withstand:

$$V_{\text{max}} = 240 \times 10^6\ \text{V/m} \times 12 \times 10^{-6}\ \text{m} = 2900\ \text{V}.$$

Energy Storage in a Capacitor

16.9

We said earlier that a major application of capacitors is their use as energy storage devices. Let us now examine this energy-storing capability. We start by considering an initially uncharged capacitor. When a charge q' is transferred from one plate to the other, the potential difference V between the plates due to this transfer is q'/C. Then, if a small additional charge Δq is transferred, the amount of work required is

$$\Delta W = V\,\Delta q.$$

As more charge is transferred, the potential difference increases with the total charge q according to

$$V = \frac{q}{C}.$$

The total work done in charging the capacitor from zero to a voltage V is the sum of the work needed to transfer each amount of charge Δq:

$$W = \sum \Delta W = \sum V\,\Delta q.$$

We can compute this sum from the area under the voltage–charge curve (Fig. 16.18). The area of this triangular region is $\frac{1}{2}Vq$. Thus the energy required to charge an initially uncharged capacitor to a potential difference V and charge q is

$$W = \tfrac{1}{2}Vq.$$

This is also the energy stored on the capacitor when it becomes charged. If we use the relation $V = q/C$, we can express this energy as

$$W = \frac{1}{2}\frac{q^2}{C},$$

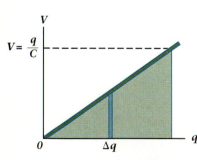

Figure 16.18   Graph of voltage versus charge for a capacitor.

or

$$W = \tfrac{1}{2}CV^2. \tag{16.13}$$

The energy stored on the capacitor may be considered as electrical potential energy, which can be released when the capacitor is discharged by connecting a conducting path between its plates. We can think of capacitors as devices for storing energy or as devices for storing electric charge.

We have seen that, when fringing can be ignored, the electric field within a parallel-plate capacitor is uniform; that is, it is the same at all points between the plates. Thus the *energy density u*, the stored energy per unit volume, should also be uniform. We can determine the energy density from the ratio of the energy stored to the volume occupied by the field,

$$u = \frac{\text{energy stored}}{\text{volume}},$$

$$u = \frac{W}{Ad} = \frac{\tfrac{1}{2}CV^2}{Ad},$$

where Ad is the volume between the plates. If we substitute the relationship $C = \kappa\epsilon_0 A/d$, the energy density becomes

$$u = \frac{1}{2} \kappa\epsilon_0 \left(\frac{V}{d}\right)^2.$$

However, V/d is the electric field within the capacitor, so

$$u = \tfrac{1}{2}\kappa\epsilon_0 E^2. \tag{16.14}$$

Although we have derived the result of Eq. (16.14) for the special case of the parallel-plate capacitor, it is true in general. That is, if an electric field **E** is present at any point in space, then we can think of that point as having a stored energy density given by Eq. (16.14).

Example 16.11

Energy storage in a capacitor.

A 20-μF capacitor in a TV set is charged to a potential difference of 100 V. (a) How much charge is stored on the capacitor? (b) How much energy is stored?

Solution   The charge stored on the capacitor is computed from the definition of capacitance as

$$q = CV = 20 \times 10^{-6}\,\text{F} \times 100\,\text{V} = 2 \times 10^{-3}\,\text{C}.$$

The energy stored may be calculated from

$$W = \tfrac{1}{2}CV^2 = \tfrac{1}{2} \times 20 \times 10^{-6}\,\text{F} \times (100\,\text{V})^2,$$
$$W = 0.10\,\text{J}.$$

Example 16.12

Energy storage and dielectric strength.

A capacitor made of polyethylene film 25 μm thick has an effective surface of 0.10 m². How much energy can be stored on the capacitor?

Solution   First we compute the capacitance, using $\kappa = 2.6$:

$$C = \frac{\kappa\epsilon_0 A}{d} = \frac{2.6 \times 8.85 \times 10^{-12} \text{ F·m}^{-1} \times 0.10 \text{ m}^2}{25 \times 10^{-6} \text{ m}},$$
$$C = 9.2 \times 10^{-8} \text{ F}.$$

The maximum voltage we can put on the capacitor is obtained from the product of the film's dielectric strength and its thickness:

$$V_{\text{max}} = 25 \times 10^6 \text{ V·m}^{-1} \times 25 \times 10^{-6} \text{ m} = 625 \text{ V}.$$

The maximum energy stored is then

$$W = \tfrac{1}{2}CV^2 = \tfrac{1}{2} \times 9.2 \times 10^{-8} \text{ F } (625 \text{ V})^2,$$
$$W = 0.018 \text{ J} = 18 \text{ mJ}.$$

SUMMARY

Useful Concepts

• Electrical potential is the potential energy per unit positive charge:

$$V = \frac{\text{PE}_{\text{elec}}}{q_0}.$$

• For a single point charge Q the electric potential is

$$V = \frac{Q}{4\pi\epsilon_0 r}.$$

• The electrical potential difference between points A and B is

$$V_{\text{AB}} = V_B - V_A = \frac{W_{\text{AB}}}{q}.$$

• A surface for which all points are at the same potential is called an equipotential surface. When all the charges have come to rest, the surface of a conductor is always an equipotential surface.

• The electric field is related to the gradient of the electric potential by

$$E = -\frac{\Delta V}{\Delta s}.$$

• A capacitor is a device for storing electric charge. The capacitance is defined to be

$$C = \frac{q}{V}.$$

• The dielectric constant of a material is

$$\kappa = \frac{C(\text{dielectric layer})}{C(\text{vacuum layer})}.$$

• The capacitance of a parallel-plate capacitor filled with material of dielectric constant κ is

$$C = \frac{\kappa\epsilon_0 A}{d}.$$

• The energy stored in a capacitor is

$$W = \tfrac{1}{2}CV^2.$$

• The energy density u in an electric field E is

$$u = \tfrac{1}{2}\kappa\epsilon_0 E^2.$$

Important New Terms

You should be able to write the definition or meaning of each of the following terms:

- electric potential
- volt
- electric potential difference
- Van de Graaff generator
- equipotential surface
- electric potential gradient

- capacitor
- capacitance
- farad
- dielectric constant
- dielectric
- dielectric strength

QUESTIONS

16.1 Sketch the equipotential lines about a charged sphere.

16.2 Sketch the equipotential lines about a charged cube located a great distance from any other conductors. How do the equipotentials look close to the cube? Far away from the cube?

16.3 Why can a Leyden jar not be charged very well if the outer coating is insulated during the charging?

16.4 How can a potential gradient, that is, an electric field, have a direction in space although the electric potential has no direction?

16.5 Why is it that the passengers in an automobile are not electrocuted if the automobile is struck by lightning?

16.6 Would you be safe inside the dome of a large Van de Graaff generator when it is operating?

16.7 Would you be safe on top of the dome of a large Van de Graaff generator when it is operating? Assume that you are in contact with the dome only. Discuss the conditions under which this behavior would be safe or unsafe.

16.8 Work is required to pull apart two charged capacitor plates. What happens to the energy expended in pulling them apart?

16.9 How does the breakdown strength of the dielectric affect the physical size of practical capacitors?

16.10 A hollow metal surface supported on an insulating stand has a large radius on one end and a small radius on the other (Fig. 16.19). What can you say about the electric potential, the electric field, and the electric charge density on the surface when it contains a net charge Q? Make a sketch of the equipotential lines about the object.

16.11 What is the purpose of the short strap frequently seen dragging along behind a gasoline tanker truck?

16.12 Explain the following advice that is often given for protection from lightning. During an electrical storm, if the hair begins to stand on end, lightning is likely to strike nearby. For safety, drop to your knees and bend forward, placing your hands on your knees, being careful not to place your hands on the ground.

16.13 Sketch the lines of force corresponding to the equipotential lines about two like charges of unequal magnitude (Fig. 16.20).

Figure 16.19   Question 16.10.

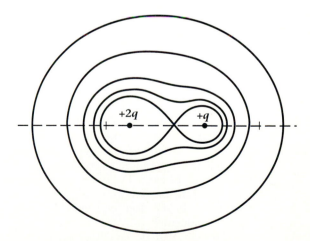

Figure 16.20   Question 16.13.

PROBLEMS

Hints for Solving Problems

Remember that electric potential is a scalar and has no direction, unlike the electric field, which is a vector. You have an electric field only where there is a gradient in the electric potential. The electric field is directed opposite to the potential gradient and has a magnitude equal to the magnitude of the potential gradient. Remember also that the lines of electric field are everywhere perpendicular to the equipotential surfaces.

Section 16.1   Electric Potential
Section 16.2   The Van de Graaff Electrostatic Generator

16.1   Two point charges q_1 and q_2 are arranged as shown in Fig. 16.21. What is the electric potential difference $V_B - V_A$ between the two points labeled A and B?

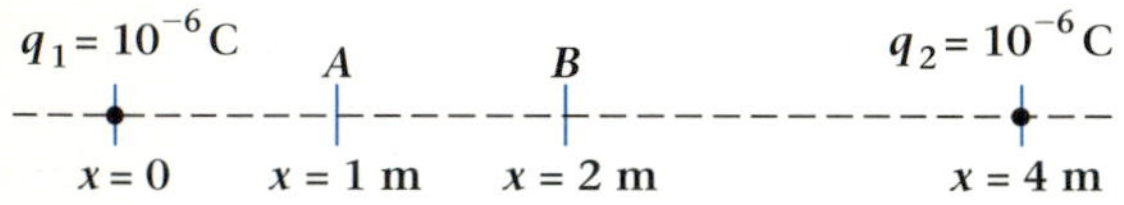

$q_1 = 10^{-6}$ C A B $q_2 = 10^{-6}$ C

$x = 0$ $x = 1$ m $x = 2$ m $x = 4$ m

Figure 16.21   Problem 16.1.

16.2   What is the electric potential at a point 0.70 m away from a point charge of 2.0 mC?

16.3   The potential 0.010 m from a point charge is 0.10 V. What is the magnitude of the point charge?

16.4   If 3.7×10^{-4} J of work is required to move 13 μC of charge from one point to another, what is the electrical potential difference between the two points?

16.5   A charge $q_1 = 0.244$ μC is initially positioned 3.00 cm directly above a fixed charge $q_2 = 21.6$ μC. The charge q_1 is moved 6.25 cm straight upward and then moved 7.50 cm horizontally away from the charge q_2. What is the change in the electrical potential energy of the movable charge q_1?

16.6   The dielectric strength of air, $E = 3.0 \times 10^6$ V/m, is the maximum field that the air can withstand before it breaks down and becomes conducting. (a) How much charge can be placed on a spherical conductor with a 10-cm radius before the field at its surface exceeds the breakdown strength of the air? (b) What would be the electric potential at the surface of this conductor?

16.7   A demonstration Van de Graaff generator has a 30-cm-diameter dome. The generator is capable of producing a potential of 150,000 V. Calculate the charge required to raise the dome to this potential. Also find the value of the electric field near the dome.

16.8   A demonstration Van de Graaff generator has a charge of 3.0 μC stored on its 30-cm-diameter dome. (a) What is the value of the electric field at the surface of the dome? (b) What is the electric field 0.50 m from the surface of the dome?

16.9   A Van de Graaff generator has a dome 50 cm in diameter. Calculate the maximum electric potential it can reach before the surrounding air breaks down.

Section 16.3   Equipotential Surfaces

16.10   From your knowledge of the behavior of the electric field and potential near a point charge, and of the fact that the surface of a conductor is an equipotential, sketch the lines of electric field and the equipotentials for a point charge q located a distance x from a large conducting plane.

16.11   A thick, conducting, spherical shell of inner radius a and outer radius b surrounds a point charge $+Q_0$ at its center (Fig. 16.22). (a) What is the surface charge residing on the inner surface of the shell? (b) What charge resides on the outer surface? (c) What is the electric field in the region $a < r < b$?

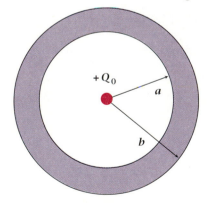

Figure 16.22   Problem 16.11.

16.12   Two isolated metal spheres are charged to the same potential V. The first sphere has a radius that is twice as large as the radius of the second sphere. (a) Which sphere carries the greater electric charge? (b) Which sphere has the greater electric field at its surface?

16.13   An isolated metal sphere of radius R carries a positive charge Q. An identical isolated metal sphere, initially uncharged, is brought into contact with the first sphere and the two are then separated. How much charge remains on the first sphere?

16.14 An isolated metal sphere of radius R carries a positive charge Q. Another isolated metal sphere, initially uncharged and of radius $R/2$, is brought into contact with the first sphere and the two are then separated. How much charge remains on the first sphere?

*Section 16.4   Electric Potential Gradient

16.15 A pair of parallel plates is separated by 1.0 mm of air. If the plates are charged to a potential difference of 75 V, what is the electric field between the plates?

16.16 An electron ($q = 1.6 \times 10^{-19}$ C) is placed in a uniform electric field of 2.0×10^{9} V/m. (a) What is the force on the electron? (b) What is the acceleration of the electron if its mass is 9.1×10^{-31} kg?

16.17 An electron (charge $q = 1.6 \times 10^{-19}$ C and mass $m = 9.1 \times 10^{-31}$ kg) is released in a vacuum between two flat, parallel metal plates that are 10.0 cm apart and are maintained at a constant electric potential difference of 750 V. If the electron is released at the negative plate, what is its speed just before it strikes the positive plate?

16.18 A pair of parallel plates is separated by 1.5 mm of air. An electric field of 3.0×10^{4} V/m exists between the plates. What is the potential difference between the plates?

16.19 To move a charged particle through an electric potential difference of 10^{-3} V requires 2×10^{-6} J of work. What is the magnitude of the charge?

16.20 (a) What is the electric potential a distance 0.30 m away from a point charge of 1.0×10^{-9} C? (b) How much work is required to bring an identical charge from very far away up to this distance of 0.30 m from the first charge?

16.21 A proton of mass $m_p = 1.67 \times 10^{-27}$ kg and charge $+e = 1.6 \times 10^{-19}$ C is accelerated from rest through an electric potential of 400 kV. What is its final velocity?

Section 16.5   Capacitors

16.22 When a charge of 120 μC is applied to the plates of a capacitor, the potential difference between the plates is 15 V. What is the capacitance?

16.23 When the plates of a radio capacitor are charged with 18×10^{-5} C, the potential difference between them is 9.0 V. What is the capacitance?

16.24 A 20-μF capacitor is charged to 15 V. How much charge is stored on the capacitor?

16.25 A charge of 3.0×10^{-3} C is stored on a 20-μF capacitor. What is the electric potential difference across the capacitor?

16.26 How much charge is stored on each plate of a 5000-μF capacitor whose plates are held at a potential difference of 400 V?

Section 16.6   The Parallel-Plate Capacitor

For Problems 16.27–16.31, assume that the plates are separated by air.

16.27 The plates of a capacitor are separated by 0.10 mm. If the area of each plate is 100 cm², what is the value of the capacitance?

16.28 A parallel-plate capacitor is made of metal plates separated by 0.10 mm. What is its capacitance if the area of the plates is 113 cm²?

16.29 Calculate the capacitance of two flat metal pie pans 23 cm in diameter separated by 2.0 mm.

16.30 A parallel-plate capacitor is made of two 50.0-cm-diameter circular plates spaced 2.50 mm apart. How much charge is stored on the capacitor when it is charged to 750 V?

16.31 A parallel-plate capacitor is designed so that the plates can be pulled apart. The capacitor is initially charged to a potential difference of 50 V when the plates are 1.0 mm apart. The plates are insulated so that the charge cannot leak off. What is the potential difference between the plates when they are pulled to a new separation 2.0 mm apart?

Section 16.7   Electric Field of a Parallel-Plate Capacitor

16.32 How much work is required to move an electron 0.010 m through a uniform field of 7.0 V/m?

16.33 A parallel-plate capacitor has plates with an area of 250 cm² separated by 1.0 mm. What is the electric field between the plates when they are charged with 0.50 μC?

16.34 A homemade capacitor is made from two metal pie pans 18.5 cm in diameter separated by 2.00 mm. The capacitor is charged to a potential of 15.7 V. How much charge is stored on each plate of the capacitor?

Section 16.8   Dielectrics

16.35 The parallel plates of a parallel-plate capacitor are separated by 1.0 mm. The gap between the plates is filled with polyethylene and the plates are charged to a potential of 80 V and insulated to keep the charge from leaking off. Then the polyethylene layer is pulled out. What is the potential difference between the plates of the capacitor?

16.36 A capacitor is made from a Teflon film that is 25 μm thick. (a) What is the effective area of the plates of this capacitor if it has a capacitance of 0.50 μF? (b) What is the maximum working voltage of the capacitor; that is, how much voltage can it stand before breakdown?

16.37 A capacitor is made from a PVDF film that is 25 μm thick. (a) If the effective area of the film and the conducting plates is 0.60 m^2, what is the capacitance of the capacitor? (b) How much voltage can the capacitor withstand without breaking down?

16.38 A manufacturer plans to make a capacitor with tantalum pentoxide as the dielectric. What must be the minimum thickness of the dielectric layer if the capacitor must withstand 25 V?

16.39 Calculate the maximum potential difference that can be maintained across a 25-μm layer of polyethylene.

16.40 What must be the effective surface area of a 10-μF aluminum oxide capacitor if the average film thickness is 2.0×10^{-8} m? Use $\kappa = 8.0$.

16.41 Show that for a given plate area, the product of capacitance and the maximum voltage of a capacitor is fixed by the type of material used for the dielectric.

16.42 A parallel-plate capacitor is made of two flat metal plates pressed against a thin slab of dielectric material. The capacitor is connected to a power supply and a potential difference of 90 V is applied to the plates. With the power supply disconnected, the dielectric material is removed and the potential difference between the plates is measured to be 500 V. What is the dielectric constant of the material that was initially used to fill the gap between the plates?

Section 16.9   Energy Storage in a Capacitor

16.43 An electrolytic capacitor is rated at 25 μF at 450 V. (a) How much charge can be stored on that capacitor? (b) How much energy can it store?

16.44 A capacitor of 0.010 μF is charged to 10 V. What is the charge and what is the energy stored on the capacitor?

16.45 How much energy is stored on a 5000-μF capacitor charged to 400 V?

16.46 A parallel-plate capacitor is made of square metal plates 30 cm $\times$ 30 cm. The plates are pressed against a glass plate 3.0 mm thick of dielectric constant $\kappa = 6.3$ that exactly fills the space between them. The capacitor is charged to a potential difference of 500 V and then insulated from the surroundings. (a) How much energy is stored on the capacitor? (b) If the piece of glass is removed, how much energy will be stored on the capacitor?

Additional Problems

16.47 Three point charges are arranged in an equilateral
• triangle of edge length 1.0 cm. What is the electric potential at the midpoint of one edge if each charge is 1.0 μC?

16.48 Four point charges $q = 3.6$ μC are arranged in a
• square of edge length 2.0 cm. What is the electric potential at the center of the square and at the midpoint of one edge?

16.49 A positive point charge of 1.0 nC is located at po-
•• sition $x = 0$. A second point charge $q = -1.0$ nC is located at $x = 0.10$ m. (a) What is the electric potential as a function of x? (b) What is the electric potential at $x = -0.050$ m?

16.50 An isolated metal sphere of radius R_1 carries an elec-
• trical charge Q. A second sphere of radius R_2 and initially uncharged is brought into contact with the first sphere. Charge is exchanged between the two and they are separated. How much charge remains on the first sphere?

16.51 Two identical point charges of $q = 1.00 \times 10^{-8}$ C are separated by a distance of 1.00 m. How much work is required to move them closer together so that they are only 0.50 m apart?

16.52 A pair of charged metal plates are separated by 2.0
• mm. An electric field of 10^5 V/m exists between the plates. An electron with mass $m_e = 9.1 \times 10^{-31}$ kg and charge $e = -1.6 \times 10^{-19}$ C is released from the negative plate and is accelerated by the field toward the positive plate. (a) What is the change in kinetic energy of the electron? (b) What is the velocity of the electron just before it strikes the positive plate?

16.53 Compute the potential gradient of a point charge q
•• by direct calculation of $\Delta V = V(r + \Delta r) - V(r)$. Show that in the limit of $\Delta r << r$ the magnitude of the gradient, $\Delta V/\Delta r$, is equal to $q/4\pi\epsilon_0 r^2$.

16.54 A large insulating sheet has a uniform static charge
• of surface charge density $\sigma = 3.54 \times 10^{-8}$ C/m^2. (a) Calculate the electric field near the sheet. (b) Calculate the separation of equipotential surfaces spaced 100 V apart.

16.55 Two hollow metal spheres are mounted on insulat-
• ing stands. Sphere 1 has a radius r_1 and sphere 2 has a radius $r_2 = 2r_1$. A charge Q is placed on sphere 1 to give it an electric potential of $+180$ V. The other sphere is uncharged at a potential of 0 V. If the two spheres are placed in contact and then separated, what will be their new potential?

16.56 The stored electrical energy of a 4000-μF capacitor
• charged to 500 V is converted to thermal energy by discharging the capacitor through a heating element submerged in 200 g of water in an insulating cup.

What is the increase in temperature of the water if the heat capacity of the heater can be ignored? Assume that all of the energy stored on the capacitor goes into heating the water.

16.57 A parallel-plate capacitor has a surface charge den-
• sity $\sigma = 8.5 \times 10^{-5}$ C/m^2 and an electric field $E = 2.5 \times 10^6$ V/m in the material between the plates. (a) Express the dielectric constant κ in terms of E and σ. (b) What is the value of the dielectric constant for the material between the plates?

16.58 Derive an expression for the work done in pulling
• out the dielectric layer in Problem 16.35. Give your answer in terms of C_0 and V_0, the initial capacitance and voltage.

16.59 A 100-μF parallel-plate capacitor is charged to 500
•• V. It is then disconnected from the charging source and insulated. (a) How much energy is stored on the capacitor? (b) If the plates are pulled apart to a separation that is twice their initial separation, what energy is stored on the capacitor? (c) If the answers are not identical, account for their difference.

16.60 A parallel-plate capacitor made of circular plates of
•• radius 25 cm separated by 0.20 cm is charged to a potential difference of 1000 V by a battery. Then a sheet of polyethylene is pushed between the plates, completely filling the gap between them. How much additional charge flows from the battery to one of the plates when the polyethylene is inserted?

ADDITIONAL READING

Crane, H. R., "How Things Work: Locating the Studs in a Wall." *The Physics Teacher,* February 1985, p. 104.

Hazen, R. M., "Perovskites." *Scientific American,* June 1988, p. 74. Perovskites are crystals having the structure of calcium titanate. They often have a permanent electric dipole moment. Synthetic materials similar to the perovskites have been found to be superconductors.

Kristjansson, L., "On the Drawing of Lines of Force and Equipotentials." *The Physics Teacher,* April 1985, p. 202.

Price, R. H., and R. J. Crowley, "The Lightning-Rod Fallacy." *American Journal of Physics,* September 1985, p. 843.

Trotter, D. M., Jr., "Capacitors." *Scientific American,* July 1988, p. 86.

Walker, J., "The Amateur Scientist: How to Map Electrically Charged Patches with Parsley, Sage, Rosemary and Thyme." *Scientific American,* April 1988, p. 114.

17

Electric Current and Resistance

17.1   Electric Current and Electromotive Force

17.2   Electric Resistance and Ohm's Law

17.3   Resistivity

17.4   Power and Energy in Electric Circuits

17.5   Short Circuits and Open Circuits

17.6   Kirchhoff's Rules

17.7   Simple Resistive Circuits

17.8   Capacitors in Combination

17.9   Internal Resistance of a Battery

*17.10 Home Power Distribution

A WORD TO THE STUDENT

Today, electric circuits are a commonplace part of our daily lives. In this chapter we use the ideas of the two previous chapters to introduce the practical subject of electric circuits. We focus on a familiar concept—energy—and apply the concept of electrical energy to simple circuits. We introduce several new terms, such as current, resistance, and electromotive force, that are necessary for describing these circuits. In the rest of the chapter, we develop several other new concepts as we apply this knowledge to other circuit combinations. We do not expect you will need to analyze complicated electronic circuits and equipment, but a familiarity with fundamental ideas of circuit operation will help you understand practical aspects of electricity, from home power distribution to basic safety from electrical shock.

In 1791 the Italian Luigi Galvani (1737–1798) announced a discovery that eventually helped lay the foundations for much of modern industry and technology. In present-day terminology we would say that he found a way to produce electric current. Galvani noticed that the muscles of a dissected frog's leg twitched when a nearby electrostatic generator was in operation. Further experiments showed that the electrostatic generator was not the cause of the twitching, but that the twitching occurred when the frog's leg was attached to a brass hook and placed on an iron plate against which the hook was pressed. He concluded, incorrectly, that the nerves and muscles of the frog produced electricity, which he had detected. Galvani's work was soon overshadowed by that of Count Alessandro Volta (1745–1827), who maintained that the electricity detected by the frog's muscles was due to contact between the dissimilar metals brass and iron. Volta was a careful investigator who was already famous for other electrical discoveries. He invented the first battery, which became known as the voltaic pile. This pile consisted of a repeating stack of disks of copper, zinc, and cardboard moistened with salt water. A stack ("battery") of twenty such units gave a strong shock and could produce sparks.

It was Volta who first recognized the crucial difference between the electricity from a battery and that from an electrostatic device such as a Leyden jar: The electricity from a battery flowed continuously. This discovery of a source of continuous electricity opened up opportunities for the discovery of effects not available with static electricity. These effects include electrochemistry, electric heating, electromagnetism, and all the applications that follow from them, including investigations into the properties of matter itself.

In this chapter we use the concepts of charge, field, and potential to study charges in motion, or current electricity. In our ordinary experience, electric currents move along closed conducting paths called circuits. To understand how circuits work, we will use the laws of conservation of charge and conservation of energy, fundamental laws that apply to all electric circuits.

Electric Current and Electromotive Force

17.1

We know from common experience that a battery is a source of electrical energy because it can turn a motor that does mechanical work. One simple energy source is the dry cell, commonly used in flashlights and portable radios. Several cells connected together form a battery, which is essentially the same as Volta's original pile of cells (Fig. 17.1a). Although the word **battery** literally means an array of cells, this word often refers to single cells as well. For example, common flashlight batteries are really single cells. Because of their combined convenience, reliability, and portability, batteries are used in a wide variety of applications, ranging from portable

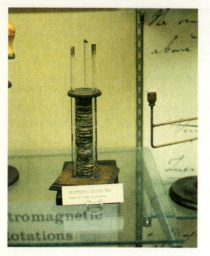

(a)

video cameras to watches and hearing aids. Several common types of batteries are shown in Fig. 17.1(b).

A flashlight with a good bulb and fresh batteries gives off a bright light as the batteries convert chemical energy to electrical energy. As we will see later, this electrical energy generates heat, which causes the filament in the bulb to become hot enough to glow. If we remove the bulb from its socket and connect it to the batteries with copper wire, it continues to glow with no perceptible change in brightness. We deduce from this observation that copper wire is a good conductor of electricity.

When we use a conducting wire to provide a continuous path from one terminal of a battery to the other, a current of electric charge passes through the wire. We define this **electric current** to be the rate at which electric charge passes through a conductor. If we use the conventional symbol I for electric current, we have

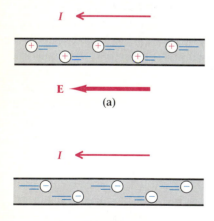

(b)

$$I \equiv \frac{\Delta q}{\Delta t}. \tag{17.1}$$

For the present, we consider only the case in which the charge always flows in the same direction and at the same rate. This is called **direct current** or **dc**. In Chapter 20 we will discuss the case in which the rate and direction change. The dimensions of electric current are charge per unit of time, and the SI unit for current is the **ampere** (abbreviated A).* The unit for charge is the coulomb, so

$$1 \text{ ampere} = \frac{1 \text{ coulomb}}{\text{second}}.$$

The source of this current is the battery. In particular, the battery provides a potential difference, or voltage, V between its terminals. The corresponding electric field causes charges to move within the wire, thus generating the electric current. (The existence of a field within a conducting wire does not contradict our statement in Chapter 16 that a field cannot exist within a conductor. There we were discussing charges at rest; here we are talking about charges in motion.) Energy is released from the battery when an amount of charge q moves through this potential difference: $W = qV$. Electrical energy can be released from a battery (or other source of electrical energy) only when the current has a complete conducting path available from one side of the potential difference to the other. We call this conducting path a complete circuit.

By convention, the direction associated with the current in a circuit is the direction in which a *positive* charge would move in the potential provided by the battery (Fig. 17.2). Thus the current in the external circuit is said to be directed from the positive terminal to the negative terminal

Figure 17.2   (a) The direction of electric current is the direction of motion of positive charge. (b) The motion of electrons is opposite to the direction associated with the current.

*The ampere is specified more completely in Section 18.6.

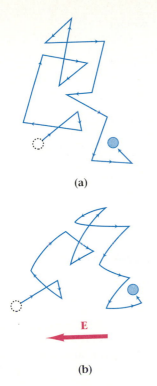

(a)

(b)

Figure 17.3 (a) Random motion of an electron through a metal when no electric field is present. After many collisions, the electron remains close to its original position. (b) Motion of the electron in a metal when an electric field is present. Notice the net drift of the electron in the direction opposite to the electric field.

of the battery. In solids, such as wires, the mobile charges that are actually free to flow, and therefore make up the current, are the negatively charged electrons. Consequently, the motion of the electrons is opposite to the conventional direction assigned to the current. However, since a flow of positive charges in one direction is equivalent to a flow of negative charges in the opposite direction, we use the convention of positive current in almost all cases. Although the current consists of the motion of discrete charges, the number of them is so large that we may ignore this granularity and consider the current to be a smooth, continuous flow of charge.

As we will see in more detail later, electrons do not flow through a wire unimpeded. If they did, the field would eventually accelerate them to very great speeds. Instead, the moving electrons interact with the relatively stationary ions in the metal. The result of the applied field is to impose a net velocity on the random motions of the electrons. In this respect they behave much like the particles of a gas, having large random velocities, but a small net velocity as they drift with a fairly constant average speed through the three-dimensional latticework of the conductor ions. This average speed of translation through the conductor is called the *drift velocity,* and ranges from a few hundredths of a millimeter per second to several centimeters per second for copper wires in ordinary situations (Fig. 17.3). The drift velocity is the rate at which electrons themselves move through a wire; it is not the rate at which electric signals move. The speed of electric signals is very high, close to the speed of light. Charge in a conductor is somewhat analogous to water in a pipe. When water begins flowing into one end of a full pipe, water begins to flow out the other end almost immediately. Similarly, when the switch is closed to complete a circuit, electrons move throughout the whole wire.

The potential difference that appears between the terminals of a battery when no current is present is called the **electromotive force** or **emf**. (This emf is *not* a force, despite its name.) The symbol for emf is $\mathcal{E}$. The emf is measured in volts, like any other potential difference.

Any device that can maintain a potential difference and supply current to an external circuit is a source of emf. Examples include batteries, solar cells, and generators. If the emf of a battery is zero, there is no current when a wire is connected across its terminals. In this case there is no potential difference to drive the charges. But if the emf is nonzero, a current is present when the terminals are connected with a conducting wire to form a complete circuit (Fig. 17.4). The greater the emf, the greater the current in the circuit.

A flashlight dry cell has an emf of 1.5 V; mercury cells sometimes used in watches and cameras have emfs of 1.35 V; a typical multicell automobile battery has an emf of approximately 12 V; and a newly charged nickel-cadmium rechargeable cell (Nicad cell) has an emf of about 1.2 V. The low emf of Nicad cells, in comparison to ordinary dry cells, explains why they cannot be used to replace dry cells for some applications.

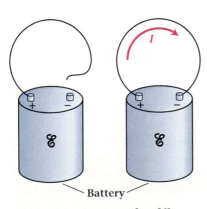

Figure 17.4 A source of emf $\mathcal{E}$ causes a current in a conductor when the circuit is completed.

Example 17.1

When you press one of the buttons on a pocket calculator the battery provides a current of 300 μA for 10 ms. (a) How much charge flows during that time? (b) How many electrons flow in that time?

Solution   (a) From the definition of current we get

$$\Delta q = I \, \Delta t.$$

The charge that flows in 10 ms is then

$$\Delta q = (300 \times 10^{-6} \, \text{A}) \times (10 \times 10^{-3} \, \text{s}) = 3.0 \times 10^{-6} \, \text{C}.$$

(b) The magnitude of the charge of an electron is 1.60×10^{-19} C, so that

$$\text{number of electrons} = \frac{\text{total charge in 0.010 s}}{\text{charge on one electron}},$$

$$\text{number of electrons} = \frac{3.0 \times 10^{-6} \, \text{C}}{1.60 \times 10^{-19} \, \text{C}} = 1.9 \times 10^{13}.$$

Electric Resistance and Ohm's Law

17.2

If you maintain an electric potential difference, or voltage V, across any conductor, an electric current occurs (Fig. 17.5). In general, the magnitude of the current depends on the potential difference. The ratio of the applied voltage to the resulting current is defined to be the **resistance** R of the material:

$$R \equiv \frac{V}{I}. \tag{17.2}$$

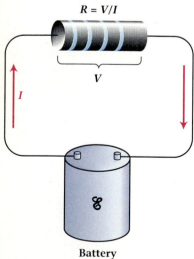

$R = V/I$

V

I

$\mathcal{E}$

Battery

Figure 17.5 An electric potential across a conductor causes a current. The conductor has a resistance defined by $R = V/I$.

You can think of resistance as the ability of a material to resist the flow of charge when it is subject to a given potential difference. An insulating material, which passes only a small current for a particular voltage V, has a large resistance. By contrast, a good conductor carries a large current at the same voltage V, and thus has a small resistance. If V is measured in volts and I in amperes, then resistance has the unit of volt per ampere, which is called an **ohm** (abbreviated Ω). We emphasize that Eq. (17.2) is the *definition* of resistance, and does not imply that the resistance measured under one set of conditions is equal to the resistance measured under different conditions.

If you change the potential difference applied *across* a conductor, the resulting current *through it* may also change. Each combination of V and I determines a particular value of resistance, as given by Eq. (17.2). However, if the change in the current is proportional to the change in the voltage, the resistance remains constant as the potential difference changes. Using symbols, we say that V and I are proportional ($V \propto I$), and the constant of proportionality is the resistance R. This relationship may be expressed as

$$V = IR, \tag{17.3}$$

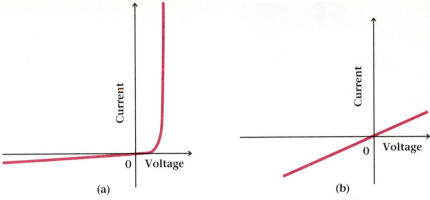

Figure 17.6 (a) Current through a typical silicon *pn*-junction diode as a function of the applied voltage. The relationship between current and voltage is not linear. (b) Current through a piece of copper wire kept at constant temperature. The relationship between current and voltage is linear.

where R is a constant. Equation (17.3) is known as **Ohm's law** in honor of Georg Simon Ohm (1787–1854), the German physicist who first demonstrated this relationship between current and potential difference in 1827. Materials whose resistance is constant over a wide range of voltages are said to obey Ohm's law. In these materials, current is proportional to the applied potential difference V and inversely proportional to the resistance R. Ohm's law is not a law of nature in the sense that conservation of momentum or the universal law of gravitation are laws of nature; rather, it is an experimental observation about the behavior of some materials under a limited range of circumstances.

Note the distinction between the definition of resistance, Eq. (17.2), and Ohm's law, Eq. (17.3). We can measure the electric resistance of a conductor by making simultaneous measurements of current and voltage. Figure 17.6(a) shows the current through a typical silicon *pn*-junction diode as a function of the applied voltage. The current increases with increased voltage, so a resistance is defined for all conditions even though it is not constant. Such a material is said to be nonohmic because the resistance is not constant. Figure 17.6(b) shows the current through a piece of copper wire kept at constant temperature. Here the relationship between current and voltage is linear and the copper is said to obey Ohm's law over this range. Indeed, Ohm's law is widely used in describing electric circuits. Nevertheless, you should remember that it may not always be applicable.

As an aid to understanding the relationship between current, potential difference, and resistance, an analogy is sometimes used. The rate at which water flows in a closed pipe is analogous to the electric current in a wire; a pump that delivers a specific pressure is analogous to a battery of a specific emf; and the electric resistance is analogous to the pipe's resistance to the flow of the water. This is an instructive analogy when you first encounter the idea of current, but it is only an analogy, and you should try to think in terms of electrical, rather than mechanical, quantities.

Example 17.2

The filament of Thomas Edison's first practical electric light conducted a current of about 0.3 A when a potential of approximately 18 V was applied. What was the approximate resistance of the filament under these conditions?

Figure 17.7   Hot-molded carbon composition resistors. The black carbon grains of the resistor are encased in an insulating phenolic jacket. Colored stripes are used to indicate the value of the resistance.

Solution   The resistance can be easily found from Eq. (17.3) to be

$$R = \frac{V}{I} = \frac{18 \text{ V}}{0.3 \text{ A}} = 60 \ \Omega.$$

Electric components manufactured especially for their resistance are called **resistors.** They are commonly available in a wide range of values from a few ohms to millions of ohms. These resistors are introduced into circuits to provide resistances that are large compared with the resistance of the wires and connectors joining together the other circuit components.

One of the most common resistor types is the carbon composition resistor (Fig. 17.7). Composition resistors are normally molded into a cylindrical shape from carbon granules compacted together with a binding resin. Resistors made this way are relatively inexpensive. They come in various sizes according to their power-handling ability, with power ratings of $\frac{1}{8}$, $\frac{1}{4}$, $\frac{1}{2}$, 1, and 2 watts being common. The limited power-dissipating ability of composition resistors is sometimes a disadvantage. When higher power ratings are required, wire-wound resistors, which have a much

PHYSICS IN PRACTICE

Superconductivity

We have seen that normal materials have electric resistance, which leads to power losses due to heating. Many electrical applications require materials with very low resistance—the lower the better. Power transmission lines, electromagnets, computer chips—all would be revolutionized by resistanceless materials. In fact, under special conditions, materials with zero resistance do exist; but making practical use of them continues to be a difficult problem.

In 1908 the Dutch physicist H. Kamerlingh Onnes (1853–1926) succeeded in liquefying helium, for which he received the 1913 Nobel Prize in physics. At atmospheric pressure, helium liquefies at 4.2 K. Because of its extremely low boiling point, liquid helium came to be used as a coolant, cooling other objects to its boiling temperature. Moreover, when the pressure above liquid helium is reduced, its temperature can be lowered to below 1 K.

It was well known at that time that the electric resistance of metals decreases with decreasing temperature, approaching a limiting, or residual, value as the temperature approaches zero (Fig. B17.1a). In 1911 one of the early experiments conducted at the newly attainable low temperatures was the investigation of the electric resistance of metals with change in temperature. When the resistance of platinum was measured at low temperature, it indeed fell to a limiting value, as expected. That limiting value was found to be related to the purity of the specimen, with purer specimens having smaller residual resistance.

However, when mercury was cooled to temperatures below 4 K, its resistance suddenly dropped to zero at a particular transition temperature T_c (Fig. B17.1b). This loss of resistance

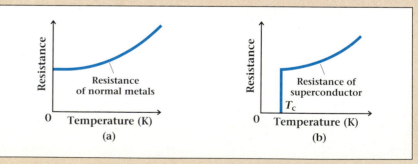

Figure B17.1   (a) As the temperature of a normal metal approaches absolute zero its resistance approaches a limiting value. (b) The resistance of a superconductor suddenly drops to zero as its temperature is lowered through the superconducting transition temperature, T_c.

higher current-carrying capability, are often used. Wire-wound resistors are commonly made from the alloys manganin and constantan because they have only small variations in resistance with changes in temperature.

Resistivity

17.3

By working with wires of different thicknesses and lengths, Ohm found that the amount of current transmitted by a wire for a given potential difference was directly proportional to the cross-sectional area of the wire and inversely proportional to its length. Ohm's observation makes sense intuitively when we consider that current is carried through a conductor by the motion of electrons. We pointed out before (Section 17.1) that electrons have a drift velocity in a current-carrying conductor, interacting with the ions of the conductor and transferring charge along in this fashion. A conductor with a larger cross-sectional area allows more charges to flow easily, so a larger current results. However, for a longer conductor

occurred even when the mercury was impure. Kamerlingh Onnes realized that the mercury had undergone a phase transition to a new state, and had become a *superconductor*. In this new superconducting state, the resistance of the mercury was truly zero.

In the years that followed, many other materials were identified as superconductors, including aluminum ($T_c = 1.2$ K), lead ($T_c = 7.2$ K), niobium ($T_c = 9.3$ K), and a number of intermetallic compounds such as niobium-tin (Nb_3Sn, $T_c = 18$ K). Materials such as niobium-tin have been used for the current-carrying windings of superconducting magnets for research and medicine (Fig. B17.2). Magnets of this type are capable of very large magnetic fields. However, to keep them cold (so that they are below their transition temperature), they must be well insulated and cooled with liquid helium.

Because of the expense and difficulty of using liquid helium as a coolant, many researchers sought super-

conductors with higher transition temperatures. In particular, they sought superconductors with transition temperatures above 77 K, which is the boiling point of liquid nitrogen, a relatively inexpensive coolant. For many years, the highest known superconducting transition temperature was 23 K, the transition temperature of

Figure B17.2 Large superconducting magnets are an integral part of magnetic resonance imaging systems used in medicine.

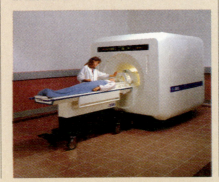

niobium-germanium (Nb_3Ge). However, in 1986 J. G. Bednorz and K. A. Müller of the IBM Zurich laboratory found an oxide compound of barium, lanthanum, and copper that became superconducting at 35 K. This discovery earned them the 1987 Nobel Prize in physics, and it set off a wave of activity in laboratories around the world by investigators searching for materials with even higher transition temperatures. By 1988, superconductors with transition temperatures as high as 125 K had been reported.

Because the new high-temperature superconductors require relatively inexpensive coolants, they hold promise for a wide range of applications. Possibilities include electric power transmission without resistive losses, new magnets, and perhaps even magnetically levitated vehicles. However, these applications require improved materials because the available high-temperature superconductors lose their superconductivity when carrying large currents.

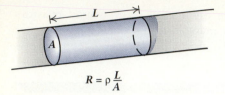

$$R = \rho \frac{L}{A}$$

Figure 17.8   The resistance of a piece of wire or other conducting material is proportional to its length L and inversely proportional to its cross-sectional area A.

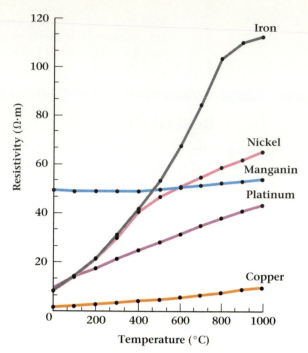

Figure 17.9   The resistivity of several metals as a function of temperature. Notice that the resistivities of all these metals are nonlinear over the temperature range shown.

TABLE 17.1 Electric resistivity of some metals	
Material	**Resistivity (Ω·m) (at 20°C)**
Conductors	
Aluminum	2.65×10^{-8}
Copper	1.72×10^{-8}
Gold	2.24×10^{-8}
Iron	9.71×10^{-8}
Nichrome	100×10^{-8}
Platinum	10.6×10^{-8}
Silver	1.59×10^{-8}
Tungsten	5.65×10^{-8}
Semiconductors	
Carbon (graphite)	1.5×10^{-5}
Germanium (pure)	5×10^{-1}
Silicon (pure)	3×10^{3}
Insulators	
Glass	$10^{7} - 10^{10}$
Quartz	7.5×10^{17}

and the same applied potential, the electric field is reduced, resulting in a reduced electron flow.

Ohm had already shown that the current was inversely proportional to the resistance (Ohm's law). When these observations are combined, we find that the resistance of a wire must be proportional to its length L and inversely proportional to its cross-sectional area A (Fig. 17.8):

$$R = \rho \frac{L}{A}. \tag{17.4}$$

The constant of proportionality ρ is the electric **resistivity**. In SI units, resistivity is given in ohm-meters (Ω·m). The resistivities of a selected list of metals are found in Table 17.1. For most metals the resistivity increases with increasing temperature (Fig. 17.9). For some materials, over narrow ranges, the change in resistivity is approximately proportional to the change in temperature.

Example 17.3

Calculating the resistance of a wire.

What is the electric resistance of an iron wire 0.50 m long with a diameter of 1.3 mm if the resistivity of iron is 9.7×10^{-8} Ω·m?

Solution   We can calculate resistance from Eq. (17.4) once we know the cross-sectional area of the wire. If d represents the diameter of the wire, then

$$A = \frac{\pi d^2}{4} = \frac{\pi (1.3 \times 10^{-3} \text{ m})^2}{4} = 1.33 \times 10^{-6} \text{ m}^2.$$

The resistance is then

$$R = \frac{\rho L}{A} = \frac{(9.7 \times 10^{-8} \text{ } \Omega \cdot \text{m})(0.50 \text{ m})}{1.3 \times 10^{-6} \text{ m}^2} = 0.037 \text{ } \Omega.$$

Example 17.4

A piece of copper wire has a cross section of 4.0 mm² and a length of 2.0 m. (a) What is the electric resistance of the wire at 20°C? (b) What is the potential difference across the wire when it carries a current of 10 A?

Solution   (a) First we calculate the resistance using Eq. (17.4). According to Table 17.1, the resistivity of copper is 1.72×10^{-8} $\Omega \cdot$m:

$$R = \frac{\rho L}{A} = \frac{1.72 \times 10^{-8} \text{ } \Omega \cdot \text{m} \times 2.0 \text{ m}}{4.0 \times 10^{-6} \text{ m}^2} = 8.6 \times 10^{-3} \text{ } \Omega.$$

(b) When the wire carries a current of 10 A, the voltage can be determined from Ohm's law:

$$V = IR = 10 \text{ A} \times 8.6 \times 10^{-3} \text{ } \Omega,$$
$$V = 8.6 \times 10^{-2} \text{ V} = 0.086 \text{ V}.$$

Power and Energy in Electric Circuits

17.4

If we connect a lamp to a pair of batteries, using good conductors whose resistance is small compared with that of the lamp itself, then we can consider the lamp as the only resistive component in the circuit. We can then treat the connecting wires as ideal conductors with no resistance. Figure 17.10(a) on page 504 illustrates this physical setup. The same situation is represented by the circuit diagram in Fig. 17.10(b). In drawing diagrams of electric circuits a special symbol (⌇⌇⌇) is used to indicate resistors. Figure 17.10(c,d,e) shows some other standard symbols.

Since the resistance of the wires is negligible, there is virtually no potential difference across them and the full potential difference of the batteries appears across the resistor. The potential difference (voltage) from points *b* to *c* in Fig. 17.10(b) is the same as the voltage from *a* to *d* because the voltage differences from *a* to *b* and from *c* to *d* are both zero. The magnitude of the electric current is constant in this circuit, the condition known as direct current or dc.

As the current passes through the lamp, charges are moving from a higher potential to a lower one. Energy is being lost from the battery and converted in the filament of the lamp into heat and light. The amount of energy released by a charge *q* as it falls through the potential *V* across the

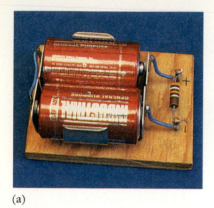

(a)

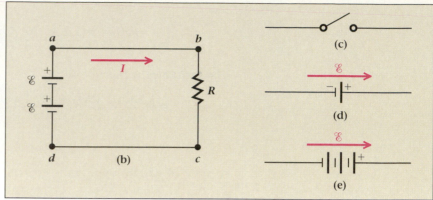

(b)   (c)   (d)   (e)

Figure 17.10   A simple resistive circuit. (a) The physical circuit. (b) A schematic diagram of the circuit, using standard symbols for the circuit components. Standard circuit symbols for (c) a switch, (d) a single-cell battery, and (e) a multicell battery. The arrow shows the direction of increasing potential for the sources of emf.

lamp is $W = qV$. Since the potential is constant, the rate at which energy is released, or the power P, is

$$P = \frac{\Delta W}{\Delta t} = \frac{\Delta(qV)}{\Delta t} = V\frac{\Delta q}{\Delta t},$$

$$\boxed{P = IV.} \tag{17.5}$$

When the potential difference is measured in volts and the current in amperes, their product gives the power in watts. We can see this from the definitions of the units:

$$(\text{ampere})(\text{volt}) = \left(\frac{\text{coulomb}}{\text{second}}\right)\left(\frac{\text{joule}}{\text{coulomb}}\right) = \frac{\text{joule}}{\text{second}} = \text{watt}.$$

In a resistor, the energy dissipated appears as thermal energy, a thermodynamically irreversible process in the sense introduced in Chapter 11. This effect is used in appliances such as electric stoves, heaters, and hair dryers (Fig. 17.11). In an incandescent lamp, the energy delivered to the filament raises its temperature so high that light is emitted. In other circuit elements the energy may take on different forms. For example, the energy may appear as mechanical work done by a motor, as sound from a loudspeaker, or as stored chemical energy in a battery when the battery is being recharged. Conversion from electrical to mechanical energy is never 100% efficient. The difference appears as heat. Anyone who has felt an operating electric motor is well aware of this, especially if the motor was drawing a lot of current.

Example 17.5

Resistance of an electric iron.

An electric iron is a good example of electric resistance. A typical iron designed to operate on a 120-V household circuit is rated at 1100 W. What is the resistance of the iron?

Solution   Since we know the voltage, we can determine the resistance from its definition if we can first find the current. The current can be

Figure 17.11   A radiant heater converts electrical energy into thermal energy.

calculated from the power relation of Eq. (17.5):*

$$I = \frac{P}{V} = \frac{1100 \text{ W}}{120 \text{ V}} = 9.17 \text{ A}.$$

When this value for the current is inserted into the definition of resistance, we obtain

$$R = \frac{V}{I} = \frac{120 \text{ V}}{9.17 \text{ A}} = 13.1 \ \Omega.$$

Example 17.6

Power used by an electric toaster.

What is the power consumption of an electric toaster if its resistance is 10.6 Ω and it operates on a household circuit?

Solution This problem is similar to Example 17.5. Here we are given the resistance (10.6 Ω) and the voltage (assume an effective household voltage of 120 V) and are asked to find the power. Using voltage and resistance, we can easily find the current:

$$I = \frac{V}{R} = \frac{120 \text{ V}}{10.6 \ \Omega} = 11.3 \text{ A}.$$

We can then combine current and voltage to get the power:

$$P = IV = (11.3 \text{ A})(120 \text{ V})$$
$$= 1360 \text{ W}.$$

In the two preceding examples we used two equations sequentially to obtain the required answer. We could have combined these equations at the outset and then substituted the values only once. For example, if we use Eq. (17.2) to eliminate the current in Eq. (17.5), we get a new equation relating power to the resistance and voltage across an electric device:

$$P = IV = \left(\frac{V}{R}\right) V,$$

$$P = \frac{V^2}{R}. \qquad (17.6)$$

This last equation could have been used from the start in both examples to calculate the answer in one step. In both cases we were given two of the three quantities, resistance, power, and voltage, and the third could have been easily calculated from Eq. (17.6). Similarly, we can eliminate the voltage from Eqs. (17.2) and (17.5) to get a relation between the

---

*The voltage in household circuits oscillates with time, a condition commonly called alternating current (ac). However, the value of the voltage is usually expressed as the dc equivalent. The effective values of oscillating currents and voltages are discussed in Chapter 20.

current through a device, resistance, and power:

$$P = IV = I(IR),$$

$$P = I^2R. \tag{17.7}$$

As we explained earlier in this section, when an electric current passes through a resistor, electric energy is irreversibly transformed to thermal energy. Equations (17.6) and (17.7) describe the rate of transfer of electrical energy to heat energy in a resistor. Both these equations are known as **Joule's law.**

Example 17.7

Application of Joule's law.

A piece of wire has a resistance of 30 Ω. How much power is dissipated in the wire if it carries a current of 0.50 A?

Solution We may use Eq. (17.7) to calculate the power directly,

$$P = (0.50 \text{ A})^2 (30 \text{ Ω}) = 7.5 \text{ A}^2 \text{ Ω}.$$

Since we have consistently used SI units, the power is in watts. Let's check the units to be sure. The ohm is one volt per ampere, so

$$A^2 \text{ Ω} = (\text{ampere})^2 \text{ ohm} = (\text{ampere})^2 \left(\frac{\text{volt}}{\text{ampere}} \right)$$

$$= (\text{ampere})(\text{volt}) = \left(\frac{\text{coulomb}}{\text{second}} \right) \left(\frac{\text{joule}}{\text{coulomb}} \right)$$

$$= \frac{\text{joule}}{\text{second}} = \text{watt}.$$

The power is indeed given in watts and is $P = 7.5 \text{ W}$.

The energy dissipated in a circuit is the product of the power and the time. Although we often speak of electric utilities as power companies, the product that we buy is energy; the power is the rate at which the energy is used. One unit of energy is the joule or watt-second. But this unit is inconveniently small for conventional power applications. In our homes, for example, we commonly use energy at the rate of kilowatts, and do so for extended periods of hours or days. As a result, a common unit for electrical energy is the kilowatt-hour (kWh). One kilowatt-hour is the energy consumed in one hour (3.6×10^3 s) by operating at a constant power of 1 kW (10^3 W),

$$1 \text{ kWh} = 3.6 \times 10^6 \text{ J}.$$

Example 17.8

Calculating electricity costs.

In the stairwell of a ten-story building there are two continuously burning 75-W safety lamps for each floor. (a) What is the total energy (in kilowatt-hours) used in one year? (b) What will it cost to use the lamps for a year if the cost of electricity is $0.078/kWh?

Solution   (a) The total power used is $2 \times 10 \times 75 \text{ W} = 1.5 \text{ kW}$. The energy, or work, is then

$$W = P \times t,$$

$$W = 1.5 \text{ kW} \times 1 \text{ year} \times \frac{365 \text{ days}}{1 \text{ year}} \times \frac{24 \text{ hours}}{\text{day}} = 1.3 \times 10^4 \text{ kWh}.$$

(b) The cost for a year of operation is

$$(1.3 \times 10^4 \text{ kWh}) \times \$0.078/\text{kWh} \approx \$1000.$$

Short Circuits and Open Circuits

17.5

A battery produces a voltage between its terminals. Either terminal can be chosen as the reference potential. In household circuits we choose the reference potential to be that of the earth or ground. That is, one side of the circuit is maintained with a zero potential difference with respect to the earth and is called the **ground potential** (Fig. 17.12a). The other side of the circuit is called the "hot" side. A current will pass through any conducting object that simultaneously contacts the hot side of the circuit and a conducting path to ground. Birds and small animals can sit or run along high-voltage wires without harm because their bodies do not make contact with the hot side of the circuit and ground at the same time.

Occasionally, electric equipment experiences a type of failure in which an alternative unwanted conducting path allows large currents. Such a path is usually called a **short circuit** because the current is not flowing through the desired circuit but instead passes through a parallel "shorter" path of lower resistance (Fig. 17.12b). Since we usually refer one side of a circuit to ground potential, we also describe the condition of a short circuit as being grounded out or simply *grounded*. The result of a short circuit is frequently a high current, since for a given potential difference, a lower resistance leads to a higher current.

Another common problem in electric circuits is the interrupted or **open circuit**. If the conducting path is interrupted or broken at any point, no current can occur in the circuit. It is just as if a switch had been opened.

Figure 17.12 (a) A circuit with one side grounded. The symbol for ground is shown. (b) Conditions for a short circuit. When the switch S is closed, the current passes through the switch instead of going through the lamp. (c) An open circuit. With the switch S open there is no current through the lamp.

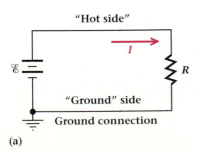

(a)

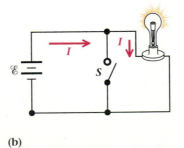

(b)

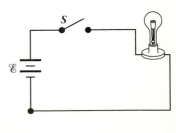

(c)

Open circuits most often occur when the conducting path is mechanically broken or burns out from an overload. The classic example of this is the failure of a string of single-wire decorative Christmas lights when a bulb burns out. All of the lights go out when one fails because no current can flow through the open circuit.

Kirchhoff's Rules

17.6

In analyzing resistive circuits, we use two fundamental ideas: conservation of energy and conservation of charge. When applied to electric circuits these laws are known as **Kirchhoff's rules**. The first rule is the law of conservation of energy, and says that *the algebraic sum of the potential differences around a closed conducting loop must be zero*. This means that if a charge is taken from some point in the circuit and moved completely around the loop to its starting point, the net work done on the charge is zero. This must be so because the charge has returned to the same energy level. This statement is called **Kirchhoff's voltage rule**, sometimes referred to as the *loop rule*.

In order to apply the loop rule we must carefully determine the signs of each potential difference. As a start, consider the potential difference across a battery or other source of emf. The direction of the emf is defined to be from the terminal of lower potential to the terminal of higher potential. Frequently the arrow and the plus and minus signs are understood and are not explicitly given on the symbol for a battery. However, the direction of the emf is always understood to be the same as that indicated in Fig. 17.10.

When a resistor is connected to the terminals of a battery (Fig. 17.13a), charge flows through the circuit. If we use the conventional direction for the current (the direction in which a positive charge would move), then the direction of current through the resistor is in the direction of decreasing electric potential (Fig. 17.13b). We can combine this observation with the definition of an emf to list the convention for finding the potential differences for the Kirchhoff voltage rule. For an imaginary traversal of a loop we use the following convention:

1. The potential difference is $+\mathscr{E}$ when a source of emf is traversed in the forward direction of the emf.
2. The potential difference is $-\mathscr{E}$ when a source of emf is traversed in the backward direction.
3. The potential difference is $-IR$ when a resistor is traversed along the direction of the current.
4. The potential difference is $+IR$ when a resistor is traversed in the direction opposite to that of the current.

The Kirchhoff voltage rule holds equally well for the loop of Fig. 17.13 or any of the loops in Fig. 17.14. The algebraic sum of the potential differences is zero around the loop *abefa* (Fig. 17.14). If the potential difference between points a and b is written as V_{ab}, then we have

$$V_{ab} + V_{be} + V_{ef} + V_{fa} = 0.$$

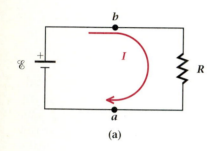

(a)

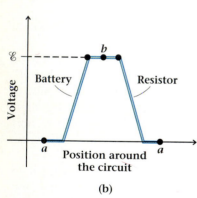

(b)

Figure 17.13   (a) A current exists in the circuit when a resistance is connected to a battery. (b) According to Kirchhoff's voltage rule, the sum of the potential changes around the loop is zero. In the diagram of voltage versus position around the circuit, sources of emf increase the potential and resistors (traversed in the direction of the current) decrease the potential.

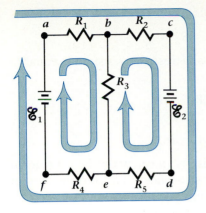

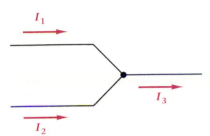

Figure 17.14 The sum of the voltages around a loop is zero. This holds for loop *abefa*, loop *abcdefa*, and loop *bcdeb*.

Had we chosen a different loop the result would still be the same:

$$V_{bc} + V_{cd} + V_{de} + V_{eb} = 0.$$

The second fundamental law that applies to circuit analysis is the conservation of charge at a circuit junction. When expressed in terms of current, this means that the sum of the currents flowing into a junction must equal the sum of the currents leaving the junction. If this were not so, then the junction would be either creating or destroying charge. Thus a single-loop circuit must have the same current in all parts of the loop. For a multiloop circuit, the *net current entering into any junction must be zero*. This last statement is known as **Kirchhoff's current rule**, sometimes called the *junction rule*. In Fig. 17.15 we may consider currents I_1 and I_2 to be positive since their directions are into the junction. The other current is taken as negative because it is directed away from the junction. Kirchhoff's current rule for this junction can be written as

$$I_1 + I_2 - I_3 = 0.$$

We will not analyze any complicated circuits requiring the solution of simultaneous equations based on Kirchhoff's rules. Instead we will refer to these rules for justifying our analysis of simple circuits in the rest of this chapter. We have included some problems at the end of the chapter involving more complicated circuits so you can try this kind of analysis on your own.

Figure 17.15 Kirchhoff's current rule: The algebraic sum of the currents into the junction is zero.

Simple Resistive Circuits

17.7

Let us return to our example of the simple circuit consisting of two batteries and a resistor (Fig. 17.16a). The potential difference of each battery is 1.5 V. When the two batteries are joined in series (that is, the positive side of one battery connected to the negative side of the other), their emfs are in the same direction in the sense of the Kirchhoff voltage rule. Thus the voltages add to provide a potential difference of 3.0 V. When a second resistor is added in series with the first (Fig. 17.16b), so that current passes through one and then the other, the total potential difference across them must add up to the three volts available from the batteries. Since there is only one source of current and only one conduction path, the current I is the same throughout the entire circuit.

The potential across the first resistor is

$$V_1 = IR_1$$

and across the second is

$$V_2 = IR_2.$$

Kirchhoff's voltage rule says that the algebraic sum of the potential dif-

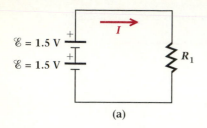

(a)

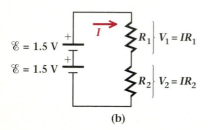

(b)

Figure 17.16　(a) A simple resistive circuit with one resistor connected across the potential of two batteries. (b) Two resistors connected in series across the same batteries. The current is the same in each resistor connected in series.

ferences across the battery and the resistors is zero:

$$\mathscr{E} + \mathscr{E} - IR_1 - IR_2 = 0.$$

Thus the magnitude of the total potential across the resistors is equal to the magnitude of the battery voltage V,

$$V = \mathscr{E} + \mathscr{E} = IR_1 + IR_2,$$

or

$$\mathscr{E} = I(R_1 + R_2) = IR_s.$$

From this equation we see that two resistors in series can be considered equivalent to a single resistor R_s, whose resistance is equal to the sum of the individual resistances. This reasoning is true not only for the two resistors in series in Fig. 17.16(b), but for any number of resistors joined in series. Thus **the equivalent resistance of resistors connected in series** is

$$R_s = R_1 + R_2 + R_3 + \cdots \qquad (17.8)$$

Example 17.9

Adding resistors in series.

Three 10-Ω resistors and one 15-Ω resistor are connected in series across a 9-V battery (Fig. 17.17). Find the voltage drop across the 15-Ω resistor.

Solution　From Eq. (17.8), the total resistance of the four resistors is 45 Ω. The current in the circuit is

$$I = \frac{V}{R_s},$$

where V is the battery voltage and R_s is the total resistance. Thus

$$I = \frac{9 \text{ V}}{45 \ \Omega} = 0.2 \text{ A}.$$

The voltage across the 15-Ω resistor is

$$V_3 = IR_3 = (0.2 \text{ A})(15 \ \Omega) = 3 \text{ V}.$$

Note that the voltage drop across this resistor would be the same regardless of how the resistors are arranged, as long as they remain connected in series across the same applied voltage.

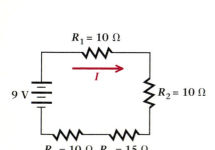

Figure 17.17　Example 17.9: Four resistors joined in series.

Two resistors can also be joined together across the same battery in parallel (Fig. 17.18). This connection allows current to flow independently through the parallel branches of the circuit, while the voltage across each branch is the full voltage of the battery. The current I_1 through resistor R_1 is

$$I_1 = \frac{V}{R_1},$$

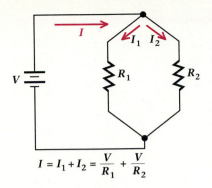

Figure 17.18 Two resistors joined in parallel. The voltage is the same across each resistor.

$$I = I_1 + I_2 = \frac{V}{R_1} + \frac{V}{R_2}$$

and the current through resistor R_2 is

$$I_2 = \frac{V}{R_2}.$$

From the conservation of charge (Kirchhoff's current rule), the total current provided by the battery is the sum of the two branch currents I_1 and I_2,

$$I = I_1 + I_2.$$

Therefore

$$I = \frac{V}{R_1} + \frac{V}{R_2},$$

or

$$I = V\left(\frac{1}{R_1} + \frac{1}{R_2}\right).$$

The total current is the potential divided by the equivalent resistance,

$$I = \frac{V}{R_p}.$$

By comparing these last two expressions for the current, we see that the equivalent resistance R_p of two resistors connected in parallel is

$$\frac{1}{R_p} = \frac{1}{R_1} + \frac{1}{R_2}.$$

This argument can be extended to show that **the equivalent resistance of resistors connected in parallel** is

$$\frac{1}{R_p} = \frac{1}{R_1} + \frac{1}{R_2} + \frac{1}{R_3} + \cdots \qquad (17.9)$$

For the special case of only two resistors, the equivalent parallel resistance may be written as

$$R_p = \frac{R_1 R_2}{R_1 + R_2}. \qquad (17.10)$$

From this last equation we observe that the equivalent resistance of two resistors joined in parallel is always less than the magnitude of the smaller resistor. To see this suppose that $R_1 < R_2$. Then

$$R_p = R_1\left(\frac{R_2}{R_1 + R_2}\right) < R_1.$$

In the special case of two identical resistors, $R_1 = R_2 = R$, and the equivalent parallel resistance becomes $R/2$.

An important point to remember concerning series and parallel connection of resistors is that for series connection the *current through* the resistors is the same, and for parallel connection the *voltage across* the resistors is the same.

Example 17.10

What is the equivalent resistance of three resistors joined in parallel if their values are 330 Ω, 100 Ω, and 220 Ω?

Solution   The combined resistance is found directly from Eq. (17.9):

$$\frac{1}{R_p} = \frac{1}{330 \ \Omega} + \frac{1}{100 \ \Omega} + \frac{1}{220 \ \Omega},$$

$$\frac{1}{R_p} = \frac{20 + 66 + 30}{6600 \ \Omega} = \frac{116}{6600 \ \Omega}.$$

The equivalent resistance is

$$R_p = \frac{6600 \ \Omega}{116} = 56.9 \ \Omega.$$

Example 17.11

A 10-Ω resistor is joined in series with a 47-Ω resistor and a 23-Ω resistor that are connected in parallel (Fig. 17.19a). What is the equivalent resistance of this combination?

Solution   We can determine the resistance of this arrangement by first reducing the parallel combination to a single equivalent resistor, and then combining that with the 10-Ω resistor, using the rule for series combination. This general procedure can be used to treat more complicated combinations than the one given here.

The combined resistance of the two connected in parallel is

$$\frac{1}{R_p} = \frac{1}{R_1} + \frac{1}{R_2},$$

$$\frac{1}{R_p} = \frac{1}{47 \ \Omega} + \frac{1}{23 \ \Omega}.$$

Solving for R_p gives

$$R_p = 15 \ \Omega.$$

This, in turn, is combined in series (Fig. 17.19b) to give

$$R = 15 \ \Omega + 10 \ \Omega,$$

$$R = 25 \ \Omega.$$

This is the value of the equivalent single resistance shown in Fig. 17.19(c).

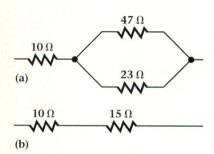

Figure 17.19   Example 17.11:
(a) Two resistors connected in parallel are connected in series to a third resistor. (b) The equivalent resistance of the parallel resistors in series with the third resistor. (c) The single resistor equivalent to the three resistors of part (a).

Capacitors in Combination

17.8

So far in this chapter, the only circuit components we have discussed are current sources and resistors. Now that we have learned to analyze these resistive circuits, let's add in the circuit component we examined in the previous chapter: the capacitor. There we learned that capacitance is proportional to the area of the plates of a parallel-plate capacitor. Thus if we

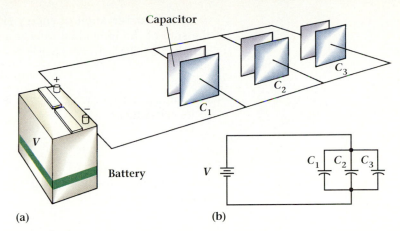

Figure 17.20 (a) Three capacitors connected in parallel. The voltage is the same across each capacitor. (b) Schematic representation of three capacitors connected in parallel.

(a)   (b)

could double the area we would double the capacitance. We can effectively do just this by joining two identical capacitors in parallel. Then, when we apply a potential across them, charge flows to each capacitor, resulting in a total stored charge that is twice the charge on each individually. Hence the capacity of the combination is twice that of one capacitor alone.

We can use this same reasoning to calculate the capacitance of any number of capacitors, of any size, joined together in parallel. Consider, for example, three capacitors in parallel (Fig. 17.20). The voltage V supplied by the battery appears across each capacitor. The charge on each capacitor depends on this voltage, and the capacitance of each is determined from the equation $C = q/V$. Thus the charge q_1 on capacitor 1 is

$$q_1 = C_1 V.$$

Similarly, the charges on the other two capacitors are

$$q_2 = C_2 V \quad \text{and} \quad q_3 = C_3 V.$$

The effective capacitance C_p of the parallel combination of capacitors is the ratio of the total charge to the applied voltage. Since the total charge is just the sum of q_1, q_2 and q_3, we have

$$C_p = \frac{(q_1 + q_2 + q_3)}{V} = \frac{q_1}{V} + \frac{q_2}{V} + \frac{q_3}{V} = C_1 + C_2 + C_3.$$

We thus arrive at an addition rule for capacitors: **The equivalent capacitance C_p of capacitors connected in parallel is**

$$C_p = C_1 + C_2 + C_3 + \cdots \tag{17.11}$$

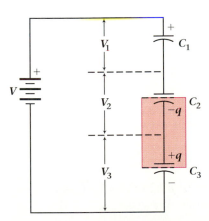

Figure 17.21 Capacitors connected in series. The magnitude of the charge is the same on each plate of each capacitor. (That is, there is no net charge in the region marked by the colored rectangle.)

Suppose instead that we join the capacitors end-to-end in series across the battery (Fig. 17.21). For this situation, as the first capacitor becomes charged, it induces an equal charge in the second capacitor, which in turn induces an equal charge in the third capacitor. Since the capacitors permit no flow of charge between their plates, there is no net charge within the region marked by the colored rectangle in Fig. 17.21. Therefore all of the capacitors must have the same charge q:

$$q_1 = q_2 = q_3 = q.$$

Now, conservation of energy (Kirchhoff's voltage rule) requires that the sum of the individual capacitor voltages be equal to the total voltage supplied by the battery. That is,

$$V = V_1 + V_2 + V_3.$$

The voltage across each capacitor is given by the ratio of the charge on it to its capacitance,

$$V_1 = \frac{q_1}{C_1}, \qquad V_2 = \frac{q_2}{C_2}, \qquad V_3 = \frac{q_3}{C_3}.$$

Replacing the individual voltages by their equivalents in terms of q/C, we find that

$$V = q\left(\frac{1}{C_1} + \frac{1}{C_2} + \frac{1}{C_3}\right).$$

This last equation has the form $V = q/C$ if we identify the equivalent series capacitance C_s as

$$\frac{1}{C_s} = \frac{1}{C_1} + \frac{1}{C_2} + \frac{1}{C_3}.$$

Thus we arrive at a second addition rule for capacitors: **The equivalent capacitance C_s of capacitors connected in series is**

$$\frac{1}{C_s} = \frac{1}{C_1} + \frac{1}{C_2} + \frac{1}{C_3} + \cdots \qquad (17.12)$$

Note that for capacitors connected in parallel, the voltage is the same across each capacitor, while for capacitors connected in series, the charge is the same on each capacitor.

Example 17.12

Equivalent capacitance of a network.

Calculate the capacitance (sometimes just called the capacity) of the network shown in Fig. 17.22.

Solution   We begin by first finding the equivalent capacitance of the three capacitors joined in parallel. Their combined capacitance is the sum of their individual capacitances,

$$C_p = C_1 + C_2 + C_3 = 5 \ \mu F + 5 \ \mu F + 10 \ \mu F = 20 \ \mu F.$$

Next we add their effective capacitance of 20 μF to the 10 μF, according to Eq. (17.12) for adding capacitors in series. We get

$$\frac{1}{C_s} = \frac{1}{20 \ \mu F} + \frac{1}{10 \ \mu F} = \frac{3}{20 \ \mu F}.$$

Upon solving this equation for C_s we have the capacitance of the network,

$$C = \tfrac{20}{3} \ \mu F = 6.7 \ \mu F.$$

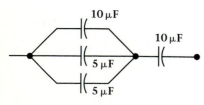

Figure 17.22   Example 17.12: A capacitor network.

Internal Resistance of a Battery

17.9

We described batteries as devices that maintain a fixed electric potential difference between two points. However, when a real battery is used to provide electrical energy, the external voltage across the terminals is less than the emf. This reduction in voltage is due to the potential drop occurring across the internal resistance of the battery itself. As more current is drawn from a battery, a greater voltage drop occurs across its internal resistance. This effect is most easily visualized by considering a real battery to consist of an ideal, resistanceless battery in series with a resistance (Fig. 17.23a). The resistance r is the internal resistance of the battery.

If the battery is connected to an external resistance R (called a load resistor), the circuit could be drawn as shown in Fig. 17.23(b), which explicitly includes the internal resistance. The current through the circuit depends on the emf $\mathcal{E}$ and the total resistance,

$$I = \frac{\mathcal{E}}{r + R}.$$

The potential difference across the terminals of the battery is the terminal potential difference, abbreviated TPD. It is the emf reduced by the voltage drop across the internal resistance r. Thus the TPD has a value

$$TPD = \mathcal{E} - Ir = \mathcal{E} - \frac{\mathcal{E}r}{r + R},$$

which reduces to

$$TPD = \frac{R}{r + R}\mathcal{E}. \tag{17.13}$$

According to this equation, when the load resistance R is small, the terminal voltage is appreciably less than the emf. However, when the load resistance is large compared with the internal resistance of the battery, the terminal voltage approximately equals the emf.

Carbon-zinc dry cells are notorious for their relatively large internal resistance, which becomes particularly noticeable as the cells age. Consequently, simply checking the output voltage of a dry cell battery with a voltmeter is no real indication of the battery's condition. The reason that this simple test is meaningless is that, as we will see later, a voltmeter has a high resistance. When a voltmeter is connected alone across a battery, it gives a reading very nearly that of the emf. But if the battery is connected first across a low resistance, say 200 Ω, the TPD measured by the voltmeter is much less than the emf. As an illustration of this, you might measure the TPD of the battery in a radio while it is operating, and see what reading you get.

The decrease in TPD associated with increased current is not limited to batteries. Any network composed of sources of emf and resistors can

(a)

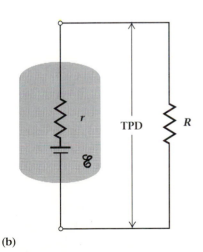

(b)

Figure 17.23 (a) A real battery can be represented as an ideal emf in series with an internal resistance. (b) A real battery in a circuit. The terminal potential difference (TPD) is less than the emf.

be represented by an equivalent series combination of a single battery and a single resistor. (This statement is known as Thévenin's theorem.) Such a circuit is fully equivalent in the way in which it influences an external (load) circuit. The output behavior of any battery, battery network, or power supply can be described by this equivalent circuit. Thus the general behavior of any voltage source is that the terminal voltage decreases as the current drawn by the load increases.

Example 17.13

A transistor radio battery has an emf of 9.0 V. When a short copper wire is connected directly across the battery terminals, a current of 4.0 A passes through the wire. What is the internal resistance of the battery? What is the terminal potential difference across a 10-Ω load?

PHYSICS IN PRACTICE

Electric Shock

Electric shock is a hazard associated with all electric appliances and equipment. The danger of electrocution is not just for the high voltages of transmission lines. Unfortunately, people have been killed by ordinary house current at 120 V or by contact with industrial equipment at 40 or 50 V.

The important measure of shock intensity is not the voltage, but the amount of current that passes through the body (Fig. B17.3). Thus any electric device using ordinary household voltages can potentially supply a fatal current. The amount of current can be determined from Ohm's law, but since the body's electric resistance varies enormously, it is not possible to give precise statements of safe or dangerous voltages. For example, the body's effective resistance depends largely upon the area of contact and the condition of the skin. But skin resistance may vary from about 500,000 Ω dry to as little as 500 Ω when wet.

The hazards of electric shock depend not only on the amount of current involved, but also on the path of the current. A current passing through your arm from fingertip to elbow may produce a painful and unpleasant shock, but the same current passing

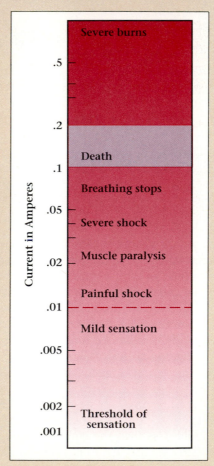

Figure B17.3   Electrical shock hazard at various values of current.

from one hand to the other through your chest may be fatal.

Electric current can damage the body in three distinct ways: (1) it may subject the body to intense heat and cause burns; (2) it may disrupt the proper functioning of the nervous system and heart; and (3) it may cause the muscles to twitch uncontrollably. Currents as low as 20 mA may cause difficulty in breathing, and at 75 mA breathing may stop completely. Currents between about 100 and 200 mA result in ventricular fibrillation of the heart, which means an uncoordinated and uncontrolled twitching of the heart muscles. The resulting loss of pumping action is fatal. At still greater currents, the heart may stop completely without going into fibrillation. Under such conditions, the chance of survival may actually be improved, since the heartbeat can be more easily restored from being stopped than from fibrillation. The defibrillators used in medical emergencies apply a large momentary voltage to the body to stop the heart and facilitate the restoration of the normal heart rhythm.

The best medicine for electric shock is prevention. Have respect for electricity at all voltages. Be cautious and follow normal safety procedures when working with electrical equipment.

Solution   Because the resistance of the wire is so small, the short circuit current is limited by the internal resistance of the battery. Thus

$$r = \frac{\mathscr{E}}{I} = \frac{9.0 \text{ V}}{4.0 \text{ A}} = 2.25 \ \Omega.$$

When the battery is placed across a 10-Ω load, the terminal potential difference is less than the emf. From Eq. (17.13),

$$\text{TPD} = \frac{R}{r + R} \mathscr{E}.$$

Here $R = 10 \ \Omega$, $r = 2.25 \ \Omega$, and $\mathscr{E} = 9.0$ V, so

$$\text{TPD} = \frac{10}{10 + 2.25} \ 9.0 \text{ V} = 7.3 \text{ V}.$$

Home Power Distribution

*17.10

One very important characteristic of parallel circuits is that the current can be interrupted in one branch without interrupting the current in the other branches. This is especially significant in a power distribution system, where many separate users are simultaneously provided with electricity from the same network. Similarly, each user would like to operate lamps and appliances independently. For this reason the power distribution circuits that are so familiar to us in our homes are parallel circuits.

In wiring a house or building, usually several separate circuits are connected in parallel to the main incoming power line. In turn, each of these circuits carries several outlets, appliances, or lamps, also in a parallel configuration (Fig. 17.24). Because of the parallel circuitry, each outlet and each lamp can be used independently.

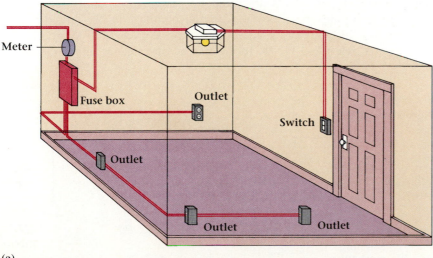

Figure 17.24   (a) Wiring scheme for a room. (b) Schematic diagram of the same room. In practice each separately fused circuit would have several outlets and lamps.

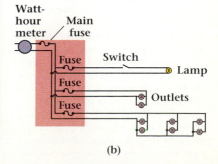

(a)   (b)

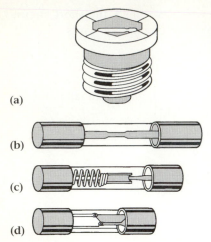

(a)

(b)

(c)

(d)

Figure 17.25 (a) A fuse of the type used in older buildings. Typical fuses of the type found in cars, amplifiers, and tape decks: (b) standard, (c) slow-blowing type, and (d) fast-acting fuse.

The wiring that carries current through the house has a low resistance, but it is not zero. Consequently, the temperature of the wires increases when the current gets large. If the amount of current is not limited somehow, the wires may overheat and become a fire hazard. This danger can be avoided by installing a safety device that interrupts the circuit and cuts off the current whenever it gets too large to be safe. The most common safety devices are fuses and circuit breakers.

A fuse consists of a metal wire or ribbon that melts at a particular current level (Fig. 17.25). For small currents, the fuse merely acts as a conductor. But when the current exceeds the rated capacity of the fuse, it overheats, melts, and leaves an air gap in the circuit. Once a fuse "blows," it has done its job of protecting the circuit and is no longer useful. It must then be replaced with a new fuse before the circuit can operate again.

In newer buildings the role of the fuse is handled by a circuit breaker (Fig. 17.26). This device mechanically interrupts the circuit in response to an overload current. The most widely used circuit breakers are thermal-magnetic devices. They use a combination of bimetallic strips and electromagnets to provide both thermal and magnetic protection from excessive currents. Unlike a fuse, a circuit breaker can be reset and used again. Each time the current exceeds the overload value, the circuit breaker trips and remains off until it is reset. Whenever a circuit breaker or fuse is blown repeatedly, the circuit should be carefully checked to find the cause of the overload. Then the source of the overload should be removed before resetting the breaker or replacing the fuse.

Household circuits are normally fused at 15 or 20 A. Exceptions are the special high-current circuits used for water heaters, ranges, and other large appliances. These circuits may carry up to 60 A and will not usually have other outlets on the same circuit. Such circuits also have larger-diameter wires to reduce their resistance.

Appliances such as toasters, frying pans, and radios are normally rated according to their power consumption in watts. Lamps, too, are commonly identified by their power rating. The result is that while the circuits are rated by current, the loads are rated by power. The connection is Eq. (17.5), which states that $P = IV$. The house voltage in the United States is nominally 120 V. Thus a 20-A circuit can deliver 2400 W at 120 V. This means that if a 1350-W frying pan is operated on the same 20-A

Figure 17.26 Schematic diagram of a circuit breaker showing only the thermal trip mechanism. (a) Normally the current path is closed. (b) When overheated, the bimetal strip bends, releasing the trip mechanism that allows the contacts to open. Additional protection is provided by a magnetically activated trip mechanism that responds more quickly to overload than does the thermal trip.

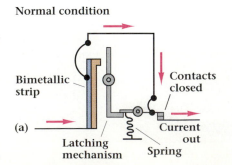

Normal condition

Bimetallic strip

Contacts closed

(a)

Latching mechanism

Spring

Current out

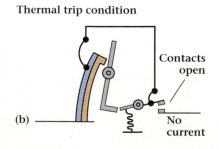

Thermal trip condition

Contacts open

(b)

No current

circuit as a 1500-W toaster, the circuit breaker will trip and shut off the current to that circuit. Occurrences similar to this are familiar to everyone.

Many electrical appliances are wired with three conductors and use a three-pronged plug on their power cord. Homes and offices are wired with three conductors and have outlets to match the three-pronged plugs. The third conductor is added for safety. In normal operation, two of the wires provide the path for the current. The third wire, which is grounded directly to the earth, is usually connected to the case of the appliance, thereby ensuring that should a short circuit occur, the case remains at ground potential.

SUMMARY

Useful Concepts

- Electric current is the rate at which electric charge flows and is defined by

$$I \equiv \frac{\Delta q}{\Delta t}.$$

- Electric resistance is defined by

$$R \equiv \frac{V}{I}.$$

- A resistive material is said to obey Ohm's law if current is proportional to the potential difference across a sample of that material:

$$V = IR$$

- The resistance of a specific piece of wire is

$$R = \rho \frac{L}{A}.$$

- The power dissipated in an electric circuit is

$$P = IV = \frac{V^2}{R} = I^2 R.$$

- The equivalent resistance R_s of a number of resistors connected in series is

$$R_s = R_1 + R_2 + R_3 + \cdots$$

- The equivalent resistance R_p of a number of resistors connected in parallel is

$$\frac{1}{R_p} = \frac{1}{R_1} + \frac{1}{R_2} + \frac{1}{R_3} + \cdots$$

- Capacitors connected in parallel have an equivalent capacitance of

$$C_p = C_1 + C_2 + C_3 + \cdots$$

- Capacitors connected in series have an equivalent capacitance given by

$$\frac{1}{C_s} = \frac{1}{C_1} + \frac{1}{C_2} + \frac{1}{C_3} + \cdots$$

- The terminal potential difference (TPD) of a battery with internal resistance r connected across an external resistance R is

$$\text{TPD} = \frac{R}{r + R} \mathscr{E}.$$

Important Terms

You should be able to write the definition or meaning of each of the following terms:

- battery
- electric current
- direct current
- ampere
- emf
- resistance
- ohm
- Ohm's law

- resistor
- resistivity
- Joule's law
- ground potential
- short circuit
- open circuit
- Kirchhoff's rules

QUESTIONS

17.1   Why are the 1.5-V D cells commonly used in flashlights so much larger than the 9-V batteries commonly used in small radios?

17.2   Under what conditions might the terminal voltage of a battery be greater than its emf?

17.3   Under what conditions would you arrange cells in series? In parallel?

17.4   What is the difference between resistance and resistivity?

17.5   Compare electric resistivity with thermal resistivity. (The thermal resistivity is the reciprocal of the thermal conductivity described in Chapter 10.) Are good electric conductors also good thermal conductors? Are good thermal conductors always good electric conductors? Give some examples.

17.6   Incandescent lamps usually burn out just as they

are turned on. Can you explain why this is so?

17.7   How does a fuse or circuit breaker act to protect electrical equipment?

17.8   Some electric appliances, notably TV sets, come with a two-pronged, polarized electric plug, which fits into an electric outlet in only one way. How do the plugs accomplish this polarization and what is the purpose of having them do so?

17.9   Figure 17.9 shows the temperature dependence of the resistivity of several metals. Explain why each of these metals is or is not a suitable choice of material as the temperature sensor for use in a resistance thermometer.

17.10   Explain why it is dangerous to reach into the back of a TV set even when it has been turned off and unplugged.

PROBLEMS

Hints for Solving Problems

The SI unit of current is the ampere (A), the SI unit of resistance is the ohm (Ω), the SI unit of power is the watt (W), and the SI unit of energy is the watt-second or joule (J). Sometimes we use the kilowatt-hour (kWh) as a unit for energy. Remember that the potential is measured *across* a resistor or other circuit component, but the current passes *through* the element. The voltage across resistors in parallel is the same; the current through resistors in series is the same. The resistivity is a property of a material itself; the resistance is a property of the material and its dimensions.

Section 17.1   Electric Current and Electromotive Force

17.1   A steady electric current of 0.50 A flows through a wire. (a) How many coulombs of charge pass through the wire per second? (b) How many per minute?

17.2   A charge of 600 C was transferred through a wire in 1.00 min. What was the average electric current during that interval?

17.3   A charge of 1200 C was transferred through a wire in 5.0 min. What was the average electric current during that interval?

17.4   How long does it take for a current of 0.80 A to transfer 67 C?

17.5   A charge corresponding to one electron for every

person in the United States passes a point in 0.01 s. What is the resulting current? (Take the population to be 250 million people. The charge on an electron is 1.60×10^{-19} C.)

Section 17.2   Electric Resistance and Ohm's Law

17.6   A 40-W electric lamp draws a current of 0.33 A when operated at a potential of 120 V. What is the resistance of the lamp?

17.7   What is the current drawn by a 500-Ω resistor when a potential difference of 25.0 V is maintained across it?

17.8   What is the potential difference across a 300-Ω resistor when a current of 3.00 A flows through it?

17.9   A three-cell flashlight draws a current of 0.50 A. What is the operating resistance of the lightbulb if each cell provides a potential of 1.5 V?

17.10   A piece of Nichrome wire passes a current of 0.75 A when a potential of 1.80 V is applied. What is the resistance of the wire?

17.11   What is the resistance of a resistor through which a charge of 8.0×10^4 C flows in one hour if the potential difference across it is 12 V?

17.12   Compute the current drawn by the 60-W lamp of Fig. 17.27 at 20-V intervals. Draw a graph of I versus V.

17.13   Compute the current drawn by the 200-Ω resistor of Fig. 17.27 at 20-V intervals and draw a graph of I versus V.

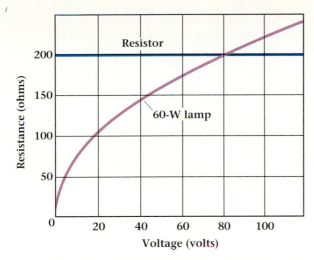

Figure 17.27 Problems 17.12 and 17.13. The re-sistance of a 60-W lamp is shown as a function of applied voltage. The horizontal line is the resistance of a 200-Ω resistor.

Section 17.3   Resistivity

17.14 A piece of 20-gauge wire one meter long has an electric resistance of 0.19 Ω. Calculate its resistivity and identify the composition of the wire from the values given in Table 17.1. (*Hint:* 20-gauge wire has a cross-sectional area of 0.5176 mm².)

17.15 Calculate the resistance of a piece of 20-gauge copper wire 2.00 m long. (*Hint:* The cross-sectional area of 20-gauge wire is 0.5176 mm².)

17.16 The light-duty extension cords that are routinely sold for household use are made from 18-gauge copper wire. Cords for higher current usage are often made from 16-gauge wire. Compare the resistances of wires 5.00 m long. (*Hint:* The cross-sectional area of 18-gauge wire is 0.8231 mm² and that of 16-gauge wire is 1.309 mm².)

17.17 When aluminum is used for electrical wire, it is common to use a larger-diameter wire than would be necessary for copper. Compare the resistance of a 16-gauge aluminum wire with that of an 18-gauge copper wire of equal length. (*Hint:* The cross-sectional area of 18-gauge wire is 0.8231 mm² and that of 16-gauge wire is 1.309 mm².)

17.18 Calculate the electric resistance of an iron rod 2.00 m long, assuming that its cross-sectional area is 0.90 mm².

17.19 Calculate the resistance of a Nichrome wire 0.50 mm in diameter and 1.50 m long.

17.20 Two conducting wires of the same material are to have the same resistance. One wire is 20 m long and 0.40 mm in diameter. If the other wire is 0.30 mm in diameter, how long should it be?

Section 17.4   Power and Energy in Electric Circuits

17.21 A flashlight lamp connected to a battery that provides 1.4 V draws a current of 0.10 A. What electrical power is used by the lamp?

17.22 (a) How much current is drawn by a 100-W lamp operating at its rated voltage of 120 V? (b) What is the resistance of the lamp when operating?

17.23 The label on a toaster reads 800 W at 120 V. How much current does it draw?

17.24 (a) What is the operating resistance of a lamp rated 40 W at 130 V? (b) How does that compare with the "cold" resistance of 31 Ω measured at very low voltage? (c) What does this calculation tell you about the relationship between resistance and temperature for the material of the filament?

17.25 A 100-Ω resistor is rated at 1.00 W maximum power capacity. (a) What is the maximum voltage that can be applied across the resistor without exceeding its maximum power rating? (b) What is the current at this voltage?

17.26 A 1500-Ω resistor is rated at 2.0 W maximum power capacity. (a) What is the maximum voltage that can be applied across the resistor without exceeding its maximum power rating? (b) What is the maximum current?

17.27 A 150-W street lamp is operated for 12 hours a day. How much energy does it take to operate the lamp for 30 days? Express your answer in kilowatt-hours and in joules.

17.28 A drilling machine operates with an electric power consumption of 840 W. How much does it cost to operate the machine continuously for eight hours if the electricity costs $0.080 per kWh?

17.29 A refrigerator is equipped with a motor that draws 100 W but operates only 25% of the time. What is the cost of operating the refrigerator for 30 days if electricity costs $0.080/kWh?

17.30 How much does it cost to operate a 100-W lamp 8 hours a day for 30 days if electricity costs $0.075/kWh?

Section 17.5   Short Circuits and Open Circuits

17.31 Two identical 100-Ω resistors are joined together as shown in Fig. 17.28. (a) What is the current through each resistor when the switch *S* is open? (b) When *S* is closed?

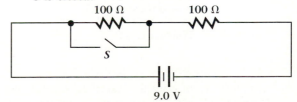

Figure 17.28 Problem 17.31.

17.32 Two identical 100-Ω resistors are joined together as shown in Fig. 17.29. (a) What is the current through each when the switch S is open? (b) What is it when S is closed?

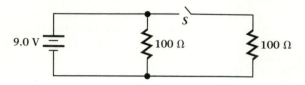

Figure 17.29   Problem 17.32.

Section 17.6   Kirchhoff's Rules
Section 17.7   Simple Resistive Circuits

17.33 Three 47-Ω resistors and one 15-Ω resistor are joined in series across a 9.0-V battery. Find the voltage drop across the 15-Ω resistor.

17.34 A 100-Ω resistor is joined in series with a 33-Ω resistor. (a) If the circuit has a current of 340 mA, what is the voltage applied? (b) What voltage drop appears across the 100-Ω resistor?

17.35 A 100-Ω resistor is joined in parallel with a 33-Ω resistor. What is the equivalent resistance?

17.36 Three resistors of 100, 45, and 33 Ω are joined together in parallel. What is the equivalent resistance?

17.37 Find the resistance of the network shown in Fig. 17.30.

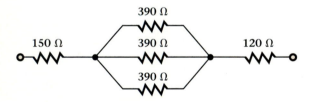

Figure 17.30   Problem 17.37.

17.38 Find the resistance of the network shown in Fig. 17.31.

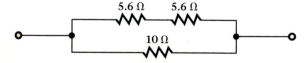

Figure 17.31   Problem 17.38.

Section 17.8   Capacitors in Combination

17.39 A radio capacitor is made of metal plates, each with an area of 6.0 cm^2. The plates are arranged to make 15 parallel-plate capacitors wired in a parallel con-

figuration. What is the capacitance of the capacitor if the separation between plates is 0.50 mm?

17.40 Capacitors of 5.0, 10, and 20 μF are joined together in series. What is their equivalent capacitance?

17.41 Three capacitors of 1.0, 1.5, and 5.0 μF are joined together in series. What is their equivalent capacitance?

17.42 Three capacitors of 1.0, 1.5, and 5.0 μF are joined together in parallel. What is their equivalent capacitance?

17.43 Three capacitors of 2.0, 5.0 and 10 μF are joined together in parallel. What is their equivalent capacitance?

17.44 Calculate the effective capacitance of the capacitor network shown in Fig. 17.32.

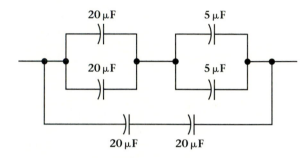

Figure 17.32   Problem 17.44.

17.45 Calculate the effective capacitance of the network shown in Fig. 17.33.

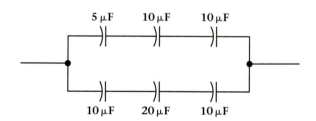

Figure 17.33   Problem 17.45.

17.46 Calculate the voltage across the 20-μF capacitor in the circuit of Fig. 17.34. The battery voltage is 9.0 V.

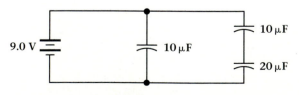

Figure 17.34   Problem 17.46.

17.47 Calculate the voltage across the 10-μF capacitor in the circuit of Fig. 17.35. The battery voltage is 9.0 V.

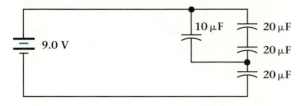

Figure 17.35   Problem 17.47.

17.48 A 15-μF capacitor is joined in series with a 25-μF capacitor and a 12-V battery. (a) What is the potential difference across each capacitor? (b) What is the charge on each capacitor? (c) How much energy is stored by each capacitor?

17.49 Suppose the two capacitors of Problem 17.48 were joined in parallel across the battery. (a) What would be the potential difference across each? (b) What would be the charge on each capacitor? (c) What would be the energy stored on each?

Section 17.9   Internal Resistance of a Battery

17.50 A 9.0-V battery delivers 170 mA to a 47-Ω load. (a) What is the internal resistance of the battery? (b) What is its terminal potential difference when joined to this load?

17.51 A certain battery has a terminal potential difference
• of 9.60 V when connected across a 200-Ω load and a terminal potential difference of 10.30 V when connected across a 300-Ω load. What is the emf of this battery?

17.52 (a) Find the terminal potential difference of the battery in Fig. 17.36, given that the internal resistance r is 375 Ω and the load resistance R_L is 1500 Ω. (b) What is the TPD if the load becomes 750 Ω? (c) How much power is delivered to the 750-Ω load?

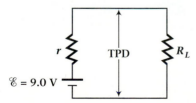

Figure 17.36   Problem 17.52.

*Section 17.10   Home Power Distribution

17.53 A single household circuit operates at 115 V and is protected by a 15-A circuit breaker. How many 100-W lamps can be operated simultaneously without tripping the circuit breaker?

17.54 A household circuit is wired for 115 V with a 15-A circuit breaker. (a) Can a 1500-W toaster, a 2200-W griddle, a 960-W iron, and a 300-W lamp be operated one at a time on this circuit? (b) Can any of them be operated simultaneously?

17.55 A particular household circuit is fused for 20 A at 115 V. (a) Can the circuit carry a 1200-W blow dryer? (b) Can it carry two dryers at the same time?

Additional Problems

17.56 Two cylindrical bars, each with diameter of 2.30 cm,
• are welded together end-to-end. One of the original bars is copper and is 0.370 m long. The other bar is iron and is 0.185 m long. What is the resistance between the ends of the welded bar at 20°C?

17.57 An elevator in a 20-story building is used to raise a
• 7000-N load 60 m. How much would it cost to raise the load if electricity costs $0.078/kWh? Assume that the elevator system is 50% efficient, that is, that the energy expended in raising the load is half of the total energy consumed.

17.58 A slide projector operating on a 120-V circuit has a
• 75-W lamp and a 1/256-hp motor fan. How long can the projector and fan run before using one cent worth of electrical energy if electricity costs $0.076/kWh?

17.59 A 120-V motor draws a current of 2.0 A while lifting
• a load at a speed of 0.65 m/s. If the efficiency for transforming electrical energy into mechanical energy is 62%, what mass is being lifted?

17.60 What is the current through the 10-Ω resistor of
• Fig. 17.37?

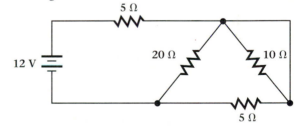

Figure 17.37   Problem 17.60.

17.61 Find the equivalent resistance of the network shown
• in Fig. 17.38.

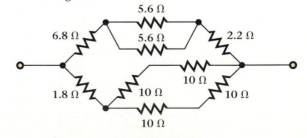

Figure 17.38   Problem 17.61.

17.62 A student has three capacitors of 10 μF each. How
• many different combinations of capacitance can he
make and what are their values?

17.63 An 1100-W iron, a 1200-W frying pan, and a lamp
• are all connected to a 120-V household circuit fused
for 20 A. (a) What is the maximum wattage lamp
that can be used simultaneously with the iron and
the frying pan? (b) What would happen if a 60-W
lamp were used? (c) A 200-W lamp?

17.64 A Wheatstone's bridge (Fig. 17.39) is used to
• measure resistance. When the bridge circuit is balanced, there is no current through the galvanometer
and the voltage from point A to point C is zero.
Verify that the current through the galvanometer
vanishes when $R_x = R_2R_3/R_1$.

17.65 The galvanometer of a Wheatstone's bridge (Fig.
• 17.39) is disconnected by opening switch S. If the
resistor values are $R_1 = 10.0$ Ω, $R_2 = 5.00$ Ω,
$R_3 = 20.0$ Ω, and $R_x = 9.00$, what is the potential
difference from point A to point C if the battery
potential is 12.0 V?

17.66 A Wheatstone's bridge similar to Fig. 17.39 is balanced for an unknown resistor R_x. What is R_x if
$R_1 = 100.0$ Ω, $R_2 = 20.0$ Ω, and $R_3 = 18.3$ Ω?

17.67 Calculate the current in the 50-Ω resistor in the
•• network of Fig. 17.40.

17.68 (a) Calculate the potential difference between points
• A and B in Fig. 17.41. (b) Calculate the power delivered to the 15-Ω resistor.

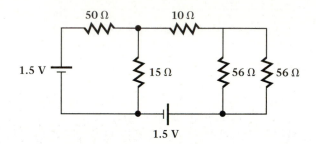

Figure 17.40 Problem 17.67.

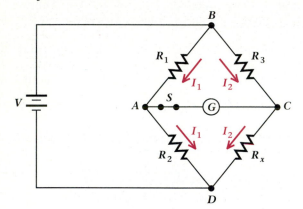

Figure 17.39 Problems 17.64, 17.65, and 17.66.

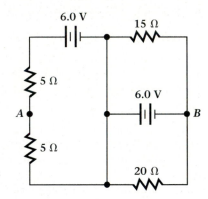

Figure 17.41 Problem 17.68.

ADDITIONAL READING

Frensley, W. R., "Gallium Arsenide Transistors." *Scientific American,* August 1987,
p. 80.

Heilblum, M., and L. T. Eastman, "Ballistic Electrons in Semiconductors." *Scientific American,* February 1987, p. 102.

Kaner, R. B., and A. G. MacDiarmid, "Plastics That Conduct Electricity." *Scientific
American,* February 1988, p. 106.

Müller, K. A., and J. G. Bednorz, "The Discovery of a Class of High-Temperature
Superconductors." *Science,* September 4, 1987, p. 1133.

Tinkham, M., "Special Issue: Superconductivity." *Physics Today,* March 1986,
p. 22.

18

Magnetism

18.1   Magnets and Magnetic Fields

18.2   Oersted's Discovery: Electric Current Produces Magnetism

18.3   Magnetic Forces on Electric Currents

18.4   Magnetic Forces on Moving Charged Particles

*18.5   The Cyclotron

18.6   Magnetic Field Due to a Current-Carrying Wire

18.7   Magnetic Force on a Current Loop

18.8   Galvanometers, Ammeters, and Voltmeters

*18.9   Ampère's Law

*18.10   Magnetic Materials

A WORD TO THE STUDENT

In this chapter we present some of the fundamental experiments and laws showing the relationship between electricity and magnetism. You should pay special attention to the nature of the force exerted on moving charges by a magnetic field. In addition, you need to understand the way in which an electric current produces a magnetic field. We will examine the magnetic field around two important current-carrying shapes: a long straight wire and a circular loop.

Together, the topics covered in Chapters 15–17 and those covered in Chapters 18–20 form the subject area of electromagnetism. The laws of electromagnetism apply to the design of such things as electric motors, meters, loudspeakers, and televisions. We will see in Chapter 19 that all the effects of electromagnetism can be described, in essence, by one unified set of four basic equations.

Magnetism has been known for thousands of years, dating back to the discovery recorded by the ancient Greeks that the naturally occurring black mineral called lodestone could attract iron objects.* Very little else was recorded until 1269, when Petrus Peregrinus de Maricourt, writing to a friend, gave a clear description of the magnetic compass, which had been developed during the eleventh and twelfth centuries. De Maricourt also recounted his own magnetic research. He observed that spherically shaped lodestones have two special points called "poles," which he located by laying bits of iron wire on the stone. The lines drawn along these wires intersected at two points that he named the *north pole* and the *south pole* in analogy with the celestial sphere. He also observed the attraction of unlike poles and repulsion of like poles. These ideas were incorporated in the work of William Gilbert, who published the results of his own experiments in 1600. Gilbert was the first to realize that the earth itself acts as a magnet, and he clearly stated the differences and similarities known at that time between electricity and magnetism.

By the beginning of the nineteenth century, a large amount of experimental information about the nature of electricity and magnetism had been accumulated. The discoveries and ideas of Gilbert, Franklin, Coulomb, Volta, and many others were well known. The similarities between electrical and magnetic attraction, the knowledge that compass needles on ships struck by lightning sometimes changed polarity, and experiments like Franklin's magnetization of needles by passing an electric discharge through them—all hinted at a possible connection between electric and magnetic behavior. But magnetism and electricity were considered to be two independent phenomena until early in 1820. In that year Hans Christian Oersted, a professor of physics at the University of Copenhagen, discovered that an electric current could indeed affect a magnet.

Today most people are familiar with magnetic compasses and permanent magnets. Electromagnets, which are perhaps less familiar, are found in such places as motors, tape recorders, and power plants. We now recognize the connection between electricity and magnetism as so fundamental that we no longer refer to the force of electricity and the force of magnetism, but instead speak of a single force of *electromagnetism*.

Magnets and Magnetic Fields

18.1

A compass needle is simply a slender magnet that is supported at its center so it can rotate freely. If there are no other magnets nearby, the needle lines up in an approximately north-south direction. The end of the needle that points toward the north is the north-seeking pole, or simply the *north pole*. The other end of the needle is the *south pole*. Larger bar magnets

---

*Lodestone is found in most parts of the world and is usually in the form of magnetite (Fe_3O_4). Large deposits were found in Asia Minor, near two ancient towns named Magnesia. These rocks became known as magnesia rocks and later as magnets.

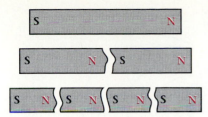

Figure 18.1 When a magnet is broken in two, each piece contains both a north and a south pole. Isolated magnetic poles (monopoles) are not observed to exist.

exhibit the same behavior. For this reason, the ends of bar magnets are often marked as north and south poles.

If two bar magnets are brought near each other, they exert forces on each other. If the north pole of one is brought near the north pole of the other, the force between them is repulsive. Similarly, if the south pole of one is brought near the south pole of the other, the force is also repulsive. But if the north pole of one is brought near the south pole of the other, the force between them is attractive. Thus we say that *like poles repel* but *unlike poles attract.*

In showing both attractive and repulsive behavior, magnetic poles are somewhat like electric charges. But don't take the analogy too far, because isolated magnetic poles do not exist. If a bar magnet is broken in two, you do not get isolated north and south poles; instead you get two new magnets, each having a north and a south pole (Fig. 18.1). If these pieces are in turn broken in two, each new piece contains both a north and a south pole. Magnets of this type are called **magnetic dipoles** because they have two poles.* Single magnetic poles, called monopoles, have not been observed. Although some theories suggest that isolated magnetic monopoles exist, no convincing experiments have verified their existence.

When a compass is brought near a magnet, the compass needle deflects. The deflection of the needle depends on the location of the compass relative to the magnet. We explain this behavior by saying that a **magnetic field** exists in the region surrounding the magnet. This magnetic field exerts a torque on the compass needle, causing it to rotate into the direction of the field. (As we shall see in the next section, we can determine the magnitude of the field from the magnitude of the torque acting on the magnetic dipole.) Thus the direction of the magnetic field (denoted by **B**) at any given point in space is defined as the direction indicated by the north pole of a compass needle when placed at that point. We can map out the lines of magnetic field around a bar magnet (or any other magnet) by placing many small compass needles in the neighborhood of the magnet and observing their orientations (Fig. 18.2a). We can then sketch the magnetic field by drawing lines tangent to the directions of the needles. Note that the field is directed away from a north pole and toward a south pole (Figs. 18.2b and 18.3).

Figure 18.2 (a) The direction of the magnetic field near a magnet is shown by the directions of small compass needles. (b) Representation of the field lines around a bar magnet. (c) Iron filings reveal the shape of the field near a bar magnet.

---

*Dipole magnets have the simplest configuration, but other arrangements of poles are possible, including quadrupole, octapole, and so on, all being multiples of dipoles.

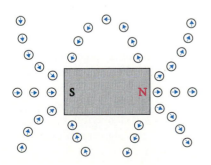

(a)

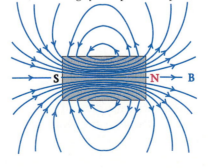

(b)

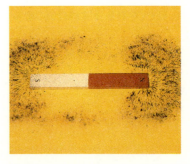

(c)

(a)

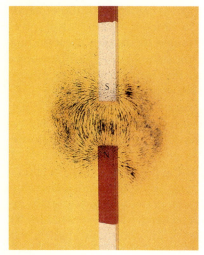

(b)

Figure 18.3   Iron filings reveal the shape of the magnetic field around (a) a horseshoe magnet and (b) two bar magnets.

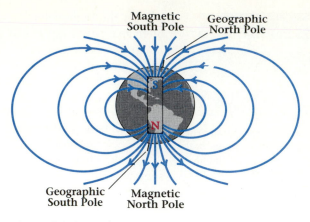

Figure 18.4   The earth behaves like a giant magnetic dipole with a south magnetic pole near the north geographic pole. The magnetic dipole is tilted approximately 11° from the earth's rotation axis.

Some objects, such as iron paper clips and nails, do not usually attract each other. But when they are brought near a magnet they become strongly attracted to it and to each other. We say they are ferromagnetic* (from the Latin word *ferrum* meaning iron). In the presence of the magnetic field they become magnetically polarized; that is, they behave as tiny magnets themselves. For this reason, when iron filings are sprinkled on a sheet of paper held close to a magnet, they orient themselves and clump together along the lines of magnetic field, revealing the shape of the field (Fig. 18.2c).

As Gilbert first pointed out, the earth itself is a magnetic dipole. It is the earth's magnetic field that causes compass needles to line up along the approximately north-south direction. However, since the north pole of a compass is attracted toward the geographic north, the magnetic pole that is located in that region is in reality a south magnetic pole (Fig. 18.4). Note also that the earth's magnetic poles do not coincide exactly with its geographic poles. For this reason, the magnetic north indicated by a compass is not quite the same as the true geographic north.

Oersted's Discovery: Electric Current Produces Magnetism

18.2

Oersted's discovery is thought to be the only case of a major scientific discovery being made during a lecture to students. While demonstrating an electrical experiment before a class of advanced students, Oersted positioned a wire near to and parallel with a compass needle. When he passed a current through the wire, the compass acted as if a magnet had been brought nearby. Oersted immediately recognized the significance of this behavior, but did not pursue the matter at the time.

*Ferromagnetism is discussed in greater detail in Section 18.10.

(a)

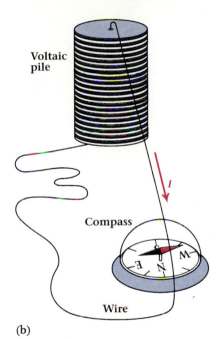

Voltaic pile

Compass

Wire

(b)

Figure 18.5 (a) Hans Christian Oersted (1777–1851). (b) Experimental setup for Oersted's experiment. A current through the wire caused the compass needle to deflect.

Three months later, after constructing a more powerful battery, Oersted extended his investigation. Using an improved experimental setup, similar to that shown in Fig. 18.5(b), Oersted conducted an extensive series of experiments. He connected a battery of some 20 copper-zinc cells to a straight section of the wire, which was held horizontally above the compass needle and parallel to it. When Oersted closed the circuit, the compass needle rotated so that the end nearest the negative side of the battery deflected to the west. The effect decreased as he moved the wire away from the compass. However, no noticeable change in the effect occurred when glass, metal, wood, water, or stone were placed between the wire and the compass. In July 1820, Oersted announced his discovery in a brief four-page pamphlet in Latin.

A flood of new research and discovery followed Oersted's announcement. Within three months, André-Marie Ampère (1775–1836) formulated Oersted's results mathematically and added some results of his own. He introduced what is now known as the "right-hand rule" for determining the direction of magnetic field due to a current.

In Oersted's experiment, the presence of an electric current in the wire produces a torque, causing the magnetic compass needle to rotate. Since the compass needle is a magnetic dipole (that is, two magnetic poles separated by a fixed distance), we can describe it with a **magnetic dipole moment μ** similar to the electric dipole moment **p** defined in Chapter 15. The deflection of the compass needle is due to the interaction of the magnetic moment with the magnetic field **B**. The behavior of a magnetic dipole in a magnetic field can be described in exactly the same way that we described the behavior of an electric dipole in an electric field. The torque on the compass needle is proportional to the product of its magnetic dipole moment **μ** and the magnetic field **B**,

$$\tau = \mu B \sin \theta, \tag{18.1}$$

where θ is the angle between **μ** and **B**, in analogy with Eq. (15.5). Here the magnetic field is caused by the current in the wire.

The magnetic field lines due to a current-carrying wire form concentric circles about the wire (Fig. 18.6, p. 530). The direction of the magnetic field about the wire is given by a right-hand rule (Fig. 18.7, p. 530): *If the thumb of the right hand is taken as the direction of current, the curled fingers point in the direction of the magnetic field about that current.* If we trace out the lines of the magnetic field, we see that they are continuous, and do not terminate on magnetic charges or poles the way electric field lines terminate on electric charges. This observation implies that we do not have single magnetic poles (monopoles) that exist independently of other poles in the same way that we have isolated positive and negative electric charges. If magnetic monopoles do not exist, then we can make a statement for magnetic fields similar to Gauss's law for electrostatics: **The net magnetic flux through any (real or imaginary) closed surface is zero.** This statement is known as **Gauss's law for magnetism.** We define the magnetic flux as the product of the magnetic field **B** and the element of area ΔA projected in the direction of the field, summed over the entire surface (Fig. 18.8, p. 530). This flux is analogous to the electric flux defined in

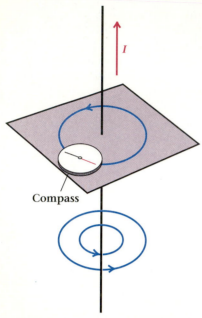

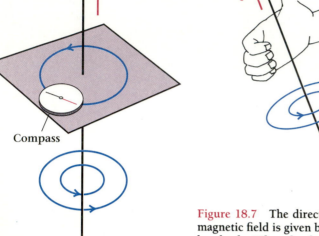

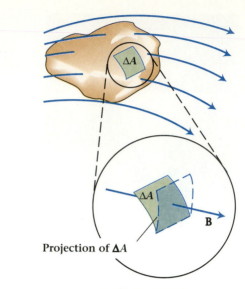

(a)

Figure 18.7   The direction of the magnetic field is given by the right-hand rule: When the thumb of the right hand points in the direction of the conventional (positive) current, the fingers curl around the wire in the direction of the magnetic field.

Figure 18.8   The magnetic flux through an element of surface **Δ**A is the product of the magnetic field **B** and the area **Δ**A projected along the direction of the field. The total magnetic flux is the sum of these products over the entire surface.

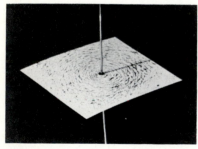

(b)

Figure 18.6   (a) Magnetic field lines around a long straight wire carrying an electric current form concentric circles about the wire. (b) Iron filings align with the magnetic field, indicating its circular symmetry.

Chapter 15. Another way of expressing Gauss's law for magnetism is to say that magnetic field lines do not start or stop at any point in space, but form closed loops. Thus any magnetic field line entering a closed surface must also leave that surface, so the net flux is zero.

As mentioned in Section 18.1, some theories of modern physics predict the existence of magnetic monopoles. Because these theories have been successful in many other areas, it is important to test them in as much detail as possible. Accordingly, several ingenious, highly sensitive experiments have been carried out in the search for magnetic monopoles. To date, no reproducible experiments have been conducted that verify the existence of monopoles. Thus the present consensus is that isolated magnetic poles do not exist.

Magnetic Forces on Electric Currents

18.3

Oersted's experiment showed that a wire carrying an electric current exerts a force on a magnet. It follows from Newton's third law that a magnet must also exert a force on a current-carrying wire. For example, consider a current-carrying wire in a region of uniform magnetic field (Fig. 18.9). We observe that the wire experiences a magnetic force that is always

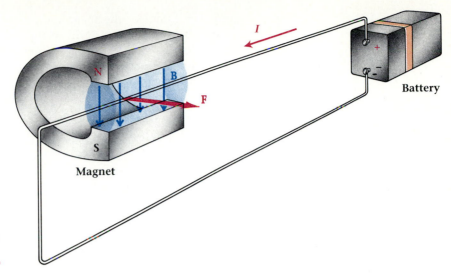

Figure 18.9 The magnetic force on a current-carrying wire in a magnetic field is perpendicular to the field **B** and to the length of the wire in the field. The force is maximum when the wire is perpendicular to the field.

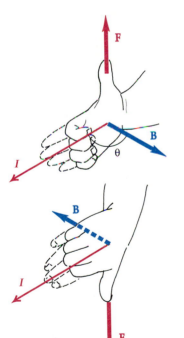

Figure 18.10 The direction of the magnetic force on a current-carrying wire is perpendicular to both the current I and the magnetic field **B**. When a right hand is held so that the fingers can be curled from the direction of I into the direction of **B**, the thumb points in the direction of the force.

perpendicular to the magnetic field and also perpendicular to the length of the wire in the field. Thus the force is perpendicular to the motion of the electric charges in the wire—that is, to the current. If we change the direction of the wire relative to the direction of the field, we find that the force is maximum when the wire is perpendicular to the field and that the force is zero when the wire is parallel to the field.

The magnitude of the force on a section of wire of length l carrying a current I in the presence of a magnetic field **B** is given by

$$F = IlB \sin \theta, \qquad (18.2)$$

where θ is the angle between the direction of the current and the direction of the field. We emphasize that the force vector is perpendicular to both the current and the field. The direction of the force is that given by another right-hand rule: *When a right hand is held so that the fingers can be curled from the direction of the current into the direction of the magnetic field, the vector **F** points in the direction of the thumb* (Fig. 18.10).

The effect of the magnetic field on a current-carrying wire is quite different from the effect of an electric field on a charge, or a gravitational field on a mass. An electric field exerts a force on a charged particle that is along the direction of the field (or opposite, if the charge is negative); a gravitational field exerts a force on a mass along the direction of the gravitational field. But a magnetic field exerts a force on a current-carrying wire that is perpendicular to both the direction of the magnetic field and the direction of the current.

If the current in Eq. (18.2) is measured in amperes, length in meters, and force in newtons, then the magnetic field has units of newtons per ampere-meter. In the SI system, this particular combination of units is given the name **tesla**, abbreviated as T.* The tesla is a relatively large

---

*The unit for magnetic field is named in honor of Nikola Tesla (1856–1943), an engineer and inventor who was largely responsible for the adoption of the alternating-current electrical distribution system that is in worldwide use today.

unit. For example, the magnitude of the earth's magnetic field near its surface is $\approx 5 \times 10^{-5}$ T. Even the field near the end of a common bar magnet is only about 10^{-2} T, while the largest fields available from superconducting magnets are on the order of 5 T.

Example 18.1

Force on a current-carrying wire in a magnetic field.

A straight wire is placed between the poles of a laboratory magnet that produces a uniform field of 1.00 T over a region that is 0.250 m wide. The wire is perpendicular to the direction of the field and, when connected to a source of electricity, carries a current. What must the current be if the field exerts a force of 9.80 N on the wire?

Solution   Equation (18.2) gives the force on the wire as

$$F = IlB \sin \theta.$$

To obtain the current required for a given force, we rearrange the equation to get

$$I = \frac{F}{lB \sin \theta}.$$

When we insert the numerical values of $F = 9.80$ N, $l = 0.250$ m, $B = 1.00$ T, and $\sin \theta = \sin 90° = 1$, we find the current:

$$I = \frac{9.80 \text{ N}}{0.250 \text{ m} \times 1.00 \text{ T} \times 1} = 39.2 \text{ A}.$$

If two parallel wires each carry a current, each wire exerts a force on the other. To understand what is happening, consider one wire as setting up a magnetic field that interacts with the current of the second wire. Figure 18.11 shows two parallel wires carrying currents I_1 and I_2 in the same direction. The direction of the magnetic field about the first wire is given by the right-hand rule (Fig. 18.7). Since the two wires are parallel, the magnetic field due to the first wire is perpendicular to the second wire and thus to current I_2. The resulting force on the second wire is directed toward the first, as shown in the figure. This agrees with the right-hand rule, since for rotation of the fingers from the direction of current I_2 toward the direction of the field **B**, the thumb points toward the first wire. Thus the force between the wires is attractive.

By Newton's third law, the first wire experiences an attractive force of the same magnitude toward the second wire. The wires are thus drawn toward each other. We would reach this same conclusion by considering the interaction of the current I_1 in the field due to current I_2. If the two currents had been directed opposite to each other, that is, antiparallel, then the force between the two wires would be repulsive and they would be pushed apart.

Experiments on current-carrying wires show that at any point, the field due to a single wire is proportional to the current I in that wire. Similarly, measurements show that the force between two parallel wires depends on the product of their currents. In fact, the force between two

Figure 18.11 Two wires carrying parallel currents attract each other. The current in each wire produces a magnetic field that interacts with the current in the other wire, resulting in an attractive force.

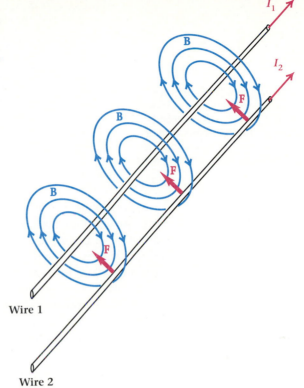

Wire 1

Wire 2

current-carrying wires is used to define the ampere in terms of mechanical units. Once the ampere has been established, then the coulomb (ampere-second) and the volt (joule/coulomb) can be set in accordance with the size of the ampere. Similarly, the unit for magnetic field, the tesla, depends on the size of the ampere. This dependence is discussed in more detail in Section 18.6.

Magnetic Forces on Moving Charged Particles

18.4

Moving charged particles constitute a current, whether they move through a wire or through free space. Thus a moving point charge q experiences a force due to a magnetic field **B** that is similar to the force on a current-carrying wire. Experiments confirm that this force is given by

$$F = qvB \sin \theta, \tag{18.3}$$

where v is the speed of the moving charge, B is the magnitude of the magnetic field, and θ is the angle between the velocity of the moving charge and the direction of the field. The force is proportional to the component of **v** perpendicular to **B**, that is, $v_\perp = v \sin \theta$ (Fig. 18.12). The component of **v** parallel to **B** does not result in a force on the charge. The direction of the force is always perpendicular to both **v** and **B**. For a

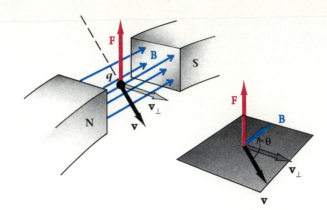

Figure 18.12 The magnitude of the force on a particle of charge q moving with a velocity v in a magnetic field **B** is $F = qvB \sin \theta$ and is directed according to the right-hand rule. If the charge is negative, the particle experiences a force of the same magnitude in the opposite direction.

positive charge, the direction of the force is given by the same right-hand rule we used to find the force on a current-carrying wire because the direction of the velocity is the same as the direction of the current. However, if the charge is negative, the force is in the direction opposite to that given by the right-hand rule. Thus the force exerted on a moving negative charge by a magnetic field is equal in magnitude but opposite in direction to the force exerted on a positive charge of the same magnitude moving with the same velocity. Note that Eq. (18.3) may be used as the defining equation for the magnetic field, because F, q, v, and θ may be measured by independent means.

We can show that Eq. (18.2), for the force on a wire segment of length l carrying a current I perpendicular to a magnetic field **B**, is equivalent to Eq. (18.3), for the force on a single charge q moving with speed v perpendicular to a magnetic field **B**. Consider a wire segment of length l that contains a total moving charge Q (Fig. 18.13). The current due to the charge Q passing a given point in the wire is

$$I = \frac{Q}{t},$$

where t is the time for an individual charge to go a distance l. This time t depends on the speed v of the charge and is $t = l/v$. Consequently, the current is

$$I = Qv/l.$$

From Eq. (18.2), the force on a length l of wire with current I is

$$F_w = IlB = (Qv/l)lB = QvB,$$

where Q is the total charge moving in that segment of wire at any instant. The total charge Q is the sum of all the individual moving charges q. Thus the force on the wire is the sum of the forces on all of the individual charges, and the force on each charge is $F = qvB$, in accord with Eq. (18.3).

Suppose we inject a positively charged particle with mass m and charge q into an otherwise empty region of uniform magnetic field **B**. If the particle's velocity **v** is perpendicular to **B**, what is the particle's subsequent motion? According to Eq. (18.3), the particle experiences a force

$$F = qvB,$$

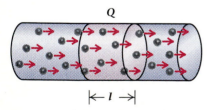

Figure 18.13 A wire segment of length l contains a total charge Q. The motion of those charges constitutes a current.

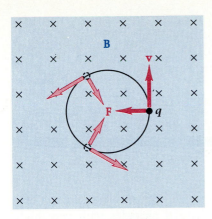

Figure 18.14   A magnetic field **B** causes a particle of charge q moving with a velocity **v** in a plane perpendicular to the field to move in a circular orbit. The ✕'s represent the tail-ends of magnetic field lines directed into the page.

since **v** is perpendicular to **B**. The particle moves in accordance with Newton's laws of motion under the influence of this force. Thus the particle undergoes an acceleration of magnitude

$$a = \frac{F}{m} = \frac{qvB}{m}.$$

Because the force is always perpendicular to both **v** and **B**, the acceleration is also perpendicular to **v** and **B**. Since the acceleration is always perpendicular to the velocity, the path of the particle is curved, as we learned in Chapter 5. In fact, since the magnitude of the acceleration is constant and **a** is perpendicular to the instantaneous velocity, the particle's path is a circle (Fig. 18.14).

The centripetal acceleration required to keep the particle in a circular orbit is provided by the interaction of the moving charge with the magnetic field. If we equate the expression for centripetal acceleration with the acceleration due to the magnetic field, we get

$$a_c = \frac{v^2}{r} = \frac{qvB}{m}.$$

This equation can be rearranged to give

$$r = \frac{mv}{qB}. \tag{18.4}$$

Thus the particle moves in a circular path whose radius r depends on the momentum mv of the particle, its charge, and the magnetic field.

Example 18.2

Motion of a proton in a magnetic field.

A proton, having a mass of 1.67×10^{-27} kg and an electric charge of 1.6×10^{-19} C, moves in a magnetic field of 5.0×10^{-3} T. What is the speed of the proton if it moves in a circular arc of radius 0.50 m and if the plane of the orbit is perpendicular to the magnetic field?

Solution   We can rearrange Eq. (18.4) to give

$$v = \frac{rqB}{m},$$

where all quantities on the right are known. Thus

$$v = \frac{0.50 \text{ m} \times 1.6 \times 10^{-19} \text{ C} \times 5.0 \times 10^{-3} \text{ T}}{1.67 \times 10^{-27} \text{ kg}}$$

$$= 2.4 \times 10^5 \text{ m/s}.$$

Since only SI units were used, we expect that the speed of the particle should be in meters per second. Let us check. The units of tesla are

$$\text{tesla} = \frac{\text{N}}{\text{A} \cdot \text{m}} = \frac{\text{kg} \cdot \text{m/s}^2}{(\text{C/s})\text{m}} = \frac{\text{kg}}{\text{C} \cdot \text{s}}.$$

The units of rqB/m become

$$\frac{\text{m·C·T}}{\text{kg}} = \frac{\text{m·C·}\dfrac{\text{kg}}{\text{C·s}}}{\text{kg}} = \text{m/s}.$$

As expected, the speed is in units of meters per second.

The Cyclotron

*18.5

Electrostatic accelerators, such as the Van de Graaff accelerator described in Chapter 16, can generate beams of charged particles with very high velocities. These beams of energetic charged particles have many uses in research, industry, and medicine. Once such a beam emerges from an accelerator, we can change its direction most easily by sending it through a region of intense magnetic field. By adjusting the magnitude of this field we can control the amount of deflection of the beam. Thus beam-steering magnets may be used to direct the beam to different targets.

Magnetic fields also play a direct role in the operation of many accelerators, including the **cyclotron**, invented by Ernest O. Lawrence (1901–1958). In 1930 Lawrence and M. S. Livingston (1905–1986) constructed an accelerator that used a magnetic field to bend the paths of the particles into nearly circular orbits. A simplified diagram of a cyclotron (Fig. 18.15) includes two semicircular cavities (called "dees" because of their shape), an ion source for producing the charged particles, an oscillating voltage source, and the magnets that produce the magnetic field.

A properly timed oscillating voltage source makes the electric potential difference between the dees change direction each time the charged particles pass through the gap between them, so that the particles encounter an accelerating voltage each time they pass through the gap. A particle with speed v moves in a circular path of radius r within one of the dees. After being accelerated across the gap it moves with a greater speed and a greater radius of curvature. However, we can show that the time required to complete a revolution in the field is constant.

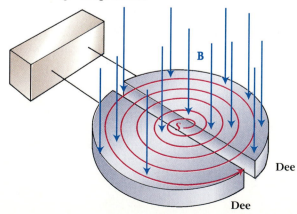

Oscillating voltage source

Figure 18.15   A simplified diagram of a cyclotron. The ion source S is located within the dees and the magnetic field is perpendicular to the plane of the dees.

The time T required to complete one orbit (the period) is the path length divided by the speed:

$$T = \frac{2\pi r}{v}.$$

Solving Eq. (18.4) for v and substituting gives us

$$T = \frac{2\pi r}{qBr/m} = \frac{2\pi m}{qB}.$$

The period needed by the charged particles to complete an orbit is constant, regardless of the values of r and v.* A constant period means that the frequency is constant. If the electric potential between the dees oscillates at this same frequency, then the charged particles accelerate each time they cross the gap. When subjected to many of these accelerations, the particles can attain very high velocities and thus possess large kinetic energies.

Other accelerators have been invented and developed since the first cyclotron. All of them use magnetic fields to guide the accelerated particles in the desired paths. New acceleration techniques have been perfected, but the circular shape of most large accelerators reflects Lawrence's original idea. Further discussion of accelerators appears in Chapter 31.

Example 18.3

Frequency of a cyclotron.

A cyclotron operating at a magnetic field of 1.6 T is used to accelerate protons (mass $= 1.67 \times 10^{-27}$ kg and charge $= 1.6 \times 10^{-19}$ C). What must be the frequency of oscillation of the accelerating voltage?

Solution   The frequency is the reciprocal of the period. Using the equation for T just given, we get

$$f = \frac{1}{T} = \frac{qB}{2\pi m} = \frac{1.6 \times 10^{-19}\ \text{C} \times 1.6\ \text{T}}{2\pi \times 1.67 \times 10^{-27}\ \text{kg}},$$

$$f = 2.4 \times 10^7\ \text{Hz} = 24\ \text{MHz}.$$

Frequencies of 24 MHz are easy to obtain with conventional electronic components.

Magnetic Field Due to a Current-Carrying Wire

18.6

We have already shown (Section 18.3) that two current-carrying wires exert forces on each other. We can use this effect to determine the relationship between a magnetic field and the current generating that field. Experiments have shown that the current I in a small segment of wire Δl

---

*At speeds approaching the speed of light, the period changes because of effects discussed in Chapter 24.

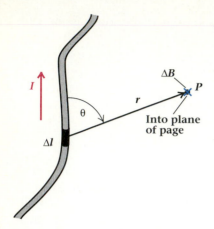

Figure 18.16 The magnetic field ΔB at point P due to a current I in a length of wire Δl is given by the Biot-Savart law, Eq. (18.5).

(Fig. 18.16) is responsible for a contribution ΔB to the magnetic field around the wire,

$$\Delta B = k' \frac{I\,\Delta l\,\sin\,\theta}{r^2}, \qquad (18.5)$$

where r is the distance from Δl to the point P at which the field is measured, θ is the angle between Δl and r, and k' is a proportionality constant. The increment of field ΔB is actually a vector, and is directed perpendicularly to both Δl and r. This expression for ΔB is known as the **law of Biot and Savart**.* It came from experimental observations and was first proposed by Biot and Savart in 1820, only a few months after Oersted's discovery.

Going from the contribution to the magnetic field ΔB given by Eq. (18.5) to an expression for the total magnetic field due to a current-carrying wire is maybe difficult mathematically, even with calculus. However, we can derive results for certain simple cases of practical importance, and the analysis of their situations may give you better insight into the physics than just a statement of formulas. So, let's consider the magnetic field due to a long straight current-carrying wire. Figure 18.17 shows a point A located a distance d from a long straight wire carrying a constant current I. The magnetic field at point A is the vector sum of all the elements $\Delta \mathbf{B}$ due to all segments Δl of the whole wire. From Eq. (18.5) and Fig. 18.16 we already know that the field $\mathbf{B}$ at point A is directed into the plane of the paper. Summing up Eq. (18.5) for the contribution to the field due to each of n segments of wire, we get

$$B = k'I \sum \frac{\Delta l_n \sin\,\theta_n}{r_n^2}.$$

If the wire is very long compared with d, the methods of calculus reduce the summation to a simple result: The magnitude of the field B at a distance

---

*Jean-Baptiste Biot (1774–1862) and Felix Savart (1791–1841) discovered this law.

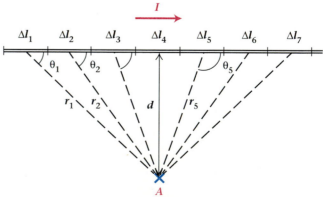

Figure 18.17 The magnetic field at A is due to contributions from all elements Δl of a long current-carrying wire.

d from a long wire is

$$B = k' \frac{2I}{d}.$$ (18.6)

This relation for the field due to a long wire was discovered experimentally by Biot and Savart a few months before they discovered Eq. (18.5). For this reason, Eq. (18.6) is also known as the Biot-Savart law.

The field due to a long wire always lies in a plane perpendicular to that wire (Fig. 18.6). If another wire of length l parallel to the first wire carries a current I_2 in the same direction, the attractive force on that wire is

$$F = I_2 l B$$

in accord with Eq. (18.2). If the two wires are a distance d apart, then according to Eq. (18.6) the force per unit length becomes

$$\frac{F}{l} = k' \frac{2I_1 l_2}{d}.$$

If the two currents are equal, this equation becomes

$$\frac{F}{l} = k' \frac{2I^2}{d}.$$

The force per unit length, which is an easy quantity to measure, depends on the separation d, the current I, and the proportionality constant k'. In the SI system we choose $k' = 10^{-7}$ N/A² so that the equation we just derived becomes the defining equation for the ampere. That is, **when the attractive force per unit length of two long parallel wires placed one meter apart and carrying the same current is 2×10^{-7} N/m, that current is defined to be exactly one ampere.**

The values of k' and of the ampere are not independent. This particular value of k' is chosen to make the ampere as just defined equal to the ampere that was in use prior to the adoption of the SI standard units. Current balances based on the principles described here provide sensitive and practical instruments for measuring currents and calibrating other current meters. After the ampere is established, we can readily set the unit of charge (the coulomb) and the unit of electric potential (the volt). The choice of the ampere for the base unit of electricity in the SI system is an operational choice based on the relative ease of measurement and observation of the forces involved and on the precision with which they can be determined.

The constant k' plays a role similar to that of the constant k that we encountered in Coulomb's law. For reasons similar to those that led us to replace k by $1/4\pi\epsilon_0$, it is common practice to introduce a constant called the magnetic permeability of free space μ_0, related to k' through the equation

$$\mu_0 = 4\pi k' = 4\pi \times 10^{-7} \text{ N/A}^2.$$ (18.7)

Hence,

$$k' = \frac{\mu_0}{4\pi}.$$

Example 18.4

Magnetic field of a long straight wire.

What is the magnitude of the magnetic field at a distance of 3.0 m from a long straight wire that carries a dc current of 15 A?

Solution   We can calculate the magnitude of the field from Eq. (18.6), where $I = 15$ A, $d = 3.0$ m, and $k' = 10^{-7}$ N/A^2:

$$B = k' \frac{2I}{d} = 10^{-7} \text{ N/A}^2 \frac{2 \times 15 \text{ A}}{3.0 \text{ m}},$$

$$B = 1.0 \times 10^{-6} \text{ N/A·m}.$$

Thus

$$B = 1.0 \times 10^{-6} \text{ T},$$

where we have recognized that the combination of units N/A·m is the tesla. The field calculated here is about 2% as large as the earth's magnetic field at its surface.

Example 18.5

Magnetic field at the center of a current-carrying loop.

Calculate the magnetic field at the center of a loop of wire of radius R carrying a current I.

Solution   We can calculate the magnetic field from the law of Biot and Savart:

$$B = \frac{\mu_0}{4\pi} I \sum \frac{\Delta l_n \sin \theta_n}{r_n^2}.$$

At the center of the loop (Fig. 18.18), all elements of the loop are equidistant at $r_n = R$. The angle between Δl_n and R is 90°, so that $\sin \theta_n = 1$. The equation for the field becomes

$$B = \frac{\mu_0 I}{4\pi} \sum \frac{\Delta l_n}{R^2},$$

$$B = \frac{\mu_0 I}{4\pi R^2} \sum \Delta l_n.$$

The sum of the Δl_n is just the circumference of the loop, which is $2\pi R$. Thus the field becomes

$$B = \frac{\mu_0 I}{4\pi R^2} 2\pi R,$$

$$B = \frac{\mu_0 I}{2R}.$$

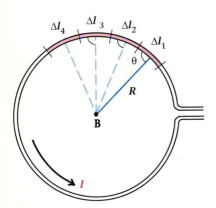

Figure 18.18   Example 18.5: A current loop of radius R carrying a current I. The magnetic field at the center of the loop is directed out of the plane of the loop, in accord with the right-hand rule.

Magnetic Force on a Current Loop

18.7

We have already discussed the effect of a magnetic field on a straight current-carrying wire. However, if the wire is not straight, but forms a loop instead, we must consider the force acting on each part of the loop. The behavior of current loops and the shapes of their magnetic fields are important for understanding electric generators, motors, and transformers, among many other electrical applications.

The simplest situation to analyze is a rectangular loop of wire in a region of uniform magnetic field **B** (Fig. 18.19a). For simplicity we have aligned the loop so that two of its edges (*a* and *c*) are perpendicular to **B**. Figure 18.19(b) shows a view of the same loop from the side. Note that edges *b* and *d* are not perpendicular to **B**. The normal (perpendicular) to the plane of the loop is represented by the vector **N**, which makes an angle θ with the magnetic field.

The force on the loop's near edge (*b*), of length *h*, is directed outward, perpendicular to **B** (Fig. 18.19a), as you can see by applying the right-hand rule (rotate *I* into **B**). Similarly, the force on the far edge (*d*) is directed inward, perpendicular to **B**. Because these two forces are equal and opposite and lie along the same line, there is no net torque on the loop due to sides *b* and *d*.

The forces on the other two edges (*a* and *c*) are also equal to each other and opposite in direction, but they are not directed along a common line, as seen in Fig. 18.19(b). The net effect is to exert a torque about a horizontal axis through the loop. The direction of the torque (clockwise in Fig. 18.19b) tends to rotate the loop so that its normal aligns along **B**.

From Eq. (18.2), the force on edge *a* is

$$F_a = IwB$$

and is directed upward. The force on edge *c* is

$$F_c = IwB$$

and is directed opposite to **F**$_a$. The torque due to **F**$_a$ about a horizontal axis through the middle of the loop is

$$\tau_a = \frac{h}{2} IwB \sin\,\theta,$$

while that due to **F**$_c$ is

$$\tau_c = \frac{h}{2} IwB \sin\,\theta.$$

Since both torques are in the same direction, the total torque is the sum of the individual torques,

$$\tau = IwhB \sin\,\theta,$$

or

$$\tau = IAB \sin\,\theta, \tag{18.8}$$

where *A* is the area of the loop and is equal to *wh*.

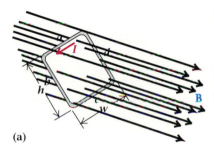

(a)

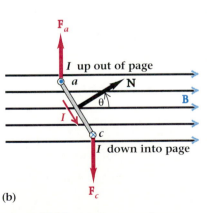

(b)

Figure 18.19   (a) A rectangular current loop in a magnetic field. (b) View of the same loop from the side, showing direction of forces along the edges *a* and *c*.

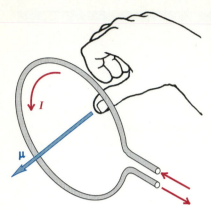

Figure 18.20 Orientation of the magnetic dipole moment of a current loop. The direction of **μ** is the same as the direction in which your thumb points if you curl the fingers of your right hand in the direction of positive current.

Equation (18.8) has the same form as Eq. (18.1), which gave the torque on a magnetic dipole due to an applied magnetic field. Thus the current loop is a magnetic dipole with magnetic moment

$$\mu = IA.$$

The direction of the magnetic moment is along the direction of **N**, the normal to the plane of the current loop. It is given by another right-hand rule (Fig. 18.20): *The direction of **μ** is the same as the direction in which your thumb points if you curl the fingers of your right hand in the direction of positive current.*

A current loop can consist of more than one single loop of wire. In this case the magnetic moment is proportional to the total number of loops or turns around a given area A. For example, if the wire makes five turns around an area A, then the magnetic moment is $5IA$. For N loops, the moment becomes

$$\mu = NIA. \tag{18.9}$$

That is, the magnetic moment depends on the number of turns of wire, the current in the wire, and the area of the loop. We chose a rectangular loop for its obvious simplicity, but our result is valid for any plane loop, no matter what its shape. The torque on a current-carrying loop is then $\tau = \mu B \sin \theta$, where μ is given by Eq. (18.9).

Example 18.6

Torque on a current loop.

A flat coil is made by wrapping wire around a 2-L bottle and then gently slipping it off. The resulting coil of 25 turns of wire has a radius of 5.5 cm. (a) What is the magnetic moment of the coil when it carries a current of 1.5 A? (b) What is the maximum torque exerted on the coil by the earth's magnetic field of 5.0×10^{-5} T when the coil conducts a current of 1.5 A?

Solution   (a) The magnetic moment may be computed from Eq. (18.9) as

$$\mu = NIA.$$

In this case $N = 25$, $I = 1.5$ A, and the area $A = \pi(5.5 \times 10^{-2} \text{ m})^2$. So

$$\mu = 25 \times 1.5 \text{ A} \times \pi(5.5 \times 10^{-2} \text{ m})^2 = 0.36 \text{ A} \cdot \text{m}^2.$$

(b) The maximum torque occurs when **μ** is perpendicular to **B**. It is

$$\tau = \mu B = 0.36 \text{ A} \cdot \text{m}^2 \times 5 \times 10^{-5} \text{ T} = 1.8 \times 10^{-5} \text{ m} \cdot \text{N}.$$

The magnetic field around a circular current loop is shown in Fig. 18.21. It is similar to the magnetic dipole field near a bar magnet (Fig. 18.2). Note that the magnetic field lines are all continuous loops having neither beginning nor end. In this respect the magnetic dipole is different

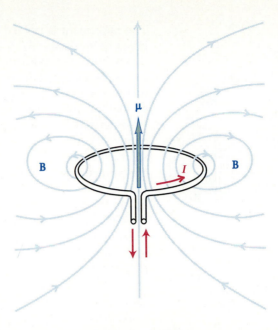

Figure 18.21 **Magnetic field due to a current loop. The magnetic moment μ points in the direction of a north magnetic pole.**

from the electric dipole, in which the field lines terminate on the electric charges.

Galvanometers, Ammeters, and Voltmeters

18.8

A practical instrument whose operation depends on the torque developed on a current-carrying coil is the **galvanometer**. Its most direct application is for sensitive measurement of electric current. Galvanometers are the basic component of many electrical meters. A particularly reliable galvanometer was built in 1880 by the French physicist Jacques Arsène d'Arsonval, who surrounded a cylindrical iron core with a movable coil placed between the poles of a magnet. When a current was passed through the coil, it produced a magnetic dipole whose interaction with the field of the permanent magnet created a torque on the coil. The coil turned through an angle that depended on the magnitude of the current causing the field. Precise measurements of the angle permitted precise measurement of the current I.

In 1888 Edward Weston used d'Arsonval's idea to construct the first portable ammeter that could be read directly from its scale. His design became standard and is still in use. Weston's design uses a horseshoe-shaped permanent magnet with iron pole pieces shaped to a cylindrical gap (Fig. 18.22). A cylindrical iron core, only slightly smaller than the gap, is fixed in the center between the pole pieces. A coil placed in the gap between the core and the pole pieces is supported on jewelled bearings. A tiny coil spring holds the coil in the zero position when no current is applied. Current is carried to the coil through the spring and the pivots.

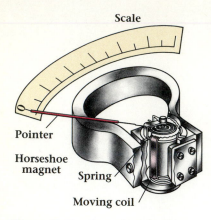

Scale

Pointer

Horseshoe magnet

Spring

Moving coil

Figure 18.22 Operation of a d'Arsonval-Weston meter movement. Current through the movable coil produces a magnetic field, resulting in a torque. The deflection of the coil is proportional to the current through it.

The movement is rugged enough for portable use while at the same time being quite sensitive.

A current-measuring instrument is usually called an **ammeter**. It measures the current passing through it. Although the heart of the ammeter is a galvanometer, the ammeter may be constructed and calibrated so that it can measure currents larger than the maximum current for full-scale deflection of the galvanometer movement. To make an ammeter, we connect a resistor in parallel with the galvanometer movement so that some of the current bypasses the galvanometer (Fig. 18.23a). By choosing the proper bypass resistor, an ammeter of any desired range can be constructed. Meters of this sort are often called by such names as milliammeter or microammeter, depending on their range.

An instrument for measuring potential difference, or voltage, is usually called a **voltmeter**. We can construct a voltmeter from a galvanometer movement by connecting a resistor in series with it to limit the current (Fig. 18.23b). Ideally, the voltmeter would measure the potential difference without allowing any current to pass through the meter. However, some current must pass in order to move the galvanometer coil. The resistor limits that current and sets the range of the meter. The maximum range of the voltmeter is the product of the current required for full scale deflection of the meter and the combined resistance of the galvanometer and the series resistor.

Example 18.7

Making a voltmeter from a galvanometer.

A student needs to make a voltmeter that measures 100 V full scale. The only galvanometer available has full-scale deflection at a current of 100 μA, and its resistance is 10 kΩ. What resistance is needed and how should it be connected to the galvanometer?

Solution To make a voltmeter from the galvanometer movement, the student needs to connect a resistor in series with the movement. For full-scale deflection at 100 V, the current through this particular meter must be 100 μA. The resistance of the voltmeter is then

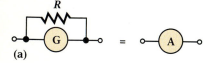

(a)

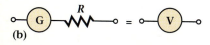

(b)

Figure 18.23 (a) A parallel connection of a resistor to a galvanometer to make an ammeter. (b) A series connection of a resistor to a galvanometer to form a voltmeter. Remember that the galvanometer itself also has a resistance.

$$R_m = \frac{V}{I} = \frac{100 \text{ V}}{100 \times 10^{-6} \text{ A}} = 10^6 \ \Omega,$$
$$R_m = 1000 \text{ k}\Omega.$$

The voltmeter resistance is the sum of the series resistance and the resistance of the galvanometer,

$$R_m = R + R_G.$$

The value of the series resistor is

$$R = R_m - R_G = 1000 \text{ k}\Omega - 10 \text{ k}\Omega,$$
$$R = 990 \text{ k}\Omega.$$

Thus the student can make a 100-V meter by connecting a 990-kΩ resistor in series with the galvanometer movement.

Ampère's Law

Soon after the discoveries of Oersted and of Biot and Savart, Ampère found a useful relation between electric currents and magnetic fields. This relation may be applied in highly symmetric problems to find the magnetic field more easily than by computing with the Biot-Savart law. In either case, the result is the same. In this respect, using Ampère's law is somewhat like finding the electric field by using Gauss's law rather than Coulomb's law. In cases that lack the proper symmetry, Ampère's law is not easily applied. It is always valid, even though the Biot-Savart law is needed sometimes for detailed calculation of the magnetic field.

According to Ampère's law, for an arbitrary closed curve around a current-carrying conductor (Fig. 18.24), the component of magnetic field tangent to the curve is related to the net current passing through the (imaginary) surface bounded by the curve. In particular, for each element Δl of the curve, we can calculate a contribution $B \, \Delta l \cos \theta$, where θ is the angle between $\mathbf{B}$ and Δl. **Ampère's law** states that the sum of these contributions around the entire loop is proportional to the net current I through the loop. That is,

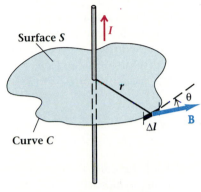

$$\sum B \, \Delta l \cos \theta = \mu_0 I. \qquad (18.10)$$

Figure 18.24 Illustration of Ampère's law. A closed curve C bounds a surface S. Ampere's law states that the sum of the contributions $B \, \Delta l \cos \theta$ around the curve is proportional to the net current I crossing the surface S.

We can easily apply Ampère's law for the case of the magnetic field about a long straight wire. We know already that the field is always perpendicular to the wire, since the lines of $\mathbf{B}$ make circles centered about the wire. Thus if we choose a circular path with the wire at its center (Fig. 18.25), then $\cos \theta = 1$ everywhere around the loop. Furthermore, B is constant around the loop, so that

$$\sum B \, \Delta l \cos \theta = B \sum \Delta l = B(2\pi r) = \mu_0 I,$$

where $\Sigma \, \Delta l$ is $2\pi r$, the circumference of the loop. The field due to the current in the wire is

$$B = \frac{\mu_0 I}{2\pi r}, \qquad (18.11)$$

which is equivalent to the expression found by Biot and Savart (Eq. 18.6) when $\mu_0/4\pi$ is substituted for k'. Notice that in this particular case, the summation over the closed path (Ampère's law) is much easier than the summation over distances to all parts of the wire (Biot-Savart law). This is not always true, but we can choose whichever technique gives us the easier calculation of the magnetic field.

Another important current configuration is that of a **solenoid**, which is a helical winding of wire that may carry a current I. The magnetic field of the solenoid is the vector sum of the fields due to each nearly circular

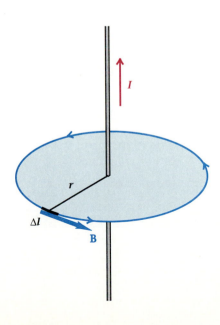

Figure 18.25 A circular closed loop located a distance r about a long straight wire. Since B is constant around the loop and is directed parallel to Δl at each point, $\theta = 0$ and $\cos \theta = 1$.

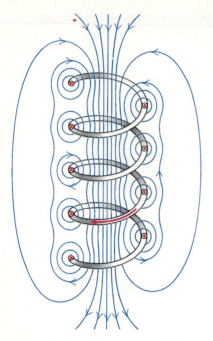

Figure 18.26   The magnetic field around a loosely wound solenoid. Some field lines leak out between the coils.

loop comprising the solenoid. Figure 18.26 depicts the field around a loosely wound solenoid. Its behavior is similar to that of a magnetic dipole. When the coil is tightly wound with adjacent wires nearly touching, the solenoid resembles a cylindrically flowing current. As the solenoid is wound tighter and made longer, the field just outside the central part of the solenoid becomes smaller. Eventually, as the length of the solenoid greatly exceeds its diameter, the field just outside the central part of the solenoid becomes negligibly small.

We can apply Ampère's law to determine the magnitude of the field inside the solenoid for the case of a very long solenoid. Figure 18.27(a) illustrates a section of an infinitely long, tightly wound solenoid. In the cutaway diagram in Fig. 18.27(b), the closed loop for computing with Ampère's law is the rectangle *abcd*. From the symmetry we expect that **B** should be perpendicular to the sides *ab* and *cd*. Outside, along *da*, the field is zero. Inside, the field along *bc* is constant. For this case, Ampère's law (Eq. 18.10) gives

$$\Sigma B \, \Delta l \cos \theta = 0_{(ab)} + Bl_{(bc)} + 0_{(cd)} + 0_{(da)}$$
$$= Bl = \mu_0 nlI,$$

where n is the number of turns per unit length of the coil. Thus nl gives the number of turns with current I that pass through the surface bounded by *abcd*. The resulting field is parallel to the axis of the solenoid, with a magnitude given by

$$B = \mu_0 nI, \tag{18.12}$$

and is independent of position within the solenoid. That is, near the center of a long solenoid the magnetic field is constant.

Example 18.8

Magnetic field in a long solenoid.

Compute the magnetic field near the center of a long solenoid having 20 turns per centimeter and carrying a current of 5.0 A.

Solution   The field is given by Eq. (18.12), with $n = 2.0 \times 10^3$/m:

$$B = (4\pi \times 10^{-7} \, \text{N/A}^2)(2 \times 10^3/\text{m})(5.0 \, \text{A}),$$
$$B = 4\pi \times 10^{-3} \, \text{N/A·m} = 0.013 \, \text{T}.$$

This field is about 250 times greater than the magnitude of the earth's field in most parts of the United States (about 5×10^{-5} T).

Figure 18.27   (a) A section of a long solenoid. If the coils are tightly wound, we can neglect any field leakage through the sides. (b) Cutaway diagram of a section of a long solenoid. The dots represent current out of the page and the crosses indicate current into the page. Applying Ampère's law to the closed path *abcda*, we can determine the magnitude of the magnetic field in a tightly wound solenoid.

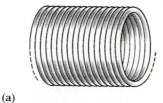

(a)

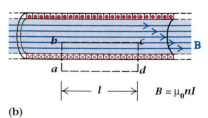

(b)

Magnetic Materials

*18.10

In general, whenever we place an object in a magnetic field, the object becomes polarized; that is, it develops a magnetic dipole moment. This polarization may be measured and used to characterize the object. We define the magnetization **M** to be the dipole moment per unit volume, a vector quantity whose direction is that of the magnetic dipole moment.

The total magnetic field inside the object is the sum of the magnetic field arising from the magnetization and the magnetic field due to electric currents inside and outside the material. The ratio of the magnetization to the external magnetic field **B** that induces it is the **magnetic susceptibility** per unit volume, χ:

$$\chi \equiv \mu_0 \frac{M}{B}.$$

(18.13)

Here B is the magnitude of the external magnetic field without the sample object being present.

The magnitude of the magnetic susceptibility depends on the inherent properties of the material. In fact, it depends on the very nature of the electronic structure of the ions that make up the material. Therefore we can use susceptibility measurements to categorize magnetic materials. Table 18.1 lists several types of common magnetic behavior. The weakest effect is *diamagnetism,* characterized by a small, negative susceptibility that is usually independent of temperature and applied field. The negative susceptibility indicates that the direction of the induced moment is opposite to the direction of the applied field. The origin of diamagnetism is a change in the atomic electron orbits about the direction of the field, resulting in a magnetic field opposite to the applied field. The diamagnetic susceptibility is small and is usually less interesting than the other types of magnetic behavior. However, diamagnetism is always present even when masked by much larger, positive effects. Note that the polarization of individual ions disappears whenever the external field is removed; that is, diamagnetic materials have no permanent dipoles.

Superconductors belong to a special class of diamagnetic materials.

TABLE 18.1
Commonly observed types of magnetic behavior

Type	Comments	Examples
Diamagnetism	Small, negative χ; independent of temperature	Benzene Silicon
Paramagnetism	Positive χ; $\chi \propto 1/T$	Aluminum
Ferromagnetism	Positive χ; complex temperature dependence below the transition temperature T_C; $\chi \propto 1/(T - T_C)$ above the transition temperature	Iron ($T_C = 1043$ K) Nickel ($T_C = 631$ K) Gadolinium ($T_C = 289$ K)

Figure 18.28 A small magnet floats above a high-temperature superconductor. The presence of the magnet induces a polarization in the superconductor that repels the magnet.

When you place a superconducting material in a magnetic field, electric currents are generated at its surface that screen out the external field so there is zero magnetic field within the material. This behavior is known as *perfect diamagnetism*. If a magnet is brought near a superconductor, the diamagnetic polarization of the superconductor causes a repulsive force between them. This effect may be used to float a magnet above a superconductor (Fig. 18.28). A practical application of this effect on a larger scale may one day lead to magnetically levitated vehicles.

Another common magnetic behavior is *paramagnetism*, which is described by a positive susceptibility that depends inversely on the absolute temperature:

$$\chi = C/T. \tag{18.14}$$

This equation is known as the **Curie law** and C is the Curie constant. Each particular paramagnetic material has a characteristic Curie constant.

Paramagnetism arises from atoms or ions with permanent magnetic dipole moments that exist independently of any applied field. The orientation of each moment is independent of the orientation of its neighbors. If no field is applied, these moments are randomly directed and there is no net magnetization of the material. However, when a magnetic field is present, the moments align preferentially along the direction of the external field. The sum of these moments causes the magnetization.

Example 18.9

Paramagnetic susceptibility at two temperatures.

Copper sulfate is paramagnetic with a susceptibility of 1.68×10^{-4} at 293 K. (In the SI system the susceptibility is dimensionless.) What is the susceptibility of copper sulfate at the temperature of liquid nitrogen (77.4 K) if it follows the Curie law?

Solution According to the Curie law (Eq. 18.14), the susceptibility depends inversely on the temperature:

$$\chi = \frac{C}{T}.$$

By dividing the susceptibility at one temperature by the susceptibility at another temperature we get the ratio

$$\frac{\chi(T_1)}{\chi(T_2)} = \frac{T_2}{T_1}.$$

So

$$\chi(77.4 \text{ K}) = \frac{293}{77.4} \chi(293 \text{ K}),$$

$$\chi(77.4 \text{ K}) = 3.79 \times 1.68 \times 10^{-4},$$

$$\chi(77.4 \text{ K}) = 6.37 \times 10^{-4}.$$

(a)

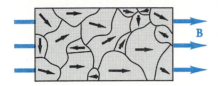

(b)

Figure 18.29   (a) Magnetic domains in an unpolarized ferromagnetic solid are random because of the random orientation of the grains within the material. (b) Under the influence of an external field, the polarization direction of the individual domains and their size may change irreversibly, creating a permanent magnet.

Other magnetic effects, distinct from diamagnetism and paramagnetism, result from the cooperative interactions of individual ionic moments. By "cooperative" we mean that the individual moments are not independent, but interact strongly. The most widely known of these magnetic phenomena is **ferromagnetism**. Ferromagnetic behavior is what the ancient Greeks observed in their lodestones and is what you observe when a magnet attracts a nail or other object. Thus materials that we loosely call "magnetic" because they are noticeably attracted by a magnet are really ferromagnets.

Ferromagnetic solids are characterized by small regions in which all of the ionic moments are aligned in the same direction. These tiny regions are called **magnetic domains**. A given ferromagnetic object (such as an iron nail) usually consists of many small crystals, or grains, and each grain contains several magnetic domains whose polarizations are in different directions (Fig. 18.29a). Thus the net magnetization of the nail is zero. When an external field is applied, the boundaries between the domains move (Fig. 18.29b). Those domains aligned close to the field direction increase in size, while those in other directions decrease. The result is a net magnetization that is very much larger than that of a paramagnet. For a small polarizing field (say 10^{-2} T), the magnetic field resulting from the magnetization of a ferromagnet is typically 10,000 to 100,000 times larger than the polarizing field. Furthermore, if the external field is large enough, the domain structure may be permanently affected, so that a net magnetization remains even when the applied field is removed. Consequently the resulting material is a permanent magnet.

The alignment of the ionic moments in the domains is characteristic of ferromagnets. This alignment is the result of a quantum-mechanical effect known as exchange, which is beyond the scope of the present discussion. However, it is often useful to think of ferromagnetism in terms of an internal field (the exchange field) that tends to align the individual ionic moments. The effect of the internal field is counteracted by the randomizing effect of thermal agitation, which tends to disorient these moments. The result is a critical temperature above which the material ceases to be ferromagnetic and becomes paramagnetic instead. The temperature that marks the transition from ferromagnetism to paramagnetism is known as the *Curie temperature* T_C. Below T_C the individual moments are all aligned in domains. Above T_C the individual moments behave independently, as in a paramagnet, and the material no longer has a permanent magnetic moment. The susceptibility above T_C follows a modified Curie law known as the *Curie-Weiss law*:

$$\chi = \frac{C}{T - T_C}.$$

(18.15)

In addition to ferromagnetism there are other types of magnetic behavior corresponding to a definite alignment of magnetic moments. One example is ferrimagnetism, in which the moments do not align in the same direction but still have a net permanent magnetization. Many ferrimagnets are insulators. They are widely used in the antennas of portable radios.

SUMMARY

Helpful Concepts

- Like magnetic poles repel, unlike poles attract.
- The simplest type of magnet is the magnetic dipole. It can be characterized by its magnetic dipole moment μ. The torque on a magnetic dipole in a magnetic field is

$$\tau = \mu B \sin \theta.$$

- Gauss's law for magnetism states that the net magnetic flux through any closed surface is zero.
- The force on a current-carrying wire in a magnetic field is

$$F = IlB \sin \theta.$$

- A charged particle moving in a magnetic field is subject to a force given by

$$F = qvB \sin \theta.$$

- A charged particle moving perpendicular to a magnetic field travels in a circle of radius

$$r = \frac{mv}{qB}.$$

- The Biot-Savart law gives the contribution to the magnetic field due to a length of current-carrying wire,

$$\Delta B = k' \frac{I \, \Delta l \sin \theta}{r^2}.$$

- The magnetic field a distance d from a long straight wire carrying a current I is

$$B = k' \frac{2I}{d} = \frac{\mu_0 I}{2\pi d}.$$

- The definition of the ampere: When the attractive force per unit length between two long parallel wires placed one meter apart and carrying the same electric current is 2×10^{-7} N/m, that current is defined to be exactly one ampere.
- A current loop has a magnetic moment given by

$$\mu = NIA.$$

- Ampère's law can be written

$$\Sigma B \, \Delta l \cos \theta = \mu_0 I,$$

where the sum is carried out around a closed loop.
- The magnetic susceptibility per unit volume is

$$\chi = \mu_0 \frac{M}{B}.$$

- The Curie law for paramagnetism is

$$\chi = \frac{C}{T}.$$

Right-Hand Rules

- The direction of the magnetic field about a wire is given by a right-hand rule: If the thumb of the right hand is taken as the direction of current, the curled fingers point in the direction of the magnetic field about that current.
- The direction of the force on a current-carrying wire in a magnetic field is also given by a right-hand rule: When a right hand is held so that the fingers can be curled from the direction of the current into the direction of the magnetic field, the force lies in the direction of the thumb.
- The direction of the magnetic moment of a current loop is given by another right-hand rule: The magnetic moment is in the direction that your thumb points when you curl the fingers of your right hand in the direction of positive current.

Important Terms

You should be able to write the definition or meaning of each of the following terms:

- magnetic dipole
- magnetic field
- magnetic dipole moment
- Gauss's law for magnetism
- tesla
- cyclotron
- law of Biot and Savart
- galvanometer
- ammeter
- voltmeter
- Ampère's law
- solenoid
- magnetic susceptibility
- Curie law
- ferromagnetism
- magnetic domains

QUESTIONS

18.1 The current-carrying wire shown in Fig. 18.6(b) is turned to lie in a horizontal plane and a horizontal card is placed on each side of the wire. When iron filings are sprinkled on the card, what pattern will form? Sketch your answer.

18.2 Sketch the pattern of iron filings that will form on the card in Fig. 18.6(b) if two closely spaced current-carrying wires pass through the card. Sketch patterns for the two cases in which the currents are in the same or in opposite directions.

18.3 Why does Gauss's law for magnetism imply that there are no magnetic monopoles?

18.4 A current-carrying wire passing through a certain region of space does not experience any force acting on it. Can we conclude that there is no magnetic field in that region?

18.5 A positively charged proton is projected toward you in a uniform magnetic field that is directed straight upward. In what direction is the proton deflected?

18.6 Equation (18.6) for the magnitude of the magnetic field near a long straight current-carrying wire becomes infinitely large when the distance d becomes zero. Does this mean that you can actually have an infinitely large magnetic field?

18.7 Some older textbooks describe the behavior of a current-carrying loop in a magnetic field in terms somewhat like this: The loop aligns itself so as to enclose as many of the lines of force of the magnetic field as possible. Explain this statement in terms of the discussion given in this text.

18.8 Suppose you make an electromagnet by winding a coil on each leg of a U-shaped iron bar and connecting the coils in series with each other across a battery. If you look directly at the ends of the pole pieces and conclude that the current flows clockwise in one coil, in what direction should it flow in the other coil?

18.9 If the magnetic field of the earth is due to a circulating current below the earth's crust, does it flow from west to east or east to west?

18.10 Discuss how you would go about constructing an ammeter and a voltmeter from a given galvanometer and whatever resistors and wires you need. What would be the effects of using large or small resistance for each?

18.11 How can a piece of iron (such as a nail) be ferromagnetic and yet not have a net magnetic dipole moment?

18.12 How many ways can you think of to give an iron rod a net permanent magnetic moment?

PROBLEMS

> **Hints for Solving Problems**
> A magnetic field exerts a torque on a magnetic dipole to rotate it into the direction of the field. A magnetic field exerts a force on a moving charge (or current) that is at right angles to the velocity of the charge and to the direction of the field. The right-hand rules for directions of fields and forces are listed in the summary.

Section 18.1   Magnets and Magnetic Fields

18.1 Sketch the magnetic field due to two identical bar magnets (dipoles) placed together (Fig. 18.30).

18.2 Sketch the magnetic field due to two identical bar magnets (dipoles) placed together (Fig. 18.31).

Section 18.2   Oersted's Discovery: Electric Current Produces Magnetism

18.3 The units of magnetic field are N/A·m. Use this fact and Eq. (18.1) to show that the units of magnetic moment are A·m^2.

18.4 A compass needle in a uniform magnetic field of 0.30 T experiences a torque of 0.050 N·m when it points in a direction 45° away from the direction of the field. What is the magnitude of the needle's magnetic moment?

18.5 A small bar magnet has an effective magnetic moment $\mu = 7.50 \times 10^6$ A·m^2. What is the maximum possible torque on the magnet due to the earth's field, $B = 5.00 \times 10^{-5}$ T?

18.6 A torque of 4.0×10^{-3} N·m is needed to hold a certain compass needle at right angles to the earth's magnetic field, $B = 5.0 \times 10^{-5}$ T. What is the magnetic moment of the compass needle?

18.7 Imagine that you are standing directly beneath a current-carrying wire and facing in the direction of the current. There is a magnetic field passing

S	N		N	S

Figure 18.30   Problem 18.1.

Figure 18.31   Problem 18.2.

through your body due to the current in the wire. In which direction does the field point?

18.8 A magnet with a magnetic moment of 6.5×10^6 A·m² experiences a torque of 8.7×10^{-3} N·m when the axis of the magnet is 27° away from the direction of a uniform magnetic field. What is the magnitude of the field?

Section 18.3 Magnetic Forces on Electric Currents

18.9 A magnet with a field $B = 0.20$ T over an effective area 10 cm in diameter exerts a force on a current-carrying wire that crosses between the pole faces at right angles to the field. What is the force on the wire if the current is 10 A?

18.10 A 0.50-m-long wire of mass $m = 1.0$ g lies on a horizontal platform in a region of uniform magnetic field $B = 1.0 \times 10^{-3}$ T. The field is perpendicular to the wire so that an upward force is exerted on the wire when a current passes through it. What is the minimum current required to lift the wire against its gravitational force?

18.11 An electric trolleycar cable carries a dc current of 400 A in a region where the vertical component of the earth's magnetic field is 5.00×10^{-6} T. What is the horizontal force on a 10.0-m section of the wire due to magnetic effects?

18.12 A wire 0.500 m long carries a current of 6.00 A and makes an angle of 45.0° with a uniform magnetic field. If the force on the wire is 0.106 N, what is the magnitude of the magnetic field?

18.13 A horizontal wire 30.0 cm long makes an angle of 45.0° with a magnetic field $B = 0.0500$ T horizontally directed to the north. What are the magnitude and the direction of the force on the wire when it carries a current of 1.00 A in the direction from southwest to northeast?

18.14 A wire 0.250 m long carries a current of 0.750 A in a region of uniform magnetic field. The wire is subject to a force of 2.50 N when held perpendicular to the direction of the field. What is the magnitude of the magnetic field?

Section 18.4 Magnetic Forces on Moving Charged Particles

18.15 A proton of charge 1.60×10^{-19} C and mass 1.67×10^{-27} kg is projected into a region of magnetic field $B = 0.140$ T. What is the acceleration of the proton if it is traveling perpendicularly to the field with a speed $v = 1.05 \times 10^6$ m/s?

18.16 A particle having an electric charge $q = 3.2 \times 10^{-19}$ C is injected into a magnetic field $B = 0.10$ T with a speed of 3.0×10^6 m/s. (a) If the velocity of the particle is perpendicular to the direction of the magnetic field, what is the force on the particle?

(b) How great an electric field would be required to exert a force with the same magnitude on the particle?

18.17 A proton of charge $+1.60 \times 10^{-19}$ C and mass 1.67×10^{-27} kg is introduced into a region of $B = 1.05$ T with an initial velocity of 1.20×10^6 m/s perpendicular to **B**. What is the radius of the proton's path?

18.18 Suppose the initial velocity of the proton in Problem 18.17 makes an angle of 45° with the direction of the magnetic field **B**. What is the subsequent motion of the proton?

18.19 A proton moving at right angles to a magnetic field of 0.100 T travels in a circular orbit of radius 5.00 cm. The proton has an electric charge $q = 1.60 \times 10^{-19}$ C and a mass of 1.67×10^{-27} kg. (a) What is the speed of the proton? (b) What is its kinetic energy?

18.20 Alpha particles from a particular radioactive source have a speed of 1.8×10^7 m/s. How large a magnetic field is required to bend the path of the particles into a circle of 0.580 m radius? An alpha particle has a mass of 6.64×10^{-27} kg and a charge of 3.20×10^{-19} C.

*Section 18.5 The Cyclotron

18.21 A laboratory cyclotron operates in a magnetic field of 1.50 T. If it is used to accelerate deuterons (mass 3.34×10^{-27} kg and charge 1.60×10^{-19} C), what must be the frequency of oscillation of the accelerating voltage?

18.22 A cyclotron used to accelerate deuterons is operated at a frequency of 9.16 MHz. What is the operating magnetic field of the cyclotron? (*Hint:* The deuteron mass $= 3.34 \times 10^{-27}$ kg; charge $= 1.60 \times 10^{-19}$ C.)

18.23 What is the cyclotron frequency of an electron of mass 9.1×10^{-31} kg and charge -1.6×10^{-19} C in a region of field $B = 1.0$ T?

18.24 What would be the operating frequency of the cyclotron of Example 18.3 if it had a magnetic field of only 0.50 T?

18.25 Derive the expression

$$\text{KE} = \frac{q^2 B^2 r^2}{2m}$$

for the kinetic energy of a particle of charge q and mass m moving at a radius r in a cyclotron with uniform magnetic field B.

18.26 At what radius must deuterons be extracted from a cyclotron with a 1.00-T magnetic field if they are to have an energy of 2.00×10^{-12} J each? (*Hint:* Use the results of Problem 18.25 and the information in Problem 18.21.)

Section 18.6   Magnetic Field Due to a Current-Carrying Wire

18.27 Use the law of Biot and Savart to show that the magnetic field at the center of a circular coil of N turns of radius r carrying a current I is given by

$$B = \frac{\mu_0 N I}{2r}.$$

18.28 What is the current in a 10-cm-radius wire loop if the magnetic field generated at the center of the loop is equal to the magnetic field of the earth at its surface? (*Hint:* See Example 18.5 and take the earth's field to be 5.0×10^{-5} T.)

18.29 Fifty turns of wire are wrapped into a coil 0.32 m in diameter and connected to a 12-V battery. If the total resistance of the coil is 0.90 Ω, what is the magnetic field at its center? (*Hint:* See Example 18.5.)

18.30 A long straight wire carries a current of 100 mA. What is the magnitude of the magnetic field 10.0 cm from the wire?

18.31 A long straight wire carries a current of 50 A. At what distance from the wire will the magnetic field become comparable to the earth's field of 5×10^{-5} T?

18.32 Two parallel wires spaced 0.50 m apart in a horizontal plane carry oppositely directed currents of 200 A. (a) Are the wires attracted or repelled? (b) What is the force per unit length on the wires?

18.33 Two parallel wires spaced a distance d apart experience an attractive force per unit length of 6.00×10^{-4} N/m. If the currents have a magnitude of 10.0 A each, what is the distance d?

18.34 Two parallel wires repel each other with a force per length of 1.00×10^{-3} N/m when spaced a distance 2.50 cm apart. If one wire has a current of 100 A, what is the current in the other wire?

Section 18.7   Magnetic Force on a Current Loop

18.35 A coil of 32 turns of wire is wound about a circular form. The average diameter of the individual loops is 28.5 cm. Calculate the magnetic moment of the coil when it carries a current of 1.75 A.

18.36 What is the magnetic moment of a 12.5-cm-diameter coil of 150 turns when it carries a current of 75 mA?

18.37 A small coil has a cross section of 3.0 cm^2 and consists of 8 turns of wire. If the loop carries a current of 100 mA, what is the torque required to hold the loop at right angles to a field of $B = 0.010$ T?

18.38 A rectangular coil 4.3 cm wide by 5.8 cm long is made from 6 turns of wire. The coil carries a current of 200 mA in a magnetic field $B = 0.055$ T. How much torque is required to hold the loop so that it makes an angle of 45° with the magnetic field?

Section 18.8   Galvanometers, Ammeters, and Voltmeters

18.39 A student needs to make a voltmeter that measures 200 V full scale. The only galvanometer available deflects full scale at a current of 10.0 mA, and its resistance is 1.00 kΩ. What resistance is needed and how should it be connected to the galvanometer?

18.40 A technician is building a voltmeter from a galvanometer movement that has a resistance of 10 kΩ and full-scale deflection at a current of 100 μA. If she places a 490-kΩ resistor in series with the galvanometer, what will be the full-scale voltage reading of the resulting meter?

18.41 The resistance of a galvanometer movement is 500 Ω, and the current required for full-scale deflection is 250 μA. What resistance is needed to convert the galvanometer to an ammeter that reads 5.0 A full scale and how should it be connected?

18.42 A lab technician needs to make an ammeter that measures 15 A full scale. The only galvanometer available deflects full scale at a current of 100 μA, and its resistance is 10 kΩ. What resistance is needed and how should it be connected to the galvanometer?

*Section 18.9   Ampère's Law

18.43 A solenoid 0.435 m long has 900 turns of wire. What is the magnetic field in the center of the solenoid when it carries a current of 2.75 A?

18.44 A 0.53-m-long solenoid consisting of 1000 turns is used to generate a magnetic field 0.15 T. How much current is required?

18.45 A large solenoid used in experiments with subnuclear particles produces an axial magnetic field of 0.800 T. The 2.00-m-long solenoid has 320 turns of wire. What current is required to produce the required field?

Additional Problems

18.46 In a region of uniform magnetic field B, how much
•   work is required to rotate a bar magnet of dipole moment μ from parallel alignment to antiparallel alignment along the magnetic field?

18.47 A slender, 150-g bar magnet 15.0 cm long has a
••   magnetic moment of 5.30×10^6 A·m^2. It is held at an angle of 45° to the direction of a uniform magnetic field. When released, it begins to rotate about its center with an initial angular acceleration of 5.09×10^{-3} rad/s^2. What is the magnitude of the magnetic field?

18.48 A freely turning compass needle in a uniform mag-
•• netic field oscillates with simple harmonic motion
when displaced from equilibrium by a small angle.
Find the equation for the oscillation frequency in
terms of the moment of inertia I, the magnetic dipole moment μ, and the magnetic field strength B.
(*Hint*: Refer to Sections 7.11 and 13.5.)

18.49 Show that the total force on a current-carrying
•• straight wire in a uniform magnetic field is independent of the length of the wire if the potential
difference across the wire is held constant. (*Hint*:
Assume Ohm's law to be valid for the wire.)

18.50 A charged particle is observed to move in a circle
• of 0.0823 m radius in a plane perpendicular to a
uniform magnetic field of 7.69×10^{-3} T. From
other measurements, the momentum of the particle
is known to be 2.03×10^{-22} kg·m/s. What is the
electric charge of the particle?

18.51 A charged particle moving through a bubble cham-
• ber travels in a circular arc of radius 0.53 m. If the
magnetic field perpendicular to the plane of the arc
is 0.10 T, what is the momentum of the particle?
(*Hint*: Assume that the charge of the particle is that
of an electron, $e = -1.6 \times 10^{-19}$ C.)

18.52 In a high-energy physics experiment a subnuclear
•• particle moves in a circular arc of 0.27 m radius
perpendicular to a magnetic field of 2.7×10^{-2} T.
The kinetic energy of the particle is determined to
be 4.1×10^{-16} J. Identify the particle from its mass.
The masses of the electron, pion, and proton are
9.1×10^{-31} kg, 2.5×10^{-28} kg, and 1.67×10^{-27} kg, respectively.

18.53 A rectangular loop of wire carries a current of
• 6.3 A. Nearby in the plane of the loop is a long
straight wire carrying a current of 5.4 A (Fig. 18.32).
What are the magnitude and the direction of the
force acting on the loop if it has the dimensions
given in the figure?

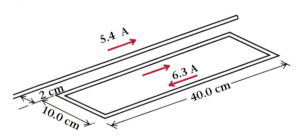

Figure 18.32   Problem 18.53.

18.54 Show that at a point x measured along the axis from
•• the center of a circular loop of radius R carrying a

current I (Fig. 18.33), the magnetic field is

$$B = \frac{\mu_0 I R^2}{2(R^2 + x^2)^{3/2}},$$

and **B** is directed along the axis.

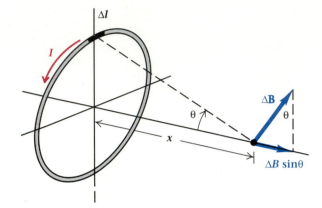

**Figure 18.33   Problem 18.54: An element of field
$\boldsymbol{\Delta}$B at x due to the current I in the element $\boldsymbol{\Delta}$l of the
loop.**

18.55 A rectangular coil 4.0 cm wide by 5.0 cm long is
• made from 10 turns of wire carrying a current of 50
mA. The coil is aligned with its magnetic moment
along the direction of a uniform applied field $B = 0.50$ T. How much work is required to rotate the
coil 180°?

18.56 A coil of area $A = 120$ cm^2 is made from 40 turns
• of wire. When a current of 2.5 A flows through the
coil, it aligns with a uniform external field **B**. If it
takes 0.65 J of energy to rotate the coil so that it is
aligned antiparallel to **B**, what is the magnitude of
the magnetic field?

18.57 A toroid is a coil wrapped around a doughnut-
•• shaped core (Fig. 18.34). (a) Use Ampère's law to
calculate the field inside the toroid along the circle
of radius r. (b) How is the field different from that
of a solenoid? (c) What is the field outside of the
toroid?

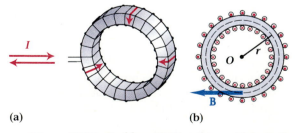

**Figure 18.34   Problem 18.57: (a) A toroidal winding.
(b) Cross section of a toroid.**

18.58 Suppose you wish to make a magnetic dipole from
•• a piece of wire of length l that can support a current I. How should you wind the coil to get the maximum dipole moment? Should you wind many small loops, thereby increasing N, or one large loop, increasing A? Reach your answer by deriving an expression for the magnetic moment as a function of I, l, and N.

18.59 Positive and negative particles are created in an ex-
• periment in which subnuclear particles are made to collide at high velocities in the center of a large empty solenoid. The design of the experimental apparatus calls for protons with speeds of 2.00×10^6 m/s perpendicular to the magnetic field to be deflected in a circular path of radius 2.00 m. If a current of 1000 A can be supplied to the solenoid, how many turns per meter should it have?

18.60 Figure 18.35 shows a cross-sectional view of a co-
• axial cable. The center conductor is surrounded by an insulating layer surrounded by an outer conductor. The current in the outer conductor is equal and opposite to the current in the inner conductor. Show that the magnetic field outside the coaxial cable due to its current is zero.

18.61 A magnetic dipole with $\mu = 5.07 \times 10^{-3}$ A·m² is
• placed in a long solenoid so that the magnetic moment makes an angle of 34.6° with the axis of the solenoid. The 50.3-cm-long solenoid is wound from 120 turns of wire and carries a current of 2.75 A. What is the torque on the dipole?

18.62 A cylindrical conductor of radius a carries a uni-
• formly distributed current I (Fig. 18.36). Show that the magnetic field within the conductor measured at a distance r from the center is given by the following equation,

$$B = \frac{\mu_0 I r}{2 \pi a^2}.$$

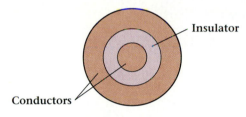

Figure 18.35 Problem 18.60.

Figure 18.36 Problem 18.62.

ADDITIONAL READING

Agassi, J., *Faraday As a Natural Philosopher*. Chicago: University of Chicago Press, 1971.

Foster, K. R., and A. W. Guy, "The Microwave Problem." *Scientific American,* September 1986, p. 32.

Franksen, O. I., *H. C. Ørsted—A Man of Two Cultures*. Birkerød, Denmark: Strandbergs Forlag, 1981. A brief biography of Ørsted that includes a translation of his paper announcing the discovery of the effects of an electric current on a compass needle.

Hoffman, K. A., "Ancient Magnetic Reversals: Clues to the Geodynamo." *Scientific American,* May 1988, p. 76.

Rubin, L. G., and P. A. Wolff, "High Magnetic Fields for Physics." *Physics Today,* August 1984, p. 24.

Runcorn, S. K., "The Moon's Ancient Magnetism." *Scientific American,* December 1987, p. 60.

19

Electromagnetic Induction

19.1 Faraday's Law

19.2 Motional Emf

19.3 Generators and Motors

19.4 The Transformer

19.5 Inductance

*19.6 Energy Storage in an Inductor

19.7 Maxwell's Equations and Electromagnetic Waves

A WORD TO THE STUDENT

In this chapter we continue to develop the laws of electromagnetism, introducing Faraday's law. Understanding Faraday's law will help you understand how electric generators and electric motors work. Faraday's law also underlies the behavior of another basic circuit element—the inductor.

We also discuss Maxwell's equations, which sum up the ideas of this chapter and the four preceding chapters. Although we do not emphasize their mathematical form, you should remember that all of electricity and magnetism can be represented by these few equations. In terms of their power and usefulness, Maxwell's equations are one of the great unifying achievements in physics. In the next chapter we bring together all the material related to electrical circuits, showing how they work together to achieve a wide range of results.

One of the greatest contributions to the rapidly growing science of electromagnetism was made by Michael Faraday (1791–1867). Following Oersted's discovery of the magnetic effects of a current and Ampère's discovery of the mutual forces between two currents, Faraday began to experiment with magnets and current-carrying wires at the Royal Institution in London. In 1831 he discovered that electric currents could be induced by magnetic fields. The same discovery was independently made at about the same time by the American Joseph Henry (1797–1878) and by the Russian Heinrich Lenz (1804–1865).

When Henry published his results in 1832, he also explained that a changing electric current in a coil can induce another current in the *same* coil. As a result, the current in a coil consists of two components, the initial current plus an induced current. This effect is known as self-inductance or, more simply, as just inductance. In 1834 Faraday also discovered self-inductance independently of Henry. This time, however, Henry received the credit for being first. Inductance, like resistance and capacitance, plays an important role in the behavior of electrical circuits.

During the middle of the nineteenth century, James Clerk Maxwell (1831–1879) showed that a complete theory of electromagnetism could be constructed from a set of only four equations. These four equations, now known as Maxwell's equations, hold a central position in the theory of electromagnetism, analogous to the central position of Newton's laws in the theory of mechanics. Maxwell's equations sum up the concepts developed in this and the previous four chapters, and often serve as the starting point for advanced texts in electromagnetism.

Applications of electromagnetism surround us in such forms as electric generating plants and their network of power lines, and as appliances that depend on motors for their operation. Our communication and information networks are all based on electromagnetism, and we can trace their roots to the work of Maxwell and his predecessors.

Faraday's Law

19.1

Faraday (Fig. 19.1a, p. 558) was investigating the effect of a current-carrying solenoid on a second solenoid wound over the first one (Fig. 19.1b). One coil was joined to a battery, while the other was connected to a galvanometer. Even a large current flowing through the first coil caused no detectable current in the other. However, Faraday did notice an effect when he connected or disconnected the circuit of the first coil. At the moment when the first circuit was closed, a sudden and slight movement occurred in the galvanometer in the second circuit. When the circuit was opened, again a momentary deflection of the galvanometer occurred, but this time in the opposite direction. Further experiments showed that the current in the second (or secondary) coil depended not on the primary current in the first coil, but on the *rate of change* of the current. This effect was greatly enhanced by winding the coils around an iron ring (Fig. 19.1c).

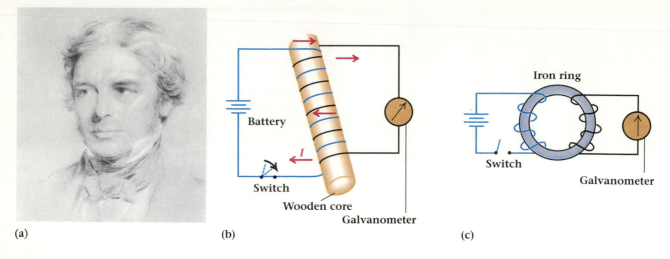

(a)   (b)   (c)

Figure 19.1   (a) Michael Faraday. (b) A simplified drawing of Faraday's experiment. Two separate coils are shown, one connected to a battery and the other connected to a galvanometer. A change in the current in the first coil produced a current in the second coil. (c) Faraday wound the coils about a ring of soft iron to enhance the effect.

Figure 19.2   (a) Moving a magnet toward a coil generates an electric current in the loop, detected by the galvanometer. (b) Moving the same end of the magnet away from the coil generates a current in the opposite direction.

Faraday continued his research and observed that, if he thrust a magnet into a coil of wire (Fig. 19.2a), a current was induced in the coil *while the magnet was moving* relative to the coil. Moving the magnet away from the coil caused the galvanometer to deflect in the opposite way (Fig. 19.2b). Thus the motion of the north pole of a bar magnet away from the coil induced a current whose direction was opposite to that caused by the motion of the north pole toward the coil. Faraday also noted that moving the magnet toward the coil had the same effect as moving the coil toward the magnet; only the relative motion was important. Furthermore, he observed that bringing the south pole of a bar magnet toward the coil caused the galvanometer to deflect opposite to the way it deflected when the north pole of the magnet was brought toward the coil. In other words, moving the south pole of a bar magnet toward the coil produced the same effect as moving the north pole away from the coil.

All these observations were finally explained by Faraday's realization that the currents observed in the secondary coil were entirely due to the changing magnetic flux within the coil. We can summarize Faraday's observations by saying that **the emf induced in a loop of wire is proportional to the rate of change of magnetic flux through the coil.** This statement is known as **Faraday's law.** We often call the emf induced by

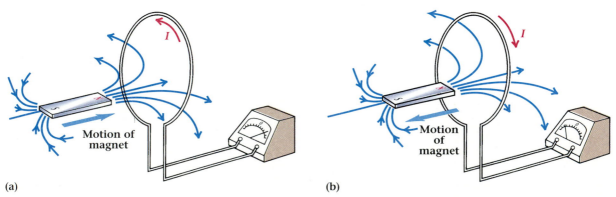

(a)   (b)

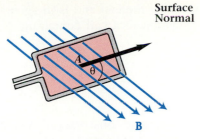

Figure 19.3 Magnetic field passing through an area A. The normal to the area makes an angle θ with the direction of the magnetic field. The magnetic flux depends on the orientation of the field to the area through the factor $\cos\theta$.

a changing magnetic flux an **induced emf**, and the current it produces is called an **induced current** or an induction current.

The amount of magnetic field that passes through a given area, such as the area enclosed by a current-carrying loop of wire, contributes to the magnetic flux, which was introduced in Section 18.2. The magnetic flux ϕ_m through an area A is the product of the magnitude of the magnetic field $|\mathbf{B}|$, the area A, and the cosine of the angle θ between the direction of the field and the direction of the normal to the area (Fig. 19.3):

$$\phi_m = BA\cos\theta. \qquad (19.1)$$

The maximum number of field lines pass through a wire loop when its area is perpendicular to the direction of the magnetic field. As the loop turns so that the angle between its plane and the field gets smaller, the effective area becomes smaller and fewer field lines can pass through it. For this reason, the definition of flux includes the effect of the angle between the area and the field lines. As before, we define the orientation of the area as the direction of its normal (a line perpendicular to the area). In defining the flux, the angle θ is the angle between the field lines and the normal line.

Shortly after Faraday's discovery, Heinrich Lenz, working in Russia, extended our understanding of induction currents by stating a related principle. Although Faraday knew how to determine the direction of an induced current, Lenz expressed this relationship more directly. **Lenz's law** states that **the direction of an induced current is such that its own magnetic field opposes the original change in the flux that induced the current.** This law is illustrated in Fig. 19.4, which shows the south pole of a bar magnet moving toward a conducting loop. The direction of the magnetic field is indicated by the small arrows. As the magnet approaches the loop, the field passing through the loop increases, thus increasing the magnetic flux through the loop. The increasing flux induces an emf in the loop, and, since the loop is closed, the emf induces a current I. Now, the direction of this induced current is such as to generate a magnetic field in the direction that opposes the change in the flux. In this case, the initial field is opposite to the direction of motion of the magnet (Fig. 19.4a), so the induced field must be in the direction of motion of the magnet (Fig. 19.4b). According to the right-hand rule, the current is clockwise as seen from the position of the magnet. If the magnet were moved away from the loop, the flux through the loop would decrease and a current would be induced in the opposite direction, whose field would still try to maintain the field through the loop constant. That is, the induced field opposes the change in external flux.

We can write Faraday's law of induction mathematically with the equation

$$\mathscr{E} = -\frac{\Delta\phi_m}{\Delta t}. \qquad (19.2)$$

Here $\mathscr{E}$ is the induced emf in one loop of wire, $\Delta\phi_m$ is the change in magnetic flux, and Δt is the time interval over which the change takes

Field due to magnet only

(a)

Induced field only

(b)

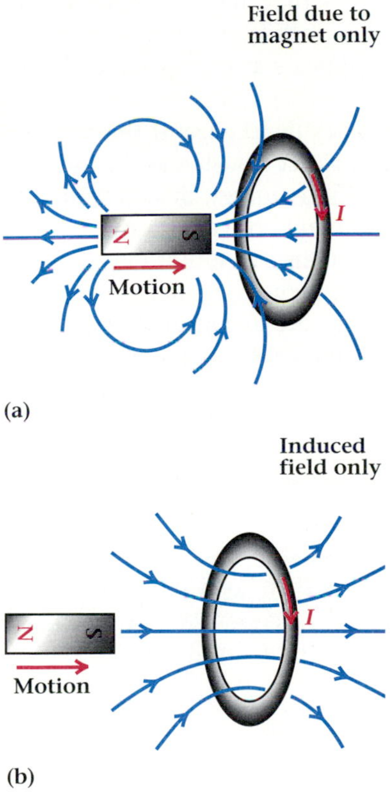

Figure 19.4 (a) The motion of a magnet toward a coil induces a current in the coil. As the south pole gets closer to the loop, the flux through the loop increases. (b) By Lenz's law, the direction of the induced current produces a magnetic field that opposes the increase in the flux.

place. The minus sign indicates the direction of the emf and is in agreement with Lenz's law. If we apply this equation to a coil containing N loops or turns of wire, then

$$\mathscr{E} = -N \frac{\Delta\phi_m}{\Delta t}. \tag{19.3}$$

Faraday's and Lenz's laws are examples of energy conservation. When a magnet is moved closer to a coil (or the coil is moved toward the magnet), by Faraday's law an emf is created that can generate an electric current if the coil is a closed circuit. By Lenz's law, the direction of this induced current is such that the resulting magnetic field exerts an opposing force on the moving magnet. In other words, if we push the magnet toward the coil, the induced current generates a magnetic field that opposes the push. If we pull the magnet away from the coil, the induced magnetic field reverses direction and opposes that, too. Thus we always encounter a resisting force and are required to do work to induce a current. The work done on the magnet-coil system appears as energy expended in the form of I^2R heating losses in the coil.

Notice that *the induced field always opposes the change*. If the induced field assisted the change, we could give the magnet a push and the induced field would pull it into the loop, releasing more energy than we initially gave it. Such behavior would violate the principle of conservation of energy. Thus Lenz's law is consistent with the conservation of energy.

Faraday's results had very practical consequences, which we describe in the rest of this chapter. They led directly to the invention of the transformer, the alternator, and the generator. These inventions, together with the electric motor, brought about the many applications of electrical energy that are so widespread today.

Example 19.1

Induced emf in a coil.

A square coil 10 cm on a side has 20 turns of wire. Initially it is at rest between the poles of a magnet whose field is 0.25 T, with the plane of the coil perpendicular to the field. What average voltage (emf) is produced on the coil if it is withdrawn completely from the field in 0.10 s?

Solution   Since the field is perpendicular to the plane of the coil, the initial flux is just the product of the field B with the area of the coil. For a square coil of side length l, the area is l^2. Thus the initial flux is

$$\phi_i = Bl^2.$$

After the coil is withdrawn from the magnet, $\phi = 0$. Thus

$$\Delta\phi = 0 - \phi_i = -\phi_i.$$

The induced emf is

$$\mathscr{E} = -N \frac{\Delta\phi_m}{\Delta t} = \frac{NBl^2}{\Delta t} = \frac{20(0.25 \text{ T})(0.10 \text{ m})^2}{0.10 \text{ s}},$$

$$\mathscr{E} = 0.50 \text{ V}.$$

Let us also check to be sure that the units are correct. The units above are, as expected,

$$\frac{\text{T} \cdot \text{m}^2}{\text{s}} = \frac{(\text{N/A} \cdot \text{m}) \cdot \text{m}^2}{\text{s}} = \frac{\text{N} \cdot \text{m}}{\text{A} \cdot \text{s}} = \frac{\text{J}}{\text{C}} = \text{V}.$$

Example 19.2

Induced current in a coil.

Compute the current through a 37-Ω resistor connected to a five-turn circular loop 10 cm in diameter, assuming that the magnetic field through the loop is changing at a rate of 0.050 T/s.

Solution If the resistance of the wire is small compared with that of the resistor, the current through the loop depends only on the emf and the resistance:

$$I = \frac{\mathscr{E}}{R}.$$

The emf is

$$\mathscr{E} = -N\frac{\Delta\phi_m}{\Delta t} = -N\pi\left(\frac{D}{2}\right)^2\frac{\Delta B}{\Delta t},$$

where D is the diameter of the loop. We obtain

$$I = \frac{\mathscr{E}}{R} = -\frac{N\pi D^2 \frac{\Delta B}{\Delta t}}{4R}$$

$$I = -\frac{5\pi(0.10 \text{ m})^2(0.050 \text{ T/s})}{4 \times 37 \ \Omega} = -53 \ \mu\text{A}.$$

The minus sign reminds us that for a particular case, we need to determine the direction of the current so that the induced magnetic field opposes the given change in flux.

Motional Emf

19.2

In our discussion of Faraday's law we examined the changing flux in a loop of wire. Now we extend our study to include the effects of a magnetic field on a straight piece of wire or other conductor that moves relative to the field. We will see that an emf is developed between the ends of the wire. To investigate the induction of emf on a moving conductor, let's consider a conducting bar of length l sliding along a stationary U-shaped conductor that is perpendicular to a magnetic field **B** (Fig. 19.5). The bar moves with a constant speed v in the x direction. As the bar moves, the area enclosed by the loop consisting of the bar and the U-shaped conductor increases. Consequently, the magnetic flux through the loop increases.

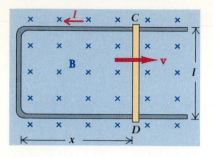

Figure 19.5 A conducting bar sliding along a stationary U-shaped conductor that is perpendicular to a magnetic field **B**. The area of the conducting loop changes, inducing an emf. The crosses indicate a magnetic field directed into the plane of the loop.

The emf developed around the loop is obtained from Faraday's law:

$$\mathscr{E} = -\frac{\Delta\phi}{\Delta t} = -\frac{\Delta(BA)}{\Delta t} = -B\frac{\Delta A}{\Delta t} = -B\frac{(l\,\Delta x)}{\Delta t} = -Bl\frac{\Delta x}{\Delta t},$$

where A is the area enclosed by the loop. Since $\Delta x/\Delta t = v$, we have

$$\mathscr{E} = -Blv.$$

Notice that the direction of the emf would cause a counterclockwise current in the loop shown in Fig. 19.5.

We should also be able to understand this result from considering the force of the magnetic field acting on the charges inside the moving bar. The development of an emf between the ends of the moving bar CD depends on the motion of the bar perpendicular to the magnetic field. The emf generated in this manner is called a *motional emf*.

We can explain motional emf as follows. In Chapter 18 we saw that a charge q moving with speed v at right angles to a magnetic field **B** experiences a force $F = qvB$. The direction of this force is perpendicular to both the direction of motion and the magnetic field. When the bar moves in the x direction (Fig. 19.5), the magnetic field exerts a force on the mobile charges along the direction of the bar. The work needed to move a positive charge q from D to C is*

$$W = Fl = qvBl.$$

The emf is an electric potential difference. Thus emf is the work per unit charge to move a charge q from D to C:

$$\mathscr{E} = \frac{W}{q} = Blv. \qquad (19.4)$$

The emf of Eq. (19.4) corresponds to the potential difference $V_C - V_D$. The direction of this emf would drive a counterclockwise current if contact were made between the bar and the U-shaped conductor.

Motional emf is due to the force of a magnetic field on moving charges and cannot be used to deduce Faraday's law. However, as we have seen, when the bar slides on the U-shaped conductor, the emf induced in the loop is the same as the motional emf.

Example 19.3

Motional emf in a moving aluminum rod.

What emf is developed between the ends of a 0.20-m aluminum rod when it is moved with a speed of 2.5 m/s between the poles of a laboratory magnet generating a magnetic field of 0.85 T (Fig. 19.6)? The rod and its motion are perpendicular to the direction of the magnetic field.

Solution   We may use Eq. (19.4) to get the emf:

$$\mathscr{E} = Blv = (0.85\text{ T})(0.20\text{ m})(2.5\text{ m/s}) = 0.43\text{ V}.$$

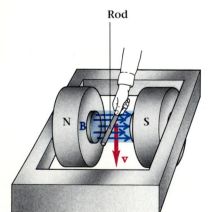

Figure 19.6 Example 19.3: An aluminum rod moves through the magnetic field of a laboratory magnet.

---

*As we explained earlier, the mobile charges in a metal rod are the negatively charged electrons, which would move in the opposite direction (from C to D). However, the resulting emf is unchanged.

Generators and Motors

19.3

Faraday's discovery of induced currents soon led to the development of practical machines for converting mechanical energy into electrical energy. Such machines, known as electric **generators**, may be classified into two main types: *dc generators*, for producing unidirectional current, and ac generators, called *alternators*, for producing alternating current.

The simplest ac generator consists of a loop of wire turning in a magnetic field (Fig. 19.7a). The wire loop (called the armature in this application) is connected to a pair of rotating contacts called slip rings, which rub against stationary contacts called brushes. As the armature rotates in the magnetic field, the flux through the coil continuously changes. This induces a changing emf, which appears at the brush contacts. As the armature rotates clockwise from part (a) to part (b) of the figure, the magnetic flux increases, reaching a maximum when the normal to the loop is parallel to the magnetic field. As the armature continues to rotate past the position of part (b), the flux decreases to zero when the normal to the loop is 90° from the direction of the field (part c). When the loop rotates past the 90° position (part d), the direction of the normal is opposite to the direction of the field, so that the flux through the armature is negative with respect to the flux in parts (a) and (b).

If the field **B** is uniform and the coil rotates with a constant angular speed ω, the magnetic flux through the coil may be expressed as

$$\phi_m = BA \cos \theta = BA \cos \omega t,$$

where A is the area of the loop (Fig. 19.8a). From Faraday's law we calculate the emf to be

$$\mathcal{E} = -\frac{\Delta \phi_m}{\Delta t} = -\frac{BA \, \Delta(\cos \omega t)}{\Delta t}.$$

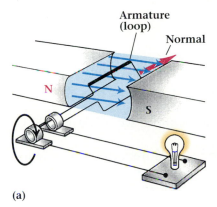

(a)

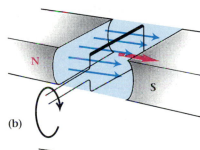

(b)

Armature (loop)

Normal

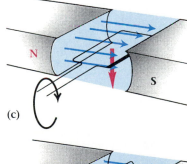

(c)

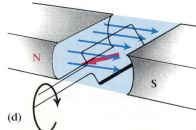

(d)

Figure 19.7 (*left*) Operation of an ac generator, called an alternator. (a) A loop of wire (the armature) rotates in a magnetic field. The induced alternating emf enters the external circuit through metal brushes that rub against rotating metal rings, called slip rings, attached to the armature. As the armature rotates clockwise to (b), the magnetic flux increases to a maximum. When the armature rotates to (c) the flux is zero. At position (d) the flux is negative relative to positions (a) and (b).

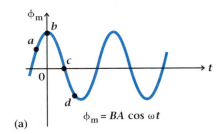

(a)

$$\phi_m = BA \cos \omega t$$

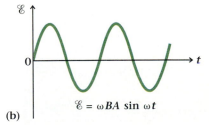

(b)

$$\mathcal{E} = \omega BA \sin \omega t$$

Figure 19.8 (*above*) (a) Graph of the magnetic flux as a function of time for an armature rotating at constant angular velocity. The points on the curve refer to the positions of the armature in parts (a–d) of Fig. 19.7. (b) Graph of the voltage output of an alternator versus time.

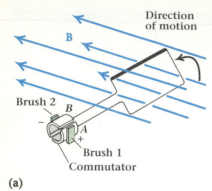

Direction of motion

B

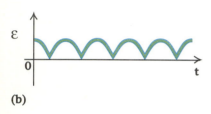

Brush 2

B

A

Brush 1

Commutator

(a)

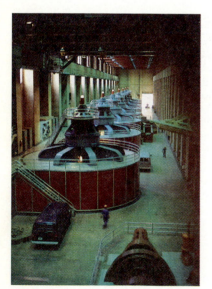

$\mathscr{E}$

0   t

(b)

Figure 19.9   A dc generator. (a) As the loop rotates in the magnetic field, an induced emf is produced between contacts A and B. As the loop rotates to a position where the polarity of the emf between the commutator contacts reverses, the commutator moves to a new position so that contact A touches brush 2 and contact B touches brush 1. Thus the polarity of the output potential from brush 1 to brush 2 is maintained. (b) Graph of the voltage output of the dc generator versus time.

Now, the change of cos ωt with the change in time is just the slope of the curve of cos ωt versus t. In Chapter 13 we saw that the rate of change of the cosine function was proportional to the sine function. Using this result in the last equation, we get

$$\mathscr{E} = \omega BA \sin \omega t. \qquad (19.5)$$

Thus the output of an alternator is a sinusoidal emf. When an alternator is connected to a closed circuit, it produces a sinusoidally **alternating current** (Fig. 19.8b). In the next chapter we will examine alternating-current circuits, which make up the majority of electric circuits you are familiar with. When we mention ac sources there, we will generally be referring to alternators.

A dc (unidirectional) generator can be made from the same rotating armature if we replace the slip rings by a commutator (Fig. 19.9a). The commutator is simply a slip ring split across the diameter, so that as the loop rotates, it changes the direction of its connection to the output circuit. During one half of a rotation, $\Delta\phi_m/\Delta t$ is positive and the induced current goes toward brush 1. During the other half of the rotation, the change in flux is negative and the induced current goes in the other direction; however, because the commutator switches its contact with the brushes, the current still goes toward brush 1. If the brush contacts are properly arranged with respect to the commutator and the position of the loop, the polarity of the emf can remain constant (Fig. 19.9b).

Practical generators are similar in principle to the single-loop generator just described. In a practical generator many turns of wire are wrapped about a rotor or armature, which turns in a magnetic field. The field may be due to permanent magnets or to electromagnets surrounding the rotor (Fig. 19.10).

As long as the external circuit is open, no current flows through the armature and it turns freely. But when the terminals of the generator are joined to a closed circuit so that current flows through the loop, a torque results that resists the motion of the loop (Fig. 19.11). Mechanical energy must be supplied to the generator to overcome this countertorque. As more current is drawn through the loop, more external torque is required to turn the generator. In other words, mechanical energy must be supplied as electrical energy is used in the external circuit.

Example 19.4

Emf of a rotating loop.

A single loop of area 0.010 m² is rotated about an axis perpendicular to the earth's magnetic field (about 5.0×10^{-5} T) at a frequency of 60 Hz. What is the peak emf generated by the loop?

Figure 19.10   A row of ac generators at the Hoover Dam hydroelectric plant.

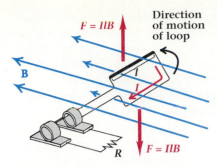

Figure 19.11   As the loop rotates counterclockwise, an emf is induced, which, according to Lenz's law, causes a current through the loop in the direction shown by the arrows. The force F of the field on the current-carrying wire is $F = IlB$, which causes a torque opposing the motion of the loop.

Solution   The peak emf is given by Eq. (19.5) with sin $\omega t = 1$:

$$\mathcal{E} = \omega BA.$$

The angular frequency $\omega = 2\pi f = 2\pi(60$ Hz$) = 377$ rad/s. The emf becomes

$$\mathcal{E} = (377 \text{ rad/s})(5.0 \times 10^{-5} \text{ T})(0.010 \text{ m}^2),$$
$$\mathcal{E} = 0.19 \times 10^{-3} \text{ V} = 0.19 \text{ mV}.$$

An **electric motor** is similar in operation and design to a generator, but works in reverse. A single-loop armature rotates in a magnetic field when a current is passed through the loop. For example, if we apply an external dc potential to the brushes of the elementary dc generator of Fig. 19.9, producing a current in the loop, the magnetic field will exert a torque on the armature, causing it to rotate. As the armature rotates through the position where the plane of the loop is perpendicular to the field, the commutator reverses the contacts, so a torque in the same direction continues to turn the armature. Here the input is electrical energy (from an external source) and the output is mechanical energy (a torque acting through an angle).

Although small motors are often built with permanent magnets, most motors are built with electromagnets providing the field for the armature. In the simplest type of motor, the armature is wired in series with this field magnet. In such a series-wound motor the magnetic field increases proportionally with increases in armature current. Since the torque depends on both the field strength and the current, we can control the speed of the motor by regulating its current. Motors of this kind are often used in power tools. In a shunt-wound motor, the field magnet is wired in parallel with the armature winding. In this case, the motor runs with a speed that is nearly independent of the load on the motor. Both series-wound and shunt-wound motors are used primarily in dc applications.

Most of the motors you are likely to use are induction motors, which have no contacts to the armature. Instead, the armature windings are wrapped around an iron core. The motor is designed so that an alternating current applied to the field electromagnet produces a rotating magnetic field, which induces a current in the armature windings. The torque on the armature due to the rotating field causes the armature to rotate. Induction motors are widely used in fans, refrigerators, and other common appliances. Induction motors can be designed to rotate synchronously with the field, which means that their rotation frequency is directly related to the frequency of the alternating current. Motors of this type are used in phonograph turntables and electric clocks.

We have already mentioned the presence of countertorque in generators. A related effect occurs in motors. As the armature moves in the field, the motion induces an emf in the armature. This emf is directed in the opposite sense to the applied voltage and is thus called the **back emf**. When a motor is first switched on, a large current surges through the armature wire. As the motor gets up to speed, the back emf opposes the applied potential. The net effect is a reduced potential across the armature and thus a smaller armature current.

You often see the effect of back emf in household circuits when you turn on a large appliance motor. For example, a momentary dimming of the lights often accompanies the start-up of an air conditioner or refrigerator motor. The dimming occurs because the terminal potential difference of the house circuit drops as the motor draws a large start-up current. As the motor comes up to speed, its current demand decreases, along with the corresponding I^2R losses in the supply circuit. When this happens, the household circuit potential rises back to its initial value.

Because the back emf reduces the current, a motor that can be safely operated under normal conditions may become overheated and burn out if its rotation is hindered or stopped. For example, an electric mixer that is unable to turn because the batter to be mixed is too thick or because a spoon is caught in its blades will draw more current than it is designed to carry. Consequently, it will overheat, and it may even burn out, if it is not switched off or disconnected from the power line.

Example 19.5

Finding the back emf of a motor.

When the motor in a window air conditioner is first turned on, it momentarily draws 40 A. Then the current quickly drops to a steady value of 13.8 A. If the motor is operated on a 120-V power source, what is the back emf generated while the motor is running?

Solution   For the purpose of this problem we will approximate a motor by a series connection of an inductive source of emf and a resistance. If we apply Kirchhoff's voltage rule (Section 17.6) to the circuit, we obtain

$$V - \mathscr{E} = IR,$$

where V is the 120 V of the power line, $\mathscr{E}$ is the back emf of the motor, I is the current in the circuit, and R is the resistance of the motor. When the air conditioner is first turned on the motor is at rest, there is no back emf ($\mathscr{E} = 0$), and the current is 40 A. The resistance can then be determined:

$$R = \frac{V}{I} = \frac{120 \text{ V}}{40 \text{ A}} = 3.0 \ \Omega.$$

When the motor is running at normal speed the current is 13.8 A, so the back emf must be

$$\mathscr{E} = V - IR = 120 \text{ V} - 13.8 \text{ A} \times 3.0 \ \Omega = (120 - 41) \text{ V} = 79 \text{ V}.$$

The Transformer

19.4

An extremely important application of electromagnetic induction is the **transformer**, a device featuring two multiturn coils of wire wound on the same iron core (Fig. 19.12). The common ferromagnetic core increases the magnetic flux due to the input current and maximizes the magnetic coupling between the coils. The transformer is used for changing ac voltages without appreciable power loss. We use it by applying an ac voltage

Figure 19.12 (a) Schematic diagram of a transformer. We can choose the number of turns on the input coil N_1 and the output coil N_2 to obtain the desired ratio of input voltage to output voltage. (b) Circuit symbol for a transformer.

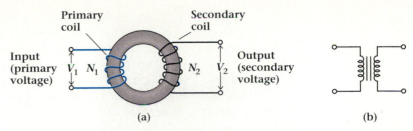

(a)   (b)

V_1 across the primary coil (input) of the transformer. The resulting current induces a changing magnetic flux through the iron core that is proportional to the number of turns N_1 in the primary coil. This changing flux travels through the core and intercepts the secondary winding, where it induces a voltage V_2 proportional to the number of turns N_2. This voltage is the output voltage of the transformer. The ratio of input to output voltages is

$$\frac{V_1}{V_2} = \frac{N_1}{N_2}. \tag{19.6}$$

By a suitable choice of N_1 and N_2, we can make the transformer step up (increase) the voltage output or step down (decrease) the output with respect to the input.

If we connect a resistive load across the secondary coil of the transformer, a current I_2 will pass through the load. This current is due to the potential difference V_2 induced by the changing flux. If we neglect any losses within the transformer, the law of conservation of energy requires that the electrical energy output of the transformer equal the electrical energy input. Since power is the rate of change of energy, this is equivalent to saying that the input and output powers are equal. The power is $P = IV$, so we can express the conservation rule as

$$I_1V_2 = I_2V_2. \tag{19.7}$$

The above argument assumes no energy loss due to Joule heating or flux leakage. In practice, these effects make the output power less than the input power. However, in a well-designed transformer the loss is a small fraction of the total power. Typical transformers have losses of about 4–8%.

The primary application of the transformer is in raising or lowering an ac voltage to a desired level. But, as seen in Eq. (19.7), this change cannot be achieved without affecting the available current. That is, for a given input voltage and current, you cannot increase the output voltage without decreasing the available output current. This is a consequence of energy conservation.

The transmission of electrical power from a generating station to the consumer often takes place over many miles of wire. In order to minimize the I^2R losses in the transmission lines, it is standard practice to use transformers to step up the voltage to very high levels (Fig. 19.13). Typical transmission-line voltage is 138 kV, but in some instances voltages as high

Figure 19.13 Commercial transformers at the Hoover Dam hydroelectric power station. You can judge the size of the transformers from the truck parked near them.

as 1 MV are used. Because power is the product of current and voltage, the current required for a given level of power is reduced when the transmission lines are operated at high voltage. At the end points of the distribution system, the voltage levels are stepped down to more modest levels. These features—the ability to shift voltage levels up and down and to minimize I^2R losses during transmission—are important reasons for the use of ac instead of dc electricity in power distribution networks.

Example 19.6

Turns ratio in a doorbell transformer.

A doorbell transformer steps down an ac voltage from 120 V to 24 V. What is the ratio of turns on the primary winding to turns on the secondary winding?

Solution   The ratio of the voltages is equal to the ratio of the number of turns. Thus

$$\frac{N_1}{N_2} = \frac{V_1}{V_2},$$

$$\frac{N_1}{N_2} = \frac{120 \text{ V}}{24 \text{ V}} = 5.$$

That is, the primary coil has five turns of wire for each turn of wire in the secondary coil.

Inductance

19.5

Let's consider a coil of N turns of wire carrying a current I. The current generates a magnetic field that passes through each loop of the coil (Fig. 19.14a). The total magnetic flux passing through the coil, which is the product of the number of turns and the flux through each turn ϕ_m, is proportional to the current carried by the coil. We can therefore write

$$N\phi_m = LI. \tag{19.8}$$

The proportionality constant L is called the **inductance** of the coil, and depends only on the shape and size of the coil. Sometimes L is referred to as self-inductance, to distinguish it from the mutual inductance between two coils.

If the current in a coil changes, as in the case of alternating current, a corresponding change in the flux occurs, given by

$$\Delta(N\phi_m) = L \, \Delta I.$$

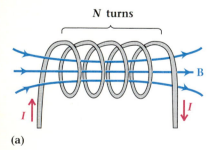

N turns

(a)

(b)

Figure 19.14   (a) A current I generates a flux ϕ passing through N turns. The proportionality constant is the inductance of the coil. When the current changes the flux changes, inducing an emf that is proportional to the change in current. (b) The circuit symbol for an inductor.

According to Faraday's law, the changing flux induces an emf given by

$$\mathcal{E} = -\frac{\Delta(N\phi_{\mathrm{m}})}{\Delta t} = -L\frac{\Delta I}{\Delta t}. \tag{19.9}$$

Thus a changing current in a coil induces an emf opposing that change, and this induced emf is proportional to the inductance of the coil.

The units of inductance are volt-seconds per ampere. This combination of units has been given the name **henry** (abbreviated H):

1 henry = 1 volt-second/ampere.

Any circuit component that exhibits inductance, like a solenoid or other wire coil, is called an **inductor**. In a circuit, an inductor may be regarded as a source of emf of $-L(\Delta i/\Delta t)$. As such, an inductor acts to resist changes in current. Along with resistors and capacitors, inductors are found as circuit elements in many applications including radios, television sets, and tape recorders. The circuit symbol used to represent the inductor is a stylized coil (Fig. 19.14b).

Example 19.7

Inductance of a solenoid.

Calculate the self-inductance L of a solenoid containing N turns of wire in a coil of length l and cross-sectional area A. Then find the value of L, given that $N = 100$ turns, $l = 3.0$ cm, and $A = 2.0$ cm^2.

Solution   The inductance L is given by Eq. (19.8) as

$$L = \frac{N\phi_{\mathrm{m}}}{I}.$$

The field of a solenoid was calculated in Chapter 18 to be

$$B = \mu_0 nI = \mu_0\left(\frac{N}{l}\right)I.$$

Since the field is uniform inside the solenoid, the corresponding flux is

$$\phi_{\mathrm{m}} = BA = \frac{\mu_0 NIA}{l}.$$

The inductance thus becomes

$$L = \left(\frac{N}{I}\right)\left(\frac{\mu_0 NIA}{l}\right) = \frac{\mu_0 N^2 A}{l}.$$

We can compute the numerical size of L from this last equation if we insert the values for N, l, A, and μ_0. Thus the inductance becomes

$$L = \frac{4\pi \times 10^{-7}\ \mathrm{N/A^2} \times (100)^2(2.0 \times 10^{-4}\ \mathrm{m^2})}{3.0 \times 10^{-2}\ \mathrm{m}},$$

$$L = 8.4 \times 10^{-5}\ \mathrm{H} = 84\ \mu\mathrm{H}.$$

Note that we calculated the inductance from purely geometrical considerations; inductance does not depend on current.

Inductors are usually made from many turns of wire wound about a supporting core. The greater the number of turns of wire, the greater the total magnetic flux for a given current, and that flux passes through more turns. So the inductance is larger for a larger number of turns. Inductors are commercially available in sizes ranging from a few microhenries to several hundred henries. Small inductors may be only a few turns of wire about a nonmagnetic support. Large inductors are made with many windings about a magnetic core. The magnetic material increases the magnetic flux passing through the coils of wire and thus increases the value of the inductance.

Inductors have resistance in addition to their inductance because the wire used in the windings necessarily has resistance. In a circuit, a real inductor contributes both inductance and resistance.

Energy Storage in an Inductor

*19.6

In Chapter 17 we found that the power delivered to a resistive load in an electric circuit was the product of current and electric potential difference across the resistor. When the load is an inductance of magnitude L, the power required to produce a current I changing at a rate $\Delta I / \Delta t$ is

$$P = \mathscr{E}I = LI\frac{\Delta I}{\Delta t}. \tag{19.10}$$

In a small time interval Δt, the circuit's source of emf delivers an amount of energy ΔW to the inductor, given by

$$\Delta W = P\,\Delta t = LI\,\Delta I.$$

The total energy required to build up the current from zero to some value I_0 is given by the sum

$$W = \Sigma\,\Delta W = \Sigma\,LI\,\Delta I.$$

This sum may be computed from the area under the graph of LI versus I, in the same way that we calculated the energy stored on a capacitor in Section 16.9. Thus the total energy is found to be

$$W = \tfrac{1}{2}LI_0^2. \tag{19.11}$$

This is the energy required to build up a current I_0 in an inductor.

If the current is held constant at I_0, no back emf is induced across the inductor and no further energy is required to sustain the current. (We are not now concerned with the energy loss due to the resistance present in a real inductor.) The energy of Eq. (19.11) has gone into creating a magnetic field surrounding the inductor. If the circuit is altered to reduce the current, the magnetic field decreases. The energy that was stored in the field provides the energy for the new emf of the inductor as it attempts to sustain the current and oppose the change.

We saw in Chapter 16 that when a capacitor was charged, energy was stored in the electric field within the capacitor. In the present case, we find that when an inductor carries a current, energy is stored in the magnetic field surrounding the inductor. In both cases, the stored energy can be released and used to do work.

Maxwell's Equations and Electromagnetic Waves

19.7

Over the last few chapters, we have studied the contributions to electromagnetic theory embodied in the laws of Coulomb, Gauss, Faraday, Ampère, and others. Each of these laws was discovered independently of the others, and each extended our understanding of electricity and magnetism in a distinct area. However, in 1865, at about the time that the American Civil War was ending, the great Scottish physicist James Clerk Maxwell expressed all these ideas and discoveries in one unified picture of electromagnetism. Maxwell showed that a complete description of electromagnetic effects could be based on a set of only four equations. These four equations, known today as **Maxwell's electromagnetic equations**, did not originate with him. But Maxwell showed conclusively that these four equations could be used to interpret and explain an impressive array of electromagnetic phenomena.

We shall not list the mathematical statements of Maxwell's equations, since their formal complexity is beyond the scope of this text. However, we can summarize them descriptively as follows:

1. **Gauss's law for electricity.** This law describes the electric field due to electric charges, and can be used to derive Coulomb's law.
2. **Gauss's law for magnetism.** According to this law, magnetic field lines are continuous and without end, and thus there are no isolated magnetic poles.
3. **Ampère's law.** Maxwell extended this law to describe the production of magnetic fields not only by electric currents but by changing electric fields as well.
4. **Faraday's law of induction.** This law describes the production of electric fields as a result of changing magnetic fields.

It was Faraday who introduced the concepts of fields and of lines of force to describe the interaction of objects separated in space. This formulation still remains a powerful way to describe electric, magnetic, and other fields. But Maxwell's precise mathematical description of these fields, and their reality, as demonstrated by many experiments, remains an important part of our understanding of electromagnetism. In addition, Maxwell showed that electric and magnetic fields in the form of waves radiate from an oscillating electric charge. The experimental confirmation of these **electromagnetic waves** did not come in Maxwell's lifetime, but was achieved by Heinrich Hertz (1857–1894) in 1887, eight years after Maxwell's death. This discovery opened the way to the wireless telegraph, radio, television, microwaves, and radar. In fact, today Maxwell's equations are considered the foundation of our understanding of electromagnetism, equal in importance to Newton's laws of motion in mechanics.

Electromagnetic waves can be generated by oscillating electric charges. When a sinusoidally varying current source is attached to a simple antenna (Fig. 19.15) a sinusoidally oscillating electromagnetic wave radiates from the antenna. Consequently, at some distance away from the antenna an electric field arises that oscillates sinusoidally with time as the wave passes.

The direction of the electric field vector in the radiated wave is determined by the orientation of the antenna. This electric field makes up only part of the electromagnetic wave, for there is also a magnetic field

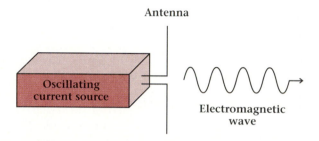

Figure 19.15 A sinusoidally oscillating electromagnetic wave radiates from an antenna driven by a sinusoidally varying current.

PHYSICS IN PRACTICE

Linear Accelerators for Radiation Therapy

Radiation therapy is a common medical tool, especially for the treatment of cancer patients. One common source of radiation is a precisely controlled beam of electrons produced with a linear accelerator, a device that accelerates charged particles from lower to higher velocities. These electron beams can be used directly or they can be used to generate highly energetic x rays. (Production of x rays is discussed in Chapters 25 and 26.)

We can construct a linear accelerator in which charged particles are accelerated in a straight line by using a series of conducting tubes connected to an alternating voltage supply (Fig. B19.1). Positively charged particles traveling along the axis of the tubes are accelerated from left to right across the gap between the tubes if the electric field between them is directed from left to right. For example, if the first tube is positive and the second tube is negative, a positively charged particle is accelerated across the gap between them. If the voltage changes to make the second tube positive and the third tube negative by the time the particle has reached the gap between them, the particle is again accelerated across the gap. If we adjust the period of the alternating voltage supply to match the transit time of the particles through each pair of adjacent tubes, the particles accelerate as they cross each gap between the tubes.

When electrons are accelerated in a linear accelerator of this type, they quickly reach very high speeds because of their small mass. Even if the voltage oscillates at radio frequencies ($\approx 10^6$ Hz), the tube lengths become excessively long due to the electron velocity.

Figure B19.1   A simple linear accelerator, consisting of conducting tubes alternately connected to opposite sides of an alternating voltage source.

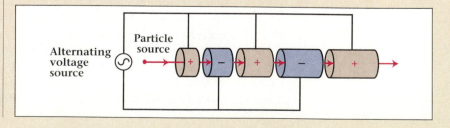

associated with the moving charges. The complete wave has a sinusoidally oscillating magnetic field coupled to the electric field.

At distances far from the antenna the wave is essentially a plane wave; that is, at a given instant of time the electric and magnetic fields are uniform over a plane perpendicular to the direction of propagation. The electric and magnetic fields are perpendicular to the direction of propagation and to each other. These fields vary sinusoidally with space and time (Fig. 19.16). The electromagnetic wave is a transverse wave because

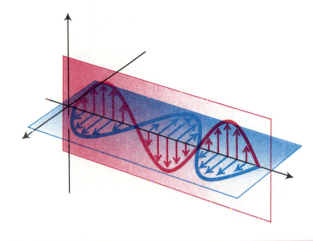

Figure 19.16 The appearance of the electric and magnetic fields in a plane electromagnetic wave. Each field varies sinusoidally with space and time. The speed of the wave is c, the speed of light.

If the source is made to oscillate at the even higher frequencies of microwave generators ($\approx 10^9$–10^{12} Hz), the wavelengths of the microwave oscillations are comparable to the dimensions of the tubes, and a system like that of Fig. B19.1 is no longer practical.

In the contemporary linear accelerator the accelerator tubes are replaced by a waveguide, that is, a hollow metal conductor through which high-frequency microwaves are propagated. This waveguide can be used to accelerate electrons, which are injected into the waveguide and travel in the same direction as the waves. An electron moving down the symmetry axis will be accelerated, depending on its position in the field. If the wave is propagating to the right, an electron moving with the same velocity, and initially placed at a position in the field where it experiences a force to the right, will be continuously accelerated. In this

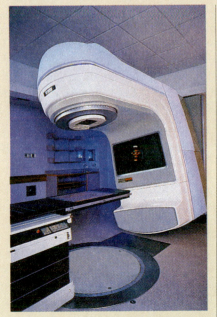

Figure B19.2 A linear accelerator for radiation therapy.

manner the electrons are moved along with the wave in much the same way that a surfer rides along on an ocean wave.

The electrons move along the waveguide with an increasing speed that approaches the speed of light. They gain energy from the electromagnetic field as they go. Electron beams with energies equal to that of electrons accelerated through a potential of 15 to 20 MV can be generated in waveguide systems less than one meter long. Linear accelerators are reliable sources of intense beams of high-energy electrons. Precise energy control is obtained by careful control of the microwave frequency. In many cases the electron beams are directed into a metal target where they produce x rays on impact. These intense beams of highly penetrating x rays are then directed at cancerous tumors to destroy their cells (Fig. B19.2).

the wave disturbance (the field) is transverse to the direction of propagation.

The electric field of broadcast television waves is horizontal because of the orientation of the transmitting antennas. This is the reason that television receiving antennas are made as they are. It is also the reason that one turns an antenna to "aim" it at the station. The maximum signal and hence the best reception occur when the elements of the antenna are aligned parallel to the direction of the electric field in the radiated wave.

Analysis of Maxwell's equations not only predicts the existence of electromagnetic waves, but also predicts the speed of propagation of the waves. The value predicted for the speed depends on the value of the constant ϵ_0 found in Coulomb's law and on the value of μ_0 found in Ampère's law. The speed of an electromagnetic wave is given by

$$c = \frac{1}{\sqrt{\mu_0 \epsilon_0}}.$$

By carefully making the appropriate electrical measurements to determine the value of these constants, the speed of electromagnetic waves was calculated. Such measurements, largely made between the time of Maxwell's first work and the end of the nineteenth century, gave results in agreement with the value for the velocity of light. The inference, later shown to be correct, was that light is an electromagnetic wave.

The properties and behavior of light, discussed in Chapters 21 through 23, can all be deduced from Maxwell's equations. These include not only wavelike behavior of the type just discussed, but the straight-line propagation that we normally observe and even the bending of the light path as it passes through a lens or prism.

Over the years, many determinations of the speed of light have been made. (Some of these methods are described in Chapter 21.) By the early 1980s the best value for the speed of light in vacuum was $c = 299{,}792{,}458 \pm 1.2$ m/s. The primary limitation of the measurement was the precision with which the length of the meter could be established. In 1983 the Seventeenth General Conference on Weights and Measures adopted a new definition of the meter, based on the best value for the speed of light:

The meter is the length of path traveled by light in vacuum during a time interval of 1/299,792,458 of a second.

By the late nineteenth century, careful study of Maxwell's theory showed it to be in conflict with the formulation of mechanics by Galileo and Newton. This created considerable consternation for the physicists of the time. Maxwell's theory gave a unique value to the speed of light, but Newtonian mechanics did not allow this uniqueness and was unable to account satisfactorily for the behavior of objects whose speeds approached the speed of light. It became painfully obvious that one of these hallowed theories had to be incorrect. The problem was resolved in 1905 when Albert Einstein described his theory of special relativity. According to relativity, Maxwell's theory of electromagnetism was correct, but Newtonian mechanics was incomplete. In Chapter 24, we will discuss Einstein's resolution of this problem and some of his unexpected results.

SUMMARY

Helpful Concepts

- Magnetic flux is defined as

$$\phi_m = BA \cos \theta.$$

- Faraday's law of induction gives the emf in a coil of N turns due to changing magnetic flux,

$$\mathscr{E} = -N \frac{\Delta \phi_m}{\Delta t}.$$

- Lenz's law states that induced current occurs in a direction that opposes the change in flux. Thus the minus sign in Faraday's law is an expression of Lenz's law.
- The ratio of input to output voltages of a transformer is equal to the ratio of turns in the primary coil to the turns in the secondary coil:

$$\frac{V_1}{V_2} = \frac{N_1}{N_2}.$$

- The voltages and currents in a transformer's primary and secondary coils are related by

$$I_1 V_1 = I_2 V_2.$$

- In a coil of N turns carrying a current I and through which a magnetic flux ϕ_m passes, the inductance is

$$L = \frac{N \phi_m}{I}.$$

- The emf induced in an inductor by a changing current is

$$\mathscr{E} = -L \frac{\Delta I}{\Delta t}.$$

- The energy stored in an inductor carrying current I_0 is

$$W = \tfrac{1}{2} L I_0^2.$$

- The laws of electromagnetism are summed up in Maxwell's equations.
- The speed of light is

$$c = 299{,}792{,}458 \text{ m/s}, \quad \text{or} \quad c \approx 3 \times 10^8 \text{ m/s}.$$

Important Terms

You should be able to write the definition or meaning of each of the following terms:

- Faraday's law
- induced emf
- induced current
- Lenz's law
- generator
- alternating current
- electric motor

- back emf
- transformer
- inductance
- henry
- inductor
- Maxwell's equations
- electromagnetic waves

QUESTIONS

19.1 Discuss several different ways of producing an emf in a wire.

19.2 A flexible wire is held in the shape of a loop perpendicular to a magnetic field. The ends of the wire are suddenly jerked so that the wire is pulled taut. Is an emf generated in the wire? Explain.

19.3 Most automobiles use alternators to provide the emf for charging the battery. How can this be done, since the battery requires a unidirectional (dc) current for charging?

19.4 The belt on a belt-driven generator breaks while the generator is being used to charge a storage battery. You observe that the generator continues to turn. What is the explanation of this observation?

19.5 A bar magnet is moved toward a circular loop of wire as shown in Fig. 19.17. In which direction does the induced current flow? In which direction will it flow if the magnet is moved away from the loop?

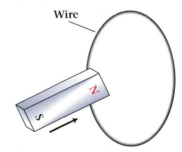

Figure 19.17 Question 19.5.

19.6 How does the inductance of a coil change when an iron core is put into it?

19.7 Two coils connected in series are originally so far apart that the magnetic field of each does not overlap the other coil. If they are brought together so that there is flux in each coil due to current in the other, can the total inductance increase? Can they be brought together in such a way that the total inductance is decreased? Explain your answers.

19.8   An airplane is flying from west to east in the northern hemisphere. Which wing tip (left or right) acquires a positive electric potential relative to the other tip as a result of the motional emf induced by the earth's magnetic field?

19.9   Explain why a spinning metal top slows down when a magnet is brought near.

19.10   Hospitals have emergency generators to provide electricity when the incoming electric service is interrupted. Usually, the emergency generators are de-signed to provide three times as much power as the hospital normally uses. Why is it necessary to have generators with so much capacity?

19.11   A coil of many turns, connected to a battery, carries a steady current. Is energy stored in this circumstance? Where? How can you test the correctness of your answer?

19.12   In the United States the standard frequency for electrical power distribution is 120 V at 60 Hz, while in many other parts of the world it is 240 V at 50 Hz. If a transformer is available to provide the proper voltage level, what kinds of electrical appliances or equipment can be used on either system and what kinds cannot be used?

PROBLEMS

Hints for Solving Problems
An emf is induced in a conducting loop when the magnetic flux through the loop changes. If the loop makes a closed circuit, an induced current is generated in a direction to create a magnetic field that opposes the change in flux. An inductor behaves as a source of emf when the current in it is changing. The direction of the inductor emf is such that it opposes the change in current.

Section 19.1   Faraday's Law

19.1   A bar magnet is rotated at a steady rate about its center (Fig. 19.18). Describe the effect this will have on a loop of wire located nearby. Sketch a graph of the voltage output of the loop as a function of the position of the magnet and as a function of time.

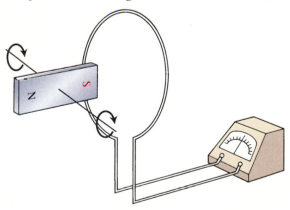

Figure 19.18   Problem 19.1.

19.2   A wire coil of 25 turns has a cross-sectional area of 10 cm². The coil is placed in a uniform field of a large magnet with $B = 0.20$ T. The coil is suddenly

rotated 90° from an orientation parallel to the field to one perpendicular to the field. If the time to flip the coil is 0.50 s, what is the average emf produced? What will be the current if the circuit of the coil is closed and the resistance is 75 Ω?

19.3   A coil contains 100 turns of wire in a loop 15 cm in diameter. The loop is placed between the poles of a large electromagnet, $B = 1.0$ T, with the plane of the loop perpendicular to the field. If the magnetic field is steadily reduced from 1.0 T to 0 T in 16 s, what is the average emf in the coil while the field is changing?

19.4   A flat coil of 300 turns is wound into a square loop 5.0 cm on a side. The coil is designed so that it can be rotated by 90° in 0.25 s. The coil is placed so that the magnetic flux is zero and it is rotated so that the flux is a maximum. If the average voltage on the coil due to the induced emf is 0.30 mV, how large is the magnetic field?

19.5   The flux in a single-loop coil of area 37 cm² steadily changes from 6.5×10^{-3} T to 9.3×10^{-3} T in 0.50 s. What emf is induced in the coil?

19.6   The plane of a square loop of wire with edge length of 8.0 cm is perpendicular to a 0.017-T magnetic field (Fig. 19.19a). What is the average emf between

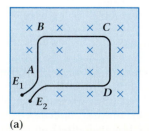
(a)

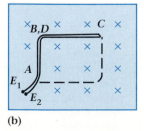
(b)

Figure 19.19   Problem 19.6.

the points E_1 and E_2 when the corner D is quickly folded about the diagonal AC so as to lie on top of B (Fig. 19.19b) if it takes 0.13 s to make the fold?

19.7 A "flip coil" of 50 mm radius and 50 turns is oriented perpendicular to a magnetic field. The coil is quickly flipped through 180° about its diameter in 0.070 s. The average emf in the coil is measured to be 2.0 V. What is the strength of the magnetic field in the region of the coil?

19.8 A copper wire (ab) is wound around a wooden rod (Fig. 19.20). Another wire (cd) is wound on top of it and connected to a resistor R. What is the direction of current in R if (a) a current flowing from a to b is increased, (b) a current flowing from a to b is decreased, (c) a current flowing from b to a is increased, and (d) a current flowing from b to a is decreased?

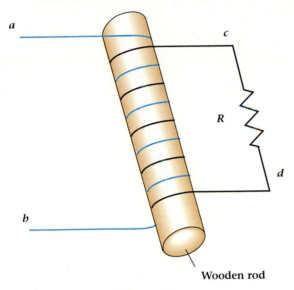

Figure 19.20   Problem 19.8.

19.9 A coil 17 cm in diameter and wound with three turns of wire is placed with the plane of the coil at right angles to a magnetic field of 3.8×10^{-2} T. What emf is induced in the coil if (a) the field is halved in 0.27 s, (b) the field is reversed in 0.27 s, (c) the coil is rotated through an angle of 90° about its diameter in 0.27 s, (d) the coil is rotated through an angle of 180° about its diameter in 0.27 s?

Section 19.2   Motional Emf

19.10 A flat-bed truck travels due south at 120 km/h in a location where the vertical component of the earth's magnetic field is 5×10^{-6} T. What is the emf induced in a 1.0-m-long copper bar held horizontally above the bed of the truck and perpendicular to its direction of motion?

19.11 A 0.57-m-long rod lies in a plane containing a magnetic field $B = 7.0 \times 10^{-3}$ T. The axis of the rod makes an angle of 60° with the field as shown (Fig. 19.21). The rod moves perpendicular to the plane at a speed of 1.03 m/s in the direction shown. (a) What is the emf between the two ends of the rod? (b) Which end is at the higher potential?

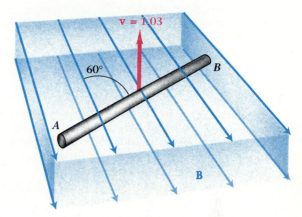

Figure 19.21   Problem 19.11.

19.12 What emf is developed between the tips of the wings of a Boeing 767 jet liner in level flight at 850 km/h in a location where the vertical component of the earth's magnetic field is 1.75×10^{-5} T? The distance between the wing tips of the 767 is 47.65 m.

Section 19.3   Generators and Motors

19.13 A 50-turn loop of wire has a cross-sectional area of 0.010 m². It is rotated about an axis perpendicular to the earth's field of 5.0×10^{-5} T. What is the peak value of the emf generated if the loop rotates at 600 rotations per minute?

19.14 What would be the peak emf of the generating coil of Problem 19.13 if it were turned in a magnetic field of 1.0 T?

19.15 A rectangular loop with an area of 120 cm² is placed in a magnetic field of 1.3×10^{-2} T and rotated about its long axis at 5.0 Hz. What is the instantaneous emf generated at the instant when the normal to the loop makes an angle $\theta = 0.27°$ with the magnetic field?

19.16 A single loop of area 0.173 m² is placed perpendicular to a magnetic field of 0.354 T. The loop is rotated about an axis lying in the plane of the loop and passing through its center. An instantaneous emf of 8.41 V is generated when the plane of the loop makes an angle of 45.0° with the field direction. What is the angular speed of the loop?

Section 19.4   The Transformer

19.17 A transformer is made by winding a primary coil of 500 turns around an iron core. A secondary winding of 50 turns is wound around the same core. If the primary voltage is 120 V, what is the output voltage on the secondary?

19.18 A transformer is made by winding a primary coil of 300 turns around an iron core. A secondary winding of 750 turns is wound about the same core. If the primary voltage is 120 V, what is the output voltage on the secondary?

19.19 A doorbell transformer has an output voltage of 12 V when connected to a 120-V household supply. What is the ratio of the number of turns on the primary coil to the number of turns on the secondary coil?

19.20 A transformer attached to a 138-kV transmission line has an output voltage of 38 kV. What is the ratio of the number of turns on the primary coil to the number of turns on the secondary coil?

19.21 A transformer attached to a 10-kV transmission line has 25 times as many primary turns as it does secondary turns. What is the output voltage?

19.22 A step-down transformer that converts 120 V ac to 10 V ac is rated for 1.5 A output current on the secondary winding. What is the input current when the output is 1.5 A?

19.23 A step-up transformer changes 115 V ac to 230 V ac. What is the magnitude of input current when the output is 2.0 A?

19.24 Two transformers are connected as shown (Fig. 19.22), where N_1 and N_2 are the number of primary and secondary turns on the transformer T_1, and N_3 and N_4 are the number of primary and secondary turns on the transformer T_2. Show that

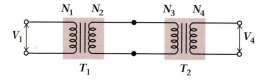

Figure 19.22 Problem 19.24.

19.25 A low-voltage outdoor lighting system uses a transformer that steps the 120-V household voltage down to 24 V for safety. The equivalent resistance of all of the low-voltage lamps is 9.6 ohms. (a) What is the current in the secondary coil? (b) What is the current in the primary coil? (c) How much power is used, neglecting losses in the transformer and line?

Section 19.5   Inductance

19.26 Show that the units of μ_0 are henries per meter.

19.27 Calculate the self-inductance of a small solenoid 5.0 cm long made of 10 turns of wire, each loop enclosing 1.0 cm². Assume that end effects are negligible.

19.28 An inductor is wound about a paper core 5.0 cm long and 12 cm² in cross section. How many turns of wire are required to make an inductance of 1.3 mH?

19.29 A 0.80-H inductor carries a current that decreases at a rate of 0.10 A/s. What is the induced emf?

19.30 A 25-mH inductor carries a current that is increasing at a rate of 1.25×10^{-2} A/s. What is the emf induced in the inductor and what is its polarity?

19.31 At what rate does the current change in a 35-mH inductor when an emf of 0.019 V develops across it?

*Section 19.6   Energy Storage in an Inductor

19.32 A 50-mH inductor carries a current of 48 mA. How much energy is stored in the field of the inductor?

19.33 A 38-mH inductor stores 4.0×10^{-5} J when carrying a dc current. What is the magnitude of that current?

19.34 A 0.8 H-inductor carries a current of 0.50 A. How much energy is stored in the field of the inductor? What is the rate of energy loss in the inductor if it has a resistance of 1.00 Ω?

Section 19.7   Maxwell's Equations and Electromagnetic Waves

19.35 Calculate the speed of light using the relationship $c = 1/\sqrt{\mu_0 \epsilon_0}$.

19.36 Imagine that there exists another universe with the same laws of physics that we know except that in that universe the force between charged particles is only 1/4000 as much as in our universe. What is the speed of light in this new universe?

19.37 A computer is designed to perform calculations in nanoseconds (ns). What is the maximum separation between any two elements in the computer if an electrical signal is to go from one to the other and back again in 1.0 ns? (*Hint:* The maximum velocity of the propagation of an electrical signal is the velocity of light.)

Additional Problems

19.38 A copper rod 25 cm long rotates about one end in
•• a plane perpendicular to a 3.0×10^{-2}-T uniform magnetic field. The rod rotates at a rate of 2.5 revolutions per second. The outer end makes contact with a stationary conducting ring (Fig. 19.23). (a) What is the emf between the conducting ring R

and the center of rotation A? (b) If a 2.3-Ω resistor is connected between A and R, how much power is required to keep the copper rod turning at the same rate? Neglect friction and neglect the resistance of the bar and the ring.

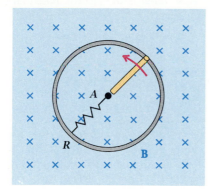

Figure 19.23   Problem 19.38.

19.39 A 1.0-m-long aluminum bar is held horizontally in
• the east-west direction and dropped from a height of 12 m at a place where the horizontal component of the earth's magnetic field is 4.0×10^{-5} T. What is the emf between the ends of the bar just before it strikes the ground?

19.40 What is the emf between the ends of a 1.0-m-long
•• light metal chain if you swing it in a horizontal circle about your head at 2.0 revolutions per second at a location where the vertical component of the earth's magnetic field is 2.5×10^{-5}T?

19.41 Two 1000-turn secondary coils are wound on the
•• same transformer, which has a 500-turn primary coil. The secondary coils are each connected to identical resistors. If the input to the primary coil is 0.150 A at 120 V, what is the current in each secondary coil? (*Hint:* Use conservation of energy.)

19.42 A resistor R is connected across the secondary coil
• of a transformer with N_1 primary turns and N_2 secondary turns. Show that the relationship between the primary current I_1 and voltage V_1 is

$$I_1 = \left(\frac{N_2}{N_1}\right)^2 \frac{V_1}{R}.$$

19.43 A transformer has a primary voltage of 3750 V and
•• a primary current of 20 A. The secondary voltage is 1200 V and the secondary current is 60.25 A. If all of the losses in the transformer are resistive losses, what will be the emergent temperature of cooling oil which flows through the transformer at 5.0 kg/min if the initial temperature of the oil is 15°C? Take the specific heat of the oil to be 3.0 kJ/kg·°C.

19.44 The direct transmission lines from a generating sta-
•• tion to a manufacturing plant are 20 km long and have a resistance of 6.2 Ω. The generators produce electrical power at 13,800 V. (a) Assuming that no step-up transformer is used at the power station, determine the rate of energy loss in the transmission lines if the potential difference at the plant is 12,000 V. (b) Suppose that transformers are to be used to step up the voltage output from the generators and to step it down again at the plant. If the power loss in the line is to be cut to 10^{-3} of the loss found in part (a), what should be the output voltage of the transformer at the generating plant? (c) If electricity cost $0.08/kWh, how much money is saved in one year by using the higher voltage for transmission?

19.45 A transformer near your house steps down higher
•• voltage to 240 V for use by heavy appliances in your home. The transformer loses 4.0% of the input electrical energy by Joule heating and other effects. The ratio of primary to secondary turns is 52. What is the current in the primary if the only load on the secondary is a 1500-W electric stove?

19.46 Show that the self-inductance of an air-core toroid
• of cross-sectional area A and mean circumferential length l wound with N turns of wire is given by

$$L = \frac{\mu_0 N^2 A}{l}.$$

Use the approximation that the magnetic flux across the cross section is just the product of the area times the value of B at the mean radius to determine the self-inductance of the toroid.

19.47 Find the inductance of an air-core toroid with a
• cross-sectional area of 6.0 cm^2 and mean circumferential length of 30 cm, wound with 200 turns of wire. (*Hint:* use the results of Problem 19.46)

19.48 Use the results of Problem 19.46 to show that the
•• energy density in the magnetic field of the toroid is given by $u = B^2/2\mu_0$. (*Hint:* The magnetic field is confined within the toroid. Calculate the energy stored and relate it to the magnetic field of a toroid. The energy density is the energy divided by the volume.)

19.49 A conducting rod of resistance R is moved along
• horizontal rails joined at one end to make a loop similar to the one in Fig. 19.5. A uniform magnetic field **B** is perpendicular to the plane of the loop and extends over the region in which the rod moves. Assume that the rod moves with constant speed v and that the rails have negligible resistance and friction. (a) Find an expression for the emf around the loop. (b) Find an expression for the emf across the rod. (c) What is the current in the circuit? (d) What force is required to keep the rod moving with constant velocity?

19.50 A rod of length l and mass m slides down parallel
•• conducting rails making an angle θ with the hori-

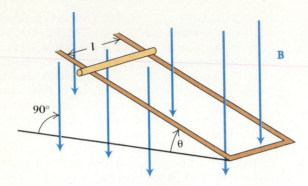

Figure 19.24   Problem 19.50.

zontal (Fig. 19.24). The rails have negligible resistance and are joined at the bottom to form a loop. A uniform magnetic field **B** is directed vertically downward in the region of the rails. What is the terminal velocity of the rod if it has an electrical resistance R? Ignore friction between rod and rails.

19.51   A conducting rod lies on two parallel, horizontal
•• rails 0.10 m apart that are connected by a resistor of 200 Ω in a region of uniform vertical magnetic field of 0.15 T (Fig. 19.25). How much work is required to move the bar along the rails at constant speed for a distance of 1.0 m in 0.25 s? Neglect friction between the bar and the rails and assume that the bar and the rails have no electrical resistance.

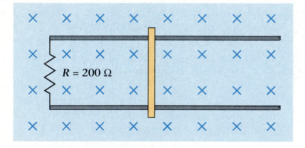

Figure 19.25   Problem 19.51.

19.52   A 500-μF capacitor is charged to 450 V. The capac-
•• itor discharges when connected across a 1.3-H inductor. What is the maximum current through the inductor? (*Hint:* Use conservation of energy and let all of the energy stored in the capacitor go into the magnetic field surrounding the inductor.)

19.53   A circular loop of wire with a diameter of 5.0 cm
•• and a total resistance of 1.6 Ω is placed inside a long solenoid with the axis of the loop parallel to the axis of the solenoid. The solenoid is 10.0 cm in diameter and has 2.0×10^4 turns per meter. At time $t = 0$ the current in the coil turns on and increases linearly with time at the rate of 0.020 A/s for 3.0 s, after

which it remains constant. How much heat (in joules) is given off by the circular wire loop during the time that current is changing in the solenoid?

19.54   At the instant of time $t = 0$ the current from A to
•• C in Fig. 19.26 is 30 A and is decreasing at the rate of 10 A/s. What is the electric potential difference between A and C at $t = 0$?

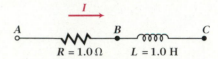

Figure 19.26   Problems 19.54 and 19.55.

19.55   For the circuit shown in Fig. 19.26, is it possible to
•• pick values of R, L, initial current from A to C, and $\Delta I/\Delta t$ so that the potential at A is the same as the potential at C?

19.56   Show that the equivalent inductance, L_S, of two in-
•• ductors added in series is

$$L_S = L_1 + L_2.$$

(*Hint:* The total potential difference across the two inductors is the sum of the individual emfs of each inductor and the rate of change of current with time is the same for both inductors.)

19.57   Show that the equivalent inductance, L_p, of two in-
•• ductors added in parallel is

$$\frac{1}{L_p} = \frac{1}{L_1} + \frac{1}{L_2}.$$

(*Hint:* Kirchhoff's law requires that the two emfs be equal and that the total current be the sum of the individual currents.)

19.58   A 480-mH inductor is connected in series with a 370-mH inductor. What is the total inductance of the combination? (*Hint:* See Problem 19.56.)

19.59   A 0.27-H inductor is connected in series with a second inductor to give a total inductance of 0.87 H. What is the inductance of the second inductor? (*Hint:* See Problem 19.56.)

19.60   A 0.83-H inductor is joined in parallel to a 0.52-H inductor. What is the inductance of the combination? (*Hint:* See Problem 19.57.)

19.61   Two inductors connected in parallel have an effective inductance of 0.037 H. If the inductance of one inductor is twice that of the other, what is the inductance of each inductor? (*Hint:* See Problem 19.57.)

19.62   Three 0.25-H inductors are to be joined together to give a maximum inductance. What is the maximum value of the combination and how should they be joined together? (*Hint:* See Problems 19.56 and 19.57.)

20

Alternating-Current Circuits

20.1 The *RL* Circuit

20.2 The *RC* Circuit

20.3 Effective Values of Alternating Current

20.4 Reactance

20.5 The *RLC* Series Circuit

20.6 Resonant Circuits

*20.7 Rectifiers

*20.8 Operational Amplifiers

A WORD TO THE STUDENT

In the previous chapters we introduced three basic electrical circuit elements: the resistor, the capacitor, and the inductor. Now we can examine their behavior in some simple alternating-current circuits.

We will not go very far in exploring the enormous variety of modern electronic circuits. Compared with the basic circuits we will examine, the circuits in today's hand-held calculators, television sets, and compact-disk players are miracles of complexity and manufacturing technology. However, all of these devices rely for their operation on the same fundamental principles, which we present in this chapter.

In the preceding chapter we introduced inductance, a third major component of electric circuits. However, since the effects of inductance are observed only with changing currents, we have waited until now to complete our study of circuits. Many common circuits in electrical and electronic equipment use alternating current, which produces a different response in circuit components than does direct current. We examine this behavior in this chapter, seeing how we can employ different combinations of resistance, inductance, and capacitance in useful ways. We also discuss two important special circuits: resonant circuits and amplifiers.

The development of our knowledge of electromagnetism, from bits of rubbed amber and lodestone in the time of the ancient Greeks to today's world filled with complex electronic equipment, is a fascinating story. We will continue to use our knowledge of electromagnetism as we study other areas of physics. In fact, much of what we have learned about physics in this century has come from experiments in which electronic instruments enable us to observe the world far beyond the reach of our normal senses. Understanding how these instruments work is an essential part of planning and interpreting experiments. Our experimental verification of contemporary physical theories rests in part on our confidence in modern electronics.

The *RL* Circuit

20.1

We have already seen that an inductor reacts to a change in current with an induced emf that opposes that change. We can analyze the behavior of an inductor in a circuit by using the Kirchhoff loop rule. Consider a circuit containing an inductor and a resistor (Fig. 20.1). The resistance R represents the total resistance in the circuit, including the resistance of the inductor. When we close the switch S, a current starts through the circuit. The inductance resists the sudden increase in current and delays its buildup to its ultimate value.

From Kirchhoff's law, the equation for potential difference around the loop is

$$V - IR - L\frac{\Delta I}{\Delta t} = 0.$$

Upon rearranging, this equation becomes

$$\frac{\Delta I}{\Delta t} = \frac{V - IR}{L} = -\frac{R}{L}(I - V/R).$$

Because V/R is constant, the change in $I - V/R$ is the same as the change in I. Thus we can write

$$\frac{\Delta(I - V/R)}{\Delta t} = -\frac{R}{L}(I - V/R).$$

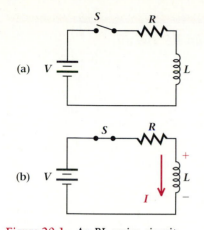

Figure 20.1 An *RL* series circuit. (a) A resistance R, a battery V, and an inductance L are joined in **series** with a switch S. (b) When the switch is closed, a current exists in the circuit.

This equation has the form of the equations we encountered in Chapter 12. In both cases, we have a rate of change of a quantity that is proportional to the value of the quantity. Thus the quantity $I - V/R$ has the form of an exponential function of time,

$$I - V/R = I_0 e^{-Rt/L},$$

which may be rearranged to give

$$I = V/R + I_0 e^{-Rt/L}.$$

We can evaluate the proportionality constant I_0 by considering what happens at time $t = 0$, when the switch is closed. There was no current prior to closing the switch and because the inductor inhibits any change in current, the current is zero *immediately* after the switch is closed. Thus at $t = 0$, $I = 0$ and the equation for the current becomes

$$0 = V/R + I_0.$$

Consequently,

$$I_0 = -V/R.$$

We can then express the current as a function of time by

$$I = \frac{V}{R}(1 - e^{-Rt/L}). \tag{20.1}$$

A graph of the current in this circuit is shown in Fig. 20.2(a). Note that for long times ($t \gg L/R$), the current in the circuit is determined by the battery and the resistance alone; that is, the current approaches $I = V/R$.

We can also obtain the potential difference across the inductor (Fig. 20.2b) from the loop equation as

$$V - IR + V_L = 0,$$

$$V_L = IR - V = -Ve^{-Rt/L}. \tag{20.2}$$

The inductor reacts to the closing of the switch by producing an emf opposite in direction to the battery voltage. This emf inhibits the current initially, but the current gradually builds up as the inductor voltage decays. The time for the current to build up to within $1/e$ of its final value is $\tau = L/R$, called the **inductive time constant.** Circuits with small inductive time constants attain their steady-state value of current quickly; circuits with large time constants take a longer time to attain maximum current.

If the switch in Fig. 20.1 is suddenly opened, the rate of change of current $\Delta I/\Delta t$ may be quite large. Since $\mathscr{E} = -L(\Delta I/\Delta t)$, the inductive voltage will also be large. This voltage, called an inductive kick, may be large enough to cause a dielectric breakdown of the air and produce arcing in the switch. When the inductance in a circuit is large, the inductive kick can present serious problems.

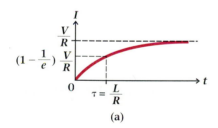

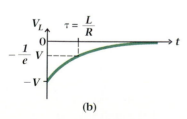

Figure 20.2 (a) Graph showing the buildup of current in the *RL* circuit of Fig. 20.1. (b) Graph of inductor voltage for the circuit of Fig. 20.1.

Example 20.1

Calculating current and voltage in an RL circuit.

An 8.0-H inductance of negligible resistance is joined in series with a resistance of 5.0 Ω. What is the time constant of this combination? If the inductor and resistor are connected in a circuit like that of Fig. 20.1 along with a 3.0-V battery, what is the current in the loop at $t = \tau$ if the switch S is closed at time $t = 0$? What is the voltage across the inductor at $t = \tau$?

Solution   The time constant of the *LR* combination is

$$\tau = \frac{L}{R} = \frac{8.0 \text{ H}}{5.0 \text{ }\Omega} = 1.6 \text{ s.}$$

At time $t = \tau$, the current may be evaluated from Eq. (20.1) as

$$I = \frac{V_0}{R} (1 - e^{-Rt/L}) = \frac{V_0}{R} (1 - e^{-1}) = \frac{V_0}{R} (0.63),$$

where V_0 is the battery voltage and R is the resistance. Thus

$$I = \frac{3.0 \text{ V}}{5.0 \text{ }\Omega} (0.63) = 0.38 \text{ A.}$$

At $t = \tau$, the current has risen to 63% of its final value. As t becomes very large compared with τ, the current rises to its final value of $V_0/R = 0.60$ A.

The voltage across the inductor is given by Eq. (20.2) as

$$V_L = -V_0 e^{-Rt/L}.$$

When $t = \tau$, the inductor voltage becomes

$$V_L = -V_0 e^{-1} = -0.37(3.0 \text{ V}) = -1.11 \text{ V.}$$

Thus, in one time constant, the potential across the inductor falls to 37% of its initial value.

The *RC* Circuit

20.2

In Chapter 16 we found that a capacitor could be used to store charge. Because the dielectric material of the capacitor is nonconducting, a capacitor does not allow a steady dc current to pass. However, a capacitor allows a time-varying current to pass in order to add or remove charge from the plates. As the current adds negative charges to one plate, equal numbers of negative charges are removed from the other plate, making it positive. The effect is as if the current passed through the capacitor.

A simple circuit for charging a capacitor consists of a battery of potential V connected in series with a resistor R and a capacitor C through a switch S (Fig. 20.3). Initially there is no charge on the capacitor and the switch is open. At time $t = 0$ the switch is closed.

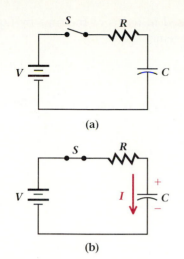

Figure 20.3   An *RC* circuit for charging a capacitor. (a) A resistance *R*, a battery *V*, and a capacitance *C* are joined in series with a switch *S*. (b) When the switch is closed, charge flows from the battery to the capacitor.

We apply Kirchhoff's voltage law to the circuit to find

$$V - IR - q/C = 0,$$

where q/C is the voltage across the capacitor. We have included the negative sign because the potential difference built up across the capacitor is opposite to that of the battery.

The current is defined to be the time rate of change of electric charge. If we substitute $\Delta q/\Delta t$ for the current in the Kirchhoff equation above, we find that

$$\frac{\Delta q}{\Delta t} = -\frac{1}{RC}(q - VC). \tag{20.3}$$

Notice the similarity between this equation and the equation for the *RL* circuit in the preceding section. Although the combination of constants is different, the equations have the same form. The solution also has the same form:

$$q = VC + q_0 e^{-t/RC}.$$

We can evaluate q_0 from the initial conditions. At $t = 0$, $q = 0$ and $q_0 = -VC$. Inserting this value of q_0 in the equation for q, we find that the charge on the capacitor grows with time according to

$$q = VC(1 - e^{-t/RC}). \tag{20.4}$$

The potential difference V_C across the capacitor is $V_C = q/C$. Upon substituting Eq. (20.4) for the charge q we find the voltage to be

$$\boxed{V_C = V(1 - e^{-t/RC}),} \tag{20.5}$$

where V is the battery potential.

The current in the circuit is $\Delta q/\Delta t$, which may be obtained from Eq. (20.3) when Eq. (20.4) is inserted for the charge q:

$$\boxed{I = \frac{V}{R}e^{-t/RC}.} \tag{20.6}$$

Figure 20.4 shows graphs of the charging current and the voltage on the capacitor as a function of time. The current in the *RC* circuit is at a maximum the instant the switch is closed and decreases exponentially with time. The characteristic time constant for the current to decay to $1/e$ of its initial value is $\tau = RC$, the **capacitive time constant.** A circuit with a large capacitive time constant charges and discharges slowly; a circuit with a small time constant charges and discharges quickly.

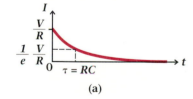

(a)

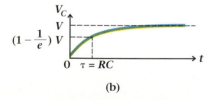

(b)

Figure 20.4   Graphs of (a) the charging current and (b) the voltage on the capacitor of Fig. 20.3. Time $t = 0$ corresponds to the closing of the switch.

Example 20.2

Charging a capacitor.

How many time constants must elapse before the voltage across a capacitor reaches 98% of its final value?

Solution   The capacitor voltage is expressed in terms of the time by Eq. (20.5) with $\tau = RC$, which, upon rearrangement, is

$$e^{-t/\tau} = \frac{V - V_C}{V}.$$

Next we insert the value of the capacitor voltage at time t, $V_C = 0.98$ V, to get

$$e^{-t/\tau} = \frac{1.00 - 0.98}{1.00} = 0.02.$$

With the aid of a calculator we find that

$$-t/\tau = \ln 0.02 = -3.9,$$

or

$$t = 3.9\tau.$$

A capacitor reaches 98% of its final voltage in 3.9 time constants.

We can discharge a charged capacitor by completing a closed circuit across its terminals (Fig. 20.5). The resistor R could be a large resistance or it could be merely the effective resistance of the conducting wire. When the switch S is closed, Kirchhoff's voltage law gives

$$RI + q/C = 0.$$

If we replace I by $\Delta q/\Delta t$, we get

$$\frac{\Delta q}{\Delta t} = -\frac{q}{RC}.$$

Again we have an equation that has an exponential function for a solution. The charge on the capacitor decreases with time according to

$$q = q_0 e^{-t/RC},$$

where q_0 is the initial charge on the capacitor at time $t = 0$. The capacitor voltage is again given by the ratio of q/C:

$$V_C = \frac{q_0}{C} e^{-t/RC},$$

$$V_C = V_0 e^{-t/RC}.$$

The current is given by

$$I = \frac{V_0}{R} e^{-t/RC}. \tag{20.7}$$

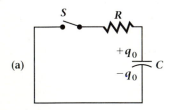

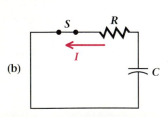

Figure 20.5 An *RC* circuit for discharging a capacitor. (a) The capacitor is charged and the switch is open. (b) A discharge current exists when the switch is closed.

The *RL* and *RC* circuits are similar in that both of them respond to an increase in applied voltage with a current that changes exponentially with time. However, they respond in different ways. Because of the induced emf, the *RL* circuit inhibits the current initially, but allows it to grow with time as the induced emf decays. The *RC* circuit permits an initial current, but as the capacitor becomes charged, that current approaches zero. As we will see later, the different behavior of these devices leads to dramatically different responses to ac signals.

Effective Values of Alternating Current

20.3

In discussing household circuits (Chapter 17), we mentioned that the instantaneous electric potential (the voltage) is not constant, but we assumed that it had a dc equivalent. In practice, the voltage in household circuits changes continuously, smoothly oscillating back and forth from positive to negative (Fig. 20.6). As a result, the currents also oscillate in a sinusoidal manner, with the same frequency as the voltage. As we discussed in Chapter 19, this alternating current, or ac, is generated naturally by a conducting loop rotating in a magnetic field. In North America, household current alternates at a rate of 60 complete oscillations per second, or 60 Hz. Much of the rest of the world uses 50-Hz electrical power.

The curve of Fig. 20.6 describes a sinusoidally oscillating source voltage. The instantaneous potential difference v between the two terminals of the source oscillates with time according to

$$v = V_0 \sin 2\pi ft,$$

where f is the frequency of the oscillation, t is the time, and V_0 is the amplitude (or peak value) of the oscillation. The sinusoidally varying voltage (or current) can be described in terms of its amplitude and frequency. However, in dealing with ac voltages and currents it is more common to use their dc equivalent values than their amplitudes.

To see how we can determine dc equivalent values of an ac circuit, consider the example of a current passing through a resistor R. For a direct current I, the power dissipated is I^2R. Since dc is constant, the instantaneous dc power is also the average power. Now if we apply a time-varying current of instantaneous magnitude i to the circuit,* the instantaneous power dissipation is still

$$p = i^2R,$$

---

*We represent the instantaneous value of a sinusoidal quantity with a lowercase letter and the amplitude, or peak value, of the quantity with an uppercase letter.

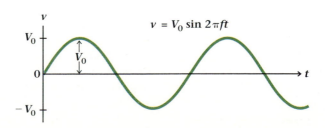

Figure 20.6   A graph of a sinusoidal voltage plotted as a function of time. The amplitude is V_0, the peak value of the signal.

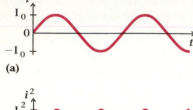

(a)

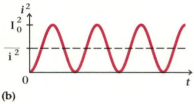

(b)

Figure 20.7   (a) A graph of a sinusoidally oscillating current i. The average of the current over a complete oscillation is zero. (b) A graph of i^2, the square of the current of part (a). The dashed line shows the average value of i^2.

as before. However, the average power is

$$\bar{p} = \overline{i^2 R} = \overline{i^2} R,$$

if R is constant.

The average power depends on the average of the square of the current, $\overline{i^2}$. A sinusoidal current reverses direction and its average over one period is zero; however, the square of the current is always positive or zero (Fig. 20.7). As a result, the time average of i^2 over one period is nonzero. The square root of $\overline{i^2}$ is called the **root-mean-square current** or the **rms current**, I_{rms}. The rms value of the current is also called the **effective value** of the current. We can then express the average power as

$$\bar{p} = I_{rms}^2 R, \tag{20.8}$$

where I_{rms} is the root-mean-square current,

$$I_{rms} = \sqrt{\overline{i^2}}. \tag{20.9}$$

Had we chosen to start with instantaneous power given by $p = iv$, then the average power would be

$$\bar{p} = I_{rms} V_{rms}, \tag{20.10}$$

where V_{rms} is the root-mean-square value of the oscillating voltage. The **rms voltage** is

$$V_{rms} = \sqrt{\overline{v^2}}. \tag{20.11}$$

We can compute the rms value of current from the average of the square of the sinusoidal current over one period. The average value of the square of the sine function over one period is $\frac{1}{2}$. So we have

$$I_{rms} = \frac{I_0}{\sqrt{2}}, \tag{20.12}$$

where I_0 is the amplitude of the oscillating current. Note that the relationship of Eq. (20.12) is appropriate only for sinusoidally oscillating currents.

The effective or rms values of voltage and current are so commonly used that whenever a numerical value for an ac signal is given without qualification, it is understood to mean the rms value. For example, the 120-V outlets ordinarily found in homes supply a constant 120 V rms. Voltmeters and ammeters used in ac circuits are usually calibrated to read rms values. In fact, rms usage is so common that subscripts are usually not included; that is, instead of V_{rms} and I_{rms} we simply use V and I.

Note that the average or effective value of the power given by Eqs. (20.8) and (20.10) is the same as that produced by dc current and voltage of magnitude equal to I and V. In other words, one ampere rms ac produces exactly the same heating of a resistor as does one ampere dc.

Example 20.3

Relating peak value to rms value.

The output of a step-down transformer is measured at 12.6 V ac when the primary side is connected to a normal household power line. What is the peak value of the output voltage?

Solution   The measured output voltage of the transformer is the effective, or rms, value. The relation between the rms value and the peak value of the voltage is similar to Eq. (20.12) for the current:

$$V_{rms} = \frac{V_0}{\sqrt{2}}.$$

So the peak value of the voltage output is

$$V_0 = \sqrt{2}V_{rms} = \sqrt{2} \times 12.6 \text{ V} = 17.8 \text{ V}.$$

Reactance

20.4

We now wish to consider the response of capacitors and inductors subjected to a sinusoidal potential or current. We know that a capacitor does not permit a direct current to pass through it, but does respond to changes in potential by allowing a charging or discharging current. Thus we expect that the response of a capacitor to an oscillating potential depends on the frequency of the oscillation.

For example, suppose we place an oscillating potential of frequency f and amplitude V_0 across a capacitor C (Fig. 20.8). We may express the potential as

$$v = V_0 \sin 2\pi ft.$$

The charge on the capacitor is proportional to its voltage through

$$q = Cv = CV_0 \sin 2\pi ft. \tag{20.13}$$

The current is obtained from the rate of change of the charge with time,

$$i = \frac{\Delta q}{\Delta t}.$$

We saw in Chapter 13 that the rate of change of a sine function was proportional to a cosine function:

$$\frac{\Delta(\sin 2\pi ft)}{\Delta t} = 2\pi f \cos 2\pi ft.$$

Thus the current is described by

$$i = 2\pi fCV_0 \cos 2\pi ft. \tag{20.14}$$

These equations for current and voltage tell us that *the current and the voltage do not reach their maxima at the same time.* We can see this behavior

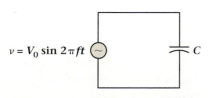

$v = V_0 \sin 2\pi ft$ C

Figure 20.8   **An alternating voltage source connected across a capacitor.**

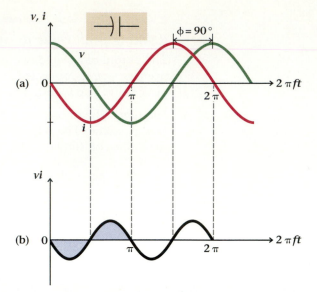

Figure 20.9   (a) The current and voltage across a capacitor do not reach their maxima at the same time. The current *leads* the voltage by one quarter of a cycle, corresponding to a phase shift of 90°. (b) The power delivered to the capacitor, given by the product *vi*, oscillates in time. The time average of the power is zero.

with the aid of Fig. 20.9(a). The current reaches its maximum one quarter of a cycle before the voltage reaches its maximum. Thus we say that the current *leads* the voltage across a capacitor by 90°. In fact, the current can be expressed as

$$i = 2\pi fCV_0 \sin (2\pi ft + 90°).$$

The difference in time between the occurrence of the maximum current and the maximum voltage is usually given in terms of a phase angle ϕ. One full cycle or period corresponds to a phase angle of 360°; thus a shift of one quarter of a cycle is equivalent to a shift of 90°. For the capacitor we say that the current is shifted 90° ahead of the voltage. We could also say that the voltage lags the current by 90°.

The instantaneous power delivered to the capacitor is vi. When this product is summed over an entire cycle the result is zero because of the 90° phase shift between v and i. Figure 20.9(b) shows the product of the v and i curves of Fig. 20.9(a). The sum over one cycle is represented by the shaded area of the curve. The area above the time axis is positive and the area below it is negative. From the symmetry of the curve we see that the average power delivered to the capacitor is zero.

The maximum value of the current varies linearly with the maximum value of the applied voltage for a capacitor, in analogy with Ohm's law for resistors. From Eq. (20.14), the maximum value of the current is

$$I_0 = 2\pi fCV_0.$$

If we compare this relationship with the current-voltage relationship for resistors, $R = V/I$, we get a proportionality factor analogous to resistance. This proportionality factor is the **capacitive reactance**, defined to be

$$X_C \equiv \frac{V_0}{I_0} = \frac{1}{2\pi fC}. \tag{20.15}$$

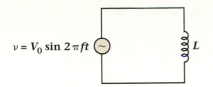

$v = V_0 \sin 2\pi ft$ L

Figure 20.10 An alternating voltage source connected across an inductor.

Notice that the reactance is frequency-dependent, as we expected, and becomes infinitely large as the frequency f approaches zero. This corresponds to the fact that capacitors do not pass a constant (dc) current. At high frequencies the reactance becomes quite small and the capacitor does little to impede the current. Thus a capacitor that serves to isolate one dc voltage from another may readily transmit ac.

An inductor also displays a linear relationship between current and voltage. As in the case of the capacitor, this relationship depends on frequency. For example, if we place a source of oscillating potential across an inductor (Fig. 20.10), an oscillating current occurs. According to Kirchhoff's voltage law, the sum of the potentials around the loop must be zero, so

$$v = V_0 \sin 2\pi ft = L \frac{\Delta i}{\Delta t}.$$

According to this equation, the potential is proportional to the slope of the curve for current. Since the current whose slope is a sine is a negative cosine, we can write the current as

$$i = -I_0 \cos 2\pi ft = I_0 \sin (2\pi ft - 90°).$$

Thus the current *lags* the voltage across an inductor by 90° (Fig. 20.11).

The instantaneous power delivered to the inductor is iv. When this product is summed over a complete cycle, the result is zero because the current and voltage are 90° apart. Thus, over a complete cycle, the inductor takes no power from the circuit. In this respect, it behaves like the capacitor.

We define the term **inductive reactance** in analogy with capacitive reactance. If we calculate the rate of change of the current using the techniques of Chapter 13, we find

$$\frac{\Delta i}{\Delta t} = 2\pi fI_0 \sin 2\pi ft.$$

The proportionality between peak current and peak voltage in an inductor is the inductive reactance X_L. We obtain the value of the reactance by inserting the expression for $\Delta i/\Delta t$ into the equation for the voltage. The result is

$$X_L \equiv \frac{V_0}{I_0} = 2\pi fL. \qquad (20.16)$$

At high frequencies the inductive reactance is large and the inductance impedes the current. This behavior corresponds to the fact that an inductor opposes changes in current. At frequencies approaching zero, the reactance is small and the inductance offers little hindrance to the current.

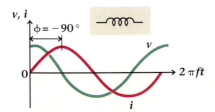

v, i

$\phi = -90°$

0 $2\pi ft$

v

i

Figure 20.11 Graphs of current and voltage across an inductor. The current *lags* the voltage by one quarter of a cycle.

Example 20.4

Capacitive reactance and voltage.

What is the reactance of a 25-μF capacitor at a frequency of 15 kHz? What is the voltage across the capacitor if the current is 5.8 mA?

Solution   The reactance can be found from Eq. (20.15):

$$X_C = \frac{1}{2\pi fC} = \frac{1}{2\pi(15 \times 10^3 \text{ Hz})(25 \times 10^{-6} \text{ F})},$$
$$X_C = 0.42 \ \Omega.$$

(Problem 20.19 asks you to show that the unit of reactance is the ohm.)
The voltage across the capacitor is the product of the current with the reactance:

$$V_C = IX_C = 5.8 \times 10^{-3} \text{ A} \times 0.42 \ \Omega = 2.4 \times 10^{-3} \text{ V} = 2.4 \text{ mV}.$$

The *RLC* Series Circuit

20.5

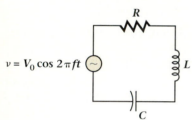

Figure 20.12   A series circuit containing resistance R, inductance L, and capacitance C. The circuit is driven by an oscillating voltage source.

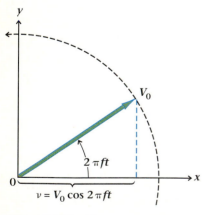

Figure 20.13   Phasor representation of an oscillating voltage $v = V_0 \cos 2\pi ft$.

Let us now tie all of these ideas together by examining the behavior of a series circuit containing resistance, inductance, and capacitance when the circuit is driven by an oscillating voltage source. This so-called *RLC* circuit (Fig. 20.12) is a fundamental circuit underlying many kinds of electronic equipment. We can understand the behavior of the circuit by considering the voltages around the loop. According to Kirchhoff's voltage law, at any instant the voltage provided by the source must equal the sum of the potential drops across the three elements *R*, *L* and *C*. That is,

$$v = v_R + v_L + v_C.$$

Each of these potentials oscillates with the frequency of the source. However, each voltage has a difference phase relationship with the current in the circuit. The voltage v_R across the resistor is in phase ($\phi = 0$) with the current. The voltage v_L across the inductor leads the current by a phase angle of 90°. The capacitive voltage v_C lags the current by 90°. As a consequence of these phase differences, we cannot simply add the voltages by summing either their peak values or their rms values.

To handle this situation, we return to the relationship we studied in Chapter 13 between simple harmonic motion and circular motion. Specifically, we represent an oscillating voltage (or current) by a **phasor**, which is a two-dimensional mathematical quantity that rotates with a constant angular frequency $2\pi f$. Phasors are added head to tail, like vectors. The direction of the voltage phasor represents the phase angle relative to a reference phase, and is not related to the spatial direction of the potential difference. For example, we represent a voltage $v = V_0 \cos 2\pi ft$ by the projection along the *x* axis of a phasor of magnitude V_0 rotating about the origin (Fig. 20.13).

The phasor diagram for the *RLC* circuit of Fig. 20.12 is drawn in Fig. 20.14(a). We choose the reference phasor to be the current in the circuit. The phasor describing the voltage across the resistor lies along the same direction as the current phasor. We see them both drawn at a particular instant of time *t*. The phasor representing the inductor voltage leads the current by 90°; its magnitude is $V_L = X_L I_0$. Similarly, the phasor for the capacitor voltage has a magnitude $V_C = X_C I_0$ and lags the current by 90°. The net voltage is the sum of these individual phasors, which we can

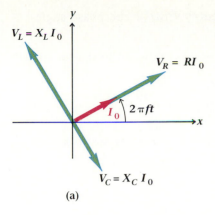

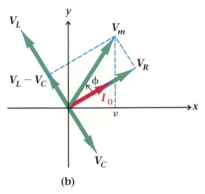

Figure 20.14 (a) Phasor diagram for the *RLC* circuit of Fig. 20.12. (b) The maximum net voltage is V_m, the phasor sum of V_L, V_C, and V_R. At any instant of time, the net voltage v is the projection along the *x* axis of V_m.

obtain as if we were adding vectors (Fig. 20.14b). Note that this sum gives us the maximum voltage V_m; the net voltage v at any instant is the projection of V_m along the *x* axis.

The voltage across the *RLC* combination has a maximum value V_m and oscillates with frequency *f*. However, it is phase-shifted by an amount ϕ relative to the current. This phase angle can be evaluated with the aid of Fig. 20.14(b). The tangent of ϕ is given by

$$\tan \phi = \frac{V_L - V_C}{V_R},$$

or, dividing each voltage by the current I_0,

$$\phi = \tan^{-1}\left(\frac{X_L - X_C}{R}\right). \tag{20.17}$$

We can also calculate the peak or maximum value of the voltage from the figure by applying the Pythagorean theorem:

$$V_m = \sqrt{V_R^2 + (V_L - V_C)^2}.$$

When the individual voltages are expressed in terms of the current I_0, the last equation becomes

$$V_m = I_0\sqrt{R^2 + (X_L - X_C)^2}.$$

The factor that relates the peak voltage to the peak current is called the **impedance** Z,

$$Z = \sqrt{R^2 + (X_L - X_C)^2}, \tag{20.18a}$$

or

$$Z = \sqrt{R^2 + \left(2\pi fL - \frac{1}{2\pi fC}\right)^2}. \tag{20.18b}$$

The impedance is also the ratio of the rms voltage to the rms current. Thus we can use it to express a generalized form of Ohm's law, applicable to ac circuits:

$$V_m = I_0 Z.$$

Note that impedance depends not only on the resistance, capacitance, and inductance of the *RLC* circuit, but also on the frequency of the applied voltage (since reactance is frequency-dependent).

Example 20.5

Analysis of an RLC circuit.

A series circuit has $R = 30\ \Omega$, $L = 0.10$ H, and $C = 100\ \mu$F. What is the peak voltage across each of these elements if a 60-Hz sinusoidal current with a peak value of 1.0 A flows in the circuit? What is the peak voltage across the combination of these elements? What is the phase angle between voltage and current?

Solution   We calculate the peak voltages one at a time. The resistor voltage is

$$V_R = RI_0 = 30 \; \Omega \times 1.0 \; \text{A} = 30 \; \text{V}.$$

The peak voltage on the inductor is

$$V_L = X_L I_0 = 2\pi f L I_0 = 2\pi \times 60 \; \text{Hz} \times 0.10 \; \text{H} \times 1.0 \; \text{A},$$
$$V_L = 37.7 \; \text{V}.$$

The peak voltage across the capacitor is

$$V_C = X_C I_0 = \frac{I_0}{2\pi f C} = \frac{1 \; \text{A}}{2\pi \times 60 \; \text{Hz} \times 10^{-4} \; \text{F}},$$
$$V_C = 26.5 \; \text{V}.$$

The voltage across the combination is found from the phasor sum of the three voltages just calculated. It is

$$V_m = \sqrt{V_R^2 + (V_L - V_C)^2} \; \text{V},$$
$$V_m = \sqrt{30^2 + (37.7 - 26.5)^2} \; \text{V} = 32 \; \text{V}.$$

Notice that while the peak voltage on the inductor is 37.7 V, the peak voltage across all three elements is only 32 V because of the phase relationships.

The phase angle between voltage and current is

$$\phi = \tan^{-1} \left(\frac{V_L - V_C}{V_R} \right) = \tan^{-1} \left(\frac{37.7 - 26.5}{30} \right),$$
$$\phi = 20.5°.$$

The voltage leads the current by 20.5°.

Resonant Circuits

20.6

Electrical systems, like mechanical systems, can display resonance. Resonant circuits have long been used in the tuning circuits of radios. To understand electrical resonance, remember that the response of a series *RLC* circuit to an applied voltage $v = V_0 \cos 2\pi f t$ depends on the frequency. The peak value of the current is

$$I_0 = \frac{V_0}{Z}.$$

Since the peak value and the rms value are related by a numerical constant, we may also express the current as

$$I_{\text{rms}} = \frac{V_{\text{rms}}}{Z}.$$

Using Eq. (20.18b) for the impedance, we can write this as

$$I_{rms} = \frac{V_{rms}}{\sqrt{R^2 + \left(2\pi f L - \dfrac{1}{2\pi f C}\right)^2}},$$

which shows the frequency dependence explicitly.

When the frequency is zero, the term $1/2\pi f C$ is infinite and the current is zero. (The capacitor blocks low frequencies.) At very high frequencies the term $2\pi f L$ becomes large and again the current approaches zero. (The inductor blocks high frequencies.) At intermediate frequencies, the current is appreciable and reaches a maximum when $2\pi f L = 1/2\pi f C$. This behavior of the circuit is called *resonance* and the frequency at which the maximum current occurs is the *natural* or *resonant frequency* f_0, given by

$$f_0 = \frac{1}{2\pi\sqrt{LC}}. \tag{20.19}$$

When the resistance is small the resonance is very sharp, leading to an appreciable current over only a narrow range of frequencies. However, when the resistance is large, the resonance becomes broad. The effect of the resistance is seen in Fig. 20.15, which shows the current as a function of frequency in an *RLC* series circuit for two different values of *R*. The lower curve corresponds to a resistance three times larger than that of the upper curve. Notice the similarities between electrical resonance and mechanical resonance (Section 13.7).

Resonance also occurs for other circuit combinations of *R*, *L*, and *C*. In particular, it occurs for the parallel combination of inductance and capacitance (Fig. 20.16). For this case, resonance is characterized by a maximum impedance, rather than a minimum. At low frequencies, the

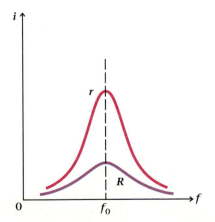

Figure 20.15 Graph of current versus frequency for a series *RLC* circuit. The lower curve corresponds to a resistance value three times larger than for the upper curve.

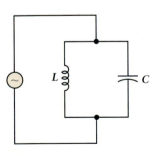

Figure 20.16 An oscillating voltage source drives a parallel combination of inductance and capacitance in a resonant circuit.

reactance of the inductor is small and a large current passes through the branch containing the inductor. At high frequencies, the capacitive reactance is low, allowing a large current in the branch containing the capacitor. At some intermediate frequency the magnitudes of the impedance in the two branches are equal, and the total current and voltage are in phase. If the resistance is small, the resonant frequency for this parallel combination of L and C is also given by Eq. (20.19).

Frequency-selective circuits (resonant circuits) are commonplace in

PHYSICS IN PRACTICE

AM, FM, and Stereo

What kind of radio station do you usually listen to—AM or FM? What's the difference? Aside from programming, there are some interesting differences between the physics of AM and FM radio.

Radio waves are electromagnetic waves generated by oscillating electric charges (Section 19.7). In the United States, every licensed commercial radio station is assigned a specific broadcast frequency by the Federal Communications Commission (FCC), in the range of 540–1640 kHz for AM and 88–108 MHz for FM. This frequency is the frequency of the carrier wave broadcast by the station at all times, and is the frequency you tune on your radio dial. Superimposed on the carrier wave is the signal wave, representing the transmitted information (music, news, etc.). Signal frequencies are limited to below 5 kHz for AM and 15 kHz for FM. The superimposing of the signal and the carrier waves, called modulation, can occur in either of two ways: amplitude modulation (AM) or frequency modulation (FM).

Amplitude modulation occurs when the two waves are brought together in a diode or transistor. The amplitude of a carrier wave of frequency f_0 is modulated by a signal wave of lower frequency f. The amplitude of the carrier waveform is made to vary by an amount $mV_0 \cos 2\pi ft$, where V_0 is the

unmodulated amplitude of the carrier wave. We can express the resulting modulated wave (Fig. B20.1c) as

$$v = V_0 (1 + m \cos 2\pi ft) \sin 2\pi f_0 t.$$

The modulation index m should always be less than unity, otherwise the

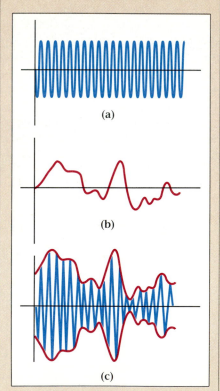

Figure B20.1 Amplitude modulation. (a) The carrier wave; (b) a non-sinusoidal audio signal; and (c) the amplitude-modulated wave.

carrier wave becomes so severely distorted that some of the signal information is destroyed and cannot be retrieved. The result of the modulation is to produce three components in the wave: a component at the carrier frequency f_0, and two additional components at the sum and difference of the carrier frequency and the signal frequency. These two additional components are known as *sidebands*. If the modulating signal is a band of frequencies, as is usual, and not just a single frequency, the term sideband seems even more appropriate (Fig. B20.2). Since we choose the carrier frequency to be large compared with the signal frequency, we can demodulate (detect) the desired signal by passing the modulated wave through a circuit that passes only the lower frequency.

Your radio receiver contains a tuner circuit that you set to resonate at the carrier frequency of the desired station by adjusting the tuning knob. The signal broadcast by the station induces a small voltage in the receiving antenna. That signal, whose frequency matches the tuner frequency, passes through the tuning circuit to be amplified. The audio information is obtained by demodulation, a process that separates the signal from the carrier wave, and the resulting electrical signal is amplified and sent to a speaker to produce the sounds you hear.

In *frequency modulation* the frequency of the carrier wave, rather than

radio, television, and other electronic applications. The tuning circuit in a radio often consists of an inductor in parallel with a variable capacitor (Fig. 20.17, p. 598). You tune the resonant frequency of the circuit to the frequency of the desired station by adjusting the value of capacitance. The radio's antenna brings many radio signals of different frequencies to the resonant circuit. Signals not at the resonant frequency of the tuning circuit pass to the ground through the low impedance of the circuit for those frequencies. Consequently, the signal voltage reaching the amplifier is very

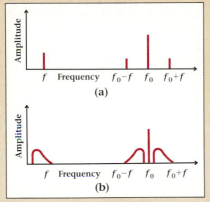

Figure B20.2 Frequency spectrum of an amplitude-modulated wave. (a) Single frequency signal. (b) Information signal of bandwidth f.

the amplitude, is varied in response to the information signal (Fig. B20.3). As in the case of amplitude modulation, sidebands are produced. FM radio has become popular in recent years for several reasons. One reason is that since the amplitude of the signal does not carry the information, noise bursts that affect the amplitude do not contribute to the demodulated signal. Thus FM has less interference noise (static) than AM. Second, FM carrier frequencies are about one hundred times greater than AM carrier frequencies. This higher frequency range allows for a greater separation between assigned carrier frequencies and thus permits a greater sideband frequency range than

for AM. Consequently, information signals can contain higher frequencies, leading to higher fidelity in the reproduced sound. Also, greater bandwidth permits simultaneous broadcast of stereo signals, as we now describe.

In standard FM broadcasting, the FCC allows each station a bandwidth of 75 kHz on either side of the carrier frequency. This bandwidth is much greater than necessary for transmitting a high-fidelity audio signal of 15 kHz bandwidth; however, stations can use this extra bandwidth for stereo broadcasting. Let us designate the two separate information channels used in stereo systems as L (for left) and R (for right). In the main audio channel the modulating signal is $L + R$, the sum of the two separate stereo signals. This permits monophonic reception without losing part of the signal. An $L - R$ signal is produced, shifted in frequency by modulation with a 38-kHz wave, and then used to modulate the FM carrier wave (Fig. B20.4). In the

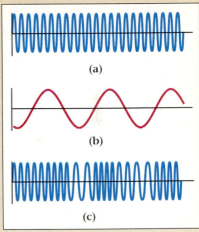

Figure B20.3 Frequency modulation. (a) The carrier wave; (b) the audio signal; and (c) the frequency-modulated wave.

stereo FM receiver, the $L + R$ and $L - R$ signals are obtained separately. Finally, the $L + R$ and $L - R$ are added to produce the L channel and are subtracted to give the R channel.

Figure B20.4 Frequency spectrum of an FM stereo sideband signal.

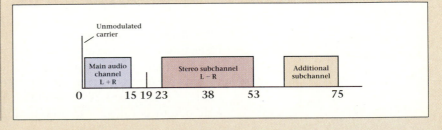

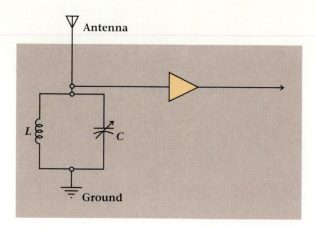

Figure 20.17 A parallel *LC* combination forms the tuning circuit for a radio receiver.

small. At the resonant frequency, the impedance of the tuning circuit is high so that an appreciable signal voltage appears across the circuit and is delivered to the input of the amplifier.

Example 20.6

Frequency of a radio tuning circuit.

A radio tuning circuit is made from a 150-μH inductor connected in parallel with a variable capacitor. If the capacitor is adjusted so that its capacitance is 169 pF, what is the resonant frequency of the circuit?

Solution   The resonant frequency is given by Eq. (20.19):

$$f_0 = \frac{1}{2\pi\sqrt{LC}} = \frac{1}{2\pi\sqrt{150 \times 10^{-6}\,\text{H} \times 169 \times 10^{-12}\,\text{F}}},$$

$$f_0 = 1000 \text{ kHz}.$$

This frequency is near the center of the frequencies allowed for AM radio.

Rectifiers

*20.7

Another simple electronic device is the **diode**, or **rectifier**, which passes current in one direction and blocks it in the opposite direction. In fact, an ideal diode would have zero resistance for one polarity of applied voltage and infinite resistance for the opposite polarity. The graph of current versus voltage would consist of two straight line segments (Fig. 20.18a). In practice, we cannot achieve this ideal situation, but real diodes can come quite close (Fig. 20.18b). The direction associated with the low resistance is known as the *forward* direction and the direction for high resistance is the *reverse* or *backward* direction (Fig. 20.19).

A major application of diodes is the rectification of alternating voltages to provide direct voltages suitable for the operation of radios, tape recorders, calculators, etc. The circuit that converts an ac input voltage to a dc output is called a dc power supply. In many applications the power

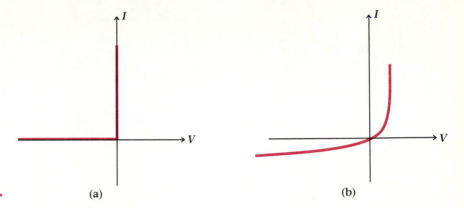

Figure 20.18 The current–voltage characteristic curves for (a) an ideal diode and (b) an actual semiconductor diode.

(a)　　　　　　　　　　(b)

supply is designed to operate directly from standard ac power lines. Such circuits range from very simple rectifier circuits to elaborately regulated power supplies that maintain a precise dc voltage level regardless of input fluctuations or changes in the load circuit.

The simplest rectifier circuit is the half-wave rectifier, consisting of a single diode in series with a transformer and a load R (Fig. 20.20a). By choosing the proper transformer, we can have the output of the power supply suit any particular need. In the half-wave rectifier shown, the transformer output is $v = V_0 \sin 2\pi ft$. During the positive half-cycle the diode conducts and delivers the positive half of the sine wave voltage to the load. During the negative half-cycle the diode blocks the current and the voltage across the load drops to zero. This leads to the half-wave rectified signal shown in Fig. 20.20(b).

The half-wave rectifier circuit uses only half of the available power, since the circuit conducts only half the time. To use more of the available power and also achieve a waveform that is more easily smoothed to a

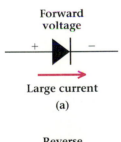

Forward voltage

+　　　−

Large current

(a)

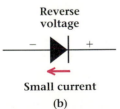

Reverse voltage

−　　　+

Small current

(b)

Figure 20.19 The circuit symbol for a diode. (a) The symbol arrow points in the direction in which a large current can pass (forward voltage). (b) Polarity for reverse voltage of a diode, resulting in a very small current in the reverse direction.

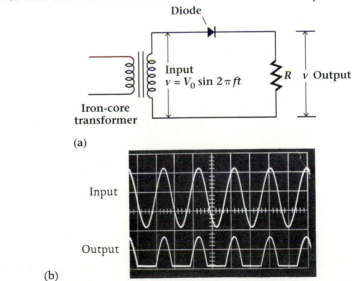

Figure 20.20 (a) A half-wave rectifier circuit. The diode conducts current only during the positive half-cycle of the oscillating voltage. (b) Input and output waveforms.

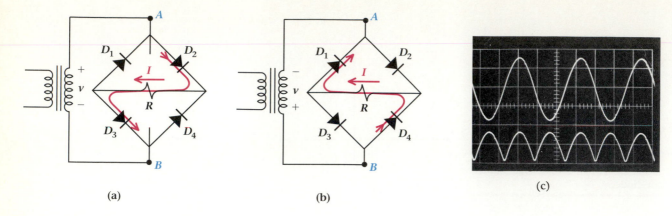

(a)　　　　　　　　(b)　　　　　　　　(c)

Figure 20.21 A full-wave rectifier circuit. The direction of the voltage across the load (and thus the current) is the same during both the (a) positive and (b) negative half cycles of the input voltage. (c) Input and output waveforms.

constant value, we need a full-wave rectifier. One circuit for achieving this is the full-wave bridge shown in Fig. 20.21(a). When point A of the diagram is positive with respect to point B, diodes D_2 and D_3 provide a path for current through the load resistor in the circuit and diodes D_1 and D_4 block the current. When the input voltage reverses polarity, D_1 and D_4 conduct while D_2 and D_3 block (Fig. 20.21b). The full rectified sine wave signal is shown in Fig. 20.21(c).

The maximum current in the bridge circuit is $I_0 = V_0/R$. Similarly, the dc or average value of the current is $I_{dc} = I = V/R$. The average values of the current and voltage can also be expressed in terms of the maximum values (Fig. 20.22). We can use graphical analysis to show that

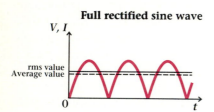

Figure 20.22. A full rectified sine wave. The lines indicate the average and effective values of the wave.

$$\bar{I} = \frac{2I_0}{\pi}, \quad \text{and} \quad \bar{V} = V_{dc} = \frac{2V_0}{\pi}. \quad \text{(full rectified sine wave)}$$

The rms value is the same as for the sine wave itself:

$$I_{rms} = \frac{I_0}{\sqrt{2}}, \quad \text{and} \quad V_{rms} = \frac{V_0}{\sqrt{2}}. \quad \text{(full rectified sine wave)}$$

The pulsating voltage (and current) of a rectifier circuit can be smoothed out to produce a nearly constant direct voltage. This is usually done with suitable arrangements of capacitors and inductors that are known as *filters* because they filter out the oscillating component of the voltage. The simplest filter circuit, the shunt-capacitor filter, consists of a capacitor connected across the output of the rectifier, in parallel with the load (Fig. 20.23a). When the voltage is first applied, the capacitor charges rapidly with rising input voltage (Fig. 20.23b). As the input voltage begins to drop, the potential difference across the diode reverses and the diode stops conducting. The charge stored on the capacitor then flows off through the load resistor. During this part of the cycle the voltage across the load resistor decreases, following the exponential decay of the RC discharge. The cycle repeats when the input voltage again gets high enough for the diode to conduct. As the product RC is made larger, the time constant of the discharge becomes larger and the waveform is smoother; that is, the ripple (or alternating part of the waveform) decreases. At the same time, the average (dc) value of the waveform increases toward the peak value of the input wave.

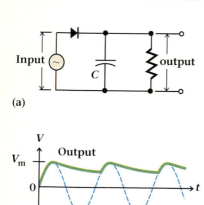

Figure 20.23 (a) A capacitively filtered half-wave rectifier. (b) The output voltage waveform.

Example 20.7

Output voltage of a full-wave rectifier.

Let's consider a practical example of a full-wave rectifier. Suppose a transformer in your radio has an rms output of 6.3 V. Assuming that no filter is used, what direct voltage is produced when the transformer is coupled to a full-wave rectifier like that of Fig. 20.21(a)?

Solution We know that for a full-wave rectifier, the dc level of the output voltage is $V_{dc} = 2V_0/\pi$. The value of V_0 is related to the rms value of the transformer voltage through $V_0 = \sqrt{2}V_{rms}$. Combining these equations, we get

$$V_{dc} = \frac{2\sqrt{2}V_{rms}}{\pi}.$$

When we insert 6.3 V for the rms voltage of the transformer, the output dc voltage becomes $V_{dc} = 5.7$ V.

Operational Amplifiers

*20.8

The invention of the transistor in 1948 by John Bardeen, Walter H. Brattain, and William Shockley stimulated research and development in physics and engineering that led to a revolution in electronic technology. Within ten years, designers and manufacturers were using transistors in their new products almost exclusively in place of vacuum tubes. By 1968 the number of transistor devices had grown many times, and a further development, integrated circuits, was becoming important. By 1978 the electronics industry had turned increasingly to integrated circuits and today they are commonplace.

Integrated circuits (IC's) are a combination of interconnected circuit elements that are inseparably mounted on a single support (Fig. 20.24). An integrated circuit may be a very complex circuit composed of many thousands of individual components, such as transistors, diodes, resistors, etc. Indeed, the pocket calculator and the electronic digital watch were made possible by the development of IC's. Pocket calculators commonly contain IC's with thousands of components on a single tiny chip. Perhaps more important is the low cost of IC's compared with the cost of building comparable circuits from individual components. Complete amplifiers are available in IC form at prices comparable with the cost of individual transistors. For these reasons, the principal component in present-day electronic circuits is the IC.

One particularly important class of integrated circuits is the **operational amplifier** (op amp), which came into prominence in the late 1960s. The behavior of these amplifiers depends on injecting some portion of the output signal back into the input, a process known as **feedback**. By using different feedback connections, we can obtain many different results from one type of amplifier. It is this versatility of operation that gives rise to the name operational amplifier.

Figure 20.24   **An integrated circuit chip.**

inverting input

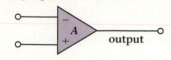

output

non-inverting input

Figure 20.25 The circuit symbol for an operational amplifier. A positive signal at the inverting input causes a negative output; a positive signal at the noninverting input causes a positive output.

An operational amplifier (Fig. 20.25) has an extremely high ratio of output voltage to input voltage, a ratio known as the circuit's *gain*. It also has a large input resistance and a small output resistance. The amplifier has two inputs, called inverting and noninverting inputs. A positive signal at the inverting input produces a negative output; a positive signal at the noninverting input produces a positive output. The output signal depends on the voltage difference between the two inputs.

In the basic *inverting amplifier* (Fig. 20.26), the noninverting (+) input is grounded. The input signal is applied to resistor R_i, the input resistor, connected to the inverting (−) input. The output at O is coupled to the same inverting input through the feedback resistor R_f. Because the signal is inverted by the amplifier, a positive input signal causes a negative output. Consequently, the potential from S to O is positive. For this reason we indicate positive current to the right in the figure. The feedback from O through R_f is negative and tends to cancel the input signal at S. Because of the large gain of the amplifier, the effect of the feedback is to drive S to the potential of the noninverting input, which in this instance is at ground potential.

Since the amplifier has high input impedance, no appreciable current flows from S into the amplifier. Applying Kirchhoff's current rule to the junction at S tells us that the currents i_{in} and i_f must be identical: $i_{in} = i_f$. The output voltage is negative with respect to S and hence to ground. It is given by

$$V_{out} = -i_f R_f.$$

Because S is at ground potential, the input current is $i_{in} = V_{in}/R_i$. The ratio of output voltage to input voltage becomes

$$\frac{V_{out}}{V_{in}} = -\frac{R_f}{R_i}. \qquad (20.20)$$

Thus the gain of the amplifier, defined to be the ratio of V_{out}/V_{in}, in this case is determined by the ratio of the two resistors.

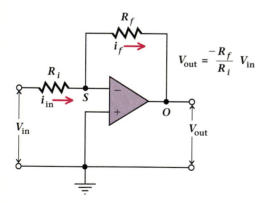

Figure 20.26 Basic configuration of an inverting operational amplifier circuit.

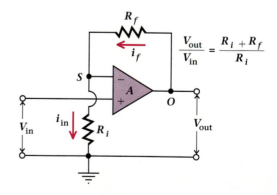

Figure 20.27 Basic configuration of a noninverting operational amplifier circuit.

The inverting amplifier circuit demonstrates an important characteristic of operational amplifiers. *Whenever the feedback circuit is closed, the inverting (−) input is driven to the potential of the noninverting (+) input.* We can use this observation to analyze and understand other operational amplifier circuits.

The basic *noninverting amplifier* is shown in Fig. 20.27. A positive signal at the noninverting input drives the output positive. The current through the series resistors R_f and R_i gives rise to a positive potential at S. Thus the potential of the inverting input (V_s) is driven to the level of the noninverting input (V_{in}). As before, there is no appreciable current into either input of the operational amplifier because of the high input impedances. Thus we have

$$i_f = i_{in} = \frac{V_s}{R_i} = \frac{V_{in}}{R_i}.$$

From Fig. 20.27 we see that the output voltage is

$$V_{out} = V_s + i_f R_f = i_f(R_i + R_f).$$

The gain is obtained from

$$\text{gain} = \frac{V_{out}}{V_{in}} = \frac{i_f(R_i + R_f)}{V_{in}}.$$

If we substitute V_{in}/R_i for i_f, we find the gain to be

$$\text{gain} = \frac{R_i + R_f}{R_i}. \tag{20.21}$$

We can make many other useful circuits by inserting other devices or networks in place of the feedback resistor R_f. Examples of operational amplifier circuits include circuits to regulate voltage or current, signal-processing circuits, signal generators, and filters.

Example 20.8

Gain of an operational amplifier.

An operational amplifier is connected as shown in Fig. 20.28 with $R_i = 2.0$ kΩ and $R_f = 10$ kΩ. What is the gain of the circuit?

Solution We see from the diagram that the amplifier is connected as an inverting amplifier. Therefore, the input and output currents are the same:

$$i_{in} = \frac{V_{in}}{R_i} = i_f = -\frac{V_{out}}{R_f}.$$

The negative sign arises because of the direction of i_f and the choice of point S as effective ground. The gain of the circuit is then

$$\text{voltage gain} = \frac{V_{out}}{V_{in}} = -\frac{R_f}{R_i} = -\frac{10 \text{ k}\Omega}{2.0 \text{ k}\Omega} = -5.0.$$

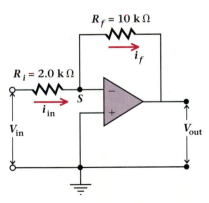

Figure 20.28 Example 20.8: An inverting amplifier circuit.

$R_f = 10$ kΩ

i_f

$R_i = 2.0$ kΩ

i_{in}

S

V_{in}

V_{out}

SUMMARY

Useful Concepts

- In an ac circuit, an inductor opposes the change in current. The current in an RL circuit containing a voltage source V is

$$I = \frac{V}{R}(1 - e^{-Rt/L}),$$

where L/R is the inductive time constant. The potential difference across the inductor is

$$V_L = -Ve^{-Rt/L}.$$

- The potential difference across the capacitor in an RC circuit containing a voltage source V is

$$V_C = V(1 - e^{-t/RC}),$$

and the current is

$$I = \frac{V}{R}e^{-t/RC},$$

where RC is the capacitive time constant.
- The rms value of a sinusoidally oscillating current with amplitude I_0 is

$$I_{\text{rms}} = \frac{I_0}{\sqrt{2}},$$

and the rms value of a sinusoidally oscillating voltage with amplitude V_0 is

$$V_{\text{rms}} = \frac{V_0}{\sqrt{2}}.$$

- The average power in a circuit with ac current and voltage is given by

$$\bar{p} = I_{\text{rms}}V_{\text{rms}} = I_{\text{rms}}^2 R.$$

- The capacitive reactance is

$$X_C = \frac{1}{2\pi fC}.$$

- The inductive reactance is

$$X_L = 2\pi fL.$$

- A phasor represents an oscillating voltage or current. Its length is proportional to the magnitude of the voltage or current and rotates with an angular frequency $2\pi f$. Phasors can represent circuit combinations of alternating voltages or currents.
- In an RLC series circuit the impedance is

$$Z = \sqrt{R^2 + (X_L - X_C)^2}.$$

- The phase angle between the voltage and the current is

$$\phi = \tan^{-1}\left(\frac{X_L - X_C}{R}\right).$$

- In an RLC series circuit the resonance frequency is given by

$$f_0 = \frac{1}{2\pi\sqrt{LC}}.$$

Important Terms

You should be able to write the meaning or definition of each of the following terms:

- inductive time constant
- capacitive time constant
- rms current
- rms voltage
- capacitive reactance
- inductive reactance

- phasor
- impedance
- diode
- rectifier
- operational amplifier
- feedback

QUESTIONS

20.1   How can an ac distribution system transmit power if the average of the current over every cycle is zero?

20.2   What would be the advantages or disadvantages of changing the frequency of household electricity to 1200 Hz rather than 60 Hz? What would they be if it were changed to 10 Hz?

20.3   Is it possible to adjust R, L, and C in some circuit so that the impedance is zero, thereby giving you free electrical energy?

20.4   If a half-wave rectifier is connected to an ordinary electric lamp in a household circuit, the lamp is observed to be dimmer than before; however, it does not blink off and on as you might expect from examining Fig. 20.20(b). Can you give two reasons for the failure to observe such flickering?

20.5   Trace the path of the current through the circuit shown in Fig. 20.29 when (a) the voltage at point A is positive with respect to point C and (b) the

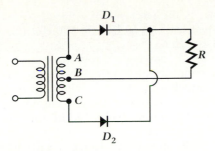

Figure 20.29 Question 20.5.

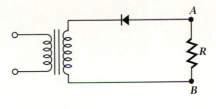

Figure 20.30 Question 20.6.

voltage at point *C* is positive with respect to *A*. Could this circuit be used as a practical full-wave rectifier?

20.6 If a sinusoidal voltage is applied to the transformer shown in Fig. 20.30, what voltage appears across the load *R*? That is, make a sketch of the voltage from *A* to *B* versus time.

20.7 Suppose a large capacitor is placed in parallel with the load resistance in the circuit of Fig. 20.30. What effect would it have on the current and voltage across *R*? Sketch a diagram of the resulting voltage waveform.

20.8 A pocket calculator containing several IC chips can easily have the equivalent of 30,000 transistors. Suppose that the manufacturer had been limited to using individual transistors. If each individual transistor occupied a volume of about 5 mm × 5 mm × 5 mm, what would be the minimum volume occupied by the transistors alone? What is the edge length of a cube of this volume?

20.9 In what ways does a filter capacitor improve the performance of a dc power supply?

20.10 How can a galvanometer movement be adapted for use in an ac meter?

PROBLEMS

Hints for Solving Problems
Remember that the time constant for an *RC* circuit is $\tau = RC$ and for an *RL* circuit is $\tau = L/R$. The current in an *RC* circuit decreases with time while that in an *RL* circuit increases with time. The effective value of a sinusoidal current or voltage is the rms value, equal to $1/\sqrt{2}$ of the peak value. In a series *RLC* circuit, the impedance is a minimum at the resonant frequency.

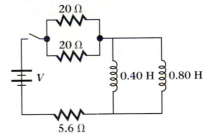

Figure 20.31 Problem 20.4.

Section 20.1 The *RL* Circuit

20.1 Show that the dimension of henries per ohm is time.

20.2 What is the inductive time constant of an *RL* circuit of $L = 0.80$ H and $R = 27$ Ω? What is the ratio of the current in the circuit at time $t = L/R$ to the current after a very long time ($t \gg L/R$)?

20.3 A series *RL* circuit containing a 0.17-H inductor has a time constant of 0.13 s. What is the resistance in the circuit?

20.4 What is the inductive time constant for the circuit shown in Fig. 20.31?

20.5 A series circuit of a resistor *R*, an inductor *L*, and a battery of emf *V* has a time constant τ. The circuit

is closed at time $t = 0$. What is the value of the current in the circuit when $t = \tau/2$? When $t = 2\tau$? When $t = 4\tau$?

20.6 The current in an *RL* circuit rises to half of its maximum value in 7.0 s. What is the inductive time constant for the circuit? If $L = 0.80$ H, what is the value of the resistance?

Section 20.2 The *RC* Circuit

20.7 Show that the dimension of resistance times capacitance is time.

20.8 What is the time constant of the circuit in Fig. 20.32?

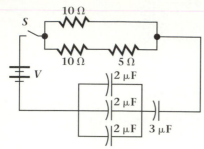

Figure 20.32   Problem 20.8.

20.9 A 200-μF capacitor is discharged through a 3.9-MΩ resistance. What is the time constant?

20.10 Two 5000-μF capacitors are joined in parallel and charged to 50 V. The capacitors discharge through a resistance of 100 kΩ. (a) What is the time constant of the circuit? (b) How much voltage will remain on the capacitors after one hour has elapsed?

20.11 A 1000-μF capacitor is discharged across a 500-kΩ resistor. What is the time constant of this discharge? If the initial voltage on the capacitor is 100 V, what is the voltage at $t = 120$ s after the circuit is closed?

Section 20.3   Effective Values of Alternating Current

20.12 An electric signal oscillates sinusoidally at a frequency of 60 Hz. What is the period of this oscillation?

20.13 The period of the alternating voltage in many European countries is 20 ms. What is the frequency of such oscillations?

20.14 What is the peak voltage of a household voltage outlet of 120 V ac?

20.15 A radio circuit carries a sinusoidal current with a peak of 17 mA. What is the effective value of the current?

20.16 Graph the function $y = \sin^2 x$ over an interval from $x = 0$ to $360°$. Verify graphically that the average value of $\sin^2 x$ over that interval is $\frac{1}{2}$.

20.17 Use the fact that $\overline{\sin^2 x}$ is $\frac{1}{2}$ over one cycle to derive the relation between the rms value of a sinusoidal oscillation and its amplitude.

20.18 Show that the average power delivered over one cycle to a resistive load R is

$$\overline{p} = \frac{V_{rms}^2}{R}.$$

Section 20.4   Reactance

20.19 Show that the dimensions of inductive and capacitive reactance are the same as the dimension of resistance.

20.20 What is the reactance of a 1.0-μF capacitor at a frequency of (a) 60 Hz? (b) 500 Hz? (c) 1000 Hz? (d) 10 kHz?

20.21 A 0.10-μF capacitor is driven by a voltage oscillating at a frequency of 10 kHz. What is the reactance of the capacitor? If the peak voltage on the capacitor is 9.0 V, what is the maximum current through it?

20.22 At what frequency will a 100-μF capacitor have a reactance of 4.0 Ω?

20.23 A 4.7-μF capacitor is driven by a voltage oscillating at 490 Hz. (a) What is the reactance of the capacitor? (b) What is the peak current if the peak voltage across the capacitor is 5.0 V?

20.24 Calculate the reactance of a 500-mH inductance operated at 2000 Hz.

20.25 An inductor has an inductance of 15 mH. What is the reactance of this inductor at frequencies of 60, 1000, and 10,000 Hz?

20.26 At what frequency will a 25-mH inductor have a reactance of 100 Ω?

20.27 A 50-mH inductance is driven by a voltage oscillating at 120 Hz. What is the peak value of the current if the maximum voltage is 10 V?

20.28 A 1.5-μF capacitor is connected to an 80-V, 60-Hz power supply. What is the effective current through the capacitor?

Section 20.5   The *RLC* Series Circuit

20.29 Calculate the amplitude of the current for a series circuit containing *R*, *L*, and *C* driven by a voltage source $v = V_0 \cos 2\pi ft$, given that the source frequency is $f = 2000$ Hz. Let $R = 50\ \Omega$, $L = 10$ mH, $C = 1\ \mu$F, and $V_0 = 10$ V.

20.30 An *RLC* series circuit operating at 1000 Hz has a
• 2.38-Ω resistor and a 23-mH inductor. What is the value of the capacitor that when added to the circuit will make the impedance of the entire circuit twice as large as the magnitude of the resistance?

20.31 Construct the phasor diagram for the situation of Problem 20.29.

20.32 A 0.8-H inductor is connected in series with a 330-Ω resistor across a 60-Hz voltage source. An ac voltmeter across the source indicates 120 V rms. (a) What current flows in the circuit? (b) What is the phase angle between current and voltage? (c) What is the power dissipation in the circuit? (*Hint:* The power is I^2R.)

20.33 The current to a loudspeaker is 350 mA (rms) and the terminal voltage is 2.8 V rms. The phase angle between current and voltage is 42°. Find the power input and the effective impedance.

20.34 A series RC circuit is driven by an alternating voltage source at a frequency of 100 Hz. (a) If $R = 330$ Ω and $C = 100\ \mu$F, what is the phase angle between current and voltage? (b) What is the phase angle between capacitor voltage and resistor voltage? (c) What is the phase angle between capacitor current and resistor current?

Section 20.6 Resonant Circuits

20.35 A series circuit of $R = 1500 \ \Omega$, $L = 350$ mH, and $C = 0.15 \ \mu$F is driven with an oscillating source of variable frequency. (a) What is the resonant frequency of the circuit? (b) What is the impedance at the resonant frequency?

20.36 A series circuit of $R = 4700 \ \Omega$, $L = 225$ mH, and $C = 1.5 \ \mu$F is driven by a voltage source oscillating at a frequency of 300 Hz. (a) Calculate the impedance at 300 Hz. (b) What is the resonant frequency? (c) What is the impedance at the resonant frequency?

20.37 A 270-μF capacitor and a 25-mH inductor are in series with a 330-Ω resistor. (a) What is the resonant frequency? (b) What is the impedance at the resonant frequency?

20.38 A series circuit containing R, L, and C is driven by a voltage source $v = V_0 \cos 2\pi f t$. If $R = 50 \ \Omega$, $L = 10$ mH, and $C = 1.0 \ \mu$F, what is the current at the resonant frequency? Let $V_0 = 10$ V.

20.39 Construct the phasor diagram for Problem 20.38.

20.40 A 0.010-μF capacitor is connected across a 2.5-mH inductor. What is the natural oscillation frequency of the combination?

20.41 A capacitor $C = 4.7 \times 10^{-9}$ F is connected across a 2.5-mH inductor. What is the natural oscillation frequency of the combination?

20.42 A 10-μF capacitor is charged to 30 V and then connected across a 0.45-H inductor. (a) What is the maximum value of the current in the circuit? (b) What is the frequency of the oscillations? (*Hint:* Use conservation of energy.)

20.43 A 15-μF capacitor is connected across a 0.50-H inductor. The capacitor is initially charged to 30 V. (a) What is the peak current in the circuit? (b) What is the frequency of the oscillations?

*Section 20.7 Rectifiers

20.44 Sketch the output signal across R_L from the full-wave rectifier of Fig. 20.21(a) as a function of time, assuming that the transformer provides a sinusoidal voltage with a peak of 5.0 V. Consider the situations: (a) for ideal diodes with no resistance, and (b) for real diodes with a forward voltage drop of 0.50 V.

20.45 A transformer in a stereo amplifier has an rms output of 15 V. (a) What is the average value of the voltage produced when the transformer is coupled to a full-wave rectifier like that of Fig. 20.21(a)? (b) What is the rms voltage?

20.46 A transformer in a TV has an rms output of 6.3 V. (a) What is the average value of the voltage produced when the transformer is coupled to a half-wave rectifier like that of Fig. 20.20(a)? (b) What is the rms voltage?

20.47 What is the average (dc) current for the situation of Problem 20.45 if the load resistance is 470 Ω?

*Section 20.8 Operational Amplifiers

20.48 An operational amplifier circuit is constructed with two resistors, $R_i = 1.0$ kΩ and $R_f = 2.0$ kΩ. (a) What is the gain of the circuit if it is operated as an inverting amplifier? (b) What is the gain if it is operated as a noninverting amplifier?

20.49 An operational amplifier circuit is constructed with $R_i = 10$ kΩ and $R_f = 5.0$ kΩ. (a) What is the gain of the inverting circuit? (b) What is the gain of the noninverting circuit?

Additional Problems

20.50 A resistance of 100 Ω and an inductance of 50 mH
• are connected in series across a 9.0-V battery. How long will it take after the circuit is completed for the current to reach 80 mA?

20.51 A 1000-Ω resistor is joined in series with a 400-mH
• inductance and a 12-V battery. (a) What is the time constant of the circuit? (b) How long will it take after the circuit is completed for the voltage across the resistor to reach 3.0 V?

20.52 Calculate the dc current through the circuit of Ques-
• tion 20.5 (Fig. 20.29), given that the load resistor is $R_L = 330 \ \Omega$. The transformer output is 12 V rms from A to C. The connection B is halfway between A and C.

20.53 (a) What is the inductive time constant of an RL
• circuit of $L = 0.80$ H and $R = 5.0 \ \Omega$? (b) How long after the switch is closed does it take for the current in the circuit to reach 99% of its ultimate value?

20.54 A 5000-Ω resistor is joined in series with a 100-μF
• capacitor. (a) What is the capacitive time constant of this combination? (b) If a 9.0-V battery is suddenly connected across the RC combination, how long will it take for the capacitor voltage to reach 8.0 V?

20.55 A resistance of 100,000 Ω is used to discharge
• a 0.47-μF capacitor that is charged to 1500 V. (a) How many time constants must elapse before the capacitor voltage falls to 10 V? (b) How much time must elapse?

20.56 A series combination of $R = 1000 \ \Omega$ and $C = 500$
• μF is connected to a 12-V battery. How long after the circuit is completed will it take for the capacitor voltage to reach 10 V?

20.57 Show that for an inductor and a resistor connected
• in a series ac circuit the power dissipated in the
resistance is given by $P = V_{rms} I_{rms} \cos \phi$, where ϕ
is the phase difference between current and voltage.

20.58 The operational amplifier circuit shown in Fig.
•• 20.33 is called a voltage follower. Analyze it as a
noninverting amplifier with $R_i = \infty$ and $R_f = 0$ to
prove that $V_{out} = V_{in}$.

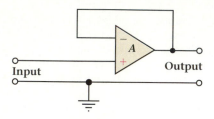

Figure 20.33 Problem 20.58.

20.59 An operational amplifier is constructed according to
• the circuit shown (Fig. 20.34). Show that the output
is given by

$$V_{out} = - (2V_1 + 3V_2 + 10V_3).$$

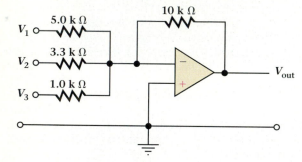

Figure 20.34 Problem 20.59.

20.60 Show that the amplifier circuit of Fig. 20.35 has an
•• output of -0.6 V when $V_{in} > 0$ and an output of
V_{max} when $V_{in} < 0$. The forward voltage drop of the
diode is 0.6 V and the maximum output of the amplifier is V_{max}.

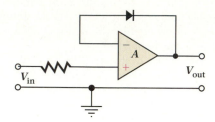

Figure 20.35 Problem 20.60.

20.61 The voltage from a square-wave oscillator varies
•• with time as shown in Fig. 20.36. Calculate (a) the
average and (b) the rms voltage for such a wave in
terms of the peak voltage V_0.

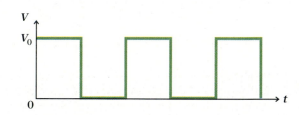

Figure 20.36 Problem 20.61.

ADDITIONAL READING

Brattain, W. H., "The Genesis of the Transistor." *The Physics Teacher,* March 1968, p. 109.

Duffy, R. J., and U. Haber-Schaim, "Establishing $V = L (dI/dt)$ Directly from Experiment." *American Journal of Physics,* February 1977, p. 170.

MacInnes, I., "Notes on the Condition for Maximum Power Transfer to the Load." *The Physics Teacher,* April 1985, p. 224.

Malmstadt, H. V., C. G. Enke, and S. R. Crouch, *Electronics and Instrumentation for Scientists.* Reading, Mass.: Addison-Wesley, 1981. A book covering all the fundamentals of modern electronic instrumentation.

Scientific American, September 1977, full issue on microelectronics.

21

Geometrical Optics

21.1 Models of Light: Rays and Waves

21.2 Reflection and Refraction

21.3 Total Internal Reflection

21.4 Thin Lenses

21.5 Locating Images by Ray Tracing

21.6 The Thin-Lens Equation

21.7 Spherical Mirrors

*21.8 Lens Aberrations

A WORD TO THE STUDENT

We saw at the end of Chapter 19 how Maxwell's equations unified electricity and magnetism into one comprehensive theory. The outstanding achievement of this theory was the prediction of electromagnetic radiation, culminating in the realization that light is a form of this radiation. Thus the study of electromagnetism leads directly to optics—the study of light.

Four chapters in this text deal with optics. In this chapter we deal briefly with the debate over the nature of light (particle versus wave). We then focus on phenomena in which light can be considered to travel in a straight line. This study is called geometrical optics and is the basis for understanding ordinary optical instruments such as cameras, microscopes, telescopes and even the human eye, all of which are discussed in Chapter 22. In Chapter 23 we will show that geometrical optics does not explain all optical observations, and we will examine instances in which light must be considered as a wave. The fourth chapter on optics is Chapter 29, which describes lasers and holography. These subjects can best be understood only after studying the intervening chapters on atomic physics and quantum mechanics.

Because of the importance of vision, people have been interested in the nature and behavior of light for many centuries. Although we now know that light behaves like an electromagnetic wave, as described by Maxwell's equations (Section 19.8), most of the fundamental laws of optics were discovered before the time of Maxwell. This is particularly true for the material of this chapter. For example, the apparent bending of a stick partially immersed at an angle into water, an effect due to the phenomenon of optical refraction, was described nearly 2000 years ago by the Greek astronomer and geographer Ptolemy (ca. 100 A.D.). Another related effect, the ability to see the sun after it has gone below the horizon, was known to Tycho Brahe, Johannes Kepler, and others in the sixteenth century.

By the end of the seventeenth century, many of the now familiar laws and relationships of optics had been stated. However, the nature of light was still not agreed upon. Some scientists, such as Robert Hooke and Christiaan Huygens, proposed that light was a wave, while Isaac Newton proposed that light consisted of particles, or corpuscles as he called them. Newton's reputation was so great that the corpuscular theory predominated for some time.

In the early 1800s, the wave theory of light began to attract renewed attention as a result of experiments by Thomas Young and theoretical work by August Jean Fresnel. By the last quarter of the century, Maxwell had predicted the existence of electromagnetic waves with a speed equal to the speed of light. Maxwell's waves were verified experimentally by Heinrich Hertz at the close of the century.

What is light in our contemporary view of physics—waves or particles? There is no simple answer to this question. In fact, as we will see in Chapter 27, quantum mechanics has shown that the difference between a wave and a particle depends to a large extent on your point of view. Perhaps the best way to think of it is that waves and particles are both simplified models of reality, and that light is a complicated phenomenon that doesn't quite fit either model alone. Both models are used extensively. Generally, we adopt whichever model provides an explanation of the optical behavior under study.

In this chapter we will examine optical behavior that does not depend on the nature of light, but only on the path it travels. This study is called geometrical optics, and includes reflection and refraction.

Models of Light: Rays and Waves

21.1

Although we usually consider light to be a wave phenomenon, that is not the only possible view. The idea that light is composed of a stream of particles is also consistent with the observation that light seems to travel in straight lines except when it encounters an interface between two opti-

Figure 21.1 A ray of light from a laser. The light travels in straight lines unless it encounters an interface between two optical media.

cal media. A beam of light from a flashlight, laser, or other source is seen to travel in a straight line (Fig. 21.1). We call this straight-line path a **ray** of light. The ray model allows us to explain in the simplest terms the formation of images by lenses and mirrors.

Furthermore, we know that light displays particlelike characteristics when it interacts with matter, as it does, for example, when sunlight falls on a leaf and photosynthesis takes place. The apparent conflict over this wave-particle duality cannot be understood with just the theories of Newton or Maxwell, but is resolved by the twentieth-century theory of quantum mechanics, discussed in Chapter 27.

We have seen in the preceding chapters how discoveries in electricity and magnetism culminated in Maxwell's formulation of the theory of electromagnetism. The realization that light was the same as the electromagnetic radiation predicted by Maxwell's equations united the study of electromagnetism and optics. The range of wavelengths that comprise visible light make up only a small portion of the entire electromagnetic spectrum (Fig. 21.2), which spans many orders of magnitude. There are no sharp dividing lines separating the various regions; there is just a continuous blending from one region to the next.

When sunlight is spread into a spectrum we see the characteristic band of visible colors from red to violet. Beyond the violet edge of the visible spectrum are frequencies of radiation that exceed that of the violet. We use the name **ultraviolet** to describe this invisible extension of the spectrum. Beyond the red end of the visible spectrum lie frequencies below those we can see. These frequencies make up the **infrared** region of the spectrum.

Figure 21.2 The electromagnetic spectrum. There are no sharp divisions between types of electromagnetic waves, nor are there sharp boundaries between different colors of visible light.

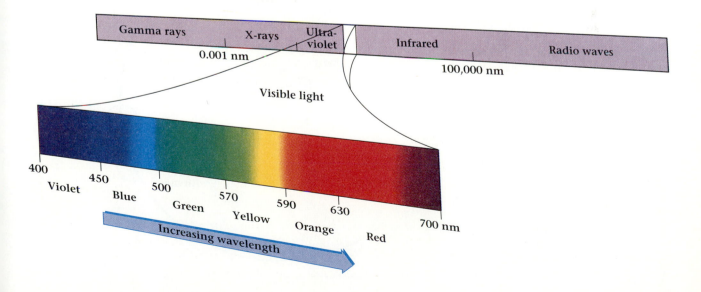

The fundamental behavior of all components of the electromagnetic spectrum is the same. They differ only in their wavelengths and frequencies and in the kinds of devices that can be used to generate and detect them. The behavior of all electromagnetic waves can be predicted from Maxwell's equations and a knowledge of the composition and shape of the lenses, reflectors, and other components involved. For example, the design of microwave antennas (Fig. 21.3) follows the same underlying principles as does the design of telescope mirrors.

Figure 21.3   Microwave antennas are curved reflectors used to focus microwaves.

BACK TO THE FUTURE

The Speed of Light

Interest in the speed of light goes back many centuries. Galileo was perhaps the first person to suggest a method for its measurement. He proposed an experiment in which two observers would be stationed on mountaintops some distance apart, each having a lantern enclosed in an opaque housing except for a hole covered by a shutter. Both observers were to begin with the shutters closed. At the start, one observer would open his shutter and begin to mark time. When the second observer saw the light from the first lantern, he was to open his own shutter. When the light from the second lantern reached the first observer, he was to note the elapsed time since his shutter was opened. The speed of light could then be determined from knowledge of the elapsed time and the distance between the mountains. We do not know whether the experiment was ever performed. As it happens, the speed of light is far too great and human reaction times are much too long for this experiment to succeed. However, the principle is sound and forms a basis for a number of successful determinations of the speed of light.

Several measurements of the speed of light have been based on astronomical observations. In 1676 the Danish astronomer Ole Roemer (1644–1710) announced his discovery of systematic variations in the time intervals, as observed on the earth, between successive disappearances of Jupiter's moons as they moved into the planet's shadow. These variations were related to the positions of the earth in its orbit (Fig. B21.1) and were associated with the variable distance from Jupiter to the earth. He found a time difference of about 22 minutes over six months, corresponding to the time needed for light to cross the diameter of the earth's

Figure B21.1   Orbits of the earth, Jupiter, and one of Jupiter's moons at two times, six months apart. Intervals between the moon's eclipses as measured on the earth depend on the relative positions of the earth and Jupiter.

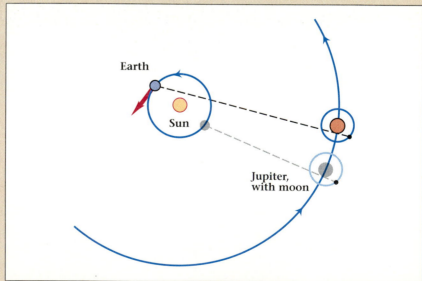

Reflection and Refraction

21.2

Suppose we shine a beam of light at a mirror (Fig. 21.4, p. 614). The light strikes the mirror at a point P and is then reflected. What is the direction of the reflected beam? You may know the answer even if you have not formulated it mathematically: The incident and reflected beams make equal angles with the mirror.

In optics it is an established convention to measure angles with respect to the normal, which is a line perpendicular to the surface, as indicated.

---

orbit. Roemer is thus credited with discovering the finite speed of light. In 1678 Huygens combined Roemer's time measurement with the distance traveled, taken from the recently estimated value of the earth–sun distance. He obtained a value for the speed of light equivalent to about 2.3×10^8 m/s.

An experiment similar in concept to that of Galileo was performed in the suburbs of Paris by Armand Hippolyte Louis Fizeau (1819–1896) in 1849. His experiment provided the first actual measurement of the speed of light performed on the earth. In order to overcome the difficulties of measuring extremely short intervals of time, he used a rotating toothed wheel and a mirror (Fig. B21.2). Light was focused on the rim of the wheel after reflection from a half-silvered mirror (M_1 in the figure). The light then passed through one of the gaps between the 720 teeth at the edge of the wheel and continued to a second mirror (M_2), where it reflected back along the same path to the wheel, passed again through a gap, and reached the mirror M_1. Some of the light reaching the mirror was transmitted to an observer.

When the wheel was rotated at certain speeds, the light passing through

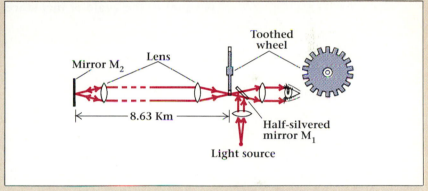

Figure B21.2 **The experimental arrangement for Fizeau's experiment to measure the speed of light.**

a gap as it proceeded to the mirror M_2 struck a tooth when it returned and the observer saw a decreased intensity. At other speeds of rotation, the light would encounter a gap on the return trip and be seen again. From knowledge of the number of gaps, the rotational speed of the wheel, and the path length of the light, Fizeau calculated a value equivalent to 3.15×10^8 m/s for the speed of light.

More recently, measurements of the speed of light have utilized its wave behavior and the knowledge that the velocity of a wave is the product of its frequency and wavelength, $v = f\lambda$. The wavelength of a highly stabilized laser was determined by high-precision techniques, while its frequency was measured by comparison with cesium-beam atomic clocks. From the values obtained, the best value for the speed of light is $299{,}792{,}458 \pm 1.2$ m/s. Because the chief limitation in these measurements was the uncertainty in the length of the meter based on previous standards, this value of the speed of light has been adopted and the meter redefined in terms of the speed of light (Section 19.7).

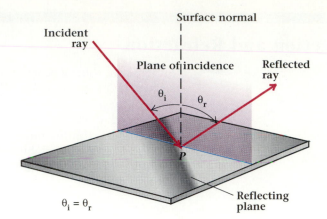

Figure 21.4   Reflection of light from a flat mirror. The incident and reflected beams make equal angles with the normal. The plane of incidence is defined by the incident and reflected rays.

The angle between the incoming ray and the normal is called the **angle of incidence** θ_i. The angle between the outgoing ray and the normal is the **angle of reflection** θ_r. Both angles are measured positive from the normal. The observed **law of reflection** is that **the angle of reflection is equal to the angle of incidence,** or

$$\theta_r \;=\; \theta_i. \tag{21.1}$$

The normal, the incident ray, and the reflected ray all lie in a plane perpendicular to the reflecting surface, known as the **plane of incidence.** The light ray does not "turn" out of the plane of incidence as it is reflected.

The law of reflection applies to both flat and curved surfaces. For a curved surface, the angle of reflection is determined by the angle of incidence between the incident ray and the surface normal at the point where the incident ray strikes the surface.

Reflection from a smooth mirrorlike surface is called *specular reflection* (Fig. 21.5a). When you look at a mirror, you do not usually see the mirror surface itself, but see the specularly reflected image of other objects instead. For example, you may look at a mirror and see an image of yourself or someone else. What happens when light is reflected from an object whose surface is not perfectly smooth? In that case the light is *diffusely reflected,* with different parts of the incident light beam scattered in different directions according to the law of reflection (Fig. 21.5b). The light reflected at each little region of the surface is reflected at an angle equal to the local angle of incidence. Most objects that we see are made visible by the diffuse reflection of light from their surfaces. The paper in this book reflects light diffusely so that you can see it from any angle. To see a light reflected in a mirror, however, requires that you be in just the right place so that the specularly reflected light can reach your eye.

Sometimes both diffuse and specular reflection occur simultaneously from the same surface. For example, when sunlight shines on a car, you can see the car from any direction around it. That is diffuse reflection. If the car is highly polished, you may also see the image of distant objects reflected from its surface. That is specular reflection.

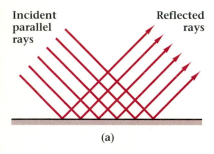

(a)

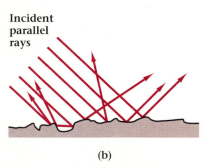

(b)

Figure 21.5   (a) Specular reflection from a smooth surface. (b) Diffuse reflection from a rough surface.

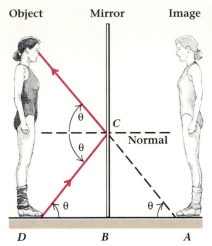

Figure 21.6 Example 21.1: A person looking into a plane mirror sees her image behind the mirror. The person and the image are both equidistant from the mirror.

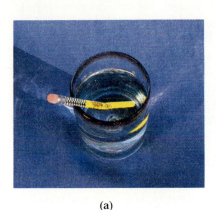

(a)

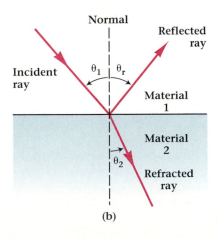

(b)

Example 21.1

How far behind a mirror does your reflection appear to be as you view yourself?

Solution   Figure 21.6 is a schematic diagram of a person looking directly into a mirror. Consider any ray of light that leaves any part of the person, is reflected in the mirror, and arrives at the person's eye. All such rays are reflected in accordance with the law of reflection. We have selected and drawn a ray in the figure from the foot to the eye. Since the brain interprets light striking the eye as having traveled in a straight-line path, the dashed line represents the path of the light as it appears to the viewer. The intersection of the dashed line with the horizontal line representing the floor locates the apparent position of the image. From the geometry we see that triangles *ABC* and *DBC* are similar and have a common side. Thus *AB* = *BD*. Therefore the apparent distance of the image behind the mirror is the same as the real distance of the person in front of the mirror.

If light strikes a *transparent* body, we observe both reflection and transmission. (A transparent body contrasts with an *opaque* body, which does not permit transmission of light through it.) The transmitted beam is bent, or *refracted,* as it crosses the surface between one medium (such as air) and another (water, for example). Ptolemy described the observation that a beam of light entering water was bent toward the normal (Fig. 21.7a). He also knew that the amount of bending depends on the angle of incidence. Later investigation showed that the amount of bending also depends on the particular materials found on either side of the interface (Fig. 21.7b). However, not until about 1600 did the Dutch mathematician Willebrod Snell (1591–1626) empirically discover the exact law of refraction. We can now derive this relationship from Maxwell's equations or from a more elementary wave picture of light (see Chapter 23). The **law of refraction**, also called **Snell's law**, can be written

$$n_1 \sin \theta_1 = n_2 \sin \theta_2, \qquad (21.2)$$

where n_1 depends only on the optical properties of material 1 and n_2 depends only on the optical properties of material 2. The constant n is called the **index of refraction** of the material. We define the index of refraction n of a medium to be the ratio of the speed of light in vacuum c to the speed of light in that medium v,

$$n = \frac{c}{v}. \qquad (21.3)$$

Figure 21.7   (a) The apparent bending of a pencil at the water surface due to the refraction of light. The magnification of the lower image results from the curvature of the container. (b) Refraction of light at a boundary between two different materials. The direction of a single ray is changed when it crosses the boundary between materials 1 and 2. Some of the light is also reflected at the surface.

TABLE 21.1
Index of refraction of some common materials ($\lambda = 589$ nm)

Material	Index of refraction (n)
Gases (at atmospheric pressure and 0°C)	
Hydrogen	1.0001
Air	1.0003
Carbon dioxide (CO_2)	1.0005
Liquids (at 20°C)	
Water	1.333
Ethyl alcohol	1.362
Glycerine	1.473
Solids (at room temperature)	
Ice	1.31
Polystyrene	1.59
Crown glass	1.50–1.62
Flint glass	1.57–1.75
Diamond	2.417
Acrylic (polymethylmethacrylate)	1.49

We can verify Snell's law for a large number of materials and use it to determine their indices of refraction. In fact, we can do this without any knowledge of the nature of light; we assume only that light travels in straight lines unless it encounters the boundary between two materials. Table 21.1 lists the indices of refraction for a number of common transparent materials. Usually the index of refraction of air can be taken as unity without introducing any significant error. In most materials the speed of light depends upon its wavelength. Generally, the index of refraction decreases slowly with increasing wavelength.

If light is initially incident at the angle θ_2 upon an interface from within material 2, Eq. (21.2) is still valid. If the refracted ray in Fig. 21.7(b) strikes a mirror so that it is reflected back along the same path, it will be refracted at the interface and emerge along the same path as the incident ray. We say that light rays are reversible. That is, if a light ray traveling in some direction takes a particular path, a light ray going in the opposite direction will follow exactly the same path.

Example 21.2

Refraction from air to glass to water.

The side of a fish tank is made of a thick parallel-sided piece of glass with an index of refraction of 1.50 (Fig. 21.8). A beam of light strikes the surface of the glass at an angle of 46° with respect to the normal. What is the direction of the beam in the water?

Solution We solve this problem by first finding the angle of refraction at the air–glass interface and then using it to determine the angle of refraction at the glass–water interface.

At the first surface, let θ_1 and θ_2 be the angles of incidence and refraction, respectively. The indices of refraction are n_1 for the air and n_2 for the glass. We can insert the values of these quantities into Snell's law

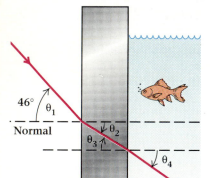

Air   Glass   Water
$n_1 = 1.00$ $n_2 = n_3 = 1.50$ $n_4 = 1.33$

46° θ_1

Normal

θ_2

θ_3

θ_4

Figure 21.8 Example 21.2: A beam of light in air is refracted as it enters a slab of glass. The beam is refracted again as it passes from the glass into water.

to find θ_2:

$$n_1 \sin \theta_1 = n_2 \sin \theta_2,$$

$$\sin \theta_2 = \left(\frac{n_1}{n_2}\right) \sin \theta_1 = \left(\frac{1.00}{1.50}\right) \sin 46°,$$

$$\sin \theta_2 = 0.480,$$

$$\theta_2 = 28.7°.$$

Thus in the glass the refracted ray makes an angle of 28.7° with the normal.
At the second surface, we use Snell's law again to get

$$\sin \theta_4 = \left(\frac{n_3}{n_4}\right) \sin \theta_3,$$

where θ_3 is the angle of incidence of the light within the glass and θ_4 is
the angle of refraction in the water. The index of refraction for the glass
is $n_3 = n_2 = 1.50$ and for the water is $n_4 = 1.33$. Since the faces of the
glass are parallel, $\theta_3 = \theta_2$ and $\sin \theta_3 = \sin \theta_2$. Upon inserting the values
for the indices of refraction and the value of $\sin \theta_2$ from our first calculation, we find that

$$\sin \theta_4 = \left(\frac{1.50}{1.33}\right) 0.480 = 0.541,$$

$$\theta_4 = 32.8°.$$

Total Internal Reflection

21.3

A particularly interesting situation arises when the light incident on an
interface comes from within the denser medium, that is, from the material
with greater index of refraction. In this case n_1 is greater than n_2. Consequently, in accord with Snell's law, θ_2 is larger than θ_1. As θ_1 increases,
θ_2 must also increase, but it reaches a limit at 90° (Fig. 21.9a, p. 618). At
this point, the refracted ray runs along the interface. The incidence angle
corresponding to a 90° transmission angle is called the **critical angle** θ_c
and is found from Snell's law where:

$$n_1 \sin \theta_c = n_2 \sin 90°,$$

$$\sin \theta_c = \frac{n_2}{n_1}. \tag{21.4}$$

When the angle of incidence is smaller than θ_c, the light is transmitted
at some θ_2 between 0 and 90°. But for angles of incidence *greater* than θ_c,
Snell's law is no longer valid and experiments show that no light penetrates into the less dense medium. Instead, when the angle of incidence
exceeds the critical angle, all the incident light is reflected back into the
denser medium by the interface between two normally transparent materials. We call this phenomenon **total internal reflection**. The light from

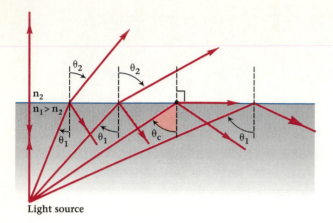

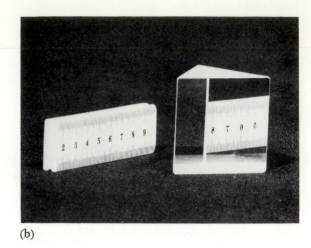

(b)

Figure 21.9 (a) When the angle of incidence exceeds the critical angle θ_c, light is totally internally reflected. (b) The image of a ruler is totally reflected from the back surface of a triangular prism.

the ruler in Fig. (21.9b) is totally reflected from the back of the prism to create the mirror image.

Example 21.3

Show that a 45°–45°–90° glass prism can deflect a light beam 90° by total internal reflection. The index of refraction of the glass is 1.52.

Solution   If the incident ray strikes the short side of the prism normally (that is, perpendicularly; Fig. 21.10), there is no refraction at the interface I_1 and the transmitted beam hits the back surface I_2, making an angle of incidence of 45°. If the critical angle is less than 45°, all of the light is reflected from the back of the prism.

To calculate θ_c, we use $n_1 = 1.52$ for the glass and $n_2 = 1.00$ for air. The critical angle is

$$\theta_c = \sin^{-1}\left(\frac{n_2}{n_1}\right) = \sin^{-1}\left(\frac{1.00}{1.52}\right),$$

$$\theta_c = 41.1°.$$

The light incident at 45° exceeds the critical angle, so the light striking I_2 is indeed totally reflected. It strikes the third interface I_3 normally, and emerges at an angle of 90° from the initial direction. Periscopes make use of two such prisms to enable the viewer to see around corners or other obstacles.

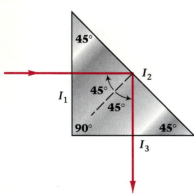

Figure 21.10 Example 21.3: Light incident normally on a 45°–45°–90° prism is totally internally reflected and emerges from the other face.

Total internal reflection can work for us in *light pipes*. If we direct light into one end of a long rod of glass or plastic, the light is totally internally reflected by the walls, bouncing back and forth until it emerges at the far end (Fig. 21.11a). If we bend the light pipe into a particular shape, the light follows the shape of the pipe and emerges only at the end.

Sometimes light pipes are made of very thin *optical fibers,* which may be grouped into a bundle (Fig. 21.11b). If the fiber ends are polished and their spatial arrangement is the same at both ends, an image may be transmitted from one end of the fiber bundle to the other (Fig. 21.12a).

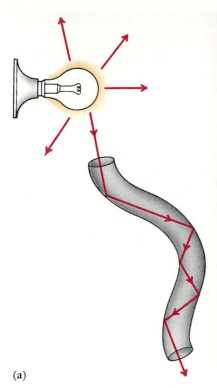

(a)

(b)

Figure 21.11 (a) Light rays trapped by total internal reflection are channeled along the length of a light pipe. (b) Light can be seen emerging from the ends of tiny optical fibers.

A fiber bundle that transmits an image is called a *coherent bundle* or an *image conduit*. Bundles of fibers that do not have exactly the same alignment at both ends transmit light but not images. They are known as *incoherent bundles,* or *light guides*. Because their fibers are so flexible, fiber optics light guides and coherent bundles are used in instruments designed to permit direct visual observation of otherwise inaccessible objects. Prime examples are endoscopes (Fig. 21.12b), which are used by physicians to examine the interior of a patient's body.

Optical fibers are also used in communication systems to transmit modulated light beams. These fibers replace metal wire, thus reducing cost and improving signal transmission.

Figure 21.12 (a) An image conduit composed of a coherent bundle of optical fibers transmits an image from one end to the other. (b) A flexible, fiber optic endoscope for observing the inside of a patient's body.

(a)

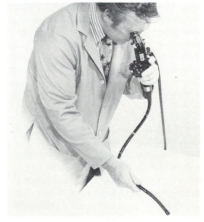

(b)

Thin Lenses

21.4

An optical lens is a piece of glass or other transparent material used to direct or control rays of light. Usually the lens surfaces are spherical, although other shapes (parabolic, cylindrical, etc.) are not uncommon. The refraction of light at the surface of a lens depends on its shape, its index of refraction, and the nature of the medium surrounding it (usually air). Because lenses can be used to produce images of objects, they form the basis of most optical instruments, from cameras and projectors to microscopes and telescopes. In fact, the eye itself functions partly as a lens.

There are only a few distinct ways of combining flat, convex, and concave surfaces to form a single lens (Fig. 21.13). If you hold a single lens like one of these at a moderate distance in front of your eye and look at something through it, you can make several observations. Consider first some lenses that are thicker in the center than at their edges, such as those in Fig. 21.13(a). These are called positive or **converging lenses**, because they refract incident parallel rays so that they converge on a focal point located on the opposite side of the lens. A distant object viewed through such a lens appears smaller and inverted if the lens is held some distance from the eye. An object held close to the same lens appears erect and enlarged. Now look at some lenses that are thinner in the center than at their edges, such as those in Fig. 21.13(b). These are called negative or **diverging lenses**, because they refract incident parallel rays so that they appear to diverge from a focal point located on the incident side of the lens. Any object viewed through such a lens always appears erect and smaller than when viewed with the unaided eye. The image remains erect no matter how far the object is from the lens or how far the lens is from the eye.

Let's take a converging lens and allow light from the sun to fall on it. Because the sun is so far away, the rays striking the lens are essentially parallel to one another. When these light rays strike the lens parallel to its axis of symmetry, the rays converge to a point called the *focal point* of the lens (Fig. 21.14). Thus we can say that any ray incident on a converging lens and parallel to its *optical axis* (that is, the symmetry axis of the lens) passes through the focal point upon leaving the lens. A lens

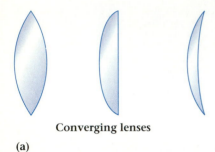

Converging lenses

(a)

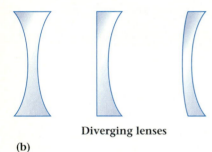

Diverging lenses

(b)

Figure 21.13 Cross-section views of the most common types of thin lenses having spherical surfaces: (a) converging lenses and (b) diverging lenses. These lenses are often referred to by their shapes (from left to right): double convex, plano-convex, positive meniscus, double concave, plano-concave, and negative meniscus.

Figure 21.14 (a) Parallel light incident from the left on a converging lens is brought to a focus at the focal point *f*. (b) Light emerging from this focal point passes through the lens and emerges parallel to the optical axis. (c) Parallel light incident from the right converges to a focal point *f′* on the left side of the lens. For a thin lens these two focal points are equidistant from the lens.

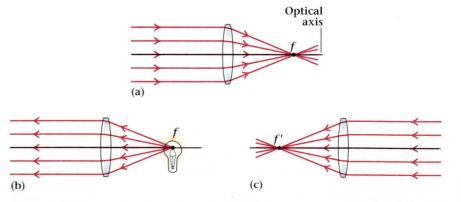

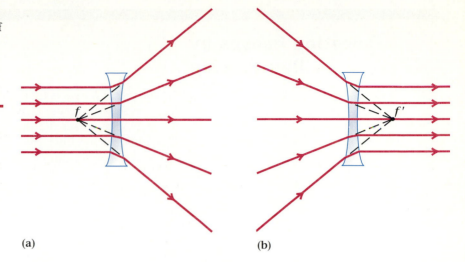

(a)   **(b)**

Figure 21.15 (a) Parallel light incident on a concave lens diverges as if it comes from the point f. (b) Light converging toward the point f' diverges so that it emerges parallel after passing through the concave lens.

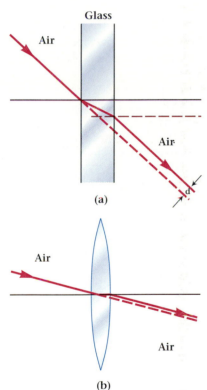

(a)

(b)

Figure 21.16 (a) A ray of light incident on a parallel-sided slab of glass is displaced an amount d, proportional to the thickness of the glass. The direction of the emerging beam is parallel to the incident beam. (b) A ray of light incident on the center of a thin lens is essentially undeviated if the angle of incidence is small.

whose thickness is small in comparison with its focal length is called a **thin lens**. The distance from the center of the thin lens to the focal point is the **focal length** of the lens.

There is a symmetry to the passage of light through a thin lens. Any light ray entering the lens from the left parallel to the optical axis (Fig. 21.14a) passes through the focal point f on the right after it leaves the lens. Conversely, any ray coming from the focal point on the right and striking the lens emerges parallel to the axis (Fig. 21.14b). Parallel light entering the same lens from the right converges to a point f' on the left of the lens (Fig. 21.14c). For a thin lens, the distances from the center of the lens to f and f' are the same, even though the lens surfaces are not identical. Thus we consider a thin lens to have two identical focal lengths, one on either side of the lens.* In the remainder of this chapter, we consider all lenses to be thin lenses. In addition, we will further assume that the incident light rays are nearly parallel to the lens axis (the *paraxial approximation*). Later, in Section 21.8, we discuss some of the ways in which real lenses differ from our idealized thin lens.

Parallel light incident on a concave lens parallel to its optical axis diverges (Fig. 21.15a). After passing through the lens, the light behaves as if it came from a point source located at f. Thus f is the focal length for this lens. Figure 21.15(b) shows that the same lens refracts light directed toward the focal point on the other side so that the rays emerge parallel to the optical axis after passing through the lens.

An ordinary piece of glass, such as a window with parallel faces, passes light without changing its direction. Although the direction of the ray does not change, it is displaced laterally (Fig. 21.16a). At the center of a lens the front and back surfaces are parallel, so to a very good approximation, a light ray striking the center of the lens goes through the lens undeviated. If the incoming ray does not make too large an angle with the axis and if the lens is thin, the offset in the ray is negligible (Fig. 21.16b).

---

*For a thick lens—that is, one for which the thickness is not small in comparison with the focal length—it is possible for f and f' to be different.

Locating Images by Ray Tracing

21.5

The major usefulness of lenses is their ability to form images of an object. The object may be self-luminous, giving off its own light (like the sun or a light bulb), or it may reflect the light that falls on it (like an apple or the page of this book). In either case, an image of the object is formed where light rays that come from points on the object intersect, or at the points from which they appear to originate. When the light rays actually intersect at the image, we call this a **real image**. A screen placed at the image point would show the image in the same way that pictures appear on a movie screen. A **virtual image** is formed at a point where the light appears to converge, or from which it appears to come. If you place a screen at the position of a virtual image, no image is observed on the screen, for the light rays do not actually intersect there. An example of images in a mirror may help to explain virtual images. If you stand one meter in front of a mirror, your image appears to lie one meter behind the mirror. However, you will not find an image on a screen placed at the image position one meter behind the mirror.

An object, such as a pencil, reflects light rays in all directions from all points of the pencil. We illustrate this in Fig. 21.17 by using arrows from the tip of the pencil to represent light rays. However, to locate the image of the pencil formed by a thin lens, you don't need to trace the light rays from all over the object. Instead, you can locate the image by tracing three particular rays from a point on the object. Figure 21.17(a) shows these rays refracted by a converging lens; Fig. 21.17(b) shows them for a diverging lens. In each case the focal length of the lens and the distance from the object to the center of the lens (the object distance) determines the distance of the image from the lens (the image distance).

In the paraxial approximation, a small planar object perpendicular to the optical axis is imaged by a thin lens into a planar region that is also perpendicular to that axis. Thus all points in the object plane are represented as points in an image plane. When the object is three-dimensional,

Figure 21.17 Graphical technique for locating image position with (a) a converging lens and (b) a diverging lens. Dashed lines indicate where light appears to come from (or go to).

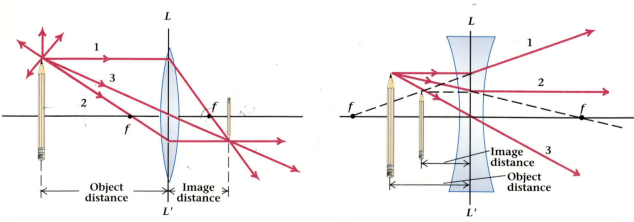

(a)   (b)

like the pencil in Fig. 21.17(a), points located in different object planes do not focus in a single image plane. For instructional purposes we restrict our discussion to planar objects and their planar images. Thus we normally represent objects and images with simple arrows perpendicular to the optical axis.

Let us state the graphical procedure to follow for locating the image from a given lens when you know the object's position. In the next section we will describe how to calculate the image height and location algebraically from this graphical ray diagram. You should refer to both parts of Fig. 21.17 as you read these steps.

A. Select an appropriate scale (meters, centimeters, etc.) and mark the position of the lens on the optical axis.
B. Draw a line LL' through the center of the lens and perpendicular to the optical axis, as shown. This line will replace the actual lens surfaces that refract the light.
C. Mark the focal points of the lens on the optical axis and locate the object on the axis, all to the same scale.
D. Draw the following rays. Any two of these rays are sufficient to locate the image, but you should always draw the third ray as a check.
   1. Draw ray 1 parallel to the optical axis from the object to the line LL' (the lens). For a converging lens, extend this ray from LL' through the focal point on the side opposite the incident light. For a diverging lens, extend the ray from LL' as though the ray came from the focal point on the same side as the incident light.
   2. Draw ray 2 from the object to the line LL', passing through a focal point. For a converging lens, draw ray 2 through the focal point on the same side as the incident light; for a diverging lens, draw ray 2 in the direction of the focal point on the opposite side. In both cases, continue the ray from LL' parallel to the optical axis.
   3. Draw ray 3 from the object to line LL' at the center of the lens and continue it undeviated.
E. If the rays converge on the side opposite the incident light, the image is real and is located at the intersection of these three rays. If the rays diverge, the image is virtual and is located where these three rays appear to originate. You can measure the position and image height directly from the scale drawing.

You could make a ray diagram for each point on the object to fully locate the image. However, if all of the object is at approximately the same distance from the lens, then you need only consider the image of one point of the object and the rest of the object is thereby located. Again, note that for a given object distance, the size and position of the image is completely determined by the focal length of the lens. Once you have determined the height of the image, you can determine the **lateral** or **linear magnification**, the ratio of the image height to the object height.

Example 21.4

A thin converging lens of focal length 6.0 cm is used to image a 35-mm slide on a screen. If the slide is 10.0 cm from the center of the lens, where should you locate the screen so that it coincides with the image?

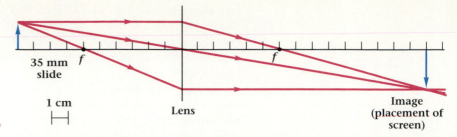

Figure 21.18 Example 21.4: Scale drawing for locating the image of a 35-mm slide.

Solution   This situation is drawn to scale in Fig. 21.18. You can measure the distance from the lens to the image position directly from the figure as 15 cm. By comparing the image height to the object height, the magnification is found to be 1.5. When drawing a scale diagram it is especially necessary to use care. You will find it helpful to draw the diagrams to a scale as large as your paper will allow in order to minimize the errors due to the finite width of the lines.

The Thin-Lens Equation

21.6

Using a geometrical ray diagram for a general case, we can derive an algebraic expression relating the object distance, the image distance, and the focal length of the lens. Applying this equation requires careful treatment of positive and negative distances from the lens, which we will discuss after presenting the basic derivation.

A general ray diagram is drawn in Fig. 21.19. We designate the object distance by o, the image distance by i, and the focal length by f. The height of the object is labeled h_o and that of the image h_i. From the similar triangles ABH and CDH we have

$$\frac{HA}{AB} = \frac{CD}{HC} \quad \text{or} \quad \frac{HA}{CD} = \frac{AB}{HC}.$$

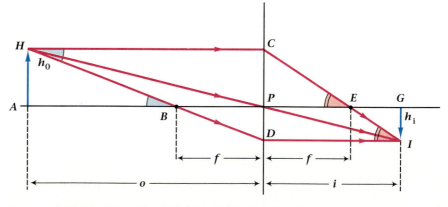

Figure 21.19   Geometry for derivation of the thin-lens formula.

From the similar triangles *CPE* and *CDI* we have

$$\frac{CP}{PE} = \frac{CD}{DI} \quad \text{or} \quad \frac{CP}{CD} = \frac{PE}{DI}.$$

But $HA = CP$, so that

$$\frac{PE}{DI} = \frac{AB}{HC}.$$

From the figure we see that $AB = o - f$, $HC = o$, $PE = f$, and $DI = i$, so that this equation becomes

$$\frac{f}{i} = \frac{o - f}{o},$$

which, upon rearranging, becomes

$$\frac{1}{i} + \frac{1}{o} = \frac{1}{f}. \tag{21.5}$$

This equation is called the **thin-lens equation** in Gaussian form.* Recall that by a thin lens we mean one for which the focal length is much greater than the thickness of the lens. For such a lens, Eq. (21.5) is sufficient to specify the position of the image.

The lateral or linear magnification of an object is given by the ratio of the image height to the height of the original object:

$$m = \frac{h_i}{h_o}. \tag{21.6}$$

From the similar triangles *HAP* and *IGP* in Fig. 21.19 we see that

$$\frac{h_i}{i} = \frac{h_o}{o},$$

so that the lateral magnification becomes

$$m = -\frac{i}{o}. \tag{21.7}$$

The minus sign is included since, as we will see, it is convenient to have a positive magnification for an erect image and a negative magnification for an inverted (upside-down) image.

In the preceding derivation, we examined the special case of a converging lens with the object placed outside the focal point. However, the thin-lens equation holds for both converging and diverging thin lenses and for any object distance, provided the following sign conventions are

---

*The behavior of a thin lens is described using other characteristic distances in the Newtonian form of the thin-lens equation. (See Problem 21.50.) The Gaussian form is more common.

maintained:

1. The sign of f is positive for a converging lens and negative for a diverging lens.
2. The sign of o is positive if the object is on the same side of the lens as the incident light (real object), negative if the object is on the other side (virtual object).*
3. The sign of i is positive (real) if the image is on the side of the lens opposite the incident light, negative (virtual) if the image is on the same side.
4. Object and image heights are positive if above the optical axis, negative if below it.

These conventions are consistent with the results of geometrical ray tracing. In fact, the use of Eq. (21.5) and the use of ray tracing techniques always give the same results. In general, it is helpful to make a scale drawing to clarify the problem in your mind and then use the equation for numerical precision. A few examples will serve to illustrate these principles and conventions.

Example 21.5

Image made with a diverging lens.

A diverging lens has a focal length of -5.0 cm. If you look through the lens at a dime held 4.0 cm from the lens, what are the location and the magnification of the image?

Solution   Figure 21.20 diagrams the situation. Solving Eq. (21.5) for the image distance, we have

$$i = \frac{fo}{o - f}.$$

The numerical values of f and o with their proper signs are $f = -5.0$ cm

---

*Virtual objects occur in compound optical systems like those described in Chapter 22.

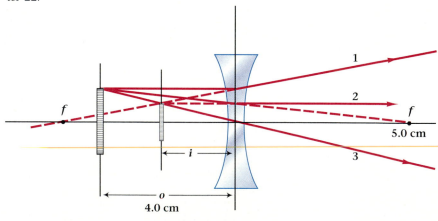

Figure 21.20   Example 21.5: Ray diagram for a diverging lens.

and $o = +4.0$ cm. The image distance is

$$i = \frac{(+4.0 \text{ cm}) \times (-5.0 \text{ cm})}{(4.0 \text{ cm}) - (-5.0 \text{ cm})},$$

$$i = -2.2 \text{ cm}.$$

The negative image distance tells us that the image is virtual. The magnification is given by

$$m = -\frac{i}{o} = -\left(\frac{-2.2 \text{ cm}}{4.0 \text{ cm}}\right) = 0.55.$$

The image is smaller than the object. The positive value for m indicates that the image is erect, as shown in the figure.

Example 21.6

Image made with a converging lens.

A penny is placed 4.0 cm in front of a converging lens whose focal length is 15 cm. Where is the image and what is its nature? Is it real or virtual, erect or inverted, and what is its magnification?

Solution   The following calculation corresponds to the diagram in Fig. 21.21. Solving Eq. (21.5) for the image distance, we have

$$i = \frac{of}{o - f}.$$

Inserting the numerical values of o and f, we get

$$i = \frac{4.0 \text{ cm} \times 15 \text{ cm}}{4.0 \text{ cm} - 15 \text{ cm}} = -5.5 \text{ cm}.$$

The image is virtual with a linear magnification of

$$m = -\frac{i}{o} = 1.4.$$

The positive value of m indicates an erect image, as seen in the figure.

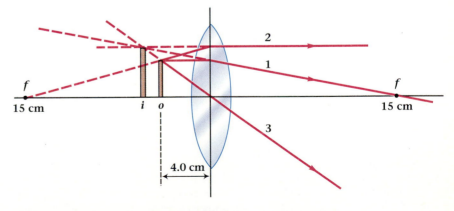

Figure 21.21   **Example 21.6: Ray diagram for a converging lens.**

We can use the thin-lens equation to reach some basic conclusions about the images obtained from lenses. For a converging lens, a real object outside the focal length produces a real, inverted image; a real object within the focal length produces a virtual, erect image. For a diverging lens, the image of a real object is always virtual and erect, regardless of the object's distance from the lens.

When lenses are used in combination, as in a compound instrument such as a microscope, the image of one lens becomes the object for a second lens. When that happens, the first image may lie on the negative side of the second lens, acting as a virtual object. For a converging lens, a virtual object always forms a real image. For a diverging lens, a virtual object within the focal length forms a real image; a virtual object beyond the focal length forms a virtual image. We will see some of this imaging in Chapter 22 on optical instruments.

Spherical Mirrors

21.7

Many optical instruments use curved mirrors as image-forming devices. Curved mirrors, like lenses, are used to focus light and create images. However, mirrors work by reflection of light, rather than by refraction, so their practical applications are different.

Curved mirrors are often spherical in shape. A concave mirror reflects light from the inner surface of a sphere; a convex mirror reflects light from the outer surface. Figure 21.22 depicts a beam of parallel light reflecting from a concave spherical mirror and converging to a single point, called the focus of the mirror. However, if the diameter of the incident beam is a large fraction of the diameter of the mirror (Fig. 21.23), the reflected rays do not all converge to the same point, and the image is somewhat blurred. This effect, called spherical aberration, is discussed in more detail in Section 21.8. For this reason we limit our discussion of spherical mirrors to those cases where the diameter of the incident light beam is small compared with the radius of curvature of the surface.

Figure 21.24 shows a concave mirror illuminated by a beam of light parallel to the optical axis. A single ray parallel to the axis strikes the mirror surface at M and is reflected through the axis at the focal point f. A line drawn from the center of curvature R to the mirror at M forms the base of an isosceles triangle RMf, with equal sides Rf and fM. In the limit of small angle θ, the distances Rf and fM are equal to one-half the radius R of the surface. The focal length f of the concave spherical mirror is then

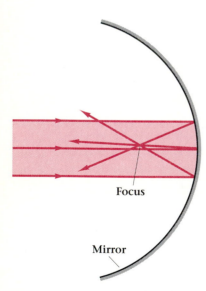

Focus

Mirror

Figure 21.22 The reflection of light rays from a curved surface obeys the law of reflection at each point. When parallel light rays strike near the center of a concave spherical mirror they are reflected to a common point, called the focus.

$$f = \frac{R}{2},$$

(21.8)

where R is the radius of curvature of the mirror.

A concave mirror focuses a parallel beam of light in a manner analogous to that of a converging lens. On the other hand, if the mirror surface is convex, the reflected light diverges, in a manner analogous to the spreading of a beam by a diverging lens. A convex mirror makes the reflected

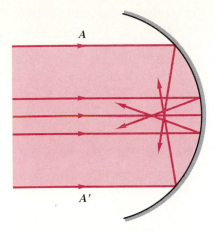

Figure 21.23 Parallel light rays *A* and *A'* strike the spherical mirror at so great a distance from the central axis that they are not reflected to the common focus of the rays incident close to the axis.

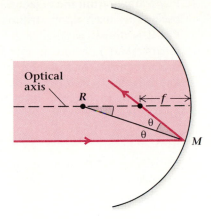

Figure 21.24 Geometric construction showing that rays near the axis of a spherical mirror converge through a point *f* located a distance *R*/2 from the mirror, where *R* is the radius of curvature.

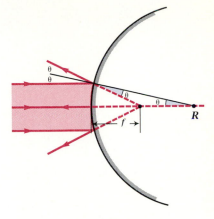

Figure 21.25 Parallel light rays diverge upon reflection from a convex spherical mirror. The diverging rays appear to come from a virtual source point located at $f = R/2$ inside the mirror.

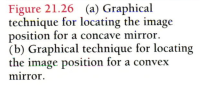

Figure 21.26 (a) Graphical technique for locating the image position for a concave mirror. (b) Graphical technique for locating the image position for a convex mirror.

beams appear to come from a focal point *f* inside the mirror surface (Fig. 21.25). For a small angle θ, this focal point is halfway between the reflecting surface and the center of curvature, so again $f = R/2$.

The geometrical procedure for locating the image produced by a spherical mirror is similar to that used for lenses. As you read these steps, follow along with the aid of Fig. 21.26.

A. Select an appropriate scale and mark the position of the object and the mirror on the optical axis, as shown.
B. Draw a line *MM'* through the center of the mirror and perpendicular to the optical axis.
C. Mark the focal point of the mirror $f = R/2$ on the optical axis and locate the object on the axis, all to the same scale.
D. Draw the following rays to locate the image. As with lenses, any two of these reflected rays are sufficient to locate the image position at their intersection, but you should always draw the third ray as a check.

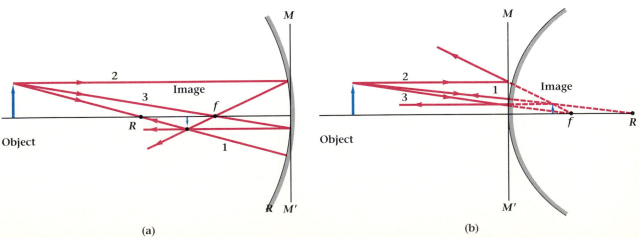

(a)   (b)

1. Ray 1 drawn from the object through the center of curvature reflects back along itself upon striking the mirror.
2. Ray 2 is drawn parallel to the optical axis from the object to the mirror (MM′), where it is reflected *through* the focal point of a concave mirror or *from* the focal point of a convex mirror.
3. Ray 3 is drawn from the object to the mirror (MM′) along a path through or toward the focal point and is reflected parallel to the optical axis.

Example 21.7

An object is placed 2.0 cm in front of a concave spherical mirror whose radius of curvature is 8.0 cm. Locate the position of the image by ray tracing. Is the image erect or inverted?

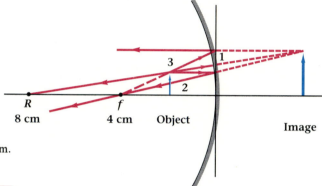

Figure 21.27   Example 21.7: Graphical ray diagram. The dashed lines indicate where light appears to come from.

Solution   The situation is drawn to scale in Fig. 21.27. Since the mirror's radius is 8.0 cm, the focal point *f* is 4.0 cm from the reflecting surface. Ray 1, drawn through the center of curvature, reflects back along itself. The extension of this reflected ray is drawn as a dashed line on the other side of the mirror. Ray 2, drawn parallel to the optical axis, reflects back through the focal point. The extension of ray 2, also shown as a dashed line, intersects the extension of the first ray at a measured distance 4.0 cm behind the mirror. Ray 3, which appears to come from the focal point, reflects back parallel to the axis. Its extension also intersects the other rays. Thus the image is located 4.0 cm behind the mirror. The image is erect and measures twice as tall as the object.

We can compute image positions for mirrors by using the same equation we used for thin lenses,

$$\frac{1}{i} + \frac{1}{o} = \frac{1}{f},$$

where *o* is the distance from the object to the mirror, *i* is the distance from the image to the mirror, and $f = R/2$ is the focal length. However, we require a new set of sign conventions for mirror problems:

1. The sign of *f* is positive if the focal point is on the same side of the mirror as the incident light, negative otherwise. (In other words, *f* is positive for a concave mirror and negative for a convex mirror.)

2. Both o and i are positive in sign if they lie on the same side of the mirror as the incident light and negative if they lie on the opposite side.

The use of the mirror equation and its sign conventions are illustrated in the following two examples. Note that the definition of magnification given earlier for lenses also applies to mirrors as well.

Example 21.8

Calculating the image from a concave spherical mirror.

An object is placed 2.0 cm in front of a concave spherical mirror whose radius of curvature is 8.0 cm. Use the mirror equation to locate the position of the image and its size. (This problem is the same as Example 21.7 except that we now find the image position and height by computation instead of by ray tracing.)

Solution First note that the focal length f is positive, according to rule 1. It is $f = R/2 = 4.0$ cm. According to rule 2 the object distance is positive. The image distance can then be computed by rearranging the mirror equation,

$$i = \frac{fo}{o - f} = \frac{(4.0 \text{ cm})(2.0 \text{ cm})}{2.0 \text{ cm} - (4.0 \text{ cm})},$$

$$i = -\frac{8.0 \text{ cm}}{2.0} = -4.0 \text{ cm}.$$

The image distance is negative, indicating that the image is virtual and lies on the opposite side of the mirror. Its magnification is

$$m = -\frac{i}{o} = -\frac{-4.0 \text{ cm}}{2.0 \text{ cm}} = +2.0.$$

The image is erect and is twice as tall as the object.

Example 21.9

Calculating the image for a convex spherical mirror.

A flea is located 3.0 cm from a convex spherical surface of radius 10 cm. Where is the image of the flea?

Solution In this case the surface is convex, so the center of curvature is behind the mirror, leading to a negative focal length $f = -5.0$ cm. Again we obtain the image distance from the mirror equation,

$$i = \frac{fo}{o - f} = \frac{(-5.0 \text{ cm})(3.0 \text{ cm})}{3.0 \text{ cm} - (-5.0 \text{ cm})} = 1.9 \text{ cm}.$$

The image is virtual and lies 1.9 cm within the mirror.

Note that a convex mirror always produces a virtual image of a real object. The image always lies within the mirror and is always erect. For a concave mirror, a real object placed outside the focal length is reflected to form a real, inverted image. An object placed within the focal length is reflected to form a virtual, erect image. In this case the image is enlarged. Shaving and make-up mirrors are convex mirrors with focal lengths larger than the normal distance from the mirror to your eye.

The most common nonspherical mirrors are concave parabolic reflectors. A parallel beam incident along the optical axis of a parabolic mirror is imaged to a point, without the complications of spherical aberration mentioned earlier. Similarly, light from a small source can be reflected into a parallel beam, as is done in searchlights. The big dish antennas that are used for microwave communication, satellite TV links, and radio telescopes are parabolic reflectors.

Lens Aberrations

*21.8

Up to now, our discussion of lenses has not taken into account some of the optical imperfections that are inherent in single lenses made of uniform material with spherical surfaces. These failures of a lens to give a perfect image are known as **aberrations**. They have nothing to do with the composition of a lens or the smoothness of its surfaces, but arise naturally from the geometry of image formation. They may be reduced to an acceptable level but, in general, the reduction of one type of aberration tends to increase another.

If we carefully reexamine our original observation that the sun's rays

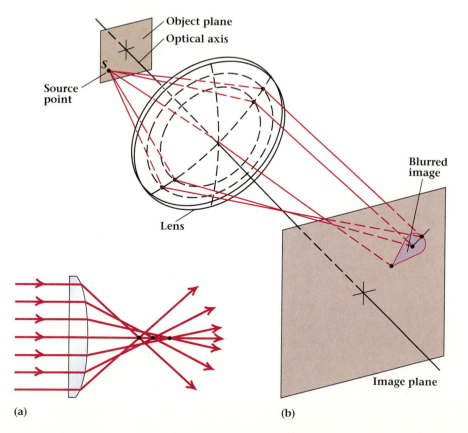

Figure 21.28   (a) Rays coming in parallel to the optical axis do not all intersect at the same point on the axis. Incident rays farther from the optical axis intersect closer to the lens. This effect is called spherical aberration. (b) Parallel rays incident at an angle to the optical axis of the lens fail to intersect at a point, an effect called coma.

(a)

(b)

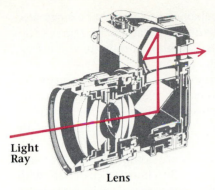

**Light
Ray**

Lens

Figure 21.29 Cross section of a
single-lens reflex camera with a
modern photographic lens. The combination of several lenses helps correct for aberrations.

all converge to a focal point after passing through a simple lens, we see
that this is only approximately so. Figure 21.28(a) shows incident rays
parallel to the optical axis of a converging lens. Careful measurement of
the refraction angles shows that incident rays farther from the optical axis
intersect slightly closer to the lens than the focal point. This effect is called
spherical aberration. Figure 21.28(b) shows incident rays at an angle to
the optical axis. Again, careful measurement shows that the refracted rays
do not meet at one point, but create a trailing blur away from the optical
axis. This effect is called *coma*. We see that what we said earlier about
thin lenses is strictly true only in the so-called paraxial approximation,
that is, only for incoming rays that are not far from the axis and that are
parallel to the axis. These limitations explain why the center of an image
is often relatively clear while the edges are considerably more distorted.

Aberrations can be reduced, but not removed entirely, by suitably
combining different spherical surfaces with appropriate radii. However, to
reduce aberrations of this kind to an acceptable level for optical instruments such as microscopes, binoculars, or cameras, each lens must be
made of several thin lenses in combination. Figure 21.29 shows such a
lens designed for a camera. The design of these lenses is quite complicated
in practice, but the basic principle used is still that of refraction as determined by Snell's law. Using computers, lens designers trace ray paths
through complicated optical systems in a short time. Snell's law is applied
at each boundary between materials with different indices of refraction.
The number of individual lenses, their composition, and the shapes of the
surfaces are chosen so as to reduce the distortion. The application of
computers to lens design problems has resulted in improved camera lenses
and other optical systems, along with reductions in their cost.

Along with the aberrations mentioned above there is an additional
complication: The index of refraction varies slightly for light of different
colors, that is, different wavelengths. This variation is most easily seen in
the action of a prism (Fig. 21.30). Incoming white light is a mixture of
all colors. When white light strikes the prism surface, it is refracted according to Snell's law, but since the indices of refraction are slightly different for the various colors, each color refracts through a different angle.
This spreading of light according to wavelength is called **dispersion**. A
similar thing happens at the second surface, resulting in a spectrum of
colors. Because the surfaces of a single lens are not parallel, it behaves

Figure 21.30 Dispersion of light
into a spectrum by a prism. The angular spread of the emerging beam
has been exaggerated. In a typical
glass prism it is much less than shown
here.

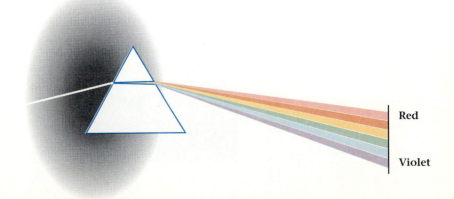

Red

Violet

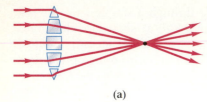

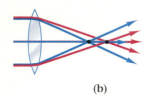

Figure 21.31   (a) A lens can be considered to be made of sections of prisms. (b) Dispersion of the light by a lens causes a spreading of the focus, with blue light converging nearer to the lens than red light does.

like a prism. Figure 21.31 shows this behavior in somewhat exaggerated form, in an effect called *chromatic aberration*. It cannot be removed by geometrical shaping of a single lens. The variation of the index of refraction is described further in Section 23.8.

Chromatic aberration can be troublesome in any optical instrument that uses lenses. To overcome this problem in telescopes, Isaac Newton invented the reflecting telescope, which uses a focusing mirror instead of an objective lens. No dispersion occurs when light is reflected from the mirror, so chromatic aberration is eliminated from that element.

As we have seen, modern camera lenses are compound lenses made with groups of simple lenses used together. By choosing materials with indices of refraction that depend in different ways on the color of the light, positive and negative lens elements can compensate for each other by having equal but opposite chromatic effects. Such lenses are said to be *achromatic*, because they exhibit only minimal chromatic aberration. The first achromatic lens was made by John Dollond, a London optician, who succeeded in building an achromatic telescope in 1758.

We can take a complicated grouping of lenses like that in Fig. 21.29 and consider the whole as if it were a single lens with a single focal point on either side. Clearly, such a lens is not literally a thin lens. Often the lens is almost as thick as its stated focal length. However, the reason for all the aberration corrections is to make the lens behave just as we said an ideal simple lens does. We can therefore deal with such compound lenses as single lenses and use the formulas developed for thin lenses to a good approximation.

Figure 21.32 illustrates lens aberrations and their correction. It shows the same scene photographed through a simple uncorrected lens and through a well-corrected lens. The blurring in the first photograph (Fig. 21.32a) is the result of spherical aberration and the inaccuracy in the color is due to chromatic aberration. A color photograph through the well-corrected lens (Fig. 21.32b) shows no visible color errors.

In recent years lensmakers have begun to depart from the traditional manufacturing techniques of grinding lenses to spherical surfaces. Newer methods allow the production of lenses with one or both surfaces neither planar nor spherical. The resulting lenses, known as *aspherics*, have less aberration than an equivalent spherical lens. Alternatively, they can be made with shorter focal lengths than those possible with spherical lenses of equal diameter and equal spherical aberration. Lenses of this type are

Figure 21.32   (a) A photograph taken using a single thin, plano-convex 50-mm lens (b) The same scene photographed through a well-corrected camera lens of the same focal length (50 mm).

(a)

(b)

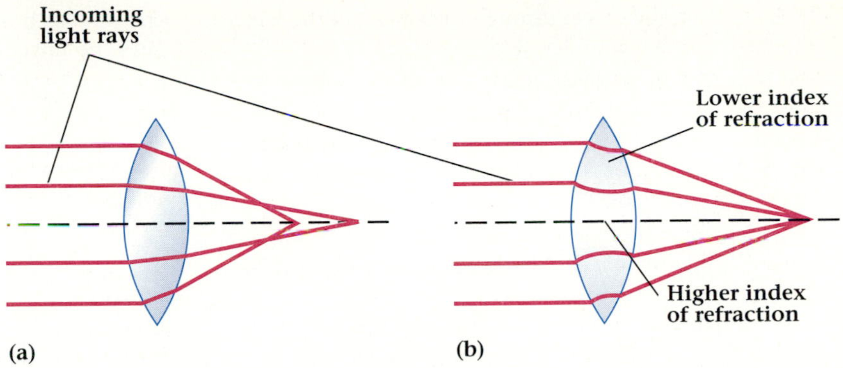

Figure 21.33 (a) An ordinary lens bends light more at the edges than near the center. (b) A gradient-index lens brings all the light to the same focus.

Incoming light rays

Lower index of refraction

Higher index of refraction

(a)   (b)

now being used in cameras and as condenser lenses in projectors. (Cameras and projectors are discussed in Section 22.3.)

One of the most recent advances in lens technology is the development of *gradient-index* lenses, in which the index of refraction decreases as a function of the radius. Light rays incident near the edge of the lens are not refracted as much as those near the center of the lens. Consequently, a gradient-index lens can be used to correct for spherical aberration (Fig. 21.33). Lenses of this type may soon be used in cameras, but at present they are limited to small-diameter rods with parallel, flat faces. The rod axis is the optical axis of the lens. Because of their nonuniform index of refraction, they can be used for imaging even though both faces are flat.

SUMMARY

Useful Concepts

- The law of reflection is

$$\theta_r = \theta_i.$$

- Snell's law of refraction is

$$n_1 \sin \theta_1 = n_2 \sin \theta_2,$$

where n_1 is the index of refraction of material 1 and n_2 is the index of refraction of material 2. The index of refraction of a medium is the ratio of the speed of light in vacuum c to the speed of light in the medium v:

$$n = \frac{c}{v}.$$

- The critical angle for total internal reflection within an optically denser medium of refractive index n_1 is

$$\sin \theta_c = \frac{n_2}{n_1}.$$

- The thin-lens (and spherical-mirror) equation is

$$\frac{1}{i} + \frac{1}{o} = \frac{1}{f}.$$

- The lateral magnification due to a thin lens is given by

$$m = -\frac{i}{o}.$$

- The focal length of a spherical mirror is

$$f = \frac{R}{2}.$$

- The location of the image of a point formed by a lens can be found from a scale diagram by drawing at least two of the following rays:

  1. A ray parallel to the optical axis from the object to the lens. For a converging lens, extend the ray through the focal point on the opposite side. For a diverging lens, extend the ray from the lens as though the ray came from the focal point on the same side.

  2. A ray from the object to the lens passing through a focal point. For a converging lens, draw the ray through the focal point on the same side as the incident light; for a diverging lens draw the ray in the direction of the focal point on the opposite side. Continue the ray parallel to the optical axis after passing through the lens.

3. A ray from the object through the center of the lens, which continues undeviated.

• The intersection of these rays (or their extensions) locates the image of the object point.

• The location of the image of a point formed by a mirror can be found from a scale diagram by drawing at least two of the following rays:

1. A ray from the object through the center of curvature, which is reflected back along itself upon striking the mirror.

2. A ray parallel to the optical axis from the object to the mirror, where it is reflected through the focal point of a concave mirror or from the focal point of a convex mirror.

3. A ray from the object to the mirror along a path passing through or toward the focal point which is reflected parallel to the optical axis.

• The intersection of these rays (or their extensions) locates the image of the object point.

Important Terms

You should be able to write the definition or meaning of each of the following terms:

• ray
• ultraviolet
• infrared
• angle of incidence
• angle of reflection
• plane of incidence
• index of refraction
• critical angle
• total internal reflection

• converging lens
• diverging lens
• thin lens
• focal length
• real image
• virtual image
• lateral magnification
• aberrations
• dispersion

QUESTIONS

21.1 If you use an ordinary electric lamp with a frosted bulb to cast a shadow of your hand on the wall, are all parts of the shadow equally dark? Why?

21.2 Explain what, if anything, is wrong with the following statement: When you look in a mirror it reverses left and right; it should therefore also reverse up and down.

21.3 Could you see the surface of a mirror that was in all ways perfectly reflecting?

21.4 Imagine that you write your initials on a piece of paper and place it flat on a table in front of a mirror on the wall. Sketch how your initials appear in the mirror.

21.5 Explain the operation of one-way mirrors. Under what conditions do they work best? Are they really one-way?

21.6 Under what conditions is it possible to immerse a transparent object into a clear liquid and not be able to see the object?

21.7 Explain how the day/night mirrors commonly found in automobiles work.

21.8 If you stand on a river bank trying to spear a fish swimming in the water, where should you aim the spear?

21.9 What happens to the focal length of a glass lens immersed in water? Consider both positive and negative lenses.

21.10 Explain the use of convex mirrors as wide-angle mirrors.

21.11 The index of refraction of the atmosphere is slightly greater than unity. Do you see the rising sun sooner or later than you would if the earth had no atmosphere? Draw a diagram illustrating your answer.

21.12 Describe the images you would see upon looking into the bowl of a silver spoon. How does the image change if you move the spoon closer to your eye? What happens if you turn it over and look at the back side?

21.13 What happens to the image formed on a screen by a lens if you cover the central portion of the lens with a small piece of opaque paper? What happens if you cover the upper half of the lens?

21.14 The following statement is found in a handbook for scuba divers: "Underwater objects seen through a flat face mask appear about 25% larger or closer." Using diagrams, explain the reasons behind this statement.

21.15 A hollow transparent container is shaped like a lens, being thicker in the center and thinner at the edges. Describe the optical behavior of the container if it is sealed and then immersed in a fish tank filled with water. How would your answer change if the container were thinner in the center and thicker at the edges?

PROBLEMS

Section 21.2   Reflection and Refraction

21.1 A woman 152 cm tall stands in front of a plane mirror attached to a vertical wall. What is the minimum length that the mirror must have for the woman to see her entire height?

21.2 A ray of light strikes a mirror and is reflected so that the angle between the incident and reflected beam is 30°. (a) If the mirror is rotated to increase the angle of incidence by 1°, what will be the new angle between the incident and reflected beam? (b) If, instead, the mirror is moved to decrease the angle of incidence by 1°, what will be the new angle between the incident and reflected beam?

21.3 A light ray is directed at the inside of a highly polished silver hemispherical bowl and is incident parallel to the symmetry axis of the bowl. If it strikes the bowl so that the angle between the incident ray and a radius of the hemisphere is 45°, in what direction will it emerge from the bowl?

21.4 A person whose eyes are 1.6 m above ground stands on one shore of a still lake and observes the reflection of a 1.4-m-tall fence on the opposite shore. His line of sight to the reflection of the top of the fence is 6.0° below the horizontal. How wide is the lake?

21.5 If you run toward a plane mirror at a speed of 2.7 m/s, how fast do you approach your image?

21.6 What is the speed of light in water, ice, and diamond? (*Hint:* Use the information in Table 21.1.)

21.7 A light beam in air enters a transparent material at an angle of 47° with respect to the normal. After refraction it makes an angle of 32.5° with respect to the normal. Identify the material on the basis of information in Table 21.1.

21.8 A rectangular glass block has opposite sides that are exactly parallel. (a) Show that light incident on one side emerges from the other side parallel to its original direction. (b) Show also that the displacement of the two beams depends linearly on the thickness of the prism.

21.9 Water in a fish tank is 38 cm deep, and a coin rests on the bottom. How far below the surface does the coin appear to be? (Assume a small angle of incidence.)

21.10 A beam of light makes an angle of 30° with the normal of a mirror made of 2.0-mm-thick glass silvered on the back. If the index of refraction of the glass is 1.5, how far is the point at which the beam leaves the glass surface (after being reflected from the silver backing) from the point at which the beam entered the glass?

21.11 A beam of light is directed at a glass ($n = 1.50$)
• cylinder in a plane perpendicular to the cylinder axis. It strikes the cylinder at an angle of 30° with respect to the normal at the point of incidence. In what direction, with respect to the original beam direction, does it leave the cylinder?

Section 21.3   Total Internal Reflection

21.12 What is the critical angle for total internal reflection in plastic (polymethylmethacrylate) at an interface with air?

21.13 What is the critical angle for a glass-to-water interface? Assume a refractive index of 1.50 for the glass.

21.14 What is the critical angle for light inside a diamond having an index of refraction $n = 2.417$?

21.15 A person lying on the bottom of a pool notices that he can see all of the above-water surroundings. The view is distorted and compressed, but everything is visible that could be seen from the surface of the water. What is the angular diameter of this view?

21.16 Suppose that the prism of Example 21.3 were made of ice ($n = 1.31$). (a) Would a ray normally incident (see Fig. 21.10) be totally reflected? (b) Design an ice prism that would totally reflect the light.

Section 21.5   Locating Images by Ray Tracing

21.17 An overhead projector with a lens of 35.6 cm focal length is used to throw an image on a screen 1.80 m away. If you consider the lens to be a thin lens, how far is the slide from the lens when the image is in focus? What is the magnification in this setup? Work this problem by ray tracing.

21.18 A diverging lens with focal length of 25 cm is used to view a postage stamp 50 cm from the lens. (a) How far from the lens does the stamp appear?

(b) What is the magnification of the image? Work this problem by ray tracing.

21.19 A hollow cube 6.0 cm on an edge is aligned with its center on the optical axis of a converging lens of focal length 20 cm. The cube is turned so that its sides are parallel to the axis of the lens. The center of the cube is 30 cm from the lens. (a) Determine the position and size of the images of the front and rear faces of the cube. (b) Sketch the image of the cube.

21.20 A small light bulb, a lens, and a piece of white cardboard are mounted on a meter stick so that an inverted image of the bulb is formed on the cardboard. The image is three times as large as the actual size of the bulb. The bulb is located 10 cm from one end of the meter stick and the cardboard is 10 cm from the other end. (a) What is the location of the lens? (b) What is the focal length of the lens? Work this problem by ray tracing.

21.21 An incandescent light bulb is marked 60 W on its upper end. A lens of 25 cm focal length, when held above the light bulb, forms an image of the "60 W" on the ceiling. If the distance from the lens to the ceiling is 1.30 m, how much larger are the letters and numbers in the image on the ceiling than on the object light bulb? Work this problem by ray tracing.

21.22 A diverging lens of 0.75-m focal length is held so that the image of a chair appears to be 0.40 m from the lens. How far from the lens is the chair? Work this problem by ray tracing.

Section 21.6   The Thin-Lens Equation

21.23 A thin lens forms an image of an overhead electric lamp on a piece of paper. The distance to the lamp from the lens is 128 cm; the distance to the paper is 16 cm. What is the focal length of the lens?

21.24 Work Problem 21.17 using the thin-lens equation.

21.25 Work Problem 21.18 using the thin-lens equation.

21.26 A small hand-held camera has a lens with a focal length of 50 mm. The camera is arranged so that by moving the lens in or out, objects from 0.50 m away to "infinity" can be brought to an exact focus on the film. How many centimeters will the lens have to be moved as the focus is changed from the closest to the farthest object?

21.27 A luminous object is placed 3.00 m from a screen. When a lens is interposed between the object and the screen at a distance of 0.50 m from the screen, an in-focus image of the object forms on the screen. It is claimed that if the object is moved 1.70 m farther from the screen and the lens moved 0.30 m farther from the screen, the image will again be in focus. Is this claim true?

21.28 An image of a flower is produced on a screen by a lens 1.27 m from the screen. The image of the flower has a linear magnification of 2.5. What is the focal length of the lens?

21.29 A screen is placed exactly one meter away from an illuminated test pattern. A lens placed between them forms a clear (focused) image of the pattern on the screen when the center of the lens is 180 mm from the test pattern. What is the focal length of the lens?

Section 21.7   Spherical Mirrors

21.30 A coin 2.0 cm in diameter is held 20 cm from a concave spherical mirror of 30 cm radius of curvature. Locate the image of the coin and its height.

21.31 A 3.0-cm-tall domino is located 100 cm from a concave spherical mirror of $R = 30$ cm. Locate the position of the image and find the magnification.

21.32 The radius of a concave spherical mirror is 20 cm. A chess piece 4.2 cm high is located at distances of (a) 12 cm and (b) 6.0 cm in front of the mirror. Find the image position for both of these object distances. Describe the images.

21.33 The radius of a convex spherical mirror is 20.0 cm. A sugar cube 1.50 cm high is located at distances of (a) 20.0 cm, (b) 12.0 cm, and (c) 6.00 cm in front of the mirror. Locate the image positions for each of these object distances.

21.34 A pencil is held perpendicular to the optical axis of a concave spherical mirror. The pencil is 87 cm from the center of the mirror and its image is found 18 cm from the mirror. (a) Find the focal length of the mirror. (b) What is the radius of curvature of the mirror? (c) What is the magnification of the pencil's image?

21.35 When you hold a concave mirror of radius R at just the right distance you see a double-size image of your eye. What is that distance?

21.36 A shiny, spherical Christmas tree ornament is 95 mm in diameter. (a) Where is the image of a child standing 2.0 m away from the ornament? (b) What is the magnification of the image? (c) Is the image inverted or erect?

21.37 A wide-angle mirror in a grocery store is made from a section of a spherical surface whose radius of curvature is 0.75 m. How far behind the mirror is the image of a person standing 2.4 m in front of it?

Additional Problems

21.38 Two plane mirrors are joined at right angles to each
• other. A small statue of Newton is placed along the diagonal between the mirrors at a small distance from the vertex. The statue is viewed from a point not on the diagonal. Determine, by the use of appropriate drawings, the total number of images of Newton that can be seen in the mirrors.

21.39 Prove that if an object is at a distance of twice the
 • focal length from a lens, the image is the same size
as the object. Show this using both an algebraic and
a geometrical method.

21.40 A lamp that gives off light in all directions is placed
 •• 2.00 m below the surface of the water in the middle
of a large swimming pool. What fraction of the light
can escape directly from the water? Do not consider
light reflected from the bottom or sides of the pool.
(*Hint:* Consider the fraction of light emerging
through part of the surface of a sphere centered on
the lamp. The surface area of a sphere of radius R
is $4\pi R^2$. The surface area of the portion of a sphere
cut off by a plane that intersects the sphere a radial
distance h from the surface is $2\pi Rh$.)

21.41 A 10-cm-long pencil embedded in a block of trans-
 •• parent ice is parallel to and 20 cm from the surface.
Assume that you place your eye 60 cm from the
surface along a line that passes through the center
of the pencil. (a) What angle does the pencil subtend at your eye? (b) What angle would the pencil
subtend at your eye if the ice were not present?
(*Hint:* Use the small-angle approximation given in
Chapter 13; that is, $\sin \theta \approx \tan \theta \approx \theta$.)

21.42 A fish tank filled with water is made of flat glass
 • walls of $n = 1.50$. What is the maximum angle of
incidence for a light ray within the water to strike
the glass wall and still emerge to the outside air?

21.43 Light is incident normally upon the long face of a
 •• symmetric prism (Fig. 21.34). What is the maximum value of α that will permit total reflection from
both rear faces for glass of $n = 1.55$?

21.44 A converging lens of 15 cm focal length is placed
 • on an optical bench that is 120 cm long. At one end
a self-luminous object is fixed; at the other end is a
fixed screen. (a) Measuring from the object end of
the bench, determine the *two* positions for which an
image will be in focus on the screen. (b) What is
the magnification in each case? Compare your results with a geometrical construction in one case.

21.45 Two thin lenses, each with $+12$ cm focal length,
 •• are placed 60 cm apart. An object 18 cm to the left
of the first lens is imaged a distance x to the right
of the second lens. Determine x.

21.46 Show that the effective focal length of two thin len-
 •• ses of focal length f_1 and f_2 placed close together is

$$\frac{1}{f} = \frac{1}{f_1} + \frac{1}{f_2},$$

when the distance d between the lenses approaches
zero.

21.47 A spherical glass ball of index 1.5 and radius R is
 • used as a burning lens. By direct calculation using
Snell's law, find the approximate point to which rays
converge. To do this, find the intersection of a ray
directed parallel to the optical axis with a ray incident on the ball a distance $R/2$ from the optical axis.
It will help to make a scale drawing.

21.48 A converging lens of 16 cm focal length is placed at
 • the center of curvature of a concave spherical mirror
whose radius is 32 cm. Locate the final image of an
object that is very far away.

21.49 The mathematical expression that relates the shape
 •• of a lens, its index of refraction, and its focal length
is called the *lensmaker's equation*. For a lens in air
it is

$$\frac{1}{f} = (n - 1) \left(\frac{1}{R_1} - \frac{1}{R_2} \right),$$

where R_1 and R_2 are the radii of curvature of the
first and second surfaces, respectively (Fig. 21.35).

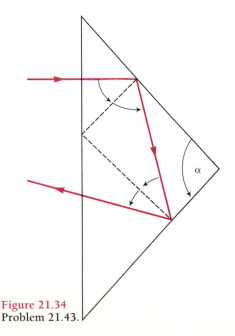

Figure 21.34
Problem 21.43.

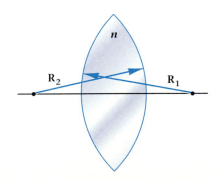

Figure 21.35   Problem 21.49

The radius is positive if the surface encountered by the light beam is convex, negative if it is concave. What is the focal length of a double convex lens made from glass with index of refraction $n = 1.52$ if $R_1 = 0.35$ m and $R_2 = -0.35$ m?

21.50 Show that you may write the thin-lens equation in
 • the form

$$f^2 = xx',$$

where x and x' are the distances of the object and image from the focal points. This equation is known as the Newtonian form of the thin-lens equation. (*Hint:* Make a drawing like Fig. 21.19 with $x = o - f$ and $x' = i - f$.)

21.51 Show that the lateral magnification of a thin lens
 • may be expressed in the form

$$m = -\frac{x'}{f},$$

where x' is the distance from the focal point to the image point, $x' = i - f$. This expression is known as the Newtonian formula for lateral magnification. (*Hint:* Make a drawing similar to Fig. 21.19 with $x' = i - f$.)

21.52 A converging lens held near a flashlight images the
 • bulb on a wall 1.60 m from the lens. With the flashlight fixed, the lens is moved closer to the wall and a sharp image is again formed when the lens is 15 cm from the wall. (a) What is the focal length of the lens? (b) How far is the flashlight bulb from the wall?

21.53 Derive the mirror equation using a geometrical dia-
 • gram. Employ a method similar to the one used in the text to derive the equation for a thin lens.

21.54 Show that the thin-lens formula for a converging
 • lens can be written as $i = (1 - m)f$.

21.55 The lateral magnification of a thin lens is $m = -i/o$.
 •• Show that the longitudinal magnification, that is, the magnification in the direction of the optical axis, is given by m^2.

21.56 Derive an expression for determining the speed of
 •• light using a Fizeau apparatus in terms of the number of teeth N, the angular velocity ω of the wheel, and the distance D between the mirrors. (*Hint:* Refer to Fig. B21.2.)

21.57 A capillary tube with an internal diameter d is made
 •• of glass of refractive index n. What is the apparent internal diameter of the capillary when viewed in air from a distance that is large compared with the diameter of the tube?

21.58 Two thin lenses, one with $+12$ cm focal length and
 •• the other with -6.0 cm focal length, are placed 18 cm apart. Locate the final image of an object 18 cm to the left of the converging lens. Give your answer with respect to the position of the diverging lens.

21.59 Three thin lenses are aligned along their optical axis.
 •• The first lens has a $+24$-cm focal length. The second lens has a -18-cm focal length and is placed 18 cm from the first lens. The third lens has a $+12$-cm focal length and is placed 9.0 cm beyond the second lens. Locate the position of the image of an object located 36 cm from the first lens.

21.60 A lens forms an image of a candle flame on a wall
 • 2.50 m away from the flame. The linear size of the image is 3.5 times the size of the flame. (a) Using the techniques of ray tracing, find the focal length of the lens. (b) Check your results by using the thin-lens equation to find the focal length.

ADDITIONAL READING

Falk, D. S., D. R. Brill, and D. G. Stork, *Seeing the Light*. New York: Harper & Row, 1986. An interesting and readable account of a wide range of topics including the eye and vision, color perception, photography, optical instruments, and holography.

Tape, W., "The Topology of Mirages." *Scientific American,* June 1985, p. 54.

Thomas, D. E., "Mirror Images." *Scientific American,* December 1980, p. 206.

Walker, J., "The Amateur Scientist: Mirrors Make a Maze So Bewildering That the Explorer Must Rely on a Map." *Scientific American*, June 1986, p. 120.

Walker, J., "The Amateur Scientist: The Kaleidoscope Now Comes Equipped with Flashing Diodes and Focusing Lenses." *Scientific American*, December 1985, p. 134.

22

Optical Instruments

22.1 The Eye

22.2 The Magnifying Glass

22.3 Cameras and Projectors

22.4 Compound Microscopes

22.5 Telescopes

*22.6 Zoom Lenses

A WORD TO THE STUDENT

In this chapter we present the basic principles underlying the operation of a variety of optical instruments, including cameras, microscopes, and telescopes. All these instruments use various arrangements of the same kinds of lenses and mirrors as you studied in the previous chapter. We do not present new theories here, but instead apply our knowledge of lenses and mirrors in new ways.

We have already talked about the importance of vision in motivating the entire study of optics. As we learn more about the nature and behavior of light, we seek to devise new optical instruments that will expand our range of vision in new ways and to new dimensions. Optical microscopes let us see objects as small as 0.1 μm in diameter. Telescopes show us galaxies so far away that their light has taken billions of years to reach us. Film and video cameras record images from visible and infrared light. Lenses that gather more light and display fewer aberrations are being designed and produced continually.

The ideas of optics and optical instruments have led to new concepts and designs for instruments that extend the range of our visual senses far beyond the limits of visible light. For example, beginning with the development of the electron microscope in the 1930s, improvements, innovations, and new methods have reached the point where we can now image individual atoms. In hospitals, acoustic waves are routinely used to generate ultrasonic images of the interior of the human body. However, before we describe the principles of optical instruments, we first examine the most fascinating and interesting of all optical instruments, the human eye.

The Eye

22.1

The human eye is one of the most familiar optical instruments, yet it is also one of the most complex. The overall mechanism of vision is extremely complicated and even now is not fully understood. Yet we can understand some of the important principles of vision by considering the optical properties of the human eye.

Figure 22.1 is a schematic drawing of a human eye, seen in cross section from above. The principal optical elements are the cornea (a transparent sheath across the front of the eye); the pupil (an opening); the iris (which adjusts the size of the pupil); a fluid called the aqueous humor in front of the lens; the lens; the ciliary muscle (which controls the shape of the lens); the gel-like vitreous humor; and the retina (the light-sensitive

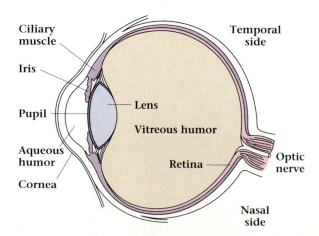

Figure 22.1   Cross section of a human eye viewed from above, showing the main optical elements. The function of each part is described in the text.

layer that lines the back of the eye). The eye cannot be treated as a thin lens for several reasons, including the facts that there are more than two refracting boundaries and that the object and image are in different optical media. The greatest relative change in the index of refraction takes place between the air ($n \approx 1.000$) and the cornea ($n \approx 1.376$). Consequently, the greatest refraction takes place at that surface also. The eye lens is actually a gradient-index lens with a larger index ($n \approx 1.406$) near the center than near the edges ($n \approx 1.386$). Because the aqueous and vitreous humors that surround the lens have indices of $n \approx 1.336$, the refraction at the lens surfaces is small compared with the refraction at the cornea.

One of the most notable optical properties of the eye is that of accommodation. The normal relaxed eye presents distant objects in focus on the retina (Fig. 22.2a). For close objects to be in focus on the retina, some optical parameters of the eye must change. If a distant object moves toward the eye, the eye changes to keep the image in focus on the retina. This effect is called **accommodation** and is accomplished primarily by contraction of the ciliary muscles, which changes the shape of the lens. When the eye focuses on nearby objects, the lens becomes thicker and the surfaces more curved (Fig. 22.2b). Because the lens has a higher index of refraction than the surrounding medium, the net effect is to shorten the focal length of the lens.

The ability of the eye to accommodate is limited. We may not move

Figure 22.2 (a) The relaxed, normal eye forms an image of distant objects on the retina. (b) For close objects, the lens of the normal eye changes shape to focus the image on the retina. (c) In a nearsighted eye, the image of a distant object forms in front of the retina. (d) A diverging lens corrects for nearsightedness. (e) In a farsighted eye, the image of a nearby object forms beyond the retina. (f) A converging lens corrects for farsightedness.

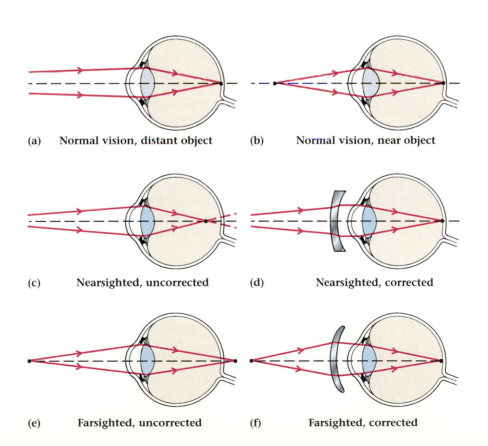

(a) Normal vision, distant object   (b) Normal vision, near object

(c) Nearsighted, uncorrected   (d) Nearsighted, corrected

(e) Farsighted, uncorrected   (f) Farsighted, corrected

an object any closer than a distance called the near point and still view it comfortably. The **near point** is the distance from the unaided eye that produces the largest retinal image without blurring. Objects closer than the near point cannot be brought into focus. The average value for the near point is about 25 cm, although there is considerable individual variation, even for people with so-called normal vision. For simplicity, we will use a value of 25 cm.

Most people do not have ideal vision. The range of visual ability that is considered normal is quite broad. Two common types of vision defects are simply related to the optical properties of the eye and can be readily corrected. They are called nearsightedness *(myopia)* and farsightedness *(hyperopia)*. A nearsighted eye is unable to accommodate over the normal range from 25 cm to infinity. Instead, there is a far point beyond which vision is not distinct. The image of a more distant object comes to a focus in front of the retina, so that only a blurred image forms on the retina (Fig. 22.2c). Clear vision of distant objects is restored by placing a diverging lens in front of the eye. This diverging lens forms an image of the distant object within the accommodation range of that particular eye (Fig. 22.2d). For a farsighted eye, the near point is farther away than the normal 25 cm (Fig. 22.2e). The image of a nearby object comes to focus behind the retina, so again the image is blurred. To see things as close as 25 cm from the eye requires the use of a converging lens, which images the object at least as far away as the actual near point of the particular eye in question (Fig. 22.2f).

Another common visual problem, **astigmatism**, occurs when the cornea or lens surfaces are not spherical. An astigmatic eye images point objects as lines. Usually the shape of an astigmatic eye can be approximated by the combination of a spherical surface with a cylindrical deformation superposed. Corrections are made for astigmatism by using a compensating cylindrical eyeglass lens.

Opticians generally express the strength of lenses used to correct visual defects in terms other than of their focal length. The common unit is the **diopter. The strength of a lens in diopters is the reciprocal of the focal length expressed in meters.** For example, a lens with a focal length of $+0.25$ m has a strength of $+4.0$ diopters. Shorter focal lengths correspond to greater dioptic strengths. The use of lens strength is especially convenient when several thin lenses are used in close proximity. Then the strength of the combination is just the sum of the strengths of the individual lenses. (See Problem 21.46.) Selecting a lens of proper strength for correcting a visual defect of the kind shown in Fig. 22.2 is usually accomplished by trying various combinations of lenses in front of the eye until the clearest vision is obtained. Then a single lens is chosen that has the same strength as the combination of trial lenses.

For people whose eyes are unable to accommodate fully at both ends of the range, lenses are available with two distinct regions of different dioptic strength. Such lenses are known as *bifocals*. They have an upper region of dioptic strength appropriate for distant vision and a lower region designed for close vision. In some cases, *trifocal* lenses are used to provide better vision at intermediate distances. The numerical example below illustrates the use of a corrective lens.

Example 22.1

Your nearsighted friend has a far point of 2.00 m. Objects farther away than the far point are not imaged clearly. What lens will provide clear vision for very distant objects?

Solution Your friend needs a lens that forms an image of a distant object at a distance from the eye equal to the far point. We only need to apply the thin-lens equation to solve the problem. For the object distance o we use infinity (for a very distant object). Because we want the image on the same side of the lens as the incident light, the image distance is negative. It is $i = -2.00$ m, the far point. From the thin-lens equation we get

$$\frac{1}{f} = \frac{1}{o} + \frac{1}{i} = \frac{1}{\infty} + \frac{1}{-2.00 \text{ m}},$$

or
$$f = -2.00 \text{ m}.$$

The proper lens has a focal length $f = -2.00$ m. It is a diverging lens. The strength of the lens in diopters is

$$\text{strength of lens} = \frac{1}{f} = \frac{1}{-2.00} = -0.50 \text{ diopters}.$$

The Magnifying Glass

22.2

Before considering optical instruments with multiple lenses, let's investigate the simplest instrument of all: the **magnifying glass**, or simple microscope. The magnifying glass is a single converging lens that, when held near the eye, gives an image whose size on the retina is larger than that observed by the unaided eye. By adjusting the distances at which the lens and object are held from the eye, the viewer can obtain maximum magnification without undue eye strain or blurring of the retinal image.

As an object is moved closer, it subtends a larger angle at the eye, so that the image produced covers a larger part of the retina (Fig. 22.3).

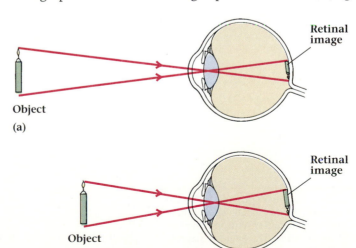

Figure 22.3 (a) The image of an object seen with the unaided eye is formed on the retina. (b) As the object comes closer, its image covers a larger part of the retina.

However, we cannot bring it any closer than the near point and still see it clearly. A magnifying glass allows us to increase the visual angle that an object subtends at the eye—that is, to form a larger image on the retina—without requiring our eye to focus closer than the near point. For example, Fig. 22.4(a) shows an object of height h placed at the near point (25 cm), where it subtends an angle θ. If the angle θ is small, we can write θ in radians as

$$\theta = \frac{h}{25},$$

where both h and the near point are expressed in centimeters. In Fig. 22.4(b), we have placed a magnifying glass (converging lens) next to the eye and moved the object within the focal length of the lens. The positions of the lens and the object have been adjusted so that the enlarged virtual image of the object falls at the eye's near point. Since the image is on the same side of the lens as the object, the image distance is negative.

What magnification have we achieved? To find out, we first calculate the object distance, using the thin-lens formula. If distances are measured in centimeters, we get the result

$$o = \frac{25f}{f + 25}.$$

From the figure we see that the angle θ' subtended by the virtual image is approximately

$$\theta' = \frac{h}{o} = \frac{h(f + 25)}{25f}.$$

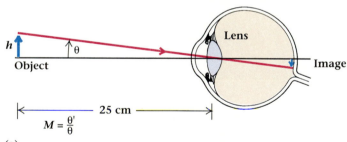

(a)

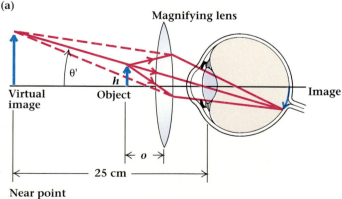

Figure 22.4   (a) An object placed at the near point of an unaided eye. (b) The same object viewed through a magnifier with the virtual image at the near point. The angular magnification M is θ'/θ.

(b)

The magnification of a magnifying glass is the **angular magnification**, defined by

$$M \equiv \frac{\theta'}{\theta},$$

(22.1)

which for this case becomes

$$M = \frac{25 \text{ cm}}{f} + 1$$

(image at near point).   (22.2)

Remember that f is measured in centimeters in this equation.

If the object is held at, or just inside, the focal point of the lens, the image forms very far away, essentially at infinity, rather than at the near point. This corresponds to the most comfortable viewing distance because the eye is relaxed. (See Problem 22.20.) The magnification in this case is given by

$$M = \frac{25 \text{ cm}}{f}$$

(image at infinity).   (22.3)

Manufacturers commonly use Eq. (22.3) to specify the magnification of a magnifying glass. Thus a magnifying glass of 10 cm focal length is marked 2.5×.

Not only are magnifying glasses used singly to enlarge such things as fine print and small objects, but they are also used to enlarge the images formed by other lenses. When they are utilized in this way, they are referred to as eyepieces. The primary function of magnifying glasses is to increase the angular size of the image and to allow viewing with a relaxed eye.

Example 22.2

Focal length of a magnifier.

A photographer has an 8× magnifier for examining negatives. What is the focal length of the magnifier lens?

Solution   The magnification of the lens is specified for producing an image at infinity. Therefore we can obtain the focal length of the magnifier lens from Eq. (22.3),

$$M = \frac{25 \text{ cm}}{f} = 8.$$

Upon rearranging, we get

$$f = \frac{25 \text{ cm}}{8} = 3.1 \text{ cm}.$$

Example 22.3

Magnification with a lens.

A biology student wishes to use a 6-cm focal-length lens as a magnifier. (a) What is the magnification of the lens when used with a relaxed eye? (b) What is the maximum magnification of the lens?

Solution   (a) For the relaxed eye, the image will be at infinity, so the magnification can be obtained with Eq. (22.3):

$$M = \frac{25 \text{ cm}}{f} = \frac{25 \text{ cm}}{6 \text{ cm}} = 4\times.$$

(b) The maximum magnification occurs when the image is at the near point. In this case the magnification is obtained from Eq. (22.2):

$$M = \frac{25 \text{ cm}}{f} + 1 = \frac{25 \text{ cm}}{6 \text{ cm}} + 1 = 5\times.$$

Cameras and Projectors

22.3

In its simplest form, a photographic camera is a light-tight box with a lens set into one side and a light-sensitive material (film) placed on the side opposite the lens (Fig. 22.5). Normally, light is prevented from entering the camera by means of a shutter, either at the lens position or just in front of the film. To take a photograph, you first adjust the lens position so that a real image of the object is in focus on the film. Then, when you press a button, the shutter momentarily opens and an image forms on the film. This image is stored in the light-sensitive material for later chemical processing, yielding a reproduction of the scene as it originally fell on the film.

Modern cameras, such as the 35-mm camera of Fig. 21.29, are basically the same as the simple camera. They have well-corrected lenses of the type discussed in Section 21.8. Many have automatic exposure control and automatic focusing, features that are made possible through the use of integrated circuit electronics. Computer-aided design of lenses and computer-aided manufacturing allow us to have inexpensive lenses of high quality that were not available for any price only twenty years ago. Pictures from space, taken with high-quality lenses, give us new information about the earth (Fig. 22.6).

Other cameras, such as motion picture cameras and television cameras, use the same basic principle. A motion picture camera takes a rapid sequence of still pictures, which are eventually projected at the same rate at which they were taken. The apparent smoothness of the motion is due to the fact that the rate at which the pictures are taken and projected, ordinarily 24 frames per second, is higher than the rate at which we can distinguish between individual images. The persistence of vision gives us the appearance of smooth continuous motion. In a television camera, although the image is detected, transmitted, and reproduced by electronic means, the optical principles are the same as for a movie or still camera.

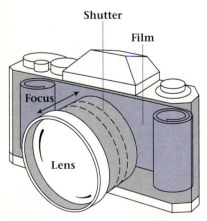

Figure 22.5   A camera is a light-tight box for holding the film and lens in proper position. It also includes a shutter to control the time during which light reaches the film.

Figure 22.6 A photograph of India and Sri Lanka from space, looking north, taken from *Gemini XI* at an altitude of 760 km.

With modern photographic films and with adequate lighting, the shutter need be open only a small fraction of a second to successfully record an image on film. Many cameras have a range of available shutter speeds, that is, lengths of time during which the shutter is open. Shutter speeds of 1/30, 1/60, 1/125, 1/250, 1/500, 1/1000, and 1/2000 second are standard. Note that each of these is approximately half as long (or, in photographic terminology, "twice as fast") as the preceding one. The faster the shutter speed, the faster the object can be moving and still produce a sharp image.

Getting the "correct exposure" in a photograph corresponds to allowing the appropriate amount of light to strike the film.* This amount is different for different types of film. The amount of light that strikes the film is determined not only by how long the shutter stays open, but by how large the effective lens opening is. Thus it is analogous to the amount of water flowing from a faucet, which depends on both the length of time the faucet stays open and the cross-sectional area of the opening. The size of the lens opening, or aperture, is often measured by what is called the *f*-value or **f-number**. This is defined to be

$$f\text{-number} = \frac{\text{focal length of lens}}{\text{diameter of lens}} = \frac{f}{d}. \qquad (22.4)$$

For example, a lens with a diameter one-half its focal length has an *f*-number of 2, which is written *f*/2. A variable-diameter opening in the camera, called an iris diaphragm (after the iris in your eye), can be adjusted to decrease the effective lens opening and therefore increase the *f*-number. By being able to adjust both shutter speed and *f*-number, the photographer can give the proper exposure to the film while still having the option of using a particular shutter speed or a particular lens opening.

Table 22.1 lists the standard *f*-number intervals. These numbers are often referred to as *f*-stops, or simply stops. Notice that the values of successive *f*-stops differ by a factor of approximately $\sqrt{2}$. Thus, if you change the *f*-number of a lens from *f*/4 to *f*/5.6, you reduce the diameter of the aperture by a factor of $1/\sqrt{2}$. The area of the aperture decreases by the square of this factor. So, if the shutter speed remains constant, the amount of light passing through the lens opening is cut in half. A change in aperture equivalent to going from one *f*-stop to the next successive one is known as a change of one full stop. It increases or decreases the light passed through the lens by a factor of 2.

TABLE 22.1
Standard full-stop *f*-numbers

f-number	(*f*-number)²
0.7	0.49
1.0	1.
1.4	1.96
2.0	4.
2.8	7.84
4	16.
5.6	31.4
8	64.
11	121.
16	256.
22	484.

Example 22.4

Setting exposure.

The light meter in your camera indicates that your film is properly exposed at a shutter speed of 1/50 s and an aperture setting of *f*/16. However, to photograph a bird flying across the park, you must shorten the exposure time to 1/200 s to avoid blurring. What should the new aperture setting be?

---

*By "amount of light" we mean the energy deposited on the film by the light.

Solution   Shortening the exposure time by a factor of 4 requires that the aperture be increased to four times its original area. This corresponds to opening the lens two full stops. Thus the lens opening should be $f/8$.

A useful consequence of the f-number method of classifying lens openings is that for the same shutter speed, lenses of different focal lengths give proper exposure at the same f-numbers. This result is due to two compensating factors. First, the amount of light that passes through the aperture is proportional to its area, and hence to the square of its diameter, d^2. Second, the light per unit area that reaches the film depends inversely on the area of the image. For the usual situation, in which the object distance is large compared with the focal length, the linear magnification is proportional to the focal length f of the lens, so the area of the image is proportional to f^2. The rate at which a photographic image is formed, or the *speed* of the lens, is then

$$\text{speed of lens} \propto \frac{d^2}{f^2} = \frac{1}{(f\text{-number})^2}. \qquad (22.5)$$

For most purposes, the speed of the lens depends only on the f-number and is independent of the particular focal length.

The film camera has an optical inverse—the slide projector. The basic components of a slide projector are shown in Fig. 22.7. The transparent slide with the likeness of the photographed object, the lens, and the image projected on the screen are the reverse of the camera used to make the slide. By choosing the proper focal-length lens, you can make the image of the slide completely fill the screen at the chosen projector–screen distance. The need to accommodate different sizes of screens and rooms has led to the use of variable-focal-length, or zoom, lenses for home projectors.

The illuminating system of a projector is equally as important as the image-forming system. It usually consists of a lamp and reflector, a con-

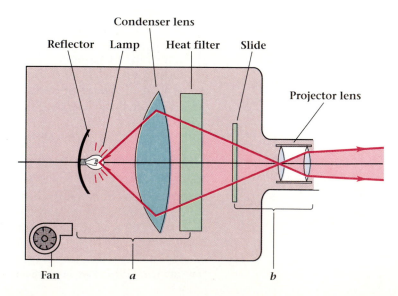

Figure 22.7 The basic optical components of a slide projector. The illuminating system is indicated by bracket *a*, and the image-forming system by bracket *b*. Most projectors also include a fan to cool the slide and prevent damage due to overheating.

densing lens, and a piece of heat-absorbing glass to protect the slide from the heat of the lamp. The reflector simply directs more light in the direction of the slide. The condenser lens is a strongly converging lens, ensuring that light passing through the edges of the slide passes through the projector lens. Thus the condenser not only increases the amount of light passing through the slide, but also increases the amount of light reaching the screen. If the condenser lens is removed, the entire slide will still be illuminated, but the image from the edges of the slide will not appear on the screen, and only the central part of the picture will be seen.

The optical principles of a movie projector are the same as those of a slide projector. However, a movie projector also contains a mechanical means of moving the film and for interrupting the light when the film is moving. The result is that the viewer sees a series of individual still pictures. As mentioned earlier, the pictures are presented at a rate of 24 per second, and the brain blends the sequence of individual images into a smoothly flowing scene. Ordinarily, a three-bladed shutter is used to block the light when the film is moving. Because the shutter makes one rotation for each frame, it also interrupts the light during the presentation of each individual picture. The shutter interrupts the light 72 times per second, a rate so fast that the eye cannot see the flicker.

Compound Microscopes

22.4

The microscope and the telescope were developed at about the same time in the early 1600s. Early naturalists soon utilized microscopes to make important discoveries. Marcello Malpighi's discovery of capillaries, Anton van Leeuwenhoek's discovery of protozoa, and Robert Hooke's beautiful drawings of magnified cells were important contributions to the understanding of biological processes. These advances would have been impossible without the microscope. The microscope and telescope have perhaps played a greater role than any other scientific instruments in establishing our current understanding of natural laws.

A typical **compound microscope** consists of a tube with a lens at each end (Fig. 22.8a, p. 652). Though in modern microscopes each lens may actually consist of a group of lenses to minimize distortion from lens aberrations, we can understand the optical principles by treating each group as a single lens. The lens close to the object being viewed is called the **objective**. The lens through which one looks is called the **eyepiece** or **ocular**.

Figure 22.8(b) is a ray diagram of a compound microscope. The objective is a lens of comparatively short focal length f_o. When an object is placed just outside the focal point of the objective, a real, enlarged image forms at the plane A. If a screen were placed at A, the image of the object would appear there. The eyepiece functions as a magnifier for viewing this image. Thus the image formed by the objective lens serves as the object for the eyepiece lens.

The image formed at plane A by the objective has a linear magnification m_o. The Newtonian expression for this magnification (see Problem

21.51) is

$$m_o \approx -\frac{L}{f_o},$$

where L is the distance from the focal point to the image plane. This image is then viewed through the eyepiece with an angular magnification

$$M_e = \frac{25 \text{ cm}}{f_e}.$$

The overall magnification of the microscope is the product of the linear magnification of the objective and the angular magnification of the eyepiece, resulting in a magnification

$$M = m_o M_e = \left(-\frac{L}{f_o}\right)\left(\frac{25 \text{ cm}}{f_e}\right). \tag{22.6}$$

Microscope objectives and eyepieces are commonly labeled according to their effective magnifications when used with a standard separation between them. Most (but not all) manufacturers design their microscopes so that the distance L, between the focal point of the objective and that of the eyepiece, is 16.0 cm. The magnification of an objective lens can then be expressed as $m_o = 16.0 \text{ cm}/f_o$. If we know the magnifications of the eyepiece and objective lens, we can use Eq. (22.6) to find the overall magnification. A $10\times$ eyepiece has an angular magnification of 10 times and a $10\times$ objective has a linear magnification of 10 times. Used together they give an overall magnification of $100\times$.

Example 22.5

Magnification of a microscope.

A laboratory microscope has a $20\times$ objective and a $10\times$ eyepiece. Determine (a) the overall magnification and (b) the focal length of the objective lens.

Solution (a) The overall magnification is the product of the two magnifications:

$$M = m_o M_e = 20 \times 10 = 200\times.$$

(b) The focal length of the objective lens may be obtained from its magnifying power,

$$m_o = \frac{16.0 \text{ cm}}{f_o}.$$

Upon rearranging, we get

$$f_o = \frac{16.0 \text{ cm}}{20} = 0.80 \text{ cm}.$$

You might expect that you could use stronger and stronger lenses to make an optical microscope as powerful as you desire, but magnification alone is not the only criterion to consider. What is really important is not

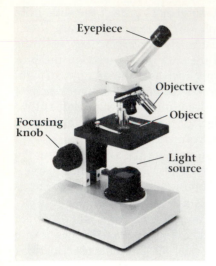

(a)

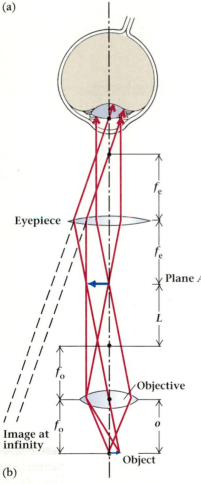

(b)

Figure 22.8 (a) A modern microscope. (b) Ray diagram of a compound microscope.

simply the size of the image, but the ability to distinguish, or resolve, two object points that are very close together. As we will see in the next chapter, the minimum distance between two object points that can be resolved in an image depends on the wavelength of the illumination and on the diameter of the lens. In optical systems, we reach fundamental, unavoidable limits resulting from the wave nature of light.

Telescopes

22.5

Though different in purpose, the telescope has a great deal in common with the microscope as an optical instrument. A telescope also consists of a long tube with an objective lens toward the object and an eyepiece lens toward the viewer. Figure 22.9(a) shows a diagram of a **refracting astronomical**, or **inverting, telescope**. In a telescope, the objective lens has a relatively long focal length and the object being viewed is far away

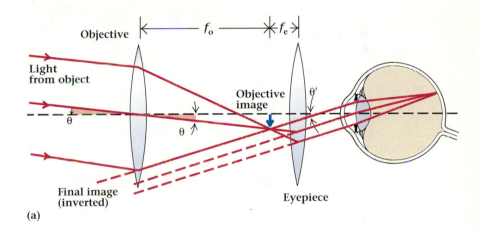

(a)

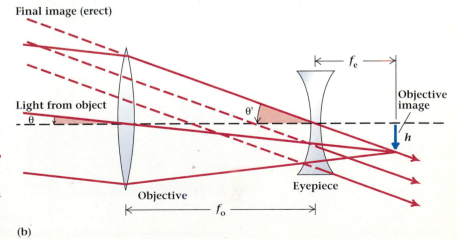

(b)

Figure 22.9 (a) Principle of operation of a refracting astronomical (or inverting) telescope. (b) Ray diagram of a Galilean telescope. The diverging eyepiece produces an erect, virtual image.

compared with this focal length. As a result, rays from the object come in nearly parallel and a real image forms near the focal point of the objective. This image would show up clearly on a screen placed at A, just as in the case of the microscope. Note, however, that in this case the image is smaller than the physical size of the object. Again we use an eyepiece to view the inverted image that has been brought to a focus at plane A by the objective lens.

We define the magnification of a telescope as the ratio of the angle θ' subtended by the object when viewed through the telescope to the angle θ subtended when the object is viewed with the unaided eye. The angle subtended at the eye is essentially the same as that subtended at the objective by the object. With this approximation, one can show from the

BACK TO THE FUTURE

Development of the Telescope

By about the year 1600, combinations of lenses were used to make telescopes in Holland. Not long afterwards, Galileo made a telescope and used it in astronomy (Fig. B22.1). It is worth noting that the earliest applications of the telescope in astronomy used its magnifying ability, as was done by Galileo. Not until much later were improvements made in observing the positions of the planets and stars. (Recall that Tycho Brahe's observations, from about 1570 to 1600, were all made without the aid of telescopes.)

Galileo first heard of the telescope in about 1609, as he tells:*

About ten months ago a report reached my ears that a certain Fleming had constructed a spyglass by means of which visible objects, though very distant from the eye of the observer, were distinctly seen as if nearby. Of this truly remarkable effect several experiences were related, to which some persons gave

*From "The Starry Messenger" in S. Drake, *Discoveries and Opinions of Galileo*. Garden City, N.Y.: Doubleday-Anchor Books, 1957.

credence while others denied them. A few days later the report was confirmed to me in a letter from a noble Frenchman at Paris, Jacques Badovere [a former pupil of Galileo], which caused me to apply myself wholeheartedly to inquire into the means by which I might arrive at the

Figure B22.1   (a) Two of Galileo's telescopes.

invention of a similar instrument. This I did shortly afterwards, my basis being the theory of refraction. First I prepared a tube of lead, at the ends of which I fitted two glass lenses, both plane on one side while on the other side one was spherically convex and the other concave. Then placing my eye near the concave lens I perceived objects satisfactorily large and near, for they appeared three times closer and nine times larger than when with the naked eye alone. Next I constructed another one, more accurate, which represented objects as enlarged more than sixty times. Finally, sparing neither labor nor expense, I succeeded in constructing for myself so excellent an instrument that objects seen by means of it appeared nearly one thousand times larger and over thirty times closer than when regarded with our natural vision.

Though Galileo was not the first to use a telescope, he was the first to use the telescope to make a systematic study of the heavens. Galileo's observations were published in 1610 in a work he called "The Starry Messenger." In it he describes his observations of the moon, giving the first evidence that the surface of the moon is not

diagram that

$$M = \frac{\theta'}{\theta} \cong -\frac{f_o}{f_e}. \tag{22.7}$$

If we replace the converging eyepiece with a diverging lens, we get an erect image. Telescopes of this type are called **Galilean telescopes** after Galileo (Fig. 22.9b). If the eyepiece were not present, incoming rays from a distant object would come to a focus essentially at the focal point of the objective lens. A real inverted image would be formed on a screen placed at this point. However, when the diverging eyepiece lens is placed within

smooth and featureless, as was previously thought and as it appears to the unaided eye. He also discovered four moons of Jupiter and made the first observations of sunspots. These observations were in dramatic conflict with the teachings of the established church concerning the "perfect, unblemished heavens" and the earth as the center of the cosmos.

The primary purpose of Galileo's telescope was to magnify, so he could better see the details of the planets. In modern astronomy one of the important properties of telescopes is their ability to gather large amounts of light and thereby let us see fainter and more distant stars. Telescopes that gather more light have larger diameters. The world's largest telescope—to be finished in 1990—is the 10-m-diameter Keck telescope on Mauna Loa volcano in Hawaii (Fig. B22.2). It has nearly twice the light-gathering power of the next largest telescope, the 6-m reflector at the Crimea Observatory in the Soviet Union. Because of the large size, it is necessary to compensate for sagging and bending of the glass mirror as the telescope changes orientation. To reduce weight, and consequently bending, the Keck telescope is made of 36 hexagonal segments. Each hexagonal segment can be individually moved

as little as 4 nm by computer-controlled motors. This new telescope will make measurements on stars so far away that the light detected from them

left approximately 10 billion years ago. By knowing the conditions in the past, we add to our knowledge of the origin of the stars and of the universe itself.

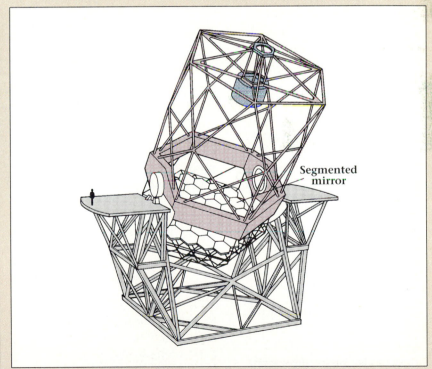

Figure B22.2 The 10-m Keck telescope in Hawaii contains a mirror made from 36 hexagonal segments.

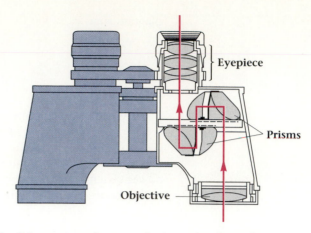

Figure 22.10   Prisms are used to erect the image in a binocular.

the focal length of the objective, so that the rays striking the eyepiece emerge from it parallel to each other, a virtual image is produced. From the diagram we see that the magnification is again given by Eq. (22.7). Note that when f_e is negative, as it is here, the magnification is positive, indicating an erect image.

Binoculars are essentially twin refracting telescopes mounted side by side (Fig. 22.10). The prisms, which give most binoculars their characteristic shape, are used to erect the image that would otherwise be inverted. (The light is totally internally reflected in the prisms.) Because the index of refraction of the prisms is greater than that of air and because they fold the light path, the tubes are shorter than the tubes of a simple telescope of the same magnification. Opera glasses differ from binoculars in that they consist of a pair of side-by-side Galilean telescopes. Because their images are already erect, no prisms are needed.

Telescopes may also be constructed with mirrors for the objective. Isaac Newton constructed such a telescope in 1668 in order to avoid the problems of chromatic aberration found in lenses. The **Newtonian telescope** has a converging (concave) mirror as its first element (Fig. 22.11). A flat mirror mounted diagonally in the middle of the tube reflects the light so that it comes to a focus just outside the tube. An eyepiece is used to view the resulting image. Other types of reflecting telescopes differ mainly in where the image is brought to focus. Modern telescopes used in astronomy are reflecting telescopes.

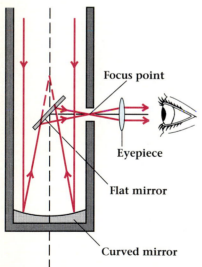

Figure 22.11   Ray diagram of a Newtonian telescope, which uses a mirror instead of a lens to focus light.

Example 22.6

Magnification of a telescope.

An astronomical telescope is used to study the moon, which subtends an angle of 0.5° at the earth's surface. The objective lens of the telescope has a focal length of 0.75 m and the eyepiece has a focal length of 0.10 m. (a) What is the angular magnification? (b) What angle does the moon's image subtend at the eye of the person looking through the telescope?

Solution   (a) The magnification is found directly from Eq. (22.7),

$$M = -\frac{f_o}{f_e} = -\frac{0.75 \text{ m}}{0.10 \text{ m}} = -7.5\times.$$

(b) The angular appearance of the moon's image can also be found from Eq. (22.7). Since $M = \theta'/\theta$, the image angle θ' is

$$\theta' = M\theta = 7.5 \times 0.5° = 3.8°.$$

Zoom Lenses

*22.6

One of the marvels of technology is the zoom lens, widely used in photography and television. The zoom lens can be changed quickly from a wide-angle lens (short focal length) to a telephoto lens (long focal length). A true **zoom lens** maintains focus throughout the entire zoom range at any focusing distance, provided you have focused sharply on an object. A **varifocal lens** must be refocused whenever you change its focal length.

Modern zoom lenses used with television cameras are available with focal-length changes of as much as 20 to 1. Lenses used on 35-mm cameras come in a range of zoom ratios from the limited 2:1 ratio of a 35–70-mm lens to the 7.5:1 ratio of a 28–210-mm lens. Thus a single lens can be adjusted to obtain the focal length needed to provide the desired composition and magnification.

To understand the operation of a varifocal lens, consider the behavior of two thin lenses of equal and opposite focal lengths as the separation between them changes (Fig. 22.12). Suppose the lenses have focal lengths of $+12$ cm and -12 cm. When the lenses are separated by 4 cm, light incident on the converging lens parallel to its axis is ultimately brought to focus 24 cm from the diverging lens. If the separation of the lenses is increased to 8 cm, the light converges only 6 cm from the diverging lens, resulting in a wider-angle field of view.

Commercial zoom lenses are much more complicated than a simple combination of two thin lenses. A typical zoom lens has 12 or more separate glass lenses arranged in four groups (Fig. 22.13). Two or more of

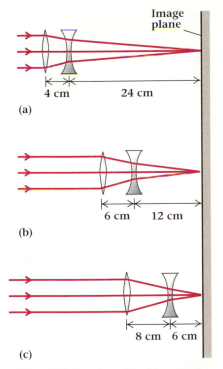

(a)

(b)

(c)

Figure 22.12 A varifocal lens made from two components of equal but opposite focal length (12 cm). The distance between the diverging lens and the effective focal point depends on the separation of the lenses: (a) 4 cm, (b) 6 cm, and (c) 8 cm.

Figure 22.13 A typical zoom lens for a 35-mm camera has 12 lenses arranged in four groups.

these groups move relative to the others in order to vary the effective focal length, all the while maintaining a constant image position so that focus is maintained. The numerous lenses in each group are needed to minimize aberrations and produce a lens that is not restricted to paraxial rays.

Because a zoom lens has so many elements, it is extremely important that each surface be coated with an antireflection layer. (See Section 23.4.) These coatings not only increase the amount of light passing through the system, but also increase the contrast and reduce the flare that results from multiple reflections from the surfaces. Some reduction in the number of elements is achieved by using aspheric lenses. In the future, the incorporation of gradient-index lenses may further reduce the number of elements, making zoom lenses lighter and cheaper.

SUMMARY

Useful Concepts

- The strength of a lens in diopters is the reciprocal of the focal length when the focal length is expressed in meters.
- A converging lens may be used as a magnifying glass. The important magnification is the angular magnification,

$$M \equiv \frac{\theta'}{\theta}.$$

- If the lens is held close to the eye, the angular magnification is

$$M = \frac{25 \text{ cm}}{f} + 1 \qquad \text{(image at the near point)}$$

or

$$M = \frac{25 \text{ cm}}{f} \qquad \text{(image at infinity)}.$$

- The latter equation is normally used to specify the magnification of a magnifying glass.
- The f-number of a lens is given by

$$f\text{-number} = \frac{\text{focal length}}{\text{diameter}}.$$

- The magnification of a compound microscope is the product of m_o, the linear magnification of the objective lens, and M_e, the angular magnification of the eyepiece.
- The magnification of a telescope is the ratio of the focal length of the objective lens to the focal length of the eyepiece:

$$M = -\frac{f_o}{f_e}.$$

- Zoom and varifocal lenses have multiple elements that move relative to one another to allow the effective focal length of the combination to be changed.

Important Terms

You should be able to write the definition or meaning of each of the following terms:

- accommodation
- near point
- astigmatism
- diopter
- magnifying glass
- angular magnification
- f-number
- compound microscope
- objective lens
- eyepiece
- ocular
- refracting astronomical telescope
- Galilean telescope
- Newtonian telescope
- zoom lens
- varifocal lens

QUESTIONS

22.1 Explain the operation of a simple periscope. What would happen to the image if the mirror at the top could be rotated around so that you were looking behind you instead of in front?

22.2 What is the advantage of having large-diameter objective lenses on telescopes or binoculars?

22.3 Many simple cameras have fixed-focus lenses; that is, the lens does not move with respect to the film position. How is it possible to take pictures in which both distant objects and nearby objects all seem to be in focus simultaneously?

22.4 Explain the operation of an overhead projector.

22.5 Explain the optical function and practical use of bifocal glasses.

22.6 A nearsighted person wishes to buy both glasses and contact lenses. The dioptic power of the prescription contact lenses is different from the power of the lenses used in the glasses. Why?

22.7 What are the advantages in the construction and operation of large reflecting telescopes compared with large refracting telescopes?

22.8 Given an additional converging lens, how can you use it to modify an inverting telescope so that the final image will be erect? You need not keep the tube length the same.

22.9 You can make a pinhole camera by taking a light-tight box, making a small hole in the center of one face, and placing a sheet of film on the opposite face inside the box. A suitable flap serves as a shutter. For such a camera, is the image erect or inverted? What is the magnification? Use diagrams to explain your answer.

22.10 Pick reasonable values for the dimensions of a 35-mm slide projector and determine appropriate values for the focal lengths and *f*-numbers of both the condenser and projection lenses. Assume that the projector is to be used at home, rather than in an auditorium. The dimensions of the film are 24.5 mm × 36.3 mm.

22.11 Archeologists frequently make records of their excavations with photographs taken from a ladder or other high support directly above the site. The following rule is often used to estimate the support height and lens focal length needed: The ratio of lens focal length to the linear size of the negative should equal the ratio of the camera height above the ground to the linear size of the excavation area. Show that this rule gives the optimum image for a lens of a given focal length.

22.12 What are some of the reasons your vision improves in bright light? What role does the narrowing of the pupils play in this?

PROBLEMS

Section 22.1 The Eye

22.1 An eyeglass lens has a strength of +5.00 diopters. What is the focal length of the lens?

22.2 What is the strength in diopters of a camera lens of 55 mm focal length?

22.3 Show that to find the equivalent strength, in diopters, of two thin lenses held close together, you add their individual dioptic powers. (*Hint:* Use the results of Problem 21.46.)

22.4 What is the equivalent focal length of a +5 diopter lens and −2 diopter lens held together? (*Hint:* Use the result of Problem 21.46.)

22.5 A farsighted eye has a near point of 150 cm. A converging lens is used to permit clear vision of a book placed 25 cm in front of the eye. Find the focal length of the lens and express its strength in diopters.

22.6 A nearsighted eye has a far point of one meter. Objects beyond one meter are not sharply focused. What lens could be used to permit clear vision for objects at infinity? Express your answer in diopters.

22.7 A myopic person wears eyeglasses with a lens strength of −2.0 diopters. Where is the far point for that person's eyes?

22.8 A farsighted person wears eyeglasses with a lens strength of +2.0 diopters. Where is the near point for that person's eyes?

22.9 One lens in a pair of eyeglasses is used to form the image of an overhead light fixture on a tabletop. The light fixture is 1.42 m above the table and the eyeglasses are 0.16 m above the table. What is the strength of the lens in diopters?

22.10 The prescription for a student's eyeglasses is changed from 0.80 diopters to 1.10 diopters. By what distance had the student's near point shifted?

Section 22.2 The Magnifying Glass

22.11 A magnifying glass enlarges an object by an angular factor of four. What is its approximate focal length?

22.12 What is the power in diopters of a 3× magnifying glass?

22.13 A lens focuses the sun's rays to a point 15 cm away. What power magnifying glass will this lens make?

22.14 A dime (diameter about 1.8 cm) is viewed through a 5× magnifying glass. Approximately what angle does it appear to subtend at the eye?

22.15 An 8× magnifying glass is placed 4.0 cm away from a postage stamp. (a) Where is the image of the stamp? (b) Is it real or virtual?

22.16 By what constant do you multiply the strength of a lens in diopters to obtain its magnification?

22.17 A magnifying glass held 13 cm in front of a television screen projects an image on a wall 3.37 m from the television screen. What magnifying power is marked on the lens?

22.18 From top to bottom, the letters on a coin subtend an angle of 2.5° when viewed through an $8 \times$ magnifying glass by a person with normal vision. What is the height of the letters on the coin?

22.19 A magnifier has two lenses of focal length 0.10 m and 0.16 m, which may be used singly or in combination. What are the possible magnifying powers? (*Hint:* Use the results of Problem 21.46.)

22.20 Show that Eq. (22.3) gives the correct magnification when the image is at infinity.

Section 22.3   Cameras and Projectors

22.21 A movie camera lens has a focal length of 17 cm and a diameter of 4.0 cm. What is its f-number?

22.22 You are taking photos with a shutter speed of 1/500 s and a lens aperture of $f/1.4$. If you then set the lens to $f/2.8$, what should be the shutter speed to give the same exposure to the film?

22.23 A camera has a lens with a focal length of 55 mm and an effective maximum lens diameter of 1.5 cm. A correct exposure of a particular scene is obtained for $f/5.6$ at 1/50 s. Can the shutter be set at 1/500 s to avoid blurring and still obtain the correct exposure? If not, what maximum shutter speed could be used?

22.24 Given that the size of the negative in a 35-mm camera is 24.5 mm $\times$ 36.3 mm, determine the approximate angular field of view for lenses of 24, 55, and 250 mm focal length.

22.25 What is the focal length of a 3.2-cm-diameter $f/2.8$ lens?

22.26 What is the diameter of a $f/5.6$ 2.3-diopter lens?

22.27 The Hale telescope on Mt. Palomar in California, which is 5 m (200 in.) in diameter, has a focal length of about 17 m. What is its f-number?

22.28 An advertisement in an astronomy magazine offers a 20-cm-diameter $f/10$ telescope mirror. What is the mirror's focal length?

22.29 In darkness, the iris of your eye opens to a diameter of about 8 mm. It is known that in darkness the eye has an aperture of about $f/2.8$. What is the approximate equivalent focal length of a thin lens in air that corresponds to these values?

22.30 On some lenses the widest aperture does not correspond to one of the standard full-stop f-numbers. A half-stop is an f-number corresponding to a lens opening with an area half-way between the full-stop values given in Table 22.1. What is the f-number of the half-stop between $f/1.4$ and $f/2.0$?

Section 22.4   Compound Microscopes

22.31 Given a 180.0-mm-long tube with a lens of 2.00 mm focal length at one end and a lens of 30.0 mm focal length at the other, where should an object be placed to use the tube and lenses as a microscope with the maximum magnification?

22.32 In a laboratory microscope, the first image of a specimen is formed inside the microscope 15 cm from the objective lens. If the specimen is 3.0 mm from the objective when the image is in focus, what is the focal length of the objective?

22.33 The maximum useful magnification for ordinary visible-light microscopes is about $2000 \times$. Given a $20 \times$ eyepiece, what should be the focal length of the objective used in combination with it to give this magnification in a standard microscope?

22.34 An eyepiece of 2.5 cm focal length and an objective of 0.32 cm focal length are used in a standard microscope. (a) What power should be marked on each element? (b) What is the overall magnification of the combination?

22.35 A new microscope comes with $5 \times$ and $10 \times$ eyepieces and a revolving "nosepiece" that contains $10 \times$, $20 \times$, and $40 \times$ objectives. (a) List all possible overall magnifications. (b) List all the magnifications that would be possible if the $5 \times$ eyepiece were replaced by a $15 \times$ eyepiece.

Section 22.5   Telescopes

22.36 An astronomical telescope is used to view the moon. If the objective has a focal length of 50 cm and the eyepiece a focal length of 3.0 cm, what is the angular magnification of the moon?

22.37 An astronomical telescope is designed with an overall magnification of $35 \times$. (a) If the objective has a focal length of 100 cm, what should be the focal length of the eyepiece? (b) How far should the eyepiece be from the objective?

22.38 The distance between the objective and eyepiece lenses of an inverting telescope with a $5 \times$ eyepiece is 55 cm. What is the telescope's overall magnification?

22.39 Binoculars denoted as 7×50 have a total magnification of 7 times and objective lenses of 50 mm diameter. What is the focal length of the eyepiece if the objective focal length is 21 cm?

22.40 A 1.90-m-tall football player is 50 m from you. If you look at him with a pair of $7 \times$ binoculars, what angle in radians does he subtend at your eye?

22.41 What is the magnifying power of a Galilean telescope with an objective of 20 cm focal length and an eyepiece of -4.0 cm focal length?

22.42 Opera glasses are usually made from two Galilean telescopes side by side. One pair is made with objective lenses of 20 cm focal length and eyepieces of −5.0 cm focal length. What should be the separation between the two lenses in these glasses if you want to view a soprano 30 m away? Refer to Fig. 22.9(b).

22.43 An upright meter-stick 53.8 m away is viewed through the 5× eyepiece of an inverting telescope. The observed image subtends an angle of 0.151 radians. (a) What is the angular magnification of the telescope? (b) What is the focal length of the objective lens?

22.44 A Galilean telescope with two lenses spaced 30 cm apart has an objective of 50 cm focal length. (a) What is the focal length of the eyepiece? (b) What is the magnification of the telescope? (*Hint:* Assume the object to be far away.)

*Section 22.6 Zoom Lenses

22.45 A simple zoom lens is made from a converging lens with a 10-cm focal length followed by a diverging lens with a −10-cm focal length. (a) Calculate the distance between the diverging lens and the image of a very distant object when the lenses are separated by 5.0 cm. (b) Calculate the distance when the lenses are separated by 8.0 cm.

22.46 A varifocal lens is made from a converging lens with a 12-cm focal length followed by a diverging lens with a −10-cm focal length. (a) Calculate the distance between the diverging lens and the image of a very distant object when the lenses are separated by 4.0 cm. (b) Calculate the distance when the lenses are separated by 8.0 cm.

Additional Problems

22.47 Each of two lenses is used to form the image of a
•   distant window on a screen. One has a diameter of 4.0 cm and a focal length of 10 cm; the second has a diameter of 6.0 cm and a focal length of 24 cm. Which lens produces a brighter image of the object?

22.48 A person who has been wearing −2.70-diopter len-
•   ses is informed that her far point has moved inward by 20% of its former value. What is the correct strength of a new lens that will correct her vision?

22.49 Use the definition of angular magnification and sim-
•   ple geometry to derive Eq. (22.7) for the magnification of a refracting astronomical telescope.

22.50 The expressions for the magnification of a simple
• • magnifier, Eqs. (22.2) and (22.3), were derived neglecting the distance between the lens and the eye. Redraw Fig. 22.4 with a nonzero spacing d between the eye and lens and show that

$$M = \frac{25}{f} + \frac{25}{s} - \frac{25\,d}{sf},$$

where s is the distance of the magnified image from the eye and f is the focal length of the magnifier. (Distances are to be measured in centimeters.)

22.51 How much more energy reaches the screen from a
•   projector with an $f/2.0$ lens than from an identical projector with an $f/3.5$ lens of the same focal length?

22.52 The Nikon N8008 camera has a shutter speed of
•   1/8000 s. (a) If the correct exposure is obtained at a full aperture of $f/1.8$ and a speed of 1/8000 s, what is the correct aperture (f-number) for an exposure of 1/500 s? (b) How far will a car traveling 130 km/h go during a 1/8000-s exposure?

22.53 A 30-cm focal-length lens is placed over a rectan-
•   gular 1.5 cm × 1.0 cm opening in the center of one side of a closed cubical box. A shutter and film are placed so that the box serves as a camera. For the purpose of exposing the film, what is the equivalent f-number of the lens in the rectangular aperture?

22.54 A "normal" lens for a particular camera is one with
•   a focal length approximately equal to the diagonal dimension of the film used. Thus for a 35-mm camera with a negative size of 24.5 mm × 36.3 mm, the "normal" lens has a focal length of about 43.8. In reality most normal 35-mm camera lenses have focal lengths nearer 50 mm. Show that for a camera with a "normal" lens, a given scene photographed from the same place will form an image that occupies the same relative proportion of the film regardless of the size of the camera. (*Hint:* Assume the object to be very far away.)

22.55 A close-up lens is mounted onto a camera lens to
• • allow focusing on closer-than-normal objects. A 55-mm focal-length lens on a particular 35-mm camera can be focused from infinity to 1.0 m from the lens. If it is desirable to have the nearest focus only 15 cm from the lens, approximately what should be the strength of the close-up attachment lens in diopters?

22.56 A 20× objective and a 5× eyepiece from a
•   standard-length (16.0 cm) microscope are placed in a microscope with a 16.5-cm spacing. What is the ratio of the overall magnification of the combination to that of the standard microscope?

22.57 A specimen is in focus under a standard microscope
• • with a 10× eyepiece and a 30× objective. The specimen is moved 0.010 mm toward the objective. How much will the eyepiece have to be moved to bring the object back into focus?

22.58 The numerical aperture, N.A., of a microscope ob-
• • jective is a measure of its light-gathering power. In air, the N.A. is defined to be the sine of half the apex angle α of the cone of light received by the lens (Fig. 22.14). A large N.A. corresponds to a large light-gathering power. (a) What is the N.A. of a

0.92-cm-diameter objective with a working distance *s* of 1.79 cm? (b) What is the approximate *f*-number of the lens? (c) Show that the light energy entering the lens is proportional to the square of the N.A.

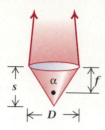

Figure 22.14   Problem 22.58.

22.59 An 8× astronomical telescope has a 30-cm focal-
• length objective lens. After looking at stars, an astronomer moves the eyepiece 1.0 cm farther away

from the objective to focus on nearer objects. What is the distance to the nearer objects?

22.60 An inverting telescope with a 5× eyepiece is fo-
• cused on a distant car. As the car approaches, the eyepiece is moved to keep the car in focus. When the car is 30 m away the eyepiece has been moved 0.437 cm from its original position. (a) What is the focal length of the objective lens? (b) What is the overall magnification of the telescope?

22.61 A telephoto lens is made with a front element of
•• positive focal length *f*, followed by a second element with a negative focal length −*f*. The spacing between the lenses is 0.5*f*. (a) Show that the image of a distant object comes to focus at a distance 1.5*f* from the front element. (b) Show that the image is the same size as one produced by a single element of focal length 2*f*. (c) What is the ratio of the equivalent focal length to the distance from the front element to the film?

ADDITIONAL READING

Blunk, W. , and A. Burkert, "Doing Physics: A Telescope Workshop." *The Physics Teacher*, February 1987, p. 111.

Darius, J., *Beyond Vision*. Oxford: Oxford University Press, 1984. A collection of over 100 scientific photographs selected for esthetic appeal and historical significance. Information inaccessible to the naked eye is shown over a range from submicroscopic to cosmic.

Drake, S., and C. T. Kowal, "Galileo's Sighting of Neptune." *Scientific American,* December 1980, p. 74 .

Huffman, A., "Measuring a Glasses Prescription." *American Journal of Physics*, April 1980, p. 309.

Koretz, J. F., and G. H. Handelman, "How the Human Eye Focuses." *Scientific American,* July 1988, p. 92.

Wave Optics

23.1   Huygens' Principle

23.2   Reflection and Refraction of Light Waves

23.3   Interference of Light

23.4   Interference in Thin Films

23.5   Diffraction by a Single Slit

23.6   Multiple-Slit Diffraction and Gratings

23.7   Resolution and the Rayleigh Criterion

23.8   Dispersion

23.9   Spectroscopes and Spectra

23.10   Polarization

*23.11   Scattering

A WORD TO THE STUDENT

You may find it helpful to review Chapter 14 on waves, especially the later sections on combining waves, before you start this chapter. We begin the study of wave optics by showing how geometrical optics, in particular Snell's law, can be derived from a wave theory of light. Then we describe the experiment by Thomas Young that first established the wave nature of light. Subsequent sections discuss other optical observations and show how they, too, can be understood from a wave-theory perspective.

The material of this chapter is important not only for the sake of understanding optics, but also because these ideas are necessary for understanding much of modern physics. This will become particularly apparent in Chapter 27, on quantum mechanics.

In the previous two chapters we saw that geometrical optics provides a lot of practical and useful information. The assumption that light travels in a straight line except at the interface between two media satisfactorily explains most of what we see. It is sufficient to allow us to design and understand the majority of optical instruments, including cameras, projectors, telescopes, and microscopes.

However, in considering geometrical optics we did not need to inquire into the nature of light itself. In this chapter we see that when the optical components become sufficiently small, wavelike properties of light become more noticeable. Additional evidence for the wave nature of light comes from such apparently diverse phenomena as the behavior of polarized sunglasses, the blue of the sky, and the red of the setting sun.

On the other hand, we will see later, in Chapter 26, that situations occur in which light clearly displays particlelike properties. As we have said before, this apparent duality between the wave and particle nature of light cannot be explained in terms of the classical mechanics and electromagnetism discussed so far in this text. A different type of theory, called quantum mechanics, is needed. It is especially useful in describing phenomena on the molecular, atomic, and smaller scales, as mentioned earlier in Section 4.10 and shown in Fig. 4.20.

In 1807 Thomas Young (1773–1829) published his *Lectures on Natural Philosophy,* containing the description of an optical experiment now referred to as Young's double-slit experiment. This demonstration of interference effects firmly established the wave theory of light on experimental grounds and provided a straightforward means for measuring the wavelengths.

As we will see later, in Chapter 29, wave optics and quantum mechanics play key roles in the operation of lasers. Furthermore, recent cosmological theories on the creation of the universe and its possible future also rely heavily on interpretations of wave optics and quantum mechanics.

Huygens' Principle

23.1

The first thing necessary in discussing a wave theory of light is a technique for describing wave motion. A simple technique was developed by the Dutch scientist Christiaan Huygens and used by him to explain reflection and refraction. We can understand his principle in terms of simple water waves. Suppose we toss a rock into a pond. When the rock hits the water, it generates circular waves that spread out from the point of impact (Fig. 23.1). As we watch the wavefront expand, we are actually following a series of points of constant phase. Every point along the wave crest all around the circle has the same phase, and as the wave expands this circle of constant phase expands. If the wave encounters a barrier, the wave is reflected; but if the barrier has an opening, part of the wave passes through.

Figure 23.1   Circular ripples spread from a disturbance at the surface of a pond.

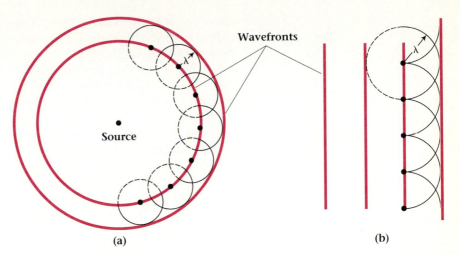

Figure 23.2   Huygens' construction of (a) a circular wave; (b) a plane wave. The radius of each wavelet, equal to the distance between successive wavefronts, is usually taken to be one wavelength.

Huygens recognized that it is possible to determine how the wave crest advances by considering each point along the wave crest to be a source point for tiny, expanding, circular wavelets, which expand with the speed of the wave. The contour of the advancing wave is the envelope tangent to these wavelets. In turn, this envelope generates source points for determining a later position of the wave. Figure 23.2 illustrates Huygens' constructions for describing circular waves and plane waves. The radius of each wavelet, equal to the distance between successive wave fronts, is taken to be one wavelength. This technique for describing the motion of waves, called **Huygens' principle**, is true for all types of waves.

We can extend Huygens' principle beyond two-dimensional waves to include three-dimensional waves, such as sound and light. The points of constant phase define a surface called the **wave surface** or **wavefront**. We can determine the shape of the wave surface as the wave advances by applying Huygens' principle as just described. Because the wavefront at any point is perpendicular to the direction in which the wave is advancing, the wavefront is perpendicular to the ray that describes the path of the light.

Reflection and Refraction of Light Waves

23.2

The law of reflection states that the angle of incidence and the angle of reflection are equal. This rule comes straight from everyday observations, as we discussed in Chapter 21. However, we can derive it geometrically with Huygens' principle. Imagine a beam of light incident at an angle θ_i on a smooth surface (Fig. 23.3a). The wavefront WF is perpendicular to the path of the beam. The angle OAF is equal to θ_i, since OA and AF are perpendicular to the lines forming θ_i.

According to Huygens' principle, we can locate the next successive wavefront by considering the expansion of spherical wavelets from the wavefront WF. In Fig. 23.3, wavelets along WF strike the surface and

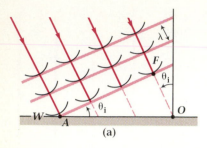

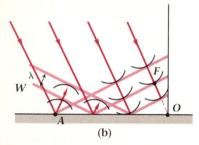

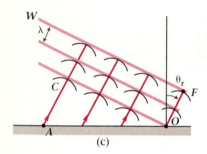

Figure 23.3 Reflection of a plane wave by a mirror surface as described with Huygens waves. Illustrations (a), (b), and (c) show the progress of the wavefronts with time.

reflect upward. The wavelet from W reaches the surface first. As it reflects up by one wavelength, the wavelet from F extends one wavelength toward the surface (Fig. 23.3b). Then, one by one, other wavelets along WF strike the surface and reflect up. The right triangle AOC in Fig. 23.3(c) is congruent to the right triangle OAF of Fig. 23.3(a), because they share the common side AO and $AC = FO$ because the wave speed remains the same. This means that the angle AOC (part c) is the same as OAF (part a). Hence, the *angle of reflection θ_r is identical to the angle of incidence θ_i*. This statement is the law of reflection given earlier in Chapter 21:

$$\theta_i = \theta_r. \tag{23.1}$$

We can also derive the law of refraction from Huygens' principle. In Chapter 21 we discussed Snell's law, the relationship between the indices of refraction and the angles of a light ray as it passes from a medium of one index to another. Let's now look at a wave derivation of Snell's law. Consider a Huygens wave incident upon an interface between medium 1 (say air) and medium 2 with a larger index of refraction (perhaps glass). The resulting behavior is like that shown in Fig. 23.4. The wave speed in air is v_i (incident light) and the wave speed in the glass is v_t (transmitted light). Since we assumed that n_2 is greater than n_1, then v_i is greater than v_t. As the wave enters into medium 2 at point A, its speed is reduced, thus shortening its wavelength (Fig. 23.4b). The wavefront transmitted in the lower medium consequently travels at a different angle with respect to the normal.

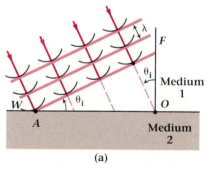

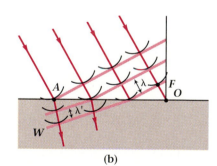

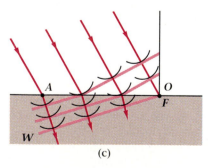

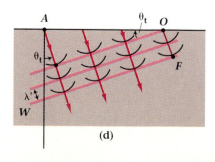

Figure 23.4 Refraction of a plane wave at a flat interface between two transparent media, with n_2 greater than n_1. Illustrations (a), (b), (c), and (d) show the progress of the wavefronts with time.

We observe that the two triangles AOF (Fig. 23.4a) and AOC (Fig. 23.4d) have one side in common. Further, we see that

$$\sin \theta_i = \frac{OF}{AO} \quad \text{and} \quad \sin \theta_t = \frac{AC}{AO}.$$

Upon dividing, we find that

$$\frac{\sin \theta_i}{\sin \theta_t} = \frac{OF}{AC}.$$

However, OF and AC are proportional to the speeds v_i and v_t of the wave. So we obtain

$$\frac{\sin \theta_i}{\sin \theta_t} = \frac{v_i t}{v_t t} = \frac{c/n_i}{c/n_t} = \frac{n_t}{n_i}.$$

This equation is the law of refraction given in Chapter 21 as

$$n_i \sin \theta_i = n_t \sin \theta_t. \tag{23.2}$$

Thus we have shown here that the observed laws of geometrical optics—the law of reflection and Snell's law—follow naturally from the assumption that light is a wave.

We have already seen that the velocity of light in a medium is not the same as the free-space velocity c, but is given by $v = c/n$. When a light wave passes from one material into another, the frequency remains constant.* The wavelength inside a material of index of refraction n is smaller than the free-space wavelength because, as we saw in Chapter 14, the velocity of a wave is the product of its wavelength and frequency, $v = f\lambda$. Thus if the velocity changes, the wavelength changes proportionately. The new wavelength inside the material of index n is $\lambda' = \lambda/n$.

Interference of Light

23.3

To understand why Thomas Young's double-slit experiment was crucial to a wave theory of light, we first need to examine the interference effects of two in-phase wave sources. Then we will show that these same effects were produced by Young's experimental setup, implying that light is a wave. Finally, we follow Young's calculations in determining the wavelength of visible light.

In Chapter 14 we observed that a wave disturbance due to two or more sources can usually be taken as the algebraic sum of the individual disturbances. If the individual sources vibrate with different frequencies, there is nothing special about the resulting disturbance. But if two or more sources vibrate at the same frequency and with a constant relative phase, interesting interference effects occur.

---

*In this respect light waves are like sound waves. Remember, from Section 14.9, that the frequency of sound waves from a guitar or violin is determined by the vibrational frequency of the string. As the waves pass into the air, their frequency does not change even though the speed of sound waves in air is different from the speed of waves in the string.

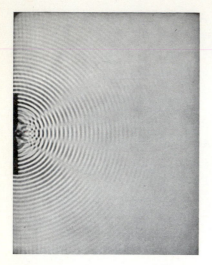

Figure 23.5 A pattern of constructive and destructive interference, produced by two in-phase sets of circular water waves.

Suppose the sources are two little bobs that vibrate up and down on the surface of a body of water. If the two bobs are joined together so that they vibrate with the same frequency, each of them causes circular water waves to spread out from the point of contact. The waves from these two sources interfere with one another. In some directions the waves combine constructively, making waves of larger amplitude. In other directions they combine destructively so that there is little or no wave amplitude. A pattern develops, as seen in Fig. 23.5. The wedge-shaped areas of sharp contrast indicate the crests (light) and troughs (dark) of strongly reinforced waves. However, in some radial directions waves from the two sources arrive exactly out of phase. Since the resulting amplitude is zero, we see no contrasting lines representing the crests and troughs; instead we see a gray region indicating no wave motion.

The condition for constructive **interference** is that waves from two sources with the same frequency arrive at the same point together with the same phase. We can see from Fig. 23.6 that the resultant wave at point P has maximum amplitude because the two contributing wave crests (and subsequently the troughs) arrive at the same time. If the two sources oscillate with exactly the same phase, the condition for maximum constructive interference is that the path lengths of the two waves must be identical or else differ by an integer multiple of the wavelength; that is,

$$D_1 - D_2 = \Delta D = n\lambda, \quad n = 0, 1, 2, 3, \ldots \quad \text{(constructive interference)}.$$

When the two waves arrive exactly out of phase, destructive interference occurs and the resulting wave is diminished. If the two sources have the same amplitude, destructive interference results in zero net amplitude. This occurs if the path lengths differ by half a wavelength or by any half-integer multiple of a wavelength; that is,

$$\Delta D = (n + \tfrac{1}{2}) \lambda, \quad n = 0, 1, 2, 3, \ldots \quad \text{(destructive interference)}.$$

If the distances D_1 and D_2 are quite large compared with the separation d between the sources, we can express the conditions for constructive and destructive interference in terms of the angle θ shown in Fig. 23.7. When the distance L between the sources and the plane containing the observation point is much greater than d, the two paths D_1 and D_2 are nearly parallel. We can approximate the difference between them by $\Delta D \cong d \sin \theta$. This result leads to two new equations for determining the maxima and minima of the resultant wave at P:

$$n\lambda = d \sin \theta, \quad n = 0, 1, 2, 3, \ldots \quad \text{(maxima)}, \quad (23.3a)$$

$$(n + \tfrac{1}{2})\lambda = d \sin \theta, \quad n = 0, 1, 2, 3, \ldots \quad \text{(minima)}. \quad (23.3b)$$

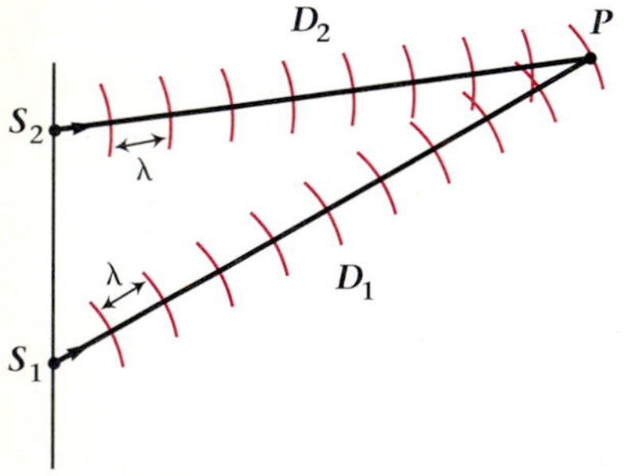

Figure 23.6 Waves from sources S_1 and S_2 reach point P in phase when $D_1 - D_2 = n\lambda$.

Equations (23.3) enable us to relate the directions of the maxima and minima with the wavelength λ, and the source separation d. These equations are true for all waves, not just water waves. In particular, they accurately describe the interference of two sound waves and, as we shall see, two light waves.

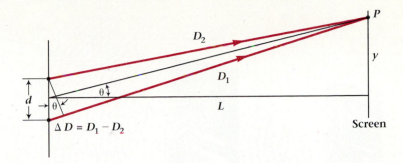

Figure 23.7 Geometrical construction for describing the interference pattern of two sources when the path length is much greater than the distance between the sources.

In Fig. 23.7, the maximum (or minimum) occurs at a point P above the center line. When D_2 is greater than D_1, the maximum (or minimum) lies below the center line. Thus the pattern of maxima and minima is symmetric about the center line. (See Fig. 23.5.) A central maximum occurs along the center line for $n = 0$ in Eq. (23.3a). For this reason it is also referred to as the zero-order maximum. The next maxima on either side of the central beam are called the first-order maxima and correspond to $n = \pm 1$. Similarly, the other peaks are labeled by their order. For example, the second-order maximum is the second maximum away from the center line and corresponds to $n = 2$ in Eq. (23.3a). Similarly, the minima given by Eq. (23.3b) are labeled in order of their position from the central maximum. A first minimum occurs to either side of the central maximum. A second minimum occurs beyond the first, and so on. Note, however, that the first minimum corresponds to $n = 0$ in Eq. (23.3b).

The interference pattern just described also arises when a single wave strikes a barrier pierced by two narrow slits. The wave passes through the slits, which act as if they were new sources of waves (Fig. 23.8). The waves emerging from the two slits must have the same frequency because they were generated from the same initial wave. Because the incident wave crests strike both slits at the same time, the waves passing through the two slits have the same phase. The slits thus act like two sources of identical frequency and phase, a condition known as **coherence**. The emerging waves produce an interference pattern exactly like the one described for two sources. This type of interference occurs for all kinds of waves and is known as double-slit interference.

Thomas Young's experimental setup (Fig. 23.9a, p. 670) allowed sunlight emerging from a small aperture to strike two very small slits. When light from the two slits fell upon a screen, dark stripes appeared, dividing the area into regularly spaced light and dark portions. Figure 22.9(b) shows the appearance of a typical double-slit pattern, made with the red light of a helium–neon laser.

Young recognized that for interference to occur, the light falling on the two slits must be coherent. The purpose of the first aperture was to ensure that the only light striking the two narrow slits came from the same source and was thus coherent. Figure 23.10 (p. 670) shows the wave diagram drawn by Young to explain the origin of the light and dark bands of the interference pattern. Young reasoned, as we did above, that the maxima (bright bands) occurred when the path lengths from the two slits

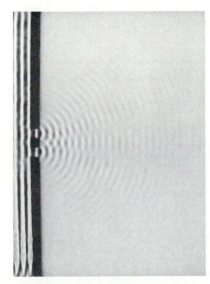

Figure 23.8 Interference pattern of water waves caused when a plane wave (left) passes through a pair of slits. Note the similarity with Fig. 23.5.

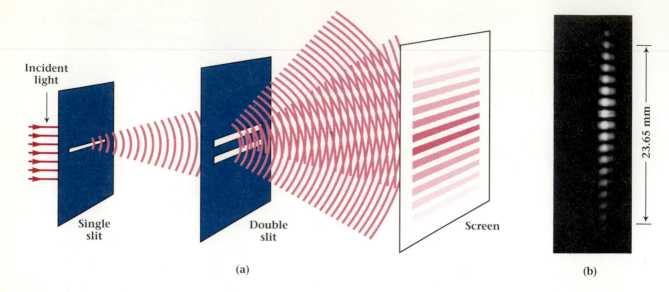

(a)

(b)

Figure 23.9 (a) The arrangement for Young's double-slit experiment. Sunlight passing through the first slit is coherent and falls on two slits close to each other. Light passing beyond the two slits produces an interference pattern on a screen. (b) The double-slit pattern of a small diameter beam of light from a helium-neon laser. The center-to-center separation of the two slits was 0.113 mm. The film was located 29.1 cm from the slits. The reference marks seen along the edge of the film were spaced 23.65 mm apart. From this data, compute the wavelength of the laser light.

to the screen differed by integral multiples of the wavelength, and that the minima (dark bands) occurred when the paths differed by an odd number of half-wavelengths. The situation is exactly the same as that described in Fig. 23.7.

Once Young had worked out this explanation of the double-slit interference pattern, he realized that by measuring the spacing between the slits, the distance L to the screen and the positions y of the maxima and minima, he could determine the wavelength of light. Young used an analysis similar to our use of Eq. (23.3) to find that the wavelengths of light range from about 400 nm in the extreme violet to about 700 nm in the extreme red.*

*Young's actual values were given as one 60-thousandth of an inch to one 36-thousandth of an inch.

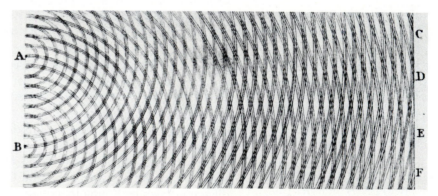

Figure 23.10 Diagram of the interference of light waves emerging from two points A and B as drawn by Thomas Young. Points C, D, E, and F indicate regions of destructive interference.

Example 23.1

Using interference to find the wavelength of light.

A narrow beam of coherent light from a laser passes through a pair of narrow slits and strikes a screen placed 1.00 m beyond the slits. The slits are spaced 0.050 mm apart. The resulting interference pattern on the screen has maxima that are 12.7 mm apart. What is the wavelength of the laser light?

Solution   The maxima are evenly spaced near the center of the double-slit pattern. For convenience, we may choose the separation between the central maximum, which lies on the symmetry axis, and the maximum nearest it (a first-order maximum) as the distance y in Fig. 23.7. This distance corresponds to a path difference of one wavelength, that is, to $n = 1$ in Eq. (23.3a). The angle θ is so small that we may approximate $\sin\theta$ as $\sin\theta \approx \tan\theta \approx y/L$. The condition for an interference maximum, Eq. (23.3a), becomes

$$\lambda = \frac{dy}{L},$$

where we have substituted y/L for $\sin\theta$ and set $n = 1$. Inserting the numerical values, we may compute the wavelength:

$$\lambda = \frac{(0.050\ \text{mm})(12.7\ \text{mm})}{1.00\ \text{m}} = \frac{(5 \times 10^{-5}\ \text{m})(1.27 \times 10^{-2}\ \text{m})}{1.00\ \text{m}}$$

$$\lambda = 6.4 \times 10^{-7}\ \text{m} = 640 \times 10^{-9}\ \text{m} = 640\ \text{nm}.$$

Interference in Thin Films

23.4

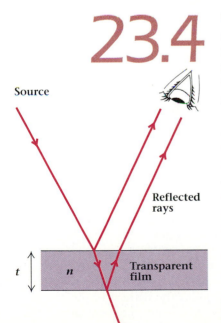

Source

Reflected rays

t n Transparent film

We can observe the interference of light in ways other than the double-slit experiment. For example, interference also occurs when light is reflected from or transmitted by a thin film of transparent material. This interference is responsible for the colors of soap bubbles and oil slicks. As we will show, it is also the basis for the nonreflecting coatings commonly used on binoculars and photographic lenses.

Figure 23.11 shows a diagram of monochromatic light incident from above on a thin transparent dielectric film of thickness t and index of refraction n that is greater than that of the surrounding medium. At the first interface the light is partly transmitted (refracted) and partly reflected. At the second interface a portion of the transmitted light is reflected and follows the path indicated in the figure. Thus an incident light beam produces two coherent reflecting beams, one from each surface of the thin film.

If light strikes the film at nearly normal incidence, the two reflected beams may interfere constructively or destructively, depending on whether they are in phase or out of phase. The path length of the beam reflected

Figure 23.11   Monochromatic light incident on a thin transparent film is reflected from both the top and bottom surfaces.

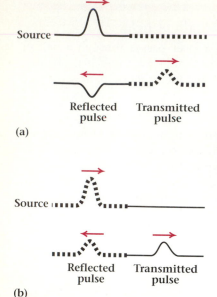

Source

Reflected pulse   Transmitted pulse

(a)

Source

Reflected pulse   Transmitted pulse

(b)

Figure 23.12 Reflection of a pulse wave at the boundary between ropes of different linear density. (a) Pulse incident in the less dense rope changes phase upon reflection. (b) Pulse incident in the denser rope does not change phase.

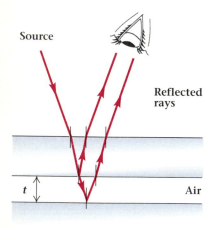

Source

Reflected rays

t Air

Figure 23.13 Reflection of light by a thin film of air between two glass plates.

from the second surface is $2t$ greater than that of the beam reflected from the first surface. If there were no other effects, the beams would interfere constructively when this path difference equaled a whole number of wavelengths, in a manner similar to our reasoning for the double-slit experiment. However, there is another effect at work here. When light passes from a medium of lesser index of refraction to one of greater index, a phase change of 180°, corresponding to $\frac{1}{2}\lambda$, takes place upon reflection. The situation is analogous to the case of pulses reflecting back from the boundary between two ropes of different density (Fig. 23.12). A pulse from a lighter rope to a heavier rope reverses its phase upon reflection, whereas a pulse from a heavier rope to a lighter one does not change phase on reflection. Similarly, light going from a medium of smaller index of refraction (air) to a larger index (oil or water) changes phase upon reflection, while light going in the opposite direction does not change phase.

Applying these ideas of phase changes to our thin film, we see that the light ray reflected from the top surface of the film undergoes a phase change of 180°. The transmitted ray does not experience any phase change during refraction at the upper surface of the film nor does it undergo any phase change as it reflects from the bottom surface of the film. There is, however, a phase change associated with the path traveled through the film. The total effect taking place here, combining the phase change upon reflection with the path difference through the film, introduces an extra difference of $\lambda/2$ into our previous conditions for constructive and destructive interference.

Inside the film, where the index of refraction is n, the wavelength λ' is smaller than the incident wavelength λ by a factor of $1/n$. For constructive interference of light at normal incidence reflected from a thin film, the path length in the film must be a half-integer multiple of the wavelength λ'. Thus the condition for constructive interference is

$$(m + \tfrac{1}{2})\lambda = 2nt, \qquad m = 0, 1, 2, 3 \ldots \qquad \text{(maxima)}. \qquad (23.4a)$$

Similarly, minima in the reflected intensity occur for

$$m\lambda = 2nt, \qquad m = 0, 1, 2, 3 \ldots \qquad \text{(minima)}. \qquad (23.4b)$$

The same considerations also hold true for light reflected from a thin film of air (of thickness t) between two media of higher index such as two glass plates (Fig. 23.13). Light reflected from the first air–glass interface does not have a change in phase (higher refractive index to lower). However, light reflected from the second air–glass interface does undergo a phase change (lower refractive index to higher). The resulting equations are the same as Eqs. (23.4), where n is the index of refraction of air.

We can see this effect when a piece of glass with a slightly curved bottom (such as a lens with a large radius of curvature) is placed on a flat glass plate and illuminated from overhead by a point source of light. When you look from above, you see a series of concentric rings. These rings, known as *Newton's rings*, arise from the interference between light reflected at the curved surface and light reflected from the underlying flat surface. Their appearance can be used to judge the flatness of the plate or the sphericity of the lens surface. If the source is not monochromatic but provides white light instead, the rings will be colored.

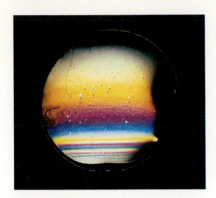

Figure 23.14 Interference pattern of a soap film suspended in a loop of wire.

We encounter interference of reflected light in many thin-film situations involving different indices of refraction. For example, a thin film of oil ($n = 1.36$) on a glass plate ($n = 1.55$) involves two phase changes, one for reflection from each interface. The conditions for constructive and destructive interference are the same as Eqs. (23.4), only now the location of maxima and minima are reversed. Rather than try to remember which combination of equations fits which thin-film situation, simply remember to count the phase changes that occur for light reflected from each interface between media of different refractive indices. Then write down the conditions for constructive interference (path difference equals integer number of wavelengths) and destructive interference (path difference equals half-integer number of wavelengths) and add a factor of one-half wavelength for each difference in phase change between reflected rays.

Another interesting case occurs when the thickness of the film changes along its length, giving rise to alternating regions of constructive and destructive interference. For example, for a soap film suspended in a loop (Fig. 23.14), the upper portion of the film is thinner than the lower portion. When the light striking the soap film of varying thickness is white, the various wavelengths of light constructively interfere at different places in the film, leading to a separation of the colors of white light. This thin-film phenomenon is also responsible for the rainbow colors visible on oil slicks. Some fish have scales with thin-film coatings that produce colors in the same way.

One of the most important applications of interference in thin films occurs when we place a thin film in contact with a third medium of still greater index of refraction (Fig. 23.15). As we have already explained, a phase change occurs upon reflection at both surfaces, since both surfaces are low-index to high-index boundaries. As a result, the condition for constructive interference in the reflected beams is

$$m\lambda = 2nt, \qquad m = 0, 1, 2, 3 \ldots \qquad \text{(maxima)}. \qquad (23.5a)$$

When monochromatic light strikes the surface at normal incidence, a phase difference occurs between the two reflected beams as a result of the difference in their paths. In this case destructive interference occurs for

$$(m + \tfrac{1}{2})\lambda = 2nt, \qquad m = 0, 1, 2, 3 \ldots \qquad \text{(minima)}. \qquad (23.5b)$$

When the thickness of the film is $\lambda/4n$, corresponding to $m = 0$, there is no reflection at that wavelength. So, if we coat a piece of glass, such as a lens, with a thin film of the right thickness, and with an index of refraction intermediate between those of air and glass, we can minimize reflection from the glass. Such a film is called an **antireflection coating**. Because reflections are reduced, more light is transmitted through lenses that have antireflection coatings.

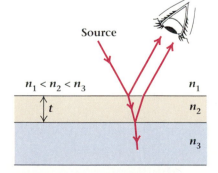

Source

$n_1 < n_2 < n_3$

t

n_1

n_2

n_3

Figure 23.15 Reflection of light from a thin transparent coating on a material of higher index of refraction.

Example 23.2

Thickness of an antireflection coating.

For a photographic lens, we want to design a simple antireflection coating that will be most effective for green light of $\lambda = 550$ nm. The coating material is magnesium fluoride, which has an index of refraction of 1.38. How thick should the coating be?

Solution   We want maximum transmission and minimum reflection of the green light. Thus we want a film of the correct thickness to create maximum destructive interference for reflected light. This condition is given above in Eq. (23.5b). If we choose $m = 0$, we get

$$t = \frac{\lambda}{4n}.$$

Substituting the numbers into the expression gives

$$t = \frac{550 \times 10^{-9} \text{ m}}{4 \times 1.38} = 99.6 \text{ nm}.$$

The film thickness chosen here, one fourth of the wavelength of light in the coating, will provide good antireflection behavior for the green light specified. However, this coating will not be as effective at either end of the visible spectrum, since the wavelengths there are quite different. Better antireflection coatings, effective over the whole visible spectrum, are built up from several layers of thin films of alternating low and high indices of refraction.

Diffraction by a Single Slit

23.5

We are all familiar with shadows thrown on a wall by an object, such as a hand, that blocks part of a beam of light, and we know that the shadow has the same geometric shape as the object. Young's double-slit experiment shows that light does not travel past objects in simple straight lines, but instead spreads out in wavefronts that can interfere with each other. This spreading out of light passing through a small aperture or around a sharp edge is called **diffraction**. Diffractive spreading is exactly what we would expect from Huygens' principle and, as we saw, was used by Young to illuminate his two slits.

Light always spreads out as it travels, but diffractive effects become noticeable only when light travels through a small enough aperture or past a sharp edge. Figure 23.16 shows a plane wave incident upon three slits

Figure 23.16   Plane waves incident on a single slit spread as they pass through. When the slit width approaches the wavelength, the waves spread as if from a point source.

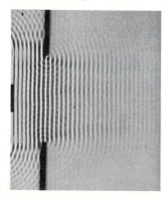

(1)

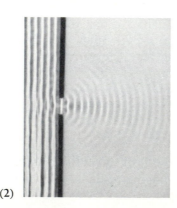

(2)

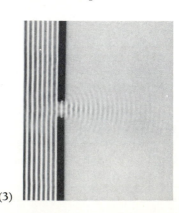

(3)

Figure 23.17 Shadow of a 3-mm-diameter ball bearing illuminated with laser light.

of different sizes. You can see that once the slit width approaches the dimension of the wavelength, the waves spread out as if from a point source. Figure 23.17 shows another example: the shadow of a ball bearing. Since the ball bearing has circular symmetry, light diffracted around its edge interferes constructively at the very center of the shadow. This bright spot in the center of a shadow was first predicted in 1819 by the French physicist Simeon Poisson as a necessary consequence of Augustin Fresnel's wave theory of light. Poisson believed the prediction—and the theory—ridiculous; but in fact, François Arago showed experimentally that the bright spot did exist, thus supporting rather than disproving the wave theory.

The pattern produced by a plane light wave illuminating a single slit depends on the size of the slit relative to the wavelength λ. When the slit is very wide compared with λ, the pattern closely resembles the geometrical shadow of the slit. As in the case of the water waves in Fig. 23.16, the pattern spreads out as the width of the slit is narrowed (Fig. 23.18).

The explanation for the pattern of single-slit diffraction is similar to the explanation for the double-slit pattern. Figure 23.19 shows the geometry for the diffraction of light by a single slit of width b. When the paths from the slit to the screen for light beams passing across the upper and lower edges of the slit differ by an integral multiple of λ, a dark region appears on the screen. This happens because light from the center of the slit is out of phase with light from the edges. (That is, the path difference is one-half wavelength.) Thus we have a condition for minima in the single-slit pattern,

$$n\lambda = b \sin \theta, \qquad n = 1, 2, 3, \ldots \qquad \text{(minima)}. \qquad (23.6)$$

Between each pair of minima is a maximum. The brightest maximum occurs right in the center and the other maxima get successively dimmer (Fig. 23.20). We can calculate the intensities and positions of these maxima, but the process requires techniques beyond the scope of this book.

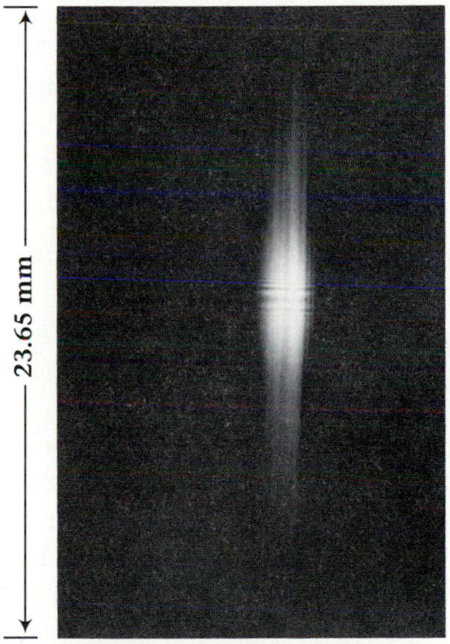

23.65 mm

Figure 23.18 Diffraction pattern of a single horizontal slit 0.05 mm wide. The photographic film was placed 29.1 cm from the slit. The reference marks along the edge of the film were 23.65 mm apart. Diffraction makes the vertical image taller than the slit width.

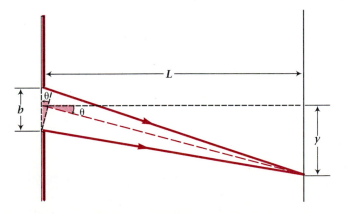

Figure 23.19 Geometry of the single-slit diffraction pattern.

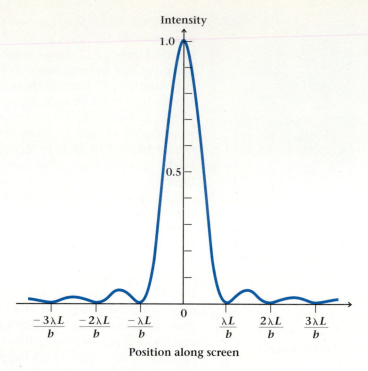

Figure 23.20   Light intensity along the screen for diffraction by a single slit. By far the brightest spot occurs at the center line of the slit.

Example 23.3

Analysis of single-slit diffraction.

A single narrow slit is illuminated with red light of wavelength 632.8 nm. A screen placed 1.60 m from the slit shows a typical single-slit diffraction pattern. The separation between the two first minima is 4.0 mm. What is the width of the slit?

Solution   Refer to Fig. 23.19, which illustrates the geometry of single-slit diffraction. The angle θ is given by

$$\theta \approx \tan \theta = \frac{y}{L},$$

where y, measured from the symmetry axis, is the distance from the middle of the central maximum to the first minimum. We are given that the total distance between the two first minima, one on either side of the center, is 4.0 mm. Thus the quantity y is half of this, or 2.0 mm. The slit-to-screen distance L is 1.60 m and is many times greater than y. We are thus able to express Eq. (23.6) in the form

$$\lambda = \frac{by}{L},$$

where we have set n equal to 1. Upon rearranging, we find that

$$b = \frac{L\lambda}{y}.$$

Substituting the appropriate values into the equation, we find that the width of the slit is

$$b = \frac{1.60 \text{ m} \times 632.8 \times 10^{-9} \text{ m}}{2.0 \text{ mm} \times \dfrac{10^{-3} \text{ m}}{\text{mm}}} = 5.1 \times 10^{-4} \text{ m} = 0.51 \text{ mm}.$$

Multiple-Slit Diffraction and Gratings

23.6

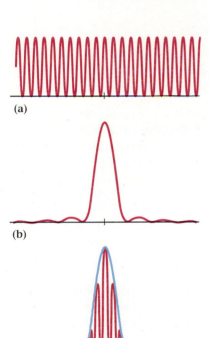

Figure 23.21 (a) Intensity pattern expected from double-slit interference without considering effects due to the width of the slits. (b) Single-slit diffraction intensity pattern. (c) Double-slit interference pattern showing the effects of the single-slit diffraction envelope. This figure is drawn for the slit separation to be three times the slit width.

By now you may be wondering about the distinction between diffraction and interference. These phenomena are inseparably linked and are not really different. The pattern that results from the diffraction of light by a single slit could be thought of as the self-interference of light passing through the slit. In double-slit interference, the light is diffracted by each slit. Thus the patterns describe both interference and diffraction at the same time. However, we generally use the term "diffraction" to describe the effect of a wave encountering an obstacle, while we use "interference" to describe the effects of combining multiple sources or parts of a wave.

In our initial treatment of the double-slit experiment, we were not concerned with the effect of the finite slit width. However, the slit width does determine the overall extent of the double-slit pattern, in that we find the interference pattern only in the region where light is diffracted by each single slit. Figure 23.21 illustrates the interference pattern of two slits superposed on the intensity pattern of a single slit. For this case we have chosen the slit width to be small compared with the spacing. Consequently, since the spreading of the patterns is inversely related to slit spacing and to slit width, the two first minima of the single-slit pattern are spread farther apart than the separation between the double-slit minima. The first minimum of the single-slit pattern occurs at an angle θ given by

$$\sin \theta = \frac{\lambda}{b} \qquad \text{(first minimum).} \qquad (23.7)$$

The double-slit maxima are separated by an angle θ' given by Eq. (23.3a):

$$\sin \theta' = \frac{\lambda}{d}.$$

Thus, illuminating two narrow slits separated by a distance d several times their width b produces a broadly spread diffraction pattern with closely spaced interference fringes.

Example 23.4

A pair of 0.30-mm-wide slits are used to generate an interference pattern. The fifth-order maximum of the double-slit pattern is not observed be-

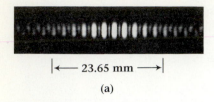

|← 23.65 mm →|

(a)

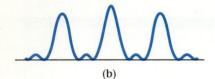

(b)

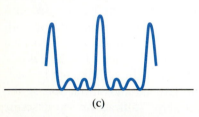

(c)

Figure 23.22   (a) Photo of a three-slit diffraction pattern. The conditions were the same as in Figure 23.9. (b) Sketch of the intensity profile due to three slits. (c) Sketch of the intensity profile due to four slits.

cause it occurs at the position of the first-order minimum in the single-slit pattern. What is the separation of the slits?

Solution   The first minimum of the single-slit pattern occurs at an angle θ given by

$$\sin \theta = \frac{\lambda}{b},$$

where b is the width of the slit and λ is the wavelength of the light.

The fifth-order maximum of the double-slit pattern occurs at angle θ' given by Eq. (23.3a) for $n = 5$,

$$\sin \theta' = \frac{5\lambda}{d}.$$

Since the single-slit minimum occurs at the same position as the fifth-order maximum for the double slit, $\theta = \theta'$ and

$$\frac{\lambda}{b} = \frac{5\lambda}{d}.$$

So

$$d = 5b = 5 \times 0.30 \text{ mm} = 1.5 \text{ mm}.$$

The separation of the slits is 1.5 mm, five times their width.

What happens to the resulting pattern if we increase the number of slits? First, if we use three equally spaced slits instead of two, the pattern looks like Fig. 23.22(a). The intensity peaks still occur at the same positions as in the double-slit pattern, since these are the same positions for which the light from all three slits arrive in phase. However, the peaks are narrower, and subsidiary maxima appear between the principal maxima (Fig. 23.22b). For two slits, no light appears at the angle where the two waves are exactly out of phase. But when three waves are present, one wave remains when the other two exactly cancel. Thus the smaller maxima in the three-slit pattern occur at the positions of the minima in the two-slit pattern.

When we add a fourth slit, another subsidiary peak occurs between the principal maxima (Fig. 23.22c). The ratio of the intensities of the smaller maxima to that of the principal maxima is even smaller in the four-slit pattern than for the three-slit pattern. In fact, as more and more slits are used, the general behavior in the resulting interference pattern is that the subsidiary peaks are suppressed and the principal peaks become narrower. If a large number of slits are used, the principal maxima can become quite sharp. An array of a large number of parallel, equally spaced slits is called a **diffraction grating.** We can analyze its resulting pattern of light intensity as due to the interference of many slits.

When white light from the sun passes through a diffraction grating it forms a spectrum, a rainbowlike pattern of colors. Spectra can also be produced by reflection from a grating. For example, you can see colors in the light reflected from the ridged surface of a black phonograph record. The grooves of the record act as a diffraction grating. Similar color patterns can be seen on a compact disk (CD), these being due to diffraction from

Figure 23.23 One of the authors (RLC) holds a sphere covered with diffraction foil. White light striking the foil is spread into many colors.

the spiraling line of dimples pressed into the disk. Brilliant colors are also produced from embossed plastic film gratings with reflective backings (Fig. 23.23).

A spectrum is produced because white light consists of a mixture of colors, each with its own wavelength, as we saw in Chapter 21. When white light passes through a grating, each wavelength is diffracted through its own characteristic angle. Thus the light disperses into a spectrum of component colors. Diffraction gratings are routinely used for spreading light into a spectrum (Section 23.9).

The angle of diffraction of each wavelength of light passed through a grating is given by the diffraction equation,

$$n\lambda = d \sin \theta, \qquad n = 0, 1, 2, 3, \ldots, \tag{23.8}$$

where d is the spacing between the centers of the slits in the grating, and we have assumed the incident light to be normal to the plane of the grating. This equation is the same one we used to describe the maxima of double-slit interference and it occurs here for exactly the same reasons.

In diffraction, larger wavelengths are deflected through larger angles. Thus diffraction spectra are formed with violet closer to the normal and red farther away. When the adjacent path lengths differ by two or more wavelengths, corresponding to $n = 2$ or more, a secondary spectrum occurs at an even greater angle of deflection. In general, light falling on a grating is diffracted into several principal maxima or orders simultaneously, in addition to the zeroth-order beam that goes straight through.

Example 23.5

White light is normally incident on a diffraction grating of 5000 lines/cm. What is the angular separation between red light ($\lambda = 650$ nm) and blue light (450 nm) in the first-order spectrum?

Solution Since the problem involves a beam incident normally upon the grating we can use Eq. (23.8). We need to determine θ for the two wavelengths given for $n = 1$. The spacing d of the grating is the reciprocal of the number of lines per centimeter, that is,

$$d = \frac{1}{5000 \text{ lines/cm}} = 2 \times 10^{-4} \text{ cm/line} = 2 \times 10^{-6} \text{ m}.$$

For the red light the angle is obtained from Eq. (23.8),

$$\sin \theta_r = \frac{\lambda_r}{d} = \frac{650 \times 10^{-9} \text{ m}}{2 \times 10^{-6} \text{ m}} = 0.325,$$

$$\theta_r = 19°.$$

At the blue end the angle is given by

$$\sin \theta_b = \frac{\lambda_b}{d} = \frac{450 \times 10^{-9} \text{ m}}{2 \times 10^{-6} \text{ m}} = 0.225,$$

$$\theta_b = 13°.$$

The angular separation between the two beams is $\Delta\theta = \theta_r - \theta_b = 6°$.

Resolution and the Rayleigh Criterion

23.7

An important prediction of the wave theory of light is that the ability of an optical instrument to produce distinct images of objects that are very close together is limited. The *resolving power,* or *resolution* as it is often called, is a measure of this ability to produce sharp images. No matter how carefully a lens is designed and made, there is a limit to its resolution. This limit is determined by the diffraction pattern of the lens.

In Section 23.5 we saw that when parallel light passes through a single slit of width b the light spreads out by diffraction. The resulting diffraction pattern consists of a bright central region bordered by alternating dark and bright bands. The angle θ between the center of the central maxima and the first minima (Fig. 23.19) is obtained from Eq. (23.6):

$$\sin \theta = \frac{\lambda}{b},$$

where λ is the wavelength of the light. For small values of the angle θ, we can approximate $\sin \theta$ by θ so that the equation becomes

$$\theta = \frac{\lambda}{b}.$$

As we will see, this angle determines the minimum separation between objects in order for them to be imaged individually.

Let's start by considering two point sources of light, P and Q, that subtend an angle α at the slit S (Fig. 23.24). The light passing through the slit is imaged on a screen by a converging lens. A ray drawn through the center of the lens from each source point determines the positions on the screen of their images P' and Q'. However, the images are not simply points, but are spread into a larger area as indicated. If the angle α is larger than the angular spread θ of each image due to diffraction, the two sources are imaged distinctly (Fig. 23.25a,b). We say the images are well resolved. If the angle α is smaller than θ, the two images overlap and cannot be resolved into two images (Fig. 23.25d). When $\alpha = \theta$ the variation in intensity across the pattern is discernible to the human eye (Fig. 23.25c). For this separation, $\alpha = \theta$, the first minimum of one diffraction

Figure 23.24 The light from two point sources P and Q is imaged on a screen after passing through a slit S. The images P' and Q' are not points, but are spread out by diffraction.

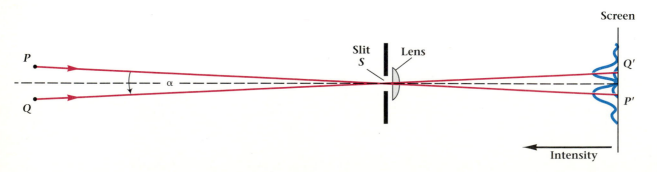

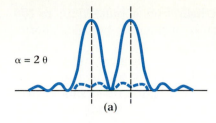

$\alpha = 2\,\theta$

(a)

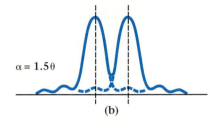

$\alpha = 1.5\theta$

(b)

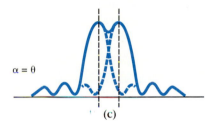

$\alpha = \theta$

(c)

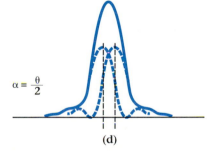

$\alpha = \dfrac{\theta}{2}$

(d)

Figure 23.25 The diffracted image of two point sources at different angular separations α. Rayleigh's criterion for just resolving the two images is shown in (c).

image falls on the central maximum of the other image. Lord Rayleigh* chose this separation as the minimum angle that could be resolved. His choice is often referred to as **Rayleigh's criterion**.

When the diffracting aperture is a circle rather than a slit, the angle of the first minimum in the diffraction pattern is somewhat larger than that given above. The angular position of the first minimum in the diffraction pattern of a circular aperture of diameter D occurs at an angle

$$\theta_{m} = 1.22\,\frac{\lambda}{D}. \tag{23.9}$$

By Rayleigh's criterion, θ_{m} is the limiting angular separation of two objects that are just resolved. This minimum separation is a theoretical lower limit. In practice, the angular separation of the two objects may have to be even greater in order to achieve distinct images. Since the limiting angle is inversely proportional to the diameter D, we can obtain improved resolution by making D larger or λ smaller.

The diameter of the objective lens (or mirror) sets a limit on the resolution of a telescope. Astronomical telescopes are made with large-diameter objectives for this reason—and for increased light-gathering ability—rather than to obtain high magnification. No matter what the magnification, a telescope's ability to resolve two nearby stars is limited by the diffraction pattern of its objective. In addition, atmospheric turbulence sets a practical limitation on the resolution of ground-based telescopes. For this reason, telescopes are now being placed in earth-orbiting satellites to put them above the image-degrading atmosphere.

Example 23.6

Resolution of binoculars.

What is the theoretical minimum angle of resolution of binoculars with 35-mm-diameter objective lenses? Take $\lambda = 550$ nm.

Solution The minimum angle of resolution is determined from Eq. (23.9),

$$\theta_{m} = 1.22\,\frac{\lambda}{D} = 1.22\,\frac{550 \times 10^{-9}\ \text{m}}{35 \times 10^{-3}\ \text{m}} = 1.9 \times 10^{-5}\ \text{radians},$$

$$\theta_{m} = 4.0\ \text{seconds of arc}.$$

This corresponds to the angular separation of the centers of two pennies held side by side and viewed from a distance of 100 m.

The resolving power of microscopes is also limited by diffraction. It is the resolution that limits the useful magnification. Increasing the magnification by the use of stronger lenses only enlarges an already blurred image. Increasing the diameter of the objective lens is not practical for microscopes, as it is for telescopes. The inherent limitation of optical

*John William Strutt, Baron Rayleigh (1842–1919) is famous for his study of waves. He was awarded the 1904 Nobel Prize in physics for his discovery of argon.

Figure 23.26   A scanning electron micrograph of a midge.

microscopes is the wavelength of visible light, from about 400 nm to 700 nm. Any improvement in resolution can be made only by using shorter wavelengths.

After the discovery of the wave properties of electrons (see Chapter 27) it was soon realized that electrons could easily be accelerated to high velocities at which their wavelengths were a thousand times shorter than the wavelengths of visible light. Furthermore, electron beams can be controlled by electrostatic and magnetic fields acting as lenses. Thus the idea of the *electron microscope* was born. The first electron microscope was built by Ernst Ruska in Germany in 1932, and improved versions became commercially available in 1938. Professor Ruska's efforts were belatedly recognized in 1986 when he received the Nobel Prize in physics.

Modern electron microscopes are commonplace in laboratories throughout the world, where they are used to form images of extremely tiny objects. The resolution of these microscopes routinely allows the observation of objects as small as 0.5 nm, and some of the most powerful microscopes permit the observation of detail as small as 0.05 nm. Thus large molecules such as DNA can be seen with electron microscopes. Figure 23.26 was taken with an electron microscope. We discuss the principle of electron microscopes in more detail in Chapter 27.

Dispersion

23.8

White light can be spread into its component colors by a glass prism as well as by diffraction. Isaac Newton showed that one prism could separate white light into a spectrum and another prism turned the opposite way could combine that light back into a white beam again. He also showed that a small portion of the spectrum could not be spread into any other colors by passing the light through a second prism. However, the spreading of white light into colors is not just a trick you can do with a prism; it is also responsible for chromatic aberration (see Section 21.8), for the attractiveness of diamonds and cut glass, and even for the beauty of the rainbow.

The production of a continuous spectrum by a prism is due to the variation of wave velocity with wavelength that commonly occurs in transparent media. Water, glass, transparent plastics, and quartz are all dispersive materials. Generally, shorter wavelengths travel with slightly smaller wave velocities than do longer wavelengths. This means that the prism's index of refraction is not constant across the visible spectrum, but decreases continuously as the wavelengths increase from violet to red. As stated in Chapter 21, the dependence of the index of refraction upon the wavelength (or color) of light is called *dispersion*. Since the index of refraction is greater for violet than for red, violet light deflects through a greater angle than does red. (See Fig. 21.30.) This effect of a prism is just the opposite of the dispersive effect of a diffraction grating, which deflects red light more than violet. The graph in Fig. 23.27 shows the typical dispersive behavior of decreasing refractive index with increasing wavelength for a transparent plastic.

Diamonds are highly regarded as gems because of their large disper-

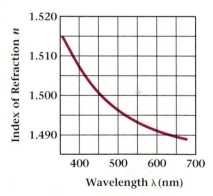

Figure 23.27   Index of refraction as a function of wavelength for polymethylmethacrylate, a transparent plastic used in lenses and other optical components.

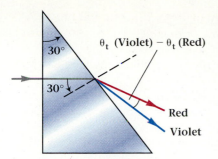

Figure 23.28 Example 23.7: Light from a hydrogen source is normally incident on a 30° glass prism. The light is dispersed as it emerges.

sion. Gems are cut and polished to refract and reflect the light incident on them. The brilliant colors you see in an otherwise colorless gem like a diamond are due to the dispersion of the light as it is refracted. Slightly rotating a properly faceted jewel causes the colors to change and flash from the different facets as the angle of incidence of the light changes.

Example 23.7

Dispersion by a glass prism.

Light from a hydrogen source is normally incident on a 30° crown glass prism (Fig. 23.28). Find the angular separation of the emerging rays of light, assuming $\lambda_{red} = 656$ nm and $\lambda_{violet} = 434$ nm. The refractive index of the glass at these two wavelengths is 1.514 (red) and 1.528 (violet).

Solution Because the light enters the prism at normal incidence, no change in direction of the ray occurs at the first interface. We see that the ray then strikes the back surface with an angle of incidence of 30°. We next calculate the angles of refracted (transmitted) rays relative to the surface normal. For red we get

$$n_i \sin \theta_i = n_t \sin \theta_t$$

$$\sin \theta_t \text{ (red)} = \frac{n_i \sin \theta_i}{n_t} = \frac{1.514 \sin 30°}{1.000} = 0.757,$$

$$\theta_t = 49.20°.$$

For violet,

$$\sin \theta_t \text{ (violet)} = \frac{1.528 \sin 30°}{1.000} = 0.764,$$

$$\theta_t = 49.82°.$$

The angular separation is θ_t (violet) $- \theta_t$ (red) $= 0.62°$. Notice that the violet light is deflected through the larger angle because the index of refraction is greater for violet than it is for red.

The formation of a rainbow by drops of water in the atmosphere is an example of dispersion. In a rainbow, water drops disperse reflected sunlight, revealing the colors of the spectrum across the sky.

To see a rainbow you must have the sun to your back while you look toward a cloud of water drops. (These drops may be in a natural cloud or in a spray or mist that you could make with a garden hose.) To see the primary rainbow (there is also a secondary bow), the angle between the direction from the sun to you and your line of sight to the cloud of mist will be about 42°. Normally we see rainbows when the sun is low in the sky and its light refracts and reflects from many drops of water in the air. The complete rainbow is a circle and can be seen as such from an airplane. Usually we see only the top portion of the circular arc, as the rest lies below the horizon, where it is not easily seen against the background and where the number of reflecting drops in our line of sight is small. For this same reason we do not see rainbows when the sun is higher than 42° above the horizon.

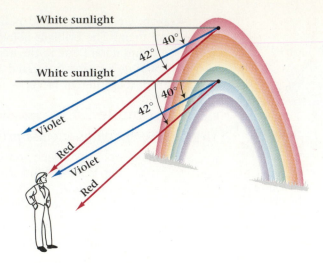

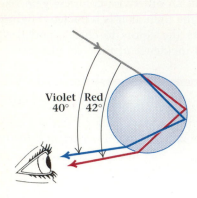

Figure 23.29 Refraction and dispersion of light by a spherical drop of water.

Figure 23.30 The eye sees the different colors of the rainbow in the light reflected and refracted from drops at different heights.

Figure 23.29 shows the refraction of light by a spherical drop of water, showing those rays responsible for the rainbow. The effect of refraction and reflection bunches the rays so that more leave the drop near the angles indicated than in other directions. As a result, the reflected and refracted light is enhanced at the angles shown. Dispersion by the water drop causes light from the sun to spread out into a spectrum. If there are many drops (Fig. 23.30), the eye receives the different colors from drops at different heights. The result is that the top of the rainbow appears red and the inner arc appears violet.

Spectroscopes and Spectra

23.9

If you look through a prism or a diffraction grating at a light source, you will see a band of overlapping images, each in a different color. If the source is an ordinary incandescent lamp with a frosted bulb, the images overlap so that most of the resulting broadened image appears white with a violet border on one side and a red one on the other. If you place a narrow slit in front of the lamp so that you can see only a thin strip of the source through the prism, the band of images has very little overlap and you see bright spectral colors across the band.

A **spectrometer** is an optical instrument designed to enhance this effect and permit analysis of spectra. A *spectroscope* (Fig. 23.31) is a spectrometer that permits direct observation of spectra. Light from the source we wish to analyze passes through a narrow slit and is focused into a parallel beam by a collimating lens. (A collimating lens is a converging lens that focuses a diverging light beam into a parallel beam.) After this beam passes through a prism, parallel beams emerge, each at a different angle according to its wavelength. A telescope focuses the parallel beams

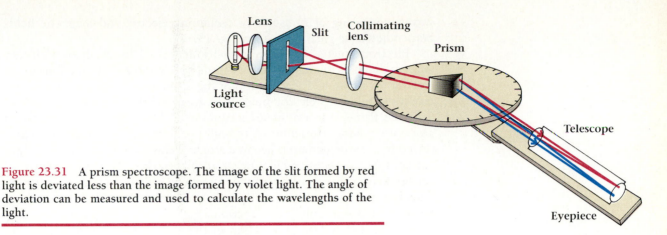

Figure 23.31 A prism spectroscope. The image of the slit formed by red light is deviated less than the image formed by violet light. The angle of deviation can be measured and used to calculate the wavelengths of the light.

and allows an observer to see an image of the slit. Most spectroscopes also have a calibrated circular scale. This scale enables the observer to measure the angle of the emerging light for each image, which is then used to determine the wavelengths of the light in the spectrum. Each image of the slit is called a spectral line. For an incandescent source, the lines merge together into a continuum. Other sources, however, give characteristic lines. As we will see in Chapter 26, knowing what wavelengths are emitted by a light source gives us information about the composition of the source.

Most modern spectrometers are made with a diffraction grating in place of a prism because the dispersion of a grating can be made much greater than that of a prism. Other instruments called *spectrographs*, which are very similar to spectroscopes, permit recording of the spectra photographically. *Spectrophotometers* are devices that convert the resultant light intensity to an electrical signal and then use a chart recorder or computer terminal to display the spectra graphically. Figure 23.32 shows the intensity spectrum taken with a spectrophotometer of light transmitted by a colored glass filter. As you can see, the filter transmits green light (540 nm) but does not pass red or blue light.

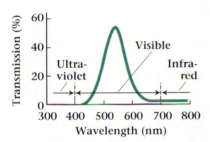

Figure 23.32 Intensity spectrum of the light transmitted by a green glass filter (Hoya #G-533).

Polarization

23.10

So far in this chapter we have discussed the optical properties of light that are due to interference and diffraction, showing how these characteristics support a wave theory of light. However, another important property of light, called polarization, is due not just to light being a wave, but to light being a transverse wave. Longitudinal waves, such as sound in air, do not exhibit polarization, but light, as well as other electromagnetic radiation, can be polarized.

Recall from Chapter 14 that the distinguishing characteristic of transverse waves is that their oscillating motion occurs in planes perpendicular to the direction in which the wave itself is moving. As we discussed in Chapter 19, electromagnetic waves, including visible light, are transverse

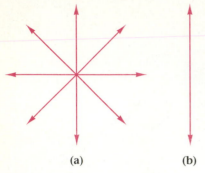

(a)   (b)

Figure 23.33   (a) Light from an ordinary source contains a mixture of
waves oscillating in different directions. (b) A plane-polarized light
beam contains only one direction
of oscillation.

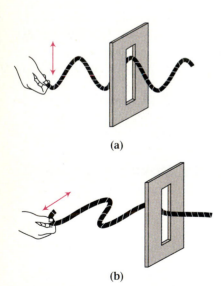

(a)

(b)

Figure 23.34   (a) Shaking a rope up
and down produces a vertically polarized wave that passes through a
vertical slit. (b) Shaking a rope from
side to side produces a horizontally
polarized wave that will not pass
through a vertical slit.

waves, consisting of sinusoidally oscillating electric and magnetic fields
perpendicular to each other (see Fig. 19.16).

Electromagnetic waves are usually generated by oscillating electric
charges, such as the oscillating current in a radio or TV antenna or an
electron in an atom. The direction of the oscillation determines the orientation of the wave's electric field. In most common light sources, such as
a candle flame, the sun, or an incandescent lamp, the many oscillating
atoms are randomly oriented. Although the electric field of each wave produced by a particular atom lies in a single plane, the overall beam of light
contains electric fields oscillating in all planes (Fig. 23.33a). However, if
the oscillating charges are confined to move in only one plane, as is the oscillating current in an antenna, then the entire electric field of the resulting
beam oscillates in only that direction (Fig. 23.33b). Such a wave is said
to be polarized and the orientation of the electric field vector is taken as
the direction of **polarization**. The complete wave has a sinusoidally oscillating magnetic field coupled to the electric field, but traditionally we
refer to the electric field to indicate the direction of polarization.

A useful mechanical analogy for polarized light is the motion of transverse waves along a rope. If you shake a rope up and down, the wave is
vertically plane-polarized. Such a wave can travel through a vertical slit
(Fig. 23.34a), but not through a horizontal slit. Similarly, if you shake a
rope side to side, the wave is confined to a horizontal plane; it can pass
through a horizontal slit but not a vertical one (Fig. 23.34b).

It is possible to produce polarized light by filtering ordinary light
through a *polarizer,* which is a material that transmits only those waves
that oscillate in a single plane. Several materials have this property, including the naturally occurring mineral tourmaline. When an ordinary
beam of light strikes a tourmaline crystal, part of the light is absorbed and
part is transmitted (Fig. 23.35a). If a second tourmaline crystal is placed
behind the first and aligned in the same orientation, light is transmitted.
If the second crystal is rotated by 90° about the axis of the light beam, no
light passes through (Fig. 23.35b). We can explain this observation in
terms of polarization. The initial light beam is unpolarized. We think of
it as containing a mixture of random polarizations. When this beam strikes
the first crystal, the tourmaline passes only light polarized along the polarization axis of the crystal. The light emerging from the first crystal is
completely polarized along the direction shown. When this light falls on
the second tourmaline crystal, all the light is absorbed, because the second
crystal has been turned so that its polarization axis is 90° away from that
of the first crystal.

Crystal polarizers like tourmaline are not used much now because of
the development of inexpensive plastic polarizers called *Polaroids*. Synthetic sheet polarizers were invented in 1928 by Edwin H. Land and improved by him ten years later. When viewed individually, these transparent
sheets have a neutral grey appearance. However, when two polarizers
overlap with their polarization directions at right angles, the region of
overlap looks black (Fig. 23.36).

If the angle between the two Polaroids is made less than 90°, the area
of overlap is lighter. It gradually gets lighter and lighter as the angle
decreases to 0°. The intensity of light that passes through two polarizers

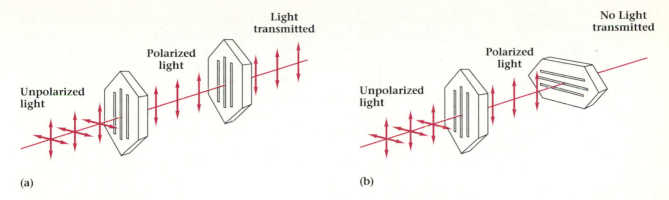

(a)   **(b)**

Figure 23.35   Polarization of light by tourmaline crystals. (a) When the polarization axes of the crystals are parallel, plane-polarized light passes through. (b) When the axes are perpendicular, no light passes through.

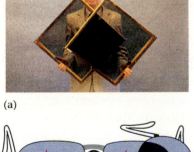

(a)

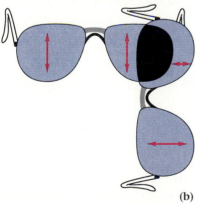

(b)

Figure 23.36   (a) One of the authors (ERJ) holding two large polarizing filters with their polarization directions at right angles. (b) You can demonstrate the same effect yourself with two pairs of polarizing sunglasses.

that are aligned with an angle θ between their polarization directions is

$$I = I_m \cos^2 \theta, \tag{23.10}$$

where I_m is the maximum amount of light transmitted when the two polarizers are aligned along the same direction of polarization. Equation (23.10) is known as *Malus's law* after its discoverer, who first observed this effect experimentally in 1809.

An ideal polarizer passes 100% of the incident light that is plane-polarized along its polarization direction and 0% of the light that is polarized at 90° to its polarization direction. Since unpolarized light is a mix of all possible polarizations, an ideal polarizer transmits only 50% of the incident unpolarized light intensity.

Example 23.8

Light transmitted by two polarizers.

Two ideal polarizers are aligned with a 30° angle between their polarization directions. If unpolarized light of intensity I_0 is incident upon them, what is the intensity of the transmitted light?

Solution   The light passing through the first polarizer is half the incident intensity, so

$$I_1 = \tfrac{1}{2}I_0.$$

The quantity I_1 becomes the I_m of Malus's law, Eq. (23.10). The intensity of the light emerging from the second polarizer is

$$I_2 = I_1 \cos^2 \theta = \tfrac{1}{2}I_0 \cos^2 30° = 0.375 \, I_0.$$

Malus also discovered that light becomes polarized upon reflection from glass windows and surfaces of water. In 1814, some six years after this discovery, David Brewster* (1781–1868) found that at a particular angle of incidence, now called the Brewster angle, polarization of the reflected light is complete, with polarization perpendicular to the plane of

---

*Brewster is also known for his invention of the kaleidoscope and for his improvements on the stereoscope invented by Sir Charles Wheatstone.

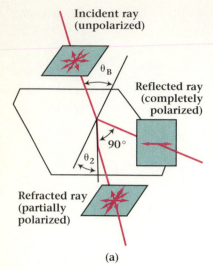

Incident ray
(unpolarized)

θ_B

Reflected ray
(completely
polarized)

90°

θ_2

Refracted ray
(partially
polarized)

(a)

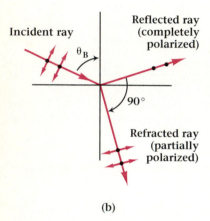

Incident ray

Reflected ray
(completely
polarized)

θ_B

90°

Refracted ray
(partially
polarized)

(b)

Figure 23.37   (a) Polarization of light by reflection at the Brewster angle. (b) The reflected light is completely polarized in a direction perpendicular to the plane of incidence. The refracted light, which is partially polarized parallel to the plane of incidence, makes an angle of 90° with the reflected ray.

incidence (Fig. 23.37). At other angles the reflected light is only partially polarized, with more light polarized in this direction than in any other.

Brewster found that maximum polarization occurs when the reflected ray and the refracted ray are at right angles to each other. By combining Brewster's observation with Snell's law, we can find a relation between a material's index of refraction and its Brewster angle. Maximum polarization occurs for $\theta_r + \theta_t = 90°$, where θ_r is the angle of reflection and θ_t is the angle of transmission (refraction). From Snell's law we can show that

$$n_i \sin \theta_i = n_t \sin \theta_t = n_t \sin (90° - \theta_r) = n_t \cos \theta_r.$$

This equation reduces to

$$n_i \sin \theta_i = n_t \cos \theta_i,$$

when we use the fact that $\theta_i = \theta_r$. At this condition of maximum polarization we can denote the angle of incidence as the Brewster angle θ_B and rewrite this equation as

$$\tan \theta_B = \frac{n_t}{n_i}. \tag{23.11}$$

This last equation is known as *Brewster's law*. Note that the Brewster angle depends on the index of refraction of the materials on both sides of the reflecting dielectric surface.

Example 23.9

Brewster's angle for an air-water interface.

At what angle of incidence is the light reflected from the surface of water completely polarized? Assume that the light is incident in air.

Solution   The index of refraction of water is 1.33. Inserting this into Brewster's law, we find that

$$\tan \theta_B = \frac{n_t}{n_i} = \frac{1.33}{1.00} = 1.33,$$

$$\theta_B = 53°.$$

Sunglasses made from Polaroid sheet are widely used because of their ability to block the glare of specularly reflected sunlight from water surfaces, or from surfaces of other smooth reflecting objects such as automobile windows. The glasses are made with a vertical axis of polarization, since glare resulting from the sun is usually horizontally polarized. You can see this by rotating such glasses by 90°. In this new position the glare light passes through because the transmission axis of the glasses is aligned with the polarization of the light. Normally, of course, the polarization direction of the glasses is crossed with that of the glare light so that the glare is extinguished.

Scattering

*23.11

We conclude this chapter by examining one more optical phenomenon that is best explained by a wave theory of light. This time, we answer an interesting question about our physical world—why is the sky blue?

In a perfectly transparent, homogeneous, and structureless medium, a beam of parallel light proceeds without broadening or lateral spreading. A beam of light in a vacuum behaves this way. But all real materials have some molecular or crystal structure and are not perfectly transparent to visible light. Furthermore, most materials contain impurities; even air has water vapor, dust, smoke, and other minute particles suspended in it. When a beam of light, such as a searchlight beam, passes through fog or dust-laden air, you can see the beam itself from the side. We say that the light has been scattered; molecules of air and its impurities have absorbed some light from the beam and then reradiated it in other directions, a process we call **scattering**. Most of the light travels in the initial beam direction, but some is scattered to the sides and even backwards.

The details of the scattering process depend on the relative size of the scattering particles compared with the wavelength of light. When the particles that cause the scattering are smaller than the wavelength of the incident light we have what is called **Rayleigh scattering**. This type of scattering was first explained by Lord Rayleigh, who showed that the scattering was proportional to the fourth power of the frequency of the light.

We can see what this means by examining a simplified version of an experiment for viewing Rayleigh scattering (Fig. 23.38). Light from a bright source of white light passes through a glass tank and strikes a screen. The collimating lens makes the beam parallel in the tank; the second lens focuses an image of the end of the tank on the screen. The tank is filled with water, into which we have mixed fine microscopic particles. (Such a mixture can be made by adding a few drops of milk to a tank of water.) From the side, the path of the beam is visible and it has a bluish tint. Blue light has a higher frequency than red light and, according to Rayleigh's f^4 law, is scattered more. In fact, it is scattered in all directions, as can be observed by walking around the tank. Because the blue end of the spectrum is scattered out of the beam, less blue and relatively more red appears as the beam proceeds. The light emerging from the right-hand end of the tank appears somewhat reddish-orange. This is not due to any change in the scattering, but happens because this end of the tank is illuminated by light that is now deficient in blue.

Figure 23.38 An experiment to demonstrate Rayleigh scattering. As the blue light is scattered out of the beam, the transmitted light appears red-orange.

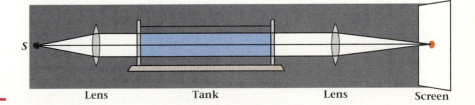

Lens   Tank   Lens   Screen

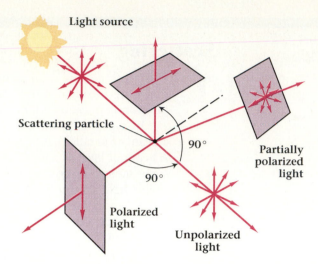

Light source

Scattering particle

90°

90°

Partially polarized light

Polarized light

Unpolarized light

Figure 23.39 Light scattered at right angles to the initial beam direction is polarized.

Light from the sun incident on the earth's atmosphere is scattered most strongly in the violet and blue region of the spectrum. In fact, the sky would appear violet except that our eyes are not very sensitive to violet. The combination of Rayleigh scattering and the sensitivity of our eyes explains the blue of the daytime sky. When the sun is on the horizon, as at sunrise or sunset, the optical path through the atmosphere is much longer than when the sun is directly overhead, so more blue is scattered out of the direct beam. Thus both the rising and setting sun appear red. These effects are due to Rayleigh scattering of the molecules of the air itself, which have dimensions of the order of a fraction of a nanometer, and not to suspended dust or vapor in the atmosphere.

Smoke rising from the end of a lighted cigarette or from around the edges of a pile of burning leaves appears blue because of Rayleigh scattering. On the other hand, exhaled smoke appears white because it contains larger water droplets. When the scattering particles are many times larger than the wavelength of light, the light is reflected and refracted as in geometrical optics, not absorbed and reradiated, so the scattered light is white. This type of scattering is called Tyndall scattering and is responsible for the appearance of fog, clouds, powders and ground glass.

Suppose we look through a polarizing filter at the light scattered out the sides of the tank in Fig. 23.38, looking at right angles to the incident beam. We find that the intensity of the light grows and diminishes as we rotate the filter. This demonstrates that the scattered light is polarized, an effect that was also predicted by Rayleigh. Because light is a transverse wave, the plane of polarization is perpendicular to the plane containing the incident light beam and the perpendicular line of sight of the viewer (Fig. 23.39). When viewed at other angles, the polarization is not complete.

If you look at the sky through a polarizing filter in a direction perpendicular to the direction of the sun's rays, you will find the sky to change from bright to dark as you rotate the filter. The polarization of the light is not 100%, but may be as much as 70% to 80% when the air is clear. Some insects, especially bees, are able to find their way over long distances by using the direction of polarization as a sort of compass.

SUMMARY

Useful Concepts

- The laws of geometrical optics, such as the law of reflection and Snell's law, can be derived from a wave interpretation of light using Huygens' principle. This principle considers each point on a wavefront to be a point source for expanding wavelets. The resulting wave is the envelope of these wavelets and, in turn, it can be considered as a source for finding a still later position of the wave.

- For Young's double-slit experiment the angular locations of the maxima and minima are given by.

$$n\lambda = d \sin \theta, \quad n = 0, 1, 2, 3, \ldots \quad \text{(maxima)},$$
$$(n + \tfrac{1}{2})\lambda = d \sin \theta, \quad n = 0, 1, 2, 3, \ldots \quad \text{(minima)},$$

  where d is the separation between the slits.

- When light is normally incident in air on a thin film of index of refraction n, the interference between the front and back surface reflections gives maxima and minima when

$$(m + \tfrac{1}{2})\lambda = 2nt, \quad m = 0, 1, 2, 3 \ldots \quad \text{(maxima)},$$
$$m\lambda = 2nt, \quad m = 0, 1, 2, 3 \ldots \quad \text{(minima)},$$

  where t is the thickness of the film and λ is the wavelength of the light.

- Antireflection coatings can be made by coating a surface with a thin film of thickness t and index of refraction n that is intermediate between the index of the surface and the index of the surrounding medium (usually air) when

$$(m + \tfrac{1}{2})\lambda = 2nt, \quad m = 0, 1, 2, 3 \ldots \quad \text{(minima)}.$$

- The spreading out of light passing through a small aperture or around a sharp edge is called diffraction. For diffraction by a single slit of width b, the angular locations of the minima are given by

$$n\lambda = b \sin \theta, \quad n = 0, 1, 2, 3, \ldots$$

- For a diffraction grating, the angle of diffraction of each wavelength is

$$n\lambda = d \sin \theta, \quad n = 0, 1, 2, 3, \ldots,$$

  where d is the spacing between slits.

- For a circular aperture of diameter D, the minimum angle of resolution as given by the Rayleigh criterion is

$$\theta_{\mathrm{m}} = 1.22 \frac{\lambda}{D}.$$

- A polarized light beam contains waves whose electric field vectors are all oriented along the same direction.

- The intensity of light passing through two polarizers is given by Malus's law:

$$I = I_{\mathrm{m}} \cos^2 \theta.$$

- The angle of maximum polarization by reflection (Brewster's law) is

$$\tan \theta_{\mathrm{B}} = \frac{n_{\mathrm{t}}}{n_{\mathrm{i}}}.$$

Important Terms

You should be able to write the definition or meaning of each of the following terms:

- Huygens' principle
- wavefront (or wave surface)
- interference
- coherence
- antireflection coating
- diffraction
- diffraction grating
- Rayleigh's criterion
- spectrometer
- polarization
- scattering
- Rayleigh scattering

QUESTIONS

23.1 Describe one method of using sound waves to carry out an experiment analogous to Young's double-slit experiment.

23.2 Can a Young's double-slit experiment be done for the case of microwaves ($\lambda \approx 1$ cm)? How?

23.3 The source of illumination for a Young's double-slit setup is white light. Suppose you cover one slit with a filter that transmits only blue light and the other slit with a filter that transmits only red light. What would you expect to see on the screen?

23.4 Suppose you have a Young's double-slit arrangement that produces an interference pattern of light on a screen. If you submerge the entire apparatus in water, what effect will that have on the pattern?

23.5 Some very high-quality lenses are said to be "diffraction limited." What does this mean?

23.6 A card is placed between a point source and a screen, but relatively close to the screen. Do you expect the card to cast a sharp shadow? If not, what do you expect and why?

23.7 Monochromatic light falls on a single slit and forms a pattern on the screen. What do you observe when the slit width is equal to the wavelength of the light? What will you observe as the slit width is increased to many times the wavelength of the light?

23.8 Can you arrange two glass plates so that a narrow beam of light reflected from the first one and incident on the second is not reflected by the second one? Explain your answer.

23.9 Maxwell showed that electromagnetic waves would not penetrate very far into conductors. Therefore conductors should not be transparent to light. Mention several cases and discuss whether Maxwell's prediction is consistent with your observations.

23.10 Is there any optical basis for the old saying regarding weather, "Red light at night, sailors delight; Red light in morning, sailors take warning"?

23.11 How is the resolution of a camera lens affected by the *f*-value of the lens?

23.12 If a salesperson claims to have Polaroid sunglasses, how can you check to see if this is correct?

23.13 What is the approximate size of the smallest things that can be successfully investigated with a microscope using visible light?

23.14 When a drop of oil is placed on the surface of a bowl of water, the surface becomes less reflecting. What can you say about the index of refraction and any other properties of the oil?

23.15 If you are given a photograph of a diffraction pattern and told that it was made by one or more slits with monochromatic light, how much could you determine about the slit arrangement that made the pattern? This is the underlying principle of holography, discussed in detail in Chapter 29.

PROBLEMS

Hints for Solving Problems

The wavelength of light is proportional to its velocity and inversely proportional to its frequency. Constructive interference occurs when two waves are in phase. In the double-slit experiment the waves are in phase when their path difference is an integer multiple of the wavelength ($n\lambda$). They are out of phase (destructive interference) when the path lengths differ by a half-integer multiple of the wavelength $(n + \frac{1}{2})\lambda$. In computing for interference from the reflection from thin films, you must also include the phase change (equivalent to a path interval of $\lambda/2$) associated with reflection from a more dense medium. Remember to measure angles of reflection and refraction (including Brewster's angle) from the normal to the surface.

Section 23.2   Reflection and Refraction of Light Waves

23.1 Show by Huygens' geometric construction that the image of an object located a distance D from a plane mirror is located a distance D on the other side of the mirror.

23.2 Using Huygens' principle, find the point at which plane waves incident on a concave spherical mirror converge. Use only that part of the wavefront closest to the axis.

23.3 Using Huygens' principle (rather than Snell's law), find the angle of refraction when light is incident on the interface between air ($n = 1.00$) and glass ($n = 1.50$) at an angle of 45°.

23.4 Light from a sodium lamp has a wavelength $\lambda = 589$ nm in empty space. What is the wavelength of this light in water? What is it in polymethylmethacrylate? (*Hint:* Refer to Table 21.1.)

23.5 A beam of light has a wavelength $\lambda = 589$ nm in air. What is its wavelength in diamond? What is it in ethyl alcohol? (*Hint:* Refer to Table 21.1.)

23.6 Light from a laser with $\lambda = 633$ nm is passed through dense flint glass with an index of refraction of 1.65. What is the speed of the light in the glass? What is the wavelength of the light in the glass?

Section 23.3   Interference of Light

23.7 At what angle with respect to the optical axis will the second-order maximum lie when a pair of slits with a separation of 0.50 mm is illuminated with green light of wavelength 550 nm?

23.8 An interference pattern is produced by illuminating double slits of center-to-center separation 0.23 mm with monochromatic light. What is the wavelength of the light if the third-order maximum beam makes an angle of 0.51° with respect to the optical axis?

23.9 A pair of narrow slits is illuminated with green light of wavelength $\lambda = 546.1$ nm. The resulting interference maxima are found to be separated by 1.10 mm on a screen 1.00 m from the slits. What is the separation of the slits?

23.10 Red light of wavelength $\lambda = 632.8$ nm is incident on a pair of slits whose separation is 0.75 mm. What is the separation of neighboring interference maxima on a screen 80 cm from the slits?

23.11 Two small loudspeakers sounding in phase at 2000
• Hz are placed side by side and 20 cm apart in a large grassy field where sound is not strongly reflected from the ground. At what minimum angle, with respect to the perpendicular bisector of the line joining the two speakers, can you stand and hear no sound?

23.12 A collimated (parallel) beam of blue light of wavelength 440 nm falls normally on a pair of slits and forms an interference pattern on a distant screen. If the light source is changed, what wavelength of visible light can be used to give an interference minimum at the same place that the first-order maximum occurred for the blue light?

23.13 Monochromatic light is incident on two narrow slits whose separation is 0.500 mm. An interference pattern forms on a screen 90.0 cm away. If adjacent interference maxima are separated by 1.06 mm, what is the wavelength of the light?

23.14 A pair of narrow slits is illuminated with red light of wavelength $\lambda = 633$ nm. The slits are separated by 0.10 mm center-to-center. (a) What is the angular separation of the interference maxima near the center of the pattern? (b) How far apart are neighboring maxima if they are observed on a wall 6.55 m away from the slits?

23.15 Red light of wavelength $\lambda = 633$ nm is incident on a pair of slits whose center-to-center separation is 0.10 mm. What is the linear separation of the maxima corresponding to $m = +7$ and $m = -7$ if they are observed on a wall 5.75 m away from the slits?

23.16 Two narrow slits 0.20 mm apart are illuminated with monochromatic light. The third-order maximum lies at a certain point on a screen some distance away. What would be the separation of the slits required to give a fifth-order maximum at the same screen point if the wavelength and the distance to the screen are unchanged?

23.17 What ratio of λ/d produces a first-order interference maximum at angle $\theta = 0.01°$?

23.18 Double slits of 0.070-mm separation are placed at the front of a lecture room that is 20 m long. The interference pattern is cast on a screen at the front by reflecting the pattern from a mirror on the back wall of the room. If the light source is a helium-neon laser ($\lambda = 633$ nm), what is the separation between the first two bright fringes?

Section 23.4 Interference in Thin Films

23.19 How thick should be the coating of a material with index of refraction 1.25 on a piece of flat glass if we wish for normally incident light of wavelength 450 nm to be transmitted with maximum efficiency?

23.20 How thick should be the coating of a material with index of refraction 1.32 on a piece of flat glass if we wish normally incident light of wavelength 560 nm to be reflected with maximum efficiency?

23.21 What wavelength in the visible region will be best reflected at near normal incidence from a glass sheet that is 400 nm thick? Take the index of refraction of the glass to be 1.50.

23.22 How thick should a sheet of mica ($n = 1.58$) be if it is to be as thin as possible and still give rise to destructive interference for reflection, given that the incident light is in the blue ($\lambda = 450$ nm) part of the spectrum?

23.23 The index of refraction of ice ($n = 1.31$) is slightly less than that of water ($n = 1.33$). Is it possible to have a sheet of ice floating on water that does not reflect the green (550 nm) part of the spectrum from the sun when it is directly overhead? If so, what is the minimum thickness of the sheet?

23.24 Determine the thickness that gives the maximum transmission and the thickness that gives the minimum transmission for light of wavelength 500 nm when a material whose index of refraction is 2.00 is deposited on glass which has an index of refraction of 1.50.

23.25 If you are below the surface of a still pond on which
• a 0.001-mm thickness of oil ($n = 1.25$) is floating, what, if any, colors will be enhanced in the sunlight from directly overhead?

Section 23.5 Diffraction by a Single Slit

23.26 Calculate the width of the central maximum in the single-slit diffraction pattern of yellow light of $\lambda = 589.0$ nm by a slit 0.25 mm wide viewed on a screen 2.00 m away. (*Hint*: Determine the separation of the two first-order minima.).

23.27 A single-slit pattern is formed when light of $\lambda = 633$ nm is passed through a narrow slit. The pattern is viewed on a screen placed one meter from the slit. What is the width of the slit if the width of the central maximum is 2.53 cm?

23.28 A single-slit pattern is formed by light passing through a narrow slit 0.050 mm wide. If the width of the central maximum is 3.8 cm on a screen 1.5 m away from the slit, what is the wavelength of the light?

23.29 A single-slit pattern is formed by light passing through a narrow slit 0.050 mm wide. If the central maximum has a width of 4.7 cm on a screen 2.0 m from the slit, what is the wavelength of the light?

Section 23.6   Multiple-Slit Diffraction and Gratings

23.30 A pair of narrow slits 0.50 mm in width is used to generate an interference pattern. The fourth-order maxima of the double-slit pattern occur at the positions of the first minima in the single-slit pattern. What is the separation of the slits?

23.31 A pair of narrow slits is used to generate an interference pattern. The third-order maxima of the double-slit pattern are missing as they occur at the position of the first minima of the single-slit pattern. What is the ratio of slit separation to slit width?

23.32 Light of wavelength $\lambda = 632.8$ nm is normally incident upon a grating of 5000 lines/cm. How many different diffraction orders can be seen in transmission?

23.33 Repeat Example 23.5 for second-order diffraction.

23.34 The first-order diffracted beam of a normally incident well-collimated 600-nm light beam emerges from a grating at an angle of 5° with respect to the normal. How many lines per centimeter does the grating have?

23.35 A long row of small loudspeakers connected to
• move in phase are fed a 8000-Hz signal. A person walking parallel to the speakers and 5.0 m away hears a decrease in the intensity every 0.50 m. How far apart are the speakers?

Section 23.7   Resolution and the Rayleigh Criterion

23.36 What is the minimum angle of resolution of an eye with a pupil diameter of 2.0 mm? This diameter corresponds to a typical human eye in bright light. Take $\lambda = 550$ nm.

23.37 Calculate the minimum theoretical angle of resolution of a telescope 5.08 m (200 in.) in diameter for light of $\lambda = 550$ nm.

23.38 Two ducks are flying close together with one leading the other by 0.60 m. Their motion is transverse to a duck hunter who sees them through binoculars with 35-mm-diameter objective lenses. How far away from the hunter are the ducks if they are just resolved as a pair? Take $\lambda = 550$ nm.

23.39 Imagine an airplane passenger looking down on two automobiles side by side on the ground. At what maximum altitude will the passenger just resolve them as two if their center-to-center separation is 2.3 m? Assume that the passenger has good eyesight with pupils narrowed to 2.0 mm because of bright light. Take $\lambda = 550$ nm.

23.40 Two small street-lights 2.00 m apart are photo-
• graphed by a camera 300 m away. The camera's 50-mm-focal-length lens has an aperture of $f/1.8$. Can the two lamps be distinguished on the photograph? Take $\lambda = 550$ nm.

23.41 A camera with a 135-mm focal-length lens is to be
• used to obtain a sharp image of a dot pattern on a wall 100 m away. The dots are small compared with the spacing between them and are regularly spaced 0.33 mm apart. What is the largest f-number that can be used if the lens performance is limited by diffraction alone? Take $\lambda = 550$ nm.

Section 23.8   Dispersion

23.42 Using the data below, make a graph of the index of refraction versus wavelength for high dispersive crown glass and heavy flint glass. Compare your curves with Fig. 23.27.

	Wavelength (nm)				
Glass	434	486	589	656	768
High dispersive crown	1.546	1.533	1.527	1.520	1.517
Heavy flint	1.675	1.664	1.650	1.644	1.638

23.43 Light is normally incident on one face of a 30° flint-glass prism. Calculate the angular separation of red light ($\lambda = 650$ nm) and violet light ($\lambda = 450$ nm) emerging from the back face. Use $n_{\text{red}} = 1.644$ and $n_{\text{violet}} = 1.675$. (*Hint:* See Fig. 23.28)

Section 23.10   Polarization

23.44 Two polarizers transmit light of intensity I_m when their polarization directions are aligned. How much light is passed when they are oriented at an angle of 40° with respect to each other?

23.45 Two ideal polarizers transmit light of intensity I_m when their polarization directions are aligned. What is the angle between their directions of polarization when the transmitted intensity is 21% of I_m?

23.46 An ideal polarizer passes 50% of the incident light intensity when the incident light is unpolarized. Unpolarized light from a particular lamp has an intensity I_0 measured with a light meter. (a) What is the intensity reading on the meter when a single ideal polarizer is inserted in the beam? (b) What is the intensity reading if a second ideal polarizer is inserted in the beam with its polarization direction aligned parallel to that of the first polarizer? (c) What is the intensity when the second polarizer is rotated by 45°?

23.47 What is the angle of incidence for maximum polarization for light reflected from the surface of glass of index 1.52?

23.48 Find the Brewster angle for light reflected at a water–glass interface. Use $n = 1.52$ for the glass.

23.49 The Brewster angle for light reflected from a piece of glass is 58.9°. (a) What is the index of refraction of the glass? (b) What kind of glass is this? (*Hint:* Refer to Table 21.1.)

23.50 Find the Brewster angle for the maximum polarization for light reflecting off a water–air interface; that is, the incident light comes from within the water. Compare this result with the result of Example 23.9.

*Section 23.11   Scattering

23.51 Make a graph of the relative intensity of Rayleigh scattering in the visible region against wavelength. Use any arbitrary value for the intensity of the scattering at the violet end of the spectrum and calculate for other wavelengths on a relative basis.

Additional Problems

23.52 A pair of narrow slits is illuminated with blue light
• of wavelength $\lambda = 434$ nm. The resulting interference maxima are separated by 1.00 mm on a screen 1.00 m from the slits. What would be the separation of the maxima if the illuminating light were red light of $\lambda = 656$ nm?

23.53 The first-order maximum formed by a double slit
• illuminated with red light of $\lambda = 650$ nm lies at a given point on a screen. (a) If the slits are illuminated with light of another wavelength, is it possible for the second-order maximum of the new wavelength to lie at the same point? (b) Is this wavelength visible?

23.54 Two glass microscope slides of 9.0 cm length are
• placed in contact. A 0.50-mm-thick spacer is placed between them at one end, forming a wedge of air. When illuminated from above with monochromatic light of $\lambda = 650$ nm, an alternating pattern of bright and dark bands is seen. What is the distance between the center of one dark band and the next?

23.55 Laser light $\lambda = 633$ nm is normally incident on a
• 0.10-mm-wide slit in a thin opaque plastic sheet floating on the surface of a swimming pool. Calculate the width of the central maximum at the bottom of the pool 4.0 m below the surface.

23.56 Loudspeaker arrays are often required to spread the
•• sound out in a large angle in the horizontal plane. Such speaker arrays are frequently made up of a vertical line of individual round speakers spaced close together and wired so as to move in phase. Suppose that it is desired that the sound be spread out over an angle of 90°. Estimate the size of the individual speakers, given that the program material has an average frequency of 4000 Hz. (*Hint:* Consider the speakers to form a vertical slit.)

23.57 In a double-slit experiment the sixth-order maxima
• are missing. The slits are illuminated with green light of $\lambda = 546$ nm and observed on a screen 4.0 m away. If the separation between maxima is 4.2 mm, what is the width of the slits?

23.58 White light is normally incident on a diffraction
• grating that has 6000 lines/cm. What is the angular separation between red light ($\lambda = 650$ nm) and blue light ($\lambda = 450$ nm) in the second order?

23.59 White light is normally incident on a 4000-lines/cm
• diffraction grating. (a) Do the first- and second-order spectra overlap? (b) Do the second- and third-order spectra overlap? (*Hint:* Take the visible spectrum to range from 400 nm for violet to 700 nm for red.)

23.60 Prove that for light incident on a diffraction grating
•• at an angle ϕ from the normal to the plane of the grating (Fig. 23.40), the equation for diffraction maxima becomes $n\lambda = d(\sin \phi + \sin \theta)$, where n is an integer, λ is the wavelength of the light, and d is the grating spacing.

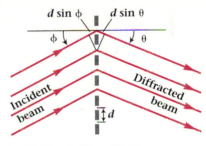

Figure 23.40.   Problem 23.60.

23.61 Light of $\lambda = 633$ nm is incident upon a transmis-
•• sion grating of 2000 lines/cm at an angle of 15° from the normal. (a) Find the angles of emergence of the two first-order diffracted beams. (b) Draw a diagram showing what occurs. (*Hint:* See Problem 23.60.)

23.62 Laser light of wavelength $\lambda = 633$ nm is incident
•• on a 5000-lines/cm diffraction grating. If the incident light makes an angle $\phi = 30°$ with the grating, what are the angles of the two first-order diffracted beams? (*Hint:* See Problem 23.60.)

23.63 A radio station broadcasting on a frequency of 1500
•• kHz generates a directional beam by using an array of four antennas that are all driven in phase. The antennas are arranged along an east-west line so that each antenna is 50 m from the next. (a) In what direction will the radiated signal be the greatest? (b) How much signal will radiate along the east-west line?

23.64 A weather satellite in stationary orbit has a camera
•• lens 75 mm in diameter. What is the minimum size object that it can resolve on the earth's surface? (*Hint:* First find the altitude of the satellite above the earth's surface. Then assume that the object diameter will subtend an angle θ_m at the camera lens. Take $\lambda = 550$ nm.)

23.65 White light is normally incident on one face of a
• 30° prism of polymethylmethacrylate. Find the approximate angular separation of red light of $\lambda = 650$ nm and blue light of $\lambda = 450$ nm as they emerge from the second face. Refer to Fig. 23.27 for the index of refraction.

23.66 If a rainbow were made by drops of liquid with an
•• index of refraction greater than that of water, say $n = 1.47$, would the bow appear higher or lower in the sky? Justify your answer and draw a diagram of the angles involved.

23.67 An amateur astronomer has a reflecting telescope of
• 20-cm (8 in.) diameter. What is the maximum distance between objects on the moon which can be resolved according to Rayleigh's criterion? (*Hint:* Take the wavelength of light to be 550 nm and the earth–moon distance to be 3.8×10^8 m.)

23.68 Two polarizers are oriented at an angle of 90° with
• each other, blocking the passage of light. When a third polarizer is added between them, light is transmitted. (a) Explain how this can occur. (b) Assuming ideal polarizers, find the angle of the third polarizer for maximum light transmission through the system.

23.69 A commercially available Polaroid sheet passes only
•• 32% of the incident unpolarized light. (a) If two of these Polaroids are aligned so that their polarization directions coincide, what percent of an incident unpolarized beam will be transmitted by the pair of polaroids? (b) What percent of the incident light will be transmitted if their polarization directions differ by 32°?

23.70 The resolution of photographic lenses is often de-
•• termined by photographing a chart with groups of parallel lines. The developed film is examined under magnification and the most closely spaced group of lines that can be resolved is noted to give the resolution of the lens in lines per millimeter. Typical resolutions for the central image of good quality 35-mm camera lenses are in the range of from 50–80 lines/mm, depending on the aperture. Because effects in addition to diffraction also limit resolution, the theoretical resolution is much better than commercial camera lenses usually achieve. The best lenses approach the "diffraction limit." Determine the diffraction limit, in lines per millimeter, of the resolution of an $f/1.8$, 50-mm focal length lens. (*Hint:* The resolution given is for the lines on the film, not at the object position. Assume a thin lens and let the object be so far away that the lens–film distance is equal to the focal length.)

ADDITIONAL READING

Babcock, H. W., "Diffraction Gratings at the Mount Wilson Observatory." *Physics Today,* July 1986, p. 34.

Nussenzveig, H. M., "The Theory of the Rainbow." *Scientific American,* April 1977, p. 116.

Walker, J., "The Amateur Scientist: Interference Patterns Made by Dust Motes on Dusty Mirrors." *Scientific American,* August 1981, p. 146.

Walker, J., "The Amateur Scientist: Music and Ammonia Vapor Excite the Color Patterns of a Soap Film." *Scientific American* , August 1987, p. 104.

Young, A. T., "Rayleigh Scattering." *Physics Today,* January 1982, p. 42.

24

Relativity

24.1	Frames of Reference
24.2	Einstein's Postulates of Special Relativity
24.3	Velocity Addition
24.4	Simultaneity
24.5	Time Dilation
24.6	Length Contraction
24.7	Mass and Energy
24.8	Relativistic Momentum
24.9	Relativistic Kinetic Energy
*24.10	The Relativistic Doppler Effect
*24.11	The Principle of Equivalence
*24.12	General Relativity

A WORD TO THE STUDENT

We present here an introduction to the topic of relativity in which we have taken particular care to include its experimental basis and confirmation. The conceptual picture of space and time that arises from relativity theory is based on only a few basic postulates. Just as you need to understand Newton's laws of motion to follow the explanations of classical mechanics, you need to pay careful attention to these basic ideas of relativity. The consequences of relativity are not usually apparent in our daily lives because we do not travel at speeds near the speed of light. However, the ideas of relativity have profound implications for our basic understanding of space and time and the structure of the universe.

Toward the end of the nineteenth century, the once separate disciplines of mechanics, optics, acoustics, thermodynamics, and electromagnetism had become interrelated through the concepts of energy and energy conservation. Physicists had developed these various subfields of physics and used them to explain many natural phenomena. In fact, physicists of that day were familiar with most of the ideas we have discussed so far in this book—ideas that came to be known as classical physics. Chief among the classical theories was Newtonian mechanics, which, more than any other science, had shaped people's view of the world. Scientists relied on mechanistic models to explain almost everything in and out of physics; thus Newton's ideas transformed society as well as physics.

Into this setting came Maxwell's extremely successful theory of electromagnetism. Its explanation of the wave nature of light contributed to the unification of physics. However, Maxwell's theory was not entirely consistent with Newtonian mechanics. Maxwell predicted that an electromagnetic wave should propagate in a vacuum with a unique speed $c = 3 \times 10^8$ m/s. This speed became a point of conflict because classical mechanics had no explanation for why light should always travel at the same characteristic speed.

In 1905 Albert Einstein devised a theory that reconciled the discrepancies between mechanics and electromagnetism. This theory of relativity challenged the Newtonian view and has affected the way people look at the world. The effects of the theory of relativity on contemporary society have been so great that "relativity" and "Albert Einstein" are among the outstanding words and images of modern science.

Frames of Reference

24.1

The concept of relative motion was introduced in Chapter 3 when we considered relative velocities. There we pointed out that before we can measure an object's velocity, we must first specify the coordinate system or reference frame within which we make our measurements. Both Galileo and Newton recognized the basic ideas of relative motion. Newton conceived of space as absolute, and he defined absolute motion as the translation of a body from one absolute place to another. However, he also recognized that all motion is necessarily measured relative to the position from which it is observed.

Newton's laws apply in any nonaccelerating frame of reference, or **inertial reference frame** (so named because in such a frame an object at rest remains at rest unless acted upon by a force). If the laws of mechanics are valid in one inertial frame, such as a railway station, they must also be valid in any other inertial frame moving with constant velocity relative to the first one, such as a train moving past. In fact, there is no way to prove that any given inertial frame is absolutely at rest. It is reasonable to say that the train is at rest and the earth is moving beneath it.

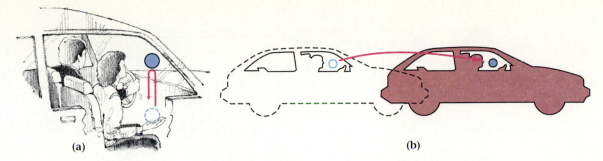

(a)   (b)

Figure 24.1   (a) Motion of a ball thrown vertically by a passenger in a car moving with constant velocity, as seen by an observer in the car. The ball's motion appears the same as if the car were not moving. (b) Motion of a ball thrown vertically by a passenger in a moving car, as seen by an outside observer at rest relative to the road. Now the ball has a forward velocity equal to that of the car.

The statement that the laws of mechanics are valid in all inertial frames is known as the principle of **Galilean relativity**. It is so named in honor of Galileo, who first discussed the motion of an object in one coordinate system as seen from another. Galilean relativity is part of our everyday experience. For example, if a passenger riding in a car moving with constant velocity pours a soft drink into a paper cup, the results are the same as if she had performed this task in her home. Similarly, if the passenger in the moving car performs a simple experiment, such as tossing a ball straight up, the ball falls straight back down, just as it would if she were standing on the ground (Fig. 24.1a). The same laws of gravity and motion apply in each inertial reference frame.

However, a person standing alongside the road occupies a different inertial frame from the car passenger. If this observer traces the path of the ball, it appears to travel in a parabolic path as a result of its horizontal velocity—the velocity of the car (Fig. 24.1b). Both observers see an event consistent with Newton's laws. To the passenger in the car, the ball has no forward velocity; to the roadside observer, the ball has a forward velocity equal to that of the car. But they both observe the same laws of mechanics and they both agree on the result—the ball is thrown up, comes back down, and is then caught.

Applying Galilean relativity to Maxwell's theory of electromagnetic waves led to contradictory results. The dominance of Newton's mechanistic ideas of motion led physicists to think that electromagnetic waves must propagate in some mechanical medium, which they called the ether. The absolute reference frame in which Maxwell's equations were valid and light would travel with speed c was thought to be a frame that was stationary or fixed with respect to the ether. Since the earth rotates about its axis and revolves about the sun, then the earth must surely be moving through the ether; if indeed the ether does exist. Galilean theory predicted that the speed of light would not be the same if viewed from different frames of reference moving with constant velocity relative to each other. Therefore the velocity of light should be different depending on whether it is measured along a direction parallel to the earth's motion through the ether or along a direction at right angles to its motion.

A crucial experiment, designed to measure how the earth's motion through the ether affected the speed of light, was first performed by the American physicist A. A. Michelson (1852–1931) in 1881. The experiment was later repeated with greater precision many more times by Michelson and E. W. Morley (1838–1923). In the Michelson-Morley experiment a beam of light was split into two parts traveling at right angles to each other (Fig. 24.2, p. 700). The beams were then reflected and recombined

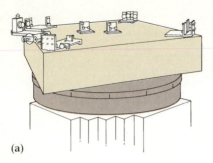

(a)

Figure 24.2   In the Michelson-Morley experiment, a light beam is divided in two by a partially silvered mirror. The two beams travel along different paths, are reflected several times, and then are recombined. The resulting interference pattern is extremely sensitive to changes in path length or wave speed in either branch. (a) The apparatus, called an interferometer. (b) Diagram of the paths of light.

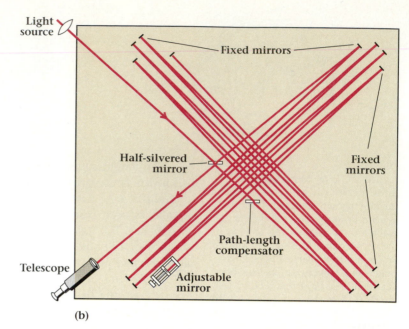

(b)

BACK TO THE FUTURE

Albert Einstein

Albert Einstein burst into the world news in 1919 in a way no other scientist ever has. That year saw a dramatic confirmation of his prediction for the amount of bending of a light ray by a gravitational field; as a result, he became *the* scientist in the public eye. Most readers did not know what the theory of relativity was. Nevertheless, its confirmation by British astronomers during a total solar eclipse in 1919, the intriguing name of the theory, and the personality of the author, all combined to create unprecedented interest in a theory with no apparent practical applications. Einstein quickly became, and still remains, one of the most widely recognized public images (Fig. B24.1). His usually rumpled appearance made him popular with photographers and his unaffected and open nature made him popular with any audience. He was so popular that

the London Palladium once tried to engage him for a three-week appearance.

As a young man, the German-born Swiss citizen spent the years 1902–1908 at the Swiss Patent Office in Bern. There he had the freedom to concentrate and work alone on basic questions, a way of working that characterized his entire career. In 1905 he published three major papers. The first was on special relativity. The second gave an explanation of Brownian motion and helped establish the reality of the molecular view of matter (Chapter 25). A third paper, about the nature of light and its interaction with matter, was a part of the foundation of quantum mechanics (Chapter 27). Later in the same year he published a fourth paper, which contained the well-known equation $E = mc^2$. Einstein went on to develop the general theory of relativity, publishing his first results in 1915. In 1921, after the confirmation of the general theory of relativity,

Figure B24.1   Albert Einstein (1879–1955), shown at his desk at the Swiss Patent Office around the time of his 1905 papers.

he received the Nobel Prize in physics for his contribution to theoretical physics and his explanation of the interaction of light with matter.

The widely told legend that young

to form an interference pattern. Changes in the interference pattern could be used to detect very minute changes in the velocity of light. Thus, by rotating the apparatus, they could tell whether the velocity of light changed with direction. Michelson's experiments gave a negative result; no changes were observed in the velocity of light. Thus there was no preferred frame of reference.

Einstein's Postulates of Special Relativity

24.2

Numerous attempts were made to reconcile the discrepancies between Galilean relativity and Maxwell's equations. They were all unsuccessful, however, until Albert Einstein proposed his special theory of relativity. Einstein's work resolved the problem raised by the Michelson-Morley experiments, although his ideas were not motivated by their experimental results. In his famous paper "On the Electrodynamics of Moving Bodies," which appeared in 1905, Einstein suggested that absolute motion had no meaning in electrodynamics and that the ether concept was superfluous.

Einstein was a poor student is untrue. He made above average marks, but did not like formal schooling, preferring to follow his own train of thought without regard for authority. He worked in his own way while at the Patent Office, his most productive period and one that he remembered as the happiest in his life.

At least as important as Einstein's inventive mind and deep knowledge of physics was his ability to concentrate on problems for as long as it took to solve them. The ideas of general relativity took eight years to develop. He so thoroughly followed his own train of thought that when he arrived at such ideas as doing away with the conventional concepts of space and time, they seemed entirely natural and unavoidable to him. Anyone with such a disposition is likely to have little regard for authority and convention, either in science or in any other area.

Einstein is invariably compared with Newton with respect to the importance of their theories and the intensity with which they worked on problems. But they had little else in common. We earlier mentioned Newton's pride, intellectual vanity, and combative personality. Einstein was just the opposite. Quotations from two scientists who knew him convey something of his character. According to Otto Frisch, "The quality that dominated his personality was a very great and genuine modesty. When anybody contradicted him he thought it over and if he found he was wrong he was delighted because he felt that he had escaped from an error and that now he knew better than before."

Robert Oppenheimer wrote,*

Einstein is also, and I think rightly, known as a man of great good will

*From Robert Oppenheimer, *Science and Synthesis* (Paris: Unesco Publications, 1967), quoted with permission.

and humanity. Indeed, if I had to think of a single word for his attitude toward human problems, I would pick the Sanscrit word *Ahinsa,* not to hurt, harmlessness. He had a deep distrust of power; he did not have that convenient and natural converse with statesmen and men of power that was quite appropriate to Rutherford and to Bohr, perhaps the two physicists of this century who most nearly rivalled him in eminence. . . .

After what you have heard, I need not say how luminous was his intelligence. He was almost wholly without sophistication and wholly without worldliness. I think that in England people would have said that he did not have much 'background', and in America that he lacked 'education'. This may throw some light on how these words are used. . . . There was always with him a wonderful purity at once childlike and profoundly stubborn.

His **special theory of relativity** was based on two simple postulates:

I.   **The laws of physics are the same in all inertial reference frames.**
II. **The speed of light in free space is independent of the motion of its source and of the motion of the observer.**

From these two postulates he formed a theory that brought mechanics and electromagnetism together, but that demanded changes in the formulation of mechanics for speeds approaching the speed of light. Special relativity is no longer considered a theory to be tested, but, like Newton's laws or Maxwell's equations, it is a physicist's everyday working tool in its appropriate domain.

The simplicity of the postulates is perhaps a little surprising, given the reputation of special relativity for producing unusual predictions. The first postulate appears to be a reasonable requirement of any theory. After all, it is natural to expect that the laws of nature do not change according to the situation. In particular, the laws of nature should not depend on relative constant motion. Perhaps the second postulate is not so intuitively apparent, but it is possible to test this statement, and we find that it does agree with experiments.

What was so unexpected was that two simple and reasonable postulates, when followed to their logical conclusion, gave such unexpected results, but results that are nevertheless confirmed by experiments. As we shall see, these results revolutionized our conceptions of time and space. Even the seemingly simple matter of how we add velocities must be reconsidered in light of these postulates.

Velocity Addition

24.3

When you run forward and throw a ball in the direction in which you are running, the ball moves faster than it would if you were standing still. For example, if you can throw a ball 30 m/s when standing still, the same throw when you are running at 3 m/s gives the ball a speed of 33 m/s. You can summarize your ordinary experience in adding velocities as

$$u = u' + v,$$

where v is your velocity with respect to the ground, u' is the velocity of the ball with respect to you, and u is the velocity of the ball with respect to the ground. This expression is called the Galilean, or classical, velocity addition formula.

Though the Galilean velocity addition formula seems to agree with our everyday experience, it violates the second postulate of special relativity, which, as we have already indicated, agrees with experimental observations. This difficulty is resolved by use of the correct velocity addition formula, which was first given by Einstein. It is

$$u = \frac{u' + v}{1 + \dfrac{u'v}{c^2}}, \tag{24.1}$$

where the symbols u, u', and v have the same meaning as before, and c is the velocity of light. Equation (24.1) is in agreement with both our everyday experience and the second postulate of relativity. When the velocities involved are small compared with the velocity of light, the relativistic velocity addition rule gives the same result that would be expected from classical mechanics. When the velocities are not small compared with the speed of light, the results of the velocity addition rule are no longer so intuitive.

Note that the velocities in Eq. (24.1) all lie in the same direction. The formula can also be used when one velocity is directed opposite to the other, if you choose one direction positive and the other negative. An appropriate generalization of Eq. (24.1) is used for motion that is not colinear.

We find that we must modify our ideas of velocity addition to bring them into agreement with the postulates of special relativity. Because velocity is the displacement divided by the time, we need to reevaluate our concepts of both time and space in light of the postulates of special relativity. Before looking more closely at space and time, we first need to ask a question that does not even occur in classical mechanics, "What do we really mean when we say that things happen simultaneously?"

Example 24.1

Speed of light from a moving source.

Is the velocity addition rule consistent with Einstein's postulate that the speed of light is the same to all observers?

Solution Suppose one observer moves with velocity v relative to another observer. The first observer shines a light beam directly ahead of him at a velocity c measured in his frame. The second observer measures the speed of the light beam as

$$u = \frac{c + v}{1 + \dfrac{cv}{c^2}} = \frac{c + v}{1 + \dfrac{v}{c}} = \frac{c(c + v)}{c + v} = c.$$

Thus the velocity addition rule does give us the desired result, as indeed it must since it was derived from Einstein's postulates.

Example 24.2

Relativistic addition of velocities.

A spaceship is moving away from the earth at a speed of $0.5c$ when it launches a satellite in the same direction with a speed of $0.5c$, as seen from the ship. What is the satellite's speed as seen from earth?

Solution Straightforward application of Eq. (24.1) with $u' = v = 0.5c$ gives

$$u = \frac{0.5c + 0.5c}{1 + \dfrac{(0.5c)(0.5c)}{c^2}} = \frac{c}{1 + 0.25} = 0.8c.$$

The satellite moves away from the earth with a speed of $0.8c$ as measured by an observer on earth—less than the result predicted by the classical addition of relative velocities.

Example 24.3

A spaceship exploring a strange planet moves away from the planet with a speed of $0.48c$ and fires a probe back toward the planet at a speed of $0.63c$ as seen from the spaceship (Fig. 24.3). What is the speed of the probe as seen from the planet?

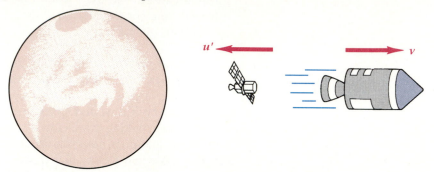

Figure 24.3 Example 24.3: A space ship, moving away at speed $0.48c$ from a strange planet, fires a probe back toward the planet at speed $0.63c$. Seen from the planet, the probe's speed is greater than the difference between these two speeds.

Solution   Again we apply the relativistic velocity addition formula, choosing the direction away from the planet as positive. Thus the velocity of the spaceship is $v = 0.48c$ and the velocity of the probe measured in the frame of the spaceship is $u' = -0.63c$. The velocity of the probe as seen from the planet is

$$u = \frac{u' + v}{1 + \dfrac{u'v}{c^2}} = \frac{-0.63c + 0.48c}{1 + \dfrac{(-0.63c)(0.48c)}{c^2}} = \frac{-0.15c}{1 - 0.30} = -0.22c.$$

To an observer on the planet, the probe approaches at a speed of $0.22c$.

Simultaneity

24.4

Relativity theory—whether special relativity or classical Galilean relativity—compares measurements made by two observers in relative motion to each other. Without any loss of generality, we can pick one observer to be stationary and the other to be in motion with constant velocity relative to the first observer. For example, suppose the "stationary" observer is on the ground and the "moving" observer is on a train. To the observer seated in the railway car, the train seems at rest and the world seems to be rushing past. You may have noticed the same thing if you've ever sat at a train window as another train opposite you starts to move. It is often difficult to know at first if you are moving and the other train is stationary, or if you are at rest and the other train is moving.

We can describe this situation by using inertial reference frames (Fig. 24.4) and showing the coordinates of the outside stationary observer and

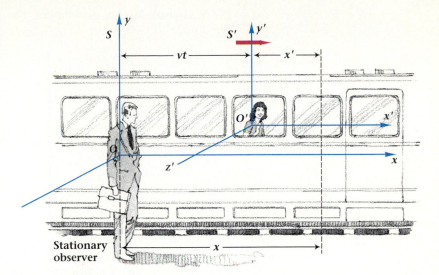

Figure 24.4   A stationary observer on the ground watches a traveler passing by on a train. The passenger's frame of reference S' moves with constant speed v along the positive x direction relative to the observer's stationary frame S.

the passenger on the train. Reference frame S corresponds to the observer on the ground; frame S' corresponds to the train passenger, moving with velocity v in the x direction. The distance x' represents a distance measured by the passenger within the train. For example, it may be the distance from the passenger to the empty seat in front of her. The distance x represents the x component of the distance from the outside observer to the empty seat, and is measured from the outside observer's position on the ground. The observer sees this distance to be increasing with time as the train moves away with speed v. We assume intuitively that time is measured the same on the train as it is on the ground. We have, therefore, a set of **transformation equations**, relating positions and times in the two systems. From the viewpoint of the stationary observer, these equations are

$$\left.\begin{array}{l} x' = x - vt, \\ y' = y, \\ z' = z, \\ t' = t. \end{array}\right\} \quad \text{(Galilean transformations)} \qquad (24.2)$$

At any one moment, an observer in the reference frame S can calculate x' in the train by measuring x and t in his own reference frame, so long as v is known. Similarly, an observer in S' can calculate x from the seat to the outside observer by measuring x' and t' in his own frame and calculating $x = x' + vt'$.

However, although transformation equations of this kind appeal to our common sense, they do not allow observers in the two systems to get the same value for the speed of light. The equations must be modified because they are not in agreement with the second postulate of relativity.

Using his two postulates of special relativity, Einstein derived the correct form of the transformation equations. They are known as the *Lorentz transformation equations* because they were first developed in 1904 by the Dutch physicist H. A. Lorentz (1853–1928) in an attempt to explain the Michelson-Morley experiment. For motion along the x direction, they are:

$$\left.\begin{array}{l} x' = \dfrac{x - vt}{\sqrt{1 - \dfrac{v^2}{c^2}}}, \\[2em] y' = y, \\ z' = z, \\[1em] t' = \dfrac{t - \dfrac{vx}{c^2}}{\sqrt{1 - \dfrac{v^2}{c^2}}}, \end{array}\right\} \quad \text{(Lorentz transformations)} \qquad (24.3)$$

where v is the velocity of S' relative to S, and c is the speed of light.

When you are in motion relative to some observer, your measurements of length and time are different from that observer's measurements, and they depend on the ratio of your relative speed to the speed of light. The Lorentz equations apply to situations similar to Fig. 24.4. An observer in the frame S uses a meter-stick (or other rule) to measure distances x, y, and z from the origin of the coordinate system and measures time t at the origin with a clock. In the same way, an observer moving with the frame S' measures distances x', y', and z' with an identical meter-stick and measures time t' with an identical clock. As we will see, their different measurements are real and are due solely to their relative motion.

To compare time measurements in two reference frames moving relative to each other, we need to establish a common reference between them. That is, we need to synchronize the clocks at some particular time. How can we do this? First, consider two observers in frame S (Fig. 24.5), measuring time with separate but identical clocks. Imagine one observer to be at the origin and the other at a distant point x. They can synchronize their clocks in the following way. A light pulse flashed by the observer at the origin at time $t = 0$ requires a time interval x/c to reach the second observer at x. When the observer sees the light flash, he sets his clock to time $t = x/c$. Thus the two clocks are synchronized. Other observers in S can set their clocks in a similar manner.

Observers in the frame S' can also measure time with identical clocks. Using the same procedure, they can synchronize their clocks with a clock at the origin of S'. Thus a clock at any point x' is set to $t' = x'/c$ when a light pulse emitted from the origin of S' at $t' = 0$ reaches the point x'.

At the instant when the origins of the two reference frames coincide, we can synchronize the clocks at the origins of both reference frames. Then the Lorentz relations tell us that when an observer in S records an event occurring at point x, y, z and time t, then an observer in frame S' records the same event at a position measured as x', y', z' and time t'.

Notice that according to Eq. (24.3), time measured in frame S' depends not only on the time t of a clock in frame S, but also on the *position* of that clock. The greater the separation of the clock in S from the origin of S, the greater the disagreement between t and t'.

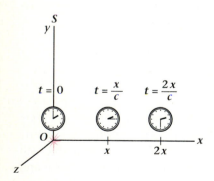

Figure 24.5 A clock at a distance x from the origin may be synchronized to a clock at the origin. When a light pulse sent from the origin at time $t = 0$ is received at x, the clock at x is set to $t = x/c$. Other clocks may be set in the same manner, where the time is related to their position divided by the speed of light.

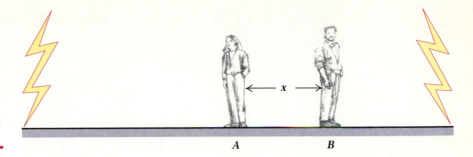

Figure 24.6 If an observer at *A* sees two lightning flashes to be simultaneous, the observer at *B* does not see them as simultaneous.

Through his insistence on the constancy of the speed of light as a necessary part of a correct relativistic relationship, Einstein revealed an unexpected phenomenon. Time itself is relative. This was a startling revelation to people accustomed to Newton's intuitive view that time is absolute and completely separate from space. In his *Principia*, Newton had written that "absolute, true and mathematical time, of itself, and from its own nature flows evenly without regard to anything external." But time and space are inseparably linked.

The fact that an observer's measurements of the space and time intervals between two events depend on the motion of the observer's reference frame forces us to reexamine our notions of simultaneity. For example, consider an observer at rest in a coordinate system *S*. He perceives two events—say two lightning flashes—as being **simultaneous** if light waves from the events reach his position at the same time (Fig. 24.6). However, since the transit time of a light signal depends on the distance from source to observer, these same two events will not appear simultaneous to a second observer stationed a distance *x* away from the first. Thus two clocks synchronized in the manner described above appear to show the same time only to an observer located equidistant between them. Everyone else sees the two clocks showing different times. (Of course, in most everyday situations, these differences are extremely small and go unnoticed.)

Now let's consider two observers in relative motion. Suppose an observer in stationary reference frame *S* sees two clocks to be synchronized in his reference frame. An observer in moving reference frame *S'* will not see these same two clocks as synchronized. Figure 24.7 (p. 708) illustrates this loss of simultaneity by a thought experiment due to Einstein. We suppose that two lightning bolts strike opposite ends of a boxcar that is moving to the right with velocity *v* relative to an observer *O* standing near the tracks halfway between *A* and *B*. A passenger *O'* riding in the boxcar observes the same two bolts. However, since the observer *O'* is moving toward the right, the light pulse from the right reaches him before the light pulse from the left. He concludes that lightning struck the front of the boxcar first. However, light pulses from left and right reach observer *O* at the same time. He concludes that the two events are simultaneous.

Thus two events that one observer sees as simultaneous may be sequential to someone else, and may perhaps even be in reverse sequence to a third viewer. However, causal relations are always maintained. That is, if event *A* causes event *B* to happen, no observer, regardless of his

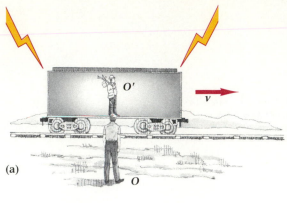

(a)

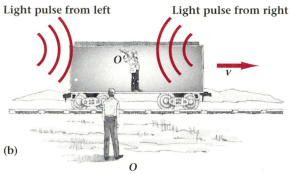

Light pulse from left   **Light pulse from right**

(b)

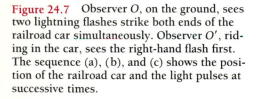

Figure 24.7   Observer O, on the ground, sees two lightning flashes strike both ends of the railroad car simultaneously. Observer O', riding in the car, sees the right-hand flash first. The sequence (a), (b), and (c) shows the position of the railroad car and the light pulses at successive times.

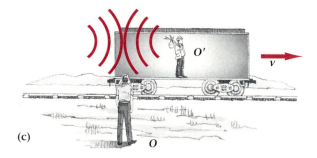

(c)

motion, will ever be able to see these two events occur in the reverse order of B before A. No observer can ever see an effect precede the event that causes it to take place.

Time Dilation

24.5

One of the most intriguing effects of special relativity is the phenomenon of **time dilation**. Because of this effect, a mechanical (or biological) clock in an inertial reference frame in motion relative to you runs more slowly than an identical clock at rest in your reference frame. As Einstein himself observed, when we move we not only change our position in space, we also change the rate at which we advance into the future. This observation is in disagreement with the Newtonian concept of absolute, evenly flowing time.

We can derive the mathematical expression of time dilation from the following thought experiment. Imagine a spaceship moving to the right with a constant speed v. A flashlamp emits a light pulse that travels upward until it hits a mirror fixed to the ceiling of the spaceship a distance d above the lamp. The mirror reflects the pulse back down to a light detector mounted beside the lamp. An observer in the moving reference frame S', at rest with respect to the spaceship, uses an electronic clock to measure the time interval Δt_0 for the light pulse to travel from the lamp to the mirror and back to the detector (Fig. 24.8a). This time interval is given by the path length traveled by the pulse divided by the speed of light,

$$\Delta t_0 = \frac{2d}{c}.$$

However, since the spaceship moves with a speed v relative to the reference frame S, an observer in S sees the light pulse travel a longer path than $2d$ in the time Δt measured in his frame (Fig. 24.8b). Because the speed of light is the same in both reference frames (Postulate I), the time interval measured in the frame S must be longer than the time interval measured in the moving frame. It is given by

$$\Delta t = \frac{2L}{c},$$

where $2L$ is the total path traveled by the light. Applying the Pythagorean theorem to Fig. 24.8(b), we obtain

$$\left(\frac{c\,\Delta t}{2}\right)^2 = d^2 + \left(\frac{v\,\Delta t}{2}\right)^2.$$

Since $2d/c$ is Δt_0 in the reference frame S', we get

$$(\Delta t)^2 = (\Delta t_0)^2 + \frac{v^2}{c^2}(\Delta t)^2.$$

Upon rearranging, we find that

$$\Delta t = \frac{\Delta t_0}{\sqrt{1 - \dfrac{v^2}{c^2}}}. \qquad (24.4)$$

Figure 24.8 (a) The time of flight of a light pulse as measured with a clock stationary in the spaceship (frame S') is $2d/c$. (b) An observer in S who sees the spaceship moving with speed v sees the light beam travel a longer path and thus measures a longer time of flight.

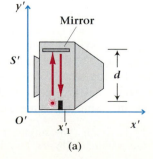

(a)

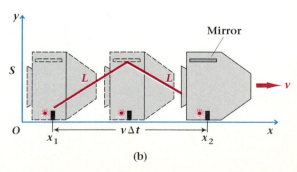

(b)

Thus an observer moving with respect to the clock sees the time interval indicated by the clock stretched out. For this reason we call this effect time dilation.

The time interval Δt_0 measured in the reference frame in which the clock is as rest is called the **proper time**. The time interval measured in any frame moving with speed v relative to the clock is greater than the proper time according to Eq. (24.4).

Example 24.4

Time dilation.

The starship *Enterprise* makes a long trip into space, traveling at a speed $v = 0.80c$ relative to the earth. The trip takes 20 years, according to a shipboard clock. How many years elapse on earth during the voyage?

Solution   We use Eq. (24.4) to determine the elapsed time. The proper time Δt_0 of the shipboard clock is 20 years. To an earthbound observer, the elapsed time is

$$\Delta t = \frac{\Delta t_0}{\sqrt{1 - \dfrac{v^2}{c^2}}} = \frac{20 \text{ years}}{\sqrt{1 - (0.80)^2}} = 33 \text{ years}.$$

The first measurement of time dilation for radioactive particles moving at high speed relative to the earth was made in 1941 by Bruno Rossi and D. B. Hall. We will describe a similar measurement made in 1963 by D. H. Frisch and J. H. Smith.* The experiment involves the detection of muons, which are subatomic particles that are produced in the upper atmosphere and that rain down toward the earth with a speed close to the speed of light. As they travel downward, some of them spontaneously decay in flight. Consequently, the number arriving at a medium altitude— say on top of a mountain—is greater than the number that survive to reach sea level. In their experiment, Frisch and Smith counted the number of muons having a narrow range of speeds near the peak of Mount Washington, New Hampshire. Then they went down to Cambridge, Massachusetts, and counted the muons that survived the trip down to sea level. The probability that a muon will decay, and thus its mean life, is determined only by forces within the muon itself. Therefore any dependence of their mean life on their speed relative to us is due only to relativity.

Let us examine their findings in detail. The Mount Washington apparatus was set up to detect and stop muons with speeds of $0.995c$. The time intervals between the arrival of a muon in the detecting apparatus and its subsequent decay were measured. Measurements from one run (Fig. 24.9) corresponded to a mean life of 2.2 μs. The average number of muons per hour arriving at this detector was 563. The time required for the muons to travel the 1907 m from the elevation of the mountain top to the laboratory in Cambridge, measured with a clock stationary with

*D. H. Frisch, and J. H. Smith, "Measurement of the Relativistic Time Dilation Using μ-Mesons." *American Journal of Physics*, May 1963, p. 342.

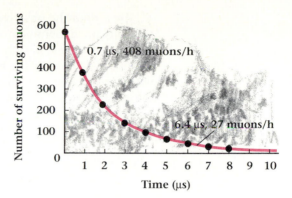

Figure 24.9 Surviving muons as a function of time. (From Frisch and Smith, with permission.)

respect to the laboratory, is

$$\Delta t = \frac{d}{v} = \frac{1907 \text{ m}}{0.995 \times 3 \times 10^8 \text{ m/s}} = 6.4 \ \mu s.$$

From Fig. 24.9 we see that after 6.4 μs elapse, only about 27 muons per hour would be expected to survive the trip down. But when detectors were set up in Cambridge to count muons of initial speed $0.995c$, they observed an average rate of 408 muons per hour. We conclude that the muons decay more slowly when they are moving rapidly relative to us than when they are at rest relative to us.

PHYSICS IN PRACTICE

The Twin Paradox

A fascinating aspect of the theory of relativity is what is often called the *twin paradox*. The problem is quite simple. Imagine a pair of identical twins, one a scientist and the other an astronaut. The astronaut departs from earth in a rocket ship to visit a nearby star. He travels at a speed of $0.99c$ relative to the earth on a round trip that lasts 35 years, according to clocks on earth.

Time passes more slowly for the moving space traveler than for his earthbound brother, who occupies a different reference frame. When the astronaut returns, he will have aged less than the scientist who stayed home:

$$\Delta t_0 = \Delta t \sqrt{1 - v^2/c^2} \approx 0.14 \ \Delta t.$$

Here Δt_0 is the proper time interval of the astronaut and Δt is the time interval of the brother on earth. The age difference between the twins depends on the duration of the trip and the speed of the spaceship. In this case, the traveler ages 5 years while his earthbound brother ages 35 years.

If we view everything from the reference frame of the astronaut, he perceives his twin brother on earth moving at $0.99c$ relative to the rocket ship. Therefore the scientist should be the one who ages more slowly. This presents the paradox: Each twin thinks he is 30 years older than his brother.

We can resolve this dilemma by recalling the postulates on which the theory is based. The first postulate states that the laws of physics are the same in all inertial frames of reference. But in the case of the twins, the two frames of reference are clearly not equivalent. The astronaut twin does not travel at constant velocity—first he must be accelerated to enormous speed, then be accelerated again as he turns around at the star, and then be accelerated a third time as he returns to earth. Thus the astronaut's frame of reference is not an inertial frame and the two views of time are not equivalent. The general theory of relativity (Section 24.12) is needed to properly treat the astronaut's motion, but the result is the same as predicted by the earthbound scientist.

The effect of time dilation is very real. The traveling twin really would return only 5 years older than when he left, while the earthbound brother would have aged 35 years. However, neither twin would have sensed anything unusual about the passage of time during those years. Everything would have seemed normal to each. Only when they came back together to the same inertial frame would the differences become apparent.

How does the observed count of 408 muons per hour correspond to relativity predictions? On the graph we see that 408 muons/h corresponds to an elapsed time of only 0.7 μs. That means that time runs slow in the frame of reference of the moving muons by a factor of 0.7/6.4 compared with time measured in the laboratory. We can check these results against the prediction of Eq. (24.4) by rearranging that equation for v and substituting for $\Delta t_0/\Delta t$:

$$v = c\sqrt{1 - (\Delta t_0/\Delta t)^2} = c\sqrt{1 - (0.7/6.4)^2} = 0.994c.$$

This calculated value for the speed of the muon in the laboratory frame is consistent with the value previously established for the experiment.

This experiment confirms time dilation and supports the special theory of relativity. Furthermore, it shows that relativistic effects can become large when the relative speeds approach the speed of light. The effects observed here are general effects of relativity and are not limited to the special case of radioactive decay.

Example 24.5

Time dilation for moving muons.

The mean life of muons at rest is 2.2 μs. What is the mean life of a beam of muons moving with a speed of 0.95c?

Solution   We use the time dilation equation with the proper time $\Delta t_0 = 2.2$ μs and $v/c = 0.95$. The result is

$$\Delta t = \frac{2.2 \times 10^{-6}\text{ s}}{\sqrt{(1 - (0.95)^2)}} = 7.0 \times 10^{-6}\text{ s} = 7.0 \ \mu\text{s}.$$

The apparent lifetime of the muons increases by about a factor of 3 over their proper mean life.

Length Contraction

24.6

We have seen that observers at rest relative to one another occupy the same reference frame and thus agree on their measurements of space and time. However, if they are moving relative to one another, their measurements of space and time do not agree. These differences are significant only at speeds approaching the speed of light, but though negligible, they exist at everyday speeds as well. With these concepts in mind, let's examine in detail the problems of measuring the length of a moving object.

Consider an observer in a frame S' moving with speed v relative to frame S (Fig. 24.10). This observer is at rest with respect to a rod moving with the system. The observer in frame S' measures the rod to have a length l_0 along the direction of motion, $l_0 = x_2' - x_1'$. This is called its **proper length**: the length measured by the observer at rest with respect to the rod. The observer also sees a marker at point P in frame S approach and then recede at speed v, covering a distance equal to the length of the rod in a time $\Delta t = l_0/v$. This time interval Δt is not a proper time, for

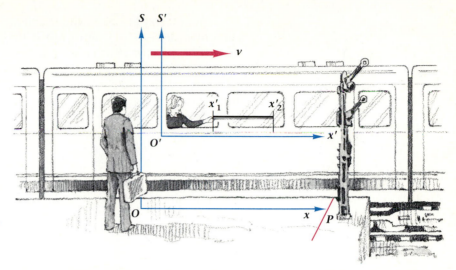

Figure 24.10 A rod at rest in system S' moves with speed v relative to the frame S. The length of the rod as measured in S' is l_0, the proper length of the rod.

the events that define it occur at different places (the marker passing first one end of the rod and then the other). Moreover, the observer must use synchronized clocks to measure Δt.

PHYSICS IN PRACTICE

The Appearance of Moving Objects

We have talked rather freely about the effects of time dilation and length contraction associated with clocks and other objects moving at great speed relative to an observer. In describing the muon experiment we explicitly calculated the effects of time dilation. However, observing systems traveling at speeds near c is no simple matter.

What would we see if we could look at an object passing by at a relativistic speed? For more than 50 years the accepted notion was that you could readily see or photograph the length contraction of a rapidly moving object. For example, it was thought that a passenger in a high-speed rocket seeing another rocket passing in the opposite direction would see the rocket shortened in the direction of motion. But in 1959 J. Terrell critically analyzed the situation and showed that length contraction would not be directly observed

with the eye. Because of the finite speed of light, the light coming from different parts of the rocket does not strike the eye at the same time. Therefore, as the rocket approaches, some of the light coming from the front of the rocket reaches the eye at the same time

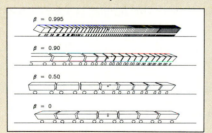

Figure B24.2 Views of a train of boxcars (each of 1 × 1 × 3 units) by an observer 10 units away at an angle 30° above the plane of the tracks. The axis of the observer's camera is directed at the center of the middle car, shown by a cross (+). The dot (·) marks the center of the near edge of the middle car in each case. (After Scott and van Driel, with permission.)

as light that left other parts of the rocket slightly earlier. The view of the rocket becomes distorted. Its exact appearance depends on the angle between the line of sight and the direction of motion, as well as on its speed. An interesting account of this effect has been published by G. D. Scott and H. J. van Driel in the *American Journal of Physics*, August 1970, p. 971 (Fig. B24.2).

Suppose we could devise a camera with a very-large-diameter lens and a properly designed shutter. If the camera were large enough—that is, if it ran the length of the rocket—and if it received only light coming perpendicularly to the path of the passing rocket, then we could photograph the Lorentz contraction.

All of this serves as a warning to be careful in the interpretation of relativistic effects and the equations that describe them. The effects of relative motion on length and time are real enough, but how they are observed is not always obvious.

An observer at rest in the frame S sees the rod move past the marker in a time $\Delta t_0 = l/v$, where l is the length of the rod as observed from S. The time measurement is made at one point with a clock at rest, so it is a proper time.

The length measured in frame S is $l = v\,\Delta t_0$, and the length measured in frame S' is $l_0 = v\,\Delta t$. These two measurements of length may be compared. The result is

$$\frac{l}{l_0} = \frac{v\,\Delta t_0}{v\,\Delta t} = \sqrt{1 - v^2/c^2},$$

where we have used Eq. (24.4) for $\Delta t_0/\Delta t$.

The observer in motion with speed v relative to the rod sees a length l given by

$$l = l_0\,\sqrt{1 - v^2/c^2}. \qquad (24.5)$$

That is, this observer measures the rod to be shorter than does the observer at rest with the rod. This effect is known as **length contraction**. It is also called Lorentz contraction.

The earlier discussion of the muon decay experiment can be reformulated in terms of observations from the muon's rest frame. From that point of view the height of the mountain appears shortened (by length contraction) so that the time to travel from top to bottom is reduced. When the calculations are made we get the same result for the number of muons reaching sea level that we found before. (See Problem 24.19.)

Example 24.6

A spaceship with a proper length of 100 m is moving with a speed $v = 0.6c$ relative to an earthbound observer. What is the apparent length of the spaceship as determined by the observer on earth?

Solution   The length is found from Eq. (24.5), where $l_0 = 100$ m and $v/c = 0.60$:

$$l = l_0\,\sqrt{1 - v^2/c^2} = 100\ \text{m}\,\sqrt{1 - (0.60)^2} = 80\ \text{m}.$$

Mass and Energy

24.7

Einstein showed that the basic physical quantities of time, distance, and velocity are not absolute, as previously thought, but are all relative to the motion of the observer. What, then, can we say about the hallmarks of classical physics, the laws of conservation of energy and momentum? We shall see that in order to maintain their validity, we have to modify our understanding of mass, energy, and momentum. First we will discuss the relationship Einstein derived between mass and energy; then, in the next sections, we will show how the postulates of special relativity affect our understanding of momentum and energy at high speeds.

One result of Maxwell's electromagnetic theory is the prediction that

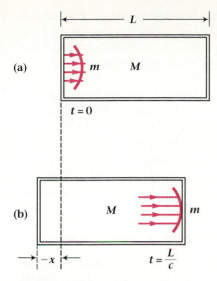

Figure 24.11 (a) At time $t = 0$ the box at rest emits a flash of light. (b) At time $t \approx L/c$ the light strikes the other end of the box and is absorbed. During this time interval the box moves a distance $-x$.

when light strikes an object, it exerts a pressure. The presence of this pressure implies that light waves carry momentum. Moreover, the theory requires that this momentum be proportional to the energy carried by the light wave. Experiments support the theory, having verified this effect in several ways. The result is that a wave carrying an amount of energy E has a momentum given by

$$p = \frac{E}{c},$$

where c is the speed of light. We shall use this result and the principle of conservation of momentum to derive a relation between mass and energy.

Let us consider a thought experiment due to Einstein. Imagine a closed box of length L and mass M. The box is initially at rest with respect to us, but is suspended so that it is completely free to move without friction. Now suppose that at some instant of time, a bank of flashbulbs at one end of the box emits a flash of light (Fig. 24.11a). Since the light carries momentum, the box must recoil with an equal but opposite momentum, of magnitude

$$p_{\text{box}} = Mv = p_{\text{light}} = \frac{E}{c},$$

where v is the recoil velocity of the box and E is the energy carried by the flash of light.

The light requires a time t to reach the other end of the box, where it is completely absorbed. During this time t, given approximately by $t \approx L/c$, the box moves a distance $-x$ as it recoils (Fig. 24.11b). Because the speed of light is so great, the box moves only a very tiny distance while the light travels the length of the box. When the light is absorbed at the other end of the box, the system returns to rest, in agreement with conservation of momentum.

Since no outside forces act on the box, its center of mass must not move during this event, even though the position of the box obviously changes. The only way this can happen is for a shift of mass to occur within the box as it moves, maintaining the position of the center of mass. Consequently there must be an equivalent mass m associated with the light, of such magnitude that while the box of mass M moves a distance $-x$, its movement about the center of mass is compensated by the light moving a distance L:

$$mL = Mx.$$

Since we know the momentum of the box, and thus its velocity, we can use this information along with the time of flight of the light to evaluate the distance the box moves:

$$x = vt = \left(\frac{E}{Mc}\right)\left(\frac{L}{c}\right).$$

Using this value for x in the previous equation and solving for the energy E, we get

$$E = mc^2. \tag{24.6}$$

This result is Einstein's famous **mass–energy relation**. For the example at hand, it says that when radiation is emitted by one end of the box, the box loses an amount of mass given by E/c^2. Similarly, when the light is absorbed at the other end, the box must gain an amount of mass given by E/c^2.

The significance of the mass–energy relation lies in its generalization of the law of conservation of energy to include mass. It tells us that any change in the energy of an object necessarily involves a change in the object's mass. Conversely, a change in an object's mass is accompanied by the absorption or emission of energy. The most dramatic example of energy released by a change in an object's mass is the release of energy in nuclear reactions, especially in fission and fusion. These topics will be discussed in detail in Chapter 28.

Example 24.7

A free neutron is unstable and spontaneously decays into a proton and an electron. The mass of the neutron is $m_n = 1.67493 \times 10^{-27}$ kg, the mass of the proton is $m_p = 1.67262 \times 10^{-27}$ kg, and the mass of the electron is $m_e = 9.1094 \times 10^{-31}$ kg. How much energy is released when the neutron decays?

Solution   The sum of the masses of the electron and the proton is less than the mass of the neutron. The difference corresponds to the energy released:

$$\Delta E = \Delta m c^2 = (m_n - m_p - m_e)c^2$$
$$\Delta E = (1.40 \times 10^{-30} \text{ kg})(3.00 \times 10^8 \text{ m/s})^2,$$
$$\Delta E = 1.26 \times 10^{-13} \text{ J}.$$

The mass that disappears in the decay of the neutron is converted to kinetic energy. Some of this kinetic energy goes into the resulting motion of the proton and electron; the rest is carried off by a massless particle called the neutrino (Chapter 28).

Relativistic Momentum

24.8

One of the most basic physical laws is the law of conservation of momentum. It is fundamental to Newtonian mechanics and is consistent with our observations. Is it also valid from the relativistic point of view?

To answer this question, let us imagine an elastic collision involving two identical balls. When measured in a frame in which the two balls are initially at rest, their masses are identical. Suppose the two balls move toward each other with sufficient speed so that relativistic effects will be obvious. Now, let the two balls collide with a glancing blow that sends each ball recoiling to some extent perpendicular to the direction of their initial relative motion. (In such a collision neither mass receives any appreciable velocity in its own reference frame along the direction of relative motion.) We can describe the collision in each of two frames in which one of the balls is at rest.

To an observer in the frame S, containing ball B, ball A at rest in the moving frame S' approaches from the negative x direction with a speed v (Fig. 24.12a). After the collision takes place, ball B in the frame S has a small velocity in the positive z direction (Fig. 24.12b). As viewed from S,

Figure 24.12   (a) Ball A in frame S' approaches ball B in S at a relative speed v along the x direction. The two balls are situated on opposite sides of the x axis so that the collision is a glancing one. (b) From the frame S, the ball B moves along the positive z axis after the collision, while ball A is deflected by a small angle. (c) From the viewpoint of frame S', ball B is initially moving and ball A is stationary. (d) In frame S', after the collision, the ball A moves along the negative z' axis while ball B moves off with a small component of velocity along the positive z' axis.

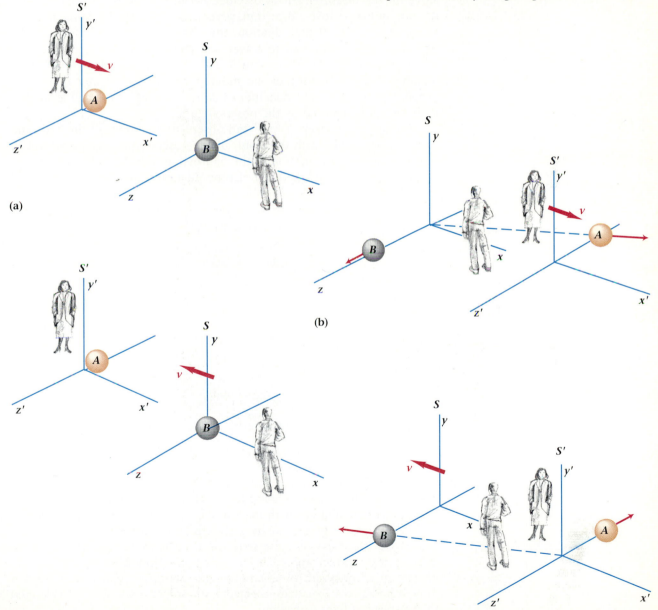

the moving ball A is deflected by a very small angle so that it has a small component of velocity in the negative z direction.

An observer in the frame S' sees the same collision differently. From her viewpoint ball A, at rest in S', is struck by ball B moving fast in the negative x' direction (Fig. 24.12c). After the glancing collision, ball A recoils slowly along the negative z' axis, while ball B speeds off in the negative x' direction, with a small component of velocity in the positive z' direction (Fig. 24.12d). Both observers determine the velocity of the mass in their reference frame and report the result to the other.

Suppose both observers watch each other make their measurements of time and distance traveled by the two balls after the collision. They will agree on the measurement of distance because there is no Lorentz contraction in the positive z direction, perpendicular to their relative motion. However, because of time dilation, the observer in S concludes that the time required for the ball to travel one meter in S' is longer than the value reported by the observer in S' by a factor of $1/\sqrt{1 - v^2/c^2}$. (Remember that the lifetime of the moving muons was greater than their proper lifetime.) Thus he concludes that the velocity of the ball is smaller than the value reported by the observer in S'.

Similarly, the observer in S' can observe the motions of the balls and compare her results with the result of the observer in S. The observer in S' concludes that the motion of the ball in S as seen by the observer in S is improperly reported, because of time dilation. Thus the observer in S' concludes that the transverse velocity of ball B is too slow.

The result is that the law of conservation of momentum holds for both observers only if we redefine momentum to be

$$p = \frac{m_0 v}{\sqrt{1 - v^2/c^2}}. \tag{24.7}$$

This equation is identical to our original, classical definition of momentum except for the factor $\sqrt{1 - v^2/c^2}$. It is possible to interpret this as meaning that an object's mass is velocity-dependent according to

$$m = \frac{m_0}{\sqrt{1 - v^2/c^2}}. \tag{24.8}$$

When the velocity v is zero, the apparent mass m is m_0, which we call the **rest mass** or proper mass of the object.

We must point out that since Eq. (24.7) arose because of the effects of time dilation, we could choose to associate the factor $\sqrt{1 - v^2/c^2}$ with the momentum, and not with mass alone. This choice is usually made in advanced courses. Instead, we have chosen here to associate the relativistic factor with the mass, because that approach is more common in introductory texts. Either approach leads to a description of rapidly moving objects in accord with observations. An interesting conclusion we can draw from Eq. (24.7) is that no object with proper mass can be accelerated to the speed of light or beyond.

Example 24.8

What is the apparent mass of a muon traveling at a speed of 0.995c relative to an observer in the laboratory?

Solution The observer in the lab finds that the muon is more massive than m_0, according to Eq. (24.8). In this case, $v/c = 0.995$ and $v^2/c^2 = 0.990$. Inserting these values, we find that

$$m = \frac{m_0}{\sqrt{1 - v^2/c^2}} = \frac{m_0}{\sqrt{1 - 0.990}} = \frac{m_0}{\sqrt{0.01}} = \frac{m_0}{0.1} = 10 \ m_0.$$

Note that when v/c is close to 1, the mass is strongly dependent on v. For example, if we had used $v/c = 0.994$ instead of 0.995, the value for m would have been only $9.1 \ m_0$.

Relativistic Kinetic Energy

24.9

In classical mechanics we defined the quantity $\frac{1}{2}mv^2$ to be the kinetic energy of an object of mass m moving with speed v. Now let's examine the modifications of this expression imposed by special relativity.

We begin with the mass-velocity relationship of Eq. (24.8). In the limit of small velocities, $v/c \ll 1$, we can expand the square root term in a binomial series so that to a very good approximation the mass becomes

$$m \approx m_0 \left(1 + \tfrac{1}{2}v^2/c^2\right).^*$$

If we multiply through by c^2, we get

$$mc^2 \approx m_0c^2 + \tfrac{1}{2}m_0v^2.$$

The left-hand term is the total energy of the object. The right-hand term consists of two parts: a term related to the rest mass of the object (m_0c^2) and a term equal to the classical kinetic energy ($\frac{1}{2}m_0v^2$). At higher speeds the equation may not look so simple, but we can still separate the total energy into a rest-mass energy and a kinetic energy. Now, suppose we redefine the object's kinetic energy to be the difference between its total energy and its rest-mass energy. Then the kinetic energy is equal to the work required to give the object a velocity v. Thus

$$\boxed{KE = mc^2 - m_0c^2.} \tag{24.9}$$

---

*The binomial series that represents $1/\sqrt{1 - x}$ is

$$\frac{1}{\sqrt{1 - x}} = 1 + \tfrac{1}{2}x + \tfrac{3}{8}x^2 + \tfrac{15}{48}x^3 + \ldots,$$

where $x < 1$. When x is very small, the series is closely approximated by the first two terms:

$$\frac{1}{\sqrt{1 - x}} \approx 1 + \tfrac{1}{2}x, \qquad x \ll 1.$$

When we insert the value of the relativistic mass into this equation, we get

$$KE = m_0 c^2 \left[\frac{1}{\sqrt{1 - v^2/c^2}} - 1 \right]. \tag{24.10}$$

When the velocity is small, this relativistic expression for kinetic energy goes smoothly to the classical value. (You can check this for yourself by applying the binomial theorem to the first term.) So long as an object's velocity is much less than the speed of light, the classical formulas are accurate enough. But when v approaches c, the classical formula for kinetic energy is no longer adequate and we must use Eq. (24.10) instead.

In some cases it is helpful to express an object's total energy in terms of its rest mass and momentum. If we square the relation $E = mc^2$, we get

$$E^2 = m^2 c^4 = m^2 c^2 (c^2 + v^2 - v^2).$$

Observing that $p = mv$ and that $m = m_0/\sqrt{1 - v^2/c^2}$, we find after some algebra that

$$E^2 = p^2 c^2 + m_0^2 c^4. \tag{24.11}$$

One of the effects mentioned in the previous section is that it becomes harder to accelerate an object as its speed approaches c. We see from the kinetic energy equation that a large amount of energy is required to accelerate an object to a speed close to c. The object behaves as if it gets more and more massive and requires more and more kinetic energy for each small increase in speed. Therefore an infinite amount of energy would be required to accelerate an object with rest mass m_0 to the speed of light. Since an infinite amount of energy is not available, the conclusion is inescapable that particles with rest mass cannot attain the speed of light.

Example 24.9

An electron is accelerated to a kinetic energy twice its rest-mass energy. How fast is it traveling?

Solution   We evaluate the speed v using Eq. (24.10) with $KE = 2m_0 c^2$:

$$KE = 2m_0 c^2 = m_0 c^2 \left[\frac{1}{\sqrt{1 - v^2/c^2}} - 1 \right].$$

Upon factoring out the $m_0 c^2$ term, we get

$$2 = \frac{1}{\sqrt{1 - v^2/c^2}} - 1.$$

This is rearranged to give

$$\sqrt{1 - v^2/c^2} = \tfrac{1}{3}.$$

Now square both sides and solve for v to give

$$v = 0.94c.$$

The Relativistic Doppler Effect

*24.10

In Chapter 14 we discussed the Doppler effect, which describes the apparent change in frequency of a sound from a source in motion relative to an observer. Now let's examine the effect of relative motion on light waves, making sure that our description is consistent with the special theory of relativity.

Imagine a starship moving with velocity v toward an observer on earth. As the starship approaches, it emits a light signal of frequency f_0 measured in the reference frame in which the rocket is at rest. Because of time dilation, the period T of the signal appears longer to the observer on earth than to the starship's captain, who measures a period T_0. Moreover, as the ship approaches the observer, the wavefronts of light are compressed from their normal wavelength, measured with a stationary source, by an amount equal to the distance traveled by the starship during one period (Fig. 24.13). The resulting wavelength is

$$\lambda = cT - vT = (c - v)T = \frac{(c - v)T_0}{\sqrt{1 - v^2/c^2}},$$

where T_0 is the proper period. Consequently, the frequency f measured by the earthbound observer is

$$f = \frac{c}{\lambda}$$

or, in terms of the proper frequency of the source, $f_0 = 1/T_0$,

$$f = \frac{f_0 \sqrt{(1 - v^2/c^2)}}{1 - v/c}.$$

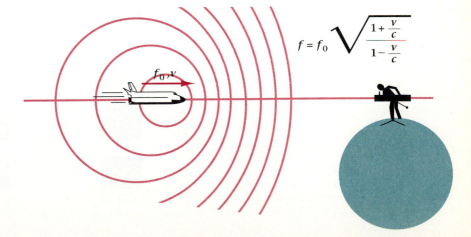

$$f = f_0 \sqrt{\frac{1 + \dfrac{v}{c}}{1 - \dfrac{v}{c}}}$$

Figure 24.13 A starship approaching the earth sends out a radio signal with frequency f_0. The wavefronts are compressed by the motion of the starship so that an observer on earth detects them at a higher frequency f.

We can rearrange this equation to give

$$f = \frac{f_0\sqrt{(1 - v/c)(1 + v/c)}}{\sqrt{(1 - v/c)(1 - v/c)}},$$

which may be simplified to

$$f = f_0 \sqrt{\frac{1 + v/c}{1 - v/c}} \qquad \text{(approaching)}. \qquad (24.12)$$

The observed frequency increases as a result of the relative motion of source and observer. If the source and observer were moving apart, then v in Eq. (24.12) would become negative and the frequency would decrease. Notice that in the relativistic case, the only velocity that matters is the *relative* velocity of the source and observer. This was not true for sound (Chapter 14).

The relativistic Doppler effect applies to any electromagnetic wave. If the waves are reflected from a moving object back toward the source, the shift in frequency can be used to determine the velocity of that object relative to the source. A practical application of this effect can be found in the use of radar to measure the speed of cars and aircraft.

Detailed examination of the light from stars and galaxies shows that, in general, the light is shifted to lower frequencies. Equivalently, we say that the light is red-shifted; that is, it is shifted to longer wavelengths. This red shift arises from the relativistic Doppler effect associated with the motion of the stars away from us. Everywhere we look in the sky, the galaxies are moving away. When we analyze their motion using the Doppler formula we find that the farther away the galaxies are from us, the faster they are receding (Fig. 24.14). Thus we live in an expanding universe.

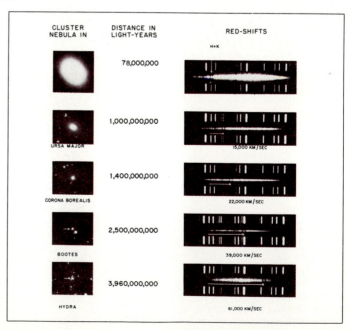

Figure 24.14 Light from more distant galaxies is red-shifted more. The spectra of light from the galaxies is compared with spectra from the laboratory. (Line spectra are discussed in Chapter 26.)

Example 24.10

A spacecraft exploring a distant galaxy is moving very rapidly toward a strange planet that emits a yellow light of frequency 5.0×10^{14} Hz, measured in the rest frame of the planet. What frequency does the spacecraft crew observe for the light if their velocity relative to the planet is $0.095c$?

Solution   A straightforward application of Eq. (24.12) determines the frequency of the light as

$$f = 5.0 \times 10^{14} \text{ Hz} \sqrt{\frac{1 + 0.095}{1 - 0.095}} = 5.5 \times 10^{14} \text{ Hz}.$$

The frequency seen by the crew is $f = 5.5 \times 10^{14}$ Hz. Since we more often refer to light in terms of wavelength, we can convert this to

$$\lambda = \frac{c}{f} = \frac{3.0 \times 10^8 \text{ m/s}}{5.5 \times 10^{14} \text{ Hz}} = 550 \text{ nm}.$$

This wavelength corresponds to green light. The yellow light is shifted to green for the observers moving toward the source at the speed $v = 0.095c$.

The Principle of Equivalence

*24.11

A few years after his first work in special relativity, Einstein advanced a new relationship between accelerated motion and gravitational force. This relation, called the **principle of equivalence**, states that *experiments performed in a uniformly accelerating reference frame, having an acceleration a with respect to an inertial frame, give exactly the same results as identical experiments carried out in an inertial frame containing a uniform gravitational field* $-a$.

To illustrate the principle of equivalence, imagine a spaceship accelerating with a constant acceleration equal to the acceleration of gravity g (Fig. 24.15, p. 724). An astronaut in the spaceship releases a ball. Because of the ball's inertia, it continues to move with the same velocity it possessed at the instant it was released. However, since the spaceship is accelerating, it overtakes the ball. From the astronaut's point of view, the ball falls toward the back of the spaceship with an acceleration $-g$. That is, from his reference frame, the astronaut cannot distinguish between the effects of the spaceship's acceleration g and the effects of a gravitational field in the ship of acceleration $-g$. Thus the principle of equivalence tells us that an accelerated motion in one direction has exactly the same effect as a gravitational field in the opposite direction.

Let us now examine the Doppler effect of a light in a stationary reference frame S as seen from an accelerating frame S'. Then, by the principle of equivalence, we shall relate our results to an inertial frame with a gravitational field.

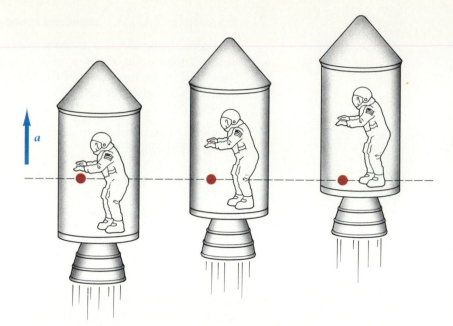

Figure 24.15 An astronaut standing in an accelerating spaceship releases a ball, which falls in the direction opposite to the rocket's acceleration.

Consider a light source of frequency f_0 stationary in the frame S at a point a distance l from the origin on the x axis (Fig. 24.16). At time $t = 0$ the origins of the two frames coincide and the frame S' begins to move in the positive x direction with constant acceleration a. Also at time $t = 0$ the light is turned on. In a time approximately equal to l/c a light pulse reaches the origin of the frame S', just as it attains a velocity $v = at = al/c$, relative to S. From the relativistic Doppler equation, the frequency of the light observed at the origin of S' is

$$f = f_0 \sqrt{\frac{1 + v/c}{1 - v/c}} = f_0 \frac{\sqrt{1 + v/c}}{\sqrt{1 - v/c}}.$$

Using the binomial theorem to evaluate both square roots for $v \ll c$, we get

$$f \approx f_0 \left(1 + \frac{v}{c}\right) = f_0 \left(1 + \frac{al}{c^2}\right). \qquad (24.13)$$

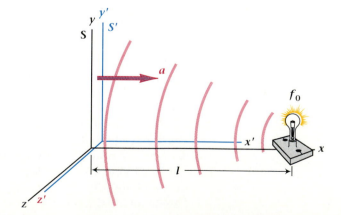

Figure 24.16 At time 0 the light at a distance l from the origin flashes and the system S' acquires a uniform acceleration a.

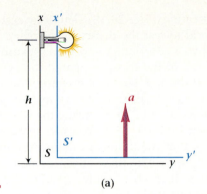

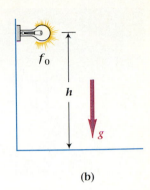

(a)   (b)

Figure 24.17   (a) An accelerating frame S' moving toward the light source. (b) The light source viewed in the equivalent stationary frame with a gravitational acceleration $-g$.

According to the principle of equivalence, we should expect to get the identical frequency in a system that is at rest relative to S but that has a gravitational field of magnitude a in the negative x direction. We can relate this to the earth and its gravitational field by considering the x direction of the frame S' to be directed vertically upward from the earth's surface (Fig. 24.17). Here the gravitational field is directed downward as required and has magnitude g near the earth's surface. We can then make the following interpretation: A transmitter of electromagnetic waves located a height h above the ground and radiating a signal at frequency f_0 is detected on the ground at a frequency f given by

$$f = f_0 \left(1 + \frac{gh}{c^2} \right).$$ (24.14)

Example 24.11

Gravitational shift of a radio frequency.

A radio signal is broadcast from a plane at an altitude of 10,000 m and received on the ground. What is the relative shift in the frequency due to gravitational effects?

Solution   The relative shift in frequency is

$$\frac{\Delta f}{f_0} \approx \frac{f - f_0}{f_0} \approx \frac{gh}{c^2} \approx 10^{-12}.$$

If light is traveling opposite to the direction of a gravitational field, its frequency is reduced. Thus light traveling from the sun to the earth has its frequency reduced as it travels in the sun's gravitational field. Since its wavelength increases toward the red end of the spectrum, this effect is often referred to as the *gravitational red shift*. This effect has been observed in the light from the sun when it is compared with light generated in a laboratory, and agrees with the predicted shift to within a few percent. More recently, experiments performed using a remarkably sensitive technique known as the Mössbauer effect show that the measured frequency shift agrees with the value predicted by the theory to within 0.1%.

An additional experiment involving the principle of equivalence and kinematic time dilation was conducted in 1971 when J. C. Hafele and

Figure 24.18 Joseph Hafele, left, and Richard Keating pose with two atomic clocks just prior to their round-the-world flight to test the theory of relativity.

R. E. Keating (Fig. 24.18) flew four cesium beam atomic clocks around the world to test the theory of relativity. They flew the clocks on regular jet flights once around the world eastward and once around westward and, at the end of the trips, compared the time on their clocks with reference clocks at the U.S. Naval Observatory. According to the theory of relativity, the airborne clocks should have lost about 40 ns during the eastward trip and gained about 275 ns during the westward trip. Their measured values are given in Table 24.1. You can see that they agree with the predicted values within the limits of the uncertainties.

Let's see where these effects come from. We have seen that a clock in motion with speed v relative to an observer runs slowly compared with a clock at rest with the observer. Now suppose a clock is flown along the equator in an airplane with ground speed v. Because of time dilation, time on the plane will run at a different rate from time on the earth's surface. To an observer at rest with respect to the fixed stars, a clock on the earth is moving because of the earth's rotation. If the airborne clock is flown eastward, he observes it moving faster than a clock on the earth. Thus we expect that time should run slower on the plane. If the clock is flown westward, the fixed observer sees it moving slower than a clock on the earth. Thus time runs faster on the plane than on the earth. But there is more. We need to include the gravitational effect. We have seen that a source of radiation with frequency f_0 at a height h is detected on the ground with a higher frequency. There is a corresponding difference in the periods of the radiation and in the time measured as a number of periods. The result is, time runs faster on the plane as it flies higher. This effect is due to gravitation and is independent of the plane's velocity. The rate of the clock is determined by both gravitational and kinematic effects.

The significance of the Hafele-Keating experiment lies in its confirmation that the time differences predicted by the theory of relativity are real—not only for muons, but also for real macroscopic objects, such as clocks. The time differences involved were small, but were well within the capability of the atomic clocks to resolve. Thus merely by traveling eastward around the earth, the experimenters aged a few nanoseconds less than their earthbound colleagues. If the speeds were greater the relative effect would be greater also. This is the result of the experiment: Time is indeed relative and is not the same for all observers.

TABLE 24.1
Measured and predicted time differences in the experiment of Hafele and Keating*

| | Δt (ns) | |
	Eastward	Westward
Measured	−59 ± 10	273 ± 7
Predicted	−40 ± 23	275 ± 21

*J. C. Hafele, and R. E. Keating, *Science*, Vol. 177, 14 July 1972, pp. 166–170. Copyright 1972 by the AAAS.

General Relativity

*24.12

In proposing his special theory of relativity, Einstein postulated that the laws of physics were the same to all observers in inertial frames of reference. The acceptance of this theory and the validity of the postulates are supported by a wide range of observations and experiments. The "specialness" of special relativity is the restriction to inertial frames of reference. To describe the behavior of accelerating systems, we need a more general theory. Such a theory was proposed by Einstein in 1916. This **general theory of relativity** creates a theoretical framework applicable to all systems, inertial or noninertial. That is, it includes accelerating systems.

The basic postulate of general relativity is that **all physical laws can be formulated so as to be valid for any observer, regardless of his motion.** When this postulate is combined with the principle of equivalence, which treats gravitational fields as equivalent to accelerations, we see that the general theory becomes more than a description of accelerated systems. It is a theory of gravitation.

The special relationship between gravitation and acceleration has intrigued physicists since the time of Isaac Newton. In Newton's second law, $F = ma$, the acceleration that an object acquires from the application of an external force depends on its mass m. Moreover, according to Newton's law of universal gravitation, the gravitational attraction of that object by another object also depends on the mass. These two entirely separate effects depend on the same quantity, the mass. Other forces do not show this behavior. For example, the electromagnetic force on a charged particle depends on the magnitude of its charge, while its acceleration depends on its mass. There seems to be something special about gravitational forces.

By using the principle of equivalence, we can transform our viewpoint to a reference frame with just the right acceleration to eliminate the gravitational field at any particular point in space. But in general, the acceleration required to eliminate the gravitational field at an arbitrary second point in space will be different. For example, consider the gravitational field of the earth. Above the North Pole the field is directed toward the center of the earth along its axis of rotation. But at some other point above the earth, say above the equator, the gravitational field is radial to the center of the earth and perpendicular to the axis of the earth's rotation. Thus, while a transformation of coordinates may appear to remove a local gravitational field at one point, it cannot eliminate all gravitational effects.

Observations such as these suggested to Einstein that the usual conception of space itself should be reexamined. Einstein proposed that instead of being always the rectilinear Euclidean space of Newton's laws, space itself might be curved (Fig. 24.19). In such a space the motion of an object could be described in terms of the geometry of the space rather than in terms of external forces.

For example, in a Euclidean inertial reference frame, a ray of light moves in a straight line. However, when this event is viewed from an accelerating frame of reference, the path of the light ray appears curved. By the principle of equivalence, we should expect the light ray to follow a curved path in a gravitational field. Since light has no mass, we cannot state the force involved in terms of Newton's laws. However, from the perspective of general relativity, it is space itself that is viewed as being

Figure 24.19   A two-dimensional surface representing a curved two-dimensional space. We draw the figure as if the distortion exists in the third dimension, but that is only to help visualize the curvature. An observer on the surface can tell that the surface is not flat because the ratio of circumference to diameter of a finite circle drawn on the surface does not equal π.

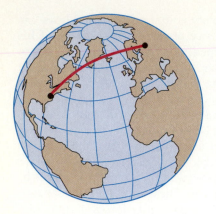

Figure 24.20 The shortest path between New York and Moscow is the great circle path (geodesic) shown.

curved or distorted by the presence of the gravitational field (that is, by the presence of mass). In such a curved space, light merely travels along a **geodesic**, which is the curved path that represents the minimum distance between two points in that space. Such behavior has an analogy in travel on the earth's surface. The shortest route between cities is not a straight line, but a curved path along a great circle around the earth (that is, along a circumferential circle, Fig. 24.20). Such a path is the geodesic on the spherical surface of the earth. This is a natural consequence of the fact that the earth is spherical and not flat.

We do not have to explain the deflection of light in terms of Newtonian forces. Instead, we say that light always follows a geodesic through space, which may itself be curved. Thus, where space is Euclidean (that is, far away from any mass), light travels in a straight line. Where space is curved (that is, near a significant mass, like a star), light travels along the geodesic in that space.

There are only a few experiments that can distinguish the predictions of the general theory of relativity from those of Newtonian mechanics. Such experiments are known as *tests* of the theory. These experiments include, among others, the deflection of light in a gravitational field; the exact path of Mercury's orbit around the sun; and the time delay of radar signals that pass near the sun. The agreement between the observations and the predictions shows that the general theory of relativity provides a suitable explanation. However, they do not prove the theory to be correct. These tests are examined briefly in the following paragraphs. Note that there are experiments, such as those mentioned in the previous section, that test the principle of equivalence but do not really test general relativity.

When light from a star passes close to the sun, it is deflected by the sun's gravitational field. This deflection causes the star to appear slightly displaced (Fig. 24.21). The displacement has been measured by photo-

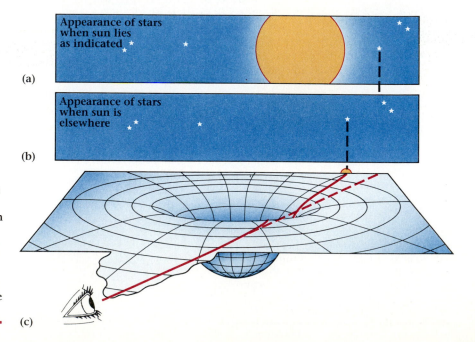

Figure 24.21 Deflection of a light beam results from curvature of space near a massive object, such as the sun. (a) The appearance of stars when the sun lies as indicated is different from their appearance when the sun lies elsewhere (b). (c) The path of the light from the star to the observer's eye follows a geodesic in the curved space, represented here by the curved surface.

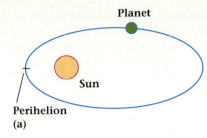

Planet

Sun

Perihelion
(a)

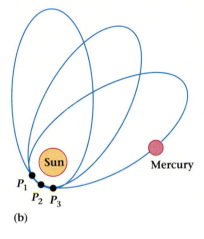

Sun

Mercury

P_1
P_2 P_3

(b)

Figure 24.22 (a) Elliptical orbit of a planet about the sun. The closest point of approach is the perihelion. (b) Exaggerated drawing of the advancing orbit of Mercury, showing the motion of the perihelion from point P_1 to P_2 to P_3.

graphing the apparent positions of stars during a solar eclipse and comparing these positions with those observed in the night sky six months later. Apparent shifts of less than 2 seconds of arc have been measured this way, in close agreement with the theoretical prediction of 1.75 seconds of arc.

As the planets travel about the sun in elliptical orbits with the sun at one focus, they affect one another so that their orbits are not perfect ellipses. In the case of the planet Mercury, its orbit moves slowly about the sun (Fig. 24.22). The point of closest approach (the perihelion) advances around the sun at a rate of nearly 500 seconds of arc per century. When all of the complex calculations of Newtonian forces are made, including the gravitational effects of Venus, the earth, Jupiter, etc., a discrepancy remains of about 43 seconds of arc per century between observation and prediction. If we apply special relativity to account for effects of mass change and time dilations, the resulting discrepancy between experiment and theory is still about 21 seconds of arc per century. But if we apply general relativity, the predicted precession is 43 seconds of arc per century, in agreement with the observations.

These tests cannot be cited as absolute proof of the general theory of relativity, since other theories could conceivably explain these results as well. However, contemporary physicists generally agree that the principle of equivalence is valid and that Einstein's general theory is consistent with experimental observations. Various details of the general theory have been modified over the years by the insights of other theoretical physicists, but the overall structure stands as one of the great general concepts of twentieth-century physics.

SUMMARY

Useful Concepts

- Special relativity is based on two postulates:

  I. The laws of physics are the same in all inertial frames of reference.
  II. The speed of light in free space is independent of the motion of its source and of the motion of the observer.

- Velocities are added according to

$$u = \frac{u' + v}{1 + \dfrac{u'v}{c^2}},$$

where u' is a velocity measured in a reference frame moving with velocity v relative to the frame in which u is measured.

- An observer at rest with respect to a clock records a proper time interval Δt_0. An observer moving with respect to the clock sees the time interval indicated by the clock stretched out to Δt:

$$\Delta t = \frac{\Delta t_0}{\sqrt{1 - \dfrac{v^2}{c^2}}}.$$

This effect is called time dilation.

- An observer in motion with speed v relative to a rod of proper length l_0 sees a length l given by

$$l = l_0 \sqrt{1 - v^2/c^2}.$$

This effect is known as length contraction.

- The mass–energy relationship is

$$E = mc^2.$$

- The momentum of an object is

$$p = \frac{m_0 v}{\sqrt{(1 - v^2/c^2)}}.$$

The behavior of an object moving with speed v may be interpreted as if the object has mass

$$m = \frac{m_0}{\sqrt{1 - v^2/c^2}}.$$

- A particle of rest mass m_0 has kinetic energy

$$KE = mc^2 - m_0 c^2$$

$$= m_0 c^2 \left[\frac{1}{\sqrt{1 - v^2/c^2}} - 1 \right].$$

- The total energy of a particle of rest mass m_0 is

$$E = p^2 c^2 + m_0^2 c^4.$$

- The Doppler effect for light is

$$f = f_0 \sqrt{\frac{1 + v/c}{1 - v/c}}.$$

- The principle of equivalence states that experiments performed in a uniformly accelerating frame of reference, having an acceleration a with respect to an inertial frame, give exactly the same results as identical experiments carried out in an inertial frame having a uniform gravitational field $-a$.
- The general theory of relativity creates a theoretical framework applicable to both inertial and noninertial systems. The basic postulate of general relativity is that all physical laws can be formulated so as to be valid for any observer, regardless of his motion.

Important Terms

You should be able to write the definition or meaning of each of the following terms:

- inertial reference frame
- Galilean relativity
- special theory of relativity
- transformation equations
- simultaneous
- time dilation
- proper time
- proper length
- length contraction
- rest mass

QUESTIONS

24.1 Passengers in an accelerating railroad car have with them a ball and a ruler but no clock. The windows are painted over, so they may not see outside. Can they, by observing what happens when they drop the ball, tell whether they are in an inertial reference frame? Could they answer the question if they were in an accelerating elevator?

24.2 Is it reasonable to expect the physical theories describing nature to be in accord with "common sense"? What does the term really mean, and how important is it likely to be in formulating and judging theories for cases in which speeds are near the speed of light?

24.3 In the Lorentz transformations for space and time, do the results depend on whether the frame S' is moving toward or away from you in S?

24.4 Is it correct to say that the Galilean transformations are wrong?

24.5 What would be the effect on our daily lives if the velocity of light were 30 m/s? How would the world differ from the way it seems now?

24.6 Is the twin paradox aptly named? Before you answer, look up the meaning of "paradox" in a dictionary large enough to give several definitions for the word.

24.7 Is it true that "matter can neither be created nor destroyed"? If not, what is the correct statement? How might this bear on the conservation of mass when balancing chemical equations?

24.8 Is the mass of a red-hot piece of metal greater than its mass when it is at room temperature? Explain your answer.

24.9 The special theory of relativity establishes an upper limit to the speed of a material body. Does this theory set an upper limit to the momentum and kinetic energy of the same body?

24.10 Twins separate, with one going to live on top of a very high mountain and the other living in a cave deep underground. Which twin ages faster?

24.11 Suppose you set your clock using time signals from the radio transmitter at the National Bureau of Standards in Fort Collins, Colorado. Approximately how long does it take for the radio signal to travel to your receiver?

PROBLEMS

Section 24.1   Frames of Reference

24.1 Two airplanes are flying side by side at the same altitude. Plane A is slowly overtaking plane B at 4 km/h. A stewardess in plane A is walking toward the rear of the plane at a speed of 2 km/h and a passenger is walking toward the front at 2 km/h. What are their speeds relative to a passenger watching them from plane B?

24.2 A train moves slowly past a station at 8 km/h. A passenger on the train runs 8 km/h down the aisle in the direction opposite to the train's velocity. (a) What is the passenger's speed relative to the station? (b) What would it be if he reversed his direction?

24.3 A ferry boat headed directly away from the river bank is traveling at 3.00 km/h with respect to the water in a river flowing at 4.00 km/h. A passenger walks diagonally toward the stern of the boat with a speed of 1.39 m/s. The passenger's path is at an angle of 53.1° toward the upstream direction from the length of the boat. What is the passenger's speed relative to the bank?

Section 24.3   Velocity Addition

24.4 A space station moving away from the earth at $0.50c$ launches a rocket toward earth with a velocity $0.70c$. What is the rocket's speed as seen from earth?

24.5 A rocketship hurtling toward Mars at a speed of $0.80c$ launches a probe toward the planet at a speed $v = 0.60c$ relative to the rocketship. How fast does the probe approach Mars as measured by an observer on Mars?

24.6 A spaceship approaching the moon at a speed $0.50c$ sends a light signal to an observer stationed on the lunar surface. (a) What is the speed of the light signal as measured by the pilot of the spaceship? (b) What is the speed of the light signal seen by the lunar observer?

24.7 A spaceship racing toward Jupiter at $0.70c$ launches a probe toward the planet at a relative speed of $0.50c$. How fast does the probe approach Jupiter as measured by an observer on Jupiter?

24.8 Two space stations approach each other at a relative speed of $0.50c$. A rocket is sent from station A to station B with a speed $0.40c$ as measured from station A. What is the approach speed as seen from station B?

24.9 A spaceship moving away from Mars at speed $v = 0.60c$ relative to an observer there is pursued by an enemy ship, traveling at $v_e = 0.70c$ relative to Mars. The pursuit ship launches a missile with relative velocity $v_m = 0.30c$ toward the first spaceship. How fast does the missile appear to move as seen from the lead spaceship?

24.10 In a colliding-beam experiment particles traveling at $0.950c$ relative to the laboratory meet in a head-on collision with other particles traveling in the opposite direction at $0.950c$ relative to the lab. What is the relative speed of approach of the particles as seen in the rest frame of one of them?

Section 24.4   Simultaneity

24.11 An astronaut on the moon wants to synchronize his clock to a time signal on earth. He receives a radio message saying that the time at the tone will be exactly 6:00. To what time should his clock be set at the instant of the tone?

24.12 An astronaut on the Martian surface, 8.96×10^{10} m from home, receives a time signal from earth. The message says that at the sound of the tone it will be exactly 5:00. When the tone sounds, she sets her clock to 5:00. When she returns home, she compares her clock to the standard clock. What does she find? Neglect any effect due to the trip home.

Section 24.5   Time Dilation

24.13 An astronaut journeying to the nearby star Lacaille 9352 at a speed of $0.85c$ relative to earth requires 20 years as measured on earth to complete his trip. How many years does he age during his journey?

24.14 An astronaut journeying to the nearby star Alpha Centauri at a speed of $0.99c$ relative to the earth requires 8.57 years as measured on earth to complete her trip. How many years would the astronaut age during this journey?

24.15 A cosmonaut journeys at high speed for two years. Upon returning he finds that eight years have elapsed on earth. What was his average speed?

24.16 Astronaut Kirk travels at high speed for three years.

Upon returning to earth he finds that six years have elapsed. What was his average speed?

24.17 The mean life of the pion is 2.6×10^{-8} s. What is the mean life of a beam of pions moving with a speed of $0.99c$ relative to you?

24.18 The mean life of muons is 2.2 μs. What is the mean life of a beam of muons moving at a speed of $0.9995c$ relative to you?

24.19 In the muon experiment described in the text, the muons appeared to have an abnormally long lifetime as a result of the time dilation associated with their velocity. (a) From the rest frame of the muons, calculate the contracted distance from the mountaintop to sea level, 1907 m in the rest frame of the earth. (b) How long does it require the muon to travel this distance at a relative speed of $0.994c$? (c) What fraction of the muons should remain after the time found in (b)? (d) Does this agree with the experimental results?

Section 24.6   Length Contraction

24.20 A spaceship with a proper length of 300 m passes near a space platform at a relative speed of $0.80c$. What is the length of the spaceship when measured in the frame of the space platform?

24.21 A spaceship moves with a speed $v = 0.90c$ relative to a space platform that has a landing strip 5000 m long. What is the apparent length of the landing strip as measured in the frame of the spaceship?

24.22 A meter-stick is accelerated to a velocity such that it is contracted to only 50 cm relative to an interested observer. How fast must the meter-stick be moving in the observer's frame of reference?

24.23 A meter-stick is accelerated to such a velocity that to an interested observer it appears only 80 cm long. How fast must the stick be moving in the observer's frame of reference?

24.24 By how much is a 100-m-long train shortened as observed by someone at rest relative to the track if the train is traveling 320 km/h?

24.25 An electron travels along a straight 10-m section of a particle accelerator at a speed of $0.99c$. If you could ride along with the electron, how long would that straight section appear to you?

Section 24.7   Mass and Energy

24.26 Suppose that all of the matter in a 100-mg grain of sand was entirely converted to energy. How much energy would be released?

24.27 An electron and a positron (a positive electron) can come together and annihilate; that is, they disappear, producing a flash of electromagnetic radiation. If each of these particles has a mass of 9.11×10^{-31} kg, what is the total energy of the radiation?

24.28 A certain nuclear reactor is capable of producing 1.0×10^9 W of power. During operation, mass is converted to energy. How much mass is converted per hour?

Section 24.8   Relativistic Momentum

24.29 A baseball of rest mass m_0 is accelerated to a speed $0.95c$ relative to the earth. What is the apparent mass of the ball measured in the rest frame of the earth?

24.30 Calculate the mass of a muon as measured in a frame in which its velocity is $0.999c$. Express your answer in terms of its rest mass.

24.31 Calculate the mass of a particle of rest mass m_0 when it is observed in a frame of reference in which its velocity is $0.990c$.

24.32 How fast must a particle travel relative to an observer for its apparent mass to be double its rest mass?

24.33 How fast must a particle travel for its apparent mass to be six times its rest mass?

Section 24.9   Relativistic Kinetic Energy

24.34 A muon travels with a kinetic energy equal to its rest-mass energy. How fast is it traveling?

24.35 A particle of mass m_0 moves with a speed of 2.0×10^8 m/s relative to an observer. What are the apparent mass and kinetic energy of the particle in the rest frame of the observer?

24.36 A particle of mass m_0 is given a kinetic energy equal to one-half its rest-mass energy. How fast must the particle be traveling?

24.37 A particle of mass m_0 travels at a speed of $0.75c$ relative to the laboratory. (a) What is its apparent mass? (b) What is its kinetic energy? (c) What is its total energy?

24.38 Calculate the velocity of a particle whose kinetic energy is 10 times its rest-mass energy.

24.39 A particle's total energy is five times its rest-mass energy. How fast is the particle traveling?

*Section 24.10   The Relativistic Doppler Effect

24.40 A radio station broadcasting on a frequency of 106 MHz is received by an astronaut headed toward the station at a speed of 6.0×10^7 m/s. To what frequency must the astronaut's radio be tuned?

24.41 Radio signals from a distant planet are received at a frequency of 106 MHz on a spaceship headed directly toward the planet at a speed of $0.30c$. What is the frequency of the radio signal as measured on that planet?

24.42 A spaceship traveling at $v = 0.35c$ is moving transverse to the line of sight of an earthbound observer. (a) If the ship's radio has a proper frequency of 105

MHz, at what frequency will it be detected by the observer? (b) What frequency would be observed if the ship were headed directly toward the earth? For transverse motion, only the shift due to time dilation holds.

24.43 The red light emitted by hydrogen has a wavelength of 656 nm measured in the laboratory. An astronomer studying a distant star observes light at a longer wavelength, which he believes is hydrogen light that has been Doppler-shifted because of the star's motion. Is the star moving toward or away from him? If the star is moving at a speed of 40,000 km/s, what is the observed wavelength of the light?

*Section 24.11   The Principle of Equivalence

24.44 A radio signal of frequency f_0 is broadcast by a station at sea level. What is the shift in frequency of that station as measured by an observer on a mountaintop 2000 m above sea level?

24.45 A satellite at an altitude of 100 km broadcasts a radio signal at 108 MHz. What is the shift in frequency at sea level due to gravitational effects? Is the shift positive or negative? (*Hint:* Take g as constant.)

*Section 24.12   General Relativity

24.46 The separation between the planet Mercury and the sun is 46×10^9 m at the perihelion. The period of revolution of Mercury about the sun is 88 days and its perihelion advances at the rate of 573 seconds of arc per century. Calculate the distance through which the perihelion of Mercury advances in one period of the planet's motion about the sun.

Additional Problems

24.47 A stunt diver jumps from the top of a 20.0-m-high
• building. After falling for 1.50 s, the diver flips a coin upward with a relative speed of 14.7 m/s. What is the initial speed of the coin with respect to the ground as it leaves the diver's hand?

24.48 Solve the Lorentz transformations, Eq. (24.3), to get
• x and t in terms of x' and t'. How do the new equations relate to the original pair?

24.49 Derive the velocity transformation, Eq. (24.1), from
•• the Lorentz transformation, Eq. (24.3).

24.50 A beam of particles traveling at $0.75c$ is observed in
• the laboratory. In a time equal to one mean life in the frame of the particles, the beam has traveled 6.12 m in the laboratory. What is the mean life of these particles?

24.51 A cosmonaut orbits the earth at an altitude of 300
• km above the mean radius of the earth. (a) Calculate the tangential velocity of the spaceship. (b) Calculate the period of one complete orbit. (c) If the cos-

monaut remains in orbit for 14 days, how much younger would he be upon his return than his twin brother who stayed behind? (*Hint:* Use the approximation that for $x << 1$, $\sqrt{1 - x} \approx 1 - x/2$.)

24.52 The Concorde airplane is 62.1 m long. (a) What
• difference will an observer on the ground find between the Concorde's length when at rest on the runway and during flight at 1700 km/h? (b) How fast would the Concorde have to fly to increase this difference by a factor of 5? (*Hint:* Use the approximation that for $x << 1$, $\sqrt{1 - x} \approx 1 - x/2$.)

24.53 Suppose that one milligram of matter was converted
• to energy and that all of the energy was used (with no waste) to operate a 200-W lamp. How long could the lamp be operated?

24.54 An astronaut moves rapidly past a galactic traffic
• signal that emits yellow light of frequency 5.0×10^{14} Hz measured in the rest frame of the signal. The astronaut looks at the signal in his rearview mirror. (a) What frequency does the astronaut observe if his velocity is $0.095c$ relative to the signal? (b) To what color does this correspond?

24.55 In a hypothetical experiment, a plane flies eastward
•• around the earth at an average supersonic speed of 600 m/s relative to the ground and an average altitude of 12 km. The flying time is 18 hours and the flight path is equatorial. Calculate the time difference that should result when a clock that was on the plane is compared with a clock that was stationary on the ground at the equator. (*Hint:* To find the difference due to special relativity choose an imaginary clock that is stationary with respect to the earth's center of mass. Then find the time on a clock moving with the earth's surface as the earth rotates about its axis, and the time on a plane moving with speed v relative to the earth. Then add the differences due to gravitation to the difference due to special relativity.)

24.56 In a hypothetical experiment, a plane flies westward
•• around the earth at an average supersonic speed of 600 m/s relative to the ground and an average altitude of 12 km. The flying time is 18 hours and the flight path is equatorial. Calculate the time difference that should result when a clock that was on the plane is compared with a clock that was stationary on the ground. (*Hint:* See the hint for Problem 24.55. Notice that the direction of the plane's velocity is negative with respect to the direction in that problem.)

24.57 Show that the relativistic expression for kinetic en-
• ergy goes smoothly to the classical expression when the velocity is small compared with the speed of light. (*Hint:* Use the binomial series to evaluate the relativistic expression.)

25

The Discovery of
Atomic Structure

25.1 Evidence of Atoms from Solids and Gases

25.2 Electrolysis and the Quantization of Charge

25.3 Avogadro's Number and the Periodic Table

25.4 The Size of Atoms

25.5 Crystals and X-Ray Diffraction

25.6 Discovery of the Electron

25.7 Radioactivity

25.8 Radioactive Decay

25.9 Discovery of the Atomic Nucleus

A WORD TO THE STUDENT

You have probably known for many years about the existence of atoms and nuclei. However, most people cannot cite much experimental evidence for them. In this chapter we discuss the experiments that form the basis for our knowledge of atoms. In doing so, we examine many fundamental facts of atomic structure and our reasons for believing in them. This will lay the groundwork for our further discussion of atomic physics in the following chapters.

In addition, we present the principal observations about natural radioactivity and the law of radioactive decay. We then show how Rutherford's scattering experiment led to our present concept of the nuclear atom. This same general technique is still used today to probe the composition and nature of the subatomic world.

In Chapter 12 we examined the properties of gases, basing our description on the kinetic theory. This model treats gases as collections of rapidly moving atoms. Using the kinetic-theory model, we found ways to relate macroscopic properties such as pressure and temperature to the microscopic behavior of the atoms. We can consider the success of the kinetic theory as evidence for the existence of atoms, but the observational evidence for atoms is much greater than simply the behavior of gases.

The atomic concept is fundamental to the modern view of physics and chemistry—but it has not always been so. Only in the twentieth century has the existence of atoms become universally accepted. The size of individual atoms is so small that we are unable to see them directly, and therefore we find evidence for them difficult to comprehend. It is not unusual to meet people who accept atoms as the objects that chemists and physicists talk about, but who cannot cite any experimental justification for believing in the existence of atoms. For these reasons we turn our attention to the development of the concept of atomicity and to the many observations that have established its validity.

Atomic theory has proved successful because it explains observations that could not otherwise be understood. As more details of atomic structure have become known, older theories of matter and its behavior have given way to newer and more comprehensive ideas. Knowledge of atomic structure has deepened our understanding of other areas of physics. For example, greater understanding of atomic structure led to better understanding of solids, which in turn led to the development of new kinds of materials, including new semiconductors, magnets, and superconductors.

In this chapter we discuss the development of atomic theory up to the early twentieth century, describing the basic facts upon which later ideas were built and which remain fundamental to contemporary understanding. In the next chapter we describe how classical physics was insufficient to explain observations of atomic structure and how the concepts introduced by Planck, Einstein, and Bohr led to a new understanding of matter.

Evidence of Atoms from Solids and Gases

25.1

An atomic theory of matter was first proposed by the Greek philosopher Leucippus about 500 B.C. and developed by his pupil Democritus. They believed that matter was made up of particles so small that they could not be divided. These ultimate particles were called *atoms,* from the Greek word *atomos,* which means indivisible. Democritus proposed that the different properties of various substances were due to differences in the nature of their atoms. His atomic theory was not widely accepted and was rejected by the dominant philosophers of the next generation. In particular, Aristotle argued that matter was continuous and therefore infinitely

Henri Becquerel is the first to observe radioactive decay; three kinds of radioactive emissions are eventually found.

1895

1896

1897

Wilhelm Roentgen discovers x rays, demonstrating they can penetrate matter.

J. J. Thompson discovers the electron, measures its mass-to-charge ratio, and establishes that electrons are basic constituents of all atoms.

divisible. Thus there was no need for atoms. This rejection by Aristotle doomed the atomic concept until the seventeenth century, when the ideas of Galileo and Newton undermined Aristotle's authority and atomism was revived.

An important stimulus to the acceptance of atomism came from studies of crystals. One of the most remarkable properties of crystals is that many possess a regular geometric shape. For example, quartz crystals are hexagonal (Fig. 25.1a). Other crystals have their own particular shapes (Fig. 25.1b). The shapes of small crystals intrigued the English scientist Robert Hooke when he examined them with his microscope. He presumed that the symmetry of the crystals reflected the regular arrangement of the tiny particles that composed the solid. He illustrated his ideas with several drawings published in 1665 (Fig. 25.1c,d). In 1690 the Dutch scientist Christiaan Huygens showed that the geometrical shape of Iceland spar (calcite) could be explained if the mineral were composed of small rounded bodies arranged as shown in Fig. 25.2.

Figure 25.1 (a) Crystals of quartz have hexagonal shapes. (b) Pyrite (FeS_2) occurs in three characteristic shapes; cubic, octahedral, and pyritohedral. (c, d) Drawings of crystals made by Robert Hooke.

(a)

(b)

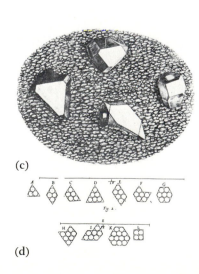

(c)

(d)

W. L. Bragg (shown)
and his father,
W. H. Bragg, measure
the wavelength of x rays
by using the structure
of crystals.

1908　　　　　　　　　　**1909**　　　　　　　　　　　　**1911**

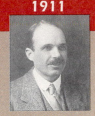

Jean Perrin determines the
value of Avogadro's number,
thereby demonstrating the
existence of atoms.

Ernest Rutherford proposes the
nuclear model of the atom,
based on his analysis of particle
scattering experiments.

(a)

(b)

Figure 25.2 (a) A calcite crystal.
(b) A model of calcite composed of
small rounded constituent particles.
(After Huygens.)

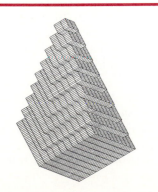

The ability of some crystals to cleave—that is, to split along definite planes, leaving smooth surfaces—led the French abbé Rene Just Haüy to a deeper understanding of crystal structure. In 1781 Haüy accidentally dropped a calcite cluster, which broke apart. Haüy noticed that the broken crystal had a single fracture along one edge. When he tried to break it in other directions, it broke into rhombohedral pieces. He later cleaved calcite crystals of other initial shapes, always finding that they could be broken to reveal the rhombohedral form.

After similar experience with other crystals, Haüy proposed that continued cleavage of crystals into smaller and smaller pieces ultimately reduces the crystal to the smallest possible unit or building block. Today we call this smallest shape the **unit cell**. A whole crystal can be formed by stacking these unit cells side by side.

Haüy showed that his model of unit cells could also be used to explain the angles between other natural faces that occur when the crystal layers are successively smaller and recede from the edge in a regular manner, like the steps of a large building (Fig. 25.3). Thus the crystallographers of the early nineteenth century became convinced that the shape of crystals was a direct result of their construction from invisible building blocks in a regular fashion. Eventually, links were established between the shapes and properties of crystals and the physical and chemical properties of their constituent atoms.

At the same time that crystallographers were discovering Haüy's ideas, chemistry was beginning to provide independent evidence for atoms. By 1800 the law of definite proportions had been demonstrated by the French chemist J. L. Proust, though it was by no means universally accepted. According to this rule, the proportions (by mass) of the elements in a chemical compound are constant. If water, consisting of the elements hydrogen and oxygen, is decomposed, the mass of the oxygen released is *always* 8 times greater than the mass of the hydrogen. Similarly, when

Figure 25.3 Haüy's interpretation of crystal facets arising from stepped layers.

oxygen and hydrogen combine to form water they *always* do so in the same 8 to 1 proportion by mass.

Inspired in part by this rule, the English chemist John Dalton around 1803 developed a quantitative atomic theory of chemistry. In addition to ideas similar to the atomic theory of Democritus, the principal postulates of Dalton's theory were that chemical reactions only separate or join atoms, and that when different atoms combine, the resulting compound always contains the same relative number of atoms. That is, Dalton's theory incorporated the observed law of definite proportions.

Dalton and others applied these ideas to understanding the problems of chemical combinations. Tables of atomic masses were compiled, which compared the mass of an element with that of hydrogen. Still, no one knew either the mass or the size of a single atom.

Soon after the publication of Dalton's atomic theory, J. L. Gay-Lussac found an empirical rule governing the behavior of gases. He rediscovered an observation by Cavendish, that hydrogen and oxygen gases combined in volume proportions of 2 to 1 to form water. He further showed that other gases reacted in volumes whose ratios were small whole numbers. This rule, known as the law of volumes, says that gases unite in simple and definite proportions by volume.

The law of definite proportions, Dalton's atomic theory, and the law of volumes were brought together by the Italian count Amadeo Avogadro (1776–1856). Avogadro suggested that equal volumes of gases (at the same temperature and pressure) contain equal numbers of molecules. This hypothesis clearly differentiates between molecules and atoms: The molecule is a combination of atoms. Avogadro had no way of knowing how many molecules were present in a given volume of gas, but he knew the number must be large.

Using his hypothesis of equal numbers of molecules in equal volumes of gases and Dalton's laws, Avogadro satisfactorily explained the law of definite proportions and the law of volumes. In his view, gases consisted of molecules, and molecules consisted of atoms; equal volumes of gases contained equal numbers of molecules; and gases reacted by exchanging atoms, thus changing the ratios of numbers of molecules (and hence volumes) by small whole numbers. Avogadro applied his hypothesis to explain Cavendish's experimental observation, correctly concluding that a molecule of water was formed from half a molecule of oxygen and two half-molecules of hydrogen.

Unfortunately, Avogadro's hypothesis was ignored by his contemporaries. Had it been accepted, chemists would have been spared half a century of confusion. However, we should remember that few facts were available in Avogadro's time that could support his hypothesis, and even these "facts" were not without controversy.

Because atoms are very small, it follows that vast numbers of them are required to make any appreciable amount of matter, such as one gram. How many atoms are in one gram of matter? This question was central to the development of atomic theory, as well as to the theory of chemical reactions. The next step in answering this question came from Faraday's experiments with electricity.

Electrolysis and the Quantization of Charge

25.2

In the early 1800s, experiments showed that chemical compounds could be separated into their constituent components by passing an electric current through them, a process called **electrolysis**. Michael Faraday discovered that the amount of substance decomposed in electrolysis is proportional to the magnitude of the electric current and to the elapsed time (Fig. 25.4). Consequently, the mass of material released or deposited on the electrodes is proportional to the electric charge that passes through the system.

Faraday also found that the mass of material deposited was proportional to its chemical equivalent mass, which is the atomic mass divided by the valence, or most common combining ratio. For example, electrolysis of sodium chloride (table salt) released amounts of sodium and chlorine in direct proportion to their atomic masses, as given by Dalton. However, electrolysis of copper chloride sufficient to release an amount of chlorine equal to its atomic mass would only deposit an amount of copper equal to half its atomic mass. Faraday would have said that copper has a valence of 2, so that in each case, electrolysis released the equivalent mass of the substance. For many elements these equivalents were the same as Dalton's atomic masses (corresponding to a valence of 1). This correlation was explained by assuming that charge is *quantized* (existing in only discrete amounts), with a unit of charge e. Each single-valent ion (the particle collected by the electrolysis) was thought to carry one unit of charge. Faraday carefully measured the total amount of charge required to deposit one gram atomic mass (one mole) of a single-valence element. (One gram atomic mass is the amount of substance, in grams, equal to its atomic mass.) This amount of charge became known as the *faraday* and its presently accepted value is 96,485 coulombs.

According to Avogadro's hypothesis, a gram atomic mass of a substance would contain a definite number N_A of molecules, called **Avogadro's number**. Thus the faraday, F, must be the product of e and the number N_A. That is,

$$F = N_A \times e. \tag{25.1}$$

Since the faraday constant is known with great precision, Eq. (25.1) is important because it can be used to determine either Avogadro's number N_A or the elementary charge e if the other is known. Faraday recognized the significance of this relationship, but was unable to measure either N_A or e independently.

Faraday did not prove that atoms exist; he could not determine their size or mass, for example. However, he used the assumption of atomicity to explain successfully the quantitative aspects of electrolysis, and he extended this assumption to include the concept of quantization of electric charge. In doing so, he gave still more credibility to the idea that atoms do, indeed, exist.

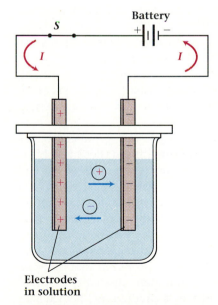

Figure 25.4 An electrolytic cell. The amount of substance that decomposes is proportional to the applied current and the elapsed time.

Example 25.1

Finding atomic mass by electrolysis.

In an electrolysis experiment a student plated 0.503 g of silver onto an electrode. During the plating, a constant current of 0.500 A was maintained for 15.0 min. Assuming singly charged ions, what is the atomic mass of the silver?

Solution   We first compute the amount of charge delivered in the electrolysis. The total charge is the product of current with time,

$$q = It = 0.500 \text{ A} \times 15.0 \text{ min} \times 60 \text{ s/min} = 450 \text{ C}.$$

The ratio of the atomic mass to the mass deposited during electrolysis is equal to the ratio of the faraday to the charge q delivered. If we let M be the atomic mass, we get

$$\frac{M}{0.503 \text{ g}} = \frac{96,485 \text{ C/mol}}{450 \text{ C}},$$

$$M = 108 \text{ g/mol}.$$

Avogadro's Number and the Periodic Table

25.3

By the middle of the nineteenth century, chemistry was in a state of confusion. Independent application of the law of definite proportions and the law of volumes without the use of Avogadro's hypothesis had led to different interpretations of reactions and different tables of atomic masses. To clear up some of this confusion, an International Chemical Congress was convened in Karlsruhe, Germany, in 1860. At this congress, Stanislao Cannizzaro recommended acceptance of Avogadro's hypothesis. Among the participants at the congress who were influenced by Cannizzaro were the German chemist Lothar Meyer (1830–1895) and the Russian chemist Dmitri Mendeleev (1834–1907).

Subsequently, Meyer and Mendeleev independently produced tables of the elements. Mendeleev's version was published first (1869) and Meyer's followed soon after. Figure 25.5 shows Mendeleev's table of the elements known at that time, as it appeared in the German translation of his work. He arranged the elements in order of increasing atomic mass, noting the periodic occurrence of elements with similar physical and chemical properties. This arrangement became known as the *periodic table* of the elements. Mendeleev left gaps in the table to fit the elements into the proper columns according to their properties, and proposed that the gaps represented elements not yet discovered. Subsequent discovery of these elements, with properties matching Mendeleev's predictions, helped win acceptance of his table among chemists. The absolute masses of atoms were still unknown, but chemists were now in agreement about their relative masses.

Avogadro's number was still unmeasured. The first reliable measurements of this quantity were published in 1908 by Jean Perrin. Perrin made

			Ti = 50	Zr = 90	? = 180
			V = 51	Nb = 94	Ta = 182
			Gr = 52	Mo = 96	W = 186
			Mn = 55	Rh = 104.4	Pt = 197.4
			Fe = 56	Ru = 104.4	Ir = 198
			Ni = Co = 59	Pd = 106.6	Os = 199
H = 1			Cu = 63.4	Ag = 108	Hg = 200
	Be = 9.4	Mg = 24	Zn = 65.2	Cd = 112	
	B = 11	Al = 27.4	? = 68	Ur = 116	Au = 197?
	C = 12	Si = 28	? = 70	Sn = 118	
	N = 14	P = 31	As = 75	Sb = 122	Bi = 210?
	O = 16	S = 32	Se = 79.4	Te = 128?	
	F = 19	Cl = 35.5	Br = 80	J = 127	
Li = 7	Na = 23	K = 39	Rb = 85.4	Cs = 133	Tl = 204
		Ca = 40	Sr = 87.6	Ba = 137	Pb = 207
		? = 45	Ce = 92		
		?Er = 56	La = 94		
		?Yt = 60	Di = 95		
		?In = 75.6	Th = 118?		

Figure 25.5 Mendeleev's periodic table of the elements as he arranged them in 1869. His predicted elements are marked in color.

quantitative measurements of **Brownian motion**, which is the continual irregular movement of minute particles suspended in a liquid. This motion was first observed in 1827 by the Scottish botanist Robert Brown. The cause of Brownian motion is molecular motion, as molecules in the liquid incessantly move and collide with one another. For a large object the net effect of collisions due to molecules of the surrounding fluids is zero because, on average, all the forces are balanced. But when the size of the object is sufficiently small, the instantaneous force is not zero. Thus a tiny particle suspended in a fluid moves wildly about with random motions. Figure 25.6 shows a drawing similar to those made by Perrin of a particle moving in a fluid. The line segments join the consecutive positions of the same particle at intervals of 30 s.

In addition, Perrin found that the suspended particles distribute themselves in the same way as do the particles of a gas. That is, the density of particles decreases exponentially with height in the same way that the

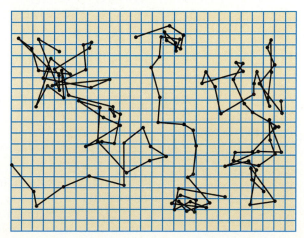

Figure 25.6 Drawings by Perrin of Brownian motion, made by connecting the consecutive positions of the same granules at intervals of 30 s.

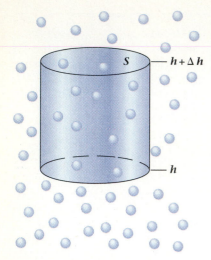

Figure 25.7 Particles with individual volume V suspended in a fluid move with random motion. The number of particles between h and $h + \Delta h$ in a column of cross section S is $nS\,\Delta h$, where n is the number of particles per unit volume.

density of the atmosphere decreases with height (Section 12.9). A pressure is associated with the motion of these particles, called osmotic pressure. It is proportional to the mean kinetic energy W of the particles. At a height h where the number of particles per unit volume is n, the pressure is

$$P(h) = \tfrac{2}{3}nW.$$

At greater height $h + \Delta h$, the number of particles changes to $n + \Delta n$, so

$$P(h + \Delta h) = \tfrac{2}{3}(n + \Delta n)W.$$

Because the system is in thermal equilibrium, the average kinetic energy W must be the same regardless of height.

Since the particles are suspended in a liquid of different density, a buoyant force acts on them. The forces on the particles due to osmotic pressure, gravitation, and bouyancy are balanced. Let the volume of each particle be taken as V, its density as ρ, and the density of the surrounding liquid as ρ'. The number of particles in the space between h and $h + \Delta h$ in a column of cross section S (Fig. 25.7) is given by $nS\,\Delta h$. The total force on the particles is the force per particle times the number of particles in the volume. The resultant of the gravitational force and the bouyant force (Section 9.3) is a downward force:

$$\text{downward force} = (nS\,\Delta h)g(\rho - \rho')V.$$

The upward force due to the osmotic pressure is

$$\text{upward force} = [P(h) - P(h + \Delta h)]S = -\tfrac{2}{3}WS\,\Delta n.$$

Upon equating these two forces, we see that

$$\frac{\Delta n}{\Delta h} = \frac{-g(\rho - \rho')Vn}{\tfrac{2}{3}W}.$$

The change with height in the number of particles per unit volume is proportional to the number of particles per unit volume. As we showed earlier (Chapter 12), equations of this type can be described with an exponential function,

$$n(h) = n(0)e^{-3g(\rho - \rho')Vh/2W}. \tag{25.2}$$

The average kinetic energy can be obtained from the ideal gas law (Section 12.6) as

$$W = \frac{3RT}{2N_A},$$

where R is the molar gas constant, T is the absolute temperature, and N_A is Avogadro's number. These two equations can be combined to give

$$N_A = \frac{RT}{g(\rho' - \rho)Vh}\ln\left[\frac{n(0)}{n(h)}\right]. \tag{25.3}$$

We now have an expression for Avogadro's number N_A in terms of measurable quantities.

In 1908 Perrin prepared a suspension of tiny particles of a yellow pigment used in water colors. He measured the density of particles and their diameter, from which he computed their volume. He then counted the number of particles at different heights in the solution. From these measurements he showed that, indeed, the number of particles per unit volume decreased exponentially with height as predicted, verifying Eq. (25.2). Perrin then used his measurements to determine the value of Avogadro's number. Perrin's initial result was a little larger than the value

$$N_A = 6.02 \times 10^{23} \text{ molecules/mole}$$

that we use today.

The atomic concept existed before Perrin's time. The kinetic-theory model of a gas, for example, had been successfully used to account for many observed gas properties. However, the success of the atomic model in kinetic theory and in chemistry was not proof of the reality of atoms and molecules. The concept might be only a convenient fiction useful for calculations. With Einstein's 1905 paper on Brownian motion, another way of subjecting the atomic hypothesis to a qualitative test was found. Perrin performed several other independent experiments on Brownian motion in addition to the one discussed here, all of them yielding similar values for Avogadro's number. Some of these experiments also tested the details of Einstein's predictions.

The agreement between the values found for Avogadro's number in independent experiments analyzed under the assumption of the reality of molecules, and the agreement with Einstein's prediction for Brownian motion, firmly established the reality of molecules and atoms. Because of Perrin's work the distinguished chemist Wilhelm Ostwald, long an opponent of the theory, wrote, "The atomic hypothesis is thus raised to the position of a scientifically well-founded theory."

The Size of Atoms

25.4

By making use of what we have just learned, especially the value of Avogadro's number, we can estimate the size of atoms. We are now in a position to determine the number of atoms that occupy a given volume. From there it is just a short step to determine the volume occupied by a single atom, that is, its size.

Example 25.2

How many copper atoms are there in a solid cube of copper which is 1.00 cm on each side?

Solution Chemical and physical measurements establish that copper has an atomic mass of 63.6 g/mol and a density of 8.96 g/cm^3. The mass of copper in the cube is

mass = density × volume = 8.96 g/cm^3 × 1.00 cm^3 = 8.96 g.

The number of moles present is given by the mass divided by the atomic mass:

$$\text{number of moles} = \frac{\text{mass}}{\text{atomic mass}} = \frac{8.96 \text{ g}}{63.6 \text{ g/mol}} = 0.141 \text{ mol}.$$

Finally, multiplying the number of moles by Avogadro's number, we get the number of atoms:

$$\text{number of atoms} = \text{number of moles} \times N_A,$$

$$\text{number of atoms} = 8.49 \times 10^{22} \text{ atoms.}$$

We see that one cubic centimeter of copper contains 8.48×10^{22} atoms, a very large number indeed. To find the volume occupied by one atom, we divide the total volume (1 cm^3) by the number of atoms, which gives an average volume per atom of 1.18×10^{-23} cm^3 or 1.18×10^{-29} m^3. If, for simplicity, we assign each atom a small cubic volume of that size, then the cube edges have a length equal to the cube root of the volume, or 2.28×10^{-10} m. Without making any assumption about the shape of atoms or how they are arranged, we may conclude that the linear dimensions of an atom are of this order of magnitude, that is,

$$\text{atomic size} \approx 10^{-10} \text{ m.}$$

Crystals and X-Ray Diffraction

25.5

Perrin's experiments proved the existence of atoms and molecules and determined the value of Avogadro's number, but other questions remained. Are atoms really indivisible? Do they have any internal structure? Today we know the answers to these questions, but the discoveries that led to these answers did not come all at once. We first discuss the discovery of x rays, which, as we will see, served as a primary tool for examining details of atomic structure.

During the late nineteenth century, improvements by Sir William Crookes in the production of vacuums led to pressures that were thousands of times smaller than those previously attainable. To investigate the behavior of electrical discharges in such a vacuum, researchers sealed electrodes into tubes and removed the air inside with a vacuum pump. Tubes of this sort became commonly known as Crookes tubes (Fig. 25.8). When a high voltage was applied across the electrodes, conduction took place within the tube and a green glow became visible in the glass at the positive end. Sometimes, fluorescent minerals were placed into the tube and were observed to glow when the current was present. The fluorescence was due to rays emanating from the cathode (the negative electrode), which were called **cathode rays**. We will see in Section 25.6 that the attempts to understand the nature of cathode rays led to the discovery of the first subatomic particle: the electron.

Among the experimenters who studied the behavior of the Crookes tube was the German scientist Wilhelm Konrad Roentgen (1845–1923). Late in 1895, Roentgen noticed that operating the tube caused a card coated with fluorescent material to glow, even though the card was outside the tube. Even more remarkable was Roentgen's observation that the card's

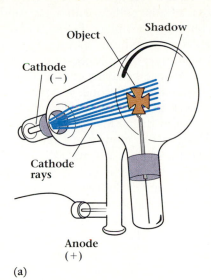

Figure 25.8 (a) A type of Crookes tube known as a Maltese Cross tube. The shadow of the cross seen on the face of the tube implies the presence of rays that emerge from the cathode and travel in straight lines. (b) The glow from the electric discharge and the shadow of the cross are visible.

(a)   (b)

fluorescence was still visible even if he completely surrounded the Crookes tube with black cardboard. In fact, the fluorescence was independent of whether the coated side or the plain side of the card faced the tube.

Roentgen concluded from these observations that some kind of unknown radiation from the Crookes tube was penetrating objects that were opaque to both visible and ultraviolet light. He named this radiation **x rays** and tested other materials to see if they were at least partially transparent to these rays. Roentgen observed that when a hand was held between a fluorescent screen and the Crookes tube, a shadow image of the bones could be seen within the lighter shadow of the hand itself (Fig. 25.9). This observation was seized immediately by the medical profession and within weeks of Roentgen's announcement, many medical researchers were experimenting with this important new discovery. In a very brief time, x rays became a routine diagnostic tool in hospitals all over the world. As a result, Roentgen became extremely famous and many people referred to x rays as Roentgen rays. In 1901 he was awarded the first Nobel Prize in physics for his discovery.

Roentgen's x rays were so intriguing that scientists began intense investigation and speculation about what they really were. The rays were not deflected by electric or magnetic fields, so they could not be streams of charged particles. They were not diffracted by diffraction gratings, so they were not waves with wavelengths close to those of light. However, there was a possibility that they could be electromagnetic waves with wavelengths very much shorter than those of visible light.

In 1912, Max von Laue (1879–1960) proposed using a crystal as a diffraction grating for x rays. As we have seen, a crystal was thought of as layer upon layer of atoms, all spaced apart at regular intervals. These intervals are of the order of 10^{-10} m, as shown in the previous section. Thus a beam of x rays striking a crystal might be diffracted, similarly to the way that light is diffracted by a grating. The experiment was carried

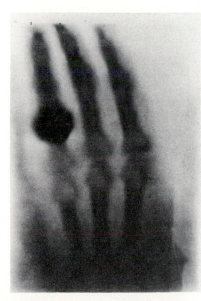

Figure 25.9 An x-ray photograph made by Roentgen at his first public lecture on his new discovery on January 23, 1896.

Figure 25.10 A diffraction pattern made by passing a beam of x rays through a small sodium chloride crystal. The symmetry of the pattern is a result of the cubic symmetry of the crystal.

out by Laue's colleagues W. Friedrich and P. Knipping, who bombarded a crystal of zinc sulfide with x rays and recorded the diffraction pattern on a photographic plate. Figure 25.10 shows a similar pattern of x rays diffracted from sodium chloride.

The diffraction of x rays was an historic event. First of all, it established the wave behavior of x rays and provided a means for measuring their wavelengths. Second, it confirmed the atomic structure of crystals and provided a means of determining the actual three-dimensional arrangement of the atoms within crystals.

Within a few months of Laue's discovery, W. L. Bragg (1890–1971), while still a student at Cambridge, worked out a simple relationship between the observed angles of x-ray diffraction and the spacing between planes of atoms in the crystal. Together with his father, W. H. Bragg, he determined the atomic crystal structure of a number of crystals.

Bragg reasoned that since x rays are very penetrating, they are only partially reflected from each layer of atoms as they pass through the crystal. Suppose the incident waves are specularly reflected from parallel planes of atoms, with each plane reflecting only a small fraction of the incident radiation. As with ordinary optical diffraction, for some angles of incidence the reflected beams are all in phase and a diffracted beam is observed (Fig. 25.11a). At other angles of incidence the waves from successive layers are not in phase and no diffraction is seen (Fig. 25.11b). For maximum intensity of the reflected beam, the waves from successive layers must be exactly in phase. Consequently, the path difference between the rays labeled 1 and 2 in Fig. 25.12 must be an integral number of wavelengths. This path difference depends on the angle of incidence θ and the separation between layers, d, and equals $2d \sin \theta$. Thus

Figure 25.11 Reflection of x rays from successive layers of atoms in a crystal. (a) The wavelets are reflected exactly in phase and reinforce each other. (b) The wavelets are reflected out of phase and cancel each other.

$$n\lambda = 2d \sin \theta, \qquad (25.4)$$

where n is an integer and λ is the x-ray wavelength. This statement is known as **Bragg's law**. The angle θ in the Bragg equation is often called

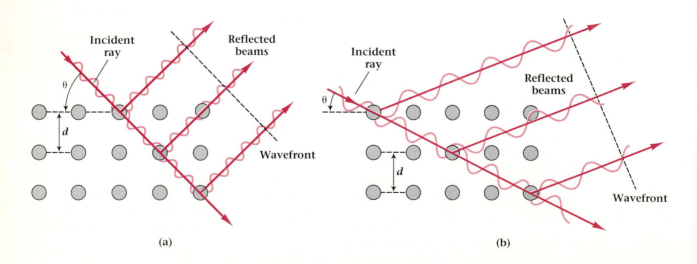

Incident ray   Reflected beams

θ

d

Wavefront

(a)

Incident ray   Reflected beams

θ

d

Wavefront

(b)

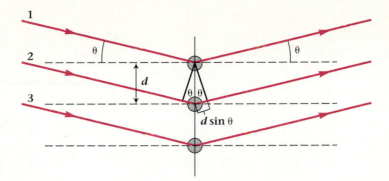

Figure 25.12 In Bragg scattering of x rays from a crystal, scattering maxima occur at angles where $2d \sin \theta$ is an integral multiple of wavelengths.

the Bragg angle. The scattering angle, which is the angle between the incident beam and the scattered beam, is 2θ.

Using Avogadro's number, we can compute the number of atoms in a crystal of known size, as in Example 25.2. From the number of atoms per unit volume, we can calculate the average spacing between atoms. This information, together with knowledge of crystal symmetry, means that we can calculate the spacing between layers, d, and then use Bragg's law to determine the wavelength of x rays. The smallest spacing between atoms in a crystal is typically about 2×10^{-10} m. From the Bragg equation we see that diffraction is not observed unless λ is smaller than $2d$. Thus the x rays used for diffraction have wavelengths smaller than the interatomic spacing in the crystal.

Example 25.3

Finding x-ray wavelength by diffraction.

In a silver bromide crystal the atoms are arranged in parallel planes 2.88×10^{-10} m apart. If diffraction is observed from these planes at a scattering angle $2\theta = 31°$, what is the wavelength of the x rays? Assume first-order diffraction, $n = 1$.

Solution Since the scattering angle is 31°, the Bragg angle is $\theta = 15.5°$. From Bragg's law we have

$$\lambda = 2d \sin \theta = 2(2.88 \times 10^{-10} \text{ m}) \sin 15.5°,$$
$$\lambda = 1.54 \times 10^{-10} \text{ m}.$$

We have discussed some of the properties of x rays, but not the details of how they are produced. According to Roentgen, x rays were "produced by the cathode rays at the glass wall of the tube." He also found that x rays were produced when other materials were struck by cathode rays. In modern terminology, x rays are produced when high-speed electrons strike matter. In a modern x-ray tube the electrons are "boiled off" a hot metal cathode and accelerated by an electric potential of 10^4 to 10^6 V in an evacuated tube (Fig. 25.13). When the electrons strike the positive target, or anode, their rapid deceleration produces x rays. Further details of x-ray production will be given in Chapter 26 after we have developed the necessary concepts.

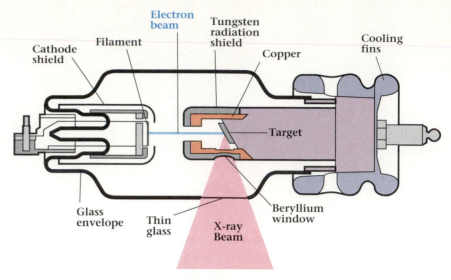

Figure 25.13   A schematic diagram of an x-ray tube with the anode shielded to reduce stray radiation. (After a design by Philips Laboratories.) Electrons leave the cathode filament and strike the anode target at high energies, producing the beam of x rays.

X rays differ from light in their frequency, which affects the way they interact with matter. The higher-frequency (and shorter-wavelength) x rays are more penetrating. Also, the penetrating power of the x rays is greater for higher accelerating potentials. Their relative transmission by matter depends on the material's composition. This range of penetrating power allows us to use x rays in a tremendous variety of applications including medical diagnosis and treatment, solid-state research, archeology, industry, and many other applications where it is desirable to "see beneath the surface" (Fig. 25.14).

Figure 25.14   (a) *A young Girl Reading*, painted by Jean-Honoré Fragonard about the year 1776. (National Gallery of Art, gift of Mrs. Mellon Bruce) (b) X-ray analysis reveals an image of a young man beneath Fragonard's painting.

(a)

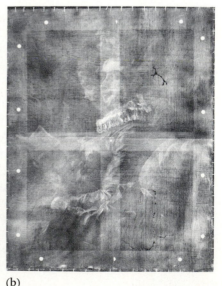

(b)

Discovery of the Electron

25.6

During the last decade of the nineteenth century, controversy abounded as to the nature of cathode rays. Were they a wave phenomenon similar to light? Or were they corpuscular radiation, consisting of tiny bits of matter? As we saw in Fig. 25.8, cathode rays travel in straight lines and cast shadows when small metal objects are inserted in their path. In addition, cathode rays are deflected by a magnetic field, unlike electromagnetic waves.

The particle theory of cathode rays was given a boost in 1895 when Perrin showed that they carried negative electric charge. Two years later, J. J. Thomson of the Cavendish Laboratory repeated Perrin's experiment and unequivocally confirmed his result. Thomson then set out to determine whether cathode rays could be deflected by an electric field. Previous attempts by others had shown no effect of electric fields on the path of cathode rays, which was a serious objection to the particle theory. For his experiment, Thomson built a special apparatus (Fig. 25.15a), which was more highly evacuated than those of earlier experimenters. When the air was removed from the tube and a potential difference applied across the electrodes, cathode rays emerged from the negatively charged cathode C and passed through slits in the grounded anodes A and B. The rays then traveled between the two parallel plates D and F and struck the end of the tube, where they produced a small, well-defined fluorescent spot.

When the two plates were connected to a battery and plate D was made negative with respect to F, the rays were deflected downward (Fig. 25.15b). When D was made positive with respect to F, the deflection was upward. Also, the magnitude of the deflection was proportional to the potential difference (and hence the electric field) between the plates. This

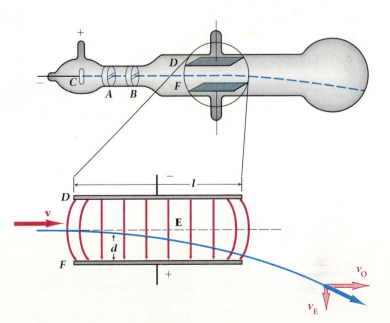

Figure 25.15 (a) Schematic diagram of Thomson's apparatus for determining the ratio e/m of an electron. (b) Motion of the cathode ray particles as they pass through the charged plates. In addition to their horizontal velocity v_0, which is constant, they acquire a transverse velocity v_E.

result lent support to the idea that cathode rays were really charged particles.

Let's follow Thomson's analysis of his experiment. Assume, as Thomson did, that the cathode rays are particles of mass m and electric charge e. A particle traveling between the plates will be transversely deflected because the electric field $\mathbf{E}$ exerts a force eE transverse to the particle's initial direction. The transverse acceleration due to the electric field is

$$a_{\mathrm{E}} = \frac{F}{m} = \frac{eE}{m}.$$

If the particle has an initial speed v_0 it will travel the length l of the plates in a time $t = l/v_0$. During that time, the particle acquires a transverse velocity v_{E} given by

$$v_{\mathrm{E}} = a_{\mathrm{E}}t = \frac{eE}{m}\frac{l}{v_0}.$$

PHYSICS IN PRACTICE

The Field Ion and Scanning Tunneling Microscopes

Atoms are extremely small. We have already seen that a reasonable estimate of atomic diameter is about 2×10^{-10} m. Thus atoms are thousands of times smaller than the wavelengths of visible light, and we cannot see atoms with our naked eyes. Even the use of an optical microscope is no help, since the wavelengths of light are so much greater than the size of the objects we wish to see.

To be able to observe the shapes of objects smaller than optical wavelengths we must use extraordinary techniques and instruments. One of the most interesting of these is the field ion microscope developed in 1951 by Professor Erwin W. Müller. Essentially, the ion microscope consists of an extremely sharp needle point raised to a high electric potential above a fluorescent screen in an evacuated enclosure (Figure B25.1). Within the evacuated chamber is a small amount of helium gas. The helium atoms become polarized by the large applied electric field and are drawn toward the positively charged needle point. As they get

very close to the tip, the electric field becomes very large, increasing the likelihood of ionization. Once an atom becomes ionized, it is accelerated away from the tip by the electric field. Its

motion lies along the direction of the field, that is, along the field lines. In this manner ions produced at a unique point on the tip are accelerated to a unique point on the screen. Thus the

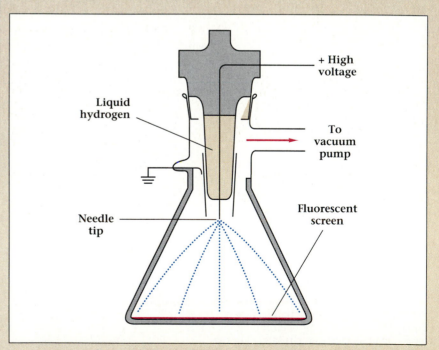

Figure B25.1   Schematic view of a field ion microscope. Dashed lines show the shape of the electric field between the needle point (object) and the fluorescent screen (image plane).

As the deflected particle leaves the region between the deflection plates and enters the field-free region, it makes an angle θ_E with the undeflected beam,

$$\theta_E \approx \tan \theta_E = \frac{v_E}{v_0} = \frac{eEl}{mv_0^2}. \tag{25.5}$$

Then Thomson deflected the cathode rays with a magnetic field, produced in the space between the electric deflection plates by a matched pair of current-carrying coils positioned outside the glass tube. Thomson shaped the coils to produce a nearly uniform magnetic field at right angles to the deflection plates and extending over the distance l. The acceleration of the particle due to just the magnetic field is

$$a_B = \frac{F}{m} = \frac{ev_0B}{m}.$$

pattern on the screen is an image of the tip of the needle.

When the ions impinge on the screen, flashes of light are recorded. If the rate of ionization is great enough, a steady pattern is observed (Fig. B25.2a). If the tip radius is small enough, we can obtain an image of the tip in which the bright spots of the pattern correspond to individual atoms on the tip's surface. Those atoms lying on the step edges of the atomic planes contribute a greater local field and thus are seen as bright spots on the screen.

Atoms sitting on top of an otherwise smooth atomic plane are also imaged brightly. Thus the ion microscope can directly reveal the atomic order in a metallic needle.

If a model of the needle tip, constructed from balls, is painted so that the balls on the edges can be easily seen, it appears as in Fig. B25.2(b). The similarity between the ball model and the microscope picture is striking.

Another microscope capable of examining surfaces on an atomic scale was developed in 1984 by Gerd Ben-

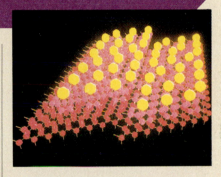

Figure B25.3 The atomic structure of a silicon crystal surface obtained from a scanning tunneling microscope. A unit cell of the atomic reorganization at the surface is highlighted with an expanded computer model.

ning and Heinrich Rohrer. The scanning tunneling microscope, as it is called, can resolve features even smaller than the size of an atom and does not require the high electric fields of the field ion microscope. A representative micrograph is shown in Fig. B25.3. A further description of the microscope and how it operates is found in Chapter 30. Although these microscopes are limited in the kinds of surfaces and materials they can image, they have produced easily understandable proof of the atomic theory: We can actually see pictures of the individual atoms.

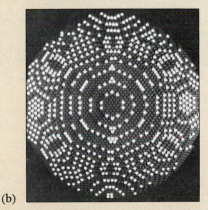

(a)   (b)

Figure B25.2 (a) Field ion micrograph of a tungsten tip with a radius of 25 nm. (b) A ball model of a 15-nm-radius tungsten tip with the protruding atoms enhanced. These are the atoms imaged by the field ion microscope. The similarity between the model and the micrograph is apparent.

The transverse velocity imparted to the cathode rays by the magnetic field is

$$v_B = a_B t = \frac{ev_0 B}{m} \frac{l}{v_0} = \frac{eBl}{m}.$$

Thus the angular deflection θ_B due to the magnetic field is

$$\theta_B \approx \tan \theta_B = \frac{eBl}{mv_0}. \tag{25.6}$$

By first deflecting the particles with the electric field alone, Thomson measured θ_E. Then by turning on and adjusting the magnetic field, he brought the beam of particles back to the undeflected position, thus making θ_B equal and opposite to θ_E. Under this condition, Eqs. (25.5) and (25.6) can be set equal to each other and we find that $v_0 = E/B$. Then, using this evaluation of v_0 in Eq. (25.6), we can determine the ratio of e/m to be

$$\frac{e}{m} = \frac{E\theta_E}{B^2 l}. \tag{25.7}$$

All the quantities on the right-hand side of this equation are measured, so in this way Thomson was able to determine the ratio of the electric charge to the mass of the cathode rays.* Today the accepted value is

$$e/m = 1.7588 \times 10^{11} \text{ C/kg}.$$

Thomson conducted his experiments with several different gases at low pressure in the evacuated tube. Each time the result was the same and was consistent with the idea that the cathode rays were charged particles. Upon considering the implication of all these experiments, he concluded that: (1) Atoms are not indivisible, because negatively charged particles had been torn from normally neutral atoms. (2) All cathode ray particles have the same charge and the same mass and are a part of all atoms. (3) If the charge e was the same size as the smallest unit of charge observed in electrochemistry, then the mass of the cathode ray particle was less than one-thousandth the mass of a hydrogen atom—the smallest mass known up to that time. Thomson first called these particles "corpuscles," but they soon became known as **electrons**, a name G. J. Stoney had coined several years earlier for the elementary electric charge found in Faraday's electrolysis experiments. Thus Thomson is considered the discoverer of the electron.

Following Perrin's measurement of Avogadro's number in 1908, the electric charge e was determined from N_A and the Faraday constant F by using Eq. (25.1). Shortly afterwards, in a series of experiments begun around 1909, the American physicist Robert A. Millikan (1865–1953) succeeded in measuring the charge of the electron with greater precision than anyone had done before him. Moreover, his work conclusively proved what Thomson had assumed, that the charge on each electron was exactly the same. In his experiment, Millikan examined the motion of single tiny

---

*Thomson actually determined the ratio m/e. It has become conventional to use the reciprocal ratio e/m in contemporary physics.

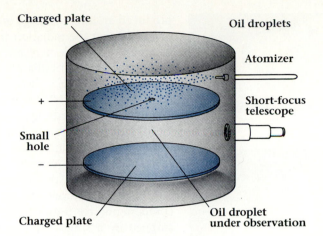

Figure 25.16   A schematic view of Millikan's oil drop experiment. Tiny drops of oil enter the space between the conducting plates through the hole in the top plate. Their motion under the simultaneous influence of the electric field between the two charged plates and the gravitational field is observed through the short-focus telescope.

drops of oil that picked up static charge from ions in the air. He suspended the charged drops in air by applying an electric field and watched them through a short-focus telescope (Fig. 25.16).

Millikan derived an expression for the charge q on the drop in terms of the large-scale measurable quantities. In a sequence of experiments he observed many drops and obtained a series of values for q. In practice, q was found to be an integral multiple of the elementary charge e. The magnitude of the elementary charge is

$$e = 1.6022 \times 10^{-19}\ \text{C}.$$

This is the charge of one electron. Using the best available value for e/m, we find that the electron's mass is

$$m_e = 9.1094 \times 10^{-31}\ \text{kg}.$$

Radioactivity

25.7

As the atomic theory became firmly established, modifications of some of its basic postulates became necessary. The discovery of the electron and of positive and negative ions showed that atoms were not indivisible. If atoms could be broken apart, what was their internal structure? How did electrons fit inside an atom? If atoms were not indivisible, then just how permanent were they? The next discovery that affected these issues came in 1896, only one year after Roentgen's announcement of x rays. This event was Henri Becquerel's discovery of radioactivity.

Becquerel was studying phosphorescent minerals that continued to emit light (fluoresce) after the cause of their excitation (usually sunlight) was removed. Some of these minerals emitted a radiation that would pass through opaque objects and darken a photographic plate. It was presumed that this radiation was associated with the phosphorescence. At one point, Becquerel developed a photographic plate that had been wrapped in heavy black paper and then left in the dark for several days beneath a piece of uranium sulfate. To his great surprise, he found that the mineral had

(a)

(b)

Figure 25.17   (a) A rock containing the uranium-bearing mineral uranophane. (b) An autoradiograph made by placing the rock on top of a photographic plate. The exposure is due to the radiation emitted by the mineral.

produced an intense image on the plate (Fig. 25.17). Becquerel soon demonstrated that the radiation was due to the presence of the uranium itself, and was independent of any phosphorescent effects. This emission of a penetrating natural radiation is called **radioactivity**.

After Becquerel's discovery, chemists searched for other radioactive materials. At first, of the many compounds tested, only those of uranium and thorium showed the property of radioactivity. However, in 1898 Marie Curie and her husband, Pierre, reported the isolation of two new radioactive elements, which they named polonium and radium. Radium turned out to be much more radioactive than either uranium or polonium.

Shortly after the discovery of radioactivity, Ernest Rutherford (1871–1937) showed that the radiation from uranium had two distinct components. He found that one component of the radiation was readily stopped by a single layer of thin aluminum foil. Rutherford named this easily absorbed component **alpha (α) rays.** The second, more penetrating component he called **beta (β) rays.** Further experiments by the French physicist Paul Villard showed that the radiation from uranium contained a third component that was even more penetrating than beta rays. This third type of radiation is called **gamma (γ) radiation.** Table 25.1 gives the relative range of penetrations of the three radiations from radium through aluminum absorbers. Table 25.2 lists the masses and charges of these radiations.

Over the course of several years, Rutherford gradually explained the nature of alpha rays. The absorption of alpha rays was proportional to the density of the absorbing material and the rate of absorption increased with distance traveled through the absorber. On the basis of this indirect evidence, Rutherford proposed that alpha rays were particles. Because they strongly ionized any gas through which they passed, Rutherford suggested they must possess electric charge. Charged particles should be deflected by a magnetic field, but such deflection was not observed initially for alpha rays. However, deflection was observed later. Rutherford explained the initial failure to notice deflection by suggesting that alpha particles were very massive. Subsequent measurements established that the alpha particles have an e/m less than 1/1000 that of electrons.

Rutherford and his associate Hans Geiger made accurate measurements of the electric charge carried by alpha rays, finding the charge of the alpha particle to be exactly twice the charge carried by a hydrogen ion. Using this information and the ratio of e/m, they showed that the mass of the alpha particle was the same as that of a helium atom. Separate experiments confirmed that alpha particles really are helium ions.

Many properties of beta rays were first observed by Becquerel. Using the experimental setup illustrated in Fig. 25.18, he found that beta rays were deflected by a magnetic field. Becquerel placed the radioactive material in a small lead container, open at the top. This container was then placed on top of a photographic plate, which was positioned in a magnetic field directed parallel to the plane of the plate. The beta rays emitted by the radioactive source were deflected by the magnetic field in a direction indicating that they carried a negative electric charge.

We have shown earlier that a particle of charge e and mass m, moving with speed v perpendicular to a magnetic field **B**, travels in a circular path

TABLE 25.1
Thickness of aluminum required to reduce radioactive intensity by one-half*

Radiation from radium	Thickness of aluminum
Alpha rays	0.0005 cm
Beta rays	0.05 cm
Gamma rays	8 cm

*E. Rutherford, *Philosophical Magazine.* Vol. 26, 1903, p. 177.

TABLE 25.2
Properties of natural radiation

Particle	Charge	Mass (in units of electron mass m_e)
Alpha (α)	$+2e$	7294
Beta (β)	$-e$	1
Gamma (γ)	0	0

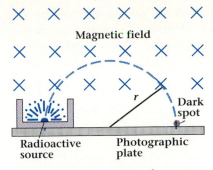

Figure 25.18 Experimental arrangement for showing the deflection of beta rays by a magnetic field. The direction of the field is into the plane of the figure.

of radius

$$r = \frac{mv}{eB}.\tag{25.8}$$

A beta particle initially directed upward from the lead box in Fig. 25.18 would be deflected by the magnetic field so that it strikes the plate a distance $2r$ from the source. With this arrangement, Becquerel found that the rays struck the plate over a range of distances, indicating a wide variation in the speeds of the beta rays. By placing thin absorbing foils over the plate, he discovered that the most easily deflected rays were also the most easily absorbed rays.

Becquerel also investigated the deflection of beta rays by electrostatic fields. By comparing the electrostatic deflection with the magnetic deflection, he was able to show that beta rays had the same ratio of e/m as cathode rays. In a series of experiments Becquerel established that beta rays are electrons, with velocities greater than those of cathode rays. These large velocities indicated that the radiation was not due to ordinary chemical behavior.

Gamma rays were found to have no charge and no mass. In this respect they are similar to x rays. Gamma rays are short-wavelength electromagnetic waves.

These three kinds of radioactive emissions represent a behavior so different from normal chemical behavior that their origins must also be different. As we will see, the discovery of the atomic nucleus (Section 25.9), which leads to a better understanding of atomic structure, reveals the source of radioactivity and completes our outline of the experimental evidence for our atomic model.

Radioactive Decay

25.8

The intensity of the radiation from uranium (or any other radioactive element) is independent of temperature, pressure, and chemical combination. It depends only on the quantity of uranium actually present. When the amount of uranium is increased, the amount of emitted radiation is proportionally increased. The measure of radioactivity is the number of radioactive emissions per second, called the **activity**. The SI unit of activity is the **becquerel** (Bq), which is one event per second.

Each radioactive atom is unstable and has a probability of decaying spontaneously into another, lighter atom with the emission of radiation. It does so independently and without any influence from its neighbors. However, when more atoms are present, the activity is proportionally increased. Mathematically, this means that the rate of decay, or activity A, is given by

$$A = \frac{\Delta N}{\Delta t} = -\lambda N,\tag{25.9}$$

where N is the number of radioactive atoms, ΔN is the number of decays in the time Δt, and λ is a proportionality constant (called the decay constant). The negative sign in Eq. (25.9) expresses the fact that the number

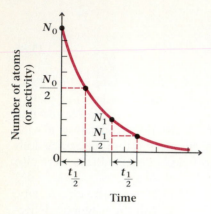

Figure 25.19 The number of atoms of a radioactive species decreases exponentially with time as the atoms decay to other species. The activity decreases in the same way because it is proportional to the number of atoms. The time interval for N_0 atoms to decrease to $N_0/2$ atoms is the same as the time for N_1 atoms to decrease to $N_1/2$ atoms. The time is the half-life $t_{1/2}$.

of radioactive atoms decreases. This equation has the same form that we have seen before in Chapter 12 and leads to an exponential relation between the number of unstable atoms and the time. If N_0 unstable atoms are present at time zero, then the number remaining at some time t is

$$N = N_0 e^{-\lambda t}. \tag{25.10}$$

A graph of Eq. (25.10) is shown in Fig. 25.19. Since the activity is proportional to the number of radioactive atoms present, the activity satisfies an exponential decay law also.

After a sufficient time has elapsed, the number of unstable atoms remaining is only one half of the initial number N_0. This time interval is known as the **half-life**. It is different for each type of radioactive atom. In each successive half-life, the number of radioactive atoms remaining is one-half the number present at the beginning of the interval. Table 25.3 shows the effect of radioactive decay on an initial sample of N_0 atoms.

The half-life is closely related to the decay constant λ. When the time interval is equal to one half-life, $t_{1/2}$, the number of atoms remaining is $N = N_0/2$ and Eq. (25.10) becomes

$$\tfrac{1}{2} = e^{-\lambda t_{1/2}}.$$

We can use this relation to obtain the half-life in terms of λ. Taking logarithms of both sides gives

$$\ln \tfrac{1}{2} = -0.693 = -\lambda t_{1/2}$$

or

$$t_{1/2} = \frac{0.693}{\lambda}.$$

Equation (25.10) can then be written

$$N = N_0 e^{-0.693 t/t_{1/2}}. \tag{25.11}$$

Notice that if the decay constant is large, the half-life is correspondingly short. Similarly, a small decay constant means a long half-life.

Example 25.4

Activity after two half-lives.

Radioactive cobalt-60 has a half-life of 5.26 years. A certain sample of cobalt-60 has an initial activity of A_0. What will its activity be 10.52 years from now?

Solution   The passage of 10.52 years is equivalent to two half-lives of cobalt-60. After one half-life the activity falls to $A_0/2$. After the second half-life it will be reduced by another factor of $\tfrac{1}{2}$ to $\tfrac{1}{2}(A_0/2) = A_0/4$.

Example 25.5

A particular sample of cobalt-60 ($t_{1/2} = 5.26$ years) has an initial activity A_0. What will its activity be after three years have passed?

Solution   We use Eq. (25.11), recognizing that both sides are proportional

TABLE 25.3
Effect of time on number of unstable atoms remaining

Time (in half-lives)	Number of unstable atoms remaining
0	N_0
1	$N_0/2$
2	$N_0/4$
3	$N_0/8$
4	$N_0/16$
5	$N_0/32$

to the activity:

$$A = A_0 e^{-0.693t/t_{1/2}}.$$

When we insert the value for the half-life and elapsed time we have

$$A = A_0 e^{-0.693(3.0/5.26)} = A_0 e^{-0.395} = 0.67A_0.$$

Two facts combine to make it possible to use the known half-life of radioactive carbon-14 as a kind of clock or calendar for determining the age of archeological objects. This process is known as radiocarbon dating. The first fact is that living plants take in carbon dioxide (CO_2) from the air. The second fact is that a small proportion of the carbon in the atmosphere is radioactive. The ratio of the amount of radioactive carbon-14 to nonradioactive carbon-12 is almost constant and is about 1.3×10^{-12}. (The distinction between the two types of carbon is not important here, but will be discussed in Chapter 28.) Radioactive carbon-14 is constantly decaying with a half-life of about 5700 years. It is constantly replenished from atmospheric nitrogen, which is converted into radioactive carbon-14 after bombardment by cosmic rays that reach the earth's atmosphere from all directions from outer space. The rates of decay and production are approximately equal.

As long as a plant is alive and growing, it takes in both radioactive and nonradioactive carbon in the form of carbon dioxide. But once the plant is cut, or is eaten by an animal or person, the intake of carbon stops and the amount of radioactive carbon begins to decrease as it decays. By determining the relative amount of radioactive to nonradioactive carbon, we can determine how long ago the object stopped living. For instance, if the relative amount of radioactive carbon in a piece of wood is one-half as much as it originally was, then one half-life of carbon-14, or about 5700 years, must have passed since the tree had been cut down.

Some uncertainty in the time estimate of radiocarbon dating occurs because the relative fraction of carbon-14 in the atmosphere changes with time over long periods. This difficulty is partly overcome by comparing radiocarbon dates with dates established by other techniques, such as tree-ring dating. Radiocarbon dating is useful until the carbon-14 remaining has become too small to measure reliably. This puts a practical limit on radiocarbon dating of about 40,000 years.

Other radioactive elements can be used to determine the age of some archeological objects and geological formations. For instance, the amounts of several different radioactive isotopes found in rocks independently indicate that the age of the earth's crust is about 5×10^9 years.

Discovery of the Atomic Nucleus

25.9

By 1909 physicists could piece together a description of the atom from assorted experimental evidence, yielding the following model: Atoms are about 10^{-10} m in diameter and are electrically neutral. They are composed of negatively charged electrons of small mass and positively charged ions,

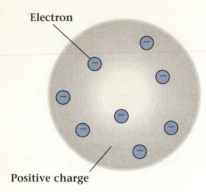

Figure 25.20   The plum-pudding model of the atom.

which contain most of the atomic mass. However, as the first decade of the century drew to a close, the arrangement of these atomic components was still unknown and, in the absence of any direct evidence, was the subject of much speculation.

Because electrons represent such a small part of the mass of the atom, one model of the atom consisted of a heavy positive charge spread uniformly over the entire atomic volume (Fig. 25.20). The electrons were imagined to be stuck in the positive charge (like plums in a pudding) in just enough numbers to exactly neutralize the positive charge. This model, advocated by J. J. Thomson, became known as the plum-pudding model.

By 1909 it was known that certain atoms, such as uranium, give off alpha and beta radiations, which penetrate thin layers of matter. Because of their enormous speeds, these particles must pass near or through any atoms that lie in their path. The deflection of an alpha particle, say, from its rectilinear path by its encounter with an atom would depend on both the magnitude and the distribution of the electric charge within the atom. Thus a powerful tool was available for obtaining information on the internal structure of the atom.

It was already established both theoretically and experimentally that a well-defined narrow beam of alpha rays was broadened by passage through air or very thin metal foil. This broadening was due to one or more collisions with atoms, which produced small-angle scattering, a well-understood phenomenon. In 1909 Hans Geiger and E. Marsden, while working with Rutherford, observed a striking new effect. A very small fraction of the alpha rays incident on a metal foil were scattered through such a large angle that they emerged from the same side of the foil from which they entered. This effect was very surprising. Because the alpha particles were comparatively heavy and were moving with high velocities, it was inconceivable that they should be scattered backward as a result of successive small angle scatterings. Also, since the alpha particles mostly penetrated the foil, it seemed surprising that there was anything to hit that could send them recoiling backwards.

Geiger and Marsden made their observations with the experimental arrangement shown schematically in Fig. 25.21. The alpha particles were directed toward a thin metal foil. A fluorescent screen was placed near the foil on the same side as the source of radiation. A thick sheet of lead was placed between source and screen to prevent the alpha particles from striking the screen directly. A microscope was brought close to the screen. Individual alpha particles that were scattered back toward the screen could be seen as tiny flashes of light where they struck the screen.

Geiger and Marsden studied the effect of using different materials for the scattering foil and found that materials of greater atomic mass scattered a greater number of alpha particles into the backward direction. Even with heavy elements for the foil, however, the backward scattering was very small. When the target foil was platinum, only about one alpha was scattered backward for every 8000 alphas incident on the foil.

Two years later, Rutherford published an explanation for the strong scattering of the alpha rays. He proposed that all of the positive charge of the atom was concentrated in a tiny **nucleus** at its center and that the compensating negative charge (due to the electrons) was uniformly dis-

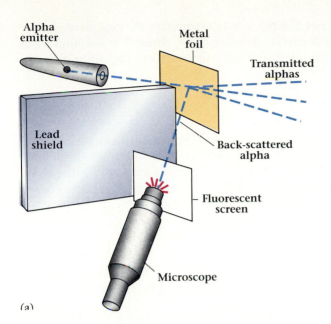

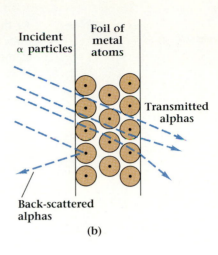

(b)

(a)

Figure 25.21 The Geiger and Marsden experiment. (a) A collimated beam of alpha particles bombards a thin gold foil; some alpha particles are scattered through a large angle and are observed by the flashes produced on a fluorescent screen. (b) The large-angle scattering occurs when the alpha particle passes very close to a nucleus.

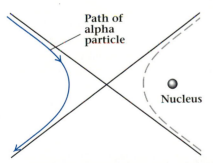

Figure 25.22 Rutherford scattering of an alpha particle by a nucleus. The colored line is the hyperbolic path of the alpha. If the force between the alpha and the nucleus were attractive, the trajectory would still be hyperbolic and would correspond to the gray line.

tributed over a sphere of radius comparable with the observed radius of an atom. He also found it necessary to assume that most of the mass of the atom was associated with the positive charge. For this model, the Coulomb force between the central positive charge and a deeply penetrating alpha particle could be many times greater than the force on the alpha due to the diffuse outer charge. When the distance of the positive alpha particle from the positive nucleus of the atom becomes small compared with the atomic radius, the repulsive force on the alpha becomes great enough to cause the large angle scattering.

Rutherford calculated the behavior of an alpha scattered by a nucleus so heavy that it could be considered to remain at rest during the collision. Using that assumption, together with the Coulomb force law, he showed that the general trajectory for the alpha was one branch of a hyperbola, with the nucleus of the atom as a focus (Fig. 25.22). (See page 13 for information on hyperbolas.) Experimental verification of Rutherford's scattering formula led to the acceptance of the nuclear model of the atom.

The heart of the technique used by Rutherford—that is, the measurement of subatomic properties by the scattering of incident particles—is still used today. Much of our knowledge of the nucleus (Chapter 28) and of its component parts (Chapter 31) comes from such experiments. It can even be said that most of our information about the physical world comes from scattering data, because to look at an object is to observe the scattering of light waves from it. We will find in Chapter 27 that light waves have particlelike properties also, so that the connection between scattering experiments is closer than it first appears.

We can estimate the size of an atomic nucleus by considering a head-on collision of an alpha particle with an atom, assuming that the Coulomb force law applies in this microscopic region. Initially, the alpha possesses a kinetic energy given by

$$KE = \tfrac{1}{2}m_\alpha v_\alpha^2.$$

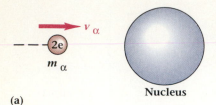

(a)

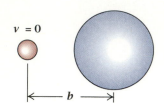

(b)

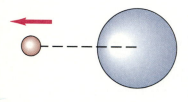

(c)

Figure 25.23 (a) Schematic diagram of an alpha particle approaching a heavy nucleus head-on. (b) At some point the alpha particle has zero velocity as it reverses direction. This point marks the distance of closest approach b, which gives us an upper limit for the radius of the nucleus. (c) Alpha particle rebounding from the nucleus.

As the alpha approaches the nucleus it encounters the repulsive electric force due to the nuclear charge, denoted by Ze, where Z is the number of elementary electric charges e present in the nucleus. The alpha's charge is $2e$ and the potential energy is

$$PE = k\,\frac{2Ze^2}{r},$$

where we have ignored the contribution to the potential due to the electrons. (The constant k is the proportionality constant from Coulomb's law, given in Chapter 15.) At the distance of closest approach b, the alpha particle's motion is reversed (Fig. 25.23). At this point, its kinetic energy is zero and its potential energy must equal its initial kinetic energy:

$$\tfrac{1}{2}m_\alpha v_\alpha^2 = \frac{k2Ze^2}{b}.$$

This equation sets an upper limit on the value of b and hence gives an estimate of the size of the nucleus. Upon rearranging, we find that

$$b = \frac{4kZe^2}{m_\alpha v_\alpha^2}. \qquad (25.12)$$

The speed of the alphas was known to Rutherford from deflection experiments and was about 1.8×10^7 m/s. Using the assumption that the atomic nucleus might be made up of alpha particles, it was thought that the number of electric charges in a gold nucleus should be about half its atomic mass of 197, say $100e$. The mass of the alpha was known to be about 6.62×10^{-27} kg. When these numbers are substituted into Eq. (25.12), we get

$$b = \frac{4(9 \times 10^9 \text{ N·m}^2/\text{C}^2)100(1.6 \times 10^{-19}\text{ C})^2}{6.62 \times 10^{-27}\text{ kg }(1.8 \times 10^7\text{ m/s})^2},$$

$$b = 4.3 \times 10^{-14}\text{ m}.$$

This value of b gives an upper limit to the nuclear radius. It is many times smaller than the value of 10^{-10} m that is typical for the atomic radius.

At the time of Rutherford's theoretical predictions there was no independent method of determining the charge on the nucleus. However, the general success of his predictions confirmed the validity of the theory. Acceptance of the nuclear theory in turn led to successful descriptions of the electronic structure of the atom, as we shall describe in the next chapter.

SUMMARY

Useful Concepts

- The observation that crystals have regular geometries and can be cleaved along smooth planes can be explained by saying that they are made up of submicroscopic components or building blocks, called atoms.
- The law of definite proportions states that the propor-

tions (by mass) of the elements in a chemical compound are constant. The law of volumes states that gases unite by volume in simple and definite proportions. These two statements led to Avogadro's law: Equal volumes of gases at the same temperature and pressure contain equal numbers of molecules.

- The experiments of Perrin on Brownian motion established the reality of atoms and led to the measurement of the charge on the electron.
- The size of atoms is of the order of 10^{-10} m.
- X rays are produced when electrons are suddenly stopped upon striking a target. The wave character of x rays is shown by the fact that they undergo diffraction. The Bragg law for the diffraction of x rays by crystalline atomic layers separated by a distance d is

$$n\lambda = 2d \sin \theta.$$

- Cathode rays were shown to be particles (electrons) by J. J. Thomson, who also measured the ratio e/m.
- The charge on the electron was measured in an experiment by R. A. Millikan. The presently accepted value for the electron charge is

$$e = 1.6022 \times 10^{-19} \text{ C}.$$

- In radioactive decay, the number of unstable atoms left after time t is

$$N = N_0 e^{-0.693t/t_{1/2}},$$

where N_0 is the number of atoms present at time 0 and $t_{1/2}$ is the half-life.
- Our model of the atom is due to Rutherford, who explained the strong backward scattering of alpha particles by proposing a small heavy positive nucleus surrounded by a diffuse sphere of electrons.

Important Terms

You should be able to write the meaning or definition of each of the following terms:

- unit cell
- electrolysis
- Avogadro's number
- Brownian motion
- cathode rays
- x rays
- Bragg's law
- electrons
- radioactivity
- alpha rays (or particles)
- beta rays (or particles)
- gamma rays
- activity
- becquerel
- half-life
- nucleus

QUESTIONS

25.1 What reasons might Hooke and Huygens have had for considering the smallest components of crystals to be spherical rather than, say, cubical?

25.2 When a small drop of some types of oil is placed on the surface of a bowl of water, it spreads out over a large area. How could this effect be used to determine an upper limit for the size of the molecules of the oil?

25.3 Explain why the law of definite proportions is a direct result of the atomicity of matter.

25.4 Is Avogadro's number a fundamental constant of nature, in the same way as the speed of light and as the charge on the electron, or is its value just happenstance?

25.5 Name some large-scale phenomena that would be observably different if Avogadro's number were 10^{20} times as large as its actual value.

25.6 What factors govern whether Brownian motion can be observed for a particle suspended in a fluid? Is it a matter of size only?

25.7 When one metal is electroplated onto another, the conducting solution used normally contains ions of the metal being deposited. Describe what happens at the anode and the cathode and in the solution.

25.8 Assume you have a radiation detector that is sensitive to alpha, beta, and gamma radiations. Describe a simple method for determining which radiations are present in a particular radioactive source.

25.9 In view of the Rutherford model of the atom, is the relative penetrating power of natural radioactivity to be expected? (See Table 25.1.) Explain.

25.10 Imagine that you could roll small hard balls at an obstacle on a smooth table. The obstacle is hidden from view, but you can see the directions of the outgoing balls. How much can you determine about the obstacle? Give several specific examples.

25.11 Which of the experiments discussed in this chapter best support the claim that all electrons have exactly the same charge?

25.12 Examine the periodic table of the elements and note those elements that, from your personal knowledge, have physical or chemical properties in common. Do these elements belong to particular rows or columns in the table?

25.13 What type of materials can be examined with the field ion microscope? What, if any, restrictions hold for the materials to be examined with the scanning tunneling microscope?

PROBLEMS

Hints for Solving Problems

Remember that in electrolysis the transfer of one faraday of electric charge corresponds to the transfer of one mole of electric charge. The amount of charge transferred is the product of current and time. In radioactive decay, the number of radioactive atoms (and thus the activity) decreases exponentially with time. If you know the number present at any time, say t, the number remaining at time Δt later is $e^{-\lambda \Delta t}$ times the number present at t, regardless of what time t you begin counting from.

Section 25.1   Evidence of Atoms from Solids and Gases

25.1 Place several pennies on a flat surface with their edges touching. Show by diagrams that if they are arranged to form a square array, they take up more surface area than when they are arranged in a hexagonal array.

25.2 When water is decomposed the weight of oxygen released is 8 times that of the hydrogen released. Oxygen and hydrogen combine in volume proportions of 1 to 2 to make water. Use this information, along with Avogadro's hypothesis, to find the ratio of the mass of a molecule of oxygen to that of a molecule of hydrogen.

25.3 Nitrogen may be combined with oxygen in weight proportions of 1.75 g nitrogen to 1.00 g oxygen to form the gas nitrous oxide, N_2O. The same elements may be combined in the ratio of 0.438 g nitrogen to 1.00 g oxygen. What is the atomic composition of the second compound?

Section 25.2   Electrolysis and the Quantization of Charge

25.4 In an electrolysis experiment a student plated silver onto an electrode, maintaining a current of 0.500 A for 40.0 min. How much mass of silver was deposited?

25.5 In an electrolysis experiment a current of one ampere is maintained for 30.0 min during which copper is deposited on the negative electrode. Copper has an atomic mass of 63.6 and its ions are divalent; that is, each ion carries two units of electric charge e. Calculate how much copper (mass) was deposited on the electrode.

25.6 In an electrolysis experiment a student plated copper from a solution of copper ions. The student did not know whether the ions were singly charged or doubly charged. After a current of 0.25 A was passed

for 40 min the electrode was removed, dried, and weighed. The mass of copper deposited was 0.396 g. Were the copper ions singly or doubly charged?

25.7 Hydrogen gas and oxygen gas are generated by passing an electric current through water. What volume of hydrogen (H_2) can be liberated in one hour in a cell that carries a current of 10 A?

25.8 Hydrogen and oxygen gas are generated by passing an electric current through water. What volume of oxygen gas (O_2) can be liberated in one hour in a cell that carries a current of 6.0 A?

25.9 Calculate the volume in liters occupied by one mole of a gas at standard temperature and pressure. (*Hint:* Remember the ideal gas law.)

Section 25.3   Avogadro's Number and the Periodic Table

25.10 In his study of Brownian motion, Perrin measured the numbers of particles at four distinct heights: 5 mm, 35 mm, 65 mm, and 95 mm. The concentrations at those heights were proportional to the numbers 100, 47, 22.6, and 12. Graph these numbers on semilogarithmic paper to see if the concentration depends exponentially on the height.

25.11 In another study, Perrin measured the concentration of particles at four levels, with each successive level 10 mm above the preceding one. The concentrations in a given area were 100, 43, 22, and 10. Graph this information on semilogarithmic paper. At what height above the first layer will the concentration drop to 50?

25.12 The density of particles n in a suspension is $n = n_0 e^{-\lambda h}$, where λ is a constant and h is the height. Show that the density falls to $n_0/2$ at the height $h = 0.693/\lambda$.

25.13 The density of particles n in a suspension is found to go as $n = n_0 e^{-\lambda h}$, where λ is a constant and h is the height. What is the value of λ if the density falls to $n_0/2$ when $h = 10 \ \mu m$?

Section 25.4   The Size of Atoms

25.14 Silver has a density of 10.5 g/cm³ and an atomic mass of 107.9 g/mol. Calculate the number of atoms in a cubic centimeter of silver.

25.15 Silicon has a density of 2.23 g/cm³ and an atomic mass of 28.09 g/mol. Calculate the number of atoms in a cubic centimeter of silicon.

25.16 Aluminum has an atomic weight of 26.98 g/mol and a density of 2.70 g/cm³. (a) How many atoms are there in a cube of aluminum 1.00 cm on an edge? (b) Calculate the average volume per atom. (c) If you treated the volume per atom as a cube, what would be the length of the cube edge associated with a single atom?

Section 25.5 Crystals and X-Ray Diffraction

25.17 X rays of wavelength $\lambda = 1.54 \times 10^{-10}$ m are aimed at a crystal with interplanar spacing of 2.88×10^{-10} m. Beams diffracted from these planes emerge for more than one angle of incidence. Determine the allowed values of 2θ.

25.18 X rays of $\lambda = 0.709 \times 10^{-10}$ m are diffracted from a set of crystal planes separated by $d = 1.44 \times 10^{-10}$ m. Calculate the allowed values of 2θ.

25.19 First-order diffraction of x rays of wavelength $\lambda = 0.709 \times 10^{-10}$ m occurs for a certain set of planes in a nickel crystal. If the scattering angle $2\theta = 23.2°$, what must be the spacing of the planes responsible for the diffraction?

25.20 X rays of $\lambda = 0.709 \times 10^{-10}$ m are diffracted in second order ($n = 2$) from a silver bromide crystal. If the scattering angle is $2\theta = 28.50°$, what is the spacing between the atomic planes?

25.21 X rays are Bragg-diffracted in third order ($n = 3$) from a cesium chloride crystal. If the wavelength of the x rays is $\lambda = 0.709 \times 10^{-10}$ m and the scattering angle is $2\theta = 29.99°$, what is the interplanar spacing?

25.22 X rays are Bragg-diffracted from a cesium chloride crystal with interplanar spacing of 4.11×10^{-10} m. In first order the scattering angle 2θ is $7.59°$. What is the wavelength of the x rays?

25.23 X rays are Bragg-diffracted from a cobalt crystal with interplanar spacing of 4.07×10^{-10} m. In first order, the scattering angle 2θ is $24.0°$. What is the wavelength of the x rays?

Section 25.6 Discovery of the Electron

25.24 An electron of mass m_e and charge e, moving with speed v perpendicular to a magnetic field **B**, travels in a circle of radius r. Show that the ratio of charge to mass of the electron is

$$\frac{e}{m_e} = \frac{v}{Br}.$$

25.25 Calculate the radius of the circular path of an electron accelerated through an electric potential of 200 V and injected into a region of magnetic field $B = 5.00 \times 10^{-3}$ T.

25.26 Electrons moving with a speed of 1.00×10^7 m/s pass through a pair of parallel plates 1.00 cm long separated by 8.00 mm (Fig. 25.15b). (a) If a potential of 100 V is applied to the plates, what is the angle of deflection of the electron beam? (b) What

is the linear deflection on a fluorescent screen placed 20.0 cm beyond the deflection plates?

25.27 Electrons in a cathode ray tube are accelerated through a potential of 19.0 kV. The electron beam then passes through a pair of deflection plates 2.00 cm long separated by 4.00 mm. What potential must be applied to the plates to achieve a 3.00-cm deflection of the beam on a fluorescent screen 30.0 cm beyond the deflection plates?

25.28 Calculate the angular deflection of an electron beam traveling at 1.0×10^7 m/s through a region of magnetic field $B = 1.0 \times 10^{-3}$ T that is 1.5 cm long.

Section 25.7 Radioactivity

25.29 Calculate the radius of curvature of the path of an alpha particle of mass 6.6×10^{-27} kg and speed 2.4×10^7 m/sec perpendicular to a magnetic field of 0.20 T.

25.30 Calculate the radius of curvature of the path of a beta particle of mass 9.1×10^{-31} kg and speed 2.0×10^7 m/sec perpendicular to a magnetic field of 0.50 T.

25.31 The hydrogen ion has a mass of 1.67×10^{-27} kg and a charge of $+e$. Compare the radius of curvature of the path of a hydrogen ion to that of an alpha particle of mass 6.6×10^{-27} kg of the same speed v in the same magnetic field **B**.

25.32 Charged particles are deflected in a circular path of 6.60 m radius in a magnetic field of 5.00×10^{-2} T. The speed of the particles is known to be 1.60×10^7 m/s. What is their ratio of charge to mass? Could the particles be alpha particles?

25.33 Alpha particles are deflected by a magnetic field of 6.40×10^{-2} T into a curved path with a radius of 4.26 m. What is the speed of the alpha particles? (*Hint:* See Table 25.2.)

Section 25.8 Radioactive Decay

25.34 Radioactive bismuth (^{210}Bi) undergoes beta decay with a 5.0-day half-life. How long will it take a sample of initial activity A to decrease to one eighth of its initial activity?

25.35 A sample of polonium has a half-life of 3.0 min and an initial activity of 10×10^6 Bq. What is its activity after 30 min have elapsed?

25.36 Radioactive ^{57}Co has a half-life of 270 days. How long will it take for a source of activity A to decrease its activity to $A/10$?

25.37 The activity of a certain sample was measured at
• intervals over a 12-h period. The data are given on page 764. Plot the decay curve on semilogarithmic graph paper with activity along the logarithmic coordinate and time along the linear coordinate. De-

termine the disintegration constant and the half-life. What would you expect the activity to have been at $t = 6$ h?

Time (h)	Activity (disintegrations/min)	Time (h)	Activity (disintegrations/min)
0	8550	7	2140
1	7015	8	1740
2	5750	9	1439
3	4720	10	1182
4	3890	11	970
5	3165	12	795

25.38 The penetration of gamma rays through copper was
• studied by measuring the number of gamma rays that pass through successive copper layers. The data recorded are given below. Plot the data on semilogarithmic graph paper to verify that the transmitted intensity is $I = I_0 e^{-mx}$, where m is a linear absorption coefficient and x is the thickness of the copper. Plot intensity along the logarithmic coordinate and thickness along the linear coordinate. Also compute the half-thickness, that is, the thickness at which an incident beam of gamma rays is diminished by one-half.

Activity (events/minute)	Thickness (cm)
21700	0
20600	0.1
18300	0.3
14500	0.6
11900	0.95
9560	1.3
6450	1.9
5050	2.2
3470	2.9

25.39 The ratio of radioactive to nonradioactive carbon in a sample of wood is found to be one-third that found in a recently cut piece of wood. Approximately how long ago was the first sample cut?

Section 25.9   Discovery of the Atomic Nucleus

25.40 What is the distance of closest approach of an alpha particle of speed $v = 1.8 \times 10^7$ m/s to a lead nucleus of $Z = 82$? (*Hint:* Assume that the lead nucleus remains at rest.)

25.41 The electric charge on a uranium nucleus is $+92e$, where e is the elementary charge. What is the distance of closest approach b of an alpha particle with initial velocity of 2.0×10^7 m/s directed straight at the uranium nucleus? (*Hint:* Assume that the uranium nucleus remains at rest.)

25.42 What is the distance of closest approach of an alpha particle of speed $v = 1.70 \times 10^7$ m/s to a gold nucleus of $Z = 79$? How close would the alpha get to a copper nucleus of $Z = 29$? (*Hint:* Assume that the gold and copper nuclei remain at rest.)

Additional Problems

25.43 Derive Eq. (25.3) from the two equations immedi-
• ately preceding it.

25.44 Calculate the radius of curvature of the path of an
• alpha particle of mass 6.6×10^{-27} kg moving with a kinetic energy of 8.7×10^{-14} J in a magnetic field of 1.00 T.

25.45 Electrons in a cathode ray tube pass through a pair
• of deflection plates 2.00 cm long separated by 4.00 mm. What electric potential must be applied to offset an angular deflection of 15° (= 0.26 radian) caused by a magnetic field $B = 1.00 \times 10^{-2}$ T?

25.46 One gram of uranium (^{238}U) contains 2.5×10^{21}
• atoms. Its half-life is 4.5×10^9 years. What is the activity of 1.0 g of uranium?

25.47 Actinium (^{227}Ac) has a half-life of 22 years. One
• gram of actinium contains 2.65×10^{21} atoms. (a) What is the activity of one gram of actinium? (b) What is the activity in curies? One curie is 3.7×10^{10} disintegrations per second and is the activity of one gram of pure radium.

25.48 A small radioactive source of gamma rays is moni-
• tored by a detector placed 5.0 cm from the source. The detector records an average count rate of 4320 counts/min. What would be the count rate if the detector were moved back to a position 10 cm from the source? (*Hint:* Assume that the radiation is emitted uniformly in all directions and that the size of the detector is fixed.)

25.49 A small radioactive source of beta rays is monitored
• by a detector placed 4.0 cm from the source. The detector records 5280 counts/min. What is the count rate when the detector is moved back to a position 9.0 cm from the source? (*Hint:* Assume that the radiation is emitted uniformly in all directions and that the size of the detector is fixed.)

25.50 A particle of mass m_0 and speed v_0 collides head-on
• with a free stationary atom of mass M. Using conservation of kinetic energy and momentum, find the recoil velocity and energy of the incident particle and of the atom when $M = 50m_0$ and when $M = 10m_0$. What can you say about Rutherford's approximation that the struck nucleus was so heavy that it did not move during the collision?

26

Origins of the Quantum Theory

26.1 Spectroscopy

26.2 Balmer's Series

26.3 Blackbody Radiation

26.4 Planck's Hypothesis

26.5 The Photoelectric Effect

26.6 Bohr's Theory of the Hydrogen Atom

26.7 Successes of the Bohr Theory

*26.8 Moseley and the Periodic Table

A WORD TO THE STUDENT

To understand atomic phenomena, we need both a model of the atom and a theory that describes the behavior of atoms. In this chapter we examine some of the experiments that led to our present understanding of the atom. These same observations also contributed to the development of quantum mechanics, which is discussed in the next chapter. The ideas of Planck, Einstein, and Bohr are especially important. However, all of the concepts introduced here need to be understood before you continue to the next chapters. You will find that ideas such as "atomic energy levels" play a fundamental role in explaining observations in areas as diverse as chemical reactions, cosmology, lasers, and integrated circuits.

TIME LINE

Johann Balmer finds a formula for the visible series of spectroscopic lines of hydrogen.

1814

1895

1900

Joseph Fraunhofer discovers absorption spectra in sunlight. He is the first to use diffraction gratings to measure spectral wavelengths.

Max Planck proposes quantization of energy to explain the shape of the blackbody radiation curve.

During the early twentieth century, several seemingly unrelated developments in physics converged to provide the basis for a surprising new model of atomic structure. One contribution was the collective observations from the study of optical spectra. Another was Planck's explanation of the distribution of the energy of light emitted from an incandescent object. Planck's proposal that the energy carried by the light depended on the frequency of the light was in turn used by Einstein to explain an observation known as the photoelectric effect.

These diverse ideas, along with Rutherford's discovery of the atomic nucleus in 1911, led Niels Bohr in 1913 to conceive a radically new atomic model. Bohr's theory successfully explained many of the puzzles that had arisen in the studies of optical spectra; they also provided an explanation of x-ray spectra. His ideas led directly to the development of modern quantum theory and its applications, including lasers and semiconductors.

Bohr's theory was highly successful, but had its limitations as well. Today we have replaced Bohr's original theory with the quantum mechanical approach to atomic structure, described in the next chapter. However, many concepts introduced by Bohr are essentially correct and helped point the way to contemporary quantum theory. His atomic model is still useful as a tool for visualizing the quantum description of atoms. This is another example of the evolution of ideas in physics, in which one successful explanation becomes modified by later ideas, yet some of the original assumptions are accepted in later theories.

Bohr's theory was a key step in another major development in physics: the overthrow of the Newtonian world view. Einstein's theory of special relativity showed that Newtonian mechanics was insufficient to explain the behavior of objects moving at speeds approaching the speed of light. Bohr's theory of the atom showed that classical physics also failed to explain the behavior of objects on the extremely small scale of atomic dimensions. Thus this chapter documents the progress of the second great revolution of modern physics, the early development of quantum mechanics.

Henry Moseley uses x-ray spectroscopy to relate atomic number to the charge of the nucleus.

1905

1913

1914

Albert Einstein uses Planck's quantum idea to explain the photoelectric effect, introducing the concept of photons of light.

Niels Bohr makes the first quantum model of the hydrogen atom, which successfully explains the wavelength of the spectral lines.

Spectroscopy

26.1

The beginning of a new understanding of atomic structure was based on observational evidence, much of it from the spectra of light emitted or absorbed by the various chemical elements. These observations were important in the development of new theories because the theoretical models had to be consistent with the observations. Initially, **spectroscopy** was the study of optical spectra, which result from dispersing light into its component colors, that is, spreading the light according to its wavelength (Section 23.9). Eventually, this idea was extended to other regions of the electromagnetic spectrum, so that today spectroscopy includes the study of spectra from microwaves, infrared rays, ultraviolet rays, and x rays, as well as visible light. Spectroscopy in all its forms plays a major role in determining the chemical composition of substances; much of modern chemistry depends on spectroscopic analysis of materials.

Since the Middle Ages, people have known that introducing various substances into a flame changes the color of the flame. For example, copper produces a green flame and potassium produces a violet one. By the end of the nineteenth century, it was known that if an electric current is passed through a low-pressure gas, the gas emits light whose color is characteristic of that particular gas. (This effect finds commercial use today in the ubiquitous neon sign.)

As we saw in Chapter 23, Newton used a prism to disperse sunlight into a spectrum of visible colors. Then, around 1750, the Scotsman Thomas Melville began investigating the spectra emitted by various substances held in a flame. He found that the distribution of wavelengths (colors) emitted was not the same, but varied from one element to another. Eventually, as the quality of prisms improved, scientists were able to catalog spectra, associating particular sets of emitted wavelengths with particular chemical elements. It became clear that since each kind of atom produces a different spectrum, the cause of spectra—whatever it is—must be connected with the nature of atomic structure.

Today we think of spectra as the fingerprints of matter. Every atom or molecule emits its own characteristic spectrum of light—no two spectra are alike. We may produce spectra by heating a gas at low pressure in an electronic discharge tube. *Emission spectra* appear as a series of bright lines, each representing a particular wavelength of light (Fig. 26.1). (Each line is actually the image of a slit, as described in Section 23.9.) *Absorption spectra* appear as a series of dark lines, not always equivalent to the bright lines of emission spectra. Each line represents a particular wavelength of radiation absorbed from an otherwise continuous spectrum of transmitted light. The cataloging of the different line patterns was an important step in being able to identify the chemical composition of substances isolated in chemical and biological research. However, the nineteenth-century theories of physics could not explain why the spectral patterns differed.

Some of the most important work in cataloging spectra was done by the lensmaker and physicist Joseph Fraunhofer (1787–1826). He discovered a series of dark lines (an absorption spectrum) present in the otherwise continuous solar spectrum. These Fraunhofer lines established the presence of individual chemical elements in the sun. But the explanation for the particular patterns of atomic spectra was still unknown.

BACK TO THE FUTURE

Fraunhofer and the Solar Spectrum

What is the sun made of? Does it consist of the same chemical elements we have on earth, or is it formed from its own kind of matter? The first clue to the answers to these questions came in 1814 from an analysis of the spectrum of sunlight by Joseph Fraunhofer. Fraunhofer was a skilled lensmaker, as well as a physicist, and invented new methods and machines for measuring, grinding, and polishing lenses and prisms. He discovered a method of accurately computing the shapes of lenses and learned how to make larger pieces of optical quality glass than his predecessors could make.

Fraunhofer wanted to determine the index of refraction of different kinds of glass for different colors of light, so that he might design more accurate achromatic lenses. During this work he discovered a pair of bright yellow lines in the spectrum of the light from an oil lamp. He found that similar lines were also present in other kinds of firelight. The bright lines always occurred at the same place in the spectrum, and so would be useful in determining refractive indices. We now know that these lines are due to the presence of the element sodium. Modern sodium vapor lamps, easily recognized by their distinctive orange color, are often used for outdoor lighting because of their efficiency.

To determine accurately the wavelengths of the yellow lines, Fraunhofer constructed a spectroscope (see Chapter 23) consisting of a narrow slit placed some distance from a flint glass prism. He observed the spectrum of the lamp through a telescope directed at the prism. He next studied sunlight to see whether the bright lines were there, too. To his surprise, he found a large number of dark lines in the otherwise

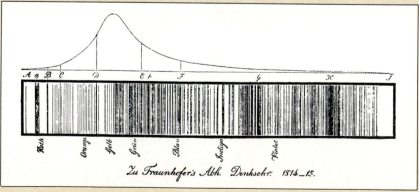

Figure B26.1   Fraunhofer's drawing of the dark lines observed in the solar spectrum.

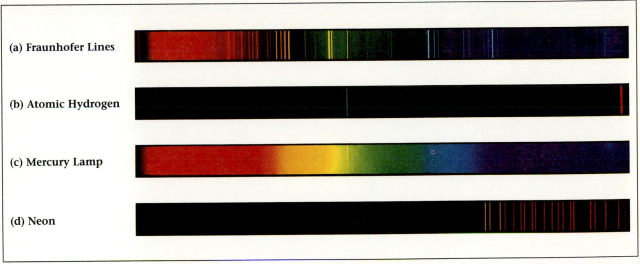

Figure 26.1 (a) Continuous spectrum of the sun showing the absorption lines; (b) emission spectrum of hydrogen; (c) emission spectrum of mercury; (d) emission spectrum of neon.

bright solar spectrum. At the exact position of the yellow lines seen in the firelight, he saw dark lines rather than bright lines. Fraunhofer's sketch of what he saw is shown in Fig. B26.1.

Fraunhofer measured the wavelengths of the more prominent lines, labeling them with the letters from A to K. The lines coinciding with the bright lines in the firelight became the D lines. Because of his contribution to the discovery of the dark lines in the solar spectrum, these lines have become known as the Fraunhofer lines.

Fraunhofer was also the first person to use diffraction gratings to observe spectra. He formed his own gratings of fine wire, closely spaced, and used them to make the first measurement of optical wavelengths. His measurement of the mean wavelength of the D lines was 588.8 nm, a value which was quite close to the presently accepted one of 589.3 nm.

By the late nineteenth century, the identification of spectra with chemical elements on the earth led to the realization that Fraunhofer's lines indicated the presence of chemical elements in the sun. For example, the fact that the D line is part of the Fraunhofer series proved the presence of sodium in the sun.

Other prominent parts of the Fraunhofer series indicate the presence of hydrogen and helium, which are known to be the major constituents of the sun. (Helium was discovered in the analysis of solar spectra before it was discovered on earth. The word helium comes from the Greek *helios*, the sun.) Later study of spectra from starlight showed that the sun and the stars are made of the same material. In short, the sun is a star.

The study of spectra has continued to be a major source of information about astronomical objects. For example, ionized gases have a slightly different spectral pattern from nonionized gases, so by analysis of spectral intensities we can estimate the percentage of ionization in a star. This, in turn, enables us to estimate the temperature of stars or interstellar dust clouds. The Doppler shift of spectral lines, de-

Figure B26.2 The Horsehead Nebula. The red light in the photograph is due to hydrogen emission; the blue light is starlight scattered from dust clouds.

scribed in Chapter 24, provides information on the speeds of objects in outer space. Our picture of the universe, on the enormous scale of galaxies as well as on the tiny scale of atoms, is built up in large part from spectroscopic observation and analysis (Fig. B26.2).

Balmer's Series

26.2

Fraunhofer's work spurred a great interest in spectroscopy, leading to the development of better techniques and instruments. By the late nineteenth century, spectroscopy had become a well-developed field of physics. The spectra of most elements had been carefully measured, and detailed tables of wavelengths were available. But still, the reasons for the existence of spectral lines were not understood.

In 1885 a Swiss schoolteacher, Johann Jacob Balmer, found a simple mathematical formula that related the wavelengths of the prominent lines in the visible and near-ultraviolet spectrum of hydrogen gas. (Hydrogen exhibits one of the simplest atomic spectra.) Balmer obtained the wavelengths of the four visible hydrogen lines by multiplying a constant number by the numerical coefficients 9/5, 4/3, 25/21, and 9/8. At first glance, these four numbers seem to have no simple relation, but Balmer realized that if the numerator and denominator of the second and fourth of these numbers are multiplied by 4, the series becomes 9/5, 16/12, 25/21, 36/32. Now the relation is more apparent: Each numerator is the square of an integer, and the denominator is 4 less than the numerator. Thus Balmer generated a formula for the wavelength λ of the hydrogen lines,

$$\lambda = 364.56 \text{ nm } \frac{n^2}{n^2 - 2^2},\qquad(26.1)$$

where n is an integer that takes on the values $n = 3, 4, 5, 6, \ldots$ Equation (26.1) is *Balmer's formula* and the lines observed in the visible spectrum of hydrogen are called the **Balmer series** (Fig. 26.2).

Using his formula, Balmer calculated the wavelengths of the nine lines (four visible and five ultraviolet) that were then known to exist in the spectrum of hydrogen. Balmer's formula was strictly empirical. That is, it was not derived from any physical model or theory of physical behavior; instead, Balmer offered his formula simply as a mathematical relationship that was consistent with observations. There was no apparent reason why it should work. Nevertheless, it provided amazingly precise computation of the wavelengths in the hydrogen spectrum. Even for the worst case,

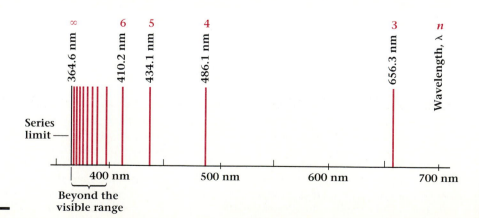

Figure 26.2 Sketch of the Balmer lines in the spectrum of atomic hydrogen.

which occurred for $n = 11$, Balmer's calculated wavelength was within 0.1% of the measured value.

In announcing his formula, Balmer suggested it might possibly be a special case of some more general formula that would apply to other series of lines in other elements. The Swedish spectroscopist J. R. Rydberg sought to find such a formula. From the large amount of data available in 1889, Rydberg found several spectral series that fit an empirical formula that he showed was equivalent to the Balmer formula. The **Rydberg formula** can be written to give the reciprocal of the wavelength of the emitted light as

$$\frac{1}{\lambda} = R \left(\frac{1}{n_1^2} - \frac{1}{n_2^2} \right), \qquad n_1 < n_2, \tag{26.2}$$

where R is the *Rydberg constant,* $10{,}973{,}731.534$ m^{-1}, and n_1 and n_2 are integers. For the Balmer series, $n_1 = 2$ and n_2 takes on values of 3, 4, 5, 6, ... As n_2 becomes very large the lines converge toward a *series limit.* Later observations by other spectroscopists confirmed additional spectral series in hydrogen corresponding to other values of n_1, including $n_1 = 1$, 3, 4, and 5.

By the year 1900, mathematical formulas were known that could provide extremely accurate calculations of the spectral lines in hydrogen. And yet, as far as atomic structure was concerned, no one had devised any model that could account for the existence of the observed spectra, nor could anyone explain why Rydberg's formula worked as well as it did.

Example 26.1

Calculate the wavelength of the line in the Balmer series with the longest wavelength.

Solution   To give the largest wavelength λ, the right-hand side of Eq. (26.2) must be as small as possible. This means that n_1 and n_2 should differ by unity; or when $n_1 = 2$, as it does for the Balmer series, then $n_2 = 3$. We then have

$$\frac{1}{\lambda} = R \left(\frac{1}{2^2} - \frac{1}{3^2} \right),$$

$$\frac{1}{\lambda} = (1.097 \times 10^7 \text{ m}^{-1}) \left(\frac{1}{2^2} - \frac{1}{3^2} \right) = 1.5236 \times 10^6 \text{ m}^{-1},$$

$$\lambda = 656 \text{ nm}.$$

This line is in the red, or long-wavelength, part of the visible spectrum.

Blackbody Radiation

26.3

About the same time that spectroscopists were wrestling with the problems of spectral series, other scientists were studying the continuous spectrum of light emitted from a hot object. The interest was in part theoretical, but it was also a practical interest growing out of the needs of the growing

TABLE 26.1
Color scale of temperatures

Color of glowing object	Approximate temperature (°C)
Incipient red	500–550
Dark red	650–750
Bright red	850–950
Yellowish red	1050–1150
Incipient white	1250–1350
White	1450–1550

public—and home—illumination industry. The eventual explanation for the shape of this spectrum departed from classical physics and pointed the way to a key aspect of the new model of the atom.

When a solid object is heated to several hundred degrees Celsius, it becomes incandescent; that is, it glows. The radiation emitted by an incandescent object forms a continuous range of wavelengths, part of which lies in the visible range. As the object's temperature increases, the relative intensities of the light across the visible spectrum change. This causes a perceptible shift in the observed color, which can be used to estimate the temperature (see Table 26.1).

An ideal object for producing incandescent light is called a **blackbody**. At low temperatures, it would absorb all of the radiation that fell on it. (A blackbody is also a perfect radiator.) Such an object might be an enclosed oven or kiln with only a tiny opening for light to get out. When the oven is cold, light from outside that passes into the opening is not reflected back (Fig. 26.3a) and so the hole looks completely black. When the oven is very hot, light emitted inside passes out through the hole. Figure 26.3(b) shows the blackbody radiation emitted by a kiln operated at a temperature of 1400 K. The light depends only on the temperature of the oven and not on the oven's composition.

Measurements of the spectra of blackbody radiation show that for any temperature T, one wavelength has a greater intensity than all others. This wavelength λ_m decreases with increasing temperature according to the rule

$$\lambda_m T = 2.90 \times 10^{-3} \text{ m·K.} \tag{26.3}$$

This rule is known as the **Wien displacement law**.

Figure 26.3 A small kiln of the type used to fire ceramics behaves nearly like an ideal blackbody cavity. (a) At a temperature of approximately 300 K the kiln is very dark inside. Light entering from outside is absorbed. (b) When the kiln is heated to 1400 K, the interior glows brightly and the light emitted inside the cavity passes out through the opening.

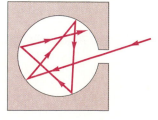
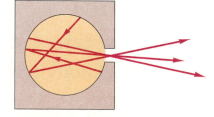

(a)   (b)

Example 26.2

Wavelength corresponding to blackbody temperature.

A photoflood lamp operates at a filament temperature of 3400 K. What is the wavelength λ_m of the peak in its blackbody spectrum?

Solution   The wavelength λ_m obtained from the Wien displacement law is

$$\lambda_m = \frac{2.90 \times 10^{-3} \text{ m·K}}{3.40 \times 10^3 \text{ K}} = 853 \text{ nm}.$$

The spectral peak lies in the near-infrared region, just beyond the visible spectrum, and most of the energy radiated is not visible.

Figure 26.4 shows the radiant energy distribution for a blackbody at several temperatures. As the temperature increases, the wavelength of the maximum intensity λ_m shifts to smaller values, according to the Wien displacement law. The total radiant energy is proportional to the area under the curve, which grows rapidly with increasing temperature. (According to the Stefan-Boltzmann law, Section 10.7, energy varies as T^4.) The radiation appears as a continuous spectrum, having neither the bright lines seen in flame spectra and in gas discharge spectra nor the dark lines seen in the light from the sun. Instead, all wavelengths are present over a wide range. (Note that the curve for 6000 K corresponds to the approximate temperature of the sun and that the peak for λ_m occurs in the visible part of the spectrum.)

Numerous attempts were made to explain the shape of the blackbody spectrum. Theories based on the well-established laws of classical electromagnetism and thermodynamics failed to reproduce the experimental measurements adequately. The explanation of blackbody radiation was simply not to be found in classical physics. The shape of the blackbody radiation curve was first derived as a result of a novel, *ad hoc* hypothesis of Max Planck (1858–1947).

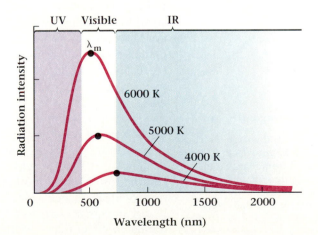

Figure 26.4   Blackbody spectra for several different temperatures. The peak of the curve (λ_m) shifts to shorter wavelengths at higher temperatures. The total amount of energy radiated is proportional to the area under the curve.

Planck's Hypothesis

26.4

In 1900 Planck reported his discovery of a formula that accurately described the shape of the blackbody spectrum for all wavelengths and temperatures. Having found an empirical formula that fit the observations, he then sought a physical reason for the success of the formula. Planck's thinking began, much like other theories of his day, by supposing the existence of identical vibrating oscillators within the walls of a hot cavity like the kiln in Fig. 26.3(b). When a large number of oscillators, such as might be found in any ordinary-sized object, exchange energy, they do so by the emission and absorption of electromagnetic radiation. Radiation is emitted when an oscillator makes a transition from one energy level to a lower one. Absorption of radiation is an inverse process in which the oscillator jumps from a lower energy level to a higher one.

At this point Planck broke with tradition. In order for the theory to fit his successful empirical formula, Planck was forced to assume that the energy of each oscillator was proportional to its frequency f. The same is true for the energy of the radiation, that is,

$$E = hf, \tag{26.4}$$

where the constant of proportionality h is a universal constant. Thus Planck postulated that vibrational energy is *quantized,* that is, limited to certain discrete quantities. Planck himself determined the value of h from previous experimental measurements of blackbody radiation. The value of h, which today is called **Planck's constant**, is

$$h = 6.626 \times 10^{-34} \text{ J·s.}$$

We should emphasize here that in formulating his theory of blackbody radiation, Planck did not draw upon any direct evidence of energy quantization, nor did his results come from extension of the classical theories. Instead, he introduced the **quantum** concept as a modification of classical ideas that brought his theory into agreement with experimental observations. (The word *quantum* has the same origin as *quantity,* and means the smallest possible unit of energy.) Physicists of the time had considerable doubt as to the validity of the quantization of electromagnetic radiation. Planck originally suspected that it was a mathematical trick that did not correspond to reality. However, his formula for blackbody radiation commanded attention because of its striking agreement with observations (Fig. 26.5).

The replacement of the view that energy flowed like a smooth unbroken stream of water, by one in which energy needed to be thought of as coming in little packets, marked the beginning of quantum mechanics and the end of a time in which all physical explanations were in terms of continuous flows or motions. However, in common with many revolutionary ideas, Planck's idea had little influence when it was first proposed. It did not gain credence until Einstein used it to explain a seemingly unrelated effect that is discussed in the next section.

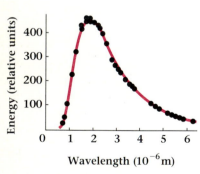

Figure 26.5 Spectral energy distribution of a blackbody at 1596 K. The line is computed from Planck's formula for blackbody radiation. The circles show observations by Coblentz, published in *Bull. Nat'l. Bureau of Standards,* **13**, 436 (1916).

Example 26.3

Wavelength of a quantum of radiation.

What is the wavelength of a quantum of radiation whose energy is 3.05×10^{-19} J?

Solution   From the Planck relation, the energy of the radiation is

$$E = hf = \frac{hc}{\lambda}.$$

The wavelength is obtained from

$$\lambda = \frac{hc}{E} = \frac{(6.626 \times 10^{-34} \text{ J·s})(3.00 \times 10^8 \text{ m/s})}{(3.05 \times 10^{-19} \text{ J})},$$

$$\lambda = 652 \text{ nm}.$$

The radiation lies in the red portion of the visible spectrum.

The Photoelectric Effect

26.5

By the year 1905, many areas of physics had undergone a transition from confidence to confusion. The existence of line spectra was still unexplained. Blackbody radiation had been successfully explained, but only on an assumption that contradicted classical physics. X rays and radioactivity had been discovered and also were unexplained. However, the next ten years saw the beginning of understanding for all these areas of physics, including an explanation of the photoelectric effect. Einstein's photoelectric theory, proposed in 1905 (the year of his special theory of relativity), shed new light on Planck's idea of the quantization of energy. Along with Rutherford's proposal in 1911 that atoms have nuclei, it led to Bohr's 1913 theory of atomic structure.

In 1887 Heinrich Hertz was studying the generation of electromagnetic waves with a spark gap. He found that electrical discharges between the electrodes in the gap were enhanced when ultraviolet light was allowed to shine on those electrodes. This enhancement was an unexpected behavior that could not be readily explained. Later investigators showed that a freshly polished zinc plate, when negatively charged, would lose its charge when exposed to ultraviolet light, while a positively charged plate showed no such effect. The conclusion was drawn that a negatively charged plate emitted negatively charged particles when illuminated with ultraviolet light. This emission of negative charges from material as a result of light falling on it is called the **photoelectric effect.**

The nature of the negatively charged particles was unknown at first, but was resolved by J. J. Thomson's discovery of the electron in 1897. The negatively charged particles emitted in the photoelectric effect were found to be electrons. Still, the phenomenon could not be explained on the basis of classical physics.

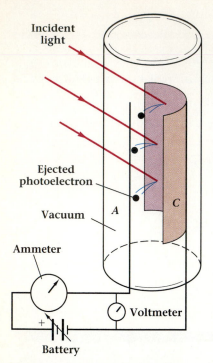

Figure 26.6 Schematic view of apparatus for observing the photoelectric effect. When light strikes the plate *C*, it ejects electrons that are collected at the wire *A*, causing a current.

To examine the photoelectric effect quantitatively, consider an evacuated quartz tube containing two electrodes (Fig. 26.6). In normal operation the thin wire *A* is held at a positive potential with respect to the curved metal plate *C*. When light of a given frequency strikes the plate *C*, it ejects electrons from its surface. These electrons (sometimes called photoelectrons) are attracted to the wire by the positive electric potential and give rise to a measurable current. If we gradually reduce the potential applied to the wire until it becomes negative with respect to the plate, some of the electrons ejected from the plate will not have enough kinetic energy to reach the wire. They will be repelled back to the plate, reducing the current. As the potential of the wire is made still more negative, the current decreases until it becomes zero at a potential called the *stopping potential* V_s. The maximum kinetic energy of the photoelectron is the energy equivalent of the stopping potential.

Experiments with this apparatus show that for light of a given frequency the magnitude of the current produced (that is, the number of photoelectrons per unit time) depends on the intensity, or brightness, of the light (Fig. 26.7). However, the stopping potential is independent of the intensity of the light and is different for different materials (Fig. 26.8a, b). Thus the maximum kinetic energy of the photoelectrons is also independent of the light intensity. Instead, the electron's maximum kinetic energy *does* depend on the *frequency* of the light falling on the cathode (Fig. 26.9).

In addition to determining the photoelectron's maximum kinetic energy, the frequency of the light enters into the photoelectric effect in

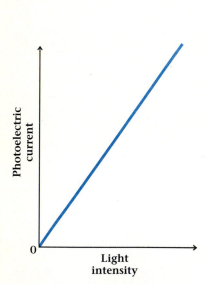

Figure 26.7 Photoelectric current plotted against the intensity of the incident light for a case in which photoelectrons are emitted.

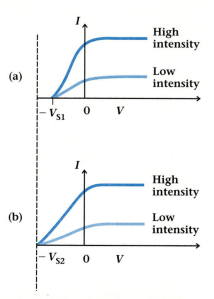

Figure 26.8 Photoelectric current plotted against the retarding potential for two different intensities of light. The stopping potentials (V_s) are different for (a) material 1 and (b) material 2.

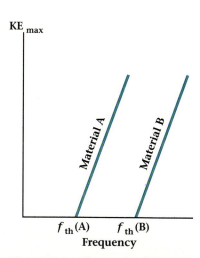

Figure 26.9 The maximum kinetic energy of photoelectrons as a function of the frequency of the incident light for two materials. Each material exhibits a threshold frequency f_{th}.

TABLE 26.2
Photoelectric work functions of selected metals

Metal	Work Function ϕ (eV)*
Cesium	2.14
Potassium	2.30
Sodium	2.75
Silver	4.74
Copper	4.94
Gold	5.31
Platinum	5.65

*1 eV = 1.6×10^{-19} J.

another way. Not all frequencies of light cause photoelectric emission from a given substance. Instead, each cathode material exhibits a **threshold frequency** f_{th}. Illumination with light of frequency less than f_{th} will not cause the ejection of photoelectrons, no matter how great the intensity of the radiation. But illumination with a frequency greater than f_{th} produces photoelectrons instantaneously, even if the light intensity is very small.

We may analyze these experimental observations by noting that they fall into two classes: those effects that can be explained by classical physics and those effects that cannot be so explained. One effect that is understandable on the basis of classical ideas is the increase in current with an increase in the intensity of light. The principal observations that have no classical physics explanation are as follows:

1. *Electrons are emitted only when the frequency of the light is above some threshold value, no matter how intense the light.* The classical expectation is that you should be able to expel electrons if you provide enough energy in the form of sufficiently intense light.
2. *The maximum kinetic energy of the emitted electrons depends on the frequency of the light.* According to classical physics, you could increase the electron's kinetic energy with any frequency of light simply by making the light more intense.
3. *The photoelectrons are emitted almost at once when the light strikes.* The expectation of classical physics is that the photoelectrons will absorb energy over a period of time as the light continues to shine, eventually gaining enough energy to escape the material.

The explanation of the photoelectric effect was given in 1905 by Albert Einstein. His theory was beautiful in its simplicity and accounted for all of the experimental observations. Einstein hypothesized that incident light consists of streams of Planck's energy quanta, called **photons**. When monochromatic light shines on the cathode in the photoelectric experiment, the photons penetrate the surface and give up their energy to the electrons of the cathode. The simplest process envisioned by Einstein was one in which the entire energy of the photon is given up to a single electron. The energetic electron then makes its way through the surface and escapes from the material. A certain minimum amount of energy must be acquired by the electron before it can escape from the material of the cathode. This energy is called the **work function** ϕ of the particular material used and is the basis of the explanation of the threshold frequency. Table 26.2 lists the photoelectric work functions for selected metals.

Einstein's idea incorporated Planck's quantum hypothesis into a statement of energy conservation. The energy of the incident photon must equal the energy needed to free the electron plus the electron's kinetic energy:

$$hf = \text{energy to free the electron} + KE.$$

The maximum kinetic energy of the photoelectrons is then the difference between the energy of the incident photon and the minimum energy ϕ necessary to free the electron from the material. That is,

$$KE_{max} = hf - \phi. \tag{26.5}$$

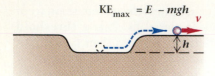

$$\text{KE}_{max} = E - mgh$$

Figure 26.10   A mechanical analogy of the photoelectric effect. If a ball in a ditch is given kinetic energy E greater than mgh, it will escape from the ditch. The maximum kinetic energy after escape is $\text{KE}_{max} = E - mgh$.

This equation is known as the **Einstein photoelectric equation.** Figure 26.10 presents a mechanical analog of the photoelectric effect.

The minimum (threshold) energy required to remove an electron from the surface with no kinetic energy is given by Eq.(26.5) with KE = 0. Thus the threshold frequency is defined in terms of the work function ϕ as

$$hf_{th} = \phi. \tag{26.6}$$

For lower frequencies, where hf is less than ϕ, no emission of photoelectrons occurs. At frequencies for which hf exceeds the work function, photoelectrons are emitted in numbers proportional to the number of incident photons (that is, the light intensity) and with kinetic energy maxima given by the Einstein equation. Subsequent experiments by Millikan verified Einstein's equation and provided an independent measurement of the Planck constant.

Einstein's successful explanation of the photoelectric effect furthered the acceptance of the idea of the quantization of radiation, even though the particlelike behavior of the photon seemed contradictory. On one hand, the light is treated like a wave having a frequency f. On the other hand, a single photon transfers all of its energy in an encounter with a single electron, a particlelike behavior. This dual behavior is characteristic of all waves and particles on the atomic scale.

In discussing photoelectrons, and afterwards as well, we find it convenient to use a unit of energy suitable to atomic-scale interactions. As we have shown in Chapter 16, whenever a charge q is moved through an electric potential difference V, an amount of work $W = qV$ is required. If the amount of charge is the electron charge, e, and the potential difference is 1 V, then the work done is

$$W = (e)(1 \text{ V}) = (1.602 \times 10^{-19} \text{ C})(1 \text{ V}) = 1.602 \times 10^{-19} \text{ J}.$$

This amount of work or energy is given the special name **electron volt,** abbreviated eV:

$$1 \text{ eV} = 1.602 \times 10^{-19} \text{ J}. \tag{26.7}$$

The electron volt is a practical unit for measuring energy on a molecular or smaller level. It is a natural unit to use in the photoelectric effect for both the work function and the electron kinetic energy.

The electron volt is a valid unit of energy regardless of how the energy is acquired. In the case at hand, the electrons gain energy from an encounter with a photon. Upon escaping from the material, the electrons have a kinetic energy that is conveniently expressed in electron volts, though it was not attained by actually moving through a particular potential difference.

Example 26.4

Energy of a photon.

What is the energy in electron volts of a photon of green light of wavelength 546 nm?

Solution   The energy of the photon is given by

$$E = hf = \frac{hc}{\lambda},$$

$$E = \frac{(6.626 \times 10^{-34} \text{ J·s})(3.00 \times 10^8 \text{ m/s})}{(546 \times 10^{-9} \text{ m})(1.60 \times 10^{-19} \text{ J/eV})} = 2.28 \text{ eV}.$$

Example 26.5

The photoelectric work function for cesium is 2.14 eV. What is the maximum kinetic energy of electrons ejected from the surface of cesium by light of wavelength $\lambda = 546$ nm?

Solution   The maximum kinetic energy is given by the Einstein equation as

$$\text{KE} = hf - \phi = \frac{hc}{\lambda} - \phi,$$

which is

$$\text{KE} = \frac{(6.626 \times 10^{-34} \text{ J·s})(3.00 \times 10^8 \text{ m/s})}{(546 \times 10^{-9} \text{ m})(1.60 \times 10^{-19} \text{ J/eV})} - 2.14 \text{ eV},$$

$$\text{KE} = 2.28 \text{ eV} - 2.14 \text{ eV} = 0.14 \text{ eV}.$$

Example 26.6

Find an expression for the energy of a photon in electron volts when the wavelength of the photon is given in nanometers.

Solution   We begin with the Planck formula,

$$E = hf = \frac{hc}{\lambda},$$

and insert the values of h, c, and the conversion factors:

$$E = \frac{(6.626 \times 10^{-34} \text{ J·s})(3.00 \times 10^8 \text{ m/s})}{\lambda(10^{-9} \text{ m/nm})} \frac{1 \text{ eV}}{1.60 \times 10^{-19} \text{ J}},$$

$$E = \frac{1240}{\lambda} \text{ eV·nm}.$$

This is a useful rule for people who work with light and x rays.

Bohr's Theory of the Hydrogen Atom

26.6

The experimental evidence of spectral lines, blackbody radiation, and the photoelectric effect were all clues to a deeper understanding of the nature and behavior of atoms, but these clues were subtle and not readily interpreted. In 1911 Rutherford added a vital contribution when he proposed

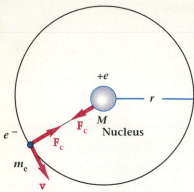

Figure 26.11 A positively charged nucleus with a single electron traveling in a closed orbit around it. The centripetal force on the electron is given by Coulomb's law.

that the atom had a tiny, positively charged nucleus that held most of the atom's mass.

Rutherford's description of a positively charged nucleus surrounded by electrons conflicted with classical electromagnetic theory. If the electrons were assumed to be at rest, no possible stable configuration could be found; the electrons would just fall in toward the nucleus. On the other hand, if they revolved in a circular orbit about the nucleus, they would undergo constant acceleration toward the center. Maxwell's equations predict that the accelerating charges would radiate away their energy and spiral into the nucleus. What is it, then, that keeps the atom from collapsing?

In 1913 the Danish physicist Niels Bohr published his answer to the question. Bohr had worked a year with J. J. Thomson and a year with Rutherford. Upon returning home to Copenhagen he combined Rutherford's concept of the nucleus with the quantum hypothesis of Planck and Einstein to develop a radically new theory of the atom. Bohr evaded the classical difficulties by observing that the existence of stable atoms implied that the laws of classical electrodynamics were not appropriate for the description of atoms and therefore must be altered.

Bohr began by considering the simplest case, that of a positively charged nucleus with a single electron traveling in a closed orbit around it (Fig. 26.11). This model corresponds to the hydrogen atom. He further assumed that the mass of the electron was negligibly small compared with the mass of the nucleus. Then he postulated that:

1. **The electron revolves about the nucleus in stationary, or stable, nonradiating orbits corresponding to fixed energy states**. Even though this postulate is not consistent with classical physics, Bohr assumed that while the electrons are in stationary orbits they may be treated with classical dynamics, including Newton's laws, Coulomb's law, and the ordinary conservation rules.

2. **The atom emits light only when the electron makes a sudden change from one energy state to another**. Bohr assumed that when radiation does take place, the relationship between the energy radiated and the frequency is given by Planck's law.

Bohr applied conservation of energy by assuming that the energy of a photon emitted by an atom is exactly equal to the energy difference between two energy levels (Fig. 26.12). Thus when an electron in an atomic energy level E_2 makes a transition to a lower energy level E_1, a photon is emitted with an energy

$$hf = E_2 - E_1. \tag{26.8}$$

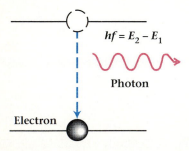

$hf = E_2 - E_1$

Photon

Electron

Figure 26.12 The energy of an emitted photon is equal to the energy difference between two energy levels, or states.

This equation is known as the Bohr frequency condition.

By using these postulates, Bohr was able to calculate the energy levels of the hydrogen atom and thereby account for the frequencies of the Balmer lines observed in the hydrogen spectra. The derivation we present here is not the one given in Bohr's original paper. In that paper he showed that, as a result of his assumptions, it follows that angular momentum (first discussed in Chapter 7) is also quantized for the motion of electrons in the hydrogen atom. The quantization of angular momentum is an extremely important concept for atomic and smaller-scale phenomena, and we will use it in succeeding chapters. By employing the quantization of angular momentum from the outset we can considerably shorten the derivation of the consequences of Bohr's assumptions.

We begin by assuming an electron of mass m traveling with speed v in a circular orbit of radius r about a stationary nucleus. For such an orbit Bohr showed that the angular momentum L of an electron is

$$L = mvr = n\frac{h}{2\pi}, \qquad n = 1, 2, 3, 4, \ldots,\qquad (26.9)$$

where h is Planck's constant and n is an integer. This equation expresses the requirement that the orbital angular momentum of an electron in a stationary state must be an integer multiple of $h/2\pi$. Thus we have a quantization rule for angular momentum: **Angular momentum is quantized in units of $h/2\pi$.**

In the Bohr model of the hydrogen atom, the electron travels in a circular orbit about a stationary nucleus of charge $+e$. For the electron to travel in a stationary, or stable, orbit, the electrostatic attraction due to the nucleus must provide the force required for circular motion. Thus the centripetal force mv^2/r is equal to the electrostatic force, given by

$$F_e = \frac{e^2}{4\pi\epsilon_0 r^2},$$

where ϵ_0 is the permittivity constant introduced in Chapter 15. When we equate the centripetal force to the electrostatic force and use the angular momentum of Eq. (26.9) to eliminate the speed, we get

$$r_n = \frac{\epsilon_0 n^2 h^2}{\pi m e^2}, \qquad n = 1, 2, 3, 4, \ldots\qquad (26.10)$$

Equation (26.10) gives the radii of the orbits in the Bohr model of the atom. Only those orbits corresponding to integral values of n are allowed (Fig. 26.13). The integer n is called the **principal quantum**

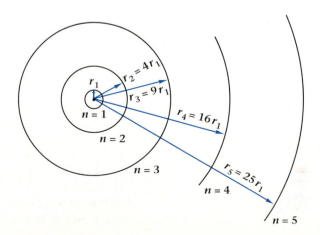

Figure 26.13 Sketch of the allowed electron orbits in Bohr's model of hydrogen. The radii increase with the square of n.

number, for it establishes an electron's orbit and, as we will see, the allowed energy for the electron.

You should not equate the Bohr radii r_n with atomic radii. The idea of electron orbits as circles with sharply defined radii contradicts experimental evidence. However, for hydrogen, good agreement exists between the predictions of Bohr's theory and experimental observations of spectra and energy levels.

Example 26.7

Radius of the smallest Bohr orbit of hydrogen.

Calculate the radius of the smallest orbit of hydrogen.

Solution   The smallest orbit corresponds to the situation in which the principal quantum number takes its smallest value, $n = 1$. From Eq. (26.10) we have

$$r_1 = \frac{1^2 \cdot h^2 \epsilon_0}{\pi m e^2},$$

$$r_1 = \frac{(6.626 \times 10^{-34} \text{ J·s})^2 (8.854 \times 10^{-12} \text{ C}^2/\text{J·m})}{\pi (9.109 \times 10^{-31} \text{ kg})(1.602 \times 10^{-19} \text{ C})^2},$$

$$r_1 = 5.29 \times 10^{-11} \text{ m} = 0.0529 \text{ nm}.$$

This number is called the *first Bohr radius of the hydrogen atom* and is comparable to the atomic radius of hydrogen as determined by other means.

The total energy E of an electron orbiting around the nucleus includes both kinetic and potential terms. The potential energy is due to the electric interaction and is negative (Chapter 16). The total energy is

$$E = \text{KE} + \text{PE} = \tfrac{1}{2}mv^2 - \frac{e^2}{4\pi\epsilon_0 r_n},$$

where only certain orbits, with radius r_n, are allowed. By substituting the value of v from the angular momentum quantization rule (Eq. 26.9) into the KE term, we get an expression for E as a function of r_n. Then by substituting in both terms for r_n from Eq. (26.10), we get

$$E = \frac{-me^4}{8\epsilon_0^2 h^2 n^2}, \qquad n = 1, 2, 3, 4, \ldots \tag{26.11}$$

The total energy is negative, which means that energy from outside is required to remove the electron from its orbit around the nucleus. Note that the smallest value of n and, therefore, the smallest value of r correspond to the lowest, or most tightly bound, energy state of hydrogen. The state of lowest energy of a system is called its **ground state.**

Example 26.8

The ground state energy of hydrogen.

Calculate the energy of the lowest state of hydrogen.

Solution   The lowest energy state occurs for $n = 1$ in Eq. (26.11),

$$E = \frac{-me^4}{8\epsilon_0^2 h^2}.$$

Upon inserting the values for m, e, ϵ_0, h, and the conversion factor, we get

$$E = \frac{-(9.109 \times 10^{-31} \text{ kg})(1.602 \times 10^{-19} \text{ C})^4}{8(8.854 \times 10^{-12} \text{ C}^2/\text{J·m})^2(6.626 \times 10^{-34} \text{ J·s})^2(1.602 \times 10^{-19} \text{ J/eV})},$$

$$E = -13.6 \text{ eV}.$$

This value of 13.6 eV is in excellent agreement with the experimentally observed value for the ionization energy of hydrogen—that is, the energy required to remove the single electron from the lowest energy state of the hydrogen atom.

The light emitted when an electron makes a transition from one energy level to another carries away energy equal to the difference between the energy levels, according to Eq. (26.8). If these two levels are characterized by two integers n_1 and n_2, we get

$$hf = E_2 - E_1 = \frac{-me^4}{8\epsilon_0^2 h^2 n_2^2} - \frac{-me^4}{8\epsilon_0^2 h^2 n_1^2},$$

where we have used Eq. (26.11) for the energy. Applying the relation between frequency and wavelength, $c = f\lambda$, we have

$$\frac{1}{\lambda} = \frac{me^4}{8\epsilon_0^2 h^3 c} \left(\frac{1}{n_1^2} - \frac{1}{n_2^2} \right). \tag{26.12}$$

We have developed the Rydberg equation. If we calculate the quantity in front of the parentheses from the fundamental constants, we find a value of $1.097 \times 10^7 \text{ m}^{-1}$, the same value determined spectroscopically for the Rydberg constant R.

Bohr's value for the combination of constants in front of the parentheses differed from the spectroscopically measured value by only 6%, an error due to the poor precision in the values for e, m, and h available to him. The difference in the two values was within the uncertainty caused by the errors in the values of the constants. His success was truly remarkable.

Successes of the Bohr Theory

26.7

Starting with the simple model of a single electron making circular orbits about a positive charge, Bohr explained the Rydberg formula. The essential assumptions he made were that the electron could exist in a stationary orbit and that radiation was emitted only when the electron made a transition from one stationary orbit to another. In addition, he used Planck's

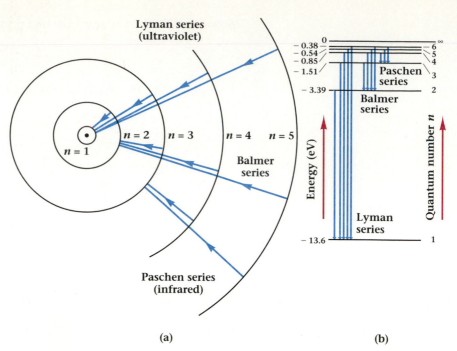

Figure 26.14 (a) Electron orbits of Bohr's model of the hydrogen atom, showing the transitions corresponding to the frequencies of spectroscopic lines in hydrogen. (b) Energy-level representation of the hydrogen atom. The downward arrows represent electronic transitions from higher energy states to lower energy states, corresponding to light emission. Upward arrows correspond to resonant absorption of light. The lengths of the arrows are proportional to the energy of the photons emitted (or absorbed).

constant to relate the frequency of the radiation to the energy difference between the orbits. A consequence of the assumptions was that angular momentum, as well as energy, was quantized.

Bohr recognized that Eq. (26.12) gave the wavelengths of the Balmer series when n_1 was set equal to 2 and n_2 was allowed to be 3, 4, 5, . . . Similarly, if $n_1 = 3$ and $n_2 = 4, 5, 6, . . .$, his formula gave the wavelengths of the infrared series observed for hydrogen by F. Paschen in 1908. Bohr then predicted that other spectral series not yet observed would be found, corresponding to $n_1 = 1$ and to n_1 greater than 3. Indeed, a series of ultraviolet lines in hydrogen was observed by T. Lyman in 1914, corresponding to $n_1 = 1$. Other series corresponding to $n_1 = 4$ and $n_1 = 5$ were discovered in the 1920s.

The interpretations of Bohr's equations can be understood with the aid of Fig. 26.14(a), which represents the electron orbits in space as given by Bohr's model. As we have seen, the radii of the orbits increase as the square of the quantum number n (Eq. 26.10). Figure 26.14(b) shows a representation of the allowed orbits in terms of their energy E. A figure like this is called an **energy-level diagram**. Vertical arrows drawn between the energy levels represent electronic transitions from one energy level to another. A downward transition corresponds to emission of light. The energy of the emitted photon is given by the energy difference between the levels. Transitions from higher energy states to the $n = 2$ level correspond to light emitted at the frequencies seen in the Balmer series.

Absorption spectra can be understood as upward transitions in Fig. 26.14. Notice that absorption is a resonant phenomenon; that is, not all frequencies of light can be absorbed. There are two requirements for absorption of light by an atom. First, only those frequencies can be absorbed that exactly correspond (by Planck's equation) to the difference between two distinct energy states. Second, the absorption can happen only if an electron is present in a lower energy level to absorb the energy and if there is a vacancy in the allowed level at an energy hf above it, into which the electron can jump. Resonant absorption accounts for the dark lines observed by Fraunhofer in the solar spectrum. These absorption lines, which are observed superimposed on the continuous spectrum of the sun, are due to absorption in the outer (cooler) layers of the sun. Atoms in these cooler layers have more electrons in lower energy levels and a larger number of higher energy levels available than do atoms in the sun's hotter interior.

Hydrogen absorption spectra do not show lines corresponding to the Balmer series observed in emission. In emission the transitions of the Balmer series correspond to electrons making transitions from higher states to the $n = 2$ state, which is 10.2 eV above the ground ($n = 1$) state. In cool hydrogen gas, the $n = 2$ state is so far above the $n = 1$ state that the electrons are primarily in the ground state. Because there are so few atoms with electrons in the $n = 2$ state, light at the Balmer frequencies cannot be resonantly absorbed, but simply passes through the gas.

Moseley and the Periodic Table

*26.8

Another successful correlation between Bohr's atomic model and experimental observations came through the study of x-ray spectra by the English physicist Henry G. J. Moseley. While Bohr was developing his atomic theory, W. H. Bragg and W. L. Bragg were carrying out experiments with their newly constructed x-ray spectrometer (Section 25.5). In 1913 they showed that an x-ray spectrum consists of two parts: a continuous part superimposed on a line spectrum similar to the emission spectra of a hot gas (Fig. 26.15). The continuous part is due to an electromagnetic effect: Electrons radiate when decelerating as they are stopped by the target of the x-ray tube. The lines are characteristic of the material in the target and have no classical explanation. The most intense line is called the K_α line.

Just a few months after Bohr's theory was published, Moseley reported a study of the x-ray spectra of the elements of atomic mass between calcium and zinc. He showed that the frequencies of all of the K_α lines were given by a single formula,

$$f = \tfrac{3}{4}\, cR\, (Z - 1)^2, \tag{26.13}$$

where c is the velocity of light and R is the Rydberg constant (Fig. 26.16). The parameter Z is an integer that was found to be the charge on the nucleus in units of the electron's charge. This equation can be rearranged

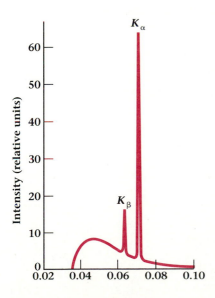

Figure 26.15   The x-ray spectrum of a molybdenum target. (Line widths are not to scale.)

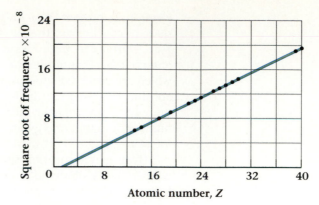

Figure 26.16   A graph of Moseley's data showing the linear relation between the square root of the frequencies of the x ray lines and atomic number.

and written in terms of the wavelength as

$$\frac{1}{\lambda} = \frac{f}{c} = R(Z - 1)^2 \left(\frac{1}{1^2} - \frac{1}{2^2}\right).$$

In other words, the K_α line has a frequency described by an equation similar to the Rydberg equation, but with an additional factor that depends on the nuclear charge.

The interpretation of Moseley's equation is straightforward in light of the Bohr theory. The equation gives the frequency of the x ray when an electron goes from the $n = 2$ state to the $n = 1$ state of an atom of nuclear charge Ze. The simple Bohr theory would have predicted a frequency proportional to Z^2 for an atom with a single electron. The origin of the $(Z - 1)^2$ factor lies in the fact that elements heavier than hydrogen contain two electrons in the lowest-energy level. When one electron is removed from the lowest level, there is still one negative charge near the nucleus. This negative charge reduces the effective positive nuclear charge to $(Z - 1)e$.

Example 26.9

Energy of the K_α x rays from chromium.

Calculate the wavelength and energy of the K_α x rays from chromium ($Z = 24$).

Solution   First we calculate the wavelength, using Moseley's law:

$$\frac{1}{\lambda} = \frac{f}{c} = \tfrac{3}{4}R(Z - 1)^2,$$

$$\lambda = \frac{4}{3R(Z - 1)^2} = \frac{4}{3(1.097 \times 10^7 \text{ m}^{-1})(23)^2} = 0.230 \text{ nm}.$$

Then we determine the energy from the wavelength:

$$E = \frac{hc}{\lambda} = \frac{(6.626 \times 10^{-34} \text{ J·s})(3.00 \times 10^8 \text{ m/s})}{(0.230 \times 10^{-9} \text{ m})},$$

$$E = (8.643 \times 10^{-16} \text{ J}) \times \frac{1 \text{ eV}}{(1.602 \times 10^{-19} \text{ J})} = 5.40 \text{ keV}.$$

Example 26.10

K_α x-ray energy for an atom of atomic number Z.

Find an expression for the energy in electron volts of the K_α line of an element of atomic number Z.

Solution   The energy may be determined from $E = hf$, which becomes

$$E = hf = \tfrac{3}{4}hcR \, (Z - 1)^2.$$

When we use the values of h, c, R, and the conversion constant we have

$$E = \tfrac{3}{4}(6.626 \times 10^{-34} \text{ J·s})(3.00 \times 10^8 \text{ m/s}) \times$$
$$(1.097 \times 10^7 \text{ m}^{-1})(Z - 1)^2,$$

$$E = \left(1.636 \times 10^{-18}(Z - 1)^2 \text{ J} \right) \frac{1 \text{ eV}}{1.602 \times 10^{-19} \text{ J}},$$

$$E = 10.2 \, (Z - 1)^2 \text{ eV}.$$

The real significance of Moseley's work lies in his characterization of the elements. His observation that the square root of an element's K_α line frequency is linearly related to its nuclear charge, Eq. (26.13), has become known as **Moseley's law**. It provides a direct way to number the elements according to their nuclear charge. The number of nuclear charges in an atom is its atomic number, which gives the sequence of elements as they occur in the periodic table (see the inside back cover of this book).

A result of this work was a better understanding of the periodic table, including several corrections and some predictions. For example, prior to this discovery, it was known that the atomic mass of nickel was less than that of cobalt. Yet nickel was placed after cobalt in most tables of the elements because of its chemical properties. Moseley's work identified cobalt as element 27 and nickel as 28, showing that nickel does properly follow cobalt. It is the atomic number that is important in constructing the periodic table, not the atomic mass. Furthermore, the atomic numbers are consecutive integers, so no other element would be expected to occur between cobalt and nickel.

The atomic number of the heaviest known naturally occurring element, uranium, was found to be 92. Thus from the simplest element, hydrogen, to the heaviest, uranium, only 92 elements could exist. By the time of Moseley's untimely death as a British soldier in World War I, most of the elements had been assigned atomic numbers on the basis of their x-ray spectra. All of the elements missing in Moseley's time were subsequently found. Although elements number 43 (technium) and 61 (promethium) do not occur naturally in the earth's crust, they have been made in the laboratory. Elements beyond uranium have also been created in the laboratory. Atomic numbers up to 109 have been reported.

Moseley's work was of primary importance because it provided an unambiguous method of numbering the elements. Furthermore, Moseley's results reinforced the new models of Rutherford and Bohr. Together these ideas form the basis of our modern view of atoms and how one atom differs from another.

SUMMARY

Useful Concepts

- The Rydberg formula for the spectral lines of hydrogen is

$$\frac{1}{\lambda} = R\left(\frac{1}{n_1^2} - \frac{1}{n_2^2}\right), \qquad n_1 < n_2.$$

For the Balmer series, $n_1 = 2$ and $n_2 = 3, 4, 5, \ldots$

- For blackbody radiation the temperature of the body and the maximum wavelength are related through the Wien displacement law,

$$\lambda_m T = 2.90 \times 10^{-3} \text{ m·K}.$$

- According to Planck's hypothesis, the energy of a quantum of radiation is determined by its frequency,

$$E = hf.$$

- The Einstein photoelectric equation gives the maximum energy of a photoelectron as

$$\text{KE}_{max} = hf - \phi.$$

- Bohr postulated that atomic electrons exist in definite energy levels, or orbits, and do not radiate (or absorb) energy except when they make a transition to another energy level or orbit. The Bohr theory of the hydrogen atom gives the Rydberg and Balmer formulas through the use of

$$hf = E_2 - E_1.$$

- The orbital angular momentum of an electron in the hydrogen atom is

$$L = n\frac{h}{2\pi}, \qquad \text{where} \quad n = 1, 2, 3, \ldots$$

- The frequency of the K_α line of an element is given by Moseley's law,

$$f = \tfrac{3}{4}\, cR(Z - 1)^2.$$

Important Terms

You should be able to write the definition or meaning of each of the following terms:

- spectroscopy
- Balmer series
- Rydberg formula
- blackbody
- Wien displacement law
- Planck's constant
- quantum
- photoelectric effect
- threshold frequency

- photon
- work function
- Einstein photoelectric equation
- electron volt
- principal quantum number
- ground state
- energy-level diagram
- Moseley's law

QUESTIONS

26.1 Stars appear to have distinct colors. Some stars look red, some yellow, and others blue. What is a possible explanation for this?

26.2 How does the gravitational red shift affect the wavelength of the Fraunhofer lines measured in the solar spectrum, compared with the same lines measured in the laboratory?

26.3 When the interior of an operating kiln reaches a uniform temperature, it is virtually impossible to see the objects that are being heated. Explain why this is so.

26.4 Do all electrons emitted in the photoelectric effect have the same kinetic energy?

26.5 What difference would it make if the electrons in an atom obeyed classical mechanics?

26.6 Explain the differences between the emission of light by an incandescent lamp, a neon sign, and a fluorescent lamp.

26.7 Can x rays be emitted by hydrogen?

26.8 How are x rays produced? Explain the origin of the line spectra and the continuous spectra. What limits the minimum size of x-ray wavelengths?

26.9 You are examining the spectrum of a particular gas that is excited in a discharge tube. You are viewing the discharge through a transparent box that contains the same gas. Under what conditions would you expect to see dark lines in the spectrum?

26.10 Why is it desirable for suntan lotions to block out the ultraviolet rays?

26.11 What kind of materials have the lowest work functions? A study of Table 26.2 and the periodic table should help you answer this.

PROBLEMS

Section 26.2   Balmer's Series

26.1 (a) What is the wavelength and color of the line in the Balmer spectrum for $n_2 = 4$? (b) What is the wavelength and color of the line for $n_2 = 5$?

26.2 Calculate the wavelength of the Balmer line corresponding to $n_2 = 7$. Why is this line not seen visually?

26.3 Find the relationship between the numerical constant in Balmer's formula (call it b) and the Rydberg constant.

26.4 Calculate the wavelengths of the first four lines in the Balmer series. These lines are in the visible spectrum.

Section 26.3   Blackbody Radiation

26.5 The peak intensity of the solar spectrum occurs for $\lambda_m = 475$ nm. What is the temperature of the surface of the sun?

26.6 A certain star has a peak intensity in its blackbody spectrum at 380 nm. What is the surface temperature of the star?

26.7 A star has an effective surface temperature of 5100 K. What is the wavelength λ_m of the peak in its blackbody spectrum?

26.8 What is the wavelength λ_m of the most intense radiation from a blackbody at room temperature (20°C)?

26.9 A photoflood lamp operates at a temperature of 3200 K. What is the wavelength of the peak in its blackbody spectrum?

26.10 Write the Wien displacement law in terms of the temperature and the frequency.

Section 26.4   Planck's Hypothesis

26.11 What is the energy in joules of a quantum of blue light of wavelength 400 nm?

26.12 What is the energy in joules of a quantum of infrared light whose wavelength is 1000 nm?

26.13 What is the wavelength of a quantum of light whose energy is 3.50×10^{-19} J?

26.14 What is the wavelength of a quantum of light having an energy of 5.00×10^{-19} J?

Section 26.5   The Photoelectric Effect

26.15 What is the Planck constant in terms of eV·s?

26.16 Calculate the wavelength of a photon whose energy is 2.00 eV.

26.17 A gamma ray from a nuclear decay has an energy of 1.20 MeV. What is its wavelength?

26.18 (a) Calculate the wavelength of an x-ray photon whose energy is 25.0 keV. (b) What is the frequency of the photon?

26.19 X rays are produced when energetic electrons collide with a material target. The radiation consists of a line spectrum plus a broad-banded continuous spectrum, which extends up to the maximum energy of the electrons. Find the shortest x-ray wavelength due to electrons accelerated through 28.7 kV.

26.20 What is the minimum voltage required in an x-ray tube to produce photons whose wavelength is 0.10 nm?

26.21 The eye is most sensitive to light of wavelength 550 nm. What is the energy of photons with that wavelength? Give your answer in electron volts.

26.22 The mean wavelength of the D lines in sodium is 589.3 nm. What is the mean photon energy of the D lines? Give your answer in electron volts.

26.23 What is the maximum kinetic energy of photoelectrons ejected from the surface of sodium by blue light of wavelength 434 nm?

26.24 What is the threshold wavelength for photoelectric emission from silver?

26.25 Light from the 253.7-nm UV line in the mercury spectrum ejects electrons from the surface of metallic sodium. (a) What is the maximum kinetic energy of these photoelectrons? (b) Can photoelectrons be ejected from sodium by the green light ($\lambda = 546$ nm) in the mercury spectrum?

26.26 (a) Calculate the work function of a surface from which the green light from mercury ($\lambda = 546.1$ nm) ejects photoelectrons with maximum kinetic energy of 0.13 eV. (b) What is the material of the surface?

Section 26.6   Bohr's Theory of the Hydrogen Atom

26.27 Calculate the second Bohr radius of the hydrogen atom, that is, the radius for $n = 2$.

26.28 Calculate the third Bohr radius of the hydrogen atom, that is, the radius for $n = 3$.

26.29 As the integer n_2 in the Rydberg equation (Eq. 26.2) approaches infinity, the spectral lines converge to a minimum wavelength known as the series limit. Calculate the series limit for the Balmer lines.

26.30 (a) What is the energy of the photon that will raise a hydrogen atom from its ground state to the $n = 3$ excited state? (b) What is the wavelength of this photon?

26.31 If you twirl a small 5.00-g wooden ball in a circle at 3.0 revolutions per second on a light 10-cm string, approximately how many units of angular momentum does it have?

Section 26.7   Successes of the Bohr Theory

26.32 Calculate the series limit for the infrared series (Paschen series) for which $n_1 = 3$.

26.33 What is the longest-wavelength photon that can ionize a neutral hydrogen atom initially in the ground, or lowest, energy state?

26.34 A free electron combines with a free hydrogen nucleus to form a hydrogen atom in the $n = 4$ level. What is the wavelength of the photon emitted during this process?

26.35 (a) What is the energy of the least energetic photon that can be absorbed by hydrogen gas at room temperature? (b) What is the wavelength of such a photon? (*Hint:* The electrons are in the lowest state at the start.)

26.36 What is the angular momentum of an electron in the third Bohr orbit?

26.37 Show that the quantity $h/2\pi$ has the proper dimensions of an angular momentum.

26.38 Evaluate the Rydberg constant from its definition using the values for m, c, e, h, and ϵ_0.

*Section 26.8   Moseley and the Periodic Table

You may need the periodic table found on the inside back cover of this book.

26.39 On the basis of Moseley's law, calculate the energy of the K_α line from (a) gadolinium ($Z = 64$) and (b) thorium ($Z = 90$).

26.40 On the basis of Moseley's law, calculate the energy of the K_α line from (a) aluminum ($Z = 13$) and (b) lead ($Z = 82$).

26.41 (a) What is the wavelength of the K_α line from tungsten ($Z = 74$)? (b) From molybdenum ($Z = 42$)?

26.42 What are the wavelengths of the K_α line from (a) nickel ($Z = 28$) and (b) iron ($Z = 26$)?

26.43 The energy of the K_α line in an unknown sample is 8.07 keV. What is the element producing these x rays?

26.44 The K_α x rays from a certain target are found to have a wavelength of 0.143 nm. What is the element of the target?

26.45 The K_α x rays from a certain target are found to have a wavelength of 0.079 nm. What is the element of the target?

26.46 What is the minimum possible voltage for an x-ray tube that produces x rays of wavelength 0.048 nm?

26.47 In an x-ray tube, electrons are accelerated through a potential of 100 kV and allowed to strike a metal target. (a) What is the maximum energy of the x rays produced? (b) What is the wavelength of these x rays?

Additional Problems

26.48 Derive the equation for the first Bohr radius of singly
• ionized helium. (*Hint:* Helium has a nuclear charge of $Z = +2e$.)

26.49 (a) Calculate the speed of an electron in the second
• Bohr orbit. (b) How does it compare with the speed of light? What does this suggest?

26.50 The spectral series corresponding to $n_1 = 4$ in hy-
• drogen is called the Brackett series in honor of its discoverer. What are the longest and shortest wavelengths in the Brackett series?

26.51 Calculate the longest and shortest wavelengths in
• the Lyman series, corresponding to $n_1 = 1$ in hydrogen.

26.52 (a) Find an expression for the momentum of a pho-
•• ton in terms of its frequency. (b) A beam of light is directed upwards at a small, horizontally held piece of aluminum foil. How many photons per second of blue light (475 nm) must strike the 0.50-g, 3.0 cm × 3.0 cm piece of foil from below in order to make it float? (c) What is the power of such a beam, and (d) is the experiment a practical one? Why? (*Hint:* Use the relationship given in Chapter 24 that the momentum and energy of light are related by $p = E/c$. Recall that the force is the rate of change of the momentum. Assume that the photons are perfectly reflected from the aluminum foil and recall what this meant for the molecules striking the wall in the kinetic theory model of a gas.)

26.53 A material has a photoelectric work function of 3.66
• eV. What is the maximum velocity of photoelectrons ejected from it by light of wavelength 250 nm?

26.54 Work out in detail each step of the derivation of
• Bohr's results between Eq. (26.9) and Eq. (26.12).

26.55 What is the ratio of the wavelength emitted by the
• $n = 3$ to $n = 2$ transition in the hydrogen atom to the Bohr radius of the $n = 2$ orbit?

26.56 (a) What energy is required to completely remove
• an electron from the $n = 3$ level of hydrogen?

(b) What is the electron's energy when removed?
(c) What frequency of radiation will be emitted if the electron falls back to its former level?

26.57 The frequency of a 0.500-mW He-Ne laser is 633
• nm. How many photons per second pass along the beam?

26.58 The human eye is quite a sensitive detector. Only a
• few photons are needed to trigger a visual stimulus. Assume that 10 photons of red (600 nm) light cause a visible flash in your eye. (a) What is the total energy deposited in your eye? (b) If a thousand times as much energy is used to raise a 2.0-μg speck of pollen, how high will it be lifted?

26.59 At what temperature will hydrogen gas have an aver-
• • age kinetic energy per molecule equal to the energy required to remove the electron from the hydrogen?

ADDITIONAL READING

Bayfield, J. E., "Rydberg Atoms." *American Scientist*, July 1983, p. 375.

Blaedel, N., *Harmony and Unity, The Life of Niels Bohr*. Madison, Wisconsin: Science Tech Publishers, 1988. A scientific biography accessible to both the scientist and the general reader.

Hansch, T. W., A. L. Schawlow, and G. W. Series, "The Spectrum of Atomic Hydrogen." *Scientific American*, March 1979, p. 94.

Heilbron, J. L., *The Dilemmas of an Upright Man: Max Planck as Spokesman for German Science*. Berkeley, Calif.: University of California Press, 1986. This biography includes Planck's scientific work and his public life. It describes the conflict between his patriotism and his moral anguish over the events in Germany in the 1930s.

27

Quantum Mechanics

27.1	Classical and Quantum Mechanics
27.2	The Compton Effect
27.3	De Broglie Waves
27.4	Schrödinger's Equation
27.5	The Uncertainty Principle
27.6	Interpretation of the Wave Function
27.7	The Particle in a Box
27.8	Tunneling or Barrier Penetration
*27.9	Wave Theory of the Hydrogen Atom
*27.10	The Zeeman Effect and Space Quantization
27.11	The Pauli Exclusion Principle
*27.12	Understanding the Periodic Table

A WORD TO THE STUDENT

The quantum theory forms the basis of our contemporary understanding of atomic, nuclear, and subnuclear processes. It is also necessary for understanding chemical reactions and, increasingly, biological processes as well. Though quantum mechanics is essentially different from the theories of classical mechanics and electromagnetism that we studied earlier, you need a clear understanding of those theories, particularly the concepts of energy and of wave motion, in order to understand quantum ideas. In this chapter we discuss several experimental results before introducing Schrödinger's wave equation, which is at the heart of the theory. Then we develop some examples of the consequences of this equation, using only trigonometric and exponential functions.

As we have seen in the preceding chapters, the search for the ultimate structure of matter led to the conclusion that matter is not infinitely divisible but is composed of tiny particles, called atoms. Subsequent discoveries demonstrated that the atom itself was not indivisible as had previously been thought. Instead, the atom has component parts: a heavy, positively charged nucleus and lighter, negatively charged electrons. Moseley's experiments with x rays showed that the nuclei of atoms possess positive charge in integral multiples of the electronic charge e.

The explanations of blackbody radiation and the photoelectric effect introduced the concept of quantization of electromagnetic radiation. These ideas, in turn, led Bohr to a remarkably successful description of the hydrogen atom, which utilized the idea of energy quantization. These developments, in the early decades of the twentieth century, showed that the classical laws of physics, which worked so successfully when applied to macroscopic objects, were not appropriate for the description of matter at the microscopic (atomic) level.

Despite the successes of the Bohr theory, it had limitations. For example, although it could predict the wavelength of the spectral lines of the hydrogen atom, the procedure did not work for all atoms. Bohr's theory was one of the great milestones in physics, but it needed further refinement and development. In this chapter we trace the development of what is commonly called quantum mechanics or, sometimes, wave mechanics. We start by setting forth some of the important differences between classical physics and quantum physics.

One caution should be added. Keep in mind that our description is a model. We hope that each improved conception is a more accurate description of reality—in the present case, of the nature of the atom. But our description is not reality itself. A model or theory does not make an electron behave in a particular way. However, if a model is complete enough and accurate enough, we may use it to understand certain features of electron behavior and perhaps to predict new effects.

Classical and Quantum Mechanics

27.1

Before we discuss the development of quantum mechanics and present some of its key results and applications, let's try to clarify some of the important differences between quantum mechanics and classical mechanics. We will expand on these ideas again later in this chapter, but here we preview some of the ways in which quantum mechanics has changed our view of the physical world.

According to classical physics, there are no limits to how accurately we can measure physical quantities. Certainly we have practical limitations on the precision and accuracy of our instruments. Also, the task of keeping track of some things—say, the speed and position of every molecule in a

TIME LINE

Louis de Broglie proposes that all moving particles have a wavelength, a property previously associated only with waves.

1923

1924

1925

A. H. Compton shows that x-ray photons display particlelike properties and obey the law of conservation of momentum.

Wolfgang Pauli states the exclusion principle: No two electrons in the same atom can have the same four quantum numbers.

gas—is beyond doing. These are practical obstacles, however. In the Newtonian view of things we can, in principle, make all the measurements needed to predict accurately the future state of the entire universe. Even special relativity did not change this perception.

Quantum mechanics changes all this. Even in principle, it is not possible to know all physical quantities simultaneously with complete accuracy. This idea has its clearest statement in the uncertainty principle (Section 27.5), which gives specific limitations to how accurately we can know certain properties of a system at the same time. This is not a limitation of our measuring apparatus; this is an inherent property of the physical world.

How, then, can we measure and calculate properties of atoms and other systems? The answer given by quantum mechanics is that we can assign probabilities for values of physical properties. We cannot say that the radius of an electron's orbit in hydrogen is exactly equal to the Bohr radius; however, we can say that the Bohr radius is the distance from the nucleus at which the electron is most likely to be found at a given time. The probability of finding an electron farther than five times the Bohr radius from the nucleus is about 1 chance in 2000. The probabilities involved are calculated using the mathematics of quantum mechanics.

The introduction of probability in quantum mechanics is one of its most significant departures from classical physics. Einstein had difficulty accepting this idea, as reflected by his statement that has been frequently paraphrased as "God does not play dice with the world." Nevertheless, quantum mechanics has had such great success in explaining observed behavior and predicting new behavior that it is universally accepted today. The quantum mechanical interpretation of matter has led to the development of such things as transistors, lasers, and the tunneling microscope, and to an increased understanding of the life cycles of stars.

The interpretation of quantum mechanical results in terms of probabilities is sometimes called the "Copenhagen interpretation," referring to the home of Niels Bohr, who pioneered this view. After having developed the first quantum model of the hydrogen atom in 1913, he went on to make significant contributions to the newer quantum theory, developing

C. J. Davisson (shown) and
L. H. Germer observe electron
diffraction, as does G. P.
Thomson in an independent
experiment, confirming that
particles act as waves.

1926

1927

Erwin Schrödinger formulates
a wave equation that describes
the behavior of particles on the
atomic scale.

Werner Heisenberg formulates his
uncertainty principle, which states that
we cannot measure position and momentum simultaneously with arbitrary
precision.

new theories and insights. Bohr, more than anyone else, provided the philosophical framework for our present understanding of quantum mechanics.

Note that the use of probability in quantum mechanics differs from our use of probability in kinetic theory, where statistics describes the behavior of large numbers of particles, each behaving in accord with Newtonian dynamics. In quantum mechanics, each individual particle does not follow Newtonian dynamics, but has probabilities associated with such properties as position, momentum, and energy.

Bohr also influenced the development of quantum mechanics by his *correspondence principle.* According to this principle, a more general theory, such as relativity or quantum mechanics, must give the same results as a more restricted theory, such as classical mechanics, where the latter is appropriate. For example, for velocities small compared with the speed of light, the special theory of relativity gives the same results as does Newtonian mechanics. Moreover, when we apply the quantization of energy and angular momentum to large orbits, say with $n = 10,000$, then an orbit corresponding to $n \pm 1$ is essentially indistinguishable from the orbit for n. In this case, we have the classical result that energy and angular momentum vary smoothly. The indistinct boundaries between the domains of applicability of the theories shown in Fig. 4.20 reflect this view.

An idea associated with the correspondence principle is that the results of a quantum mechanical calculation should give the classical result if Planck's constant h is decreased to zero. This expectation is reasonable because the size of h determines the size of quanta, which is zero in classical physics. You may test the quantum mechanical results in this chapter by applying this principle.

Another important difference between classical and quantum mechanics has to do with the role of the observer, that is, the person making the measurements in any experiment. It is a basic idea of quantum mechanics that making a measurement unavoidably affects the object being measured. As a result, we can no longer look at the laws of physics as furnishing a perfectly objective description of the physical universe; instead we must consider our role in making these observations.

795

In classical mechanics most of the ideas have clear intuitive meanings that are extrapolated from our own experience. In contrast, many of the ideas of quantum mechanics are not intuitively clear and do not correspond to anything in our prior experience. Nevertheless, quantum mechanics is an extremely successful theory, without which little or no progress could be made in most modern fields of research.

The Compton Effect

27.2

Roentgen's discovery of x rays in 1895 opened the way for numerous advances in physics. For example, the discovery of x-ray diffraction and its explanation (Section 25.5) led to knowledge of the detailed atomic structure of crystalline solids. The study of x-ray spectra helped confirm Bohr's theory and led to a clearer understanding of nuclear charge in the elements.

In 1923 the American physicist Arthur Holly Compton (1892–1962) carried out experiments on the scattering of x rays by matter. The interpretation of these experiments had an impact on the developing quantum theory that was comparable to the impact of Einstein's theory of the photoelectric effect.

According to classical theory, as a wave such as an x ray penetrates matter, it affects every electron in the region through which it travels. The observed scattering of the x ray is due to the combined effect of all these electrons. On the other hand, in the photoelectric effect a single quantum of light—a photon—causes the ejection of a single electron. Thus Compton reasoned that from the quantum point of view, a single "x-ray photon" should interact with a single electron. In doing so, the photon could scatter by imparting energy and momentum to the electron.

The deflection of an x ray can occur only with a change in its momentum. Consequently, by conservation of momentum, the struck electron must recoil with a momentum equal to the x ray's change in momentum. The energy of the scattered photon must equal the energy of the incident photon minus the recoil kinetic energy of the electron. Since the scattered photon necessarily has less energy than the incident photon, then according to the Planck relation $E = hc/\lambda$, the scattered photon has a correspondingly increased wavelength. Compton's experiment confirmed this change in wavelength.

We can analyze this collision using conservation laws just as we analyzed elastic collisions in Chapter 8. Imagine an x ray of frequency f scattered by a stationary free electron* of mass m (Fig. 27.1a). Since the x ray moves at the speed of light, the momentum of the incident photon is

$$p_\gamma = \frac{E}{c} = \frac{hf}{c} = \frac{h}{\lambda}.$$

*The assumption of a stationary electron is reasonable because the x ray moves with a velocity c. The assumption of a free electron is valid if the recoil energy is very much greater than the energy binding the electron to the atom in the scattering material.

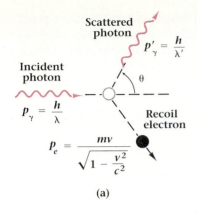

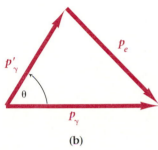

Figure 27.1 Compton scattering of an x-ray photon by an electron. **(a)** Diagram of the collision. **(b)** Momentum vector diagram for a photon of energy hf scattered by a free electron. The three momentum vectors form a triangle.

Similarly, the momentum of the scattered photon is

$$p_{\gamma}' = \frac{E'}{c} = \frac{h}{\lambda'},$$

and its direction makes an angle θ with the incident beam. The momentum of the electron recoiling with speed v is

$$p_e = \frac{mv}{\sqrt{1 - v^2/c^2}},$$

where we use the relativistic formula for momentum developed in Chapter 24. With Compton's assumption that we can treat the x-ray photon as a particle, then the direction of the electron's momentum is determined by conservation of momentum. Figure 27.1(b) shows the momentum diagram for the collision of Fig. 27.1(a). Conservation of momentum requires the three momentum vectors to be coplanar. Furthermore, p_{γ}' plus p_e must equal p_{γ}. Thus the three momentum vectors will form a triangle as shown in Fig. 27.1(b). The statement of momentum conservation can be put into an equation by using the law of cosines (see Appendix A, page 931):

$$p_e^2 = p_{\gamma}^2 + p_{\gamma}'^2 - 2p_{\gamma}p_{\gamma}' \cos \theta.$$

Expressing the momenta in terms of the preceding equations, we get

$$\left(\frac{mv}{\sqrt{1 - v^2/c^2}}\right)^2 = \frac{h^2}{\lambda^2} + \frac{h^2}{\lambda'^2} - \frac{2h^2 \cos \theta}{\lambda\lambda'} \qquad \text{(momentum)}.$$

According to the law of energy conservation, the energy of the incident photon equals the total energy of the scattered photon and the recoil electron. Thus

$$\frac{hc}{\lambda} = \frac{hc}{\lambda'} + mc^2 \left(\frac{1}{\sqrt{1 - v^2/c^2}} - 1\right) \qquad \text{(energy)}.$$

We now have two equations for the two unknown quantities λ' and v, similar to our earlier analysis of elastic collisions. We obtain the scattered wavelength λ' by solving these equations simultaneously. The result is

$$\lambda' - \lambda = \frac{h}{mc} (1 - \cos \theta). \qquad (27.1)$$

The wavelength of the scattered photon is greater than that of the incident photon, as expected. The scattered wavelength is angular-dependent and is greatest for scattering in the backward direction ($\theta = 180°$). The quantity h/mc is known as the **Compton wavelength** of the electron and has the value 2.426×10^{-12} m.

The Compton effect is very important because it points directly to the wave–particle duality of light. Planck's idea of quantized radiation did not fit the classical wave description, and Einstein's explanation of the photoelectric effect supported Planck's quantum hypothesis. But Compton's discovery goes even further. His analysis treats the photon as a particle, having momentum and energy, that collides with another particle, the electron. On one hand we treat the event as a two-body collision problem; on the other hand we quantize the photon energy in terms of its wavelength. The same x rays that are scattered like particles by the electrons in the target material are diffracted by the crystal in the spectrometer that measures their wavelength.

Example 27.1

Energy loss in Compton scattering.

A 17.2-keV x ray from molybdenum is Compton-scattered through an angle of 90°. What is the energy of the x ray after scattering?

Solution   For the incident x-ray photon, the wavelength may be calculated using the rule of thumb given in Example 26.6:

$$\lambda = \frac{1240 \text{ eV·nm}}{E} = \frac{1240 \text{ eV·nm}}{17.2 \text{ keV} \times 10^3 \text{ eV/keV}} = 0.0721 \text{ nm.}$$

The wavelength of the scattered x ray is found from Eq. (27.1) to be

$$\lambda' = \lambda + \frac{h}{mc} (1 - \cos \theta).$$

For $\theta = 90°$, $\cos \theta = 0$. The quantity h/mc has the value 0.002426 nm. So

$$\lambda' = 0.0721 \text{ nm} + 0.0024 \text{ nm} = 0.0745 \text{ nm.}$$

We can compute the energy of the scattered x ray from its wavelength:

$$E = \frac{1240 \text{ eV·nm}}{0.0745 \text{ nm}} = 16,600 \text{ eV} = 16.6 \text{ keV.}$$

The missing energy, 0.6 keV, is carried off by the recoil electron.

De Broglie Waves

27.3

By 1920 the dual nature of electromagnetic radiation was generally accepted, although it was not understood. As physicists gradually accumulated experimental data, they became accustomed to using a particle theory or a wave theory of light, depending on the nature of the particular experiment to be interpreted. Into this setting came Louis de Broglie, a graduate student at the University of Paris. De Broglie was strongly influenced by conversations about the photoelectric effect, which gave particle properties to waves. He began thinking about the reverse effect: that particles might have some wavelike properties.

In 1924, using arguments from both electrodynamics and relativity, de Broglie proposed that a particle of mass m traveling with speed v should have associated with it a wavelength λ, given by

$$\lambda = \frac{h}{mv} = \frac{h}{p}, \tag{27.2}$$

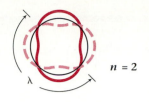

$n = 2$

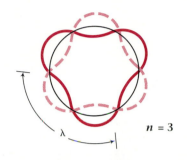

$n = 3$

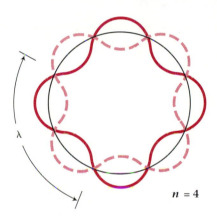

$n = 4$

Figure 27.2 Standing waves on a circle. The circumference is shown successively equal to two, three, and four wavelengths.

where $p = mv$, the momentum of the particle. (This proposal was part of de Broglie's doctoral dissertation.) The wavelength given by Eq. (27.2) is known as the **de Broglie wavelength**. The de Broglie equation (Eq. 27.2) also gives the correct momentum for a photon of wavelength λ.

The idea of de Broglie waves provided an alternative way for deriving Bohr's condition for quantization of atomic electron orbits. If we describe the electron as a wave, its stable orbits are those that meet the conditions for a standing wave (Fig. 27.2). For a standing wave to be set up, an integral number of wavelengths must coincide exactly with the circumference of the orbit. Thus we get $n\lambda = 2\pi r$, where n is an integer. When the de Broglie relation is used for the wavelength, we get $nh/p = 2\pi r$, which, upon rearrangement, gives the quantization condition for angular momentum:

$$L = pr = n\frac{h}{2\pi}. \tag{27.3}$$

Thus, by associating a wavelength with the electron, the quantization of atomic orbits evolves as a natural consequence of the behavior of the electron. Bohr's principal quantum number n becomes the number of integral wavelengths that fit into each orbit. You may note a similarity of the conditions for a standing wave in the circular orbit and the resonant conditions for a standing wave on a string.

Example 27.2

De Broglie wavelengths of electrons and falling dust.

In a research laboratory, electrons are accelerated to speeds of 6.0×10^6 m/s. Nearby, at the same laboratory, a 1.0×10^{-9} kg speck of dust falls through the air at a speed of 0.020 m/s. Calculate the de Broglie wavelength in both cases.

Solution   For the electron,

$$\lambda_e = \frac{h}{p} = \frac{h}{mv} = \frac{(6.63 \times 10^{-34} \text{ J·s})}{(9.11 \times 10^{-31} \text{ kg})(6.0 \times 10^6 \text{ m/s})}$$

$$\lambda_e = 1.2 \times 10^{-10} \text{ m}.$$

For the dust speck,

$$\lambda_{\text{dust}} = \frac{h}{p} = \frac{h}{mv} = \frac{(6.63 \times 10^{-34} \text{ J·s})}{(1.0 \times 10^{-9} \text{ kg})(0.020 \text{ m/s})}$$

$$\lambda_{\text{dust}} = 3.3 \times 10^{-23} \text{ m}.$$

The de Broglie wavelength of the electron is the same order of magnitude as the distance between atomic planes in a crystal. From our knowledge of diffraction and interference, we expect to be able to detect the wavelike behavior of electrons by their diffraction from crystals. The momentum of the speck of dust is very small for ordinary matter, but is enormously larger than the momentum of the electron just described. Consequently, the de Broglie wavelength of the dust speck is many orders of magnitude smaller than that of the electron, even smaller than nuclear dimensions (10^{-15} m). It is so small that we do not observe its wavelike behavior.

Example 27.3

Find an expression for the de Broglie wavelength, in nanometers, of an electron when the kinetic energy of the electron is given in terms of its accelerating potential difference V.

Solution   If the electron is moving slowly enough, we can use the classical expression for kinetic energy,

$$KE = \tfrac{1}{2}mv^2 = \frac{p^2}{2m}.$$

Then the wavelength becomes

$$\lambda = \frac{h}{p} = \frac{h}{\sqrt{2m \cdot KE}}.$$

PHYSICS IN PRACTICE

Electron Microscopes

The first electron microscope was built only a few years after the discovery that high-speed electrons had wavelengths many times smaller than the wavelengths of light. Because of the smaller wavelengths, electron microscopes are capable of much higher resolution than optical microscopes. The beams of electrons can be focused by suitably shaped electric or magnetic fields. For example, the magnetic field of a solenoid acts like a converging lens for electrons (Fig. B27.1). In transmission electron microscopes (TEM), an energetic beam of electrons focused by a condenser lens illuminates the speci-

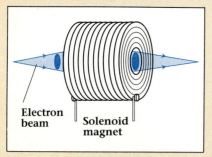

Figure B27.1   The field of a solenoid magnet acts as a converging lens to focus a beam of electrons.

Figure B27.2   Comparison of optical and electron microscopes. The components of the microscopes are arranged side by side to show the similarity of the two systems.

men being examined and passes through to an objective lens, which forms an intermediate image (Fig. B27.2). A projector lens then magnifies a portion of the intermediate image to form the final image, which may be viewed on a fluorescent screen or recorded on a photographic plate. Just as in the case of an optical microscope, the overall magnification is the product of the magnifications of the objective and projector lenses, and with the proper choice of lens currents can be as high as 200,000×. The critical limitation is the resolution. Modern instruments are capable of resolving objects smaller than 0.5 nm.

Scanning electron microscopes (SEM), which came into wide usage in the 1970s and 1980s, operate on a different principle from the TEM. In the scanning electron microscope a finely focused electron beam scans across the surface of the specimen being examined. As the beam scans across the specimen, the incident electrons (pri-

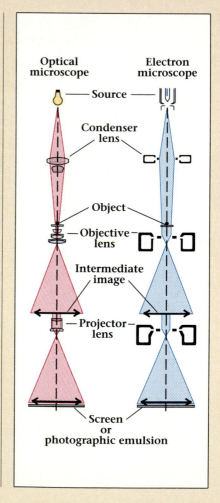

If an electron is accelerated through an electric potential difference V, its kinetic energy is $KE = eV$ and the wavelength becomes

$$\lambda = \frac{h}{\sqrt{2meV}}.$$

Inserting numerical values for h, m, and e and measuring V in volts, we get

$$\lambda = \sqrt{\frac{1.50}{V}} \text{ nm.}$$

This equation is a very handy formula for computing electron wavelengths at low energies, that is, at low voltages. At energies above a few kiloelectron volts, the electron velocity becomes so great that its momen-

maries) knock out other electrons (secondaries) that come from the area on the specimen where the primary beam is focused (Fig. B27.3). These secondary electrons are collected at an electrode. The intensity of the secondary current changes as the primary beam sweeps across the specimen, since more secondaries are generated when the beam strikes a sharply curved edge or sloping surface than when it strikes a flat surface. The intensity information is used to generate a televisionlike image on a cathode ray

Figure B27.4   A false-color scanning electron micrograph of an array of infrared detector diodes. The orange mesas of p-type semiconductor material are isolated from the red valleys of n-type material (see Chapter 30). The blue bumps of metallic indium are used to make electrical connection to the detectors. Individual diodes are spaced 100 μm apart, center to center.

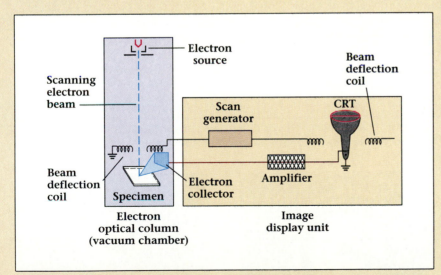

Figure B27.3   Schematic view of a scanning electron microscope. The tightly focused primary beam scans across the specimen, releasing secondary electrons. These secondary electrons are collected and their current is used to control the intensity of the electron beam in a cathode ray tube (CRT) that is scanned synchronously with the primary beam. An image forms on the face of the CRT.

tube, which gives an impression of three-dimensional surface relief (Fig. B27.4). The size of the beam (typically less than 10 nm) limits the resolution of the instrument. In both types of electron microscopes the extremely small de Broglie wavelengths of high-speed electrons permit imaging with high resolution.

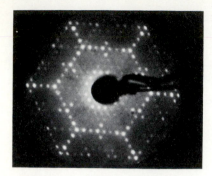

Figure 27.3   A diffraction pattern of low-energy electrons (approximately 48 eV) from the surface of silicon.

tum must be calculated relativistically. As a result, the wavelength is reduced from the value predicted classically. For example, at 50,000 eV the relativistic correction is about 2.5%.

Three years after de Broglie asserted that particles of matter could possess wavelike properties, the diffraction of electrons from the surface of a solid crystal was experimentally observed by C. J. Davisson and L. H. Germer of the Bell Telephone Laboratory. In 1927 they reported their investigation of the angular distribution of electrons scattered from nickel. With careful analysis, they showed that the electron beam was scattered by the surface atoms of the nickel in the exact manner predicted for the diffraction of waves according to Bragg's formula, with a wavelength given by the de Broglie equation. A striking characteristic of diffracted electron beams is their similarity to the Laue pattern resulting from the same crystal when the incident beam is a beam of x rays (Fig. 27.3).

Also in 1927 G. P. Thomson, the son of J. J. Thomson, reported his experiments, in which a beam of energetic electrons was diffracted by a thin foil. Thomson found patterns that resembled the x-ray patterns made with powdered (polycrystalline) samples. This kind of diffraction, by many randomly oriented crystalline grains, produces rings (Fig. 27.4). If the wavelength of the electrons is changed by changing their incident energy, the diameters of the diffraction rings change proportionally, as expected from Bragg's equation. These experiments by Davisson and Germer and by Thomson proved that de Broglie waves are not simply mathematical conveniences, but have observable physical effects. Just as Compton showed that waves could act as particles, Davisson and Germer showed that particles could act as waves.

The concept of particles as waves and waves as particles is incompatible with the ideas of classical physics. However, this wave–particle duality, as it is sometimes called, is an integral part of quantum mechanics. To help explain this seeming contradiction, Bohr postulated the *principle of complementarity*. It states that it is neither possible nor necessary to choose between the wave and particle descriptions, but that both are essential for a complete description of nature.

Figure 27.4   Transmission electron diffraction patterns produced from a gold foil. Ring diameters decrease with increasing beam voltage: (a) 60 kV, (b) 80 kV, (c) 100 kV.

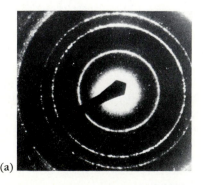

(a)

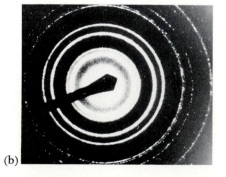

(b)

(c)

Schrödinger's Equation

27.4

The discovery of the wavelike behavior of electrons created the need for a wave theory describing the behavior of matter on the atomic scale. Such a theory was proposed by the Austrian physicist Erwin Schrödinger (1887–1961) only two years after de Broglie formulated the idea of particle waves.

Schrödinger reasoned that if an electron behaves as a wave, then it should be possible to mathematically describe the electron's behavior in space and time by a wave equation. In 1926, from his knowledge of the general theory of wave phenomena and of de Broglie's work, Schrödinger deduced the fundamental equation that now bears his name. We can write the **Schrödinger equation** in the form

$$\frac{\Delta(\Delta\psi/\Delta x)}{\Delta x} + \frac{8\pi^2 m}{h^2}(E - V)\psi = 0, \tag{27.4}$$

where E is the total energy of the particle, V is its potential energy, m is its mass, and h is Planck's constant.* The symbol ψ (the Greek letter psi) is called the **wave function**. The Schrödinger equation as written here is for only one dimension; the complete equation is written to include three dimensions.

What does ψ, the symbol for the wave function, represent? Its mathematical role in the Schrödinger equation is analogous to the amplitude of a wave described by classical mechanics (Chapter 14), such as a transverse wave propagating along a string. What do we mean by the amplitude in Schrödinger's wave equation?

We cannot measure a property of an electron that directly corresponds to the "amplitude" in Schrödinger's equation. Nonetheless, we can use Schrödinger's equation to calculate wave functions and predict the results of experiments. We will show several examples of such calculations later in this chapter. We will also discuss the significance of the square of the wave function, which relates not to a physical property of the particle, but to the probability of a given behavior. However, the wave function itself is not a physically observable quantity. From a classical point of view, this is a somewhat unsatisfactory situation. In quantum mechanics we calculate wave functions in various circumstances (for example, multielectron atoms, liquids, gases, etc.) and then, from the results, we predict the properties and behavior of the corresponding particles. Quantum mechanics allows us to obtain results and predictions that our experimental observations confirm.

With his wave equation, Schrödinger obtained the wavelength of a free electron, in agreement with de Broglie's formula. For a free electron the potential $V = 0$ in Eq. (27.4). The wave equation then predicts that the electron wave can take on any wavelength $\lambda = h/p$ associated with

*As it stands, Eq. (27.4) is in a form called a difference equation. It becomes a differential equation in the limit that Δx goes to zero.

any energy $E = p^2/2m$, in the limit of nonrelativistic velocities. Since classical physics allows a free particle to have any energy whatsoever, all energies are allowed for free particles and the results of the wave equation are identical with those of classical physics. But, as we shall see below, differences do occur between the two theories when the particle is no longer free.

The Uncertainty Principle

27.5

We can write the equation for a traveling harmonic wave, such as we described in Chapter 14, as

$$\psi(x, t) = \psi_0 \sin 2\pi \left(\frac{x}{\lambda} - \frac{t}{T} \right). \tag{27.5}$$

Such a wave extends infinitely far in both the positive and negative directions (Fig. 27.5a). Here ψ_0 is the amplitude of the wave and T is its period. The wave is characterized by a single wavelength λ_0 (Fig. 27.5b). However, if we expect a wave to represent an electron or other particle, then it seems unreasonable to have the wave extending to infinity. Instead, the wave should be localized in the vicinity of the electron. Indeed, we see the need for localizing a wave to a small region of space even in the case of light waves. For example, in the photoelectric effect and in Compton scattering, the photon interacts with a single electron, not with all of them.

Fortunately, there is a relatively simple answer to our dilemma. Any complicated wave can be built up from a superposition of harmonic waves of different wavelengths. For example, if we add together two waves ψ_1 and ψ_2 of slightly different wavelengths, a new wave results (Fig. 27.6). Notice that this new wave does not have an appreciable amplitude at all positions, but only over certain repetitive ranges of x. If we add a third wave ψ_3, the waves combine with even greater amplitude in some regions and less in others (Fig. 27.7). By combining a large number of waves of different wavelengths, we can build up a localized wave disturbance like the one shown in Fig. 27.8(a). We call this kind of resultant wave a **wave packet.** It has an appreciable amplitude only in a small region of space Δx, which corresponds to the location of the particle it represents. Because its component waves are traveling waves, the wave packet moves with time. We identify the velocity of the wave packet with the velocity of the particle.

To construct a wave packet like that of Fig. 27.8(a), the resultant wave amplitude must be vanishingly small everywhere outside of the small region Δx. An infinite number of pure harmonic waves are needed to completely eliminate the resultant amplitude beyond Δx. However, not all of these pure harmonic waves have large amplitudes. Figure 27.8(b) shows the distribution of amplitudes versus wavelength for the component waves. Most of the waves with appreciable amplitude lie in a wavelength range $\Delta \lambda$ about the average wavelength λ_0. The distance Δx is some multiple n of the average wavelength; that is, $\Delta x = n \, \Delta \lambda_0$. We use this fact to prove an important relation between position and momentum.

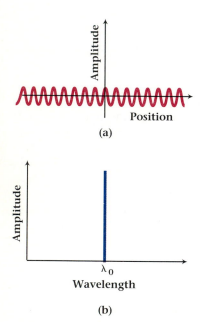

Figure 27.5 (a) A harmonic wave of wavelength λ_0 extending infinitely far in the x direction. (b) The wavelength spectrum of a harmonic wave contains only one component at $\lambda = \lambda_0$.

Figure 27.6 (a) Two harmonic waves ψ_1 and ψ_2, of wavelengths λ_1 and λ_2, interfere to produce a new wave $\psi_1 + \psi_2$. (b) The wavelength spectrum of the resultant wave $\psi_1 + \psi_2$.

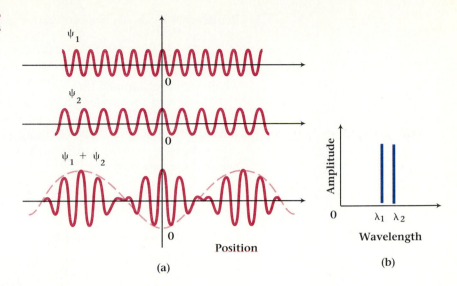

(a)

(b)

Figure 27.7 The addition of wave ψ_3 to $\psi_1 + \psi_2$ gives a new wave.

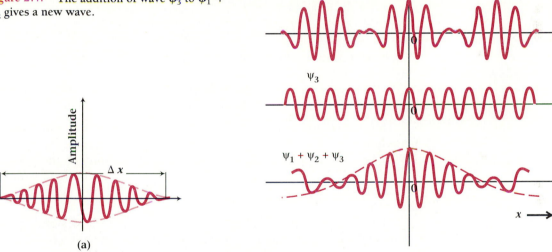

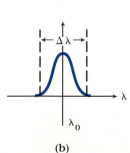

(a)

(b)

Figure 27.8 (a) A wave packet of average wavelength λ_0 and spatial extent Δx. (b) The wavelength spectrum of the wave packet of part (a).

The number of wavelengths of average wavelength λ_0 within the region Δx is $n = \Delta x / \lambda_0$. Outside of Δx the total amplitude is zero and the component waves interfere destructively to cancel one another. That is, some of them must be exactly out of phase with the average wave λ_0. Waves of different wavelengths that are exactly in phase at the center of the wave packet will destructively interfere somewhere else. Some of the waves within the spread $\Delta\lambda$ have wavelengths less than λ_0 that are small enough so that at least $n + 1$ of the wavelengths can fit into the length Δx. Thus for a wavelength $\lambda' = \lambda_0 - \Delta\lambda$, the number of wavelengths contained in Δx will be

$$\frac{\Delta x}{\lambda'} = \frac{\Delta x}{\lambda_0 - \Delta\lambda} \geq n + 1.$$

Replacing n by $\Delta x/\lambda_0$, we can rearrange this inequality to yield

$$\frac{\Delta x \, \Delta \lambda}{\lambda_0^2} \geq 1, \tag{27.6}$$

where we have assumed that $\Delta \lambda \ll \lambda_0$.

The spread in wavelength can be associated with the spread in momenta through de Broglie's equation. For example, consider two waves of wavelengths λ_1 and λ_2. The de Broglie relation gives the corresponding momenta as $p_1 = h/\lambda_1$ and $p_2 = h/\lambda_2$. Thus the difference in momenta between the two waves is

$$\Delta p = p_1 - p_2 = h \left(\frac{1}{\lambda_1} - \frac{1}{\lambda_2} \right).$$

Finding the common denominator for the terms in parentheses gives

$$\Delta p = \frac{h(\lambda_2 - \lambda_1)}{\lambda_1 \lambda_2}.$$

If the difference $\lambda_2 - \lambda_1 = \Delta \lambda$ is small, then the product $\lambda_1 \lambda_2$ may be safely approximated by λ_0^2, where λ_0 is the average wavelength. Thus

$$\frac{\Delta p}{h} = \frac{\Delta \lambda}{\lambda_0^2}.$$

When this result is substituted for $\Delta \lambda / \lambda_0^2$ in Eq. (27.6), we get

$$\boxed{\Delta x \, \Delta p \geq h.} \tag{27.7}$$

This inequality is the mathematical statement of the **uncertainty principle**,* which was first advanced in 1927 by Werner Heisenberg (1901–1976). The quantity Δx represents the uncertainty in the position of the particle. In other words, we know the particle is confined within a region Δx. Similarly, Δp represents the uncertainty in the momentum of the particle.

The uncertainty principle limits how well we can know simultaneous values of the position and linear momentum of a particle or wave. It means that we cannot measure both the position and the related momentum simultaneously with arbitrarily great precision. For example, suppose that we know the momentum of a certain free electron precisely, but do not know its position. Then according to the uncertainty principle, any attempt to measure its position will alter its momentum. Consequently, after any experiment to determine its position, our knowledge of its momentum is limited by Eq. (27.7).

In more concrete terms, consider the act of measuring the position of an electron. You might imagine trying to see the electron through a microscope, which would involve a photon of light coming from a source, bouncing off the electron, and then traveling through the microscope to

---

*Heisenberg's original derivation gave the result $\Delta x \, \Delta p \geq h/2\pi$. This more precise form of the uncertainty principle is most frequently used. However, in this chapter we will use Eq. (27.7).

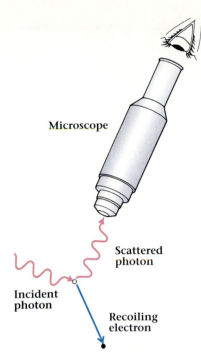

Microscope

Scattered photon

Incident photon

Recoiling electron

Figure 27.9 A thought experiment for viewing an electron with a microscope. The photon that carries the information of the position of the electron imparts momentum to it, so that you cannot locate the electron without disturbing its momentum.

your eye (Fig. 27.9). The photon imparts momentum $p \approx h/\lambda$ to the electron, and so you cannot locate the electron without changing its momentum. Now, how accurately can a photon measure the position of the electron? Only as accurately as the wavelength of the photon, as we saw back in Chapter 23. If we use a photon of shorter wavelength, like an x-ray photon, we will know the position more precisely, but the more energetic photon will impart more momentum to the electron. Consequently, the product of Δx and Δp always remains of the order of Planck's constant.

When we consider motion in three dimensions, three uncertainty relations are required, one for each of the three perpendicular directions. However, because of the independent nature of motion in three directions, there is no uncertainty relation involving other products of these quantities. For example, a measurement of the x position of an object does not limit our knowledge of the object's y position or its momentum in the y direction.

There is another version of the uncertainty principle that relates energy and time. Suppose we try to measure the frequency of a wave (perhaps a photon) by counting the number of cycles that occur in a time interval Δt. The longer we make Δt, the better our measurement of the frequency. Since we have a minimum uncertainty of one count per time interval, we have a spread or uncertainty in frequency of $\Delta f \geq 1/\Delta t$. An uncertainty in frequency implies a corresponding uncertainty in energy, since $\Delta E = h \, \Delta f$. These two equations may be combined to give the uncertainty relations between time and energy:

$$\Delta E \, \Delta t \geq h. \tag{27.8}$$

There is an uncertainty in the energy of a photon (or any particle or wave) that is related to the time required to define the energy.

The uncertainty principle helped overthrow the ideas of strict determinism in physics. Regardless of our measuring instruments, we cannot predict the motion of an electron beyond an unavoidable uncertainty. According to quantum mechanics, this is the way nature is.

Example 27.4

An experiment determines that an electron is within 0.10 mm of a particular point. If we try to measure its velocity, what will be the minimum uncertainty?

Solution Using the Heisenberg relation, we get

$$\Delta v_x = \frac{\Delta p_x}{m} \geq \frac{h}{m \, \Delta x},$$

$$\Delta v_x \geq \frac{6.63 \times 10^{-34} \text{ J} \cdot \text{s}}{(9.1 \times 10^{-31} \text{ kg})(1.0 \times 10^{-4} \text{ m})},$$

$$\Delta v_x \geq 7.3 \text{ m/s}.$$

The uncertainty in the electron's velocity is 7.3 m/s. Thus we can predict its velocity only to within 7.3 m/s. Locating the electron at one position affects our ability to know where it will be at later times.

Example 27.5

Uncertainty in the motion of a grain of sand.

A grain of sand with a mass of 1.00×10^{-3} g appears to be at rest on a smooth surface. We locate its position to within 0.010 mm. What velocity limit is implied by our measurement of its position?

Solution   As before,

$$\Delta v_x = \frac{\Delta p_x}{m} \geq \frac{h}{m \, \Delta x},$$

$$\Delta v_x \geq \frac{6.63 \times 10^{-34} \, \text{J·s}}{(1.0 \times 10^{-6} \, \text{kg})(1.0 \times 10^{-5} \, \text{m})},$$

$$\Delta v_x \geq 6.6 \times 10^{-23} \, \text{m/s}.$$

The uncertainty in momentum leads to an uncertainty in velocity of 6.6×10^{-23} m/s. This velocity is so small that we do not observe it. The grain of sand may still be considered at rest, as our experience says it should.

Example 27.6

Energy of an electron confined to nuclear dimensions.

Calculate the approximate energy needed to confine an electron to within a space of the order of nuclear dimensions, about 10^{-14} m.

Solution   We can estimate the momentum of the electron from the uncertainty relationship, $\Delta p \, \Delta x \geq h$. If the electron is confined to within $\Delta x = 10^{-14}$ m, then the uncertainty in momentum is

$$\Delta p \geq \frac{6.63 \times 10^{-34} \, \text{J·s}}{1 \times 10^{-14} \, \text{m}} = 6.63 \times 10^{-20} \, \text{J·s/m}.$$

The momentum must be at least as great as the uncertainty in momentum Δp. The minimum energy is obtained from

$$E^2 = c^2 p^2 + m_0^2 c^4,$$

$$E^2 = (3.00 \times 10^8 \, \text{m/s})^2 (6.63 \times 10^{-20} \, \text{J·s/m})^2$$
$$+ (9.11 \times 10^{-31} \, \text{kg})^2 (3.00 \times 10^8 \, \text{m/s})^4.$$

The energy is

$$E = 2 \times 10^{-11} \, \text{J} \approx 100 \, \text{MeV}.$$

Thus the energy of an electron confined within a nucleus would be of the order of 100 MeV. The observed energies of the electrons in beta decay, however, are of the order of 2 or 3 MeV. Because of this very large discrepancy, we believe that the nucleus is not composed of, and does not contain, electrons, even though electrons are emitted during nuclear beta decay. The resolution to this seeming paradox will be discussed in Chapter 28.

Interpretation of the Wave Function

27.6

The Schrödinger wave theory was very successful for calculating quantum behavior; we shall examine some of these successes soon. However, there was still no physical interpretation of the wave function ψ. Initially, Schrödinger tried to associate the wave amplitude with a distribution of electric charge in space, but this idea was soon found unacceptable. Shortly after the publication of Schrödinger's work, the German physicist Max Born (1882–1970) began to apply quantum mechanics to the analysis of collisions between particles. As an outgrowth of this work, Born proposed a surprising interpretation of the physical significance of the wave function.

Born recognized that the quantity ψ could not be measured directly. Instead, he concluded that the square of the absolute value of the wave function $|\psi|^2$ is physically meaningful. Born interpreted this quantity as the **probability density** for the electron (or other particle). That is, the probability or likelihood of finding an electron in a particular small volume of space ΔV is given by

$$P = |\psi|^2 \, \Delta V. \tag{27.9}$$

Thus the particle is most likely to be found where the wave function is largest and is not likely to be found where the wave function is small.

Furthermore, Born pointed out that in collision problems, for example, quantum mechanics does not describe precisely the state of the particles following the collision, as classical mechanics does. Instead, the quantum theory gives the probabilities of various possible outcomes. In other words, quantum mechanics is a statistical theory. It does not precisely describe a single event, but rather it describes the distribution of outcomes of a large number of identical events. The wave function carries statistical information about the outcome of any particular experimental measurement designed to provide information about the particle.

Suppose, for example, we imagine a beam of electrons traveling along the positive x direction (Fig. 27.10, p. 810). If the beam passes through a narrow slit whose length is perpendicular to the y direction, we expect the beam to spread out in the y direction by diffraction. (This discussion is equally valid for photons as well as electrons.) If the diffracted electron beam strikes a fluorescent screen, we expect an intensity pattern of fluorescent light like that shown.

Now suppose that we limit the beam to such a small number of electrons that they pass through the slit one at a time. We could then observe scintillations on the screen due to individual electrons. We would no longer see an interference pattern such as that of Fig. 27.10. Instead we would see only the tiny flashes of light indicating where each electron hit the screen (Fig. 27.11a, p. 810). We would be helpless to predict where successive electrons would strike the screen. However, we could calculate the *probability* that an electron would strike any particular position on the screen. If we then waited long enough, until a large number of electrons

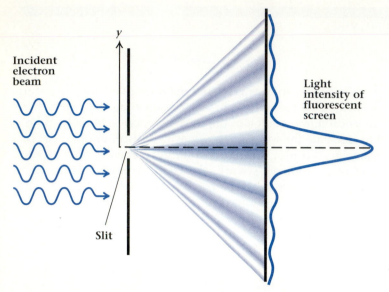

Figure 27.10   Diffraction pattern observed on a fluorescent screen for an electron beam diffracted by a narrow slit.

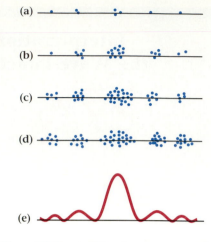

Figure 27.11   A diffraction pattern can be formed from a succession of single-particle events. (a) Few events; (b), (c), (d), more events. (e) Theoretical pattern for the sum of a large number of single events, matching the diffraction pattern of Fig. 27.10.

had passed through the slit and struck the screen (Fig. 27.11b,c,d), the distribution of the observed individual impacts would take on the shape of the theoretical probability curve for electrons striking the screen (Fig. 27.11e) This curve is identical with the diffraction pattern calculated from simple wave theory.

Here again the wave–particle duality appears. The electrons (or photons) are diffracted as waves by the slit. They travel to the screen (or other detector) where they are recorded as individual, particlelike events. Our theory cannot predict exactly where any particular electron will strike, but it can describe the probability of the electron's striking the screen at various points. The sum of a large number of events conforms closely with the probabilistic prediction given by the theory.

We can extend our discussion of single-slit diffraction to include Young's double-slit experiment. As in the case of the single slit, the same general diffraction pattern results from either photons (light) or electrons. You are already familiar with the details of the double-slit experiment with light (Chapter 23). Now let us consider the quantum mechanical interpretation of that experiment. In accordance with Bohr's correspondence principle, quantum mechanics predicts that for high-intensity light (that is, large numbers of photons), the pattern will be the same as predicted by the wave theory (Fig. 27.12a). Just as in the case of the single slit, we cannot calculate where a given photon will strike the screen, but we can calculate the probability that it will strike a particular place on the screen. The resulting probability distribution has the same pattern as that given by the classical theory. These results have been confirmed by experiments in which the light intensity was so low that only one photon at a time passed through the slits.

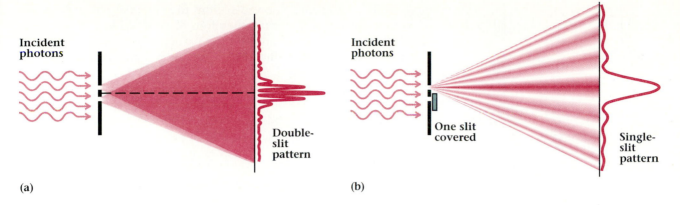

(a)

(b)

Figure 27.12 (a) Quantum mechanics predicts the intensity distribution for light diffracted by a pair of narrow slits. (b) If one of the slits is covered, the resulting pattern is the usual single-slit pattern.

An interesting point arises when you ask what happens, not at the screen, but at the slits themselves. Considering incident photons one at a time, we may cover either of the slits and, after a sufficiently long time, observe the buildup of a single-slit pattern (Fig. 27.12b). However, without changing anything else, if we uncover both slits, the pattern built up after a long time will be the usual double-slit one, easily distinguished from that of a single slit. The probability that a photon will arrive at a given place on the screen depends on whether one slit or two slits are open.

If you regard the photons strictly as particles, then individually they must go through one slit or the other, not both. If they go through only one slit, how do they "know" if the other slit is open or closed, so that they generate the appropriate pattern?

The answer to this seeming paradox is not in the description of the experiment, but in our attempt to model the photon as being either a particle or a wave. The solution frequently offered, that the photon (or electron) interferes with itself, is difficult to understand. One way to resolve the difficulty is to say that the wave function ψ must include information about the apparatus, as well as about the particle itself. The wave functions for the cases of one slit only and of both slits are different because the apparatus is physically different. With this view of the experiment there should be no surprise that the diffraction patterns satisfy the correspondence principle.

This experiment points out one of the underlying problems with quantum mechanics. The mathematical rules for predicting results are clear, but interpreting what happens in terms of a physical model is sometimes difficult.

The Particle in a Box

27.7

One of the natural consequences of wave mechanics is the quantization of energy. For example, the stationary electronic states of atoms are intimately related to the wavelike nature of electrons. To illustrate how quantum mechanics reveals this relationship, we examine the case of a particle bound within a one-dimensional box. For simplicity, we assume that the particle is confined between rigid, unyielding walls separated by a distance

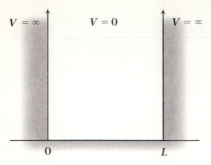

$V = \infty$ $V = 0$ $V = \infty$

0   L

L (Fig. 27.13). Beyond this allowed region, the potential energy V is infinite and the wave function must therefore be zero. Within the box the potential V may be taken as zero and the particle has a wave function ψ and energy E. We often call this kind of situation a potential well, since it resembles a well sunk into the ground. Although this is a somewhat artificial situation, it serves as a good example of a simple quantum mechanical system. With it we obtain quantization of energy.

Within the box, the Schrödinger equation for the particle becomes

$$\frac{\Delta(\Delta\psi/\Delta x)}{\Delta x} + \frac{8\pi^2 m}{h^2} E\psi = 0. \tag{27.10}$$

The condition of the rigid boundaries demands that the wave function should vanish for $x = 0$ and for $x = L$ because the probability density $|\psi|^2$ is zero outside the box and, hence, is also zero right at the boundaries. These conditions are the same as the boundary conditions for a wave along a string fixed at the ends. Wave solutions that are consistent with these boundary conditions are

$$\psi_n(x) = \psi_0 \sin(n\pi x/L), \qquad n = 1, 2, 3, \ldots, \quad 0 < x < L, \tag{27.11}$$
$$\psi_n(x) = 0, \qquad\qquad\qquad\qquad \text{everywhere else.}$$

The integer n is the quantum number that describes the state of the system. The wave functions $\psi_n(x)$ are zero at the boundaries $x = 0$ and $x = L$ as required. The first four standing wave solutions are shown in Fig. 27.14.

We can now use these wave functions in the Schrödinger equation, Eq. (27.10), to find the energy associated with each quantum state (or wave function). We may conclude from our discussion of harmonic motion in Chapter 14 that the rate of change of a cosine function is a negative sine function and that the rate of change of a sine function is a cosine function. Using this information, we can calculate the first term in Eq. (27.10) to be

$$\frac{\Delta(\Delta\psi_n/\Delta x)}{\Delta x} = -\left(\frac{n\pi}{L}\right)^2 \psi_n.$$

Equating the coefficients of ψ_n and solving for E_n, we find a value for the energy of

$$E_n = \frac{h^2}{8mL^2}n^2, \qquad n = 1, 2, 3, \ldots \tag{27.12}$$

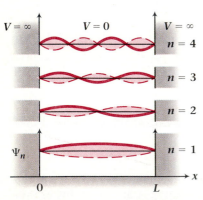

$V = \infty$ $V = 0$ $V = \infty$

$n = 4$

$n = 3$

$n = 2$

Ψ_n $n = 1$

0   L x

Figure 27.14 The first four standing wave solutions describing the wave function of a particle in a one-dimensional square potential well.

The energy levels are quantized according to the square of the quantum number n. Figure 27.15 shows the first four levels. Particles in a one-dimensional potential well can only exist in states of definite energy E_n, described by the wave functions ψ_n. These states are labeled by the quantum number n. All other energy values are excluded from consideration because the wave functions corresponding to such energies do not satisfy the boundary conditions and are consequently forbidden.

The energy E_n given above is entirely kinetic energy because the potential energy is zero. An important consequence of the wave theory is that zero kinetic energy is not allowed. If we choose $n = 0$, so that $E = $

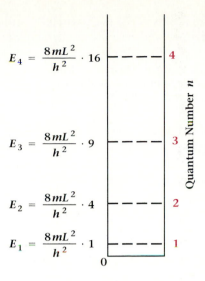

$$E_4 = \frac{8mL^2}{h^2} \cdot 16 \quad 4$$

$$E_3 = \frac{8mL^2}{h^2} \cdot 9 \quad 3$$

$$E_2 = \frac{8mL^2}{h^2} \cdot 4 \quad 2$$

$$E_1 = \frac{8mL^2}{h^2} \cdot 1 \quad 1$$

Figure 27.15 Left: the first four energy levels of a one-dimensional potential well of width L; right: the corresponding quantum numbers.

0, we find that the wave function vanishes everywhere, since sin $0 = 0$. Thus there is no wave function corresponding to zero energy. The lowest energy corresponds to $n = 1$ and is not zero. This result is typical for quantum systems in that, even for a temperature of absolute zero, where classical motions cease, a quantum system still possesses a small amount of kinetic energy. This residual energy is called the **zero-point energy**.

The zero-point energy is observed in solid-state, atomic, and nuclear physics. It is especially apparent in the low-temperature behavior of helium. Classical thermodynamics predicts that the kinetic energy of molecules goes to zero as the absolute temperature goes to zero. This prediction, however, does not take into account the quantum mechanical prediction of zero-point energy. For the case of helium, the zero-point energy is so great that at atmospheric pressure helium does not liquefy even when cooled to below 1 K. Instead, helium remains a liquid at temperatures far below the freezing points of all other substances.

Tunneling or Barrier Penetration

27.8

A startling prediction of quantum mechanics is the tunneling of particles through a potential-energy barrier. This behavior is unknown in classical physics, where particles cannot penetrate such barriers and can cross them only if the particles have sufficient kinetic energy. The quantum mechanical case is analogous to a football player running headlong into a solid brick wall and passing right through to the other side without making a hole in the wall or going over the top. Nevertheless, tunneling is a widespread occurrence throughout the atomic and subatomic world.

In our preceding discussion of the particle in a box we demanded that the wave function vanish at the boundary. But suppose we relax that requirement, realizing that in actual situations the potential outside is not truly infinite. Then the wave function need not vanish exactly at the boundary. We need only insist that the wave function inside and outside the box merge smoothly at the boundary (Fig. 27.16).

In the region where the potential V exceeds the energy E, the Schrödinger equation becomes

$$\frac{\Delta(\Delta\psi/\Delta x)}{\Delta x} = \frac{8\pi^2 m}{h^2}(V - E)\psi,$$

and the term on the right-hand side is positive. Wave solutions are no longer possible. However, the equation does have exponential solutions of the form

$$\psi(x) = \psi_0 e^{-\alpha x}, \qquad (27.13)$$

where α is a constant and x is measured from the boundary into the region

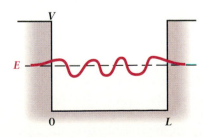

Figure 27.16 Wave function for a particle in a potential well of finite height V. The function is oscillatory within the well, but decreases exponentially outside the well.

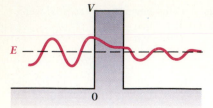

Figure 27.17 Potential energy and wave functions for a finite potential barrier. The presence of ψ on the right-hand side indicates a reduced but finite probability of finding the particle on that side of the barrier.

of high potential. In the allowed region $0 < x < L$, the particle is still described by an oscillatory function of sines and cosines. In the region beyond the boundary, the wave function for the particle decreases exponentially with distance into the barrier (Fig. 27.16). There is thus a finite though small probability of finding the particle outside the allowed region. This probability diminishes with increasing potential V or increasing distance x into the barrier.

Now suppose the barrier has a finite thickness (Fig. 27.17). Then the amplitude of the wave function does not decrease completely to zero across the width of the barrier. Instead, a wave solution exists on the opposite side of the barrier, but with a reduced amplitude. The probability for the electron (or other particle) appearing on the other side of the barrier is nonzero. This effect is known as **barrier penetration** or **tunneling**.

Tunneling of particles is not predicted in classical mechanics. A particle confined in a gravitational potential cannot escape unless its total energy E exceeds the potential energy barrier V. Quantum mechanically, a particle such as an electron can tunnel out of a finite potential well even though its total energy E is less than that of the potential barrier. Tunneling effects are important in the radioactive decay of nuclei, in semiconductors, and in superconductors.

Wave Theory of the Hydrogen Atom

*27.9

If we extend our simple example of the particle moving in a rigid box to two dimensions, we find that the wave functions must satisfy boundary requirements in both dimensions simultaneously. However, since the two motions are perpendicular, they are independent. Mathematically, this means the solutions to the Schrödinger equation must contain two independent quantum numbers. If we choose the two dimensions of the box as x and y, then the wave must satisfy the condition for standing waves in each direction.

For motion in three dimensions, we need three quantum numbers to specify the complete wave function. Thus we could describe a particle in a three-dimensional box in terms of the energy states corresponding to various values of the three quantum numbers n_x, n_y, and n_z.

When we apply wave mechanics to the description of a real three-dimensional atomic system, such as the hydrogen atom, it is more natural and more convenient to use a system of spherical coordinates. We still have three quantum numbers although they are no longer n_x, n_y, and n_z. We shall not attempt here to solve the Schrödinger equation for the hydrogen atom because of the mathematical complexity. However, we will discuss the three quantum numbers and point out some results from the quantum mechanical treatment of the hydrogen atom. As we will see in the rest of this chapter, most of these results, with some modification and extension, underlie the quantum mechanical description of all other atoms.

The electron wave functions that are solutions of the Schrödinger equation for the hydrogen atom are specified by three quantum numbers, one for each of the three coordinates. The first of these is the *principal quantum number n*, which is associated with the radial part of the wave function. This quantum number plays much the same role as the quantum number introduced by Bohr. The energy levels of the atom are roughly arranged into levels or shells associated with the principal quantum number *n*, which takes on integer values 1, 2, 3, 4, . . . just as in the Bohr theory. Frequently, we designate each shell by a capital letter:

Principal quantum number *n*: 1 2 3 4 . . .

Letter designation: *K L M N* . . .

The second quantum number is the azimuthal or *orbital quantum number l*, which characterizes the orbital angular momentum. This quantum number can take on integer values of 0, 1, 2, . . . up to $n - 1$. Each of the principal atomic shells is composed of subshells, characterized by their orbital quantum number. These subshells are labeled with lower-case letters as follows:

Orbital quantum number *l*: 0 1 2 3 4 5 . . .

Subshell letter designation: *s p d f g h* . . .

Finally, there is the **magnetic quantum number** m_l, which describes the magnitude of the component of the angular momentum along a particular spatial direction. This quantum number can be any integer value between $-l$ and $+l$, including 0. Each subshell consists of several states, each of which can accommodate two electrons.

To summarize, the energy levels of the hydrogen atom are characterized by three quantum numbers, which can assume integer values according to the following scheme:

$$n = 1, 2, 3, 4, \ldots ,$$
$$l = 0, 1, 2, 3, \ldots , n - 1, \tag{27.14}$$
$$m_l = 0, \pm 1, \pm 2, \pm 3, \ldots , +l.$$

Table 27.1 illustrates the labeling of the first few levels.

When no external fields are present, the energy associated with each wave function depends only on *n* and is given by

$$E_n = -\frac{me^4}{8\epsilon_0^2 h^2 n^2}.$$

TABLE 27.1
Designations of the lowest atomic energy levels

Principal quantum number n	Shell designation	Orbital quantum number l	Subshell designation
1	K	0	1s
2	L	0	2s
		1	2p
3	M	0	3s
		1	3p
		2	3d

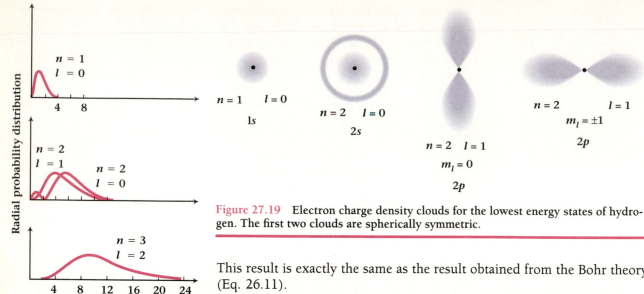

Figure 27.18   Radial probability distributions for several wave functions for hydrogen.

Figure 27.19   Electron charge density clouds for the lowest energy states of hydrogen. The first two clouds are spherically symmetric.

This result is exactly the same as the result obtained from the Bohr theory (Eq. 26.11).

When we compute the radial probability densities from the wave functions, using $|\psi|^2$, they have the forms shown graphically in Fig. 27.18. The probability density for the lowest-energy wave function, labeled by $n = 1$ and $l = 0$, has a maximum at $r = 0.53 \times 10^{-10}$ m, which is the same radial distance from the nucleus as the first Bohr orbit. Similarly, other wave functions for n and $l = n - 1$ have a probability maximum corresponding to the radius of the appropriate Bohr orbit.

When the atom is in a definite quantum state, the wave function is constant in time, just as it was for the earlier example of the particle in a box . Therefore $|\psi|^2$ is also constant. Thus the Schrödinger description of the atom replaces the Bohr model of a point electron racing in eternal orbit about the nucleus with a more abstract model of a standing wave, whose amplitude squared measures the probability that the electron is at any particular distance from the nucleus. It is often convenient to imagine the atom as consisting of a point nucleus surrounded by a diffuse cloud of electric charge of density ρ, given by

$$\rho = -e\,|\psi|^2.$$

In many cases this model can help us understand the behavior of an atom with its surroundings. Figure 27.19 shows a few examples of the charge density clouds for several wave functions.

The Zeeman Effect and Space Quantization

*27.10

We have shown that the principal quantum number n has a clear physical meaning in quantum mechanics, corresponding to distinct energy levels in the atom. What physical characteristics of the atom are associated with the quantum numbers l and m_l? Several experiments conducted before the development of Schrödinger's theory help provide the answer.

After the initial success of the Bohr model of the atom and prior to the development of wave mechanics, Arnold Sommerfeld (1868–1951) extended the Bohr theory to include elliptical orbits. According to his model, stationary elliptical orbits were permitted when the ratio of the major to minor axes of the ellipse was a ratio of integers, n/l, where n was the principal quantum number and $l = 0, 1, 2, \ldots, n - 1$ was the azimuthal quantum number, which labeled the magnitude of the angular momentum of the orbit. These two quantum numbers turned out to be the same as those that arose from wave mechanics.

Sommerfeld then argued that a charged electron moving about an orbit is equivalent to a small current loop and therefore behaves like a tiny magnet. When no external magnetic field is present, the spatial orientations of the orbital magnetic moments are all equivalent. But when an external magnetic field is present, the orientation of the orbital magnetic moment contributes to the energy of the atom according to

$$E_{\mathrm{m}} = -\mu B \cos \theta,$$

where μ is the magnitude of the magnetic moment, B is the magnitude of the applied magnetic field, and θ is the angle between them. (In Chapter 15 we discussed a similar equation for the potential energy of an electric dipole in an electric field.)

According to Sommerfeld's theory, these current loops can only orient themselves in directions such that the component of the orbital angular momentum along the direction of the field is an integer multiple of $h/2\pi$ (Fig. 27.20). This effect is known as **space quantization**. To keep track of the possible orientations, Sommerfeld introduced the magnetic quantum number m_l, where

$$m_l = 0, \pm 1, \pm 2, \ldots, \pm l.$$

Notice that the quantum number m_l of the Sommerfeld theory is also found in the Schrödinger wave theory (Eq. 27.14).

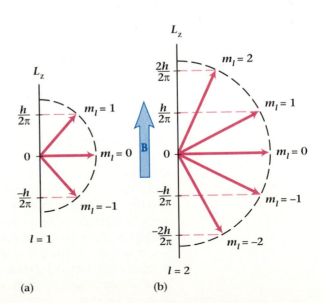

Figure 27.20   **Projections of the angular momentum vector along the direction of an applied field B. The projections are quantized in units of $h/2\pi$. (a) $l = 1$; (b) $l = 2$.**

The relation between an electron's orbital magnetic moment and its angular momentum can be found from classical electrodynamics. The magnetic moment μ of a current loop enclosing an area A and carrying a current i is

$$\mu = iA.$$

Here the current is the charge $-e$ divided by its period $T = 2\pi r/v$, where v is the tangential speed of the electron. Thus

$$i = -\frac{e}{T} = -\frac{ev}{2\pi r}.$$

The area is πr^2, so that

$$\mu = -\frac{ev}{2\pi r}(\pi r^2) = -\frac{evr}{2}.$$

But the definition of angular momentum is $L = mvr$, where m is the mass of the electron. Thus the magnetic moment becomes

$$\mu = \frac{-e}{2m}L. \tag{27.15}$$

We can replace the angular momentum L with its quantized form $l\,(h/2\pi)$:

$$\mu = \frac{-eh}{4\pi m}l.$$

We define $eh/4\pi m$ as the **Bohr magneton** μ_B, which has the value 9.274×10^{-24} J/T. In terms of μ_B, the orbital magnetic moment is

$$\mu = -l\mu_B.$$

In the presence of an applied magnetic field **B**, an electronic energy level with orbital angular momentum labeled by l and having a projection m_l along the field direction is shifted in energy by an amount

$$E_m = m_l\mu_B B. \tag{27.16}$$

This separation in energy of otherwise identical energy levels provides a semiclassical explanation for the **Zeeman effect,** which is the splitting of spectral lines into separate components by applying a magnetic field to the atom. The explanation for this effect is not found in the Bohr theory.

For the case of an electronic transition from an $l = 1$ state to an $l = 0$ state, light is emitted. The presence of an external magnetic field splits the expected single spectral line into three closely spaced lines, called a Zeeman triplet (Fig. 27.21). With no applied field, the various m_l states correspond to the same energy. When the magnetic field is applied, the initially identical levels separate according to Eq. (27.16). Transitions from these slightly different states lead to the distinct lines seen in the Zeeman spectra. Classically, only a broadening of the original line would occur, since a classical magnet has no restrictions on the orientations it can assume with respect to the applied field.

While the explanation of the Zeeman effect in terms of space quantization of orbital angular momentum is satisfying, it did not unequivocally confirm space quantization. An experiment designed to directly test space

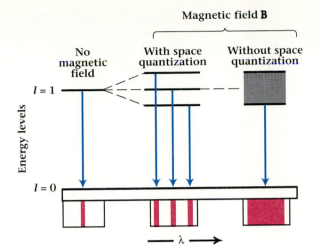

Figure 27.21 In the Zeeman effect, a single spectral line due to a transition from an $l = 1$ state to an $l = 0$ state is split into three lines by an applied magnetic field because the different m_l levels have different energies. Without space quantization, a broadened line would be expected.

quantization was proposed by Otto Stern in 1921 and was subsequently carried out by Stern and W. Gerlach in late 1921 and early 1922.

An electric dipole in a nonuniform electric field experiences a net translational force because the forces acting on the positive and negative charges are unequal. A magnetic dipole in a nonuniform magnetic field also experiences a translational force. Suppose, then, that we direct a narrow beam of atoms, each possessing a magnetic moment, between the poles of a magnet shaped to produce a strong magnetic field gradient perpendicular to the path of the beam. Those atoms with a component of their magnetic moment along the direction of the field experience a deflection. If the component of the magnetic moment lies along the direction of the field, the atoms are deflected one way; if it is opposite to the field, they are deflected oppositely.

In their experiment, Stern and Gerlach generated a beam of silver atoms by evaporation from a heated oven (Fig. 27.22). The atoms escaped

Figure 27.22 Schematic arrangement of the apparatus for the Stern-Gerlach experiments. A beam of atoms was directed along the positive x direction between the specially shaped magnetic pole pieces, which produced a strong gradient in B_z. Splitting of the atomic beam was observed at the glass plate.

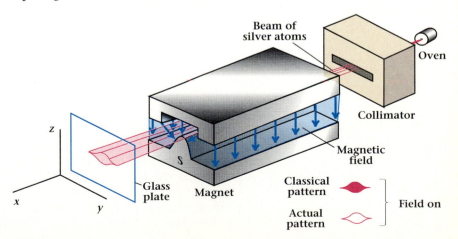

through a small aperture into a region of high vacuum. The emerging beam was first shaped by a series of apertures to a narrow cross section and then directed down the path between the pole pieces of the magnet. After passing through the magnet, the silver atoms struck a glass plate where, after several hours exposure, they accumulated to form a thin film that was then darkened by chemical treatment. The resulting exposures all had one prominent characteristic: The atomic beam was split into two separate beams by the magnetic field and no atoms passed through undeflected. This experiment was considered another triumph of the early quantum theory and was taken as direct experimental verification of space quantization. If there were no quantization of the magnetic moments, the Stern-Gerlach experiment would give a smeared pattern instead of the two distinct lines observed. The existence of the two distinct beams is proof of two (and only two) unique orientations of the atomic magnetic moment relative to the direction of the applied magnetic field.

The Pauli Exclusion Principle

27.11

In 1925, before publication of Schrödinger's wave equation, Wolfgang Pauli (1900–1958) recognized that the shell structure of atoms could be explained naturally if each electron state was labeled by four quantum numbers and if each individual state could be occupied by only a single electron. Pauli combined the four quantum numbers according to the then current atomic model to give an understanding of the periodic table and of the multiple line structures observed in optical spectra.

As part of this development, Pauli formulated the now famous **exclusion principle: No two electrons in an atom may occupy the same quantum state; that is, no two electrons in an atom may have the same four quantum numbers.** If an electron in an atom exists in a quantum state labeled by a set of four distinct quantum numbers, then that state is occupied and all other electrons in the atom are excluded from that particular state.

In studying Pauli's work on the exclusion principle, Samuel Goudsmit and George Uhlenbeck were struck by the absence of any concrete picture or physical meaning to connect with the fourth quantum number. It was introduced as just part of the formalism of Pauli's atomic model. Goudsmit and Uhlenbeck were familiar with the idea that each quantum number corresponded to a particular coordinate. The prevailing idea in 1925 of a point electron allowed for only three coordinates and, hence, only three quantum numbers.

Goudsmit showed that the four quantum numbers used by Pauli could be replaced by the quantum numbers n, l, m_l, and m_s. The first three quantum numbers are the same three already discussed. The quantum number m_s was observed to take on only the values $+\frac{1}{2}$ and $-\frac{1}{2}$. Goudsmit and Uhlenbeck proposed that this fourth quantum number is associated with an additional property—the angular momentum of the electron or, as it has become known, the electron's **spin**. According to their idea, the electron possesses an intrinsic angular momentum $s = \frac{1}{2}$ whose projection m_s along a given direction could take on only the value of $m_s = +\frac{1}{2}$ or $m_s = -\frac{1}{2}$.* They also postulated that the ratio of the intrinsic magnetic moment to the intrinsic angular momentum had to be twice as great as for orbital motion. That is,

$$\mu_s = 2\,\frac{e}{2m}\,\frac{h}{2\pi}\,m_s = 2\mu_B m_s. \tag{27.17}$$

Although they did not realize it at the time of their 1922 experiment, Stern and Gerlach were really observing the intrinsic magnetic moment of the electron and not the effects of orbital motion. Had they used other atoms their results could have been quite different. Nevertheless, their experiment and many different experiments performed since that time established the fact that, in the presence of a magnetic field, atomic magnetic moments arising from both spin and orbital contributions are quantized.

Example 27.7

Energy difference between two states of electron spin.

Calculate the energy difference between two electron states of $m_s = \pm\frac{1}{2}$ in an applied magnetic field of $B = 0.50$ T.

Solution   By analogy with Eq. (27.16), we can determine the energy of each state from

$$E = \mu_s B = 2\mu_B m_s B.$$

The quantum number m_s takes on values of $\pm\frac{1}{2}$. Thus

$$\Delta E = E_+ - E_- = 2\mu_B B\left[\tfrac{1}{2} - (-\tfrac{1}{2})\right] = 2\mu_B B,$$

$$\Delta E = 2 \times 9.27 \times 10^{-24}\,\text{J/T} \times 0.50\,\text{T},$$

$$\Delta E = 9.27 \times 10^{-24}\,\text{J} = 5.8 \times 10^{-5}\,\text{eV}.$$

This energy is nearly 30,000 times smaller than the energy corresponding to the least energetic Balmer transition in hydrogen.

Understanding the Periodic Table

*27.12

With the help of the Pauli exclusion principle, we can bring together the information about quantum states in the atom and their associated quantum numbers to finally understand the electronic structure of the elements and their orderly arrangement in the periodic table. We visualize the lowest energy level or ground state of an atom by considering a simple model of energy levels and placing electrons into those levels. In the ground state, we place electrons into the lowest electronic energy levels and fill

---

*An intrinsic property is a permanent property of the body itself, such as mass or charge. Quantities that depend on the body's circumstances, such as velocity, kinetic energy, and ordinary angular momentum, are not intrinsic properties.

TABLE 27.2
Quantum numbers of the electron states for the first three principal energy levels

Shell	n	l	m_l	m_s	Number of electrons In subshell	Number of electrons In shell
K	1	0	0	$+\frac{1}{2}$		
	1	0	0	$-\frac{1}{2}$		$\}2$
L	2	0	0	$+\frac{1}{2}$		
	2	0	0	$-\frac{1}{2}$	$\}2$	
	2	1	$+1$	$+\frac{1}{2}$		
	2	1	$+1$	$-\frac{1}{2}$		
	2	1	0	$+\frac{1}{2}$		$\}8$
	2	1	0	$-\frac{1}{2}$	$\}6$	
	2	1	-1	$+\frac{1}{2}$		
	2	1	-1	$-\frac{1}{2}$		
M	3	0	0	$+\frac{1}{2}$		
	3	0	0	$-\frac{1}{2}$	$\}2$	
	3	1	$+1$	$+\frac{1}{2}$		
	3	1	$+1$	$-\frac{1}{2}$		
	3	1	0	$+\frac{1}{2}$		
	3	1	0	$-\frac{1}{2}$	$\}6$	
	3	1	-1	$+\frac{1}{2}$		
	3	1	-1	$-\frac{1}{2}$		
	3	2	$+2$	$+\frac{1}{2}$		$\}18$
	3	2	$+2$	$-\frac{1}{2}$		
	3	2	$+1$	$+\frac{1}{2}$		
	3	2	$+1$	$-\frac{1}{2}$		
	3	2	0	$+\frac{1}{2}$		
	3	2	0	$-\frac{1}{2}$	$\}10$	
	3	2	-1	$+\frac{1}{2}$		
	3	2	-1	$-\frac{1}{2}$		
	3	2	-2	$+\frac{1}{2}$		
	3	2	-2	$-\frac{1}{2}$		

successive levels until all electrons are accounted for. The remaining energy levels of higher energy are unoccupied. An illustration of such a scheme is shown in Table 27.2.

The first element ($Z = 1$) in the periodic table is hydrogen, which has only one electron. The second element, helium ($Z = 2$), has two electrons, corresponding to the first two states in Table 27.2. Two electrons exhaust all possibilities for energy states with $n = 1$. Together they make up the first shell of electrons about the nucleus, which is called the K shell.

The element for $Z = 3$ is lithium. Two of its electrons occupy the K shell, while the third occupies a state for which $n = 2$, beginning the L shell. This configuration of one electron outside a closed shell is similar to hydrogen and accounts for the position of lithium in the periodic table. The fourth element, beryllium, has four electrons, which in the ground state fill the K shell and the subshell of $n = 2$, $l = 0$. The L shell can be considered to be made up of two subshells with $l = 0$ and $l = 1$. The maximum number of electrons in the L shell is 8, corresponding to

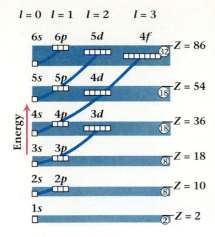

$l = 0$ $l = 1$ $l = 2$ $l = 3$

6s 6p 5d 4f

$Z = 86$ ③②

5s 5p 4d

$Z = 54$ ⑱

4s 4p 3d

$Z = 36$ ⑱

3s 3p

$Z = 18$ ⑧

2s 2p

$Z = 10$ ⑧

1s

$Z = 2$ ②

Energy

Figure 27.23 Schematic diagram of the electronic states of atoms. Each small square represents two electron states (one each for $m_s = \pm\frac{1}{2}$). Separated groups of squares correspond to subshells. The total number of electrons corresponding to a closed shell is shown in the circle at the right. The number at the far right is the atomic number of the element for which the shells are filled. States having the same value of n are connected with the heavy line.

the element neon ($Z = 10$). See Fig. 27.23. In the periodic table, neon is placed below the element helium, which also has a closed shell of electrons.

The periodicity of chemical behavior of the elements was the basis for Mendeleev's original periodic table (Chapter 25). The additional ordering of these elements according to atomic number Z came as a result of Moseley's x-ray experiments. Finally, with the development of quantum mechanics, the electronic structure of the atoms explained the periodicity of the elements.

Further details about the electronic structure of atoms are beyond the scope of this text. However, we should point out that some important effects have not been included. The energies of the quantum states do not always follow the simple order our treatment implies. For example, subshells of a given value of n sometimes have higher energies than some subshells with a higher value of n. This phenomenon accounts for the positions of the transition metals, which occupy a large fraction of the periodic table. We have also ignored the effect of coupling interactions between the spin and the orbital motion of the electrons. The principal point to establish here is that the behavior of the elements and their arrangement into the periodic table is well explained within the framework of quantum mechanics.

SUMMARY

Useful Concepts

- The change in wavelength of Compton-scattered photons is

$$\lambda - \lambda' = \frac{h}{mc}(1 - \cos\theta).$$

- The de Broglie wavelength of a particle is

$$\lambda = \frac{h}{p}.$$

- Nature is most adequately described on a microscopic scale by the Schrödinger wave equation. The one-dimensional Schrödinger wave equation can be written

$$\frac{\Delta(\Delta\psi/\Delta x)}{\Delta x} + \frac{8\pi^2 m}{h^2}(E - V)\psi = 0.$$

- The probability P of finding a particle in a volume of space ΔV is given by

$$P = |\psi|^2 \Delta V.$$

- The Heisenberg uncertainty relations can be written in the form

$$\Delta x \, \Delta p_x \geq h$$

and

$$\Delta E \, \Delta t \geq h.$$

- The energy levels of particles confined to one-dimensional motion within a fixed distance L are

$$E_n = \frac{h^2}{8mL^2}n^2, \qquad n = 1, 2, 3, \ldots$$

- The Pauli exclusion principle states that no two electrons in any atom may have the same quantum numbers.

You should be able to write the definition or meaning of each of the following terms:

- Compton wavelength
- de Broglie wavelength
- Schrödinger equation
- wave function
- wave packet
- uncertainty principle
- probability density
- zero-point energy

- barrier penetration or tunneling
- space quantization
- Bohr magneton
- Zeeman effect
- exclusion principle
- spin

QUESTIONS

27.1 How does the Compton effect differ from the photoelectric effect?

27.2 Why does the Compton effect occur for x rays but not for visible light?

27.3 Estimate what the de Broglie wavelength of a baseball thrown hard by a pitcher would be if Planck's constant had a value of 1.0 J·s.

27.4 What is your approximate de Broglie wavelength when you run at 6 m/s? How does that compare with the size of a hydrogen atom? What is the physical significance of such a wavelength?

27.5 Is a baseball diffracted when thrown through a picket fence? Explain.

27.6 Does the uncertainty principle have any noticeable effect in large-scale phenomena, such as steering a fast-moving car or shooting a rocket to the moon? Explain.

27.7 How does the uncertainty principle affect our understanding of the principle of conservation of energy?

27.8 How is the interpretation of the Schrödinger wave amplitude similar to the interpretation of the amplitude of an electromagnetic wave?

27.9 Decide, on the basis of some numerical estimates, whether the quantized energy states of a person confined to a room can be observationally distinguished from the smooth range allowed by classical physics.

27.10 Is the probability that a book lying on a table will tunnel or penetrate through the table exactly zero or is it just extremely small? Could you estimate an upper limit to this probability, based on an estimate of the number of tables in the world and the number of tunneling events reported each year?

27.11 Explain the effects on tunneling of large potentials ($E << V$) and of wide barriers ($x >> 0$).

27.12 Why is m_l called the magnetic quantum number?

27.13 How would the atomic shell structure be affected if electrons did not obey the Pauli exclusion principle?

27.14 What are the quantum numbers for the outermost electron of a sodium atom in the ground state? What would be the quantum numbers if the outermost electron were raised to the first excited state?

27.15 How would atomic shell structures be affected if electrons did not obey the Pauli exclusion principle?

PROBLEMS

Hints for Solving Problems
Remember that for speeds at which relativistic effects are important you must use the relativistic relationships for kinetic energy and momentum. For a given energy a lighter particle goes faster than a heavier one. The uncertainty principle can be used to estimate the value of a quantity: For instance, the momentum of a particle in the nucleus can be estimated because its uncertainty in momentum cannot be larger than its momentum.

Section 27.2   The Compton Effect

27.1 Compute the value of the Compton wavelength of (a) an electron (mass 9.11×10^{-31} kg), (b) a proton (mass 1.67×10^{-27} kg), and (c) a muon (mass 1.88×10^{-28} kg).

27.2 A photon whose wavelength is 0.070 nm is Compton-scattered through an angle of 60°. (a) What are the wavelength and the energy of the scattered photon? (b) How much energy is given to the recoil electron?

27.3 A 100-keV x ray is Compton-scattered through an angle of 90°. What is the energy of the x ray after scattering?

27.4 A 0.025-nm x ray is scattered through an angle of 60°. What is the frequency of the emergent x ray?

27.5 A 17.2-keV x ray from molybdenum is Compton-scattered through an angle of 120°. (a) What is the energy of the scattered x ray? (b) What is the kinetic energy of the recoil electron?

27.6 An x ray is scattered through an angle of 47°. What
• is the fractional change in its energy if the incident wavelength is ten times the Compton wavelength of the electron?

Section 27.3   De Broglie Waves

27.7 What is the de Broglie wavelength of a particle moving at a speed of 1.00×10^6 m/s if it is (a) an electron, (b) a proton? (*Hint:* $m_e = 9.11 \times 10^{-31}$ kg, $m_p = 1.67 \times 10^{-27}$ kg.)

27.8 Compute the de Broglie wavelength of a baseball of mass 0.145 kg moving at a speed of 50 m/s.

27.9 The de Broglie wavelength of an electron is 6.0×10^{-9} m. What is the electron's speed?

27.10 The de Broglie wavelength of a particle moving with a speed of $0.100c$ is 1.32×10^{-14} m. Is the particle an electron or a proton?

27.11 Compute the de Broglie wavelength of an electron accelerated from rest through a potential of 1000 V.

27.12 Electrons are accelerated to 30 GeV for experiments at the DESY laboratory in Hamburg, BRD. What is the de Broglie wavelength of these electrons?

27.13 Compute the de Broglie wavelength of the earth in its orbit.

27.14 What is the de Broglie wavelength of a 100-MeV
• proton?

27.15 What is the kinetic energy of an electron of wavelength 0.30 nm?

27.16 What is the wavelength of an electron whose total
• energy is 0.87 MeV?

27.17 A 0.25-kg ball is swung on a string in a 0.75-m-radius horizontal circle once every 1.0 s. (a) How many de Broglie wavelengths of the moving ball fit into the circle? (b) By approximately what factor would Planck's constant have to change if the answer to part (a) were 1000 de Broglie wavelengths?

27.18 Calculate the Bragg angle θ for 400-eV electrons diffracted from the planes of a nickel crystal of spacing $d = 0.203$ nm.

27.19 Calculate the Bragg angle θ for 300-eV electrons diffracted from the atomic planes of silver spaced at $d = 0.236$ nm.

27.20 In the Davisson and Germer experiment, 54-eV electrons were diffracted from nickel so that the angle between the incident and the diffracted beams was 50°. What is the spacing of the crystal planes?

Section 27.5   The Uncertainty Principle

27.21 A slowly moving electron is localized to be within 0.50 mm of a particular point. What is the minimum uncertainty with which its velocity can be simultaneously determined?

27.22 An electron is localized to within a distance of 1.0×10^{-10} m (approximately the diameter of a hydrogen atom). (a) Treat this as a one-dimensional problem and determine the uncertainty in the electron's momentum. (b) What is the kinetic energy associated with this momentum?

27.23 In radiating a photon of green light ($\lambda = 520$ nm) an atom radiates for 1.00×10^{-9} s. What is the energy of the photon? What is the uncertainty in its energy?

27.24 A proton has a kinetic energy of 10.0 MeV. If the x component of its momentum is measured with an uncertainty of 1.00%, what is the uncertainty in its position along the x axis?

27.25 Photons of energy 14.4 keV can be produced with a relative uncertainty in energy of one part in 10^{11}. What is the uncertainty in the lifetime of the state that emits such photons?

27.26 What is the uncertainty in momentum of an electron with de Broglie wavelength $\lambda = 5.00 \times 10^{-2}$ nm if the uncertainty in wavelength is $\Delta\lambda = 1.00 \times 10^{-3}$ nm?

27.27 An electron with de Broglie wavelength $\lambda = 0.10$ nm has an uncertainty in momentum of $\Delta p = 6.6 \times 10^{-25}$ kg·m/s. What is the uncertainty in the electron's wavelength?

27.28 An experiment measures the energy of a photon to one part in 10^7. What is the minimum uncertainty in locating the photon if it is (a) an x ray of $\lambda = 0.071$ nm, (b) visible light of $\lambda = 550$ nm, (c) an AM broadcast wave of $\lambda = 300$ m?

Section 27.7   The Particle in a Box

27.29 Imagine an electron confined in a one-dimensional potential box of width $L = 1.0$ nm. Compute the energy of the first three energy levels.

27.30 Imagine an electron confined in a one-dimensional potential box of width $L = 0.14$ nm. Compute the energy of the first three energy levels.

27.31 Imagine a proton of mass 1.67×10^{-27} kg confined in a one-dimensional potential box of width $L = 1.0 \times 10^{-14}$ m, roughly the size of an atomic nucleus. Compute the energy of the first three energy levels.

27.32 The ground-state energy of an electron confined along the x axis is 6.03×10^{-18} J. The first excited state ($n = 2$) energy is 2.41×10^{-17} J. What is the approximate length of the region to which the electron is confined?

27.33 A particle confined along the x axis to the dimensions of a nucleus ($\approx 1.0 \times 10^{-14}$ m) has a ground-state energy of 3.28×10^{-13} J. The second excited state ($n = 3$) energy is 2.96×10^{-12} J. What is the particle's mass?

Section 27.8   Tunneling or Barrier Penetration

27.34 A wave function within a barrier is described by $\psi(x) = \psi_0 e^{-\alpha x}$. What is the ratio of the value of $\psi(x)$ for $x = 1/\alpha$ to that for $x = 10/\alpha$?

27.35 A wave function within a barrier is described by $\psi(x) = \psi_0 e^{-\alpha x}$. What is the probability density for finding the wave at $x = 3/\alpha$? Give your answer as a multiple of ψ_0^2.

*Section 27.9   Wave Theory of the Hydrogen Atom

27.36 Compute the first six energy levels for the two-
• dimensional square potential box of edge length L. Give the quantum numbers n_x and n_y for each level.

27.37 Compute the first six energy levels for the two-
• dimensional rectangular potential box of dimensions $L_x = L$ and $L_y = 2L$. Give the quantum numbers n_x and n_y for each level.

27.38 A given atomic level is characterized by principal quantum number $n = 3$. How many orbital subshells are possible?

27.39 How many orbital subshells are associated with the L electronic shell?

27.40 How many magnetic substates are possible in a p subshell?

27.41 How many magnetic substates are possible in an f subshell?

*Section 27.10   The Zeeman Effect and Space Quantization

27.42 Find the shift in energy of a particular orbital angular momentum level having a value $m_l = 1$ due to the presence of an applied magnetic field of 1.00 T. Express this energy in units of electron volts.

27.43 Calculate the value of the Bohr magneton from its definition. Give your answer in SI units.

27.44 In the presence of a magnetic field **B**, an electronic p state splits into three energy levels corresponding to $m_l = 0, +1, -1$. Calculate the energy separation between adjacent levels.

Section 27.11   The Pauli Exclusion Principle

27.45 Find the energy difference between states of $m_s = +\frac{1}{2}$ and $m_s = -\frac{1}{2}$ in an applied magnetic field of $B = 0.10$ T. Express this energy in electron volts.

25.46 (a) Find the energy difference between electron states of $m_s = +\frac{1}{2}$ and $m_s = -\frac{1}{2}$ in an applied magnetic field of $B = 0.80$ T. Express this energy

in electron volts. (b) What is the wavelength of a photon having this same energy?

*Section 27.12   Understanding the Periodic Table

27.47 Extend Table 27.2 to the N shell for which $n = 4$.

27.48 Predict the electronic states for sodium ($Z = 11$) using Table 27.2.

27.49 Predict the electronic states for boron ($Z = 5$) using Table 27.2.

27.50 Predict the electronic states for neon ($Z = 10$) using Table 27.2.

Additional Problems

27.51 Perform the algebraic steps necessary for eliminating
•• v and f from the equations for conservation of momentum and energy for the Compton effect, and thereby obtain Eq. (27.1).

27.52 What is the maximum recoil velocity of a free elec-
• tron that is scattered by an x ray of wavelength $\lambda = 7.1 \times 10^{-11}$ m?

27.53 An x-ray photon of wavelength 0.0023 nm loses
• 0.024 of its initial energy during a Compton scattering. Through what angle is it scattered?

27.54 (a) What is the maximum change in wavelength
• of a photon that undergoes Compton scattering? (b) What is the ratio of this maximum change in wavelength to the wavelength of a 50-kV x ray?

27.55 Prove that the emergent photon in Compton scat-
•• tering has a frequency that lies between f and approximately $mc^2/2h$ when the incident x ray is very energetic. (*Hint:* By very energetic we mean a photon for which $f \gg mc^2/h$.)

27.56 Find the expression requested in Example 27.3 for
•• the case in which relativistic effects must be taken into account.

27.57 Diffraction rings of the type shown in Fig. 27.4 are
• made by electrons diffracted from a polycrystalline sample located a distance L from a fluorescent screen. Prove that the ring diameter is given by

$$D = \frac{2L\lambda}{d}$$

for electrons of kiloelectronvolt energies for which the Bragg angle θ is small and the wavelength λ is much less than the spacing d between atomic planes.

27.58 Show that the uncertainty principle can be ex-
• pressed in the form

$$\Delta L \cdot \Delta \theta \geq h,$$

where ΔL is the uncertainty in the angular momentum and $\Delta \theta$ is the uncertainty in the angular position. (*Hint:* Use a particle moving in a circle.)

27.59 Show that when the uncertainty in the position of a particle is equal to its de Broglie wavelength, the uncertainty in its momentum is equal to its momentum.

27.60 An atom is excited to an energy level 2.4 eV above
• its ground state, where it remains an average of 2.0×10^{-8} s before making a transition to the ground state. What is the ratio of the spread in the frequency of the light emitted to the average frequency of this transition?

27.61 What is the average de Broglie wavelength of oxygen
• molecules in the air at a temperature of 27°C? (*Hint:* Use the results of the kinetic theory of gases. The mass of an oxygen molecule is 5.31×10^{-26} kg.)

27.62 A particle of mass m is confined to a one-dimensional box of length Δx. What is its lowest possible energy?

27.63 The probability that a particle will penetrate a bar-
•• rier of thickness Δx is 0.0010. What is the probability of penetrating a barrier that is twice as thick? Use the wave function of Eq. (27.13) and assume that everything else is the same.

27.64 Derive the expression analogous to Eq. (27.12) for
•• the energy of the three-dimensional rectangular potential box of edge lengths L_x, L_y, and L_z.

27.65 In the presence of a magnetic field of $B = 0.10$ T,
• an electronic d state is split into sublevels of different energies. What is the energy separation between the highest and lowest sublevels?

25.66 An electronic state is characterized only by its in-
• trinsic magnetic moment, given by Eq. (27.17). In a magnetic field **B** the state is split into two states and the system resonantly absorbs microwave photons of $\lambda = 2.00$ cm. What is the magnitude of **B**?

ADDITIONAL READING

Miller, A. I., "Werner Heisenberg and the Beginning of Nuclear Physics." *Physics Today*, November 1985, p. 60.

Murdoch, D. R., *Niels Bohr's Philosophy of Physics*. New York: Cambridge University Press, 1987. Describes the historical background of Bohr's ideas; traces the origins of complementarity and its meaning.

Pykett, I. L., "NMR Imaging in Medicine." *Scientific American*, May 1982, p. 78.

Shimony, A., "The Reality of the Quantum World." *Scientific American,* January 1988, p. 46.

Tonomura, A., J. Endo, T. Matsuda, T. Kawasaki, and H. Ezawa, "Demonstration of Single-Electron Buildup of an Interference Pattern." *American Journal of Physics,* February 1989, p. 117.

28

The Nucleus

28.1	Chadwick's Discovery of the Neutron
28.2	Composition of the Nucleus
28.3	Nuclear Size, Forces, and Binding Energy
28.4	Radioactive Decay Versus Nuclear Stability
28.5	Natural Radioactive Decay Series
28.6	Models for Alpha, Beta, and Gamma Decay
*28.7	Detectors of Radiation
*28.8	Radiation Measurement and Biological Effects
28.9	Induced Transmutation and Reactions
28.10	Nuclear Fission
28.11	Nuclear Fusion

A WORD TO THE STUDENT

In the preceding three chapters we have shown how our picture of the atom has evolved from a crude, indivisible bit of matter to a complex structure best understood in terms of quantum mechanics. Now we consider the composition and properties of the atomic nucleus itself. We begin with the discovery of the neutron, then go on to describe the composition of nuclei in terms of neutrons and protons; we also discuss of the size of nuclei and the forces that hold them together. We next examine the conditions for nuclear stability and the natural radioactive processes of alpha, beta, and gamma decay. Finally, we consider nuclear reactions, fission, and fusion—topics of current commercial and political, as well as scientific, interest.

We began our investigation of the atom and its constituent parts in Chapter 25. There we discussed the properties of the atom, evidence for the existence of the atomic nucleus, and the types of particles emitted when a nucleus undergoes radioactive decay. Now we examine the nucleus itself. We do not introduce these ideas in the order in which they were discovered; instead we have tried to organize this chapter so as best to explain both the properties of the nucleus and the experimental evidence for those properties.

Our understanding of the forces and structure within the nucleus is still incomplete, but present research is advancing along exciting paths. Today's discoveries are of interest to the public as well as to scientists, for they have led to a wealth of practical devices and techniques. Many people think immediately of nuclear weapons and nuclear power, but knowledge of the nucleus has also revolutionized many areas of medicine and biological science. Radioactive isotopes have been important tools for many years in diagnosing disease, and radioactive emission still contributes to the war against cancer.

Chadwick's Discovery of the Neutron

28.1

After Rutherford's discovery of the nucleus in 1911, physicists slowly realized that the nucleus itself must be made of smaller constituent parts. Initially, it seemed reasonable to postulate that the nuclei of atoms heavier than hydrogen consisted of hydrogen nuclei, called **protons**, and electrons. The protons would provide the mass and the positive charge, while the electrons, which were presumed to be in the nucleus, would neutralize some of the charge. This theory could satisfactorily account for a nucleus of atomic mass A and atomic number Z by assuming that the nucleus contained A protons and $A - Z$ electrons. The proton-electron model of the nucleus gave the correct charge and approximately the correct mass for any nucleus. Furthermore, this model accounted for the emission of alpha particles, which were assumed to be composed of four protons and two electrons, and for the emission of beta particles (electrons). According to this theory, gamma radiation was emitted from an excited nuclear state in a manner analogous to the emission of light from an excited state of the electrons of an atom.

However, the proton-electron model of the atom had several difficulties. For instance, if you take account of the observed spin of particles emitted during decay processes, the proton-electron model of the nucleus does not conserve angular momentum. This fact alone was a strong reason for discarding or drastically revising the theory. Furthermore, as we showed in the previous chapter (Example 27.6), the energy needed to confine an electron to a region comparable to the size of the nucleus greatly exceeds the electron energies observed in beta decay. For these reasons, Rutherford suggested as early as 1920 that the nucleus should contain a neutral particle of approximately the same mass as the proton. The search for such a particle was made difficult by the fact that the radiation-detecting devices of the time were sensitive only to charged particles. Not until 1932 did James Chadwick (1891–1974) conclusively demonstrate the existence of such a neutral particle.

In Chadwick's experimental apparatus, alpha particles were allowed to bombard a beryllium target (Fig. 28.1). The beryllium then emitted an unknown radiation, which was unaffected by electric or magnetic fields and thus was electrically neutral. This radiation struck a paraffin sheet. Chadwick used paraffin because it contains a large proportion of hydrogen. The paraffin ejected protons (hydrogen nuclei) when hit by the unknown radiation, and these protons were detected by an ionization chamber.

The maximum energy of the protons emerging from the paraffin was measured to be 5.7 MeV. Then Chadwick showed, by an analysis like that for Compton scattering, that if the radiation that knocked these protons out of the paraffin were electromagnetic radiation, such as x rays or gamma rays, it must have an energy of about 55 MeV. This was a surprisingly large amount of energy, much greater than ordinary nuclear decay energies. On the other hand, if the unknown radiation were a neutral particle with no electric charge but with a specific mass, then its interaction with the protons in paraffin could be analyzed by applying the laws of conservation of momentum and energy to a collision between two particles, just as we did back in Chapter 8. (See Problem 28.53.) The unknown particle's velocity and mass could not both be determined in one experiment, so Chadwick repeated the experiment, substituting a nitrogen target for the paraffin.

We denote the unknown incident particle with the subscript U, the hydrogen atoms ejected by the paraffin with the subscript H, and the nitrogen by the subscript N. Then by applying the laws of conservation of momentum and energy to a head-on collision we obtain the following

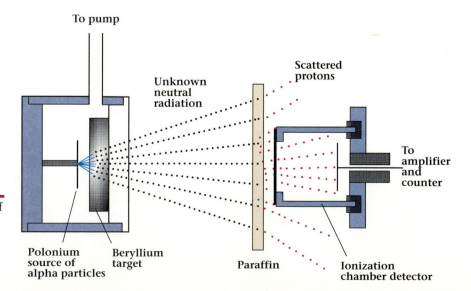

Figure 28.1 A schematic diagram of Chadwick's experimental apparatus. Chadwick discovered neutrons by analyzing their interaction with hydrogen-rich paraffin, in which protons were ejected.

To pump

Unknown neutral radiation

Scattered protons

To amplifier and counter

Polonium source of alpha particles

Beryllium target

Paraffin

Ionization chamber detector

relationships:

$$v'_H = \frac{2m_U}{m_U + 1m_H} v_U \quad \text{and} \quad v'_N = \frac{2m_U}{m_U + 14m_H} v_U. \quad (28.1)$$

The quantity v_U indicates the speed of the unknown particle before the collision, and the quantities v'_H and v'_N indicate recoil speeds after the collision. Also we have substituted $14m_H$ for m_N. Dividing the first of these equations by the other and inserting the measured values of the final speeds, we get

$$\frac{m_U + 14m_H}{m_U + 1m_H} = \frac{v'_H}{v'_N} = \frac{3.3 \times 10^7 \text{ m/s}}{4.7 \times 10^6 \text{ m/s}} = 7.0$$

Upon solving for m_U, we find

$$m_U = 1.15m_H.$$

The mass of the unknown particle is approximately the same as that of the proton. Within the precision of Chadwick's experiment, the masses of the two particles were considered the same.

Later, more refined measurements showed the mass to be even closer to the proton mass than indicated here. If the mass of the unknown particle is the same as the mass of the proton, then, assuming an elastic collision, the energy of the incoming particle is transferred to the outgoing target particle, just as for billiard ball collisions. The point is that the model of a neutral particle with the mass of a proton does not give rise to the large energies required if the unknown radiation were electromagnetic. Instead, the radiation could have an energy of 5.7 MeV, well within the range of nuclear decay energies, and still produce the observed effects.

Chadwick's analysis thereby demonstrated the existence of a neutral particle, which was given the name **neutron**. This name had been proposed twelve years earlier by Rutherford for a particle thought to consist of a proton and an electron. Note that Chadwick's analysis demonstrated the neutron's existence without directly observing neutrons. Instead, he measured the effects of neutron bombardment on target atoms and applied conservation of energy and momentum to determine the neutron's properties. This is an example of how contemporary physics research uses models and conservation laws to determine new results. Since Chadwick's discovery, the masses of the neutron and proton have been carefully measured in many independent experiments; they are listed in Table 28.1.

TABLE 28.1
Mass and charge of some subnuclear particles

Name	Symbol	Electric charge	Mass (m_e)	Mass (MeV)	Mass (kg)
Photon (gamma)	γ	0	0	0	0
Electron	e^- (β^-)	$-e$	1	0.5110	9.109×10^{-31}
Positron	e^+ (β^+)	$+e$	1	0.5110	9.109×10^{-31}
Proton	p	$+e$	1836	938.272	1.673×10^{-27}
Neutron	n	0	1839	939.566	1.675×10^{-27}

Composition of the Nucleus

28.2

Because protons and neutrons both appear in the nucleus, they are collectively called **nucleons.** According to today's accepted proton-neutron model of the nucleus, the number of protons in a nucleus is Z, the *atomic number* of the element, and the number of protons plus the number of neutrons is A, the element's *atomic mass number*. This leads to the following way of designating the nucleus of an element X, where X stands for the chemical symbol for the element:

$$_Z^A X. \tag{28.2}$$

For example, the nucleus of hydrogen (one proton) is $_1^1 H$, the nucleus of helium (two protons and two neutrons) is $_2^4 He$, and so on. The number of neutrons in a nucleus is simply $A - Z$.

Nuclei that have the same number of protons (Z) but different numbers of neutrons (different A) are called **isotopes**. Isotopes of the same element have the same chemical properties because they have the same number and arrangement of electrons. (Remember, the number of electrons in a neutral atom equals its number of protons.) However, isotopes may not have the same nuclear properties. For instance, one isotope of an element may be stable and one or more others may be radioactive. This is the case for carbon, whose isotope with $A = 12$ is stable, but whose isotope with $A = 14$ is radioactive, as we saw in Chapter 25. The naturally occurring elements have about 280 stable isotopes. However, there are more than 800 known radioactive isotopes, both naturally occurring and laboratory produced. Figure 28.2 shows a chart of the known stable isotopes. Note that for elements of low atomic numbers, the number of protons and neutrons in stable nuclei is about the same. As we go to higher atomic numbers, stable nuclei have a small excess of neutrons, yet the general tendency remains for a stable nucleus to have about half protons and half neutrons.

Very often an element has more than one stable isotope. For example, stable boron occurs in the form $_5^{10} B$, with five protons and five neutrons, and in the form $_5^{11} B$, with five protons and six neutrons. The isotope $_5^{10} B$ has a relative atomic abundance of 19.8% and the isotope $_5^{11} B$ has a relative atomic abundance of 80.2%. An ordinary chemical sample contains a mixture of the two isotopes in the same proportions as their natural abundance, although in some cases a slight variation occurs, depending on the origin of the sample. An element's *chemical atomic mass* (sometimes called *atomic weight*) is a weighted average of its stable isotopes, counting each isotope according to its natural abundance.

Example 28.1

Calculate the expected atomic mass of boron from knowledge of its two naturally occurring isotopes and their relative abundance.

Solution   We just mentioned that a sample of natural boron contains

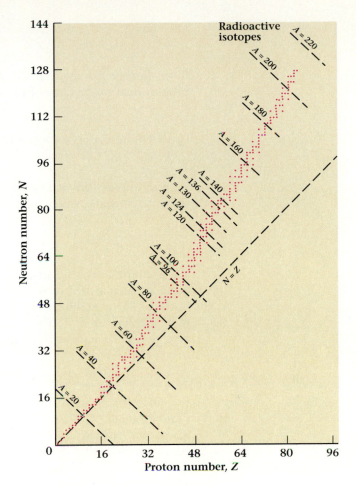

Figure 28.2 Chart of the known stable isotopes. Note how close they lie to the line $N = Z$, indicating equal numbers of neutrons and protons.

isotopes $^{10}_{5}B$ and $^{11}_{5}B$ in relative abundances of 19.8% and 80.2% respectively. Thus 19.8% of the sample has atomic mass of approximately 10, and 80.2% of the sample has atomic mass of approximately 11. The averaged mass is then

$$10 \times 0.198 = 1.98$$
$$11 \times 0.802 = \underline{8.82}$$
$$10.80$$

The sum 10.8 represents the approximate average atomic mass of the combined sample. Compare the result of this simple calculation with the value given in the periodic table on the inside back cover.

Nuclear Size, Forces, and Binding Energy

28.3

Let us make a simple model and assume that that the nucleons fill a nucleus in a manner similar to the way marbles fill a bag. Then we can write that the volume V of the nucleus is proportional to the number of nucleons A,

$$V \propto A.$$

If we further assume that the nucleus is spherical, then the volume is proportional to the cube of the radius,

$$r^3 \propto A.$$

Upon taking the cube root of each side, we get

$$r \propto A^{1/3}.$$

We can express this relationship as an equality by introducing a proportionality constant, which we call R_0:

$$r = R_0 A^{1/3}. \tag{28.3}$$

Scattering experiments, both with charged particles and with neutral particles such as neutrons, have been used to determine the radius of the nucleus. The experiments generally agree with the conclusion that nuclear radii increase as $A^{1/3}$. The exact value of R_0 depends somewhat on the type of experiment used to measure it. A typical value is

$$R_0 \approx 1.2 \times 10^{-15} \text{ m.}$$

This value inserted in Eq. (28.3) predicts a value of $r = 7 \times 10^{-15}$ m for gold, which has 197 nucleons. (Compare this value with the upper limit estimate of 40×10^{-15} m made in Chapter 25 on the basis of Rutherford scattering.) You should keep in mind that this is an extremely simple model and only roughly represents reality.

The forces holding the nucleus together must be extremely strong. This is evident from the fact that the positively charged protons remain confined to the small volume of the nucleus. If the nuclear force were weaker, the electrostatic (Coulomb) repulsion of the positively charged protons would predominate and the nucleus would fly apart or, at the very least, the nucleus would occupy a much larger volume.

Example 28.2

Calculating nuclear radii.

What is the approximate ratio of the radius of $^{252}_{98}\text{Cf}$ to the radius of $^{56}_{26}\text{Fe}$?

Solution We can use Eq. (28.3) to get

$$\frac{r_{\text{Cf}}}{r_{\text{Fe}}} = \left(\frac{252}{56}\right)^{1/3} = (4.5)^{1/3} = 1.65.$$

Many experiments have shown that the attractive force between nucleons—whether between protons and protons, neutrons and neutrons, or neutrons and protons—is the same, once we subtract the effect of the electrostatic Coulomb repulsion between protons. We call this fundamental attractive force the *strong nuclear force*. The strong nuclear force be-

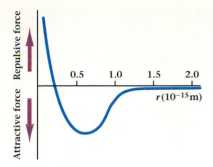

Figure 28.3 General shape of the force between any two nucleons as a function of the distance between them.

tween nucleons in a nucleus is approximately 10^2 times as strong as the Coulomb repulsion between protons, and approximately 10^{39} times as strong as the attractive gravitational force between nucleons. The general form of the strong force between individual nucleons as a function of their separation is sketched in Fig. 28.3. A word of caution is necessary here: This curve does *not* represent the nuclear force in the same precise way that a graph of the gravitational or electrical force represents those forces. We have drawn it only to help gain some insight into the nature of the nucleon–nucleon interaction. For example, the force between nucleons is not necessarily directed along the line joining them. We can, however, argue for the general shape of the curve on several points. We do not observe nuclear forces at distances greater than the size of the nucleus. Thus there should be essentially zero force at distances greater than about R_0. The nucleus is tightly bound, so the force must be large and attractive at distances less than the nuclear radius. However, the nucleus does have a finite size, which depends on the number of particles in the nucleus. This observation indicates that two or more nucleons cannot occupy the same space. Therefore there must also be a strong repulsive force at very short distances.

Strictly speaking, at distances of nuclear dimensions we must use only quantum-mechanical considerations. In quantum mechanics we do not speak of the force between two particles, but instead describe the energy of their interaction. We have greatly oversimplified the picture of the nuclear force in order to gain some understanding of nuclear behavior. The forces among many nucleons are not easily described and are quite different from the forces we have studied so far. The nature of the nuclear force is not fully understood in any simple way, even though a great deal is known about it. Attempts to solve this problem have led to many interesting ideas and results, some of which are discussed in Chapter 31.

Even without knowing the exact nature of the nuclear force we can, in part, understand the way a nucleus is bound together by considering its energy. Let us use the Einstein mass–energy relationship $E = mc^2$. The masses of the proton and neutron are well known, as are the nuclear masses of several hundred nuclear species. Table 28.2 (p. 836) lists the masses of several light atoms. Nuclear and atomic masses are often measured in terms of atomic mass units* u. Through the use of the Einstein energy equation we can show that 1 u is equivalent to 931.5 MeV. In nuclear physics, masses are frequently given in units of MeV.

Let us compute the total mass of the constituent parts of a helium atom, consisting of two neutrons, two protons, and two electrons. Rather than adding together the masses of the six particles individually, we will add the mass of two hydrogen atoms to the mass of two neutrons. We can do this because the energy that binds the electrons to the atoms is small, a few electron volts, in comparison with the mass of the particles involved. (Remember that energy and mass are connected through $E = mc^2$.) The result is a somewhat shorter calculation. Using the values

---

*The unified atomic mass unit u, also abbreviated as amu, is equal to 1/12 the mass of an atom of $^{12}_{6}C$; 1 u = 1.66054×10^{-27} kg.

	Mass	Mass	Mass
Symbol	(u)	(MeV)	$(10^{-27}$ kg)
p	1.007276	938.272	1.67262
n	1.008665	939.566	1.67493
H	1.007825	938.783	1.67353
$^2_1$H (D)	2.014102	1876.125	3.34450
$^3_1$H (T)	3.016049	2809.433	5.00827
$^3_2$He	3.016029	2809.414	5.00824
$^4_2$He	4.002603	3728.402	6.64648
$^5_2$He	5.012220	4668.854	8.32299
$^5_3$Li	5.012538	4669.151	8.32352
$^6_3$Li	6.015121	5603.051	9.98835
$^7_3$Li	7.016003	6535.367	11.65036
$^9_4$Be	9.012182	8394.796	14.96509

TABLE 28.2
Atomic mass of some light elements*
The atomic mass is given for the neutral atom, including its electrons.

*The masses of the proton and neutron have been included as an aid in working problems.

from Table 28.2, we have

$$2m_H = 2(1.007825) = 2.015650 \text{ u}$$
$$+\ 2m_n = 2(1.008665) = \underline{2.017330 \text{ u}}$$
$$4.032980 \text{ u}.$$

The mass of the $^4_2$He atom is 4.002603 u, a value that is significantly smaller than the sum of the masses of the constituent particles. Using the Einstein relation, we may say that the mass energy of the components is greater than the mass energy of the atom. This difference in energy is what holds the nucleus together. It is called the **binding energy** of the nucleus. The binding energy is the difference in mass energy of the constituent particles when bound together in the nucleus and when considered as separate, isolated particles. Therefore it is the energy needed to separate the component parts of the nucleus infinitely far apart. For $^4_2$He, the mass difference is

$$\Delta m = (2m_H + 2m_n) - m(^4_2\text{He}) = 0.030377 \text{ u}.$$

The binding energy is

$$\Delta E = \Delta mc^2 = (0.030377 \text{ u}) (931.5 \text{ MeV/u}),$$
$$\Delta E = 28.3 \text{ MeV}.$$

The binding energy of $^4_2$He is rather large for a light nucleus, corresponding to the observation that the alpha particle is very stable. We can calculate the binding energy of any nucleus whose mass and composition are known, in the same way that we calculated the binding energy of the alpha particle. Figure 28.4 shows the binding energy per nucleon for a large number of nuclei. Notice that the binding energy per nucleon is greatest for nuclei with A between 40 and 80. The general shape of this

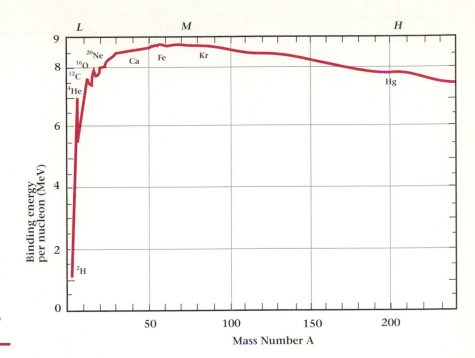

Figure 28.4 Graph of the binding energy per nucleon (in MeV) versus the atomic mass number A.

curve is important for analyzing some kinds of nuclear reactions. We will discuss its significance further in Sections 28.10 and 28.11.

Radioactive Decay Versus Nuclear Stability

28.4

As we saw in Chapter 25, some nuclei are unstable and decay, most commonly by emitting alpha, beta, or gamma particles. Such decays must be consistent with the law of conservation of energy. However, before considering the energies involved in nuclear decay, we will first describe the notation used to discuss decay and introduce some of the other conservation laws that determine the outcome of a nuclear decay.

We can write equations for the radioactive decay of nuclei in a manner analogous to the way in which chemical equations are written. The radiation that Becquerel first observed in 1896 was the emission of alpha particles by the decay of uranium into thorium. The equation for this decay is written

$$^{238}_{92}\text{U} \rightarrow \,^{234}_{90}\text{Th} + \,^{4}_{2}\text{He}.$$

The rules that govern the "balancing" of such equations are *conservation of charge* and *conservation of the total number of nucleons*. Notice that the total number of nucleons (the sum of the protons and neutrons) is the same before and after the decay. As we have seen, this does not mean that mass is conserved in nuclear reactions. We refer to the original nucleus as the *parent nucleus* and the decay product as the *daughter nucleus*.

Example 28.3

Radioactive decay by alpha emission.

The unstable nucleus $^{147}_{62}\text{Sm}$ decays by the emission of alpha particles. What is the daughter nucleus?

Solution   Using X to stand for the unknown element, we write

$$^{147}_{62}\text{Sm} \rightarrow \text{X} + {}^4_2\text{He}.$$

The total charge is 62 units on the left side of the equation. Because the alpha particle (helium nucleus) has two units of charge, we assign 60 units of charge to the nucleus X. There are 147 nucleons in the parent nucleus and the alpha particle has four nucleons. Thus the daughter nucleus must have 143 nucleons to conserve the total number of nucleons. We have

$$^{147}_{62}\text{Sm} \rightarrow {}^{143}_{60}\text{X} + {}^4_2\text{He}.$$

We may discover the name of element X by consulting the periodic table of the elements. The element with atomic number 60 is neodymium (Nd). We insert the symbol Nd in the above equation to give

$$^{147}_{62}\text{Sm} \rightarrow {}^{143}_{60}\text{Nd} + {}^4_2\text{He}.$$

Example 28.4

Radioactive decay by beta emission.

Write the the equation for the beta decay of $^{87}_{37}\text{Rb}$.

Solution   We start with conservation of nuclear charge and conservation of nucleon number. Energy and momentum are also conserved in a nuclear decay. Taking them into consideration would show that an additional uncharged particle of negligible mass is also one of the decay products in beta decay. This particle is called the anti-neutrino $\overline{\nu}$ and is discussed in Section 28.6. Writing an X for the unknown nucleus, we have

$$^{87}_{37}\text{Rb} \rightarrow \text{X} + {}^{\ 0}_{-1}\beta + \overline{\nu}.$$

In this case the charge of the daughter nucleus must be greater than that of the parent, even though the total number of nucleons is the same. The result is equivalent to the conversion of one neutron into a proton and an electron. From the periodic table we identify the daughter nucleus as strontium (Sr):

$$^{87}_{37}\text{Rb} \rightarrow {}^{87}_{38}\text{Sr} + {}^{\ 0}_{-1}\beta + \overline{\nu}.$$

The emission of negative electrons from a nucleus—which contains only positive and neutral particles—also presents a conceptual problem. We will discuss the explanation in Section 28.6. Here we only point out that the transformation within the nucleus of a neutron into a proton requires the emission of a negative charge.

We can use energy considerations to investigate the important question of stability: Under what conditions can a nucleus spontaneously decay into particular products and under what conditions is it stable? We answer the question by considering the relative mass energies of the parent nucleus and the decay products. For the decay to take place the mass energy of the parent nucleus must be greater than the total mass energy of the decay products. If the decay products have more mass than the parent nucleus, additional energy is required from some other source in order to accomplish the reaction. Thus radioactive decay is possible only if

$$m(\text{parent}) > m(\text{daughter}) + m(\text{decay particle}).$$

The difference between the initial and final mass energies is called the **Q value**. Thus, using $E = mc^2$, we have

$$Q = [m(\text{parent}) - m(\text{daughter}) - m(\text{decay particle})]c^2. \qquad (28.4)$$

A positive Q value indicates that a spontaneous reaction may occur, but a negative Q value indicates that a spontaneous reaction cannot take place. The positive Q value is available to the decay products as kinetic energy.

Example 28.5

Determine whether $^{210}_{84}\text{Po}$ can decay by emission of an alpha particle, and if so, find the kinetic energy released in the process.

Solution   The reaction is

$$^{210}_{84}\text{Po} \rightarrow {}^{206}_{82}\text{Pb} + {}^{4}_{2}\text{He}.$$

The observed masses are

$$m(^{210}_{84}\text{Po}) = 209.98285 \text{ u},$$
$$m(^{206}_{82}\text{Pb}) = 205.97440 \text{ u},$$
$$m(^{4}_{2}\text{He}) = 4.00260 \text{ u}.$$

The total mass of the decay products is

$$m(^{206}_{82}\text{Pb}) + m(^{4}_{2}\text{He}) = 209.97700 \text{ u}.$$

This is less than the mass of $^{210}_{84}\text{Po}$, so the decay may occur.
 The total kinetic energy available is the Q value:

$$\text{KE} = Q = m(^{210}_{84}\text{Po}) - [m(^{206}_{82}\text{Pb}) + m(^{4}_{2}\text{He})],$$
$$\text{KE} = 209.98285 \text{ u} - 209.97700 \text{ u} = 0.00585 \text{ u},$$
$$\text{KE} = 0.00585 \text{ u} \times 931.5 \text{ MeV/u} = 5.45 \text{ MeV}.$$

This kinetic energy is shared between the alpha particle and the daughter nucleus.

Natural Radioactive
Decay Series

28.5

In the years following the discovery of radioactivity, scientists found many radioactive elements, most of them relatively heavy. Eventually, they realized that these radioactive elements formed decay chains or series. That is, when one nucleus decays into another, the resulting daughter nucleus

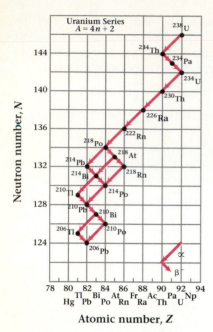

Figure 28.5 Plot of the number of neutrons $N = A - Z$ versus Z for the uranium series. Alpha decays proceed toward the lower left; beta decays proceed toward the lower right.

is also radioactive and subsequently decays, and so on. The end product of each decay chain is a stable isotope of lead ($Z = 82$).

All known naturally radioactive nuclei can be grouped into one of three decay series, which originate with long-lived heavy elements. Each series is named after the longest-lived member of the series: uranium, thorium, and actinium. A fourth series, the neptunium series, consists of isotopes that no longer exist naturally in the earth's crust; it terminates in $^{209}_{83}\text{Bi}$. We can categorize each radioactive nucleus into one of these four series according to its mass number A. If n is an integer, then the series are labeled as follows:

Thorium series: $A = 4n$ Uranium series: $A = 4n + 2$
Neptunium series: $A = 4n + 1$ Actinium series: $A = 4n + 3$

A radioactive decay that begins with a member of one of these series produces daughter nuclei that are also members of the same series. Subsequent decay products remain in the same series because alpha decay involves a mass number change of four, while beta and gamma decay involve no changes in the mass number A.

We can display these decay series by plotting the number of neutrons N ($N = A - Z$) against the number of protons Z for each nucleus in the series (Fig. 28.5). An alpha decay leads to the lower left of the graph, toward lower N and lower Z. A beta decay leads to the lower right, toward lower N but higher Z. Starting in the upper right-hand corner in each case, you can trace out the sequential decays as the process proceeds. Note that some radioactive nuclei have alternative decay modes, decaying by either alpha emission or beta emission.

Models for Alpha, Beta, and Gamma Decay

28.6

We have seen that many naturally occurring nuclei are unstable and decay to other nuclei, with the emission of radioactivity. You may recall from our discussion of radioactive half-life (Section 25.8) that the time elapsed before any individual nucleus undergoes decay is not subject to exact prediction, but is subject to statistical laws.

The details of the decay process are complex, but we will examine some of the major features, using our knowledge of conservation laws and quantum mechanics. In doing this we will learn some important facts about nuclear structure and the nature of the subatomic particles themselves. We first classified nuclear decay processes into three types, corresponding to the emission of alpha, beta, or gamma radiation. Now let's look at the major features of each process.

Alpha Decay

A characteristic feature observed in alpha decay is that all alpha particles from a given type of nuclear reaction have essentially the same kinetic energy. This observation allows us to consider alpha decay as a simple

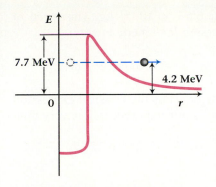

Figure 28.6 The potential-energy diagram of an alpha particle in a $^{235}_{92}$U nucleus. The observed kinetic energy (4.2 MeV) of the emitted alpha particle is not large enough to overcome the nuclear Coulomb potential barrier, which is greater than 7 MeV. However, the alpha particle can escape by quantum mechanical barrier penetration.

(a)

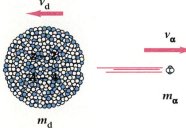

(b)

Figure 28.7 The decay of (a) an unstable nucleus Z, A into (b) a daughter nucleus $Z - 2$, $A - 4$ by the emission of an alpha particle. The momentum of the daughter nucleus is equal in magnitude and opposite in direction to the momentum of the alpha particle.

breakup of a single parent nucleus into two pieces: the daughter nucleus and the alpha particle.

One approach to understanding alpha decay is to consider the alpha particle to be formed inside the nucleus before it is emitted. This seems reasonable in view of the energetic stability of the combination of two neutrons and two protons in one bound state (an alpha particle). However, the observed energies of emitted alpha particles are not great enough for them to have escaped the nucleus by overcoming the potential barrier at the nuclear surface (Fig. 28.6). The explanation of how the alpha particle gets out of the nucleus lies in quantum mechanical tunneling. As discussed in the previous chapter, even though the barrier height is greater than the particle's energy, the particle can still get through. The probability of tunneling through the barrier depends on the energy of the particle and the height of the barrier. After working out the details, one gets a prediction that agrees with the experimental observations: The more energetic alpha particles come from isotopes with shorter half-lives.

If the parent nucleus is initially at rest, then conservation of momentum requires that the alpha particle and the daughter nucleus move apart with equal but opposite momenta (Fig. 28.7). Their kinetic energies are provided by the Q value for the decay,

$$Q = \tfrac{1}{2}m_\alpha v_\alpha^2 + \tfrac{1}{2}m_d v_d^2,$$

where subscripts α and d refer to the alpha particle and the daughter nucleus, respectively. Any decay involves a specific energy Q, and in such a two-body breakup it is uniquely shared by the two decay products. For a daughter nucleus of mass number A, the masses m_α and m_d are essentially 4 and A, respectively, in atomic mass units. It can be shown that the kinetic energy of the emitted alpha particle is approximately

$$KE_\alpha \approx \frac{A}{A + 4} Q. \tag{28.5}$$

The only naturally occurring nuclei that undergo spontaneous alpha decay are those with $Z > 82$. For these heavy nuclei the fraction $A/(A + 4)$ is approximately unity, and the alpha particle takes away almost all of the reaction energy Q.

Example 28.6

Alpha decay of radium-223.

Radium-223 decays via alpha emission to radon ($^{219}_{86}$Rn) with an energy release of $Q = 5.864$ MeV. Find the approximate kinetic energy of the alpha and its speed.

Solution From Eq. (28.5), the approximate kinetic energy of the alpha particle is

$$KE_\alpha \approx \frac{A}{A + 4} Q = \frac{219}{223} (5.864 \text{ MeV}),$$

$$KE_\alpha \approx 5.76 \text{ MeV} = 9.23 \times 10^{-13} \text{ J}.$$

The mass of the alpha is 4.00 u or 6.65×10^{-27} kg. The speed of the alpha is determined from the expression for kinetic energy,

$$v_\alpha = \sqrt{\frac{2KE_\alpha}{m}} = \sqrt{\frac{2(9.23 \times 10^{-13} \text{ J})}{6.65 \times 10^{-27} \text{ kg}}} = 1.67 \times 10^7 \text{ m/s}.$$

This speed is about 5% of the speed of light. Thus the classical expression for energy is valid and relativistic theory is not required, unless very high precision is called for.

Beta Decay

In 1932 Carl W. Anderson (b. 1905) discovered the **positron** or **beta plus** particle (β^+) in cosmic rays. Subsequent experiments confirmed that the positron is similar to an ordinary negative electron except for its charge. A positron has the same mass as an electron but carries one unit (e) of positive, rather than negative, electric charge. For clarity, the (negative) electron is often called the **beta minus** (β^-) particle. Positron emission has been observed in some nuclear decays. In a related decay process, known as **electron capture**, the nucleus absorbs an orbital electron from the atom. Any process in which the nucleus spontaneously emits or absorbs an electron or positron is called *beta decay*.

If we assume that beta decay is a breakup process in which the end products are a nucleus and an electron, then we should be able to calculate the final energy of the electron by using the laws of conservation of energy and momentum. Furthermore, because all nuclei of a given species have the same mass and all electrons have the same mass, the outgoing electrons should all have the same kinetic energy. Such behavior would be similar to alpha decay. However, this is not what we observe experimentally. Figure 28.8 shows the results of measurements made of the energies of electrons in beta decay. You should note two things about the data:

1. The electrons do not all have the same energy.
2. The energy predicted on the basis of conservation of energy and momentum for a two-body final state (marked by the arrow) is the upper limit for the observed energies.

This type of result has been obtained for both β^- and β^+ decay. The large discrepancy between what was predicted and what is observed gave rise to serious problems. In the early 1930s it became apparent that either conservation of momentum had to be abandoned or some revision of the theory was necessary.

In 1934 Wolfgang Pauli proposed a solution that was so radical as to be almost unbelievable. However, it became widely accepted because it restored conservation of momentum to full standing. Pauli suggested that an additional particle is emitted in beta decay. This particle must be electrically neutral because charge is already conserved without it. The mass of the particle must be zero or near zero, because some electrons in beta decay do have energies almost as high as the predicted electron energy. Other considerations indicated that the particle should interact only weakly with matter and that it is not electromagnetic in nature (that is, it

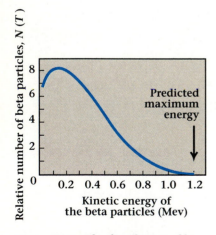

Figure 28.8 The distribution of kinetic energies of beta particles from the decay of bismuth-210. The arrow indicates the energy that the electron would have if the decay were a two-body process, leaving only the daughter nucleus and the electron.

cannot be a photon). The Italian physicist Enrico Fermi (1901–1954) named this particle the **neutrino**, which means "little neutral one." The symbol for the neutrino is ν (nu). Neutrinos were not observed by interaction with matter until 1953, when they were first detected by F. Reines and C. L. Cowan.

Fermi used the neutrino hypothesis to develop a theory of beta decay. His theory accurately predicts the shape of the beta decay energy curves like that shown in Fig. 28.8. Close examination of the data in the region of the upper energy limits of these curves reveals that the mass of the neutrino is zero or very nearly so. According to Fermi's theory, any beta decay process is characterized by the change of a neutron into a proton or vice versa, with the simultaneous emission of a neutrino. Thus the fundamental beta decay processes are

$$n \rightarrow p + \beta^- + \bar{\nu} \qquad \text{(beta emission)}, \qquad (28.6a)$$
$$p \rightarrow n + \beta^+ + \nu \qquad \text{(positron emission)}, \qquad (28.6b)$$
$$p + e^- \rightarrow n + \nu \qquad \text{(electron capture)}. \qquad (28.6c)$$

The line over the Greek letter ν (pronounced nu-bar) denotes an antineutrino. The distinction between neutrino and antineutrino need not concern us here. (Particles and antiparticles are discussed in Chapter 31.)

As a simple example of Eq. (28.6a), the decay of free neutrons into protons occurs with a half-life of about fourteen minutes. Such a decay is energetically favored, for it has a positive Q value. The decay of an isolated proton into a neutron is not allowed, since the neutron mass exceeds the proton mass.

Example 28.7

Calculate the Q value in the beta decay of a free neutron.

Solution If we assume that the antineutrino has no mass, we need only consider the mass of the proton, neutron, and electron:

$$Q = m_n - m_p - m_e.$$

Using the values from Tables 28.1, we find that

$$Q = 939.566 \text{ MeV} - 938.272 \text{ MeV} - 0.511 \text{ MeV} = 0.783 \text{ MeV}.$$

In the beta decay of a free neutron, the upper limit of the kinetic energy of the beta particle is 0.783 MeV.

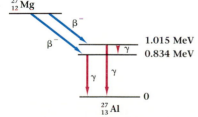

Figure 28.9 Nuclear energy-level diagram for gamma decay of excited $^{27}_{13}$Al following the beta decay of $^{27}_{12}$Mg. The $^{27}_{12}$Mg decays to either of the excited $^{27}_{13}$Al states, which then eventually decay to the ground state by emitting gamma-ray photons, in analogy to the emission of photons of light by an excited atom.

Gamma Decay

Frequently the daughter nucleus resulting from alpha or beta decay is left in an excited energy state. These energy states are energy levels of the nucleus, analogous to the electronic energy levels of an atom. Nuclei in excited states can give up their excess energy by emitting electromagnetic radiation. That is, an excited nucleus can return to its lowest or ground state through the emission of a photon (Fig. 28.9). The photons associated with nuclear energy-level changes are called gamma rays. Typical energies of gamma rays are in the range from tens of kilo electron volts to a few mega electron volts.

Gamma rays and x rays are both very energetic photons. The distinction between them has to do with their origins: X rays come from electronic excitations and gamma rays come from nuclear excitations. In general the gamma rays from nuclear decay are more energetic than the x rays from most x-ray machines, though there is some overlap (Fig. 21.2).

Detectors of Radiation

*28.7

So far we have discussed the general ideas involved in nuclear stability and nuclear decay. Later we will discuss the units used for radiation measurement and the biological effects of radiation. However, first let's pause to look at the instruments we use to detect the products of nuclear decay and see what we can learn from them.

By observing the flashes, or scintillations, made when radiation strikes a fluorescent screen, physicists gathered much of the early data concerning nuclear physics. However, looking at flashes through a microscope in a darkened room was not easy. (The story is told that Rutherford picked his assistants in part on the basis of how sharp their eyesight was for this task.) Two assistants in Rutherford's laboratory, Hans Geiger and W. Müller, developed a different technique for detecting particles. They invented what was called the Geiger-Müller tube, but is now more frequently referred to as simply a *Geiger tube*. This tube is the detecting element in a Geiger counter.

A typical Geiger tube consists of a cylindrical conducting tube with a wire electrode placed along the axis (Fig. 28.10). The ends of the tube are covered and the tube is filled with a special gas mixture, often argon with a small admixture of polyatomic gases, at a pressure below atmospheric pressure. A dc voltage is applied between the wire and the tube. This voltage must be higher than some threshold value, depending on the gas and the geometry of the tube, but is not so high that discharge takes place between the wire and the tube.

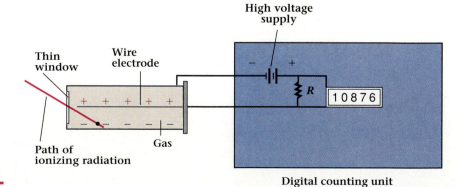

Figure 28.10 Schematic diagram of a Geiger counter. A charged particle entering the tube ionizes atoms of the gas, releasing a shower of electrons to the wire. The counter keeps track of each resulting current pulse.

When a charged particle passes through the gas in the tube, it ionizes, or knocks some electrons out of, an atom of gas. The strong electric field between the tube and wire accelerates the freed electrons and gives them enough energy to ionize other atoms, freeing yet more electrons. This multiplication of charge allows a large number of electrons to be collected at the positive wire electrode. A large current then flows through the resistor *R*, causing a large potential drop across the resistor and thereby reducing the potential difference between the central wire and the tube. This effect extinguishes the discharge, and the potential of the central wire returns to its original value. The tube is then ready to detect the next charged particle. The positive gas ions are attracted to the negative tube wall, but are prevented from cascading as the electrons do by the presence of the polyatomic gas. Each time a current pulse flows through the resistor, it is counted by an electronic counting circuit, which displays the number of counts on some numerical indicator. Frequently an audio circuit is included, which gives a "click" for each ionizing particle.

The Geiger tube may be provided with thin walls and with especially thin end windows for the detection of low-energy alpha and beta rays. X rays and gamma rays may produce an electron-positron pair in the walls and give rise to particles that will be counted. However, the efficiency for detection of x rays and gamma rays is quite low.

Although the Geiger tube, or some variation of it, is useful for many purposes, another frequently used radiation detector is a modern descendant of Rutherford's fluorescent scintillating screen. A *scintillation counter* (Fig. 28.11) contains a material that emits a flash of light when radiation passes through it. A typical material for use as a scintillator is a single crystal of sodium iodide with a small amount of thallium; however, many plastics and even some liquids are commonly used. The scintillator is coated with a light-tight layer everywhere except where it is in optical contact with the front of a photomultiplier tube. The tube detects weak flashes of light.

The photomultiplier is a vacuum tube containing several highly sensitive electrodes. The potential of each electrode increases with its position along the length of the tube. When a photon from the scintillator strikes

Figure 28.11 Schematic diagram of a scintillation counter. An incoming particle causes the scintillator to emit photons, which cause an electrode to emit photoelectrons. The chain of electrodes magnifies this effect into a sizable current pulse.

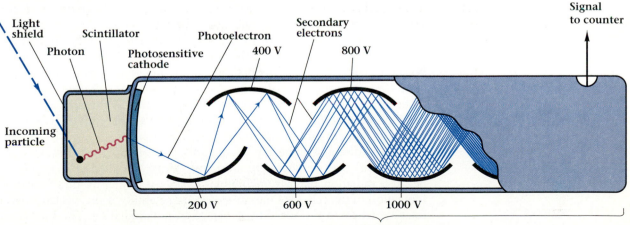

Figure 28.12   A neutrino-induced reaction in a freon-filled bubble chamber. The tracks are tiny bubbles generated by moving charged particles. The particles' paths are curved because of the magnetic field present.

the electrode at the top of the tube, an electron is ejected by the photo-electric effect. This electron is then accelerated toward the next electrode, arriving with enough kinetic energy to release several secondary electrons from it. These electrons, in turn, accelerate toward the next electrode along the tube, triggering an avalanche effect. The end result is a sizable electric pulse, which is registered by an electronic counter. For many scintillation materials, the output of the photomultiplier tube is proportional to the energy of the incident radiation.

One advantage of scintillation counters is that they can be made very sensitive to the detection of x rays and gamma rays, radiation for which the Geiger counter has an efficiency of only a few percent. In addition, they can be ready to record a new event in as little as 10^{-9} s, while a typical Geiger tube requires of the order of 10^{-4} s.

Another modern type of detector is the solid state or semiconductor detector. (The physics of semiconductors is discussed in Chapter 30.) It has the advantages of being quite small, having a fast response, and not requiring a high-voltage power supply. Solid state detectors are frequently useful when good energy resolution is required. However, they are not very efficient detectors; almost all incident particles pass right through the sensitive region of the device.

There are also several kinds of "track-visualization" devices. These instruments record the path in space of an individual particle. They include photographic film, the very first detector used by Roentgen in the discovery of radiation, and bubble chambers, where the path through a liquid is made visible by the formation of bubbles (Fig. 28.12).

Radiation Measurement and Biological Effects

*28.8

Everyone knows that radioactivity can be dangerous to your health. But before we can examine these biological effects, we need to establish units for measuring how much radioactivity a body is exposed to and how much of this radioactivity a body absorbs. The study of radioactivity measurement, both physical and biological, is called dosimetry.

As we mentioned previously, the number of radioactive disintegrations per unit time is the *activity* of the substance. The SI unit of activity is the *becquerel* (Bq), which is defined as one disintegration per second. The *curie* (Ci), an earlier unit of activity that is sometimes still used, is defined to be 3.7×10^{10} disintegrations per second; this is the activity of one gram of pure radium. The curie turns out to be a large unit of activity, so mCi and μCi are more commonly used.

When discussing the effects of radiation, other quantities are often more useful than activity. The ionizing ability of radiation is important because radiation damage in biological cells is primarily due to excess ionization in the cell. One measure of ionizing ability is the electric charge released when air is ionized by x rays or gamma radiation. The unit is the *roentgen* (R), defined to be 2.58×10^{-4} C/kg of air. However, since the

TABLE 28.3
Quality factor for several types of radiation

Radiation type	Quality factor
X rays, gamma rays, beta particles	1
Neutrons, energy < 10 keV	3
Neutrons, energy > 10 keV	10
Protons	10
Alpha particles, fission fragments, recoil nuclei	20

TABLE 28.4
Recommended maximum permissible doses, excluding intentional medical exposures

Type of exposure	Maximum permissible dose (mSv/y)
Radiation workers:	
Whole body, gonads, or lenses of eyes	50
Skin of whole body	300
Hands and feet	750
General population:	
Whole body	5
Gonads	1.7

roentgen is defined only for x rays and gamma rays in air, it is not widely used today.

An important quantity for measuring physical effects due to absorption of radiation in matter is called the absorbed dose D, or simply, the **dose**. The dose is defined as the energy deposited per unit mass by absorbed radiation. The SI unit of dose is the *gray* (Gy), defined as 1 J/kg. The gray can be used for measuring the energy absorbed from any type of radiation in any material. Another common unit for dose is the *rad* (radiation absorbed dose): One rad equals 0.01 Gy.

Radiation gives rise to two types of biological effects: genetic and somatic. Genetic effects cause mutations in the reproductive cells of an organism and so affect subsequent generations. Since genetic damage occurs only when the reproductive cells are irradiated, the gonads should be shielded if possible any time you are being x-rayed. The younger you are, the more important this is.

Somatic effects (sometimes called "radiation sickness") harm individuals directly, primarily by ionizing molecules in cells. The extent of the damage depends not only on the type of radiation, but on the part of the body irradiated and on the individual's age. Again, generally speaking, the younger you are, the more hazardous the radiation. In fact, radiation doses before birth are the most dangerous, and pregnant women often require special precautions against radiation exposure. Some somatic effects include reddening of the skin, loss of hair, ulceration, fibrosis of the lungs, formation of holes in tissues, reduction of white blood cells, and induction of cataracts in the eyes. Perhaps one of the most feared effects of radiation is the occurrence of cancer on the skin and in various other organs of the body.

The extent of damage to biological organisms due to radiation depends on the particular type of radiation as well as on the energy deposited. Thus the dose, which measures only the energy deposited per unit mass, is not sufficient for indicating damage to living tissue. Instead, we use the **dose equivalent** H, which measures the product of the dose (in grays) times a dimensionless number, called the **quality factor** Q, which takes account of the biological effect of each radiation. The result is an equivalent dose, measured in the SI unit called the sievert (Sv). The equivalent dose is also measured in *rem* (roentgen equivalent man), where 1 rem = 0.01 Sv:

$$H \text{ (in Sv)} = D \text{ (in Gy)} \times Q. \qquad (28.7)$$

Table 28.3 lists the quality factors for several types of radiation. The quality factor is a judgment based on experiments and experience and is not the direct result of experimentation alone.

To guard against the health hazards of radiation, the U.S. Nuclear Regulatory Commission (NRC) has established upper limits of acceptable equivalent doses for human exposure to radiation. Table 28.4 gives the present recommended maximum permissible dose for radiation workers and for the general population. The NRC reviews and revises these standards periodically. We should also point out that other countries establish their own standards for maximum permissible equivalent dose, often different from U.S. standards.

TABLE 28.5 Estimated annual effective dose equivalent to individuals in the United States*	
Source	**Average annual effective dose equivalent in the U.S. population† (mSv)**
Natural radiation:	
Cosmic rays	0.280
Terrestrial	0.280
In the body	0.390
Inhaled radon	≈2.000
Radiation from human activity:	
Occupational	0.009
Nuclear fuel cycle	0.0005
Consumer products	0.05–0.13
Environmental sources	0.0006
Medical	
Diagnostic x rays	0.39
Nuclear medicine	0.14
Rounded total	3.6

*Taken with permission from "Ionizing Radiation Exposure of the Population of the United States." NCRP Report No. 93, (National Council on Radiation Protection and Measurements, Bethesda, Md.,1987).
†The effective dose equivalent is essentially the dose equivalent when the whole body is irradiated uniformly.

In our daily environment we are subject to radiation exposure from two natural sources: cosmic rays and naturally occurring radioactive isotopes on the earth's surface. This exposure is known as background radiation and varies somewhat from place to place. For example, cosmic ray exposure is much greater at high elevations than at sea level. Additionally, there are several common sources of radiation that are the result of human activity, some of which are listed in Table 28.5. Figure 28.13 shows the relative contributions of the major sources of radiation to the population of the United States.

Although there is some uncertainty in the exact value, it is clear that radon, a colorless odorless gas, is a major contributor to the equivalent dose in the United States. Radon is part of the natural decay chain of the uranium found in some rocks, and it seeps into houses and buildings through cracks in the foundations. The inhaled radon decays to alpha emitters in the lungs and causes damage and tumor growth. The Environmental Protection Agency estimates that radon causes between 5,000 and 20,000 of the 140,000 annual lung cancer deaths in the United States.

The amount of radon released in the soil varies markedly from place to place because of the underlying geological formations, so that not every area receives the same dose. Furthermore, how the foundation of a building is sealed and how the building is ventilated make a difference. You can now buy inexpensive radon monitors for your home, and the Environmental Protection Agency publishes several helpful booklets.

There are two basic methods of shielding against any radiation: distance and mass. The simpler is distance. Most radiation is emitted equally

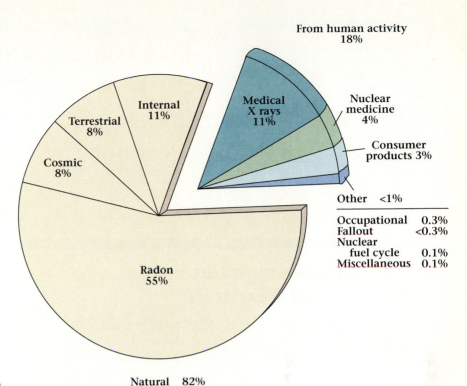

Figure 28.13   The percentage contribution of various radiation sources to the total average effective dose equivalent in the U. S. population.

in all directions from its source, so its intensity decreases as $1/r^2$. For example, increasing your distance from a radioactive source by three times reduces your received dose by a factor of nine. The other means of shielding is to place absorbing material around the source. In most cases, for a given thickness of shielding, the more massive it is, the more effective it is. Thus lead is best for most cases. However, lead is sometimes prohibitively expensive and thicker pieces of other material, such as special types of concrete, are used. Because neutrons interact with matter differently from other radiation, shielding against them is best done with a material such as cadmium, which has a high probability of capturing a neutron, or with material containing light nuclei, such as hydrogen, which can absorb the neutron's kinetic energy. Thus water and paraffin are among the materials commonly used for neutron shielding.

Example 28.8

Dose equivalent of a dental x ray.

A patient receives a dose equivalent of 1.0 mSv in 0.2 kg of tissue from a dental x-ray machine operating at 90 keV. (a) What is the total energy deposited in the patient? (b) How many x-ray photons contribute to the dose if you assume that each photon gives up all its energy?

Solution   (a) To find the total energy we must first find the dose (in grays). X rays have a quality factor of 1, so we get

$$D = \frac{H}{Q} = \frac{1.0 \text{ mSv}}{1} = 1.0 \times 10^{-3} \text{ Gy}$$

$$D = 1.0 \times 10^{-3} \text{ J/kg}.$$

The total energy absorbed E_T is

$$E_T = (1.0 \times 10^{-3} \text{ J/kg}) \times (0.20 \text{ kg}) = 2.0 \times 10^{-4} \text{ J}.$$

(b) Each x-ray photon has an energy E_γ of

$$E_\gamma = 90{,}000 \text{ eV} \times 1.60 \times 10^{-19} \text{ J/eV} = 1.44 \times 10^{-14} \text{ J}.$$

The number of photons N is

$$N = \frac{\text{total energy}}{\text{energy per photon}} = \frac{E_T}{E_\gamma} = \frac{2.0 \times 10^{-4} \text{ J}}{1.44 \times 10^{-14} \text{ J/photon}},$$

$$N = 1.4 \times 10^{10} \text{ photons}.$$

Induced Transmutation and Reactions

28.9

In 1902 Rutherford and Soddy proposed that when a radioactive element decayed, the remaining material was a different chemical element. This view was accepted only after several years of experimental verification. With the discovery of radioactivity the dream of the alchemists had come true: One element was changed into another. Yet it seemed that there was no way to control this transmutation process. No way was known of speeding up or slowing down the rates of radioactive decay or of causing nuclei to decay if they did not already do so naturally.

However, in 1919 Rutherford made the remarkable discovery that nuclear transmutation can be induced, or caused to happen. Rutherford's apparatus is shown in Fig. 28.14. A radioactive source was placed at D within a chamber that could be either evacuated or filled with some gas through the tubes A. In the first part of the experiment the chamber was filled with hydrogen. When alpha particles from the radioactive source collided head-on with the protons of hydrogen gas, the protons acquired a speed greater than that of the alpha particles. The faster, lighter protons were more penetrating than the alpha particles. When the hydrogen was replaced by air, which contains nitrogen, a surprising effect was observed. Rutherford expected that the head-on collisions of alphas with any of the heavier components of the air would yield slowly moving particles. Instead, he observed rapidly moving particles. He concluded that these rap-

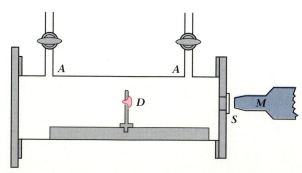

Figure 28.14   Rutherford's apparatus used in the discovery of induced transmutation. D = radiation source, A = gas inlet and outlet, S = zinc sulfide screen, M = microscope.

idly moving particles were protons and that the only explanation was that the nitrogen was disintegrated in the collision with the rapidly moving alpha particle. Rutherford described the reaction as

$$_{2}^{4}\text{He} + {}_{7}^{14}\text{N} \rightarrow {}_{8}^{17}\text{O} + {}_{1}^{1}\text{H}.$$

The resultant product must be ${}_{8}^{17}\text{O}$ in order to conserve both charge and nucleon number.

Other elements were soon investigated to see if similar results could be obtained. Two reactions that were found early were

$$_{2}^{4}\text{He} + {}_{4}^{9}\text{Be} \rightarrow {}_{6}^{12}\text{C} + {}_{0}^{1}\text{n}$$

and

$$_{2}^{4}\text{He} + {}_{5}^{11}\text{B} \rightarrow {}_{7}^{14}\text{N} + {}_{0}^{1}\text{n}.$$

In both cases alpha particles were used as the bombarding projectiles. However, the alpha particles from naturally radioactive materials have only enough energy to overcome the repulsive Coulomb potential barrier (Fig. 28.6) of a limited number of nuclei. The need to experiment with particles of higher energy led to the development of what are called particle accelerators. An early accelerator is shown in Fig. 28.15; more recent ones are discussed in Chapter 31.

The first reaction observed with machine-accelerated nuclear particles was produced by the bombardment of lithium foil with protons:

$$_{1}^{1}\text{H} + {}_{3}^{7}\text{Li} \rightarrow {}_{2}^{4}\text{He} + {}_{2}^{4}\text{He}.$$

Now we can bombard almost any nucleus with almost any other one. We may write a general nuclear reaction in the form

$$A + B \rightarrow C + D. \tag{28.8}$$

The total energy of the particles on the left in Eq. (28.8) must be the same as the total energy of the particles on the right. By "total energy" we mean the sum of the kinetic energies and the rest-mass energies. There is no potential-energy term to be included because the short-range nuclear forces act only during the brief time of the interaction itself.

A given reaction, such as the general one in Eq. (28.8), is characterized by the masses of the nuclei involved. We can calculate the difference in the initial and final masses, and therefore the Q value, for any reaction we care to consider. If the Q value is positive, the kinetic energy of the products is greater than that of the reactants. If the Q value is negative, we must supply energy to the reactant nuclei for the reaction to actually take place.

For a reaction with a negative Q value, the reactant nuclei must have a kinetic energy equal to or greater than the magnitude of the Q value. If the target particle (B) is at rest, the kinetic energy of the incoming particle (A) must be high enough to supply the energy needed for the reaction. Otherwise the reaction will not occur. However, all of the incoming kinetic energy is not available for the reaction, as some must be transferred to the other particles in order to conserve momentum. The lowest energy for the incoming particle at which the reaction will take place is called the **threshold energy**. We can calculate the relationship between the threshold energy KE_{th} and the absolute magnitude of the Q value for the reaction,

Figure 28.15   One of Van de Graaff's early electrostatic generators. The first electrostatic generator announced by Van de Graaff in 1931 was capable of producing a potential difference of 1,500,000 V. A later machine was capable of reliably accelerating electrons or ions to energies of 2.75 MeV.

using conservation of energy and momentum. The result of the somewhat lengthy calculation is

$$KE_{th} = (1 + m_A/m_B)|Q|. \tag{28.9}$$

We must emphasize that among the principal tools used in investigating nuclear reactions are the conservation rules of momentum and energy. Reactions like those we have just described are two-body interactions, in which two particles come in and two particles go out. Their collision is handled in the same manner as are the collisions between larger bodies (Chapter 8). However, here we also use the Einstein mass–energy relationship. Examples of threshold energy calculations follow.

Example 28.9

Calculating threshold energy.

Determine the minimum kinetic energy a proton must have to make the following reaction occur:

$$p + {}^7_3Li \rightarrow {}^4_2He + {}^4_2He.$$

Solution   The appropriate masses are atomic masses, so we use the mass of ${}^1_1H$ for the incident proton:

$$m({}^1_1H) = 1.007825 \text{ u},$$

$$m({}^7_3Li) = 7.016003 \text{ u},$$

$$m({}^4_2He) = 4.002603 \text{ u};$$

then

$$Q = \text{initial mass energies} - \text{final mass energies}$$
$$= [m({}^1_1H) + m({}^7_3Li) - 2m({}^4_2He)]c^2$$
$$= (0.018622 \text{ u})(931.5 \text{ MeV/u}),$$
$$Q = 17.3 \text{ MeV}.$$

The Q value for the reaction is positive, which means that energy is given off in the proposed reaction. Thus there is no minimum threshold energy. If, on the other hand, the Q value had been negative, the reaction would not proceed by itself and kinetic energy would be required.

Example 28.10

Determine the threshold energy of the alpha particle in the reaction

$${}^4_2He + {}^{14}_7N \rightarrow {}^{17}_8O + {}^1_1H.$$

Solution   The appropriate masses are

$$m({}^4_2He) = 4.002603 \text{ u},$$

$$m({}^{14}_7N) = 14.003074 \text{ u},$$

$$m({}^{17}_8O) = 16.999130 \text{ u},$$

$$m({}^1_1H) = 1.007825 \text{ u};$$

then

$$Q = \text{initial mass} - \text{final mass} = -0.001278 \text{ u} = -1.190 \text{ MeV}.$$

The Q value is negative so kinetic energy must be supplied. The threshold, or minimum energy required, is

$$KE_{th} = [1 + m(^4_2\text{He})/m(^{14}_7\text{N})]|Q|,$$

$$KE_{th} = [1 + (4.0026)/(14.003)] \, (1.190 \text{ MeV}),$$

$$KE_{th} = 1.53 \text{ MeV}.$$

Thus if the target $^{14}_7\text{N}$ nucleus is initially at rest, the alpha particle must have a kinetic energy of at least 1.53 MeV for the reaction to take place.

Nuclear Fission

28.10

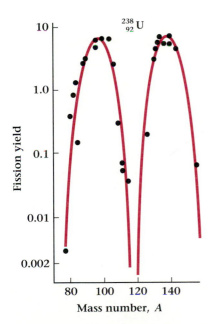

Figure 28.16 Fission yields from $^{238}_{92}$U. The fission yield is the relative number of times, here plotted in arbitrary units, that a nucleus of atomic number A is a product of fission.

In 1939 Otto Hahn and Fritz Strassman published their work on a special kind of nuclear reaction of both theoretical and practical interest. It was already known that many products of artificially induced transmutations were themselves radioactive. Hahn and Strassman, along with Lise Meitner, had bombarded uranium with neutrons. Their experiments were similar to other experiments being done at that time in an effort to create elements of atomic number greater than 92 (uranium).

The products of the reaction were radioactive, but there was great difficulty in identifying the actual nuclei. Finally, a series of extremely careful chemical analyses by Hahn showed that the reaction products contained both $^{139}_{56}$Ba and $^{140}_{57}$La, two materials much lighter than uranium. Meitner and O. R. Frisch, working in Sweden, soon proposed the explanation that the uranium nucleus had split into two nuclei of smaller mass. The reaction, as we now understand, can lead to several such splittings, one of which is

$$^{238}_{92}\text{U} + ^1_0\text{n} \rightarrow ^{139}_{56}\text{Ba} + ^{97}_{36}\text{Kr} + 3^1_0\text{n}.$$

Such a process is called **fission.** It is different from other nuclear reactions in that the unstable nucleus splits into two relatively heavy parts plus a few neutrons. Furthermore, the same products do not always result from the same initial bombardment. For instance, the following reaction is also observed:

$$^{238}_{92}\text{U} + ^1_0\text{n} \rightarrow ^{140}_{57}\text{La} + ^{97}_{35}\text{Br} + 2^1_0\text{n}.$$

This explains why Hahn found both barium and lanthanum among the reaction products. Many other reactions are observed as well. Figure 28.16 shows the relative distribution of the products of the fission of $^{238}_{92}$U by neutrons. Most fission fragments are highly excited and go through many decays before reaching a stable isotope. Other products accompany the nuclei, as shown in Fig. 28.17. Several neutrons are usually released in the fission process. For $^{238}_{92}$U the average number of neutrons released per fission is 2.5. Other isotopes of uranium, as well as other heavy nuclei, can undergo fission.

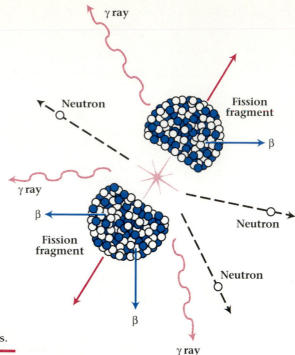

Figure 28.17   Schematic drawing of fission fragments.

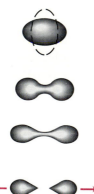

Figure 28.18   Oscillation and breakup of a liquid drop. This process can serve as a model for the fission of a large unstable nucleus.

Nuclei that undergo fission are of such a large size that we can describe the breakup in terms of a model, treating the nucleus as a liquid drop. A liquid drop that is disturbed oscillates and, if the disturbance is great enough, eventually breaks apart (Fig. 28.18). The fission process, in which the neutron provides the initial disturbance, is analogous to this behavior.

Fission processes have the possibility of being of some practical use because they have positive Q values; that is, the kinetic energy of the products is greater than the kinetic energy of the initial reactants. The positive Q value occurs because the heavier elements have a smaller binding energy per nucleon than do the medium-mass elements (Fig. 28.4). The heavy elements can then break up into two components that are more tightly bound and therefore have less total mass energy. On the average, about 200 MeV of energy is released per fission. Compare this value with the average energy in chemical reactions, which is less than 1 eV per molecule.

The large energy produced per fission would be of no practical use if it were not for the fact that each fission releases an average of 2.5 neutrons. If we place the fissioning nuclei in the proper configuration, these neutrons can cause other fissions. The neutrons from those fissions can cause still other fissions, each fission releasing more neutrons and more energy. Such a process is called a **chain reaction** (Fig. 28.19). If the process occurs in an uncontrolled fashion, the result is an explosion. If the process is carried out in a manner that limits the number of neutrons causing successive fission events, then the heat due to the kinetic energy can be extracted and made to do useful work, as in nuclear-fueled power plants. In either case, the products of the fission process are extremely radioactive. The

Figure 28.19   Schematic representation of a chain reaction. Here in each step the number of fissioning nuclei increases. In a controlled process this number is made to stay constant by absorbing the excess neutrons in a moderator.

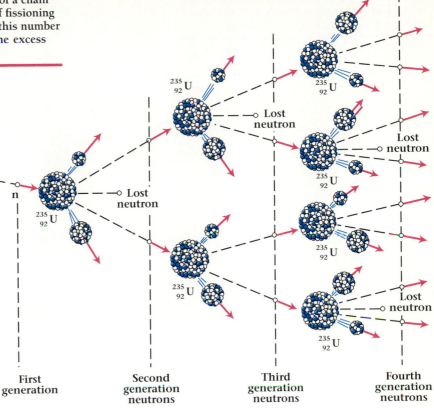

First generation   **Second generation neutrons**   **Third generation neutrons**   **Fourth generation neutrons**

products are mostly long-lived gamma-ray emitters, which offer considerable health hazard.

Figure 28.20 (p. 856) shows the basic components of a fission power plant. The first successful self-sustaining, controlled nuclear chain reaction was accomplished by Enrico Fermi and his colleagues at the University of Chicago in 1942. Although the design of the reactor in Fig. 28.20 differs from Fermi's original reactor, all reactors have several things in common.

Most reactors use uranium for their nuclear fuel. Natural uranium is mostly $^{238}_{92}U$ (about 99.3%) with a small amount of $^{235}_{92}U$ (about 0.7%). It is the $^{235}_{92}U$ isotope that plays the important role in the chain reaction.* Because the amount of $^{235}_{92}U$ in natural uranium is so small, the uranium used for nuclear fuel is enriched to about 3.5% of this isotope.

The probability that an incident neutron will cause $^{235}_{92}U$ to fission is significant only for low-energy neutrons. The fast neutrons released by fission must be slowed down so they can cause other fissions to occur. This is accomplished by a *moderator,* sometimes carbon, ordinary water, or heavy water (D_2O). The neutrons lose kinetic energy by collisions with the moderator. The neutrons eventually slow down until their kinetic

---

*The $^{238}_{92}U$ isotope does not readily fission when struck by a neutron; instead it captures the neutron to become $^{239}_{92}U$, which, after two beta decays, becomes $^{239}_{94}Pu$. The isotope $^{235}_{92}U$ fissions readily when struck by a slowly moving neutron.

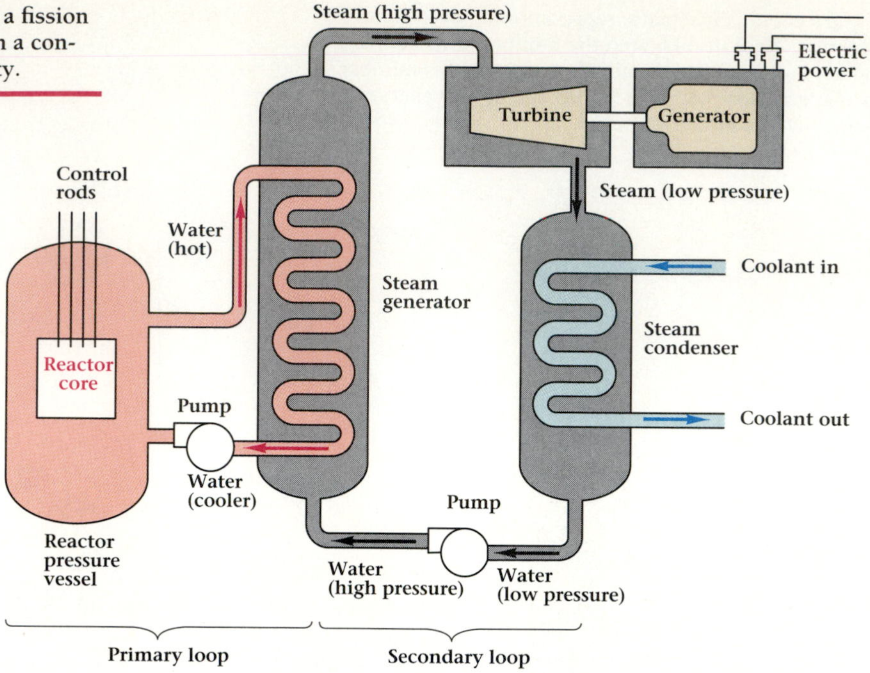

Figure 28.20   The basic components of a fission power plant. The primary loop is kept in a containment structure to seal in radioactivity.

energies are of the order of kT, where T is the temperature of the moderator. Slow neutrons of these energies are called *thermal neutrons*.

The placement of the nuclear fuel, which is normally made into the form of long rods, must be carefully designed. The uranium and the moderator must be arranged so that for every fission at least one neutron goes on to cause another fission. There also must be a way to control the rate at which fissions occur. It is necessary to be able to start the chain reaction slowly, regulate it when in progress, and stop it when desired. This control is achieved with control rods of neutron-absorbing material such as cadmium. When the control rods are fully inserted among the fuel rods, so many neutrons are absorbed that the chain reaction is blocked. The rods are slowly withdrawn to allow the reaction to begin and to build up to the desired level for a self-sustaining chain reaction.

The kinetic energy of the fission fragments and neutrons is converted to thermal energy. Thus the reactor is a large heat source capable of extremely high temperatures. The thermal energy is removed by a liquid coolant that flows through pipes in the reactor core. This primary coolant passes through a heat exchanger where it gives up heat to a secondary coolant, which may be used to drive a turbine that turns an electric generator. Thus the energy of fission can be used to generate electricity.

Nuclear reactors are reasonably efficient sources of electrical energy, but they have at least two principal problems. The first problem is with the safety of the reactor itself. The chain reaction must not be allowed to get out of control so that the fuel melts or pressures build up and cause a break in the surrounding containment vessel, because such a break could release radioactive materials into the environment. The solutions to this problem are well within current technology, but require large expenditures

Figure 28.21 Cerenkov radiation is seen as the blue glow in the water surrounding spent fuel as it is removed from the High Flux Isotope Reactor.

of money and effort. A more serious and long-term problem involves the disposal of the radioactive waste from the spent fuel. The solution of this problem is less well agreed upon, and to a great extent involves political and economic decisions, as well as scientific ones.

Special nuclear reactors are designed for research. The High Flux Isotope Reactor (HFIR) at the Oak Ridge National Laboratory is able to produce an extremely high flux of neutrons in its core. These neutrons bombard various target materials to produce elements heavier than uranium. Beams of thermal neutrons can also be used for neutron diffraction studies of magnetic solids.

HFIR, a water-cooled and water-moderated reactor operating at 85 MW, produces intense radiation. A blue glow in the water surrounding the radioactive core is caused by high-speed beta particles emitted in the decay of fission products (Fig. 28.21). This glow, called **Cerenkov radiation**, is analogous to the shock wave caused by an object moving faster than the speed of sound. Here the shock wave is electromagnetic, caused by charged particles exceeding the speed of light in the water. (The speed of the particles is less than c, the speed of light in a vacuum, in accord with the theory of relativity.)

Nuclear Fusion

28.11

Inspection of Fig. 28.4 suggests another energy-releasing process. When undergoing fission, heavy nuclei release energy in going from H to M on the graph. We should also be able to obtain energy by combining two elements in the region L, forming nuclei of greater binding energy per nucleon in the region M. The process of combining two nuclei to form a heavier one is called **fusion**. An example is

$$_1^2H + {}_1^2H \rightarrow {}_0^1n + {}_2^3He,$$

where $_1^2H$ represents the naturally occurring heavy isotope of hydrogen, known as deuterium (D). Deuterium is a naturally occurring component of water and can be extracted from the sea in large amounts. This reaction releases approximately 4 MeV, which is about the same energy per nucleon as released by fission. An attractive feature of this reaction is that the product $_2^3He$ is stable and not radioactive.

In order for energy to be released in such a fusion reaction, the two deuterium nuclei must get close enough for the nuclear forces to act. At distances larger than the range of the nuclear force, the repulsive Coulomb force dominates. (Recall Fig. 28.6, which depicts the nuclear Coulomb potential experienced by an alpha particle.) The energy necessary to accomplish the reaction is the energy needed to overcome the Coulomb potential energy barrier at a distance of about 10^{-14} m, the approximate

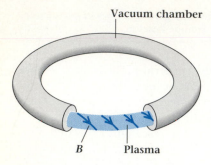

Vacuum chamber

B   **Plasma**

Figure 28.22 Schematic diagram of the toroidal vacuum chamber of a tokamak fusion chamber. The plasma is confined by the helical magnetic field.

distance at which the reaction will occur. This energy is

$$E = k\frac{e^2}{r} = \frac{(9 \times 10^9 \text{ N·m}^2/\text{C}^2)(1.6 \times 10^{-19}\text{C})^2}{10^{-14} \text{ m}},$$

$$E = (2.3 \times 10^{-14} \text{ J})(1 \text{ MeV}/1.6 \times 10^{-13} \text{ J}) = 0.14 \text{ MeV}.$$

For the fusion reaction to be self-sustaining, its energy must be released in the vicinity of other deuterium nuclei so that they can subsequently interact. We can determine the required temperature of a deuterium gas in which the particles would have an average kinetic energy of 0.14 MeV from

$$\tfrac{3}{2}kT = \overline{\text{KE}}.$$

The temperature is

$$T = \tfrac{2}{3}\overline{\text{KE}}/k \approx 1 \times 10^9 \text{ K}.$$

The atoms in a gas at such high temperatures are completely ionized. Such a high-temperature gas of positive and negative charged particles is called a *plasma*.

The energy released from fusion is what keeps the sun and other stars hot. The extremely high temperatures and densities, in turn, provide the conditions necessary for a self-sustaining reaction. Even though the reaction will actually take place at temperatures that are lower by a factor of 100, the principal barrier to practical self-sustaining fusion reactions in the laboratory is clear; the high-density fusion fuel must be maintained at an extremely high temperature.

One of the most likely ways of accomplishing a self-sustaining fusion reaction on earth is to compress a high-temperature plasma using magnetic fields. Two methods of doing this are currently under study: the tokamak and the magnetic mirror configuration.

The tokamak, first devised in the Soviet Union, has a toroidal shaped chamber with a complex arrangement of coils to generate a magnetic field that spirals around the torus (Fig. 28.22). The circulating plasma is compressed and kept away from the walls by the magnetic field. Research using the tokamak principle is underway in the United States, Great Britain, Japan, and the USSR.

An alternative method of containing the plasma uses a differently shaped magnetic field that reflects the charged particles back and forth between "magnetic mirrors" (Fig. 28.23). Charged particles move along helical paths in the region where the field is nearly uniform. In the region where the field lines come closer together the field has a radial component. As the particles move into that region they encounter a force that retards their motion into the high-field region. Eventually their motion toward the high-field region is reversed and they are reflected back toward the region of lower field.

In addition to the magnetic confinement methods, other approaches to achieving fusion-produced power are possible. One approach is to bombard small solid pellets containing fusionable material with a burst of high intensity from many lasers at once. This pulsed technique attempts to achieve the required temperatures and densities for a sufficiently long time. Another, entirely different approach, is to use muons as part of a

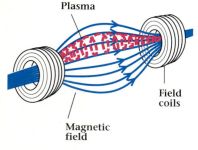

Plasma

Field coils

Magnetic field

Figure 28.23 The magnetic mirror principle of plasma confinement.

lower-temperature fusion reaction. Muons are subatomic particles (discussed in Chapter 31), which are not stable and therefore must be produced by an accelerator.

There is hope that fusion reactors will become practical energy sources, especially since the deuterium fuel is relatively inexpensive and the direct waste products are not radioactive. Furthermore, there is no possibility of the fusion reaction running out of control and exploding. However, the early expectations of the 1950s and 1960s that nuclear fusion would provide practical power before the end of the century were too optimistic. It seems probable now that another 25–50 years of research and development will be required before fusion power becomes a practical reality. Despite their virtues, fusion reactors will not be the perfect source of energy. They will be extremely expensive to build, they will be extremely radioactive as a result of their high flux of neutrons, and they will be sources of thermal pollution.

SUMMARY

Useful Concepts

- A nucleus of the element X with charge Z and total number of nucleons A is written

$$^A_Z X.$$

- The radius of a nucleus of mass number A is given approximately by

$$r = R_0 A^{1/3}.$$

- Nuclear reaction equations at low energies can be "balanced" using conservation of charge and conservation of nucleon number.
- The Q value of a decay reaction is written

$$Q = [m(\text{parent}) - m(\text{daughter}) - m(\text{decay particle})]c^2.$$

- In alpha decay processes, the kinetic energy of the emitted alpha is

$$\text{KE}_\alpha \approx \left[\frac{A}{A+4}\right]Q.$$

- The fundamental beta decay processes are:

$$n \rightarrow p + \beta^- + \bar{\nu} \quad \text{(beta emission)},$$

$$p \rightarrow n + \beta^+ + \nu \quad \text{(positron emission)},$$
$$p + e^- \rightarrow n + \nu \quad \text{(electron capture)}.$$

- The threshold energy for a nuclear reaction is

$$\text{KE}_{\text{th}} = (1 + m_A/m_B)|Q|.$$

- The release of energy in both fission and fusion reactions is accomplished through a change in the binding energy per nucleon.

Important Terms

You should be able to write the definition or meaning of each of the following terms:

- proton
- neutron
- nucleon
- isotope
- binding energy
- Q value
- positron
- beta plus
- beta minus
- electron capture
- neutrino
- dose
- dose equivalent
- quality factor
- threshold energy
- fission
- chain reaction
- fusion
- Cerenkov radiation

QUESTIONS

28.1 What are the properties of a nucleus that is likely to undergo alpha decay; beta decay; gamma decay?

28.2 What would the distribution of electron kinetic energies be like if no neutrinos were emitted in beta decay? What would it look like if a very massive neutral particle were always emitted?

28.3 Why is the decay rate of radioactive nuclei not changed by chemical environment, heating, or pressure?

28.4 Why is radon still found in nature? Its half-life is 1600 years and the estimated age of the universe is five billion years.

28.5 How can isotopes of an element be separated, since they have the same chemical properties?

28.6 An inventor claims to have invented a device with an input power of a few watts and an output power of many times the input power. He says it is not to be considered a "perpetual motion machine," because it converts mass into energy. The machine gives off no radiation and leaves no radioactive by-products. What tests would you suggest to test his claim, and what questions would you ask based on your knowledge of electricity, relativity, and nuclear physics?

28.7 Can you suggest any reasons why natural radioactivity contains helium nuclei but neither hydrogen nor lithium nuclei? (*Hint:* One key can be found in the binding energy of the decay products.)

28.8 What could be some of the reasons why stable nuclei have about the same number of protons and neutrons, with an excess number of neutrons as A increases?

28.9 A Geiger counter, or any other detection device, registers a finite, though perhaps small, count rate even when no clearly radioactive source is present. Name as many possible sources of this "background" radiation as you can.

28.10 What are some of the advantages and disadvantages of fusion power generation?

28.11 Over which of the radiation sources that affect you do you have any control? By completely eliminating all nonnatural sources, by what fraction could the average U.S. citizen reduce the received dose of radiation?

PROBLEMS

Hints for Solving Problems

Remember that the atomic number Z is the number of protons in the nucleus. The number of neutrons is $A - Z$, where A is the atomic mass number. The Q value of a reaction is the difference between the initial mass and the final mass. It may be expressed as an energy. A positive Q value means that a reaction can take place spontaneously. If the Q value is negative, then energy must be supplied from outside to get the reaction to take place.

Section 28.1   Chadwick's Discovery of the Neutron

28.1 If Chadwick had used paraffin and carbon targets instead of paraffin and a nitrogen-containing substance in his neutron-scattering experiments, what value would he have obtained for the ratio of the final velocities for the two cases? Assume the mass of the neutron and proton to be the same.

28.2 Using the fact that Chadwick's unknown particle is now known to be a neutron of mass 1, find the expected ratio of the speeds of the outgoing protons from the bombardment of paraffin and of the recoiling nitrogen ions in Chadwick's experiment.

28.3 Imagine an experiment similar to Chadwick's in which magnesium, with a relative mass of 24, is used instead of the nitrogen target. What would be the expected ratio of the recoil velocities of the protons relative to the magnesium ions?

Section 28.2   Composition of the Nucleus

28.4 How many protons and how many neutrons are needed to make the following elements: $^{126}_{53}I$, $^{56}_{26}Fe$, and $^{207}_{82}Pb$?

28.5 How many protons and how many neutrons are in the nucleus of each atom: $^{200}_{80}Hg$, $^{16}_{8}O$, and $^{232}_{90}Th$?

28.6 Nuclei with the same number of nucleons but with the number of protons and neutrons interchanged are called mirror nuclei and are useful for studying nuclear forces. How many protons and how many neutrons would be needed to make the following mirror nuclei: $^{39}_{19}K$, $^{39}_{20}Ca$, and $^{23}_{11}Na$, $^{23}_{12}Mg$?

28.7 Identify the element in which a given nucleus has an atomic mass number of 29 and one more neutron than protons.

28.8 A nucleus of neon contains ten neutrons. What is the atomic mass number, A, of this nucleus?

28.9 Natural chlorine occurs as a mixture of two isotopes. The isotope $^{35}_{17}Cl$ has a relative abundance of 75.5% and the isotope $^{37}_{17}Cl$ has a relative abundance of 24.5%. Calculate the atomic mass of a natural sample of chlorine.

28.10 Natural magnesium occurs as a mixture of three isotopes: $^{24}_{12}Mg$ at 78.7%, $^{25}_{12}Mg$ at 10.1%, and $^{26}_{12}Mg$ at 11.2%. Calculate the atomic mass of a natural sample of magnesium.

28.11 Natural copper occurs as a mixture of two isotopes. The isotope $^{63}_{29}Cu$ has a relative abundance of 69%

while the isotope $^{65}_{29}$Cu has a relative abundance of 31%. What is the approximate atomic mass of natural copper?

Section 28.3 Nuclear Size, Forces, and Binding Energy

28.12 What is the approximate radius of the nucleus of $^{235}_{92}$U?

28.13 What is the approximate volume of the nucleus of $^{138}_{56}$Ba?

28.14 Compare the approximate nuclear radius of $^{232}_{90}$Th with that of $^{57}_{26}$Fe.

28.15 What is the approximate mass number of a nucleus whose radius is measured to be 6.0×10^{-15} m?

28.16 Use the mass of the proton and Eq. (28.3) to estimate the density of nuclear matter.

28.17 What is the ratio of the strength of the gravitational force to the strength of the Coulomb force between two protons separated by a distance of the order of nuclear dimensions (10^{-14} m)?

28.18 Show that 1 u = 931.5 MeV. (*Hint:* 1 u = 1.66054 $\times 10^{-27}$ kg, 1.6022×10^{-19} J = 1 eV.)

28.19 Using Fig. 28.4, estimate the approximate total binding energy of $^{20}_{10}$Ne and $^{200}_{80}$Hg.

28.20 Using Fig. 28.4, estimate the approximate total binding energy of $^{58}_{28}$Ni and $^{232}_{90}$Th.

28.21 Calculate the binding energy of $^5_2$He, using the data in Table 28.2.

28.22 Calculate the binding energy of $^9_4$Be, using the data in Table 28.2.

28.23 Using the data in Table 28.2 and Section 28.3, calculate the ratio of binding energy per nucleon of $^4_2$He to that of $^6_3$Li.

Section 28.4 Radioactive Decay Versus Nuclear Stability

28.24 Identify the daughter nucleus in the decay of $^{234}_{92}$U by alpha particle emission.

28.25 Identify the nucleus designated by X in each of the following reactions: (a) $^{226}_{88}$Ra $\rightarrow$ X + α, (b) $^{233}_{91}$Pa $\rightarrow$ X + β^-, (c) $^{59}_{26}$Fe $\rightarrow$ X + γ.

28.26 (a) What nucleus results from the negative beta decay of $^{66}_{29}$Cu? (b) What nucleus results from the positive beta decay of $^{38}_{19}$K?

28.27 When $^6_3$Li is bombarded with neutrons, tritium ($^3_1$H) is emitted. What is the remaining nucleus?

28.28 What is the name of the element that results from the alpha decay of polonium?

28.29 What particle is emitted in the decay of $^{14}_6$C to $^{14}_7$N?

28.30 What nucleus decays by negative beta emission to $^{90}_{39}$Y?

28.31 May the following reaction take place naturally? $^9_4$Be $\rightarrow$ α + $^5_2$He.

28.32 (a) Determine whether $^{226}_{88}$Ra can decay by the emission of an alpha particle, and if so, find the kinetic energy released in the process. (b) What is the kinetic energy of the alpha particle? The observed masses are $m(^{226}_{88}$Ra$) = 226.02540$ u and $m(^{222}_{86}$Rn$) = 222.01757$ u.

28.33 (a) Determine the kinetic energy available to products of alpha decay of $^{235}_{92}$U. (b) Also find the kinetic energy of the alpha particle. The observed masses are $m(^{235}_{92}$U$) = 235.04392$ u and $m(^{231}_{90}$Th$) = 231.03630$ u.

Section 28.5 Natural Radioactive Decay Series

28.34 The isotope $^{234}_{92}$U undergoes five successive alpha decays. Identify the daughter nucleus at each decay step.

28.35 Which uranium isotope belongs to the neptunium decay series? To the actinium decay series?

28.36 (a) In which decay series do you find the isotope $^{234}_{91}$Pa? (b) In which do you find $^{231}_{91}$Pa?

Section 28.6 Models for Alpha, Beta, and Gamma Decay

28.37 Show that an alpha particle with a kinetic energy of 5 MeV need not be treated as a relativistic particle.

28.38 Calculate the Q value of the decay $^{218}_{84}$Po $\rightarrow$ $^{214}_{82}$Pb + α, given that the measured kinetic energy of the alpha is 5.998 MeV.

28.39 Bismuth-212 decays via alpha emission to $^{208}_{81}$Tl with an energy release of $Q = 6.170$ MeV. Find the approximate kinetic energy of the alpha and its speed.

28.40 Calculate the Q value for the beta decay of tritium ($^3_1$H).

28.41 The atomic mass m_a of a given isotope ^A_ZX can be considered to be the sum of the nuclear mass m_N and the mass of Z electrons, that is,

$$m_a \,(^A_Z\text{X}) = m_N(^A_Z\text{X}) + Zm_e.$$

Show that nucleus ^A_ZX can decay via emission of a negative beta particle if

$$m_a(^A_Z\text{X}) > m_a(_{Z+1}^{\ A}\text{Y}).$$

*Section 28.7 Detectors of Radiation

28.42 A small radioactive source is monitored by a detector 10 cm away. The sensitive part of the detector is a circle of 1.50-cm radius and faces directly at the source. If the average count rate of the detector is 4320 counts per min, what is the approximate activity of the sample in becquerels?

*Section 28.8   Radiation Measurement and Biological Effects

28.43 A patient undergoing therapy is to receive an equivalent dose of 5.0 Sv. What dose (in grays) is this of (a) x rays, (b) fast neutrons?

28.44 A 90-keV x-ray equivalent dose of 1.0 Sv is absorbed by 2.0 kg of tissue in a person's leg. How many photons are absorbed?

Section 28.9   Induced Transmutation and Reactions

28.45 Determine the identity of the nucleus X in the reaction $^{13}_6C + n \rightarrow X + \gamma$.

28.46 Identify the particle Y in the reaction $^2_1H + Y \rightarrow ^4_2He + n$.

28.47 What element is formed in the reaction $^{10}_5B + ^4_2He \rightarrow X$?

28.48 Calculate the threshold energy of the proton for the reaction $^1_1H + ^{14}_6C \rightarrow ^{14}_7N + ^1_0n$. (*Hint:* $m(^{14}_6C) = 14.003241$ u, $m(^{14}_7N) = 14.003074$ u.)

Section 28.10   Nuclear Fission

28.49 How much mass is lost in the fission of the nuclear fuel in a 1000-MW power plant in one year?

28.50 What is the kinetic energy in MeV of a room-temperature thermal neutron? By thermal neutron we mean one with kinetic energy kT. (*Hint:* Take room temperature to be 300 K.)

Section 28.11   Nuclear Fusion

28.51 Calculate the energy released in the fusion reaction $^2_1H + ^2_1H \rightarrow ^1_0n + ^3_2He$.

28.52 Calculate the energy released in the reaction $^3_2He + ^3_2He \rightarrow ^4_2He + ^1_1H + ^1_1H$.

Additional Problems

28.53 Derive Eq. (28.1) from the principles of conservation of energy and momentum. Assume head-on collisions.
•

28.54 A nucleus has 126 neutrons and a volume of approximately 1.51×10^{-42} m³. Assume that Eq. (28.3) and the value given for R_0 are exactly correct and identify the element.
•

28.55 Determine the velocity of the alpha particle in Example 28.5.
•

28.56 Using conservation of energy and momentum, show that the kinetic energy of an alpha particle in a decay
•

with a given Q value is

$$KE_\alpha = \frac{Q}{1 + m_\alpha/m_d},$$

where m_α is the mass of the alpha particle and m_d is the mass of the daughter nucleus.

28.57 Using conservation of energy and momentum, derive an expression for the kinetic energy of recoil of the daughter nucleus of mass number A resulting from an alpha decay with a given Q value.
•

28.58 A beta particle emitted from a radioactive nucleus moves in a circular path perpendicular to a uniform magnetic field of 0.37 T. The kinetic energy of the particle is 0.66 MeV. What is the radius of the circle? (*Hint:* First decide if the speed is relativistic.)
• •

28.59 Show that nucleus ^A_ZX can decay by emission of a positron only if
•

$$m_a(^A_ZX) > m_a(_{Z-1}^A Y) + 2m_e,$$

where m_a is the atomic mass and m_e is the mass of the positron.

28.60 Radioactive materials used for pharmaceutical purposes are usually given in amounts of a few millicuries. What would be the mass of radioactive iodine (^{131}I) having activity of 20 mCi? (*Hint:* The half-life of ^{131}I is 8 days and its atomic mass is 131. You may want to refer to Section 25.8.)
•

28.61 According to a newspaper account a nuclear power plant had an accidental release of 33,000 curies of tritium. How many grams of tritium were released? (*Hint:* The half-life of tritium is 12.3 years. You may want to refer to Section 25.8.)
•

28.62 The elimination of radioactive material from the human body follows an exponential law so that the amount of material remaining in the body decreases with a characteristic half-life t_{body}. The radioactive material has a natural half-life t_{rad} of its own. Show that the effective biological half-life of the radioactivity in the body is given by
• •

$$\frac{1}{t_{biol}} = \frac{1}{t_{body}} + \frac{1}{t_{rad}}.$$

28.63 An alpha particle collides head-on with a stationary proton in an elastic collision. Show that for an initial alpha speed v_α, the struck proton will move off with a velocity $1.6\ v_\alpha$.
•

28.64 (a) Estimate the kinetic energy needed to bring two ^3_2He nuclei close enough for a reaction to occur ($\approx 10^{-14}$ m). (b) What is the equivalent temperature?
•

29

Lasers, Holography, and Color

29.1   Stimulated Emission
        of Light

29.2   Lasers

29.3   The Helium-Neon Laser

29.4   Properties of Laser Light

29.5   Holography

29.6   Light and Color

29.7   Color by Addition and
        Subtraction

A WORD TO THE STUDENT

We have now discussed enough atomic physics to understand some of the exciting contemporary developments in physics. In this chapter we examine modern optics. We begin by describing how a laser works. Then we discuss one of the most fascinating applications of lasers: holography. Holography is a type of three-dimensional photography that can be understood only on the basis of the wave nature of light. The concluding topic, color, has assumed great importance in our modern era of full-color communication. You will find that all of this material relies on your understanding of classical optics, atomic physics, and quantum physics.

Throughout much of our discussion of optics in Chapters 21–23 we assumed monochromatic light emitted coherently (in phase). In reality, this is a very rare kind of light. Most light, whether from the sun, from fluorescent lamps, or from incandescent bulbs, comes from large numbers of atoms randomly radiating photons. However, with the invention of the laser in 1960, a practical source of coherent light became available, and physicists took a renewed interest in optics. The coherence of laser light made possible many new experiments. For example, laser light can be focused into beams of extremely high intensity, a feature that has led to basic discoveries concerning the nature of the interaction between light and matter. It has also led to practical applications such as drills for microscopic holes in the hardest substances and surgical instruments for delicate and precise operations.

One of the better-known applications of lasers is the production of three-dimensional images, called holograms. Holography is a rapidly growing field, with scientific and commercial applications. In this chapter we present the basic ideas of holography, in which the modern tool of laser light is combined with the classical principles of diffraction.

The development of the laser would not have been possible without an understanding of quantum mechanics and atomic physics. Similarly, our understanding of other optical phenomena, such as color and vision, has also advanced through our knowledge of quantum physics. The importance of color and how we see it has grown along with the explosive growth of information in the late twentieth century. The use of color enhances our ability to assimilate information rapidly, as false-color images have demonstrated. Color has become an important tool for presenting information in all areas of life.

Stimulated Emission of Light

29.1

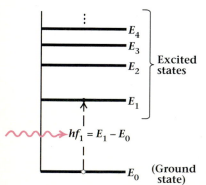

We have seen in earlier chapters that the electronic energy levels of an atom determine the frequencies of the light emitted or absorbed by the atom. We have also seen that the likelihood of an atom existing in a particular energy state decreases with the energy of that state. It can be shown that the probability that an atom exists in an energy state E is proportional to $e^{-E/kT}$, called the Boltzmann factor, where E is the energy, k is Boltzmann's constant, and T is the temperature. Thus, for example, a cool (room temperature) gas exists predominantly in the lowest or ground state. When the energy difference between the first excited state E_1 and the ground state E_0 is large compared with kT, almost all of the gas atoms will be in the state E_0 (Fig. 29.1). Illuminating such a gas with light of energy $hf_1 = E_1 - E_0$ resonantly excites the atoms from the ground state E_0 to the excited state E_1, with the consequent absorption of

Figure 29.1   Energy levels of an atom. When light of frequency f_1 is resonantly absorbed, the atom goes from ground state E_0 to the excited state E_1.

(a)

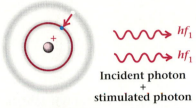

(b)

Figure 29.2 **Stimulated emission of light. (a) A photon of resonant energy hf_1 incident upon an excited atom. (b) The stimulated atom emits a photon of energy hf_1 in phase with the continuing photon.**

the light. For each atom excited from E_0 to E_1, a photon of energy hf_1 is absorbed.

Atoms in excited states are no longer in thermodynamic equilibrium with their surroundings. They are unstable and after a short time return to the ground state, emitting a photon during this process. The average lifetime of an excited atomic state is designated by τ. The emission of photons that occurs this way—through the random de-excitation of excited atoms—is known as **spontaneous emission**.

In 1917 Albert Einstein showed that incident light of the proper frequency could trigger the emission of light by excited atoms. This process, called **stimulated emission**, takes place when the frequency of the incident photon equals the resonant frequency of the transition between energy levels. Thus, when light of frequency f_1 is incident on an atom with an electron in energy level E_1 (Fig. 29.2), a photon of energy hf_1 can stimulate a transition to the ground state, with the simultaneous emission of another photon, also of frequency f_1. In such a case, instead of being absorbed the incident photon continues its travel, accompanied now by a second photon of the same frequency and phase.

In 1958, some forty years after Einstein's theoretical discovery of stimulated emission, Charles H. Townes and Arthur L. Schawlow published a paper which discussed how this principle could be used in a practical light source.* Two years later, T. H. Maiman succeeded in operating the first light source of this type. This device soon became known as a laser, an acronym of Light Amplification by Stimulated Emission of Radiation.

Lasers

29.2

To use stimulated emission in a practical laser, we must satisfy several basic conditions. First, we need an **active medium**, that is, a material containing atoms or molecules that can be made to emit light. Secondly, we need a method for adding energy to this medium in order to promote a sufficient number of its atoms to an excited state. The process of exciting the laser medium is called **pumping**. Finally, we need a way of confining the light so that we can trigger many stimulated emissions before the light escapes from the medium, thus building up an intense output beam. We achieve this result by placing the active medium in an **optical resonator**, which consists of mirrors that reflect the light back and forth. Light emitted along the axis is reflected, causing many additional stimulated emissions, while light directed off the axis is quickly lost from the system.

As an example of a laser system, let us consider a ruby laser similar to the first laser built by Maiman (Fig. 29.3). The active medium of this laser is a small ruby rod. Ruby is a crystalline form of aluminum oxide containing small amounts of chromium, which give it a characteristic pink color that is due to the absorption of green and blue light. The absorption bands arise from closely spaced energy levels, called energy bands, which

---

*Townes had already demonstrated the stimulated emission of radio frequency radiation (microwaves) in 1954. Two Russian physicists, Nikolai Basov and Alexander Prokhorov, independently described the laser theoretically.

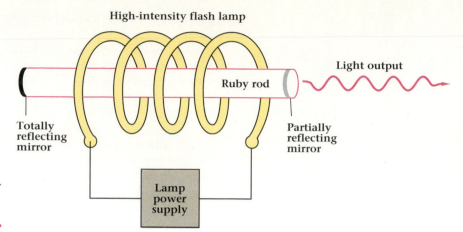

Figure 29.3 Schematic representation of a ruby laser, containing a ruby rod with mirrored ends and a flash lamp.

are shown in the simplified energy-level diagram for chromium ions in ruby (Fig. 29.4). The presence of the broad energy bands permits excitation of the chromium ions by exposure to bright white light from a flash lamp. Pumping of the active medium by the absorption of light is known as *optical pumping*.

Both of the absorption bands in ruby de-excite by nonradiative transitions, in which the excitation energy is given up to the crystal instead of being carried off by photons. These transitions lead to the two closely spaced states labeled E_1. These energy states are also unstable and give up their energy by emitting light at wavelengths of 694.3 and 692.7 nm. The laser transition corresponds to the 694.3-nm line, which is dominant in the presence of stimulated emission.

Normally, for atoms in thermodynamic equilibrium, upper energy levels are less densely populated than lower ones. In this condition, absorption of photons dominates stimulated emission. That is, because there are more ground state ions available to be excited than there are excited ions to be stimulated to emission, there is more absorption than emission. In contrast, when more ions exist in an upper state than in a lower state, we have an **inverted population**. Laser action cannot take place unless we can generate an inverted population of atoms in the active medium.

In a ruby laser, the flash lamp gives out sudden bursts of light, pumping the chromium ions into their excited states. The excited states quickly de-excite, populating the state E_1. Although E_1 is also unstable, its lifetime is long compared with those of the absorption bands. Thus the pumping leads to an inverted population in state E_1. (Such a state is called a **metastable state**.) When more ions are in the excited state E_1 than in the ground state E_0, the probability that a single photon will cause stimulated emission exceeds the probability that the photon will be absorbed. As a result, the system emits a brief flash of light—laser light. The ruby laser gives out a pulse of laser light each time the flash lamp is activated. Such a laser is called, for obvious reasons, a *pulsed laser*.

The ends of the ruby crystal are polished and silvered to form a pair of parallel mirrors—the optical resonator. One mirror is totally reflecting, while the other mirror is partially reflecting and partially transmitting. The mirrors reflect the light back and forth through the crystal many times

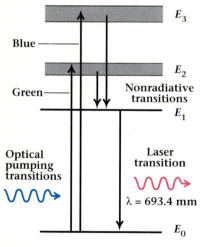

Figure 29.4 Energy-level diagram for chromium ions in ruby. Absorption bands E_2 and E_3 are excited by optical pumping from the ground state E_0. Fast nonradiative transitions to E_1 create the inverted population needed for laser transition from E_1 to E_0.

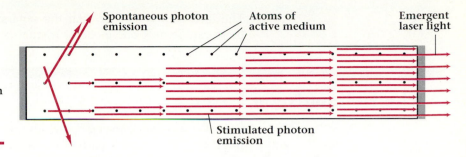

Spontaneous photon emission

Atoms of active medium

Emergent laser light

Stimulated photon emission

Figure 29.5 As light travels through the active medium the intensity increases because of stimulated emission.

before it escapes through the partially transmitting mirror. This process of extending the path of the light through the active medium increases the probability of obtaining stimulated emissions caused by the photons that are traveling parallel to the rod's axis. Photons emitted in other directions, the products of spontaneous emission, quickly leave the rod and play no role in the laser action. The result is an avalanche effect, as each emitted photon causes the stimulated emission of other photons, all traveling parallel to the laser axis (Fig. 29.5). The beam that finally emerges along the axis has a relatively small angular divergence.

The reflecting mirrors not only maintain a high light intensity within the medium, but also serve as the optical resonator that tunes the laser to a very sharp frequency. The narrow band of frequencies associated with the natural width of energy level E_1 is further restricted to the resonant frequencies of the resonant cavity (Fig. 29.6). Thus light emitted by the laser is much more sharply defined than we would otherwise expect. In this way the laser becomes a source of essentially monochromatic light.

We can determine the separation in frequency of the resonator modes in the same way we found resonant frequencies in Chapter 14. For constructive interference within the resonator, the path length between mirrors must be an integral number of half-wavelengths of the light. So, for mirrors separated by a length L, the resonant frequencies are

$$f = \frac{nc}{2L}, \tag{29.1}$$

where n is an integer and c is the speed of light. For a frequency f' corresponding to $n' = n + 1$ we see that

$$\Delta f = f' - f = \frac{c}{2L}. \tag{29.2}$$

The spacing of the resonator modes is illustrated in Fig. 29.6.

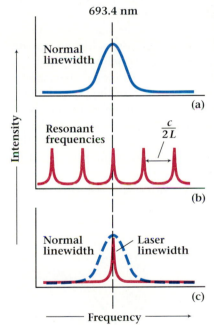

693.4 nm

Normal linewidth

(a)

Resonant frequencies

$\frac{c}{2L}$

(b)

Normal linewidth · Laser linewidth

(c)

Frequency

Intensity

Figure 29.6 Effect of the laser cavity on the frequency. (a) The natural spread of frequencies in the emission line. (b) The resonant frequencies of the optical resonator created by the mirrors. These frequencies are separated by an amount $c/2L$, where c is the speed of light and L is the separation between the mirrors. (c) The product of the linewidth and the cavity resonance determines the laser output, here shown with only one resonant frequency.

Example 29.1

A laser has mirrors 10 cm apart. If the natural width of the emission line for the laser transition is $\Delta f_0 = 10^8$ Hz, can the laser have more than one frequency?

Solution The separation between the modes is found from Eq. (29.2),

$$\Delta f = \frac{c}{2L} = \frac{3.0 \times 10^8 \text{ m/s}}{2 \times 0.10 \text{ m}} = 1.5 \times 10^9 \text{ Hz}.$$

The linewidth is smaller than the separation of the cavity modes. Thus there can be only one mode or frequency for this laser.

The Helium-Neon Laser

29.3

Pulsed lasers can deliver a large amount of energy in a very short time interval. Such a laser is particularly appropriate for applications requiring high power in the laser beam, such as drilling or welding. On the other hand, other applications may require a laser that operates continuously. Such a laser is known as a *continuous wave* or *cw laser*.

One of the best-known cw lasers is the helium-neon laser, which was first proposed in 1959 by Ali Javan of Bell Telephone Laboratories. This laser uses a low-pressure mixture of 90% helium and 10% neon gas for the active medium. An electric discharge causes the gas to glow in essentially the same process that occurs in neon signs (Fig. 29.7). However, by using a mixture of gases, we can achieve an inverted population in the neon atoms, thus leading to the possibility of laser action.

Figure 29.8 shows the energy-level diagram for the helium-neon system. The helium atoms are easily excited by electron collisions in the electric discharge. Two of the lowest excited states of helium, labeled 2^1S and 2^3S, are sufficiently long lived that they are considered metastable. These two states are at almost the same energy as two of the excited states of the neon atoms, the $4S$ and $5S$ states. Because the energies match so closely, a helium atom can readily transfer its excitation energy to a neon atom during a collision between atoms in the gas. After the collision, the helium atom returns to its ground state while the neon atom is raised to one of its excited states. This process of pumping by collision creates an inverted population distribution in the neon. Atoms in these two excited

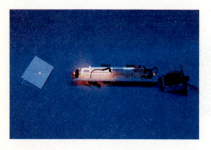

Figure 29.7   A helium-neon laser in operation.

Figure 29.8   Energy-level diagram of the helium-neon laser. The laser transitions are labeled with their wavelengths.

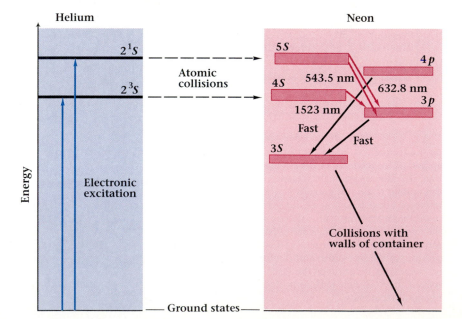

levels can then radiate energy by stimulated emission to the states labeled 4*p* and 3*p* in Fig. 29.8, which quickly de-excite to still lower states, maintaining an inverted population in the excited upper levels. Thus the laser action is continuous.

In order to select one laser wavelength over others, it is common to use special mirrors that reflect only near the wavelength of the desired transition. By using mirrors that reflect at 632.8 nm but transmit in the infrared, manufacturers construct helium-neon lasers that emit red light at 632.8 nm. Low-power helium-neon lasers of this type are relatively inexpensive. For this reason they are used by many high school and college students. Other mirrors permit the construction of helium-neon lasers operating in the green at 543.5 nm and in the infrared at 1523 nm.

Ruby lasers and helium-neon lasers are only two of a growing number of types of lasers. Other gases such as argon and krypton are also used as laser media, as are metal vapors and molecular gases such as carbon dioxide. Other materials are used in the same manner as ruby in optically pumped lasers. Still other lasers are made from semiconductor diode junctions. Dye lasers, containing fluorescent dyes as the active medium, can be made to emit a continuously variable wavelength. Thus we can obtain laser light at almost any wavelength desired.

Properties of Laser Light

29.4

When light from a laser passes through a prism or diffraction grating, no noticeable dispersion occurs—an indication that the laser light is monochromatic. In contrast, sunlight is dispersed by a grating into a continuous spectrum of colors, indicating that the sunlight actually consists of a mixture of many different wavelengths. Related effects due to interference of white light often go unnoticed, since the minima and maxima for each separate wavelength occur at different positions and the pattern is washed out. However, monochromatic laser light produces interference patterns that are quite pronounced.

When laser light reflects off a surface, its unusual character immediately becomes visible. The reflected light seems to sparkle with bright regions separated by dark areas. This so-called *speckle pattern* is a characteristic of laser light that is due to coherence. What you see is actually an interference pattern in the reflected light. As you move about, the speckle pattern also moves. Whether it moves in a direction that is the same as, or opposite to, your motion depends on whether you are near-sighted or far-sighted. A near-sighted observer sees the pattern move in the opposite direction to his motion; a far-sighted observer sees the pattern move in the same direction. You may wish to try this for yourself.

In our discussion of interference in Chapter 23, we noted that Thomas Young's double-slit experiment was crucial in establishing wave behavior of light. In Young's experiment, white light from the sun fell on two narrow, closely spaced slits. The light passing through the slits formed an interference pattern of colored stripes on a screen. We can limit the white light to predominantly one color by first passing it through a prism, which spreads the beam into a spectrum, and then directing the spectrum to a

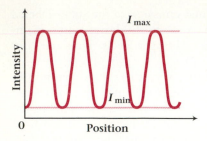

Figure 29.9 Intensity profile of an interference pattern. The visibility is defined in terms of the maximum and minimum intensities, I_{max} and I_{min}.

narrow slit. We can use the relatively narrow band of wavelengths that pass the slit to generate a double-slit interference pattern. This pattern is more distinct than that made by white light and appears as a set of alternating bright and dark bands.

Now, suppose we illuminate the slits with light from a laser. Since only one wavelength is present, the pattern becomes even more pronounced. If the illumination falling on the two slits is uniform, then we expect that where the waves interfere destructively the effect should be total; that is, we should observe no light intensity. Then we can take the visibility of the interference lines, or fringes, as a quantitative measure of how well the light interferes with itself. That is, the fringe visibility measures the *coherence* of the light.

We define the fringe visibility V to be the ratio

$$V \equiv \frac{I_{max} - I_{min}}{I_{max} + I_{min}}, \qquad (29.3)$$

where I_{max} is the maximum intensity in the pattern and I_{min} is the minimum intensity (Fig. 29.9). If $I_{max} = I_{min}$, no fringes are seen, the visibility is zero, and the light is completely incoherent. If $I_{min} = 0$, the visibility is 1 and the light is completely coherent. Finally, partial coherence results for V between 0 and 1. If the illumination of the slits is equal, the degree of coherence is identical with the fringe visibility. If the illumination is unequal, then the interference pattern has reduced visibility even for perfectly coherent light. In that case the coherence and the fringe visibility are still proportional. The coherence γ is then defined as

$$\gamma \equiv \frac{I_1 + I_2}{2\sqrt{I_1 I_2}} V, \qquad (29.4)$$

where I_1 and I_2 are the intensities due to each slit independently.

Example 29.2

A double slit, evenly illuminated by a light source, produces interference fringes that are barely visible. Measurement of the intensities reveals that $I_{max} = 2I_{min}$. What is the degree of coherence of the light source?

Solution   The fringe visibility is

$$V = \frac{I_{max} - I_{min}}{I_{max} + I_{min}} = \frac{2I_{min} - I_{min}}{2I_{min} + I_{min}} = \frac{1}{3} = 0.33.$$

Since the slits are evenly illuminated, the degree of coherence and the visibility are identical: $\gamma = 0.33$.

As light is made more nearly monochromatic, its coherence increases. Thus two light waves of the same wavelength traveling in the same direction maintain a constant phase between them (Fig. 29.10a). On the other hand, two waves of different wavelengths change their relative phase as

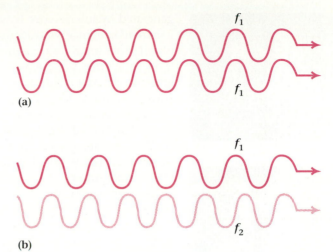

Figure 29.10 (a) Two waves of the same wavelength retain a constant relative phase as they travel through space. (b) Two waves of different wavelength have a changing relative phase as they travel through space.

they move through space (Fig. 29.10b). As these waves pass a fixed point in space, they go in and out of phase with time. As a result, their time-averaged effect is one of incoherence. A high degree of coherence can occur only if the light is very nearly monochromatic. This coherence is referred to as **temporal coherence**. But even waves of exactly the same wavelength will not be coherent unless their phases are constant in time.

Besides temporal coherence, there is a degree of coherence arising from the spatial extent of the light source, which is called **spatial coherence**. We can associate spatial coherence with the coherence of light from different positions on the wavefront. This complication was taken into account by Young, who passed light through a single pinhole before allowing it to fall on the double slit. In consequence, the light falling on the double slit came from a very small region of the initial wavefront. Light from farther along the wavefront was not sufficiently coherent to produce the interference pattern. Laser light has a high degree of both spatial and temporal coherence, which accounts for its remarkable diffraction and interference effects.

Holography

29.5

Most common optical devices, such as cameras, projectors, and binoculars, use lenses to refract light and form images. As we showed in Chapter 22, we can explain the formation of images by applying the techniques of ray tracing to determine their position and size. However, we have firmly established that light is a wave phenomenon. In fact, only by treating light as a wave can we explain many observed effects. Therefore we should expect the wave nature of light to be important in understanding image formation. Indeed, it is very important, as our discussion of resolution in Section 23.7 made clear. Further investigation of image formation by waves led to one of the exciting areas of modern optics: holography.

In 1947 the British physicist Dennis Gabor began experiments to overcome problems of image formation that were due to aberrations. Gabor recognized that for an object illuminated with coherent light, the pattern

(a)

(b)

(c)

(d)

Figure 29.11   (a,b) Two photographs of the same hologram taken at slightly different vertical positions. Note the effect of parallax in the relative positions of the lens and ruler. (c,d) Two photographs of the same hologram taken from the same position but with the camera focused at different distances. (c) The camera was focused on the standing figure. (d) The camera was focused on the rim of the magnifying glass.

generated by interference between the wave scattered off the object and a coherent reference wave contains all of the visual information about the object. This interference pattern can be recorded photographically. Then, if the reference wave alone is used to illuminate the developed photographic film, it is scattered by the film in such a way as to generate a new wave, essentially identical to the original wave reflected from the object. The result is the formation of an image without the use of a lens.

Gabor coined the word **hologram** for this photographic recording of the interference pattern produced by the combination of reference and object beams. This word is derived from the Greek words *holos*, meaning whole, and *gramme*, meaning what is written or drawn. The making and study of holograms is called **holography.** Gabor successfully demonstrated the effect in 1948, more than a decade before the invention of the laser. In the ten years immediately following his discovery, holography received little attention and its development was hampered because of the lack of a suitable source of coherent light. After the invention of the laser, Emmett N. Leith used coherent laser light to create images in the manner first demonstrated by Gabor. With Leith's introduction of the laser and use of a reference beam at an angle to the object beam, holography became a practical method of photographically recording optical information.

The images created from holograms contain more information than images produced with ordinary photographic techniques. Holographic images are three-dimensional and have both depth and parallax. As you move your head when viewing a holographic image it seems as if the original objects are present within the image. Figure 29.11 shows four images reconstructed from the same hologram. Parts (a) and (b) show the effects of parallax when viewing the holographic image from different angles. Parts (c) and (d) illustrate the depth in the image, revealed when the camera is focused on different portions of the image. Both these effects are missing in ordinary photographs. Even matched stereo pairs of photographs do not contain the visual information of a hologram.

The production of a hologram requires two coherent beams of light (Fig. 29.12). These two beams are generated from a single laser beam by a half-silvered mirror or other beam-splitting device. One beam, called the reference beam, reflects directly on the photographic film. Mirrors direct the other beam to the object whose image is to be recorded. Light scattered from the object to the photographic film is called the object beam. No lenses are used to image the object onto the film; instead, the object and reference beam both illuminate the film. Since these beams are coherent, they interfere and the photographic film records that interference pattern.

You can see a holographic image of the object by illuminating the developed film with a laser beam similar to the reference beam. In general, two images are formed (Fig. 29.13). If the reference beam used in making the hologram was a beam of parallel light and the reconstructing beam is also parallel, then one of these two images is a virtual image. You can observe the virtual image by looking through the hologram as indicated in the figure. The three-dimensional image appears behind the film. The second image is real and can be formed on a screen. If you are farther away from the film than the image is, your eyes can focus the rays diverging from the real image.

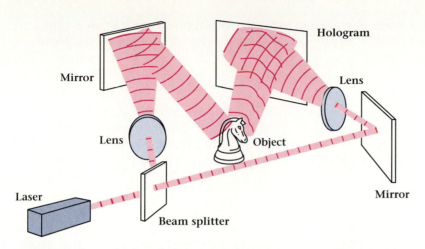

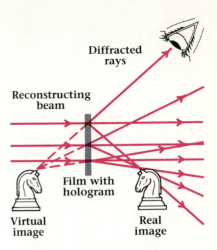

Figure 29.12   Producing a hologram. Light from the laser is split into two beams. The reference beam, here shown to be parallel, is reflected directly at the photographic film. The other beam is scattered off the object onto the film.

Figure 29.13   Reconstructing a holographic image. The virtual image may be seen with the unaided eye.

Notice that in producing the hologram, light from each point on the object is scattered over the entire photographic film. This behavior is in sharp contrast to the one-to-one correspondence between points on the object and points on the image in ordinary photography. (See, for example, Fig. 21.17.) Thus each tiny area of the hologram contains information about the entire object. Even if a large part of the hologram is destroyed, we can still reconstruct the entire image from the part that remains. There is no way to reconstruct part of an ordinary photograph that has been destroyed.

Whether the holographic image is real or virtual depends on both the initial reference beam and the reconstructing beam. These beams need not be parallel, as shown in Fig. 29.13, but can be diverging or converging. Therefore it is possible to make a hologram and reconstruct it so that either of these two images is real or virtual.

To analyze holography, we must think in terms of diffraction and interference. For simplicity, let's consider the interference of two beams of parallel coherent light striking a photographic film (Fig. 29.14). The

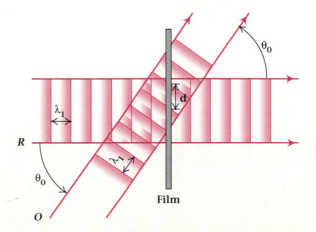

Figure 29.14   Interference pattern formed in the plane of the film by two coherent light beams R and O. The spacing d is given by $d = \lambda_1 / \sin \theta_0$.

object beam O makes an angle θ_0 with the reference beam R. The two beams combine to produce an interference pattern, with spacing d between maxima given by

$$d = \lambda_1/\sin \theta_0, \qquad (29.5)$$

where λ_1 is the wavelength of the light. A stationary interference pattern occurs only if the two beams are coherent. Furthermore, the relative positions of the light sources, the film, and any other components (such as beam splitters and mirrors) must be stationary and free from vibrations; otherwise the interference pattern in the plane of the film is averaged or

PHYSICS IN PRACTICE

White-Light Holograms

So far we have been describing transmission holograms, which produce sharp images only when illuminated with light that is nearly monochromatic. If the light contains a band of wavelengths, as in white light, each wavelength creates its own holographic image. These individual images overlap, resulting in a blurred effect. Also we need the light to come from a point source. If a large or diffuse source is used, each source point produces its own image, again resulting in a smeared image, even if the light is monochromatic. These requirements are equivalent to demanding that the light be both spatially and temporally coherent.

One type of hologram that can be viewed in white light is a *reflection hologram*. In a reflection hologram the

Figure B29.1   Making a white-light reflection hologram. (a) The reference beam of light passes through the photographic plate to also act as illumination for the object. (b) When illuminated with white light similar to the original reference beam, the developed hologram scatters light backward as if it came from the object. For each angle of incidence, the viewer sees only a narrow range of wavelengths, resulting in a sharp image.

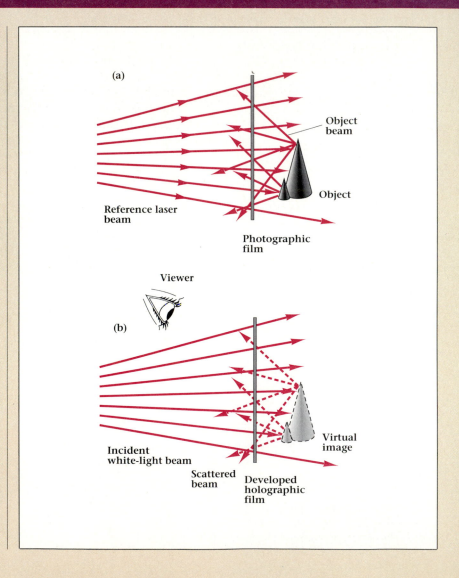

washed out. Vibrations with amplitudes as small as $\lambda/4$ are sufficient to prevent recording of the interference pattern.

When the film is developed, it becomes a linear diffraction grating. However, instead of displaying alternating opaque and transparent lines, the film varies in darkness sinusoidally across the width of the film. Such a film is known as a sinusoidal diffraction grating. The spacing d between successive lines on the film is given by Eq. (29.5). When a beam of parallel light of wavelength λ_2 is perpendicularly incident on this linear grating, we expect first-order diffraction at the angle θ given by

$$\sin \theta = \lambda_2/d.$$

object and reference beams are brought together from opposite sides of the photographic film (Fig. B29.1a). The resulting interference pattern has structure perpendicular to the plane of the film. If the thickness of the emulsion is greater than about 15 μm, then the interference pattern recorded in the film is truly three-dimensional, with twenty or more layers within the emulsion. The structure of the resulting hologram is analogous to a thin crystal twenty or thirty atomic layers thick. When the developed film is illuminated by a white light beam in the same direction as the original reference beam, the hologram diffracts some of the light backward (Fig. B29.1b). The resulting wave is similar to the original object wave. For each particular angle of incidence, only a narrow range of wavelengths is reflected into the viewer's eye. This diffraction is analogous to the effect of Bragg scattering of x rays. The hologram automatically selects the proper wavelength for each angle of incidence. Thus we can view reflection holograms with white light, provided it comes from a point source.

Another type of white-light hologram is the *rainbow hologram*, invented by Steven Benton of the Polaroid Corporation in 1969. The rainbow hologram may be viewed with transmitted light from an incandescent lamp. The different wavelengths are dispersed so that each forms only a small portion of the total image, resulting in a rainbow-like change of color across the image. This effect is achieved in a two-step process. First we make a transmission hologram of the object in the usual way. Then we make a second hologram, using the real image from the first hologram as the object. A narrow horizontal slit is placed next to the first hologram during this process. This eliminates the vertical parallax in the second hologram, but the second hologram does record the image of the entire object.

When the second hologram is illuminated with a white light source each wavelength forms an image of the slit. Each of these images occurs at a different vertical angle, as expected from the laws of diffraction. When looking for the image of the object, an observer actually looks through one of the colored images of the slit. The original object appears with horizontal parallax. As the observer moves up and down, the image remains constant, that is, shows no vertical parallax—but it changes color depending on the vertical position.

In recent years, holograms have been made using a special photographic emulsion called photoresist. Instead of variations in opacity, as in normal photographic emulsion, the developed photoresist has variations in

Figure B29.2 Photograph of an embossed rainbow hologram.

thickness in response to the intensity of the light incident upon it. By using electrolysis, we can deposit a thin layer of nickel on the developed photoresist, making a three-dimensional mold. Duplicates of the interference pattern, made from the mold, can be stamped into plastic and coated with a thin layer of aluminum. The aluminum functions like a mirror, reflecting incident white light back through the interference layers to produce the holographic image. The result is called an *embossed hologram*. Because they can be reproduced inexpensively, embossed holograms are finding their way onto magazine covers, buttons, and novelties (Fig. B29.2). In 1984 credit card manufacturers began incorporating embossed holograms into their cards in an effort to thwart counterfeiting.

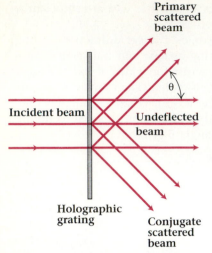

Figure 29.15 A light beam incident on a holographic grating, formed as in Fig. 29.14, is broken into three beams.

Upon inserting Eq. (29.5) into this equation, we find that

$$\sin \theta = \frac{\lambda_2}{\lambda_1} \sin \theta_0.$$

If the reconstructing wavelength λ_2 and the recording wavelength λ_1 are the same, then $\theta = \theta_0$ and one of the diffracted beams is directed in exactly the same direction as the original object beam. The other diffracted beam is scattered on the opposite side of the normal from the direction of the original object beam (Fig. 29.15). In addition, some of the light passes through undeflected, just as in any other grating. However, in contrast with the gratings described in Chapter 23, the sinusoidal grating does not produce any higher-order beams. That is, because of the sinusoidal variation of the intensity in the grating, only zero- and first-order beams are generated.

If the object beam of Fig. 29.14 had been a diverging beam from a point source, the pattern recorded on the film would have been more complex than that of the linear sinusoidal grating. However, it would still have some of the same properties. As before, only zero- and first-order diffraction would occur. One of the first-order beams would be an expanding wave, emerging in the same direction as the original object wave. This scattered wavefront would have the same shape as the original object wave. For this reason holography has also been called wavefront reconstruction. The other first-order beam is scattered on the other side of the zero-order beam, as shown in Fig. 29.14. But this wave has a converging wavefront that comes to a point, forming a real image.

We can consider a hologram of a complex object as a superposition of point source holograms from all possible source points on the object. Even in this case it is still true that only zero- and first-order diffracted beams are generated during holographic reconstruction. One of these first-order beams, called the primary beam, recreates the waveform as it was scattered from the object. This beam produces the virtual image indicated in Fig. 29.13. The other first-order diffracted beam is called the conjugate beam, which converges to form the real image in Fig. 29.13.

We can apply Eq. (29.5) to compute the resolution requirements for photographic film used in holography. The inverse of the line spacing d gives the number of lines per unit length—the **spatial frequency** of the interference pattern. If the film is to record this pattern, it must be capable of recording lines of at least this spatial frequency. For example, suppose that the angle between the incident beams is 45°. If the coherent light source is a helium-neon laser with $\lambda = 632.8$ nm, then the spatial frequency F of the interference pattern would be

$$F = \frac{1}{d} = \frac{\sin 45°}{632.8 \text{ nm}} = 1117/\text{mm}.$$

Thus in order to record the interference pattern and serve as a hologram, the film must be capable of recording more than 1117 lines/mm. Most ordinary photographic films have resolutions an order of magnitude less than this and consequently are unsuitable for holography. Fortunately, special holographic films with resolutions in excess of 2000 lines/mm are available.

Example 29.3

Angular field of view of a hologram.

A hologram is made with film whose resolution is 1000 lines/mm. Light from a helium-neon laser, $\lambda = 632.8$ nm, is used to record and view the hologram. The reference beam makes normal incidence with the hologram. What is the maximum angular field of view of this hologram?

Solution The spatial frequency due to light incident at an angle θ from the normal is

$$F = \frac{\sin \theta}{\lambda}.$$

The highest resolution of the film is 1000 lines/mm, so this must be the maximum spatial frequency. Substituting into the equation, we get

$$\sin \theta = F\lambda = \frac{1000}{10^{-3} \text{ m}} 632.8 \times 10^{-9} \text{ m} = 0.6328,$$

$$\theta = 39.3°.$$

The field of view is limited to a cone making an angle of 39.3° from the reference beam.

Light and Color

29.6

Isaac Newton's first scientific paper, published in 1672, described his experiments with light and color. Newton passed a beam of sunlight through a prism, spreading the light into a spectrum of colors. A second prism turned the opposite way converged the spectrum back into a narrow beam of white light. Later development of the diffraction grating led to measurements of the wavelengths of light. Although each color merges smoothly into the next across the spectrum, it is possible to assign approximate wavelength ranges to each color (Table 29.1).

Although white light contains all the colors of the spectrum, they need not be present with the same intensity. For example, sunlight, often used as a reference standard for white light, does not have the same intensity at all wavelengths. Instead, it has the intensity distribution shown in Fig. 29.16. The intensity of sunlight reaching the earth's surface peaks near a wavelength of 475 nm and falls off sharply in the ultraviolet. A comparison

TABLE 29.1
Colors in the visible spectrum

Color	Wavelength range (nm)
Red	630–700
Orange	590–630
Yellow	570–590
Green	500–570
Blue	450–500
Violet	400–450

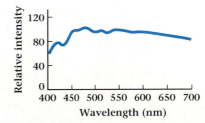

Figure 29.16 Spectral distribution of the intensity of sunlight at the earth's surface at noon.

TABLE 29.2
Approximate color temperature of some common light sources

Source	Color temperature (K)
Mercury arc	6000
Daylight	5500
High-intensity carbon arc	5500
Cool white fluorescent	4200
Incandescent tungsten (photoflood 3400)	3400
Incandescent tungsten (photoflood 3200)	3200
Warm white fluorescent	3000
Incandescent tungsten (100W)	2900
Incandescent tungsten (40W)	2650
High-pressure sodium arc	2200

of the solar data with the Planck theory of blackbody radiation can be used to determine the sun's temperature, giving a result of about 6000 K.

Incandescent lamps are good approximations to blackbody radiators and give off white light; however, this light is deficient in the blue and violet. The spectrum is skewed toward the red because of the relatively low operating temperature of incandescent lamps, about 2900 K. We describe the spectral distribution of the light in terms of the corresponding temperature of a blackbody, termed the *color temperature*. At a color temperature of 2900 K only 3% of the energy dissipated in the lamp emerges as visible light. Special high-temperature lamps designed for photography and television usually operate at one of two reference temperatures, 3200 K or 3400 K. They produce considerably more blue than do ordinary household lamps. Table 29.2 shows some of the characteristics of a few common light sources.

Example 29.4

Peak-intensity wavelength of a photoflood lamp.

Where is the peak in the spectral distribution of light from a photoflood lamp operating at 3400 K?

Solution We can compute the wavelength with the greatest intensity from the Wien displacement law (Chapter 26):

$$\lambda_m T = 2.90 \times 10^{-3} \text{ m·K}.$$

Here T is the temperature of our light source, treated as a blackbody radiator. Thus

$$\lambda_m = \frac{2.90 \times 10^{-3} \text{ m·K}}{3400 \text{ K}} = 8.53 \times 10^{-7} \text{ m} = 853 \text{ nm}.$$

The peak of the spectral distribution lies beyond the visible range, in the infrared region of the electromagnetic spectrum.

The spectral distribution of a light source depends on more than just the operating temperature. As we saw in Chapters 26 and 27, characteristic emission spectra of elements are due to electronic transitions between atomic energy levels. Thus the light from each element has its own characteristic set of emission wavelengths and intensities, which give it a characteristic color. An electric discharge in hydrogen produces light of the Balmer series. We perceive this mixture of wavelengths as pink. Sodium vapor lamps produce a characteristic yellow light that is due to the intensity of the D lines. Similarly, neon lamps produce a red-orange light.

The fluorescent lamp is a gas-discharge tube containing mercury vapor and a coating of fluorescent powder on the inside of the tube. Fluorescent materials absorb light at some wavelengths and then radiate light at longer wavelengths. Generally, both the absorption and emission states are broad bands rather than sharply defined energy levels of the type observed in free atoms. When the mercury vapor in a fluorescent lamp is excited by an electric discharge, it emits its characteristic spectral radiation. Part of this radiation extends beyond the visible range into the ultraviolet. The fluorescent material absorbs the ultraviolet radiation and then radiates light in the visible spectrum. Fluorescent lamps of this type are considerably more efficient than incandescent lamps, converting about 20% of the electrical energy into visible light.

Objects can absorb, reflect, or transmit light, and some objects can do a combination of all three processes. Thus in addition to color that depends on the detailed nature of the source, light may take on a particular color as it passes through a material that selectively absorbs some wavelengths. For example, a red piece of glass is a filter that selectively passes red light, but absorbs shorter wavelengths. Similarly, a blue filter transmits blue but not green, yellow, or red. Thus color may be produced by selective absorption.

Color is also produced by reflection as well as transmission. The perceived color of most objects is due to the selective reflection of light. Thus a red object reflects red light but absorbs green and blue. A red apple appears a vivid red when illuminated with red light, but it looks dark under blue light because it does not reflect blue light well.

We should caution you that while we can characterize light according to its spectral distribution, this is *not* the same as describing its color. Color is a *visual* sensation. The biological mechanisms of vision play an important role in our determination of color. Experiments by Edwin Land have demonstrated convincingly that the sensation of color depends on more than the spectral intensities of light reaching our eyes.

Color by Addition and Subtraction

29.7

The visible spectrum contains a continuous range of color, varying smoothly from violet on one end to deep red on the other. However, we can approximate the spectrum using only three separate bands of color, representing equal intervals of wavelength. These bands are called the three **additive primary colors** of light: red, green, and blue. By combining these three colors in various combinations, we can create a wide range of other colors. For instance, the combination of red and green light produces yellow. The addition of blue light to red light gives magenta, while blue added to green gives cyan (Fig. 29.17). If we mix the three primary colors in various proportions, we can obtain a full range of other colors, including purple, brown, and orange. If we add all three primaries with equal intensity, the result is white light. The production of colors in this manner is called *color mixing by addition.*

Color television generates color pictures through the additive mixing of color. If you look closely at the front of a television picture tube, you will see a pattern of closely spaced dots arranged in groups of three.* Each dot contains a fluorescent material that glows when excited by the electron beam. The system uses three different materials to produce red, green, and blue light. By controlling the intensity of the light from the separate dots, a wide range of colors can be produced (Fig. 29.18).

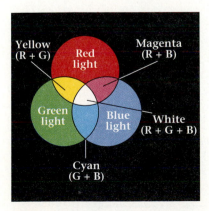

Figure 29.17 **The primary colors of light, also called the additive primaries, can combine to generate a full range of colors.**

---

*Some manufacturers use closely spaced lines instead of the dots described here.

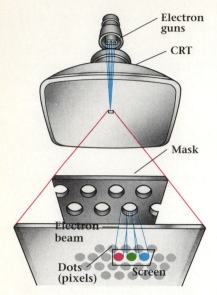

Figure 29.18.   Three different phosphors generate tiny dots of primary colored light on the face of a television screen. The colors blend together to make a full range of colors.

We can also produce color by subtraction. In this process we generate colors by selectively filtering light, thus removing certain wavelengths from the beam. For example, a red filter removes green and blue from white light, resulting in a beam of red. Similarly, a green filter transmits green light and a blue filter transmits blue light. However, we cannot combine such filters to produce other colors. Since the red filter passes only red, the combination of red plus green filter or red plus blue filter allows no light to pass (Fig. 29.19). Thus red, green, and blue filters are not appropriate for producing other colors by subtraction.

We use the term *complementary colors* for any pair of colors of light that can be combined to produce white light. Thus the color complementary to red is cyan, the color complementary to blue is yellow, and the color complementary to green is magenta. These three colors—cyan, magenta, and yellow—are the **subtractive primary colors**. Filters of these colors *can* be used in succession to produce other colors. For example, passing white light successively through yellow and magenta filters produces red light (Fig. 29.20). The yellow filter removes blue light while the magenta filter removes the green. The remaining light is red.

Color photography relies on the subtractive method of color production. Photographic films and papers are made with three layers of photosensitive materials, each of which responds to one of the primary colors of light. When the film is developed, dye images in one of the subtractive

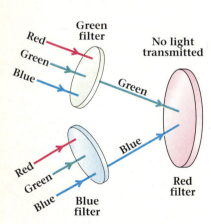

Figure 29.19   Filters in the color of the additive primaries block all other colors of light, so they cannot be used to generate other colors.

Figure 29.20   Use of subtractive color filters. (a) Cyan, magenta, and yellow filters block the passage of their complementary colors—respectively, red, green, and blue. (b) By using two subtractive filters in succession, we can block the passage of two of the additive primaries and allow the other to pass.

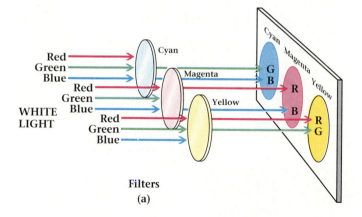

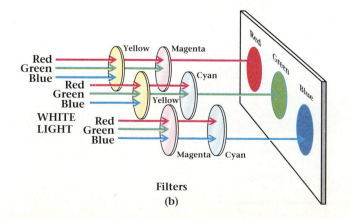

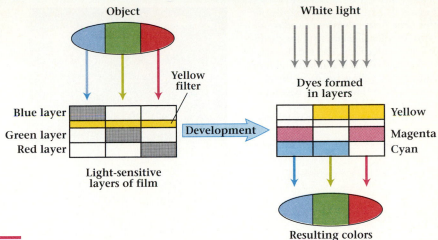

Figure 29.21   Color formation in color slide film.

primaries form in each layer (Fig. 29.21). The varying densities of these filters control the color of the light that passes through. These same subtractive primaries are also used in color printing.

Although both addition and subtraction of colors give good results, as observed in color television and photography, these methods cannot accurately reproduce pure spectral colors. To a television camera, a spectral red of some particular wavelength is detected the same as any other spectral red. The resulting red displayed on the screen is necessarily limited to whatever red light is characteristic of the particular phosphor placed in the tube by the manufacturer. Why, then, does color television (and photography) work so well? The answer is that most of the light that we normally see consists of mixtures of wavelengths and not just spectral lines. Most colors are due to broad-banded absorptions or reflections. A yellow lemon photographs as yellow because it reflects both green and red light over a broad range. Since most colors are broad banded, the limitations in color reproductions usually go unnoticed.

We can use our knowledge of color to see beyond the visible range with the help of false-color infrared photography (Fig. 29.22). Special

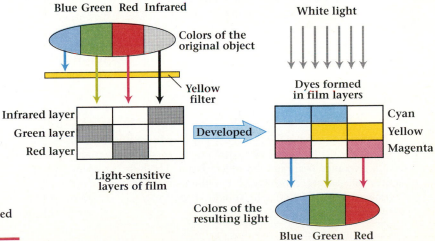

Figure 29.22   Color formation in infrared color film.

Figure 29.23 Two photographs of the same flowers with (a) Koda-chrome color film and (b) Ektachrome infrared color film. The red flowers reflect both the red and infrared light, causing the yellow image in the infrared photo. The green leaves reflect both green and infrared, causing them to look magenta in the infrared photo.

color films are designed with sensitivity that extends into the near-infrared region of the spectrum. They differ from usual color films in that the three image layers are sensitized to green, red, and infrared instead of blue, green, and red light. During exposure, a yellow filter is used on the camera to block blue light, to which the layers are also sensitive. When processed, the green-sensitive layer has a magenta image, the red-sensitive layer has a yellow image, and the infrared-sensitive layer has a cyan image. These layers, when combined, filter the light to produce an image in which the original infrared, red, and green light appears as red, green, and blue. Leaves appear magenta in such photos because they reflect both infrared and green light, which the film translates into red and blue that combine to give magenta (Fig. 29.23). False-color infrared images from high altitudes are used in studying the earth's surface for land use and for vegetation and water coverage.

SUMMARY

Useful Concepts

- To create a laser you need an active medium of atoms or molecules with an inverted population of excited states enclosed in an optical resonator. The laser light is generated by the stimulated emission of photons by the excited medium. Laser light is characterized by its monochromaticity and its coherence.

- The fringe visibility is

$$V \equiv \frac{I_{max} - I_{min}}{I_{max} + I_{min}}.$$

- The coherence is

$$\gamma \equiv \frac{I_1 + I_2}{2\sqrt{I_1 I_2}} V.$$

- Holograms are made by recording the interference pattern of a reference beam and an object beam. The holographic images are reproduced by illuminating the hologram with a beam of light similar to the original reference beam.

- A full range of colors can be produced by the addition of primary colors of light (red, green, and blue). Colors can also be produced by passing white light through a combination of subtractive filters (the subtractive primaries of yellow, magenta, and cyan).

Important Terms

You should be able to write the definition or meaning of each of the following terms:

- spontaneous emission
- stimulated emission
- laser
- active medium
- pumping
- optical resonator
- inverted population
- metastable state
- temporal coherence
- spatial coherence
- hologram
- holography
- spatial frequency
- additive primary colors
- subtractive primary colors

QUESTIONS

29.1 A photographer has a photoflood lamp whose temperature is too low for the film available. If she uses a color filter to correct the light balance to better match the film, what color should it be?

29.2 Where is the interference pattern observed as laser speckle?

29.3 What effect would there be if both mirrors of a laser were partly reflecting?

29.4 Give a semiclassical explanation for stimulated emission. Does it explain why the radiation is in the same direction as the incident light?

29.5 Sketch a setup to make a hologram. Identify all components shown in your drawing. What special requirements must be met to make a hologram?

29.6 When a hologram is cut in two, can a complete image of the object still be reconstructed from each piece? What differences will there be in the images from the two pieces?

29.7 Why do colored fabrics sometimes appear different when viewed in sunlight, as compared with their appearance under incandescent light or fluorescent light?

29.8 Examine a Polaroid color slide with a microscope. Is the color reproduction in the image an additive or subtractive process?

29.9 How can a color television reproduce yellow light when the available phosphors are limited to red, green, and blue?

29.10 Why does uneven illumination falling on a pair of slits affect the fringe visibility?

29.11 Explain why helium is necessary for the operation of the helium-neon laser.

29.12 Explain the photographer's rule that a colored filter lightens the image of objects of its own color and darkens the image of those of the complementary color.

PROBLEMS

Hints for Solving Problems
The degree of coherence of light is the same as the visibility of the double-slit pattern when the slits are evenly illuminated. The maximum spatial frequency that a hologram can have is limited by the resolution of the film. The additive primary colors of light are red, green, and blue. Their complementary colors, cyan, magenta, and yellow, are the subtractive primaries.

Section 29.2   Lasers

29.1 Calculate the number of half-wavelengths in the laser cavity of a helium-neon laser whose wavelength is exactly 632.8 nm and whose mirrors are 30.0 cm apart.

29.2 Calculate the number of half-wavelengths in a ruby laser operating at 694.3 nm, given that the ruby rod, silvered on the ends, is exactly 10.0 cm long. (*Hint:* The index of refraction of ruby is 1.69.)

29.3 Calculate the separation between modes in a helium-neon laser whose cavity length is 35.5 cm.

29.4 Calculate the separation between modes in a short laser whose cavity length is only 5.00 cm. Assume that the index of refraction is 1.00.

29.5 What is the energy of the transition giving rise to laser light at 694.3 nm? Give your answer in eV.

Section 29.3   The Helium-Neon Laser

29.6 What is the energy separation of the states that give rise to the 632.8-nm laser line in the helium-neon laser? What is the energy separation of the states that give rise to the infrared line at 1150 nm? Give your answers in electron volts.

29.7 The helium-neon laser has a line with a wavelength of 543 nm. What is the energy, in electron volts, of photons at that wavelength? What color is the light?

Section 29.4   Properties of Laser Light

29.8 A double-slit fringe pattern is observed to have intensity maxima that are three times greater than the intensity minima. What is the visibility of these fringes?

29.9 An interference pattern produces fringes with a visibility of 0.40. What is the ratio of I_{max}/I_{min} for this pattern?

29.10 A certain laser produces a fringe visibility of 0.50. What is the ratio of I_{max}/I_{min} for these fringes?

29.11 A double-slit interference pattern is observed for which the intensity maxima are ten times as great as the minima. Assuming even illumination on the slits, what is the degree of coherence of the incident light?

29.12 The interference pattern of a double slit has intensity maxima that are 7.7 times greater than the minima. It is known that the intensity due to one slit alone is twice as great as the intensity due to the other slit. What is the coherence of the incident light?

29.13 Completely coherent laser light is used to illuminate a double slit. The intensity of the light reaching one slit is five times as great as the light reaching the other one. What will be the visibility of the resulting fringes?

29.14 A partially coherent light source, $\gamma = 0.80$, illuminates a double slit unevenly so that one slit receives 1.5 times more light than the other one. Compute the visibility of the fringes and the ratio I_{max}/I_{min} in the interference pattern.

29.15 The interference pattern of a double slit has intensity maxima that are 5.5 times greater than the minima. It is known that the intensity due to one slit alone is three times as great as the intensity due to the other slit. What is the coherence of the light source?

Section 29.5   Holography

29.16 In making a particular hologram, the maximum angle between the reference and object beams is 30°. What is the minimum resolution required of the holographic film if the wavelength of the light used is 693.4 nm?

29.17 A hologram is made with film whose resolution is 1500 lines/mm. Light from an argon laser, $\lambda = 488$ nm, was used in recording and viewing the hologram. The reference beam strikes the hologram at normal incidence. What is the angular field of view of this hologram?

29.18 A holographic image subtends an angle of 34° at the center of the holographic film when illuminated with a parallel beam of light of wavelength 488 nm. If the illumination is changed to light from a helium-neon laser at 632.8 nm, what angle will be subtended by the image?

29.19 Two beams of parallel laser light with $\lambda = 632.8$ nm are brought together on a photographic plate to form a sinusoidal holographic grating. One beam strikes the plate at normal incidence. The other beam makes an angle of 28° with the normal. (See Fig. 29.14.) The processed hologram is illuminated with light of wavelength 546.1 nm. If this light strikes the hologram at normal incidence, how many beams emerge and at what angles?

29.20 A holographic grating of 1500 lines/mm is used as the dispersing element in a spectroscope. In the first-order spectrum, what is the angular separation of the sodium D lines, $\lambda = 589.0$ and 589.6 nm? Assume normal incidence.

29.21 A grating spectroscope is used to study the emission spectrum from hydrogen. A grating of 1500 lines/mm is illuminated at normal incidence. Determine the angles at which the Balmer lines are found. (*Hint*: The wavelengths are 656.3, 486.1, 434.0, and 410.2 nm.)

Section 29.6   Light and Color

29.22 Radiation from the sun has its maximum intensity at a wavelength of 475 nm. Compute the temperature of the sun using the Wien law and compare it with the value of 6000 K given by the Planck law.

29.23 The blackbody spectrum radiated by a certain lamp has a maximum around the wavelength 967 nm. What is the temperature of this lamp?

29.24 A large room is illuminated with 40 fluorescent lights, each rated at 40 W. How many 100-W incandescent lamps would be required to produce the same level of illumination in the room? Assume that the efficiency of the fluorescent lamps is 20%, while that of the incandescents is 3%.

29.25 A new office building is designed to be illuminated with fluorescent lamps that are 20% efficient in converting electrical energy into light. By what fraction would the energy costs for illumination be increased if the building were illuminated at the same level by incandescent lamps of 3.5% efficiency?

Section 29.7   Color by Addition and Subtraction

29.26 White light passes through a cyan filter, which is, in turn, followed by a second filter. What color emerges if the second filter is (a) yellow, (b) magenta, (c) blue, (d) violet, (e) orange?

29.27 Simultaneous excitation in equal intensities of the red and green phosphors on a TV screen produces a yellow color. (a) What color results from full-intensity red and half-intensity green? (b) What color results from full-intensity blue and half-intensity red?

29.28 White light passes through two overlapping, full-strength color filters. What is the color of light that emerges if the two filters are (a) red and green, (b) magenta and yellow, (c) red and yellow, (d) magenta and cyan?

29.29 White light passes through a yellow filter, which is, in turn, followed by a second filter. What color light emerges if the second filter is (a) green, (b) cyan, (c) magenta, (d) blue?

29.30 The principal lines in the emission spectrum of mercury lie at wavelengths of 365.5, 435.8, 546.1, 577.0, and 579.1 nm. Characterize these wavelengths according to their color.

29.31 The Balmer lines in the hydrogen spectrum are wavelengths of 410.2, 434.0, 486.1, and 656.3 nm. What are the colors of these lines?

29.32 Violet light (λ = 410 nm) and red light (λ = 615 nm) are shined simultaneously on a pair of slits with separation of 0.50 mm. The diffraction pattern is formed on a screen 1.0 m away. Describe the location and color of the first five bright fringes.

Additional Problems

29.33 An advertisement for a commercial helium-neon
• laser says that the wavelength is 633 nm and the bandwidth of the line is 2 × 10⁹ Hz. This bandwidth has the effect of smearing out the interference pattern that would be due to an absolutely monochromatic source. For a double-slit experiment using this laser, determine the position of the first maximum in the pattern from a double slit with a spacing of 0.20 mm that is 2.50 m from a screen. By how many millimeters is the position of the first maximum smeared out by the bandwidth of the laser?

ADDITIONAL READING

Bromberg, J. L., "The Birth of the Laser." *Physics Today,* October 1988, p. 26.

Deutsch, T. F., "Medical Applications of Lasers." *Physics Today,* October 1988, p. 56.

McAlister, H. A., "Seeing Stars with Speckle Interferometry." *American Scientist,* March-April 1988, p. 166.

Walker, J., "The Amateur Scientist: How to Stop Worrying About Vibration and Make Holograms Viewable in White Light." *Scientific American,* May 1989, p. 134.

30

Condensed Matter

A WORD TO THE STUDENT

One of the early triumphs of quantum theory was its application to understanding the behavior of solids. The success of solid state theory arose in part from the choice of an appropriate model. The development of such a model forms the main subject of this chapter. We will be especially concerned with the electrical behavior of solids. We also give special attention to the study of semiconductors, which is basic to the understanding of transistors and integrated circuits.

30.1	Types of Condensed Matter
30.2	The Free-Electron Model of Metals
30.3	Electrical Conductivity and Ohm's Law
30.4	Band Theory of Solids
30.5	Pure Semiconductors
30.6	The Hall Effect
30.7	Impure Semiconductors
30.8	The *pn* Junction
*30.9	Solar Cells and Light-Emitting Diodes

In Chapter 25 we examined some of the early research on crystal structure, which provided evidence for the atomic theory of matter. Although the early crystallographers made progress in categorizing different types of crystal structures, they did not get very far in understanding the behavior of solids. From our present viewpoint, this limited progress is not too surprising. After all, as we pointed out in Chapter 10, atoms in solids are much closer together than atoms in gases. Thus the kinetic-theory model of a gas that works well in explaining the gas laws is inappropriate for a solid or a liquid. The kinetic theory of gases assumes Newtonian mechanics, but a description of solids or liquids must be based on quantum mechanics. The development of quantum mechanics, in the early part of the twentieth century, and the development of x-ray diffraction techniques were key steps along the way to our contemporary understanding of solids.

Our presentation of condensed-matter physics is necessarily brief and introductory. We focus primarily on crystalline solids. However, without getting into the mathematical complexities of quantum mechanics, we can still examine many of the important concepts needed to understand solids. We will develop a descriptive model of the electronic structure of solids, called band theory. This model has been especially successful in explaining the behavior of metals and semiconductors.

Types of Condensed Matter

30.1

The average separation between molecules in a liquid is comparable to their diameters. Individual molecules are free to move about, but they are constrained to move so that the average separation between near neighbors remains constant. As a result, a given mass of liquid is virtually incompressible and has a definite volume, although its shape can change to match the shape of its container.

In solids the interatomic separations are also comparable to atomic diameters. But, unlike those in liquids, the atoms in a solid are not free to move about. Instead, atoms of a solid are rigidly fixed and do not move except for small oscillations about their average positions. Thus the solid has not only a definite volume but a definite shape as well.

As we mentioned in Chapter 25, when the atoms of a solid are arranged in a regular manner, forming a three-dimensional array or lattice, we say that the solid is **crystalline**. We can construct this array by repeating a basic unit pattern in three dimensions, in much the same way that bricks stacked in an orderly fashion form a neat pile. A schematic illustration of a two-dimensional lattice is shown in Fig. 30.1(a). The basic repeating unit, or *unit cell*, of this rectangular lattice is shown by the

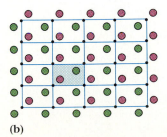

(a)

(b)

Figure 30.1 Schematic representation of two-dimensional crystals. The basic repeating unit, the unit cell, is shown by the dashed lines. The points in (a) represent the lattice. (b) A crystal is formed when atoms or molecules are associated with each lattice point.

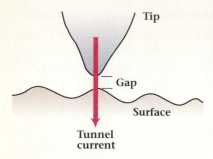

Figure 30.2 The tunneling of electrons across a narrow vacuum gap between two conductors is constant for a constant separation between them.

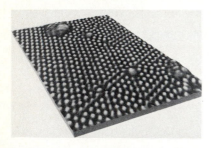

Figure 30.3 Tunneling microscope picture of germanium, showing the location of atoms on the surface. Impurity areas are present in the picture.

dashed lines. The entire lattice is built up by adding more unit cells in the adjacent space. In Fig. 30.1(a) we have drawn the unit cell by connecting neighboring points of the lattice with the dashed line. If we associate an atom or molecule with each lattice point, the result is a crystal. If two atoms are associated with each lattice point of Fig. 30.1(a), the resulting crystal is more complex, but the shape of the unit cell is unchanged. Figure 30.1(b) shows a rectangular crystal structure in which two atoms are associated with each lattice point. The periodicity of the atoms in the crystal lattice affects all of the physical properties of crystals.

The scanning tunneling microscope, which we first mentioned in Chapter 25, reveals three-dimensional images of solid surfaces that show vertical position differences as small as 0.01 nm and horizontal position differences as small as 0.6 nm. This microscopy technique utilizes the quantum mechanical tunneling of electrons between two conducting solids separated by a narrow layer of insulator or vacuum.

The tunneling microscope takes advantage of the strong exponential dependence of the tunneling current on the width of the insulating gap between the two conducting solids. When a sharp probe tip (Fig. 30.2) is scanned across the surface to be investigated, the vertical position of the tip is changed to keep the tunnel current constant. In this manner the tip-to-surface distance is kept constant so that the probe follows the surface contour as it scans across. By monitoring the vertical position of the probe, a two-dimensional representation of the surface contours can be made for each scan. A three-dimensional image may be obtained from a sequence of such scans. Figure 30.3 shows one such image of a surface. The regular positions of the surface atoms are easily seen.

When the regularity of the pattern extends over the entire crystal, it is called a single crystal or monocrystal. (Examples of single crystals were shown in Figs. 25.1 and 25.2.) However, frequently the periodicity is disrupted so that the entire solid is made up of numerous individual crystalline subunits, called grains. Such material is said to be a **polycrystalline solid.** Figure 30.4(a) shows a possible arrangement of atoms near a grain boundary dividing two otherwise perfect crystalline regions within a solid. Figure 30.4(b) is a micrograph of a metal, revealing the individual crystalline grains.

Figure 30.4 (a) Schematic drawing of a grain boundary between two otherwise perfect crystalline grains. (b) Optical micrograph of polycrystalline shows crystal grains.

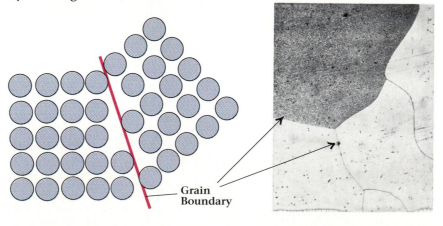

(a)   (b)

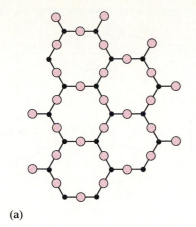

(a)

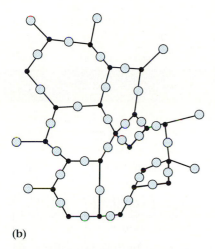

(b)

Figure 30.5 (a) Two-dimensional analog of a crystal. (b) Two-dimensional analog of a glass.

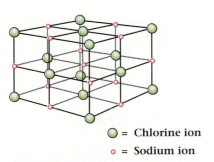

○ = Chlorine ion

○ = Sodium ion

Figure 30.6 The crystal structure of sodium chloride. The sodium ions are depicted as smaller than the chlorine ions. The cubic unit cell is shown. This cell is face-centered cubic.

Not all solids have the long-range order of crystals. **Amorphous solids** possess only a short-range order, similar to that of liquids. Glass is one example. The arrangement of molecules in a glass is much like that of a liquid in which the atomic positions have become rigid. Schematic representations of crystalline and glassy solids (Fig. 30.5) show an orderly arrangement between neighboring atoms (short-range order) in a glass, but no long-range periodicity of the atomic positions.

A class of material in which the regular periodic arrangement exists in only one or two dimensions is known as a **liquid crystal.** Liquid crystals can flow, form droplets, and maintain long-range orientational and spatial order. Because of these special properties, some physicists refer to liquid crystals as a fourth state of matter. Liquid crystals usually consist of large molecules whose geometric shape forces the periodic structure.

In the remainder of this chapter we will examine only crystalline solids. In most instances we will consider the solid to be a perfect single crystal. However, remember that real crystals are never perfect and the presence of imperfections alters the properties of the solid. These imperfections are especially noticeable in the semiconductors, as we shall see below.

The electronic forces binding the atoms in a crystal are not all the same, and differences arise as a result of the geometric arrangement of charges. We often use specialized terms to categorize different crystal bindings, depending on the dominant type of interaction. Thus we speak of ionic, covalent, and metallic crystals, which correspond to chemical bonding of those three types.

In an ionic crystal, such as ordinary table salt (sodium chloride, NaCl), each sodium atom gives up its outermost electron, becoming a positively charged ion. Each chlorine atom absorbs one of these electrons to become a negatively charged ion. The ions then arrange themselves in what is called a face-centered cubic lattice (Fig. 30.6). In this crystalline form the energy of the system is reduced well below the energy of the free ions. If this were not so, the crystal would not be so stable. The large binding energy is a direct result of the fact that each ion is surrounded by six nearest neighbors of opposite charge, thus reducing the electrostatic potential energy.

Since the electrostatic potential gets lower (more negative) as the separation between ions decreases, you might conclude that the crystal would collapse until it disappeared. However, we know that this does not happen. As the ions are brought closer and closer together, their electronic distributions begin to overlap. At these distances a repulsive force associated with the Pauli exclusion principle arises (Fig. 30.7, p. 890). Other ionic crystals have different crystal structures with different nearest-neighbor distances. For example, cesium chloride, CsCl, has the structure shown in Fig. 30.8 (p. 890).

Covalent crystals result from atoms that exhibit covalent chemical bonding, in which electrons are shared by two closely spaced atoms. Carbon (in the form of diamond), silicon, and germanium all have the same type of crystal structure, known as the diamond structure, in which each atom is bonded to four nearest neighbors in a tetrahedral arrangement (Fig. 30.9, p. 891). There appears to be a continuous range of crystal types between the extremes of the purely ionic and purely covalent limits. Often

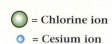

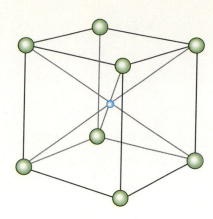

Figure 30.7   Potential energy per molecule of an ionic crystal as a function of nearest-neighbor distance.

Figure 30.8   The crystal structure of cesium chloride. The space lattice is simple cubic with one cesium ion on the corner of the cube and one chlorine ion in the center.

= Chlorine ion

= Cesium ion

PHYSICS IN PRACTICE

Liquid Crystal Displays

Look around you, wherever you are. Chances are you are not too far from a liquid crystal display (LCD). Adapted in the 1970s for use as display devices in digital watches (Fig. B30.1), they are now used in making digital meters, fever thermometers, and screens for laptop computers. LCDs are even used to make the display screens for pocket televisions. In all of these applications the alignment of molecules within the liquid crystal, and hence the optical

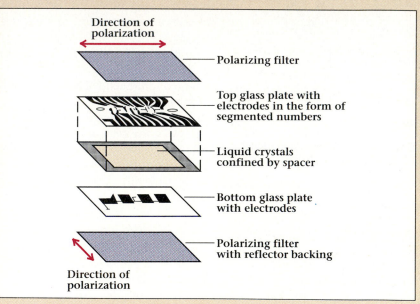

Figure B30.2   An exploded view of a liquid crystal display.

Direction of polarization

Polarizing filter

Top glass plate with electrodes in the form of segmented numbers

Liquid crystals confined by spacer

Bottom glass plate with electrodes

Polarizing filter with reflector backing

Direction of polarization

Figure B30.1   An electric digital watch with a liquid crystal display.

behavior of the crystal, is changed by applying an electric field. Since very little current is required to activate the LCDs, they are excellent choices for all kinds of battery-operated electronic equipment.

A typical LCD display is made from liquid crystal materials that have rod-like molecules with electric dipole moments along the molecular axes. A thin layer of this material is sandwiched between transparent electrodes and polarizing filters (Fig. B30.2). In the structure shown, the polarizers are

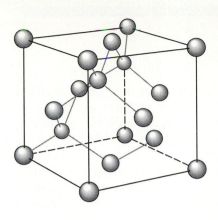

the bonding is something intermediate and is characterized for convenience as being mainly ionic or mainly covalent.

Metallic crystals are characterized by high electrical and thermal conductivity and a lustrous appearance. The bonding in a metal may be likened to a form of covalent bonding in which there are more potential bonding sites than there are electrons to be shared. So the shared electrons may move through the crystal, traveling from one bonding site to another. Since these outer electrons are free to move through the solid, it is better to describe them as belonging to the entire solid rather than to particular ions. We consider their behavior further in the next section.

Figure 30.9 The diamond structure. The edges of the cubic unit cell are shown. Each atom is tetrahedrally bonded to four nearest neighbors.

aligned at right angles to each other. The two surfaces that contact the liquid crystals are treated so that the molecules align with their axes parallel to the surfaces, but with a 90° rotation from the top to bottom surface when no voltage is applied between the electrodes (Fig. B30.3a). Light incident on the display is linearly polarized by the first polarizer. As the light passes through the layer of liquid crystal, the plane of polarization rotates by 90°. The light passes through the second polarizer, strikes a reflector and is reflected back through the system.

When a voltage is applied to the liquid crystal material, the molecules rotate to align with the resulting electric field (Fig. B30.3b). The plane of polarization of the incident linearly polarized light no longer rotates as the light traverses the liquid crystal material. Consequently, the second polarizer absorbs this light. If a segmented display is used, the number segments to which the voltage is applied will appear dark on a light background. If the orientation of the two polarizers is made parallel, then the display is reversed, with light numbers appearing on a dark background.

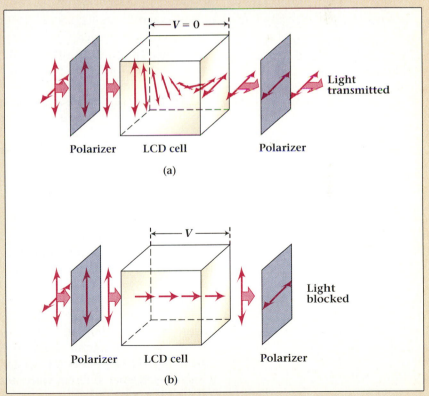

Figure B30.3 The action of an LCD device on polarized light passing through it. (a) The plane of polarization is rotated when no voltage is applied to the cell. (b) When a voltage is applied across the cell, the molecules of the liquid crystal rotate so that light passes through with no change in its polarization.

The Free-Electron Model of Metals

30.2

One of the best-known characteristics of metals is their large electrical conductivity. Since most metals are solid at room temperature, their ions are stationary in the crystal lattice except for small-amplitude oscillations about their equilibrium positions. Thus the large electrical conductivity is not due to motion of these positive ions but results from the motion of some of the negative electrons, which are not tightly bound to the ions.

Let us construct a model of a metallic crystal that accounts for these observations. For simplicity, we consider a metal from the first column of the periodic table of the elements, such as sodium. Elements from the first column characteristically have one electron outside of a completely filled electron shell. This outermost electron is only weakly bound to the ion core. In fact, in the solid phase these outermost electrons are no longer bound to any particular ion, but are free to move throughout the metallic crystal. They belong to the whole crystal rather than to an individual ion. In this model, known as the free-electron model of metals, we do not worry about details of interactions of the electrons with each other or with the ions. Instead, we treat them as a gas of particles confined to the volume of the crystal.

We know, however, that we cannot describe the electrons using Newtonian mechanics, as we did for the classical kinetic model of gases. Instead, we must formulate our model in terms of quantum mechanics. We have already laid the groundwork for this model in our discussion in Chapter 27 of the particle in a box. In the present case, however, we need to treat the problem in three dimensions. Suppose that the electrons are confined to a cube of edge length L. The wave function that satisfies the Schrödinger equation for the three-dimensional box with rigid walls is

$$\Psi_n(x,\,y,\,z) \; = \; \psi_0 \sin \frac{n_x \pi x}{L} \sin \frac{n_y \pi y}{L} \sin \frac{n_z \pi z}{L},$$

where n_x, n_y, and n_z are integers. These integers are the quantum numbers that describe the wave function in the x, y, and z directions.

In applying quantum mechanics to a three-dimensional potential, we assumed fixed boundary conditions. However, we have already established that the essence of the crystal structure is its periodicity. Thus the environment of each metallic ion must be identical to that of any other similar ion in the crystal. This fact leads us to consider other boundary conditions. Instead of demanding that the electron wave function vanish at the boundaries, as we required earlier, we need only require that the wave function be the same at points where the potential is the same. As a result we get a somewhat different wave function, one that is a periodic traveling wave with a period L. The resulting expression for the electron energy levels is

$$E_n \; = \; \frac{h^2}{2mV^{2/3}}\, n^2, \tag{30.1}$$

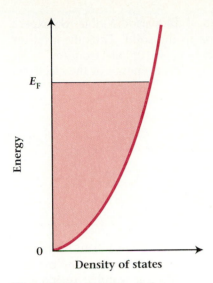

Figure 30.10 Density of electron states (as abscissa) plotted against electron energy (as ordinate). The shaded area represents the occupied states for temperature near 0 K. The cutoff energy is the Fermi energy E_F. Only electrons with energy near E_F interact with applied fields.

where $n^2 = n_x^2 + n_y^2 + n_z^2$ and V is the volume L^3. In this case we are not restricted to positive integers for the quantum numbers n_x, n_y, and n_z as before. Instead they may be positive or negative integers or even zero:

$$n_x, n_y, n_z = 0, \pm 1, \pm 2, \pm 3, \dots$$

Each energy state described by a particular set of quantum numbers n_x, n_y, n_z can hold no more than two electrons, one each with spin quantum numbers of $+\frac{1}{2}$ and $-\frac{1}{2}$ as required by the Pauli exclusion principle. Of course, a given energy state could be empty. At very low temperatures we would expect the lower energy states to be filled, beginning with the lowest state and continuing into successively higher states until all of the free electrons were accounted for. The last filled level would define a cutoff energy. The levels above this cutoff would be vacant (Fig. 30.10). The cutoff energy is known as the **Fermi energy** E_F. The Fermi energy* depends on the electron density, that is, on the number of free electrons N divided by the volume V, and is given by

$$E_F = \frac{h^2}{2m}\left(\frac{3N}{8\pi V}\right)^{2/3}. \tag{30.2}$$

For good conductors, the free-electron density N/V is of the order of 5×10^{28} electrons/m³. Fermi energies corresponding to electron densities of this magnitude are of the order of 5 eV. Note that although we have described the lowest or ground state of the system, individual electrons possess kinetic energies from 0 to E_F. At temperatures on the order of ordinary room temperatures, where $T \approx 300$ K, the kinetic energy of an electron near the Fermi energy far exceeds the energy (kT) associated with thermal processes. (At room temperature $kT \approx 1/40$ eV.) Consequently, the Fermi energy of a metal is essentially unaffected by changes in temperature of tens of kelvins, and properties such as electrical conductivity change slowly with change in temperature.

Example 30.1

Calculate the Fermi energy in lithium.

Solution We begin by calculating the number of lithium atoms per unit volume in the solid. We can determine this number from the mass density, the atomic mass, and Avogadro's number. Lithium has a density of 0.53 g/cm³ and an atomic mass of 6.94 g/mol. Dividing the density by the atomic mass and multiplying the result by Avogadro's number, we get

$$\frac{\text{density} \times N_A}{\text{atomic mass}} = \frac{(0.53 \text{ g/cm}^3)(6.02 \times 10^{23} \text{ atoms/mol})}{6.94 \text{ g/mol}},$$

$$\text{atomic density} = 4.6 \times 10^{22} \text{ atoms/cm}^3 = 4.6 \times 10^{28} \text{ atoms/m}^3.$$

*The Fermi energy is named after the Italian physicist Enrico Fermi (1901–1954), whose pioneer efforts produced the first nuclear chain reaction. Fermi is also well known for his work on the statistical (thermal) behavior of particles that obey the Pauli principle. These particles, including electrons and protons among others, are collectively known as fermions.

Since lithium belongs to the first column of the periodic table, we expect each atom to contribute one free electron to the solid. Thus the electron density is the same as the atomic density, 4.6×10^{28} electrons/m³.

Now we can use Eq. (30.2) to find the Fermi energy of the electrons in lithium:

$$E_F = \frac{h^2}{2m} \left(\frac{3N}{8\pi V} \right)^{2/3}.$$

Using the free-electron density just determined for N/V, we get

$$E_F = \frac{(6.63 \times 10^{-34} \text{ J·s})^2}{2(9.1 \times 10^{-31} \text{ kg})} \left(\frac{3 \times 4.6 \times 10^{28}/\text{m}^3}{8\pi} \right)^{2/3},$$

$$E_F = 7.52 \times 10^{-19} \text{ J} = 4.7 \text{ eV}.$$

We can summarize our model of a metal as follows. The outermost electrons of the metal atoms are not bound tightly to the ions but are bound only by the potential function of the whole solid. Nonetheless, subject to the rules of quantum mechanics and the Pauli exclusion principle, the electrons are characterized by wave functions that are like free-electron wave functions, except for the restriction on the allowed wavelengths and, hence, energies. In the absence of electric fields applied to the metal, there is no electric current in the crystal even though the electrons are in motion. For each electron characterized by a momentum $\mathbf{p}_i$ there is another electron of equal and opposite momentum $-\mathbf{p}_i$. The net transfer of electric charge within the metal is zero.

The presence of an applied electric field within the metal changes the momentum distribution of the electrons so that the average momentum is nonzero and an electric current flows. The momentum distribution can change only if energy levels are available to which the electrons can be promoted by the field. These available levels are present in metals; indeed, they are the chief characteristic of metals, as we shall see below. The effect of the applied field is to give each electron an additional momentum $\Delta\mathbf{p}$. The entire electron gas then has a net momentum $N\Delta\mathbf{p}$. Associated with this net momentum is an average drift velocity $\mathbf{v}$ of the electron gas. (Recall our discussion of drift velocity in Chapter 17.)

When an electric field is applied to a metal, the current reaches a finite steady-state value. The electrons are no longer accelerated by the field and there is some limit to the magnitude of their drift velocity. This limit arises because the electrons are not really free. They do interact with defects, impurities, and thermal distortions of their host crystal. The resulting collisions cause changes in the electron momentum that limit the magnitude of the drift velocity.

We can reexamine the photoelectric effect to see what it can tell us about metals. In Chapter 26, we found that the relation between the maximum kinetic energy of photoelectrons and the incident light frequency f is given by the Einstein equation:

$$\text{KE}_{\text{max}} = hf - \phi.$$

The work function ϕ represents the minimum energy required to just remove an electron from the surface of the metal.

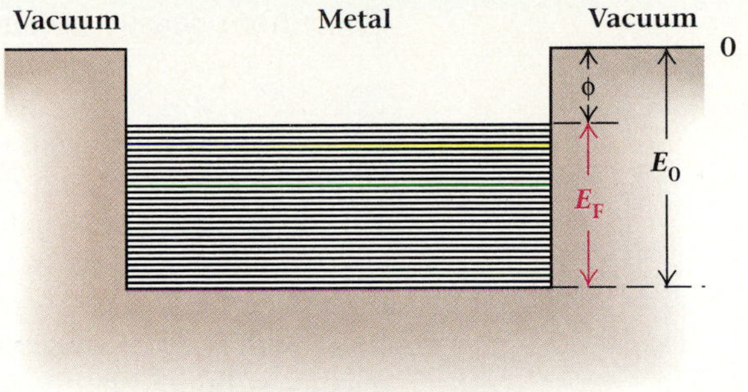

Vacuum   Metal   Vacuum

E_0

E_F

Figure 30.11 The potential well of a metal. The energy levels are filled for $E < E_F$ and empty for $E > E_F$. The photoelectric work function ϕ is the energy needed to take an electron from the Fermi energy E_F in the metal to the top of the potential well.

We can now interpret the photoelectric effect to mean that in the metal the highest occupied electronic energy levels lie at an energy ϕ below the top of the potential well of the metallic crystal. Furthermore, the potential well describing the crystal potential must be finite. Our schematic picture of the metallic potential is revised along the lines of Fig. 30.11, where we see the work function ϕ as the energy required to take an electron from the Fermi energy E_F to a state of zero kinetic energy in the vacuum. The difference between the zero kinetic energy state in the vacuum and the bottom of the conduction band is $E_0 = E_F + \phi$.

Electrical Conductivity and Ohm's Law

30.3

Let us now reexamine electrical conductivity in more detail than we did in Chapter 17. We need to introduce some new quantities that will help us to understand the nature of metals and semiconductors.

We learned earlier that when we apply an electric potential difference to the ends of a conducting wire, a current flows through the wire. This current is due to the net drift velocity of the electrons in the wire. A related quantity is the current density **j**, which we define to be the vector whose magnitude is given by the current divided by the cross-sectional area A of the conductor through which it passes,

$$j = I/A. \tag{30.3}$$

The direction of the current density vector is always the same as the direction of the applied electric field that drives the current. It is, therefore, the direction in which a positive charge would move under the influence of the field.

To understand the relation between current density and drift velocity, consider a segment of a uniform conducting wire (Fig. 30.12). In a time t, electrons moving with drift velocity **v** travel a distance $l = vt$. Thus in a time t, all the electrons in the volume lA pass through cross section A. If the electron density is n, then the current is given by the electric charge in the volume lA divided by the time,

$$I = \frac{n(-e)lA}{t} = -nevA.$$

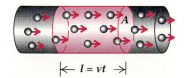

$\leftarrow l = vt \rightarrow$

Figure 30.12 Section of a conducting wire. The arrows indicate only the drift motion.

The current density becomes

$$\mathbf{j} = -ne\mathbf{v}. \tag{30.4}$$

In the solid the electrons suffer collisions with impurities and defects in the crystals, including distortions of the potential that result from thermal vibrations. These collisions occur with a characteristic time τ and serve to limit the drift velocity of the electrons as they are accelerated by an applied field. The presence of an electric field causes the electrons to move with an acceleration $\mathbf{a} = -e\mathbf{E}/m$. They acquire an average drift velocity $\mathbf{v}$, limited by the average collision time, $\mathbf{v} = \mathbf{a}\tau$. When these two equations are combined with Eq. (30.4) for current density, we get an expression for current density in terms of the field,

$$\mathbf{j} = \frac{ne^2\tau}{m}\mathbf{E} = \sigma\mathbf{E}. \tag{30.5}$$

The quantity σ is the **electrical conductivity**.

Electrical conductivity is an intrinsic characteristic of the material from which the wire is made, and is independent of the shape and size of the wire. The SI units of conductivity are $(\text{ohm}\cdot\text{m})^{-1}$. A large conductivity means that a given applied field can produce a large current density in the material. Good conductors like copper and silver have large conductivities.

The reciprocal of the conductivity is the resistivity ρ. Thus a large conductivity corresponds to a small resistivity. (Table 17.1 lists the resistivities of a number of materials.) The resistivity is

$$\rho = \frac{1}{\sigma} = \frac{E}{j}.$$

For a particular conductor of length l and cross section A, the electric field E within the conductors depends on the applied voltage across it through $E = V/l$. We can then express the resistivity as

$$\rho = \frac{V/l}{i/A},$$

which can be rearranged to give

$$\frac{V}{i} = \frac{\rho l}{A}.$$

However, we earlier defined the ratio of voltage to current to be the electrical resistance R. Thus the resistance is

$$R = \rho\frac{l}{A}.$$

This is the same expression for resistance in terms of resistivity that we had in Chapter 17. If the resistivity depends only on the microscopic properties of the material and not on the applied field, then the resistance is independent of voltage and the material obeys Ohm's law.

Another quantity related to the drift velocity is the **mobility**, defined as the magnitude of the drift velocity per unit field,

$$\mu = v/E. \tag{30.6}$$

The mobility changes slowly with temperature and, for constant temperature, is constant for a given crystal. By combining Eqs. (30.4) and (30.5), we find that the mobility is primarily determined by the electron collision processes through the collision time τ, since

$$\mu = \frac{e\tau}{m}.$$

As a matter of convention, μ is always taken to be a positive quantity.

The conductivity can also be expressed in terms of the mobility through

$$\sigma = ne\mu. \tag{30.7}$$

This equation for conductivity allows us to compute the conductivity from knowledge of the electron density n and the mobility. If there are collections of electrons with different mobilities, we can obtain the total conductivity by adding the contributions due to the different mobilities. We find this approach especially useful when dealing with semiconductors.

Example 30.2

Calculating electron mobility in lithium.

The electrical conductivity of lithium is $1.08 \times 10^7 \ \Omega^{-1}m^{-1}$. Calculate the mobility of electrons in lithium using the electron density found in Example 30.1.

Solution We can find the mobility from the relation

$$\sigma = ne\mu,$$

which can be rearranged to get

$$\mu = \frac{\sigma}{ne} = \frac{1.08 \times 10^7 \ \Omega^{-1} \ m^{-1}}{4.6 \times 10^{28} \ m^{-3} \times 1.6 \times 10^{-19} \ C},$$

$$\mu = 1.47 \times 10^{-3} \ m^2 \ V^{-1} \ s^{-1}.$$

The mobility in lithium is nearly 100 times smaller than the mobility of electrons in silicon.

Band Theory of Solids

30.4

In the preceding sections we developed a model for metals, consisting of nearly free electrons contained within a uniform potential well that extends over the entire crystal. We now examine a somewhat more realistic model in which the potential varies periodically with distance through the crystal. The period of the potential is the interatomic spacing between the

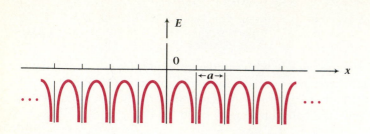

Figure 30.13   Potential energy versus position along a one-dimensional lattice of evenly spaced nuclei separated by a distance a.

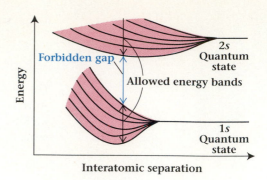

Figure 30.14   Dependence of energy levels on interatomic separation.

ions that make up the crystal. One such potential is illustrated in Fig. 30.13.

We can gain some insight by imagining a solid with interatomic separations so large that the electronic energy levels of the crystal are just those of the isolated atoms. This would create an enormous number of states with the same energy. The number of independent states of a particular energy is the product of the number of atoms in the crystal and the number of distinct states of that energy in an isolated atom. Thus a crystal of N widely spaced atoms that possess m states of energy E will have Nm independent states belonging to the same energy level E.

If we reduce the interatomic spacing in our imaginary crystal, the mutual interaction of the atoms causes a shift in the energy levels, so that the states no longer have the same energy. The result is that each original energy level is split into a large number of distinct levels (Fig. 30.14). The total number of distinct levels is Nm, which is the total number of states that belong to the same basic atomic level. For a real crystal with a mass of one milligram, $N \approx 10^{19}$. Thus the number of levels in any macroscopic sample is very great. These levels occur within an energy spread on the order of one electron volt. Thus the separation between levels is about 10^{-19} eV. For this reason we often speak of the electronic levels of a crystal as a continuous band of states. However, the band is not truly continuous; it contains a very large but finite number of distinct levels.

It is helpful to illustrate the energy bands with a picture like that of Fig. 30.15, which combines the crystal potential of Fig. 30.13 with information on the electronic bands obtained from Fig. 30.14. The low-lying bands correspond to atomiclike wave functions that have appreciable amplitude only near the ion cores. The higher energy states are represented as being more evenly spread throughout the crystal. The nonlocalization of the higher-lying states means that although the electrons are bound to the entire solid, they are not associated with particular atoms. The relatively free motion of some of the electrons through the metal crystal is due to this nonlocalization.

The effect of the periodic crystal potential on the electronic energy levels is to group the higher states into **allowed bands**, separated by regions of energy in which no electron states are allowed, known as

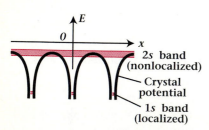

Figure 30.15   Energy bands superposed on a diagram of potential energy versus distance in a crystal.

forbidden zones or gaps. An electron within an allowed band can be described by a wave packet in a manner similar to the treatment of the free-electron model. Each of these wave packets corresponds to a particular electron momentum.

Because of the symmetry of the crystal and, hence, of the energy bands, when all the energy states are filled there is no net momentum in the electron system. Consequently, a completely filled band carries no electric current because the net velocity of the electrons in a filled band is always zero. In addition, the restriction imposed by the forbidden zones means that there are no accessible energy states nearby. Thus there can be no current in a material with filled energy bands even in the presence of an applied electric field. In other words, we have just described the electronic structure of an insulator.

But how do we describe a metal with the band model? Remember that a set of N regularly spaced potential wells can give rise to a band of N independent states. If each atomic energy level corresponds to two states associated with the electron spin, then the energy band contains $2N$ independent states. Also, we have seen that a completely filled band does not conduct a current any better than an empty band. This fact suggests that the best conductors are those with a half-filled allowed band, such as the alkali metals of the first column in the periodic table. For example, sodium has a single electron above a core of filled atomic levels. Thus the energy band corresponding to this atomic level in metallic sodium is only half filled (Fig. 30.16a). In fact, sodium does have a high conductivity and is considered a good metal, as would be expected from the band model.

At this point, you might expect that elements in the second column of the periodic table, such as beryllium and magnesium, would be insulators, because each atom contributes two electrons to the highest occupied band. Thus the band would be filled, as in the case of the insulator previously described. However, elements from the second column are not insulators, but are metals with conductivities nearly as good as those of the first column. Their conductivities arise from the overlap of a higher-energy band with an otherwise filled band (Fig. 30.16b). The noble metals (gold and silver) contribute one electron per atom. They are indeed good metals, as might be predicted from this model.

In some materials the bands do not overlap. In these crystals, if the number of electrons is just sufficient to fill several of these energy bands,

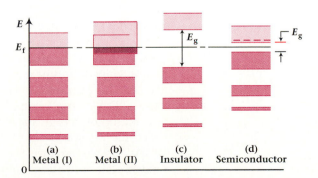

Figure 30.16 Band diagrams for (a) type I metal (partially filled band), (b) type II metal (band overlap), (c) insulator, (d) semiconductor.

TABLE 30.1	
Energy gaps of some common semiconductors at room temperature	
Material	Energy gap (eV)
Indium antimonide (InSb)	0.16
Germanium (Ge)	0.67
Silicon (Si)	1.14
Gallium arsenide (GaAs)	1.4
Cadmium sulfide (CdS)	2.42
Zinc sulfide (ZnS)	3.6

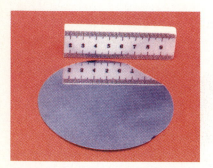

Figure 30.17   A silicon crystal wafer of the type used in the manufacture of integrated circuits.

the material may be an insulator. In particular, an insulator occurs when the energy gap E_g of the forbidden zone, separating the uppermost filled band from the nearest empty band of allowed states, is so large that thermal or photoexcitation of electrons across the gap is insignificant (Fig. 30.16c). An example of such a material is diamond, which has a large energy gap (5.33 eV) between the filled energy band and the next allowed band.

When the energy gap E_g is sufficiently small (Fig. 30.16d), thermal excitation may become significant and the crystal will no longer be an insulator. Instead, it becomes a **semiconductor**, having a conductivity intermediate between those of an insulator and a metal. In addition, its conductivity may increase in the presence of light as a result of photoexcitation of electrons from the uppermost filled band to the next higher allowed band. Silicon and germanium have crystal structures like that of diamond, but their band gap energies are much lower: 1.14 eV in silicon and 0.67 eV in germanium. They are both semiconductors. Table 30.1 lists the gap energies of some other semiconductors.

Silicon and germanium crystals both look metallic (Fig. 30.17). We explain their appearance by noting that the photon energy in visible light is sufficient to cause photoexcitation of electrons from the filled band to the next higher band, called the conduction band. Thus the presence of light generates a high density of free electrons within the bands, which creates the metallic luster. Diamond, on the other hand, is transparent to visible light because the photon energy is smaller than the band gap energy.

When semiconductor materials are kept in the dark, random thermal excitations promote electrons into the conduction band. As the temperature of the material is lowered, the thermal excitations diminish and the number of electrons within the conduction band decreases. As the temperature gets closer to absolute zero, the conductivity of the semiconductor material gets progressively poorer until it becomes an insulator.

Pure Semiconductors

30.5

Semiconductors are just what their name implies; that is, they conduct to some extent, with conductivities intermediate between those of good conductors and good insulators. The conductivities of semiconductors range from about 10^{-6} to $10^4/\Omega\cdot m$ (see Table 17.1). The electrical properties of semiconductors are greatly influenced by impurities, present in amounts as small as one part per million or even less. Semiconductors that are very pure and whose behavior is not dominated by impurity atoms are called **pure (intrinsic) semiconductors**.

At temperatures near absolute zero, an intrinsic semiconductor is characterized by a completely empty energy band, separated by a small energy gap (≈ 1 eV) from a completely filled band (Fig. 30.18). The filled band is made up of energy states of the valence electrons of the atoms and is called the *valence band*.

As the temperature rises, some electrons are excited across the forbidden zone into the empty upper band. Then a continuous process of

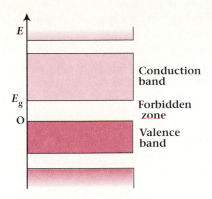

Figure 30.18 Energy band diagram near the valence band in a semiconductor. The top of the valence band is arbitrarily chosen to be the zero level.

electron transfer occurs, both into the upper band from below and out of the upper band into the valence band. At any given temperature the number of electrons in the upper band changes until the rates of transfer into and out of the band become equal. Since the electrons in this nearly empty band have many available states nearby, they can carry a current. For this reason, the upper band is called the *conduction band*.

The transfer of electrons out of the valence band leaves empty states within that band, allowing for a nonzero distribution of momentum within the band. Thus the valence band can also carry a current. The effective number of charge carriers is equal to the number of empty electron states. Because of the effects of the crystal potential on the momentum of the electrons, we can treat these empty states as positive charge carriers called *holes*. Although the actual mechanism of charge transport is due to electrons, in a nearly filled band we commonly describe the transport in terms of the holes.

In an intrinsic semiconductor, all of the electrons excited into the conduction band leave holes in the valence band: Their numbers must be exactly equal. If we let n be the electron density in the conduction band and p be the hole density in the valence band, then $n = p$ in an intrinsic material. In determining the overall conductivity of the semiconductor, we must add the contributions due to both kinds of carriers. Thus we write

$$\sigma = |e| \, (n\mu_n + p\mu_p), \tag{30.8}$$

where μ_n and μ_p represent the electron and hole mobilities, respectively. In the following section we describe an experimental technique that allows us to measure the density, as well as the sign, of the charge carriers.

The Hall Effect

30.6

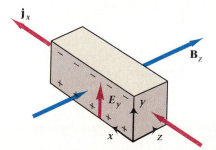

Figure 30.19 Hall effect geometry: current in the *x* direction, magnetic field in the *z* direction, and Hall field in the *y* direction.

Before we consider the behavior of impure semiconductors, let's examine one experiment that will help us establish the reality of charge transport by holes. In 1879 E. H. Hall observed that a voltage could be generated across a current-carrying conductor by placing the conductor in a magnetic field. When the direction of the current is perpendicular to the direction of the magnetic field, a voltage develops across the conductor at right angles to both the current and the magnetic field. This voltage is known as the Hall voltage, and its occurrence is called the **Hall effect**.

Let's examine the geometry of the Hall effect more closely (Fig. 30.19). An electric field applied to the conducting material in the *x* direction gives rise to an electric current density $\mathbf{j}_x$. As the charge carriers move through the material they are deflected by the magnetic field $\mathbf{B}_z$. The result is a net separation of the charge, with an accumulation of negative charge on one edge and positive charge on the other. The separated charges establish an electric field whose effect is to oppose the magnetic deflection on the moving charges. Thus in the steady state no current exists in the *y* direction. Subsequent charges pass through without deflection, since the opposing forces are balanced.

A charge e moving in the x direction in Fig. 30.19 experiences a force in the negative y direction due to the magnetic field, given by $F_m = ev_x B_z$. This deflection causes a separation of charge in the material, which generates an electric field $\mathbf{E}$ in the positive y direction. This field, called the Hall field, exerts a force $F_e = eE_y$ on the charge carriers. Since there is no steady-state current in the y direction, the magnitude of these two forces must be equal. Thus $F_e = F_m$ or

$$eE_y = ev_x B_z.$$

Using the relation for the current density, $j_x = nev_x$, we can rewrite this equation as

$$eE_y = \frac{j_x B_z}{n}.$$

Upon rearrangement, the density of charge carriers is given by

$$n = \frac{j_x B_z}{eE_y}. \tag{30.9}$$

Thus by measuring the current density j_x, the magnetic field B_z, and the Hall field E_y, it is possible to determine the density of mobile charge carriers within the material.

The direction of the Hall field depends on the sign of the charge carrier. Thus the Hall effect is a relatively simple means for determining the sign of charge carriers in a material. For alkali metals and noble metals, the experimentally determined values of n are close to the values calculated from free-electron theory and the direction of the Hall field is that expected for the motion of negative charges. For many other materials the agreement with free-electron theory is poorer. In some cases, like metallic bismuth, the sign of the Hall field is reversed, indicating an effective positive charge on the current carriers. Thus the conductivity of bismuth is described in terms of the motion of positively charged holes.

The transport of electric charge through metals and semiconductors is due to the motion of the electrons. However, when the energy bands are more than half filled, these materials behave as if the charge were being carried by positive carriers, the holes. The Hall effect is one of several experiments that are most simply interpreted in terms of electrons and holes. As you will see in the following section, the concept of charge transport by holes is a great help in understanding the behavior of semiconductors.

Example 30.3

Measuring carrier density with the Hall effect.

A rectangular ribbon of silver, having dimensions $l_z = l_y = 1.0$ mm, is suspended in a magnetic field $B_z = 1.0$ T. When the current through the ribbon is 1.0 A, the Hall field is 1.1×10^{-4} V/m. What is the density of free carriers in the silver?

Solution The density of charge carriers is readily computed from Eq. (30.9). The current density j_x is the current divided by the cross-sectional area of the ribbon, which is $l_z l_y = 1.0$ mm $\times$ 1.0 mm $= 1.0 \times 10^{-6}$ m^2. Thus

$$n = \frac{j_x B_z}{e E_y} = \frac{(1.0 \text{ A}/1.0 \times 10^{-6} \text{ m}^2)(1.0 \text{ T})}{(1.6 \times 10^{-19} \text{ C})(1.1 \times 10^{-4} \text{ V/m})} = 5.7 \times 10^{28} \text{ m}^{-3}.$$

Hall probes are Hall-effect devices used for measuring magnetic fields. When a current passes through a small semiconductor crystal in the presence of a magnetic field, a voltage develops just as we have already seen. The magnitude of this voltage depends on the magnitude of the magnetic field and the orientation of the probe relative to the direction of the field. Hall probes are routinely used to measure fields from 10^{-4} T to 10 T.

Example 30.4

Measuring magnetic field with a Hall probe.

A Hall probe used to measure a magnetic field is constructed from a crystal of indium arsenide with dimension $l_y = 2.5$ mm (Fig. 30.19) and a carrier density of $4.2 \times 10^{22}/\text{m}^3$. If the voltage due to the Hall field across the crystal is 8.9 μV when the current density j_x is 160 A/m^2, what is the magnitude of the magnetic field transverse to the crystal, that is, B_z?

Solution Equation (30.9) can be rewritten to give the magnetic field component B_z

$$B_z = \frac{neE_y}{j_x} = \frac{neV_y}{j_x l_y}.$$

When the numerical values are inserted we get

$$B_z = \frac{(4.2 \times 10^{22}/\text{m}^3)(1.6 \times 10^{-19} \text{ C})(8.9 \times 10^{-6} \text{ V})}{160 \text{ A/m}^3 \times 2.5 \times 10^{-3} \text{ m}} = 0.15 \text{ T}.$$

Impure Semiconductors

30.7

Impure (or **extrinsic**) **semiconductors** are made from pure semiconducting material by adding small numbers of impurity atoms of higher or lower valence. In silicon and germanium the addition of impurities with valence higher than four results in materials whose conductivities are dominated by electron carriers. Addition of atoms with fewer than four valence electrons leads to materials in which the dominant carriers are holes.

The controlled addition of impurities allows us to control the conductivity, as well as the sign of the majority of the carriers. The reason that we can do so is that the concentration of carriers in a pure semicon-

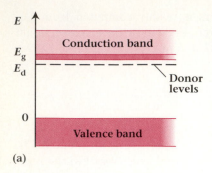

(a)

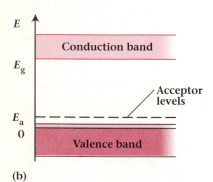

(b)

Figure 30.20 Detail of the band gap in an impurity semiconductor. (a) Donor impurity levels, lying near the bottom of the conduction band, contribute electrons into the band. (b) Acceptor levels, lying near the top of the valence band, trap electrons from the band, leaving holes free to move through the crystal.

ductor at room temperature is quite small compared with the density of atoms. For example, the carrier concentration in pure silicon is about 1.5×10^{10} cm^{-3}. The addition of impurity concentrations on the order of parts per million can increase the number of available charge carriers by a factor of a million. Thus the conductivity of a semiconductor material is extremely sensitive to the impurity concentrations.

Addition of an impurity atom with five valence electrons, such as arsenic, into a crystal of silicon results in a set of new electronic energy levels localized at the impurity atom. The new localized level corresponding to the conduction band state of the host crystal actually lies slightly below the bottom of the conduction band, in the forbidden zone. Such impurity levels usually lie only a small fraction of an electron volt below the band edge (Fig. 30.20a). The energy required to transfer an electron from an impurity level to the conduction band is so small that most impurity sites are empty (or ionized) at room temperature. Because of the ready availability of these electrons for transitions to empty states in the conduction band, this kind of "doped" semiconductor is one in which the negative carriers dominate; that is, $n \gg p$, where n is the free-electron density and p is the density of holes. Impurity atoms that increase the number of negative carriers are called *donor impurities* because they donate electrons to the band. The resulting material is called a negative or **n-type semiconductor**.

In a similar manner, impurities with only three valence electrons, such as gallium, lead to localized electron states only slightly above the valence band (Fig. 30.20b). These states would be empty at a temperature of 0 K. However, at room temperature, thermal excitations promote electrons out of the valence band and into these impurity states. This results in an excess of holes, or positive carriers. These impurities are known as *acceptor impurities* because they accept or trap electrons out of the valence band. The resulting materials have a predominance of positive carriers and are known as **p-type semiconductors**.

The *pn* Junction

30.8

Many contemporary electronic devices owe their existence to junctions where *n*-type and *p*-type semiconductor materials are joined together. Junctions of this type, known as **pn junctions,** form highly efficient rectifiers. These junctions have impurity concentrations that vary in such a way that the semiconductor crystal goes from *n*-type to *p*-type across a relatively narrow transition region. The change from *n*-type to *p*-type material may be abrupt, as when regions of virtually constant impurity concentrations contact each other, or the change may be graded so that the impurity concentration varies gradually across the transition region. In either case the crystal structure is maintained across the transition region.

As an example of a *pn* junction, let's imagine an abrupt junction between *p*-type material containing N_a acceptor impurities per unit volume and an *n*-type material containing N_d donor impurities per unit volume

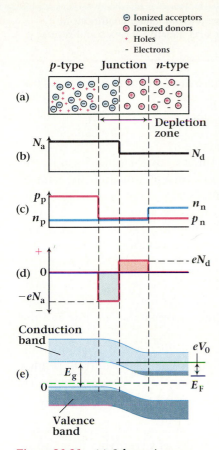

⊖ Ionized acceptors
⊙ Ionized donors
+ Holes
- Electrons

Figure 30.21 (a) Schematic representation of a pn junction in equilibrium, (b) impurity concentration, (c) mobile charge concentration, (d) charge density, and (e) energy band diagram showing the effect of the junction potential V_0.

(Fig. 30.21). Since many practical junctions are asymmetric in their impurity concentrations, we let $N_a > N_d$. At equilibrium, with zero applied voltage, electrons and holes diffuse across the junction and recombine. As a result, they create a region around the junction in which the electron and hole concentrations are greatly reduced, a region known as the *depletion zone*. There is a dipolar charge layer in the depletion zone because of the ionized impurities that are no longer neutralized by the mobile charges. This dipolar layer alters the electrostatic potential near the junction, creating an electric field that opposes the flow of electrons out of the n-type material and the flow of holes out of the p-type material.

The diffusion of electrons from the n side to the p side is limited by the electric potential V_0 resulting from the dipolar charge layer. Diffusion of electrons in the other direction, from the p side to the n side, is limited by their availability because of the small density of free electrons in the p-type material. In the same way, the potential inhibits the diffusion of holes from the p side to the n side. For this reason we refer to it as a potential barrier.

A pn junction diode consists of a pn junction with two nonrectifying contacts applied, one to the p material and one to the n material. With no external voltage applied there is an electric potential across the junction region (Fig. 30.21e). At equilibrium that potential difference is V_0 and the net current through the junction is zero.

We can determine the current–voltage relationship for a diode by considering the currents across the junction. Considering first the electrons, we allow for motion of electrons both ways across the junction. For a density n_n of electrons in the n-type material, the diffusion current of electrons across the potential barrier and into the p-type material is

$$I_f = Cn_n e^{-eV_0/kT}.$$

The factor $e^{-eV_0/kT}$ gives the probability that an electron in the n-type material has a kinetic energy eV_0 and thus is able to get over the potential barrier. C is a constant that depends on the mobility, the junction area, and the electronic charge.

There is also a current resulting from the diffusion of electrons out of the p-type material. When these electrons diffuse to the edge of the depletion zone they are swept across by the junction potential, generating a current

$$I_s = Cn_p,$$

where n_p is the density of electrons in the p-type material. At equilibrium these two opposing currents must be equal for the net current to be zero. Then

$$n_p = n_n e^{-eV_0/kT}. \tag{30.10}$$

When we apply a potential difference V in the forward direction, the potential barrier is lowered (Fig. 30.22) and the current I_f becomes

$$I_f = Cn_n e^{-e(V-V_0)/kT}.$$

The current I_s is unaffected by the added potential, since all electrons

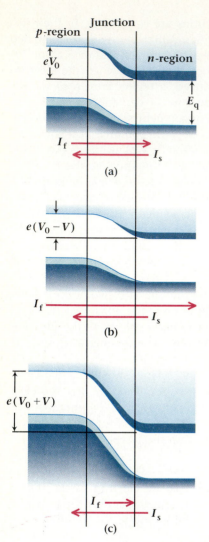

Figure 30.22   Schematic representation of a *pn* junction under conditions of (a) zero bias, (b) forward bias, and (c) reverse bias. Relative forward and reverse currents are indicated.

reaching the junction from the *p* side still contribute to the current. Thus the net current becomes

$$I_n = I_f - I_s = I_s(e^{eV/kT} - 1),$$

where we have used Eq. (30.10) to eliminate n_n.

The current I_p due to the motion of holes has the same form as the net current due to the electrons. The total current, which is the sum of the electron current and the hole current, is

$$I = I_n + I_p = I_0(e^{eV/kT} - 1), \qquad (30.11)$$

where I_0 represents the maximum current in the reverse direction, known as the *saturation current*. Equation (30.11) is known as the *rectifier equation*. Note that for strongly doped materials, the densities of holes p_n in the *n*-type material and electrons n_p in the *p*-type material are low and the saturation current is necessarily small.

When an electric potential difference is applied across a *pn* junction, making the *p* side positive with respect to the *n* side, a large current results, corresponding to a positive V in the rectifier equation. When the potential difference is reversed, corresponding to a negative V in Eq. (30.11), there is very little current.

The importance of the *pn* junction in semiconductor devices cannot be overemphasized. Whether used individually as a rectifying junction in a diode or in combination to make transistors or other devices, *pn* junctions play a paramount role in contemporary electronics. The integrated circuit memory chips that are used in modern digital computers consist of millions of *pn* junctons assembled together on a single silicon surface no larger than a nickel.

Solar Cells and Light-Emitting Diodes

*30.9

The absorption of light in semiconductors generates electron-hole pairs in excess of the number already present by virtue of thermal agitation. This effect, called photoabsorption, was mentioned earlier in connection with the photoexcitation of electrons from the valence band to the conduction band. The increased conductivity associated with the excess carriers is called photoconductivity. If the protective material surrounding a diode, transistor, or other semiconductor device has a transparent window, that device may then be controlled with light. Devices of this kind are known as *optoelectronic* devices.

A *photodiode* is a *pn*-junction diode fabricated so that light can fall on

Figure 30.23 A digital voltmeter display made with light-emitting diodes.

the region near the junction. When the diode is operated with a reverse bias, a large change in its resistance can be caused by photoabsorption. Photons of energy greater than the energy gap E_g generate excess electron-hole pairs, which contribute to an increased reverse current proportional to the incident light intensity. A *solar cell* is a photodiode with zero applied voltage. It produces a current that is due to carriers generated near the junction and carried across by the junction field. Under illumination, the solar cell can be used as a source of dc current to replace batteries.

In addition to the numerous light-activated devices, there are other devices that emit radiation. This radiation, associated with the direct recombination of electrons from the conduction band with holes in the valence band, is seen in *light-emitting diodes* (LEDs). Because the photon energy is essentially that of the band gap E_g, most LEDs emit light in the infrared or red portion of the spectrum (Fig. 30.23). However, green and yellow LEDs are available.

In normal operation an LED emits light when a forward voltage of about 1.5 to 2.0 V is applied. The light output is roughly proportional to the current in the diode. Because an LED operates in the forward conducting mode, it requires much more power to operate than does an LCD. For this reason, a pocket calculator with an LED display requires larger batteries than the equivalent calculator with a liquid crystal display.

Laser diodes are also available. They are light-emitting diodes that have been especially prepared with polished faces transverse to the depletion zone that act like the mirrors normally used for a laser. Most LEDs and laser diodes are made from alloys of the intermetallic compounds gallium arsenide and gallium phosphide because they have band gaps of the appropriate size for the emission of visible light.

Laser diodes are used in compact disc digital audio systems because they can be brought to ultrasharp focus (Fig. 30.24). A sharply focused beam, less than 1 μm in diameter, is necessary for detecting the audio

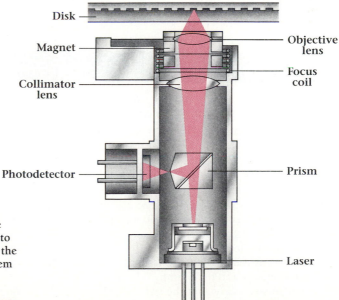

Figure 30.24 Compact disc optical system. The laser beam passes through a semireflecting prism to the lens system, which keeps it in sharp focus on the disc. Reflected light returns through the lens system to the prism, where it is deflected into the photodetector.

Disk

Magnet

Collimator lens

Photodetector

Objective lens

Focus coil

Prism

Laser

information, which is stored as tiny pits only 0.5 μm wide, 0.1 μm deep, and approximately 1 to 3 μm long in a spiral track with a pitch of only 1.6 μm. The pits are pressed into a flat, reflective surface. The beam optics direct the beam to the surface of the disc, where it is reflected and sent to a photodetector. When the beam encounters a pit, it is scattered so that very little light reaches the detector. The sequence of "on" and "off" pulses of light reflected from the disc are converted to digital electrical signals, which are sent to a computer for decoding to reproduce the original sound that was encoded on the disc in the form of the pits.

SUMMARY

Helpful Concepts

- The Fermi energy of electrons in a metal is given by

$$E_F = \frac{h^2}{2m}\left(\frac{3N}{8\pi V}\right)^{2/3}.$$

- The current density is

$$j = I/A = nev.$$

- The mobility of the charge carriers in a conductor is defined by

$$\mu = v/E.$$

- The conductivity is related to the mobility by

$$\sigma = ne\mu.$$

- In the Hall effect the density of charge carriers is

$$n = \frac{j_x B_z}{eE_y}.$$

- The current in a diode is given by the rectifier equation

$$I = I_0(e^{eV/kT} - 1).$$

Important Terms

You should be able to write the definition or meaning of each of the following terms:

- crystalline solid
- polycrystalline solid
- amorphous solid
- liquid crystal
- Fermi energy
- electrical conductivity
- mobility
- allowed bands
- forbidden zones
- semiconductor
- pure (intrinsic) semiconductors
- Hall effect
- impure (extrinsic) semiconductors
- n-type semiconductor
- p-type semiconductor
- pn junction

QUESTIONS

30.1 Carefully distinguish between crystals and glasses.

30.2 What keeps the atoms in a crystal apart?

30.3 What are the forces that hold atoms together in solids? In liquids?

30.4 Why is iron a solid at room temperature but mercury a liquid?

30.5 What gives metals their lustrous appearance?

30.6 What affects the distribution of kinetic energies of the electrons emitted in a photoelectric process when the energy of the photons exceeds the work function by a few electron volts?

30.7 Estimate the drift velocity of the electrons in a wire carrying current to a 100-W lamp. Why does the lamp come on so quickly after the switch is closed?

30.8 The band structures of diamond and silicon are quite similar. However, the band gaps between the valence and conduction bands are different, being 5.33 eV for diamond and 1.14 eV for silicon. How can this lead to a simple explanation of why silicon has a metallic appearance and why diamond is transparent?

30.9 Compare semiconductors and insulators. How are they alike and how are they different?

30.10 What evidence is there for the conduction of electricity by holes? Do the holes really exist?

30.11 Why was resonance absorption so readily seen in atomic spectroscopy and yet so difficult to find in nuclear gamma radiation?

PROBLEMS

Section 30.1   Types of Condensed Matter

30.1 Estimate the spacing between nearest-neighbor ions in a sodium chloride crystal from knowledge of the molecular weight and the density $\rho = 2.17$ g/cm^3. Compare your result with the spacing of 0.281 nm determined experimentally by x-ray diffraction.

30.2 The interatomic distance along the cube edge in cesium chloride is 0.411 nm. What is the interatomic distance between the nearest-neighbor cesium and chlorine ions? (*Hint:* All distances are measured center to center.)

30.3 The nearest-neighbor distance in diamond is one-fourth the corner-to-corner distance along the body diagonal of the cubic unit cell. Show that the near-neighbor spacing is $(\sqrt{3}/4)a$, where a is the length of the edge of the unit cell.

30.4 The cesium chloride unit cell has an edge length of 0.411 nm. What is the spacing between a cesium ion and its nearest cesium neighbor?

Section 30.2   The Free-Electron Model of Metals

30.5 Calculate the free-electron density in sodium, which has a mass density of 0.97 g/cm^3 and an atomic mass of 23.0 g/mol.

30.6 Calculate the free-electron density in potassium. Assume the mass density to be 0.82 g/cm^3 and the atomic mass to be 39.1 g/mol.

30.7 Calculate the equivalent temperature T_F of the electrons at the Fermi surface of lithium. (*Hint:* T_F is defined by $E_F = kT_F$. Use the result of Example 30.1.)

30.8 The photoelectric work function of silver is 4.74 eV. The Fermi energy is 5.51 eV. (a) What is the least energetic photon required to remove an electron from the bottom of the potential well? (b) What is the wavelength of this photon? (c) What is the maximum kinetic energy for photoelectrons ejected by photons of this wavelength?

30.9 Metallic sodium has a photoelectric work function

$\phi = 2.75$ eV and a Fermi energy $E_F = 3.10$ eV. (a) What is the least energetic photon required to remove an electron from the bottom of the potential well? (b) What is the wavelength of this photon?

30.10 The threshold wavelength for photoelectric emission from tin is 340 nm. (a) What is the value of the work function? (b) What is the color of the threshold light? (c) Will red light from a helium-neon laser generate photoelectrons from tin?

Section 30.3   Electrical Conductivity and Ohm's Law

30.11 Estimate the average collision time for electrons in silicon, given that the mobility is 1300 cm^2/V·s.

30.12 Estimate the average collision time for electrons in lithium from the mobility which was calculated in Example 30.2.

30.13 What is the drift velocity of electrons in lithium in a region of $E = 10^{-3}$ V/cm? (*Hint:* Use the result of Example 30.2.)

30.14 Compute the average electron mobility in gold, given that the electrical resistivity is 2.44×10^{-8} Ω·m and the electron density is 5.90×10^{28} electrons/m^3.

30.15 Compute the average electron mobility in silver, given that the electrical resistivity is 1.59×10^{-8} Ω·m and the electron density is 5.85×10^{28} electrons/m^3.

30.16 (a) Calculate the mobility of electrons in gold, given that the electron density is 5.90×10^{28} m^{-3} and the resistivity is 2.4×10^{-8} Ω·m. (b) What is the average collision time of electrons in gold?

30.17 The density of free electrons in silver is 5.76×10^{28} m^{-3} and the electrical conductivity is 6.17×10^7 Ω^{-1}m^{-1}. (a) What is the electronic mobility μ? (b) What is the collision time τ?

Section 30.4   Band Theory of Solids

30.18 Calculate the wavelength of the radiation just sufficient to cause electron transitions from the valence band to the conduction band in germanium. To what kind of radiation does this correspond? (*Hint:* $E_g = 0.67$ eV.)

30.19 Calculate the wavelength of the radiation just sufficient to cause electron transitions from the valence band to the conduction band in diamond. To what kind of radiation does this correspond? (*Hint:* $E_g = 5.33$ eV.)

30.20 Silicon can be used as a window for infrared radiation. What is the shortest wavelength that will pass through silicon without causing electron excitation from valence band to conduction band?

Section 30.5   Pure Semiconductors

30.21 Calculate the conductivity of pure germanium at a temperature where the density of electrons is $n = p = 2.4 \times 10^{19}$ m^{-3}, given that the mobilities are $\mu_n = 3900$ cm^2/V·s and $\mu_p = 1900$ cm^2/V·s.

30.22 Calculate the conductivity of intrinsic gallium arsenide at a temperature where the density of carriers is $n = p = 9.1 \times 10^{12}$ m^{-3}, given that the mobilities are $\mu_n = 8600$ cm^2/V·s and $\mu_p = 250$ cm^2/V·s.

30.23 The mobility of electrons in silicon is 1350 cm^2/V·s. (a) Calculate the drift velocity of an electron in silicon moving in a field of 100 V/cm. (b) Calculate the final velocity of a free electron accelerated through a potential of 100 V.

Section 30.6   The Hall Effect

30.24 Verify that the direction of E_y in Fig. 30.19 is correct for positive charge carriers. In what direction is E_y when the current density j_x is due to negative carriers?

30.25 A ribbon of copper having dimensions $l_z = 2.0$ mm and $l_y = 5.0$ mm is suspended in a magnetic field $B_z = 1.25$ T. When the current through the ribbon is 10 A, the Hall field is $E_y = 9.3 \times 10^{-5}$ V/m. What is the density of free carriers in this material?

30.26 A ribbon of n-type silicon having dimensions $l_z = 0.2$ cm and $l_y = 0.5$ cm is suspended in a magnetic field $B_z = 1.0$ T. The free carrier density in this material is 4.6×10^{22} m^{-3}. What is the Hall field E_y when a current of 125 mA is passed through the sample? What is the Hall voltage across the sample?

Section 30.7   Impure Semiconductors

30.27 Repeat Problem 30.21 under conditions that $n = 7.2 \times 10^{19}$ m^{-3} and $p = 8 \times 10^{18}$ m^{-3}.

30.28 Calculate the conductivity of n-type gallium arsenide for which $n = 5.1 \times 10^{18}$ m^{-3} and $p = 1.6 \times 10^{7}$ m^{-3}. The mobilities are $\mu_n = 8600$ cm^2/V·s and $\mu_p = 250$ cm^2/V·s.

Section 30.8   The pn Junction

30.29 Calculate the junction potential V_0 of a silicon pn junction at room temperature, given that the materials are doped so that the concentration of electrons in the n-type material is 1.0×10^{14} cm^{-3} while the concentration of electrons in the p-type material is 1.0×10^{5} cm^{-3}.

30.30 Calculate the junction potential V_0 of a silicon pn junction at room temperature (300 K), given that the materials are doped so that the concentration of electrons in the n-type material is 5.0×10^{14} cm^{-3}, while the concentration of electrons in the p-type material is 1.0×10^{5} cm^{-3}.

30.31 Using the rectifier equation, calculate the current through a diode at reverse voltages of 0.10 V and 1.0 V and forward voltage of 0.30 V, given that the saturation current is 1.0×10^{-7} A. Assume that $T = 300$ K.

30.32 A diode has a saturation current of 0.10 μA. What current would it carry at a forward voltage of 0.40 V if it exactly obeyed the rectifier equation?

*Section 30.9   Solar Cells and Light-Emitting Diodes

30.33 An LED is made from material with a band gap of 1.4 eV. At what wavelength will it radiate?

30.34 Light is emitted by an LED in a process in which an electron makes a direct transition from the conduction band to the valence band, where it recombines with a hole. If the light has a wavelength of 650 nm, what is the energy of the band gap?

30.35 Suppose that you have solar cells that are 10% efficient; that is, they convert 10% of incident sunlight into electrical energy. (a) Assuming an incident solar flux of 1.0 kW/m^2, how large a collector is needed to provide a home with electrical power at a rate of 40 kW? (b) How large a collector is needed if the sunlight makes an incident angle of 45° with the surface of the collector?

Additional Problems

30.36 The nearest-neighbor spacing in sodium chloride is
• 0.281 nm. What is the spacing between closest neighbors of the same kind; that is, what is the shortest Na–Na distance?

30.37 Calculate the distance between nearest neighbors in
• the diamond crystal in terms of the edge length of the cubic unit cell, a_0.

30.38 Calculate the Fermi energy for the electrons in ce-
• sium, which has a free-electron density of 0.86×10^{22} cm^{-3}.

30.39 Calculate the Fermi energy for the electrons in ru-
• bidium, which has a free-electron density of 1.08×10^{22} cm^{-3}.

30.40 Calculate the equivalent temperature T_F of the elec-
• trons at the Fermi surface of cesium, which has a free-electron density of 0.86×10^{22} cm^{-3}. (*Hint:* First calculate E_F. Then use the definition of T_F from $E_F = kT_F$.)

30.41 A copper wire 10 cm long has a cross section of 0.50
• mm^2. (a) Calculate the resistance of the wire, given that copper has a resistivity of 1.69×10^{-8} Ω·m. (b) Calculate the mobility of the electrons, given that the electron density in copper is 8.5×10^{28} m^{-3}. (c) What is the drift velocity of the electrons when the electric potential difference across the length of the wire is one volt?

30.42 Current to a 100-W lamp is carried by a wire with
• a cross section of 1.00 mm². The lamp operates on
24V-dc. (a) Calculate the drift velocity of the electrons in the wire, assuming that it is made of copper
with a carrier density of $8.5 \times 10^{28}/m^3$. (b) How
long will it take a particular electron to get from the
switch to the lamp if the length of wire from the
switch to the lamp is 3.0 m long?

30.43 A silicon diode has a saturation current of 0.10 μA.
• At what voltage would it carry a forward current of
10 mA if the diode exactly obeyed the rectifier equation?

ADDITIONAL READING

Döhler, G. H., "Solid State Superlattices." *Scientific American,* November 1983,
p. 144.

Golovchenko, J. A., "The Tunneling Microscope: A New Look at the Atomic
World." *Science,* 4 April 1986, p. 48.

Hamakawa, Y., "Photovoltaic Power." *Scientific American,* April 1987, p. 86.

Nelson, D. R., "Quasicrystals." *Scientific American,* August 1986, p. 42.

Quate, C. F., "Vacuum Tunneling: A New Technique for Microscopy." *Physics
Today,* August 1986, p. 26.

31

Elementary Particle Physics

31.1 Particles and
      Antiparticles

31.2 Pions and the Strong
      Nuclear Force

31.3 More and More Particles

*31.4 Accelerators and
      Detectors

31.5 Classification of
      Elementary Particles

31.6 The Quark Model
      of Matter

31.7 Unified Theories

31.8 Cosmology

A WORD TO THE STUDENT

We have discussed many different models of the physical world, from the kinetic theory of gases, through the atomic theory of matter, to subnuclear structure. In this chapter we inquire about the nature of the fundamental particles that make up the atoms and nuclei themselves, investigating questions about the forces between them and their possible internal structure. The basic ideas we present are known as the quark model of matter.

So far, most of the topics covered in this text are those for which the underlying principles have been known for some time. That is not the case for the material in this chapter. Although we have achieved considerable understanding, the behavior of the fundamental building blocks of matter remains one of the contemporary frontiers of physics research. At the end of this chapter we briefly introduce a fascinating aspect of this topic: the connections between fundamental particles and the beginning of the universe.

After Chadwick's discovery of the neutron in 1932, it seemed that physicists had found all the subnuclear particles needed as the elementary building blocks for atoms. In appropriate combinations, the proton, neutron, and electron could be used to build up every known element, from hydrogen to the most complex atom. Atoms, in turn, made up molecules and molecules made up ordinary matter. Such a scheme accounted nicely for the masses and sizes of atoms and, in conjunction with quantum mechanics, explained their chemical properties. Thus the proton, neutron, and electron, along with the photon, were called **elementary,** or **fundamental, particles**.

Soon, however, this seemingly simple and satisfactory picture was complicated by new discoveries. During the period when Schrödinger was developing his wave formulation of quantum mechanics, other people discovered several new elementary particles. Since then, literally hundreds of these seemingly elementary particles have been discovered. Attempts to understand the role played by these new particles continues to be one of the most exciting and dynamic areas of modern physics.

The study of elementary particles and their interactions, often known as high-energy physics, has created a number of new terms and classifications that we have not used before. We will try to keep the introduction of new vocabulary to a manageable level. We will focus on those topics now recognized as central to our current understanding, and will avoid the many interesting side issues that have arisen in recent years. As the material in this chapter shows, contemporary physics continues to place great importance on conservation laws, and progress in understanding fundamental particles has frequently come by identifying conserved quantities and trying to explain their origin.

Particles and Antiparticles

31.1

One of those involved in the early work on quantum mechanics was Paul A. M. Dirac (1902–1984). At the time, Dirac was Lucasian Professor of Mathematics at Cambridge, a post formerly held by Isaac Newton. One of Dirac's particular concerns was to incorporate relativity into quantum mechanics. To do so required that space and time be treated on an equal footing in quantum theory, as it already was in relativity.

With considerable insight and mathematical elegance, Dirac formulated a relativistically correct quantum wave equation for the electron. The solutions to this equation contained several unexpected results. First, the solutions indicated that electrons have spin, a fact already known from experiment. The second surprising prediction was that electrons could exist in both positive *and* negative energy states. The positive energy states correspond to the ordinary electrons that we already know about. Since energy states correspond to a particle's charge, Dirac thought at first that the negative electron states, which would have positive charge, corresponded to protons. But this could not be the case, in part because the proton is almost two thousand times as massive as the electron. So in

(a)

Figure 31.1   (a) Electron-positron pairs are produced in a bubble chamber. The tracks curve in opposite directions because they represent the paths of oppositely charged particles moving in a magnetic field. The nearly straight tracks are left by the charged particles of the incident beam. (b) A schematic view. The dashed lines represent the paths of photons, which, like all neutral particles, leave no tracks in a bubble chamber. In both examples, a photon is converted into an electron-positron pair. In the lower event a recoil electron is knocked out of a hydrogen atom.

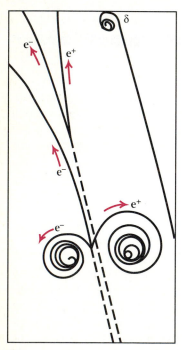

(b)

1930 Dirac postulated the existence of positively charged particles having the same mass as the electron. His prediction was confirmed in 1932, the same year that Chadwick discovered the neutron, when Carl Anderson discovered in the cosmic rays a positively charged particle with the mass of an electron. These particles are now called **positrons**.

One major consequence of Dirac's work was the realization that to any particle, not just the electron, there corresponds a similar particle of the same mass but with opposite charge. These oppositely charged particles are called **antiparticles**. The positron was the first antiparticle discovered. Later, in 1955, the existence of the antiproton was confirmed by Emilio Segrè and Owen Chamberlain. Further discoveries led to the conclusion that neutral particles also have antiparticles and that some, though not all, neutral particles are their own antiparticles.

The concept of antiparticles is well established in physics today. Symbolically, we denote the antiparticle of a given particle by placing a bar over the letter representing the particle. Thus a proton is represented by p and the antiproton is denoted by $\bar{p}$ (pee-bar).

When a particle and its antiparticle come together at rest, they annihilate each other. In doing so they produce gamma rays whose total energy equals the total available rest-mass energy. The annihilation of particles and antiparticles has given rise to a large number of interesting science fiction stories. However, producing an antiparticle in the first place requires an amount of energy equal to its rest mass, a relatively large amount of energy on a nuclear scale. Furthermore, antiparticles are produced only in particle-antiparticle pairs. The dream of making antimatter remains unfulfilled except in the sense of laboratory experiments (Fig. 31.1).

Example 31.1

A particle known as the B meson has a mass slightly more than $5\frac{1}{2}$ times that of a proton. The neutral state, B^0 (bee-naught), can be produced as one of a particle-antiparticle pair. What is the minimum energy needed to produce a B^0-$\overline{B^0}$ pair? The mass of the B^0 is 9.40×10^{-27} kg.

Solution   Because particle and antiparticle have the same mass, we need a minimum energy equivalent to twice the mass of one of the particles. This energy can be determined using a conversion factor based on $E = mc^2$ (see Problem 31.1):

$$1 \text{ MeV is equivalent to } 1.78 \times 10^{-30} \text{ kg.}$$

The total energy is then

$$2(9.41 \times 10^{-27} \text{ kg})(1 \text{ MeV}/1.78 \times 10^{-30} \text{ kg}) = 10,600 \text{ MeV} = 10.6 \text{ GeV.}$$

This is the minimum energy that an accelerator would have to provide in order to produce a pair of neutral B mesons.

Pions and the Strong Nuclear Force

31.2

Knowing the building blocks of an atom is not the same as knowing what holds them together. The Coulomb electrostatic force binds the negative electrons to the positive nucleus, but what force holds the nucleus together? The Coulomb force between protons in the nucleus is repulsive, and the gravitational force of attraction between nucleons is much weaker than this repulsive force. So, as we saw in Chapter 28, there must be a third nuclear force that is quite strong.

In 1932 the Japanese physicist Hideki Yukawa (1907–1981) proposed a mathematical form for the strong nuclear force. His potential energy function (Fig. 31.2) has the mathematical form

$$PE = -K \frac{e^{-\alpha r}}{r},$$

(31.1)

where the constant α has the dimensions of reciprocal length and K is a positive constant that must be determined from experiments. The force corresponding to the Yukawa potential has a much shorter range than does the Coulomb force. It has appreciable magnitude only over a distance comparable to nuclear dimensions, that is, about 10^{-15} m.

Yukawa predicted the approximate mass of a new elementary particle associated with the nuclear force field, basing his argument on the Heisenberg uncertainty principle. Our modern understanding of forces—and fields—is that every fundamental force has a particle associated with it. Although we did not put it in just those terms, we have already seen one example of this type of connection. The electromagnetic force between charged particles is described by Maxwell's equations, of which Coulomb's law of electrostatic force is a part. However, Maxwell's laws also predict the existence of electromagnetic waves, which we also interpret as being photons. Thus we associate the photon with the electromagnetic force. We say that the photon is the field particle of the electromagnetic force, or that the photon carries, or mediates, the electromagnetic force. We can think of the force between charged bodies as corresponding to an exchange of photons between the bodies. In an analogous fashion, we can describe the strong nuclear force between two nucleons as mediated by a new particle.

According to Yukawa's theory, the nuclear force field can be represented by a cloud of these new particles that surround a nucleon and that are quickly emitted and reabsorbed by the nucleon. However, if we consider the attraction between nucleons to be mediated by a particle with mass, as Yukawa proposed, then a problem arises with the conservation of energy. The appearance of a new particle with a finite mass seems at first to violate the law of energy conservation; after all, the particle's rest

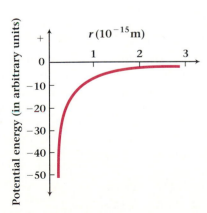

Figure 31.2 The shape of the Yukawa strong nuclear potential, drawn for the case $1/\alpha = 1.4 \times 10^{-15}$ m.

mass must come from somewhere. Nevertheless, if we think of each mediating particle as being emitted from a nucleon and then reabsorbed almost immediately, energy fails to be conserved over only a short period of time. The amount by which the conservation of energy law is uncertain, ΔE, and the time during which the particle exists outside the nucleon, Δt, are related through the uncertainty relationship

$$\Delta E \; \Delta t \approx \frac{h}{2\pi}.$$

Note that we have used the more precise form of the Heisenberg uncertainty principle.

The length of the particle's path away from a nucleon is of the same order of magnitude as the range of the nuclear force, R_0. If we assume that the particles move with the speed of light, then the uncertainty in time is the ratio of the distance to the speed of light. If the uncertainty in the energy is taken to be the particle's mass energy, the above equation becomes

$$mc^2 \frac{R_0}{c} \approx \frac{h}{2\pi},$$

or

$$m \approx \frac{h}{2\pi R_0 c}. \tag{31.2}$$

Inserting the values h, c, and R_0, we find that

$$m \approx 2.93 \times 10^{-28} \text{ kg}$$

or

$$m \approx 320 \; m_e.$$

This is not a precise calculation of the particle's mass, only an order-of-magnitude calculation. We would not expect to observe particles of this kind free in nature because they should interact strongly with all nuclear matter. However, it should be possible to produce them in collisions between nucleons when the incoming nucleon has enough kinetic energy to provide the required mass energy.

Within two years of Yukawa's prediction, researchers observed a charged particle of approximately the right mass. However, subsequent investigation showed that the new particle's interaction with matter was so weak that it could not possibly be the Yukawa particle. This new particle was named the mu meson or **muon**,* represented by the symbol μ. Muons are observed with positive or negative charge, both forms having the same mass. The intrinsic angular momentum, or spin, of the muon is $\frac{1}{2}$ in units of $h/2\pi$. This is the same value as the spin of the electron, introduced in Chapter 27. The spin is an important property for the classification of elementary particles and for understanding their reactions.

---

*The term *meson* comes from the Greek *mesos,* which means "intermediate." Those members of the meson group of particles that were discovered first have masses between the mass of an electron and a proton. Other mesons—discovered later—have masses outside this range. Particle classification is discussed later in the chapter.

In 1947 C. F. Powell (1903–1969) and his coworkers finally discovered the strongly interacting Yukawa particle while doing experiments involving cosmic rays at high altitudes. This particle was named the π meson, or **pion**, and has a mass somewhat greater than the mass of the muon. Pions occur in positive, negative, and uncharged forms, all with approximately the same mass. The spin of the pion is zero.

Charged pions are unstable and primarily decay with a half-life of about 2.8×10^{-8} s by the process

$$\pi^+ \rightarrow \mu^+ + \nu$$
$$\downarrow$$
$$e^+ + \nu + \bar{\nu}.$$

This notation means that the μ^+ (positively charged muon) subsequently decays into a positron and a neutrino-antineutrino pair. The π^- decays, by a similar process, to a μ^- and a neutrino, and then the μ^- decays into an electron and a neutrino-antineutrino pair. The neutral pion (π^0) decays primarily into two photons, with a half-life of about 0.8×10^{-16} s.

More and More Particles

31.3

In the same year that C. F. Powell and his collaborators discovered the pion, investigators observed other particles that were about a thousand times as massive as an electron. After several years of further research, it became apparent that these were actually two kinds of particles: **hyperons**, whose decay products always included a proton, and **kaons**, or K mesons, whose decay products were mesons only. Like pions, the hyperons and kaons seemed to be produced as a result of the strong nuclear force between two protons colliding with high energy. However, neither of these newly found particles seemed to be elementary building blocks of ordinary matter, nor did they seem necessary for explaining the nuclear force.

The discovery of the muon, pion, hyperon, and kaon created an interest in higher-energy accelerators, capable of enough energy to produce other new particles. Following the construction of such accelerators, other new particles were indeed found, a result that stimulated another increase in the energy of the accelerators. This leapfrog development eventually led to the discovery of a large number of "elementary" particles, some of which are listed in Table 31.1. (We will explain the classification terms lepton, meson, and baryon in Section 31.5.) We have not listed all the properties of the particles, nor even shown how the unstable ones decay. A few typical accelerator reactions that give rise to some of these elementary particles are as follows (Fig. 31.3):

$$p + p \rightarrow p + p + \pi^+ + \pi^-$$
$$p + p \rightarrow p + n + \pi^+$$
$$p + p \rightarrow p + \Lambda^0 + K^+$$
$$\pi^- + p \rightarrow \pi^- + p + \pi^0$$
$$\pi^- + p \rightarrow \pi^- + p + \pi^+ + \pi^-$$
$$\pi^- + p \rightarrow \pi^- + n + \pi^+$$

(31.3)

TABLE 31.1
Partial list of elementary particles

Classification	Name	Symbol	Charge	Mass (MeV/c²)	Lifetime (s)
	Photon	γ	0	0	∞
Leptons	Neutrino	$\nu_e, \bar{\nu}_e$ $\nu_\mu, \bar{\nu}_\mu$ $\nu_\tau, \bar{\nu}_\tau$	0 0 0	0 0 0	∞ ∞ ∞
	Electron	$e^\pm$	$\pm e$	0.5110	∞
	Muon	$\mu^\pm$	$\pm e$	105.7	2.20×10^{-6}
	Tau	$\tau^\pm$	$\pm e$	1784	3.04×10^{-13}
Mesons	Pion	$\pi^\pm$ π^0	$\pm e$ 0	139.6 135.0	2.60×10^{-8} 0.87×10^{-16}
	Kaon	$K^\pm$ $K^0 \, (\bar{K}^0)$	$\pm e$ 0	493.7 497.7	1.24×10^{-8} 0.89×10^{-10}
	D meson	$D^\pm$ $D^0 (\bar{D}^0)$	$\pm e$ 0	1869 1865	10.7×10^{-13} 4.3×10^{-13}
Baryons	Proton	p, $\bar{\text{p}}$	$\pm e$	938.3	$>10^{39}$
	Neutron	n, $\bar{\text{n}}$	0	939.6	896
	Lambda	$\Lambda^0, \bar{\Lambda}^0$	0	1116	2.63×10^{-10}
	Sigma	$\Sigma^+, \bar{\Sigma}^+$ $\Sigma^0, \bar{\Sigma}^0$ $\Sigma^-, \bar{\Sigma}^-$	$\pm e$ 0 $\pm e$	1189 1193 1197	0.80×10^{-10} 7.4×10^{-20} 1.48×10^{-10}
	Xi	$\Xi^0, \bar{\Xi}^0$ $\Xi^+, \bar{\Xi}^+$	0 $\pm e$	1315 1321	2.90×10^{-10} 1.64×10^{-10}
	Omega	$\Omega^-, \bar{\Omega}^-$	$\pm e$	1672	0.82×10^{-10}

(a)

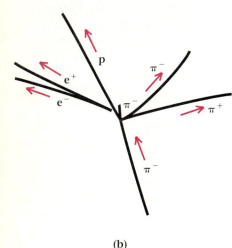

(b)

Figure 31.3 (a) A bubble chamber photo of the reaction $\pi^- + p \rightarrow \pi^- + p + \pi^- + \pi^+$. (b) A schematic diagram of the reaction. One of the emergent π^- interacts with another proton to give two neutral particles, a π^0 and a neutron. The π^0 decays into two gamma rays. One of the gamma rays creates the electron-positron pair.

The usefulness of high-energy accelerators can be seen from two standpoints. First, to produce a particle of a given mass m, the accelerated particle must have at least enough kinetic energy so that mc^2 of energy is available for the reaction. Thus, since the first accelerators were of lower energy than later accelerators, they could produce only lower mass particles, such as pions.

A second way of looking at the advantage of higher energies is to consider the accelerator as a microscope, enabling us to probe deeper into the structure of matter than we can do using only visible light. We have seen that a microscope's resolution, or its ability to discern small details, increases as the wavelength of light decreases. The electron microscope was constructed to take advantage of this effect. The de Broglie relationship between a particle's wavelength λ and momentum p is $\lambda = h/p$, where h is Planck's constant. In an electron microscope, electrons are accelerated to such high speeds that their momentum is large enough to make λ much less than the wavelength of visible light. As a result, the electron microscope has much greater resolution than the light microscope. The greater energies of modern particle accelerators, in the range of tens of GeV and even many TeV, provide still greater resolution for probing the structure of the nucleus and of the nucleons themselves. It is the decrease in wavelength with increased momentum and energy, and the resultant increase in the resolution and fineness of detail that can be probed, that makes the terms high-energy physics and elementary particle physics equivalent.

Accelerators and Detectors

*31.4

As we probe deeper into the structure of matter, we require finer and finer probes; that is, we need projectiles of shorter wavelength or, equivalently, higher momentum. Construction of accelerators of increasingly higher energies, and of devices to detect the particles emerging from these high-energy interactions, utilizes our most advanced technology. Reciprocally, the research and development put into the design of particle physics facilities has advanced the development of technology used in other fields.

The Van de Graaff accelerator and the cyclotron have features that prevent them from being scaled up to extremely high energies. The Van de Graaff accelerator is limited by the breakdown potential of its insulators. The basic principle of the cyclotron, that of constant period for the outwardly spiraling particles, becomes invalid at higher energies. As the particle's speed increases, relativistic effects become important and the conditions for constant period fail.

Several alternative schemes to overcome these difficulties have been used in constructing accelerators of increasingly higher energies. Linear accelerators use pulses of radio waves to push particles to higher speeds. Circular machines use the same general principles of acceleration, but guide the particles through nearly circular orbits, sending them through the same accelerating sections many times (Fig. 31.4a). Magnets placed around the periphery of the donut-shaped beam tube determine the path the particles travel (Fig. 31.4b).

Conservation of momentum requires that for each particle collision, the sum of the momenta of the outgoing particles must equal the momenta of the incoming particles. This requirement limits the energy available for reactions when a fixed target (zero initial momentum) is bombarded with the beam. The outgoing particles necessarily carry away some kinetic energy. This means that the energy available for reactions or production of new particles is not the energy of the incoming particle, but is the incoming energy minus the outgoing energy. For highly relativistic velocities the available energy depends on the square root of the energy of the incoming particle.

We can overcome some of the difficulties of using fixed targets by using colliding beams. Two beams of particles with the same momentum are shot directly at each other, so that the net momentum of the particle collision is zero and all of the energy of both beams is available for the interaction. Colliding beams are usually achieved with storage rings, in which particles circulate in opposite directions in each of two donut-

(a)

(b)

Figure 31.4 (a) Aerial view of the site of HERA at DESY in Hamburg, FRG. HERA is the world's only electron-proton colliding beam facility. (b) Picture inside the accelerator tunnel, showing the bending and focusing magnets around the beam pipe.

Figure 31.5 A drawing comparing the size of the SSC with Manhattan Island.

(a)

(b)

shaped tubes. The tubes intersect, creating places where the countercirculating bunches of particles can collide. In some cases, particles and antiparticles circulate in the same storage ring.

Even for storage rings, an increase in beam energy means an increase in accelerator size. This requirement is largely due to the relationship between a particle's momentum and the radius of curvature of its path in a magnetic field. The magnetic field of any particular accelerator is limited by the strength of the magnets available. Confinement of faster-moving particles requires a larger circle.

The upper limit of practical accelerators and storage rings is the proposed superconducting supercollider (SSC). The design calls for counter-circulating bunches of protons and antiprotons with energies of 20 TeV each, making available an energy of 40 TeV for the interactions. This energy is equivalent to about 40,000 proton masses. The machine will make use of the strong magnetic fields generated by superconducting magnets. The physical extent of the SSC is extremely large (Fig. 31.5); the circumference of the ring is projected to be approximately 80 km. Because such a large machine must be kept in careful alignment, the geological formation over which it is built must meet stringent requirements for stability. For this and other reasons, a site about 30 miles south of Dallas, Texas, was selected in 1989 to be the location of this facility. Though plans for the construction and use of the accelerator are going forward, the time for starting construction has not been fixed and, to a large extent, that date depends on the state of the national economy.

Experiments in particle physics require not only accelerators to provide particles for the interactions, but also detectors to determine what happens in the interactions. At high energies much of the initial energy may be converted into mass, and typically many particles emerge from the collision between two particles. In a colliding-beam experiment the net momentum is zero and the particles emerge in all directions from the interaction point. To detect and measure the particles, it is necessary to surround the interaction region completely with detectors. The objective is to determine the angle at which each particle leaves the interaction region, the charge and energy of the particle, and, when possible, its identity. This is accomplished by using layers of different types of detecting devices, often including a magnetic field as part of the detector. Figure 31.6 shows a detector for colliding-beam experiments and an event recorded by it.

Neither the analysis of data nor the actual running of an experiment can be accomplished without the aid of powerful and fast computers. The many thousands of pieces of information recorded for each event and the large number of components to be operated and monitored make computers as much a part of the detector as are the gas-filled counting tubes and the scintillation counters.

Figure 31.6 (a) Participants from six different countries assemble the ARGUS detector at the German national laboratory DESY, in Hamburg. (b) The emerging particles from a single electron-positron annihilation at 10 GeV, recorded in the central tracking chamber of the ARGUS detector. A B^0-$\overline{B}^0$ pair produced in the decay of an upsilon meson is identified from the tracks of its decay products.

Classification of Elementary Particles

31.5

The large number of elementary particles makes it important to have a classification scheme based on similar properties, as described in Table 31.1. Photons, which interact only by the electromagnetic force, occupy a category of their own. All other particles are classified as either hadrons or leptons. **Hadrons**, after the Greek word *hadros,* which means "strong," take part in nuclear interactions through the strong nuclear force. **Leptons**, named from a Greek word meaning "light," interact by the weak nuclear force. (Hadrons also experience the weak nuclear force, but the strong force has a greater range and magnitude, and so predominates in determining hadron behavior.)

Hadrons are further subdivided into two groups: the previously described *mesons* and the *baryons* (from the Greek word for "heavy"). The best known baryons are the proton and neutron. Baryons can be produced only in baryon-antibaryon pairs. Each baryon is assigned a baryon number of $+1$ and each antibaryon is assigned a baryon number of -1. In nuclear reactions, baryons obey a conservation rule: The sum of the baryon numbers after a reaction is the same as it was before the reaction. Mesons do not obey this rule: They have a baryon number of zero and may be produced singly. Examination of Table 31.1 shows that mesons are no longer confined to the mass range between the electron and the proton.

Example 31.2

Can the following reactions and decays take place? (a) $p + \bar{p} \rightarrow n + \pi^0 + \pi^- + \pi^+$, (b) $p + p \rightarrow p + n + \pi^+$, (c) $\Lambda \rightarrow p + \pi^-$.

Solution   We will test the proposed reactions to see if both baryon number and charge are conserved. If not, then the reaction cannot take place. If both baryon number and charge are conserved, there are other conserved quantities which must also be examined. However, in order to keep our presentation of the main ideas of particle physics as simple and to the point as possible, we have not introduced all of the conserved quantities. In this example, and in the problems, the other conservation rules have already been satisfied.

(a) The proton has a baryon number of $+1$ and the antiproton has a baryon number of -1, making the baryon number of the initial system 0. The right-hand side has a neutron, with a baryon number of $+1$, and three pions, with baryon number 0. Therefore the reaction cannot occur, even though charge is conserved, because it would mean that baryon number was not conserved.

(b) The total baryon number on the left-hand side is 2, and the charge, in units of the elementary charge e, is equal to 2. Looking at the right-hand side, we see that the only charged particles are p and π^+, for a total of $+2$, indicating that charge is conserved. The proton and neutron both have $+1$ baryon number and the pions have 0 baryon number. Therefore baryon number is conserved also, and the reaction can, and does, occur.

**TABLE 31.2
The fundamental forces
of nature**

Name	Relative strength	Associated particle
Strong nuclear	1	Meson
Electromagnetic	10^{-2}	Photon
Weak nuclear	10^{-13}	$W^{\pm}$, Z^0
Gravitational	10^{-40}	Graviton

(c) Table 31.1 lists the Λ as a neutral baryon. Examination of the decay products shows conservation of both charge and baryon number, so the reaction occurs.

The best-known lepton is the electron. The electron, the muon, and the tau, together with their respective neutrinos, constitute the known leptons. Since the leptons do not experience the strong nuclear force, their behavior is determined by the weak nuclear force, in addition to the much weaker gravitational force and the powerful electrostatic (Coulomb) force in the case of charged leptons. In fact, these same forces are experienced by all particles. Beta decay is associated primarily with the weak nuclear force. The weak force is not only weaker than the strong force by several orders of magnitude, it is also shorter in range. This indicates that any particle involved in mediating the weak force must be more massive than the pion, which is involved in the strong nuclear force.

Table 31.2 lists the four fundamental forces of nature in order of decreasing strength, along with the particle (or quantum) that is associated with each force. Experiments have not yet verified the particle associated with the gravitational field, the graviton. The particles associated with the weak nuclear field, the $W^{\pm}$ and the Z^0, were detected in 1983 by a group headed by Carlo Rubbia.

Other properties are associated with elementary particles in addition to those mentioned here. However, in order to outline the current understanding of the basic composition of matter in briefest terms, we will not deal with these properties. Many other parameters are necessary for a full understanding of particle physics, but are beyond the scope of this book.

Electrons and the other leptons seem to be pointlike particles. That is, they appear to have no observable spatial extent or structure. On the other hand, hadrons seem to have an internal structure. In the next section we discuss a model for this structure.

The Quark Model of Matter

31.6

The large number of known hadrons (over 200) presents a formidable barrier to understanding the basic composition of matter in any simple way. One approach toward understanding them would be some scheme or technique for classifying the properties of hadrons. The most successful of such schemes, called the **quark** model, was introduced in 1963 by Murray Gell-Mann and George Zweig. The quark model proposes that each hadron consists of a proper combination of a few elementary components called quarks. The properties of quarks may seem somewhat unusual; for example, they carry an electric charge that is a fraction of the elementary unit of charge. At first many people thought that the quark model did not represent physical reality, that it was only an extremely good mathematical arrangement. But experimental evidence has made clear that quarks are real constituents of matter. Before describing this evidence, we must introduce the elementary properties of quarks.

TABLE 31.3
Some properties of the
u, d, and s quarks

Quark	Charge	Spin	Baryon number
u	$\frac{2}{3}$	$\frac{1}{2}$	$\frac{1}{3}$
d	$-\frac{1}{3}$	$\frac{1}{2}$	$\frac{1}{3}$
s	$-\frac{1}{3}$	$\frac{1}{2}$	$\frac{1}{3}$

TABLE 31.4
Quark composition of
several particles

Particle	Quark Composition
π^-	$\bar{u}d$
K^+	$u\bar{s}$
K^-	$\bar{u}s$
p	uud
n	udd

The earliest form of the quark theory postulated three quarks, all of fractional charge, called the up (u), down (d), and strange (s) quarks (Table 31.3). These names are said to represent the different "flavors" of quarks. The quarks also have a spin quantum number of $\frac{1}{2}$ and a baryon number of $\frac{1}{3}$. Each quark has its own antiquark, which differs from it in the sign of the charge and baryon number.* To construct hadrons from quarks, there is a very simple rule: Mesons are composed of quark-antiquark pairs, and baryons consist of three quarks in proper combination. For instance, we can represent a π^+ meson by a u quark and an anti-d quark ($\bar{d}$), written as $u\bar{d}$. The proton is made up of a uud combination. Table 31.4 shows some other combinations.

Example 31.3

Can the following combinations of quarks occur in nature? (a) ud, (b) $\bar{u}d$, (c) uuu.

Solution   There are at least two tests that you can apply to see if a given combination of quarks is allowed. The first is based on the fact that all elementary particles have either a whole number or zero charge in units of the elementary charge e. Therefore the quark content has the same restriction. Likewise the quark composition of a particle must give an integer or zero for the baryon number.

(a) You should already recognize this as impossible because it is not a quark-antiquark pair. From Table 31.3, it is clear that the total charge of the proposed combination is $-\frac{1}{3}$, and the baryon number is $\frac{2}{3}$. So this is not a combination that occurs.

(b) For this combination, remembering that antiquarks have charge and baryon number opposite to the quarks, we get a charge of -1 and a baryon number of 0. The combination is possible.

(c) The total charge of the combination is 2 and the baryon number is 2. This is a possible combination also; in fact, it is the quark content of the Δ^{++}.

To verify the unusual properties of quarks, experimental searches for fractionally charged particles have been made. So far, none of them has demonstrated the existence of a *free* quark in nature. However, other experiments have pointed conclusively to the existence of quarks that are bound together.

It is possible to probe the structure of a hadron, say the proton, using electrons. The experiments are similar in many ways to Rutherford's scattering experiments to probe atomic structure. Very rapidly moving electrons have de Broglie wavelengths that are much smaller than the dimensions of a proton; thus the scattering of these electrons from protons tells us much about proton structure. Such "deep inelastic" scattering experiments reveal that the proton has an internal structure made up of other

---

*Other properties, or quantum numbers, which we have not mentioned may also be different. These frequently bear names such as "charm," "strangeness," and "color." Terms like *flavor* and *color* do not have their usual meaning and are only fanciful names for certain quantum numbers.

particles. The properties of these other particles identify them as quarks. Thus experiments have demonstrated the existence of quarks in the same way Rutherford's experiments showed the existence of the nucleus.

To explain the fact that free quarks are not observed, the quark theory proposes an attractive force between quarks that is small or zero when the distance between quarks is of the order of nucleon dimensions, but that becomes extremely large when their separation is increased. This behavior, which is known as quark confinement, is quite different from that of other fundamental forces, which decrease with distance. These assumptions have been incorporated into theories that successfully explain quark confinement and the properties of hadrons.

The three quarks listed in Table 31.3 were sufficient for the original quark model. However, the subsequent discovery of additional particles has required the introduction of additional quarks. These quarks—the *c* or charm quark and the *b* or bottom quark—bring the number of experimentally justified quarks to five. The five quarks *u, d, s, c,* and *b* have been experimentally confirmed and the existence of the *t*, or top, quark can be inferred from indirect evidence. You might conclude that the number of quarks is increasing to some large number, analogous to the large number of hadrons; however, from analysis of the neutrinos that reached the earth from the explosion of supernova 1987A, it appears that the number of flavors is not more than about fourteen. Recent laboratory evidence indicates a limit of eight, and there is a widespread belief at this time that there are only six flavors.

It now appears that quarks and leptons are the most elementary components of nature. Various combinations of these point particles make up all known constituents of ordinary matter, as well as those particles seen only in high-energy interactions. This classification is shown diagrammatically in Fig. 31.7. The ordinary world—that is, the world of neutrons,

Figure 31.7   The constituents of matter as we now know them. All matter, in the most fundamental sense, consists of various combinations of quarks and leptons.

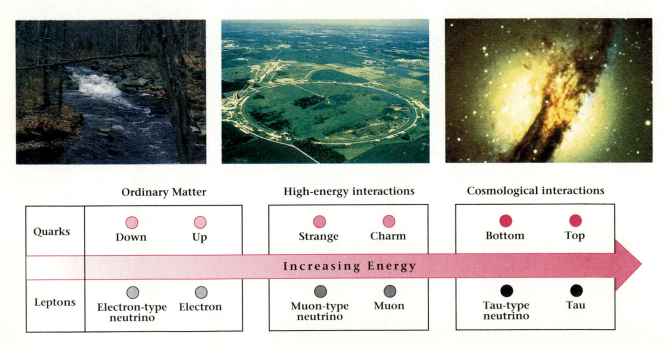

	Ordinary Matter		**High-energy interactions**		**Cosmological interactions**	
Quarks	Down	Up	Strange	Charm	Bottom	Top
			Increasing Energy			
Leptons	Electron-type neutrino	Electron	Muon-type neutrino	Muon	Tau-type neutrino	Tau

protons, and electrons—is composed from the quartet of quarks and leptons at the lowest energy level. To explain the existence of particles observed at higher energies, such as the Λ^0, we need the next higher energy level of quarks and leptons. Finally, the highest level corresponds to the great energies observed in galactic events or in very-high-energy accelerators. Observations have confirmed all of the particle quartets from the two lower levels. The only confirmed quark from the highest-energy quartet is the b quark. The present vigorous experimental search for the t quark is, in part, prompted by the symmetries seen in Fig. 31.7.

Unified Theories

31.7

The recognition of only four basic forces in nature has naturally led to the question, Are any or all of these forces perhaps related, or might they even be the same? The idea of explaining all natural forces with one comprehensive theory, called a **unified theory**, had its modern beginning with Einstein. During his later years he tried, without success, to combine his theory of gravitation with theories about the other long-range force, electromagnetism. To this date gravitation still has not been incorporated with any of the other forces, but attempts to unify the other three have met with considerable success.

In 1967 S. L. Glashow, A. Salam, and S. Weinberg introduced a theory unifying the description of the weak and electromagnetic forces. These two forces have several things in common, one being that they act on quarks and leptons equally. Since quarks and leptons are the basic constituents of matter in the quark model, then perhaps it is not surprising that one theory can describe both forces. The Glashow-Salam-Weinberg model, called the *standard model,* has successfully described a large number of observations. In a manner similar to the prediction of electromagnetic waves (photons) by Maxwell's theory of electromagnetism and the prediction of the pion by Yukawa's theory of the strong nuclear force, this theory of the "electroweak force" predicts the existence of several particles rather than one. These particles are named the $W^\pm$, the Z^0, and the Higgs particle. The most striking verification of this theory was the subsequent confirmation of the existence of the $W^\pm$ and Z^0. As of this writing, the Higgs particle has not yet been found, but there is hope that it will be observed within the next several years.

A further step of unification would be to combine the strong, weak, and electromagnetic forces in one theory. Such theories are called grand unification theories and go by the acronym GUTs. Experiments have shown that the strengths of the three forces, though different at ordinary energies and at energies available today with accelerators, tend to become more nearly the same as the energy becomes higher. Some calculations indicate that the strengths of the strong, weak, and electromagnetic forces become equal at about 10^{15} GeV, an incredible energy that is far beyond the 10^2–10^3-GeV range of present-day accelerators (Fig. 31.8). Nevertheless, GUTs have been constructed that give results in agreement with experiments.

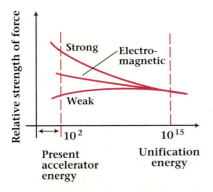

Figure 31.8 The strengths of the strong, weak, and electromagnetic forces as a function of energy.

Figure 31.9   A diver inside the IMB water scintillation detector used in a proton decay experiment.

The simplest and most reasonable grand unification theories give rise to one surprising result: the instability of the proton. They predict that protons may decay by one of several modes, such as

$$p \rightarrow \pi^0 + e^+$$

or

$$p \rightarrow \gamma + e^+.$$

Such reactions could not happen with a short half-life, or the universe as we know it would have vanished long ago—if, indeed, it could ever have been created. However, it is possible to calculate the approximate proton lifetime from GUTs. The calculations predict a proton lifetime of the order of 10^{30} years, perhaps somewhat longer. Since astronomers believe the approximate lifetime of the universe is of the order of 10^{10} years, the finite lifetime of the proton presents no real danger. Nevertheless, it is still an important prediction that is being tested.

Instead of looking at one proton for 10^{30} years, one collects a large number of protons (many times Avogadro's number of hydrogen atoms) and observes them for a few years. Such experiments often involve placing large volumes of hydrogen-containing liquids deep underground in mines or tunnels to shield them against spurious reactions due to cosmic rays (Fig. 31.9). In 1989 about three such experiments were underway or being prepared in various countries. To date, in contrast with theoretical predictions, no one has found conclusive evidence for proton decay, but a lower limit of about 5×10^{32} years has been placed on the experimentally determined proton lifetime. If further experiments push the observed lifetime much farther, some revision of existing theories may be required.

Cosmology

31.8

Physics and astronomy have been closely connected ever since Newton showed that planetary orbits are determined by the same laws of mechanics and gravitation that operate on the earth. In recent years physicists and astronomers have demonstrated a connection between the very small scale of elementary particles and quarks and the very large scale of the entire universe. This connection arises through the study of the earliest origins of the universe. We can actually observe aspects of the early universe because most stars are so distant from us that their light takes many thousands and even millions of years to reach the earth (Fig. 31.10). Thus studying the properties of far-away stars is the same as studying their state in the distant past.

As the result of many different types of observations and model-building, most astronomers now agree that the universe does not exist in some sort of steady state. Instead, it originated at some particular point in time in the distant past. This model of the early universe, called the **big bang** model, is supported by considerable evidence indicating that the age of the universe is about 10 to 20 billion years.

One important experimental observation supporting the big bang theory is the fact that all the galaxies are moving away from one another. We can draw the conclusion that at some time in the past the universe was

Figure 31.10 The distant galaxy NGC5236. The light from the galaxy takes approximately 10 million years to reach the earth.

much denser than at present—so dense that at one early time, shortly after the big bang event, the density of the entire universe was comparable to or greater than the density of nuclear matter. Here is the connection between the study of the origins of the universe and the study of the elementary properties of matter. Understanding the nature of elementary particles sheds light on the behavior of the early universe. Until the quark model and modern unified theories had been developed, there was no way in which physics could help explain how matter as we observe it today could have developed from the big bang model.

Steven Weinberg, one of the authors of the Glashow-Salam-Weinberg unified theory, has written in his book, *The First Three Minutes:**

> I remember that during the time that I was a student and then began my own research (on other problems) in the 1950's, the study of the early universe was widely regarded as not the sort of thing to which a respectable scientist would devote his time. Nor was this judgment unreasonable. Throughout most of the history of modern physics and astronomy, there simply has not existed an adequate observational and theoretical foundation on which to build a history of the early universe.
>
> Now, in just the past decade, all this has changed. A theory of the early universe has become so widely accepted that astronomers often call it "the standard model." It is more or less the same as what is sometimes called the "big bang" theory, but supplemented with a much more specific recipe for the contents of the universe. . . .

The theory outlines the early history of the universe as follows. The universe began in an explosion—not an ordinary explosion, but a tremendous burst of energy filling all of space. This energy corresponds to an extremely high temperature, of the order of 10^{32} K. At this high energy, no difference would exist between the strengths of the fundamental forces. However, as the universe cooled, with a consequent decrease of energy, the strengths of the four forces separated until they became as we now observe them in nature. The entire separation process was completed by the end of about 10^{-4} s.

During this initial phase of the universe, the temperature, or energy density, was so high that ordinary matter, such as molecules, atoms, and even nucleons, could not exist. Instead, shortly after the big bang, the universe was a soup of photons (radiation), leptons, antileptons, quarks, and antiquarks, though not necessarily in equal numbers. Because the density of the soup was high and the forces between quarks are small when they are close together, the particles all moved very much like free particles. But as the soup cooled, it began to "condense" into elementary particles, including protons and neutrons. As the temperature dropped further these nucleons condensed into nuclei. Further condensation gave rise to molecules and matter in bulk.

This brief outline skips over several interesting and important points. However, two things are especially worth noting. First, although some details of the big bang model are still uncertain, the same sort of physical

*S. Weinberg, *The First Three Minutes: A Modern View of the Origin of the Universe.* New York: Bantam Books, 1979.

theories that successfully describe the subnuclear world are also useful in describing the early history of the universe. Second, these particle physics theories have helped cosmologists deal with one of the more puzzling questions about the origin of the universe: If the initial big bang resulted in equal amounts of matter and antimatter, as our present understanding indicates, why does the universe today not have equal amounts of matter and antimatter? Astronomers have studied this question extensively, but no evidence exists for large amounts of antimatter located far away from us in the universe. One of the most impressive successes of grand unification has been the prediction that, even though the universe starts with a symmetry between matter and antimatter, the end result is a preponderance of one over the other. The theory predicts that among the first particles produced in the big bang were particles named the X-bosons. The X and its antiparticle the $\overline{X}$ are produced with equal probability, but their decay rates are not the same. As a result the average baryon number for all of the eventual decay products is not zero. The average baryon number is closer to the baryon number of the more rapidly decaying X (matter) than to that of the $\overline{X}$ (antimatter). This effect would have tipped the balance of matter in the universe during the first 10^{-35} s, even before the quark stage had been reached.

SUMMARY

Useful Concepts
- The positron was the first antiparticle found, and all particles have antiparticles.
- Yukawa's expression for the nuclear potential energy is

$$\text{PE} = -K \frac{e^{-\alpha r}}{r}.$$

- The use of colliding beams of particles in an accelerator allows all the energy of the particles to be available for interaction.
- According to the quark model of matter, mesons can be represented as a quark-antiquark pair; baryons can be represented as a combination of three quarks.

- Unified theories try to combine all of the forces of nature into one type of force.
- The quark theory has helped explain features of the big bang model of the origin of the universe.

Important Terms
You should be able to write the definition or meaning of each of the following terms:

- elementary particle
- positron
- antiparticle
- muon
- pion
- hyperon

- kaon
- hadron
- lepton
- quark
- unified theories
- big bang

QUESTIONS

31.1 Would you believe someone who claimed to have a laboratory flask full of antiwater? (Antiwater is the same as ordinary water except that each ordinary particle is replaced by its antiparticle.)

31.2 If all the particles in the universe were changed into their antiparticles overnight, could you tell that anything had happened?

31.3 What criteria can be used for classifying whether particles are fundamental?

31.4 A π^0 meson never decays into a single photon. Would doing so violate any conservation laws? If so, which ones?

31.5 Discuss the similarities and differences between the photon and the neutrino.

31.6 What is meant by the statement that "a muon is just a fat electron"?

31.7 A free neutron decays into a proton plus an electron plus an antineutrino. Why do we not also observe its decay into a proton-antiproton pair? Why does it not decay into an electron-positron pair?

31.8 Numerous experiments have searched for single particles with a charge that is a fraction of the electron charge. If such particles are never found, would you conclude that the quark model must be incorrect?

31.9 Defend or attack the following statement, giving reasons or historical analogies: "Theories and investigations into the structure of the proton and the neutron and into the first few minutes of the existence

of the universe are great intellectual achievements, but have no practical use."

31.10 What criteria must be satisfied for one elementary particle to decay into another?

31.11 Beams of negative pions have been used to replace x rays in the treatment of cancer. What might be some of the advantages of using pions rather than x rays?

31.12 The weakest of the four forces in nature, gravitation, seems to play the most important role in our daily lives. Is this really true? Explain.

31.13 Assuming that most cosmic rays are charged particles, give some reasons why you would expect the intensity to vary with position on the earth and time of day.

PROBLEMS

Hints for Solving Problems

For elementary particle interactions the most important thing is to take into account conserved quantities. This means not only using conservation of momentum and energy (keeping in mind relativity), but also using such ideas as conservation of charge and baryon number.

Section 31.1   Particles and Antiparticles

31.1 Show that the conversion factor between energy and mass can be put in the form: 1.000 MeV is equivalent to 1.783×10^{-30} kg.

31.2 How much energy is required to produce an electron-positron pair?

31.3 How much energy is released when a proton and antiproton annihilate at rest?

31.4 An electron and a positron come together at rest and then annihilate. Two photons are given off in the process. What can you say about the energy and direction of the photons?

31.5 How many proton-antiproton pairs could be produced if all of the kinetic energy gained by a 0.250 kg apple dropped from a height of 0.50 m went into producing the particles?

31.6 How much energy would be required to produce a person-antiperson pair with individual masses of 60 kg? How long would the Hoover dam hydroelectric plant have to operate to produce this much energy? (See Table 6.2.)

31.7 An electron and a positron collide head on and annihilate. Each has a kinetic energy of $\frac{1}{2}m_e c^2$ with respect to a stationary observer. What is the total energy released in the collision?

Section 31.2   Pions and the Strong Nuclear Force

31.8 Plot the Yukawa potential and the Coulomb poten-
  • tial and note any differences that you observe.

31.9 What would be the approximate mass of the Yukawa particle if the range of the nuclear force were twice as great as it is and all of the fundamental constants were unchanged?

31.10 What value of R_0 in Eq. (31.2) corresponds to the actual mass of the π meson?

31.11 The neutral pion decays most frequently into two photons. Show that when the decaying pion is at rest, the photons must have the same energy.

31.12 In the decay of a charged pion into a charged muon, what is the maximum kinetic energy available to the decay products?

Section 31.3   More and More Particles

31.13 On the average, how far will a neutral K meson go before it decays if its speed is 0.90 c? Answer the same question for the charged kaons.

31.14 Approximately how many nuclear diameters ($A \approx 100$) will a π^0 meson traverse at a speed of $0.95c$ before it decays?

*Section 31.4   Accelerators and Detectors

31.15 What momentum must a particle have for its wavelength to be of the order of nuclear dimensions?

31.16 What is the wavelength of a 10-GeV electron? How does this compare with the dimensions of a proton? (*Hint:* Use the formula for the radius of a nucleus. Also see the hint for Problem 31.20.)

31.17 The proton beam of a small accelerator is directed at a thin aluminum foil. If the beam current is 1.0 μA, how many protons strike the foil per second?

31.18 Estimate the mass of a piece of typing paper 0.5 mm square. Assume that all of the kinetic energy of a 20-TeV proton goes into lifting the paper straight up. Could you detect this change in elevation?

31.19 What is the energy in TeV of a 0.250-kg apple that has fallen through a distance of 1.0 m?

31.20 What is the wavelength of a 20-TeV proton? How does this compare with R_0? (*Hint:* For extremely relativistic particles, the mass term in the expression for the total energy may be neglected.)

Section 31.5   Classification of Elementary Particles

31.21 Which of the following reactions or decays cannot occur and why? (a) $p + p \rightarrow p + n + \pi^+$, (b) $\pi^+ + p \rightarrow n + \pi^+ + \pi^-$, (c) $n \rightarrow e^- + e^+ + \nu$, (d) $p + p \rightarrow \mu^+ + \mu^-$.

31.22 Which of the following reactions can take place? What conservation rules are violated in those that are not allowed? (a) $\Sigma^0 \rightarrow \pi^0 + \pi^0 + \gamma$, (b) $K^+ \rightarrow \pi^- + e^+ + e^+$, (c) $\Lambda^0 \rightarrow p + \pi^+ + \pi^-$.

31.23 The upsilon meson has a mass of 9460 MeV and a spread in mass of 52 keV. What is the approximate lifetime of this particle? (*Hint:* Use the Heisenberg uncertainty relationship.)

Section 31.6   The Quark Model of Matter

31.24 Verify that the quark composition of the particles listed in Table 31.4 gives rise to the correct charge and baryon number.

31.25 Write out the quark composition of the π^+.

31.26 Which of the following quark combinations can occur? Explain your answer. (a) *udd*, (b) *ūud*, (c) *dsu*, (d) *dū*.

31.27 What is the charge of a particle made up of (a) $c\bar{c}$ (the ψ meson) and (b) *uds* (the Λ hyperon)?

Additional Problems

31.28 The vacuum in the beam pipe of a storage ring is about 1.5×10^{-6} Pa. (a) Approximately how many air molecules are there per cubic meter when the temperature is 15°C? (b) To what height above the surface of the earth does this density correspond? For simplicity assume the atmosphere to be entirely composed of oxygen. (*Hint:* Use the barometric formula.)

31.29 In the reaction

$$\pi^- + p \rightarrow \Lambda^0 + K^0,$$

what is the rest-mass difference between the initial and final products? Does the reaction release energy or require it?

31.30 A Λ^0 at rest decays into a π^- and a proton. (a) How much energy is available as kinetic energy to the decay products? (b) How much momentum does each particle have, in units of MeV/c?

31.31 A hydrogenlike atom can be made by replacing the electron of normal hydrogen with a muon. (a) What wavelength radiation corresponds to the three Balmer spectral lines? (b) In what part of the electromagnetic spectrum do they lie?

31.32 One estimate of the lower limit of the proton lifetime is 2.5×10^{32} years. Some of the experiments to search for proton decay use a large volume of water. (a) How many water molecules contain 2.5×10^{32} protons? (b) What mass of water is this and what volume does it occupy?

ADDITIONAL READING

Jackson, J. D., M. Tigner, and S. Wojcicki, "The Superconducting Supercollider." *Scientific American,* March 1986, p. 66.

LoSecca, J. M., F. Reines, and D. Sinclair, "The Search for Proton Decay." *Scientific American*, June 1985, p. 54.

Schram, D., and G. Steigman, "Particle Accelerators Test Cosmological Theory." *Scientific American,* June 1988, p. 66.

Weinberg, S., *The First Three Minutes: A Modern View of the Origin of the Universe.* New York: Bantam Books, 1979.

Zee, A., *Fearful Symmetry: The Search for Beauty in Modern Physics.* New York: MacMillan, 1986. The story of how the search for basic symmetries has brought a clearer picture of nature.

Appendices

APPENDIX A: Formulas from Geometry and Trigonometry

Area and Volume

Area of a rectangle of sides a and b:	$A = ab.$
Area of a triangle with base b and height h:	$A = \dfrac{hb}{2}.$
Area of a circle of radius r:	$A = \pi r^2.$
Area of a sector of a circle when θ is the angle between the radii (in radians):	$A = r^2\theta/2.$
Area of an ellipse with semiaxes a and b:	$A = \pi ab.$
Area of the surface of a sphere of radius r:	$A = 4\pi r^2.$
Area of the curved surface of a spherical segment of height h, radius of sphere r:	$A = 2\pi rh.$
Volume of a sphere of radius r:	$V = \dfrac{4}{3}\pi r^3.$
Volume of a cylinder of radius r and height h:	$V = \pi r^2 h.$

Trigonometric Relationships

$\sin^2 x + \cos^2 x = 1.$

$\sin (x \pm y) = \sin x \cos y \pm \cos x \sin y.$

$\cos (x \pm y) = \cos x \cos y \mp \sin x \sin y.$

$\tan (x \pm y) = \dfrac{\tan x \pm \tan y}{1 \mp \tan x \tan y}.$

$\sin 2x = 2 \sin x \cos x.$

$\cos 2x = \cos^2 x - \sin^2 x.$

The law of cosines: in any triangle with angles A, B, and C and opposite sides a, b, and c, respectively,

$$a^2 = b^2 + c^2 - 2bc \cos A.$$

APPENDIX B: The International System of Units

Definitions of the SI base units.

1. unit of length (meter)   The 17th General Conference on Weights and Measures (CGPM, 1983) abolished the former definition of the meter and adopted a new definition which reads: *The meter is the length of the path traveled by light in vacuum during a time interval of 1/299 792 458 of a second.* The old prototype of the meter, which was legalized by the 1st CGPM in 1889, is still kept at the International Bureau of Weights and Measures (BIPM).

2. unit of mass (kilogram)   The 1st CGPM (1889) legalized the international prototype of the kilogram and declared: *This prototype shall henceforth be considered to be the unit of mass.* This international prototype made of platinum-iridium is kept at the BIPM under conditions specified by the 1st CGPM in 1889.

3. unit of time (second)   The second was defined originally as 1/86 400 of the mean solar day. Because a more precise definition was needed, the 13th CGPM (1967) replaced the astronomical definition of the second by the following: *The second is the duration of 9 192 631 770 periods of the radiation corresponding to the transition between the two hyperfine levels of the ground state of the cesium-133 atom.*

4. unit of electric current (ampere)   The 9th CGPM (1948) adopted the ampere for the unit of electric current, with the following definition: *The ampere is that constant current which, if maintained in two straight parallel conductors of infinite length, of negligible circular cross section, and placed 1 meter apart in vacuum, would produce between these conductors a force equal to 2×10^{-7} newton per meter of length.*

5. unit of thermodynamic temperature (kelvin)   The 13th CGPM (1967) adopted the name kelvin (symbol K), and defined the unit as follows: *The kelvin, unit of thermodynamic temperature, is the fraction 1/273.16 of the thermodynamic temperature of the triple point of water.* In addition to the thermodynamic temperature (symbol T), expressed in kelvins, use is also made of Celsius temperature (symbol t) defined by the equation $t = T - T_0$, where $T_0 = 273.15$ K by definition.

6. unit of amount of substance (mole)   The following is the definition of the mole, adopted by the 14th CGPM (1971): *The mole is the amount of substance of a system that contains as many elementary entities as there are atoms in 0.012 kilogram of carbon 12. When the mole is used, the elementary entities must be specified and may be atoms, molecules, ions, electrons, other particles, or specified groups of such particles.*

7. unit of luminous intensity (candela)   The unit based on flame or incandescent filament standards before 1948 was replaced initially by a unit based on a Planckian radiator (a blackbody) at the temperature of freezing platinum. Because of the difficulties, and new experimental techniques, the 16th CGPM (1979) adopted the following definition: *The candela is the luminous intensity, in a given direction, of a source that emits monochromatic radiation of frequency 540×10^{12} hertz and that has a radiant intensity in that direction of (1/683) watt per steradian.*

APPENDIX C: Alphabetical List of Elements

Name	Symbol	Atomic Number	Name	Symbol	Atomic Number	Name	Symbol	Atomic Number
Actinium	Ac	89	Helium	He	2	Radium	Ra	88
Aluminum	Al	13	Holmium	Ho	67	Radon	Rn	86
Americium	Am	95	Hydrogen	H	1	Rhenium	Re	75
Antimony	Sb	51	Indium	In	49	Rhodium	Rh	45
Argon	Ar	18	Iodine	I	53	Rubidium	Rb	37
Arsenic	As	33	Iridium	Ir	77	Ruthenium	Ru	44
Astatine	At	85	Iron	Fe	26	Samarium	Sm	62
Barium	Ba	56	Krypton	Kr	36	Scandium	Sc	21
Berkelium	Bk	97	Lanthanum	La	57	Selenium	Se	34
Beryllium	Be	4	Lawrencium	Lr	103	Silicon	Si	14
Bismuth	Bi	83	Lead	Pb	82	Silver	Ag	47
Boron	B	5	Lithium	Li	3	Sodium	Na	11
Bromine	Br	35	Lutetium	Lu	71	Strontium	Sr	38
Cadmium	Cd	48	Magnesium	Mg	12	Sulfur	S	16
Calcium	Ca	20	Manganese	Mn	25	Tantalum	Ta	73
Californium	Cf	98	Mendelevium	Md	101	Technetium	Tc	43
Carbon	C	6	Mercury	Hg	80	Tellurium	Te	52
Cerium	Ce	58	Molybdenum	Mo	42	Terbium	Tb	65
Cesium	Cs	55	Neodymium	Nd	60	Thallium	Tl	81
Chlorine	Cl	17	Neon	Ne	10	Thorium	Th	90
Chromium	Cr	24	Neptunium	Np	93	Thulium	Tm	69
Cobalt	Co	27	Nickel	Ni	28	Tin	Sn	50
Copper	Cu	29	Niobium	Nb	41	Titanium	Ti	22
Curium	Cm	96	Nitrogen	N	7	Tungsten	W	74
Dysprosium	Dy	66	Nobelium	No	102	(Unnilhexium)	(Unh)	106
Einsteinium	Es	99	Osmium	Os	76	(Unnilpentium)	(Unp)	105
Erbium	Er	68	Oxygen	O	8	(Unnilquadium)	(Unq)	104
Europium	Eu	63	Palladium	Pd	46	(Unnilseptium)	(Uns)	107
Fermium	Fm	100	Phosphorus	P	15	Uranium	U	92
Fluorine	F	9	Platinum	Pt	78	Vanadium	V	23
Francium	Fr	87	Plutonium	Pu	94	Xenon	Xe	54
Gadolinium	Gd	64	Polonium	Po	84	Ytterbium	Yb	70
Gallium	Ga	31	Potassium	K	19	Yttrium	Y	39
Germanium	Ge	32	Praseodymium	Pr	59	Zinc	Zn	30
Gold	Au	79	Promethium	Pm	61	Zirconium	Zr	40
Hafnium	Hf	72	Protactinium	Pa	91			

Answers to odd-numbered problems

Chapter 1

1.1 1.25 m   **1.3** 7.2 ft   **1.5** 2.0 m
1.7 (a) 3.5 rad/s, (b) 4.7 rad/s   **1.9** 0.549°
1.11 1 year^2/AU^3 or 3.0×10^{-19} s^2/m^3 **1.13** untrue
1.15 18.25 min/year   **1.17** 2.3×10^4 cm^3
1.19 100 cm^2 **1.21** 3.59×10^5 cm^3 **1.23** 2.16×10^4
1.25 391   **1.27** 2.9×10^5 m^3 **1.29** 6×10^{10} gallons
1.31 2.3 km $= 2.3 \times 10^5$ cm   **1.33** 53 min   **1.35** 30 cm
1.37 173 cm   **1.39** 74.6 miles/h   **1.41** $326.27
1.43 4.0×10^{-2} km^2 **1.45** 0.92 km
1.47 Wheel cannot be made.   **1.51** less accurate
1.53 3.15 Gs   **1.55** 9.41 L/100 km   **1.61** 0.354

Chapter 2

2.1 (a) 88 km/h, (b) 25 m/s   **2.3** 60.9 km/h   **2.5** 1.3 h
2.7 68 km/h   **2.9** 7.5 s   **2.11** 160 min
2.13 120 m ccw from the start   **2.15** -53 m/s
2.17 12th floor   **2.21** (a) 30 km/h, (b) 90 km/h
2.23 100 m   **2.25** 10 m/s   **2.27** (a) 3.8 m/s^2, (b) 2.4 s
2.29 3.4 m/s^2 **2.31** 3.2 s   **2.33** 80 ft/s, 55 mi/h
2.35 First mark at 0.40 m, fourth mark at 6.4 m
2.37 (a) 13 m/s^2, (b) 1.3   **2.39** 66.6 km/h, 133 km/h
2.41 (a) 144° W of N, (b) 72° E of N
2.43 (a) 36 ft/s, (b) 432 ft, (c) 18 ft/s   **2.45** 1.18 s
2.47 (a) 0.071 s, (b) 0.025 s   **2.49** -0.4 cm/s, -2 cm/s

2.51 (a) $v_2 = 2v_1$, (b) $a = \dfrac{2v_1{}^2}{R}$ **2.53** 20 m/s

2.55 39 m   **2.57** Yes, the stone catches the parachute.
2.59 (a) 1900 km/h, (b) 11 km/h, (c) 170

Chapter 3

3.1 (a) $C = 10$ blocks, (b) $\theta = 121°$ from the East.
3.3 (a) 54 units at 306°, (b) 28 units at 177°
3.5 (a) 5.8 km, (b) 121° from E
3.7 (a) 56 units at an angle of 121°   **3.9** 28 lb Northwest
3.11 (a) 160 km/h, (b) 160 km/h,
(c) -10 km/h and -170 km/h   **3.13** 30 min, 6.5 km
3.15 15 at angle of 37°   **3.17** (a) ± 17, (b) $\pm 112°$
3.19 (a) 13 at $-8.7°$, (b) 17 at 114°   **3.21** 10 at 96°

3.23 (a) 26°, (b) 8 km/h   **3.25** 15 at 70°   **3.27** 3.9 m
3.29 3.1 m/s   **3.31** 47.1 m/s   **3.33** 2.9 m/s

3.35 77 m   **3.37** $t = \dfrac{v_o \sin\theta + \sqrt{v_o^2 \sin^2\theta + 2gy_o}}{g}$

3.39 $\frac{1}{4} R \tan\theta$
3.41 $A = 50$ m, $B = 100$ m, $\theta = 63°$ from north
3.43 67 km at angle $-57°$ **3.45** 6%
3.47 (a) 6.12 m, (b) 5.94 m   **3.49** 151 m
3.51 (a) 3.81 cm, (b) 3.80 cm
3.53 $\sin\theta \approx \theta$ to within 0.5% for θ less than about 10°
3.55 (a) 56 cm, (b) 26.6°, (c) 56°   **3.57** 18 m/s
3.59 9.8 m

Chapter 4

4.1 (a) 2.0 N, (b) 0.10 m/s^2 **4.3** 3.33×10^4 N
4.5 8.2 m   **4.7** (a) 667 N, (b) 68.1 kg   **4.9** 88.2 N
4.11 (a) 1.9 kg, (b) 19 N   **4.13** 0.1%
4.15 (a) 780 N, (b) 940 N, (c) 780 N, (d) 660 N,(e) 780 N
4.17 (a) Reading increases momentarily, then returns to
initial reading. (b) Same as initial reading. (c) Same as (a).
4.19 780 N   **4.21** 1.4×10^4 N
4.23 (a) 2.0 m/s^2, (b) 1.0 N   **4.25** 3.3×10^3 N
4.27 180 N   **4.29** 1.36 m/s^2
4.31 (a) 3.7 m/s^2, (b) 31 N   **4.33** 0.58   **4.35** 30°
4.37 0.59 kg   **4.39** 1000 N   **4.41** 49°
4.43 Equilibrium only if $m_1 = m_2$.
4.45 $a_{Enterprise} = 8a_{alien}$ **4.47** 210 N
4.49 (a) 100 N, (b) 1100 N, (c) 1500 N   **4.51** 3.3 m/s^2
4.53 9.7 m/s^2
4.55 (a) 30°, (b) 0.65 m/s^2 down the incline
4.57 (a) 0.64 m/s^2, (b) $T_1 = 24$ N and $T_2 = 37$ N
4.59 $T_1 = 16.4$ N, $T_2 = 18.5$ N
4.61 (a) 4.9 m/s^2, (b) 320 N

Chapter 5

5.1 (a) 6.28 rad/s, (b) 12.6 m/s, (c) 79.0 m/s^2
5.3 6.2 m/s^2 **5.5** (a) 3.49 rad/s, (b) 4.71 rad/s
5.7 (a) 0.75 Hz, (b) 1.33 s   **5.9** (a) 4.0 Hz, (b) 25 rad/s
5.11 (a) 1.73×10^4 m/s, (b) 4.11×10^{-5} rad/s,

(c) 0.712 m/s^2 **5.13** (a) 0.62 Hz **5.15** 11.6 N
5.17 (a) 0.043 Hz, (b) 23 s, (c) 100 m **5.19** 49 m/s
5.21 0.900 **5.23** (a) $2.59 \times 10^{-5} \text{ m}^3$, (b) 2.96 cm
5.25 17 kg
5.27 (a) $2.70 \times 10^{-3} \text{ m/s}^2$,(b) $3.33 \ 10^{-5} \text{ m/s}^2$
5.29 1.39×10^{21} N **5.31** 8.49 h or 3.05×10^4 s
5.33 224 d **5.35** 1.90×10^{27} kg
5.37 (a) 1.2 g, (b) 720 kg/m^3 **5.39** 40 N
5.41 $1.2 \times 10^{-10} \text{ m/s}^2$ **5.43** -13.2 N
5.45 (a) 21.4 rad/s, (b) 3.41 Hz **5.47** 0.60 Hz
5.49 (a) 2.87×10^{-5}, (b) 553 m **5.51** 2.50×10^6 m
5.53 1.71% **5.55** 26 m/s
5.57 (a) 6.02×10^{24} kg,(b) $5.53 \times 10^3 \text{ kg/m}^3$
5.59 6.23×10^3 N **5.61** (a) 0.94 g, (b) 2.94 g
5.63 52.5 m

Chapter 6

6.1 1.6×10^3 J **6.3** 56 J **6.5** 6.7 m/s
6.7 (a) 1.3, (b) 1.7 **6.9** 3.412×10^3 Btu **6.11** 2100
6.13 (a) 3.6×10^5 J, (b) 5.4×10^5 J **6.15** 38 m/s
6.17 8.9 km/h **6.19** (a) 196 J, (b) 145 ft-lb
6.21 9.8 J for each path **6.23** -6.25×10^7 J
6.25 3.12×10^7 J **6.27** 1.48×10^4 J **6.29** 35 J
6.31 (a) 5.9×10^4 J, (b) 63 m/s **6.33** 7.0 m/s
6.35 3.7 m/s **6.37** 199 **6.39** 7.68×10^{-3} hp
6.41 340 kg **6.43** 1000 W **6.45** (a) 120 J, (b) 61 J
6.47 4.0 J
6.49 (a) 1.63×10^{15} W, (b) 1.4×10^{10} m, (c) 0.15%
6.51 (a) 3.7×10^4 J, (b) 7.4×10^4 J
6.53 -3.82×10^{28} J **6.55** (a) 19 J, (b) 4.2 rev/s
6.57 1.3 m/s **6.59** 0.074 m
6.61 At B, 11.7 m/s; at C, 8.9 m/s; at D, 9.9 m/s; 1.6 m/s^2
6.67 3 cm

Chapter 7

7.1 1.75×10^4 kg·m/s **7.3** 5.0×10^{-21} kg·m/s
7.5 1.8 ms **7.7** 4.1×10^3 N **7.9** -2.4 m/s
7.11 0.37 m/s **7.13** 0.25 m
7.15 2.78 m/s at an angle of 70° from the initial direction
of the 1300 kg auto
7.17 7 m/s at 1.56° from the direction of the train
7.19 3.3×10^7 N **7.21** 40 N **7.23** 1.6×10^{-4} N·m
7.25 (a) 330 N, (b) 270 N **7.29** 330 N **7.31** 6.7 m/s
7.33 (a) 0.40 rad/s², (b) 4.0 rad/s **7.35** 8 rad/s
7.37 (a) 1.4 rad/s², (b) 8.3 rad/s, (c) 31 m/s²
7.39 $m_1 = 0.72$ kg, $m_2 = 2.16$ kg
7.41 (a) 0.46 m/s, (b) 0.54 m/s, (c) 0.50 m/s **7.43** 0.94
7.45 28 km/h in the x direction

7.47 (a) 3.7 m/s², (b) 5.9 m/s² **7.49** 2.4 kg
7.51 1200 N **7.53** (a) 1010 N, (b) 793 N
7.55 (a) 200 N, (b) 700 N
7.59 (a) 0.047 kg·m², (b) 0.68 m **7.61** 0.18 rev/s
7.63 (b) $-\frac{1}{2} G \dfrac{Mm}{r}$

Chapter 8

8.1 -0.20 **8.3** $v_1 = -v_o, v_2 = +v_o$
8.5 $v_1 = -2$ m/s, $v_2 = +2$ m/s
8.7 $v_1 = -2.0$ m/s, $v_2 = 3.0$ m/s **8.9** 1280 J
8.11 3/8 **8.13** $v'_1 = 0.47$ m/s, $v'_2 = 0.17$ m/s
8.15 45° **8.17** Hoop, 5.4 m/s; disk, 6.3 m/s **8.19** 23 m
8.21 $\sqrt{\dfrac{10}{7} gh}$ **8.25** 59 m
8.27 (a) 30 ft, 58 ft, 96 ft, 144 ft **8.29** 13 m, 100 m
8.31 1.11×10^4 m/s **8.33** 1.2×10^4 m/s **8.35** 0.93
8.37 1.05 **8.39** 0.375 **8.41** 4.8×10^6 J
8.43 $\dfrac{8}{9} \left(\dfrac{v_o^2}{g} \right) - \dfrac{7}{9} \text{h}$
8.45 128 km/h, which exceeds the speed limit
8.47 (b) $0.088 \, v_{esc}$ **8.49** $-135°$ **8.51** 2.5 cm
8.53 $v_1 = v'_1 + v'_2, v_1^2 = v'^2_1 + v'^2_2$
8.55 (a) $\sqrt{\dfrac{4}{27} gh}$, (b) $\sqrt{\dfrac{4}{3} gh}$ **8.57** 2.0 N·m
8.59 0.100 g **8.61** 1.3 m/s **8.63** 0.03

Chapter 9

9.1 240 kPa **9.3** 0.0735 m² **9.5** 2170 kPa
9.7 17 MPa **9.9** 10.1 km **9.11** 11.1 kPa
9.13 10000 N **9.15** 147 kPa **9.17** 18.5 cm
9.19 (a) 1.6 kg, (b) 1.83 kg
9.21 (a) 2.99×10^{-4} m³, (b) 2.05 kg
9.23 (a) Density is $12.5 \times 10^3 \text{ kg/m}^3$.
Crown is not solid gold.
(b) 23% **9.25** 556 m³ **9.27** 8.0 cm
9.29 (a) 2.2 kg, (b) 1.5 g **9.31** 4.5×10^{-2} N **9.33** 4
9.35 2.4 cm **9.37** $p = 4.2 \times 10^5$ Pa, $v = 12.5$ m/s
9.39 4.4 kPa **9.41** reduced by 19%
9.43 8.1×10^{-4} m³/s **9.45** 0.15 m/s
9.47 (a) 65 Pa·s, (b) 7.2 Pa·s **9.49** 0.018 m/s
9.51 1.14×10^5 N, 6.40×10^4 N, consumption $\propto v^2$
9.53 $a = g - F_{\text{drag}}/m$ **9.55** 4 cm **9.57** 2 m/s
9.59 30.6 kPa **9.61** 85.6°C **9.63** 6.6 mm
9.65 (a) 40 m/s, (b) 19 m **9.67** 1.1 W **9.69** 19 Hz
9.73 0.150 mm

Chapter 10

10.1 (a) 136°F, (b) −129°F **10.3** −40°

10.5 $T(°F) = \frac{9}{5}T(K) - 459.67$

10.7 77.35 K = −320.4°F **10.9** $2.9 \times 10^5/°C$
10.11 1.51 m **10.13** 1.6 L **10.15** 2.1 cm
10.17 0.4 mm, up **10.19** 30 cm^3 **10.21** 252 cal
10.23 0.36 °C/s **10.25** 5.7°C **10.27** 5.9×10^4 cal
10.29 103 g **10.31** 73°C **10.33** 1.2°C
10.35 4.18×10^5 J **10.37** 3.65×10^5 cal
10.39 1.4 MJ **10.41** 265 km **10.43** 8.8 kW
10.45 (a) 116 W, (b) 249 W **10.47** 4.0 W
10.49 459 W/m^2, 5.67×10^4 W/m^2, 4.59×10^6 W/m^2
10.51 5980 K **10.55** (a) 3.33×10^{-4}, (b) 14.4 s
10.57 22.1°C **10.59** 11.4 $ft^2 \cdot h \cdot °F/Btu$
10.63 5.25 cal/g°C **10.65** 48.3 g
10.67 All the water remains. **10.69** 1.72°C

Chapter 11

11.1 22.9 kJ **11.3** (a) 33.4 kJ, (b) larger
11.5 impossible **11.7** 7.8 J **11.9** (a) 250 J, (b) 300 J
11.11 260 J **11.13** −150 J **11.15** Better to lower T_C.
11.17 (a) 26 J, (b) 9 J **11.19** 1.27 J **11.21** 0.80
11.23 0.22 K **11.25** (a) 0.502, (b) 0.424
11.27 (a) 0.175, (b) 0.308 **11.29** (a) 0.264, (b) 3410 MW
11.31 1 **11.33** −32°C **11.35** (a) 21, (b) 8.9
11.37 122 J/K **11.39** 606 J/K **11.41** 2.6×10^{-3} J/K
11.43 8.8 MW/K
11.45 (a) 3060 J/K, (b) −2840 J/K, (c) 220 J/K

11.47 $W = P_o V_o + \frac{3}{2}P_o b V_o^2$ **11.49** 42 J

11.53 (a) 14.8 J/K, (b) 11.3 J/K
11.55 (a) 3000 MW, (b) 2000 MW, (c) 8°C
11.57 (a) 97 W, (b) 0.33 W/K
11.59 56% more heat released

Chapter 12

12.1 8.17 m **12.3** 5.2×10^{19} N **12.5** 590 N
12.7 3 atm **12.9** 1.26×10^5 Pa **12.11** 0.94 L
12.13 68 cm from the closed end
12.15 (a) 546 K, (b) 746 K, (c) 2546 K **12.17** 0.039 mol
12.19 (a) 40.1 g/mol, (b) argon **12.21** 1.7×10^{-3} m^3
12.23 1.92 kg/s **12.25** 1.5 kg/m^3 **12.27** 1.05
12.29 6.21×10^{-21} J **12.31** (a) 1.24×10^{-24} kg, (b) No
12.33 3740 J **12.35** (a) 3120 J, (b) 1870 J
12.37 (a) 26 K, (b) 16 K **12.39** (a) 0.750, (b) 0.749
12.43 0.50 P_o **12.45** 1.21 **12.47** 2.16×10^4 N
12.49 (a) 7.3×10^{26} molecules, (b) 340 N,
(c) 3.3×10^{-9} m, (d) 0.035 m^3 **12.51** 1.6 cm
12.53 (a) 1.71×10^4 J/mol, (b) 1.02×10^4 J/mol
12.55 5.02 **12.57** 117 kPa **A12.5** 49/min
A12.7 19.34

Chapter 13

13.1 100 N **13.3** 23 cm **13.5** 66.7 N/m
13.7 0.575 m **13.9** 19.6 m/s^2 **13.11** 2.8 s
13.13 (a) 6.0 m/s^2, (b) −4.2 m/s^2, (c) 10.6 cm
13.15 0.087 m **13.17** 64 N/m **13.19** 1.2×10^{-2} J
13.21 (a) 0.067 m, (b) 2.3×10^{-2} J
13.23 (a) 0.318 Hz, (b) 3.14 s
13.25 (a) 40 cm/s, (b) 0.80 Hz **13.27** 0.99 m
13.29 (a) 1.55 Hz, (b) 0.645 s
13.31 (a) 0.140 m, (b) 0.994 m **13.33** (a) 4.0 s, (b) 3.5 s
13.35 9.783 m/s^2
13.37 St. Thomas, 122.3 cm; New York, 122.5 cm; London,
122.6 cm; Spitzbergen, 122.9 cm
13.41 (a) 3.0×10^{-3}/s, (b) 0.069 **13.45** 2:1

13.47 3.92 m/s^2 **13.51** $F = -\frac{4}{3}G\pi\rho mr = -kr$

13.53 5.9 J
13.55 $PE_{spring} = \frac{1}{2}$ gravitational energy lost. Difference
appears as heat.
13.57 (a) $a = -(g/L)x$, (b) $T = 2\pi\sqrt{L/g}$ **13.59** 0.76 Hz
13.61 4.92 s **13.63** 2.0 m

Chapter 14

14.1 6.7 Hz **14.3** $y_o \sin(kx - \omega t)$
14.5 (a) 3.23 m, (b) 560 m, (c) 25.2 m
14.7 4.3×10^{14} Hz to 7.5×10^{14} Hz
14.9 The child must triple the frequency.
14.11 352.5 m/s
14.13 (a) 10000 Hz, 1000 Hz, 100 Hz, 10 Hz,
(b) All heard but the 10 Hz. **14.15** 2.0 cm, 680 cm
14.17 $y_1/y_2 = r_2/r_1$ **14.19** 0.050 m **14.21** 10^8
14.23 37 dB **14.25** 3.16×10^6
14.27 6.3×10^{-6} W/m^2 **14.29** (a) 916 Hz, (b) 724 Hz
14.31 (a) 44 Hz, (b) You will not hear it. **14.33** 14.5°
14.35 28° **14.37** 330 m/s
14.39 333 Hz, 667 Hz, 1000 Hz **14.41** 1.00 m
14.43 0.35 **14.45** 1.01 g/m **14.47** 4
14.49 (a) 307 Hz, (b) Fourth harmonic, (c) 77 Hz
14.51 (a) 340 Hz, (b) 170 Hz **14.53** 1.7 Hz
14.55 0.9 mm **14.57** 293 Hz **14.59** (a) Yes, (b) No

14.61 $y_o \sin\left(\frac{2\pi x}{0.64 \text{ m}} - \frac{2\pi t}{2.1 \text{ s}} - \frac{\pi}{2}\right)$ **14.63** 1524 Hz

14.65 800 Hz$(1 + 0.04 \cos 6\pi t)$ 14.67 0.018 Hz
14.69 (a) 10, (b) 360 crests, (c) 120 crests, (d) 96
14.71 $(1.8 \times 10^{-3})\lambda\Delta T$ 14.73 -9.5 dB
14.75 (a) $f_o/(1 + gt/v)$, (b) $f_o(1 - gt/v)$
14.77 $y = 2y_o\cos(2\pi x/\lambda)\sin(2\pi t/T)$ 14.79 0.11 m/s^2
14.81 86.5 dB

Chapter 15

15.1 3.2 N, attractive   15.3 9.5 m   15.5 36
15.7 (a) 320 N, attractive, (b) 40 N, repulsive
15.9 -8.2×10^{-8} N   15.11 1.7 km
15.13 3.68×10^{-9} N   15.15 $\left(\sqrt{2} + \dfrac{1}{2}\right) k \dfrac{q^2}{L^2}$,

diagonally outward
15.17 End charges: 0.10 N, outward, Center charge: zero
15.19 -8.4 N, toward the origin   15.21 1.21×10^6 N/C
15.23 -2.45×10^{-8} C   15.27 3.29 μC
15.29 (a) 8.00×10^{-16} N, (b) 3.2×10^{-17} J
15.31 1.50×10^{-2} N at 53.1°   15.35 $E = Q/4\pi\epsilon_o r^2$
15.37 $\sigma/2\epsilon_o$
15.39 (a) $-\rho qr/3\epsilon_o$, (b) $r_o\cos \omega t$, where $\omega = \sqrt{\rho q/3\epsilon_o m}$
15.41 1.6×10^{-26} J   15.43 4.33 mm   15.45 100 N/C
15.47 9.7×10^{-9} C   15.49 0.119 m
15.51 (a) 8.22×10^{-8} N, (b) 2.19×10^6 m/s,
(c) 6.57×10^{15} Hz
15.53 (a) 5.4×10^6 N/C in the $-x$ direction,
(b) 1.7×10^5 N/C in the $-x$ direction
15.55 $\tan^{-1}\left(\dfrac{q\sigma}{2\epsilon_o mg}\right)$ 15.57 (a) $\rho R^3/3\epsilon_o r^2$, (b) $\rho r/3\epsilon_o$
15.59 $qQ/4\pi\epsilon_o r^2$ 15.61 3.13×10^4 N/C

Chapter 16

16.1 -3 kV   16.3 0.11×10^{-12} C   16.5 -1.18 V
16.7 2.5×10^{-6} C, 1.00×10^6 V/m   16.9 7.5×10^5 V
16.11 (a) $-Q_o$, (b) $+Q_o$, (c) 0   16.13 $Q/2$
16.15 7.5×10^4 V/m   16.17 1.6×10^7 m/s
16.19 2×10^{-3} C   16.21 8.8×10^6 m/s   16.23 20 μF
16.25 150 V   16.27 8.9×10^{-10} F   16.29 0.18 nF
16.31 100 V   16.33 2.3×10^6 V/m   16.35 208 V
16.37 (a) 2.3 μF, (b) 5 kV   16.39 625 V   16.41 $\kappa\epsilon_o AE$
16.43 (a) 1.1×10^{-2} C, (b) 2.5 J   16.45 400 J
16.47 4.6×10^6 V
16.49 (a) 9.0 V$\left(\dfrac{1}{|x|} + \dfrac{1}{|x - 0.10|}\right)$, (b) 120 V
16.51 9.0×10^{-7} J   16.53 $q/4\pi\epsilon_o r^2$ 16.55 60 V
16.57 (a) $\sigma/\epsilon_o E$, (b) 3.8
16.59 (a) 12.5 J, (b) 25 J, (c) Work must be supplied to
separate the plates.

Chapter 17

17.1 0.50 C, 30 C   17.3 4.0 A   17.5 4 nA
17.7 50 mA   17.9 9.0 Ω 17.11 0.54 Ω
17.13 0.10 A, 0.20 A, 0.30 A, 0.40 A, 0.50 A, 0.60 A
17.15 6.65×10^{-2} Ω 17.17 $R_{Al(16)} = 0.969 R_{Cu(18)}$
17.19 7.6 Ω 17.21 0.14 W   17.23 6.67 A
17.25 (a) 10.0 V, (b) 0.100 A
17.27 54 kWh, 1.94×10^8 J   17.29 \$1.44
17.31 (a) 0.045 A, (b) zero in left resistor,
0.090 A in right resistor   17.33 0.87 V   17.35 25 Ω
17.37 400 Ω 17.39 160 pF   17.41 0.54 μF
17.43 17 μF   17.45 6.5 μF   17.47 4.5 V
17.49 (a) 12 V, (b) 180 μC, 300 μC,
(c) 1.1×10^{-3} J, 1.8×10^{-3} J   17.51 12.1 V   17.53 17
17.55 (a) Yes, (b) No   17.57 1.8¢   17.59 23 kg
17.61 5.9 Ω
17.63 (a) 100 W, (b) no problem, (c) overload the circuit
17.65 $R_x = R_2R_3/R_1$ 17.67 16 mA

Chapter 18

18.3 A·m^2 18.5 375 N·m   18.7 to your left
18.9 0.20 N   18.11 0.020 N   18.13 0.0106 N up
18.15 1.41×10^{13} m/s^2 18.17 1.19 cm
18.19 (a) 4.79×10^5 m/s, (b) 1.92×10^{-16} J
18.21 11.4 MHz   18.23 28 GHz   18.29 2.6 mT
18.31 20 cm   18.33 3.33 cm   18.35 3.57 A·m^2
18.37 2.4×10^{-6} N·m   18.39 19000 Ω in series
18.41 2.5×10^{-2} Ω in parallel with galvanometer
18.43 7.15 mT   18.45 4000 A   18.47 3.82×10^{-13} T
18.49 VBA/ρ 18.51 8.5×10^{-21} kg·m/s
18.53 1.1×10^{-4} N, attractive   18.55 1.0×10^{-3} J
18.57 (a) $\mu_o NI/2\pi r$, (b) varies with position, (c) zero
18.59 8.31 turns/m   18.61 4.18×10^{-6} N·m

Chapter 19

19.3 0.11 V   19.5 21 μV   19.7 0.18 T
19.9 (a) 4.8 mV, (b) 19 mV, (c) 9.6 mV, (d) 19 mV
19.11 (a) 3.6 mV, (b) The end with higher potential is B.
19.13 1.6 mV   19.15 2.3×10^{-5} V   19.17 12 V
19.19 10   19.21 400 V   19.23 4.0 A
19.25 (a) 2.5 A, (b) 0.50 A, (c) 60 W   19.27 0.25 μH
19.29 80 mV   19.31 0.54 A/s   19.33 45 mA
19.35 2.998×10^8 m/s   19.37 0.15 m   19.39 0.61 mV
19.41 37.5 mA   19.43 25.8°C   19.45 0.125 A
19.47 100 μH
19.49 (a) Blv, (b) Blv, (c) Blv/R, (d) B^2l^2v/R 19.51 4.5 J
19.53 1.8×10^{-12} J   19.55 Yes   19.59 0.60 H
19.61 0.11 H

Chapter 20

20.3 $1.3 \ \Omega$ **20.5** $0.39 \ V/R, 0.86 \ V/R, 0.98 \ V/R$

20.9 $780 \ s$ **20.11** $500 \ s, 79 \ V$ **20.13** $50 \ Hz$

20.15 $12 \ mA$ **20.17** $V_o/\sqrt{2}$ **20.21** $160 \ \Omega, 56 \ mA$

20.23 (a) $69 \ \Omega$, (b) $72 \ mA$ **20.25** $5.7 \ k\Omega, 94 \ k\Omega, 940 \ k\Omega$

20.27 $0.27 \ A$ **20.29** $0.15 \ A$ **20.33** $0.73 \ W, 8.0 \ \Omega$

20.35 (a) $690 \ Hz$, (b) $1500 \ \Omega$ **20.37** (a) $61 \ Hz$, (b) $330 \ \Omega$

20.41 $46 \ kHz$ **20.43** (a) $0.16 \ A$, (b) $58 \ Hz$

20.45 (a) $13.5 \ V$, (b) $15 \ V$ **20.47** $6.0 \ mA$

20.49 (a) -0.5, (b) 1.5 **20.51** (a) $0.40 \ ms$, (b) $0.55 \ s$

20.53 (a) $0.16 \ s$, (b) $0.74 \ s$ **20.55** (a) 5τ, (b) $0.24 \ s$

20.61 (a) $V_o/2$, (b) $V_o/\sqrt{2}$

Chapter 21

21.1 $76 \ cm$ **21.3** parallel to the symmetry axis

21.5 $5.4 \ m/s$ **21.7** ethyl alcohol   **21.9** $29 \ cm$

21.11 $21°$ **21.13** $62.5°$ **21.15** $97.2°$

21.17 $44 \ cm, -4.1$

21.19 (a) front, $77 \ cm$, $17 \ cm$ high, rear, $51 \ cm$, $9 \ cm$ high

21.21 -4.2 **21.23** $14 \ cm$ **21.25** (a) $-17 \ cm$, (b) 0.33

21.27 The claim cannot be true.   **21.29** $148 \ mm$

21.31 $17.6 \ cm, -0.18$

21.33 (a) $-6.67 \ cm$, (b) $-5.45 \ cm$, (c) $-3.75 \ cm$

21.35 $3R/2$ **21.37** $-0.32 \ m$

21.41 (a) $0.133 \ rad$, (b) $0.125 \ rad$ **21.43** $80.4°$

21.45 $24 \ cm$

21.47 $0.39 \ R$ from the back surface of the sphere

21.49 $0.34 \ m$ **21.57** nd

21.59 $18 \ cm$ to the right of the third lens

Chapter 22

22.1 $20.0 \ cm$ **22.5** 3.3 diopters   **22.7** $50 \ cm$

22.9 7.0 diopters   **22.11** $6.3 \ cm$ **22.13** $1.7\times$

22.15 (a) $14 \ cm$, (b) real   **22.17** $2\times$

22.19 $2.5\times, 1.6\times, 4.0\times$ **22.21** $f/4.3$ **22.23** $1/115 \ s$

22.25 $9.0 \ cm$ **22.27** $f/3.4$ **22.29** $2.2 \ cm$

22.31 $2.03 \ mm$ from the objective lens   **22.33** $1.6 \ mm$

22.35 (a) $50\times, 100\times, 200\times, 400\times$,

(b) $100\times, 150\times, 200\times, 300\times, 400\times, 600\times$

22.37 (a) $2.9 \ cm$, (b) $103 \ cm$ **22.39** $3 \ cm$

22.41 $+5.0\times$ **22.43** (a) $8.12\times$, (b) $40.6 \ cm$

22.45 (a) $+10 \ cm$, (b) $+2.5 \ cm$ **22.47** smaller lens

22.51 3.1 times as much, or 210% more   **22.53** $f/22$

22.55 5.7 diopters   **22.57** $7 \ mm$ **22.59** $9.3 \ m$

22.61 (a) $1.5 \ f$, (c) 1.33

Chapter 23

23.3 $28°$ **23.5** (a) $244 \ nm$, (b) $432 \ nm$ **23.7** $0.13°$

23.9 $0.496 \ mm$ **23.11** $25°$ **23.13** $589 \ nm$

23.15 $51 \ cm$ **23.17** 1.75×10^{-4} **23.19** $90 \ nm$

23.21 $480 \ nm$ **23.23** Yes, $105 \ nm$

23.25 $714 \ nm$, far red; $556 \ nm$, green; $455 \ nm$, blue

23.27 $50 \ \mu m$ **23.29** $590 \ nm$ **23.31** 3 **23.33** $14.8°$

23.35 $0.43 \ m$ **23.37** $1.32 \times 10^{-7} \ rad$ **23.39** $6.9 \ km$

23.41 $f/0.66$ **23.43** $1.59°$ **23.45** $63°$ **23.47** $56.7°$

23.49 (a) 1.66, (b) flint glass

23.53 (a) Yes, (b) not visible   **23.55** $3.8 \ cm$

23.57 $87 \ \mu m$ **23.59** (a) No overlap, (b) Yes

23.61 (a) $-7.6°, -22.7°$ **23.63** (a) Along a north-south line, (b) no signal   **23.65** $0.48°$ **23.67** $1.3 \ km$

23.69 (a) 20%, (b) 15%

Chapter 24

24.1 $2 \ km/h, 6 \ km/h$ **24.3** 0 **24.5** $0.95c$

24.7 $0.89c$ **24.9** $0.45c$ **24.11** $6:00:01.28$

24.13 $10.5 \ y$ **24.15** $0.97c$ **24.17** $1.8 \times 10^{-7} \ s$

24.19 (a) $209 \ m$, (b) $0.70 \ \mu s$, (c) 0.72, (d) Yes

24.21 $2200 \ m$ **24.23** $0.60c$ **24.25** $1.4 \ m$

24.27 $1.64 \times 10^{-13} \ J$ **24.29** $3.2 \ m_0$ **24.31** $7.09 \ m_0$

24.33 $0.986c$ **24.35** $1.34 \ m_0, 0.34 \ m_0 c^2$

24.37 (a) $1.5 \ m_0$, (b) $0.5 \ m_0 c^2$, (c) $1.5 \ m_0 c^2$ **24.39** $0.98c$

24.41 $78 \ MHz$ **24.43** moving away, $750 \ nm$

24.45 $1.2 \times 10^{-3} \ Hz$ **24.47** 0

24.51 (a) $7.73 \times 10^3 \ m/s$, (b) $5240 \ s$, (c) $400 \ \mu s$

24.53 14 years   **24.55** $-245 \ ns$

Chapter 25

25.3 NO_2 **25.5** $0.593 \ g$ **25.7** $4.2 \ L$ **25.9** $22.41 \ L$

25.11 $9 \ mm$ **25.13** $6.93 \times 10^4/m$

25.15 $4.78 \times 10^{22}/cm^3$ **25.17** $31.0°, 64.7°, 107°$

25.19 $1.76 \times 10^{-10} \ m$ **25.21** $4.11 \times 10^{-10} \ m$

25.23 $1.69 \times 10^{-10} \ m$ **25.25** $9.54 \times 10^{-3} \ m$

25.27 $760 \ V$ **25.29** $2.5 \ m$ **25.31** 0.51

25.33 $1.31 \times 10^7 \ m/s$ **25.35** $9.8 \times 10^3 \ Bq$

25.37 $t_{1/2} = 3.5 \ h, \lambda = 0.20/h$, 2600 disintegrations/min

25.39 $9050 \ y$ **25.41** $3.2 \times 10^{-14} \ m$ **25.45** $5.4 \ kV$

25.47 (a) $2.6 \times 10^{12} \ Bq$, (b) $70 \ Ci$ **25.49** $1040/min$

Chapter 26

26.1 (a) $486.08 \ nm$, blue, (b) $434.00 \ nm$, violet

26.3 $b = 4/R$ **26.5** $6110 \ K$ **26.7** $569 \ nm$

26.9 $910 \ nm$ **26.11** $4.97 \times 10^{-19} \ J$ **26.13** $568 \ nm$

26.15 $4.136 \times 10^{-15} \ eV·s$ **26.17** $1.04 \times 10^{-12} \ m$

26.19 $4.32 \times 10^{-11} \ m$ **26.21** $2.26 \ eV$ **26.23** $0.11 \ eV$

26.25 (a) $2.14 \ eV$, (b) No   **26.27** $2.12 \times 10^{-10} \ m$

26.29 364.6 nm   26.31 9.8×10^{31} 26.33 91.2 nm
26.35 (a) 10.2 eV, (b) 122 nm
26.39 (a) 40.5 keV, (b) 80.8 keV
26.41 (a) 2.28×10^{-11} m, (b) 7.23×10^{-11} m
26.43 copper   26.45 zirconium
26.47 (a) 100 keV, (b) 1.24×10^{-11} m
26.49 (a) 1.09×10^{6} m/s, (b) 3.65×10^{-3}
26.51 122 nm, 91.2 nm   26.53 6.78×10^{5} m/s
26.55 3.09×10^{3} 26.57 1.59×10^{15}/s
26.59 1.05×10^{5} K

Chapter 27

27.1 (a) 2.43×10^{-12} m, (b) 1.32×10^{-15} m,
(c) 1.18×10^{-14} m   27.3 83.6 keV
27.5 (a) 16.4 keV, (b) 0.8 keV
27.7 (a) 0.727 nm, (b) 3.97×10^{-13} m
27.9 1.2×10^{5} m/s   27.11 3.88×10^{-11} m
27.13 3.7×10^{-63} m   27.15 16.7 eV
27.17 8.4×10^{33} 27.19 8.62°   27.21 1.5 m/s
27.23 2.4 eV, 4.1×10^{-6} eV   27.25 28.8 ns
27.27 1.0×10^{-11} m
27.29 6.0×10^{-20} J, 2.4×10^{-19} J, 5.4×10^{-19} J
27.31 3.3×10^{-13} J, 1.3×10^{-12} J, 3.0×10^{-12} J
27.33 1.7×10^{-27} kg   27.35 $2.48 \times 10^{-3} \, \psi_0^2$
27.37 Listed as n_x, n_y, and $(8mL^2/h^2)E$: 1, 1, 1.25; 1, 2, 2;
1, 3, 3.25; 1, 4, 5; 2, 1, 4.25; 2, 3, 6.25   27.39 2 subshells
27.41 7   27.43 9.274×10^{-24} J/T
27.45 1.16×10^{-5} eV
27.49 Two 1s states, two 2s states, one 2p state
27.53 12.7°   27.61 2.58×10^{-11} m   27.63 1.0×10^{-6}
27.65 3.7×10^{-24} J

Chapter 28

28.1 6.5   28.3 12.5
28.5 80 p, 120 n; 8 p, 8 n; 90 p, 142 n   28.7 $^{29}_{14}$Si
28.9 35.5   28.11 63.6   28.13 1.0×10^{-42} m³
28.15 125   28.17 8.08×10^{-37}
28.19 164 MeV, 1520 MeV   28.21 26.388 MeV
28.23 1.343   28.25 $^{222}_{86}$Rn, $^{233}_{92}$U, $^{59}_{26}$Fe   28.27 $^{4}_{2}$He
28.29 electron   28.31 No
28.33 (a) 4.67 MeV, (b) 4.59 MeV   28.35 $^{233}_{92}$U, $^{235}_{92}$U
28.39 6.05 MeV, 1.71×10^{7} m/s

28.43 (a) 5.0 Gy, (b) 0.50 Gy   28.45 $^{14}_{6}$C   28.47 $^{14}_{7}$N
28.49 0.35 kg   28.51 3.27 MeV   28.55 1.60×10^{7} m/s
28.57 $4Q/(A + 4)$ 28.61 3.4 g   28.63 $1.6 \, v_\alpha$

Chapter 29

29.1 9.48×10^{5} 29.3 423 MHz   29.5 1.786 eV
29.7 2.28 eV, green   29.9 2.3   29.11 0.82
29.13 0.75   29.15 0.80   29.17 47.1°
29.19 0°, ± 23.9°   29.21 79.88°, 46.82°, 40.62°, 37.97°
29.23 3000 K   29.25 5.7   29.27 (a) Orange, (b) purple
29.29 (a) Green, (b) green, (c) red, (d) no light emerges
29.31 violet, violet, blue, red   29.33 3.3×10^{-5} mm

Chapter 30

30.1 0.282 nm   30.3 $\dfrac{\sqrt{3}}{4}$ a
30.5 2.54×10^{28} electrons/m³   30.7 5.5×10^{4} K
30.9 (a) 5.84 eV, (b) 212 nm   30.11 7.40×10^{-15} s
30.13 1.47×10^{-4} m/s   30.15 6.72×10^{-3} m²/V·s
30.17 3.81×10^{-14} s   30.19 2.33 nm
30.21 2.23/Ω·m
30.23 (a) 1.35 km/s, (b) 5.93×10^{6} m/s
30.25 8.4×10^{28}/m³   30.27 4.74/Ω·m   30.29 0.54 V
30.31 −0.098 μA, −0.1 μA, +11 mA   30.33 890 nm
30.35 (a) 400 m², (b) 570 m²   30.37 $\dfrac{\sqrt{3}}{4}$ a_0
30.39 1.79 eV
30.41 (a) 3.4×10^{-3} Ω, (b) 4.35×10^{-3} m²/V·s,
(c) 4.35×10^{-2} m/s   30.43 0.30 V

Chapter 31

31.1 1.783×10^{-30} kg   31.3 1877 MeV
31.5 4.1×10^{9} pair   31.7 1.533 MeV   31.9 160 m_e
31.13 7.68 m   31.15 7×10^{-20} kg·m⁻¹·s⁻¹
31.17 6.3×10^{12} protons/s   31.19 1.5×10^{7} TeV
31.21 (c) Does not conserve baryon number,
(d) does not conserve baryon number or charge
31.23 1.3×10^{-20} s   31.25 u$\bar{\text{d}}$
31.27 (a) zero, (b) zero
31.29 −536 MeV, requires energy
31.31 (a) 3.17 nm, 2.35 nm, 2.10 nm, (b) x-ray region

Photo Credits

Chapter One

B1.1, The Bettmann Archives; B1.2(a + b), University of Toronto; B1.3, The Bettmann Archives; B1.4, Courtesy Intelsat

Chapter Two

2.15, Michael Sparks Keegan; 2.19, Photo Courtesy of United Airlines; 2.22(a + b), "The Role of Music in Galileo's Experiment," Stillman Drake, Scientific American, June 1975; 2.23(a + b), NASA Langley Research Center; B2.1, The Bettmann Archives; 2.24, Courtesy Six Flags over Georgia; B2.2, Dr. Harold Edgerton, MIT

Chapter Three

3.23, Jones & Childers

Chapter Four

B4.1, Bettmann Archives; B4.2, Royal Mail Stamps and Philately; 4.2(a), 4.3(a), Jones & Childers; 4.7, NASA; 4.8, Thomas Cooper Library/US; STC4.1, © Time Inc/Del Molkey; STC4.2, Sgt. Tom Shriver; 4.13, Photri; B4.3, Courtesy Darlington Raceway; B4.4, Jones & Childers; 4.15, Brit Rail Travel International; 4.22, North Wind Picture Archives

Chapter Five

5.2, Photri; 5.5, Courtesy Six Flags over Georgia; 5.10(c), Courtesy International Equipment Co.; B5.1, Thomas Cooper Library/USC; B5.2, from S.C. Geological Survey MS-21, "Simple Bouquer Anomaly Map of South Carolina"; STC5.1, Photo Researchers, Inc.; 5.22, Courtesy Six Flags over Georgia

Chapter Six

6.8, FPG International; 6.16, Photri; STC6.1, NASA; STC6.2, Harold & Erica Van Pelt Photographers/K. Proctor Collection

Chapter Seven

7.1, Dr. Harold Edgerton, MIT; 7.18, The Bettmann Archives; 7.23, © Time Inc./Heinz Kluetmeier; 7.26, Agence Nature/NHPA

Chapter Eight

8.2, Dr. Harold Edgerton, MIT; 8.7, PSSC PHYSICS, 2nd Edition, 1965; D.C. Heath & Co. with Educational Development Center, Newton, MA.; B8.1, Stuart Johnson

Chapter Nine

9.11, Courtesy Goodyear Tire & Rubber Co.; 9.14, Dr. Harold Edgerton, MIT; B9.3, *The Pathway for Oxygen,* Ewald R. Weibel, Harvard University Press reprinted with permission; B9.7, © Lee Marshall 1988; 9.23, FPG International; 9.24, Photography by Thomas Corke and Hassan Nagib, Fluid Dynamic Research Center, Illinois Institute of Tech; 9.25(c), F.N.M. Brown

Chapter Ten

10.1, State of Alaska, Division of Tourism; B10.1, Courtesy of Omega Engineering, Inc., An Omega Technologies Co.; 10.3(b), Meredith Nightingale; 10.7, Courtesy of the Triborough Bridge & Tunnel Authority

Chapter Eleven

11.4, Thomas Cooper Library/USC; 11.6(a), 11.6(b), Photri; 11.6(c), G E Aircraft Engines; 11.13(a), General Motors Corp.; 11.13(b), FPG International; 11.13(c), Courtesy Paul Gipe & Associates; 11.13(d), Jones & Childers; 11.14, Tennessee Valley Authority

Chapter Twelve

12.3, Burndy Library; B12.1, Bettmann Archives; B12.2, FPG International

Chapter Fourteen

14.8, Pam Clark, Brigham & Women's Hospital; 14.9(a), FPG International; 14.9(b), Meredith Nightingale; 14.9(c), Motown Studio A—Courtesy Motown Historical Museum, Detroit, MI; 14.12, Courtesy United States Army; 14.16, Wide World Photos, Inc.

Chapter Fifteen

15.1, FPG International; 15.11(b + c), 15.13(b), Photos by Harold M. Waage; 15.20, Photo Deutsches Museum Munchen

Chapter Sixteen

16.4(b), Fischer Scientific; 16.5, High Voltage Engineering Corp.; B16.2, The Bettmann Archives; 16.10(a), Jones & Childers; 16.12(b), Photos by Harold M. Waage

Chapter Seventeen

17.1(a + b), Jones & Childers; 17.7, Allen-Bradley Company; B17.2, Photo Courtesy of GE Medical Systems; 17.10(a), Jones & Childers; 17.11, Meredith Nightingale

Chapter Eighteen

18.2(c), 18.3(a + b), Jones & Childers; 18.5(a), Danmarks Tekniskes Museum; 18.6(b), Jones & Childers; 18.28, Ken Gatherum—Boeing Computer Services Richland, Inc.

Chapter Nineteen

19.1(a), By Courtesy of the Director of the Royal Institute; 19.10, 19.13, Jones & Childers; B19.2, Photo Courtesy of Varian Associates

Chapter Twenty

20.20(b), 20.21(c), Jones & Childers; 20.24, Courtesy of AT&T Archives, a unit of AT&T Library network

Chapter Twenty-One

21.1, Courtesy of AT&T Archives, a unit of AT&T Library network; 21.3, New England Satellite Systems, Inc.; 21.7(a), Meredith Nightingale; 21.9(b), Jones & Childers; 21.11(b), Courtesy of SPEX Industries, manufacturer of light-measurement equipment; 21.12(a), Meredith Nightingale; 21.12(b), American Cystoscope; 21.13(a + b), Bausch & Lomb; 21.29, Eugene Hecht, *Optics,* 2nd ed. © 1987, Addison-Wesley Publishing Co., Inc., Reading, MA. Fig 5.110. Reprinted with permission of the publisher. 21.32(a + b), Jones & Childers

Chapter Twenty-Two

22.6, NASA; 22.8(a), Central Scientific Company; B22.1, Scala/Art Resource; 22.13, Tokina Corp./Courtesy DISCOVER

Chapter Twenty-Three

23.1, Jones & Childers; 23.5, 23.8, EDC; 23.9(b), Jones & Childers; 23.10, Thomas Cooper Library/USC; 23.14, Jones & Childers; 23.16(a–c), EDC; 23.17, Eugene Hecht, *Optics,* © 1987, Addison-Wesley Publishing Co., Inc., Reading, MA. Fig 10.56 on page 374. Reprinted with permission. 23.18, 23.23, Jones & Childers; 23.26, N. Watabe; 23.35, Jones & Childers

Chapter Twenty-Four

B24.1, By permission of the Hebrew University of Jerusalem/Courtesy AIP Neil Bohr Library; B24.2, reprinted from the American Journal of Physics; 24.14, Palomar Observatory, CAL Tech; 24.18, Wide World Photos

Chapter Twenty-Five

TL25.1, Culver Pictures; TL25.2, TL25.3, The Bettmann Archives; TL25.4, AIP Niels Bohr Library; TL25.5, Cambridge University Library/Courtesy AIP Niels Bohr Library; TL25.6, AIP Niels Bohr Library, W. F. Meggers Collection; 25.1(a + b), Jones & Childers; 25.1(c + d), Thomas Cooper Library/USC; 25.2(a), Jones & Childers; 25.2(b), 25.6, Thomas Cooper Library/USC; 25.8(b), Jones & Childers; 25.9, Reynolds Historical Library; 25.10, USC Chemistry x-ray Laboratory; 25.14(a + b) National Gallery in Washington; B25.2(a + b), Erwin Muller; B25.3, Courtesy of AT&T Archives; 25.17(a + b), Jones & Childers

Chapter Twenty-Six

TL26.1, Culver Pictures; TL26.2, AIP Niels Bohr Library, W. F. Meggers Collection; TL26.3, Maison Albert Schweitzer/Courtesy AIP Niels Bohr Library; TL26.4, By permission of the Hebrew University of Jerusalem/Courtesy of Neils Bohr Library; TL26.5, AIP Niels Bohr Library, Meggers Gallery of Nobel Laureates; TL26.6, University of Oxford, Museum of History of Science/Courtesy of Niels Bohr Library; 26.1, Bausch & Lomb; B26.2, Anglo-Australian Telescope Board © 1984 by David Malin; 26.3(a + b), Jones & Childers

Chapter Twenty-Seven

TL27.1, TL27.2, AIP Niels Bohr Library; TL27.3, AIP Niels Bohr Library, W. F. Meggers Collection; TL27.4, AIP Niels Bohr Library, Francis Simon Collection; TL27.5, AIP Niels Bohr Library, Bainbridge Collection; TL27.6, AIP Niels Bohr Library; B27.4, Courtesy of Amber Engineering, Inc., Goleta, CA; 27.3, Omnicron Associates; 27.4(a–c), N. Watabe

Chapter Twenty-Eight

28.12, University of Tennessee, Knoxville; 28.15, Smithsonian Institute; 28.21, Oak Ridge National Laboratory Photo

Chapter Twenty-Nine

29.7, 29.11(a–d), B29.2, 29.23(a + b), Jones & Childers

Chapter Thirty

30.3, AT&T Bell Labs & Oak Ridge National Lab; 30.4(b), S. -W. Chen, F. -Y. Shiau and Y. A. Chang; B30.1, 30.17, 30.23, Jones & Childers

Chapter Thirty-One

31.1(a + b), 31.2(a + b), Lawrence Berkeley Laboratory, University of California; 31.4(a + b), 31.6(a + b), Deutsches Elektronen-Synchrontron, Hamburg; 31.7(a), Pam Clark; 31.7(b), Fermi National Lab; 31.7(c), N. O. A. O./USA; 31.9, Joe Stancampiano & Karl Luttrell, IMB Collaboration, © National Geographic Society; 31.10, N. O. A. O./USA

Index

The abbreviation *def.* indicates a page where a term is defined; the abbreviation *tab.* indicates a table.

A

Aberrations, lens, 632–634, 683
Abscissa, 38 *def.*
Absolute zero, 299, 354
Absorbed dose, 847
Absorption spectra, 768
ac (alternating current), 505, 587–601
    effective value, 587
Acceleration, 43–46
    angular, 217
    average, 44 *def.*
    centripetal, 130
    constant, 46–50
    and force, 98
    of gravity, 51, 100
    instantaneous, 45 *def.*
    of mass on a spring, 385
    in simple harmonic motion, 386–387
    vector, 73
Accelerator, 919–920
    circular, 919
    colliding beams, 919
    cyclotron, 536-537
    fixed target, 919
    Van de Graaff, 471, 472
    linear, 572–573, 919
    as a microscope, 918
    storage ring, 919
Acceptor impurity, 904
Accommodation, 643
Achromatic lens, 634
Acoustics, 407
    of rooms, 416–417
Action and reaction, 102–103
Active medium, 865

Activity, 755, 846
Adiabatic process, 322
Aerodynamic drag, 283–284
Air column, vibrating, 428
Air drag, or resistance, 283–284
Air track, 106
Airfoil, 278
Airplane flight, 278
Alpha decay, model for, 840
Alpha emission, 838
Alpha particles, 754
Alternator, 563
AM, FM, and stereo, 596–597
Amber, 438
Ammeter, 543–544
Amorphous solid, 889
Ampere, Andre-Marie, 529, 545
Ampere (unit), 442, 496, 539 *def.*
Ampere's law, 545, 571
Amplifier,
    inverting, 602
    noninverting, 603
Amplitude in simple harmonic motion, 387 *def.*
Amplitude modulation, 596
Aneroid barometer, 269
Angle,
    critical, 617
    of incidence, 614
    radian measure, 5
    of reflection, 614
Angular acceleration, 217
    and torque, 217
Angular frequency, 133 *def.*
Angular kinematic equations, 218 *tab.*
Angular magnification, 647

Angular momentum, 211–213, 212 *def.*
    conservation of, 213–216
    quantization, 781
Angular velocity, 133 *def.*
Antenna, 572–574
    microwave, 612
Antibaryon, 921
Antinode, 425 *def.*
Antiparticles, 913–914
    annihilation of, 914
    production of, 914
Antiprotons, 914, 921
Antireflection coating, 673
Appearance of moving objects, 712
Archimedes' principle, 269–270
Area under a curve, 42
ARGUS detector, 920
Aristarchos of Samos, 7
Aristotle, 32, 385
Aspheric lens, 634, 658
Astigmatism, 644
Astronomical data, 9 *tab.*, and back end-paper
Astronomical telescope, 653–655
Atmosphere (atm), 351 *def.*
Atmospheric pressure, 261, 349–350
    variation with altitude, 368–369
Atomic clock, 14
Atomic concept, 349
Atomic mass, chemical, 832
Atomic mass of light elements, 836 *tab.*
Atomic mass unit, 835 *def.*
Atomic nucleus, discovery, 757–760
    see also nuclear
Atomic number, 832

Atomic size, 743–747
Atomic theory of
    Bohr, 779–783
    Dalton, 738
Atoms, 735–738, 743–744, 757–760
Audio frequencies, 414
Automobile
    braking, 240–243
    engine, 330–331
    tire friction, 112–113
    tire pressure, 358
Average acceleration, 44 def.
Average power, ac, 588
Average speed, 33
Average velocity, 34–38, 35 def.
Avogadro, Amadeo, 738, 739
Avogadro's number, 356, 739,
    740–743
Axioms, or laws of motion, 119

B
Back emf, 565–566
Background radiation, 848
Balance, torsional, 441
Balloons and gas laws, 352
Balmer, Johann, 766
Balmer's formula, 770
Balmer's series, 770,
    Bohr's theory of, 783–794
Band theory of solids, 897–900
Banking of curves, 138–139
Bar magnet, 527
Bardeen, John, 601
Barometer, 268
Barometric formula, 368–371
Barrier penetration, 813–814
    and alpha decay, 841
Baryon, 921
    baryon number, 921
Base unit, 18
Baseball, spinning, 286
Battery, 495,496
    internal resistance of, 515
Beats, 429
Becquerel, Henri, 736, 753
Becquerel (unit), 755, 846
Bednorz, J. G., 501
Bernoulli's equation, 276
Beta decay 838
    model for, 842
    spectrum, 842
Beta minus, 842
Beta particles, 754–755

Beta plus, 842
Bifocals, 644
Big bang model, 926–927
Bimetallic strip, 302
Binding energy, 836
    per nucleon, 836–837
Binoculars, 656
    resolution of, 681
Biological effects of radiation, 847
Biot, Jean-Baptiste, 538
Biot-Savart law, 538
Black hole, 247
Blackbody, 772 def.
    radiation, 771–773
    spectrum, 772–773
Blood pressure, 266–267
Bohr, Niels, 735, 766, 767, 779–785,
    794–795, 816
Bohr frequency condition, 780
Bohr magneton, 818
Bohr theory successes, 783–785
Bohr's postulates, 780
Bohr's theory of the hydrogen atom,
    779–783
Boiling, 308
Boltzmann, Ludwig, 339
Boltzmann
    constant, 364
    and entropy, 339
    factor, 864
Boyle's law, 351
Boyle, Robert, 351
Bragg, W. L., 737, 746, 785
Bragg's law, 746
Brahe, Tycho, 3–4, 6–7, 10, 14, 15,
    610, 654
Braking a car, 240–243
Brattain, Walter H., 601
Breakdown strength of dielectrics,
    484 tab.
Brewster, David, 687
Brewster's law, 688
Bridge, full-wave, 600
British thermal unit, 304
Brownian motion, 700, 741
Bubble chamber, 846, 914, 918
Buoyant force, 269

C
Calcite, 736
Calendar, Gregorian and Julian, 11
Caloric, 303
Calorie (unit), 304

Calorimeter, 307
Calorimetry, 305
Camera, 633, 648–650
Capacitance, 478 def.
    parallel plate capacitor, 479
Capacitive reactance, 590
Capacitive time constant, 585
Capacitors, 477–478, 477 def.
    charging, 585
    and dielectrics, 483-485
    energy stored in, 486–487
    in parallel, 513
    in series, 514
Capillary action, 273
Carbon-14, 757
Carnot, Sadi, 327
Carnot
    cycle, 326, 328–330
    efficiency, 332
    heat engine, 328
Cathode rays, 744
Cavendish balance, 144
Cavendish, Henry, 143–145, 352,
    738
Cell, 495, 497, 515
Celsius temperature scale, 296
Center of gravity, 141, 209
Center of mass, 209
Centrifugal force, 137
Centrifuge, 137
Centripetal
    acceleration, 130
    force, 135–137
Cerenkov radiation, 857
Chadwick, James, 830
Chadwick's experiment, 830–831
Chain reaction, 854–856
Changing mass, 203–204
Charge, electric, 439
    conservation of, 440
    density, linear, 455
    density, surface, 482
    isolated conductor, 452
    on some objects, 443 tab.
Charging a capacitor, 585
Charging by induction, 440–441
Charles, J. A. C., 352–353, 354–355
Charles' law, 354–355
Chromatic aberration, 634, 682
Circuit breaker, 518
Circuit diagram, 503–504
Circuit symbol for
    ac voltage source, 589

amplifier, 602
battery, 504
capacitor, 478
diode, 599
ground, 507
inductor, 568
rectifier, 599
resistor, 503, 504
switch, 504
transformer, 567
Circuits,
    alternating current, 587–601
    direct current, 496
    household, 517
    integrated, 601
    and Kirchhoff's rules, 508–509
    open, 507
    power in, 503–507
    RC, 584
    RL, 582
    RLC, 592
    resonant, 594–596
    short, 507
Circular motion, uniform, 129
    and simple harmonic motion, 386
Classical and quantum mechanics,
    793
Clausius, Rudolf, 335, 336
Clocks and watches, 14
Coefficient, sound-absorption, 417
    tab.
Coefficient of performance of a
    refrigerator, 333 def.
    heat pump, 334 def.
Coefficient of thermal expansion,
    linear, 300 def., 300 tab.
    volume, 302 def., 300 tab.
Coherence, optical, 869–871
Coherent bundle, 619
Colliding beams, 919
Collimating lens, 684
Collisions,
    one dimensional, 196, 231–235
    and conservation laws, 231-238
    elastic, 229, 230
    inelastic, 196, 230
    two and three dimensional, 200,
        235–238
Color
    by addition and subtraction,
        879–882
    filters, 880
    television, 879-880

temperature, 878 def., tab.
    of the visible spectrum, 877 tab.
Coma, 633
Communications satellite, 11
Compact-disc audio system, 907
Compass needle as a magnet, 526
Complementary colors, 880
Components of vectors, 67
Compression, longitudinal waves,
    412
Compton, A. H., 794, 796
Compton effect, 796–798
Concave lens, 620
Concave mirror, 628
Concorde, 58
Condenser, electrical, 477
Condensing lens, 651
Conduction, thermal, 310–311
Conduction band, 901
Conductivity and Ohm's law, 895
Conductivity, electrical, 896–897,
    901
Conductor, electrical, 439 def.
Conductor, isolated, 452
Confinement, quark, 925
Conic sections, 12–13
Conservation laws, 177
Conservation of charge, 440, 837
Conservation of mechanical energy,
    175, 177
Conservation of momentum,
    angular, 213–216
    and energy, 231
    linear, 195-196
    and Newton's laws, 195–196
    in one dimension, 195–196
    in two and three dimensions,
        200–203
Conservation of nucleon number,
    837
Conservative force, 175
Constant acceleration, 46 def.
Constant, spring, 383
Constants, fundamental, back
    end-papers
Constructive interference, 425
Contact force, 95, 438
Continuity equation, 275
Continuous spectra, 768–769
Continuous-wave laser, 868
Control rods, 856
Convection, 311
Conversion, temperature, 297

Conversions, unit, 24
Convex lens, 620
Convex mirror, 628
Copenhagen interpretation, 794
Copernicus, Nicholas, 7
Correspondence principle, 795
Cosine of an angle, 87 def.
Cosmic rays, 757, 842, 848–849,
    917, 926
Cosmology, 926–928
Coulomb, Charles, 441
Coulomb (unit), 442
Coulomb's balance, 441
Coulomb's law, 441–442,467
Couple, 208 def.
Crimea Observatory, 655
Critical angle, 617
Critical damping, 396
Crookes' tube, 744–745
Crystal diffraction, 744–748
Crystalline solid, 889
Crystals, 736, 887
    covalent, 889
    germanium, 900
    ionic, 889
    liquid, 889
    metallic, 889, 891
    silicon, 900
Curie, Marie and Pierre, 754
Curie (unit), 846
Curie law, 548
Curie temperature, 549
Curie-Weiss law, 549
Current balance, 539
Current density, 895, 896, 901
Current, electric, 496 def., 507
    direction, 496
    effective value, 588
    hazards of, 516
    induced, 559
    loop as a dipole, 542
    magnetic field of, 537–541
    magnetic force on, 531
    root-mean-square, 588
Current-carrying wires, force be-
    tween, 539
Cycle, Carnot, 328–330
Cyclotron, 536–537
    frequency, 537

D
D'Arsonval-Weston meter, 544
Dalton, John, 738

Damped harmonic motion, 395–397
Dark spectral lines, 768
Daughter nucleus, 837, 839
Davisson, C. J., 795, 802
dc 496
    generator, 563–564
    power supply, 598
De Broglie, Louis, 794, 798
De Broglie wavelength, 799
Decay constant, 755–756
Decay, radioactive, 755–757
Decibel, 416 *def.*
Definite proportions, law of, 737
Delta (Δ), 39 *def.*
Densities, 262 *tab.*
Density, 140 *def.*
    of charge carriers, 902
    of the earth, 143–145
    of water, 303
Depletion zone, 905
Destructive interference, 425
Detectors,
    ARGUS, 920
    in elementary particle physics,
        919–920
    of radiation, 844–846
Dewar flask, 313
Diagram, indicator, 327
Diamagnetism, 547
Diastolic pressure, 267
Dielectric, dipole model of, 485
Dielectric constant, 483 *def.*, 484 *tab.*
Dielectric films, optical properties,
    671–674
Dielectric strength, 484 *tab.*,
    485–486, 488
Dielectrics and capacitors, 483–485
Diffraction grating, 678–679
    holographic sinusoidal, 873–876
Diffraction, 674 *def.*
    double slit, 668–670
    electron, 802, 809–811
    and interference, 677
    multiple-slit, 677–678
    single slit, 674–676
Dimensions of a quantity, 20
Diode, 499, 596, 598–600
Diode, laser, 907
Diopter, 644 *def.*
Dipole, electric, 457–460
    alignment in a field, 458–459,
        484–485
    moment, 458 *def.*

Dipole, magnetic, 527
    moment, 529, 542
Dirac, Paul A. M., 913
Direct current, 496
    generator, 563–564
    power supply, 598
Dispersion, 633, 682
Displacement, 35 *def.*
    in simple harmonic motion, 386
Donor impurities, 904
Doppler effect, relativistic, 721–723
    red shift, 722, 725
Doppler effect, sound, 419–421
Doppler shift, 420–421
Dose equivalent, 847–848
Dose, 847
Double slit diffraction, 668–670
Drag coefficient, 284, 284 *tab.*
Drag, aerodynamic, 283–284
Drift velocity, 497, 895–897
Dry cell, 495, 497, 515
Dynamic equilibrium, 115 *def.*
Dynamics, 32

E

Earth's magnetic field, 532
Earth, as dipole, 528
Earth, density of, 143–145
Edison, Thomas A., 499
Effective dose equivalent, 848
Effective value of ac current,
    587–588
Efficiency, thermal, 332 *def.*
Einstein, Albert, 698, 700–701, 743,
    767, 777–778, 794, 865, 925
Elastic collision, 229, 230
    in one dimension, 231–235
    in two dimensions, 235–238
Electric charge, 439
Electric conductivity and Ohm's law,
    895
Electric conductivity, 896–897
Electric conductor, 439 *def.*
Electric current, 496 *def. See also*
    Current, electric
Electric dipole, 457–460
    moment, 456, 458 *def.*
    potential energy, 459
Electric field, 446 def
    energy density, 487
    and potential gradient, 475
    superposition, 447–449
Electric field, pictorial, 447–449

Electric flux, 450–451
Electric insulator, 439 *def.*
Electric motor, 565
Electric potential, 467–470, 469 *def.*
    energy, 468
    difference, 469 *def.*
    gradient and field, 475
Electric resistance, 498–503
Electric shock, 516, 477
Electrically neutral, 440
Electricity, 438
Electricity, Gauss's law for, 571
Electrolysis, 739
Electromagnetic induction, 558–559
Electromagnetic spectrum, 611
Electromagnetic wave, 571–574,
    596–597
    antenna production of, 572
    spectrum, 611
    speed, 574
Electromagnetism, 526, 528
Electromotive force, 497 *def. See also*
    emf
Electron capture, 842
Electron charge density of an atom,
    816
Electron microscope, 682, 800–801
Electron volt (unit), 778 *def.*
Electron, 439,
    charge, 442, 753
    charge to mass ratio, 752
    diffraction, 802, 809–811
    discovery of, 749–752
    mass of, 753
Electron-positron pair, 914
Electroscope, 440–441
Electrostatic force, 441
Electrostatic generator, Van de
    Graaff, 471
Electrostatics, 438 *def.*
Electroweak force, 925
Elementary charge, 442
Elementary particles, 913
    classification of, 921, 918 *tab.*
Elements, alphabetical list, Appendix
    C
Ellipse, 8–9, 12–13
emf, 497 *def*
    back, 565–566
    induced, 558–559
    in inductor, 569
    motional, 561–562
Emission spectra, 768–769

Endoscope, 619
Energy, 157, 162 *def.*
    binding, 836
    conservation of mechanical, 175, 177
    conversion, 340
    Fermi, 893
    and first law of thermodynamics, 322
    gravitational potential, 166
    of a harmonic oscillator, 389
    and Hooke's law, 161, 389
    internal, 321
    kinetic, 163 *def.*
    and mass, 714–716
    of a particle in a box, 812
    potential, 165
    quantization of, 774
    relativistic kinetic, 719
    rotational, 173 *def.*
    sources, alternative, 341
    transfer, thermal, 303
    transfer by wave, 411
    units, 305 *tab.*
    values, 163 *tab.*
    and work, 162
Energy, electromagnetic,
    in capacitor, 486–487
    electrical, 468, 503-507
    of an electric dipole, 459
    in electric field, 487
    in inductor, 570
    in magnetic field, 570
Energy and mass in relativity, 714–715
Energy bands, 879–901, 904–905
Energy levels, atomic, 864
    in helium-neon, 868
    in ruby, 866
Energy levels, nuclear, 843
Energy of a particle in a box, 812
Energy, binding, 836
Energy-level diagram, 784
Engine,
    Carnot heat, 328
    Diesel, 328
    gasoline, 330–331
    Otto, 328
    steam, 320, 326
Entropy, 336–340
    and probability, 339
    of the universe, 338
Environment, thermal, 322

Equal tempered scale, 427
Equation of continuity, 275
Equation of state, 356
Equations of motion, kinematic, 49 *tab.*, 218 *tab.*
Equilibrium,
    dynamic, 115 *def.*
    first condition for, 208
    rotational, 208
    second condition for, 208
    static, 115–119, 115 *def.*
    thermal, 295
    translational, 115
Equipotential lines and field lines, 473
Equipotential surface, 473–474, 473 *def.*
Equivalence principle, 723–726
Equivalent dose, 847–848
Eratosthenes, 29
Escape velocity, 246
Estimates and order of magnitude, 18–20
Exclusion principle, 820–821
Expansion, thermal, 300–303
Exponential function, 377–380, 583, 585–586
Eye, human, 642–645
Eyepiece, 651

F
f-number, 649 *def.* 649 *table*
Fahrenheit, Gabriel, 298
Fahrenheit temperature scale, 297
False color infrared film, 882
Faraday, Michael 452, 478, 485, 496, 557–558, 739
Farad (unit), 478 *def.*
Faraday (unit), 739 *def.*
Faraday cage, 452
Faraday's law of induction, 558–559, 571
Farsighted eye, 643–644
Feedback, 601
Fermi, Enrico, 843, 855
Fermi beta decay theory, 843
Fermi energy, 893
Ferrimagnetism, 549
Ferromagnetism, 528, 549
Fiber, optical, 618
Field in a parallel plate capacitor, 480–482

Field lines, 148–149
    electric, 446, 448, 449
    and equipotential lines, 473
    gravitational, 150
    magnetic, 527, 528–530
Field,
    electric, 446 *def.*
    gravitational, 147
    magnetic, 527
    superposition of electric, 447–449
Field-ion microscope, 750
Filters,
    color, 880
    electrical, 600
First condition for equilibrium, 208
First law of thermodynamics, 322–326
Fission, nuclear, 853
    fragments, 854
    power plant, 855–856
Fixed-target accelerators, 919
Fizeau, Armand Hippolyte Louis, 613
Flow,
    laminar, 275, 285
    turbulent, 276, 285
Fluid, 260 *def.*
    flow, 274–281, 284–285
Fluorescent lamp, 878
Flux, 450
    electric, 450–451
    magnetic, 530, 559
FM radio, 596
Focal length, 621 *def.*
    of spherical mirror, 628
    of thin lens, 639
Focal point, 620 *def.*
Forbidden zones, 899
Force, 94–95 *def.*
    and acceleration, 98
    buoyant, 269 *def.*
    centripetal, 135 *def.*
    conservative, 175 *def.*
    constant, spring, 383 *def.*
    contact, 95
    between current-carrying wires, 539
    electromagnetic, 95, 96, 922
    electrostatic, 439, 441
    electroweak, 95, 925
    external, 195
    frictional, 110–111
    fundamental, 95, 922, 922 *tab.*
    gravitational, 96, 922

Force (*cont.*)
  and impulse, 192
  inverse square, 92, 140, 441
  net, 94 *def.*
  nonconservative, 175 *def.*
  normal, 101
  magnetic on currents, 531
  on moving charge, 533
  restoring, 383
  strong, 95–96, 834–835, 922
  superposition of electric, 444–445
  and torque, 206
  weak, 95–96, 921–922, 925
  and work 157
Forced harmonic motion, 397–399
Fractional charge, 443
Franklin, Benjamin, 352, 439, 452,
  477
Fraunhofer, 766, 768–769
Fraunhofer dark lines, 768
Free body diagrams, 104–106,
  108–109
Free electron density, 893
Free electron model of metals,
  892–895
Free fall, Galileo and, 50–54
Freely falling body, 50
Frequency, 132 *def.*
  angular, 133
  cyclotron, 537
  fundamental , 425 *def.*
  harmonic, 425 *def.* natural oscilla-
    tor, 391
  resonant, 398
Frequency modulation, 596
Fresnel, August Jean, 610, 675
Friction, 110–111
  coefficients of, 111 *tab.*
Fringe visibility, 870
Fuel rods, 856
Full-wave rectifier, 600, 601
Fundamental constants, back end-pa-
  pers
Fundamental forces, 95
Fundamental frequency, 425 *def.*
Fundamental particles, 913
Fundamental unit, 18
Fuse, 518
Fusion, nuclear, 857–859
  power, 859

G

Gain of an amplifier, 603
Gal (unit), 145 *def.*

Galilean relativity, 699
Galilean telescope, 653, 655
Galilean transformations, 705
Galileo, 32, 52–53, 91, 296, 349,
  612, 654
Galvani, Luigi, 495
Galvanometer, 543–544
Gamma decay, model for, 843
Gamma radiation, 754–755
Gamma rays and x rays, 844
Gas constant R, 356
Gas law, 356
Gas, state of matter, 294
  ideal, 356, 360
  kinetic energy of, 361–364
  molecular speeds, 368–371
  specific heats of, 365
  van der Waals, 367
Gasoline engine, 330
Gauge pressure, 261
Gauss's law for magnetism, 529–530,
  571
Gauss's law, electricity, (qualitative)
  450–452
  (quantitative) 453–457
Gaussian surface, 451
Gay-Lussac, J. L., 352, 738
Gay-Lussac law, 355, 354–355
Geiger counter, 844–845
General relativity, 726–729
Generators, 563 *def.*
Genetic effects of radiation, 847
Geodesic, 728
Geometrical formulas, Appendix A
Geosynchronous satellite, 11,
  153
Germanium crystals, 900
Germer, L. H., 795, 802
Gilbert, William, 438, 526
Glashow-Salam-Weinberg model,
  925, 927
Glassy solid, 889
Gradient-index lens, 635, 658
Grain boundary, 888
Gram atomic mass, 739 *def.*
Grand unification theories, 925
Graphical interpretation of velocity,
  38–39
Grating, diffraction, 678–679
Gravitation, universal law of,
  140
Gravitational constant, G, 141
Gravitational field strength, 147
Gravitational force, 95, 100

Gravitational potential energy,
  general, 168–171
  near the earth, 165–168
Gravitational red shift, 725
Gravity,
  acceleration of, 51, 100
  center of, 141, 209
Gravity meter, 145
Gray (unit), 847
Gregorian calendar, 11
Ground state of hydrogen, 782
Ground, 470, 507
Grounding, 441
Guericke, Otto von, 351

H

Hadrons, 921 *def.*
Hafele-Keating experiment, 725–726
Half-life, biological, 862
Half-life, radioactive, 756
Half-wave rectifier, 599
Hall effect, 901–903
Halley, Edmund, 92
Harmonic frequency, 425 *def.*
Harmonic motion, damped, 395
Harmonic motion, simple, 385
Harmonic oscillator, 385–393
  energy, 389
  period, 390
Harmonic wave, 408
  equation, 409
Hauksbee's electrostatic machine,
  438
Heat capacity, 306 *def.*
Heat capacity of gas, molar, 365–367,
  366 *tab.*
  at constant pressure, 366
  at constant volume, 365
Heat death of the universe, 340
Heat engine, 323, 328
Heat flow, 303
  and first law, 322
Heat gained or lost, sign convention,
  307
Heat of
  fusion, 308 *def.*
  transformation, 308 *def.*, 309 *tab.*
  vaporization, 308 *def.*
Heat pump, 334–335
Heat reservoir, 328
Heat sink, 328
Heat transfer, 310–313
Heat, internal energy and tempera-
  ture, 321–326

Heat, mechanical equivalent of, 303
Heisenberg, Werner, 795, 806
Heisenberg's uncertainty principle, 804–807
Helium, 769
Helium-neon laser, 868–869
Hemispheres, Magdeburg, 351
Henry, Joseph, 557
Henry (unit), 569 *def.*
Hertz, Heinrich, 572, 610, 775
Hertz (unit), 132 *def.*
Higgs particle, 925
High Flux Isotope Reactor, 857
Holes (in semiconductor), 901
Hologram, 872
    production of, 872–873
    reconstruction of, 872–873
    white light, 874–875
    reflection, 874
    rainbow, 875
Holography, 864, 871–877
Home power distribution, 517–518
Hooke, Robert, 92, 351, 407, 610, 736
Hooke's law, 160, 382
Hoover dam, 340, 564, 567
Horsepower (unit), 182 *def.*
Horseshoe magnet, 528
Huygens, Christiaan, 191, 230, 407, 610, 664, 736
Huygens' principle, 664–665
Hydraulic lift, 265
Hydrodynamics, 260
Hydrogen atom,
    Bohr's theory of, 779–783
    wave theory of, 816–817
Hydrostatics, 260
Hyperbola, 12–13
Hyperons, 917
Hyperopia, 644

I
Ice cream churn, 310
Iceberg, 270
Iceland spar, 736, 737
Ideal gas law, 356
    deviations from, 365
Ideal gas, 360 *def.*
    internal energy, 363
Ideal heat engine, 328
Image conduit, 619
Image distance, 624

Image,
    real, 622 *def.*, 872–873
    virtual, 622 *def.*, 872–873
Impedance, 593 *def.*
Impulse, 192
Impure semiconductors, 903
Incandescent lamp, 878
Incidence, angle of, 614
Inclined plane, 107
Index of refraction, 615 *def.*, 616 *table*
Indicator diagram, 327
Induced current, 559
    in a coil, 561
Induced emf, 558–559
    in a coil, 560
Inductance, 568–569
    of a solenoid, 569
Induction, charging by, 440–441
Induction, Faraday's law, 558–559
Induction motor, 565
Inductive reactance, 591
    time constant, 583
Inductor, 569
    energy stored in, 570
    source of emf, 569
Inelastic collisions, 196, 230
Inertia, 96, 99
    moment of, 173 *def.*, 174 *tab.*
Inertial reference frame, 119, 698–699
Infrared, 299, 611, 773, 784
    color film, 882
Infrasonic frequencies, 414
Instability of the proton, 926
Instantaneous acceleration, 45 *def.*
Instantaneous velocity, 40 *def.*
Insulator, electrical, 439 *def.*
Integrated circuits, 601
Intensity, wave, 411
Intensity level, sound, 416
Intensity of sound as a function of distance, 413
Interference of light, 667–671
    and diffraction, 677
    double slits, 668
    in thin films, 671–674
Interference pattern, 873
Interference, constructive, 425
    destructive, 425
Interferometer, 700
Internal energy, 321 *def.*, 322–324
    and first law, 322

    of an ideal gas, 363
    temperature and heat, 321–326
Internal forces, 195
Internal reflection, total, 617, 618
Internal resistance, 515
International system of units, SI, 18–19, Appendix B
Inverse-square law,
    electrostatics, 441
    gravitation, 92, 140
Inverted population, 866
Inverting amplifier, 602
Ionization chamber, 830
Iris, 642
Isobaric process, 323–324
Isochoric process, 323–324
Isothermal process, 323
Isotherms, 368
Isotope, 832 *def.*
    atomic mass, 836 *tab.*
    stable, 832–833
    radioactive, 832

J
Joule, James Prescot, 304
Joule (unit), 158 *def.*
Joule's law of electrical power, 505–506
Junction, pn, 499, 904–905
Jupiter, moons of, 612, 655

K
Kamerlingh-Ohnes, H., 500
Kaons, 917
Keck telescope, 655
Kelvin, Lord William, 299
Kelvin (unit), 299–300
Kelvin temperature scale, 299
Kepler, Johannes, 7–8, 10–11, 91
Kepler's laws, 6–11, 142
    and conservation of angular momentum, 216
Kilocalorie (unit), 304
Kilogram (unit), 99
Kilowatt-hour (unit), 305 *tab.*, 506 *def.*
Kinematic equations, 49 *tab.*, 218 *tab.*
Kinematics, 32
    two dimensions, 72–75
Kinetic energy and Q value, 841, 862
Kinetic energy, 163 *def.*
    relativistic, 719–720
    rotational, 173 *def.*

Kinetic energy (*cont.*)
  translational, 164 *def.*
  and work, 164
Kinetic friction, 111
Kinetic theory of gases, 359–365
  model, assumptions of, 360
  definition of temperature, 363
Kirchhoff's rules, 508–509
  loop rule, 582, 585, 592
  voltage rule, 508

L

Laminar flow, 275–276, 285
Lamp,
  incandescent, 878
  fluorescent, 878
Land, Edwin H., 686
Laplace, P. S., 247
Laser, 611, 865–869
  diode, 907
  helium-neon, 868–869
  holography and, 864, 872
  ruby, 865–868
Laser-induced fusion, 858
Latent heat of transformation, 308
Lateral magnification, 623, 625
Laue, Max von, 744
Law of Charles and Gay-Lussac, 355
Law of definite proportions, 737
Law of volumes, 738
Lawrence, Ernest O., 536
Laws of thermodynamics,
  first, 322–326
  second, 335–336
  second law and entropy, 336–340
  zeroth, 321
LCD, 890–891
Leaning Tower of Pisa, 76
LED, 907
Leg, model of, 392
Length contraction, 712–714
Lens
  aberrations, 632
  achromatic, 634
  aspheric, 634, 658
  of camera, 648
  collimating, 684
  combinations, 618
  concave, 621
  condensing, 651
  converging, 620, 627
  convex, 620
  diverging, 620, 626

  of eye, 642
  eyepiece, 651
  focal length of, 621
  gradient-index, 635, 658
  power of, 644
  speed of, 650
  thick, 621
  thin, 620–621
  varifocal, 657
  zoom, 657–658
Lensmakers' equation, 639
Lenz, Heinrich, 557, 559
Lenz's law, 559
Leptons, 921 *def.*, 924
Lever arm, 106
Leyden jar, 476–477
Lift pump, 350
Light
  and color, 877–879
  deflection by gravity, 728
  dispersion of, 633, 682
  guide, 619
  infrared, 611
  laser, properties of, 869–871
  models of, 610–611
  pipes, 618
  polarization of, 685–688
  reflection of, 613–614
  refraction of, 615
  spectrum of, 611
  speed of, 612, 613
  ultraviolet, 611
  visible, 610–611
  as wave, 610, 664–667
Light-emitting diode, 906, 907
Lightning, 438
Lightning rod, 477
Line spectra, 685, 768–769
Linear accelerator, 572–573, 919
Linear charge density, 455
Linear expansion, thermal, 300–303,
  300 *tab.*
Linear momentum, 97 *def.*, 192 *def.*
  conservation of, 195–196
Lines of force,
  electric, 446
  gravitational, 150
Lines of magnetic field, 527
Liquid crystal, 889
  display, 890–891
Liquid drop model of the nucleus,
  854
Liquid, state of matter, 294

Livingston, M. S., 536
Lodestone, 526, 549
Logarithms, 380
Longitudinal waves, 412
Loop rule, 508, 582, 585, 592
Lorentz transformations, 705–706
Lungs and surface tension, 272

M

Mach number, 422 *def.*
Macroscopic, 362 *def.*
Magdeburg hemispheres, 351
Magnet, 526–528
  bar, 527
  horseshoe, 528
Magnetic behavior, 547 *tab.*
Magnetic dipole, 527
  current loop, 542
  dipole moment, 529
  earth as a, 528
  torque on, 529
Magnetic domains, 549
Magnetic field, 527
  at the center of a loop, 540
  due to a current, 537-541
  due to a current loop, 543
  direction of, 527, 529
  of earth, 528, 532
  energy stored in, 570
  field lines, 527, 528, 530
  path of a charged particle in,
    533–536
  right hand rules, for, 530, 531,
    542, 550
  of a solenoid, 546
  due to a long straight wire, 540
Magnetic flux, 530, 559 *def.*
Magnetic force on
  current carrying wire, 531
  current loop, 540
  moving charge, 533
Magnetic mirror, 858
Magnetic monopole, 530
Magnetic permeability, 539
Magnetic polarization, 528, 549
Magnetic quantum number, 815
Magnetic susceptibility, 547 *def.*
Magnetism, Gauss's law of, 529, 571
Magneton, Bohr, 818
Magnification,
  angular, 647
  lateral or linear, 623, 625
  of a magnifying glass, 646–647

of a microscope, 652
of a telescope, 654–655
Magnifying glass, 645–648
Magnitude of vector, 64, 67–68
Malus's law, 687
Manometer, 268
Mass and energy in relativity, 714–715
Mass, 99
and inertia, 99
and weight, 100–102
Mass, molecular, 356
Mathematical formulas, Appendix A
Maxwell, James Clerk, 368, 571, 610
Maxwell's equations, 571, 611
and electromagnetic waves, 571–574
and relativity, 699
Maxwell-Boltzmann distribution, 370
Measurement, 3, 15–18
Mechanical energy, 177 *def.*
conservation of, 177
Mechanical equivalent of heat, 304–305
Mechanics, 32 *def.*
Newtonian, 91
Mendeleev, Dmitri, 740
Meson, 916, 921
Metals, band theory of, 897–900
Metastable state, 866
Meter (unit), 21, 574 *def.*, Appendix B *def.*
Metric prefixes, 21 *tab.*
Michelson-Morley experiment, 699–700
Microscope,
field-ion, 750
optical, 651–653
resolution of optical, 681
scanning electron, 800–801
scanning tunneling, 751, 888
transmission electron, 800
Microscopic, 362 *def.*
Microwave, 456, 573
antenna, 612
oven, 456–457
Millikan, Robert A., 752
Millikan's oil-drop experiment, 753
Mirror,
plane, 614, 615
spherical, 628–632
Mobility, 896 *def.*, 897, 901
Model, 2, 7, 15, 229, 359–360, 392

Models of light, 610–611
Moderator, 855
Mole (unit), 356 *def.*
Molecular mass, 356, 357 *tab.*
Molecules, 360
Moment of inertia, 173 *def.* 174 *tab.*
Momentum, angular, 212 *def.*
conservation, of 213–216
Momentum, linear, 97 *def.*, 192 *def.*
conservation of 195–196
Momentum, relativistic, 716–719
Monochromatic light, 671, 672, 673
Monopole, magnetic, 527, 530
Moon and tides, 248–249
Moseley, Henry, G. J., 767, 785
and the periodic table, 785–787
Moseley's law, 785–787
Motion
circular, 129
damped harmonic, 395–397
of fluids, 274–281, 284–285
of free fall, 50–54
harmonic, 385–389
linear, 32–50
Newton's laws of, 96–104
periodic, 382
projectile, 75–81
relative, 65–67, 698–699
rotational, 172–175, 208–219
uniform accelerated, 46–50, 73
of waves, 407–410
Motional emf, 561–562
Motor, electric, 565
Moving charge, force on, 533
Muon, 916
time dilation of, 710
Myopia, 644

N

n-type semiconductor, 904
Natural abundance, 832
Natural decay series, 839–840
Natural frequency of an oscillator, 391
Near point, 644
Nearsighted eye, 643–645
Negative charge, 439–440
Net force, 94
Neutral, electrically, 440
Neutrino, 842–843, 917, 924, 918 *tab.*
Neutron, 829–831, 832, 913, 918 *tab.*, 927

Newton, Sir Isaac, 91, 92–93, 191, 610
Newton (unit), 99 *def.*
Newton's law of gravitation, 140, 169
Newton's laws,
applications of, 104–110, 360
and conservation of momentum, 195-196
first law, 96–97
second law 97–100
third law 102–104
Newton's rings, 672
Newtonian telescope, 656
Nicad cell, 497
Node, (def) 425 *def.*
Noninverting amplifier, 603
Normal, 101 *def.*, 613
North pole, 526
Nova, 3–4, 6–7
Nuclear
binding energy, 835–836
fission, 853
force, 834–835
fusion, 857-859
reactions, 850–851
reactor, 855–857
size, 760, 833–834
stability, 837–839
transmutations, 859
Nucleon, 832 *def.*
Nucleus
daughter, 837, 839
discovery of, 757–760
half-life of, 756
liquid-drop model of, 854
models of, 758, 829
parent , 837, 839
radioactive decay, 755–757, 837–839
size of, 760, 833–834

O

Object distance, 624
Objective, 651
Observer in quantum mechanics, 795
Octave, 427
Ocular, 651
Oersted, Hans C., 526, 528–529
Ohm (unit), 498, *def.*
Ohm's law, 499, 502
and electrical conductivity, 895–896
Opaque body, 615

Open circuit, 507
Opera glasses, 656
Operational amplifier, 601–603
Optical
  axis, 620 *def.*
  fibers, 618
  pumping, 866
  resonator, 865, 867
Optoelectric devices, 906
Orbital quantum number, 815
der of magnitude calculations, 18–20
Order to disorder, 339
Ordinate, 38 *def.*
Origins of the universe, 926
Oscillations, conditions for self-
    sustaining, 382
Oscillator, simple harmonic, 388
Oscillatory motion, 385–388
Otto cycle, 330–331
Overdamped harmonic motion, 396
Overtone, 425 *def.*

P
π-meson, 917
p-type semiconductor, 904
Pair production, 914,
Parabola, 12–13, 77, 119
Parallel connection
  of capacitors, 513
  of resistors, 511
Parallel plate capacitor,
  capacitance, 478–480
  field, 480–483
Paramagnetism, 548, 549
Paraxial approximation, 621, 622
Parent nucleus, 837, 839
Particle in a box, 811–812
  energy of, 812
Particle model of light, 610–611
Particles,
  elementary, 913
  fundamental, 913
Pascal (unit), 261 *def.*
Pascal's principle, 265
Pauli, Wolfgang, 794, 842
Pauli exclusion principle, 820–821
Pendulum, simple, 391
Perfectly inelastic collisions, 196
Period, 131 *def.*
  of circular motion, 131–132
  of cyclotron orbit, 537
  of a harmonic oscillator, 390
  natural oscillator, 391

of planets, 9 *tab.*
of a satellite, 146–147
of a simple pendulum, 394
Periodic motion, 382
Periodic table, 740, 785–787
  understanding of, 821–822
Permanent magnet, 549
Permeability of free space, 539
Permittivity of free space, 442, 480
Perpetual motion, first kind, 326
Perrin, Jean, 737, 740
Phase angle
  in simple harmonic motion, 388
  in ac circuits, 590
Phase change on reflection, 672
Phases of matter, 294
Phasor, 592
Photodiode, 906–907
Photoelectric effect, 775–779,
    894–895
  Einstein's explanation, 777-778
  mechanical analog, 778
Photographic films, 881–882
  color slides, 881
  false color infrared, 881-882
  resolution of, 876
Photomultiplier tube, 845
Photon
  Compton effect, 796–798
  energy of, 774
  exchange, 922
  spontaneous emission, 865
  stimulated emission, 864–865, 867
Physical constants, back end-papers
Physics, origin of term, 2
Pion, 917
Pipe, vibrations in, 452
Pitch of a sound, 414
Planck, Max, 335, 766, 774
Planck's constant, 774
Planck's formula, 774
Planck's hypothesis, 774
Plane of incidence, 614
Plane waves, 413
Plane, inclined, 107
Planetary orbits and periods, 9 *tab.*
Plasma, 858
Plum-pudding model, 758
pn junction, 499, 904–905
Poiseuille's law, 280
Polar molecules, 484–485
Polarization, 685–688
  mechanical analogy, 686

by scattering, 690
of skylight, 690
Polarizer, 686, 891
Polaroids, 686
Polycrystalline solid, 888
Positron, 842, 914, 917
Postulates of special relativity, 702
Potential, electrical, 467–470
  of a point charge, 469
  gradient, 475
Potential barrier, nuclear, 841
Potential difference, 498
Potential energy, 166
  diagram, 178, 180–181
  difference, 167
  electric, 468
  general form of gravitational,
      168–171
  gravitational, 166
  and Hooke's law, 161, 389
  near the earth, 165–168
  origin of the term, 157
Power transmitted by wave, 411
Power, 182 *def.*
Power, electrical, 503–507
  distribution, home, 517–518
  effective, 588
  supply, dc, 598
  and thermal pollution, 340–342
Powers of ten, 18 *tab.*, 337 *tab.*
Prefixes, SI units, 21 *tab.*
Pressure, 260 *def.*
  due to atmosphere, 261, 349–350
  of a gas, 351–353, 361
  gauge, 261
  units, 261 *tab.*
Primary colors, 879–880
Principal quantum number, 781–782,
    815
Principia, 91, 92–93, 94, 707
Principle of equivalence, 723–726
Principle of superposition, 424
Prism, 633, 639
Probability density, 809, 816
Probability in quantum mechanics,
    794
Problem solving guidelines, 22–23
Projectile motion, 75–78, 78–81
Projector, 650–651
Proper
  length, 712 *def.*
  mass, 718 *def.*
  time, 710 *def.*

Proton, 829 *def.*
  instability, 926
Proton-electron model of the nucleus, 829
Pulse, transverse, 407
Pulsed laser, 866
Pump, lift, 349
Pumping, optical, 865
PV diagram, 327
Pyrometer, 299
Pythagorean theorem, 88

Q

Q value, 838, 841, 843, 851
Quality factor, 847 *tab.*
Quantization of
  angular momentum, 781
  energy, 774
  charge, 443, 739
  space, 817–820
Quantized, 774 *def.*
Quantum, 774
Quantum and classical mechanics, 793–796
Quantum mechanical observer, 795
Quantum numbers, 815
Quark
  model, 443, 922–925
  confinement, 925

R

R value of insulation, 311
R, universal gas constant, 356
Rad, (unit), 847
Radian, 5 *def.*
Radiation,
  background, 848
  biological effects of, 847
  blackbody, 771–773
  detectors, 844–846
  gamma, 754–755
  genetic effects of, 847
  microwave, 456
  somatic effects, 847
  thermal, 311, 312–313
Radiation (thermal), 311
Radiation therapy, 572, 848–849
Radio waves, 456
Radioactive decay, 755–757, 837–839
  law, 756
Radioactivity, 753–755, 754 *def.*
Radiocarbon dating, 757

Radon, 848–849
Rail, continuous-welded, 302
Rainbow, 683–684
Rainbow hologram, 875
Range of a projectile, 78–81
Rarefaction, longitudinal wave, 412
Ray of light, 611
Rays, cosmic, 757, 842, 848–849, 917, 926
Ray-tracing, 622–624
Rayleigh, Baron (John W. Strutt), 681, 689
Rayleigh scattering, 689–690
Rayleigh's criterion, 680
RC circuit, 584
Reactance, 589–591
  capacitive, 590
  inductive, 591
Reactions, nuclear, 850–851
Reactor, 855–857
Recommended maximum dose, 847 *tab.*
Rectifier equation, 906
Rectifier, 598–601
  full wave, 600
  half-wave, 599
Red shift, 722, 725
Reference frame, 119 *def.*
  inertial, 119, 698–699
Reflection of wave pulse, 423
Reflection, optical, 613–614
  diffuse, 614
  Huygens' principle, 665-666
  law of, 614
  specular, 614
  total internal, 617, 618
Refraction, 615 *def.*
  Huygens' principle, 666–667
  index of, 615 *def.*, 616 *tab.*
  law of, 615
Refrigerator, 333–335
Relative velocity, 65–67, 70, 698–699
Relativity
  Galilean, 699
  general theory, 726–729
  and length contraction, 712–714
  postulates of, 701–702
  simultaneity and, 704–708
  special theory, 701–702
  and time dilation, 708–710
  and velocity addition, 702–703
Reservoir, heat, 328

Resistance, electrical, 498, 896
Resistivity, electrical, 501–503, 502 *tab.*, 896
Resistor, carbon composition, 500
Resistors
  in parallel, 511
  in series, 510
Resolution of vectors, 67–72
Resolving power, 680
Resonance curve, 398, 595
Resonance, 398 *def.*
Resonant circuits, 594–596
Resonant frequency, 398 *def.*
  of vibrating air columns, 428–429
  of a vibrating string, 426
Rest mass, 718
Restoring force, 383
Resultant, 62 *def.*
Reversible process, 325, 336
Reynold's number, 284
Right-hand-rule for the direction of
  angular momentum, 212
  a dipole current loop, 542
  the magnetic force on a wire, 531
  the magnetic field about current, 529
  torque, 207
RL circuit, 582–584
RLC series circuit, 592
Rocket propulsion, 204
Roemer, Ole, 298, 612
Roentgen, Wilhelm Konrad, 736, 744
Roentgen (unit), 846
Roentgen rays, 744
Rolling bodies, 238–240
Room acoustics, 416–417
Root-mean-square (rms)
  current, 588
  speed, 362
  voltage, 588
Rotational equilibrium, 208
Rotational kinetic energy, 172, 173 *def.*
Rotational motion, 172–175, 208–219
Rounding off, 16–17
Ruby laser, 865–868
Rumford, Count (Benjamin Thompson), 304
Running and walking, 391
Rutherford, E., 758–760, 830, 850
Rutherford's discovery of transmutation, 850–851

Rydberg constant, 771
Rydberg formula, 771

S

Sabin (unit), 416 *def.*
Satellite, period, 146
Saturation current, 906
Scalar, 61 *def.*
Scale, musical, 427
Scanning tunneling microscope, 751, 888
Scattering
  of alpha rays, 758
  angle, 237
  Compton, 796–798
  optical, 689-690
Schrödinger, Erwin, 795, 803
Schrödinger's equation, 803–804, 892
Schwarzschild radius, 248
Scientific notation, 17
Scintillation counter, 845
Scuba diver, 267
Second (unit), 15 *def.*
Second condition for equilibrium, 208
Second law of thermodynamics, 335–336
  and entropy, 336–340
Semiconductor radiation detector, 846
Semiconductor, 439, 900 *def.*
  extrinsic, 903–904
  impure, 903-904
  intrinsic, 900–901
  n-type, 904
  p-type, 904
  p-n junction of, 904–905
  pure, 900–901
Series resonance, 594–595
Shielding, radiation, 848
Shock waves, 421–422
Shock, electric, 477, 516
Shockley, William, 601
Short circuit, 507, 519
Shutter speed, 649
SI prefixes, 21 *tab.*
SI units, 20–21, 21 *tab.*, Appendix B
Sievert, (unit), 847
Significant figures, 16
Silicon crystals, 900
Simple harmonic motion, 382, 385 *def.*

Simple harmonic oscillator, 385–392, 388 *def.*
Simple pendulum, 391,394
Simultaneity, 704–708, 707 *def.*
Sine of an angle, 87
Single-slit diffraction, 674–676
Sink, heat, 328
Sinusoidal curve, 382
Sky, color of, 690
Skylight, polarization of, 690
Slope of a line, 39 *def.*
Smoke, color of, 690
Snell's law, 615
  from Huygens' principle, 667
Soap film, 673
Solar cell, 906–908
Solar spectrum, 768–769
Solenoid, 545 *def.*
  inductance of, 569
  magnetic field, of 546
Solid, state of matter, 294
Solid, 887
  amorphous, 889
  band theory of, 897–900
  crystalline, 887glassy, 889
  polycrystalline, 888
Somatic effects of radiation, 847
Sonic boom, 422
Sonic frequencies, 414
Sound, 412–415
  beats, 429
  Doppler effect, 419–421
  intensity of, 413, 416, 418
  measurement, 415
  pitch, 414
  speed, 413, 414 *tab.*
  ultrasonic, 414
Sound-absorption coefficient, 417
South pole, 526
Space quantization, 817
Spatial coherence, 871
Spatial frequency, 876–877
Special theory of relativity, 701–702
Specific heat, 306 *def.*, 306 *tab.*
  molar, of gases, 365–367
Speckle pattern, 869
Spectra, 685, 767–768
Spectrograph, 685
Spectrometer, 684
Spectroscopes, 684–685
Spectroscopy, 767–769, 767 *def.*
Spectrum
  absorption, 768

blackbody, 772–773
colors of the visible, 877
electromagnetic, 611, 633
emission, 768–769
line, 685, 768–769
solar, 768–769
Specular reflection, 614
Speed, 32–34
  average, 33 *def.*,
  constant, 33
  of molecules in a gas, 368–371
  root-mean-square, 362
  of running, 393
  of sound, 413–414
  of walking, 393
  of a wave, 409–410
Speed of a lens, 650
Speed of light, 574, 612, 613, 699
  and index of refraction, 615
Speed of sound, 414 *tab.*
  as a function of temperature, 413
Spherical aberration, 633
Spherical mirrors, 628–632
Spherical waves, 413
Sphygmomanometer, 267
Spin magnetic quantum number, 815
Spontaneous emission, 865
Spring constant, 383
St. Elmo's fire, 438
Stable isotopes, 832–833
Standard model, 927
Standing waves on a string, 424–428
State variables and coordinates, 325
Static equilibrium, 115–119, 115 *def.*
Static friction, 111
Statics, 116 *def.*
Statistical mechanics, 339
Steam engine, 320, 326
Stefan-Boltzmann constant, 312
Stefan-Boltzmann law, 312
Stereo broadcasting, 597
Stern and Gerlach experiment, 819
Stimulated emission, 864–865, 867
Stokes's law, 282
Stopping potential, 776
Storage ring, 919
Streamline, 275
Stringed instruments, 427
Strong nuclear force, 915, 834–835, 922
Stud finder, 485
Sunlight, spectral distribution, 877
Sunset and sunrise, color, 690

Superconducting supercollider (SSC), 920
Superconductor, 500, 548
Supernova 1987 A, 7, 924
Superposition principle, 424
  for electric fields, 447–449
  for electric forces, 444–445
Surface charge density, 482
Surface tension, 271 *def.*, 271 *tab.*
Susceptibility, 548
System, thermodynamic, 321–322
Systolic pressure, 267

T
Tangent of an angle, 87
Telescopes, 653–656
  astronomical, 653–655
  Galilean, 653, 655
  Newtonian, 656
  resolution of, 681
Television, color, 879–880
Temperature conversion, 297, 299
Temperature scales,
  Celsius, 296
  Fahrenheit, 297
  Kelvin, 299
Temperature, 295 *def.*
  Curie, 549
  internal energy and heat, 321–326
  kinetic theory definition of, 363
Temporal coherence, 871
Tension, 104 *def.*, 106
Terminal potential difference, 515
Terminal velocity, 282–283
Tesla (unit), 531 *def.*
Thermal conductivity, 311, 311 *tab.*
Thermal contact, 320
Thermal efficiency, 332
Thermal equilibrium, 295, 300
Thermal expansion, 300–303
  linear coefficient, 300 *def.*, 300 *tab.*
  volume coefficient, 302 *def.*, 300 *tab.*
Thermal neutrons, 856
Thermal pollution, 340–342
Thermodynamic system, 321
Thermodynamics,
first law, 322–326
  second law, 335–336
  second law and entropy, 336–340
  zeroth law, 321
Thermometer, 295, 302
Thermometry, 295

Thermostat, 302
Thevenin's theorem, 516
Thick lens, 621
Thin lens, 620–621, 621 *def.*
  in combination, 639
  focal length, 639
  Gaussian equation, 625
  Newtonian equation, 640
Thin-film interference, 671–674
Thomson, G. P., 795, 802
Thomson, J. J., 736, 749, 758, 775
Thomson's experiment, 749–752
Threshold energy of a nuclear reaction, 851–852
Threshold frequency, 777
Thrust of rocket, 204
Tides, 248–249
Time, 14, 15, 36, 705, 708–711, 725–726
Time constant,
  capacitive, 585
  inductive, 583
Time dilation, 708–710
Time-reversal symmetry, 336
Timekeepers, 14
Tire pressure gauge, 261
Tires, automobile 112–113
Tokamak, 858
Torque, 206 *def.*
  and angular acceleration, 217
  on a current loop, 541-542
  on an electric dipole, 458
  on magnetic dipole, 529
Torricelli, Evangelista, 349
Torricellian tube, 350
Torricellian vacuum, 350
Torsional balance, 441
Total internal reflection, 617–619, 617 *def.*
Townes, Charles H., 865
Transformer, 566–568
Translational equilibrium, 115
Transmutations, nuclear, 850
Transparent body, 615
Transverse wave, 412
  pulse, 407
Traveling wave, 408–409
  equation for, 409
Trifocals, 644
Trigonometry, 87–89, Appendix A
Tunneling, 813–814
  in alpha decay, 841
  microscope, 888

Turbulent flow, 276, 285
  and Reynolds number, 284–285
Twin paradox, 711
Tycho Brahe, 3–4, 6–7, 10, 14, 15, 610, 654
Tyndall scattering, 690

U
Ultrasonic frequencies and waves, 414
Ultrasonic imaging, 415
Ultraviolet, 611
Uncertainty principle, 804–807
Underdamped harmonic motion, 396
Unified theories, 925–926
Uniform circular motion, 129
Unit cell, 737 *def.*, 887
Unit conversions, 24
Universal gas constant R, 356
Universal gravitation, law of, 140
Universal gravitational constant, G, 141

V
Vacuum flask, 313
Valence band, 900
Van de Graaff generator, 471, 472, 851
Van der Waals' equation, 365
Varifocal lens, 657
Vector, 61 *def.*
  addition, 61–65, 68–70
  components, 67
  magnitude of, 64
  resolution, 67–72
  subtraction, 63–64, 65, 68–70
  symbols for, 61
Velocity, 34 *def.*
  addition in relativity, 702–703
  angular, 133 *def.*
  average, 34–38, 72
  drift, 497, 895–897
  escape, 246
  graphical interpretation of, 38–39
  instantaneous, 40
  kinematic equations for, 218
  relative, 65–67, 70
  root mean square, 362
  in simple harmonic motion, 387
  terminal, 282
  of a wave, 408, 410
Venturi meter, 278
Verrazano-Narrows Bridge, 301

Vibrating air column, 428
Vibrating string, 426
Viscosity, 279, 280 *tab.*
Visibility, fringe, 870
Visible spectrum, 611, 678–679, 682, 877–879
  colors of, 877 *tab.*
Volt (unit), 469 *def.*
Volta, Count Allesandro, 495, 496
Voltage, root-mean-square, 588
Voltaic pile, 497, 529
Voltmeter, 515
  from a galvanometer, 544
Volumes, law of, 738

W
Walking and running, 391
Water molecules, 456
Water, density of, 303
Watt, James, 320, 327
Watt (unit), 182 *def.*, 504
Watt-second, 506
Wave , 408 *def.*
  deBroglie, 799
  diffraction of, 674
  electromagnetic, 408, 571–574, 596–587
  energy transfer, 411
  harmonic, 408
  intensity, 411
  interference of, 667
  longitudinal, 412
  model of light, 610–611

packet, 804–805
plane, 413
power transmitted by, 411
radio, 456
reflection of, 423, 613
refraction of, 615
shock, 421
sound, 412–415
speed, 409–410
spherical, 413
standing, 424–428
on a string, 424–428
transverse, 412
traveling, 408–409
Wave equation, Schrödinger's, 803–804, 892
Wave function, 803 *def.*
  interpretation of, 809–811
Wave pulse,
  reflection,423
  transverse, 407
Wave theory of the hydrogen atom, 814–816
Wave-particle duality, 611, 810
Wavefronts, 413
  in Huygens' construction, 665
Waveguide, 573
Wavelength, 408 *def.*
  Compton, 797 *def.*
  de Broglie, 799 *def.*
  in a material, 667
Weak force, 95–96, 921–922, 925
Weight and mass, 100–102, 103

Weightlessness, 104
Wien displacement law, 772
Work function, 777 *def.*
Work, 157 *def.* in Carnot cycle, 330
  and first law, 322
  done in stretching a spring, 161
  thermal, 323
Work-energy theorem, 164
Wren, Christopher, 92, 191
Wright brothers, 58

X
X rays, 744–745, 572–573
  Bragg scattering of, 746
  Compton scattering, 796–798
  diffraction, 745–748
  Laue pattern of, 745
  production, 747–748
  spectrum, 785–787
X-ray tube, 748

Y
Young, Thomas, 407, 610, 644
Young's double-slit experiment, 667–670, 811
Yukawa particle, 916
Yukawa potential, 915

Z
Zeeman effect, 816–819
Zeppelin, 353
Zero, absolute, 299, 354
Zero-point energy, 813
Zeroth law of thermodynamics, 321
Zoom lenses, 657–658

PERIODIC TABLE OF THE ELEMENTS[a]

Key:

1	Atomic number
H	Symbol
1.00794	Atomic weight

Noble gases — 0: **He** 2, 4.00260

Period	IA	IIA	IIIB	IVB	VB	VIB	VIIB	VIII			IB	IIB	IIIA	IVA	VA	VIA	VIIA
1	1 **H** 1.00794																
2	3 **Li** 6.941	4 **Be** 9.01218											5 **B** 10.811	6 **C** 12.011	7 **N** 14.0067	8 **O** 15.9994	9 **F** 18.998403
3	11 **Na** 22.98977	12 **Mg** 24.3050											13 **Al** 26.98154	14 **Si** 28.0855	15 **P** 30.97376	16 **S** 32.066	17 **Cl** 35.4527
4	19 **K** 39.0983	20 **Ca** 40.078	21 **Sc** 44.95591	22 **Ti** 47.88	23 **V** 50.9415	24 **Cr** 51.9961	25 **Mn** 54.9380	26 **Fe** 55.847	27 **Co** 58.93320	28 **Ni** 58.69	29 **Cu** 63.546	30 **Zn** 65.39	31 **Ga** 69.723	32 **Ge** 72.61	33 **As** 74.92159	34 **Se** 78.96	35 **Br** 79.904
5	37 **Rb** 85.4678	38 **Sr** 87.62	39 **Y** 88.90585	40 **Zr** 91.224	41 **Nb** 92.90638	42 **Mo** 95.94	43 **Tc** 98.9072	44 **Ru** 101.07	45 **Rh** 102.90550	46 **Pd** 106.42	47 **Ag** 107.8682	48 **Cd** 112.411	49 **In** 114.82	50 **Sn** 118.710	51 **Sb** 121.75	52 **Te** 127.60	53 **I** 126.90447
6	55 **Cs** 132.90543	56 **Ba** 137.327	57 ***La** 138.9055	72 **Hf** 178.49	73 **Ta** 180.9479	74 **W** 183.85	75 **Re** 186.207	76 **Os** 190.2	77 **Ir** 192.22	78 **Pt** 195.08	79 **Au** 196.96654	80 **Hg** 200.59	81 **Tl** 204.3833	82 **Pb** 207.2	83 **Bi** 208.98037	84 **Po** 208.9824	85 **At** 209.9871
7	87 **Fr** 223.0197	88 **Ra** 226.0254	89 **†Ac** 227.0278	104 **Unq** 261.11	105 **Unp** 262.114	106 **Unh** 263.118	107 **Uns** 262.12										

Noble gases (group 0):
10 **Ne** 20.1797; 18 **Ar** 39.948; 36 **Kr** 83.80; 54 **Xe** 131.29; 86 **Rn** 222.0176

*Lanthanide series:

| 58 **Ce** 140.115 | 59 **Pr** 140.90765 | 60 **Nd** 144.24 | 61 **Pm** 144.9127 | 62 **Sm** 150.36 | 63 **Eu** 151.965 | 64 **Gd** 157.25 | 65 **Tb** 158.92534 | 66 **Dy** 162.50 | 67 **Ho** 164.93032 | 68 **Er** 167.26 | 69 **Tm** 168.93421 | 70 **Yb** 173.04 | 71 **Lu** 174.967 |

†Actinide series:

| 90 **Th** 232.0381 | 91 **Pa** 231.0359 | 92 **U** 238.0289 | 93 **Np** 237.0482 | 94 **Pu** 244.0642 | 95 **Am** 243.0614 | 96 **Cm** 247.0703 | 97 **Bk** 247.0703 | 98 **Cf** 242.0587 | 99 **Es** 252.083 | 100 **Fm** 257.0951 | 101 **Md** 258.10 | 102 **No** 259.1009 | 103 **Lr** 260.105 |

[a]Based on 1985 IUPAC values.